T0329341

GEOCHEMISTRY OF EARTH SURFACE SYSTEMS

FROM THE TREATISE ON GEOCHEMISTRY

Editors

H. D. Holland

University of Pennsylvania, Philadelphia, PA, USA

K. K. Turekian

Yale University, New Haven, CT, USA

ELSEVIER

AMSTERDAM • BOSTON • HEIDELBERG • LONDON • NEW YORK • OXFORD
PARIS • SAN DIEGO • SAN FRANCISCO • SINGAPORE • SYDNEY • TOKYO
Academic Press is an imprint of Elsevier

ACADEMIC
PRESS

Academic Press is an imprint of Elsevier
32 Jamestown Road, London NW1 7BY, UK
Radarweg 29, PO Box 211, 1000 AE Amsterdam, The Netherlands
30 Corporate Drive, Suite 400, Burlington, MA 01803, USA
525 B Street, Suite 1900, San Diego, CA 92101-4495, USA

First edition 2011

British Library Cataloguing in Publication Data
A catalogue record for this book is available from the British Library

Library of Congress Cataloging-in-Publication Data
A catalog record for this book is available from the Library of Congress

ISBN: 978-0-08-096706-6

For information on all Academic Press publications
visit our website at elsevierdirect.com

Printed and bound in Italy

11 12 13 14 10 9 8 7 6 5 4 3 2 1

Working together to grow
libraries in developing countries
www.elsevier.com | www.bookaid.org | www.sabre.org
ELSEVIER   BOOK AID International   Sabre Foundation

CONTENTS

Contents   iii

Introduction   v

Contributors   vii

1   Volcanic Degassing   1
    C. OPPENHEIMER, *University of Cambridge, UK*

2   Hydrothermal Processes   45
    C. R. GERMAN, *Southampton Oceanography Centre, Southampton, UK*
    K. L. VON DAMM*, *University of New Hampshire, Durham, NH, USA*

3   The Contemporary Carbon Cycle   87
    R. A. HOUGHTON, *Woods Hole Research Center, MA, USA*

4   The Global Sulfur Cycle   129
    P. BRIMBLECOMBE, *University of East Anglia, Norwich, UK*

5   The History of Planetary Degassing as Recorded by Noble Gases   167
    D. PORCELLI, *University of Oxford, UK*
    K. K. TUREKIAN, *Yale University, New Haven, CT, USA*

6   Natural Weathering Rates of Silicate Minerals   205
    A. F. WHITE, *US Geological Survey, Menlo Park, CA, USA*

7   Soil Formation   241
    R. AMUNDSON, *University of California, Berkeley, CA, USA*

8   Global Occurrence of Major Elements in Rivers   277
    M. MEYBECK, *University of Paris VI, CNRS, Paris, France*

9   Trace Elements in River Waters   293
    J. GAILLARDET, *Institut de Physique du Globe de Paris, France*
    J. VIERS and B. DUPRÉ, *Laboratoire des Mécanismes et Transferts en Géologie, Toulouse, France*

*Deceased

10   The Geologic History of the Carbon Cycle   341
     E. T. SUNDQUIST and K. VISSER, *US Geological Survey, Woods Hole, MA, USA*

11   Organic Matter in the Contemporary Ocean   389
     T. I. EGLINTON and D. J. REPETA, *Woods Hole Oceanographic Institution, MA, USA*

12   The Biological Pump   425
     C. L. DE LA ROCHA, *Alfred Wegener Institute for Polar and Marine Research,*
     Bremerhaven, Germany

13   The Biological Pump in the Past   453
     D. M. SIGMAN, *Princeton University, NJ, USA*
     G. H. HAUG, *Geoforschungszentrum Potsdam, Germany*

14   The Oceanic CaCO$_3$ Cycle   491
     W. S. BROECKER, *Columbia University, Palisades, NY, USA*

15   The Global Oxygen Cycle   511
     S. T. PETSCH, *University of Massachusetts, Amherst, MA, USA*

16   The Global Nitrogen Cycle   551
     J. N. GALLOWAY, *University of Virginia, Charlottesville, VA, USA*

17   Evolution of Sedimentary Rocks   579
     J. VEIZER, *Ruhr University, Bochum, Germany and*
     University of Ottawa, ON, Canada
     F. T. MACKENZIE, *University of Hawaii, Honolulu, HI, USA*

18   Generation of Mobile Components during Subduction of Oceanic Crust   617
     M. W. SCHMIDT, *ETH Zürich, Switzerland*
     S. POLI, *Universita di Milano, Italy*

Appendix 1   643

Appendix 2   644

Appendix 3   648

Appendix 4   649

Index   651

Introduction

It is now common parlance to refer to the Earth as a physical–chemical–biological system; but this is a relatively recent innovation. Until about 1950, geologists generally used the word 'system' to define geologic periods, or the several crystal classes, or the physical chemistry of multicomponent mixtures. Long before then, the word 'system' in its current usage had been a part of the vocabulary of a number of fields, including communication and control engineering. In fact, the first significant paper on feedback mechanisms in control systems was written by Clerk Maxwell in 1868. Its subject was governors (mechanical, not political). In the 1930s and 1940s, Norbert Wiener of the Massachusetts Institute of Technology and Arturo Rosenblueth of the Harvard Medical School applied communication engineering design to physiology, converted the field into a branch of statistical mechanics, and christened it Cybernetics. Wiener's book 'Cybernetics, or Control and Communication in the Animal and the Machine' (1948) introduced the subject to a wide audience, which, happily, included geochemists.

Independently, Barth (1952) took a step toward the use of systems analysis in geochemistry by using the river flux of the elements to the oceans to calculate their 'period of passage' through the oceans. In 1958 his results were largely confirmed by Goldberg and Arrhenius. In 1961 Goldberg raised important questions about the operation of the ocean as a geochemical system. These were answered in part by Broecker *et al*'s chapter in the same volume as Goldberg's. In their 1975 book, '*Chemical Cycles and the Global Environment: Assessing Human Influences*,' Garrels, Mackenzie, and Hunt introduced the concept of residence times, then the notion of fluxes and reservoirs, and finally, a description of the Earth system. In 1978, I chose to begin my book '*The Chemistry of the Atmosphere and Oceans*' with a chapter on the Earth as a chemical system. By then the new nomenclature had clearly arrived. Its adoption signaled more than a change in vocabulary. There was a greatly expanded application of systems analysis to the Earth Sciences. Extensive studies of reaction rates and reaction mechanisms were added to analyses of the thermodynamics of Earth materials. The consequences of positive and negative feedbacks, singularities, tipping points, chaos theory, and the linking of geochemical cycles were explored. All this led to considerable progress in our ability to understand the workings of near-surface systems, where temperatures are generally low and many reactions are slow, even on geologic time scales. The need to understand and to predict the consequences of mankind's perturbations of the Earth's surface and near-surface environments has lent a strong sense of urgency and greatly increased funding for studies of nonsteady state systems.

The Earth's near-surface systems have always been subjected to inputs of mass and energy. The intensity and the frequency of these inputs have varied widely. Many inputs have been cyclic, some have been singular, and several have evolved more or less unidirectionally during the course of Earth history. Volcanoes (Chapter 1) and oceanic hydrothermal vents (Chapter 2) have been the major sources of volatiles for near-surface systems, although these inputs are now dwarfed by the effects of fossil fuel burning and other human activities (Chapters 3 and 4). Volcanic rocks have been the most important terrestrial and marine resurfacing agents, and magmatic intrusions have continually underplated the Earth's surface. Sporadically, meteorites and possibly comets have been quantitatively important sources of mass and devastation, particularly during the earliest parts of Earth history.

Sunlight has been by far the largest source of energy for the Earth's near-surface environments, although these have been shaped to a surprisingly large extent by the small supply of energy from the

Earth's interior. The internal energy is responsible for the large supply of volatiles that are now and always have been added to the Earth's near-surface systems. Most of these volatiles have been 'digested' and removed to more permanent quarters on geologically short time scales. The heavier rare gases are an exception. Helium escapes into interplanetary space on a time scale of about 10^6 years, but most of the Ne, Ar, Kr, and Xe that has escaped from the Earth's interior has accumulated in the atmosphere (Chapter 5).

Water, the most abundant of the volcanic volatiles, rapidly becomes a participant in the hydrologic cycle and enters the oceans, where it has accumulated to form the largest of the Earth's near-surface systems. It is a critical actor in weathering (Chapter 6), in the formation of soils (Chapter 7), and in the release of the dissolved constituents of rivers that become the major solute inputs to the oceans (Chapters 8 and 9). Together with carbon and nitrogen, water cycles rapidly through the biosphere. Without these elements the Earth would not be a habitable planet. Next to water, CO_2 is the most abundant of the degassed volatiles. It is an important greenhouse gas, the most important of the weathering acids (Chapter 6), and an essential participant in photosynthesis (Chapter 3). In the oceans it is the basis of marine photosynthesis, and hence the source of much of the organic matter in the oceans (Chapter 11). Organic matter and marine $CaCO_3$ are cycled through the oceans by the marine biological pump (Chapters 12 and 13) and via the marine $CaCO_3$ cycle (Chapter 14). Small residues become parts of marine sediments. Photosynthesis via photosystem II generates O_2, the only major atmospheric constituent that is not a devolatilized gas (Chapter 15). Its history is linked both to the evolution of volcanic gases, particularly their H_2, CO_2, and SO_2 content (Holland, 2009), and to the evolution of the biosphere.

Of all the biologically important gases emitted by volcanoes, N_2 is the only constituent whose near-surface chemistry is similar to that of the rare gases. It has accumulated in the atmosphere largely because all of the common compounds of NO_3^-, NO_2^-, and NH_4^+ are highly soluble (Chapter 16). By contrast, the removal of carbon from the near-surface reservoirs is and has been speeded considerably by the precipitation of $CaCO_3$ and $CaMg(CO_3)_2$. Similarly the removal of sulfur has been speeded by the precipitation of FeS_2 and $CaSO_4$. These minerals and residual organic matter, together with the terrestrial detritus delivered to the oceans and the marine volcanic inputs, have combined to form most of the large variety of marine sediments and their diagenetic products, the sedimentary rocks (Chapter 17). The metamorphism and the melting of sedimentary rocks in subduction zones have closed the geochemical cycle. They have been accompanied by the release of volatiles. These, together with the primordial volatiles, have been the major volatile volcanic and hydrothermal inputs to the Earth surface systems (Chapter 18). The members of the biosphere have both responded to and molded their environment. The molding has become particularly intense since the arrival of mankind.

                                                                                                H.D. Holland

REFERENCES

Barth T. W. F. (1952) *Theoretical Petrology*. New York, NY: Wiley.

Broecker W. S., Gerard R. D., Ewing M., and Heezen G. C. (1961) Geochemistry and physics of ocean circulation. In: Sears M. (ed.) *Oceanography*, pp. 301–322. Washington, DC: American Association for the Advancement of Science, Publication No. 67.

Garrels R. M., Mackenzie F. T., and Hunt C. (1975) *Chemical Cycles and the Global Environment: Assessing Human Influences*. Los Altos, CA: William Kaufmann.

Goldberg E. D. (1961) Chemistry in the oceans. In: Sears M. (ed.) *Oceanography*, pp. 583–597. Washington, DC: American Association for the Advancement of Science, Publication No. 67.

Goldberg E. D. and Arrhenius G. O. S. (1958) Chemistry of Pacific pelagic sediments. *Geochimica et Cosmochimica Acta* **13**, 153–212.

Holland H. D. (1978) *The Chemistry of the Atmosphere and Oceans*. New York, NY: Wiley-Interscience.

Holland H. D. (2009) Why the atmosphere became oxygenated: A proposal. *Geochimica et Cosmochimica Acta* **73**, 5241–5255.

Maxwell J. C. (1868) On Governors. *Proceedings of the Royal Society (London)* **16**, 270–283.

Wiener N. (1948) *Cybernetics, or Control and Communication in the Animal and the Machine*. Cambridge: The MIT Press.

CONTRIBUTORS

R. Amundson
University of California, Berkeley, CA, USA

P. Brimblecombe
University of East Anglia, Norwich, UK

W. S. Broecker
Columbia University, Palisades, NY, USA

C. L. De La Rocha
Alfred Wegener Institute for Polar and Marine Research, Bremerhaven, Germany

B. Dupré
Laboratoire des Mécanismes et Transferts en Géologie, Toulouse, France

T. I. Eglinton
Woods Hole Oceanographic Institution, MA, USA

J. Gaillardet
Institut de Physique du Globe de Paris, France

J. N. Galloway
University of Virginia, Charlottesville, VA, USA

C. R. German
Southampton Oceanography Centre, Southampton, UK

G. H. Haug
Geoforschungszentrum Potsdam, Germany

R. A. Houghton
Woods Hole Research Center, MA, USA

F. T. Mackenzie
University of Hawaii, Honolulu, HI, USA

M. Meybeck
University of Paris VI, CNRS, Paris, France

C. Oppenheimer
University of Cambridge, UK

S. T. Petsch
University of Massachusetts, Amherst, MA, USA

S. Poli
Universita di Milano, Italy

D. Porcelli
University of Oxford, UK

D. J. Repeta
Woods Hole Oceanographic Institution, MA, USA

M. W. Schmidt
ETH Zürich, Switzerland

D. M. Sigman
Princeton University, NJ, USA

E. T. Sundquist
US Geological Survey, Woods Hole, MA, USA

K. K. Turekian
Yale University, New Haven, CT, USA

J. Veizer
Ruhr University, Bochum, Germany and
University of Ottawa, ON, Canada

J. Viers
Laboratoire des Mécanismes et Transferts en Géologie, Toulouse, France

K. Visser
US Geological Survey, Woods Hole, MA, USA

K. L. Von Damm*
University of New Hampshire, Durham, NH, USA

A. F. White
US Geological Survey, Menlo Park, CA, USA

*Deceased

1

Volcanic Degassing

C. Oppenheimer

University of Cambridge, UK

NOMENCLATURE		2
1.1 INTRODUCTION		2
	1.1.1 Earth Outgassing, Atmospheric Evolution and Global Climate	2
	1.1.2 Magma Evolution and Dynamics, and Volcanic Eruptions	4
	1.1.3 Volcanic Hazards and Volcano Monitoring	4
1.2 ORIGIN, SPECIATION, AND ABUNDANCE OF VOLATILES		6
	1.2.1 Sources and Abundances of Volatiles in Magmas	6
	1.2.2 Solubility and Speciation of Volatiles	8
1.3 DEGASSING		9
	1.3.1 Saturation	9
	1.3.2 Supersaturation and Nucleation	10
	1.3.3 Bubble Growth and Magma Ascent	10
	1.3.4 Bubble Coalescence	10
	1.3.5 Gas Separation	11
	1.3.6 Fragmentation	12
	1.3.7 Excess Degassing	12
1.4 EMISSIONS		13
	1.4.1 Styles of Surface Emissions	13
	1.4.1.1 Noneruptive emissions	14
	1.4.1.2 Eruptive emissions	16
	1.4.2 Chemical Composition of Volcanic Gases	16
	1.4.3 Measurement of Volatiles	16
	1.4.3.1 In situ sampling and analysis	17
	1.4.3.2 Portable remote sensing systems	18
	1.4.3.3 Satellite remote sensing	19
	1.4.3.4 Petrological methods	19
	1.4.3.5 Ice cores	20
	1.4.3.6 Application of geochemical surveillance to volcano monitoring	21
1.5 FLUXES OF VOLCANIC VOLATILES TO THE ATMOSPHERE		23
	1.5.1 Sulfur	24
	1.5.2 Carbon and Water	26
	1.5.3 Halogens	26
	1.5.4 Trace Metals	27
1.6 IMPACTS		27
	1.6.1 Stratospheric Chemistry and Radiative Impacts of Volcanic Plumes	28
	1.6.1.1 Formation of stratospheric sulfate aerosol veil	28
	1.6.1.2 Impacts on ozone chemistry	28
	1.6.1.3 Optical and radiative effects	30
	1.6.2 Climatic Impacts of Major Volcanic Eruptions	31
	1.6.2.1 Intermediate to silicic eruptions	31
	1.6.2.2 Mafic eruptions	33

1.6.3   *Tropospheric Chemistry of Volcanic Plumes*   33
1.6.4   *Impacts of Volcanic Volatiles on Vegetation and Soils*   34
1.6.5   *Impacts of Volcanic Pollution on Animal and Human Health*   35
    1.6.5.1   *Gaseous and particulate sulfur and air quality*   35
    1.6.5.2   *Fluorine*   35
    1.6.5.3   *Carbon dioxide*   35
1.7   CONCLUSIONS AND FUTURE DIRECTIONS   36
ACKNOWLEDGMENTS   36
REFERENCES   37

NOMENCLATURE

r_{eff}	effective aerosol radius
D	diffusion coefficient of given volatile in melt
I	solar irradiance at given level in atmosphere
I_o	solar irradiance at top of the atmosphere
K_R	constant in vesicularity equation (2)
N	constant in solubility equation (1)
P	pressure
Pe	Peclet number
R	bubble radius
s	constant in solubility equation (1)
V_g/V_1	ratio between the volumes of gas and melt at given pressure
X_{H_2O}	water solubility in melt
$X_{H_2O}^r$	residual melt water content at given pressure
$X_{H_2O}^0$	initial melt water content at saturation pressure
η	melt dynamic viscosity
τ	aerosol optical depth
τ_d	timescale of volatile diffusion
τ_η	timescale of viscous relaxation
ΔP	oversaturation pressure
Θ	solar zenith angle

Nature only reveals her secrets if we ask the right questions and listen, and listening in geochemistry means sampling, analyzing, plotting. (Werner Giggenbach, 1992b)

1.1   INTRODUCTION

Humans have long marveled at the odorous and colorful manifestations of volcanic emissions, and, in some cases, have harnessed them for their economic value (Figure 1). Moreover, the degassing of magma that is responsible for them is one of the key processes influencing the timing and nature of volcanic eruptions, and the emissions of these volatiles to the atmosphere can have profound effects on the atmospheric and terrestrial environment, and climate, at timescales ranging from a few years to >1 Myr, and spatial scales from local to global (Oppenheimer *et al.*, 2003a). Even more fundamental are the relationships between the history of planetary outgassing, differentiation of the Earth's interior, chemistry of the atmosphere and hydrosphere, and the origin and evolution of life (e.g., Kelley *et al.*, 2002).

This chapter focuses on the origins, composition and flux, and the environmental impacts of volcanic volatile emissions. This introductory section sets the scene by considering the general context and significance of volcanic degassing. Several chapters in this volume interface with this one on volcanic degassing, and in particular the reader is referred to the chapters on hydrothermal systems and ore formation.

1.1.1   Earth Outgassing, Atmospheric Evolution and Global Climate

Volcanic emissions have occurred throughout Earth history, and have provided the inventory of volatile elements that take part in the major geochemical cycles involving the lithosphere, hydrosphere, atmosphere, and biosphere (Sections 1.2 and 1.5; Holland, 1984; Arthur, 2000; see Chapter 5). The mantle is an important reservoir for volatiles, and its concentration of carbon, sulfur, hydrogen, oxygen, and halogens has changed through Earth history as a result of differentiation. Anhydrous minerals such as olivine, pyroxene, and garnet can hold structurally bound OH^-, while molecular water is present in amphibole, phologopite, and apatite. Carbon is present in carbonate minerals or in elemental form (e.g., diamond and graphite), and sulfur in sulfide minerals. Volatiles probably also exist in the mantle in intergranular films. Since the volatile species are incompatible, they partition into the melt phase during partial melting of the mantle. In this way, magmagenesis plays a key role in transferring volatiles between the mantle and the crust. Magma evolution then partitions volatiles between the crust and the atmosphere/hydrosphere via degassing and eruption (with important feedbacks on magma differentiation), and plate recycling ensures a return flux of a proportion of the volatiles back to the mantle.

Figure 1 "The burning valley called Vulcan's Cave near Naples" or Solfatara (Campi Flegrei), from Bankes's New System of Geography ($\sim$1800).

Major changes have occurred in atmospheric composition and in greenhouse gas forcing over Earth history, in part coupled to the evolution of life, and interacting with changes in solar flux and planetary albedo to control global climate. Over timescales exceeding 1 Myr, the carbon cycle operates as a climate thermostat on the Earth. The Archean atmosphere was anoxic, even after the onset of oxygenic photosynthesis $\sim$2.7 Gyr ago. This has been attributed to an excess of reductants (e.g., CH_4 and H_2) able to scavenge out O_2 accumulating in the atmosphere (Holland, 2002; Catling and Kasting, 2003). Bacterial consumption of hydrogen also contributed methane, such that the Archean atmosphere may have contained as much as hundreds or thousands of ppm of CH_4 (Catling *et al.*, 2001). The greenhouse forcing associated with these high concentrations of atmospheric methane may account for the absence of prolonged freezing of the Earth that should otherwise have arisen from the reduced solar luminosity (30% lower than today) early in the Earths history (the "early faint Sun paradox"; Sagan and Chyba (1997)).

Hydrogen escape from the Earth's atmosphere has been proposed as a mechanism for inexorable oxidation of the mantle, tilting the redox balance of volcanic emissions today in favor of oxidized gases (Kasting *et al.*, 1993; Catling and Kasting, 2003). If correct, this shifting redox balance could

have led to the rise of $O_2 \sim$ 2.3 Gyr ago (Catling *et al.*, 2001). However, the mechanisms for oxygenation resulting from methane-induced hydrogen escape remain controversial (e.g., Towe, 2002). Furthermore, geochemical evidence argues strongly against any substantial increase in the oxidation state of the mantle over the past 3.5 Gyr—the Archean mantle was already oxidized (Delano, 2001; Canil, 2002).

Major changes in atmospheric composition have, in turn, been held responsible for so-called "snowball Earth" events, characterized by low-latitude glaciation and effective shutdown of the hydrological cycle, $\sim$2.3 Gyr (coinciding with the rise of oxygen), and 750 Myr and 600 Myr ago (e.g., Kirschvink, 1992; Hoffman *et al.*, 1998; Kirschvink *et al.*, 2000; Hoffman and Schrag, 2002). Loss of CH_4 accompanying the rise of oxygen could explain the onset of the glacial conditions that characterize these events, with the "snowball" being ultimately thawed by the action of the carbonate–silicate inorganic cycle: under the glacial conditions, weathering rates would fall and CO_2 from volcanism would accumulate in the atmosphere, providing a long-term climatic recovery via greenhouse feedbacks. Between the Paleoproterozoic and Neoproterozoic "snowball" events, climate was warm and very stable, possibly due to methane emissions from biogenic sources (e.g., Catling *et al.*, 2001; Catling and Kasting, 2003).

On short timescales, years to decades, possibly centuries, individual volcanic eruptions, or conceivably a burst of eruptions from several volcanoes, are capable of perturbing global climate by the release of sulfur gases into the upper atmosphere. These oxidize to form a veil of minute sulfuric acid particles that can girdle the globe at heights of 20 km or more, scattering back into space some of the sunlight that would ordinarily penetrate the tropopause, and heat the lower atmosphere and Earth's surface. The complex spatial and temporal patterns of radiative forcing that result are reflected in regional climate change, and globally averaged surface temperature decreases. The 1991 eruption of Mt. Pinatubo in the Philippines has, as of early 2000s, provided the clearest evidence of these interactions, and an outstanding opportunity for climatologists to test both radiative and dynamical aspects of general circulation models (Sections 1.6.1 and 1.6.2). Such work has contributed significantly to the development of our understanding of the importance of aerosols in climate change, and has helped to unravel the natural variability from anthropogenic forcing.

1.1.2 Magma Evolution and Dynamics, and Volcanic Eruptions

The partition of magmatic volatiles from the melt to the gas phase, and their subsequent separation—collectively referred to as degassing—exert fundamental controls on magma overpressure, viscosity and density, and thereby on the chemical evolution, storage, and transport (notably ascent rate) of magmas, and the style, magnitude, and duration of eruptions (Section 1.3; Huppert and Woods, 2002). Viscosity is a critical factor and is strongly dependent on temperature, melt composition (especially silica content (Figure 2(a)) and dissolved volatile content (Figure 2(b)), which control polymerization of the melt), applied stress, bubble fraction, and crystal content. It is also weakly dependent on pressure, with a general trend of decreasing viscosity at higher pressure. Kinetic factors, including cooling rate and diffusion, are also important.

The expansion of bubbles of gas accompanying decompression of magmas plays a key role in over-pressuring of magmatic systems, and hence in eruption triggering. It also provides the energy capable of propelling pyroclasts several kilometers into the atmosphere. For example, from the ideal gas law, 5 wt.% of dissolved water in 1 m^3 of melt would occupy $\sim$700 m^3 in the vapor phase at atmospheric pressure. Such a volatile rich magma may experience explosive fragmentation on eruption if the exsolving gas cannot escape fast enough. Alternatively, if the gas is able to segregate with relative ease from the melt prior to, or

during, eruption, as is generally the case with lower viscosity magmas, then lava lakes or flows are more likely to form, accompanied perhaps by strombolian or hawaiian style activity. Degassing can result in orders of magnitude changes in magma viscosity (Figure 2(b)), because: (i) the presence of bubbles has a strong effect on rheology (that is difficult to model because of steep viscosity gradients in the melt close to bubble walls, and changing bubble size distributions); (ii) it may induce crystallization; and (iii) dissolved water helps to polymerize silicate melts (Lange, 1994). Degassing, therefore, strongly influences rheological behavior, and can initiate feedbacks between magma dynamics and eruptive style that have been documented at several volcanoes (e.g., Voight et al., 1999).

Vesiculation also has a profound effect on the permeability of magmas, and can reach the point where bubbles are sufficiently interconnected to permit gas loss from deeper to shallower levels in a conduit, or through the conduit walls. In this way, even highly viscous silicic and intermediate magmas can degas non-explosively, as seen in the case of lava dome eruptions such as that of Soufrière Hills Volcano (Montserrat). Understanding the rheological consequences of bubble growth, crystallization, magma ascent, and gas loss is clearly critical to understanding volcanic behavior.

1.1.3 Volcanic Hazards and Volcano Monitoring

Volcanic gases have been described as "telegrams from the Earth's interior" (Matsuo, 1975)—if the messages can be intercepted and interpreted, they can be used to aid in forecasting and prediction of volcanic activity. To this end, many investigators have developed and applied methods to measure the gas mixtures and emission rates from volcanoes, and to understand what controls these parameters (Section 1.4). While volcano surveillance efforts still largely concentrate on seismological and geodetic approaches, gas geochemistry is widely recognized as an important and highly desirable component of multidisciplinary monitoring, and increasingly sophisticated remote sensing techniques are becoming available to measure volcanic volatile emissions. These include a range of spectroscopic instruments that can be deployed from the ground, aircraft, and satellites. Improved forecasting of volcano behavior offers immediate benefits in mitigating the risks of eruptions to society.

In addition to offering a means for assessing eruption hazards, volcanic gas and aerosol emissions can pose a direct hazard (Sections 1.6.3–1.6.5). At local and regional scales, several gas species and aerosols (including sulfate and fine

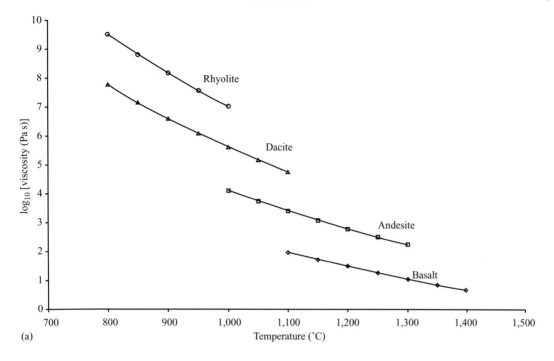

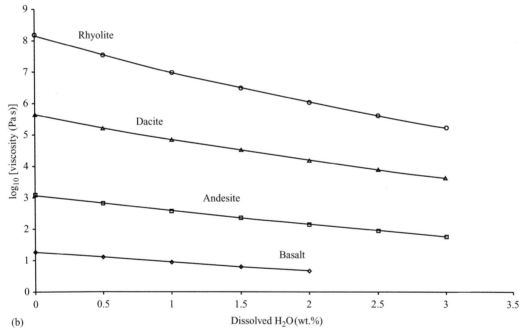

Figure 2 Viscosities at 1 bar pressure for natural melts as a function of: (a) temperature and (b) dissolved water content based on model of Shaw (1972) (see Spera (2000) for a review).

ash) can affect the health of humans and animals. In 1984 and 1986, catastrophic releases of CO_2 from two volcanic crater lakes in Cameroon claimed several thousand lives in nearby villages (Le Guern and Tazieff, 1989). Acid gas species and their aerosol products can be transported over ranges of hundreds or thousands of kilometers, and can have various impacts on respiratory and cardiovascular health. Fluorine deposited to the ground on ash has been responsible for large losses of grazing animals during a number of eruptions (e.g., Cronin *et al.*, 2003). Emissions of radioactive species, principally radon, have also given rise to speculation over potential health risks. Several components of volcanic emissions, including heavy metals as well as the acid species, can damage vegetation, though in some cases, leachable components of ashfalls can provide beneficial nutrients to terrestrial and aquatic ecosystems.

1.2 ORIGIN, SPECIATION, AND ABUNDANCE OF VOLATILES

"Next to nothing is known about the sources of the volatile components of magmas or how they are distributed and transported between the mantle and shallow levels of the crust." This is how Williams and McBirney (1979) open the chapter on volcanic gases in their influential textbook *Volcanology*. Though there is much still to learn, thanks to recent investigations spanning experimental phase petrology, analysis of glass inclusions, isotope geochemistry, thermodynamical and fluid dynamical modeling, and satellite and ground-based remote sensing, this pessimistic view can be replaced today by a much more encouraging outlook of our understanding of the origins, emissions, and impacts of volcanic gases (see Carroll and Holloway, 1994). The benefits of this are significant, not least because of the fundamental relationships between volcanic degassing, crystallization, magma dynamics, eruption style, mineral deposition, and atmospheric and climate change. This section reviews the sources and distribution of volatiles in magmas. The means by which they escape to the atmosphere are then explored in Section 1.3.

1.2.1 Sources and Abundances of Volatiles in Magmas

The origin of magmatic volatiles lies in the Earth's interior but, as a result of global tectonics, various species can be recycled through the lithosphere–hydrosphere–atmosphere system, and sourced more recently from the mantle, subducted slab, crust or hydrosphere. Interactions between magma bodies and hydrothermal fluids (with meteoric and/or seawater contributions) can also contribute to the complement of volatiles in magmas. The relative contributions from these different sources for a given volatile species are difficult to ascertain, for an individual volcano, for an arc, or for the whole Earth, but general trends exist between magmatic volatile content and tectonic environment (Table 1). These reflect magmatic differentiation in the crust, and concentration of incompatible volatiles in the melt, as well as the contributions of slab, depleted vs. undepleted mantle, and crustal assimilation (e.g., decarbonation and dehydration of crustal rocks).

The principal volatile species in magma is water, usually followed in abundance by carbon dioxide, hydrogen sulfide, sulfur dioxide, hydrogen chloride, and hydrogen (Section 1.4.2). Many other trace species have been identified in high-temperature volcanic gas discharges, including heavy metals. Analyses of trace elements and isotopic composition of volcanic rocks and fluids have provided some of the clearest indications of the source of volatiles in magmas. In particular, the emissions of arc volcanoes are thought to derive the larger part of their volatile budgets from the subducted slab, either through dehydration reactions occurring in the altered oceanic crust, release of pore water contained in sediments, or dehydration of pelagic clay minerals.

The abundance of slab-derived components is partly revealed by straightforward comparisons of N_2/He and CO_2/He ratios. For nonarc basalts, these are typically in the range 0–200 and $(2–6) \times 10^4$, respectively, whereas andesites are characterized by ranges of 1000–10,000 and $(6.5–9) \times 10^4$, respectively (Giggenbach, 1992a, b). Of the magmatic component in andesitic fluids, more than 95% is estimated to be sourced from the slab. Evidence for this is provided by oxygen-isotopic composition. Giggenbach (1992a) has promoted the idea of a common "andesitic water" feeding arc volcanoes because of characteristic $\delta^{18}O$ (around +10‰) values. The consistency in isotopic composition strongly suggests that this fluid is ultimately derived from seawater, and Giggenbach identified water released from pelagic sediments as the likely dominant source, though others believe that altered oceanic crust provides more water. According to his view, "arc magmatism is effectively driven by the inability of the mantle to accommodate the subducted water. Andesitic magmas represent simply the most suitable vehicle for the 'unwelcome' water to travel through the crust back to the surface, andesitic volcanoes 'vent holes' for excess subducted volatiles" (Giggenbach, 1992a). In other words, he suggested that most of the water dissolved in andesitic magmas derived from recycled seawater, along with high proportions of carbon, nitrogen, and chlorine (see Giggenbach, 1992a, figure 4). It is worth qualifying, though, that most andesites are thought to be the products of differentiation in the crust of basalts generated by partial melting of the mantle wedge.

Further work to characterize volatile sources has been carried out by Fischer *et al.* (1998). They analyzed carbon-, helium-, and nitrogen-isotopic ratios in emissions from Kudriavy, a basaltic andesite volcano in the Kurile arc (Table 2). Combined with measurements of sulfur dioxide flux from the volcano (see Section 1.4.3.2 for details on how this can be done), they estimated that 84% of the 2,700 mol yr^{-1} flux of helium was mantle derived (the remainder being radiogenic, and derived from the crust), and only 12% of the CO_2, and almost none of the N_2, was sourced from the mantle. For CO_2, they estimated

Table 1 Summary of typical pre-eruptive volatile abundances of magmas in different tectonic settings. Note that volatiles can be contained in solution, in crystallized phases, and in a separate vapor phase. The values reported in the table are generally for the melt phase (dissolved) though true bulk volatile abundances for a magma would sum all these potential reservoirs. In particular, some magmas show abundant evidence for a substantial fluid phase co-existing with the melt prior to eruption, which is often not represented in melt inclusion-based estimates of pre-eruptive volatiles.[a]

MORB (e.g., Mid Atlantic Ridge, East Pacific Rise)

H_2O	<0.4–0.5 wt.%, typically 0.1–0.2 wt.%; enriched mid-ocean-ridge basalt (E-MORB) up to 1.5 wt.%
CO_2	50–400 ppm; typically saturated at eruption (gas phase almost pure CO_2) leading to vesiculation
S	800–1,500 ppm; immiscible Fe–S–O liquids indicate saturation at eruption
Cl	20–50 ppm in most primitive MORB, occasionally much higher due to assimilation of hydrothermally altered rocks
F	100–600 ppm

Ocean island basalts (OIB) (e.g., Kīlauea, Galapagos, Réunion)

H_2O	0.2–1 wt.%; e.g., Hawai'i 0.4–0.9 wt.% (Dixon *et al.*, 1991)
CO_2	2,000–6,500 ppm for Hawai'i
S	up to 3,000 ppm; 200–1,900 ppm for Hawai'i
Cl	comparable to MORB; Kīlauea estimate around 90 ppm
F	comparable to MORB; Kīlauea estimate around 35 ppm

Arc basalt (island arc basalts and continental margin basalts, e.g., Cerro Negro, Marianas)

H_2O	up to 4–6 wt.% (e.g., see Roggensack *et al.* (1997), on Cerro Negro), largely sourced from subducted slab; crustal assimilation another potential source, especially for arcs built on continental crust

Back arc basin basalt (BABB) (e.g., Lau Basin)

H_2O	1–3 wt.%; generally speaking, intermediate between MORB and island arc basalts

Andesites (e.g., Soufrière Hills Volcano)

Note that it is particularly difficult to quantify pre-eruptive volatile contents of andesites because most are erupted subaerially (i.e., at atmospheric pressure) after significant degassing has taken place, and contain abundant phenocrysts (e.g., >30 wt.%) such that liquid compositions are more silicic (often rhyolitic) than bulk rock. Also, good host minerals for melt inclusions (e.g., olivine and quartz) are rare, and mineral disequilibria hamper experimental work.

H_2O	>3 wt.%
CO_2	10–1,200 ppm
S	<1,000 ppm; typically 200–400 ppm
Cl	can be high, e.g., 1,500 ppm not unusual; (5,000 ppm or more in phonolites)
F	<500 ppm

Dacites and rhyolites (e.g., Mount St. Helens 1980, Pinatubo, 1991, Bishop Tuff)

H_2O	typically 3–7 wt.%; e.g., 4.6 wt.% dissolved for Mount St. Helens 1980, 6–7 wt.% for Pinatubo 1991; there is strong evidence for vertical gradients in both dissolved and exsolved H_2O and CO_2 in pre-eruptive magmas (e.g., Wallace *et al.*, 1995; Wallace, 2001)
CO_2	often below detection limits
S	typically <200 ppm (75 ppm for Pinatubo, 1991) but melt often saturated with sulfide (pyrrhotite) or sulfate (anhydrite) crystalline phases
Cl	600–2,700 ppm in metaluminous dacites and rhyolites; 6,700 ppm in trachytes; 9,000 ppm in peralkaline rhyolites (pantellerites); 1,100 ppm for Pinatubo 1991
F	200–1,500 ppm in metaluminous dacites and rhyolites; up to 1.5 wt.% in peralkaline rhyolites (pantellerites)

---

[a] Assimilated from the detailed reviews of Johnson *et al.* (1993) and Wallace and Anderson (2000). See also Signorelli and Carroll (2000, 2002) for Cl solubility data.

Table 2 Percentage (molar) contribution to estimated fluxes of volatiles to the atmosphere from the high temperature (up to 920 °C) fumaroles of Kudriavy volcano, Kuril islands.

	Mantle	*Slab inorganic*	*Slab organic*	*Crust*	*Flux* (mol yr^{-1})
CO_2	12%	67%	21%	0	4.32×10^8
^3He	100%	0	0	0	2.5×10^{-2}
^4He	84%	0	0	16%	2.7×10^3
N_2	2%	0	98%	0	5.1×10^6
SO_2					3.47×10^8
HCl					1.06×10^8

Source: Fischer *et al.* (1998).

that most (67%) was derived from inorganic carbonate sediments and hydrothermal veins in altered oceanic crust in the slab. Subducted organic nitrogen can account for all the observed nitrogen flux. More recently, Fischer *et al.* (2002) have examined the fate of nitrogen subducted along different segments of the Central American volcanic arc. They demonstrated efficient transfer of shallow marine sedimentary nitrogen carried in subduction zones back to the atmosphere via arc volcanism, thereby limiting recycling of nitrogen to the deep mantle.

1.2.2 Solubility and Speciation of Volatiles

Fundamental controls on the solubility of volatiles in melt include pressure and temperature, and the presence of nonvolatile phases (in the case of sulfur, these could be sulfates or sulfides; in the case of chlorine, metal chlorides). As a first approximation, the solubility of water in silicate melt, X_{H_2O}, is roughly proportional to the square root of pressure, P (Burnham, 1979):

$$X_{H_2O} = nP^s \qquad (1)$$

where n and s for water have values of around 0.34 and 0.54, respectively.

Analytical work on natural and synthetic melts has helped to establish an understanding of the speciation of volatiles in both the melt and vapor phase (e.g., Stolper, 1982, 1989; Silver *et al.*,

1990; Ihinger *et al.*, 1999). Dissolved water exists in the melt in the form of OH^- groups or as H_2O molecules that are structurally bound to the aluminosilicate network of the melt (McMilan, 1994). Likewise, CO_2 dissolves as molecular CO_2 and CO_3^{2-} (Fine and Stolper, 1985; Blank and Brooker, 1994). Speciation is a function of structure of the silicate melt and oxidation state of the magma (availability of cations). In experiments, OH^- reacts strongly with the silicate melt, lowering viscosity. Molecular water is less reactive and does not disrupt the polymerization of a melt as much. Water speciation thereby has an important effect on melt viscosity. The interactions of multiple phases (e.g., CO_2 and H_2O) are not well understood but are important (Papale, 1999, Papale and Polacci, 1999; Moretti *et al.*, 2003). Newman and Lowenstern (2002) have developed a simple-to-use code to model H_2O–CO_2-melt equilibrium for rhyolite and basalt systems (Figure 3).

Sulfur solubility is complicated due to multiple valancies including S_2, H_2S, SO_2, SO_3 (in the gas phase), and nonvolatile solid phases (e.g., pyrrhotite and anhydrite) or liquid phases. Redox equilibria, therefore, play a major role in determining speciation of sulfur. Predominant phases, though, are thought to be H_2S and SO_2 in the gas phase, and S^{2-} and SO_4^{2-} in the melt (Carroll and Webster, 1994). Métrich *et al.* (2002) have shown that sulfite (SO_3^{2-}) is also an important species in arc basalt melts.

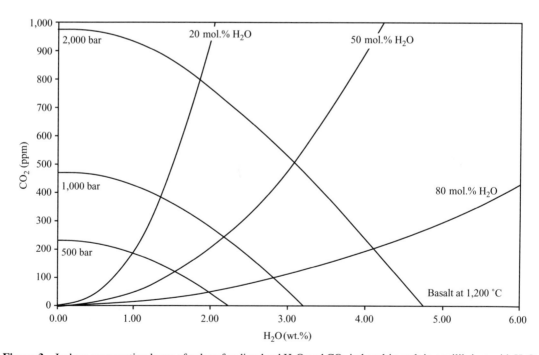

Figure 3 Isobars representing locus of values for dissolved H_2O and CO_2 in basaltic melt in equilibrium with H_2O–CO_2 vapor at 1,200 °C and selected pressures. Similarly, isopleths represent locus of basaltic melt compositions in equilibrium with given vapor compositions (20 mol.%, 50 mol.%, and 80 mol.% H_2O) at 1,200 °C (source Newman and Lowenstern, 2002). See also: http://wrgis.wr.usgs.gov/docs/geologic/jlwnstrn/other/software_jbl.html.

Fluorine and chlorine speciation and solubility in magmas are complex and interdependent to a degree, though some general processes are recognized, including an inverse relationship between chlorine solubility and pressure in water-saturated systems, direct proportionality of phosphorus and chlorine solubility in water-poor, brine saturated systems, and a typically strong dependence of chlorine solubility on melt structure and composition (Carroll and Webster, 1994; Signorelli and Carroll, 2000, 2002; Webster et al., 1999, 2001). Further experimental studies on speciation are required in order to develop predictive models of halogen solubility as a function of melt composition and physical parameters. These could prove highly fruitful in the interpretation of observed halogen content through time in volcanic gas emissions, with applications in volcano monitoring and hazard assessment (Villemant et al., 2003).

1.3 DEGASSING

Volcanic degassing begins with the expulsion or *exsolution* of volatiles from the melt and formation of gas bubbles. This is the first stage in *vesiculation*; the subsequent stages of bubble evolution are described in Sections 1.3.3–1.3.6. Vesiculated magma has very different physical properties to bubble-free melt, and exsolution can provide the trigger for magma intrusion and eruption. The term "degassing" refers to vesiculation, followed by separation of the gas phase from the melt. This section examines, in essence, how volatiles segregate from magma and ultimately reach the atmosphere. This field is of tremendous importance in linking eruptive behavior to magma dynamics (e.g., Eichelberger et al., 1986; Eichelberger, 1995; Jaupart and Allègre, 1991; Sparks, 1997), and throws up many puzzles, such as the "excess sulfur" conundrum (Section 1.3.7).

1.3.1 Saturation

Because the pressure, temperature, and composition (redox state) of the magma change through time, volatiles in solution may reach and exceed their saturation points and enter the vapor phase. In the broadest terms, H_2O and CO_2 exsolution are effectively controlled by the phase diagram, and trace elements by partition coefficients. Because of its low solubility (roughly two orders of magnitude less than H_2O), CO_2 plays an important role in vapor saturation at crustal pressures (Anderson, 1975; Newman and Lowenstern, 2002). Formation of a CO_2- and H_2O-rich vapor phase in a magma body can result in strong partitioning of other species, including sulfur and halogens, into it. For example, Keppler (1999) showed from experimentally determined fluid–melt partition coefficients for sulfur that development and accumulation of a hydrous vapor phase could efficiently extract most of the dissolved sulfur in the melt. He went on to suggest that this could explain the high sulfur yield of the 1991 eruption of Mt. Pinatubo (though his experimental work was on a calcium- and iron-free haplogranite melt rather than actual Pinatubo dacite; Scaillet et al. (1998) and (Scaillet and Pichavant, 2003).

Pressure exerts a first-order control on the solubility of volatile species (Equation (1)) because of the great increase in molar volume going from the melt to vapor phase. Therefore, decompression, which can accompany magma ascent, or failure of confining rock, is a key process leading to saturation of volatiles and exsolution. Decompression also promotes undercooling, driving crystallization, and exsolution.

Magmatic systems have sufficiently long lifetimes that magma mixing occurs between more and less evolved batches of melt. Assimilation of the country rocks that confine magma reservoirs may also provide a source of volatiles. Juxtaposition, mingling, or mixing of different magmas can induce degassing by supplying fresh volatiles, e.g., from a mafic intrusion into an intermediate or silicic magma reservoir. When a dense, mafic magma is injected into an intermediate or silicic reservoir, the induced cooling and crystallization of the mafic melt causes volatile exsolution, which lowers its bulk density (due to the presence of bubbles), potentially leading to overturn of the mafic and silicic magmas. Alternatively, the bubbles may rise and form a foam at the interface between the two magmas. What actually happens will depend on the viscosity of the mafic magma with respect to the ascent rate of the bubbles forming within it (Hammer and Rutherford, 2002).

Magma mixing may also promote degassing by affecting the stability of volatile species in both the melt and solid phase. For example, Kress (1997) invoked mixing of comparatively reduced, sulfide-bearing basalt with oxidized, anhydrite-bearing dacitic magma as responsible for sulfur degassing prior to the 1991 Pinatubo eruption. Sulfur solubility reaches a minimum at the precise oxidation state that he calculated to result from mixing these two end-members, promoting exsolution. In addition, he showed that anhydrite and pyrrhotite stability would decrease, also liberating sulfur into the vapor phase. This could have induced magma ascent, decompression, and further degassing, leading to eruption. His argument provides a mechanism for exsolving the sulfur-rich vapor phase that Westrich and Gerlach (1992) and Wallace and Gerlach (1994), among others, suggested had formed prior to the eruption.

Crystallization (of anhydrous or water-poor phases) due to decompression or substantial cooling increases the fraction of volatiles dissolved in melts. This can also lead to oversaturation and exsolution, a process termed "second boiling." In the case of closed-system, isobaric crystallization of a volatile-saturated silicic magma, because CO_2 is less soluble than H_2O, it will preferentially exsolve. Residual melts formed during progressive crystallization, therefore, contain decreasing amounts of dissolved CO_2 and increasing amounts of H_2O. Degassing and crystallization are, therefore, very closely linked—the one can induce the other—and both have profound effects on the viscosity of magma. As the crystal fraction increases to ~40 vol.%, bulk viscosity also increases dramatically. In the case of erupting intermediate and silicic magmas, extensive microlite crystallization within the eruption conduit (e.g., Cashman, 1992; Hammer *et al.*, 1999; Blundy and Cashman, 2001) is thought to play an important role in magma dynamics by dramatically increasing bulk viscosity (Lejeune and Richet, 1995; Sparks, 1997), thereby strongly influencing rheology and transport.

1.3.2   Supersaturation and Nucleation

When the melt becomes saturated or oversaturated in a given volatile phase in solution, the subsequent fate of the phase in question depends on the kinetics of bubble nucleation, and the growth of bubbles. Because there is an energy cost in creating a bubble–melt interface, a degree of supersaturation in the melt is typically required to overcome it. The energy gained by moving the volatile into the new gas bubble is proportional to the volume of the bubble nucleus. This means that there is a critical size of nucleus for which the energy terms balance. Above this size, bubble nucleation is spontaneous. Since water is the principal volatile species, considerable work has gone into understanding its diffusion in silicate melts (see Watson, 1994).

Much of the theoretical work on bubble formation has considered homogeneous nucleation but, in real magmas, the presence of crystalline phases can be important in providing nucleation sites for bubbles, with both the sizes and compositions of those phases playing a role in the exsolution (Navon and Lyakhovsky, 1998). Because the surface energy associated with a gas–crystal interface can be lower than a gas–melt interface, the degree of oversaturation required for heterogeneous nucleation can be much lower than for homogeneous nucleation. Pre-existing bubbles and other inhomogeneities in the fluid may also play a role in promoting heterogeneous nucleation. For arc magmas, which appear to be mostly fluid saturated prior to eruption, the importance of heterogeneous versus homogeneous nucleation for eruption regimes is probably limited.

1.3.3   Bubble Growth and Magma Ascent

The growth of bubbles is controlled by the rates at which volatiles in the melt can diffuse towards the bubbles, and the opposing viscous forces. Near a bubble, volatiles are depleted such that melt viscosity increases dramatically, and diffusivities drop, making it harder for volatiles to diffuse through and grow the bubble. These opposing factors are described by the nondimensional Peclet number (*Pe*), which is the ratio of the characteristic timescales of volatile diffusion ($\tau_d = r^2/D$, where r is the bubble radius and D the diffusion coefficient of the volatile in the melt) and of viscous relaxation ($\tau_\eta = \eta/\Delta P$ where η is the melt dynamic viscosity and ΔP the oversaturation pressure, i.e., $Pe = \tau_d/\tau_\eta$; Dingwell (1998) and Navon and Lyakhovsky (1998)). For $Pe > 1$ (i.e., mafic melts) bubble growth is controlled by volatile diffusion, and for $Pe \leq 1$ (i.e., viscous intermediate and silicic melts), viscous resistance dominates, resulting in bubble overpressure. The higher viscosity of the immediate shell of magma surrounding bubbles can impart a significant control on τ_η and should be taken into account when modeling conduit flow dynamics. The coupling between viscosity and diffusivity in melts has also been explored in detail by Lensky *et al.* (2001, 2002) and Blower *et al.* (2001). When magma is rising, growth models for bubbles also have to consider the falling pressure, adding a decompression timescale to bubble nucleation (Lensky *et al.*, 2003).

1.3.4   Bubble Coalescence

When bubbles have formed in magma, their subsequent evolution and role depend on their buoyancy, and the magma transport and rheology. If bubbles nucleate through decompression and diffusion, their size may be insufficient for them to move relative to the magma. If the magma ascends, the bubbles grow; as the vesicularity of the magma increases, bubbles can interact and coalesce. At high gas volume fractions, foams may develop.

Foams are typically unstable and reduce gravitational and surface energy by collapse. The failure of bubble walls is promoted by thinning as capillary and gravitational forces drain melt from the films between adjacent bubbles. Capillary forces dominate over gravity when the vesicularity exceeds 74% (Mader, 1998). Once bubbles have coalesced the new, larger bubbles tend to relax and gain a spherical shape. A fundamental control on gas separation, once a bubble forms, is the melt viscosity. Other factors being equal, hot mafic magmas permit more efficient separation of gas from melt than cooler intermediate and silicic magmas. However, the fate of bubbles will also depend on several other factors, including the motion of the magma itself (e.g., ascent or descent in a conduit, or convection in the chamber), the degree of coalescence between bubbles, the flow regime (and extent of loss of gas through conduit walls).

Bubbly magma has very different rheological and physical properties from those of dense, non-vesiculated or slightly vesiculated magma. Thus, as degassing affects magma rheology and overpressure, these, in turn, influence degassing, resulting in strong feedbacks that ultimately impact eruption style (Sparks, 2003).

1.3.5 Gas Separation

A very strong control on degassing is the degree to which gas–melt separation proceeds to the point that the gas leaves its host magma. Two end-member states can be envisaged—closed-system degassing in which the exsolved vapor does not leave the melt, and open-system degassing in which the separated gas phase produced at each stage in the magma body is expelled into country rock (possibly through a hydrothermal system) or directly, via the magmatic plumbing system, to the atmosphere (in the case of a lava–air interface such as characterizes a number of persistently active volcanoes like Stromboli and Mount Etna, where bubbles can rise much faster than the melt—Section 1.4.1.2). These states exert a strong influence on partitioning of volatiles into the vapor phase, and result in differing isotopic shifts (e.g., between ^{16}O and ^{18}O) between melt and vapor. Magma ascent rate will also determine the extent to which degassing is incomplete (high ascent rates) or is in equilibrium (low ascent rates).

Two equations roughly describe the evolution of the gas phase in ascending magmas for a closed system. Water solubility and exsolution are described by (1), and vesiculation by the ideal gas law:

$$V_g/V_l = K_R\left(X_{H_2O}^0 - X_{H_2O}^r\right)/P \qquad (2)$$

where the term on the left of the equation is the volumetric ratio of gas to melt (vesicularity) at pressure P, $X_{H_2O}^0$ is the initial melt water content at saturation pressure, $X_{H_2O}^r$ the residual melt water content at pressure P, and K_R is a constant. In the case of open-system degassing, the evolution of the water content is still controlled by water solubility (1), but the vesicles are no longer maintained by the internal pressure of bubbles, and they collapse. In this case, the bulk evolution of the system is closer to a distillation process than an equilibrium process—each small quantity of fluid that is produced is expelled from the melt. Models for open-system degassing, therefore,

incorporate the Rayleigh distillation law, but complications arise because of degassing-induced crystallization, particularly in intermediate and silicic systems, and where there is a competition between extraction of the vapor phase and mineral phases during vapor exsolution (Villemant and Boudon, 1998, 1999; Villemant *et al.*, 2003). Extensive microlite formation (of anhydrous minerals like plagioclase) acts to increase the volatile fraction in the residual melt, promoting further exsolution (Sparks, 1997). When crystallization is advanced, it impedes the expansion of bubbles, enhancing gas pressures during decompression. In intermediate and silicic systems, the highest overpressures are generally reached in the uppermost part of the conduit (Melnik and Sparks, 1999).

Theoretical models and experiments (e.g., Eichelberger *et al.*, 1986; Klug and Cashman, 1996) have indicated that magmas become permeable when the gas volume fraction reaches between 30% and 60%. In reality, deformation of bubbles may be necessary to achieve permeability at low vesicularities. In this case, elongation of bubbles would be flow parallel, and hence conduit parallel, promoting vertical permeability but not the horizontal permeability needed for degassing to take place through conduit walls. However, Jaupart (1998) suggests that fracturing in overpressured magma in the conduit may be a more likely mechanism for exporting gas into the country rock, or possibly nonlaminar flow in the conduit, and sufficient turbulence to bring significant quantities of melt into contact with the conduit walls on ascent. Studies of a fossil eruption conduit at Mule Creek, New Mexico, revealed a fractured conduit wall, and low vesicularity of the lava close to the conduit walls, indicating efficient gas loss (Jaupart, 1998). Massol and Jaupart (1999) found from models of conduit flow that significant horizontal pressure gradients can develop due to larger bubble overpressures developing in faster rising magma at the center of a conduit, exporting gas to the conduit walls given sufficient interconnection of bubbles. Permeable magma filling a conduit can permit gas contained in a deeper reservoir to flow to the surface, as has been suggested to account for high SO_2 fluxes from Soufrière Hills Volcano, Montserrat, during a hiatus in its ongoing eruption (Edmonds *et al.*, 2001; Oppenheimer *et al.*, 2002a). Crystallization can strongly promote high degrees of permeability in conduit magma because the bubbles that form are effectively squeezed along and between crystal boundaries, helping to establish an interconnected framework that promotes gas loss (Melnik and Sparks, 1999).

One of the best-understood low-viscosity magmatic plumbing systems is that of Kilauea. Gerlach and Graeber (1985) have shown how

Table 3 Representative compositional analyses obtained by direct sampling of hot gas vents at different volcanoes.

Volcano	Mt. St. Helens, USA, 1980	Mt. St. Augustine, USA, 1979	Momotombo, Nicaragua, 1980	Kilauea, USA, 1918	Kilauea, USA, 1983	Erta 'Ale, Ethiopia, 1974	Oldoinyo Lengai, Tanzania, 1999
Magma type and tectonic association	Dacite (continental margin subduction)	Andesite (island arc)	Basalt (continental margin)	OIB summit (Type I gas)	OIB rift (Type II gas)	Transitional MORB (incipient plate boundary)	Carbonatite (continental intraplate)
Temperature (°C)	802	648	820	1,170	1,010	1,130	600
H_2O (mol.%)	91.58	97.23	97.11	37.09	79.8	77.24	75.6
H_2 (mol.%)	0.8542	0.381	0.7	0.49	0.9025	1.39	
CO_2 (mol.%)	6.924	1.90	1.44	48.90	3.15	11.26	24.4
CO (mol.%)	0.06	0.0035	0.0096	1.51	0.0592	0.44	0.0787
SO_2 (mol.%)	0.2089	0.006	0.50	11.84	14.9	8.34	0.0197
H_2S (mol.%)	0.3553	0.057	0.23	0.04	0.622	0.68	
S_2 (mol.%)	0.0039		0.0003	0.02	0.309	0.21	
HCl (mol.%)		0.365	2.89	0.08	0.1	0.42	
HF (mol.%)		0.056	0.259		0.19		
OCS (mol.%)	0.0008				0.0013		

[a] Data all for thermodynamically calculated equilibrium compositions from Symonds *et al.* (1994) except for Oldoinyo Lengai (from Oppenheimer *et al.*, 2002b).

observed gas emissions (analyses for Kilauea in Table 3) indicate the differentiation of volatiles between the summit magma chamber and the sites of flank eruptions fed by intrusions from the main reservoir. Volatile-rich, mantle-derived magma arrives in the shallow summit chamber, and is saturated in CO_2. The exsolved CO_2-rich, H_2O-poor fluid degasses through summit lava lakes (when present) or through the hydrothermal system (type I gas). The magma, now equilibrated in the summit chamber, passes laterally via dikes that may feed subaerial or submarine eruption sites along the rift systems. These emit a distinct type II gas, with reversed H_2O and CO_2 contents compared with type I gas, reflecting the second degassing stage of the now CO_2-depleted magma. Degassing-driven fractionation of sulfur and halogens has been recently documented at Mount Etna, and has been related to open-system degassing and geometries and branching of the magmatic conduits (Aiuppa *et al.*, 2002; Burton *et al.*, 2003).

1.3.6 Fragmentation

When the product of viscosity and strain rate exceeds some critical value, and the strength of the magma is exceeded, fragmentation of magma occurs, driving an explosive eruption. In the case, at least, of rhyolitic systems, this occurs at the "glass transition" (Dingwell, 1998). Fragmentation can occur as the bubbly flow in a volcanic conduit accelerates and disintegrates, or due to sudden decompression of already vesiculated conduit magma, for instance resulting from a lava dome collapse and propagation of a rarefraction wave down the conduit (Cashman *et al.*, 2000; Melnik and Sparks, 2002a). Variations in, and pressurization feedback between, magma ascent rate, crystallization, and open- versus closed-system degassing may account for the rapid transitions between explosive and effusive eruption style that seem to characterize a number of intermediate and silicic eruptions (e.g., Melnik and Sparks, 1999, 2002b; Slezin, 1995, 2003; Barmin *et al.*, 2002; Sparks, 2003). Although volcanic systems are unlikely ever to be strictly deterministic, such nonlinearities have important implications for predicting volcanic behavior during periods when certain modes and timescales of behavior are dominant. Also, explosive eruptions are obviously capable of releasing very large quantities of volatiles almost instantaneously (Sections 1.4.1.2 and 1.6).

1.3.7 Excess Degassing

Numerous studies in the 1980s and 1990s identified the conundrum of "excess sulfur" or more generally "excess degassing" in which measured gas yields could not be provided by syneruptive degassing of observed quantities of erupted melt (Francis *et al.*, 1993). However, one way that excess degassing has been identified is really just an artifact of the measurement technique. For example, eruptions like Pinatubo 1991 and El Chichón 1985 produced far more sulfur than could be accounted for by petrological estimates of the volatile yields of the eruption (Section 1.4.3.4). However, in these cases, the excess sulfur was observed because the petrological method was inappropriate. Estimates of pre-eruptive volatile contents based on analyses of melt inclusions were invalid, because large amounts of volatiles had already exsolved prior to entrapment of the inclusions. Several investigations have yielded strong evidence for the existence of pre-eruptive vapor phases, e.g., for Pinatubo 1991 (e.g., Wallace and Gerlach, 1994).

Another way of resolving the excess degassing issue is to realize that when the sum of the partial pressures of dissolved volatiles equals the local confining pressure, then a separate multicomponent gas phase will exist in equilibrium with the melt. This gas can decouple from melt at many levels in the crust and potentially leak to the surface. A large body of unerupted melt can then be the source of volatiles released into the atmosphere. The degassing only seems excessive, because the mass of unerupted magma that supplied the volatiles is not being taken into account. Whether the unerupted melt exsolves gas and delivers it to the erupting magma syneruptively, or whether it has supplied vapor to the shallow parts of the pre-eruptive magma reservoir over periods of decades, centuries, or more, is uncertain, but several arguments favor the latter explanation (Wallace, 2001).

Extreme examples of excess degassing can be seen in the emissions from volcanoes such as Mt. Etna and Stromboli (Francis *et al.*, 1993; Allard *et al.*, 1994). Allard (1997) proposed that during the period 1975–1995, the sulfur observed to be degassing from Mt. Etna derived from 3.5 km^3 to 5.9 km^3 of magma, but only 10–20% of this was actually erupted. The remainder probably accretes as part of the plutonic complex within the sedimentary basement beneath the volcano. Allard *et al.* (1994) estimated an even more extreme ratio of eruptive-to-intrusive magma for Stromboli. They estimated that the observed SO_2 flux from 1980 to 1993 implied degassing of 0.01–0.02 $km^3 \, yr^{-1}$ of magma, exceeding by a factor of 100–200 the volume of material actually erupted. It should be borne in mind, however, that the extent to which degassing appears excessive will depend on the time period to which observations pertain. Volcanoes can substantially catch up on erupted mass with major events outside the observation period.

Several interesting models have been proposed to account for excess degassing, some of which show how shallow degassing drives the convection in the volcanic conduit, permitting the emptying of large fractions of chamber volatiles without major eruption. Kazahaya *et al.* (1994) and Stevenson and Blake (1998) developed similar models based on poiseuille flow, and applied them to explain the behavior of quite diverse volcanoes, including Izu-Oshima (Japan) and Stromboli (Italy) (both basaltic), Sakurajima (Japan) (andesitic), and Mount St. Helens (USA) (dacitic). Stevenson and Blake (1998) suggested that conduit convection involves concentric flow of upwelling and downwelling magma. They showed that where convective overturn controls the gas supply, the gas flux is a function of the density difference between gas-rich and degassed magma, the conduit radius, and the magma viscosity (which is itself a function of gas mass fraction). Thus, efficient degassing induces a negative feedback by increasing viscosity of the downwelling magma, and thereby slowing the supply of gas-rich magma from the reservoir. Evidence for significant mixing of magma degassed at shallow depths with "fresh" volatile-rich magma is provided for Kilauea from analyses of volatile concentrations in samples dredged from the submarine Puna ridge at depths of as much as 5.5 km. Dixon *et al.* (1991) found ranges of CO_2, H_2O, and S contents in the tholeiitic glasses that could only be explained by deep recycling of magma that had already lost its gas at shallow depths, possibly even subaerially in a lava lake.

1.4 EMISSIONS

Volcanic emissions of gases and aerosol to the atmosphere take many different forms—from geothermal/hydrothermal manifestations, to massive syn-eruptive releases, such as that of Mount Pinatubo in 1991. This section illustrates this spectrum, highlighting the importance of understanding the fluxes and composition of gases, and their time–space distribution, for assessing and predicting the atmospheric, environmental, and climatic impacts of volcanic degassing. It describes also the techniques available to analyze volatile emissions, and to interpret them in the context of volcano surveillance efforts.

1.4.1 Styles of Surface Emissions

The manifestations of volatile release from volcanoes vary tremendously—from the diffuse leaks of CO_2 on both active and dormant volcanoes to the highly concentrated, sporadic injections of water and acid gases into the upper atmosphere by major explosive eruptions (Figure 4).

1.4.1.1 Noneruptive emissions

Noneruptive volcanic emissions are most apparent when they are exhaled from discrete vents, either in gas, liquid, or gas/liquid state. However, significant fluxes of gases can be discharged over wide areas in a much more distributed fashion. Broadly speaking, these are the manifestations of magmatic–hydrothermal systems, whose constituents are derived, in varying proportions, from magma, the crust, and groundwaters of assorted provenance.

In the case of *hot springs*, the steam and gas flux is subordinate to the liquid water flux. *Geysers* are spectacular examples of hot springs (Figure 4(a)). When gaseous emissions predominate, the term *fumarole* is usually applied. Emission temperatures of fumaroles, therefore, typically exceed the local boiling point. Long-lived fumarole fields are sometimes termed *solfataras* or *soufrieres* (Figure 4(b)), their longevity giving rise to substantial alteration of host rock and deposition of sublimates. Solfatara in the Campi Flegrei north of Napoli is the classic example (Figure 1). Fumarole

(a)

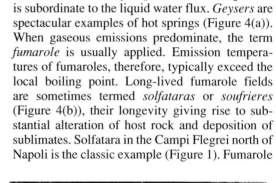

(c)

(b)

(d)

(e)

Figure 4 (Continued).

Figure 4 Illustrations of emission styles from volcanoes: (a) Old Faithful geyser in Yellowstone National Park, USA; (b) fumaroles at Kawah Ijen (Indonesia)—the pipes visible between the steam are used to condense sulfur, which is collected by miners working in the crater; (c) diffuse degassing of CO_2 produced this tree-kill zone at Mammoth Mountain, USA; (d) Kawah Ijen's crater lake—such lakes are typically enriched in acid species and have elevated temperatures; (e) open-vent degassing from Masaya (Nicaragua)—this volcano emits high fluxes of sulfur and halogens directly to the atmosphere from the surface of a magma-filled conduit or "open vent" opening on to the floor of the crater; (f) when low viscosity lava fills a crater, a lava lake forms, such as this long-lived example at Erta 'Ale (Ethiopia; Oppenheimer and Yirgu, 2002)—these can efficiently degas large volumes of subsurface magma by convective circulation; (g) slugs of gas bursting through an open vent generate strombolian eruptions, as demonstrated here at Stromboli volcano (Italy); (h) the lava dome of Soufrière Hills Volcano (Montserrat), which has sustained a considerable gas flux since its emergence in 1995 (Edmonds *et al.*, 2001, 2003a,b,c); (i) dome collapses (as seen here at Soufrière Hills Volcano) can degas lava as it fragments, fuelling eruption plumes—this is at the low-scale end of a spectrum of explosive degassing behavior that spans up to releases of well in excess of 10^7 kg s^{-1} of volatiles.

emissions are very often composed of magmatic gases and hydrothermal gases (which result from complex interactions between magmatic fluids, meteoric water, seawater, and rock).

CO_2-rich emissions at temperatures below the boiling point of water are sometimes referred to as *mofettes* when they are localized. More diffuse emissions of CO_2 (and radon) can also occur over wide areas, reflecting the exsolution of these gases from magma bodies at depth (Figure 4(c)). Such emissions can present a hazard close to the ground and in depressions, especially in calm atmospheric conditions, as CO_2 concentrations can exceed several percent (Section 1.6.5.3).

Numerous volcanoes discharge acid gases that are condensed in crater lakes (Figure 4(d)). The lake water derives, therefore, both from volcanic and meteoric inputs. Such lakes provide valuable opportunities to evaluate the heat and mass budgets of the host volcanoes, but also represent particular hazards since even small eruptions can displace large quantities of water over crater rims, triggering catastrophic lahars (Varekamp and Rowe, 2000). The black smokers associated with active oceanic ridges are another well-known manifestation of subaqueous volatile discharge.

1.4.1.2 Eruptive emissions

When magma reaches the surface, it can release gases directly into the atmosphere. Magma-filled conduits (often referred to as *open vents*; Figure 4(e)), lava lakes (Figure 4(f)), and lava domes (Figure 4(g)) can all discharge large amounts of gas. If they do this nonexplosively, the efficient segregation of gas typically reflects low-viscosity magmas in which coalescence of bubbles permits them to rise faster than the magma, or the development of significant permeability by vesiculation in more viscous intermediate or silicic magmas. Gas fluxes from open vents can be very high (Section 1.5). Various circumstances can lead to explosive fragmentation of magma, witnessed in a wide spectrum of eruptive behavior (Figures 4(h) and (i)) including discrete strombolian explosions, lava fountains, and sustained eruption columns typified by the plinian eruptions of Mt. St. Helens (1980) or Mt. Pinatubo (1991). In the intermediate and silicic cases, as bubbly magma ascends in the conduit, progressive vesiculation and inhibited gas loss can culminate in fragmentation of the accelerating mixture into a gas–particle mixture. The eruption of this mixture increases the pressure drop between the vent and chamber, typically raising the eruption intensity (mass discharge rate of magma at the vent) further. In this way, significant fractions of magma reservoirs may be erupted in a matter of hours.

1.4.2 Chemical Composition of Volcanic Gases

The major and trace element composition of volcanic gases varies widely (Table 3). In general terms, the composition of volcanic gas represents the complex sources, histories, and processes of magma generation, mixing and ascent (e.g., variable contributions from the mantle wedge or slab

in the case of arc magmas; time-dependent vapor–melt separation as a function of evolving magma composition and physical conditions), and the interactions of the gas phase after it separates from the host magma, for example, with rocks and fluids in the crust (Giggenbach, 1996). Degassing is a continuous process, and the distribution of volatiles between vapor and melt phases varies strongly as a function of depth and time. Magma bodies and their plumbing systems have ample time to evolve and can have significant vertical extents. The latter point implies that deep degassing, shallow degassing, and anything in between, can contribute to the mixture of magmatic fluids observed at the surface. Superimposed on these magmatic complexities are the effects of shallow re-equilibration of fluids due to cooling and dilution by groundwater (e.g., meteoric water, seawater, or hydrothermal fluids). Isotopic ratios of various fluid species can be used to model magmatic and hydrothermal processes. In general terms, isotopes of carbon, hydrogen, and nitrogen are good tracers of magmatic processes, while oxygen isotopes are useful for tracing the source (see Valley and Cole, 2001, for reviews).

1.4.3 Measurement of Volatiles

The conventional way to measure volcanic emissions is by direct sampling, either by *in situ* collection of samples from fumarole vents and active lava bodies using "Giggenbach bottles," filter packs and condensing systems, or within atmospheric plumes (sometimes from aircraft) using various kinds of sampling apparatus and on-board analyzers. A range of spectroscopic, gravimetric, isotopic, and chromatographic techniques is available to determine chemical concentrations in real time or subsequently in the laboratory (Symonds *et al.*, 1994). While such direct sampling can deliver very detailed and precise analyses, it is difficult to sustain routine surveillance in this way, and to compete with seismological and geodetic monitoring techniques in terms of sampling rate. However, an expanding array of remote sensing methods is available to provide the data streams needed to characterize the composition and fluxes of volatiles from volcanoes. In particular, both field-based and satellite spectroscopic methods are increasingly in use to measure gas composition remotely.

This section briefly reviews the principal measurement techniques (Figure 5), including direct sampling, ground-based, or airborne ultraviolet spectroscopy (correlation spectrometer and successors), ground-based infrared spectroscopy (Fourier transform spectroscopy and other infrared spectroscopic analysers), and spaceborne methods, including the important role of the total ozone mapping spectrometer (TOMS), and similar instruments, for measuring the SO_2 emissions from major eruptions. It also covers petrological approaches to characterizing volatile abundances and emissions.

1.4.3.1 *In situ sampling and analysis*

Conventional analyses of volcanic gases have been made by collection of samples directly from fumarole vents using evacuated bottles and caustic solutions, and subsequent laboratory analysis. The classic "Giggenbach" bottle consists of an evacuated glass vessel partially filled with NaOH (Figure 5(a); Giggenbach, 1975; Giggenbach and Matsuo, 1991). On the volcano, the gas stream is allowed to bubble through the solution via tubing inserted into the volcanic vent. Acid species condense according to reactions such as

$$CO_2 + 2OH^-_{(aq)} = CO^{2-}_{3(aq)} + H_2O \qquad (3)$$

(a)

(b)

(c)

Figure 5 (Continued).

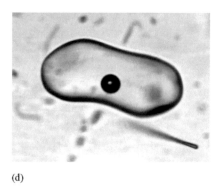

(d)   (e)

Figure 5   Techniques for measuring volcanic volatiles: (a) direct sampling of fumarole vent at Ol Doinyo Lengai Volcano (Tanzania) using Giggenbach bottle; (b) ground-based remote sensing techniques—shown here in the flanks of Mount Etna are the ultraviolet sensing COSPEC instrument (for SO_2 measurements) and a Fourier transform infrared spectrometer (for SO_2, HCl, and HF measurements); (c) satellite-based TOMS observations of the Mt. Pinatubo stratospheric SO_2 emission (courtesy of Simon Carn, TOMS Volcanic Emissions Group); (d) glassy melt inclusion in quartz and its associated shrinkage bubble, from the ignimbrite unit of the 1912 eruption at the Valley of Ten Thousand Smokes, Alaska—such inclusions can provide valuable information on volatile contents (micrograph and caption courtesy of Jake Lowenstern (Volcano Hazards Team, US Geological Survey); and (e) ice core chemical stratigraphy has been used to estimate sulfur yields from major eruptions (for review see Zielinski, 2000)—this section of the North GRIP core from Greenland shows the 10.14 kyr BP (ice stratigraphic age) Icelandic Saksunarvatn ash layer (thin horizon in the center of the core shown) (courtesy of Sune Olander Rasmussen, Trine Ebbensgaard and Sigfus Johnsen, Glaciology Group, University of Copenhagen; Dahl-Jensen *et al.*, 2002).

$$4SO_2 + 7OH^-_{(aq)} = 3SO^{2-}_{4(aq)} + HS^-_{(aq)} \\ + 3H_2O$$ (4)

and are analyzed by ion chromatography. The remaining species collect in the headspace and are usually analyzed by gas chromatography. Base-treated filters can also be used to trap acid species, and can be deployed around crater rims and in the vicinity of gas sources. Such filter-based methods have been extended to characterization of aerosol size distribution and chemistry (e.g., Vié le Sage, 1983; Allen *et al.*, 2000, 2002). The volatiles scavenged out of eruption clouds by ash particles, which then sediment to the ground, can also be studied analytically by leaching samples with distilled water (e.g., Edmonds *et al.*, 2003a).

Although such approaches offer very high sensitivity, measurements can be difficult and often dangerous to obtain, and there can be problems of postcollection reactions. Also, the inevitable delays in obtaining results limit their value in volcanic crises. Remote sensing methods can overcome these difficulties, and, importantly, several are able to constrain gas fluxes, which are hard to derive by point sampling.

1.4.3.2   Portable remote sensing systems

Since the early 1970s, ground-based, optical remote-sensing techniques have been increasingly used for volcanic gas and aerosol monitoring (Figure 5(b)). In particular, the correlation spectrometer (COSPEC), which operates in the ultraviolet region of the spectrum, using scattered

skylight as a source, has been used routinely by volcano observatories worldwide to measure SO_2 fluxes (Stoiber *et al.*, 1983). COSPEC has seen active service in numerous volcanic crises, crucially helping to ascertain whether or not new magma pathways are opening up to shallow levels beneath a volcano. More recently, a much smaller ultraviolet spectrometer (the size of a pack of cards) has been applied to measurements of SO_2 flux (Galle *et al.*, 2003; McGonigle and Oppenheimer, 2003; McGonigle *et al.*, 2002, 2003). The device is set to revolutionize ground-based sensing of volcanic gas emissions because of its extreme portability, ease of operation, and suitability for automated scanning measurements at high temporal resolution (Edmonds *et al.*, 2003b; McGonigle *et al.*, 2002, 2003).

The reasons for focusing measurements on SO_2 are that: (i) the background levels of this gas are very low in the atmosphere (typical volume mixing ratios are $\ll 1$ ppb) while mixing ratios in volcanic plumes can easily exceed 1 ppm, and (ii) several strong absorption bands for SO_2 are found in the ultraviolet and infrared regions of the spectrum. In other words, the spectroscopic observation of this species is straightforward, despite its comparatively low abundance in volcanic emissions. In contrast, H_2O or CO_2, which are the principal components of volcanic gases, are difficult to measure accurately because of the high, and in the case of H_2O, rapidly varying, atmospheric background concentrations. However, it remains highly desirable, for the purposes of volcano surveillance, to be able to measure other components of the gas emission, and to be able to follow these through time. To this end, broadband infrared measurements have been carried out using Fourier transform spectrometers. These are capable of simultaneously sensing several gas molecules of interest, including HCl, HF, CO_2, CO, OCS, SiF_4, and H_2O, as well as SO_2 (Oppenheimer *et al.*, 1998a; Burton *et al.*, 2000; Horrocks *et al.*, 2001). Measurements can even be made from moving vehicles in order to build up cross-sections of plume burdens of SO_2 and halogens that can then be integrated and multiplied by plume speed to yield estimates of the fluxes of these species (Duffell *et al.*, 2001).

Aerosol concentrations and size distributions can be investigated remotely using sun-photometry. Characterization of volcanic aerosol is important in studies of plume chemistry, atmospheric radiation, and the environmental and health impacts of particle emissions. Watson and Oppenheimer (2000, 2001) used a portable sun-photometer to observe tropospheric aerosol emitted by Mt. Etna. They found distinct aerosol optical signatures for the several plumes emitted from Etna's different summit craters, and apparent coagulation of particles as the plume aged. More recently, Porter *et al.* (2002) have obtained sun-photometer and pulsed lidar data for the plume from Pu'u O'o vent on Kilauea, Hawaii, from a moving vehicle in order to build profiles of sulfate concentration.

1.4.3.3 Satellite remote sensing

The larger releases of volcanic volatiles to the atmosphere defy synoptic measurements from the ground. Major advances in our understanding of explosive volcanism and its impact on the atmosphere and climate have been achieved thanks to satellite observations. Again, sulfur dioxide is the most readily measured species, and a number of spaceborne sensors operating in the ultraviolet (electronic lines), infrared (roto-vibrational lines), and microwave (rotational lines) have been utilized for measurements of this gas. TOMS operating in the ultraviolet region has provided the most comprehensive database on the release of SO_2 to the upper atmosphere (Section 1.5.1; Figure 5(c); Krueger *et al.*, 1995). It has detected many of the larger silicic and intermediate composition explosive eruptions that have taken place since 1979, and some mafic eruptions, prominently including those from Nyiragongo and Nyamuragira in the Great Lakes region of central Africa (Carn *et al.*, 2003). TOMS provided early estimates of the initial sulfur yield to the stratosphere of the 1991 Mt. Pinatubo eruption ($\sim$20 Tg of SO_2), but other instruments, including NASA's Microwave Limb Sounder (MLS), were able to track the SO_2 clouds as they were depleted over the following weeks (Read *et al.*, 1993).

Recently, Prata *et al.* (2002) have pointed out the suitability of an infrared system—the high-resolution infrared radiation sounder (HIRS)—carried by NOAA satellites, for measuring volcanic SO_2 above altitudes of $\sim$5 km. This could prove valuable in filling in gaps in the existing TOMS record of volcanic emissions due to instrument problems (Carn *et al.*, 2003). Hitherto, satellites have only really been capable of measuring the larger releases of SO_2 during eruptions. However, two newer infrared instruments with higher spatial resolution—the moderate resolution imaging spectrometer (MODIS) and advanced spaceborne thermal emission and reflection radiometer (ASTER)—both carried on-board NASA's Terra satellite (launched in late 1999)—show considerable promise for monitoring smaller yields of SO_2 to the troposphere. Researchers at Michigan Technological University have used imagery from both sensors to estimate sulfur dioxide fluxes from volcanoes in Guatemala, suggesting that these and future similar sensors could be used for routine volcano monitoring purposes (W. I. Rose, personal communication).

1.4.3.4 Petrological methods

There are various "petrological" approaches to determining volatile concentrations in magma (melt, vapor phase, and crystals) or volatile yields to the atmosphere. Most of the data on pre-eruptive volatile abundances reported in Table 1, for example, have been obtained by analyses of quenched glasses. These include the chilled glassy rinds of lavas or pyroclasts, and melt inclusions trapped inside phenocrysts (Figure 5(d)). In some cases, differencing melt inclusion and matrix glass concentrations for specific volatile species permits estimates of degassing yield, if it can be assumed that the former represents pre-eruptive concentrations, and the latter degassed melt. Other approaches to characterizing volatile abundance in magmas include the application of experimentally determined phase equilibria (for both natural and synthetic melts), and thermodynamic calculations based on mineral compositions (see Johnson *et al.* (1993) and Wallace and Anderson (2000) for overviews of these methods).

Petrological estimates of eruption yields of sulfur and, to a lesser extent, halogens have received much attention, partly because they offer a means to assess the volatile yield for historic and ancient eruptions. The search for suitable host phenocrysts requires careful microscopic examination of thin sections and even more careful analysis of the inclusions themselves, which can measure only a few microns across. While the method appears to work well for some systems (e.g., Thordarson *et al.*, 2003), it is not universally applicable. For example, the crystals may not always be leak-proof containers for their high-pressure melt samples, and inclusions can interact with their host minerals. Most problematic, however, is the possibility of volatile exsolution into a fluid phase *prior* to crystal growth and entrapment of the melt, leading to the inclusion recording a lower volatile content than that which the melt started with. This very likely explains why the petrological technique singularly fails to explain the sulfur release during Mount Pinatubo's eruption in 1991 (Section 1.3.7). The sulfur contents in glass and inclusions are more or less the same, which would suggest a sulfur-free eruption plume. Instead, the eruption is known to have released some 20 Tg of sulfur dioxide.

Wallace *et al.* (1995) and Wallace (2001) have explored this issue in detail, and Scaillet *et al.* (1998, 2003) and Scaillet and Pichavant (2003) have advanced petrological modeling approaches to circumvent the problem by estimating the volatile contents contained in the vapor phase prior to eruption. Another problem with petrological estimates of volatile yields is that they scale linearly with the eruption magnitude, which is often only poorly constrained from the rock record. This is especially true in cases where tephra dispersal is very widespread, perhaps largely at sea, and where

substantial burial, erosion or redeposition limit efforts to identify original thicknesses of sediment in the field.

1.4.3.5 Ice cores

The cryosphere provides an important repository of volcanic volatiles from past eruptions in the form of sulfate layers deposited within a few years of eruption (Hammer *et al.*, 1980; Zielinski *et al.*, 1994; Zielinski, 2000). Ice core stratigraphy is used to date the eruption year, and in some cases, tephra particles can be fingerprinted chemically to the products of known eruptions. The most productive cores for identifying volcanic markers have come from Greenland and Antarctica, including the Greenland Ice Sheet Project 2 (GISP2) and Greenland Icecore Project (GRIP and NorthGRIP) efforts (Figure 5(e)).

Zielinski *et al.* (1996a) have published a 110 kyr record of volcanism as recorded by the GISP2 core, providing one of the most intriguing records of palaeovolcanism during the Late Quaternary (Figure 6). By determining the flux of sulfate that formed the layer (in kg m^{-2}), it is possible to estimate the total atmospheric mass of sulfur by calibrating against fallout from

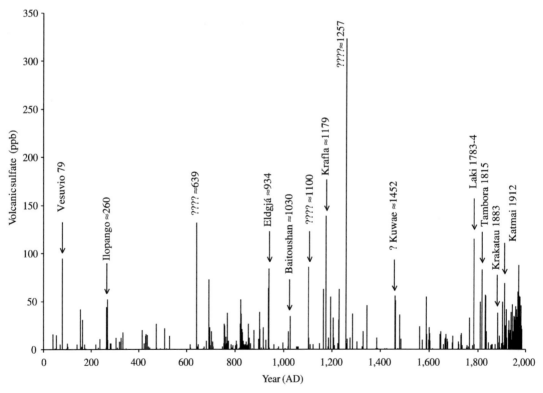

Figure 6 GISP2 volcanic sulfate markers for the past 2,000 yr based on statistical analysis (Zielinski *et al.*, 1994). Several large anomalies have not been traced to the responsible volcanoes, including the prominent AD 640 and 1259 peaks. Data provided by the National Snow and Ice Data Center, University of Colorado at Boulder, and the WDC-A for Paleoclimatology, NGDC, Boulder, CO, USA.

sources of known magnitude (Hammer *et al.*, 1980; Zielinski, 1995). The most widely used calibration is based on β-activity measurements of the layers of radioactive fallout in the ice cores that resulted from nuclear weapons tests in the atmosphere conducted in the 1950s and 1960s. Regressions can be obtained between the flux of fallout, say for the Greenland ice core, and the known yield of the explosions, and then applied to the volcanic markers. This, of course, assumes that the volcanic sulfate was transported by comparable atmospheric dynamics, from similar latitude, and so on, which is difficult to gauge for ancient eruptions (usually, the responsible volcano is unidentified). Very low deposition rates, and post-depositional effects, such as densification, diffusion, and aeolian deflation or redeposition, and other potential sources of sulfur (including marine sources), also pose difficulties in interpreting sulfate layers in the ice cores. Nevertheless, multiple estimates of sulfur yields of eruptions such as Tambora 1815, whose sulfate marker is found in both Arctic and Antarctic cores, are reasonably consistent with each other, and with estimates obtained by other methods (e.g., based on astronomical observations of atmospheric optical thickness), lending some confidence in the approach.

1.4.3.6 Application of geochemical surveillance to volcano monitoring

Time-series chemical and isotopic measurements, and flux observations can be interpreted with respect to volcano behavior, and the interrelationships between degassing, eruptive character, and other geophysical and geodetic parameters. Gas geochemistry, therefore, plays an important role in volcanic hazard assessment. Temporal changes in gas chemistry and flux, in particular, are widely regarded as potential indicators of future volcanic activity, and many volcano observatories worldwide carry out some kind of geochemical surveillance as part of their overall monitoring efforts. The basic tasks are to identify volatile sources, magmatic-hydrothermal system interactions, the dynamics of degassing, and changes in these through time. Unfortunately, interpretation of the observations is far from straightforward because of the multiple intensive parameters that control magmatic volatile content (mantle melting, slab contributions, wall–rock assimilation, etc.), exsolution and gas separation of different volatile species from magma, and the subsequent chemical and physical interactions of the exsolved fluids, for example, with crustal rocks and hydrothermal fluids, as they ascend to the surface (Figure 7). Some general observations on the interpretation of the principal volcanic gas species are made in Table 4.

In the broadest terms, Matuso (1960) recognized that the composition of volcanic gas reflects the balance of contributions from magmatic degassing and from the crust. An oft-quoted sequence of volatile release with decreasing pressure, based largely on experimental observations of solubility in the melt, is

$$C \Rightarrow S \Rightarrow Cl \Rightarrow H_2O \Rightarrow F$$

However, this greatly oversimplifies the exsolution behavior of real magmas (which, of course, vary enormously in chemical composition, viscosity, etc.) and their degassing histories, and is of limited use in trying to interpret measurements of, for example, increasing Cl/S ratio or decreasing SO_2 flux of fumarolic emissions. Also, in general

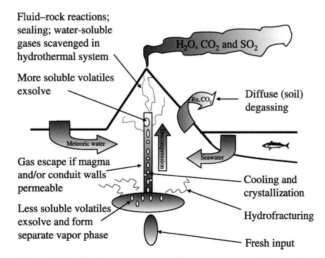

Figure 7 Potential physical and chemical processes occurring in a magmatic–hydrothermal system, including the influence of magma dynamics in the chamber-conduit plumbing system, and interactions between magmatic fluids and the crust. These can strongly modulate the speciation and flux of various magmatic components emitted into the atmosphere, complicating the interpretation of geochemical measurements of surface emissions.

Table 4   General observations on the information content of principal volcanic gas species.

H_2O	Several origins in volcanic discharges—principally magmatic, hydrothermal, meteoric, seawater (volcanic islands)—hence difficult to interpret water contents without measurements of isotopic composition.
CO_2	Predominantly magmatic origin, and comparatively inert in hydrothermal systems and the atmosphere. Fluxes can be measured by some spectrometric techniques.
SO_2	Common in high temperature volcanic gases and a good indicator of magmatic degassing. Fluxes readily measured by remote sensing methods.
H_2S	Typical of lower temperature volcanic vents and indicative of hydrothermal contributions.
HCl	Typically magmatic but potentially sourced by hydrolysis/volatilisation of chloride compounds or dissociation of brines. Readily scavenged by hydrothermal systems so changes in HCl emission can reflect both magmatic and hydrothermal processes.
He	Total He content and $^3He/^4He$ ratios are good tracers of mantle contribution and mixing with atmospheric He. Unaffected by secondary processes.
Ar	A good tracer of atmospheric and air-saturated meteoric water mixing in volcanic gas samples.
H_2	Hydrogen content increases with increasing temperature and provides the basis for gas geothermometers.
N_2	Either of atmospheric origin or from subducted slab (in which case, N_2/Ar usually >83, the value for air).
CH_4	Nonmagmatic component generally formed in hydrothermal environment and therefore a good indicator of hydrothermal system. Also favored magmatic species according to control of calcite-anhydrite buffer $(CaCO_3 + H_2S + CH_4 + 5H_2O = CaSO_4 + 2CO_2 + 8H_2)$ or the H_2S/SO_2 and rock buffers: $2Fe_3O_4 + 3SiO_2 + CO_2 + 2H_2S = 3Fe_2SiO_4 + CH_4 + 2SO_2.$
CO	Controlled by temperature and redox potential and a useful indicator of these parameters.

Source: Giggenbach *et al.* (2001).

terms, the relative proportions of magmatic carbon, sulfur, and halogen compounds are dictated by the pressure- and redox-controlled partition coefficients between vapor and melt, the mixing or mingling of different magma sources, and the dynamics of degassing. For example, important redox equilibria affecting volcanic gas composition include

$$SO_2 + 3H_2 = H_2S + 2H_2O \qquad (5)$$

$$CO_2 + H_2 = CO + H_2O \qquad (6)$$

$$CO_2 + 4H_2 = CH_4 + 2H_2O \qquad (7)$$

Reaction (5) is most likely to represent the major gas redox buffer because of the comparable abundances of the two sulfur-bearing gas species. The other effective geochemical buffer is the FeO–Fe_2O_3 "rock buffer," which can affect the redox equilibria above in magma, and as gas and wall–rock interact.

Although thermodynamical codes have been developed to restore observed analyses of gas composition to equilibrium compositions (e.g., Gerlach, 1993; Symonds *et al.*, 1994), unravelling these complex and highly nonlinear processes from a patchy record of surface observations, and identifying precise magmatic and hydrothermal processes, remain real challenges. Indeed, a casual perusal of the literature on gas geochemical monitoring will reveal conflicting interpretations of ostensibly similar observations—for instance, decreasing SO_2 fluxes could be ascribed to: (i) depletion of volatiles in a magma body, or (ii) a decrease in the permeability of the plumbing system. Process (i) might be taken to indicate diminished eruption likelihood, while (ii), perhaps

induced by sealing of bubble networks by hydrothermal precipitation, which would increase overpressure (e.g., Edmonds *et al.*, 2003c), could increase the chance of an eruption. Thus, the same observation can be interpreted in different ways with contradictory hazard implications.

Nevertheless, high SO_2 fluxes remain a reliable indicator of the presence of magma during new episodes of unrest at volcanoes (e.g., seismicity, changes in fumarole emissions), and help to discriminate between magmatic, and tectonic or hydrothermal causes of unrest. In particular, COSPEC measurements are generally recognized as having contributed significantly to hazard assessment prior to the 1991 eruption of Mt. Pinatubo (Daag *et al.*, 1996; Hoff, 1992). Immediately prior to the eruption the measured SO_2 flux increased by an order of magnitude over two weeks, mirroring seismic unrest. These observations were interpreted as evidence of shallow intrusion of magma, increasing the estimates of the probability of an impending eruption. More recently, COSPEC observations have supported conclusions of magmatic unrest prior to eruptions of Soufrière Hills Volcano (Montserrat), Popocatèpetl (Mexico), and Tungurahua (Ecuador) (Dan Miller, USGS, personal communication, 2002). For ongoing eruptions, combining observations of gas emissions with melt inclusion constraints on melt volatile contents can offer valuable insights into the sources of gases and mechanisms of degassing. For example, Edmonds *et al.* (2001) used both kinds of information to conclude that, at Soufrière Hills Volcano, the HCl content of the summit plume has been largely supported by shallow, crystallization-induced degassing of the andesite approaching and forming the lava dome, while SO_2 fluxes have been sustained, even through noneruptive periods, by degassing of a deep, possibly mafic, source.

Decreasing SO_2 fluxes accompanying diminishing posteruptive activity have also been observed at many volcanoes, including Mt. St. Helens from 1980 to 1988 (McGee, 1992), where a decline in CO_2 flux and increase in H_2O flux were also observed following the 1980 eruption. This decrease in the CO_2 and SO_2 emissions suggested that the magma reservoir was not being replenished, consistent with the decreased eruption rates (see also Stevenson, 1993; Stevenson and Blake, 1998). The increase in water vapor emission was interpreted as the result of groundwater permeating the conduit system (Gerlach and Casadevall, 1986).

Because of its deep exsolution, and comparative chemical inertness in the crust and hydrothermal systems, and atmosphere, CO_2 flux measurements are seen as a particularly valuable indicator of magmatic degassing and unrest. Infrared analyzers have been used to determine CO_2 fluxes from the ground and in atmospheric plumes (e.g., Gerlach *et al.*, 1997; McGee and Gerlach, 1998) but the available methods are time consuming (ground surveys) and expensive (where aircraft deployments are required), with the possible exception of open-path Fourier transform infrared spectroscopy combined with a method for determining SO_2 fluxes (Burton *et al.*, 2000).

Villemant *et al.* (2003) have examined in detail the implications of closed-system versus open-system degassing (Section 1.3.5) on the compositional evolution of the gas phase, focusing on halogens, since their behavior is not particularly affected by the oxidation state of the magma (e.g., Symonds and Reed, 1993), and their concentrations are relatively easy to measure in lava and tephra samples. Open-system degassing is more efficient at extracting halogens from the melt than closed-system degassing. HCl is less bound to the silicate chains in melts than HF, and is therefore more efficiently extracted by a water-vapor phase. HCl/HF ratios in an open system are therefore predicted to fall with increased degassing of water. In other words, gas–rock interaction, cooling, and decompression should not affect halogen ratios strongly. However, kinetic effects can also be significant in halogen partitioning during shallow degassing (e.g., Signorelli and Carroll, 2001), and scrubbing in hydrothermal systems (Oppenheimer, 1996) can dissolve acid species from the gas phase, stripping fumarole gases of their halogen content (e.g., Symonds *et al.*, 2001). Duffell *et al.* (2003) observed an increase in SO_2/HCl ratios prior to a minor phreatic eruption of Masaya volcano, Nicaragua, which they suggested could have reflected hydrothermal scavenging of HCl. Increased magma–water interaction may have

played a role in triggering the eruption, indicating the potential predictive value of routine monitoring of sulfur and halogen ratios. Second boiling was thought to explain a *decrease* in S/Cl ratios in gases emitted from the Showa–Shinzan lava dome, several years after its emplacement, as the slow crystallization of the thick lava body delayed the eventual exsolution of chlorine from the melt (Symonds *et al.*, 1996).

Isotopic signatures can be of great value in identifying magmatic components of volcanic fluids. In particular, the ratio of $^3He/^4He$ is useful in distinguishing the contribution from primordial (mantle) ^3He from that of radiogenic (crustal) ^4He (e.g., Sorey *et al.*, 1993, 1998; Tedesco, 1995). Nitrogen-isotopic composition is a useful tracer of sedimentary input into arc volcanic gases (Section 1.2.1; Fischer *et al.*, 2002). $\delta^{18}O$ and δD compositions of water are important indicators of mixing of magmatic, hydrothermal, marine, and meteoric sources (Section 1.2.1). Stable isotopes of carbon and sulfur can also be used to identify contributions of magmatic volatiles, assuming that isotopic compositions of surface reservoirs are constrained. For example, $\delta^{13}C_{CO_2}$ and $\delta^{34}S_{SO_2}$ measurements of gas samples collected from the lava lake of Erta 'Ale volcano (Ethiopia) indicated the mantle origin of emitted carbon and sulfur ($\delta^{13}C \approx -4‰$ and $\delta^{34}S \approx 0‰$), whereas hot fumaroles on arc volcanoes display more variable $\delta^{13}C$ $-12‰$ to $+2.5‰$) and $\delta^{34}S$ ($0‰$ to $+10‰$) indicative of contamination by slab and crustal sources (summarized by Delmelle and Stix (2000)). Very little work has been carried out on chlorine-isotopic distribution in volcanic gases, and indeed, the mantle $\delta^{37}Cl$ value is arguably still not well constrained, but improved analytical techniques could open up considerable potential to study chlorine distribution and cycling. Eggenkamp (1994) reported variations in $\delta^{37}Cl$ between about $-2.5‰$ for sublimates and $+10‰$ for gases from Lewotolo volcano (Indonesia), which were thought to indicate surficial and deep processes, respectively, demonstrating significant chlorine-isotopic fractionation in volcanic systems.

1.5 FLUXES OF VOLCANIC VOLATILES TO THE ATMOSPHERE

Inventories of the spatial and temporal distribution of volcanic emissions to the atmosphere are important for studies of tropospheric and stratospheric chemistry, and the Earth's radiation budget. This section summarizes estimates of global fluxes and distribution of several key species. As pointed out in the IPCC Report (2001), referring in this case to the volcanic sulfur budget, "...estimates are highly uncertain because only very few of the potential sources have ever been measured and the variability between sources and between different stages of activity of the sources is considerable." Crucial parameters for climate studies include the quantity and height of entrainment of sulfur and other species into the atmosphere (large explosive eruptions, e.g., Pinatubo are uniquely capable of more or less instantaneous injections of sulfur into the upper atmosphere), and zonal and seasonal controls on atmospheric circulation (Section 1.6).

The more soluble volatile components of volcanic plumes (e.g., halogens) can be removed from eruption clouds rapidly by adsorption on to tephra, or deposition in hydrometeors. The estimates of atmospheric burdens of volcanic volatiles in eruption clouds, therefore, depend on the time after eruption that the measurements were made, and can, in general, be lower than the total volatile release. Surprisingly, little work has been undertaken to quantify the total volatile budgets of eruptions (see De Hoog *et al.*, 2001, for an exception).

This section focuses on subaerial emissions. Submarine fluxes of volatiles are also important in understanding global geochemical cycles but are rather poorly constrained. They will only be referred to briefly.

1.5.1 Sulfur

There is considerable interest in volcanic emissions of sulfur compounds because of the role of atmospheric sulfur chemistry in atmospheric radiation and climate, the hydrological cycle, acid precipitation, and air quality (see Chapter 4). Early theories on the climatic effects of eruptions considered that ash particles were responsible for raising the planetary albedo, but it is now clear that even fine tephra sediment rapidly from the

atmosphere, and that the main protagonist in volcanic forcing of climate is the sulfate aerosol formed by oxidation of sulfur gases released to the upper atmosphere.

Most estimates of the volcanic source strength of sulfur to the atmosphere are based on compilations of COSPEC and related observations of lesser emissions from individual volcanoes (many exhibiting long-term degassing; Tables 5 and 6), and TOMS measurements of the larger, near instantaneous, and mostly explosive releases of SO_2 to the upper troposphere and stratosphere (Table 7). These data sources are patchy and go back to the 1970s only, and the statistical distribution of emissions is not well constrained, though some volcanoes appear responsible for substantial fractions of the total volcanic source strength. These include Mt. Etna, which is exceptional not just within Europe but globally as one of the most

Table 6 Measured annual emissions from two continuously degassing volcanoes: Mt. Etna (Italy) and Merapi (Indonesia).

Species (units)	Mt. Etna	Merapi
H_2O (Tg yr^{-1})	55	7.7
CO_2 (Tg yr^{-1})	13	0.88
SO_2 (Tg yr^{-1})	1.5	0.15
H_2S (Gg yr^{-1})	100	73
HCl (Gg yr^{-1})	110	33
HF (Gg yr^{-1})	5.5	1.1
Pb (Mg yr^{-1})	130	2.6
Hg (Mg yr^{-1})	27	0.6
Cd (Mg yr^{-1})	10	0.07
Ag (Mg yr^{-1})	3.3	0.32
Au (kg yr^{-1})	700	0.6

Courtesy of Patrick Allard.

Table 5 The top ten SO_2 emitters among "continuously erupting" volcanoes (Andres and Kasgnoc, 1998). Note that several of these volcanoes have very rarely had their SO_2 emissions measured (including those in Melanesia—four volcanoes in the top ten; and Láscar), and SO_2 fluxes from an individual volcano can and do vary on timescales of months or years (e.g., Mount Etna's SO_2 emission fell to around 0.3–0.4 Tg for the 12 months after July 2001).

Volcano	SO_2 flux (Tg yr^{-1})
Etna, Italy	1.5
Bagana, PNG	1.2
Láscar, Chile	0.88
Ruíz, Colombia	0.69
Sakurajima, Japan	0.69
Manam. PNG	0.34
Yasur, Vanuatu	0.33
Kilauea East Rift, Hawaii	0.29
Masaya, Nicaragua	0.29
Stromboli, Italy	0.27

Table 7 Top 10 recent stratospheric releases of SO_2, based on TOMS data. Typical errors on TOMS estimates are 30%. See Bluth *et al.* (1993, 1997) and Krueger *et al.* (1995) for details of methods.

Volcano	Eruption date(s)	TOMS SO_2 (Tg)
Pinatubo	June 12–15, 1991	20.2[a]
El Chichón	March 28–April 4, 1982	8.1[a]
Sierra Negra	November 13, 1979	4.5
Cerro Hudson	August 8–15, 1991	4.0[a]
Nyamuragira	December 25, 1981	3.3
Mauna Loa	March 25, 1984	2.0
Galunggung	April 5–September 19, 1982	1.73[a]
Alaid	April 27, 1981	1.1
Nyamuragira	October 17, 1998	1.1
Wolf	August 28, 1982	1.1

Courtesy of Simon Carn (TOMS group, NASA).
[a] Cumulative totals of several eruptive episodes. The largest single release was that of Pinatubo on June 15, 1991 (20 Tg of SO_2).

prodigious sources of volcanic gases to the troposphere (Tables 5 and 6). Its average SO_2 emission rate ($\sim$4,000–5,000 Mg d^{-1}) is similar to the total industrial sulfur flux from France (Allard *et al.*, 1991) and must result in substantial elevations in tropospheric sulfate in southern Italy (Graf *et al.*, 1998). It has been suggested that these emissions have caused pollution events in mainland Italy, and even that they have been responsible for deterioration of Roman monuments (Camuffo and Enzi, 1995). Etna also pumps an astonishing estimated 700 kg of gold into the atmosphere every year (Table 6).

There is a reasonable consensus regarding the magnitude of annual volcanic source strengths of sulfur, though difficulties arise in time-averaging the sporadic but large magnitude releases to the stratosphere from explosive eruptions, and in

extrapolating field data for a comparatively small number of observed tropospheric volcanic plumes to the global volcano population. The most widely used global data set is that compiled for the global emissions inventory activity (GEIA) by Andres and Kasgnoc (1998). This arrives at a global annual flux of sulfur from all sources that exceeds 10.4 Tg (Table 8). More recently, Halmer *et al.* (2002) have estimated the global volcanic SO_2 emission to the atmosphere as 15–21 Tg yr^{-1} for the period 1972–2000. Their figures for all sulfur species add considerable uncertainty to the total volcanic sulfur flux (9–46 Tg of sulfur) mainly because of a very large uncertainty in the H_2S emission (1.4–35 Tg of sulfur). For comparison, the IPCC (2001) estimates of annual emissions of other sources of sulfur include anthropogenic (76 Tg), biomass burning (2.2 Tg), and dimethyl sulfide (DMS, 25 Tg, mainly from the oceans). Most of the volcanic source strength is from the continuous degassing of many volcanoes worldwide (Table 8; Sections 1.4.1.1–1.4.1.2). Using the TOMS data set, Pyle *et al.* (1996) estimated the medium-term ($\sim$10^2 yr) annual flux of volcanic sulfur to the stratosphere to be $\sim$1 Tg (range of 0.3–3 Tg). This does not take into account the larger releases of sulfur that are indicated by historic and prehistoric eruptions such as that of Toba 74 kyr BP, and Tambora 1815 AD (Table 9).

The sulfur gases released by volcanoes are either deposited at the Earth's surface or are oxidized to form sulfate aerosol. An important point that has emerged from recent work is the disproportionate contribution of volcanic sulfur emissions to the global atmospheric sulfate budget compared with other sources of sulfur including anthropogenic and oceanic emissions. Episodic,

Table 8 Estimated volcanic emissions of sulfur species to the atmosphere between the early 1970s and 1997.

Emission	SO_2 (Mg d^{-1})	SO_2 (Tg yr^{-1})	SO_2 (Tg yr^{-1})
Sustained SO_2	26,200	9.6	8
Sporadic SO_2	200	0.07	0.04
TOMS SO_2	10,100	3.7	1.9
Total SO_2	36,500	13.4	6.7
Other S species[a]			3.7
Total volcanic S			10.4

Source: Andres and Kasgnoc (1998).
[a] Includes OCS, H_2S, H_2SO_4, etc., partitioned between sustained, sporadic and TOMS in the same proportions indicated in the first three rows.

Table 9 Estimates of sulfur yield from selected major historic and prehistoric eruptions (magnitude >10^{13} kg).

Eruption year and volcano	Magnitude (kg)[a]	Sulfur yield (Tg of S)[b]	Northern hemisphere summer cooling (K)[c]
$\approx$74 kyr BP	7$\times$10^{15}	35–3,300	>1
$\approx$AD 181 Taupo	7.7$\times$10^{13}	?6.5	0.4
$\approx$AD 1028 Baitoushan	5.8$\times$10^{13}	>2	0.5
$\approx$1257 Unknown	10^{14}–10^{15}?	>100	?
$\approx$1452 Kuwae	>8$\times$10^{13}	?40	0.5
1600 Huaynaputina	2.1$\times$10^{13}	23–55[d]	0.8
1815 Tambora	1.4$\times$10^{14}	28	0.5
1883 Krakatau	3.0$\times$10^{13}	15	0.3
1902 Santa Maria	2.2$\times$10^{13}	11[e]	Not detected
1912 Katmai	2.5$\times$10^{13}	10	0.4
1991 Pinatubo	1.3–1.8$\times$10^{13}	10.1	<0.5

[a] Total eruption magnitude for multiple phases of eruption and combining plinian and phoenix cloud ashfall and associated pyroclastic flow deposits where applicable, data mainly from Carey and Sigurdsson (1989), Chesner and Rose (1991), Monzier *et al.* (1994), Holasek *et al.* (1996), Pyle (2000), Horn and Schmincke (2000), Adams *et al.* (2001), and Oppenheimer (2003b).
[b] Stratospheric sulfur yield from Table 6, and from Zielinski (1995), de Silva and Zielinski (1998).
[c] Estimated northern hemisphere summertime temperature anomaly derived from tree-ring chronologies reported by Briffa *et al.* (1998) for eruptions before Mt. St. Helens (note that other records do indicate a$\approx$0.2 K northern hemisphere summer cooling in 1903).
[d] Costa *et al.* (2003).
[e] Since this estimate is based on ice core sulfate deposition, it may reflect the cumulative aerosol fallout of other notable 1902 eruptions, i.e., Mont Pelée (Martinique) and Soufrière (St. Vincent) as well as the Santa Maria event.

large magnitude eruptions are the principal perturbation to stratospheric aerosol levels (e.g., the 30 Mt of sulfate injected by the 1991 Pinatubo eruption). In the troposphere, the picture is less clear, but modeling suggests that up to 40% of the global tropospheric sulfate burden may be volcanogenic (Graf *et al.*, 1997), though Stevenson *et al.* (2003a) and Chin and Jacob (1996) obtained lower figures (14% and 18%, respectively), in part due to use of a lower volcanic sulfur source strength. In any case, these figures all exceed the fraction of the sulfur source to the atmosphere that is volcanogenic (around 10%) because of the generally higher altitudes of entrainment of volcanic sulfur compared with biogenic (DMS) or anthropogenic sources, and hence the longer residence time of volcanic SO_2 compared with other sources. This is largely because of lower deposition rates, and results in more conversion of SO_2 to sulfate, and a longer residence time of the higher altitude aerosol. Sulfate aerosol plays a significant role in the Earth's radiation budget, because it may both backscatter incoming shortwave solar radiation and absorb outgoing longwave radiation, the competition of these processes depending strongly on particle size. In addition, sulfate aerosol can have a secondary, and possibly more profound, radiative effect by promoting cloud condensation or modification of the microphysical properties and longevity of existing clouds (Graf *et al.*, 1997). Changes in this "background" emission in time and space could represent an important forcing that has yet to be characterized.

1.5.2   Carbon and Water

As mentioned in Section 1.4.3.2, SO_2 is the most readily measured volcanic volatile in the atmosphere. Thus, most flux estimates of other components have been based on measuring their ratios to SO_2 (often obtained by *in situ* sampling methods), and multiplying them by the SO_2 flux measured using COSPEC. Other approaches include scaling estimates of ^3He flux against measured $C/^3He$ ratios, and extrapolation of the few available direct observations of CO_2 flux. Arthur (2000) provides a review of current estimates of global subaerial volcanic CO_2 flux, which range from 15 Tg yr^{-1} to 130 Tg yr^{-1}. This is swamped by the anthropogenic source of carbon to the atmosphere and is not considered globally significant on short timescales, though CO_2 emissions are important as a local hazard (Section 1.6.5). At Kilauea, a "hot spot" volcano, the exhaled CO_2 is derived directly from the mantle (Gerlach and Taylor, 1990), but carbon-isotope studies show that at arc volcanoes most of the CO_2 is recycled from subducted organic and carbonate sediments (Section 1.2.1; Fischer *et al.*,

1998). Estimates of the CO_2 outgassing to the oceans by mid-ocean ridge basalt (MORB) volcanism are mostly in the range 90–350 Tg yr^{-1}.

While a few estimates of water budgets have been attempted for individual volcanoes and eruptions (e.g., 3 Tg yr^{-1} for White Island, New Zealand (Rose *et al.*, 1986); 13 Tg yr^{-1} for Masaya, Nicragua (Burton *et al.*, 2000)), meaningful estimates of the global volcanic flux of water to the atmosphere are unavailable as of early 2000s. Interpretation of water emission rates from volcanoes is complicated by the likelihood, in many cases, that a substantial fraction of the emitted water has been derived from groundwaters (e.g., Taran *et al.*, 1995).

1.5.3   Halogens

In contrast to carbon, volcanic emissions of halogens are thought to be substantial compared with other sources. However, the fluxes remain very poorly constrained. They are important as they play a critical role in atmospheric chemistry—in boundary layer ozone depletion, tropospheric hydrocarbon oxidation, the oxidizing capacity of the troposphere, and stratospheric ozone depletion (Section 1.6.1.2). Realistic estimates of the anthropogenic and natural emissions of halogen species are essential in order to model and assess such processes accurately. Very little work exists to improve upon the emission inventory for volcanic sources elaborated by Cadle (1975, 1980). The more recent work indicates greater than one to two orders of magnitude of uncertainty in halogen fluxes (Table 10).

Measurements of gas samples collected at Augustine Volcano, Alaska, indicate typical orders of magnitude concentrations of different halogen species in volcanic emissions (Table 11). With peak SO_2 emission rates of nearly 280 kg s^{-1} during minor eruptive episodes at Augustine, even the HBr emission corresponds to ~0.5 kg s^{-1}. These emissions are significant because they are, at many volcanoes, released directly into the free troposphere, where they can initiate and catalyze a large range of processes. Again, Mt. Etna is a

Table 10   Estimated annual mean global emissions of HCl, HF, and HBr from volcanoes.

Volcanic source	HCl (Tg)	HF (Tg)	HBr (Gg)
Cadle (1980)	7.8	0.4	78
Symonds *et al.* (1988)	0.4–11	0.06–6	
Halmer *et al.* (2002)[a]	1.2–170	0.7–8.6	2.6–43.2

[a] For period 1972–2000.

prodigious source of halogens, continuously emitting $\sim$2 kg s^{-1} of HF, and >8 kg s^{-1} of HCl (Francis *et al.*, 1998).

Bureau *et al.* (2000) considered the input of bromine to the atmosphere from explosive eruptions, based on experimental data for synthetic (albite) melts (Table 12). Their results suggested even stronger partitioning of both bromine and iodine into the fluid phase compared with chlorine. Assuming a Cl/Br mass ratio of $\sim$300 based on measurements of a range of volcanic rocks, and scaling against estimated chlorine emissions for various explosive eruptions, they suggested that volcanoes could be a significant source for stratospheric bromine. They pointed out, however, that

Table 11 Volcanic halogens at Mt. Augustine, all figures as mole fractions.

HCl	6.0×10^{-2}
HF	5.4×10^{-4}
SiF$_4$	8.0×10^{-5}
NaCl	1.4×10^{-5}
KCl	8.7×10^{-6}
HBr	6.8×10^{-6}
SiOF$_2$	2.1×10^{-6}
FeCl$_2$	1.6×10^{-6}

Source: Andres and Rose (1995).

Table 12 Estimated chlorine and bromine yields of eruptions.

Eruption	Cl yield (Tg)	Br yield (Gg)
Toba 74 kyr BP	400–4,000	1,460–33,700
Tambora 1815	216	790–1,820
Krakatau 1883	3.75	14–31
Mt. St. Helens 1980	0.67	2.4–5.6
El Chichón 1982	0.04	0.15–0.4
Mt. Pinatubo 1991	>3	>11–25

Source: Bureau *et al.* (2000).

the behavior of bromine in eruption columns is not known at present, so it is not clear what fraction of bromine degassed would actually cross the tropopause and remain in the stratosphere.

1.5.4 Trace Metals

Volcanoes are an important source of trace metals to the atmosphere—for some species (e.g., arsenic, cadmium, copper, lead, and selenium) they may be the principal natural source (e.g., Nriagu, 1998). Measurements of fluxes are complicated because of rapid condensation of the vapor phases carrying the trace metals, but estimates have been attempted, based on scaling of metal/sulfur ratios by known sulfur fluxes or by using the flux of a radioactive volatile metal (^{210}Po) as the normalizing factor (e.g., Lambert *et al.*, 1988). These methods have provided metal flux estimates for individual volcanoes (e.g., Erebus (Zreda-Gostynska *et al.*, 1997), Stromboli (Allard *et al.*, 2000), Etna (Gauthier and Le Cloarec, 1998), Vulcano (Cheynet *et al.*, 2000)), and estimates of global source strengths (e.g., Nriagu, 1989; Hinkley *et al.*, 1999; Table 13).

1.6 IMPACTS

The atmospheric, climatic, environmental, and health effects of volcanic volatile emissions depend on several factors but importantly on fluxes of sulfur and halogens. As discussed in Section 1.5.1, intermittent explosive eruptions can pump >10^{10} kg of sulfur into the stratosphere, against a background of continuous fumarolic and "open-vent" emission into the troposphere. The episodic, large explosive eruptions are the principal perturbation to stratospheric aerosol levels (e.g., 30 Mt of sulfate due to the 1991 eruption of Pinatubo), and can result in global climate

Table 13 Comparison of volcanic, total natural and anthropogenic fluxes to the atmosphere of selected species.

Element	Volcanic (Gg yr^{-1})	Total natural (Gg yr^{-1})	Anthropogenic (Gg yr^{-1})
Al	13,280[a]	48,900[b]	7,200[b]
Co	0.96[c]	6.1[c]	4.4[b]
Cu	1.0[d], 4.7[c], 15[e], 22[f]	28[c]	35[c]
Zn	4.8[c], 7.2[d], 8.5[f]	45[c]	132[c]
Pb	0.9[d], 1.7[c], 2.5[e], 4.1[f]	12[c]	332[c]
As	1.9[c]	12[c]	19[c]
Se	0.3[f], 0.5[c]	9.3[c]	6.3[c]
Mo	0.2[c]	3.0[c]	3.3[c]
Cd	0.4[c]	1.3[c]	7.6[c]

Source: Mather *et al.* (2003a).
[a] Symonds *et al.* (1988).
[b] Lantzy and Mackenzie (1979).
[c] Nriagu (1989).
[d] Hinkley *et al.* (1999).
[e] Lambert *et al.* 1988.
[f] Le Cloarec and Marty (1991).

forcing and stratospheric ozone depletion (Sections 1.6.1 and 1.6.2). Large lava eruptions, such as that of Laki in Iceland in 1783–1784, which released $\approx$120 Tg of SO_2, 7.0 Tg of HCl, and 15 Tg of HF into the upper troposphere–lower stratosphere region, have resulted in major pollution episodes responsible for regional-scale extreme weather, damage to farming and agriculture, and elevated human morbidity and mortality (Thordarson et al., 1996; Thordarson and Self, 2002). Individual continuously degassing volcanoes such as Mt. Etna and Masaya volcano (Nicaragua) can also represent major polluters. The adverse environmental and health impacts observed downwind of many degassing volcanoes are widely recognized if poorly investigated (e.g., Baxter et al., 1982; Delmelle et al., 2001, 2002; Delmelle, 2003). Tropospheric volcanic emissions and their impacts are discussed in Sections 1.6.3–1.6.5.

1.6.1 Stratospheric Chemistry and Radiative Impacts of Volcanic Plumes

On April 2, 1991, steam explosions were observed on a little known volcano called Mount Pinatubo on the island of Luzon, the Philippines. Within 11 weeks, on the afternoon of June 15, and following a crescendo in activity, the volcano erupted $(1.3–1.8) \times 10^{13}$ kg (8.4–10.4 km^3 bulk volume) of pumice. It devastated a 400 km^2 area and mantled much of Southeast Asia with ash. This was the second largest magnitude eruption of the twentieth century after that of Katmai in Alaska in 1912. The tropopause was at $\sim$17 km altitude at the time of the eruption and was punctured both by the central eruption column fed directly by the vent and by co-ignimbrite or "phoenix" clouds that lofted above immense pyroclastic currents moving down the west and south flanks of the volcano. The eruption intensity peaked between around 13:40 h and 16:40 h local time on June 15, based on infrasonic records from Japan, with vent exit velocities of $\sim$280 m s^{-1}, and discharge rates estimated at 1.6×10^9 kg s^{-1}. A wide variety of observations, especially from satellite instruments, have quantified the eruption's emissions and their impacts on the atmosphere and climate. The eruption serves as a benchmark in our understanding of the impacts of major eruptions on the Earth system, and is therefore discussed in detail below.

1.6.1.1 Formation of stratospheric sulfate aerosol veil

The stratospheric umbrella cloud formed during the Pinatubo eruption attained a vertical thickness of 10–15 km, extending from the tropopause up to $\sim$35 km above sea level. Weather satellites tracked the cloud for two days, after which other spaceborne instruments, including TOMS, were able to continue monitoring dispersal of the plume. It took 22 days for the cloud to circumnavigate the globe. Its estimated initial SO_2 yield, $\sim$17–20 Tg, remains the largest measured (Bluth et al., 1992; Read et al., 1993). The amount of SO_2 decreased daily through oxidation by $\cdot$OH radicals (Coffey, 1996):

$$SO_2 + \cdot OH \rightarrow HOSO_2 \qquad (8)$$

Various chemical pathways then promoted the formation of sulfuric acid (H_2SO_4):

$$HOSO_2 + O_2 \rightarrow SO_3 + HO_2 \qquad (9)$$

$$SO_3 + H_2O \rightarrow H_2SO_4 \qquad (10)$$

This scheme conserves HO_x and results in a predicted e-folding lifetime of SO_2 in the stratosphere of 38 days (i.e., the time taken for the abundance of SO_2 to drop by 1/e of its starting amount), only slightly longer than the observed time of 33–35 days (Figure 8; Read et al., 1993). The aerosol consisted of around 25% water and 75% sulfuric acid by weight.

The aerosol cloud was tracked and measured by several instruments, including the Stratospheric Aerosol and Gas Experiment (SAGE) satellite (McCormick et al., 1995). Independent estimates of the total mass of aerosol generated ($\sim$30 Tg) are slightly higher than the amount expected from the initial SO_2 load (Baran et al., 1993; Baran and Foot, 1994). The cloud was so thick at its peak that no sunlight was transmitted and there are missing data from the record. Gradually, the aerosol sedimented back to the surface, and, by the end of 1993, only $\sim$5 Tg of aerosol remained airborne. The observed e-folding time was $\sim$12 months, comparable to that measured after the previous major stratospheric aerosol perturbation due to the 1982 eruption of El Chichón in Mexico.

1.6.1.2 Impacts on ozone chemistry

Following the Pinatubo eruption, global stratospheric ozone levels began to show a strong downturn (McCormick et al., 1995). Ozone levels decreased 6–8% in the tropics in the first months after the eruption. These figures for the total vertical column of ozone hide local depletions of up to 20% at altitudes of 24–25 km. By mid-1992, total column ozone was lower than at any time in the preceding 12 yr, reaching a low point in April 1993 when the global deficit was $\sim$6% compared with the average. Losses were greatest in the northern hemisphere. For example, total ozone above the USA dropped 10% below average with the strongest depletion observed between 13 km and 33 km altitude. These decreases have been attributed to complex chemical reactions

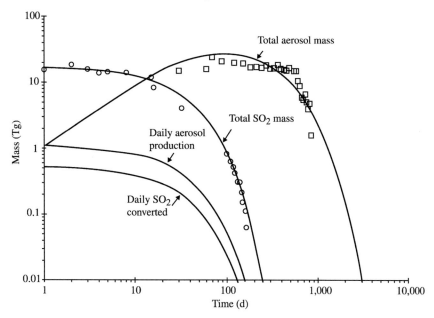

Figure 8 Oxidation of SO_2 to SO_4 in the Pinatubo stratospheric cloud. The "total SO_2 mass" curve indicates the loss of SO_2 by oxidation to sulfate aerosol according to an initial loading of 17 Tg of SO_2 (after Read *et al.* (1993), and 3 Tg lower than the TOMS-only estimate of Bluth *et al.* (1992)) and a 33 d e-folding time. The circles indicate satellite measurements of stratospheric SO_2 burden from Read *et al.* (1993). The "total aerosol mass" curve is obtained by modeling the aerosol mass generated by sulfate oxidation and an e-folding time for aerosol loss of 1 yr. The squares show estimates of the stratospheric aerosol mass from Baran *et al.* (1993). Daily rates of SO_2 conversion and aerosol production are shown by the two other curves (labeled).

exploiting the presence of the aerosol surface. These result in a shift of stratospheric chlorine from stable compounds (HCl and $ClONO_2$) into more reactive ones (e.g., HOCl) that can destroy ozone (Coffey, 1996). First, the aerosol surface promotes reactions that remove gaseous NO_x compounds. For example, N_2O_5 is removed as follows:

$$N_2O_5 + H_2O \rightarrow 2HNO_3 \quad (11)$$

$$N_2O_5 + HCl \rightarrow ClONO + HNO_3 \quad (12)$$

In the months following the eruption, increases in HNO_3 were detected (e.g., Koike *et al.*, 1994) providing compelling evidence for the transfer of nitrogen from comparatively reactive (NO_x) to more inert species (HNO_3). Depletion of N_2O_5, in turn, reduces NO_2 levels. Airborne measurements obtained above the USA and Mexico after the eruptions show strong depletions in NO_2 levels (Coffey and Mankin, 1993). Remarkably, by January 1992, seven months after the eruption, NO_2 concentrations had halved. This is important because NO_2 ordinarily captures reactive chlorine monoxide, forming the inert species $ClONO_2$:

$$ClO + NO_2 \rightarrow ClONO_2 \quad (13)$$

In other words, if stratospheric NO_x levels are lowered, chlorine monoxide levels increase, increasing rates of ozone destruction.

Although enhanced stratospheric HCl levels were reported after the El Chichón eruption (with up to 40% increases over background levels observed within the cloud), it is generally considered that most volcanic HCl is "scrubbed" out of eruption columns as they ascend through the troposphere, due to its high solubility. Pinatubo emitted an estimated 3 Tg of chlorine, but measurements indicate that it was rapidly removed, perhaps assisted by the estimated 250–300 Tg of steam erupted, and the comparable amount of tropospheric moisture entrained into the eruption column. From analyses of the sulfate marker in the Greenland ice core for the 1815 Tambora eruption, Delmas *et al.* (1992) found no evidence for changes in atmospheric chlorine associated with this much larger event, suggesting that the massive amount of chlorine thought to have been released by the eruption (Table 12) was rapidly and efficiently scavenged by the troposphere as the eruption clouds ascended.

Although simple eruption-column physical and chemical models have indicated that halogens are readily scavenged in eruption clouds by hydrometeors (Tabazadeh and Turco, 1993), more recent, and more sophisticated, models have suggested that limited scavenging by hydrometeors in a dry troposphere could permit substantial quantities of HCl and HBr to reach the stratosphere (Textor *et al.*, 2003). The observation of chlorine-rich fallout layers in the GISP2 Greenland ice core linked to another great eruption, that of Mt. Mazama (USA) in ≈5677 BC (Zdanowicz *et al.*,

1999), lends weight to the idea that there are some atmospheric environmental conditions that permit transport of substantial quantities of halogens across the tropopause.

1.6.1.3 Optical and radiative effects

The widespread dispersal of Pinatubo aerosol led to many spectacular optical effects in the atmosphere, including vividly colored sunsets and sunrises, crepuscular rays and a hazy, whitish appearance to the Sun. These phenomena occurred as the aerosol veil absorbed and scattered sunlight. An objective measure of the fraction of solar energy removed as it travels down through an aerosol layer is the optical depth, τ, defined by the relationship between the initial irradiance, I_0, at the top of the layer, and the radiation received at the bottom of the layer, I:

$$I = I_0 \exp(-\tau/\cos\theta) \qquad (14)$$

where θ is the solar zenith angle (measured between the vertical and the Sun). τ depends on the concentration and thickness (path length) of the aerosol as well as its absorbing or scattering properties. In the months following the Pinatubo eruption, optical depths in the stratosphere were the highest ever recorded by modern techniques, peaking at $\sim$0.5 in the visible region in August–September 1991.

The appearance of these optical effects of Pinatubo's aerosol veil necessarily implies that it was interrupting the transmission of visible solar radiation to the Earth's surface. In detail, the radiative effects of stratospheric aerosol veils are highly complex, because they can consist of variable proportions of minute glassy ash fragments and sulfuric acid droplets of different compositions, sizes, and shapes (and hence optical properties). The various components also accumulate and sediment out at different rates according to their masses and aerodynamic properties, so that any effects on the Earth's radiation budget can be expected to change through time. As a result, the effective radius of the aerosol veil should decrease in the months after an eruption. Measurements following the El Chichón and Pinatubo eruptions agree with this picture (McCormick *et al.*, 1995). The particles scatter incoming solar ultraviolet and visible radiation, directing some back into space but also sideways and forwards. The aerosol can also absorb radiation—short wavelength from the Sun or long wavelength from the Earth—and warm up. The net effect on the Earth's radiation budget is not straightforward to determine.

Observations by the Earth Radiation Budget Experiment (ERBE) satellite sensor revealed significant increases in the reflectivity of the Earth (Minnis *et al.*, 1993). By August 1991, the backscattering of solar radiation by the aerosol had

increased the global albedo to $\sim$0.25, some 5 SDs above the 5 yr mean of 0.236. Corresponding reductions of the direct solar beam were $\sim$25–30%. ERBE was also able to show that the albedo increase was not uniform across the Earth but was most pronounced in normally low-albedo, cloud-free regions, including the Australian deserts and the Sahara, and in typically high-albedo regions associated with convective cloud systems in the tropics such as the Congo Basin and around New Guinea. The latter observation is initially puzzling, because over regions that have a naturally high albedo, the percentage increase due to volcanic aerosol in the stratosphere is small. It appears that transport of Pinatubo aerosol across the tropopause seeded clouds, or modified the optical properties of existing upper tropospheric cirrus clouds. Satellite observations support this interpretation since they indicate a correlation between Pinatubo aerosol and increased cirrus clouds, persisting for more than 3 yr after the eruptions. The enhancement was especially noticeable in mid-latitudes $\sim$6 months after Pinatubo's eruption, consistent with the likely time lag before accumulation and initial sedimentation into the troposphere of the sulfate aerosol. The mid-latitudes are dominant sites for transfer between stratosphere and troposphere across tropospheric folds associated with the jet stream and cyclonic systems.

The albedo changes corresponded to radiative forcing, as observed by ERBE (Minnis *et al.*, 1993). In July 1991, the short-wave flux increased dramatically over the tropics. The corresponding net flux decreased by 8 W m^{-2} in August 1991, twice the magnitude of any other monthly anomaly. A similar but weaker trend was observed between 40°S and 40°N, with the net forcing for August 1991 amounting to -4.3 W m^{-2}, nearly 3 SDs from the 5 yr mean. Unfortunately, ERBE was not operating pole-ward of 40° latitude but even if there were no aerosol forcing at higher latitudes, then the globally averaged volcanic forcing still amounts to -2.7 W m^{-2}. This should represent the minimum global forcing, because enhanced stratospheric aerosols were observed at higher latitudes by mid-August 1991. The net radiation flux anomalies seen by ERBE remain the largest that have been observed by satellites. These results represented the first unambiguous, direct observations of eruption-induced radiative forcing (Stenchikov *et al.*, 1998).

Stratospheric aerosols can also have a "greenhouse" effect on the Earth if they are effective in absorbing upwelling long-wave radiation. For this warming effect to outbalance the cooling due to increased albedo, the aerosol's effective radius, r_{eff} (its mean radius weighted by surface area), should exceed 2 μm. Prior to Pinatubo, the r_{eff} of stratospheric aerosol was $\sim$0.2 μm. Pinatubo's aerosol pushed the stratospheric r_{eff} up to more than 0.5 μm compatible with the observed negative forcing.

1.6.2 Climatic Impacts of Major Volcanic Eruptions

The impacts of volcanism on climate are complex in both space and time (Robock, 2000). They can also depend on complex feedback mechanisms; for example, the lower global mean surface temperatures following the 1991 Pinatubo eruption resulted in less evaporation, and so less atmospheric heating due to water vapor (Soden *et al.*, 2002). Best understood now are the short-term climatic perturbations following high sulfur-yield explosive eruptions like Pinatubo. Scaling up to the largest known explosive eruptions (e.g., the Younger Toba Tuff, 74 kyr BP) poses considerable difficulties, not least because of uncertainties in the microphysics and chemistry of very large releases of sulfur to the atmosphere. No large mafic eruption has occurred in modern times, but there is ample documentary evidence to implicate the Laki 1783–1784 eruption in strong climatic effects, at least at regional scale. On timescales up to 10^6 yr, large igneous provinces (see Chapter 18) may be responsible for climate change, though even less is understood about the volatile yields or dynamics of such eruptions. At timescales $>10^6$–10^7 yr, variations in the rate of oceanic crust growth are associated with major global climate change via their control on eustatic sea level and atmospheric concentrations of CO_2, and hence the carbonate–silicate cycle (Section 1.1.1). This section focuses on the volcanic forcing of climate on the shorter timescales (1–10^3 yr).

1.6.2.1 Intermediate to silicic eruptions

Pinatubo's negative radiative forcing exceeded for two years the positive forcings due to anthropogenic greenhouse gases (carbon dioxide, methane, CFCs and N_2O; McCormick *et al.* (1995)). The net drop in global mean tropospheric temperature in 1992 was ∼0.2 K compared with the baseline from 1958 to 1991. This may not sound much, but it is a globally averaged figure that hides much larger regional and temporal variations, with pockets of abnormally strong surface heating as well as cooling. For example, the Siberian winter was 5 K warmer, and the North Atlantic was 5 K cooler than average. Furthermore, the actual global troposphere temperature decrease due to Pinatubo is closer to 0.4 K if a correction is made for the tropospheric warming associated with the 1992 El Niño-Southern Oscillation event.

The satellite-borne Microwave Sounding Unit (MSU) records lower stratospheric temperatures.

Global mean increases of up to 1.4 K were apparent following both the Pinatubo and El Chichón eruptions due to the local heating of the volcanic aerosol (Parker *et al.*, 1996). Interestingly, for the Pinatubo case, the temperature anomaly decreased as aerosol sedimented back into the troposphere, and, by early 1993, below average, lower stratospheric temperatures were observed. This could be due to cooling coincident with the destruction of stratospheric ozone (Section 1.6.2.2).

Kelly *et al.* (1996) examined surface temperature anomalies following the Pinatubo eruption from worldwide observatory records of surface air temperature and nighttime marine air temperature obtained from ships and buoys. The observed global cooling was initially rapid but punctuated by a warming trend, predominantly over land, between January and March 1992. Cooling resumed, and by 1 yr after the eruption amounted to ∼0.5 K. In 1992, the USA experienced its third coldest and wettest summers in 77 years. The Mississippi flooded its banks spectacularly while drought desiccated the Sahel. There were further relative warmings in early 1993 and 1995, and the mid-year cooling reduced each year. Globally averaged sea-surface temperatures indicated a slightly lower maximum cooling ∼0.4 K, which can be accounted for by the high thermal inertia of the oceans. These trends are more or less mirrored by mean lower tropospheric temperatures determined by the MSU and have been fitted quite successfully by climate models (e.g., Hansen *et al.*, 1997). Indeed, the Pinatubo case has provided an important test of the capabilities of climate models for the simulation of atmospheric radiation and dynamics.

The winter warmth after the Pinatubo eruption was concentrated over Scandinavia and Siberia and central North America. These temperature anomalies were associated with marked departures in sea-level pressure patterns in the first northern winter. There was a pole-ward shift and strengthening of North Atlantic westerlies at ∼60° N, associated with corresponding shifts in the positions and strengths of the Iceland Low and Azores High. These effects have been modeled as a result of changes to the atmospheric circulation around the Arctic (the Arctic Oscillation; Thompson and Wallace (1998)) arising from the differential heating effects of the volcanic aerosol in the upper atmosphere (Graf *et al.*, 1993; Kirchner *et al.*, 1999; Stenchikov *et al.*, 2002). Strong stratospheric heating in the tropics due to the presence of volcanic aerosol establishes a steeper meridional thermal gradient at the tropopause, enhancing geostrophic winds, and the polar vortex. At the same time, surface cooling in the subtropics leads to weaker planetary waves, and surface warming of mid-high latitudes, reducing tropospheric temperature gradients. This also helps to strengthen the polar vortex, and the combined effects amplify the Arctic Oscillation, leading to winter warming. The stronger northern hemisphere response, compared with the southern hemisphere, reflects the greater landmass area.

One apparent impact of the volcanic forcing induced by the Pinatubo eruption was a reduction in the melt area of the Greenland Ice Sheet observed by passive microwave satellite imagery (Abdalati and Steffen, 1997). Melting and freezing are of great significance in ice sheet thermal dynamics, because they represent strong positive feedback mechanisms. When snow melts its albedo drops from about 0.9 to 0.7. Thus, it absorbs even more incident radiation, warms up more, and melts more. The reverse situation occurs as wet snow freezes. The combination of Greenland's high albedo and size (its ice sheet covers 1.75×10^6 km^2) accounts for a strong cooling contribution to Arctic climate. Superimposed on a (poorly understood) 3 yr cycle in melt extent was a longer-term increasing trend in melt area of over 4% per year up to 1991. However, then, in 1992, there was a drop of almost 1.5×10^5 km^2 in the melt extent. This pattern is apparent also in coastal temperature records for Greenland, which indicate around a 2 K cooling between 1991 and 1992. The calculated decrease of melting extent in 1992 amounts to a decrease in absorption of solar radiation by the ice sheet of $\sim 10^{13}$ W. If Pinatubo were responsible for this cooling of Greenland, then it has important implications for the amplification of volcanic aerosol forcing by ice-albedo feedbacks, and the potential enhancement of regional and global cooling following large eruptions. Meanwhile, cooling of the Red Sea is thought to have enhanced mixing in the water column, bringing nutrients to the surface and stimulating algal blooms (Genin et al., 1995). These, in turn, resulted in coral mortality.

The largest known historic eruption, that of Tambora in 1815, yielded an estimated 60 Tg of sulfur to the stratosphere (based mostly on ice core sulfate deposition records; reviewed by Oppenheimer (2003a)). This is ~ 6 times the release of Pinatubo. Considerable efforts have been expended to quantify the climatic effects of this eruption, and have spanned investigations of surface temperature measurements and proxy climate indicators such as tree rings (Harington, 1992). Anomalously cold weather hit the northeastern USA, maritime provinces of Canada, and Europe in the boreal summer of 1816, which came to be known as the "year without a summer" in these regions. Widespread crop failures occurred in Europe and North America in 1816, and the eruption has been implicated in accelerated emigration from New England, widespread outbreaks of epidemic typhus, and what Post (1977) termed "the last great subsistence crisis in the western world."

Briffa and Jones (1992) reconstructed 1816 weather across Europe from contemporary meteorological observations and tree-ring data. They showed that the summer temperatures across much of Europe were 1–2 K cooler than the average for the period 1810–1819, and up to 3 K cooler than the mean for 1951–1970. Rainfall was also high across much of Europe in the summer of 1816. More recent dendrochronological studies have confirmed the distribution of these cool summers on both sides of the North Atlantic (Briffa et al., 1998). These reconstructions of northern hemisphere summer temperatures indicate 1816 as one of the very coldest of the past six centuries, second only to 1601 (the year after the eruption of Huaynaputina in Peru). The estimated mean northern hemisphere (land and marine) surface temperature anomalies in the summers of 1816, 1817, and 1818 are -0.51 K, -0.44 K, and -0.29 K, respectively. Despite Tambora's substantially greater sulfur yield to the atmosphere, it does not appear to have had a correspondingly larger effect on global surface temperatures, highlighting the nonlinear scaling between volatile yield and climatic impact.

Assessing the impacts of the largest known Quaternary "super-eruption," that of the Younger Toba Tuff (YTT), Sumatra, at 74 kyr BP, is highly challenging given uncertainties in the key parameters for the eruption (intensity, height, and magnitude), and amounts of gaseous sulfur species released (Oppenheimer, 2002). At the high end of impact claims, Rampino and Self (1992) have argued that Toba caused a "volcanic winter"—a global mean surface temperature drop of 3–5 K (Table 12). However, there are considerable challenges in modeling the climatic impacts of such large eruptions, in particular arising from uncertainties in cloud microphysics. Pinto et al. (1989) argued that the chemical and physical processes occurring in very large volcanic clouds may act in a "self-limiting" way. They applied simplified aerosol photochemical and microphysical models to show that, for a 100 Tg injection of SO_2, condensation and coagulation produce larger aerosol particles, which are less effective at scattering incoming sunlight, and also sediment more rapidly. Combined, these effects lessen the expected magnitude and duration of climate forcing expected from super-eruptions, certainly compared with linear extrapolations of observed climate response following eruptions like Pinatubo 1991.

Zielinski et al. (1996b) reported finding the Toba sulfate marker in the GISP2 core. Their high temporal resolution measurements showed that while the Toba anomaly coincides with a 1 kyr cool period between interstadials 19 and 20, it is separated by the 2 kyr long, and dramatic warming event of interstadial 19, from the prolonged (9 kyr) major glacial which began ~67.5 kyr BP. This undermines earlier suggestions that the YTT eruption played a role in initiating the last glaciation, though leaves open the possibility that it is implicated in the ~1 kyr cold period prior to interstadial 19.

Efforts to model the climatic impacts of supereruptions were reported at a Chapman Conference on Santorini in 2002. Jones and Stott (2002) showed that simulations of a 2,000 Tg injection of SO_2 in the tropical stratosphere (100 × Pinatubo) resulted in a 5–7 yr forcing with a peak of -60 W m^{-2} (global annual mean at the tropopause), corresponding to a peak global mean monthly temperature anomaly near the surface of -10 K, remaining about -2 K for 5–7 yr, and remaining around -0.5 K for up to 50 yr due to the thermal inertia of the oceans. Fifteen months after the eruption, the maximum cooling in parts of Africa and North America exceeded 20 K in the model, resulting in a large increase in land and sea ice, and reduced evaporation and precipitation. Interestingly, the snow and ice cover were neither extensive enough nor persistent enough in the model to cause an ice-albedo effect and an ice age transition.

1.6.2.2 Mafic eruptions

At first consideration, it might seem that mafic eruptions, typically more effusive in character, would not have much impact on hemispheric- to global-scale climate because of their lower-altitude sulfur emission. There are some compensating factors, however. First, mafic magma tends to contain high sulfur contents, and high-intensity lava eruptions can propel fire fountains to heights in excess of 1 km altitude, and can persist for days, weeks, months, and even years, compared with the several hour bursts of an eruption like Pinatubo's. Also, there are some rare instances of basaltic plinian eruptions, e.g., Masaya, >20 and ≈6.5 kyr BP (Williams, 1983), Mt. Etna, 122 BC (Coltelli *et al.*, 1998), and Tarawera, 1886 AD (Walker *et al.*, 1984), that may have combined high eruption intensities and plume heights with high sulfur yields.

Interest in the potential global impacts of effusive volcanism has been fuelled by the coincidence of some of the greatest mass extinctions that punctuate the fossil record with the massive outpourings of lava during flood basalt episodes (Rampino and Stothers, 1988). Estimated sulfur yields of such provinces are certainly very high (e.g., Thordarson and Self, 1996), and Ar–Ar dates have now demonstrated that the immense volumes of lava (of order 10^6 km^3) are erupted in comparatively short time periods (~1 Myr, e.g., the 30 Myr old Ethiopian Plateau basalts; Hofmann *et al.* (1997)), suggesting sustained high annual fluxes of sulfur and other volatiles to the atmosphere.

On a vastly smaller scale, the Laki fissure eruption on Iceland in 1783–1784 (which emitted ~15 km^3 of magma) yielded an estimated 122 Tg of SO_2, distributed between the troposphere and stratosphere (Thordarson and Self, 2002). Fire fountains reached an estimated 1,400 m in height and are believed to have fed convective columns to up to 10–13 km in height (the upper troposphere–lower stratosphere region). Recent chemistry-transport models for a Laki-like eruption have indicated that a surprisingly high fraction (60–70%) of the SO_2 in such a release may be deposited at the surface before it is oxidized to sulfate aerosol in the atmosphere (Stevenson *et al.*, 2003b). This suggests that assumptions of complete conversion of SO_2 to aerosol might result in overestimation of the radiative effects of the volcanic clouds. However, such results have important implications for SO_2 air quality over distances of up to thousands of kilometers from the source.

Various meteorological and proxy records do indicate regional climate anomalies in 1783–1784 (e.g., Briffa *et al.*, 1998), but the patterns are not yet well understood or modeled. More generally, it is not clear how readily volcanic plumes generated by fissure and flood basalt eruptions can reach, and entrain sulfur into, the stratosphere, and considerable work is still required to understand the circumstances by which effusive activity can compete in the climate stakes with explosive eruptions.

1.6.3 Tropospheric Chemistry of Volcanic Plumes

Thanks largely to the work following Pinatubo, there is now a good understanding of the stratospheric chemistry of volcanic eruption clouds, at least for emissions on this scale (Section 1.6.1). In contrast, the tropospheric chemistry of volcanic plumes is rather poorly known. This partly stems from a wide spectrum of emissions and emission styles (e.g., continuous degassing, minor eruptions), rapid transport of some components to the Earth's surface, and the greater concentration and variability of H_2O in the troposphere. Reactive sulfur, chlorine, and fluorine compounds may be present in both the gas and particle phase in volcanic plumes, and are typically co-emitted with many other volatile species, including water vapor, and with silicate ash particles (Figure 9).

Once in the troposphere, physical and chemical processes convert gaseous SO_2 to sulfate, while gaseous HF and HCl establish equilibria with aqueous phase H^+, F^-, and Cl^-. SO_2 reacts with ·OH radicals during daylight, giving an SO_2 lifetime of around a week in the troposphere (substantially shorter than in the stratosphere). In volcanic

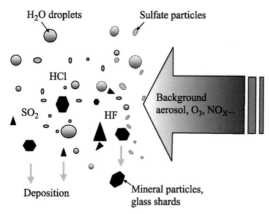

H₂O droplets   Sulfate particles

HCl

SO₂   HF

Background
aerosol, O₃, NOₓ...

Deposition   Mineral particles,
                    glass shards

Figure 9 Volcanic clouds are typically composed of gases and particles and diluted by the background atmosphere. Various chemical and physical processes and transformations acting during plume transport further modify plume composition. The chemical and physical form of plume components, their spatial and temporal distribution, and their deposition are therefore strongly controlled by atmospheric chemistry and transport of the plume.

plumes, aqueous phase oxidation of SO_2 by H_2O_2 is likely to be limited by low H_2O_2 availability. Ozone is another potential oxidant, but its effectiveness decreases rapidly as the aerosol acidifies due to dissociation of the acid gases. Observations indicate a wide variation in the lifetime of volcanic SO_2 in the troposphere (e.g., Oppenheimer et al., 1998b).

Increasing interest in the impacts of tropospheric volcanic emissions on terrestrial and aquatic ecology and on human and animal health is driving more research into the tropospheric chemistry and transport of volcanic plumes. This will be of particular importance in understanding the long-range pollution impacts of volcanic clouds (that might be expected from future eruptions like that of Laki 1783–1784).

1.6.4 Impacts of Volcanic Volatiles on Vegetation and Soils

Volcanic volatiles emitted into the atmosphere are ultimately wet or dry deposited at the Earth's surface. As discussed in Section 1.6.3, the chemical and physical form in which they are deposited, and the spatial and temporal distribution of deposition are strongly controlled by atmospheric chemistry and transport of the volcanic plume. Various components of volcanic emissions (including acid species and heavy metals) can be taken up by plants, and can have both harmful and beneficial effects. The detrimental effects are generally either mediated through acidification of soils (by dry or wet deposition) or by direct fumigation of foliage (e.g., respiration of acid gases

through stomata). Rarely, the diffuse emissions of CO_2 described in Section 1.4.1.1 can damage plant communities. The gas crisis at Mammoth Mountain, California resulted in an extensive tree kill zone (Figure 4(c); Farrar et al., 1995).

Chemical burning of leaves and flowers of vegetation downwind of the crater is a common observation during degassing crises of Masaya volcano (Figure 4(e)), with substantial economic impact from the loss of coffee crops (Delmelle et al., 2001; Delmelle, 2003). SO_2 can cause direct damage to plants once it is taken up by the foliage (Smith, 1990). The response, however, can be very variable, depending on dosage, atmospheric conditions, and leaf type (e.g., Linzon et al., 1979; Winner and Mooney, 1980; Smith, 1990). In general, chronic exposures to SO_2 concentrations of a few tens or hundreds ppb are sufficient to affect plant ecosystems, decrease agricultural productivity, and cause visible foliar chlorosis and necrosis (Winner and Mooney, 1980).

However, other gas species (e.g., HF and HCl) can be important, as well as the extent of soil acidification due to wet and dry deposition. The impacts of sulfate deposition on soils have been investigated widely in the context of anthropogenic pollution, indicating that the SO_4^{2-} retention capacity of soils varies widely and can be expected to dictate the ecosystem disturbance of volcanogenic sulfur deposition. Anion sorption in soils is directly related to the soil mineralogy and hence to the soil parent material. Delmelle et al. (2001) estimated that the amount of SO_2 and HCl dry-deposited within 44 km of Masaya volcano generates an equivalent H^+ flux ranging from <1 mg m^{-2} d^{-1} to 30 mg m^{-2} d^{-1}. Sustained acid loading at these rates can severely impact soil chemistry, reflected downwind of Masaya in low pH and depressed base-saturation contents of soils (Parnell, 1986).

Several studies have indicated benefits of fresh ashfall in supplying nutrients. For example, analyses of leachates of fresh ashfall from the modest 2000 eruption of Hekla in Iceland suggest that ash fallout can actually fertilize the oceans by supplying macronutrients and "bioactive" trace metals (Frogner et al., 2001). Cronin et al. (1998) also documented beneficial additions of soluble sulfur and selenium to agricultural soils during the 1995–1996 eruption of Ruapehu, New Zealand.

In the case of explosive eruptions such as Pinatubo, 1991, there is some evidence that despite the increase in planetary albedo and resulting global surface cooling, photosynthesis could be encouraged in some regions by reduction of the direct solar irradiance at the surface (which, when very intense, results in some varieties turning or closing their leaves to avoid damage) and increase of the diffuse flux (which can penetrate more of the canopy than direct light). Following Pinatubo, measurements at the Mauna Loa Observatory, Hawaii, indicate a decrease in the direct flux from about 520 W m^{-2} to 400 W m^{-2}, but an increase in the diffuse flux from 40 W m^{-2} to 140 W m^{-2}. Increased photosynthesis helps to explain a slowing of the rate of CO_2 increase in the atmosphere in the 2yr following the eruption (Gu *et al.*, 2002).

1.6.5 Impacts of Volcanic Pollution on Animal and Human Health

Several volcanic volatile species are harmful on contact with the skin, if taken into the lungs, or ingested (Williams-Jones and Rymer, 2000). In particular, sulfur species (in both gaseous and aerosol form) can affect respiratory and cardiovascular health in humans, fluorine can contaminate pasture, leading to poisoning of grazing animals, and catastrophic releases of volcanic CO_2 have resulted in several disasters in recent times.

1.6.5.1 Gaseous and particulate sulfur and air quality

Sulfur dioxide has well-known effects on humans—sometimes called "tear gas," its utility in dispersing rioters is regularly witnessed on television news bulletins. Downwind of Masaya Volcano, SO_2 levels exceed background levels at the surface across an area of 1,250 km^2, and an estimated 50,000 people are exposed to concentrations of SO_2 exceeding WHO air quality standards (Baxter *et al.*, 1982; Baxter, 2000; Delmelle *et al.*, 2001, 2002). Symptoms of respiratory illness are commonly reported anecdotally, though formal epidemiological studies have yet to be undertaken. Measurements in Mexico City have indicated the direct influence of Popocatépetl Volcano's emissions on urban air quality: Raga *et al.* (1999) found SO_2 quadrupled, and sulfate concentrations doubled in the city when affected by the volcanic plume. This hints at the potential impacts of future eruptions like Laki, 1783–1784, with long-range transport of SO_2 and aerosol reaching European cities, where air quality standards are already surpassed at certain times of the year. Air quality in Hawaii is reportedly affected by "vog" (volcanic fog) associated with SO_2 and sulfate aerosol from Kilauea's plume, and "laze" (lava haze), composed of HCl-rich droplets formed when active lavas enter the sea (Sutton and Elias, 1993; Mannino *et al.*, 1996).

More recent attention has focused on the abundance of very fine (submicron) and very acidic (pH$\sim$1) aerosol emitted from volcanoes (Allen *et al.*, 2002; Mather *et al.*, 2003b). The few reports available of the human health effects of fine sulfuric acid aerosol are confined to industrial incidents, e.g., the case of a community exposed to emissions from a titanium dioxide plant in Japan in the 1960s, in which some 600 individuals living within 5 km of the plant reported asthmatic symptoms (Kitagawa, 1984).

1.6.5.2 Fluorine

Several recent eruptions have been remarkable for the quantities of fluorine distributed on tephra. HF appears to be readily adsorbed on to ash surfaces, and is efficiently scavenged out of the gas-particle plume as it erupts and dilutes in the atmosphere. The sedimentation of tephra thereby carries significant quantities of volatiles to the Earth's surface. Grazing animals not only consume contaminated foliage, they can also directly ingest large quantities of tephra (an average cow on an average pasture ingests over a kg of soil per day). Much of the fluorine is contained in bio-available compounds such as NaF and CaF$_2$ and can lead rapidly to skeletal deformity, bone lesions, and deformation of teeth. When bones become saturated in fluorine, soft tissues become flooded with fluorine, leading to death. Many animals were lost during the eruptions of Hekla (1970; Óskarsson, 1980), Lonquimay (Chile, 1989–1990; Araya *et al.*, 1990, 1993), and Ruapehu (New Zealand, 1995–1996; Cronin *et al.*, 2003). During the 1783–1784 eruption of Laki, 50% of the livestock on Iceland perished, many probably as a result of fluorine poisoning (Thorarinsson, 1979). It is possible that chronic fluorosis may even have affected parts of the human population through contamination of drinking water.

1.6.5.3 Carbon dioxide

Due to its low solubility in magma, CO_2 is able to exsolve from magma even at high pressures. Diffuse carbon dioxide emissions can be dangerous, particularly when there is limited air circulation (e.g., in basements and excavations). Deaths from CO_2 asphyxia are not uncommon—two people died on the flanks of Cosiguina Volcano (Nicaragua) in 1999 while descending a water well in order to carry out repairs. More catastrophic releases have occurred in recent times, twice in Cameroun (at Lakes Monoun and Nyos in 1984 and 1986, respectively; Le Guern and Tazieff (1989) and Sigurdsson *et al.* (1987)), and on the Dieng Plateau (Java; Le Guern *et al.*, 1982). The Nyos disaster claimed $\sim$1,800 lives, and is generally thought to have resulted from sudden overturn of CO_2-rich waters near the base of the lake, possibly triggered by a landslide into the lake. As the CO_2-rich water ascended and decompressed, it released CO_2 into the gas phase. A controlled degassing project has been initiated at Nyos to pump the deep waters to the surface continuously, allowing them to lose their CO_2, and preventing the buildup that could lead to a future catastrophic overturn.

1.7 CONCLUSIONS AND FUTURE DIRECTIONS

Volcanic volatile emissions play vital roles in the major geochemical cycles of the Earth system, and have done so throughout Earth history. Their short-term (months/years) impacts on the atmosphere, climate, and environment are strongly controlled by fluxes and emission altitudes of sulfur gases (principally SO_2 or H_2S depending on oxidation state of the magma, and hydrothermal influences), which form sulfate aerosol. Halogen emissions may also be important. Episodic explosive eruptions are the principal perturbation to stratospheric aerosol. In the troposphere, the picture is less clear but recent analyses suggest that up to 40% of the global tropospheric sulfate burden may be volcanogenic. Sulfate aerosol influences the Earth's radiation budget by scattering and absorption of short-wave and long-wave radiation, and by seeding or modifying clouds. When they are brought to the boundary layer and Earth's surface, volcanic sulfur and halogens can result in profound environmental and health impacts.

The global impacts of only one large eruption (magnitude $>10^{13}$ kg) have been studied with any instrumental detail—the 1991 eruption of Mt. Pinatubo. More than a decade after the eruption, new findings are still being published on the climatic, environmental, and ecological consequences of the volatile emissions from this eruption. Despite the tremendous insights afforded by this event, it represents a very small sample of the range of volcanic eruption styles, geographic locations, atmospheric conditions, etc., that could combine to produce significant perturbations to atmospheric radiation and dynamics. This begs the question: how ready is the scientific community to record the next major climate-forcing eruption, and what measures can be taken now to ensure optimal observations of such an event? More generally, substantial further work is required to constrain the temporal and spatial distribution of gas and particle emissions (including sulfur, halogens, and trace metal species) to the atmosphere from all erupting and dormant volcanoes.

In recent years, immense progress has also been made in understanding the physics of volcanic plumes (e.g., Sparks *et al.*, 1997). This has been extremely influential for many reasons, not least since it provides a link between observable phenomena (e.g., height of an eruption column) and critical processes (e.g., the switch between lava effusions and explosive eruption), and hence a quantitative basis for volcanic hazard assessment and mitigation. The time is ripe for comparable investigations of the *chemistry* of volcanic plumes, which would substantially improve our understanding of the environmental, atmospheric, climatic, and health impacts of volcanism. Apart from studies of Pinatubo's stratospheric cloud, such work barely exists. In addition to improved observational data on the spatial and temporal distributions of volcanic volatiles to the atmosphere, further studies are required to characterize the physical and chemical interactions of gases and particles in the atmosphere. This will be essential for the realistic application of numerical models describing the transport and chemical evolution of plumes, and will contribute to a better understanding of volcanogenic pollution and improvements in options for risk mitigation.

Surveillance of gas composition and flux are essential for interpretation of volcanic activity, since the nature of degassing exerts a strong control on eruption style, and is closely associated with volcano seismicity and ground deformation. New optical remote sensing techniques are emerging for the monitoring of volcanic emissions such as the miniaturized ultraviolet spectrometers described by Galle *et al.* (2003). Proliferation of such technologies will also help in efforts to improve the global database of volcanic volatile source distribution. Unfortunately, the modeling frameworks for interpretation of geochemical data remain poorly developed, limiting the application of such data in hazard assessment. Advances in this area will benefit from development and validation of comprehensive physico-chemical models for volcanic degassing based on the integration of results from experiments on the controls on distribution of volatiles in synthetic and natural melts, analysis of dissolved volatiles preserved in melt inclusions, and observed volcanic gas geochemistry. Ultimately, such models can be applied to integrated geophysical, geodetic, and geochemical monitoring data to support eruption forecasting.

Research on volcanic degassing is a rich and diverse field affording many opportunities for interdisciplinary and multidisciplinary efforts at many scales. Broadly, the investigator of volcanic volatiles may follow the path from which the emissions came, to probe the chemistry, dynamics, and evolution of the source regions, storage zones, and volcanic plumbing systems; or consider the interactions of the emissions with the atmosphere, environment, and climate.

ACKNOWLEDGMENTS

The author gratefully acknowledge recent research support from the European Commission 5th Framework programs "MULTIMO" and

"DORSIVA," the UK Natural Environment Research Council, grants GR9/4655 and GR29250, the Italian Gruppo Nazionale per la Vulcanologia grant "Development of an integrated spectroscopic system for remote and continuous monitoring of volcanic gas," and NASA grant NAG5-10640. The author is very grateful to Bruno Scaillet, Oded Navon, and David Catling for comments on the original manuscript, and Simon Carn for the TOMS information (Table 7 and Figure 5(c)), Jake Lowenstern for Figure 5(d), and Sune Olander Rasmussen, Trine Ebbensgaard, and Sigfus Johnsen for Figure 5(e). The author thanks all the emissions enthusiasts he has worked with in recent years for their generous collaboration and friendship, including Marie Edmonds, Lisa Horrocks, Andrew McGonigle, David Pyle, Mike Burton, Tamsin Mather, Matt Watson, Hayley Duffell, Bo Galle, Claire Witham, Bill Rose, Frank Tittel, Dirk Richter, Andrew Allen, Peter Baxter, Steve Sparks, Jenni Barclay, Pierre Delmelle, Patrick Allard, Bruno Scaillet, Benoît Villemant, Pete Mouginis-Mark, Keith Horton, John Porter, and Roy Harrison. Lastly, I remember Peter Francis, who set me on sulfurous journeys a decade ago.

REFERENCES

Abdalati W. and Steffen K. (1997) The apparent effects of the Pinatubo eruption on the Greenland ice sheet melt extent. *Geophys. Res. Lett.* **24**, 1795–1797.

Adams N. K., de Silva S. L., Self S., Salas G., Schubring S., Permenter J. L., and Arbesman K. (2001) The physical volcanology of the 1600 eruption of Huaynaputina, southern Peru. *Bull. Volcanol.* **62**, 493–518.

Aiuppa A., Federico C., Paonita A., Pecoraino G., and Valenza M. (2002) S, Cl, and F degassing as an indicator of volcanic dynamics: the 2001 eruption of Mount Etna. *Geophys. Res. Lett.* **29**, DOI 10.1029/2002GL015032.

Allard P. (1997) Endogenous magma degassing and storage at Mount Etna. *Geophys. Res. Lett.* **24**, 2219–2222.

Allard P., Carbonelle J., Metrich N., and Zettwoog P. (1991) Eruptive and diffuse emissions of carbon dioxide from Etna volcano. *Nature* **351**, 38–391.

Allard P., Carbonnelle J., Métrich N., Loyer H., and Zettwoog P. (1994) Sulphur output and magma degassing budget of Stromboli volcano. *Nature* **368**, 326–330.

Allard P., Aiuppa A., Loyer H., Carrot F., Gaudry A., Pinte G., Michel A., and Dongarrà G. (2000) Acid gas and metal emission rates during long-lived basalt degassing at Stromboli volcano. *Geophys. Res. Lett.* **27**, 1207–1210.

Allen A. G., Baxter P. J., and Ottley C. J. (2000) Gas and particle emissions from Soufrière Hills volcano, Montserrat WI: characterization and health hazard assessment. *Bull. Volcanol.* **62**(1), 6–17.

Allen A. G., Oppenheimer C., Ferm M., Baxter P. J., Horrocks L., Galle B., McGonigle A. J. S., and Duffell H. J. (2002) Primary sulphate aerosol and associated emissions from Masaya volcano, Nicaragua. *J. Geophys. Res.* (4), Doi 10.1029/2002JD0 (4 December 2002).

Anderson A. T. (1975) Some basaltic and andesitic gases. *Rev. Geophys. Space Phys.* **13**, 37–55.

Andres R. J. and Kasgnoc A. D. (1998) A time-averaged inventory of subaerial volcanic sulfur emissions. *J. Geophys. Res.* **103**, 25251–25261.

Andres R. J. and Rose W. I. (1995) Remote sensing spectroscopy of volcanic plumes and clouds. In *Monitoring Active Volcanoes: Strategies, Procedures and Techniques* (eds. B. McGuire, C. Kilburn, and J. Murray). UCL Press, London, pp. 301–314.

Araya O., Wittwer F., Villa A., and Ducom C. (1990) Bovine fluorosis following volcanic activity in the southern Andes. *Vet. Rec.* **126**, 641–642.

Araya O., Wittwer F., and Villa A. (1993) Evolution of fluoride concentrations in cattle and grass following a volcanic eruption. *Vet. Hum. Toxicol.* **35**, 437–440.

Arthur M. A. (2000) Volcanic contributions to the carbon and sulfur geochemical cycles and global change. In *Encyclopedia of Volcanoes* (eds. H. Sigurdsson, B. F. Houghton, S. R. McNutt, H. Rymer, and J. Stix). Academic Press, San Diego, pp. 1046–1056.

Baran A. and Foot J. S. (1994) New application of the operational sound HIRS in determining a climatology of acid aerosol from the Pinatubo eruption. *J. Geophys. Res.* **99**, 25673–25679.

Baran A. J., Foot J. S., and Dibben P. C. (1993) Satellite detection of volcanic sulfuric acid aerosol. *Geophys. Res. Lett.* **20**, 1799–1801.

Barmin A., Melnik O., and Sparks R. S. J. (2002) Periodic behaviour in lava dome eruptions. *Earth Planet. Sci. Lett.* **199**, 173–184.

Baxter P. J. (2000) Impacts of eruptions on human health. In *Encyclopedia of Volcanoes* (eds. H. Sigurdsson, B. F. Houghton, S. R. McNutt, H. Rymer, and J. Stix). Academic Press, San Diego, pp. 1035–1043.

Baxter P. J., Stoiber R. E., and Williams S. N. (1982) Volcanic gases and health: Masaya volcano, Nicaragua. *Lancet* **2**, 150–151.

Blank J. G. and Brooker R. A. (1994) Experimental studies of carbon dioxide in silicate melts: solubility, speciation, and stable carbon isotope behavior. *Rev. Mineral.* **30**, 157–186.

Blower J. D., Mader H. M., and Wilson S. D. R. (2001) Coupling of viscous and diffusive controls on bubble growth during explosive volcanic eruptions. *Earth Planet. Sci. Lett.* **193**, 47–56.

Blundy J. and Cashman K. V. (2001) Ascent-driven crystallization of dacite magmas of Mount St Helens, 1980–1986. *Contrib. Mineral. Petrol.* **140**, 631–650.

Bluth G. J. S., Doiron S. D., Schnetzler C. C., Krueger A. J., and Walter L. S. (1992) Global tracking of the SO_2 clouds from the June, 1991 Mount Pinatubo eruptions. *Geophys. Res. Lett.* **19**, 151–154.

Bluth G. J. S., Schnetzler C. C., Krueger A. J., and Walter L. S. (1993) The contribution of explosive volcanism to global atmospheric sulphur dioxide concentrations. *Nature* **366**, 327–329.

Bluth G. J. S., Rose W. I., Sprod I. E., and Krueger A. J. (1997) Stratospheric loading of sulfur from explosive volcanic eruptions. *J. Geol.* **105**, 671–684.

Briffa K. R. and Jones P. D. (1992) The climate of Europe during the 1810s with special reference to 1816. In *The Year without a Summer? World Climate in 1816* (ed. C. R. Harington). Canadian Museum of Nature, Ottawa, pp. 372–391.

Briffa K. R., Jones P. D., Schweingruber F. H., and Osborn T. J. (1998) Influence of volcanic eruptions on northern hemisphere summer temperature over the past 600 years. *Nature* **393**, 450–455.

Bureau H., Keppler H., and Métrich N. (2000) Volcanic degassing of bromine and iodine: experimental fluid/melt partitioning data and applications to stratospheric chemistry. *Earth Planet. Sci. Lett.* **183**, 51–60.

Burnham C. W. (1979) The importance of volatile constituents. In *The Evolution of Igneous Rocks: Fiftieth Anniversary Perspectives.* (ed. H. S. Yoder, Jr.). Princeton University Press, Princeton, NJ.

Burton M., Allard P., Murè F., and Oppenheimer C. (2003) FTIR remote sensing of fractional magma degassing at Mt. Etna, Sicily. In *Volcanic Degassing,* Geological Society of London Special Publication 213 (eds. C. Oppenheimer, D. M. Pyle, and J. Barclay). Geological Society of London, pp. 281–293.

Burton M. R., Oppenheimer C., Horrocks L. A., and Francis P. W. (2000) Remote sensing of CO_2 and H_2O emission rates from Masaya volcano, Nicaragua. *Geology* **28**, 915–918.

Cadle R. D. (1975) Volcanic emissions of halides and sulfur compounds to the troposphere and stratosphere. *J. Geophys. Res.* **80**, 1650–1652.

Cadle R. D. (1980) A comparison of volcanic with other fluxes of atmospheric trace gas constituents. *Rev. Geophys. Space Phys.* **18**, 746–752.

Camuffo D. and Enzi S. (1995) Impact of clouds of volcanic aerosols in Italy during the last seven centuries. *Nat. Hazards* **11**(2), 135–161.

Canil D. (2002) Vanadium in peridotites, mantle redox and tectonic environments: Archean to present. *Earth Planet. Sci. Lett.* **195**, 75–90.

Carey S. and Sigurdsson H. (1989) The intensity of plinian eruptions. *Bull. Volcanol.* **51**, 28–40.

Carn S. A., Krueger A. J., Bluth G. J. S., Schaefer S. J., Krotkov N. A., Watson I. M., and Datta S. (2003) Volcanic eruption detection by the Total Ozone Mapping Spectrometer (TOMS) instruments: a 22-year record of sulfur dioxide and ash emissions. In *Volcanic Degassing.* Geological Society of London Special Publication 213 (eds. C. Oppenheimer, D. M. Pyle, and J. Barclay). Geological Society of London, 177–202.

Carroll M. R. and Holloway J. R. (eds.) (1994). *Volatiles in Magmas.* Reviews in Mineralogy, 30. Mineralogical Society of America, Washington, DC, 517pp.

Carroll M. R. and Webster J. D. (1994) Solubilities of sulfur, noble gases, nitrogen, chlorine, and fluorine in magmas. *Rev. Mineral.* **30**, 187–230.

Cashman K. V. (1992) Groundmass crystallization of Mount St. Helens dacite, 1980–1986—a tool for interpreting shallow magmatic processes. *Contrib. Mineral. Petrol.* **109**, 431–449.

Cashman K. V., Sturtevant B., Papale P., and Navon O. (2000) Magmatic fragmentation. In *Encyclopedia of Volcanoes* (eds. H. Sigurdsson, B. F. Houghton, S. R. McNutt, H. Rymer, and J. Stix). Academic Press, San Diego, CA, pp. 421–430.

Catling D. and Kasting J. F. (2003) Planetary atmospheres and life. In *The Emerging Science of Astrobiology* (eds. J. Baross and W. Sullivan). Cambridge University Press, (in press).

Catling D. C., Zahnle K. J., and McKay C. P. (2001) Biogenic methane, hydrogen escape, and the irreversible oxidation of the early earth. *Science* **393**, 839–843.

Chesner C. A. and Rose W. I. (1991) Stratigraphy of the Toba tuffs and the evolution of the Toba caldera complex, Sumatra, Indonesia. *Bull. Volcanol.* **53**, 343–356.

Cheynet B., Dall'Aglio M., Garavelli A., Grasso M. F., and Vurro F. (2000) Trace elements from fumaroles at Vulcano Island (Italy): rates of transport and a thermochemical model. *J. Volcanol. Geotherm. Res.,* **95**, 273–283.

Chin M. and Jacob D. J. (1996) Anthropogenic and natural contributions to tropospheric sulfate: a global model analysis. *J. Geophys. Res.* **101**, 18691–18699.

Coffey M. T. (1996) Observations of the impact of volcanic activity on stratospheric chemistry. *J. Geophys. Res.* **101**, 6767–6780.

Coffey M. T. and Mankin W. G. (1993) Observations of the loss of stratospheric NO_2 following volcanic eruptions. *Geophys. Res. Lett.* **29**, 2873–2876.

Coltelli M., Del Carlo P., and Vezzoli L. (1998) Discovery of a plinian basaltic eruption of Roman age at Etna volcano, Italy. *Geology* **26**, 1095–1098.

Costa F., Scaillet B., Gourgaud A. (2003). Massive atmospheric sulfur loading of the AD 1600 Huaynaputina

eruption and implications for petrologicl sulfur estimates. *Geophys. Res. Lett.* **30**(2), 10.102912002GL016402.

Cronin S. J., Hedley M. J., Neall V. E., and Smith G. (1998) Agronomic impact of tephra fallout from 1995 and 1996 Ruapehu volcano eruptions, New Zealand. *Environ. Geol.* **34**, 21–30.

Cronin S. J., Neall V. E., Lecointre J. A., Hedley M. J., and Loganathan P. (2003) Environmental hazards of fluoride in volcanic ash: a case study from Ruapehu volcano, New Zealand. *J. Volcanol. Geotherm. Res.* **121**, 217–191.

Daag A. S., Tubianosa B. S., Newhall C. G., Tuñgol N. M., Javier D., Dolan M. T., Delos Reyes P. J., Arboleda R. A., Martinez M. L., and Regalado T. M. (1996) Monitoring sulfur dioxide emission at Mount Pinatubo. In *Fire and Mud: Eruptions and Lahars of Mount Pinatubo Philippines* (eds. C. G. Newhall and R. S. Punongbayan). Philippine Institute of Volcanology and Seismology, Quezon City/University of Washington Press, Seattle, pp. 409–434.

Dahl-Jensen D., Gundestrup N. S., Miller H., Watanable O., Johnson S. J., Steffensen J. P., Clausen H. B., Svensson A., and Larsen L. B. (2002) The NorthGRIP deep drilling program. *Ann. Glaciol.* **35**, 1–4.

De Hoog J. C. M., Koetsier G. W., Bronto S., Sriwana T., and van Bergen M. J. (2001) Sulphur and chlorine degassing from primitive arc magmas: temporal changes during the 1982–1983 eruptions of Galunggung (West Java, Indonesia). *J. Volcanol. Geotherm. Res.* **108**, 55–83.

Delano J. W. (2001) Redox history of the earth's interior since ~3900 Ma: implications for prebiotic molecules. *Origins Life Evol. Biosphere* **31**, 311–341.

Delmas R. J., Kirchner S., Palais J. M., and Petit J. R. (1992) 1,000 years of explosive volcanism recorded at the South Pole. *Tellus* **44B**, 335–350.

Delmelle P. (2003) Environmental impacts of tropospheric volcanic gas plumes. In *Volcanic Degassing,* Geological Society of London Special Publicationb 213. (eds. C. Oppenheimer, D. M. Pyle, and J. Barclay). Geological Society of London, pp. 381–399.

Delmelle P. and Stix J. (2000) Volcanic gases. In *Encyclopedia of Volcanoes* (eds. H. Sigurdsson, B. F. Houghton, S. R. McNutt, H. Rymer, and J. Stix). Academic Press, San Diego, CA, pp. 803–815.

Delmelle P., Stix J., Bourque C. P. A., Baxter P. J., Garcia-Alvarez J., and Barquero J. (2001) Dry deposition and heavy acid loading in the vicinity of Masaya volcano, a major sulfur and chlorine source in Nicaragua. *Environ. Sci. Technol.* **35**, 1289–1293.

Delmelle P., Stix J., Baxter P. J., Garcia-Alvarez J., and Barquero J. (2002) Atmospheric dispersion, environmental effects and potential health hazard associated with the low-altitude gas plume of Masaya volcano, Nicaragua. *Bull. Volcanol.* **64**, 423–434.

Dingwell D. (1998) Recent experimental progress in the physical description of silicic magma relevant to explosive volcanism. In *The Physics of Explosive Volcanic Eruptions* Geology Society of London Special Publication 145 (eds. J. S. Gilbert and R. S. J. Sparks). Geological Society of London, pp. 9–26.

de Silva S. L. and Zielinski G. A. (1998) Global influence of the AD 1600 eruption of Huaynaputina Peru. *Nature* **393**, 455–458.

Dixon J. E., Clague D. A., and Stolper E. M. (1991) Degassing history of water, sulfur, and carbon in submarine lavas from Kilauea volcano, Hawaii. *J. Geol.* **99**, 371–394.

Duffell H., Oppenheimer C., and Burton M. (2001) Volcanic gas emission rates measured by solar occultation spectroscopy. *Geophys. Res. Lett.* **28**, 3131–3134.

Duffell H. J., Oppenheimer C., Pyle D., Galle B., McGonigle A. J. S., and Burton M. R. (2003) Geochemical precursors to a minor explosive eruption at Masaya volcano, Nicaragua. *J. Volcanol. Geotherm. Res* **126**, 327–339.

Edmonds M., Pyle D., and Oppenheimer C. (2001) A model for degassing at the Soufrière Hills Volcano, Montserrat,

West Indies, based on geochemical data. *Earth Planet. Sci. Lett.* **186**, 159–173.

Edmonds M., Oppenheimer C., Pyle D. M., and Herd R. A. (2003a) Rainwater and ash leachate analysis as a proxy for plume chemistry at Soufrière Hills Volcano, Montserrat. In *Volcanic Degassing, Geological Society of London Special Publication 213.* (eds. C. Oppenheimer, D. M. Pyle, and J. Barclay). Geological Society of London, pp. 203–218.

Edmonds M., Herd R. A., Galle B., and Oppenheimer C. (2003b) Automated, high time-resolution measurements of SO_2 flux at Soufrière Hills Volcano, Montserrat. *Bull. Volcanol.* D.o.i. 10.1007/s00445-003-0286-x.

Edmonds M., Oppenheimer C., Pyle D. M., Herd R. A., and Thompson G. (2003c) SO_2 emissions from Soufrière Hills Volcano and their relationship to conduit permeability, hydrothermal interaction and degassing regime. *J. Volcanol. Geotherm. Res.* **124**, 23–43.

Eggenkamp H.G.M. (1994). The geochemistry of chlorine isotopes. PhD Thesis, University of Utrecht, The Netherlands, 150pp (unpublished).

Eichelberger J. C. (1995) Silicic volcanism: ascent of viscous magmas from crustal reservoirs. *Ann. Rev. Earth Plant. Sci. Lett.* **23**, 41–63.

Eichelberger J. C., Carrigan C. R., Westrich H. R., and Price R. H. (1986) Non-explosive silicic volcanism. *Nature* **323**, 598–602.

Farrar C. D., Sorey M. L., Evans W. C., Howle J. F., Kerr B. D., Kennedy B. M., King C.-Y., and Southon J. R. (1995) Forest-killing diffuse CO_2 emissions at Mammoth Mountain as a sign of magmatic unrest. *Nature* **376**, 675–678.

Fine G. J. and Stolper E. M. (1985) The speciation of carbon dioxide in sodium aluminosilicate glasses. *Contrib. Mineral. Petrol.* **91**, 105–121.

Fischer T. P., Giggenbach W. F., Sano Y., and Williams S. N. (1998) Fluxes and sources of volatiles discharged from Kudryavy, a subduction zone volcano, Kurile Islands. *Earth Planet. Sci. Lett.* **160**, 81–96.

Fischer T. P., Hilton D. R., Zimmer M. M., Shaw A. M., Sharp Z. D., and Walker J. A. (2002) Subduction and recycling of nitrogen along the Central American margin. *Science* **297**, 1154–1157.

Francis P., Burton M., and Oppenheimer C. (1998) Remote measurements of volcanic gas compositions by solar FTIR spectroscopy. *Nature* **396**, 567–570.

Francis P. W., Oppenheimer C., and Stevenson D. (1993) Endogenous growth of persistently active volcanoes. *Nature* **366**, 554–557.

Frogner P., Gislason S. R., and Oskarsson N. (2001) Fertilizing potential of volcanic ash in ocean surface water. *Geology* **29**, 487–490.

Galle B., Oppenheimer C., Geyer A., McGonigle A., Edmonds M., and Horrocks L. A. (2003) A miniaturised ultraviolet spectrometer for remote sensing of SO_2 fluxes: a new tool for volcano surveillance. *J. Volcanol. Geotherm. Res.* **119**, 241–254.

Gauthier P. J. and Le Cloarec M.-F. (1998) Variability of alkali and heavy metal fluxes released by Mt. Etna volcano, Sicily, between 1991 and 1995. *J. Volcanol. Geotherm. Res.* **81**, 311–326.

Genin A., Lazar B., and Brenner S. (1995) Vertical mixing and coral death in the Red Sea following the eruption of Mt. Pinatubo. *Nature* **377**, 507–510.

Gerlach T. M. (1993) Thermodynamic evaluation and restoration of volcanic gas analyses; an example based on modern collection and analytical methods. *Geochem. J.* **27**, 305–322.

Gerlach T. M. and Casadevall T. J. (1986) Fumarole emissions at Mount St Helens volcano, June 1980 to October 1981: degassing of a magma-hydrothermal system. *J. Volcanol. Geotherm. Res.* **28**, 141–160.

Gerlach T. M. and Graeber E. J. (1985) Volatile budget of Kilauea volcano. *Nature* **313**, 273–277.

Gerlach T. M. and Taylor B. E. (1990) Carbon isotope constraints on degassing of carbon dioxide from Kilauea Volcano. *Geochim. Cosmochim.* Acta **54**, 2051–2058.

Gerlach T. M., Delgado H., McGee K. A., Doukas M. P., Venegas J. J., and Cardenas L. (1997) Application of the LI-COR CO_2 analyzer to volcanic plumes: a case study, volcan Popocatépetl, Mexico, June 7 and 10, 1995. *J. Geophys. Res.* **102**(B4), 8005–8019.

Giggenbach W. F. (1975) A simple method for the collection and analysis of volcanic gas samples. *Bull. Volcanol.* **39**, 132–145.

Giggenbach W. F. (1992a) Isotopic shifts in waters from geothermal and volcanic systems along convergent plate boundaries and their origin. *Earth Planet. Sci. Lett.* **113**, 495–510.

Giggenbach W. F. (1992b) Magma degassing and mineral deposition in hydrothermal systems along convergent plate boundaries. *Econ. Geol.* **87**, 1927–1944.

Giggenbach W. F. (1996) Chemical composition of volcanic gases. In *Monitoring and Mitigation of Volcano Hazards* (eds. R. Scarpa and R. I. Tilling). Springer, Berlin, pp. 221–256.

Giggenbach W. F. and Matsuo S. (1991) Evaluation of results from second and third IAVCEI field workshops on volcanic gases, Mt. Usu, Japan and White Island, New Zealand. *Appl. Geochem.* **6**, 125–141.

Giggenbach W. F., Tedesco D., Sulistiyo Y., Caprai A., Cioni R., Favara R., Fischer T. P., Hirabayashi J.-I., Korzhinsky M., Martini M., Menyailov I., and Shinohara H. (2001) Evaluation of results from the fourth and fifth IAVCEI field workshops on volcanic gases, Vulcano island, Italy and Java, Indonesia. *J. Volcanol. Geotherm. Res.* **108**, 157–172.

Graf H. F., Kirchner I., Robock A., and Schult I. (1993) Pinatubo eruption winter climate effects: model versus observations. *Clim. Dyn.* **9**, 81–93.

Graf H.-F., Feichter J., and Langmann B. (1997) Volcanic sulfur emissions: estimates of source strength and its contribution to the global sulfate distribution. *J. Geophys. Res.* **102**, 10727–10738.

Graf H.-F., Langmann B., and Feichter J. (1998) The contribution of earth degassing to the atmospheric sulfur budget. *Chem. Geol.* **147**, 131–145.

Gu L., Baldocchi D., Verma S. B., Black T. A., Vesala T., Falge E. M., and Dowty P. R. (2002) Advantages of diffuse radiation for terrestrial ecosystem productivity. *J. Geophys. Res.* **107**, 1–23.

Halmer M. M., Schmincke H.-U., and Graf H.-F. (2002) The annual volcanic gas input into the atmosphere, in particular into the stratosphere: a global data set for the past 100 years. *J. Volcanol. Geotherm. Res.,* **115**, 511–528.

Hammer C. U., Clausen H. B., and Dansgaard W. (1980) Greenland ice sheet evidence of post-glacial volcanism and its climatic impact. *Nature* **288**, 230–235.

Hammer J. E. and Rutherford M. J. (2002) An experimental study of the kinetics of decompression-induced crystallization in silicic melt. *J. Geophys. Res.* **107**(131), 10.1029/2001JB000281.

Hammer J. E., Cashman K. V., Hoblitt R. P., and Newman S. (1999) Degassing and microlite crystallization during pre-climactic events of the 1991 eruption of Mt. Pinatubo, Phillipines. *Bull. Volcanol.* **60**, 355–380.

Hansen J., Sato M., Ruedy R., Lacis A., Asamoah K., Beckford K., Borenstein S., Brown E., Cairns B., Carlson B., Curran B., de Castro S., Druyan L., Etwarrow P., Ferede T., Fox M., Gaffen D., Glascoe J., Gordon H., Hollandsworth S., Jiang X., Johnson C., Lawrence N., Lean J., Lerner J., Lo K., Logan J., Luckett A., McCormick M. P., McPeters R., Miller R., Minnis P., Ramberran I., Russell G., Russell P., Stone P., Tegen I., Thomas S., Thomason L., Thompson A., Wilder J., Willson R., and Zawodny J. (1997) Forcings and chaos in interannual to decadal climate change. *J. Geophys. Res.* **102**, 25679–25720.

Harington C. R. (ed.) (1992). *The Year without a Summer? World Climate in (1816).* Canadian Museum of Nature, Ottawa, 576 pp.

Hinkley T. K., Lamothe P. J., Wilson S. A., Finnegan D. L., and Gerlach T. M. (1999) Metal emissions from Kilauea, and a suggested revision of the estimated worldwide metal output by quiescent degassing of volcanoes. *Earth Planet. Sci. Lett.* **170**, 315–325.

Hoff R. M. (1992) Differential SO_2 column measurements of the Mt. Pinatubo volcanic plume. *Geophys. Res. Lett.* **19**, 175–178.

Hoffman P. F. and Schrag D. P. (2002) The snowball earth hypothesis: testing the limits of global change. *Terra Nova* **14**, 129–155.

Hoffman P. F., Kaufman A. J., Halverson G. P., and Schrag D. P. (1998) A neoproterozoic snowball earth. *Science* **281**, 1342–1346.

Hofmann C., Courtillot V., Feraud G., Rochette P., Yirgu G., Ketefo E., and Pik R. (1997) Timing of the Ethiopian flood basalt event and implications for plume birth and global change. *Nature* **389**, 838–841.

Holasek R. E., Self S., and Woods A. W. (1996) Satellite observations and interpretation of the 1991 Mount Pinatubo eruption plumes. *J. Geophys. Res.* **101**, 27635–27655.

Holland H. D. (1984) *The Chemical Evolution of the Atmosphere and Oceans.* Princeton University Press, Princeton, 598 pp.

Holland H. D. (2002) Volcanic gases, black smokers, and the great oxidation event. *Geochim. Cosmochim. Acta* **66**, 3811–3826.

Horn S. and Schmincke H.-U. (2000) Volatile emission during the eruption of Baitoushan volcano (China/North Korea) ca. 969 AD. *Bull. Volcanol.* **61**, 537–555.

Horrocks L. A., Oppenheimer C., Burton M. R., Duffell H. R., Davies N. M., Martin N. A., and Bell W. (2001) Open-path Fourier transform infrared spectroscopy of SO_2: an empirical error budget analysis, with implications for volcano monitoring. *J. Geophys. Res.* **106**, 27647–27659.

Huppert H. E. and Woods A. W. (2002) The role of volatiles in magma chamber dynamics. *Nature* **420**, 493–495.

Ihinger P. D., Zhang Y., and Stolper E. M. (1999) The speciation of dissolved water in rhyolitic melt. *Geochim. Cosmochim. Acta* **63**, 3567–3578.

Intergovernmental Panel on Climate Change Working Group (IPCC Report) (2001) *Climate Change 2001: The Scientific Basis*, Contribution of Working Group I to the Third Assessment Report of the Intergovernmental Panel on Climate Change (J. T. Houghton, Y. Ding, D. J. Griggs, M. Noguer, P. J. van der Linden, D. Xiaosu).

Jaupart C. (1998) Gas loss through conduit walls during eruption. In *The Physics of Explosive Volcanic Eruptions* Geological Society of London Special Publication 145 (eds. J. S. Gilbert and R. S. J. Sparks). Geological Society of London, pp. 73–90.

Jaupart C. and Allegre C. J. (1991) Gas content, eruption rate and instabilities of eruption regime in silicic volcanoes. *Earth Planet. Sci. Lett.* **102**, 413–429.

Johnson M. C., Anderson A. T. Jr., and Rutherford M. J. (1993) Pre-eruptive volatile contents of magmas. In *Volatiles in Magmas,* Reviews in Mineral 30 (eds. M. R. Carroll and J. R. Holloway). Mineralogical Society of America, Washington, DC, pp. 281–330.

Jones G.S., and Stott P.A., (2002). Simulation of climate response to a super eruption. In *American Geophysical Union Chapman Conference on Volcanism and the Earth's Atmosphere, Santorini 17–21 June, (2002).* Abstract volume, 45p.

Kasting J. F., Eggler D. H., and Raeburn S. P. (1993) Mantle redox evolution and the oxidation state of the Archean atmosphere. *J. Geol.* **101**, 245–257.

Kazahaya K., Shinohara H., and Saito G. (1994) Excessive degassing of Izu-Oshima volcano: magma convection in a conduit. *Bull. Volcanol.* **56**, 207–216.

Kelley D. S., Baross J. A., and Delaney J. R. (2002) Volcanoes, fluids, and life at mid-ocean ridge spreading centers. *Ann. Rev. Earth Planet. Sci.* **30**, 385–491.

Kelly P. M., Jones P. D., and Pengqun J. L. A. (1996) The spatial response of the climate system to explosive volcanic eruptions. *Int. J. Climatol.* **16**, 537–550.

Keppler H. (1999) Experimental evidence for the source of excess sulfur in explosive volcanic eruptions. *Science* **284**, 1652–1654.

Kirchner I., Stenchikov G., Graf H.-F., Robock A., and Antuna J. (1999) Climate model simulation of winter warming and summer cooling following the 1991 Mount Pinatubo volcanic eruption. *J. Geophys. Res.* **104**, 19039–19055.

Kirschvink J. L. (1992) Late proterozoic low-latitude glaciation: the snowball earth. In *The Proterozoic Biosphere* (eds. J. W. Schopf and C. Klein). Cambridge University Press, New York, pp. 51–52.

Kirschvink J. L., Gaidos E. J., Bertani L. E., Beukes N. J., Gutzmer J., Maepa L. N., and Steinberger R. E. (2000) Paleoproterozoic snowball earth: extreme climatic and geochemical global change and its biological consequences. *Proc. Natl Acad. Sci.* **97**, 1400–1405.

Kitagawa T. (1984) Cause analysis of the Yokkaichi asthma episode in Japan. *J. Air Pollut. Control Assoc.* **34**, 743–746.

Klug C. and Cashman K. V. (1996) Permeability development in vesiculating magmas: implications for fragmentation. *Bull. Volcanol.* **58**, 87–100.

Koike M., Jones N. B., Matthews W. A., Johnston P. V., McKenzie R. L., Kinnison D., and Rodriguez J. (1994) Impact of pinatubo aerosols on the partitioning between NO_2 and HNO_3. *Geophys. Res. Lett.* **21**, 597–600.

Kress V. C. (1997) Magma mixing as a source for Pinatubo sulphur. *Nature* **389**, 591–593.

Krueger A. J., Walter L. S., Bhartia P. K., Schnetzler C. C., Krotkov N. A., Sprod I., and Bluth G. J. S. (1995) Volcanic sulfur dioxide measurements from the Total Ozone Mapping Spectrometer (TOMS) instruments. *J. Geophys. Res.* **100**, 14057–14076.

Lambert G., Le Cloarec M.-F., and Pennisi M. (1988) Volcanic output of SO_2 and trace metals: a new approach. *Geochim. Cosmochim. Acta* **52**, 39–42.

Lange R. A. (1994) The effect of H_2O, CO_2, and F on the density and viscosity of silicate melts. *Rev. Mineral.* **30**, 331–369.

Lantzy R. J. and Mackenzie F. T. (1979) Atmospheric trace metals: global cycles and assessment of man's impact. *Geochim. Cosmochim. Acta* **43**, 511–525.

Le Cloarec M.-F. and Marty B. (1991) Volatile fluxes from volcanoes. *Terra Nova* **3**, 17–27.

Le Guern, F., Tazieff, H. (1989) Lake Nyos. *J. Volcanol. Geotherm. Res.* **39** (special issue).

Le Guern F., Tazieff H., and Faivre-Perret R. (1982) An example of health hazard: people killed by gas during a phreatic eruption, Dieng Plateau (Java, Indonesia), February 20th (1979). *Bull. Volcanol.* **45**, 153–156.

Lejeune A.-M. and Richet P. (1995) Rheology of crystal bearing silicate melts: an experimental study at high viscosities. *J. Geophys. Res.* **100**, 4215–4229.

Lensky N., Lyakhovsky V., and Navon O. (2001) Radial variations of melt viscosity around growing bubbles and gas overpressure in vesiculating magmas. *Earth Planet. Sci. Lett.* **186**, 1–6.

Lensky N., Lyakhovsky V., and Navon O. (2002) Expansion dynamics of volatile-saturated fluid and bulk viscosity of bubbly magmas. *J. Fluid Mech.* **460**, 39–56.

Lensky N., Navon O., and Lyakhovsky V. (2003) Bubble growth during decompression of magma: experimental and theoretical investigation. *J. Volcanol. Geotherm. Res.* (in review).

Linzon S. N., Temple P. J., and Pearson R. G. (1979) Sulfur concentrations in plant foliage and related effects. *J Air Pollut. Control Assoc.* **29**, 520–525.

Mader H. M. (1998) Conduit flow and fragmentation. In *The Physics of Explosive Volcanic Eruptions,* Geological Society of London Special Publication 145 (eds. J. S. Gilbert and R. S. J. Sparks), Geological Society of London, pp. 51–71.

Mannino D. M., Ruben S., Holschuh F. C., Holschuh T. C., Wilson M. D., and Holschuh T. (1996) Emergency department visits and hospitalizations for respiratory disease on the island of Hawaii, 1981 to 1991. *Hawaii Med. J.* **55**(3), 48–53.

Massol H. and Jaupart C. (1999) The generation of gas overpressure in volcanic eruptions. *Earth Planet. Sci. Lett.* **166**, 57–70.

Mather T. A., Pyle D. M., and Oppenheimer C. (2003a) Volcanic aerosol in the troposphere. In *Volcanism and the Earth's Atmosphere* (eds. A. Robock and C. Oppenheimer). American Geophysical Union Monograph (in press).

Mather T. A., Allen A. G., Oppenheimer C., Pyle D. M., and McGonigle A. J. S. (2003b) Size-resolved particle compositions of the tropospheric plume of Masaya volcano, Nicaragua. *J. Atmos. Chem.* (in press).

Matuso S. (1960) On the origin of volcanic gases. *J. Earth Sci. Nagoya Univ.* **8**, 222–245.

Matsuo S. (1975) Chemistry of volcanic gases. *Kazan (Bull. Volcanol. Soc. Japan)* **20**, 319–329.

McCormick M. P., Thomason L. W., and Trepte C. R. (1995) Atmospheric effects of the Mt. Pinatubo eruption. *Nature* **373**, 399–404.

McGee K. A. (1992) The structure, dynamics and chemical composition of non-eruptive plumes from Mt. St. Helens, 1980–88. *J. Volcanol. Geotherm. Res.* **51**, 269–282.

McGee K. A. and Gerlach T. M. (1998) Annual cycle of magmatic CO_2 in a tree-kill soil at Mammoth mountain, California: implications for soil acidification. *Geology* **26**, 463–466.

McGonigle A. J. S. and Oppenheimer C. (2003) Optical sensing of volcanic gas and aerosol. In *Volcanic Degassing*, Geological Society of London Special Publication, 213 (eds. C. Oppenheimer, D. M. Pyle, and J. Barclay). Geological Society of London pp. 149–168.

McGonigle A. J. S., Oppenheimer C., Galle B., Mather T., and Pyle D. (2002) Walking traverse and scanning DOAS measurements of volcanic gas emission rates. *Geophys. Res. Lett.* Doi 10.1029/2002GL015827 (October 26, 2002).

McGonigle A. J. S., Oppenheimer C., Hayes A. R., Galle B., Edmonds M., Caltabiano T., Salerno G., Burton M., and Mather T. A. (2003) Sulphur dioxide fluxes from Mount Etna, Vulcano, and Stromboli measured with an automated scanning ultraviolet spectrometer. *J. Geophys. Res.* **180**, D.o.i. 10.1029/2002JB002261 (30 September 2003).

McMilan P. F. (1994) Water solubility and speciation models. *Rev. Mineral.* **30**, 131–156.

Melnik O. and Sparks R. S. J. (1999) Nonlinear dynamics of lava extrusion. *Nature* **402**, 37–41.

Melnik O. and Sparks R. S. J. (2002a) Modelling of conduit flow dynamics during explosive activity at Soufrière Hills Volcano, Montserrat. In *The Eruption of Soufrière Hills Volcano, Montserrat, from 1995 to 1999.* Geological Society of London Memoir 21 (eds. T. H. Druitt and P. Kokelaar). Geological Society of London, pp. 307–317.

Melnik O. and Sparks R. S. J. (2002b) Dynamics of magma ascent and lava extrusion at Soufrière Hills Volcano, Montserrat. In *The Eruption of Soufrière Hills Volcano, Montserrat, from 1995 to 1999.* Geological Society of London Memoir 21 (eds. T. H. Druitt and P. Kokelaar). Geological Society of London, pp. 153–171.

Métrich N., Bonnin-Mosbah M., Susini J., Menez B., and Galoisy L. (2002) Presence of sulfite (S^{IV}) in arc magmas: implications for volcanic sulfur emissions. *Geophys. Res. Lett.* **29** DOI 10.1029/2001GL014607.

Minnis P., Harrison E. F., Stowe L. L., Gibson G. G., Ddenn F. M., Doelling D. R., and Smith W. L. Jr., (1993) Radiative climate forcing by the Mount Pinatubo eruption. *Science* **259**, 1411–1415.

Monzier M., Robin C., and Eissen J.-P. (1994) Kuwae (1425 AD); the forgotten caldera. *J. Volcanol. Geotherm. Res.* **59**, 207–218.

Moretti R., Papale P., and Ottonello G. (2003) A model for the saturation of C–O–H–S fluids in silicate melts. In *Volcanic Degassing,* Geological Society of London Special Publication 213 (eds. C. Oppenheimer, D. M. Pyle, and J. Barclay). Geological Society of London, , pp. 81–101.

Navon and Lyakhovsky (1998) Vesiculation processes in silicic magmas. In *The Physics of Explosive Volcanic Eruptions,* Geological Society of London Special Publication 145 (eds. J. S. Gilbert and R. S. J. Sparks). Geological Society of London, , pp. 27–50.

Newman S. and Lowenstern J. B. (2002) VolatileCalc: a silicate melt-H_2O–CO_2 solution model written in visual basic for excel. *Comput. Geosci.* **28**, 597–604.

Nriagu J. O. (1989) A global assessment of natural sources of atmospheric trace metals. *Nature* **338**, 47–49.

Oppenheimer C. (1996) On the role of hydrothermal systems in the transfer of volcanic sulfur to the atmosphere. *Geophys. Res. Lett.* **23**, 2057–2060.

Oppenheimer C. (2002) Limited global change due to largest known Quaternary eruption, Toba $\approx$74 kyr BP? *Quat. Sci. Rev.* **21**, 1593–1609.

Oppenheimer C. (2003a) Climatic, environmental, and human consequences of the largest known historic eruption: Tambora volcano (Indonesia) 1815. *Prog. Phys. Geogr.* **27**, 230–259.

Oppenheimer C. (2003b) Ice core and palaeoclimatic evidence for the great volcanic eruption of (1257). *Int. J. Climatol.,* **23**, 417–426.

Oppenheimer C. and Yirgu G. (2002) Imaging of an active lava lake: Erta 'Ale volcano. *Ethiopia. Int. J. Remote Sensing* **23**, 4777–4782.

Oppenheimer C., Francis P., Burton M., Maciejewski A., and Boardman L. (1998a) Remote measurement of volcanic gases by Fourier transform infrared spectroscopy. *Appl. Phys.* **B67**, 505–515.

Oppenheimer C., Francis P., and Stix J. (1998b) Depletion rates of sulfur dioxide in tropospheric volcanic plumes. *Geophys. Res. Lett.* **25**, 2671–2674.

Oppenheimer C., Edmonds M., Francis P., and Burton M. R. (2002a) Variation in HCl/SO_2 gas ratios observed by Fourier transform spectroscopy at Soufrière Hills Volcano, Montserrat. In *The Eruption of Soufrière Hills Volcano, Montserrat, from 1995 to 1999.* Geological Society of London Memoir 21 (eds. T. H. Druitt and P. Kokelaar). Geological Society of London, pp. 621–639.

Oppenheimer C., Burton M. R., Durieux J., and Pyle D. M. (2002b) Open-path Fourier transform spectroscopy of gas emissions from a carbonatite volcano: Oldoinyo Lengai, Tanzania. *Opt. Lasers Eng.* **37**, 203–214.

Oppenheimer C., Pyle D. M., and Barclay J. (eds.) (2003) *Volcanic Degassing.* Geological Society of London Special Publication 213, 420pp.

Óskarsson N. (1980) The interaction between volcanic gases and tephra, fluorine adhering to tephra of the 1970 Hekla eruption. *J. Volcanol. Geotherm. Res.* **8**, 251–266.

Papale P. (1999) Modelling of the solubility of a two-component H_2O + CO_2 fluid in silicate liquid. *Am. Mineral.* **84**, 477–492.

Papale P. and Polacci M. (1999) Role of carbon dioxide in the dynamics of magma ascent in explosive eruptions. *Bull. Volcanol.* **60**, 583–594.

Parker D. E., Wilson H., Jones P. D., Christy J. R., and Folland C. K. (1996) The impact of Mount Pinatubo on world-wide temperatures. *Int. J. Climatol.* **16**, 197–487.

Parnell R. A. (1986) Processes of soil acidification in tropical Durandepts, Nicaragua. *Soil Sci.* **42**, 43–55.

Pinto J. P., Turco R. P., and Toon O. B. (1989) Self-limiting physical and chemical effects in volcanic eruption clouds. *J. Geophys. Res.* **94**, 11165–11174.

Porter J. N., Horton K., Mouginis-Mark P., Lienert B., Lau E., Sutton A. J., Elias T., and Oppenheimer C. (2002) Sun photometer and lidar measurements of the plume from the Hawaii Kilauea volcano Pu'u O'o vent: estimates of aerosol flux rates and SO$_2$ lifetime. *Geophys. Res. Lett.* Doi 10.1029/2002GL014744 (23 August 2002).

Post J. D. (1977) *The Last Great Subsistence Crisis in the Western World.* The John Hopkins University Press, Baltimore, 240 pp.

Prata, A. J., Self, S., Rose, W. I., O'Brien, D. M., (2002). Global, long-term sulphur dioxide measurements from TOVS data: a new tool for studying explosive volcanism and climate. In *American Geophysical Union Chapman Conference on Volcanism and the Earth's Atmosphere, Santorini June 17–21, 2002.* Abstract volume, 33.

Pyle D. M. (2000) Sizes of volcanic eruptions. In *Encyclopedia of Volcanoes* (eds. H. Sigurdsson, B. F. Houghton, S. R. McNutt, H. Rymer, and J. Stix). Academic Press, San Diego, CA, pp. 263–269.

Pyle D. M., Beattie P. D., and Bluth G. J. S. (1996) Sulphur emissions to the stratosphere from explosive volcanic eruptions. *Bull. Volcanol.* 57, 663–671.

Raga G. B., Kok G. L., and Baumgardner D. (1999) Evidence for volcanic influence on Mexico City aerosols. *Geophys. Res. Lett.* 26, 1149–1152.

Rampino M. R. and Self S. (1992) Volcanic winter and accelerated glaciation following the Toba super-eruption. *Nature* 359, 50–52.

Rampino M. R. and Stothers R. B. (1988) Flood basalt volcanism during the past 250 million years. *Science* 241, 663–668.

Read W. G., Froidevaux L., and Waters J. W. (1993) Microwave limb sounder measurements of stratospheric SO$_2$ from the Mt. Pinatubo eruption. *Geophys. Res. Lett.* 20, 1299–1302.

Robock A. (2000) Volcanic eruptions and climate. *Rev. Geophys.* 38(2), 191–219.

Roggensack K. R., Hervig R. L., McKnight S. B., and Williams S. N. (1997) Explosive basaltic volcanism from Cerro Negro volcano: influence of volatiles on eruption style. *Science* 277, 1639–1642.

Rose W. I., Chuan R. L., Giggenbach W. F., Kyle P. R., and Symonds R. B. (1986) Rates of sulfur dioxide and particle emissions from White Island volcano, New Zealand, and an estimate of the total flux of major gaseous species. *Bull. Volcanol.* 48, 181–188.

Sagan C. and Chyba C. (1997) The early faint Sun paradox: organic shielding of ultraviolet-labile greenhouse gases. *Science* 276, 1217–1221.

Scaillet B. and Pichavant M. (2003) Experimental constraints on volatile abundances in arc magmas and their implications for degassing processes. In *Volcanic Degassing.* Geological Society of London Special Publication 213 (eds. C. Oppenheimer, D. M. Pyle, and J. Barclay). Geological Society of London, pp. 23–52.

Scaillet B., Clemente B., Evans B. W., and Pichavant M. (1998) Redox control of sulfur degassing in silicic magmas. *J. Geophys. Res.* 103, 23937–23949.

Scaillet B., Luhr J., and Carroll M. (2003) Petrological and volcanological constraints on volcanic sulfur emissions to the atmosphere. In *Volcanism and the Earth's Atmosphere* (eds. A. Robock and C. Oppenheimer). American Geophysical Union Monograph (in press).

Shaw H. R. (1972) Viscosities of magmatic silicate liquids: an empirical method of prediction. *Am. J. Sci.* 272, 870–893.

Signorelli S. and Carroll M. R. (2000) Solubility and fluid-melt partitioning of Cl in hydrous phonolitic melts. *Geochim. Cosmochim. Acta* 64, 2851–2862.

Signorelli S. and Carroll M. R. (2001) Experimental constraints on the origin of chlorine emissions at the Soufriere Hills Volcano. Montserrat. *Bull. Volcanol.* 62, 431–440.

Signorelli S. and Carroll M. R. (2002) Experimental study of Cl solubility in hydrous alkaline melts: constraints on the theoretical maximum amount of Cl in trachytic and phonolitic melts. *Contrib. Mineral. Petrol.* 143, 209–218.

Sigurdsson H., Devine J. D., Tchoua F. M., Presser T. S., Pringle M. K. W., and Evans W. C. (1987) Origin of the lethal has burst from Lake Monoun, Cameroon. *J. Volcanol. Geotherm. Res.* 31, 1–16.

Silver L. A., Ihinger P. D., and Stolper E. M. (1990) The influence of bulk composition on the speciation of water in silicate glasses. *Contrib. Mineral. Petrol.* 104, 142–162.

Slezin Y. B. (1995) Principal regimes of volcanic eruptions. *Volcanol. Seismol.* 17, 193–205.

Slezin Y. B. (2003) The mechanism of volcanic eruption (steady state approach). *J. Volcanol. Geotherm. Res.* 122, 7–50.

Smith W. H. (1990) *Air Pollution and Forests: Interaction between Air Contaminants and Forest Ecosystems.* Springer, New York, 618pp.

Soden B. J., Wetherald R. T., Stenchikov G. L., and Robock A. (2002) Global cooling following the eruption of Mt. Pinatubo: a test of climate feedback by water vapor. *Science* 296, 727–730.

Sorey M. L., Kennedy B. M., Evans W. C., Farrar C. D., and Suemnicht G. A. (1993) Helium isotope and gas discharge variations associated with crustal unrest in Long Valley caldera, California. *J. Geophys. Res.* 98, 15871–15889.

Sorey M. L., Evans W. C., Kennedy B. M., Farrar C. D., Hainsworth L. J., and Hausback B. (1998) Carbon dioxide and helium emissions from a reservoir of magmatic gas beneath Mammoth Mountain, California. *J. Geophys. Res.* 103, 15303–15323.

Sparks R. S. J. (1997) Causes and consequences of pressurisation in lava dome eruptions. *Earth Planet. Sci. Lett.* 150, 177–189.

Sparks R. S. J. (2003) Dynamics of magma degassing. In *Volcanic Degassing,* Geological Society of London Special Publication 213 (eds. C. Oppenheimer, D. M. Pyle, and J. Barclay). Geological Society of London, pp. 5–22.

Sparks R. S. J., Bursik M. I., Carey S. N., Gilbert J. S., Glaze L., Sigurdsson H., and Woods A. W. (1997) *Volcanic Plumes.* Wiley, Chichester, England, 557pp.

Spera F. J. (2000) Physical properties of magma. In *Encyclopedia of Volcanoes* (eds. H. Sigurdsson, B. F. Houghton, S. R. McNutt, H. Rymer, and J. Stix). Academic Press, San Diego, CA, pp. 171–190.

Stenchikov G. L., Kirchner I., Robock A., Graf H.-F., Antuna J. C., Grainger R. G., Lambert A., and Thomason L. (1998) Radiative forcing from the 1991 Mount Pinatubo volcanic eruption. *J. Geophys. Res.* 103, 13837–13857.

Stenchikov G., Robock A., Ramaswamy V., Schwarzkopf M. D., Hamilton K., and Ramachandran S. (2002) Arctic Oscillation response to the 1991 Mount Pinatubo eruption: effects of volcanic aerosols and ozone depletion. *J. Geophys. Res.* Doi 10.1029/2002JD002090 (28 December 2002).

Stevenson D. S. (1993) Physical models of fumarolic flow. *J. Volcanol. Geotherm. Res.* 57, 139–156.

Stevenson D. S. and Blake S. (1998) Modelling the dynamics and thermodynamics of volcanic degassing. *Bull. Volcanol.* 60, 307–317.

Stevenson D. S., Johnson C. E., Collins W. J., and Derwent R. G. (2003a) The tropospheric sulphur cycle and the role of volcanic SO$_2$. In *Volcanic Degassing.* Geological Society of London Special Publication 213 (eds. C. Oppenheimer, D. M. Pyle, and J. Barclay). Geological Society of London, pp. 295–305.

Stevenson D. S., Johnson C. E., Highwood E. J., Ganei V., Collins W. J., and Derwent R. G. (2003) Atmospheric impact of the 1783–1784 Laki eruption: Part 1. Chemistry modelling, *Atmos. Chem. Phys.* 3, 487–507.

Stoiber R. E., Malinconico L. L., Jr., and Williams S. N. (1983) Use of the correlation spectrometer at volcanoes. In *Forecasting Volcanic Events* (eds. H. Tazieff and J.-C. Sabroux). Elsevier, Amsterdam, pp. 425–444.

Stolper E. (1989) Temperature dependence of the speciation of water in rhyolitic melts and glasses. *Am. Mineral.* 74, 1247–1257.

Stolper E. M. (1982) The speciation of water in silicate melts. *Geochim. Cosmochim. Acta* **46**, 2609–2620.

Sutton A. J. and Elias T. (1993) Volcanic gases create air pollution in the Island of Hawaii. *Earthquakes Volcanoes* **24**, 178–196.

Symonds R. B. and Reed M. H. (1993) Calculation of multi-component chemical equilibria in gas–solid–liquid systems: calculation methods, thermochemical data and applications to studies of high-temperature volcanic gases with examples from Mount St. Helens. *Am. J. Sci.* **293**, 758–864.

Symonds R. B., Rose W. I., and Reed M. H. (1988) Contribution of Cl- and F-bearing gases to the atmosphere by volcanoes. *Nature* **334**, 415–418.

Symonds R. B., Rose W. I., Bluth G. J. S., and Gerlach T. M. (1994) Volcanic gas studies: methods, results and applications. In *Volatiles in Magmas,* Revienes in Mineralogy, 30 (eds. M. R. Carroll and J. R. Hollaway). Mineralogical Society of America, Washingon, DC, pp. 1–66.

Symonds R. B., Mizutani Y., and Briggs P. H. (1996) Long-term geochemical surveillance of fumaroles at Showa-Shinzan dome, Usu volcano, Japan. *J. Volcanol. Geotherm. Res.* **73**, 177–211.

Symonds R. B., Gerlach T. M., and Reed M. H. (2001) Magmatic gas scrubbing: implications for volcano monitoring. *J. Volcanol. Geotherm. Res.* **108**, 303–341.

Tabazadeh A. and Turco R. P. (1993) Stratospheric chlorine injection by volcanic eruptions: HCl scavenging and implications for ozone. *Science* **260**, 1082–1086.

Taran Y. A., Hedenquist J. W., Korzhinsky M. A., Tkachenko S. I., and Shmulovich K. I. (1995) Geochemistry of magmatic gases from Kudryavy volcano, Iturup, Kuril Islands. *Geochim. Cosmochim. Acta* **59**, 1749–1761.

Tedesco D. (1995) Monitoring fluids and gases at active volcanoes. In *Monitoring Active Volcanoes* (eds. B. McGuire and J. Murray). University College London Press, London, pp. 315–345.

Textor C., Sachs P. M., Graf H.-F., and Hansteen T. (2003) The scavenging of sulphur and halogen gases in a plinian volcanic plume similar to the Laacher see eruption 12,900 yr BP. In *Volcanic Degassing,* Geological Society of London Special Publication 213 (eds. C. Oppenheimer, D. M. Pyle, and J. Barclay). Geological Society of London, pp. 307–328.

Thompson D. W. J. and Wallace J. M. (1998) The Arctic Oscillation signature in the wintertime geopotential height and temperature fields. *Geophys. Res. Lett.* **25**, 1297–1300.

Thorarinsson S. (1979) On the damage caused by volcanic eruptions with special reference to tephra and gases. In *Volcanic Activity and Human Ecology* (eds. P. D. Sheets and D. K. Grayson). Academic Press, New York, pp. 125–159.

Thordarson T. and Self S. (1996) Sulfur, chlorine, and fluorine degassing and atmospheric loading by the Roza eruption, Columbia River Basalt group, Washington, USA. *J. Volcanol. Geotherm. Res.* **74**, 49–73.

Thordarson T. and Self S. (2002) Atmospheric and environmental effects of the 1783–84 Laki eruption: a-review and reassessment. *J. Geophys. Res.* **108** (DI), 4011.

Thordarson T., Self S., Oskarsson N., and Hulsebosch T. (1996) Sulfur, chlorine, and flourine degassing and atmospheric loading by the 1783–1784 AD Laki (Skaftar fires) eruption in Iceland. *Bull. Volcanol.* **58**, 205–225.

Thordarson T., Self S., Miller D. J., Larsen G., and Vilmundardóttir E. G. (2003) Sulphur release from flood lava eruptions in the Veidivötn, Grímsvötn, and Katla volcanic systems, Iceland. In *Volcanic Degassing,* Geological Society of London Special Publication 213, (eds. C. Oppenheimer, D. M. Pyle, and J. Barclay). Geological Society of London, pp. 103–121.

Towe K. M. (2002) The problematic rise of Archean oxygen. *Science* **295**, 1419.

Valley J. W. and Cole D. (eds.) (2001) Stable isotope geochemistry. Reviews in Mineralogy, 43, Mineralogical Society of America, Washington, DC, 662pp.

Varekamp J. C. and Rowe G. L., Jr. (eds.) (2000) Crater Lakes. In *Journal of Volcanology and Geothermal Research,* Special Issue 97, Elsevier, 508pp.

Vié le Sage R. (1983) Chemistry of the volcanic aerosol. In *Forecasting Volcanic Events* (eds. H. Tazieff and J.-C. Sabroux). Elsevier, Amsterdam, pp. 445–474.

Villemant B. and Boudon G. (1998) Transition between dome-building and plinian eruptive styles: H_2O and Cl degassing behaviour. *Nature* **392**, 65–69.

Villemant B. and Boudon G. (1999) H_2O and halogen (F, Cl, Br) behaviour during shallow magma degassing processes. *Earth Planet. Sci. Lett.* **168**, 271–286.

Villemant B., Boudon G., Nougrigat S., Poteaux S., and Michel A. (2003) H_2O and halogens in volcanic clasts: tracers of degassing processes during plinian and dome-forming eruptions. In *Volcanic Degassing,* Geological Society of London Special Publication 213 (eds. C. Oppenheimer, D. M. Pyle, and J. Barclay). Geological Society of London, pp. 63–79.

Voight B. V., Sparks R. S. J., Miller A. D., Stewart R. C., Hoblitt R. P., Clarke R. P., Ewart J., Aspinall W., Baptie B., Druit T. H., Herd R., Jackson P., Lockhart A. B., Loughlin S. C., Luckett R., Lynch L., McMahon J., Norton G. E., Robertson R., Watson I. M., and Young S. R. (1999) Magma flow instability and cyclic activity at Soufriere Hills volcano, Montserrat BWI. *Science* **283**, 1138–1142.

Walker G. P. L., Self S., and Wilson L. (1984) Tarawera 1886, New Zealand, a basaltic plinian eruption. *J. Volcanol. Geotherm. Res.* **21**, 61–78.

Wallace P. and Anderson A. T. (2000) Volatiles in magmas. In *Encyclopedia of Volcanoes* (eds. H. Sigurdsson, B. F. Houghton, S. R. McNutt, H. Rymer, and J. Stix). Academic Press, San Diego, CA, pp. 149–170.

Wallace P. J. (2001) Volcanic SO_2 emissions and the abundance and distribution of exsolved gas in magma bodies. *J. Volcanol. Geotherm. Res.* **108**, 85–106.

Wallace P. J. and Gerlach T. M. (1994) Magmatic vapor source for the sulfur dioxide released during volcanic eruptions: evidence from Mount Pinatubo. *Science* **265**, 497–499.

Wallace P. J., Anderson A. T., Jr., and Davis A. M. (1995) Quantification of pre-eruptive exsolved gas contents in silicic magmas. *Nature* **377**, 612–616.

Watson B. E. (1994) Diffusion in volatile-bearing magmas. *Rev. Mineral.* **30**, 371–411.

Watson I. M. and Oppenheimer C. (2000) Particle size distributions of Mt. Etna's aerosol plume constrained by sunphotometry. *J. Geophys. Res. Atmos.* **105**, 9823–9830.

Watson I. M. and Oppenheimer C. (2001) Particle-size distributions of ash-rich volcanic plumes determined by sun photometry. *Atmos. Environ.* **35**, 3561–3572.

Webster J. D., Kinzler R. J., and Mathez E. A. (1999) Chloride and water solubility in basalt and andesite melts and implications for magma degassing. *Geochim. Cosmochim. Acta* **63**, 729–738.

Webster J. D., Raia F., De Vivo B., and Rolandi G. (2001) The behavior of chlorine and sulfur during differentiation of the Mt. Somma-Vesuvius magmatic system. *Mineral. Petrol.* **73**, 177–200.

Westrich H. R. and Gerlach T. M. (1992) Magmatic gas source for the stratospheric SO_2 cloud from the June 15, 1991 eruption of Mount Pinatubo. *Geology* **20**, 867–870.

Williams S. N. (1983) Plinian airfall deposits of basaltic composition. *Geology* **11**, 211–214.

Williams H. and McBirney A. R. (1979) *Volcanology*. Freeman, Cooper and Company, San Francisco, CA, 397pp.

Williams-Jones G. and Rymer H. (2000) Hazards of volcanic gases. In *Encyclopedia of Volcanoes* (eds. H. Sigurdsson, B. F. Houghton, S. R. McNutt, H. Rymer, and J. Stix). Academic Press, San Diego, CA, pp. 997–1013.

Winner W. E. and Mooney H. A. (1980) Responses of Hawaiian plants to volcanic sulfur dioxide: stomatal behavior and foliar injury. *Science* **210**, 789–791.

Zdanowicz C. M., Zielinski G. A., and Germani M. S. (1999) Mt. Mazama eruption: calendrical age verified and atmospheric impact assessed. *Geology* **27**, 621–624.

Zielinski G. A. (1995) Stratospheric loading and optical depth estimates of explosive volcanism over the last 2100 years derived from the Greenland ice sheet project 2 ice core. *J. Geophys. Res.* **100**, 20937–20955.

Zielinski G. A. (2000) Use of paleo-records in determining variability within the volcanism-climate system. *Quat. Sci. Rev.* **19**, 417–438.

Zielinski G. A., Mayewski O. A., Meeker L. D., Whitlow S., Twickler M. S., Morrison M., Meese D. A., Gow A. J., and

Alley R. B. (1994) Record of volcanism since 7000BC from the GISP2 Greenland ice core and implications for the volcano-climate system. *Science* **264**, 948–952.

Zielinski G. A., Mayewski P. A., Meeker L. D., Whitlow S., and Twickler M. S. (1996a) A 110,000-yr record of explosive volcanism from the GISP2 (Greenland) ice core. *Quat. Res.* **45**, 109–118.

Zielinski G. A., Mayewski P. A., Meeker L. D., Whitlow S., Twickler M. S., and Taylor K. (1996b) Potential atmospheric impact of the Toba mega-eruption ∼71,000 years ago. *Geophys. Res. Lett.* **23**, 837–840.

Zreda-Gostynska G., Kyle P. R., Finnegan D. L., and Prestbo K. M. (1997) Volcanic gas emissions from Mount Erebus and their impact on the Antarctic environment. *J Geophys. Res.* **102**, 15039–15055.

Geochemistry of Earth Surface Systems
ISBN: 978-0-08-096706-6

pp. 1–44

2

Hydrothermal Processes

C. R. German

Southampton Oceanography Centre, Southampton, UK

and

K. L. Von Damm*

University of New Hampshire, Durham, NH, USA

2.1	INTRODUCTION	46
	2.1.1 What is Hydrothermal Circulation?	46
	2.1.2 Where Does Hydrothermal Circulation Occur?	48
	2.1.3 Why Should Hydrothermal Fluxes Be Considered Important?	50
2.2	VENT-FLUID GEOCHEMISTRY	51
	2.2.1 Why are Vent-fluid Compositions of Interest?	51
	2.2.2 Processes Affecting Vent-fluid Compositions	52
	2.2.3 Compositions of Hydrothermal Vent Fluids	56
	2.2.3.1 Major-element chemistry	56
	2.2.3.2 Trace-metal chemistry	59
	2.2.3.3 Gas chemistry of hydrothermal fluids	61
	2.2.3.4 Nutrient chemistry	61
	2.2.3.5 Organic geochemistry of hydrothermal vent fluids	62
	2.2.4 Geographic Variations in Vent-fluid Compositions	62
	2.2.4.1 The role of the substrate	62
	2.2.4.2 The role of temperature and pressure	62
	2.2.4.3 The role of spreading rate	63
	2.2.4.4 The role of the plumbing system	64
	2.2.5 Temporal Variability in Vent-fluid Compositions	64
2.3	THE *NET* IMPACT OF HYDROTHERMAL ACTIVITY	66
2.4	NEAR-VENT DEPOSITS	67
	2.4.1 Alteration and Mineralization of the Upper Ocean Crust	67
	2.4.2 Near-vent Hydrothermal Deposits	68
2.5	HYDROTHERMAL PLUME PROCESSES	69
	2.5.1 Dynamics of Hydrothermal Plumes	69
	2.5.2 Modification of Gross Geochemical Fluxes	70
	2.5.2.1 Dissolved noble gases	70
	2.5.2.2 Dissolved reduced gases (H_2S, H_2, CH_4)	71
	2.5.2.3 Iron and manganese geochemistry in hydrothermal plumes	72
	2.5.2.4 Co-precipitation and uptake with iron in buoyant and nonbuoyant plumes	73
	2.5.2.5 Hydrothermal scavenging by Fe-oxyhydroxides	74
	2.5.3 Physical Controls on Dispersing Plumes	75
	2.5.4 Biogeochemical Interactions in Dispersing Hydrothermal Plumes	76
	2.5.5 Impact of Hydrothermal Plumes Upon Ocean Geochemical Cycles	77

*Deceased

2.6 HYDROTHERMAL SEDIMENTS   77
    2.6.1 Near-vent Sediments   77
    2.6.2 Deposition from Hydrothermal Plumes   78
    2.6.3 Hydrothermal Sediments in Paleoceanography   78
    2.6.4 Hydrothermal Sediments and Boundary Scavenging   79
2.7 CONCLUSION   79
REFERENCES   80

2.1 INTRODUCTION

2.1.1 What is Hydrothermal Circulation?

Hydrothermal circulation occurs when seawater percolates downward through fractured ocean crust along the volcanic mid-ocean ridge (MOR) system. The seawater is first heated and then undergoes chemical modification through reaction with the host rock as it continues downward, reaching maximum temperatures that can exceed 400 °C. At these temperatures the fluids become extremely buoyant and rise rapidly back to the seafloor where they are expelled into the overlying water column. Seafloor hydrothermal circulation plays a significant role in the cycling of energy and mass between the solid earth and the oceans; the first identification of submarine hydrothermal venting and their accompanying chemosynthetically based communities in the late 1970s remains one of the most exciting discoveries in modern science. The existence of some form of hydrothermal circulation had been predicted almost as soon as the significance of ridges themselves was first recognized, with the emergence of plate tectonic theory. Magma wells up from the Earth's interior along "spreading centers" or "MORs" to produce fresh ocean crust at a rate of $\sim$20 km^3 yr^{-1}, forming new seafloor at a rate of $\sim$3.3 km^2 yr^{-1} (Parsons, 1981; White et al., 1992). The young oceanic lithosphere formed in this way cools as it moves away from the ridge crest. Although much of this cooling occurs by upward conduction of heat through the lithosphere, early heat-flow studies quickly established that a significant proportion of the total heat flux must also occur via some additional convective process (Figure 1), i.e., through circulation of cold seawater within the upper ocean crust (Anderson and Silbeck, 1981).

The first geochemical evidence for the existence of hydrothermal vents on the ocean floor came in the mid-1960s when investigations in the Red Sea revealed deep basins filled with hot, salty water (40–60 °C) and underlain by thick layers of metal-rich sediment (Degens and Ross, 1969). Because the Red Sea represents a young, rifting, ocean basin it was speculated that the phenomenon observed there might also prevail along other young MOR spreading centers. An analysis of core-top sediments from throughout the world's oceans (Figure 2) revealed that such metalliferous sediments did, indeed, appear to be concentrated along the newly recognized global ridge crest (Boström et al., 1969). Another early indication of hydrothermal activity came from the detection of plumes of excess ^{3}He in the Pacific Ocean Basin (Clarke et al., 1969)—notably the > 2,000 km wide section in the South Pacific (Lupton and Craig, 1981)—because ^{3}He present in the deep ocean could only be sourced through some form of active degassing of the Earth's interior, at the seafloor.

One area where early heat-flow studies suggested hydrothermal activity was likely to occur was along the Galapagos Spreading Center in the eastern equatorial Pacific Ocean (Anderson and Hobart, 1976). In 1977, scientists diving at this location found hydrothermal fluids discharging chemically altered seawater from young volcanic seafloor at elevated temperatures up to 17 °C (Edmond et al., 1979). Two years later, the first high-temperature (380 $\pm$ 30 °C) vent fluids were found at 21° N on the East Pacific Rise (EPR) (Spiess et al., 1980)—with fluid compositions remarkably close to those predicted from the lower-temperature Galapagos findings (Edmond et al., 1979). Since that time, hydrothermal activity has been found at more than 40 locations throughout the Pacific, North Atlantic, and Indian

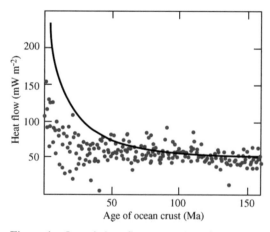

Figure 1 Oceanic heat flow versus age of ocean crust. Data from the Pacific, Atlantic, and Indian oceans, averaged over 2 Ma intervals (circles) depart from the theoretical cooling curve (solid line) indicating convective cooling of young ocean crust by circulating seawater (after C. A. Stein and S. Stein, 1994).

Oceans (e.g., Van Dover *et al.*, 2002) with further evidence—from characteristic chemical anomalies in the ocean water column—of its occurrence in even the most remote and slowly spreading ocean basins (Figure 3), from the polar seas of the Southern Ocean (German *et al.*, 2000; Klinkhammer *et al.*, 2001) to the extremes of the ice-covered Arctic (Edmonds *et al.*, 2003).

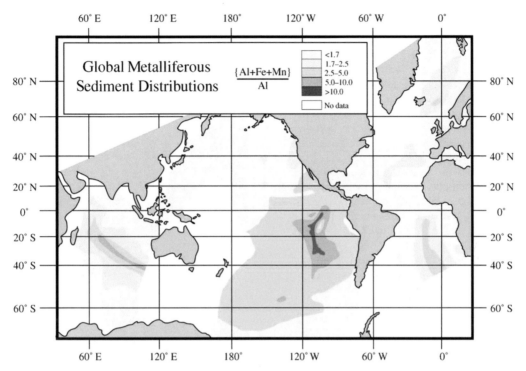

Figure 2 Global map of the (Al+Fe+Mn):Al ratio for surficial marine sediments. Highest ratios mimic the trend of the global MOR axis (after Boström *et al.*, 1969).

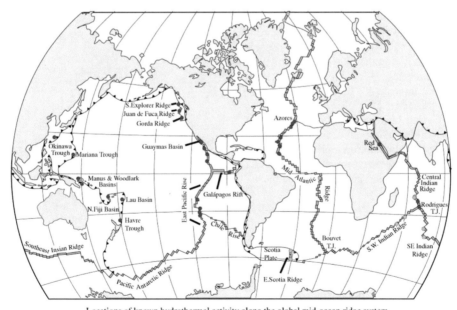

Locations of known hydrothermal activity along the global mid-ocean ridge system
● = known active sites ○ = active sites indicated by midwater chemical anomalies

Figure 3 Schematic map of the global ridge crest showing the major ridge sections along which active hydrothermal vents have already been found (red circles) or are known to exist from the detection of characteristic chemical signals in the overlying water column (orange circles). Full details of all known hydrothermally active sites and plume signals are maintained at the InterRidge web-site: http://triton.ori.u-tokyo.ac.jp/~intridge/wg-gdha.htm

The most spectacular manifestation of seafloor hydrothermal circulation is, without doubt, the high-temperature (>400 °C) "black smokers" that expel fluids from the seafloor along all parts of the global ocean ridge crest. In addition to being visually compelling, vent fluids also exhibit important enrichments and depletions when compared to ambient seawater. Many of the dissolved chemicals released from the Earth's interior during venting precipitate upon mixing with the cold, overlying seawater, generating thick columns of black metal-sulfide and oxide mineral-rich smoke—hence the colloquial name for these vents: "black smokers" (Figure 4). In spite of their common appearance, high-temperature hydrothermal vent fluids actually exhibit a wide range of temperatures and chemical compositions, which are determined by subsurface reaction conditions. Despite their spectacular appearance, however, high-temperature vents may only represent a small fraction—perhaps as little as 10%—of the total hydrothermal heat flux close to ridge axes. A range of studies—most notably along the Juan de Fuca Ridge (JdFR) in the NE Pacific Ocean (Rona and Trivett, 1992; Schultz *et al.*, 1992; Ginster *et al.*, 1994) have suggested that, instead, axial hydrothermal circulation may be dominated by much lower-temperature diffuse flow exiting the seafloor at temperatures comparable to those first observed at the Galapagos vent sites in 1977. The relative importance of high- and low-temperature hydrothermal circulation to overall ocean chemistry remains a topic of active debate.

While most studies of seafloor hydrothermal systems have focused on the currently active plate boundary (~0–1 Ma crust), pooled heat-flow data from throughout the world's ocean basins (Figure 1) indicate that convective heat loss from the oceanic lithosphere actually continues in crust from 0–65 Ma in age (Stein *et al.*, 1995). Indeed, most recent estimates would indicate that hydrothermal circulation through this older (1–65 Ma) section, termed "flank fluxes," may be responsible for some 70% or more of the total hydrothermal heat loss associated with spreading-plate boundaries—either in the form of warm (20–65 °C) altered seawater, or as cooler water, which is only much more subtly chemically altered (Mottl, 2003).

When considering the impact of hydrothermal circulation upon the chemical composition of the oceans and their underlying sediments, however, attention returns—for many elements—to the high-temperature "black smoker" systems. Only here do many species escape from the seafloor in high abundance. When they do, the buoyancy of the high-temperature fluids carries them hundreds of meters up into the overlying water column as they mix and eventually form nonbuoyant plumes containing a wide variety of both dissolved chemicals and freshly precipitated mineral phases. The processes active within these dispersing

(a)

(b)

Figure 4 (a) Photograph of a "black smoker" hydrothermal vent emitting hot (>400 °C) fluid at a depth of 2,834 m into the base of the oceanic water column at the Brandon vent site, southern EPR. The vent is instrumented with a recording temperature probe. (b) Diffuse flow hydrothermal fluids have temperatures that are generally <35 °C and, therefore, may host animal communities. This diffuse flow site at a depth of 2,500 m on the EPR at 9°50′ N is populated by *Riftia* tubeworms, mussels, crabs, and other organisms.

hydrothermal plumes play a major role in determining the net impact of hydrothermal circulation upon the oceans and marine geochemistry.

2.1.2 Where Does Hydrothermal Circulation Occur?

Hydrothermal circulation occurs predominantly along the global MOR crest, a near-continuous volcanic chain that extends over ~6 × 10⁴ km

(Figure 3). Starting in the Arctic basin this ridge system extends south through the Norwegian-Greenland Sea as far as Iceland and then continues southward as the Mid-Atlantic Ridge (MAR), passing through the Azores and onward into the far South Atlantic, where it reaches the Bouvet Triple Junction, near 50° S. To the west, a major transform fault connects this triple junction to the Sandwich and Scotia plates that are separated by the East Scotia Ridge (an isolated back-arc spreading center). These plates are also bound to north and south by two major transform faults that extend further west between South America and the Antarctic Peninsula before connecting to the South Chile Trench. To the east of the Bouvet Triple Junction lies the SW Indian Ridge, which runs east and north as far as the Rodrigues Triple Junction (~25° S, 70° E), where the ridge crest splits in two. One branch, the Central Indian Ridge, extends north through the western Indian Ocean and Gulf of Aden ending as the incipient ocean basin that is the Red Sea (Section 2.1.1). The other branch of the global ridge crest branches south east from the Rodrigues Triple Junction to form the SE Indian and Pacific-Antarctic Ridges which extend across the entire southern Indian Ocean past Australasia and on across the southern Pacific Ocean as far as ~120° W, where the ridge again strikes north. The ridge here, the EPR, extends from ~55° S to ~30° N but is intersected near 30° S by the Chile Rise, which connects to the South Chile Trench. Further north, near the equator, the Galapagos Spreading Center meets the EPR at another triple junction. The EPR (and, hence, the truly continuous portion of the global ridge crest, extending back through the Indian and Atlantic Oceans) finally ends where it runs "on-land" at the northern end of the Gulf of California. There, the ridge crest is offset to the NW by a transform zone, more commonly known as the San Andreas Fault, which continues offshore once more, off northern California at ~40° N, to form the Gorda, Juan de Fuca, and Explorer Ridges—all of which hug the NE Pacific/N. American margin up to ~55° N. Submarine hydrothermal activity is also known to be associated with the back-arc spreading centers formed behind ocean–ocean subduction zones which occur predominantly around the northern and western margins of the Pacific Ocean, from the Aleutians via the Japanese archipelago and Indonesia all the way south to New Zealand. In addition to ridge-crest hydrothermal venting, similar circulation also occurs associated with hot-spot related intraplate volcanism—most prominently in the central and western Pacific Ocean (e.g., Hawaii, Samoa, Society Islands), but these sites are much less extensive, laterally, than ridge crests and back-arc spreading centers, combined. A continuously updated map of reported hydrothermal vent sites is maintained by the Inter-Ridge community as a Vents Database (http://triton.ori.u-tokyo.ac.jp/~intridge/wg-gdha.htm).

As described earlier, the first sites of hydrothermal venting to be discovered were located along the intermediate to fast spreading Galapagos Spreading Center (6 cm yr^{-1}) and northern EPR (6–15 cm yr^{-1}). A hypothesis, not an unreasonable one, influenced heavily by these early observations but only formalized nearly 20 years later (Baker *et al.*, 1996) proposed that the incidence of hydrothermal venting along any unit length of ridge crest should correlate positively with spreading-rate because the latter is intrinsically linked to the magmatic heat flux at that location. Thus, the faster the spreading rate the more abundant the hydrothermal activity, with the most abundant venting expected (and found: Charlou *et al.*, 1996; Feely *et al.*, 1996; Ishibashi *et al.*, 1997) along the superfast spreading southern EPR (17–19° S), where ridge-spreading rate is among the fastest known (>14 cm yr^{-1}). Evidence for reasonably widespread venting has also been found most recently along some of the slowest-spreading sections of the global ridge crest, both in the SW Indian Ocean (German *et al.*, 1998a; Bach *et al.*, 2002) and in the Greenland/Arctic Basins (Connelly *et al.*, 2002; Edmonds *et al.*, 2003). Most explorations so far, however, have focused upon ridge crests closest to nations with major oceanographic research fleets and in the low- to mid-latitudes, where weather conditions are most favorable toward use of key research tools such as submersibles and deep-tow vehicles. Consequently, numerous active vent sites are known along the NE Pacific ridge crests, in the western Pacific back-arc basins and along the northern MAR (Figure 3), while other parts of the global MOR system remain largely unexplored (e.g., southern MAR, Central Indian Ridge, SE Indian Ridge, Pacific – Antarctic Ridge).

Reinforcing how little of the seafloor is well explored, as recently as December 2000 an entirely new form of seafloor hydrothermal activity, in a previously unexplored geologic setting was discovered (Kelley *et al.*, 2001). Geologists diving at the Atlantis fracture zone, which offsets part of the MAR near 30° N, found moderate-temperature fluids (40–75 °C) exiting from tall (up to 20 m) chimneys, formed predominantly from calcite [$CaCO_3$], aragonite [$CaCO_3$], and brucite [$Mg(OH)_2$]. These compositions are quite unlike previously documented hydrothermal vent fluids (Section 2.2), yet their geologic setting is one that may recur frequently along slow and very slow spreading ridges (e.g., Gracia *et al.*, 1999, 2000, Parson *et al.*, 2000; Sauter *et al.*, 2002). The Lost City vent site may, therefore, represent a new and important form of hydrothermal vent input to the oceans, which has hitherto been overlooked.

2.1.3   Why Should Hydrothermal Fluxes Be Considered Important?

Since hydrothermal systems were first discovered on the seafloor, determining the magnitude of their flux to the ocean and, hence, their importance in controlling ocean chemistry has been the overriding question that numerous authors have tried to assess (Edmond *et al.*, 1979, 1982; Staudigel and Hart, 1983; Von Damm *et al.*, 1985a; C. A. Stein and S. Stein, 1994; Elderfield and Schultz, 1996; Schultz and Elderfield, 1997; Mottl, 2003). Of the total heat flux from the interior of the Earth ($\sim$43 TW) $\sim$32 TW is associated with cooling through oceanic crust and, of this, some 34% is estimated to occur in the form of hydrothermal circulation through ocean crust up to 65 Ma in age (C. A. Stein and S. Stein, 1994). The heat supply that drives this circulation is of two parts: magmatic heat, which is actively *emplaced* close to the ridge axis during crustal formation, and heat that is *conducted* into the crust from cooling lithospheric mantle, which extends out beneath the ridge flanks.

At the ridge axis, the magmatic heat available from crustal formation can be summarized as (i) heat released from the crystallization of basaltic magma at emplacement temperatures (latent heat), and (ii) heat mined from the solidified crust during cooling from emplacement temperatures to hydrothermal temperatures, assumed by Mottl (2003) to be $1175 \pm 25°C$ and $350 \pm 25°C$, respectively. For an average crustal thickness of $\sim$6 km (White *et al.*, 1992) the mass of magma emplaced per annum is estimated at 6×10^{16} g yr^{-1} and the maximum heat available from crystallization of this basaltic magma and cooling to hydrothermal temperatures is 2.8 ± 0.3 TW (Elderfield and Schultz, 1996; Mottl, 2003). If all this heat were transported as high-temperature hydrothermal fluids expelled from the seafloor at 350 °C and 350 bar, this heat flux would equate to a volume flux of $5-7 \times 10^{16}$ g yr^{-1}. It should be noted, however, that the heat capacity (c_p) of a 3.2% NaCl solution becomes extremely sensitive to increasing temperature under hydrothermal conditions of temperature and pressure, as the critical point is approached. Thus, for example, at 350 bar, a moderate increase in temperature near 400 °C could cause an increase in c_p approaching an order of magnitude resulting in a concomitant drop in the water flux required to transport this much heat (Bischoff and Rosenbauer, 1985; Elderfield and Schultz, 1996).

Of course, high-temperature hydrothermal fluids may not be entirely responsible for the transport of all the axial hydrothermal heat flux. Elderfield and Schultz (1996) considered a uniform distribution, on the global scale, in which only 10% of the total axial hydrothermal flux occurred as "focused" flow (heat flux $= 0.2-0.4$ TW; volume flux $= 0.3-0.6 \times 10^{16}$ g yr^{-1}). In those calculations, the remainder of the axial heat flux was assumed to be transported by a much larger volume flux of lower-temperature fluid ($280-560 \times 10^{16}$ g yr^{-1} at $\sim$5 °C). But how might such diffuse flow manifest itself? Should diffuse fluid be considered as diluted high-temperature vent fluid, conductively heated seawater, or some combination of the above? Where might such diffuse fluxes occur? Even if the axial hydrothermal heat flux were only restricted to 0–0.1 Ma crust, the associated fluid flow might still extend over the range of kilometers from the axis on medium-fast ridges—i.e., out onto young ridge flanks. For slow and ultraslow spreading ridges (e.g., the MAR) by contrast, all 0–0.1 Ma and, indeed 0–1 Ma crustal circulation would occur within the confines of the axial rift valley (order 10 km wide). The partitioning of "axial" and "near-axial" hydrothermal flow, on fast and slow ridges and between "focused" and "diffuse" flow, remains very poorly constrained in the majority of MOR settings and is an area of active debate.

On older oceanic crust (1–65 Ma) hydrothermal circulation is driven by upward conduction of heat from cooling of the underlying lithospheric mantle. Heat fluxes associated with this process are estimated at 7 ± 2 TW (Mottl, 2003). These values are significantly greater than the total heat fluxes associated with axial and near-axis circulation combined, and represent as much as $75-80\%$ of Earth's total *hydrothermal* heat flux, >20% of the total *oceanic* heat flux and >15% of the Earth's *entire* heat flux. Mottl and Wheat (1994) chose to subdivide the fluid circulation associated with this heat into two components, warm (>20 °C) and cool (<20 °C) fluids, which exhibit large and small changes in the composition of the circulating seawater, respectively. Constraints from the magnesium mass balance of the oceans suggest that the cool (less altered) fluids carry some 88% of the total *flank* heat flux, representing a cool-fluid water flux (for 5–20 °C fluid temperatures) of $1-4 \times 10^{19}$ g yr^{-1} (Mottl, 2003).

To put these volume fluxes in context, the maximum flux of cool (<20 °C) hydrothermal fluids, calculated above is almost identical to the global riverine flux estimate of $3.7-4.2 \times 10^{19}$ g yr^{-1} (Palmer and Edmond, 1989). The flux of high-temperature fluids close to the ridge axis, by contrast, is $\sim$1,000-fold lower. Nevertheless, for an ocean volume of $\sim$1.4 $\times 10^{24}$ g, this still yields a (geologically short) oceanic residence time, with respect to high-temperature circulation, of $\sim$20–30 Ma—and the hydrothermal fluxes will be important for those elements which exhibit high-temperature fluid concentrations more than 1,000-fold greater than river waters. Furthermore, high-temperature fluids emitted from "black smoker" hydrothermal systems typically entrain large volumes of ambient seawater during the formation of buoyant and neutrally buoyant plumes (Section 2.5) with typical dilution ratios of $\sim 10^4$:1 (e.g., Helfrich and Speer, 1995). If 50% of the fluids circulating at high temperature through young ocean crust are entrained into hydrothermal plumes then the total water flux through hydrothermal plumes would be approximately one order of magnitude greater than all other hydrothermal fluxes *and* the global riverine flux to the oceans (Table 1). The associated residence time of the global ocean, with respect to cycling through hydrothermal plume entrainment, would be just 4–8 kyr, i.e., directly comparable to the mixing time of the global deep-ocean conveyor (~ 1.5 kyr; Broecker and Peng, 1982). From that perspective, therefore, we can anticipate that hydrothermal circulation should play an important role in the marine geochemistry of any tracer which exhibits a residence time greater than ~ 1–10 kyr in the open ocean.

The rest of the chapter is organized as follows. In Section 2.2 we discuss the chemical composition of hydrothermal fluids, why they are important, what factors control their compositions, and how these compositions vary, both in space, from one location to another, and in time. Next (Section 2.3) we identify that the fluxes established thus far represent gross fluxes into and out of the ocean crust associated with high-temperature venting. We then examine the other source and sink terms associated with hydrothermal circulation, including alteration of the oceanic crust, formation of hydrothermal mineral deposits, interactions/uptake within hydrothermal plumes and settling into deep-sea sediments. Each of these "fates" for hydrothermal material is then considered in more detail. Section 2.4 provides a detailed discussion of near-vent deposits, including the formation of

polymetallic sulfides and other minerals, as well as near-vent sediments. In Section 2.5 we present a detailed description of the processes associated with hydrothermal plumes, including a brief explanation of basic plume dynamics, a discussion of how plume processes modify the gross flux from high-temperature venting and further discussions of how plume chemistry can be both determined by, and influence, physical oceanographic, and biological interactions. Section 2.6 discusses the fate of hydrothermal products and concentrates on ridge-flank metalliferous sediments, including their potential for paleoceanographic investigations and role in "boundary scavenging" processes. We conclude (Section 2.7) by identifying some of the unresolved questions associated with hydrothermal circulation that are most in need of further investigation.

2.2 VENT-FLUID GEOCHEMISTRY

2.2.1 Why are Vent-fluid Compositions of Interest?

The compositions of vent fluids found on the global MOR system are of interest for several reasons; how and why those compositions vary has important implications. The overarching question, as mentioned in Section 2.1.3, is to determine how the fluids emitted from these systems influence and control ocean chemistry, on both short and long timescales. This question is very difficult to address in a quantitative manner because, in addition to all the heat flux and related water flux uncertainties discussed in Section 2.1, it also requires an understanding of the range of chemical variation in these systems and an understanding of the mechanisms and variables that control vent-fluid chemistries and temperatures. Essentially every hydrothermal vent that is discovered has a different composition (e.g., Von Damm, 1995) and we now know that these compositions often vary profoundly on short (minutes to years)

Table 1 Overview of hydrothermal fluxes: heat and water volume: data from Elderfield and Schultz (1996) and Mottl (2003).

(I) Summary of global heat fluxes		
Heat flux from the Earth's interior	43 TW	
Heat flux associated with ocean crust	32 TW	
Seafloor hydrothermal heat flux	11 TW	
(II) Global hydrothermal fluxes: heat and water		
	Heat flux (TW)	Water flux (10^{16} g yr^{-1})
Axial flow (0 – 1 Ma)		
All flow at 350 °C	2.8 ± 0.3	5.6 ± 0.6
10% @350 °C/90% @5 °C	2.8 ± 0.3	420 ± 140
Hydrothermal plumes (50%)		$28,000 \pm 3,000$
Off-axis flow (1 – 65 Ma)	7 ± 2	$1,000 - 4,000$
Global riverine flux		$3,700 - 4,200$

timescales. Hence, the flux question remains a difficult one to answer. Vent fluid compositions also act as sensitive and unique indicators of processes occurring within young oceanic crust and at present, this same information cannot be obtained from any other source. The "window" that vent fluids provide into subsurface crustal processes is especially important because we can not yet drill young oceanic crust, due to its unconsolidated nature, unless it is sediment covered. The chemical compositions of the fluids exiting at the seafloor provide an integrated record of the reactions and the pressure and temperature (P–T) conditions these fluids have experienced during their transit through the crust. Vent fluids can provide information on the depth of fluid circulation (hence, information on the depth to the heat source), as well as information on the residence time of fluids within the oceanic crust at certain temperatures. Because the dissolved chemicals in hydrothermal fluids provide energy sources for microbial communities living within the oceanic crust, vent-fluid chemistries can also provide information on whether such communities are active at a given location. Vent fluids may also lead to the formation of metal-rich sulfide and sulfate deposits at the seafloor. Although the mineral deposits found are not economic themselves, they have provided important insights into how

metals and sulfide can be transported in the same fluids and, thus, how economically viable mineral deposits are formed. Seafloor deposits also have the potential to provide an integrated history of hydrothermal activity at sites where actively venting fluids have ceased to flow.

2.2.2 Processes Affecting Vent-fluid Compositions

In all known cases the starting fluid for a submarine hydrothermal system is predominantly, if not entirely, seawater, which is then modified by processes occurring within the oceanic crust. Four factors have been identified: the two most important are (i) phase separation and (ii) water–rock interaction; the importance of (iii) biological processes and (iv) magmatic degassing has yet to be established.

Water–rock interaction and phase separation are processes that are inextricably linked. As water passes through the hydrothermal system it will react with the rock and/or sediment substrate that is present (Figure 5). These reactions begin in the downflow zone, and continue throughout. When vent fluids exit at the seafloor, what we observe represents the net result of all the reactions that have occurred along the entire hydrothermal flow path. Because the kinetics of most

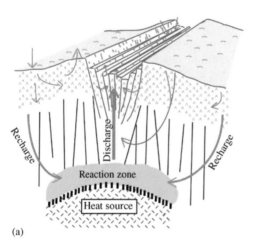
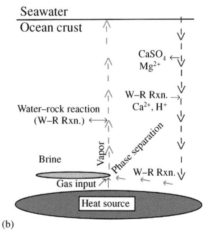

(a)   (b)

Figure 5 (a) Schematic illustration of the three key stages of submarine hydrothermal circulation through young ocean crust (after Alt, 1995). Seawater enters the crust in widespread "recharge" zones and reacts under increasing conditions of temperature and pressure as it penetrates downward. Maximum temperatures and pressures are experienced in the "reaction zone," close to the (magmatic or hot-rock) "heat source" before buoyant plumes rise rapidly back toward the seafloor—the "discharge" zone. (b) Schematic of processes controlling the composition of hydrothermal vent fluid, as it is modified from starting seawater (after Von Damm, 1995). During recharge, fluids are heated progressively. Above ~130 °C anhydrite ($CaSO_4$) precipitates and, as a result of water–rock reaction, additional calcium (Ca^{2+}) is leached from the rock in order to precipitate most of the sulfate (SO_4^{2-}) derived from seawater. Magnesium (Mg^{2+}) is also lost to the rock and protons (H^+) are added. As the fluid continues downward and up the temperature gradient, water–rock interactions continue and phase separation may occur. At at least two sites on the global MOR system, direct degassing of the magma must be occurring, because very high levels of gas (especially CO_2, and helium) are observed in the hydrothermal fluids. The buoyant fluids then rise to the seafloor. In most cases the fluids have undergone phase separation, and in at least some cases storage of the liquid or brine phase has occurred which has been observed to vent in later years from the same sulfide structure (Von Damm *et al.*, 1997). See Figure 6 for additional discussion of phase separation.

reactions are faster at higher temperatures, it is assumed that much of the reaction occurs in the "reaction zone." Phase separation may also occur at more than one location during the fluid's passage through the crust, and may continue as the P–T conditions acting on the fluid change as it rises through the oceanic crust, back toward the seafloor. However, without a direct view into any of the active seafloor hydrothermal systems, for simplicity of discussion, and because we lack better constraints, we often view the system as one of: (i) water–rock reaction on the downflow leg; (ii) phase separation and water–rock reaction in the "reaction zone"; (iii) additional water–rock reaction after the phase separation "event" (Figure 5). Unless confronted with clear inconsistencies in the (chemical) data that invalidate this approach, we usually employ this simple "flow-through" as our working conceptual model. Even though we are unable to rigorously constrain the complexities for any given system, it is always important to remember that the true system is likely far more complex than any model we employ.

In water–rock reactions, chemical species are both gained and lost from the fluids. In terms of differences from the major-element chemistry of seawater, magnesium and SO_4 are lost, and the pH is lowered so substantially that all the alkalinity is titrated. The large quantities of silicon, iron, and manganese that are frequently gained may be sufficient for these to become "major elements" in hydrothermal fluids. For example, silicon and iron can exceed the concentrations of calcium and potassium, two major elements in seawater. Much of the dissolved SO_4 in seawater is lost on the downflow leg of the hydrothermal system as $CaSO_4$ (anhydrite) precipitates at temperatures of $\sim$130 °C—just by heating seawater. Because there is more dissolved SO_4 than calcium in seawater, on a molar basis, additional calcium would have to be leached from the host rock if more than $\sim$33% of all the available seawater sulfate were to be precipitated in this way. In fact, it is now recognized that at least some dissolved SO_4 must persist down into the reaction zone, based on the inferred redox state at depth (see later discussion). Some seawater SO_4 is also reduced to H_2S, substantial quantities of which may be found in hydrothermal fluids at any temperature, based on information from sulfur isotopes (Shanks, 2001). The magnesium is lost by the formation of Mg–OH silicates. This results in the generation of H^+, which accounts for the low pH and titration of the alkalinity. Sodium can also be lost from the fluids due to Na–Ca replacement reactions in plagioclase feldspars, known as albitization. Potassium (and the other alkalis) are also involved in similar types of reaction that can also generate acidity. Large quantities of iron, manganese, and silicon are also leached out of the rocks and into the fluids.

An element that is relatively conservative through water–rock reaction is chlorine in the form of the anion chloride. Chloride is key in hydrothermal fluids, because with the precipitation and/or reduction of SO_4 and the titration of HCO_3^-/CO_3^{2-}, chloride becomes the overwhelming and almost only anion (Br is usually present in the seawater proportion to chloride). Chloride becomes a key component, therefore, because almost all of the cations in hydrothermal fluids are present as chloro-complexes; thus, the levels of chloride in a fluid effectively determine the total concentration of cationic species that can be present. A fundamental aspect of seawater is that the major ions are present in relatively constant ratios—this forms the basis of the definition of salinity (see Volume Editor's Introduction). Because these constant proportions are not maintained in vent fluids and because chloride is the predominant anion, discussions of vent fluids are best discussed in terms of their *chlorinity*, not their *salinity*.

Although small variations in chloride may be caused by rock hydration/dehydration, there are almost no mineralogic sinks for chloride in these systems. Therefore, the main process that effects changes in the chloride concentrations in the vent fluids is phase separation (Figure 6). Phase separation is a ubiquitous process in seafloor hydrothermal systems. Essentially no hydrothermal fluids are found with chlorinities equal to the local ambient seawater value. To phase separate seawater at typical intermediate-to-fast spreading MOR depths of $\sim$2,500 m requires temperatures $\geqq$ 389 °C (Bischoff, 1991). This sets a minimum temperature that fluids must have reached, therefore, during their transit through the oceanic crust. The greater the depth, the higher the temperature required for phase separation to occur. Known vent systems occur at depths of 800–3,600 m, requiring maximum temperatures in the range 297–433 °C to phase separate seawater. Seawater, being a two-component system, $H_2O+NaCl$ (to a first approximation) exhibits different phase separation behavior from pure water. The critical point for seawater is 407 °C and 298 bar (Bischoff and Rosenbauer, 1985) compared to 374 °C and 220 bar for pure water. For the salt solution, the two-phase curve does not stop at the critical point but, instead, continues beyond it. As a solution crosses the two-phase curve, it will separate into two phases, one with chlorinities greater than starting seawater, and the other with chlorinities less than starting seawater. If the fluid reaches the two-phase curve at temperature and pressure conditions *less* than the critical point, subcritical phase separation (also called boiling) will occur, with the generation of a low chlorinity "vapor" phase. This phase contains salt, the amount of which will vary depending on where the two-phase curve was intersected (Bischoff and Rosenbauer, 1987).

What is conceptually more difficult to grasp, is that when a fluid intersects the two-phase curve at *P–T* conditions *greater* than the critical point, the process is called supercritical phase separation (or condensation). In this case a small amount of a relatively high chlorinity liquid phase is condensed out of the fluid. Both sub- and super-critical phase separation occur in seafloor hydro-thermal systems. To complete the phase relations in this system, halite may also precipitate (Figure 6). There is evidence that halite forms, and subse-quently redissolves, in some seafloor hydrothermal systems (Oosting and Von Damm, 1996; Berndt and Seyfried, 1997; Butterfield *et al.*, 1997; Von Damm, 2000). The *P–T* conditions at which the fluid intersects the two-phase curve, will determine the relative compositions of the two phases, as well as their relative amounts. Throughout this discussion, we have assumed the starting fluid under-going phase separation is seawater (or, rather, an

NaCl equivalent, because the initial magnesium and SO_4 will already be lost by this stage). If the NaCl content is different, the phase relations in this system change, forming a family of curves or sur-faces that are a function of the NaCl content, as well as pressure and temperature. The critical point is also a function of the salt content, and hence is really a critical curve in *P–T–x* (*x* referring to composition) space.

As phase separation occurs, substantially changing the chloride content of vent fluids (values from <6% to ~200% of the seawater concentration have been observed), other chemi-cal species will change in concert. It has been shown, both experimentally as well as in the field, that most of the cations (and usually bro-mide) maintain their element-to-Cl ratios during the phase separation process (Berndt and Seyfried, 1990; Von Damm, 2000; Von Damm *et al.*, 2003), i.e., most elements are conservative with respect

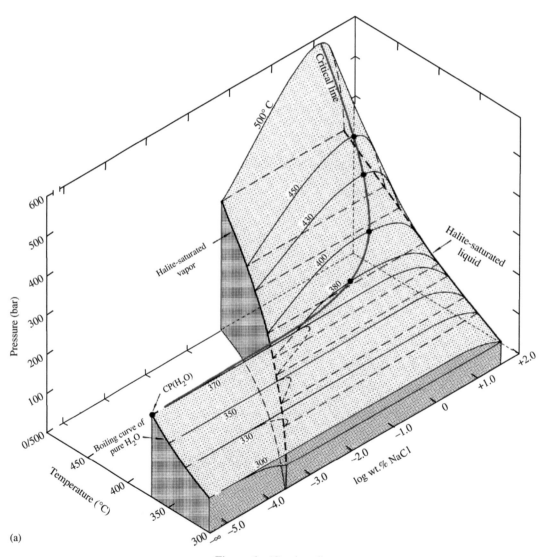

(a)

Figure 6 (Continued).

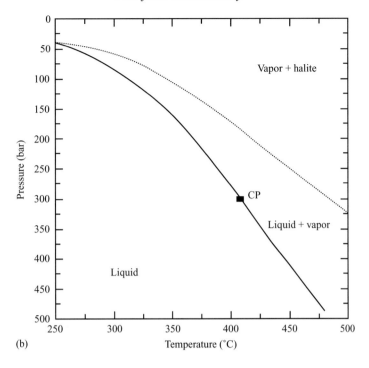

(b)

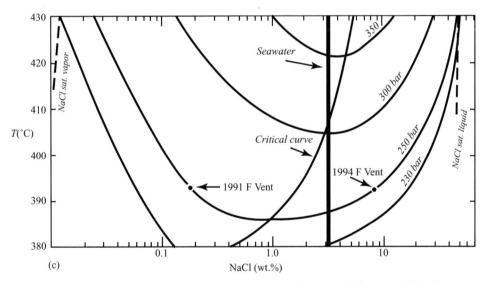

(c)

Figure 6 Phase relations in the NaCl–H$_2$O system. (a) The amount of salt (NaCl) in the NaCl–H$_2$O system varies as both a function of temperature and pressure. Bischoff and Pitzer (1989) constructed this figure of the three dimensional relationships between pressure (P), temperature (T), and composition (x) in the system based on previous literature data and new experiments, in order to better determine the phase relationships for seafloor hydrothermal systems. The P–T–x relationships define a 3D space, but more commonly various projections are shown. (b) The P–T properties for seawater including the two phase curve (solid-line) separating the liquid stability field from the liquid + vapor field, and indicating the location of the critical point (CP) at 407 °C and 298 bar. Halite can also be stable in this system and the region where halite + vapor is stable is shown, separated from the liquid + vapor stability field by the dotted line. This figure is essentially a "slice" from (a) and is a commonly used figure to show the phase relations for seawater (after Von Damm *et al.*, 1995). (c) A "slice" of (a) can also be made to better demonstrate the relationships in the system in T–x space. Here isobars show the composition of the conjugate vapor and brine (liquid) phases formed by the phase separation of seawater. This figure can be used to not only determine salt contents of the conjugate phases, but also their relative amounts. On this figure the compositions of the vapor and liquid phases sampled from "F" vent in 1991 and 1994, respectively, are shown (after Von Damm *et al.*, 1997).

to chloride. Exceptions do occur, however—
primarily for those chemical species not present
as chlorocomplexes. Dissolved gases (e.g., CO_2,
CH_4, He, H_2, H_2S) are preferentially retained in
the low chlorinity or vapor phase, and boron,
which is present as a hydroxyl complex, is relatively unaffected by phase separation (Bray and
Von Damm, 2003a). Bromide, as viewed through
the Br/Cl ratio, is sometimes seen to be fractionated from chloride; this occurs whenever halite is
formed or dissolves because bromide is preferentially excluded from the halite structure (Oosting
and Von Damm, 1996; Von Damm, 2000). Fluids
that have deposited halite, therefore, will have a
high Br/Cl ratio, while fluids that have dissolved
halite will have a low Br/Cl ratio, relative to
seawater. It is because of the ubiquity of phase
separation that vent-fluid compositions are often
now viewed or expressed as ratios with respect to
chloride, rather than as absolute concentrations.
This normalization to chloride *must* be used
when trying to evaluate net gains and losses of
chemical species as seawater traverses the hydrothermal circulation cell, to correct for the fractionation caused by phase separation.

Aside from early eruptive fluids (discussed
later), the chemical composition of most high-
temperature fluids (Figure 7) appears to be controlled by equilibrium or steady state with the
rock assemblage. (Equilibrium requires the assemblage to be at its lowest energy state, but the actual
phases present may be metastable, in which case
they are not at true thermodynamic equilibrium
but, rather, have achieved a steady-state condition.) When vent-fluid data are modeled with
geochemical modeling codes using modern thermodynamic databases, the results suggest equilibrium, or at least steady state, has been achieved.
The models cannot be rigorously applied to many
of the data, however, because the fluids are often
close to the critical point and in that region the
thermodynamic data are not as well constrained.
Based on results from both geochemical modeling
codes and elemental ratios, current data indicate
that not only the major elements, but also many
minor elements (e.g., rubidium, caesium, lithium,
and strontium) are controlled by equilibrium, or
steady-state, conditions between the fluids and
their host-rocks (Bray and Von Damm, 2003b).
The rare earth elements (REE) in vent fluids
present one such example. REE distributions in
hydrothermal fluids are light-REE enriched and
exhibit strong positive europium anomalies, apparently quite unrelated to host-rock MORB compositions (Figure 8). However, Klinkhammer
et al. (1994) have shown that when these same
REE concentrations are plotted versus their ionic
radii, the fluid trends not only become linear
but also show the same fractionation trend exhibited by plagioclase during magma segregation,

indicating that vent-fluid REE concentrations
may be intrinsically linked to the high-temperature alteration of this particular mineral.

Two other processes are known to influence the
chemistry of seafloor vent fluids: biological processes and magmatic degassing. Evidence for
"magmatic degassing" has been identified at two
sites along the global MOR system—at 9°50′ N
and at 32° S on the EPR (M. D. Lilley, personal
communication; Lupton *et al.*, 1999a). These sites
have very high levels of CO_2, and very high He/
heat ratios. The interpretation is that we are seeing
areas with recent magma resupply within the crust
and degassing of the lavas, resulting in very high
gas levels in the hydrothermal fluids found at
these sites. We do not know the spatial–temporal
variation of this process, hence, we cannot yet
evaluate its overall importance. Presumably,
every site on the global MOR system undergoes
similar processes episodically. What is not known,
however, is the frequency of recurrence at any one
site. Consequently, the importance of fluxes due
to this degassing process, versus more "steady-
state" venting, cannot currently be assessed. At
9°50′ N, high gas contents have now been
observed for almost a decade; no signature of
volatile-metal enrichment has been observed in
conjunction with these high gas contents (Von
Damm, 2003).

The fourth process influencing vent-fluid compositions is biological change, which can take the
form of either consumption or production of various chemical species. As the current known limit
to life on Earth is ∼120 °C (e.g., Holland and
Baross, 2003) this process can only affect fluids
at temperatures lower than this threshold. This
implies that high-temperature vents should not be
subject to these effects whereas they may occur in
both lower-temperature axial diffuse flow and
beneath ridge flanks. From observations at the
times of seafloor eruptions and/or diking events,
it is known that there are microbial communities
living within the oceanic crust (e.g., Haymon *et al.*,
1993). Their signatures can be seen clearly in at
least some low-temperature fluids, as noted in
particular by changes in the H_2, CH_4, and H_2S
contents of those fluids (Von Damm and Lilley,
2003). Hence, biological influences have been
observed; how widespread this is, which elements
are affected, and what the overall impact on chemical fluxes may be all remain to be resolved.

2.2.3 Compositions of Hydrothermal Vent Fluids

2.2.3.1 Major-element chemistry

The known compositional ranges of vent fluids
are summarized in Figure 9 and Table 2. Because
no two vents yet discovered have exactly the same

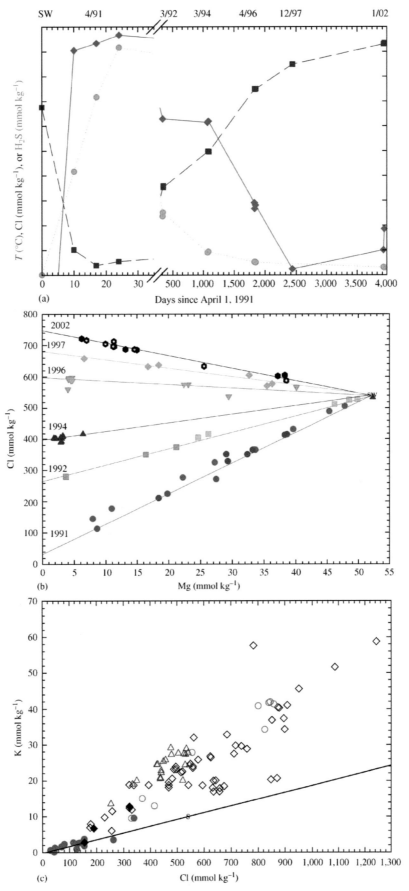

Figure 7 (Continued).

composition, these ranges often change with each new site. As discussed in Section 2.2.2, vent fluids are modified seawater characterized by the loss of magnesium, SO_4, and alkalinity and the gain of many metals, especially on a chloride normalized basis.

Vent fluids are acidic, but not as acid as may first appear from pH values measured at 25 °C and 1 atm (i.e., in shipboard laboratories). The cation H^+ in vent fluids is also present as a chloro-complex with the extent of complexation increasing as P and T increase. At the higher *in situ* conditions of P and T experienced at the seafloor, therefore, much of the H^+ is incorporated into the HCl–aqueous complex; hence, the activity of H^+ is reduced and the *in situ* pH is substantially higher than what is measured at laboratory temperatures. The K_w for water also changes as a function of P and T such that neutral pH is not necessarily 7 at other P–T conditions. For most vent fluids, the *in situ* pH is 1–2 pH units more acid than neutral, not the ~4 units of acidity that the measured (25 °C, 1 atm) data appear to imply. Most high-temperature vent fluids have (25 °C measured) pH values of 3.3 ± 0.5 but a few are more acidic whilst several are less acid. If fluids are more acid than pH 3.3 ± 0.5, it is often an indicator that metal sulfides have precipitated below the seafloor because such reactions produce protons. Two mechanisms are known that can cause fluids to be less acidic than the norm: (i) cases

where the rock substrate appears to be more highly altered—the rock cannot buffer the solutions to as low a pH; (ii) when organic matter is present, ammonium is often present and the NH_3/NH_4^+ couple serves to buffer the pH to a higher level.

Vent fluids are very reducing, as evidenced by the presence of H_2S rather than SO_4, as well as H_2, CH_4 and copious amounts of Fe^{2+} and Mn^{2+}. In

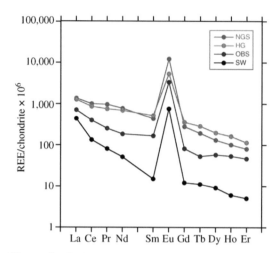

Figure 8   End-member REE concentrations in vent fluids from four different "black smokers" at the 21° N vent site, EPR, normalized to chondrite (REE data from Klinkhammer *et al.*, 1994). NGS=*National Geographic Smoker*; HG=*Hanging Gardens*; OBS=*Ocean Bottom Seismometer*; SW=*South West* vent.

Figure 7   Compositional data for vent fluids. (a) Time series data from "A" vent for chloride and H_2S concentrations and measured temperature (T). When time-series data are available, this type of figure, demonstrating the change in fluid composition in a single vent over time, is becoming more common. The data plotted are referred to as the "end-member" data (data from Von Damm *et al.*, 1995 and unpublished). Points on the y-axis are values for ambient seawater. Note the low chlorinity (vapor phase) fluids venting initially from A vent; over time the chloride content has increased and the fluids sampled more recently, in 2002, are the liquid (brine) phase. As is expected, the concentration of H_2S, a gas that will be partitioned preferentially into the vapor phase, is anticorrelated with the chloride concentration. The vertical axis has 10 divisions with the following ranges: T (°C) 200–405 (ambient seawater is 2 °C); Cl (mmol kg^{-1}) 0–800 (ambient is 540 mmol kg^{-1}); H_2S (mmol kg^{-1}) 0–120 (ambient is 0 mmol kg^{-1}). (b) Whenever vent fluids are sampled, varying amounts of ambient seawater are entrained into the sampling devices. Vent fluids contain 0 mmol kg^{-1}, while ambient seawater contains 52.2 mmol kg^{-1}. Therefore, if actual sample data are plotted as properties versus magnesium, least squares linear regression fits can be made to the data. The calculated "end-member" concentration for a given species, which represents the "pure" hydrothermal fluid is then taken as the point where that line intercepts the y-axis (i.e., the calculated value at Mg=0 mmol kg^{-1}). These plots versus magnesium are therefore mostly sampling artifacts and are referred to as "mixing" diagrams. While these types of plots were originally used to illustrate vent-fluid data, they have been largely superceded by figures such as (a) or (c). This figure (b) shows the data used to construct the time series represented in (a). Note the different lines for the different years. In some years samples were collected on more than one date. All samples for a given year are shown by the same shape, the different colors within a year grouping indicate different sample dates. In some years the chemical composition varied from day-to-day, but for simplicity a single line is shown for each year in which samples were collected (Von Damm *et al.*, 1995 and unpublished data). (c) As the chloride content of a vent fluid is a major control on the overall composition of the vent composition, most of the cations vary as a function of the chloride-content. Variations in the chloride content are a result of phase separation. This shows the relationship between the potassium (K) and chloride content in vent fluids in the global database, as of 2000. The line is the ratio of K/Cl in ambient seawater. Closed circles are from 9–10° N EPR following the 1991 eruption, open circles are other 9–10° N data not affected by the eruptive events; triangles are from sites where vents occur on enriched oceanic crust, diamonds are from bare-basalt (MORB) hosted sites, filled diamonds are other sites impacted by volcanic eruptions. Data sources and additional discussion in Von Damm (2000).

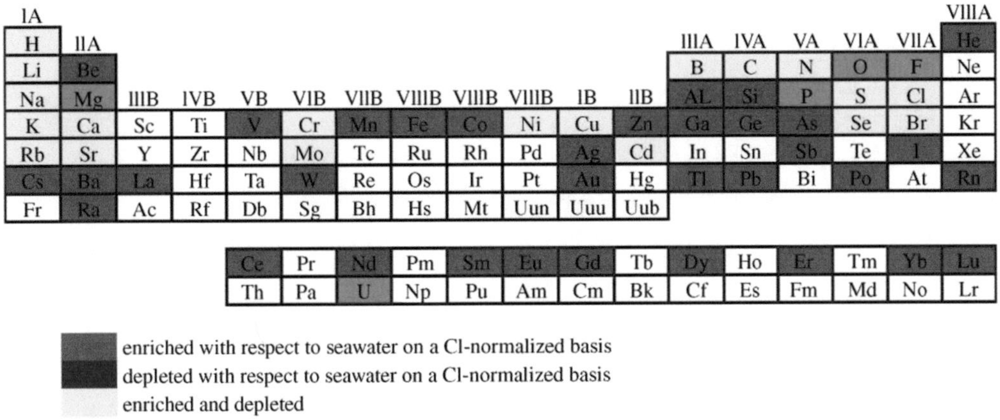

Figure 9 Periodic table of the elements showing the elements that are enriched in hydrothermal vent fluids relative to seawater (red), depleted (blue) and those which have been shown to exhibit both depletions and enrichments in different hydrothermal fluids (yellow) relative to seawater. All data are normalized to the chloride content of seawater in order to evaluate true gains and losses relative to the starting seawater concentrations.

rare cases, there can be more H_2S and/or H_2 than chloride on a molar basis and it is the prevailing high acidity that dictates that H_2S rather than $HS-$ or S^{2-} is the predominant form in high-temperature vent fluids. Free H_2 is derived as a result of water–rock reaction and there is substantially more H_2 than O_2 in these fluids. Therefore, although redox calculations are typically given in terms of the $\log f_{O_2}$, the redox state is best calculated based on the H_2/H_2O couple for seafloor vent fluids. This can then be expressed in terms of the $\log f_{O_2}$. The K for the reaction:

$$H_2 + \frac{1}{2}O_2 = H_2O$$

also changes as a function of temperature and pressure. Another way to determine how reducing vent fluids are is by comparing them to various mineralogic buffers such as pyrite–pyrrhotite–magnetite (PPM) or hematite–magnetite–pyrite (HMP). Most vent fluids lie between these two extremes, but there is some systematic variation (Seyfried and Ding, 1995). The observation that vent fluids are more oxic than would be expected based on the PPM buffer, provides one line of evidence that the reaction zone is not as reducing as initially predicted, consistent with at least some dissolved seawater SO_4 penetrating into the deeper parts of the system rather than being quantitatively removed by anhydrite precipitation within shallower levels of the downflow limb.

Lower-temperature ($<100\,^\circ$C) vent fluids found right at the axis are in most known cases a dilution of some amount of high-temperature fluids with seawater, or a low-temperature fluid with a composition close to seawater. There is some evidence for an "intermediate" fluid, perhaps most analogous to a crustal "ground water," with temperatures of $\sim150\,^\circ$C within the oceanic crust.

Evidence for the latter is found in some Ocean Drilling Program data (Magenheim *et al.*, 1992), some high-temperature vent fluids from 9°50′ N on the EPR (Ravizza *et al.*, 2001), and some very unusual $\sim$90 °C fluids from the southern EPR (O'Grady, 2001). To conclude, the major-element composition of high-temperature vent fluids can be described as acidic, reducing, metal-rich NaCl solutions whilst lower-temperature fluids are typically a dilution of this same material with seawater. The few exceptions to this will be discussed below.

2.2.3.2 Trace-metal chemistry

Compared to the number of vent fluids sampled and analyzed for their major-element data, relatively little trace-metal data exist. This is because when hot, acidic vent fluids mix with seawater, or even just cool within submersible- or ROV-deployed sampling bottles, they become supersaturated with respect to many solid phases and, thus, precipitate. Once this occurs, everything in the sampling apparatus must be treated as one sample: a budget can only be constructed by integrating these different fractions back together. In the difficult sampling environment found at high-temperature vent sites, pieces of chimney structure are also sometimes entrained into the sampling apparatus. It is necessary, therefore, to be able to discriminate between particles that have precipitated from solution in the sampling bottle and "contaminating" particles that are extraneous to the sample. In addition, water samples are often subdivided into different fractions, aboard ship, making accurate budget reconstructions difficult if not impossible to complete. It is because of these difficulties that there are few robust analyses of many trace metals, especially those that

Table 2 Ranges in chemical composition for all known vent fluids.

Chemical species	Units	Seawater	Overall range	Slow[a] (0–2.5 cm yr^{-1})	Intermediate[a] >2.5–6	Fast[a] >6–12	Ultrafast[a] >12	Sediment covered[b]	Ultramafic hosted[c]	Arc, back-arc[d]
T	°C	2	>2–405	40–369	13–382	8–403	16–405	100–315	40–364	278–334
pH	25 °C, 1 atm	7.8	2.0–9.8	2.5–4.85	2.8–4.5	2.45–>6.63	2.96–5.53	5.1–5.9	2.7–9.8	2.0–5.0
Alkalinity	meq kg^{-1}	2.4	–3.75–10.6	–3.4–0.31	–3.75–0.66	–2.69–<2.27	–1.36–0.915	1.45–10.6		–0.20–3.51
Cl	mmol kg^{-1}	540	30.5–1,245	357–675	176–1,245	30.5–902	113–1,090	412–668	515–756	255–790
SO$_4$	mmol kg^{-1}	28	<0–<28	–3.5–1.9	–25–0.763	>–8.76	–0.502–9.53	0	<12.9	0
H$_2$S	mmol kg^{-1}	0	0–110	0.5–5.9	0–19.5	0–110	0–35	1.10–5.98	0.064–1.0	2.0–13.1
Si	mmol kg^{-1}	0.03–0.18	<24	7.7–22	11–24	2.73–22.0	8.69–21.3	5.60–13.8	6.4–8.2	10.8–14.5
Li	umol kg^{-1}	26	4.04–5,800	238–1,035	160–2,350	4.04–1,620	248–1,200	370–1,290	245–345	200–5,800
Na	mmol kg^{-1}	464	10.6–983	312–584	148–924	10.6–983	109–886	315–560	479–553	210–590
K	mmol kg^{-1}	10.1	–1.17–79.0	17–28.8	6.98–58.7	–1.17–51	2.2–44.8	13.5–49.2	20.2–22	10.5–79.0
Rb	umol kg^{-1}	1.3	0.156–360	9.4–40.4	22.9–59	0.156–31.1	0.39–6.8	22.5–105	28–37.1	8.8–360
Cs	nmol kg^{-1}	2	2.3–7,700	100–285	168–364	2.3–264		1,000–7,700	331–385	
Be	nmol kg^{-1}	0	10–91		10–37		0	12–91		
Mg	mmol kg^{-1}	52.2	0	0	0	0	0	0	<19	0
Ca	mmol kg^{-1}	10.2	–1.31–109	9.9–43	9.75–109	–1.31–106	4.02–65.5	26.6–81.0	21.0–67	6.5–89.0
Sr	umol kg^{-1}	87	–29–387	42.9–133	0.0–348	–29–387	10.7–190	160–257	138–203	20–300
Ba	umol kg^{-1}	0.14	1.64–100	<52.2	>8–>46		1.64–18.6	>12	>45–79	5.9–100
Mn	umol kg^{-1}	<0.001	10–7,100	59–1,000	140–4,480	62.7–3,300	20.6–2,750	10–236	330–2,350	12–7,100
Fe	mmol kg^{-1}	<0.001	0.007–25.0	0.0241–5.590	0.009–18.7	0.007–12.1	0.038–14.7	0–0.18	2.5–25.0	13–2,500
Cu	umol kg^{-1}	0.007	0–162	0–150	0.1–142	0.18–97.3	2.6–150	<0.02–1.1	27–162	0.003–34
Zn	umol kg^{-1}	0.012	0–3,000	0–400	2.2–600	13–411	1.9–740	0.1–40.0	29–185	7.6–3,000
Co	umol kg^{-1}	0.00003	<0.005–14.1	0.130–0.422	0.022–0.227			<0.005	11.8–14.1	
Ni	umol kg^{-1}	0.012							2.2–3.6	
Ag	nmol kg^{-1}	0.02	<1–230	75–146	<1–38			<1–230	11–47	
Cd	nmol kg^{-1}	1.0	0–180		0–180			<10–46	63–178	
Pb	nmol kg^{-1}	0.01	<20–3,900	221–376	183–360			<20–652	86–169	36–3,900
B	umol kg^{-1}	415	356–3,410	356–480	465–1,874	430–617	400–499	<2,160		470–3,410
Al	umol kg^{-1}	0.02	0.1–18.7	1.03–13.9	4.0–5.2	0.1–18	9.3–18.7	0.9–7.9	1.9–4	4.9–17.0
Br	umol kg^{-1}	840	29.0–1,910	666–1,066	250–1789	29.0–1370	216–1910	770–1,180		306–1,045
F	umol kg^{-1}	68	<38.8	16.1–38.8	"0"					
CO$_2$	mmol kg^{-1}	0.0003		3.56–39.9	<5.7	<200	8.4–22		130–2,200	14.4–200
CH$_4$	umol kg^{-1}	0		150–2,150	52		7–133			
NH$_4$	mmol kg^{-1}	0	<15.6	<0.06	<0.65			5.6–15.6		
H$_2$	umol kg^{-1}	0.0003	<38,000	1.1–727	<0.45	<38,000	40–1300		250–13,000	

[a] These omit sedimented covered and um hosted.
[b] Includes: Guaymas, Escanaba, Middle Valley.
[c] Includes Rainbow, Lost City, kvd unpublished data for Logatchev.
[d] Compilation from Ishibashi and Urabe (1995).

precipitate as, or co-precipitate with, metal sulfide phases. Some general statements can, however, be made. In high-temperature vent fluids, most metals are enriched relative to seawater, sometimes by 7–8 orders of magnitude (as is sometimes true for iron). At least some data exist demonstrating the enrichment of all of vanadium, cobalt, nickel, copper, zinc, arsenic, selenium, aluminum, silver, cadmium, antimony, caesium, barium, tungsten, gold, thallium, lead, and REE relative to seawater. Data also exist showing that molybdenum and uranium are often lower than their seawater concentrations. These trace-metal data have been shown to vary with substrate and the relative enrichments of many of these trace metals varies significantly between MOR hydrothermal systems, those located in back arcs, and those with a significant sedimentary component. Even fewer trace-metal data exist for low-temperature "diffuse" fluids. The original work on the $<20\,^{\circ}\mathrm{C}$ GSC fluids (Edmond *et al.*, 1979) showed these fluids to be a mix of high-temperature fluids with seawater, with many of the transition metals present at *less than* their seawater concentrations due to precipitation and removal below the seafloor. Essentially the same results were obtained by James and Elderfield (1996) using the MEDUSA system to sample diffuse-flow fluids at TAG (26° N, MAR).

2.2.3.3 Gas chemistry of hydrothermal fluids

In general, concentrations of dissolved gases are highest in the lowest-chlorinity fluids, which represent the vapor phase. However, there are exceptions to this rule, and gas concentrations vary significantly between vents, even at a single location. In the lowest chlorinity and hottest fluids, H_2S may well be the dominant gas. However, because H_2S levels are controlled by metal-sulfide mineral solubility, this H_2S is often lost via precipitation. While the first vent sites discovered contained less than twice the CO_2 present in seawater (Welhan and Craig, 1983), more vents have higher levels of CO_2 than is commonly realized. Few MOR vent fluids have CO_2 levels less than or equal to the total CO_2 levels present in seawater (~ 2.5 mmol kg^{-1}). Instead, many fluids have concentrations approaching an order of magnitude more CO_2 than seawater; the highest approach two orders of magnitude more CO_2 than seawater, but these highest levels are uncommon (M. D. Lilley, personal communication). Back-arc systems more commonly have higher levels of CO_2 in their vent fluids, but concentrations two orders of magnitude greater than seawater are, again, close to the upper maximum of what has been sampled so far (Ishibashi and Urabe, 1995). CH_4 is much less abundant than CO_2 in most systems. Vent-fluid CH_4 concentrations are typically higher in sedimented systems and in systems hosted in ultramafic rocks, when compared to bare basaltic vent sites. CH_4 is also enriched in low-temperature vent fluids when compared to concentrations predicted from simple seawater/vent-fluid mixing (Von Damm and Lilley, 2003). Longer-chain organic molecules have also been reported from some sites, usually at even lower abundances than CO_2 and/or CH_4 (Evans *et al.*, 1988). The concentrations of H_2 gas in vent fluids vary substantially, over two orders of magnitude (M. D. Lilley, personal communication). Again the highest levels are usually observed in vapor phase fluids, especially those sampled immediately after volcanic eruptions or diking events. Relatively high values (several mmol kg^{-1}) have also been reported from sites hosted by ultramafic rocks. Of the noble gases, helium, especially ^{3}He, is most enriched in vent fluids. ^{3}He can be used as a conservative tracer in vent fluids, because its entire source in vent fluids is primordial, from within the Earth (see Section 2.5). Radon, a product of radioactive decay in the uranium series, is also greatly enriched in vent fluids (e.g., Kadko and Moore, 1988; Rudnicki and Elderfield, 1992). Less data are available for the other noble gases, at least some of which appear to be relatively conservative compared to their concentration in starting seawater (Kennedy, 1988).

2.2.3.4 Nutrient chemistry

The concentrations of nutrients available in seawater control biological productivity. Consequently, the dissolved nutrient concentrations in natural waters are always of great interest. Compared to deep-ocean seawater, the PO_4 contents of vent fluids are significantly lower, but are not zero. Much work remains to be done on the distribution of nitrogen species, and the nitrogen cycle in general, in vent fluids. Generally, in basalt-hosted systems, the nitrate + nitrite content is also lower than local deepwaters but, again, is not zero. Ammonium in these systems typically measures less than 10 μmol kg^{-1} but, in some systems in which no sediment cover is present, values of 10s to even 100s of μmol kg^{-1} are sometimes observed (cf. Von Damm, 1995). In Guaymas Basin, in contrast, ammonium concentrations as high as 15 mmol kg^{-1} have been measured (Von Damm *et al.*, 1985b). N_2 is also present. Silica concentrations are extremely high, due to interaction with host rocks at high temperature, at depth. Of course, in these systems, it could be debated what actually constitutes a "nutrient." For example, both dissolved H_2 and H_2S (as well as numerous other reduced species), represent important primary energy sources for the chemosynthetic communities invariably found at hydrothermal vent sites.

2.2.3.5 *Organic geochemistry of hydrothermal vent fluids*

Studies of the organic chemistry of vent fluids are truly in their infancy and little field data exist (Holm and Charlou, 2001). There are predictions of what should be present based on experimental work (e.g., Berndt *et al.*, 1996; Cruse and Seewald, 2001) and thermodynamic modeling (e.g., McCollom and Shock, 1997, 1998). These results await confirmation "in the field." Significant data on the organic geochemistry are uniquely available for the Guaymas Basin hydrothermal system (e.g., Simoneit, 1991), which underlies a very highly productive area of the ocean, and is hosted in an organic-rich sediment-filled basin.

2.2.4 Geographic Variations in Vent-fluid Compositions

2.2.4.1 *The role of the substrate*

There are systematic reasons for some of the variations observed in vent-fluid compositions. One of the most profound is the involvement of sedimentary material in the hydrothermal circulation cell, as seen at sediment-covered ridges (Von Damm *et al.*, 1985b; Campbell *et al.*, 1994; Butterfield *et al.*, 1994). The exact differences this imposes depend upon the nature of the sedimentary material involved: the source/nature of the aluminosilicate material, the proportions and type of organic matter it contains and the proportion and type of animal tests present, calcareous and/or siliceous. In the known sediment-hosted systems (Guaymas Basin, EPR; Escanaba Trough, Gorda Ridge; Middle Valley, JdFR; and perhaps the Red Sea) basalts are intercalated with the sediments or else underlie them. Hence, in these systems, reactions with basalt are overprinted by those with the sediments. In most cases, depending on the exact nature as well as thickness of the sedimentary cover, this causes a rise in the pH, which results in the precipitation of metal sulfides before the fluids reach the seafloor. The presence of carbonate and/or organic matter also buffers the pH to significantly higher levels (at least pH 5 at 25 °C and 1 atm).

The chemical composition of most seafloor vent fluids can be explained by reaction of unaltered basalt with seawater. However, in some cases the best explanation for the fluid chemistry is that the fluids have reacted with basalt that has already been highly altered (Von Damm *et al.*, 1998). Two indicators for this are higher pH values (pH~4 versus pH~3.3 at 25 °C), as well as lower K/Na molar ratios and lower concentrations of the REEs. In the last several years, several vent sites have been sampled that cannot be explained by these mechanisms. At some locations, vent fluids must be generated by reaction

of seawater with ultramafic rocks (Douville *et al.*, 2002). These fluids can also have major variations from each other, depending on the temperature regime. In high-temperature fluids that have reacted with an ultramafic substrate, silicon contents are generally lower than in basalt-hosted fluids; H_2, calcium, and iron contents are generally higher, but these fluids remain acidic (Douville *et al.*, 2002). Not much is yet known about the seafloor fluids that are generated from lower-temperature ultramafic hydrothermal circulation. In the one example studied thus far (Lost City) the measured pH is greater than that in seawater, and fluid compositions are clearly controlled by a quite distinct set of serpentinization-related reactions (Kelley *et al.*, 2001). An illustration of this fundamental difference is given by magnesium which is quantitatively stripped from "black smoker" hydrothermal fluids but exhibits ~20–40% of seawater concentrations (9–19 mM) in the Lost City vents, leading to the unusual magnesium-rich mineralization observed at this site (see later). The seafloor fluids from Lost City are remarkably similar to those found in continental hydrothermal systems hosted in ultramafic environments (Barnes *et al.*, 1972). In back-arc spreading centers such as those found in the western Pacific, andesitic rock types are common and profound differences in vent-fluid compositions arise (Fouquet *et al.*, 1991; Ishibashi and Urabe, 1995, Gamo *et al.*, 1997). These fluids can be both more acidic and more oxidizing than is typical and the relative enrichments of transition metals and volatile species in these fluids are quite distinct from what is observed in basalt-hosted systems.

Major differences in substrate are, therefore, reflected in the compositions of vent fluids. Insufficient trace-metal data for vent fluids exist, however, to discern more subtle substrate differences, e.g., between EMORB and NMORB on non-hotspot influenced ridges. Where the ridge axis is influenced by hot-spot volcanism, some differences may be seen in the fluid compositions, as, for example, the high barium in the Lucky Strike vent fluids, but in this case most of the fluid characteristics (e.g., potassium concentrations) do not show evidence for an enriched substrate (Langmuir *et al.*, 1997; Von Damm *et al.*, 1998).

2.2.4.2 *The role of temperature and pressure*

Temperature, of course, plays a major role in determining vent-fluid compositions. Pressure is often thought of as less important than temperature, but the relative importance of the two depends on the exact *P–T* conditions of the fluids. Because of the controls that pressure and temperature conditions exert on the thermodynamics as well as the physical properties of the fluids, the two effects cannot be discussed completely independently from each other. Not only do *P–T* conditions govern phase separation, as discussed above, they control transport in the fluids and mineral dissolution and precipitation reactions. Temperature, especially, plays a role in the quantities of elements that are leached from the host rocks. When temperature decreases, as it often does due to conductive cooling as fluids rise through the oceanic crust, most minerals become less soluble. Due to these decreasing mineral solubilities, transition metals and sulfide, in particular, may be lost from the ascending fluids. *P–T* conditions in the fluids also control the strength of the aqueous complexes. In general, as *P* and *T* rise, aqueous species become more associated. Because of the properties of water at the critical point (the dielectric constant goes to zero), all the species must become associated, as there can be no charged species in solution at the critical point. Therefore, transport of species can increase markedly as the critical point is approached because there will be smaller amounts of the (charged) species present in solution which are needed if mineral solubility products are to be exceeded (Von Damm *et al.*, 2003). It is in this critical point region that small changes in pressure can be particularly significant—for example, as a fluid is rising in the upflow ("discharge") limb of a hydrothermal cell (Figure 5). Because most vent-fluid compositions are controlled by equilibrium or steady state, and because the equilibrium constants for these reactions change as a function of pressure and temperature, *P–T* conditions will ultimately control all vent-fluid compositions. One problem associated with modeling vent fluids and trying to understand the controls on their compositions is that we really do not know the temperature in the "reaction zone." Basaltic lavas are emplaced at temperatures of 1,100–1,200 °C, but rocks must be brittle to retain fractures that allow fluid flow, and this brittle–ductile transition lies in the range 500–600 °C. A commonplace statement is that the reaction zone temperature is ~450 °C, but we do not really know this value with any accuracy, nor how variable it may be from one location to another. At the seafloor, we have sampled fluids with exit temperatures as high as ~405 °C. Hence, in addition to the constraints provided by the recognition that at least subcritical phase separation is pervasive (see above) we can further determine that (i) reaction zone temperatures must exceed 405 °C, at least in some cases, and (ii) that in cases where evidence for supercritical phase separation has been determined (e.g., Butterfield *et al.*, 1994; Von Damm *et al.*, 1998) temperatures must exceed 407 °C within the oceanic crust.

The pressure conditions at any hydrothermal field are largely controlled by the depth of the overlying water column. Pressure is most critical in terms of phase separation and vent fluids are particularly sensitive to small changes in pressure when close to the critical point. It is in this region, close to the critical point, when fluids are very expanded (i.e., at very low density) that small changes in pressure can cause significant changes in vent-fluid composition.

2.2.4.3 *The role of spreading rate*

When one considers tables of vent-fluid chemical data, one cannot separate vents from ultrafast- versus slow-spreading ridges (Table 2); the range of chemical compositions from each of these two end-member types of spreading regime overlap. There has been much debate in the marine geological literature whether rates of magma supply, rather than spreading rate, should more correctly be applied when defining ridge types (e.g., "magma-starved" versus "magma-rich" sections of ridge crest). While any one individual segment of ridge crest undoubtedly passes through different stages of a volcanic–tectonic cycle, regardless of spreading rate (e.g., Parson *et al.*, 1993; Haymon, 1996), it is generally the case that slow-spreading ridges are relatively magma-starved whilst fast spreading ridges are more typically magma-rich. Consequently, we continue to rely upon (readily quantified) spreading rate (De Mets *et al.*, 1994) as a proxy for magma supply. To a first approximation, ridge systems in the Atlantic Ocean are slow-spreading, while fast-spreading ridges are only found in the Pacific. The Pacific contains ridges that spread at rates from ~15 cm yr^{-1} full opening rate to a minimum of ~2.4 cm yr^{-1} on parts of the Gorda Ridge (comparable to the northern MAR). In the Indian Ocean, ridge spreading varies between intermediate (~6 cm yr^{-1}, CIR and SEIR) and very slow rates (<2.0 cm yr^{-1}, SWIR). In the Arctic Ocean the spreading rate is the slowest known, decreasing to <1.0 cm yr^{-1} from west to east as the Siberian shelf is approached. Discussion of vent-fluid compositions from different oceans, therefore, often approximates closely to variations in vent-fluid chemistries at different spreading rates. Although tables of the ranges of vent-fluid chemistries do not show distinct differences between ocean basins, some important differences do become apparent when those data are modeled. (NB: Although there is now firm evidence for hydrothermal activity in the Arctic, those systems have not yet been sampled for vent fluids; similarly, in the Indian Ocean, only two sites have recently been discovered). Consequently, meaningful comparisons can only readily be made, at present, between Atlantic and Pacific vent-fluid compositions. In systems on the slow spreading MAR, the calculated f_{O_2} of hydrothermal fluids is higher, suggesting that these systems are more oxidizing. Also, for example, both TAG and Lucky Strike vent fluids contain relatively little potassium compared to sodium (Edmond *et al.*, 1995; Von Damm *et al.*, 1998). Boron is also low in some of the Atlantic sites, especially at TAG and Logatchev (You *et al.*, 1994; Bray and Von Damm, 2003a). The explanation for these observations is that on the slow-spreading MAR, hydrothermal activity is active for a much longer period of time at any given site (also reflected in the relative sizes of the hydrothermal deposits formed: Humphris *et al.*, 1995; Fouquet *et al.*, submitted). Consequently, MAR vent fluids become more oxic because more dissolved seawater SO_4 has penetrated as deep as the reaction zone; the rocks within the hydrothermal flow cell have been more completely leached and altered. Because of the more pronounced tectonic (rather than volcanic) activity that is associated with slow spreading ridges, rock types that are normally found at greater depths within the oceanic crust can be exposed at or near the surface. Thus, hydrothermal sites have been located along the MARs that are hosted in ultramafic rocks: the Rainbow, Logatchev, and Lost City sites. No ultramafic-hosted systems are expected to occur, by contrast, along the fast-spreading ridges of the EPR.

2.2.4.4 *The role of the plumbing system*

Another fundamental difference observed in the nature of hydrothermal systems at fast- and slow-spreading ridges concerns intra-field differences in vent-fluid compositions. (Note that the terms "vent field" and "vent area" are often used interchangeably, have no specific size classifications, and may be used differently by different authors. In our usage, "vent area" is smaller, referring to a cluster of vents within 100 m or so, while a "vent field" may stretch for a kilometer or more along the ridge axis—but this is by no means a standard definition.) On a slow spreading ridge, such as the MAR, all of the fluids venting, for example, at the TAG site, can be shown to have a common source fluid that may have undergone some change in composition due to near surface processes such as mixing and/or conductive cooling. Many of the fluids can also be related to each other at the Lucky Strike site. In contrast, on fast spreading ridges, vents that may be within a few tens of meters of each other, clearly have different source fluids. A plausible explanation for these differences would be that vents on slow spreading ridges are fed from greater depths than those on fast-spreading ridges, with emitted fluids channeled upward from the subsurface along fault planes or other major tectonic fractures. Hence, in at least some cases, hydrothermal activity found on slow spreading ridges may be located wherever fluids have been preferentially channeled. Active vent sites on slow spreading ridges also appear to achieve greater longevity, based on the size of the sulfide structures and mounds they have produced. Fluids on fast spreading ridges, by contrast, are fed by much shallower heat sources and the conduits for these fluids appear to be much more localized, resulting in the very pronounced chemical differences often observed between immediately adjacent vent structures. Clearly, the plumbing systems at fast and slow spreading ridges must be characterized by significant, fundamental differences.

2.2.5 Temporal Variability in Vent-fluid Compositions

The MOR is, in effect, one extremely long, continuous submarine volcano. While volcanoes are commonly held to be very dynamic features, however, little temporal variability was observed for more than the first decade of work on hydrothermal systems. Indeed, a tendency arose not to view the MOR as an active volcano, at least on the timescales that were being worked on. This perspective was changed dramatically in the early 1990s. Together with evidence for recent volcanic eruptions at several sites, profound temporal variability in vent-fluid chemistries, temperatures, and styles of venting were also observed (Figure 7). In one case, the changes observed at a single vent almost span the full range of known compositions reported from throughout the globe. These temporal variations in hydrothermal venting reflect changes in the nature of the underlying heat source. The intrusion of a basaltic dike into the upper ocean crust, which may or may not be accompanied by volcanic extrusion at the seafloor, has been colloquially termed "the quantum unit of ocean accretion." These dikes are of the order 1 m wide, 10 km long, and can extend hundreds of meters upward through the upper crust toward the ocean floor. These shallow-emplaced and relatively small, transient, heat sources provide most, if not all of the heat that supports venting immediately following magmatic emplacement. Over timescales of as little as a year, however, an individual dike will have largely cooled and the heat source driving any continuing vent activity deepens. An immediate result is a decrease in measured exit temperatures for the vent fluids, because more heat is now lost, conductively, as the fluids rise from deeper within the oceanic crust. Vent-fluid compositions change, too, because the conditions of phase separation change; so, too, do the subsurface path length and residence time, such that the likelihood that circulating fluids reach equilibrium or steady state with the ocean crust also vary. Detailed time-series studies at sites perturbed by magmatic emplacements have shown that it is the vapor phase which vents first, in the earliest stages after a magmatic/volcanic event, while the high-chlorinity liquid phase is expelled somewhat later. In the best documented case study available, "brine" fluids were actively venting at a location some three years after the vapor phase had been expelled from the same individual hydrothermal chimney; at other event-affected vent sites, similar evolutions in vent-fluid composition have been observed over somewhat longer timescales.

The temporal variability that has now been observed has revolutionized our ideas about the functioning of hydrothermal systems and the time-scales over which processes occur on the deep ocean floor. It is no exaggeration to state that processes we thought would take 100–1,000 yr, have been seen to occur in <10 yr. The majority of magmatic intrusions/eruptions that have been detected have been along the JdFR (Cleft Segment, Co-axial Segment, and Axial Volcano) where acoustic monitoring of the T-phase signal that accompanies magma migration in the upper crust has provided real-time data for events in progress and allowed "rapid response" cruises to be organized at these sites, within days to weeks. We also have good evidence for two volcanic events on the ultrafast spreading southern EPR, but the best-studied eruption site, to date, has been at 9°45–52′ N on the EPR. Serendipitously, submersible studies began at 9–10° N EPR less than one month after volcanic eruption at this site (Haymon *et al.*, 1993; Rubin *et al.*, 1994). Profound chemical changes (more than a factor of two in some cases) were noted at some of these vents during a period of less than a month (Von Damm *et al.*, 1995; Von Damm, 2000). Subsequently, it has become clear that very rapid changes occur within an initial one-year period which are related to changes in the conditions of phase separation and water rock reaction. These, in turn, are presumed to reflect responses to the mining of heat from the dike-intrusion, including lengthening of the reaction path and increases in the residence time of the fluids within the crust. At none of the other eruptive sites has it been possible to complete comparable direct sampling of vent fluids within this earliest "post-event" time period. It is now clear that the first fluid to be expelled is the vapor phase (whether formed as a result of sub- or supercritical phase separation), probably because of its lower density. What happens next, however, is less clear. In several cases, the "brine" (liquid) phase has been emitted next. In some vents this has occurred as a gradual progression to higher-chlorinity fluids; in other vents, the transition appears to occur more as a step function—although those observations may be aliased by the episodic nature of the sampling programs involved. What is most certainly the case at 9° N EPR, however, is that following initial vapor-phase expulsion, some

vents have progressed to venting fluids with chlorinities greater than seawater much faster (≤3 yr), than others (∼10 yr), and several have never made the transition. Further, in some parts of the eruptive area, vapor phase fluids are still the predominant type of fluid being emitted more than a decade after the eruption event. Fluids exiting from several of the vents have begun to decrease in chlorinity again, without ever having reached seawater concentrations. Conversely, other systems (most notably those from the Cleft segment) have been emitting vent fluids with chlorinities approximately twice that of seawater for more than a decade. If one wanted to sustain an argument that hydrothermal systems followed a vapor-to-brine phase evolution as a system ages (e.g., Butterfield *et al.*, 1997), it would be difficult to reconcile the observation that systems that are presumed to be relatively "older" (e.g., those on the northern Gorda Ridge) vent vapor-phase fluids, only (Von Damm *et al.*, 2001). Finally, one vent on the southern EPR, in an area with no evidence for a recent magmatic event, is emitting fluids which are phase separating "real time," with vapor exiting from the top of the structure, and brine from the bottom, simultaneously (Von Damm *et al.*, 2003). Fundamentally, there is a chloride-mass balance problem at many known hydrothermal sites. For example, at 21° N EPR, where high-temperature venting was first discovered and an active system is now known to have persisted for at least 23 years (based on sampling expeditions from 1979 and 2002) *only* low-chlorinity fluids are now being emitted (Von Damm *et al.*, 2002). Clearly there must be some additional storage and/or transport of higher-chlorinity fluids within the underlying crust. Our understanding of such systems is, at best, poor. What *is* clear is that pronounced temporal variability occurs at many vent sites, most notably at those that have been affected by magmatic events. There is also evidence for changes accompanying seismic events that are not related to magma-migration, but most likely related to cracking within the upper ocean crust (Sohn *et al.*, 1998; Von Damm and Lilley, 2003).

In marked contrast to those sites where volcanic eruptions and/or intrusions (diking events) have been detected, there are several other sites that have been sampled repeatedly over timescales of about two decades where no magmatic activity is known to have occurred. At some of these sites, chemical variations observed over the entire sampling period fall within the analytical error of our measurement techniques. The longest such time-series is for hydrothermal venting at 21° N on the EPR, where black smokers were first discovered in 1979, and where there has been remarkably little change in the composition of at least some of the vent fluids. Similarly, the Guaymas Basin vent site was first sampled in January 1982 (Von Damm *et al.*, 1985b), South Cleft on the JdF ridge in 1984 (Von Damm and Bischoff, 1987), and TAG on the MAR in 1986 (Campbell *et al.*, 1988). All these sites have exhibited very stable vent-fluid chemistries, although only TAG would be considered as a "slow spreading" vent site. Nevertheless, it is the TAG vent site that has shown perhaps the most remarkable stability in its vent-fluid compositions; these have remained invariant for more than a decade, even after perturbation from the drilling of a series of 5 ODP holes direct into the top of the active sulfide mound (Humphris *et al.*, 1995; Edmonds *et al.*, 1996).

Accounting for temporal variability (or lack thereof) when calculating hydrothermal fluxes is, clearly, problematic. It is very difficult to calculate the volume of fluid exiting from a hydrothermal system accurately. Many of the differences from seawater are most pronounced in the early eruptive period ($\sim$1 yr), which is also the time when fluid temperatures are hottest (Von Damm, 2000). Visual observations suggest that this is a time of voluminous fluid flow, which is not unexpected given that an enhanced heat source will have recently been emplaced directly at the seafloor in the case of an eruption, or, in the case of a dike intrusion, at shallow depths beneath. The upper oceanic crust exhibits high porosity, filled with ambient seawater. At eruption/intrusion, this seawater will be heated rapidly, its density will decrease, the resultant fluid will quickly rise, and large volumes of unreacted, cooler seawater will be drawn in and quickly expelled. It is not unreasonable to assume, therefore, that the water flux through a hydrothermal system may be at its largest during this initial period, just when chemical compositions are most extreme (Von Damm *et al.*, 1995; Von Damm, 2000). The key to the problem, therefore, probably lies in determining how much time a hydrothermal system spends in its "waxing" (immediate post-eruptive) stage when compared to the time spent at "steady state" (e.g., as observed at 21° N EPR) and in a "waning" period, together with an evaluation of the relative heat, water, and chemical fluxes associated with each of these different stages. If fluid fluxes and chemical anomalies are greatest in the immediate post-eruptive period, for example, the initial 12 months of any vent-field eruption may be geochemically more significant than a further 20 years of "steady-state" emission. At fast-spreading ridges, new eruptions might even occur faster than such a vent "lifecycle" can be completed. Alternately, the converse may hold true: early-stage eruptions may prove relatively insignificant over the full lifecycle of a prolonged, unperturbed hydrothermal site.

To advance our understanding of the chemical variability of vent fluids, it will be important to continue to find new sites that may be at evolutionary stages not previously observed. Equally, it will be important to continue studies of temporal variability at known sites: both those that have varied in the past and those that have appeared to be stable over the time intervals at which they have been sampled. Understanding the mechanisms and physical processes that control these vent-fluid compositions are key to calculating hydrothermal fluxes.

2.3 THE *NET* IMPACT OF HYDROTHERMAL ACTIVITY

It is important to remember that the gross chemical flux associated with expelled vent fluids (Section 2.2) is not identical to the net flux from hydrothermal systems. Subsurface hydrothermal circulation can have a net negative flux for some chemicals, the most obvious being magnesium which is almost quantitatively removed from the starting seawater and is added instead to the oceanic crust. Another example of such removal is uranium, which is also completely removed from seawater during hydrothermal circulation and then recycled into the upper mantle through subduction of altered oceanic crust. Even where, compared to starting seawater, there is no concentration gain or loss, an element may nevertheless undergo almost complete isotopic exchange within the oceanic crust indicating that none of the substance originally present in the starting seawater has passed conservatively through the hydrothermal system—the most obvious example being that of strontium. We present a brief summary of ocean crust mineralization in Section 2.4.1, but a more detailed treatment of ocean crust alteration is given elsewhere.

For the remainder of this chapter we concentrate, instead, upon the fate of hydrothermal discharge once it reaches the seafloor. Much of this material, transported in dissolved or gaseous form in warm or hot fluids, does not remain in solution but forms solid phases as fluids cool and/or mix with colder, more alkaline seawater. These products, whose formation may be mediated as well as modified by a range of biogeochemical processes, occur from the ridge axis out into the deep ocean basins. Massive sulfide as well as silicate, oxide, and sometimes carbonate deposits formed from high-temperature fluids are progressively altered by high-temperature metasomatism, as well as low-temperature oxidation and mass wasting—much of which may be biologically mediated. Various low-temperature deposits may also be formed, again often catalyzed by biological activity. In addition to these near-vent hydrothermal products, abundant fine-grained particles are formed in hydrothermal plumes, which subsequently settle to the seafloor to form metalliferous sediments, both close to vent sites and across ridge flanks into adjacent ocean basins. The post-depositional fates for these near- and far-field deposits remain poorly understood. Sulfide deposits, for example, may undergo extensive diagenesis and dissolution, leading to further release of dissolved chemicals into the deep ocean. Conversely, oxidized hydrothermal products may remain well-preserved in the sedimentary record and only be recycled via subduction back into the Earth's interior. On ridges where volcanic eruptions are frequent, both relatively fresh and more oxidized deposits may be covered over by subsequent lava flows (on the timescale of a decade) and, thus, become assimilated into the oceanic crust, isolated from the overlying water- and sediment-columns. In the following sections we discuss the fates of various hydrothermal "products" in order of their distance from the vent-source: near-vent deposits (Section 2.4); hydrothermal plumes (Section 2.5); and hydrothermal sediments (Section 2.6).

2.4 NEAR-VENT DEPOSITS

2.4.1 Alteration and Mineralization of the Upper Ocean Crust

Hydrothermal circulation causes extensive alteration of the upper ocean crust, reflected both in mineralization of the crust and in changes to physical properties of the basement (Alt, 1995). The direction and extent of chemical and isotopic exchange between seawater and oceanic crust depends on variations in temperature and fluid penetration and, thus, vary strongly as a function of depth. Extensive mineralization of the upper ocean crust can occur where metals leached from large volumes of altered crust become concentrated at, or close beneath, the sediment–water interface (Hannington *et al.*, 1995, Herzig and Hannington, 2000). As we have seen (Section 2.2), hydrothermal circulation within the ocean crust can be subdivided into three major components—the *recharge, reaction*, and *discharge* zones (Figure 5).

Recharge zones, which are broad and diffuse, represent areas where seawater is heated and undergoes reactions with the crust as it penetrates, generally downwards. (It is important to remember, however, that except where hydrothermal systems are sediment covered, the location of the recharge zone has never been established; debate continues, for example, whether recharge occurs *along* or *across* axis.) Important reactions in the recharge zone, at progressively increasing temperatures, include: low-temperature oxidation, whereby iron oxyhydroxides replace olivines and primary sulfides; fixation of alkalis (potassium, rubidium, and caesium) in celadonite and nontronite (ferric mica and smectite, respectively) and

fixation of magnesium, as smectite ($<200\ ^\circ$C) and chlorite ($>200\ ^\circ$C). At temperatures exceeding $\sim$130–150 $^\circ$C, two other key reactions occur: formation of anhydrite ($CaSO_4$) and mobilization of the alkalis (potassium, rubidium, lithium) (Alt, 1995). Upon subduction, the altered mineralogy and composition of the ocean crust can lead to the development of chemical and isotopic heterogeneities, both in the mantle and in the composition of island arc volcanic rocks. This subject is discussed in greater detail by Staudigel.

The *reaction* zone represents the highest pressure and temperature (likely $>400\ ^\circ$C) conditions reached by hydrothermal fluids during their subsurface circulation; it is here that hydrothermal vent fluids are believed to acquire much of their chemical signatures (Section 2.2). As they become buoyant, these fluids then rise rapidly back to the seafloor through *discharge* zones. The deep roots of hydrothermal discharge zones have only ever been observed at the base of ophiolite sequences (e.g., Nehlig *et al.*, 1994). Here, fluid inclusions and losses of metals and sulfur from the rocks indicate alteration temperatures of 350–440 $^\circ$C (Alt and Bach, 2003) in reasonable agreement with vent-fluid observations (Section 2.2). Submersible investigations and towed camera surveys of the modern seafloor have allowed surficial hydrothermal deposits to be observed in some detail (see next section). By contrast, the alteration pipes and "stockwork" zones that are believed to form the "roots" underlying all seafloor hydrothermal deposits (Figure 10) and which are considered to be quantitatively important in global

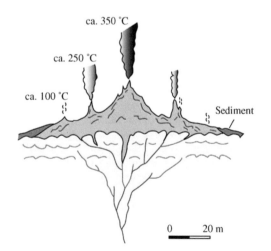

Figure 10 Schematic illustration of an idealized hydrothermal sulfide mound including: branching stockwork zone beneath the mound; emission of both high-temperature (350 °C) and lower-temperature (100–250 °C) fluids from the top of the mound together with more ephemeral diffuse flow; and deposition by mass-wasting of an apron of sulfidic sediments around the edges of the mound.

geochemical cycles (e.g., Peucker-Ehrenbrink *et al.*, 1994) remain relatively inaccessible. Direct sampling of the stockwork beneath active seafloor hydrothermal systems has only been achieved on very rare occasions, e.g., through direct sampling from ODP drilling (Humphris *et al.*, 1995) or through fault-face exposure at the seabed (e.g., Fouquet *et al.*, 2003). Because of their relative inaccessibility, even compared to all other aspects of deep-sea hydrothermal circulation, study of sub-seafloor crustal mineralization remains best studied in ancient sulfide deposits preserved in the geologic record on-land (Hannington *et al.*, 1995).

2.4.2 Near-vent Hydrothermal Deposits

The first discoveries of hydrothermal vent fields (e.g., Galapagos; EPR, 21° N) revealed three distinctive types of mineralization: (i) massive sulfide mounds deposited from focused high-temperature fluid flow, (ii) accumulations of Fe–Mn oxyhydroxides and silicates from low-temperature diffuse discharge, and (iii) fine-grained particles precipitated from hydrothermal plumes. Subsequently, a wide range of mineral deposits have been identified that are the result of hydrothermal discharge, both along the global ridge crest and in other tectonic settings (Koski *et al.*, 2003). Of course, massive sulfides only contain a fraction of the total dissolved load released from the seafloor. Much of this flux is delivered to ridge flanks via dispersion in buoyant and non-buoyant hydrothermal plumes (Section 2.5). In addition, discoveries such as the carbonate-rich "Lost City" deposits (Kelley *et al.*, 2001), silica-rich deposits formed in the Blanco Fracture Zone (Hein *et al.*, 1999), and metal-bearing fluids on the flanks of the JdFR (Mottl *et al.*, 1998) remind us that there is still much to learn about the formation of hydrothermal mineral deposits.

Haymon (1983) proposed the first model for how a black smoker chimney forms (Figure 11). The first step is the formation of an anhydrite (CaSO$_4$) framework due to the heating of seawater, and mixing of vent fluids with that seawater. The anhydrite walls then protect subsequent hydrothermal fluids from being mixed so extensively with seawater, as well as providing a template onto which sulfide minerals can precipitate as those fluids cool within the anhydrite structure. As the temperature and chemical compositions within the chimney walls evolve, a zonation of metal sulfide minerals develops, with more copper-rich phases being formed towards the interior, zinc-rich phases towards the exterior, and iron-rich phases ubiquitous. This model is directly analogous to the concept of an "intensifying hydrothermal system" developed by Eldridge *et al.* (1983), in which initial deposition of a fine-grained mineral carapace restricts mixing

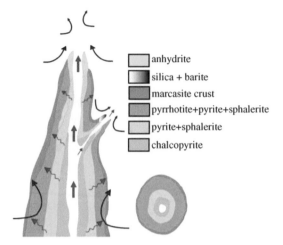

Figure 11 Schematic diagram showing mineral zonation in cross-section and in plan view for a typical black smoker chimney. Arrows indicate direction of inferred fluid flow (after Haymon, 1983).

of hydrothermal fluid and seawater at the site of discharge. Subsequently, less-dilute, higher-temperature (copper-rich) fluids interact with the sulfides within this carapace to precipitate chalcopyrite and mobilize more soluble, lower-temperature metals such as lead and zinc toward the outer, cooler parts of the deposit. Thus, it is the steep temperature and chemical gradients, caused by both mixing and diffusion, which account for the variations in wall mineralogy and Cu–Zn zonation observed in both chimneys and larger deposits. These processes, initially proposed as part of a conceptual model, have subsequently been demonstrated more rigorously by quantitative geochemical modeling of hydrothermal fluids and deposits (Tivey, 1995).

Drilling during Ocean Drilling Program Leg 158 revealed that similar internal variations can also occur on much larger scales—e.g., across the entire TAG mound (Humphris *et al.*, 1995; Petersen *et al.*, 2000). That work revealed the core of the mound to be dominated by chalcopyrite-bearing massive pyrite, pyrite–anhydrite and pyrite–silica breccias whilst the mound top and margins contained little or no chalcopyrite but more sphalerite and higher concentrations of metals soluble at lower temperatures (e.g., zinc, gold). The geochemical modeling results of Tivey (1995) point to a mechanism of entrainment of seawater into the focused upflow zone within the mound which would, almost simultaneously: (i) induce the precipitation of anhydrite, chalcopyrite, pyrite, and quartz; (ii) decrease the pH of the fluid; and (iii) mobilize zinc and other metals. When combined, the processes outlined above—zone refining and the entrainment of seawater into active sulfide deposits—appear to credibly explain mineralogical and chemical features observed both in modern hydrothermal systems (e.g., the TAG mound) and in "Cyprus-type" deposits found in many ophiolites of orogenic belts (Hannington *et al.*, 1998).

A quite different form of hydrothermal deposit has also been located, on the slow-spreading MAR. The Lost City vent site (Kelley *et al.*, 2001) occurs near 30° N on the MAR away from the more recently erupted volcanic ridge axis. Instead, it is situated high up on a tectonic massif where faulting has exposed variably altered peridotite and gabbro (Blackman *et al.*, 1998). The Lost City field hosts at least 30 active and inactive spires, extending up to 60 m in height, on a terrace that is underlain by diverse mafic and ultramafic lithologies. Cliffs adjacent to this terrace also host abundant white hydrothermal alteration both as flanges and peridotite mineralization, which is directly akin to deposits reported from Alpine ophiolites (Früh-Green *et al.*, 1990). The Lost City chimneys emit fluids up to 75 °C which have a very high pH (9.0–9.8) and compositions which are rich in H_2S, CH_4, and H_2—consistent with serpentinization reactions (Section 2.2)—but low in dissolved silica and metal contents (Kelley *et al.*, 2001). Consistent with this, the chimneys of the Lost City field are composed predominantly of magnesium and calcium-rich carbonate and hydroxide minerals, notably calcite, brucite, and aragonite.

In addition to the sulfide- and carbonate-dominated deposits described above, mounds and chimneys composed of Fe- and Mn-oxyhydroxides and silicate minerals also occur at tectonically diverse rift zones, from MORs such as the Galapagos Spreading Center to back-arc systems such as the Woodlark Basin (Corliss *et al.*, 1978; Binns *et al.*, 1993). Unlike polymetallic sulfides, Fe–Mn oxide-rich, low-temperature deposits should be chemically stable on the ocean floor. Certainly, metalliferous sediments in ophiolites—often referred to as "umbers"—have long been identified as submarine hydrothermal deposits formed in ancient ocean ridge settings. These types of ophiolite deposit may be intimately linked to the Fe–Mn–Si oxide "mound" deposits formed on pelagic ooze near the Galapagos Spreading Center (Maris and Bender, 1982). It has proven difficult, however, to determine the precise temporal and genetic relationship of umbers to massive sulfides, not least because no gradation of Fe–Mn–Si oxide to sulfide mineralization has yet been reported from ophiolitic terranes. The genetic relationship between sulfide and oxide facies deposits formed at modern hydrothermal sites also remains enigmatic. Fe–Mn–Si oxide deposits may simply represent "failed" massive sulfides. Alternately, it may well be that there are important aspects of, for example, axial versus off-axis plumbing systems (e.g., porosity, permeability, chemical variations caused by phase separation) or controls on the sulfur budget of hydrothermal systems that

remain inadequately understood. What seems certain is that the three-dimensional problem of hydrothermal deposit formation (indeed, 4D if one includes temporal evolution) cannot be solved from seafloor observations alone. Instead, what is required is a continuing program of seafloor drilling coupled with analogue studies of hydrothermal deposits preserved on land.

2.5 HYDROTHERMAL PLUME PROCESSES

2.5.1 Dynamics of Hydrothermal Plumes

Hydrothermal plumes form wherever buoyant hydrothermal fluids enter the ocean. They represent an important dispersal mechanism for the thermal and chemical fluxes delivered to the oceans while the processes active within these plumes serve to modify the gross fluxes from venting, significantly. Plumes are of further interest to geochemists because they can be exploited in the detection and location of new hydrothermal fields *and* for the calculation of total integrated fluxes from any particular vent field. To biologists, hydrothermal plumes represent an effective transport mechanism for dispersing vent fauna, aiding gene-flow between adjacent vent sites along the global ridge crest (e.g., Mullineaux and France, 1995). In certain circumstances the heat and energy released into hydrothermal plumes could act as a driving force for mid-depth oceanic circulation (Helfrich and Speer, 1995).

Present day understanding of the dynamics of hydrothermal plumes is heavily influenced by the theoretical work of Morton *et al.* (1956) and Turner (1973). When high-temperature vent fluids are expelled into the base of the much colder, stratified, oceanic water column they are buoyant and begin to rise. Shear flow between the rising fluid and the ambient water column produces turbulence and vortices or eddies are formed which are readily visible in both still and video-imaging of active hydrothermal vents. These eddies or vortices act to entrain material from the ambient water column, resulting in a continuous dilution of the original vent fluid as the plume rises. Because the oceans exhibit stable density-stratification, this mixing causes progressive dilution of the buoyant plume with water which is denser than both the initial vent fluid and the overlying water column into which the plume is rising. Thus, the plume becomes progressively less buoyant as it rises and it eventually reaches some finite maximum height above the seafloor, beyond which it cannot rise (Figure 12). The first, rising stage of hydrothermal plume evolution is termed the *buoyant plume*. The later stage, where plume rise has ceased and hydrothermal effluent begins to be dispersed laterally, is termed the *nonbuoyant*

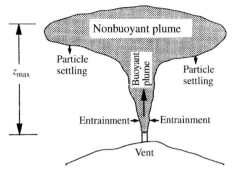

Figure 12 Sketch of the hydrothermal plume rising above an active hydrothermal vent, illustrating entrainment of ambient seawater into the buoyant hydrothermal plume, establishment of a nonbuoyant plume at height z_{max} (deeper than the maximum height of rise actually attained due to momentum overshoot) and particle settling from beneath the dispersing nonbuoyant plume (after Helfrich and Speer, 1995).

plume also referred to in earlier literature as the *neutrally buoyant plume*.

The exact height reached by any hydrothermal plume is a complex function involving key properties of both the source vent fluids and the water column into which they are injected—notably the initial buoyancy of the former and the degree of stratification of the latter. A theoretical approach to calculating the maximum height-of-rise that can be attained by any hydrothermal plume is given by Turner (1973) with the equation:

$$z_{max} = 3.76F_0^{1/4}N^{-3/4} \tag{1}$$

where F_0 and N represent parameters termed the buoyancy flux and the Brunt–Väisälä frequency, respectively. The concept of the buoyancy flux, F_0 (units $cm^4 s^{-3}$) can best be understood from an explanation that the product $F_0\rho_0$ represents the total weight deficiency produced at the vent source per unit time (units $g\ cm\ s^{-3}$). The Brunt–Väisälä frequency, also termed the buoyancy frequency, N (units s^{-1}) is defined as:

$$N^2 = -(g/\rho_0)\mathrm{d}\rho/\mathrm{d}z \tag{2}$$

where g is the acceleration due to gravity, ρ_0 is the background density at the seafloor and $\mathrm{d}\rho/\mathrm{d}z$ is the ambient vertical density gradient. In practice, buoyant hydrothermal plumes always exceed this theoretical maximum because, as they reach the level z_{max} the plume retains some finite positive vertical velocity. This leads to "momentum overshoot" (Turner, 1973) and "doming" directly above the plume source before this (now negatively buoyant) dome collapses back to the level of zero buoyancy (Figure 12).

Note also the very weak dependence of emplacement height (z_{max}) upon the buoyancy flux or heat flux of any given vent source. A doubling of z_{max} for any plume, for example,

could only be achieved by a 16-fold increase in the heat flux provided by its vent source. By contrast, the ambient water column with which the buoyant plume becomes progressively more diluted exerts a significant influence because the volumes entrained are nontrivial. For a typical plume with $F_0=10^{-2}\ m^4\ s^{-3}$, $N=10^{-3}\ s^{-1}$ the entrainment flux is of the order $10^2\ m^3\ s^{-1}$ (e.g., Helfrich and Speer, 1995) resulting in very rapid dilution of the primary vent fluid (10^2–10^3:1) within the first 5–10 m of plume rise and even greater dilution ($\sim 10^4$:1) by the time of emplacement within the nonbuoyant, spreading hydrothermal plume (Feely *et al.*, 1994). Similarly, the *time* of rise for a buoyant hydrothermal plume, is entirely dependent on the background buoyancy frequency, N (Middleton, 1979):

$$\tau = \pi N^{-1} \tag{3}$$

which, for a typical value of $N=10^{-3}\ s^{-1}$, yields a plume rise-time of ≤ 1 h.

2.5.2 Modification of Gross Geochemical Fluxes

Hydrothermal plumes represent a significant dispersal mechanism for chemicals released from seafloor venting to the oceans. Consequently, it is important to understand the physical processes that control this dispersion (Section 2.5.1). It is also important to recognize that hydrothermal plumes represent non-steady-state fluids whose chemical compositions evolve with age (Figure 13). Processes active in hydrothermal plumes can lead to significant modification of *gross* hydrothermal fluxes (cf. Edmond *et al.*, 1979; German *et al.*, 1991b) and, in the extreme, can even reverse the sign of *net* flux to/from the ocean (e.g., German *et al.*, 1990, 1991a).

2.5.2.1 *Dissolved noble gases*

For one group of tracers, however, inert marine geochemical behavior dictates that they do undergo conservative dilution and dispersion within hydrothermal plumes. Perhaps the simplest example of such behavior is primordial dissolved ^{3}He, which is trapped in the Earth's interior and only released to the deep ocean through processes linked to volcanic activity—notably, submarine hydrothermal vents. As we have seen, previously, end-member vent fluids typically undergo $\sim$10,000-fold dilution prior to emplacement in a nonbuoyant hydrothermal plume. Nevertheless, because of the large enrichments of dissolved ^{3}He in hydrothermal fluids when compared to the low background levels in seawater, pronounced enrichments of dissolved ^{3}He relative to ^{4}He can be traced over great distances in the deep ocean. Perhaps the most famous example of such

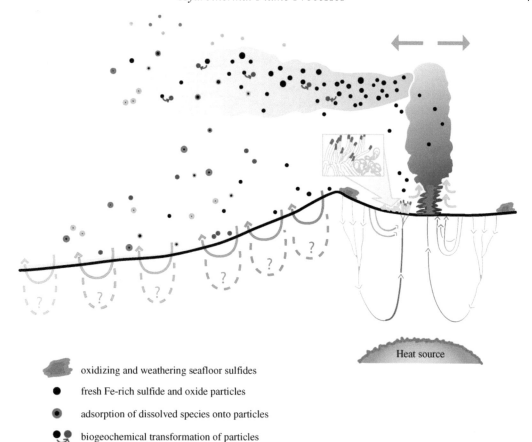

oxidizing and weathering seafloor sulfides

● fresh Fe-rich sulfide and oxide particles

◉ adsorption of dissolved species onto particles

●◉ biogeochemical transformation of particles

Figure 13 Schematic representation of an MOR hydrothermal system and its effects on the overlying water column. Circulation of seawater occurs within the oceanic crust, and so far three types of fluids have been identified and are illustrated here: high-temperature vent fluids that have likely reacted at >400 °C; high- temperature fluids that have then mixed with seawater close to the seafloor; fluids that have reacted at "intermediate" temperatures, perhaps ~150 °C. When the fluids exit the seafloor, either as diffuse flow (where animal communities may live) or as "black smokers," the water they emit rises and the hydrothermal plume then spreads out at its appropriate density level. Within the plume, sorption of aqueous oxyanions may occur onto the vent-derived particles (e.g., phosphate, vanadium, arsenic) making the plumes a sink for these elements; biogeochemical transformations also occur. These particles eventually rain-out, forming metalliferous sediments on the seafloor. While hydrothermal circulation is known to occur far out onto the flanks of the ridges, little is known about the depth to which it extends or its overall chemical composition because few sites of active ridge-flank venting have yet been identified and sampled (Von Damm, unpublished).

behavior is the pronounced ^{3}He plume identified by Lupton and Craig (1981) dispersing over ~2,000 km across the southern Pacific Ocean, west of the southern EPR (Figure 14). A more recent example of the same phenomenon, however, is the large-scale ^{3}He anomaly reported by Rüth *et al.* (2000) providing the first firm evidence for high-temperature hydrothermal venting anywhere along the southern MAR. Rn-222, a radioactive isotope of the noble element radon, is also enriched in hydrothermal fluids; while it shares the advantages of being a conservative tracer with ^{3}He, it also provides the additional advantage of acting as a "clock" for hydrothermal plume processes because it decays with a half-life of 3.83 d. Kadko *et al.* (1990) used the fractionation of ^{222}Rn/^{3}He ratios in a dispersing non-buoyant hydrothermal plume above the JdFR to estimate rates of dispersion or "ages" at different locations "down-plume" and, thus, deduce rates of uptake and/or removal of various nonconservative plume components (e.g., H$_2$, CH$_4$, manganese, particles). A complication to that approach arises, however, with the recognition that—in at least some localities—maximum plume-height ^{222}Rn/^{3}He ratios exceed pristine high-temperature vent-fluid values; clearly, entrainment from near-vent diffuse flow can act as an important additional source of dissolved ^{222}Rn entering ascending hydrothermal plumes (Rudnicki and Elderfield, 1992).

2.5.2.2 *Dissolved reduced gases (H$_2$S, H$_2$, CH$_4$)*

The next group of tracers that are important in hydrothermal plumes are the reduced dissolved

East Pacific Rise

$\delta\,^3$He (%)

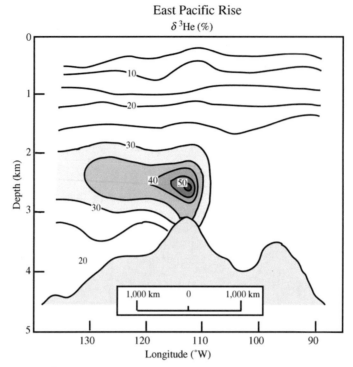

Figure 14 Distribution of δ^3He across the Pacific Ocean at 15° S. This plume corresponds to the lobe of metalliferous sediment observed at the same latitude extending westward from the crest of the EPR (Figure 2) (after Lupton and Craig, 1981).

gases, H_2S, H_2, and CH_4. As we have already seen, dissolved H_2S is commonly the most abundant dissolved reduced gas in high-temperature vent fluids (Section 2.2). Typically, however, any dissolved H_2S released to the oceans undergoes rapid reaction, either through precipitation of polymetallic sulfide phases close to the seafloor and in buoyant plumes, or through oxidation in the water column. As of the early 2000s, there has only been one report of measurable dissolved H_2S at nonbuoyant plume height anywhere in the deep ocean (Radford-Knöery et al., 2001). That study revealed maximum dissolved H_2S concentrations of ≤ 2 nM, representing a 5×10^5-fold decrease from vent-fluid concentrations (Douville et al., 2002) with complete oxidative removal occurring in just 4–5 h, within $\sim$1–10 km of the vent site. Dissolved H_2, although not commonly abundant in such high concentrations in vent fluids (Section 2.2), exhibits similarly short-lived behavior within hydrothermal plumes. Kadko et al. (1990) and German et al. (1994) have reported maximum plume-height dissolved H_2 concentrations of 12 nM and 32 nM above the Main Endeavour vent site, JdFR and above the Steinahóll vent site, Reykjanes Ridge. From use of the $^{222}Rn/^3He$ "clock," Kadko et al. (1990) estimated an apparent "oxidative-removal" half-life for dissolved H_2 of $\sim$10 h.

The most abundant and widely reported dissolved reduced gas in hydrothermal plumes is

methane which is released into the oceans from both high- and low-temperature venting and the serpentinization of ultramafic rocks (e.g., Charlou et al., 1998). Vent-fluid concentrations are significantly enriched over seawater values (10–2,000 µmol kg^{-1} versus $<$5 nmol kg^{-1}) but the behavior of dissolved methane, once released, appears variable from one location to another: e.g., rapid removal of dissolved CH_4 was observed in the Main Endeavour plume (half-life=10 d; Kadko et al., 1990) yet near-conservative behavior for the same tracer has been reported for the Rainbow hydrothermal plume, MAR, over distances up to 50 km from the vent site (Charlou et al., 1998; German et al., 1998b). Possible reasons for these significant variations are discussed later (Section 2.5.4).

2.5.2.3 *Iron and manganese geochemistry in hydrothermal plumes*

The two metals most enriched in hydrothermal vent fluids are iron and manganese. These elements are present in a reduced dissolved form (Fe^{2+}, Mn^{2+}) in end-member vent fluids yet are most stable as oxidized Fe(III) and Mn(IV) precipitates in the open ocean. Consequently, the dissolved concentrations of iron and manganese in vent fluids are enriched $\sim$10^6:1 over open-ocean values (e.g., Landing and Bruland, 1987; Von Damm, 1995; Statham et al., 1998). When these metal-laden fluids first enter the ocean, two important processes occur. First, the fluids are instantaneously cooled from >350 °C to ≤30 °C. This quenching of a hot saturated solution leads to precipitation of a range of metal sulfide phases that are rich in iron but not manganese because the latter does not readily form sulfide minerals. In addition, turbulent mixing between the sulfide-bearing vent fluid and the entrained, oxidizing, water column leads to a range of redox reactions resulting in the rapid precipitation of high concentrations of suspended iron oxyhydroxide particulate material. The dissolved manganese within the hydrothermal plume, by contrast, typically exhibits much more sluggish oxidation kinetics and remains predominantly in dissolved form at the time of emplacement in the nonbuoyant plume. Because iron and manganese are so enriched in primary vent fluids, nonbuoyant plumes typically exhibit total (dissolved and particulate) iron and manganese concentrations, which are ~100-fold higher than ambient water column values immediately following nonbuoyant plume emplacement. Consequently, iron and manganese, together with CH_4 and ^3He (above), act as extremely powerful tracers with which to identify the presence of hydrothermal activity from water-column investigations. The fate of iron in hydrothermal plumes is of particular interest because it is the geochemical cycling of this element, more than any other, which controls the fate of much of the hydrothermal flux from seafloor venting to the oceans (e.g., Lilley *et al.*, 1995).

Because of their turbulent nature, buoyant hydrothermal plumes have continued to elude detailed geochemical investigations. One approach has been to conduct direct sampling using manned submersibles or ROVs (e.g., Rudnicki and Elderfield, 1993; Feely *et al.*, 1994). An alternate, and indirect, method is to investigate the geochemistry of precipitates collected both rising in, and sinking from, buoyant hydrothermal plumes using near-vent sediment traps (e.g., Cowen *et al.*, 2001; German *et al.*, 2002). From direct observations it is apparent that up to 50% of the total dissolved iron emitted from hydrothermal vents is precipitated rapidly from buoyant hydrothermal plumes (e.g., Mottl and McConachy, 1990; Rudnicki and Elderfield, 1993) forming polymetallic sulfide phases which dominate (>90%) the iron flux to the near-vent seabed (Cowen *et al.*, 2001; German *et al.*, 2002). The remaining dissolved iron within the buoyant and nonbuoyant hydrothermal plume undergoes oxidative precipitation. In the well-ventilated N. Atlantic Ocean, very rapid Fe(II) oxidation is observed with a half-life for oxidative removal from solution of just 2–3 min (Rudnicki and Elderfield, 1993). In the NE Pacific, by contrast, corresponding half-times of up to 32 h have been reported from JdFR hydrothermal plumes (Chin *et al.*, 1994; Massoth *et al.*, 1994). Field and Sherrell (2000) have predicted that the oxidation rate for dissolved Fe^{2+} in hydrothermal plumes should decrease along the path of the thermohaline circulation, reflecting the progressively decreasing pH and dissolved oxygen content of the deep ocean (Millero *et al.*, 1987):

$$-d[Fe(II)]/dt = k[OH^-]^2[O_2][Fe(II)] \quad (4)$$

The first Fe(II) incubation experiments conducted within the Kairei hydrothermal plume, Central Indian Ridge, are consistent with that prediction, yielding an Fe(II) oxidation half-time of ~2 h (Statham *et al.*, 2003, submitted).

2.5.2.4 Co-precipitation and uptake with iron in buoyant and nonbuoyant plumes

There is significant co-precipitation of other metals enriched in hydrothermal fluids, along with iron, to form buoyant plume polymetallic sulfides. Notable among these are copper, zinc, and lead. Common accompanying phases, which also sink rapidly from buoyant plumes, are barite and anhydrite (barium and calcium sulfates) and amorphous silica (e.g., Lilley *et al.*, 1995). In nonbuoyant hydrothermal plumes, where Fe- and (to a lesser extent) Mn-oxyhydroxides predominate, even closer relationships are observed between particulate iron concentrations and numerous other tracers. To a first approximation, three differing iron-related behaviors have been identified (German *et al.*, 1991b; Rudnicki and Elderfield, 1993): (i) co-precipitation; (ii) fixed molar ratios to iron; and (iii) oxyhydroxide scavenging (Figure 15). The first is that already alluded to above and loosely termed "chalcophile" behavior—namely preferential co-precipitation with iron as sulfide phases followed by preferential settling from the nonbuoyant plume. Such elements exhibit a generally positive correlation with iron for plume particle concentrations but with highest $X:Fe$ ratios closest to the vent site ($X=Cu$, Zn, Pb) and much lower values farther afield.

The second group are particularly interesting. These are elements that establish fixed $X:Fe$ ratios in nonbuoyant hydrothermal plumes which do not vary with dilution or dispersal distance "down-plume" (Figure 15). Elements that have been shown to exhibit such "linear" behavior include potassium, vanadium, arsenic, chromium, and uranium (e.g., Trocine and Trefry, 1988; Feely *et al.*, 1990, 1991; German *et al.*, 1991a,b). Hydrothermal vent fluids are not particularly enriched in any of these elements, which typically occur as rather stable "oxyanion" dissolved species in seawater. Consequently, this uptake must represent a significant removal flux, for at

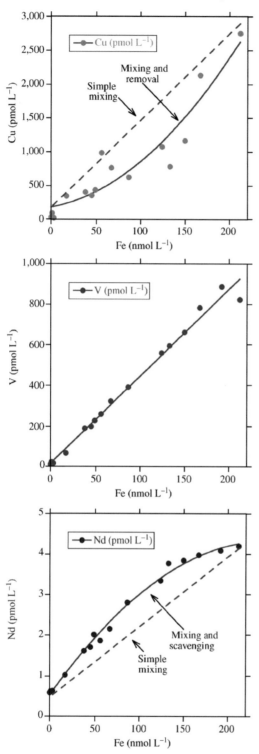

Figure 15 Plots of particulate copper, vanadium, and neodymium concentrations versus particulate iron for suspended particulate material filtered *in situ* from the TAG hydrothermal mound, MAR, 26° N (data from German *et al.*, 1990, 1991b). Note generally positive correlations with particulate Fe concentration for all three tracers but with additional negative (Cu) or positive (Nd) departure for sulfide-forming and scavenged elements, respectively.

least some of these elements, from the deep ocean. The P:Fe ratios observed throughout all Pacific hydrothermal plumes are rather similar (P:Fe=0.17–0.21; Feely *et al.*, 1998) and distinctly higher than the value observed in Atlantic hydrothermal plumes (P:Fe=0.06–0.12). This has led to speculation that plume P:Fe ratios may reflect the ambient dissolved PO_4^{3-} concentration of the host water column and, thus, may offer potential as a long-term tracer of past ocean circulation, if preserved faithfully in metalliferous marine sediments (Feely *et al.*, 1998).

2.5.2.5 Hydrothermal scavenging by Fe-oxyhydroxides

A final group of elements identified from hydrothermal plume particle investigations are of particular importance: "particle-reactive" tracers. Perhaps the best examples are the REE although other tracers that show similar behavior include beryllium, yttrium, thorium, and protactinium (German *et al.*, 1990, 1991a,b, 1993; Sherrell *et al.*, 1999). These tracers, like the two preceding groups, exhibit generally positive correlations with particulate iron concentrations in hydrothermal plumes (Figure 15). Unlike the "oxyanion" group, however, these tracers do not show constant X:Fe ratios within the nonbuoyant plume. Instead, highest values (e.g., for REE:Fe) are observed at increasing distances away from the vent site, rather than immediately above the active source (German *et al.*, 1990; Sherrell *et al.*, 1999). One possible interpretation of this phenomenon is that the Fe-oxyhydroxide particles within young nonbuoyant plumes are undersaturated with respect to surface adsorption and that continuous "scavenging" of dissolved, particle-reactive species occurs as the particles disperse through the water column (German *et al.*, 1990, 1991a,b). An alternate hypothesis (Sherrell *et al.*, 1999) argues, instead, for two-stage equilibration within a nonbuoyant hydrothermal plume: close to the vent source, a maximum in (e.g.,) particulate REE concentrations is reached, limited by equilibration at fixed distribution constants between the high particulate iron concentrations present and the finite dissolved tracer (e.g., REE) concentrations present in plume-water. As the plume disperses and undergoes dilution, however, particulate iron concentrations also decrease; re-equilibration between these particles and the diluting "pristine" ambient seawater, at the same fixed distribution constants, would then account for the higher REE/Fe ratios observed at lower particulate iron concentrations. More work is required to resolve which of these interpretations (kinetic versus equilibrium) more accurately reflects the processes active within hydrothermal plumes. What is beyond dispute concerning these particle-reactive tracers is that their uptake fluxes, in association with hydrothermal Fe-oxyhydroxide precipitates, far exceeds their dissolved fluxes entering the oceans from hydrothermal vents. Thus, hydrothermal plumes act as *sinks* rather than *sources* for these elements, even causing local depletions relative to the ambient water column concentrations (e.g., Klinkhammer *et al.*, 1983). Thus, even for those "particle-reactive" elements which are greatly enriched in vent fluids over seawater concentrations (e.g., REE), the processes active within hydrothermal plumes dictate that hydrothermal circulation causes a net *removal* of these tracers not just relative to the vent fluids themselves, but also from the oceanic water column (German *et al.*, 1990; Rudnicki and Elderfield, 1993). In the extreme, these processes can be sufficient to cause geochemical fractionations as pronounced as those caused by "boundary scavenging" in high-productivity ocean margin environments (cf.Anderson *et al.*, 1990; German *et al.*, 1997).

Thus far, we have treated the processes active in hydrothermal plumes as inorganic geochemical processes. However we know this is not strictly the case: microbial processes are well known to mediate key chemical reactions in hydrothermal plumes (Winn *et al.*, 1995; Cowen and German, 2003) and more recently the role of larger organisms such as zooplankton has also been noted (e.g., Burd *et al.*, 1992; Herring and Dixon, 1998). The biological modification of plume processes is discussed more fully below (Section 2.5.4).

2.5.3 Physical Controls on Dispersing Plumes

Physical processes associated with hydrothermal plumes may affect their impact upon ocean geochemistry; because of the fundamentally different hydrographic controls in the Pacific versus Atlantic Oceans, plume dispersion varies between these two oceans. In the Pacific Ocean, where deep waters are warmer and saltier than overlying water masses, nonbuoyant hydrothermal plumes which have entrained local deep water are typically warmer and more saline at the point of emplacement than that part of the water column into which they intrude (e.g., Lupton *et al.*, 1985). The opposite has been observed in the Atlantic Ocean where deep waters tend to be colder and less saline than the overlying water column. Consequently, for example, the TAG nonbuoyant plume is anomalously cold and fresh when compared to the background waters into which it intrudes, 300–400 m above the seafloor (Speer and Rona, 1989).

Of perhaps more significance, geochemically, are the physical processes which affect plume-dispersion *after* emplacement at nonbuoyant plume height. Here, topography at the ridge crest exerts particular influence. Along slow and ultra-slow spreading ridges (e.g., MAR, SW Indian Ridge, Central Indian Ridge) nonbuoyant plumes are typically emplaced within the confining bathymetry of the rift-valley walls (order 1,000 m). Along faster spreading ridges such as the EPR, by contrast, buoyant plumes typically rise clear of the confining topography (order 100 m, only). Within rift-valley "corridors" plume dispersion is highly dependent upon along-axis current flow. At the TAG hydrothermal field (MAR 26° N), for example, residual currents are dominated by tidal excursions (Rudnicki *et al.*, 1994). A net effect of these relatively "stagnant" conditions is that plume material trapped within the vicinity of the vent site appears to be recycled multiple times through the TAG buoyant and nonbuoyant plume, enhancing the scavenging effect upon "particle-reactive" tracers (e.g., thorium isotopes) within the local water column (German and Sparks, 1993). At the Rainbow vent site (MAR 36° N) by contrast, much stronger prevailing currents ($\sim$10 cm s^{-1}) are observed and a more unidirectional, topographically controlled flow is observed (German *et al.*, 1998b). Failure to appreciate the potential complexities of such dispersion precludes the "informed" sampling required to resolve processes of geochemical evolution within a dispersing hydrothermal plume. Nor should it be assumed that such topographic steering is entirely a local effect confined to slow spreading ridges' rift-valleys. In recent work, Speer *et al.* (2003) used a numerical simulation of ocean circulation to estimate dispersion along and away from the global ridge crest. A series of starting points were considered along the entire ridge system, 200 m above the seafloor and at spacings of 30–100 km along-axis; trajectories were then calculated over a 10 yr integrated period. With few exceptions (e.g., major fracture zones) the net effect reported was that these dispersal trajectories tend to be constrained by the overall form of the ridge and flow parallel to the ridge axis over great distances (Speer *et al.*, 2003).

The processes described above are relevant to established "chronic" hydrothermal plumes. Important exceptions (only identified rarely–so far) are Event (or "*Mega-*") Plumes. One interpretation of these features is as follows: however a hydrothermal system may evolve, it must first displace a large volume of cold seawater from pore spaces within the upper ocean crust. Initial flushing of this system must be rapid, especially on fast ridges that are extending at rates in excess of ten centimeters per year. In circumstances where there is frequent recurrence, either of intrusions of magma close beneath the seafloor or dike-fed eruptions at the seabed, seafloor venting may commence with a rapid exhalation of a large volume of hydrothermal fluid to generate an "event" plume high up in the overlying water column (e.g., Baker *et al.*, 1995; Lupton *et al.*, 1999b). Alternately, Palmer and Ernst (1998, 2000) have argued that cooling of pillow basalts, erupted at ~1,200 °C on the seafloor and the most common form of submarine volcanic extrusion, is responsible for the formation of these same "event" plume features. Whichever eruption-related process causes their formation, an important question that follows is: how much hydrothermal flux is overlooked if we fail to intercept "event" plumes at the onset of venting at any given location? To address this problem, Baker *et al.* (1998) estimated that the event plume triggered by dike-intrusion at the co-axial vent field (JdFR) contributed less than 10% of the total flux of heat and chemicals released during the ~3 yr life span of that vent. If widely applicable, those deliberations suggest that event plumes can safely be ignored when calculating global geochemical fluxes (Hein *et al.*, 2003). They remain of great interest to microbiologists, however, as a potential "window" into the deep, hot biosphere (e.g., Summit and Baross, 1998).

2.5.4 Biogeochemical Interactions in Dispersing Hydrothermal Plumes

Recognition of the predominantly along-axis flow of water masses above MORs as a result of topographic steering has renewed speculation that hydrothermal plumes may represent important vectors for gene-flow along the global ridge crest, transporting both chemicals and vent larvae alike, from one adjacent vent site to another (e.g., Van Dover *et al.*, 2002). If such is the case, however, it is also to be expected that a range of biogeochemical processes should also be active within nonbuoyant hydrothermal plumes. One particularly good example of such a process is the microbial oxidation of dissolved manganese. In the restricted circulation regime of the Guaymas Basin, formation of particulate manganese is dominated by bacteria and the dissolved manganese residence time is less than a week (Campbell *et al.*, 1988). Similarly, uptake of dissolved manganese in the Cleft Segment plume (JdFR) is bacterially dominated, albeit with much longer residence times, estimated at 0.5 yr to >2 yr (Winn *et al.*, 1995). Distributions of dissolved CH_4 and H_2 in hydrothermal plumes have also been shown to be controlled by bacterially mediated oxidation (de Angelis *et al.*, 1993) with apparent residence times which vary widely for CH_4 (7–177 d) but are much shorter for dissolved H_2 (<1 d).

The release of dissolved H_2 and CH_4 into hydrothermal plumes provides suitable substrates for both primary (chemolithoautotrophic) and secondary (heterotrophic) production within dispersing nonbuoyant hydrothermal plumes. Although the sinking organic carbon flux from hydrothermal plumes (Roth and Dymond, 1989; German *et al.*, 2002) may be less than 1% of the total oceanic photosynthetic production (Winn *et al.*, 1995) hydrothermal production of organic carbon is probably restricted to a corridor extending no more than ~10 km to either side of the ridge axis. Consequently, microbial activity within hydrothermal plumes may have a pronounced local effect—perhaps 5–10-fold greater than the photosynthetic-flux driven from the overlying upper ocean (Cowen and German, 2003). Although the detailed microbiology of hydrothermal plumes remains poorly studied, as of the early 2000s, bacterial counts from the Rainbow plume have identified both an increase in total cell concentrations at plume-height compared to background (Figure 16) and, further, that 50–75% of the microbial cells identified in that work were particle-attached compared to typical values of just 15% for the open ocean. Detailed molecular biological analysis of those particle-associated microbes have also revealed that the majority (66%) are archeal in nature rather than bacterial (German *et al.*, 2003); in the open ocean, by contrast, it is the bacteria which typically dominate (Fuhrmann *et al.*, 1993; Mullins *et al.*, 1995). It is tempting to speculate that these preliminary data may provide testament to a long-established (even on geological timescales) chemical-microbial symbiosis at hydrothermal vents.

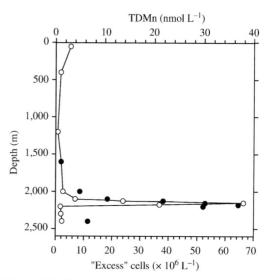

Figure 16 Plot showing co-registered enrichments of TDMn concentrations and "excess" cell counts (after subtraction of a "background" N. Atlantic cell-count profile) in the Rainbow nonbuoyant hydrothermal plume near 36°14' N, MAR (source German *et al.*, 2003).

2.5.5 Impact of Hydrothermal Plumes Upon Ocean Geochemical Cycles

Hydrothermal plumes form by the entrainment of large volumes of ambient ocean water into rising buoyant plumes driven by the release of vent fluids at the seafloor. The effect of this dilution is such that the entire volume of the oceans is cycled through buoyant and nonbuoyant hydrothermal plumes, on average, every few thousand years—a timescale comparable to that for mixing of the entire deep ocean.

Close to the vent source, rapid precipitation of a range of polymetallic sulfide, sulfate, and oxide phases leads to a strong modification of the gross dissolved metal flux from the seafloor. Independent estimates by Mottl and McConachy (1990) and German *et al.* (1991b), from buoyant and nonbuoyant plume investigations in the Pacific and Atlantic Oceans respectively, concluded that perhaps only manganese and calcium achieved a significant dissolved metal flux from hydrothermal venting to the oceans. (For comparison, the following 27 elements exhibited no evidence for a significant dissolved hydrothermal flux: iron, beryllium, aluminum, magnesium, chromium, vanadium, cobalt, copper, zinc, arsenic, yttrium, molybdenum, silver, cadmium, tin, lanthanum, cerium, praseodymium, neodymium, samarium, europium, gadolinium, terbium, holmium, erbium, lead, and uranium.) In addition to rapid co-precipitation and deposition of vent-sourced metals to the local sediments (e.g., Dymond and Roth, 1988; German *et al.*, 2002), hydrothermal plumes are also the site of additional uptake of other dissolved tracers from the water column—most notably large dissolved "oxyanion" species (phosphorus, vanadium, chromium, arsenic, uranium) and particle-reactive species (beryllium, yttrium, REE, thorium, and protactinium). For many of these tracers, hydrothermal plume removal fluxes are as great as, or at least significant when compared to, riverine input fluxes to the oceans (Table 3). What remains less certain, however, is the extent to which these particle-associated species are subsequently retained within the hydrothermal sediment record.

2.6 HYDROTHERMAL SEDIMENTS

Deep-sea metalliferous sediments were first documented in the reports of the Challenger Expedition, 1873–1876 (Murray and Renard, 1891), but it took almost a century to recognize that such metalliferous material was concentrated along all the world's ridge crests (Figure 2). Boström *et al.* (1969) attributed these distributions to some form of "volcanic emanations"; the accuracy of those predictions was confirmed some ten years later with the discovery of ridge-crest venting (Corliss *et al.*, 1978; Spiess *et al.*, 1980) although metalliferous sediments had already been found in association with hot brine pools in the Red Sea (Degens and Ross, 1969). Following the discovery of active venting, it has become recognized that hydrothermal sediments can be classified into two types: those derived from plume fall-out (including the majority of metalliferous sediments reported from ridge flanks) and those derived from mass wasting close to active vent sites (see, e.g., Mills and Elderfield, 1995).

2.6.1 Near-vent Sediments

Near-vent metalliferous sediments form from the physical degradation of hydrothermal deposits themselves, a process which begins as soon as deposition has occurred. Whilst there is ample evidence for extensive mass wasting in ancient volcanogenic sulfide deposits, only limited attention has been paid, to date, to this aspect of modern hydrothermal systems. Indeed, much of our understanding comes from a series of detailed investigations from a single site, the TAG hydrothermal field at 26° N on the MAR. It has been shown, for example, that at least some of the weathered sulfide debris at TAG is produced

Table 3 Global removal fluxes from the deep ocean into hydrothermal plumes (after Elderfield and Schultz, 1996).

Element	Hydrothermal input (mol yr^{-1})	Plume removal (mol yr^{-1})	Riverine input (mol yr^{-1})
Cr	0	4.8×10^7	6.3×10^8
V	0	4.3×10^8	5.9×10^8
As	$(0.01–3.6) \times 10^7$	1.8×10^8	8.8×10^8
P	-4.5×10^7	1.1×10^{10}	3.3×10^{10}
U	-3.8×10^5	4.3×10^7	3.6×10^7
Mo	0	1.9×10^6	2.0×10^8
Be		1.7×10^6	3.7×10^7
Ce	9.1×10^5	1.0×10^6	1.9×10^7
Nd	5.3×10^5	6.3×10^6	9.2×10^6
Lu	2.1×10^3	0.6×10^5	1.9×10^5

from collapse of the mound itself. This collapse is believed to arise from waxing and waning of hydrothermal circulation which, in turn, leads to episodic dissolution of large volumes of anhydrite within the mound (e.g., Humphris *et al.*, 1995; James and Elderfield, 1996). The mass-wasting process at TAG generates an apron of hydrothermal detritus with oxidized sulfides deposited up to 60 m out, away from the flanks of the hydrothermal mound.

Similar ponds of metalliferous sediment are observed close to other inactive sulfide structures throughout the TAG area (Rona *et al.*, 1993). Metz *et al.* (1988) characterized the metalliferous sediment in a core raised from a sediment pond close to one such deposit, ~2 km NNE of the active TAG mound. That core consisted of alternating dark red-brown layers of weathered sulfide debris and lighter calcareous ooze. Traces of pyrite, chalcopyrite, and sphalerite, together with elevated transition-metal concentrations were found in the dark red-brown layers, confirming the presence of clastic sulfide debris. Subsequently, German *et al.* (1993) investigated a short-core raised from the outer limit of the apron of "stained" hydrothermal sediment surrounding the TAG mound itself. That core penetrated through 7 cm of metal-rich degraded sulfide material into pelagic carbonate ooze. The upper "mass-wasting" layer was characterized by high transition-metal contents, just as observed by Metz *et al.* (1988), but also exhibited REE patterns similar to vent fluids (see earlier) and high uranium contents attributed to uptake from seawater during oxidation of sulfides (German *et al.*, 1993). Lead isotopic compositions in sulfidic sediments from both sites were indistinguishable from local MORB, vent fluids and chimneys (German *et al.*, 1993, Mills *et al.*, 1993). By contrast, the underlying/intercalated carbonate/ calcareous ooze layers from each core exhibited lead isotope, REE, and U–Th compositions which much more closely reflected input of Fe-oxyhydroxide particulate material from nonbuoyant hydrothermal plumes (see below).

2.6.2 Deposition from Hydrothermal Plumes

Speer *et al.* (2003) have modeled deep-water circulation above the global ridge crest and concluded that this circulation is dominated by topographically steered flow along-axis. Escape of dispersed material into adjacent deep basins is predicted to be minimal, except in key areas where pronounced across-axis circulation occurs. If this model proves to be generally valid, the majority of hydrothermal material released into nonbuoyant hydrothermal plumes should not be dispersed more than ~100 km off-axis. Instead, most hydrothermal material should settle out in a near-continuous rain of metalliferous sediment along the length of the global ridge crest. Significant off-axis dispersion is only predicted (i) close to the equator (~5° N to 5° S), (ii) where the ridge intersects boundary currents or regions of deepwater formation, and (iii) in the Antarctic Circumpolar Current (Speer *et al.*, 2003). One good example of strong across-axis flow is at the equatorial MAR where pronounced eastward flow of both Antarctic Bottom Water and lower North Atlantic Deep Water has been reported, passing through the Romanche and Chain Fracture Zones (Mercier and Speer, 1998).

Another location where the large-scale off-axis dispersion modeled by Speer *et al.* (2003) has already been well documented is on the southern EPR (Figure 14). There, metalliferous sediment enrichments underlie the pronounced dissolved ^{3}He plume which extends westward across the southern Pacific Ocean at ~15° S (cf. Boström *et al.*, 1969; Lupton and Craig, 1981). Much of our understanding of ridge-flank metalliferous sediments comes from a large-scale study carried out at this latitude (19° S) by Leg 92 of the Deep Sea Drilling Project (DSDP). That work targeted sediments underlying the westward-trending plume to investigate both temporal and spatial variability in hydrothermal output at this latitude (Lyle *et al.*, 1987). A series of holes were drilled extending westward from the ridge axis into 5–28 Ma crust; the recovered cores comprised mixtures of biogenic carbonate and Fe–Mn oxyhydroxides. One important result of that work was the demonstration, based on lead isotopic analyses, that even the most distal sediments, collected at a range of >1,000 km from the ridge axis, contained 20–30% mantle-derived lead (Barrett *et al.*, 1987). In contrast, analysis of the same samples indicated that REE distributions in the metalliferous sediments were dominated by a seawater source (Ruhlin and Owen, 1986). This is entirely consistent with what has subsequently been demonstrated for hydrothermal plumes (see Section 2.5, above) with the caveat that REE/Fe ratios in DSDP Leg 92 sediments are everywhere higher than the highest REE/Fe ratios yet measured in modern nonbuoyant hydrothermal plume particles (German *et al.*, 1990; Sherrell *et al.*, 1999).

2.6.3 Hydrothermal Sediments in Paleoceanography

Phosphorus and vanadium, which are typically present in seawater as dissolved oxyanion species, have been shown to exhibit systematic plume-particle P:Fe and V:Fe variations which differ from one ocean basin to another (e.g., Trefry and Metz, 1989; Feely *et al.*, 1990). This has led to the hypothesis (Feely *et al.*, 1998) that (i) plume P:Fe and V:Fe ratios may be directly linked to local deep-ocean dissolved phosphate concentrations and (ii) ridge-flank metalliferous sediments, preserved under oxic diagenesis, might faithfully record temporal variations in plume-particle P:Fe and/or V:Fe ratios. Encouragingly, a study of slowly accumulating ($\sim$0.5 cm kyr^{-1}) sediments from the west flank of the JdFR has revealed that V:Fe ratios in the hydrothermal component from that core appear faithfully to record local plume-particle V:Fe ratios for the past $\sim$200 kyr (German *et al.*, 1997; Feely *et al.*, 1998). More recently, however, Schaller *et al.* (2000) have shown that while cores from the flanks of the southern EPR (10° S) also exhibit V:Fe ratios that mimic modern plume-values, in sediments dating back to 60–70 kyr, the complementary P:Fe and As:Fe ratios in these samples are quite different from contemporaneous nonbuoyant plume values. These variations have been attributed to differences in the intensity of hydrothermal iron oxide formation between different hydrothermal plumes and/or significant uptake/release of phosphorus and arsenic, following deposition (Schaller *et al.*, 2000).

Unlike vanadium, REE/Fe ratios recorded in even the most recent metalliferous sediments are much higher than those in suspended hydrothermal plume particles (German *et al.*, 1990, 1997; Sherrell *et al.*, 1999). Further, hydrothermal sediments' REE/Fe ratios increase systematically with distance away from the paleo-ridge crest (Ruhlin and Owen, 1986; Olivarez and Owen, 1989). This indicates that the REE may continue to be taken up from seawater, at and near the sediment–water interface, long after the particles settle from the plume to the seabed. Because increased uptake of dissolved REE from seawater should also be accompanied by continuing fractionation across the REE series (e.g., Rudnicki and Elderfield, 1993) reconstruction of deep-water REE patterns from preserved metalliferous sediment records remain problematic. Much more tractable, however, is the exploitation of these same sample types for isotopic reconstructions.

Because seawater uptake dominates the REE content of metalliferous sediment, neodymium isotopic analysis of metalliferous carbonate can provide a reliable proxy for contemporaneous seawater, away from input of near-vent sulfide detritus (Mills *et al.*, 1993). Osmium also exhibits a similar behavior and seawater dominates the isotopic composition of metalliferous sediments even close to active vent sites (Ravizza *et al.*, 1996). Consequently, analysis of preserved metalliferous carbonate sediments has proven extremely useful in determining the past osmium isotopic composition of the oceans, both from modern marine sediments (e.g., Ravizza, 1993; Peucker-Ehrenbrink *et al.*, 1995) and those preserved in ophiolites (e.g., Ravizza *et al.*, 2001). Only in sediments close to an ultramafic-hosted hydrothermal system, have perturbations from a purely seawater osmium isotopic composition been observed (Cave *et al.*, 2003, in press).

2.6.4 Hydrothermal Sediments and Boundary Scavenging

It has been known for sometime that sediments underlying areas of high particle settling flux exhibit pronounced fractionations between particle-reactive tracers. Both ^{231}Pa and ^{10}Be, for example, exhibit pronounced enrichments relative to ^{230}Th, in ocean margin environments, when compared to sediments underlying mid-ocean gyres (e.g., Bacon, 1988; Anderson *et al.*, 1990; Lao *et al.*, 1992). Comparable fractionations between these three radiotracers (^{230}Th, ^{231}Pa, and ^{10}Be) have also been identified in sediments underlying hydrothermal plumes (German *et al.*, 1993; Bourlès *et al.*, 1994; German *et al.*, 1997). For example, a metalliferous sediment core raised from the flanks of the JdFR exhibited characteristic hydrothermal lead-isotopic and REE/Fe compositions, together with high ^{10}Be/^{230}Th ratios indicative of net focusing relative to the open ocean (German *et al.*, 1997). The degree of fractionation observed was high, even compared to high-productivity ocean-margin environments (Anderson *et al.*, 1990; Lao *et al.*, 1992), presumably due to intense scavenging onto hydrothermal Fe-oxyhydroxides. Of course, the observation that REE and thorium are scavenged into ridge-crest metalliferous sediments is not new; sediments from the EPR near 17° S, with mantle lead, excess ^{230}Th and seawater-derived REE compositions were reported more than thirty years ago by Bender *et al.* (1971). More recently, however, examination of ridge crest sediments and near-vent sediment-traps has revealed that the settling flux of scavenged tracers (e.g., ^{230}Th) from hydrothermal plumes is higher than can be sustained by *in situ* production in the overlying water column alone (German *et al.*, 2002). Thus, uptake onto Fe-oxyhydroxide material in hydrothermal plumes and sediments may act as a special form of deep-ocean "boundary scavenging" leading to the net focusing and deposition of these dissolved tracers in ridge-flank metalliferous sediments.

2.7 CONCLUSION

The field of deep-sea hydrothermal research is young; it was only in the mid-1970s when it was first discovered, anywhere in the oceans. To synthesize current understanding of its impact on marine geochemistry, therefore, could be considered akin to explaining the significance of rivers to ocean chemistry in the early part of the last century. This chapter has aimed to provide a brief synopsis of the current state of the art, but much more surely remains to be learnt. There are three key questions that will continue to focus efforts within this vigorous research field:

(i) What are the geological processes that control submarine hydrothermal venting? How might these have varied during the course of Earth's history?

(ii) To what extent do geochemical and biological processes interact to regulate hydrothermal fluxes to the ocean? How might past-ocean processes have differed from the present-day ones?

(iii) What are the timescales relevant to hydrothermal processes? Whilst some long-term proxies do exist (sufide deposits, metalliferous sediments) for active processes, we do not have any time-series records longer than 25 years!

REFERENCES

Alt J. C. (1995) Subseafloor processes in mid-ocean ridge hydrothermal systems. *Geophys. Monogr. (AGU)* **91**, 85–114.

Alt J. C. and Bach W. (2003) Alteration of oceanic crust: subsurface rock–water interactions. In *Energy and Mass Transfer in Marine Hydrothermal Systems* (eds. P. Halbach, V. Tunnicliffe, and J. Hein). DUP, Berlin, pp. 7–28.

Anderson R. N. and Hobart M. A. (1976) The relationship between heat flow, sediment thickness and age in the eastern Pacific. *J. Geophys. Res.* **81**, 2968–2989.

Anderson R. N. and Silbeck J. N. (1981) Oceanic heat flow. In *The Oceanic Lithosphere, The Seas* (ed. C. Emiliani), Wiley, New York, vol. 7, pp. 489–523.

Anderson R. F., Lao Y., Broecker W. S., Trumbore S. E., Hofmann H. J., and Wolfli W. (1990) Boundary scavenging in the Pacific Ocean: a comparison of [10]Be and [231]Pa. *Earth Planet. Sci. Lett.* **96**, 287–304.

Bach W., Banerjee N. R., Dick H. J. B., and Baker E. T. (2002) Discovery of ancient and active hydrothermal systems along the ultra-slow spreading southwest Indian ridge 10 degrees–16 degrees E. *Geophys. Geochem. Geosys.* **3**, paper 10.1029/2001GC000279.

Bacon M. P. (1988) Tracers of chemical scavenging in the ocean: boundary effects and largescale chemical fractionation. *Phil. Trans. Roy. Soc. London Ser. A* **325**, 147–160.

Baker E. T. (1998) Patterns of event and chronic hydrothermal venting following a magmatic intrusion: new perspectives from the 1996 Gorda ridge eruption. *Deep-Sea Res. II* **45**, 2599–2618.

Baker E. T., Massoth G. J., Feely R. A., Embley R. W., Thomson R. E., and Burd B. J. (1995) Hydrothermal event plumes from the CoAxial seafloor eruption site, Juan de Fuca ridge. *Geophys. Res. Lett.* **22**, 147–150.

Baker E. T., Chen Y. J., and Phipps Morgan J. (1996) The relationship between near-axis hydrothermal cooling and the spreading rate of mid-ocean ridges. *Earth Planet. Sci. Lett.* **142**, 137–145.

Baker E. T., Massoth G. J., Feely R. A., Cannon G. A., and Thomson R. E. (1998) The rise and fall of the CoAxial hydrothermal site, 1993–1996. *J. Geophys. Res.* **103**, 9791–9806.

Barnes I., Rapp J. T., O'Neill J. R., Sheppard R. A., and Gude A. J. (1972) Metamorphic assemblages and the flow of metamorphic fluid in four instances of serpentenization. *Contrib. Mineral. Petrol.* **35**, 263–276.

Barrett T. J., Taylor P. N., and Lugowski J. (1987) Metalliferous sediments from DSDP leg 92, the East Pacific rise transect. *Geochim. Cosmochim. Acta* **51**, 2241–2253.

Bender M., Broecker W., Gornitz V., Middel U., Kay R., Sun S., and Biscaye P. (1971) Geochemistry of three cores from the East Pacific Rise. *Earth Planet. Sci. Lett.* **12**, 425–433.

Berndt M. E. and Seyfried W. E. (1990) Boron, bromine, and other trace elements as clues to the fate of chlorine in mid-ocean ridge vent fluids. *Geochim. Cosmochim. Acta* **54**, 2235–2245.

Berndt M. E. and Seyfried W. E. (1997) Calibration of Br/Cl fractionation during sub-critical phase separation of seawater: possible halite at 9 to 10°N East Pacific Rise. *Geochim. Cosmochim. Acta* **61**, 2849–2854.

Berndt M. E., Allen D. E., and Seyfried W. E. (1996) Reduction of CO_2 during serpentinization of olivine at 300 degrees C and 500 bar. *Geology* **24**, 351–354.

Binns R. A., Scott S. D., Bogdanov Y. A., Lisitzin A. P., Gordeev V. V., Gurvcich E. G., Finlayson E. J., Boyd T., Dotter L. E., Wheller G. E., and Muravyev K. G. (1993) Hydrothermal oxide and gold-rich sulfate deposits of Franklin seamount, Western Woodlark Basin, Papua New Guinea. *Econ. Geol.* **88**, 2122–2153.

Bischoff J. L. (1991) Densities of liquids and vapors in boiling $NaCl–H_2O$ solutions: a PVTX summary from 300° to 500°C. *Am. J. Sci.* **291**, 309–338.

Bischoff J. L. and Pitzer K. S. (1989) Liquid-vapor relations for the system $NaCl–H_2O$: summary of the P–T–x surface from 300deg to 500deg C. *Am. J. Sci.* **289**, 217–248.

Bischoff J. L. and Rosenbauer R. J. (1985) An empirical equation of state for hydrothermal seawater (3.2% NaCl). *Am. J. Sci.* **285**, 725–763.

Bischoff J. L. and Rosenbauer R. J. (1987) Phase separation in seafloor systems—an experimental study of the effects on metal transport. *Am. J. Sci.* **287**, 953–978.

Blackman D. K., Cann J. R., Janssen B., and Smith D. K. (1998) Origin of extensional core complexes: evidence from the Mid-Atlantic Ridge at Atlantis fracture zone. *J. Geophys. Res.* **103**, 21315–21333.

Boström K., Peterson M. N. A., Joensuu O., and Fisher D. E. (1969) Aluminium-poor ferromanganoan sediments on active ocean ridges. *J. Geophys. Res.* **74**, 3261–3270.

Bourlès D. L., Brown E. T., German C. R., Measures C. I., Edmond J. M., Raisbeck G. M., and Yiou F. (1994) Hydrothermal influence on oceanic beryllium. *Earth Planet. Sci. Lett.* **122**, 143–157.

Bray A. M. and Von Damm K. L. (2003a) The role of phase separation and water–rock reactions in controlling the boron content of mid-ocean ridge hydrothermal vent fluids. *Geochim. Cosmochim. Acta* (in revision).

Bray A. M. and Von Damm K. L. (2003b) Controls on the alkali metal composition of mid-ocean ridge hydrothermal fluids: constraints from the 9–10°N East Pacific Rise time series. *Geochim. Cosmochim. Acta* (in revision).

Broecker W. S. and Peng T. H. (1982) *Tracers in the Sea*. Eldigio Press, Columbia University, New York, 690 pp.

Burd B. J., Thomson R. E., and Jamieson G. S. (1992) Composition of a deep scattering layer overlying a mid-ocean ridge hydrothermal plume. *Mar. Biol.* **113**, 517–526.

Butterfield D. A., McDuff R. E., Mottl M. J., Lilley M. D., Lupton J. E., and Massoth G. J. (1994) Gradients in the composition of hydrothermal fluids from the endeavour segment vent field: phase separation and brine loss. *J. Geophys. Res.* **99**, 9561–9583.

Butterfield D. A., Jonasson I. R., Massoth G. J., Feely R. A., Roe K. K., Embley R. E., Holden J. F., McDuff R. E., Lilley M. D., and Delaney J. R. (1997) Seafloor eruptions and evolution of hydrothermal fluid chemistry. *Phil. Trans. Roy. Soc. London* **A355**, 369–386.

Campbell A. C., Palmer M. R., Klinkhammer G. P., Bowers T. S., Edmond J. M., Lawrence J. R., Casey J. F., Thomson G., Humphris S., Rona P., and Karson J. A. (1988)

Chemistry of hot springs on the Mid-Atlantic Ridge. *Nature* **335**, 514–519.

Campbell A. C., German C. R., Palmer M. R., Gamo T., and Edmond J. M. (1994) Chemistry of hydrothermal fluids from the Escanaba Trough, Gorda Ridge. In *Geologic, Hydrothermal and Biologic Studies at Escanaba Trough, Gorda Ridge.* US Geol. Surv. Bull. 2022 (eds. J. L. Morton, R. A. Zierenberg, and C. A. Reiss). Offshore, Northern California, pp. 201–221.

Cave R. R., Ravizza G. E., German C. R., Thomson J., and Nesbitt R. W. (2003) Deposition of osmium and other platinum-group elements beneath the ultramafic-hosted rainbow hydrothermal plume. *Earth Planet. Sci. Lett.* **210**, 65–79.

Charlou J.-L., Fouquet Y., Donval J. P., Auzende J. M., Jean Baptiste P., Stievenard M., and Michel S. (1996) Mineral and gas chemistry of hydrothermal fluids on an ultrafast spreading ridge: East Pacific Rise, 17° to 19°S (NAUDUR cruise, 1993) phase separation processes controlled by volcanic and tectonic activity. *J. Geophys. Res.* **101**, 15899–15919.

Charlou J. L., Fouquet Y., Bougault H., Donval J. P., Etoubleau J., Jean-Baptiste P., Dapoigny A., Appriou P., and Rona P. A. (1998) Intense CH_4 plumes generated by serpentinization of ultramafic rocks at the intersection of the 15 degrees 20' N. *Geochim. Cosmochim. Acta* **62**, 2323–2333.

Chin C. S., Coale K. H., Elrod V. A., Johnson K. S., Johnson G. J., and Baker E. T. (1994) *In situ* observations of dissolved iron and manganese in hydrothermal vent plumes, Juan de Fuca Ridge. *J. Geophys. Res.* **99**, 4969–4984.

Clarke W. B., Beg M. A., and Craig H. (1969) Excess ^{3}He in the sea: evidence for terrestrial primordial helium. *Earth Planet. Sci. Lett.* **6**, 213–220.

Connelly D. P., German C. R., Egorov A., Pimenov N. V., and Dohsik H. (2002) Hydrothermal plumes overlying the ultra-slow spreading Knipovich Ridge, 72–78°N. *EOS Trans. AGU (abstr.)* **83**, T11A-1229.

Corliss J. B., Lyle M., and Dymond J. (1978) The chemistry of hydrothermal mounds near the Galapagos rift. *Earth Planet. Sci. Lett.* **40**, 12–24.

Cowen J. P. and German C. R. (2003) Biogeochemical cycling in hydrothermal plumes. In *Energy and Mass Transfer in Marine Hydrothermal Systems* (eds. P. Halbach, V. Tunnicliffe, and J. Hein). DUP, Berlin, pp. 303–316.

Cowen J. P., Bertram M. A., Wakeham S. G., Thomson R. E., Lavelle J. W., Baker E. T., and Feely R. A. (2001) Ascending and descending particle flux from hydrothermal plumes at endeavour segment, Juan de Fuca Ridge. *Deep-Sea Res. I* **48**, 1093–1120.

Cruse A. M. and Seewald J. S. (2001) Metal mobility in sediment-covered ridge-crest hydrothermal systems: experimental and theoretical constraints. *Geochim. Cosmochim. Acta* **65**, 3233–3247.

de Angelis M. A., Lilley M. D., Olson E. J., and Baross J. A. (1993) Methane oxidation in deep-sea hydrothermal plumes of the endeavour segment of the Juan de Fuca Ridge. *Deep-Sea Res.* **40**, 1169–1186.

Degens E. T. and Ross D. A. (eds.) (1969) *Hot Brines and Heavy Metal Deposits in the Red Sea.* Springer, New York.

De Mets C., Gordon R. G., Argus D. F., and Stein S. (1994) Effect of recent revisions to the geomagnetic reversal time-scale on estimates of current plate motions. *Geophys. Res. Lett.* **21**, 2191–2194.

Douville E., Charlou J. L., Oelkers E. H., Bienvenu P., Colon C. F. J., Donval J. P., Fouquet Y., Prieur D., and Appriou P. (2002) The rainbow vent fluids (36 degrees 14' N, MAR): the influence of ultramafic rocks and phase separation on trace metal content in Mid-Atlantic Ridge hydrothermal fluids. *Chem. Geol.* **184**, 37–48.

Dymond J. and Roth S. (1988) Plume dispersed hydrothermal particles—a time-series record of settling flux from the

endeavour ridge using moored sensors. *Geochim. Cosmochim. Acta* **52**, 2525–2536.

Edmond J. M., Measures C. I., McDuff R. E., Chan L. H., Collier R., Grant B., Gordon L. I., and Corliss J. B. (1979) Ridge crest hydrothermal activity and the balances of the major and minor elements in the ocean: the Galapagos data. *Earth Planet. Sci. Lett.* **46**, 1–18.

Edmond J. M., Von Damm K. L., McDuff R. E., and Measures C. I. (1982) Chemistry of hot springs on the East Pacific Rise and their effluent dispersal. *Nature* **297**, 187–191.

Edmond J. M., Campbell A. C., Palmer M. R., Klinkhammer G. P., German C. R., Edmonds H. N., Elderfield H., Thompson G., and Rona P. (1995) Time-series studies of vent-fluids from the TAG and MARK sites (1986, 1990) Mid-Atlantic Ridge: a new solution chemistry model and a mechanism for Cu/Zn zonation in massive sulphide ore-bodies. In *Hydrothermal Vents and Processes* Geol. Soc. Spec. Publ. 87 (eds. L. M. Parson, C. L. Walker, and D. R. Dixon), The Geological Society Publishing House, Bath, UK, pp. 77–86.

Edmonds H. N., German C. R., Green D. R. H., Huh Y., Gamo T., and Edmond J. M. (1996) Continuation of the hydrothermal fluid chemistry time series at TAG and the effects of ODP drilling. *Geophys. Res. Lett.* **23**, 3487–3489.

Edmonds H. N., Michael P. J., Baker E. T., Connelly D. P., Snow J. E., Langmuir C. H., Dick H. J. B., German C. R., and Graham D. W. (2003) Discovery of abundant hydrothermal venting on the ultraslow-spreading Gakkel Ridge in the Arctic Ocean. *Nature* **421**, 252–256.

Elderfield H. and Schultz A. (1996) Mid-ocean ridge hydrothermal fluxes and the chemical composition of the ocean. *Ann. Rev. Earth Planet. Sci.* **24**, 191–224.

Eldridge C. S., Barton P. B., and Ohmoto H. (1983) Mineral textures and their bearing on formation of the Kuroko ore-bodies. *Econ. Monogr.* **5**, 241–281.

Evans W. C., White L. D., and Rapp J. B. (1988) Geochemistry of some gases in hydrothermal fluids from the southern Juan de Fuca Ridge. *J. Geophys. Res.* **93**, 15305–15313.

Feely R. A., Massoth G. J., Baker E. T., Cowen J. P., Lamb M. F., and Krogslund K. A. (1990) The effect of hydrothermal processes on mid-water phosphorous distributions in the northeast Pacific. *Earth Planet. Sci. Lett.* **96**, 305–318.

Feely R. A., Trefry J. H., Massoth G. J., and Metz S. (1991) A comparison of the scavenging of phosphorous and arsenic from seawater by hydrothermal iron hydroxides in the Atlantic and Pacific oceans. *Deep-Sea Res.* **38**, 617–623, 1991.

Feely R. A., Massoth G. J., Trefry J. H., Baker E. T., Paulson A. J., and Lebon G. T. (1994) Composition and sedimentation of hydrothermal plume particles from North Cleft Segment, Juan de Fuca Ridge. *J. Geophys. Res.* **99**, 4985–5006.

Feely R. A., Baker E. T., Marumo K., Urabe T., Ishibashi J., Gendron J., Lebon G. T., and Okamura K. (1996) Hydrothermal plume particles and dissolved phosphate over the superfast-spreading southern East Pacific Rise. *Geochim. Cosmochim. Acta* **60**, 2297–2323.

Feely R. A., Trefry J. H., Lebon G. T., and German C. R. (1998) The relationship between P/Fe and V/Fe in hydrothermal precipitates and dissolved phosphate in seawater. *Geophys. Res. Lett.* **25**, 2253–2256.

Field M. P. and Sherrell R. M. (2000) Dissolved and particulate Fe in a hydrothermal plume at 9°45' N, East Pacific Rise: slow Fe(II) oxidation kinetics in Pacific plumes. *Geochim. Cosmochim. Acta* **64**, 619–628.

Fouquet Y., Von Stackelberg U., Charlou J. L., Donval J. P., Erzinger J., Foucher J. P., Herzig P., Mühe R., Soakai S., Wiedicke M., and Whitechurch H. (1991) Hydrothermal activity and metallogenesis in the Lau back-arc basin. *Nature* **349**, 778–781.

Fouquet Y., Henry K., Bayon G., Cambon P., Barriga F., Costa I., Ondreas H., Parson L., Ribeiro A., and Relvas G. (2003) The Rainbow hydrothermal field: geological setting, mineralogical and chemical composition of sulfide deposits (MAR 36°14' N). *Earth Planet. Sci. Lett.* (submitted).

Früh-Green G. L., Weissert H., and Bernoulli D. (1990) A multiple fluid history recorded in alpine ophiolites. *J. Geol. Soc. London* **147**, 959–970.

Fuhrmann J. A., McCallum K., and Davis A. A. (1993) Phylogenetic diversity of subsurface marine microbial communities from the Atlantic and Pacific oceans. *Appl. Environ. Microbiol.* **59**, 1294–1302.

Gamo T., Okamura K., Charlou J. L., Urabe T., Auzende J. M., Ishibashi J., Shitashima K., and Chiba H. (1997) Acidic and sulfate-rich hydrothermal fluids from the Manus back-arc basin, Papua New Guinea. *Geology* **25**, 139–142.

German C. R. and Sparks R. S. J. (1993) Particle recycling in the TAG hydrothermal plume. *Earth Planet. Sci. Lett.* **116**, 129–134.

German C. R., Klinkhammer G. P., Edmond J. M., Mitra A., and Elderfield H. (1990) Hydrothermal scavenging of rare earth elements in the ocean. *Nature* **316**, 516–518.

German C. R., Fleer A. P., Bacon M. P., and Edmond J. M. (1991a) Hydrothermal scavenging at the Mid-Atlantic Ridge: radionuclide distributions. *Earth Planet. Sci. Lett.* **105**, 170–181.

German C. R., Campbell A. C., and Edmond J. M. (1991b) Hydrothermal scavenging at the Mid-Atlantic Ridge: modification of trace element dissolved fluxes. *Earth Planet. Sci. Lett.* **107**, 101–114.

German C. R., Higgs N. C., Thomson J., Mills R., Elderfield H., Blusztajn J., Fleer A. P., and Bacon M. P. (1993) A geochemical study of metalliferous sediment from the TAG hydrothermal mound, 26°08' N, Mid-Atlantic Ridge. *J. Geophys. Res.* **98**, 9683–9692.

German C. R., Briem J., Chin C., Danielsen M., Holland S., James R., Jónsdottir A., Ludford E., Moser C., Ólafsson J., Palmer M. R., and Rudnicki M. D. (1994) Hydrothermal activity on the Reykjanes ridge: the steinahóll vent-field at 63°06' N. *Earth Planet. Sci. Lett.* **121**, 647–654.

German C. R., Bourlès D. L., Brown E. T., Hergt J., Colley S., Higgs N. C., Ludford E. M., Nelsen T. A., Feely R. A., Raisbeck G., and Yiou F. (1997) Hydrothermal scavenging on the Juan de Fuca Ridge: Th-230(xs), Be-10 and REE in ridge-flank sediments. *Geochim. Cosmochim. Acta* **61**, 4067–4078.

German C. R., Baker E. T., Mevel C. A., Tamaki K., and the FUJI scientific team (1998a) Hydrothermal activity along the south-west Indian ridge. *Nature* **395**, 490–493.

German C. R., Richards K. J., Rudnicki M. D., Lam M. M., Charlou J. L., and FLAME scientific party (1998b) Topographic control of a dispersing hydrothermal plume. *Earth Planet. Sci. Lett.* **156**, 267–273.

German C. R., Livermore R. A., Baker E. T., Bruguier N. I., Connelly D. P., Cunningham A. P., Morris P., Rouse I. P., Statham P. J., and Tyler P. A. (2000) Hydrothermal plumes above the East Scotia Ridge: an isolated high-latitude back-arc spreading centre. *Earth Planet. Sci. Lett.* **184**, 241–250.

German C. R., Colley S., Palmer M. R., Khripounoff A., and Klinkhammer G. P. (2002) Hydrothermal sediment trap fluxes: 13°N, East Pacific Rise. *Deep Sea Res. I* **49**, 1921–1940.

German C. R., Thursherr A. M., Radford-Kröery J., Charlou J. L., Jean-Baptiste P., Edmonds H. N., Patching J. W., and the FLAME 1 & II science teams (2003) Hydrothermal fluxes from the Rainbow vent-site, Mid-Atlantic Ridge: new constraints on global ocean vent-fluxes. *Nature* (submitted).

Ginster U., Mottl M. J., and VonHerzen R. P. (1994) Heat-flux from black smokers on the endeavor and Cleft segments, Juan de Fuca Ridge. *J. Geophys. Res.* **99**, 4937–4950.

Gracia E., Bideau D., Hekinian R., and Lagabrielle Y. (1999) Detailed geological mapping of two contrasting second-order segments of the Mid-Atlantic Ridge between oceanographer and Hayes fracture zones (33 degrees 30' N-35 degrees N). *J. Geophys. Res.* **104**, 22903–22921.

Gracia E., Charlou J. L., Radford-Knoery J., and Parson L. M. (2000) Non-transform offsets along the Mid-Atlantic Ridge south of the Azores (38 degrees N-34 degrees N): ultramafic exposures and hosting of hydrothermal vents. *Earth Planet. Sci. Lett.* **177**, 89–103.

Hannington M. D., Jonasson I., Herzig P., and Petersen S. (1995) Physical and chemical processes of seafloor mineralisation at mid-ocean ridges. *Geophys. Monogr. (AGU)* **91**, 115–157.

Hannington M. D., Galley A. G., Herzig P. M., and Petersen S. (1998) Comparison of the TAG mound and stock work complex with Cyprus-type massive sulfide deposits. In *Proceedings of the Ocean Drilling Program, Scientific Results.* **158**, pp. 389–415.

Haymon R. M. (1983) Growth history of hydrothermal black smoker chimneys. *Nature* **301**, 695–698.

Haymon R. M. (1996) The response of ridge-crest hydrothermal systems to segmented, episodic magma supply. In *Tectonic, Magmatic, Hydrothermal, and Biological Segmentation of Mid-ocean Ridges*, Geol. Soc. Spec. Publ. 118 (eds. C. J. McLeod, P. A. Tyler, and C. L. Walker). Geological Society Publishing House, Bath, UK, pp. 157–168.

Haymon R., Fornari D., Von Damm K., Lilley M., Perfit M., Edmond J., Shanks W., Lutz R. A., Grebmeier J. M., Carbotte S., Wright D., McLaughlin E., Smith M., Beedle N., and Olson E. (1993) Volcanic eruption of the mid-ocean ridge along the EPR crest at 9°45–52' N: I. Direct submersible observations of seafloor phenomena associated with an eruption event in April 1991. *Earth Planet. Sci. Lett.* **119**, 85–101.

Hein J. R., Koski R. A., Embley R. W., Reid J., and Chang S.-W. (1999) Diffuse-flow hydrothermal field in an oceanic fracture zone setting, northeast Pacific: deposit composition. *Explor. Mining Geol.* **8**, 299–322.

Hein J. R., Baker E. T., Cowen J. P., German C. R., Holzbecher E., Koski R. A., Mottl M. J., Pimenov N. V., Scott S. D., and Thurnherr A. M. (2003) How important are material and chemical fluxes from hydrothermal circulation to the ocean? In *Energy and Mass Transfer in Marine Hydrothermal Systems* (eds. P. Halbach, V. Tunnicliffe, and J. Hein). DUP, Berlin, pp. 337–355.

Helfrich K. R. and Speer K. G. (1995) Ocean hydrothermal circulation: mesoscale and basin-scale flow. *Geophys. Monogr. (AGU)* **91**, 347–356.

Herring P. J. and Dixon D. R. (1998) Extensive deep-sea dispersal of postlarval shrimp from a hydrothermal vent. *Deep-Sea Res.* **45**, 2105–2118.

Herzig P. M. and Hannington M. D. (2000) Input from the deep: hot vents and cold seeps. In *Marine Geochemistry* (eds. H. D. Schultz and M. Zabel). Springer, Heidelberg, pp. 397–416.

Holland M. E. and Baross J. A. (2003) Limits to life in hydrothermal systems. In *Energy and Mass Transfer in Marine Hydrothermal Systems* (eds. P. Halbach, V. Tunnicliffe, and J. Hein). DUP, Berlin, pp. 235–248.

Holm N. G. and Charlou J. L. (2001) Initial indications of abiotic formation of hydrocarbons in the rainbow ultramafic hydrothermal system, Mid-Atlantic Ridge. *Earth Planet. Sci. Lett.* **191**, 1–8.

Humphris S. E., Herzig P. M., Miller D. J., Alt J. C., Becker K., Brown D., Brügmann G., Chiba H., Fouquet Y., Gemmel J. B., Guerin G., Hannington M. D., Holm N. G., Honnorez J. J., Iturrino G. J., Knott R., Ludwig R., Nakamura K., Petersen S., Reysenbach A.-L., Rona P. A., Smith S., Sturz A. A., Tivey M. K., and Zhao X. (1995) The internal structure of an active sea-floor massive sulphide deposit. *Nature* **377**, 713–716.

Ishibashi J.-I. and Urabe T. (1995) Hydrothermal activity related to arc-backarc magmatism in the western Pacific.

In *Backarc Basins: Tectonics and Magmatism* (ed. B. Taylor). Plenum, NY, pp. 451–495.

Ishibashi J., Wakita H., Okamura K., Nakayama E., Feely R. A., Lebon G. T., Baker E. T., and Marumo K. (1997) Hydrothermal methane and manganese variation in the plume over the superfast-spreading southern East Pacific Rise. *Geochim. Cosmochim. Acta* **61**, 485–500.

James R. H. and Elderfield H. (1996) Chemistry of ore-forming fluids and mineral formation rates in an active hydrothermal sulfide deposit on the Mid-Atlantic Ridge. *Geology* **24**, 1147–1150.

Kadko D. and Moore W. (1988) Radiochemical constraints on the crustal residence time of submarine hydrothermal fluids—endeavour ridge. *Geochim. Cosmochim. Acta* **52**, 659–668.

Kadko D. C., Rosenberg N. D., Lupton J. E., Collier R. W., and Lilley M. D. (1990) Chemical reaction rates and entrainment within the endeavour ridge hydrothermal plume. *Earth Planet. Sci. Lett.* **99**, 315–335.

Kelley D. S., Karson J. A., Blackman D. K., Früh-Green G. L., Butterfield D. A., Lilley M. D., Olson E. J., Schrenk M. O., Roe K. K., Lebon G. T., Rivizzigno P., the AT3-60 shipboard party (2001) An off-axis hydrothermal vent field near the Mid-Atlantic Ridge at 30°N. *Nature* **412**, 145–149.

Kennedy B. M. (1988) Noble gases in vent water from the Juan de Fuca Ridge. *Geochim. Cosmochim. Acta* **52**, 1929–1935.

Klinkhammer G., Elderfield H., and Hudson A. (1983) Rare earth elements in seawater near hydrothermal vents. *Nature* **305**, 185–188.

Klinkhammer G. P., Elderfield H., Edmond J. M., and Mitra A. (1994) Geochemical implications of rare earth element patterns in hydrothermal fluids from mid-ocean ridges. *Geochim. Cosmochim. Acta* **58**, 5105–5113.

Klinkhammer G. P., Chin C. S., Keller R. A., Dahlmann A., Sahling H., Sarthou G., Petersen S., and Smith F. (2001) Discovery of new hydrothermal vent sites in Bransfield strait, Antarctica. *Earth Planet. Sci. Lett.* **193**, 395–407.

Koski R. A., German C. R., and Hein J. R. (2003) Fate of hydrothermal products from mid-ocean ridge hydrothermal systems: near-field to global perspectives. In *Energy and Mass Transfer in Marine Hydrothermal Systems* (eds. P. Halbach, V. Tunnicliffe, and J. Hein). DUP, Berlin, pp. 317–335.

Landing W. M. and Bruland K. W. (1987) The contrasting biogeochemistry of iron and manganese in the Pacific Ocean. *Geochim. Cosmochim. Acta* **51**, 29–43.

Langmuir C., Humphris S., Fornari D., Van Dover C., Von Damm K., Tivey M. K., Colodner D., Charlou J.-L., Desonie D., Wilson C., Fouquet Y., Klinkhammer G., and Bougault H. (1997) Description and significance of hydrothermal vents near a mantle hot spot: the lucky strike vent field at 37°N on the Mid-Atlantic Ridge. *Earth Planet. Sci. Lett.* **148**, 69–92.

Lao Y., Anderson R. F., Broecker W. S., Trumbore S. E., Hoffman H. J., and Wölfli W. (1992) Transport and burial rates of ^{10}Be and ^{231}Pa in the Pacific ocean during the Holocene period. *Earth Planet. Sci. Lett.* **113**, 173–189.

Lilley M. D., Feely R. A., and Trefry J. H. (1995) Chemical and biochemical transformation in hydrothermal plumes. *Geophys. Monogr. (AGU)* **91**, 369–391.

Lupton J. E. and Craig H. (1981) A major ^{3}He source on the East Pacific Rise. *Science* **214**, 13–18.

Lupton J. E., Delaney J. R., Johnson H. P., and Tivey M. K. (1985) Entrainment and vertical transport of deep ocean water by buoyant hydrothermal plumes. *Nature* **316**, 621–623.

Lupton J. E., Butterfield D., Lilley M., Ishibashi J., Hey D., and Evans L. (1999a) Gas chemistry of hydrothermal fluids along the East Pacific Rise, 5°S to 32°S. *EOS Trans. AGU (abstr.)* **80**, F1099.

Lupton J. E., Baker E. T., and Massoth G. J. (1999b) Helium, heat and the generation of hydrothermal event plumes at mid-ocean ridges. *Earth Planet. Sci. Lett.* **171**, 343–350.

Lyle M., Leinen M., Owen R. M., and Rea D. K. (1987) Late tertiary history of hydrothermal deposition at the East Pacific Rise, 19°S—correlation to volcano-tectonic events. *Geophys. Res. Lett.* **14**, 595–598.

Magenheim A. J., Bayhurst G., Alt J. C., and Gieskes J. M. (1992) ODP leg 137, borehole fluid chemistry in hole 504B. *Geophys. Res. Lett.* **19**, 521–524.

Maris C. R. P. and Bender M. L. (1982) Upwelling of hydrothermal solutions through ridge flank sediments shown by pore-water profiles. *Science* **216**, 623–626.

Massoth G. J., Baker E. T., Lupton J. E., Feely R. A., Butterfield D. A., VonDamm K. L., Roe K. K., and LeBon G. T. (1994) Temporal and spatial variability of hydrothermal manganese and iron at Cleft segment, Juan de Fuca Ridge. *J. Geophys. Res.* **99**, 4905–4923.

McCollom T. M. and Shock E. L. (1997) Geochemical constraints on chemolithoautotrophic metabolism by microorganisms in seafloor hydrothermal systems. *Geochim. Cosmochim. Acta* **61**, 4375–4391.

McCollom T. M. and Shock E. L. (1998) Fluid-rock interactions in the lower oceanic crust: thermodynamic models of hydrothermal alteration. *J. Geophys. Res.* **103**, 547–575.

Mercier H. and Speer K. G. (1998) Transport of bottom water in the Romanche Fracture Zone and the Chain Fracture Zone. *J. Phys. Oc.* **28**, 779–790.

Metz S., Trefry J. H., and Nelsen T. A. (1988) History and geochemistry of a metalliferous sediment core from the Mid-Atlantic Ridge at 26°N. *Geochim. Cosmochim. Acta* **52**, 2369–2378.

Middleton J. H. (1979) Times of rise for turbulent forced plumes. *Tellus* **31**, 82–88.

Millero F. J., Sotolongo S., and Izaguirre M. (1987) The oxidation kinetics of Fe(II) in seawater. *Geochim. Cosmochim. Acta* **51**, 793–801.

Mills R. A. and Elderfield H. (1995) Hydrothermal activity and the geochemistry of metalliferous sediment. *Geophys. Monogr. (AGU)* **91**, 392–407.

Mills R. A., Elderfield H., and Thomson J. (1993) A dual origin for the hydrothermal component in a metalliferous sediment core from the Mid-Atlantic Ridge. *J. Geophys. Res.* **98**, 9671–9678.

Mottl M. J. (2003) Partitioning of energy and mass fluxes between mid-Ocean ridge axes and flanks at high and low temperature. In *Energy and Mass Transfer in Marine Hydrothermal Systems* (eds. P. Halbach, V. Tunnicliffe, and J. Hein). DUP, Berlin, pp. 271–286.

Mottl M. J. and McConachy T. F. (1990) Chemical processes in buoyant hydrothermal plumes on the East Pacific Rise near 21°N. *Geochim. Cosmochim. Acta* **54**, 1911–1927.

Mottl M. J. and Wheat C. G. (1994) Hydrothermal circulation through mid-ocean ridge flanks: fluxes of heat and magnesium. *Geochim. Cosmochim. Acta* **58**, 2225–2237.

Mottl M. J., Wheat G., Baker E., Becker N., Davis E., Feely R., Grehan A., Kadko D., Lilley M., Massoth G., Moyer C., and Sansome F. (1998) Warm springs discovered on 3.5 Ma oceanic crust, eastern flank of the Juan de Fuca Ridge. *Geology* **26**, 51–54.

Morton B. R., Taylor G. I., and Turner J. S. (1956) Turbulent gravitational convection from maintained and instantaneous sources. *Proc. Roy. Soc. London Ser. A.* **234**, 1–23.

Mullineaux L. S., and France S. C. (1956) Disposal mechanisms of deep-sea hydrothermal vent fauna. *Geophys. Monogr. (AGU)* **91**, 408–424.

Mullins T. D., Britschgi T. B., Krest R. L., and Giovannoni S. J. (1995) Genetic comparisons reveal the same unknown bacterial lineages in Atlantic and Pacific bacterioplankton communities. *Limnol. Oceanogr.* **40**, 148–158.

Murray J. and Renard A. F. (1891) *Deep-sea Deposits*. Report "Challenger" Expedition (1873–1876), London.

Nehlig P., Juteau T., Bendel V., and Cotten J. (1994) The root zones of oceanic hydrothermal systems—constraints from the Samail ophiolite (Oman). *J. Geophys. Res.* **99**, 4703–4713.

O'Grady K. M. (2001) The geochemical controls on hydrothermal vent fluid chemistry from two areas on the ultrafast spreading southern East Pacific Rise. MSc Thesis, University of New Hampshire, p. 134.

Olivarez A. M. and Owen R. M. (1989) REE/Fe variation in hydrothermal sediments: implications for the REE content of seawater. *Geochim. Cosmochim. Acta* **53**, 757–762.

Oosting S. E. and Von Damm K. L. (1996) Bromide/chloride fractionation in seafloor hydrothermal fluids from 9–10°N East Pacific Rise. *Earth Planet. Sci. Lett.* **144**, 133–145.

Palmer M. R. and Edmond J. M. (1989) The strontium isotope budget of the modern ocean. *Earth Planet. Sci. Lett.* **92**, 11–26.

Palmer M. R. and Ernst G. G. J. (1998) Generation of hydrothermal megaplumes by cooling of pillow basalts at mid-ocean ridges. *Nature* **393**, 643–647.

Palmer M. R. and Ernst G. G. J. (2000) Comment on Lupton *et al.* (1999b). *Earth Planet. Sci. Lett.* **180**, 215–218.

Parson L. M., Murton B. J., Searle R. C., Booth D., Evans J., Field P., Keetin J., Laughton A., McAllister E., Millard N., Redbourne L., Rouse I., Shor A., Smith D., Spencer S., Summerhayes C., and Walker C. (1993) En echelon axial volcanic ridges at the Reykjanes ridge: a life cycle of volcanism and tectonics. *Earth Planet. Sci. Lett.* **117**, 73–87.

Parson L., Gracia E., Coller D., German C. R., and Needham H. D. (2000) Second order segmentation—the relationship between volcanism and tectonism at the MAR, 38°N–35°40' N. *Earth Planet. Sci. Lett.* **178**, 231–251.

Parsons B. (1981) The rates of plate creation and consumption. *Geophys. J. Roy. Astron. Soc.* **67**, 437–448.

Petersen S., Herzig P. M., and Hannington M. D. (2000) Third dimension of a presently forming VMS deposit: TAG hydrothermal mound, Mid-Atlantic Ridge, 26°N. *Mineralium Deposita* **35**, 233–259.

Peucker-Ehrenbrink B., Hofmann A. W., and Hart S. R. (1994) Hydrothermal lead transfer from mantle to continental crust—the role of metalliferous sediments. *Earth Planet. Sci. Lett.* **125**, 129–142.

Peucker-Ehrenbrink B., Ravizza G., and Hofmann A. W. (1995) The marine Os-187/Os-186 record of the past 180 million years. *Earth Planet. Sci. Lett.* **130**, 155–167.

Radford-Knoery J., German C. R., Charlou J.-L., Donval J.-P., and Fouquet Y. (2001) Distribution and behaviour of dissolved hydrogen sulfide in hydrothermal plumes. *Limnol. Oceanogr.* **46**, 461–464.

Ravizza G. (1993) Variations of the 187Os/186Os ratio of seawater over the past 28 million years as inferred from metalliferous carbonates. *Earth Planet. Sci. Lett.* **118**, 335–348.

Ravizza G., Martin C. E., German C. R., and Thompson G. (1996) Os isotopes as tracers in seafloor hydrothermal systems: a survey of metalliferous deposits from the TAG hydrothermal area, 26°N Mid-Atlantic Ridge. *Earth Planet. Sci. Lett.* **138**, 105–119.

Ravizza G., Blusztajn J., Von Damm K. L., Bray A. M., Bach W., and Hart S. R. (2001) Sr isotope variations in vent fluids from 9°46–54' N EPR: evidence of a non-zero-Mg fluid component at Biovent. *Geochim. Cosmochim. Acta* **65**, 729–739.

Rona P. A. and Trivett D. A. (1992) Discrete and diffuse heat transfer at ASHES vent field, axial volcano, Juan de Fuca Ridge. *Earth Planet. Sci. Lett.* **109**, 57–71.

Rona P. A., Bogdanov Y. A., Gurvich E. G., Rimskikorsakov N. A., Sagalevitch A. M., Hannington M. D., and Thompson G. (1993) Relict hydrothermal zones in the TAG hydrothermal field, Mid-Atlantic Ridge 26°N 45°W. *J. Geophys. Res.* **98**, 9715–9730.

Roth S. E. and Dymond J. (1989) Transport and settling of organic material in a deep-sea hydrothermal plume—evidence from particle-flux measurements. *Deep Sea Res.* **36**, 1237–1254.

Rubin K. H., MacDougall J. D., and Perfit M. R. (1994) $^{210}Po/^{210}Pb$ dating of recent volcanic eruptions on the seafloor. *Nature* **468**, 841–844.

Rudnicki M. D. and Elderfield H. (1992) Helium, radon and manganese at the TAG and SnakePit hydrothermal vent fields 26° and 23°N, Mid-Atlantic Ridge. *Earth Planet. Sci. Lett.* **113**, 307–321.

Rudnicki M. D. and Elderfield H. (1993) A chemical model of the buoyant and neutrally buoyant plume above the TAG vent field, 26 degrees N, Mid-Atlantic Ridge. *Geochim. Cosmochim. Acta* **57**, 2939–2957.

Rudnicki M. D., James R. H., and Elderfield H. (1994) Near-field variability of the TAG nonbuoyant plume 26°N Mid-Atlantic Ridge. *Earth Planet. Sci. Lett.* **127**, 1–10.

Ruhlin D. E. and Owen R. M. (1986) The rare earth element geochemistry of hydrothermal sediments from the East Pacific Rise: examination of a seawater scavenging mechanism. *Geochim. Cosmochim. Acta* **50**, 393–400.

Rüth C., Well R., and Roether W. (2000) Primordial ^3He in South Atlantic deep waters from sources on the Mid-Atlantic Ridge. *Deep-Sea Res.* **47**, 1059–1075.

Sauter D., Parson L., Mendel V., Rommevaux-Jestin C., Gomez O., Briais A., Mevel C., and Tamaki K. (2002) TOBI sidescan sonar imagery of the very slow-spreading southwest Indian Ridge: evidence for along-axis magma distribution. *Earth Planet. Sci. Lett.* **199**, 81–95.

Schaller T., Morford J., Emerson S. R., and Feely R. A. (2000) Oxyanions in metalliferous sediments: tracers for paleoseawater metal concentrations? *Geochim. Cosmochim. Acta* **64**, 2243–2254.

Schultz A. and Elderfield H. (1997) Controls on the physics and chemistry of seafloor hydrothermal circulation. *Phil. Trans. Roy. Soc. London A* **355**, 387–425.

Schultz A., Delaney J. R., and McDuff R. E. (1992) On the partitioning of heat-flux between diffuse and point-source sea-floor venting. *J. Geophys. Res.* **97**, 12299–12314.

Seyfried W. E. and Ding K. (1995) Phase equilibria in subseafloor hydrothermal systems: a review of the role of redox, temperature, pH and dissolved Cl on the chemistry of hot spring fluids at mid-ocean ridges. *Geophys. Monogr. (AGU)* **91**, 248–272.

Shanks W. C., III (2001) Stable isotopes in seafloor hydrothermal systems: vent fluids, hydrothermal deposits, hydrothermal alteration, and microbial processes. In *Stable Isotope Geochemistry, Rev. Mineral. Geochem. 43* (eds. J. W. Valley and D. R. Cole). Mineralogical Society of America, , pp. 469–525.

Sherrell R. M., Field M. P., and Ravizza G. (1999) Uptake and fractionation of rare earth elements on hydrothermal plume particles at 9°45' N, East Pacific Rise. *Geochim. Cosmochim. Acta* **63**, 1709–1722.

Simoneit B. R. T. (1991) Hydrothermal effects on recent diatomaceous sediments in Guaymas Basin—generation, migration, and deposition of petroleum. In *AAPG Memoir 47: The Gulf and Peninsular Province of the Californias*, American Association of Petroleum Geologists, Tulsa, OK, chap. 38, pp. 793–825.

Sohn R. A., Fornari D. J., Von Damm K. L., Hildebrand J. A., and Webb S. C. (1998) Seismic and hydrothermal evidence for a cracking event on the East Pacific Rise at 9°50' N. *Nature* **396**, 159–161.

Speer K. G. and Rona P. A. (1989) A model of an Atlantic and Pacific hydrothermal plume. *J. Geophys. Res.* **94**, 6213–6220.

Speer K. G., Maltrud M., and Thurnherr A. (2003) A global view of dispersion on the mid-oceanic ridge. In *Energy and Mass Transfer in Marine Hydrothermal Systems* (eds. P. Halbach, V. Tunnicliffe, and J. Hein). DUP, Berlin, pp. 287–302.

Spiess F. N., Ken C. M., Atwater T., Ballard R., Carranza A., Cordoba D., Cox C., Diaz Garcia V. M., Francheteau J., Guerrero J., Hawkins J., Haymon R., Hessler R., Juteau T., Kastner M., Larson R., Luyendyk B., Macdongall J. D., Miller S., Normark W., Orcutt J., and Rangin C. (1980) East Pacific Rise: hot springs and geophysical experiments. *Science* **207**, 1421–1433.

Statham P. J., Yeats P. A., and Landing W. M. (1998) Manganese in the eastern Atlantic Ocean: processes influencing deep and surface water distributions. *Mar. Chem.* **61**, 55–68.

Statham P. J., Connelly D. P., and German C. R. (2003) Fe(II) oxidation in Indian Ocean hydrothermal plumes. *Nature* (submitted).

Staudigel H. and Hart S. R. (1983) Alteration of basaltic glass—mechanisms and significance for oceanic-crust seawater budget. *Geochim. Cosmochim. Acta* **47**, 337–350.

Stein C. A. and Stein S. (1994) Constraints on hydrothermal heat flux through the oceanic lithosphere from global heat flow. *J. Geophys. Res.* **99**, 3081–3095.

Stein C. A., Stein S., and Pelayo A. M. (1995) Heat flow and hydrothermal circulation. Geophys. *Monogr. (AGU)* **91**, 425–445.

Summit M. and Baross J. A. (1998) Thermophilic subseafloor microorganisms from the 1996 north Gorda ridge eruption. *Deep-Sea Res.* **45**, 2751–2766.

Tivey M. K. (1995) The influence of hydrothermal fluid composition and advection rates on black smoker chimney mineralogy—insights from modelling transport and reaction. *Geochim. Cosmochim. Acta* **59**, 1933–1949.

Trefry J. H. and Metz S. (1989) Role of hydrothermal precipitates in the geochemical cycling of vanadium. *Nature* **342**, 531–533.

Trocine R. P. and Trefry J. H. (1988) Distribution and chemistry of suspended particles from an active hydrothermal vent site on the Mid-Atlantic Ridge at 26°N. *Earth Planet. Sci. Lett.* **88**, 1–15.

Turner J. S. (1973) *Buoyancy Effects in Fluids. Cambridge University Press*, 368pp.

Van Dover C. L., German C. R., Speer K. G., Parson L. M., and Vrijenhoek R. C. (2002) Evolution and biogeography of deep-sea vent and seep invertebrates. *Science* **295**, 1253–1257.

Von Damm K. L. (1995) Controls on the chemistry and temporal variability of seafloor hydrothermal fluids. *Geophys. Monogr. (AGU)* **91**, 222–247.

Von Damm K. L. (2000) Chemistry of hydrothermal vent fluids from 9–10°N, East Pacific Rise: time zero the immediate post-eruptive period. *J. Geophys. Res.* **105**, 11203–11222.

Von Damm K. L. (2003) Evolution of the hydrothermal system at East Pacific Rise 9°50′ N: geochemical evidence for changes in the upper oceanic crust. *Geophys. Monogr. (AGU)* (submitted).

Von Damm K. L. and Bischoff J. L. (1987) Chemistry of hydrothermal solutions from the southern Juan de Fuca Ridge. *J. Geophys. Res.* **92**, 11334–11346.

Von Damm K. L. and Lilley M. D. (2003) Diffuse flow hydrothermal fluids from 9°50′ N East Pacific Rise: origin, evolution and biogeochemical controls. *Geophys. Monogr. (AGU)* (in press).

Von Damm K. L., Edmond J. M., Grant B., Measures C. I., Walden B., and Weiss R. F. (1985a) Chemistry of submarine hydrothermal solutions at 21°N, East Pacific Rise. *Geochim. Cosmochim. Acta* **49**, 2197–2220.

Von Damm K. L., Edmond J. M., Measures C. I., and Grant B. (1985b) Chemistry of submarine hydrothermal solutions at Guaymas Basin, Gulf of California. *Geochim. Cosmochim. Acta* **49**, 2221–2237.

Von Damm K. L., Oosting S. E., Kozlowski R., Buttermore L. G., Colodner D. C., Edmonds H. N., Edmond J. M., and Grebmeier J. M. (1995) Evolution of East Pacific Rise hydrothermal vent fluids following a volcanic eruption. *Nature* **375**, 47–50.

Von Damm K. L., Buttermore L. G., Oosting S. E., Bray A. M., Fornari D. J., Lilley M. D., and Shanks W. C., III (1997) Direct observation of the evolution of a seafloor black smoker from vapor to brine. *Earth Planet. Sci. Lett.* **149**, 101–112.

Von Damm K. L., Bray A. M., Buttermore L. G., and Oosting S. E. (1998) The geochemical relationships between vent fluids from the lucky strike vent field, Mid-Atlantic Ridge. *Earth Planet. Sci. Lett.* **160**, 521–536.

Von Damm K. L., Gallant R. M., Hall J. M., Loveless J., Merchant E., and Scientific party of R/V Knorr KN162-13. (2001) The Edmond hydrothermal field: pushing the envelope on MOR brines. *EOS Trans. AGU (abstr.)* **82**, F646.

Von Damm K. L., Parker C. M., Gallant R. M., Loveless J. P., and the AdVenture 9 Science Party. (2002) Chemical evolution of hydrothermal fluids from EPR 21°N: 23 years later in a phase separating world. *EOS Trans. AGU (abstr.)* **83**, V61B-1365.

Von Damm K. L., Lilley M. D., Shanks W. C., III, Brockington M., Bray A. M., O'Grady K. M., Olson E., Graham A., Proskurowski G., and the SouEPR Science Party. (2003) Extraordinary phase separation and segregation in vent fluids from the southern East Pacific Rise. *Earth Planet. Sci. Lett.* **206**, 365–378.

Welhan J. and Craig H. (1983) Methane, hydrogen, and helium in hydrothermal fluids at 21°N on the East Pacific Rise. In *Hydrothermal Processes at Seafloor Spreading Centres* (eds. P. A. Rona, K. Boström, L. Laubier, L. Laubier, and K. L. Smith, Jr). NATO Conference Series IV: 12, Plenum, New York, pp. 391–409.

White R. S., McKenzie D., and O'Nions R. K. (1992) Oceanic crustal thickness from seismic measurements and rare earth element inversions. *J. Geophys. Res.* **97**, 19683–19715.

Winn C. D., Cowen J. P., and Karl D. M. (1995) Microbiology of hydrothermal plumes. In *Microbiology of Deep-sea Hydrothermal Vent Habitats* (ed. D. M. Karl), CRC, Boca Raton.

You C.-F., Butterfield D. A., Spivack A. J., Gieskes J. M., Gamo T., and Campbell A. J. (1994) Boron and halide systematics in submarine hydrothermal systems: effects of phase separation and sedimentary contributions. *Earth Planet. Sci. Lett.* **123**, 227–238.

Geochemistry of Earth Surface Systems
ISBN: 978-0-08-096706-6

pp. 45–86

3

The Contemporary Carbon Cycle

R. A. Houghton

Woods Hole Research Center, MA, USA

3.1	INTRODUCTION	88
3.2	MAJOR RESERVOIRS AND NATURAL FLUXES OF CARBON	88
	3.2.1 Reservoirs	88
	3.2.1.1 The atmosphere	89
	3.2.1.2 Terrestrial ecosystems: vegetation and soils	90
	3.2.1.3 The oceans	90
	3.2.1.4 Fossil fuels	91
	3.2.2 The Natural Flows of Carbon	91
	3.2.2.1 Between land and atmosphere	91
	3.2.2.2 Between oceans and atmosphere	92
	3.2.2.3 Between land and oceans	94
3.3	CHANGES IN THE STOCKS AND FLUXES OF CARBON AS A RESULT OF HUMAN ACTIVITIES	94
	3.3.1 Changes Over the Period 1850–2000	94
	3.3.1.1 Emissions of carbon from combustion of fossil fuels	94
	3.3.1.2 The increase in atmospheric CO_2	95
	3.3.1.3 Net uptake of carbon by the oceans	97
	3.3.1.4 Land: net exchange of carbon between terrestrial ecosystems and the atmosphere	98
	3.3.1.5 Land: changes in land use	99
	3.3.1.6 Land: a residual flux of carbon	100
	3.3.2 Changes Over the Period 1980–2000	101
	3.3.2.1 The global carbon budget	101
	3.3.2.2 Regional distribution of sources and sinks of carbon: the northern mid-latitudes	106
	3.3.2.3 Regional distribution of sources and sinks of carbon: the tropics	108
	3.3.2.4 Summary: synthesis of the results of different methods	110
3.4	MECHANISMS THOUGHT TO BE RESPONSIBLE FOR CURRENT SINKS OF CARBON	111
	3.4.1 Terrestrial Mechanisms	111
	3.4.1.1 Physiological or metabolic factors that enhance rates of growth and carbon accumulation	112
	3.4.1.2 Demographic or disturbance mechanisms	117
	3.4.2 Oceanic Mechanisms	118
	3.4.2.1 Physical and chemical mechanisms	118
	3.4.2.2 Biological feedback/processes	119
3.5	THE FUTURE: DELIBERATE SEQUESTERING OF CARBON (OR REDUCTION OF SOURCES)	120
	3.5.1 Terrestrial	120
	3.5.2 Oceanic	120
	3.5.3 Geologic	121
3.6	CONCLUSION	121
	REFERENCES	121

3.1 INTRODUCTION

The global carbon cycle refers to the exchanges of carbon within and between four major reservoirs: the atmosphere, the oceans, land, and fossil fuels. Carbon may be transferred from one reservoir to another in seconds (e.g., the fixation of atmospheric CO_2 into sugar through photosynthesis) or over millennia (e.g., the accumulation of fossil carbon (coal, oil, gas) through deposition and diagenesis of organic matter). This chapter emphasizes the exchanges that are important over years to decades and includes those occurring over the scale of months to a few centuries. The focus will be on the years 1980–2000 but our considerations will broadly include the years ~1850–2100. Chapter 10, deals with longer-term processes that involve rates of carbon exchange that are small on an annual timescale (weathering, vulcanism, sedimentation, and diagenesis).

The carbon cycle is important for at least three reasons. First, carbon forms the structure of all life on the planet, making up ~50% of the dry weight of living things. Second, the cycling of carbon approximates the flows of energy around the Earth, the metabolism of natural, human, and industrial systems. Plants transform radiant energy into chemical energy in the form of sugars, starches, and other forms of organic matter; this energy, whether in living organisms or dead organic matter, supports food chains in natural ecosystems as well as human ecosystems, not the least of which are industrial societies habituated (addicted?) to fossil forms of energy for heating, transportation, and generation of electricity. The increased use of fossil fuels has led to a third reason for interest in the carbon cycle. Carbon, in the form of carbon dioxide (CO_2) and methane (CH_4), forms two of the most important greenhouse gases. These gases contribute to a natural greenhouse effect that has kept the planet warm enough to evolve and support life (without the greenhouse effect the Earth's average temperature would be $-33\,°C$). Additions of greenhouse gases to the atmosphere from industrial activity, however, are increasing the concentrations of these gases, enhancing the greenhouse effect, and starting to warm the Earth.

The rate and extent of the warming depend, in part, on the global carbon cycle. If the rate at which the oceans remove CO_2 from the atmosphere were faster, e.g., concentrations of CO_2 would have increased less over the last century. If the processes removing carbon from the atmosphere and storing it on land were to diminish, concentrations of CO_2 would increase more rapidly than projected on the basis of recent history. The processes responsible for adding carbon to, and withdrawing it from, the atmosphere are not well enough understood to predict future levels of CO_2 with great accuracy. These processes are a part of the global carbon cycle.

Some of the processes that add carbon to the atmosphere or remove it, such as the combustion of fossil fuels and the establishment of tree plantations, are under direct human control. Others, such as the accumulation of carbon in the oceans or on land as a result of changes in global climate (i.e., feedbacks between the global carbon cycle and climate), are not under direct human control except through controlling rates of greenhouse gas emissions and, hence, climatic change. Because CO_2 has been more important than all of the other greenhouse gases under human control, combined, and is expected to continue so in the future, understanding the global carbon cycle is a vital part of managing global climate.

This chapter addresses, first, the reservoirs and natural flows of carbon on the earth. It then addresses the sources of carbon to the atmosphere from human uses of land and energy and the sinks of carbon on land and in the oceans that have kept the atmospheric accumulation of CO_2 lower than it would otherwise have been. The chapter describes changes in the distribution of carbon among the atmosphere, oceans, and terrestrial ecosystems over the past 150 years as a result of human-induced emissions of carbon. The processes responsible for sinks of carbon on land and in the sea are reviewed from the perspective of feedbacks, and the chapter concludes with some prospects for the future.

Earlier comprehensive summaries of the global carbon cycle include studies by Bolin *et al.* (1979, 1986), Woodwell and Pecan (1973), Bolin (1981), NRC (1983), Sundquist and Broecker (1985), and Trabalka (1985). More recently, the Intergovernmental Panel on Climate Change (IPCC) has summarized information on the carbon cycle in the context of climate change (Watson *et al.*, 1990; Schimel *et al.*, 1996; Prentice *et al.*, 2001). The basic aspects of the global carbon cycle have been understood for decades, but other aspects, such as the partitioning of the carbon sink between land and ocean, are being re-evaluated continuously with new data and analyses. The rate at which new publications revise estimates of these carbon sinks and re-evaluate the mechanisms that control the magnitude of the sinks suggests that portions of this review will be out of date by the time of publication.

3.2 MAJOR RESERVOIRS AND NATURAL FLUXES OF CARBON

3.2.1 Reservoirs

The contemporary global carbon cycle is shown in simplified form in Figure 1. The four major reservoirs important in the time frame of decades to centuries are the atmosphere, oceans, reserves of fossil fuels, and terrestrial ecosystems,

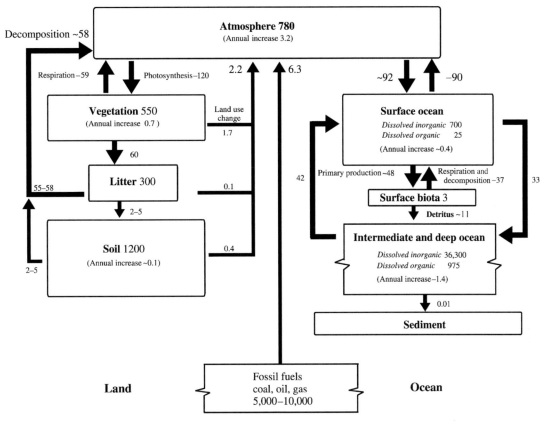

Figure 1 The contemporary global carbon cycle. Units are Pg C or Pg C yr^{-1}.

including vegetation and soils. The world's oceans contain ~50 times more carbon than either the atmosphere or the world's terrestrial vegetation, and thus shifts in the abundance of carbon among the major reservoirs will have a much greater significance for the terrestrial biota and for the atmosphere than they will for the oceans.

3.2.1.1 The atmosphere

Most of the atmosphere is made up of either nitrogen (78%) or oxygen (21%). In contrast, the concentration of CO_2 in the atmosphere is only ~0.04%. The concentrations of CO_2 in air can be measured to within one tenth of 0.1 ppmv, or 0.00001%. In the year 2000 the globally averaged concentration was ~0.0368%, or 368 ppmv, equivalent to ~780 Pg C (1 Pg = 1 petagram = 10^{15} g = 10^9 t) (Table 1).

The atmosphere is completely mixed in about a year, so any monitoring station free of local contamination will show approximately the same year-to-year increase in CO_2. There are at least 77 stations worldwide, where weekly flask samples of air are collected, analyzed for CO_2 and other constituents, and where the resulting data are integrated into a consistent global data set (Masarie and Tans, 1995; Cooperative Atmospheric Data

Table 1 Stocks and flows of carbon.

Carbon stocks (Pg C)	
Atmosphere	780
Land	2,000
Vegetation	500
Soil	1,500
Ocean	39,000
Surface	700
Deep	38,000
Fossil fuel reserves	10,000
Annual flows (Pg C yr^{-1})	
Atmosphere-oceans	90
Atmosphere-land	120
Net annual exchanges (Pg C yr^{-1})	
Fossil fuels	6
Land-use change	2
Atmospheric increase	3
Oceanic uptake	2
Other terrestrial uptake	3

Integration Project—Carbon Dioxide, 1997). The stations generally show the same year-to-year increase in concentration but vary with respect to absolute concentration, seasonal variability, and other characteristics useful for investigating the global circulation of carbon.

Most of the carbon in the atmosphere is CO_2, but small amounts of carbon exist in concentrations of CH_4, carbon monoxide (CO), and non-methane hydrocarbons. These trace gases are important because they modify the chemical and/or the radiative properties of the Earth's atmosphere. Methane is present at ~1.7 ppm, two orders of magnitude more dilute than CO_2. Methane is a reduced form of carbon, is much less stable than CO_2, and has an average residence time in the atmosphere of 5–10 years. Carbon monoxide has an atmospheric residence time of only a few months. Its low concentration, ~0.1 ppm, and its short residence time result from its chemical reactivity with OH radicals. Carbon monoxide is not a greenhouse gas, but its chemical reactivity affects the abundances of ozone and methane which are greenhouse gases. Non-methane hydrocarbons, another unstable form of carbon in the atmosphere, are present in even smaller concentrations. The oxidation of these biogenic trace gases is believed to be a major source of atmospheric CO, and, hence, these non-methane hydrocarbons also affect indirectly the Earth's radiative balance.

3.2.1.2 Terrestrial ecosystems: vegetation and soils

Carbon accounts for only ~0.27% of the mass of elements in the Earth's crust (Kempe, 1979), yet it is the basis for life on Earth. The amount of carbon contained in the living vegetation of terrestrial ecosystems (550 ± 100 Pg) is somewhat less than that present in the atmosphere (780 Pg). Soils contain 2–3 times that amount (1,500–2,000 Pg C) in the top meter (Table 2) and as much as 2,300 Pg in the top 3 m (Jobbágy and Jackson, 2000). Most terrestrial carbon is stored in the vegetation and soils of the world's forests. Forests cover ~30% of the land surface and hold ~75% of the living organic carbon. When soils are included in the inventory, forests hold almost half of the carbon of the world's terrestrial ecosystems. The soils of woodlands, grasslands, tundra, wetlands, and agricultural lands store most of the rest of the terrestrial organic carbon.

3.2.1.3 The oceans

The total amount of dissolved inorganic carbon (DIC) in the world's oceans is ~3.7×10^4 Pg, and the amount of organic carbon is ~1,000 Pg. Thus, the world's oceans contain ~50 times more carbon than the atmosphere and 70 times more than the world's terrestrial vegetation. Most of this oceanic carbon is in intermediate and deep waters; only 700–1,000 Pg C are in the surface layers of the ocean, that part of the ocean in direct contact with the atmosphere. There are also 6,000 Pg C in

Table 2 Area, carbon in living biomass, and net primary productivity of major terrestrial biomes.

Biome	Area (10⁹ ha)		Global carbon stocks (PgC)						Carbon stocks (MgC ha⁻¹)				NPP (PgC yr⁻¹)	
			WGBU			MRS IGBP			WBGU		MRS IGPB			
	WBGU	MRS	Plants	Soil	Total	Plants	Soil	Total	Plants	Soil	Plants	Soil	Ajay	MRS
Tropical forests	17.6	17.5	212	216	428	340	214	553	120	123	194	122	13.7	21.9
Temperate forests	1.04	1.04	59	100	159	139	153	292	57	96	134	147	6.5	8.1
Boreal forests	1.37	1.37	88	471	559	57	338	395	64	344	42	247	3.2	2.6
Tropical savannas and grasslands	2.25	2.76	66	264	330	79	247	326	29	117	29	90	17.7	14.9
Temperate grasslands and shrublands	1.25	1.78	9	295	304	23	176	199	7	236	13	99	5.3	7.0
Deserts and semi-deserts	4.55	2.77	8	191	199	10	159	169	2	42	4	57	1.4	3.5
Tundra	0.95	0.56	6	121	127	2	115	117	6	127	4	206	1.0	0.5
Croplands	1.60	1.35	3	128	131	4	165	169	2	80	3	122	6.8	4.1
Wetlands	0.35		15	225	240				43	643			4.3	
Total	15.12	14.93	466	2,011	2,477	654	1,567	2,221					59.9	62.6

Source: Prentice *et al.* (2001).

Table 3 The distribution of 1,000 CO_2 molecules in the atmosphere–ocean.

Atmosphere	15
Ocean	985
$\quad CO_2$	5
$\quad HCO_3^-$	875
$\quad CO_3^{2-}$	105
Total	1,000

Source: Sarmiento (1993).

reactive ocean sediments (Sundquist, 1986), but the turnover of sediments is slow, and they are not generally considered as part of the active, or short-term, carbon cycle, although they are important in determining the long-term concentration of CO_2 in the atmosphere and oceans.

Carbon dioxide behaves unlike other gases in the ocean. Most gases are not very soluble in water and are predominantly in the atmosphere. For example, only ∼1% of the world's oxygen is in the oceans; 99% exists in the atmosphere. Because of the chemistry of seawater, however, the distribution of carbon between air and sea is reversed: 98.5% of the carbon in the ocean–atmosphere systems is in the sea. Although this inorganic carbon is dissolved, less than 1% of it is in the form of dissolved CO_2 (p_{CO_2}); most of the inorganic carbon is in the form of bicarbonate and carbonate ions (Table 3).

About 1,000 PgC in the oceans (out of the total of 3.8×10^4 Pg) is organic carbon. Carbon in living organisms amounts to ∼3 Pg in the sea, in comparison to ∼550 Pg on land. The mass of animal life in the oceans is almost the same as on land, however, pointing to the very different trophic structures in the two environments. The ocean's plants are microscopic. They have a high productivity, but the production does not accumulate. Most is either grazed or decomposed in the surface waters. Only a fraction (∼25%) sinks into the deeper ocean. In contrast, terrestrial plants accumulate large amounts of carbon in long-lasting structures (trees). The distribution of organic carbon between living and dead forms of carbon is also very different on land and in the sea. The ratio is ∼1:3 on land and ∼1:300 in the sea.

3.2.1.4 Fossil fuels

The common sources of energy used by industrial societies are another form of organic matter, so-called fossil fuels. Coal, oil, and natural gas are the residuals of organic matter formed millions of years ago by green plants. The material escaped oxidation, became buried in the Earth, and over time was transformed to a (fossil) form. The energy stored in the chemical bonds of fossil fuels is released during combustion just as the energy stored in carbohydrates, proteins, and fats is released during respiration.

The difference between the two forms of organic matter (fossil and nonfossil), from the perspective of the global carbon cycle, is the rate at which they are cycled. The annual rate of formation of fossil carbon is at least 1,000 times slower than rates of photosynthesis and respiration. The formation of fossil fuels is part of a carbon cycle that operates over millions of years, and the processes that govern the behavior of this long-term system (sedimentation, weathering, vulcanism, seafloor spreading) are much slower from those that govern the behavior of the short-term system. Sedimentation of organic and inorganic carbon in the sea, e.g., is $\sim 0.2\,\mathrm{PgCyr}^{-1}$. In contrast, hundreds of petagrams of carbon are cycled annually among the reservoirs of the short-term, or active, carbon cycle. This short-term system operates over periods of seconds to centuries. When young (nonfossil) organic matter is added to or removed from the atmosphere, the total amount of carbon in the active system is unchanged. It is merely redistributed among reservoirs. When fossil fuels are oxidized, however, the CO_2 released represents a net increase in the amount of carbon in the active system.

The amount of carbon stored in recoverable reserves of coal, oil, and gas is estimated to be 5,000–10,000 PgC, larger than any other reservoir except the deep sea, and ∼10 times the carbon content of the atmosphere. Until ∼1850s this reservoir of carbon was not a significant part of the short-term cycle of carbon. The industrial revolution changed that.

3.2.2 The Natural Flows of Carbon

Carbon dioxide is chemically stable and has an average residence time in the atmosphere of about four years before it enters either the oceans or terrestrial ecosystems.

3.2.2.1 Between land and atmosphere

The inorganic form of carbon in the atmosphere (CO_2) is fixed into organic matter by green plants using energy from the Sun in the process of photosynthesis, as follows:

$$6CO_2 + 6H_2O \rightleftharpoons C_6H_{12}O_6 + 6O_2$$

The reduction of CO_2 to glucose ($C_6H_{12}O_6$) stores some of the Sun's energy in the chemical bonds of the organic matter formed. Glucose, cellulose, carbohydrates, protein, and fats are all forms of organic matter, or reduced carbon. They all embody energy and are nearly all derived ultimately from photosynthesis.

The reaction above also goes in the opposite direction during the oxidation of organic matter. Oxidation occurs during the two, seemingly dissimilar but chemically identical, processes of respiration and combustion. During either process the chemical energy stored in organic matter is released. Respiration is the biotic process that yields energy from organic matter, energy required for growth and maintenance. All living organisms oxidize organic matter; only plants and some microbes are capable of reducing CO_2 to produce organic matter.

Approximately 45–50% of the dry weight of organic matter is carbon. The organic carbon of terrestrial ecosystems exists in many forms, including living leaves and roots, animals, microbes, wood, decaying leaves, and soil humus. The turnover of these materials varies from less than one year to more than 1,000 years. In terms of carbon, the world's terrestrial biota is almost entirely vegetation; animals (including humans) account for less than 0.1% of the carbon in living organisms.

Each year the atmosphere exchanges $\sim$120 Pg C with terrestrial ecosystems through photosynthesis and respiration (Figure 1 and Table 1). The uptake of carbon through photosynthesis is gross primary production (GPP). At least half of this production is respired by the plants, themselves (autotrophic respiration (Rs_a)), leaving a net primary production (NPP) of $\sim$60 Pg C yr^{-1}. Recent estimates of global terrestrial NPP vary between 56.4 Pg C yr^{-1} and 62.6 Pg C yr^{-1} (Ajtay *et al.*, 1979; Field *et al.*, 1998; Saugier *et al.*, 2001). The annual production of organic matter is what fuels the nonplant world, providing food, feed, fiber, and fuel for natural and human systems. Thus, most of the NPP is consumed by animals or respired by decomposer organisms in the soil (heterotrophic respiration (Rs_h)). A smaller amount ($\sim$4 Pg C yr^{-1} globally) is oxidized through fires. The sum of autotrophic and heterotrophic respiration is total respiration or ecosystem respiration (Rs_e). In steady state the net flux of carbon between terrestrial ecosystems and the atmosphere (net ecosystem production (NEP)) is approximately zero, but year-to-year variations in photosynthesis and respiration (including fires) may depart from this long-term balance by as much as 5–6 Pg C yr^{-1}. The annual global exchanges may be summarized as follows:

$$NPP = GPP - Rs_a$$
$$(\sim 60 = 120 - 60 \text{ Pg C yr}^{-1})$$

$$NEP = GPP - Rs_a - Rs_h$$
$$(\sim 0 = 120 - 60 - 60 \text{ Pg C yr}^{-1})$$

$$NEP = NPP - Rs_h$$
$$(\sim 0 = 60 - 60 \text{ Pg C yr}^{-1})$$

Photosynthesis and respiration are not evenly distributed either in space or over the course of a year. About half of terrestrial photosynthesis occurs in the tropics where the conditions are generally favorable for growth, and where a large proportion of the Earth's land area exists (Table 2). Direct evidence for the importance of terrestrial metabolism (photosynthesis and respiration) can be seen in the effect it has on the atmospheric concentration of CO_2 (Figure 2(a)). The most striking feature of the figure is the regular sawtooth pattern. This pattern repeats itself annually. The cause of the oscillation is the metabolism of terrestrial ecosystems. The highest concentrations occur at the end of each winter, following the season in which respiration has exceeded photosynthesis and thereby caused a net release of CO_2 to the atmosphere. Lowest concentrations occur at the end of each summer, following the season in which photosynthesis has exceeded respiration and drawn CO_2 out of the atmosphere. The latitudinal variability in the amplitude of this oscillation suggests that it is driven largely by northern temperate and boreal ecosystems: the highest amplitudes (up to $\sim$16 ppmv) are in the northern hemisphere with the largest land area. The phase of the amplitude is reversed in the southern hemisphere, corresponding to seasonal terrestrial metabolism there. Despite the high rates of production and respiration in the tropics, equatorial regions are thought to contribute little to this oscillation. Although there is a strong seasonality in precipitation throughout much of the tropics, the seasonal changes in moisture affect photosynthesis and respiration almost equally and thus the two processes remain largely in phase with little or no net flux of CO_2.

3.2.2.2 Between oceans and atmosphere

There is $\sim$50 times more carbon in the ocean than in the atmosphere, and it is the amount of DIC in the ocean that determines the atmospheric concentration of CO_2. In the long term (millennia) the most important process determining the exchanges of carbon between the oceans and the atmosphere is the chemical equilibrium of dissolved CO_2, bicarbonate, and carbonate in the ocean. The rate at which the oceans take up or release carbon is slow on a century timescale, however, because of lags in circulation and changes in the availability of calcium ions. The carbon chemistry of seawater is discussed in more detail in the next section.

Two additional processes besides carbon chemistry keep the atmospheric CO_2 lower than it otherwise would be. One process is referred to as the solubility pump and the other as the biological pump. The solubility pump is based on the fact that CO_2 is more soluble in cold waters. In the ocean, CO_2 is $\sim$2 times more soluble in the cold mid-depth and deep waters than it is in the warm surface waters near the equator. Because sinking of cold surface waters in Arctic and Antarctic regions forms these mid-depth and deep waters, the formation of these waters with high CO_2

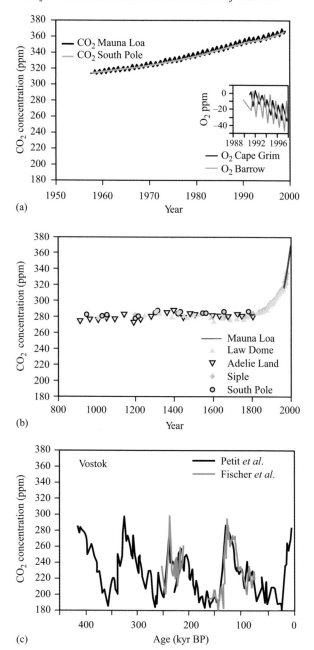

Figure 2 Concentration of CO_2 in the atmosphere: (a) over the last 42 years, (b) over the last 1,000 years, and (c) over the last $\sim4\times10^5$ years (Prentice *et al.*, 2001) (reproduced by permission of Intergovernmental Panel on Climate Change from *Climate Change 2001: The Scientific Basis*, **2001**, pp. 183–237).

keeps the CO_2 concentration of the atmosphere lower than the average concentration of surface waters.

The biological pump also transfers surface carbon to the intermediate and deep ocean. Not all of the organic matter produced by phytoplankton is respired in the surface waters where it is produced; some sinks out of the photic zone to deeper water. Eventually, this organic matter is decomposed at depth and reaches the surface again through ocean circulation. The net effect of the sinking of organic matter is to enrich the deeper waters

relative to surface waters and thus to reduce the CO_2 concentration of the atmosphere. Marine photosynthesis and the sinking of organic matter out of the surface water are estimated to keep the concentration of CO_2 in air $\sim30\%$ of what it would be in their absence.

Together the two pumps keep the DIC concentration of the surface waters $\sim10\%$ lower than at depth. Ocean models that simulate both carbon chemistry and oceanic circulation show that the concentration of CO_2 in the atmosphere (280 ppmv pre-industrially) would have been

720 ppmv if both pumps were turned off (Sarmiento, 1993).

There is another biological pump, called the carbonate pump, but its effect in reducing the concentration of CO_2 in the atmosphere is small. Some forms of phytoplankton have $CaCO_3$ shells that, in sinking, transfer carbon from the surface to deeper water, just as the biological pump transfers organic carbon to depth. The precipitation of $CaCO_3$ in the surface waters, however, increases the partial pressure of CO_2, and the evasion of this CO_2 to the atmosphere offsets the sinking of carbonate carbon.

Although ocean chemistry determines the CO_2 concentration of the atmosphere in the long term and the solubility and biological pumps act to modify this long-term equilibrium, short-term exchanges of carbon between ocean and atmosphere result from the diffusion of CO_2 across the air–sea interface. The diffusive exchanges transfer $\sim$90 Pg C yr^{-1} across the air–sea interface in both directions (Figure 1). The transfer has been estimated by two different methods. One method is based on the fact that the transfer rate of naturally produced ^{14}C into the oceans should balance the decay of ^{14}C within the oceans. Both the production rate of ^{14}C in the atmosphere and the inventory of ^{14}C in the oceans are known with enough certainty to yield an average rate of transfer of $\sim$100 Pg Cyr^{-1}, into and out of the ocean.

The second method is based on the amount of radon gas in the surface ocean. Radon gas is generated by the decay of ^{226}Ra. The concentration of the parent ^{226}Ra and its half-life allow calculation of the expected radon gas concentration in the surface water. The observed concentration is $\sim$70% of expected, so 30% of the radon must be transferred to the atmosphere during its mean lifetime of six days. Correcting for differences in the diffusivity of radon and CO_2 allows an estimation of the transfer rate for CO_2. The transfer rates given by the ^{14}C method and the radon method agree within $\sim$10%.

The net exchange of CO_2 across the air–sea interface varies latitudinally, largely as a function of the partial pressure of CO_2 in surface waters, which, in turn, is affected by temperature, upwelling or downwelling, and biological production. Cold, high-latitude waters take up carbon, while warm, lower-latitude waters tend to release carbon (outgassing of CO_2 from tropical gyres). Although the latitudinal pattern in net exchange is consistent with temperature, the dominant reason for the exchange is upwelling (in the tropics) and downwelling, or deep-water formation (at high latitudes).

The annual rate of photosynthesis in the world oceans is estimated to be $\sim$48 PgC (Table 4) (Longhurst *et al.*, 1995). About 25% of the primary production sinks from the photic zone to deeper water (Falkowski *et al.*, 1998; Laws

Table 4 Annual net primary production of the ocean.

Domain or ecosystem	NPP (Pg C yr^{-1})
Trade winds domain (tropical and subtropical)	13.0
Westerly winds domain (temperate)	16.3
Polar domain	6.4
Coastal domain	10.7
Salt marshes, estuaries, and macrophytes	1.2
Coral reefs	0.7
Total	48.3

Source: Longhurst *et al.* (1995).

et al., 2000). The gross flows of carbon between the surface ocean and the intermediate and deep ocean are estimated to be $\sim$40 Pg Cyr^{-1}, in part from the sinking of organic production (11 Pg C yr^{-1}) and in part from physical mixing (33 Pg C yr^{-1}) (Figure 1).

3.2.2.3 Between land and oceans

Most of the carbon taken up or lost by terrestrial ecosystems and the ocean is exchanged with the atmosphere, but a small flux of carbon from land to the ocean bypasses the atmosphere. The river input of inorganic carbon to the oceans (0.4 Pg C yr^{-1}) is almost balanced in steady state by a loss of carbon to carbonate sediments (0.2 Pg Cyr^{-1}) and a release of CO_2 to the atmosphere (0.1 Pg C yr^{-1}) (Sarmiento and Sundquist, 1992). The riverine flux of organic carbon is 0.3–0.5 Pg C yr^{-1}, and thus, the total flux from land to sea is 0.4–0.7 Pg C yr^{-1}.

3.3 CHANGES IN THE STOCKS AND FLUXES OF CARBON AS A RESULT OF HUMAN ACTIVITIES

3.3.1 Changes Over the Period 1850–2000

3.3.1.1 Emissions of carbon from combustion of fossil fuels

The CO_2 released annually from the combustion of fossil fuels (coal, oil, and gas) is calculated from records of fuel production compiled internationally (Marland *et al.*, 1998). Emissions of CO_2 from the production of cement and gas flaring add small amounts to the total industrial emissions, which have generally increased exponentially since $\sim$1750. Temporary interruptions in the trend occurred during the two World Wars, following the increase in oil prices in 1973 and 1979, and following the collapse of the former Soviet Union in 1992 (Figure 3). Between 1751 and 2000, the total emissions of carbon are estimated to have been $\sim$275 PgC, essentially all of it since 1860. Annual emissions averaged

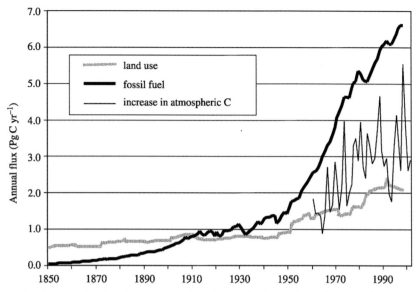

Figure 3 Annual emissions of carbon from combustion of fossil fuels and from changes in land use, and the annual increase in atmospheric CO_2 (in Pg C) since ~1750 (interannual variation in the growth rate of atmospheric CO_2 is greater than variation in emissions).

$5.4 \, Pg \, C \, yr^{-1}$ during the 1980s and $6.3 \, Pg \, C \, yr^{-1}$ during the 1990s. Estimates are thought to be known globally to within 20% before 1950 and to within 6% since 1950 (Keeling, 1973; Andres *et al.*, 1999).

The proportions of coal, oil, and gas production have changed through time. Coal was the major contributor to atmospheric CO_2 until the late 1960s, when the production of oil first exceeded that of coal. Rates of oil and gas consumption grew rapidly until 1973. After that their relative rates of growth declined dramatically, such that emissions of carbon from coal were, again, as large as those from oil during the second half of the 1980s and in the last years of the twentieth century.

The relative contributions of different world regions to the annual emissions of fossil fuel carbon have also changed. In 1925, the US, Western Europe, Japan, and Australia were responsible for ~88% of the world's fossil fuel CO_2 emissions. By 1950 the fraction contributed by these countries had decreased to 71%, and by 1980 to 48%. The annual rate of growth in the use of fossil fuels in developed countries varied between 0.5% and 1.4% in the 1970s. In contrast, the annual rate of growth in developing nations was 6.3% during this period. The share of the world's total fossil fuel used by the developing countries has grown from 6% in 1925, to 10% in 1950, to ~20% in 1980. By 2020, the developing world may be using more than half of the world's fossil fuels annually (Goldemberg *et al.*, 1985). They may then be the major source of both fossil fuel and terrestrial CO_2 to the atmosphere (Section 3.3.1.5).

Annual emissions of CO_2 from fossil fuel combustion are small relative to the natural flows of carbon through terrestrial photosynthesis and respiration (~$120 \, Pg \, C \, yr^{-1}$) and relative to the gross exchanges between oceans and atmosphere (~$90 \, Pg \, C \, yr^{-1}$) (Figure 1). Nevertheless, these anthropogenic emissions are the major contributor to increasing concentrations of CO_2 in the atmosphere. They represent a transfer of carbon from the slow carbon cycle (see Chapter 10) to the active carbon cycle.

3.3.1.2 The increase in atmospheric CO_2

Numerous measurements of atmospheric CO_2 concentrations were made in the nineteenth century (Fraser *et al.*, 1986), and Callendar (1938) estimated from these early measurements that the amount of CO_2 had increased by 6% between 1900 and 1935. Because of geographical and seasonal variations in the concentrations of CO_2, however, no reliable measure of the rate of increase was possible until after 1957 when the first continuous monitoring of CO_2 concentrations was begun at Mauna Loa, Hawaii, and at the South Pole (Keeling *et al.*, 2001). In 1958 the average concentration of CO_2 in air at Mauna Loa was ~315 ppm. In the year 2000 the concentration had reached ~368 ppm, yielding an average rate of increase of ~$1 \, ppm \, yr^{-1}$ since 1958. However, in recent decades the rate of increase in the atmosphere has been ~$1.5 \, ppm \, yr^{-1}$ (~$3 \, Pg \, C \, yr^{-1}$).

During the early 1980s, scientists developed instruments that could measure the concentration of atmospheric CO_2 in bubbles of air trapped in glacial ice. Ice cores from Greenland and Antarctica show that the pre-industrial concentration of CO_2 was between 275 ppm and 285 ppm (Neftel *et al.*, 1985; Raynaud and Barnola, 1985;

Etheridge *et al.*, 1996) (Figure 2(b)). The increase between 1700 and 2000, therefore, has been ~85 ppm, equivalent to ~175 PgC, or 30% of the pre-industrial level.

Over the last 1,000 years the concentration of CO_2 in the atmosphere has varied by less than 10 ppmv (Figure 2(b)). However, over the last 4.2×10^5 years (four glacial cycles), the concentration of CO_2 has consistently varied from ~180 ppm during glacial periods to ~280 ppm during interglacial periods (Figure 2(c)). The correlation between CO_2 concentration and the surface temperature of the Earth is evidence for the greenhouse effect of CO_2, first advanced almost a century ago by the Swedish climatologist Arrhenius (1896). As a greenhouse gas, CO_2 is more transparent to the Sun's energy entering the Earth's atmosphere than it is to the re-radiated heat energy leaving the Earth. Higher concentrations of CO_2 in the atmosphere cause a warmer Earth and lower concentrations a cooler one. There have been abrupt changes in global temperature that were not associated with a change in CO_2 concentrations (Smith *et al.*, 1999), but never in the last 4.2×10^5 years have concentrations of CO_2 changed without a discernible change in temperature (Falkowski *et al.*, 2000). The glacial–interglacial difference of 100 ppm corresponds to a temperature difference of ~10°C. The change reflects temperature changes in the upper troposphere and in the region of the ice core (Vostoc, Antarctica) and may not represent a global average. Today's CO_2 concentration of 368 ppm represents a large departure from the last 4.2×10^5 years, although the expected increase in temperature has not yet occurred.

It is impossible to say that the increase in atmospheric CO_2 is entirely the result of human activities, but the evidence is compelling. First, the known sources of carbon are more than adequate to explain the observed increase in the atmosphere. Balancing the global carbon budget requires additional carbon sinks, not an unexplained source of carbon (see Section 3.3.1.4). Since 1850, ~275 PgC have been released from the combustion of fossil fuels and another 155 PgC were released as a result of net changes in land use, i.e., from the net effects of deforestation and reforestation (Section 3.3.1.5). The observed increase in atmospheric carbon was only 175 PgC (40% of total emissions) over this 150-year period (Table 5).

Table 5 The global carbon budget for the period 1850 to 2000 (units are Pg C).

Fossil fuel emissions	275
Atmospheric increase	175
Oceanic uptake	140
Net terrestrial source	40
Land-use net source	155
Residual terrestrial sink	115

Second, for several thousand years preceding 1850 (approximately the start of the industrial revolution), the concentration of CO_2 varied by less than 10 ppmv (Etheridge *et al.*, 1996) (Figure 2(b)). Since 1850, concentrations have increased by 85 ppmv (~30%). The timing of the increase is coincident with the annual emissions of carbon from combustion of fossil fuels and the net emissions from land-use change (Figure 3).

Third, the latitudinal gradient in CO_2 concentrations is highest at northern mid-latitudes, consistent with the fact that most of the emissions of fossil fuel are located in northern mid-latitudes. Although atmospheric transport is rapid, the signal of fossil fuel combustion is discernible.

Fourth, the rate of increase of carbon in the atmosphere and the distribution of carbon isotopes and other biogeochemical tracers are consistent with scientific understanding of the sources and sinks of carbon from fossil fuels, land, and the oceans. For example, while the concentration of CO_2 has increased over the period 1850–2000, the ^{14}C content of the CO_2 has decreased. The decrease is what would be expected if the CO_2 added to the system were fossil carbon depleted in ^{14}C through radioactive decay.

Concentrations of other carbon containing gases have also increased in the last two centuries. The increase in the concentration of CH_4 has been more than 100% in the last 100 years, from background levels of less than 0.8 ppm to a value of ~1.75 ppm in 2000 (Prather and Ehhalt, 2001). The temporal pattern of the increase is similar to that of CO_2. There was no apparent trend for the 1,000 years before 1700. Between 1700 and 1900 the annual rate of increase was ~1.5 ppbv, accelerating to 15 ppb yr^{-1} in the 1980s. Since 1985, however, the annual growth rate of CH_4 (unlike CO_2) has declined. The concentration is still increasing, but not as rapidly. It is unclear whether sources have declined or whether atmospheric sinks have increased.

Methane is released from anaerobic environments, such as the sediments of wetlands, peatlands, and rice paddies and the guts of ruminants. The major sources of increased CH_4 concentrations are uncertain but are thought to include the expansion of paddy rice, the increase in the world's population of ruminants, and leaks from drilling and transport of CH_4 (Prather and Ehhalt, 2001). Atmospheric CH_4 budgets are more difficult to construct than CO_2 budgets, because increased concentrations of CH_4 occur not only from increased sources from the Earth's surface but from decreased destruction (by OH radicals) in the atmosphere as well. The increase in atmospheric CH_4 has been more significant for the greenhouse effect than it has for the carbon budget. The doubling of CH_4 concentrations since 1700 has amounted to only ~1 ppm, in comparison to the CO_2 increase of almost 90 ppm. Alternatively, CH_4 is, molecule for molecule, ~15 times more effective than CO_2 as a greenhouse gas. Its atmospheric lifetime is only 8–10 years, however.

Carbon monoxide is not a greenhouse gas, but its chemical effects on the OH radical affect the destruction of CH_4 and the formation of ozone. Because the concentration of CO is low and its lifetime is short, its atmospheric budget is less well understood than budgets for CO_2 and CH_4. Nevertheless, CO seems to have been increasing in the atmosphere until the late 1980s (Prather and Ehhalt, 2001). Its contribution to the carbon cycle is very small.

3.3.1.3 Net uptake of carbon by the oceans

As discussed above, the chemistry of carbon in seawater is such that less than 1% of the carbon exists as dissolved CO_2. More than 99% of the DIC exists as bicarbonate and carbonate anions (Table 3). The chemical equilibrium among these three forms of DIC is responsible for the high solubility of CO_2 in the oceans. It also sets up a buffer for changes in oceanic carbon. The buffer factor (or Revelle factor), ξ, is defined as follows:

$$\xi = \frac{\Delta p_{CO_2}/p_{CO_2}}{\Delta \sum CO_2 / \sum CO_2}$$

where p_{CO_2} is the partial pressure of CO_2 (the atmospheric concentration of CO_2 at equilibrium with that of seawater), ΣCO_2 is total inorganic carbon (DIC), and Δ refers to the change in the variable. The buffer factor varies with temperature, but globally averages ~10. It indicates that p_{CO_2} is sensitive to small changes in DIC: a change in the partial pressure of CO_2 (p_{CO_2}) is ~10 times the change in total CO_2. The significance of this is that the storage capacity of the ocean for excess atmospheric CO_2 is a factor of ~10 lower than might be expected by comparing reservoir sizes (Table 1). The oceans will not take up 98% of the carbon released through human activity, but only ~85% of it. The increase in atmospheric CO_2 concentration by ~30% since 1850s has been associated with a change of only ~3% in DIC of the surface waters. The other important aspect of the buffer factor is that it increases as DIC increases. The ocean will become increasingly resistant to taking up carbon (see Section 3.4.2.1).

Although the oceans determine the concentration of CO_2 in the atmosphere in the long term, in the short term, lags introduced by other processes besides chemistry allow a temporary disequilibrium. Two processes that delay the transfer of anthropogenic carbon into the ocean are: (i) the transfer of CO_2 across the air–sea interface and (ii) the mixing of water masses within the sea. The rate of transfer of CO_2 across the air–sea interface

was discussed above (Section 3.2.2.2). This transfer is believed to have reduced the oceanic absorption of CO_2 by ~10% (Broecker *et al.*, 1979).

The more important process in slowing the oceanic uptake of CO_2 is the rate of vertical mixing within the oceans. The mixing of ocean waters is determined from measured profiles of natural ^{14}C, bomb-produced ^{14}C, bomb-produced tritium, and other tracers. Profiles of these tracers were obtained during extensive oceanographic surveys: one called Geochemical Ocean Sections (GEOSECS) carried out between 1972 and 1978), a second called Transient Tracers in the Ocean (TTO) carried out in 1981, and a third called the Joint Global Ocean Flux Study (JGOSFS) carried out in the 1990s. The surveys measured profiles of carbon, oxygen, radioisotopes, and other tracers along transects in the Atlantic and Pacific Oceans. The differences between the profiles over time have been used to calculate directly the penetration of anthropogenic CO_2 into the oceans (e.g., Gruber *et al.*, 1996, described below). As of 1980, the oceans are thought to have absorbed only ~40% of the emissions (20–47%, depending on the model used; Bolin, 1986).

Direct measurement of changes in the amount of carbon in the world's oceans is difficult for two reasons: first, the oceans are not mixed as rapidly as the atmosphere, so that spatial and temporal heterogeneity is large; and, second, the background concentration of dissolved carbon in seawater is large relative to the change, so measurement of the change requires very accurate methods. Nevertheless, direct measurement of the uptake of anthropogenic carbon is possible, in theory if not practically, by two approaches. The first approach is based on measurement of changes in the oceanic inventory of carbon and the second is based on measurement of the transfer of CO_2 across the air–sea interface.

Measurement of an increase in oceanic carbon is complicated by the background concentration and the natural variability of carbon concentrations in seawater. The total uptake of anthropogenic carbon in the surface waters of the ocean is calculated by models to have been ~40 μmol kg^{-1} of water. Annual changes would, of course, be much smaller than 40 μmol kg^{-1}, as would the increase in DIC concentrations in deeper waters, where less anthropogenic carbon has penetrated. By comparison, the background concentration of DIC in surface waters is 2,000 μmol kg^{-1}. Furthermore, the seasonal variability at one site off Bermuda was 30 μmol kg^{-1}. Against this background and variability, direct measurement of change is a challenge. Analytical techniques add to uncertainties, although current techniques are capable of a precision of 1.5 μmol kg^{-1} within a laboratory and 4 μmol kg^{-1} between laboratories (Sarmiento, 1993).

A second method for directly measuring carbon uptake by the oceans, measurement of the air–sea exchange, is also made difficult by spatial and temporal variability. The approach measures the concentration of CO_2 in the air and in the surface mixed layer. The difference defines the gradient, which, together with a model that relates the exchange coefficient to wind speed, enables the rate of exchange to be calculated. An average air–sea difference (gradient) of 8 ppm, globally, is equivalent to an oceanic uptake of $2\,Pg\,C\,yr^{-1}$ (Sarmiento, 1993), but the natural variability is greater than 10 ppm. Furthermore, the gas transfer coefficient is also uncertain within a factor of 2 (Broecker, 2001).

Because of the difficulty in measuring either changes in the ocean's inventory of carbon or the exchange of carbon across the air–sea interface, the uptake of anthropogenic carbon by the oceans is calculated with models that simulate the chemistry of carbon in seawater, the air–sea exchanges of CO_2, and oceanic circulation.

Ocean carbon models. Models of the ocean carbon cycle include three processes that affect the uptake and redistribution of carbon within the ocean: the air–sea transfer of CO_2, the chemistry of CO_2 in seawater, and the circulation or mixing of the ocean's water masses.

Three tracers have been used to constrain models. One tracer is CO_2 itself. The difference between current distribution of CO_2 in the ocean and the distribution expected without anthropogenic emissions yields an estimate of oceanic uptake (Gruber *et al.*, 1996). The approach is based on changes that occur in the chemistry of seawater as it ages. With age, the organic matter present in surface waters decays, increasing the concentration of CO_2 and various nutrients, and decreasing the concentration of O_2. The hard parts $(CaCO_3)$ of marine organisms also decay with time, increasing the alkalinity of the water. From data on the concentrations of CO_2, O_2, and alkalinity throughout the oceans, it is theoretically possible to calculate the increased abundance of carbon in the ocean as a result of the increased concentration in the atmosphere. The approach is based on the assumption that the surface waters were in equilibrium with the atmosphere when they sank, or, at least, that the extent of disequilibrium is known. The approach is sensitive to seasonal variation in the CO_2 concentration in these surface waters.

A second tracer is bomb ^{14}C. The distribution of bomb ^{14}C in the oceans (Broecker *et al.*, 1995), together with an estimate of the transfer of $^{14}CO_2$ across the air–sea interface (Wanninkhof, 1992) (taking into account the fact that $^{14}CO_2$ equilibrates ~10 times more slowly than CO_2 across this interface), yields a constraint on uptake. A third constraint is based on the penetration of CFCs into the oceans (Orr and Dutay, 1999; McNeil *et al.*, 2003).

Ocean carbon models calculate changes in the oceanic carbon inventory. When these changes, together with changes in the atmospheric carbon inventory (from atmospheric and ice core CO_2 data), are subtracted from the emissions of carbon from fossil fuels, the result is an estimate of the net annual terrestrial flux of carbon.

Most current models of the ocean reproduce the major features of oceanic carbon: the vertical gradient in DIC, the seasonal and latitudinal patterns of p_{CO_2} in surface waters, and the interannual variability in p_{CO_2} observed during El Niños (Prentice *et al.*, 2001). However, ocean models do not capture the spatial distribution of ^{14}C at depth (Orr *et al.*, 2001), and they do not show an interhemispheric transport of carbon that is suggested from atmospheric CO_2 measurements (Stephens *et al.*, 1998). The models also have a tight biological coupling between carbon and nutrients, which seems not to have existed in the past and may not exist in the future. The issue is addressed below in Section 3.4.2.2.

3.3.1.4 Land: net exchange of carbon between terrestrial ecosystems and the atmosphere

Direct measurement of change in the amount of carbon held in the world's vegetation and soils may be more difficult than measurement of change in the oceans, because the land surface is not mixed. Not only are the background levels high (~550 Pg C in vegetation and ~1,500 in soils), but the spatial heterogeneity is greater on land than in the ocean. Thus, measurement of annual changes even as large as $3\,Pg\,C\,yr^{-1}$, in background levels 100 times greater, would require a very large sampling approach. Change may be measured over short intervals of a year or so in individual ecosystems by measuring fluxes of carbon, as, e.g., with the eddy flux technique (Goulden *et al.*, 1996), but, again, the results must be scaled up from 1 km^2 to the ecosystem, landscape, region, and globe.

Global changes in terrestrial carbon were initially estimated by difference, i.e., by estimates of change in the other three reservoirs. Because the global mass of carbon is conserved, when three terms of the global carbon budget are known, the fourth can be determined by difference. For the period 1850–2000, three of the terms (275 Pg C released from fossil fuels, 175 Pg C accumulated in the atmosphere, and 140 Pg C taken up by the oceans) define a net terrestrial uptake of 40 Pg C (Table 5). Temporal variations in these terrestrial sources and sinks can also be determined through inverse calculations with ocean carbon models (see Section 3.3.1.3). In inverse mode, models calculate the annual sources and sinks of carbon (output) necessary to produce observed concentrations of CO_2 in the atmosphere (input). Then, subtracting known fossil fuel sources from the calculated sources and sinks yields a residual flux of carbon, presumably terrestrial, because the other terms have been accounted for (the atmosphere and fossil fuels directly, the oceans indirectly). One such inverse calculation or deconvolution (Joos *et al.*, 1999b) suggests that terrestrial ecosystems were a net source of carbon until ~1940 and then became a small net sink. Only in the early 1990s was the net terrestrial sink greater than 0.5 Pg C yr^{-1} (Figure 4).

3.3.1.5 Land: changes in land use

At least a portion of terrestrial sources and sinks can be determined more directly from the large changes in vegetation and soil carbon that result from changes in land use, such as the conversion of forests to cleared lands. Changes in the use of land affect the amount of carbon stored in vegetation and soils and, hence, affect the flux of carbon between land and the atmosphere. The amount of carbon released to the atmosphere or accumulated on land depends not only on the magnitude and types of changes in land use, but also on the amounts of carbon held in different ecosystems. For example, the conversion of grassland to pasture may release no carbon to the atmosphere because the stocks of carbon are unchanged. The net release or accumulation of

carbon also depends on time lags introduced by the rates of decay of organic matter, the rates of oxidation of wood products, and the rates of regrowth of forests following harvest or following abandonment of agriculture land. Calculation of the net terrestrial flux of carbon requires knowledge of these rates in different ecosystems under different types of land use. Because there are several important forms of land use and many types of ecosystems in different parts of the world, and because short-term variations in the magnitude of the flux are important, computation of the annual flux requires a computer model.

Changes in terrestrial carbon calculated from changes in land use. Bookkeeping models (Houghton *et al.*, 1983; Hall and Uhlig, 1991; Houghton and Hackler, 1995) have been used to calculate net sources and sinks of carbon resulting from land-use change in all the world's regions. Calculations are based on two types of data: rates of land-use change and per hectare changes in carbon stocks that follow a change in land use. Changes in land use are defined broadly to include the clearing of lands for cultivation and pastures, the abandonment of these agricultural lands, the harvest of wood, reforestation, afforestation, and shifting cultivation. Some analyses have included wildfire because active policies of fire exclusion and fire suppression have affected carbon storage (Houghton *et al.*, 1999).

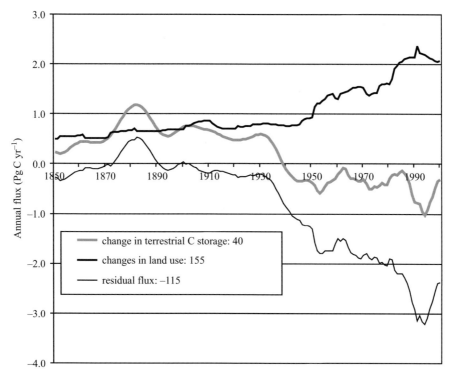

Figure 4 The net annual flux of carbon to or from terrestrial ecosystems (from inverse calculations with an ocean model (Joos *et al.*, 1999b), the flux of carbon from changes in land use (from Houghton, 2003), and the difference between the net flux and the flux from land-use change (i.e., the residual terrestrial sink). Positive values indicate a source of carbon from land and negative values indicate a terrestrial sink.

Bookkeeping models used to calculate fluxes of carbon from changes in land use track the carbon in living vegetation, dead plant material, wood products, and soils for each hectare of land cultivated, harvested, or reforested. Rates of land-use change are generally obtained from agricultural and forestry statistics, historical accounts, and national handbooks. Carbon stocks and changes in them following disturbance and growth are obtained from field studies. The data and assumptions used in the calculations are more fully documented in Houghton (1999) and Houghton and Hackler (2001).

The calculated flux is not the net flux of carbon between terrestrial ecosystems and the atmosphere, because the analysis does not consider ecosystems undisturbed by direct human activity. Rates of decay and rates of regrowth are defined in the model for different types of ecosystems and different types of land-use change, but they do not vary through time in response to changes in climate or concentrations of CO_2. The processes explicitly included in the model are the ecological processes of disturbance and recovery, not the physiological processes of photosynthesis and respiration.

The worldwide trend in land use over the last 300 years has been to reduce the area of forests, increase the area of agricultural lands, and, therefore, reduce the amount of carbon on land. Although some changes in land use increase the carbon stored on land, the net change for the 150-year period 1850–2000 is estimated to have released 156 Pg C (Houghton, 2003). An independent comparison of 1990 land cover with maps of natural vegetation suggests that another 58–75 Pg C (or ~30% of the total loss) were lost before 1850 (DeFries et al., 1999).

The net annual fluxes of carbon to the atmosphere from terrestrial ecosystems (and fossil fuels) are shown in Figure 3. The estimates of the net flux from land before 1800 are relatively less reliable, because early estimates of land-use change are often incomplete. However, the absolute errors for the early years are small because the fluxes themselves were small. There were no worldwide economic or cultural developments in the eighteenth century that would have caused changes in land use of the magnitude that began in the nineteenth century and accelerated to the present day. The net annual biotic flux of carbon to the atmosphere before 1800 was probably less than 0.5 Pg and probably less than 1 Pg C until ~1950.

It was not until the middle of the last century that the annual emissions of carbon from combustion of fossil fuels exceeded the net terrestrial source from land-use change. Since then the fossil fuel contribution has predominated, although both fluxes have accelerated in recent decades with the intensification of industrial activity and the expansion of agricultural area.

The major releases of terrestrial carbon result from the oxidation of vegetation and soils associated with the expansion of cultivated land. The harvest of forests for fuelwood and timber is less important because the release of carbon to the atmosphere from the oxidation of wood products is likely to be balanced by the storage of carbon in regrowing forests. The balance will occur only as long as the forests harvested are allowed to regrow, however. If wood harvest leads to permanent deforestation, the process will release carbon to the atmosphere.

In recent decades the net release of carbon from changes in land use has been almost entirely from the tropics, while the emissions of CO_2 from fossil fuels were almost entirely from outside the tropics. The highest biotic releases were not always from tropical countries. The release of terrestrial carbon from the tropics is a relatively recent phenomenon, post-1945. In the nineteenth century the major sources were from the industrialized regions—North America, Europe, and the Soviet Union—and from those regions with the greatest numbers of people—South Asia and China.

3.3.1.6 Land: a residual flux of carbon

The amount of carbon calculated to have been released from changes in land use since the early 1850s (156 Pg C) (Houghton, 2003) is much larger than the amount calculated to have been released using inverse calculations with global carbon models (40 Pg C) (Joos et al., 1999b) (Section 3.3.1.4). Moreover, the net source of CO_2 from changes in land use has generally increased over the past century, while the inversion approach suggests, on the contrary, that the largest releases of carbon from land were before 1930, and that since 1940 terrestrial ecosystems have been a small net sink (Figure 4).

The difference between these two estimates is greater than the errors in either one or both of the analyses, and might indicate a flux of carbon from processes not related to land-use change. The approach based on land-use change includes only the sources and sinks of carbon directly attributable to human activity; ecosystems not directly modified by human activity are left out of the analysis (assumed neither to accumulate nor release carbon). The approach based on inverse analyses with atmospheric data, in contrast, includes all ecosystems and all processes affecting carbon storage. It yields a net terrestrial flux of carbon. The difference between the two approaches thus suggests a generally increasing terrestrial sink for carbon attributable to factors other than land-use change. Ecosystems not directly cut or cleared could be accumulating or releasing carbon in response to small variations in climate, to increased concentrations of CO_2 in air, to increased availability of nitrogen or other nutrients, or to increased levels of toxins in air and soil resulting from industrialization. It is also possible that management practices not considered in analyses of land-use change may have increased the storage of carbon on lands that have been affected by land-use change. These possibilities will be discussed in more detail below (Section 3.4.1). Interestingly, the two estimates (land-use change and inverse modeling) are generally in agreement before 1935 (Figure 4), suggesting that before that date the net flux of carbon from terrestrial ecosystems was largely the result of changes in land use. Only after 1935 have changes in land use underestimated the net terrestrial carbon sink. By the mid-1990s this annual residual sink had grown to $\sim 3 \, \text{Pg} \, \text{C} \, \text{yr}^{-1}$.

3.3.2 Changes Over the Period 1980–2000

The period 1980–2000 deserves special attention not because the carbon cycle is qualitatively different over this period, but because scientists have been able to understand it better. Since 1980 new types of measurements and sophisticated methods of analysis have enabled better estimates of the uptake of carbon by the world's oceans and terrestrial ecosystems. The following section addresses the results of these analyses, first at the global level, and then at a regional level. Attention focuses on the two outstanding questions that have concerned scientists investigating the global carbon cycle since the first carbon budgets were constructed in the late 1960s (SCEP, 1970): (i) How much of the carbon released to the atmosphere from combustion of fossil fuels and changes in land use is taken up by the oceans and by terrestrial ecosystems? (ii) What are the mechanisms responsible for the uptake of carbon? The mechanisms for a carbon sink in terrestrial ecosystems have received considerable attention, in part because different mechanisms have different implications for future rates of CO_2 growth (and hence future global warming).

The previous section addressed the major reservoirs of the global carbon cycle, one at a time. This section addresses the methods used to determine changes in the amount of carbon held on land and in the sea, the two reservoirs for which changes in carbon are less well known. In contrast, the atmospheric increase in CO_2 and the emissions from fossil fuels are well documented. The order in which methods are presented is arbitrary. To set the stage, top-down (i.e., atmospherically based) approaches are described first, followed by bottom–up (ground-based) approaches (Table 6). Although the results of different methods often differ, the methods are not entirely comparable. Rather, they are complementary, and discrepancies sometimes suggest mechanisms responsible for transfers of carbon (Houghton, 2003; House *et al.*, 2003). The results from each method are presented first, and then they are added to an accumulating picture of the global carbon cycle. Again, the emphasis is on, first, the fluxes of carbon to and from terrestrial ecosystems and the ocean and, second, the mechanisms responsible for the terrestrial carbon sink.

3.3.2.1 The global carbon budget

(i) Inferring changes in terrestrial and oceanic carbon from atmospheric concentrations of CO_2 and O_2. According to the most recent assessment of climate change by the IPCC, the world's terrestrial ecosystems were a net sink averaging close to zero ($0.2 \, \text{Pg} \, \text{C} \, \text{yr}^{-1}$) during the 1980s and a significantly larger sink ($1.4 \, \text{Pg} \, \text{C} \, \text{yr}^{-1}$) during the 1990s (Prentice *et al.*, 2001). The large increase during the 1990s is difficult to explain. Surprisingly, the oceanic uptake of carbon was greater in the 1980s than the 1990s. The reverse would have been expected because atmospheric concentrations of CO_2 were higher in the 1990s. The estimates of terrestrial and oceanic uptake were based on changes in atmospheric CO_2 and O_2 and contained

Table 6 Characteristics of methods use to estimate terrestrial sinks.

	Geographic limitations	*Temporal resolution*	*Attribution of mechanism(s)*	*Precision*
Inverse modeling: oceanic data	No geographic resolution	Annual	No	Moderate
Land-use models	Data limitations in some regions	Annual	Yes	Moderate
Inverse modeling: atmospheric data	Poor in tropics	Monthly to annual	No	High: North–South Low: East–West
Forest inventories	Nearly nonexistent in the tropics	5–10 years	Yes (age classes)	High for biomass; variable for soil carbon
CO_2 flux	Site specific (a few km²); difficult to scale up	Hourly to annual	No	Some problems with windless conditions
Physiologically based models	None	Hourly to annual	Yes	Variable; difficult to validate

a small adjustment for the outgassing of O_2 from the oceans.

One approach for distinguishing terrestrial from oceanic sinks of carbon is based on atmospheric concentrations of CO_2 and O_2. CO_2 is released and O_2 taken up when fossil fuels are burned and when forests are burned. On land, CO_2 and O_2 are tightly coupled. In the oceans they are not, because O_2 is not very soluble in seawater. Thus, CO_2 is taken up by the oceans without any change in the atmospheric concentration of O_2. Because of this differential response of oceans and land, changes in atmospheric O_2 relative to CO_2 can be used to distinguish between oceanic and terrestrial sinks of carbon (Keeling and Shertz, 1992; Keeling et al., 1996b; Battle et al., 2000). Over intervals as short as a few years, slight variations in the seasonality of oceanic production and decay may appear as a change in oceanic O_2, but these variations cancel out over

many years, making the method robust over multiyear intervals (Battle et al., 2000).

Figure 5 shows how the method works. The individual points show average annual global CO_2/O_2 concentrations over the years 1990–2000. Changes in the concentrations expected from fossil fuel combustion (approximately 1:1) during this interval are drawn, starting in 1990. The departure of these two sets of data confirms that carbon has accumulated somewhere besides the atmosphere. The oceans are assumed not to be changing with respect to O_2, so the line for the oceanic sink is horizontal. The line for the terrestrial sink is approximately parallel to the line for fossil fuel, and drawn through 2000. The intersection of the terrestrial and the oceanic lines thus defines the terrestrial and oceanic sinks. According to the IPCC (Prentice et al., 2001), these sinks averaged $1.4\,\mathrm{Pg\,C\,yr^{-1}}$ and $1.7\,\mathrm{Pg\,C\,yr^{-1}}$, respectively, for the 1990s. The estimate also included a

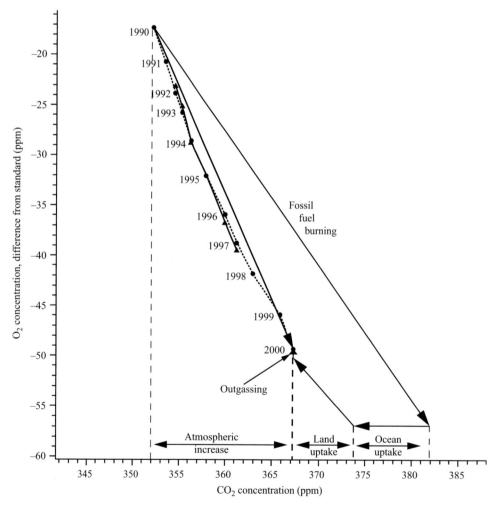

Figure 5   Terrestrial and oceanic sinks of carbon deduced from changes in atmospheric concentrations of CO_2 and O_2 (Prentice et al., 2001) (reproduced by permission of the Intergovernmental Panel on Climate Change from *Climate Change 2001: The Scientific Basis*, **2001**, pp. 183–237).

small correction for outgassing of O_2 from the ocean (in effect, recognizing that the ocean is not neutral with respect to O_2).

Recent analyses suggest that such outgassing is significantly larger than initially estimated (Bopp *et al.*, 2002; Keeling and Garcia, 2002; Plattner *et al.*, 2002). The observed decadal variability in ocean temperatures (Levitus *et al.*, 2000) suggests a warming-caused reduction in the transport rate of O_2 to deeper waters and, hence, an increased outgassing of O_2. The direct effect of the warming on O_2 solubility is estimated to have accounted for only 10% of the loss of O_2 (Plattner *et al.*, 2002). The revised estimates of O_2 outgassing change the partitioning of the carbon sink between land and ocean. The revision increases the oceanic carbon sink of the 1990s relative to that of the 1980s (average sinks of $1.7 \, \text{Pg} \, \text{C} \, \text{yr}^{-1}$ and $2.4 \, \text{Pg} \, \text{C} \, \text{yr}^{-1}$, respectively, for the 1980s and 1990s). The revised estimates are more consistent with estimates from ocean models (Orr, 2000) and from analyses based on $^{13}\text{C}/^{12}\text{C}$ ratios of atmospheric CO_2 (Joos *et al.*, 1999b; Keeling *et al.*, 2001). The revised estimate for land (a net sink of $0.7 \, \text{Pg} \, \text{C} \, \text{yr}^{-1}$ during the 1990s) (Table 7) is half of that given by the IPCC (Prentice *et al.*, 2001). The decadal change in the terrestrial sink is also much smaller (from $0.4 \, \text{Pg} \, \text{C} \, \text{yr}^{-1}$ to $0.7 \, \text{Pg} \, \text{C} \, \text{yr}^{-1}$ instead of from $0.2 \, \text{Pg} \, \text{C} \, \text{yr}^{-1}$ to $1.4 \, \text{Pg} \, \text{C} \, \text{yr}^{-1}$).

(ii) Sources and sinks inferred from inverse modeling with atmospheric transport models and atmospheric concentrations of CO_2, $^{13}CO_2$, and O_2. A second top-down method for determining oceanic and terrestrial sinks is based on spatial and temporal variations in concentrations of atmospheric CO_2 obtained through a network of flask air samples (Masarie and Tans, 1995; Cooperative Atmospheric Data Integration Project—Carbon Dioxide, 1997). Together with models of atmospheric transport, these variations are used to infer the geographic distribution of sources and sinks of carbon through a technique called inverse modeling.

Variations in the carbon isotope of CO_2 may also be used to distinguish terrestrial sources and sinks from oceanic ones. The ^{13}C isotope is slightly heavier than the ^{12}C isotope and is discriminated against during photosynthesis. Thus, trees have a lighter isotopic ratio ($-22 \, \text{ppt}$ to $-27 \, \text{ppt}$) than does air ($-7 \, \text{ppt}$) (ratios are expressed relative to a standard). The burning of forests (and fossil fuels) releases a disproportionate share of the lighter isotope, reducing the isotopic ratio of $^{13}\text{C}/^{12}\text{C}$ in air. In contrast, diffusion of CO_2 across the air–sea interface does not result in appreciable discrimination, so variations in the isotopic composition of CO_2 suggest terrestrial and fossil fuels fluxes of carbon, rather than oceanic.

Spatial and temporal variations in the concentrations of CO_2, $^{13}CO_2$, and O_2 are used with models of atmospheric transport to infer (through inverse calculations) sources and sinks of carbon at the Earth's surface. The results are dependent upon the model of atmospheric transport (Figure 6; Ciais *et al.*, 2000).

The interpretation of variations in ^{13}C is complicated. One complication results from isotopic disequilibria in carbon pools (Battle *et al.*, 2000). Disequilibria occur because the $\delta^{13}\text{C}$ taken up by plants, e.g., is representative of the $\delta^{13}\text{C}$ currently in the atmosphere (allowing for discrimination), but the $\delta^{13}\text{C}$ of CO_2 released through decay represents not the $\delta^{13}\text{C}$ of the current atmosphere but of an atmosphere several decades ago. As long as the $\delta^{13}\text{C}$ of the atmosphere is changing, the $\delta^{13}\text{C}$ in pools will reflect a mixture of earlier and current conditions. Uncertainties in the turnover of various carbon pools add uncertainty to interpretation of the $\delta^{13}\text{C}$ signal. Another complication results from unknown year-to-year variations in the photosynthesis of C_3 and C_4 plants (because these two types of plants discriminate differently against the heavier isotope). C_4 plants discriminate less than C_3 plants and leave a signal that looks oceanic, thus confounding the separation of land and ocean exchanges. These uncertainties of the $\delta^{13}\text{C}$ approach are most troublesome over long periods (Battle *et al.*, 2000); the approach is more reliable for reconstructing interannual variations in sources and sinks of carbon.

An important distinction exists between global approaches (e.g., O_2, above) and regional inverse approaches, such as implemented with ^{13}C. In the

Table 7 The global carbon budget ($\text{Pg} \, \text{C} \, \text{yr}^{-1}$).

	1980s	*1990s*
Fossil fuel emissions[a]	5.4 ± 0.3	6.3 ± 0.4
Atmospheric increase[a]	3.3 ± 0.1	3.2 ± 0.2
Oceanic uptake[b]	$-1.7 + 0.6 \, (-1.9 \pm 0.6)$	$-2.4 + 0.7 \, (-1.7 \pm 0.5)$
Net terrestrial flux[b]	$-0.4 + 0.7 \, (-0.2 \pm 0.7)$	$-0.7 + 0.8 \, (-1.4 \pm 0.7)$
Land-use change[c]	2.0 ± 0.8	2.2 ± 0.8
Residual terrestrial flux	$-2.4 + 1.1 \, (-2.2 \pm 1.1)$	$-2.9 + 1.1 \, (-3.6 \pm 1.1)$

[a] Source: Prentice *et al.* (2001).
[b] Source: Plattner *et al.* (2002) (values in parentheses are from Prentice *et al.*, 2001).
[c] Houghton (2003).

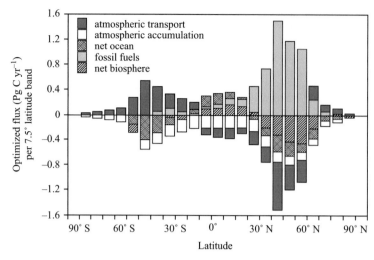

Figure 6 Terrestrial and oceanic sources and sinks of carbon inferred from inverse calculations with an atmospheric transport model and spatial and temporal variations in CO_2 concentrations. The net fluxes inferred over each region have been averaged into 7.5°-wide latitude strips (Ciais *et al.*, 2000) (reproduced by permission of the Ecological Society of America from *Ecol. Appl.*, **2000**, *10*, 1574–1589).

global top-down approach, changes in terrestrial or oceanic carbon *storage* are calculated. In contrast, the regional inverse method yields *fluxes* of carbon between the land or ocean surface and the atmosphere. These fluxes of carbon include both natural and anthropogenic components. Horizontal exchange between regions must be taken into account to estimate changes in storage. For example, the fluxes will not accurately reflect changes in the amount of carbon on land or in the sea if some of the carbon fixed by terrestrial plants is transported by rivers to the ocean and respired there (Sarmiento and Sundquist, 1992; Tans *et al.*, 1995; Aumont *et al.*, 2001).

An example of inverse calculations is the analysis by Tans *et al.* (1990). The concentration of CO_2 near the Earth's surface is ~3 ppm higher over the northern mid-latitudes than over the southern hemisphere. The "bulge" in concentration over northern mid-latitudes is consistent with the emissions of carbon from fossil fuel combustion at these latitudes. The extent of the bulge is also affected by the rate of atmospheric mixing. High rates of mixing would dilute the bulge; low rates would enhance it. By using the latitudinal gradient in CO_2 and the latitudinal distribution of fossil fuel emissions, together with a model of atmospheric transport, Tans *et al.* (1990) determined that the bulge was smaller than expected on the basis of atmospheric transport alone. Thus, carbon is being removed from the atmosphere by the land and oceans at northern mid-latitudes. Tans *et al.* estimated removal rates averaging between 2.4 Pg C yr^{-1} and 3.5 Pg C yr^{-1} for the years 1981–1987. From p_{CO_2} measurements in surface waters, Tans *et al.* calculated that the northern mid-latitude oceans were taking up only

0.2–0.4 Pg C yr^{-1}, and thus, by difference, northern mid-latitude lands were responsible for the rest, a sink of 2.0–3.4 Pg C yr^{-1}. The range resulted from uncertainties in atmospheric transport and the limited distribution of CO_2 sampling stations. Almost no stations exist over tropical continents. Thus, Tans *et al.* (1990) could not constrain the magnitude of a tropical land source or sink, but they could determine the magnitude of the northern sink relative to a tropical source. A large tropical source, as might be expected from deforestation, implied a large northern sink; a small tropical source implied a smaller northern sink.

The analysis by Tans *et al.* (1990) caused quite a stir because their estimate for oceanic uptake was only 0.3–0.8 Pg C yr^{-1}, while analyses based on ocean models yielded estimates of 2.0 ± 0.8 Pg C yr^{-1}. The discrepancy was subsequently reconciled (Sarmiento and Sundquist, 1992) by accounting for the effect of skin temperature on the calculated air–sea exchange, the effect of atmospheric transport and oxidation of CO on the carbon budget, and the effect of riverine transport of carbon on changes in carbon storage (see below). All of the adjustments increased the estimated oceanic uptake of carbon to values obtained by ocean models and lowered the estimate of the mid-latitude terrestrial sink.

Similar inverse approaches, using not only CO_2 concentrations but also spatial variations in O_2 and $^{13}CO_2$ to distinguish oceanic from terrestrial fluxes, have been carried out by several groups since 1990. An intercomparison of 16 atmospheric transport models (the TransCom 3 project) by Gurney *et al.* (2002) suggests average oceanic and terrestrial sinks of 1.3 Pg C yr^{-1} and 1.4 Pg C yr^{-1}, respectively, for the period 1992–1996.

The mean global terrestrial sink of $1.4\,Pg\,C\,yr^{-1}$ for the years 1992–1996 is higher than that obtained from changes in O_2 and CO_2 ($0.7\,Pg\,C\,yr^{-1}$) (Plattner *et al.*, 2002). However, the estimate from inverse modeling has to be adjusted to account for terrestrial sources and sinks of carbon that are not "seen" by the atmosphere. For example, the fluxes inferred from atmospheric data will not accurately reflect changes in the amount of carbon on land or in the sea if some of the carbon fixed by terrestrial plants or used in weathering minerals is transported by rivers to the ocean and respired and released to the atmosphere there. Under such circumstances, the atmosphere sees a terrestrial sink and an oceanic source, while the storage of carbon on land and in the sea may not have changed. Several studies have tried to adjust atmospherically based carbon budgets by accounting for the river transport of carbon. Sarmiento and Sundquist (1992) estimated a pre-industrial net export by rivers of 0.4–$0.7\,Pg\,C\,yr^{-1}$, balanced by a net terrestrial uptake of carbon through photosynthesis and weathering. Aumont *et al.* (2001) obtained a global estimate of $0.6\,Pg\,C\,yr^{-1}$. Adjusting the net terrestrial sink obtained through inverse calculations ($1.4\,Pg\,C\,yr^{-1}$) by $0.6\,Pg\,C\,yr^{-1}$ yields a result ($0.8\,Pg\,C\,yr^{-1}$) similar to the estimate obtained through changes in the concentrations of O_2 and CO_2 (Table 8). The two top-down methods based on atmospheric measurements yield similar global estimates of a net terrestrial sink (~0.7 (±0.8)$\,Pg\,C\,yr^{-1}$ for the 1990s).

(iii) Land-use change. Another method, independent of those based on atmospheric data and models, that has been used to estimate terrestrial sources and sinks of carbon, globally, is a method based on changes in land use (see Section 3.3.1.5). This is a ground-based or bottom-up approach. Changes in land use suggest that deforestation, reforestation, cultivation, and logging were responsible for a carbon source, globally, that averaged $2.0\,Pg\,C\,yr^{-1}$ during the 1980s and $2.2\,Pg\,C\,yr^{-1}$

during the 1990s (Houghton, 2003). The approach includes emissions of carbon from the decay of dead plant material, soil, and wood products and sinks of carbon in regrowing ecosystems, including both vegetation and soil. Analyses account for delayed sources and sinks of carbon that result from decay and regrowth following a change in land use.

Other recent analyses of land-use change give results that bound the results of this summary, although differences in the processes and regions included make comparisons somewhat misleading. An estimate by Fearnside (2000) of a $2.4\,Pg\,C\,yr^{-1}$ source includes only the tropics. A source of $0.8\,Pg\,Cyr^{-1}$ estimated by McGuire *et al.* (2001) includes changes in global cropland area but does not include either the harvest of wood or the clearing of forests for pastures, both of which contributed to the net global source. The average annual release of carbon attributed by Houghton (2003) to changes in the area of croplands ($1.2\,Pg\,C\,yr^{-1}$ for the 1980s) is higher than the estimate found by McGuire *et al.* ($0.8\,Pg\,C\,yr^{-1}$).

The calculated *source* of 2.2 (±0.8) $Pg\,C\,yr^{-1}$ for the 1990s (Houghton, 2003) is very different from the global net terrestrial *sink* determined from top-down analyses ($0.7\,Pg\,C\,yr^{-1}$) (Table 8). Are the methods biased? Biases in the inverse calculations may be in either direction. Because of the "rectifier effect" (the seasonal covariance between the terrestrial carbon flux and atmospheric transport), inverse calculations are thought to underestimate the magnitude of a northern mid-latitude sink (Denning *et al.*, 1995). However, if the near-surface concentrations of atmospheric CO_2 in northern mid-latitude regions are naturally lower than those in the southern hemisphere, the apparent sink in the north may not be anthropogenic, as usually assumed. Rather, the anthropogenic sink would be less than $0.5\,Pg\,C\,yr^{-1}$ (Taylor and Orr, 2000).

In contrast to the unknown bias of atmospheric methods, analyses based on land-use change are deliberately biased. These analyses consider only those changes in terrestrial carbon resulting directly

Table 8 Estimates of the annual terrestrial flux of carbon ($Pg\,C\,yr^{-1}$) in the 1990s according to different methods. Negative values indicate a terrestrial sink.

	O_2 and CO_2	Inverse calculations CO_2, $^{13}CO_2$, O_2	Forest inventories	Land-use change
Globe	−0.7 (±0.8)[a]	−0.8 (±0.8)[b]		2.2 (±0.6)[c]
Northern mid-latitudes		−2.1 (±0.8)[d]	−0.6 to −1.3[e]	−0.03 (±0.5)[c]
Tropics		1.5 (±1.2)[f]	−0.6 (±0.3)[g]	0.5 to 3.0[h]

[a] Plattner *et al.* (2002).
[b] −1.4 (±0.8) from Gurney *et al.* (2002) reduced by 0.6 to account for river transport (Aumont *et al.*, 2001).
[c] Houghton, 2003.
[d] −2.4 from Gurney *et al.* (2002) reduced by 0.3 to account for river transport (Aumont *et al.*, 2001).
[e] −0.65 in forests (Goodale *et al.*, 2002) and another 0.0–0.65 assumed for nonforests (see text).
[f] 1.2 from Gurney *et al.* (2002) increased by 0.3 to account for river transport (Aumont *et al.*, 2001).
[g] Undisturbed forests: −0.6 from Phillips *et al.* (1998) (challenged by Clark, 2002).
[h] 0.9 (range 0.5–1.4) from DeFries *et al.* (2002) 1.3 from Achard *et al.* (2002) adjusted for soils and degradation (see text) 2.2 (±0.8) from Houghton (2003). 2.4 from Fearnside (2000).

from human activity (conversion and modification of terrestrial ecosystems). There may be other sources and sinks of carbon not related to land-use change (such as caused by CO_2 fertilization, changes in climate, or management) that are captured by other methods but ignored in analyses of land-use change. In other words, the flux of carbon from changes in land use is not necessarily the same as the net terrestrial flux from all terrestrial processes.

If the net terrestrial flux of carbon during the 1990s was $0.7\,\mathrm{Pg\,C\,yr^{-1}}$, and $2.2\,\mathrm{Pg\,C\,yr^{-1}}$ were emitted as a result of changes in land use, then $2.9\,\mathrm{Pg\,C\,yr^{-1}}$ must have accumulated on land for reasons not related to land-use change. This residual terrestrial sink was discussed above (Table 7 and Figure 4). That the residual terrestrial sink exists at all suggests that processes other than land-use change are affecting the storage of carbon on land. Recall, however, that the residual sink is calculated by difference; if the emissions from land-use change are overestimated, the residual sink will also be high.

3.3.2.2   Regional distribution of sources and sinks of carbon: the northern mid-latitudes

Insights into the magnitude of carbon sources and sinks and the mechanisms responsible for the residual terrestrial carbon sink may be obtained from a consideration of tropical and extratropical regions separately. Inverse calculations show the tropics to be a moderate source, largely oceanic as a result of CO_2 outgassing in upwelling regions. Some of the tropical source is also terrestrial. Estimates vary greatly depending on the models of atmospheric transport and the years included in the analyses. The net global oceanic sink of $1.3\,\mathrm{Pg\,C\,yr^{-1}}$ for the period 1992–1996 is distributed in northern ($1.2\,\mathrm{Pg\,C\,yr^{-1}}$) and southern oceans ($0.8\,\mathrm{Pg\,C\,yr^{-1}}$), with a net source from tropical gyres ($0.5\,\mathrm{Pg\,C\,yr^{-1}}$) (Gurney et al., 2002).

The net terrestrial sink of $\sim0.7\,\mathrm{Pg\,C\,yr^{-1}}$ is not evenly distributed either. The comparison by Gurney et al. (2002) showed net terrestrial sinks of $2.4\pm0.8\,\mathrm{Pg\,C\,yr^{-1}}$ and $0.2\,\mathrm{Pg\,C\,yr^{-1}}$ for northern and southern mid-latitude lands, respectively, offset to some degree by a net tropical land source of $1.2\pm1.2\,\mathrm{Pg\,C\,yr^{-1}}$. Errors are larger for the tropics than the nontropics because of the lack of sampling stations and the more complex atmospheric circulation there.

River transport and subsequent oceanic release of terrestrial carbon are thought to overestimate the magnitude of the atmospherically derived northern terrestrial sink by $0.3\,\mathrm{Pg\,C\,yr^{-1}}$ and underestimate the tropical source (or overestimate its sink) by the same magnitude (Aumont et al., 2001). Thus, the northern terrestrial sink becomes

$2.1\,\mathrm{Pg\,C\,yr^{-1}}$, while the tropical terrestrial source becomes $1.5\,\mathrm{Pg\,C\,yr^{-1}}$ (Table 8).

Inverse calculations have also been used to infer east–west differences in the distribution of sources and sinks of carbon. Such calculations are more difficult because east–west gradients in CO_2 concentration are an order of magnitude smaller than north–south gradients. Some estimates placed most of the northern sink in North America (Fan et al., 1998); others placed most of it in Eurasia (Bousquet et al., 1999a,b). More recent analyses suggest a sink in both North America and Eurasia, roughly in proportion to land area (Schimel et al., 2001; Gurney et al., 2002). The analyses also suggest that higher-latitude boreal forests are small sources rather than sinks of carbon during some years.

The types of land use determining fluxes of carbon are substantially different inside and outside the tropics (Table 9). As of early 2000s, the fluxes of carbon to and from northern lands are dominated by rotational processes, e.g., logging and subsequent regrowth. Changes in the area of forests are small. The losses of carbon from decay of wood products and slash (woody debris generated as a result of harvest) are largely offset by the accumulation of carbon in regrowing forests (reforestation and regrowth following harvest). Thus, the net flux of carbon from changes in land use is small: a source of $0.06\,\mathrm{Pg\,C\,yr^{-1}}$ during the 1980s changing to a sink of $0.02\,\mathrm{Pg\,C\,yr^{-1}}$ during the 1990s. Both the US and Europe are estimated to have been carbon sinks as a result of land-use change.

Inferring changes in terrestrial carbon storage from analysis of forest inventories. An independent estimate of carbon sources and sinks in northern mid-latitudinal lands may be obtained from forest inventories. Most countries in the northern mid-latitudes conduct periodic inventories of the growing stocks in forests. Sampling is designed to yield estimates of total growing stocks (volumes of merchantable wood) that are estimated with 95% confidence to within 1–5% (Powell et al., 1993; Köhl and Päivinen, 1997; Shvidenko and Nilsson, 1997). Because annual changes due to growth and mortality are small relative to the total stocks, estimates of wood volumes are relatively less precise. A study in the southeastern US determined that regional growing stocks (m^3) were known with 95% confidence to within 1.1%, while changes in the stocks ($\mathrm{m}^3\,\mathrm{yr}^{-1}$) were known to within 39.7% (Phillips et al., 2000). Allometric regressions are used to convert growing stocks (the wood contained in the boles of trees) to carbon, including all parts of the tree (roots, stumps, branches, and foliage as well as bole), nonmerchantable and small trees and non-tree vegetation. Other measurements provide estimates of the carbon in the forest floor (litter)

Table 9 Estimates of the annual sources (+) and sinks (−) of carbon resulting from different types of land-use change and management during the 1990s ($Pg\,C\,yr^{-1}$).

Activity	Tropical regions	Temperate and boreal zones	Globe
Deforestation	2.110[a]	0.130	2.240
Afforestation	−0.100	−0.080[b]	−0.190
Reforestation (agricultural abandonment)	0[a]	−0.060	−0.060
Harvest/management	0.190	0.120	0.310
Products	0.200	0.390	0.590
Slash	0.420	0.420	0.840
Regrowth	−0.430	−0.690	−1.120
Fire suppression[c]	0	−0.030	−0.030
Nonforests			
Agricultural soils[d]	0	0.020	0.020
Woody encroachment[c]	0	−0.060	−0.060
Total	2.200	0.040	2.240

[a] Only the net effect of shifting cultivation is included here. The gross fluxes from repeated clearing and abandonment are not included.
[b] Areas of plantation forests are not generally reported in developed countries. This estimates includes only China's plantations.
[c] Probably an underestimate. The estimate is for the US only, and similar values may apply in South America, Australia, and elsewhere.
[d] These values include loss of soil carbon resulting from cultivation of new lands; they do not include accumulations of carbon that may have resulted from recent agricultural practices.

and soil. The precision of the estimates for these other pools of carbon is less than that for the growing stocks. An uncertainty analysis for 140×10^6 ha of US forests suggested an uncertainty of $0.028\,Pg\,C\,yr^{-1}$ (Heath and Smith, 2000). The strength of forest inventories is that they provide direct estimates of wood volumes on more than one million plots throughout northern mid-latitude forests, often inventoried on 5–10-year repeat cycles. Some inventories also provide estimates of growth rates and estimates of mortality from various causes, i.e., fires, insects, and harvests. One recent synthesis of these forest inventories, after converting wood volumes to total biomass and accounting for the fate of harvested products and changes in pools of woody debris, forest floor, and soils, found a net northern mid-latitude terrestrial sink of between $0.6\,Pg\,C\,yr^{-1}$ and $0.7\,Pg\,C\,yr^{-1}$ for the years around 1990 (Goodale *et al.*, 2002). The estimate is ~30% of the sink inferred from atmospheric data corrected for river transport (Table 8). Some of the difference may be explained if nonforest ecosystems throughout the region are also accumulating carbon. Inventories of nonforest lands are generally lacking, but in the US, at least, nonforests are estimated to account for 40–70% of the net terrestrial carbon sink (Houghton *et al.*, 1999; Pacala *et al.*, 2001).

It is also possible that the accumulation of carbon below ground, not directly measured in forest inventories, was underestimated and thus might account for the difference in estimates. However, the few studies that have measured the accumulation of carbon in forest soils have consistently found soils to account for only a small fraction (5–15%) of measured ecosystem sinks (Gaudinski *et al.*, 2000; Barford *et al.*, 2001;

Schlesinger and Lichter, 2001). Thus, despite the fact that the world's soils hold 2–3 times more carbon than biomass, there is no evidence, as of early 2000s, that they account for much of a terrestrial sink.

The discrepancy between estimates obtained from forest inventories and inverse calculations might also be explained by differences in the dates of measurements. The northern sink of $2.1\,Pg\,C\,yr^{-1}$ from Gurney *et al.* ($-2.4 + 0.3$ for riverine transport) is for 1992–1996 and would probably have been lower (and closer to the forest inventory-based estimate) if averaged over the entire decade (see other estimates in Prentice *et al.*, 2001). Top-down measurements based on atmospheric data are sensitive to large year-to-year variations in the growth rate of CO_2 concentrations.

Both forest inventories and inverse calculations with atmospheric data show terrestrial ecosystems to be a significant carbon sink, while changes in land use show a sink near zero. Either the analyses of land-use change are incomplete, or other mechanisms besides land-use change must be responsible for the observed sink, or some combination of both. With respect to the difference between forest inventories and land-use change, a regional comparison suggests that the recovery of forests from land-use change (abandoned farmlands, logging, fire suppression) may either overestimate or underestimate the sinks measured in forest inventories (Table 10). In Canada and Russia, the carbon sink calculated for forests recovering from harvests (land-use change) is greater than the measured sink. The difference could be error, but it is consistent with the fact that fires and insect damage increased in these regions during the 1980s and thus converted

Table 10 Annual net changes in the living vegetation of forests (Pg C yr^{-1}) in northern mid-latitude regions around the year 1990. Negative values indicate an increase in carbon stocks (i.e., a terrestrial sink).

Region	Land-use change[a]	Forest inventory[b]	Sink from land-use change relative to inventoried sink
Canada	−0.025	0.040	0.065 (larger)
Russia	−0.055	0.040	0.095 (larger)
USA	−0.035	−0.110	0.075 (smaller)
China	0.075	−0.040	0.115 (smaller)
Europe	−0.020	−0.090	0.070 (smaller)
Total	−0.060	−0.160	0.100 (smaller)

[a] Houghton (2003).
[b] From Goodale *et al.* (2002).

some of the boreal forests from sinks to sources (Kurz and Apps, 1999). These sources would not be counted in the analysis of land-use change, because natural disturbances were ignored. In time, recovery from these natural disturbances will increase the sink above that calculated on the basis of harvests alone, but as of early 2000s the sources from fire and insect damage exceed the net flux associated with harvest and regrowth.

In the three other regions (Table 10), changes in land use yield a sink that is smaller than measured in forest inventories. If the results are not simply a reflection of error, the failure of past changes in land use to explain the measured sink suggests that factors not considered in the analysis have enhanced the storage of carbon in the forests of the US, Europe, and China. Such factors include past natural disturbances, more subtle forms of management than recovery from harvest and agricultural abandonment (and fire suppression in the US), and environmental changes that may have enhanced forest growth. It is unclear whether the differences between estimates (changes in land use and forest inventories) are real or the result of errors and omissions. The differences are small, generally less than 0.1 Pg C yr^{-1} in any region. The likely errors and omissions in analyses of land-use change include uncertain rates of forest growth, natural disturbances, and many types of forest management (Spiecker *et al.*, 1996).

3.3.2.3 *Regional distribution of sources and sinks of carbon: the tropics*

How do different methods compare in the tropics? Inverse calculations show that tropical lands were a net source of carbon, 1.2 ± 1.2 Pg C yr^{-1} for the period 1992–1996 (Gurney *et al.*, 2002). Accounting for the effects of rivers (Aumont *et al.*, 2001) suggests a source of 1.5 (± 1.2) Pg C yr^{-1} (Table 8).

Forest inventories for large areas of the tropics are rare, although repeated measurements of permanent plots throughout the tropics suggest that undisturbed tropical forests are accumulating carbon, at least in the neotropics (Phillips *et al.*, 1998).

The number of such plots was too small in tropical African or Asian forests to demonstrate a change in carbon accumulation, but assuming the plots in the neotropics are representative of undisturbed forests in that region suggests a sink of 0.62 (± 0.30) Pg C yr^{-1} for mature humid neotropical forests (Phillips *et al.*, 1998). The finding of a net sink has been challenged, however, on the basis of systematic errors in measurement. Clark (2002) notes that many of the measurements of diameter included buttresses and other protuberances, while the allometric regressions used to estimate biomass were based on above-buttress relationships. Furthermore, these stem protuberances display disproportionate rates of radial growth. Finally, some of the plots were on floodplains where primary forests accumulate carbon. When plots with buttresses were excluded (and when recent floodplain (secondary) forests were excluded as well), the net increment was not statistically different from zero (Clark, 2002). Phillips *et al.* (2002) counter that the errors are minor, but the results remain contentious.

Thus, the two methods most powerful in constraining the northern net sink (inverse analyses and forest inventories) are weak or lacking in the tropics (Table 14), and the carbon balance of the tropics is less certain.

Direct measurement of CO$_2$ flux. The flux of CO$_2$ between an ecosystem and the atmosphere can be calculated directly by measuring the covariance between concentrations of CO$_2$ and vertical wind speed (Goulden *et al.*, 1996). The approach is being applied at ∼150 sites in North America, South America, Asia, and Europe. The advantage of the approach is that it includes an integrated measure for the whole ecosystem, not only the wood or the soil. The method is ideal for determining the short-term response of ecosystems to diurnal, seasonal, and interannual variations of such variables as temperature, soil moisture, and cloudiness. If measurements are made over an entire year or over a significant number of days in each season, an annual carbon balance can be determined. The results of such measured fluxes have been demonstrated in at least one ecosystem to be in agreement with independent measurements of change in the major components of the ecosystem (Barford, 2001).

As NEP is often small relative to the gross fluxes of photosynthesis and ecosystem respiration, the net flux is sometimes less than the error of measurement. More important than error is bias, and the approach is vulnerable to bias because both the fluxes of CO_2 and the micrometeorological conditions are systematically different day and night. Wind speeds below $17\ cm\ s^{-1}$ in a temperate zone forest, e.g., resulted in an underestimate of nighttime respiration (Barford *et al.*, 2001). A similar relationship between nighttime wind speed and respiration in forests in the Brazilian Amazon suggests that the assumption that lateral transport is unimportant may have been invalid (Miller *et al.*, in press).

Although the approach works well where micrometeorological conditions are met, the footprint for the measured flux is generally less than $1\ km^2$, and it is difficult to extrapolate the measured flux to large regions. Accurate extrapolations require a distribution of tower sites representative of different flux patches, but such patches are difficult to determine *a priori*. The simple extrapolation of an annual sink of $1\ Mg\ C\ ha\ yr^{-1}$ (based on 55 days of measurement) for a tropical forest in Brazil to all moist forests in the Brazilian Amazon gave an estimated sink of $\sim 1\ Pg\ C\ yr^{-1}$ (Grace *et al.*, 1995). In contrast, a more sophisticated extrapolation based on a spatial model of CO_2 flux showed a basin-wide estimate averaging only $0.3\ Pg\ C\ yr^{-1}$ (Tian *et al.*, 1998). The modeled flux agreed with the measured flux in the location of the site; spatial differences resulted from variations in modeled soil moisture throughout the basin.

Initially, support for an accumulation of carbon in undisturbed tropical forests came from measurements of CO_2 flux by eddy correlation (Grace *et al.*, 1995; Malhi *et al.*, 1998). Results showed large sinks of carbon in undisturbed forests, that, if scaled up to the entire tropics, yielded sinks in the range of $3.9-10.3\ Pg\ C\ yr^{-1}$ (Malhi *et al.*, 2001), much larger than the sources of carbon from deforestation. Tropical lands seemed to be a large net carbon sink. Recent analyses raise doubts about these initial results.

When flux measurements are corrected for calm conditions, the net carbon balance may be nearly neutral. One of the studies in an old-growth forest in the Tapajós National Forest, Pará, Brazil, showed a small net CO_2 source (Saleska *et al.*, in press). The results in that forest were supported by measurements of biomass (forest inventory) (Rice *et al.*, in press). Living trees were accumulating carbon, but the decay of downed wood released more, for a small net source. Both fluxes suggest that the stand was recovering from a disturbance several years earlier.

The observation that the rivers and streams of the Amazon are a strong source for CO_2 (Richey *et al.*, 2002) may help balance the large sinks measured in some upland sites. However, the riverine source is included in inverse calculations based on atmospheric data and does not change those estimates of a net terrestrial source (Gurney *et al.*, 2002).

Changes in land use in the tropics are clearly a source of carbon to the atmosphere, although the magnitude is uncertain (Detwiler and Hall, 1988; Fearnside, 2000; Houghton, 1999, 2003). The tropics are characterized by high rates of deforestation, and this conversion of forests to nonforests involves a large loss of carbon. Although rotational processes of land use, such as logging, are just as common in the tropics as in temperate zones (even more so because shifting cultivation is common in the tropics), the sinks of carbon in regrowing forests are dwarfed in the tropics by the large releases of carbon resulting from permanent deforestation.

Comparisons of results from different methods (Table 8) suggest at least two, mutually exclusive, interpretations for the net terrestrial source of carbon from the tropics. One interpretation is that a large release of carbon from land-use change (Fearnside, 2000; Houghton, 2003) is partially offset by a large sink in undisturbed forests (Malhi *et al.*, 1998; Phillips *et al.*, 1998, 2002). The other interpretation is that the source from deforestation is smaller (see below), and that the net flux from undisturbed forests is nearly zero (Rice *et al.*, in press; Saleska *et al.*, in press). Under the first interpretation, some sort of growth enhancement (or past natural disturbance) is required to explain the large current sink in undisturbed forests. Under the second, the entire net flux of carbon may be explained by changes in land use, but the source from land-use change is smaller than estimated by Fearnside (2000) or Houghton (2003).

A third possibility, that the net tropical source from land is larger than indicated by inverse calculations (uncertain in the tropics), is constrained by the magnitude of the net sink in northern mid-latitudes. The latitudinal gradient in CO_2 concentrations constrains the difference between the northern sink and tropical source more than it constrains the absolute fluxes. The tropical source can only be larger than indicated by inverse calculations if the northern mid-latitude sink is also larger. As discussed above, the northern mid-latitude sink is thought to be in the range of $1-2.6\ Pg\ C\ yr^{-1}$, but the estimates are based on the assumption that the pre-industrial north–south gradient in CO_2 concentrations was zero (similar concentrations at all latitudes). No data exist for the pre-industrial north–south gradient in CO_2 concentrations, but following Keeling *et al.* (1989), Tayor and Orr extrapolated the current CO_2 gradient to a zero fossil fuel release and found a negative gradient (lower concentrations in the north). They interpreted this negative gradient as the pre-industrial gradient, and their interpretation would suggest a northern sink larger than generally believed. In contrast, Conway and Tans (1999) interpret the extrapolated zero fossil fuel gradient as representing the current sources and sinks of carbon in response to fossil fuel emissions and other human activities, such as present and past land-use change. Most investigators of the carbon cycle favor this interpretation.

The second interpretation of existing estimates (a modest source of carbon from deforestation and little or no sink in undisturbed forests) is supported by satellite-based estimates of tropical deforestation. The high estimates of Fearnside (2000) and Houghton (2003) were based on rates of deforestation reported by the FAO (2001). If these rates of deforestation are high, the estimates of the carbon source are also high. Two new studies of tropical deforestation (Achard et al., 2002; DeFries et al., 2002) report lower rates than the FAO and lower emissions of carbon than Fearnside or Houghton. The study by Achard et al. (2002) found rates 23% lower than the FAO for the 1990s (Table 11). Their analysis used high resolution satellite data over a 6.5% sample of tropical humid forests, stratified by "deforestation hot-spot areas" defined by experts. In addition to observing 5.8×10^6 ha of outright deforestation in the tropical humid forests, Achard et al. also observed 2.3×10^6 ha of degradation. Their estimated carbon flux, including changes in the area of dry forests as well as humid ones, was 0.96 Pg C yr^{-1}. The estimate is probably low because it did not include the losses of soil carbon that often occur with cultivation or the losses of carbon from degradation (reduction of biomass within forests). Soils and degradation accounted for 12% and 26%, respectively, of Houghton's (2003) estimated flux of carbon for tropical Asia and America and would yield a total flux of 1.3 Pg C yr^{-1} if the same percentages were applied to the estimate by Achard et al.

A second estimate of tropical deforestation (DeFries et al., 2002) was based on coarse resolution satellite data (8 km), calibrated with high-resolution satellite data to identify percent tree cover and to account for small clearings that would be missed with the coarse resolution data. The results yielded estimates of deforestation that were, on average, 54% lower than those reported by the FAO (Table 11). According to DeFries et al., the estimated net flux of carbon for the 1990s was 0.9 (range 0.5–1.4) Pg C yr^{-1}.

If the tropical deforestation rates obtained by Archard et al. and DeFries et al. were similar, there would be little doubt that the FAO estimates are high. However, the estimates are as different from each other as they are from those of the FAO (Table 11). Absolute differences between the two studies are difficult to evaluate because Achard et al. considered only humid tropical forests, whereas DeFries et al. considered all tropical forests. The greatest differences are in tropical Africa, where the percent tree cover mapped by DeFries et al. is most unreliable because of the large areas of savanna. Both studies suggest that the FAO estimates of tropical deforestation are high, but the rates are still in question (Fearnside and Laurance, 2003; Eva et al., 2003). The tropical emissions of carbon estimated by the two studies (after adjustments for degradation and soils) are about half of Houghton's estimate: 1.3 Pg C yr^{-1} and 0.9 Pg C yr^{-1}, as opposed to 2.2 Pg C yr^{-1} (Table 8).

3.3.2.4 *Summary: synthesis of the results of different methods*

Top-down methods show consistently that terrestrial ecosystems, globally, were a small net sink in the 1980s and 1990s. The sink was in northern mid-latitudes, partially offset by a tropical source. The northern sink was distributed over both North America and Eurasia roughly in proportion to land area. The magnitudes of terrestrial sinks obtained through inverse calculations are larger (or the sources smaller) than those obtained from bottom-up analyses (land-use change and forest inventories). Is there a bias in the atmospheric

Table 11 Annual rate of change in tropical forest area[a] for the 1990s.

	Tropical humid forests			*All tropical forests*		
	FAO (2001) $(10^6$ ha $yr^{-1})$	*Achard et al. (2002)*		*FAO (2001)* $(10^6$ ha $yr^{-1})$	*DeFries et al. (2002)*	
		10^6 ha yr^{-1}	*% lower than FAO*		10^6 ha yr^{-1}	*% lower than FAO*
America	2.7	2.2	18	4.4	3.179	28
Asia	2.5	2.0	20	2.4	2.008	16
Africa	1.2	0.7	42	5.2	0.376	93
All tropics	6.4	4.9	23	12.0	5.563	54

[a] The net change in forest area is not the rate of deforestation but, rather, the rate of deforestation minus the rate of afforestation.

analyses? Or are there sinks not included in the bottom-up analyses?

For the northern mid-latitudes, when estimates of change in nonforests (poorly known) are added to the results of forest inventories, the net sink barely overlaps with estimates determined from inverse calculations. Changes in land use yield smaller estimates of a sink. It is not clear how much of the discrepancy is the result of omissions of management practices and natural disturbances from analyses of land-use change, and how much is the result of environmentally enhanced rates of tree growth. In other words, how much of the carbon sink in forests can be explained by age structure (i.e., previous disturbances and management), and how much by enhanced rates of carbon storage? The question is important for predicting future concentrations of atmospheric CO_2 (see below).

In the tropics, the uncertainties are similar but also greater because inverse calculations are more poorly constrained and because forest inventories are lacking. Existing evidence suggests two possibilities. Either large emissions of carbon from land-use change are somewhat offset by large carbon sinks in undisturbed forests, or lower releases of carbon from land-use change explain the entire net terrestrial flux, with essentially no requirement for an additional sink. The first alternative (large sources and large sinks) is most consistent with the argument that factors other than land-use change are responsible for observed carbon sinks (i.e., management or environmentally enhanced rates of growth). The second alternative is most consistent with the findings of Caspersen *et al.* (2000) that there is little enhanced growth. Overall, in both northern and tropical regions changes in land use exert a dominant influence on the flux of carbon, and it is unclear whether other factors have been important in either region. These conclusions question the assumption used in predictions of climatic change, the assumption that the current terrestrial carbon sink will increase in the future (see below).

3.4 MECHANISMS THOUGHT TO BE RESPONSIBLE FOR CURRENT SINKS OF CARBON

3.4.1 Terrestrial Mechanisms

Distinguishing between regrowth and enhanced growth in the current terrestrial sink is important. If regrowth is dominant, the current sink may be expected to diminish as forests age (Hurtt *et al.*, 2002). If enhanced growth is important, the magnitude of the carbon sink may be expected to increase in the future. Carbon cycle models used to calculate future concentrations of atmospheric CO_2 from emissions scenarios assume the latter (that the current terrestrial sink will increase) (Prentice *et al.*, 2001). These calculated concentrations are then used in general circulation models to project future rates of climatic change. If the current terrestrial sink is largely the result of regrowth, rather than enhanced growth, future projections of climate may underestimate the extent and rate of climatic change.

The issue of enhanced growth versus regrowth can be illustrated with studies from the US. Houghton *et al.* (1999) estimated a terrestrial carbon sink of 0.15–0.35 Pg C yr^{-1} for the US, attributable to changes in land use. Pacala *et al.* (2001) revised the estimate upwards by including additional processes, but in so doing they included sinks not necessarily resulting from land-use change. Their estimate for the uptake of carbon by forests, e.g., was the uptake measured by forest inventories. The measured uptake might result from previous land use (regrowth), but it might also result from environmentally enhanced growth, e.g., CO_2 fertilization (Figure 7). If all of the accumulation of carbon in US forests were the result of recovery from past land-use practices (i.e., no enhanced growth), then the measured uptake should equal the flux calculated on the basis of land-use change. The residual flux would be zero. The study by Caspersen *et al.* (2000) suggests

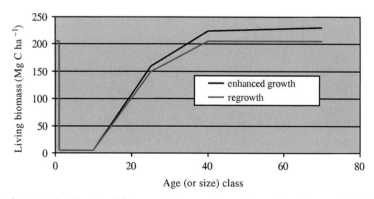

Figure 7 Idealized curves showing the difference between enhanced growth and regrowth in the accumulation of carbon in forest biomass.

Table 12 Estimated rates of carbon accumulation in the US (PgC yr^{-1} in 1990).

	Pacala et al.[a] (2001)		Houghton[b] et al. (1999)	Houghton[b] (2003)	Goodale et al. (2002)
	Low	High			
Forest trees	−0.11	−0.15	−0.072[c]	−0.046[d]	−0.11
Other forest organic matter	−0.03	−0.15	0.010	0.010	−0.11
Cropland soils	0.00	−0.04	−0.138	0.00	NE
Woody encroachment	−0.12	−0.13	−0.122	−0.061	NE
Wood products	−0.03	−0.07	−0.027	−0.027	−0.06
Sediments	−0.01	−0.04	NE	NE	NE
Total sink	−0.30	−0.58	−0.35	−0.11	−0.28
% of total sink neither in forests nor wood products	43%	36%	74%	55%	NE

NE is "not estimated". Negative values indicate an accumulation of carbon on land.
[a] Pacala *et al.* (2001) also included the import/export imbalance of food and wood products and river exports. As these would create corresponding sources outside the US, they are ignored here.
[b] Includes only the direct effects of human activity (i.e., land-use change and some management).
[c] 0.020 Pg C yr^{-1} in forests and 0.052 Pg C yr^{-1} in the thickening of western pine woodlands as a result of early fire suppression.
[d] 0.020 Pg C yr^{-1} in forests and 0.026 Pg C yr^{-1} in the thickening of western pine woodlands as a result of early fire suppression.

that such an attribution is warranted because they found that 98% of forest growth in five US states could be attributed to regrowth rather than enhanced growth. However, the analysis by Houghton *et al.* (1999) found that past changes in land use accounted for only 20–30% of the observed accumulation of carbon in trees. The uptake calculated for forests recovering from agricultural abandonment, fire suppression, and earlier harvests was only 20–30% of the uptake measured by forest inventories (∼40% if the uptake attributed to woodland "thickening" (0.26 Pg C yr^{-1}; Houghton, 2003) is included (Table 12)). The results are inconsistent with those of Caspersen *et al.* (2000). Houghton's analysis requires a significant growth enhancement to account for the observed accumulation of carbon in trees; the analysis by Caspersen *et al.* suggests little enhancement.

Both analyses merit closer scrutiny. Joos *et al.* (2002) have pointed out, e.g., that the relationship between forest age and wood volume (or biomass) is too variable to constrain the enhancement of growth to between 0.001% and 0.01% per year, as Caspersen *et al.* claimed. An enhancement of 0.1% per year fits the data as well. Furthermore, even a small enhancement of 0.1% per year in NPP yields a significant sink (∼2 Pg C yr^{-1}) if it applies globally (Joos *et al.*, 2002). Thus, Caspersen *et al.* may have underestimated the sink attributable to enhanced growth.

However, Houghton's analysis of land-use change (Houghton *et al.*, 1999; Houghton, 2003) most likely underestimates the sink attributable to regrowth. Houghton did not consider forest management practices other than harvest and subsequent regrowth. Nor did he include natural disturbances, which in boreal forests are more important than logging in determining the current

age structure and, hence, rate of carbon accumulation (Kurz and Apps, 1999). Forests might now be recovering from an earlier disturbance. A third reason why the sink may have been underestimated is that Houghton used net changes in agricultural area to obtain rates of agricultural abandonment. In contrast, rates of clearing and abandonment are often simultaneous and thus create larger areas of regrowing forests than would be predicted from net changes in agricultural area. It is unclear how much of the carbon sink in the US can be attributed to changes in land use and management, and how much can be attributed to enhanced rates of growth.

The mechanisms responsible for the current terrestrial sink fall into two broad categories (Table 13 and Figure 7): (i) enhanced growth from physiological or metabolic factors that affect rates of photosynthesis, respiration, growth, and decay and (ii) regrowth from past disturbances, changes in land use, or management, affecting the mortality of forest stands, the age structure of forests, and hence their rates of carbon accumulation. What evidence do we have that these mechanisms are important? Consider, first, enhanced rates of growth.

3.4.1.1 Physiological or metabolic factors that enhance rates of growth and carbon accumulation

CO$_2$ fertilization. Numerous reviews of the direct and indirect effects of CO$_2$ on photosynthesis and plant growth have appeared in the literature (Curtis, 1996; Koch and Mooney, 1996; Mooney *et al.*, 1999; Körner, 2000), and only a very brief review is given here. Horticulturalists have long known that annual plants respond to higher levels

Table 13 Proposed mechanisms for terrestrial carbon sinks[a].

Metabolic or physiological mechanisms
CO_2 fertilization
N fertilization
Tropospheric ozone, acid deposition
Changes in climate (temperature, moisture)
Ecosystem mechanisms
Large-scale regrowth of forests following human disturbance (includes recovery from logging and agricultural abandonment)[b]
Large-scale regrowth of forests following natural disturbance[b]
Fire suppression and woody encroachment[b]
Decreased deforestation[b]
Improved agricultural practices[b]
Erosion and re-deposition of sediment
Wood products and landfills[b]

[a] Some of these mechanisms enhance growth; some reduce decomposition. In some cases these same mechanisms may also yield sources of carbon to the atmosphere.
[b] Mechanisms included in analyses of land-use change (although not necessarily in all regions).

Table 14 Increases observed for a 100% increase in CO_2 concentrations.

Increased rates
60% increase in *photosynthesis* of young trees
33% average increase in net primary productivity (*NPP*) of crops
25% increase in *NPP* of a young pine forest
Increased stocks
14% average increase in *biomass* of grasslands and crops
~0% increase in the carbon content of mature forests

of CO_2 with increased rates of growth, and the concentration of CO_2 in greenhouses is often deliberately increased to make use of this effect. Similarly, experiments have shown that most C_3 plants (all trees, most crops, and vegetation from cold regions) respond to elevated concentrations of CO_2 with increased rates of photosynthesis and increased rates of growth.

Despite the observed stimulative effects of CO_2 on photosynthesis and plant growth, it is not clear that the effects will result in an increased storage of carbon in the world's ecosystems. One reason is that the measured effects of CO_2 have generally been short term, while over longer intervals the effects are often reduced or absent. For example, plants often acclimate to higher concentrations of CO_2 so that their rates of photosynthesis and growth return to the rates observed before the concentration was raised (Tissue and Oechel, 1987; Oren *et al.*, 2001).

Another reason why the experimental results may not apply to ecosystems is that most experiments with elevated CO_2 have been conducted with crops, annual plants, or tree seedlings. The few studies conducted at higher levels of integration or complexity, such as with mature trees and whole ecosystems, including soils as well as vegetation, suggest much reduced responses. Table 14 summarizes the results of experiments at different levels of integration. Arranged in this way (from biochemical processes to ecosystem processes), observations suggest that as the level of complexity, or the number of interacting processes, increases, the effects of CO_2 fertilization are reduced. This dampening of effects across ever-increasing levels of complexity has been noted since scientists first

began to consider the effects of CO_2 on carbon storage (Lemon, 1977).

In other words, a CO_2-enhanced increase in photosynthesis is many steps removed from an increase in carbon storage. An increase in NPP is expected to lead to increased carbon storage until the carbon lost from the detritus pool comes into a new equilibrium with the higher input of NPP. But, if the increased NPP is largely labile (easily decomposed), then it may be decomposed rapidly with little net carbon storage (Davidson and Hirsch, 2001). Results from a loblolly pine forest in North Carolina suggest a very small increase in carbon storage. Elevated CO_2 increased litter production (with a turnover time of about three years) but did not increase carbon accumulation deeper in the soil layer (Schlesinger and Lichter, 2001). Alternatively, the observation that microbes seemed to switch from old organic matter to new organic matter after CO_2 fertilization of a grassland suggests that the loss of carbon may be delayed in older, more refractory pools of soil organic matter (Cardon *et al.*, 2001).

The central question is whether natural ecosystems will accumulate carbon as a result of elevated CO_2, and whether the accumulation will persist. Few CO_2 fertilization experiments have been carried out for more than a few years in whole ecosystems, but where they have, the results generally show an initial CO_2-induced increment in biomass that diminishes after a few years. The diminution of the initial CO_2-induced effect occurred after two years in an arctic tundra (Oechel *et al.*, 1994) and after three years in a rapidly growing loblolly pine forest (Oren *et al.*, 2001). Other forests may behave differently, but the North Carolina forest was chosen in part because CO_2 fertilization, if it occurs anywhere, is likely to occur in a rapidly growing forest. The longest CO_2 fertilization experiment, in a brackish wetland on the Chesapeake Bay, has shown an enhanced net uptake of carbon even after 12 years, but the expected accumulation of carbon at the site has not been observed (Drake *et al.*, 1996).

Nitrogen fertilization. Human activity has increased the abundance of biologically active forms of nitrogen (NO_x and NH_4), largely through the production of fertilizers, the cultivation of legumes that fix atmospheric nitrogen, and the use of internal combustion engines. Because the availability of nitrogen is thought to limit NPP in temperate-zone ecosystems, the addition of nitrogen through human activities is expected to increase NPP and, hence, terrestrial carbon storage (Peterson and Melillo, 1985; Schimel *et al.*, 1996; Holland *et al.*, 1997). Based on stoichiometric relations between carbon and nitrogen, many physiologically based models predict that added nitrogen should lead to an accumulation of carbon in biomass. But the extent to which this accumulation occurs in nature is unclear. Adding nitrogen to forests does increase NPP (Bergh *et al.*, 1999). It may also modify soil organic matter and increase its residence time (Fog, 1988; Bryant *et al.*, 1998). But nitrogen deposited in an ecosystem may also be immobilized in soils (Nadelhoffer *et al.*, 1999) or lost from the ecosystem, becoming largely unavailable in either case (Davidson, 1995).

There is also evidence that additions of nitrogen above some level may saturate the ecosystem, causing: (i) increased nitrification and nitrate leaching, with associated acidification of soils and surface waters; (ii) cation depletion and nutrient imbalances; and (iii) reduced productivity (Aber *et al.*, 1998; Fenn *et al.*, 1998). Experimental nitrogen additions have had varied effects on wood production and growing stocks. Woody biomass production increased in response to nitrogen additions to two New England hardwood sites, although increased mortality at one site led to a net decrease in the stock of woody biomass (Magill *et al.*, 1997, 2000). Several studies have shown that chronic exposure to elevated nitrogen inputs can inhibit forest growth, especially in evergreen species (Tamm *et al.*, 1995; Makipaa, 1995). Fertilization decreased rates of wood production in high-elevation spruce-fir in Vermont (McNulty *et al.*, 1996) and in a heavily fertilized red pine plantation (Magill *et al.*, 2000). The long-term effects of nitrogen deposition on forest production and carbon balance remain uncertain. Furthermore, because much of the nitrogen deposited on land is in the form of acid precipitation, it is difficult to distinguish the fertilization effects of nitrogen from the adverse effects of acidity (see below).

Atmospheric chemistry. Other factors besides nitrogen saturation may have negative effects on NPP, thus reducing the uptake of carbon in ecosystems and perhaps changing them from sinks to sources of carbon. Two factors that have received attention are tropospheric ozone and sulfur (acid rain). Experimental studies show leaf injury and reduced growth in crops and trees exposed to ozone. At the level of the ecosystem, elevated levels of ozone have been associated with reduced forest growth in North America (Mclaughlin and

Percy, 2000) and Europe (Braun *et al.*, 2000). Acidification of soil as a result of deposition of NO_3^- and SO_4^{2-} in precipitation depletes the soils of available plant nutrients (Ca^{2+}, Mg^{2+}, K^+), increases the mobility and toxicity of aluminum, and increases the amount of nitrogen and sulfur stored in forest soils (Driscoll *et al.*, 2001). The loss of plant nutrients raises concerns about the long-term health and productivity of forests in the northeastern US, Europe, and southern China.

Although the effects of tropospheric ozone and sulfur generally reduce NPP, their actual or potential effects on carbon stocks are not known. The pollutants could potentially increase carbon stocks if they reduce decomposition of organic matter more than they reduce NPP.

Climatic variability and climatic change. Year-to-year differences in the growth rate of CO_2 in the atmosphere are large (Figure 3). The annual rate of increase ranged from 1.9 Pg C in 1992 to 6.0 Pg C in 1998 (Prentice *et al.*, 2001; see also Conway *et al.*, 1994). In 1998 the net global sink (ocean and land) was nearly 0 Pg C, while the average combined sink in the previous eight years was ~3.5 Pg C yr^{-1} (Tans *et al.*, 2001). The terrestrial sink is generally twice as variable as the oceanic sink (Bousquet *et al.*, 2000). This temporal variability in terrestrial fluxes is probably caused by the effect of climate on carbon pools with short lifetimes (foliage, plant litter, soil microbes) through variations in photosynthesis, respiration, and possibly fire (Schimel *et al.*, 2001). Measurements in terrestrial ecosystems suggest that respiration, rather than photosynthesis, is the major contributor to variability (Valentini *et al.*, 2000). Annual respiration was almost twice as variable as photosynthesis over a five-year period in the Harvard Forest (Goulden *et al.*, 1996). Respiration is also more sensitive than photosynthesis to changes in both temperature and moisture. For example, during a dry year at the Harvard Forest, both photosynthesis and respiration were reduced, but the reduction in respiration was greater, yielding a greater than average net uptake of carbon for the year (Goulden *et al.*, 1996). A tropical forest in the Brazilian Amazon behaved similarly (Saleska *et al.*, in press).

The greater sensitivity of respiration to climatic variations is also observed at the global scale. An analysis of satellite data over the US, together with an ecosystem model, shows that the variability in NPP is considerably less than the variability in the growth rate of atmospheric CO_2 inferred from inverse modeling, suggesting that the cause of the year-to-year variability in carbon fluxes is largely from varying rates of respiration rather than photosynthesis (Hicke *et al.*, 2002). Also, global NPP was remarkably constant over the three-year transition from El Niño to La Niña (Behrenfeld *et al.*, 2001). Myneni *et al.* (1995)

found a positive correlation between annual "greenness," derived from satellites, and the growth rate of CO_2. Greener years presumably had more photosynthesis and higher GPP, but they also had proportionately more respiration, thus yielding a net release of carbon from land (or reduced uptake) despite increased greenness.

Climatic factors influence terrestrial carbon storage through effects on photosynthesis, respiration, growth, and decay. However, prediction of future terrestrial sinks resulting from climate change requires an understanding of not only plant and microbial physiology, but the regional aspects of future climate change, as well. The important aspects of climate are: (i) temperature, including the length of the growing season; (ii) moisture; and (iii) solar radiation and clouds. Although year-to-year variations in the growth rate of CO_2 are probably the result of terrestrial responses to climatic variability, longer-term changes in carbon storage involve acclimation and other physiological adjustments that generally reduce short-term responses.

In cold ecosystems, such as those in high latitudes (tundra and taiga), an increase in temperature might be expected to increase NPP and, perhaps, carbon storage (although the effects might be indirect through increased rates of nitrogen mineralization; Jarvis and Linder, 2000). Satellite records of "greenness" over the boreal zone and temperate Europe show a lengthening of the growing season (Myneni *et al.*, 1997), suggesting greater growth and carbon storage. Measurements of CO_2 flux in these ecosystems do not consistently show a net uptake of carbon in response to warm temperatures (Oechel *et al.*, 1993; Goulden *et al.*, 1998), however, presumably because warmer soils release more carbon than plants take up. Increased temperatures in boreal forests may also reduce plant growth if the higher temperatures are associated with drier conditions (Barber *et al.*, 2000; Lloyd and Fastie, 2002). The same is true in the tropics, especially as the risk of fires increases with drought (Nepstad *et al.*, 1999; Page *et al.*, 2002). A warming-enhanced increase in rates of respiration and decay may already have begun to release carbon to the atmosphere (Woodwell, 1983; Raich and Schlesinger, 1992; Houghton *et al.*, 1998).

The results of short-term experiments may be misleading, however, because of acclimation or because the more easily decomposed material is respired rapidly. The long-term, or equilibrium, effects of climate on carbon storage can be inferred from the fact that cool, wet habitats store more carbon in soils than hot, dry habitats (Post *et al.*, 1982). The transient effects of climatic change on carbon storage, however, are difficult to predict, in large part because of uncertainty in predicting regional and temporal changes in temperature and moisture (extremes as well as means) and rates of climatic change, but also from incomplete understanding of how such changes affect fires, disease, pests, and species migration rates.

In the short term of seasons to a few years, variations in terrestrial carbon storage are most likely driven by variations in climate (temperature, moisture, light, length of growing season). Carbon dioxide fertilization and nitrogen deposition, in contrast, are unlikely to change abruptly. Interannual variations in the emissions of carbon from land-use change are also likely to be small ($<0.2\,\mathrm{Pg\,C\,yr^{-1}}$) because socioeconomic changes in different regions generally offset each other, and because the releases and uptake of carbon associated with a land-use change lag the change in land use itself and thus spread the emissions over time (Houghton, 2000). Figure 8 shows the annual net emissions of carbon from deforestation and reforestation in the Brazilian Amazon relative to the annual fluxes observed in the growth rates of trees and modeled on the basis of physiological responses to climatic variation. Clearly, metabolic responses to climatic variations are more important in the short term than interannual variations in rates of land-use change.

Understanding short-term variations in atmospheric CO_2 may not be adequate for predicting longer-term trends, however. Organisms and populations acclimate and adapt in ways that generally diminish short-term responses. Just as increased rates of photosynthesis in response to elevated levels of CO_2 often, but not always, decline within months or years (Tissue and Oechel, 1987), the same diminished response has been observed for higher temperatures (Luo *et al.*, 2001). Thus, over decades and centuries the factors most important in influencing concentrations of atmospheric CO_2 (fossil fuel emissions, land-use change, oceanic uptake) are probably different from those factors important in determining the short-term variations in atmospheric CO_2 (Houghton, 2000). Long-term changes in climate, as opposed to climatic variability, may eventually lead to long-term changes in carbon storage, but probably not at the rates suggested by short-term experiments.

One further observation is discussed here. Over the last decades the amplitude of the seasonal oscillation of CO_2 concentration increased by ~20% at Mauna Loa, Hawaii, and by ~40% at Point Barrow, Alaska (Keeling *et al.*, 1996a). This winter–summer oscillation in concentrations seems to be largely the result of terrestrial metabolism in northern mid-latitudes. The increase in amplitude suggests that the rate of processing of carbon may be increasing. Increased rates of summer photosynthesis, increased rates of winter respiration, or both would increase the amplitude of the oscillation, but it is difficult to ascertain which has contributed most. Furthermore, the increase in

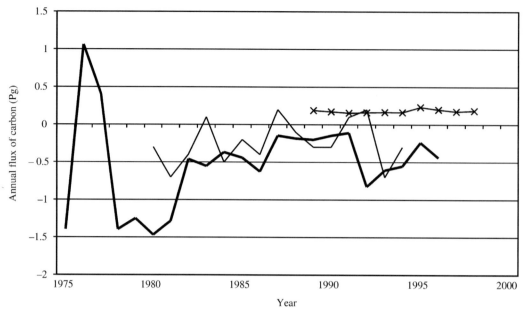

Figure 8   Net annual sources (+) and sinks (−) of carbon for the Brazilian Amazon, as determined by three different methods: (×) land-use change (Houghton *et al.*, 2000) (reproduced by permission of the American Geophysical Union from *J. Geophys. Res.*, **2000**, *105*, 20121–20130); (—) tree growth (Phillips *et al.*, 1998); and (—) modeled ecosystem metabolism (Tian *et al.*, 1998).

the amplitude does not, by itself, indicate an increasing terrestrial sink. In fact, the increase in amplitude is too large to be attributed to CO_2 fertilization or to a temperature-caused increase in winter respiration (Houghton, 1987; Randerson *et al.*, 1997). It is consistent with the observation that growing seasons have been starting earlier over the last decades (Randerson *et al.*, 1999). The trend has been observed in the temperature data, in decreasing snow cover (Folland and Karl, 2001), and in the satellite record of vegetation activity (Myneni *et al.*, 1997).

Synergies among physiological "mechanisms." The factors influencing carbon storage often interact nonadditively. For example, higher concentrations of CO_2 in air enable plants to acquire the same amount of carbon with a smaller loss of water through their leaves. This increased water-use efficiency reduces the effects of drought. Higher levels of CO_2 may also alleviate other stresses of plants, such as temperature and ozone. The observation that NPP is increased relatively more in "low productivity" years suggests that the indirect effects of CO_2 in ameliorating stress may be more important than the direct effects of CO_2 on photosynthesis (Luo *et al.*, 1999).

Another example of synergistic effects is the observation that the combination of nitrogen fertilizer and elevated CO_2 concentration may have a greater effect on the growth of biomass in a growing forest than the expected additive effect (Oren *et al.*, 2001). The relative increase was greater in a nutritionally poor site. The synergy between

nitrogen and CO_2 was different in a grassland, however (Hu *et al.*, 2001). There, elevated CO_2 increased plant uptake of nitrogen, increased NPP, and increased the carbon available for microbes; but it reduced microbial decomposition, presumably because the utilization of nitrogen by plants reduced its availability for microbes. The net effect of the reduced decomposition was an increase in the accumulation of carbon in soil.

Relatively few experiments have included more than one environmental variable at a time. A recent experiment involving combinations of four variables shows the importance of such work. Shaw *et al.* (2003) exposed an annual grassland community in California to increased temperature, precipitation, nitrogen deposition, and atmospheric CO_2 concentration. Alone, each of the treatments increased NPP in the third year of treatment. Across all multifactor treatments, however, elevated CO_2 decreased the positive effects of the other treatments. That is, elevated CO_2 increased productivity under "poor" growing conditions, but reduced it under favorable growing conditions. The most likely explanation is that some soil nutrient became limiting, either because of increased microbial activity or decreased root allocation (Shaw *et al.*, 2003).

The expense of such multifactor experiments has led scientists to use process-based ecosystem models (see the discussion of "terrestrial carbon models" below) to predict the response of terrestrial ecosystems to future climates. When predicting the effects of CO_2 alone, six global biogeochemical models showed a global terrestrial sink that began in the early part of the twentieth century and increased (with one exception) towards the year 2100 (Cramer *et al.*, 2001). The maximum sink varied from $\sim$4 Pg C yr^{-1} to $\sim$10 Pg C yr^{-1}. Adding changes in climate (predicted by the Hadley Centre) to these models reduced the future sink (with one exception), and in one case reduced the sink to zero near the year 2100.

Terrestrial carbon models. A number of ecosystem models have been developed to calculate gross and net fluxes of carbon from environmentally induced changes in plant or microbial metabolism, such as photosynthesis, plant respiration, decomposition, and heterotrophic respiration (Cramer *et al.*, 2001; McGuire *et al.*, 2001). For example, six global models yielded net terrestrial sinks of carbon ranging between 1.5 Pg C yr^{-1} and 4.0 Pg C yr^{-1} for the year 2000 (Cramer *et al.*, 2001). The differences among models became larger as environmental conditions departed from existing conditions. The magnitude of the terrestrial carbon flux projected for the year 2100 varied between a source of 0.5 Pg C yr^{-1} and a sink of 7 Pg C yr^{-1}. Other physiologically based models, including the effects of climate on plant distribution as well as growth, projected a net source from land as tropical forests were replaced with savannas (White *et al.*, 1999; Cox *et al.*, 2000).

The advantage of such models is that they allow the effects of different mechanisms to be distinguished. However, they may not include all of the important processes affecting changes in carbon stocks. To date, e.g., few process-based terrestrial models have included changes in land use.

Although some processes, such as photosynthesis, are well enough understood for predicting responses to multiple factors, other processes, such as biomass allocation, phenology, and the replacement of one species by another, are not. Even if the physiological mechanisms and their interactions were well understood and incorporated into the models, other nonphysiological factors that affect carbon storage (e.g., fires, storms, insects, and disease) are not considered in the present generation of models. Furthermore, the factors influencing short-term changes in terrestrial carbon storage may not be the ones responsible for long-term changes (Houghton, 2000) (see next section). The variability among model predictions suggests that they are not reliable enough to demonstrate the mechanisms responsible for the current modest terrestrial sink (Cramer *et al.*, 2001; Knorr and Heimann, 2001).

3.4.1.2 Demographic or disturbance mechanisms

Terrestrial sinks also result from the recovery (growth) of ecosystems disturbed in the past. The processes responsible for regrowth include physiological and metabolic processes, but they also involve higher-order or more integrated processes, such as succession, growth, and aging. Forests accumulate carbon as they grow. Regrowth is initiated either by disturbances or by the planting of trees on open land. Disturbances may be either natural (insects, disease, some fires) or human induced (management and changes in land use, including fire management). Climatic effects—e.g., droughts, storms, or fires—thus affect terrestrial carbon storage not only through physiological or metabolic effects on plant growth and respiration, but also through effects on stand demography and growth.

In some regions of the world—e.g., the US and Europe—past changes in land use are responsible for an existing sink (Houghton *et al.*, 1999; Caspersen *et al.*, 2000; Houghton, 2003). Processes include the accumulation of carbon in forests as a result of fire suppression, the growth of forests on lands abandoned from agriculture, and the growth of forests earlier harvested. In tropical regions carbon accumulates in forests that are in the fallow period of shifting cultivation. All regions, even countries with high rates of deforestation, have sinks of carbon in recovering forests, but often these sinks are offset by large emissions (Table 9). The sinks in tropical regions as a result of logging are nearly the same in magnitude as those outside the tropics.

Sinks of carbon are not limited to forests. Some analyses of the US (Houghton *et al.*, 1999; Pacala *et al.*, 2001) show that a number of processes in nonforest ecosystems may also be responsible for carbon sinks. Processes include the encroachment of woody vegetation into formerly herbaceous ecosystems, the accumulation of carbon in agricultural soils as a result of conservation tillage or other practices, exportation of wood and food, and the riverine export of carbon from land to the sea (Table 12). At least a portion of these last two processes (import/export of food and wood and river export) represents an export of carbon from the US (an apparent sink) but not a global sink because these exports presumably become sources somewhere else (either in ocean waters or in another country).

Which terrestrial mechanisms are important? Until recently, the most common explanations for the residual carbon sink in the 1980s and 1990s were factors that affect the physiology of plants and microbes: CO_2 fertilization, nitrogen deposition, and climatic variability (see Table 13). Several findings have started to shift the explanation to include management practices and disturbances that affect the age structure or demography of ecosystems. For example, the suggestion that CO_2 fertilization may be less important in forests than in short-term greenhouse experiments (Oren *et al.*, 2001) was discussed above. Second, physiological models quantifying the effects of CO_2 fertilization and climate change on the growth of US forests could account for only a small fraction of the carbon accumulation observed in those forests (Schimel *et al.*, 2000). The authors acknowledged that past changes in land use were likely to be important. Third, and most importantly, 98% of recent accumulations of carbon in US forests can be explained on the basis of the age structure of trees without requiring growth enhancement due to CO_2 or nitrogen fertilization (Caspersen *et al.*, 2000). Either the physiological effects of CO_2, nitrogen, and climate have been unimportant or their effects have been offset by unknown influences. Finally, the estimates of sinks in the US (Houghton *et al.*, 1999; Pacala *et al.*, 2001; Table 12) are based, to a large extent, on changes in land use and management, and not on physiological models of plant and soil metabolism.

To date, investigations of these two different classes of mechanisms have been largely independent. The effects of changing environmental conditions have been largely ignored in analyses of land-use change (see Section 3.3.1.5), and physiological models have generally ignored changes in land use (see Section 3.4.1.1).

As of early 2000s, the importance of different mechanisms in explaining known terrestrial carbon sinks remains unclear. Management and past disturbances seem to be the dominant mechanisms for a sink in mid-latitudes, but they are unlikely to explain a large carbon sink in the tropics (if one exists). Recovery from past disturbances is unlikely to explain a large carbon sink in the tropics, because both the area of forests and the stocks of carbon within forests have been declining. Rates of human-caused disturbance have been accelerating. Clearly there are tropical forests recovering from natural disturbances, but there is no evidence that the frequency of disturbances changed during the last century, and thus no evidence to suggest that the sink in recovering forests is larger or smaller today than in previous centuries. The lack of systematic forest inventories over large areas in the tropics precludes a more definitive test of where forests are accumulating carbon and where they are losing it.

Enhanced rates of plant growth cannot be ruled out as an explanation for apparent sinks in either the tropics or mid-latitude lands, but it is possible that the current sink is entirely the result of recovery from earlier disturbances, anthropogenic and natural.

How will the magnitude of the current terrestrial sink change in the future? Identifying the mechanisms responsible for past and current carbon sinks is important because some mechanisms are more likely than others to persist into the future. As discussed above, physiologically based models predict that CO_2 fertilization will increase

the global terrestrial sink over the next 100 years (Cramer *et al.*, 2001). Including the effects of projected climate change reduces the magnitude of projected sinks in many models but turns the current sink into a future global source in models that include the longer-term effects of climate on plant distribution (White *et al.*, 1999; Cox *et al.*, 2000). Thus, although increased levels of CO_2 are thought to increase carbon storage in forests, the effect of warmer temperatures may replace forests with savannas and grasslands, and, in the process, release carbon to the atmosphere. Future changes in natural systems are difficult to predict.

To the extent the current terrestrial sink is a result of regrowth (changes in age structure), the future terrestrial sink is more constrained. First, the net effect of continued land-use change is likely to release carbon, rather than store it. Second, forests that might have accumulated carbon in recent decades (whatever the cause) will cease to function as sinks if they are turned into croplands. Third, the current sink in regrowing forests will diminish as forests mature (Hurtt *et al.*, 2002).

Despite the recent evidence that changes in land use are more important in explaining the current terrestrial carbon sink than physiological responses to environmental changes in CO_2, nitrogen, or climate, most projections of future rates of climatic change are based on the assumption that the current terrestrial sink will not only continue, but grow in proportion to concentrations of CO_2. Positive biotic feedbacks and changes in land use are not included in the general circulation models (GCMs) used to predict future rates of climate change. The GCMs include physical feedbacks such as water vapor, clouds, snow, and polar ice, but not biotic feedbacks (Woodwell and Mackenzie, 1995). Thus, unless negative feedbacks in the biosphere become more important in the future, through physiological or other processes, these climate projections underestimate the rate and extent of climatic change. If the terrestrial sink were to diminish in the next decades, concentrations of CO_2 by the year 2100 might be hundreds of ppm higher than commonly projected.

3.4.2 Oceanic Mechanisms

3.4.2.1 *Physical and chemical mechanisms*

Increasing the concentration of CO_2 in the atmosphere is expected to affect the rate of oceanic uptake of carbon through at least eight mechanisms, half of them physical or chemical, and half of them biological. Most of the mechanisms reduce the short-term uptake of carbon by the oceans.

The buffer factor. The oceanic buffer factor (or Revelle factor), by which the concentration of CO_2 in the atmosphere is determined, increases as the concentration of CO_2 increases. The buffer factor is discussed above in Section 3.3.1.3. Here, it is sufficient to describe the chemical equation for the dissolution of CO_2 in seawater.

$$2HCO_3 + Ca \rightleftharpoons CaCO_3 + CO_2 + H_2O$$

Every molecule of CO_2 entering the oceans consumes a molecule of carbonate as the CO_2 is converted to bicarbonate. Thus, as CO_2 enters the ocean, the concentration of carbonate ions decreases, and further additions of CO_2 remain as dissolved CO_2 rather than being converted to HCO_3^-. The ocean becomes less effective in taking up additional CO_2. The effect is large. The change in DIC for a 100 ppm increase above 280 ppm (pre-industrial) was 40% larger than a 100 ppm increase would be today. The change in DIC for a 100 ppm increase above 750 ppm will be 60% lower than it would be today (Prentice *et al.*, 2001). Thus, the fraction of added CO_2 going into the ocean decreases and the fraction remaining in the atmosphere increases as concentrations continue to increase.

Warming. The solubility of CO_2 in seawater decreases with temperature. Raising the ocean temperature 1°C increases the equilibrium p_{CO_2} in seawater by 10–20 ppm, thus increasing the atmospheric concentration by that much as well. This mechanism is a positive feedback to a global warming.

Vertical mixing and stratification. If the warming of the oceans takes place in the surface layers first, the warming would be expected to increase the stability of the water column. As discussed in Section 3.3.1.3, the bottleneck for oceanic uptake of CO_2 is largely the rate at which the surface oceans exchange CO_2 with the intermediate and deeper waters. Greater stability of the water column, as a result of warming, might constrict this bottleneck further. Similarly, if the warming of the Earth's surface is greater at the poles than at the equator, the latitudinal gradient in surface ocean temperature will be reduced; and because that thermal gradient plays a role in the intensity of atmospheric mixing, a smaller gradient might be expected to subdue mixing and increase stagnation. Alternatively, the increased intensity of the hydrologic cycle expected for a warmer Earth will probably increase the intensity of storms and might, thereby, increase oceanic mixing. Interactions between oceanic stability and biological production might also change the ocean's carbon cycle, with consequences for the oceanic uptake of carbon that are difficult to predict (Sarmiento *et al.*, 1998; Matear and Hirst, 1999).

One aspect of the ocean's circulation that seems particularly vulnerable to climate change is the thermohaline circulation, which is related to the formation of North Atlantic Deep Water (NADW). Increased warming of surface waters may intensify the hydrologic cycle, leading to a reduced salinity in the sea surface at high latitudes, a reduction (even collapse) of NADW formation, reduction in the surface-to-deep transport of anthropogenic carbon, and thus a higher rate of CO_2 growth in the atmosphere. In a model simulation, modest rates of warming reduced the rate of oceanic uptake of carbon, but the reduced uptake was largely compensated by changes in the marine biological cycle (Joos *et al.*, 1999a). For higher rates of global warming, however, the NADW formation collapsed and the concentration of CO_2 in the atmosphere was 22% (and global temperature 0.6 °C) higher than expected in the absence of this feedback.

Rate of CO_2 emission. High rates of CO_2 emissions will increase the atmosphere–ocean gradient in CO_2 concentrations. Although this gradient drives the uptake of carbon by surface waters, if the rate of CO_2 emissions is greater than the rate of CO_2 uptake, the fraction of emitted CO_2 remaining in the atmosphere will be higher. Under the business-as-usual scenario for future CO_2 emissions, rates of emissions increase by more than a factor of 3, from approximately $6\,Pg\,C\,yr^{-1}$ in the 1990s to $20\,Pg\,C\,yr^{-1}$ by the end of the twenty-first century.

3.4.2.2 Biological feedback/processes

Changes in biological processes may offset some of the physical and chemical effects described above (Sarmiento *et al.*, 1998; Joos *et al.*, 1999a), but the understanding of these processes is incomplete, and the net effects far from predictable. Potential effects fall into four categories (Falkowski *et al.*, 1998).

(i) Addition of nutrients limiting primary production. Nutrient enrichment experiments and observations of nutrient distributions throughout the oceans suggest that marine primary productivity is often limited by the availability of fixed inorganic nitrogen. As most of the nitrogen for marine production comes from upwelling, physical changes in ocean circulation might also affect oceanic primary production and, hence, the biological pump. Some nitrogen is made available through nitrogen fixation, however, and some is lost through denitrification, both of which are biological processes, limited by trace nutrients and the concentration of oxygen. The two processes are not coupled, however, and differential changes in either one would affect the inventory of fixed nitrogen in the ocean.

(ii) Enhanced utilization of nutrients. One of the mysteries of ocean biology today is the observation of "high nutrient, low chlorophyll (HNLC) regions." That is, why does primary production in major regions of the surface ocean stop before all of the available nitrogen and phosphorous have been used up? It is possible that grazing pressures keep phytoplankton populations from consuming

the available nitrogen and phosphorous, and any reduction in grazing pressures might increase the export of organic matter from the surface. Another possibility that has received considerable attention is that iron may limit production (Martin, 1990). In fact, deliberate iron fertilization of the ocean has received serious attention as a way of reducing atmospheric CO_2 (see Section 3.5.2, below). Iron might also become more available naturally as a result of increased human eutrophication of coastal waters, or it might be less available as a result of a warmer (more strongly stratified) ocean or reduced transport of dust (Falkowski et al., 1998). The aeolian transport of iron in dust is a major source of iron for the open ocean, and dust could either increase or decrease in the future, depending on changes in the distribution of precipitation.

(iii) Changes in the elemental ratios of organic matter in the ocean. The elemental ratio of $C : N : P$ in marine organic particles has long been recognized as conservative (Falkowski et al., 1998). The extent to which the ratios can depart from observed concentrations is not known, yet variations could reduce the limitation of nitrogen and thus act in the same manner as the addition of nitrogen in affecting production, export, and thus oceanic uptake of CO_2.

(iv) Increases in the organic carbon/carbonate ratio of export production. The biological and carbonate pumps are described above (Section 3.2.2.2). Both pumps transport carbon out of the surface waters, and the subsequent decay at depth is responsible for the higher concentration of carbon in the intermediate and deep ocean. The formation of carbonate shells in the surface waters has the additional effect of increasing the p_{CO_2} in these waters, thus negating the export of the carbonate shells out of the surface. Any increase in the organic carbon/carbonate ratio of export production would enhance the efficiency of the biological pump.

3.5 THE FUTURE: DELIBERATE SEQUESTERING OF CARBON (OR REDUCTION OF SOURCES)

Section 3.4 addressed the factors thought to be influencing current terrestrial and oceanic sinks, and how they might change in the future. It is possible, of course, that CO_2 fertilization will become more important in the future as concentrations of CO_2 increase. Multiyear, whole ecosystems experiments with elevated CO_2 do not uniformly support this possibility, but higher concentrations of CO_2, together with nitrogen deposition or increases in moisture, might yet be important. Rather than wait for a more definitive answer, a more cautious approach to the future, besides reducing emissions of CO_2, would consider strategies for withdrawing

carbon from the atmosphere through management. Three general options for sequestering carbon have received attention: terrestrial, oceanic, and geological management.

3.5.1 Terrestrial

Even if CO_2 fertilization and other environment effects turn out to be unimportant in enhancing terrestrial carbon storage, terrestrial sinks can still be counted on to offset carbon emissions or to reduce atmospheric concentrations of CO_2. Increasing the amount of carbon held on land might be achieved through at least six management options (Houghton, 1996; Kohlmaier et al., 1998): (i) a reduction in the rate of deforestation (current rates of deforestation in the tropics are responsible for an annual release of $1–2 Pg C$ (Section 3.3.2.3); (ii) an increase in the area of forests (afforestation); (iii) an increase in the stocks of carbon within existing forests; (iv) an increase in the use of wood (including increased efficiency of wood harvest and use); (v) the substitution of wood fuels for fossil fuels; and (vi) the substitution of wood for more energy-intensive materials, such as aluminum, concrete, and steel. Estimates of the amount of carbon that might be sequestered on land over the 55-year period 1995–2050 range between $60 Pg C$ and $87 Pg C$ ($1–2 Pg C yr^{-1}$ on average) (Brown, 1996). Additional carbon might also be sequestered in agricultural soils through conservation tillage and other agricultural management practices, and in grassland soils (Sampson and Scholes, 2000). An optimistic assessment, considering all types of ecosystems over the Earth, estimated a potential for storing $5–10 Pg C yr^{-1}$ over a period of 25–50 years (DOE, 1999).

The amount of carbon potentially sequestered is small relative to projected emissions of CO_2 from business-as-usual energy practices, and thus the terrestrial options for sequestering carbon should be viewed as temporary, "buying time" for the development and implementation of longer-lasting measures for reducing fossil fuel emissions (Watson et al., 2000).

3.5.2 Oceanic

Schemes for increasing the storage of carbon in the oceans include stimulation of primary production with iron fertilization and direct injection of CO_2 at depth. As pointed out in Section 3.4.2.2., there are large areas of the ocean with high nutrient, low chlorophyll, concentrations. One explanation is that marine production is limited by the micronutrient iron. Adding iron to these regions might thus increase the ocean's biological pump, thereby reducing atmospheric CO_2 (Martin, 1990; Falkowski et al., 1998). Mesoscale fertilization experiments have been carried out (Boyd *et al.*, 2000), but the effects of large-scale iron fertilization of the ocean are not known (Chisholm, 2000).

The direct injection of concentrated CO_2 (probably in liquid form) below the thermocline or on the seafloor might sequester carbon for hundreds of years (Herzog *et al.*, 2000). The gas might be dissolved within the water column or held in solid, ice-like CO_2 hydrates. The possibility is receiving attention in several national and international experiments (DOE, 1999). Large uncertainties exist in understanding the formation and stability of CO_2 hydrates, the effect of the concentrated CO_2 on ocean ecosystems, and the permanence of the sequestration.

3.5.3 Geologic

CO_2 may be able to be sequestered in geological formations, such as active and depleted oil and gas reservoirs, coalbeds, and deep saline aquifers. Such formations are widespread and have the potential to sequester large amounts of CO_2 (Herzog *et al.*, 2000). A model project is underway in the North Sea off the coast of Norway. The Sleipner offshore oil and natural gas field contains a gas mixture of natural gas and CO_2 (9%). Because the Norwegian government taxes emissions of CO_2 in excess of 2.5%, companies have the incentive to separate CO_2 from the natural gas and pump it into an aquifer 1,000 m under the sea. Although the potential for sequestering carbon in geological formations is large, technical and economic aspects of an operational program require considerable research.

3.6 CONCLUSION

We are conducting a great geochemical experiment, unlike anything in human history and unlikely to be repeated again on Earth. "Within a few centuries we are returning to the atmosphere and oceans the concentrated organic carbon stored in sedimentary rocks over hundreds of millions of years" (Revelle and Suess, 1957). During the last 150 years ($\sim$1850–2000), there has been a 30% increase in the amount of carbon in the atmosphere. Although most of this carbon has come from the combustion of fossil fuels, an estimated 150–160 Pg C have been lost during this time from terrestrial ecosystems as a result of human management (another 58–75 Pg C were lost before 1850). The global carbon balance suggests that other terrestrial ecosystems have accumulated $\sim$115 Pg C since about 1930, at a steadily increasing rate. The annual net fluxes of carbon appear small relative to the sizes of the reservoirs, but the fluxes have been accelerating. Fifty percent of the carbon mobilized over the last 300 years ($\sim$1700–2000) was mobilized in the last 30–40 of these years (Houghton and Skole, 1990) (Figure 3). The major

drivers of the geochemical experiment are reasonably well known. However, the results are uncertain, and there is no control. Furthermore, the experiment would take a long time to stop (or reverse) if the results turned out to be deleterious.

In an attempt to put some bounds on the experiment, in 1992 the nations of the world adopted the United Nations Framework Convention on Climate Change, which has as its objective "stabilization of greenhouse gas concentrations in the atmosphere at a level that would prevent dangerous anthropogenic interference with the climate system" (UNFCCC, 1992). The Convention's soft commitment suggested that the emissions of greenhouse gases from industrial nations in 2000 be no higher than the emissions in 1990. This commitment has been achieved, although more by accident than as a result of deliberate changes in policy. The "stabilization" resulted from reduced emissions from Russia, as a result of economic downturn, balanced by increased emissions almost everywhere else. In the US, e.g., emissions were 18% higher in the year 2000 than they had been in 1990. The near-zero increase in industrial nations' emissions between 1990 and 2000 does not suggest that the stabilization will last.

Ironically, even if the annual rate of global emissions were to be stabilized, concentrations of the gases would continue to increase. Stabilization of concentrations at early 2000's levels, e.g., would require reductions of 60% or more in the emission of long-lived gases, such as CO_2. The 5% average reduction in 1990 emissions by 2010, agreed to by the industrialized countries in the Kyoto Protocol (higher than 5% for the participating countries now that the US is no longer participating), falls far short of stabilizing atmospheric concentrations. Such a stabilization will require nothing less than a switch from fossil fuels to renewable forms of energy (solar, wind, hydropower, biomass), a switch that would have salubrious economic, political, security, and health consequences quite apart from limiting climatic change. Nevertheless, the geophysical experiment seems likely to continue for at least the near future, matched by a sociopolitical experiment of similar proportions, dealing with the consequences of either mitigation or not enough mitigation.

REFERENCES

Aber J. D., McDowell W. H., Nadelhoffer K. J., Magill A. H., Berntson G., Kamakea M., McNulty S., Currie W., Rustad L., and Fernandex I. (1998) Nitrogen saturation in temperate forest ecosystems. *BioScience* **48**, 921–934.

Achard F., Eva H. D., Stibig H.-J., Mayaux P., Gallego J., Richards T., and Malingreau J.-P. (2002) Determination of deforestation rates of the world's humid tropical forests. *Science* **297**, 999–1002.

Ajtay G. L., Ketner P., and Duvigneaud P. (1979) Terrestrial primary production and phytomass. In *The Global Carbon*

Cycle (eds. B. Bolin, E. T. Degens, S. Kempe, and P. Ketner). Wiley, New York, pp. 129–182.

Andres R. J., Fielding D. J., Marland G., Boden T. A., Kumar N., and Kearney A. T. (1999) Carbon dioxide emissions from fossil-fuel use, 1751–1950. *Tellus* **51B**, 759–765.

Arrhenius S. (1896) On the influence of carbonic acid in the air upon the temperature of the ground. *Phil. Magazine J. Sci.* **41**, 237–276.

Aumont O., Orr J. C., Monfray P., Ludwig W., Amiotte-Suchet P., and Probst J.-L. (2001) Riverine-driven interhemispheric transport of carbon. *Global Biogeochem. Cycles* **15**, 393–405.

Barber V., Juday G. P., and Finney B. (2000) Reduced growth of Alaskan white spruce in the twentieth century from temperature-induced drought stress. *Nature* **405**, 668–673.

Barford C. C., Wofsy S. C., Goulden M. L., Munger J. W., Hammond Pyle E., Urbanski S. P., Hutyr L., Saleska S. R., Fitzjarrald D., and Moore K. (2001) Factors controlling long- and short-term sequestration of atmospheric CO_2 in a mid-latitude forest. *Science* **294**, 1688–1691.

Battle M., Bender M., Tans P. P., White J. W. C., Ellis J. T., Conway T., and Francey R. J. (2000) Global carbon sinks and their variability, inferred from atmospheric O_2 and $\delta^{13}C$. *Science* **287**, 2467–2470.

Behrenfeld M. J., Randerson J. T., McClain C. R., Feldman G. C., Los S. O., Tucker C. J., Falkowski P. G., Field C. B., Frouin R., Esaias W. W., Kolber D. D., and Pollack N. H. (2001) Biospheric primary production during an ENSO transition. *Science* **291**, 2594–2597.

Bergh J., Linder S., Lundmark T., and Elfving B. (1999) The effect of water and nutrient availability on the productivity of Norway spruce in northern and southern Sweden. *Forest Ecol. Manage.* **119**, 51–62.

Bolin B. (ed.) (1981) *Carbon Cycle Modelling*. Wiley, New York.

Bolin B. (1986) How much CO_2 will remain in the atmosphere? In *The Greenhouse Effect, Climatic Change, and Ecosystems* (eds. B. Bolin, B. R. Doos, J. Jager, and R. A. Warrick). Wiley, Chichester, England, pp. 93–155.

Bolin B., Degens E. T., Kempe S., and Ketner P. (eds.) (1979) *The Global Carbon Cycle*. Wiley, New York.

Bolin B., Doos B. R., Jager J., and Warrick R. A. (1986) *The Greenhouse Effect, Climatic Change, and Ecosystems*. Wiley, New York.

Bopp L., Le Quéré C., Heimann M., and Manning A. C. (2002) Climate-induced oceanic oxygen fluxes: implications for the contemporary carbon budget. *Global Biogeochem. Cycles* **16**, doi: 10.1029/2001GB001445.

Bousquet P., Ciais P., Peylin P., Ramonet M., and Monfray P. (1999a) Inverse modeling of annual atmospheric CO_2 sources and sinks: 1. Method and control inversion. *J. Geophys. Res.* **104**, 26161–26178.

Bousquet P., Peylin P., Ciais P., Ramonet M., and Monfray P. (1999b) Inverse modeling of annual atmospheric CO_2 sources and sinks: 2. Sensitivity study. *J. Geophys. Res.* **104**, 26179–26193.

Bousquet P., Peylin P., Ciais P., Le Quérè C., Friedlingstein P., and Tans P. P. (2000) Regional changes in carbon dioxide fluxes of land and oceans since 1980. *Science* **290**, 1342–1346.

Boyd P. W., Watson A. J., Cliff S., Law C. S., Abraham E. R., Trulli T., Murdoch R., Bakker D. C. E., Bowie A. R., Buesseler K. O., Chang H., Charette M., Croot P., Downing K., Frew R., Gall M., Hadfield M., Hall J., Harvey M., Jameson G., Laroche J., Liddicoat M., Ling R., Maldonado M. T., Mckay R. M., Nodder S., Pickmere S., Pridmore R., Rintoul S., Safi K., Sutton P., Strzepek R. T., Tanneberger K., Turner S., Waite A., and Zeldis J. (2000) A mesoscale phytoplankton bloom in the polar Southern Ocean stimulated by iron fertilization. *Nature* **407**, 695–702.

Braun S., Rihm B., Schindler C., and Fluckiger W. (2000) Growth of mature beech in relation to ozone and nitrogen deposition: an epidemiological approach. *Water Air Soil Pollut.* **116**, 356–364.

Broecker W. S. (2001) A Ewing Symposium on the contemporary carbon cycle. *Global Biogeochem. Cycles* **15**, 1031–1032.

Broecker W. S., Takahashi T., Simpson H. H., and Peng T.-H. (1979) Fate of fossil fuel carbon dioxide and the global carbon budget. *Science* **206**, 409–418.

Broecker W. S., Sutherland S., Smethie W., Peng T. H., and Ostlund G. (1995) Oceanic radiocarbon: separation of the natural and bomb components. *Global Biogeochem. Cycles* **9**, 263–288.

Brown S. (1996) Management of forests for mitigation of greenhouse gas emissions. In *Climatic Change 1995. Impacts, Adaptations and Mitigation of Climate Change: Scientific-technical Analyses* (eds. J. T. Houghton, G. J. Jenkins, and J. J. Ephraums). Cambridge University Press, Cambridge, pp. 773–797.

Bryant D. M., Holland E. A., Seastedt T. R., and Walker M. D. (1998) Analysis of litter decomposition in an alpine tundra. *Canadian J. Botany* **76**, 1295–1304.

Callendar G. S. (1938) The artificial production of carbon dioxide and its influence on temperature. *Quarterly J. Roy. Meteorol. Soc.* **64**, 223–240.

Cardon Z. G., Hungate B. A., Cambardella C. A., Chapin F. S., Field C. B., Holland E. A., and Mooney H. A. (2001) Contrasting effects of elevated CO_2 on old and new soil carbon pools. *Soil Biol. Biochem.* **33**, 365–373.

Caspersen J. P., Pacala S. W., Jenkins J. C., Hurtt G. C., Moorcroft P. R., and Birdsey R. A. (2000) Contributions of land-use history to carbon accumulation in US forests. *Science* **290**, 1148–1151.

Chisholm S. W. (2000) Stirring times in the Southern Ocean. *Nature* **407**, 685–687.

Ciais P., Peylin P., and Bousquet P. (2000) Regional biospheric carbon fluxes as inferred from atmospheric CO_2 measurements. *Ecol. Appl.* **10**, 1574–1589.

Clark D. (2002) Are tropical forests an important carbon sink? Reanalysis of the long-term plot data. *Ecol. Appl.* **12**, 3–7.

Conway T. J. and Tans P. P. (1999) Development of the CO_2 latitude gradient in recent decades. *Global Biogeochem. Cycles* **13**, 821–826.

Conway T. J., Tans P. P., Waterman L. S., Thoning K. W., Kitzis D. R., Masarie K. A., and Zhang N. (1994) Evidence for interannual variability of the carbon cycle from the National Oceanic and Atmospheric Administration/Climate Monitoring and Diagnostics Laboratory Global Air Sampling Network. *J. Geophys. Res.* **99**, 22831–22855.

Cooperative Atmospheric Data Integration Project—Carbon Dioxide (1997) *GLOBALVIEW—CO_2*. National Oceanic and Atmospheric Administration, Boulder, CO (CD-ROM).

Cox P. M., Betts R. A., Jones C. D., Spall S. A., and Totterdell I. J. (2000) Acceleration of global warming due to carbon-cycle feedbacks in a coupled climate model. *Nature* **408**, 184–187.

Cramer W., Bondeau A., Woodward F. I., Prentice I. C., Betts R. A., Brovkin V., Cox P. M., Fisher V., Foley J. A., Friend A. D., Kucharik C., Lomas M. R., Ramankutty N., Sitch S., Smith B., White A., and Young Molling C. (2001) Global response of terrestrial ecosystem structure and function to CO_2 and climate change: results from six dynamic global vegetation models. *Global Change Biol.* **7**, 357–373.

Curtis P. S. (1996) A meta-analysis of leaf gas exchange and nitrogen in trees grown under elevated carbon dioxide. *Plant Cell Environ.* **19**, 127–137.

Davidson E. A. (1995) Linkages between carbon and nitrogen cycling and their implications for storage of carbon in terrestrial ecosystems. In *Biotic Feedbacks in the Global Climatic System: Will the Warming Feed the Warming?* (eds. G. M. Woodwell and F. T. Mackenzie). Oxford University Press, New York, pp. 219–230.

Davidson E. A. and Hirsch A. I. (2001) Fertile forest experiments. *Nature* **411**, 431–433.

DeFries R. S., Field C. B., Fung I., Collatz G. J., and Bounoua L. (1999) Combining satellite data and biogeochemical models to estimate global effects of human-induced land cover change on carbon emissions and primary productivity. *Global Biogeochem. Cycles* **13**, 803–815.

DeFries R. S., Houghton R. A., Hansen M. C., Field C. B., Skole D., and Townshend J. (2002) Carbon emissions from tropical deforestation and regrowth based on satellite observations for the 1980s and 90s. *Proc. Natl. Acad. Sci.* **99**, 14256–14261.

Denning A. S., Fung I. Y., and Randall D. A. (1995) Latitudinal gradient of atmospheric CO_2 due to seasonal exchange with land biota. *Nature* **376**, 240–243.

Department of Energy (DOE) (1999) *Carbon Sequestration Research and Development*. National Technical Information Service, Springfield, Virginia (www.ornl.gov/carbon_sequestration/).

Detwiler R. P. and Hall C. A. S. (1988) Tropical forests and the global carbon cycle. *Science* **239**, 42–47.

Drake B. G., Muche M. S., Peresta G., Gonzalez-Meler M. A., and Matamala R. (1996) Acclimation of photosynthesis, respiration and ecosystem carbon flux of a wetland on Chesapeake Bay, Maryland to elevated atmospheric CO_2 concentration. *Plant Soil* **187**, 111–118.

Driscoll C. T., Lawrence G. B., Bulger A. J., Butler T. J., Cronan C. S., Eagar C., Lambert K. F., Likens G. E., Stoddard J. L., and Weathers K. C. (2001) Acidic deposition in the northeastern United States: sources and inputs, ecosystem effects, and management strategies. *BioScience* **51**, 180–198.

Etheridge D. M., Steele L. P., Langenfelds R. L., Francey R. J., Barnola J. M., and Morgan V. I. (1996) Natural and anthropogenic changes in atmospheric CO_2 over the last 1000 years from air in Antarctic ice and firn. *J. Geophys. Res.* **101**, 4115–4128.

Eva H. D., Achard F., Stibig H. J., and Mayaux P. (2003) Response to comment on Achard *et al.* (2002). *Science* **299**, 1015b.

Falkowski P. G., Barber R. T., and Smetacek V. (1998) Biogeochemical controls and feedbacks on ocean primary production. *Science* **281**, 200–206.

Falkowski P., Scholes R. J., Boyle E., Canadell J., Canfield D., Elser J., Gruber N., Hibbard K., Högberg P., Linder S., Mackenzie F. T., Moore B., Pedersen T., Rosenthal Y., Seitzinger S., Smetacek V., and Steffen W. (2000) The global carbon cycle: a test of our knowledge of earth as a system. *Science* **290**, 291–296.

Fan S., Gloor M., Mahlman J., Pacala S., Sarmiento J., Takahashi T., and Tans P. (1998) A large terrestrial carbon sink in North America implied by atmospheric and oceanic CO_2 data and models. *Science* **282**, 442–446.

FAO (2001) *Global Forest Resources Assessment 2000*. Main Report, FAO Forestry Paper 140, Rome.

Fearnside P. M. (2000) Global warming and tropical land-use change: greenhouse gas emissions from biomass burning, decomposition and soils in forest conversion, shifting cultivation and secondary vegetation. *Climat. Change* **46**, 115–158.

Fearnside P. M. and Laurance W. F. (2003) Comment on Achard *et al.* (2002). *Science* **299**, 1015a.

Fenn M. E., Poth M. A., Aber J. D., Baron J. S., Bormann B. T., Johnson D. W., Lemly A. D., McNulty S. G., Ryan D. F., and Stottlemyer R. (1998) Nitrogen excess in North American ecosystems: predisposing factors, ecosystem responses and management strategies. *Ecol. Appl.* **8**, 706–733.

Field C. B., Behrenfeld M. J., Randerson J. T., and Falkowski P. G. (1998) Primary production of the biosphere: integrating terrestrial and oceanic components. *Science* **281**, 237–240.

Fog K. (1988) The effect of added nitrogen on the rate of decomposition of organic matter. *Biol. Rev. Cambridge Phil. Soc.* **63**, 433–462.

Folland C. K. and Karl T. R. (2001) Observed climate variability and change. In *Climate Change 2001: The Scientific Basis. Contribution of Working Group I to the 3rd Assessment Report of the Intergovernmental Panel on Climate Change* (eds. J. T. Houghton, Y. Ding, D. J. Griggs, M. Noguer, P. J. van der Linden, X. Dai, K. Maskell, and C. A. Johnson). Cambridge University Press, Cambridge, UK and New York, pp. 99–181.

Fraser P. J., Elliott W. P., and Waterman L. S. (1986) Atmospheric CO_2 record from direct chemical measurements during the 19th century. In *The Changing Carbon Cycle. A Global Analysis* (eds. J. R. Trabalka and D. E. Reichle). Springer, New York, , pp. 66–88.

Gaudinski J. B., Trumbore S. E., Davidson E. A., and Zheng S. (2000) Soil carbon cycling in a temperate forest: radiocarbon-based estimates of residence times, sequestration rates and partitioning of fluxes. *Biogeochemistry* **51**, 33–69.

Goldemberg J., Johansson T. B., Reddy A. K. N., and Williams R. H. (1985) An end-use oriented global energy strategy. *Ann. Rev. Energy* **10**, 613–688.

Goodale C. L., Apps M. J., Birdsey R. A., Field C. B., Heath L. S., Houghton R. A., Jenkins J. C., Kohlmaier G. H., Kurz W., Liu S., Nabuurs G.-J., Nilsson S., and Shvidenko A. Z. (2002) Forest carbon sinks in the northern hemisphere. *Ecol. Appl.* **12**, 891–899.

Goulden M. L., Munger J. W., Fan S.-M., Daube B. C., and Wofsy S. C. (1996) Exchange of carbon dioxide by a deciduous forest: response to interannual climate variability. *Science* **271**, 1576–1578.

Goulden M. L., Wofsy S. C., Harden J. W., Trumbore S. E., Crill P. M., Gower S. T., Fries T., Daube B. C., Fau S., Sulton D. J., Bazzaz A., and Munger J. W. (1998) Sensitivity of boreal forest carbon balance to soil thaw. *Science* **279**, 214–217.

Grace J., Lloyd J., McIntyre J., Miranda A. C., Meir P., Miranda H. S., Nobre C., Moncrieff J., Massheder J., Malhi Y., Wright I., and Gash J. (1995) Carbon dioxide uptake by an undisturbed tropical rain forest in southwest Amazonia, 1992 to 1993. *Science* **270**, 778–780.

Gruber N., Sarmiento J. L., and Stocker T. F. (1996) An improved method for detecting anthropogenic CO_2 in the oceans. *Global Biogeochem. Cycles* **10**, 809–837.

Gurney K. R., Law R. M., Denning A. S., Rayner P. J., Baker D., Bousquet P., Bruhwiler L., Chen Y. H., Ciais P., Fan S., Fung I. Y., Gloor M., Heimann M., Higuchi K., John J., Maki T., Maksyutov S., Masarie K., Peylin P., Prather M., Pak B. C., Randerson J., Sarmiento J., Taguchi S., Takahashi T., and Yuen C. W. (2002) Towards robust regional estimates of CO_2 sources and sinks using atmospheric transport models. *Nature* **415**, 626–630.

Hall C. A. S. and Uhlig J. (1991) Refining estimates of carbon released from tropical land-use change. *Canadian J. Forest Res.* **21**, 118–131.

Heath L. S. and Smith J. E. (2000) An assessment of uncertainty in forest carbon budget projections. *Environ. Sci. Policy* **3**, 73–82.

Herzog H., Eliasson B., and Kaarstad O. (2000) Capturing greenhouse gases. *Sci. Am.* **282**(2), 72–79.

Hicke J. A., Asner G. P., Randerson J. T., Tucker C., Los S., Birdsey R., Jenkins J. C., Field C., and Holland E. (2002) Satellite-derived increases in net primary productivity across North America, 1982–1998. *Geophys. Res. Lett.*, 10.1029/2001GL013578.

Holland E. A., Braswell B. H., Lamarque J.-F., Townsend A., Sulzman J., Muller J.-F., Dentener F., Brasseur G., Levy H., Penner J. E., and Roelots G.-J. (1997) Variations in the predicted spatial distribution of atmospheric nitrogen deposition and their impact on carbon uptake by terrestrial ecosystems. *J. Geophys. Res.* **102**, 15849–15866.

Houghton R. A. (1987) Biotic changes consistent with the increased seasonal amplitude of atmospheric CO_2 concentrations. *J. Geophys. Res.* **92**, 4223–4230.

Houghton R. A. (1996) Converting terrestrial ecosystems from sources to sinks of carbon. *Ambio* **25**, 267–272.

Houghton R. A. (1999) The annual net flux of carbon to the atmosphere from changes in land use 1850–1990. *Tellus* **51B**, 298–313.

Houghton R. A. (2000) Interannual variability in the global carbon cycle. *J. Geophys. Res.* **105**, 20121–20130.

Houghton R. A. (2003) Revised estimates of the annual net flux of carbon to the atmosphere from changes in land use and land management 1850–2000. *Tellus* **55B**, 378–390.

Houghton R. A. and Hackler J. L. (1995) *Continental Scale Estimates of the Biotic Carbon Flux from Land Cover Change: 1850–1980.* ORNL/CDIAC-79, NDP-050, Oak Ridge National Laboratory, Oak Ridge, TN, 144pp.

Houghton R. A. and Hackler J. L. (2001) *Carbon Flux to the Atmosphere from Land-use Changes: 1850–1990.* ORNL/CDIAC-131, NDP-050/R1, US Department of Energy, Oak Ridge National Laboratory, Carbon Dioxide Information Analysis Center, Oak Ridge, TN.

Houghton R. A. and Skole D. L. (1990) Carbon. In *The Earth as Transformed by Human Action* (eds. B. L. Turner, W. C. Clark, R. W. Kates, J. F. Richards, J. T. Mathews, and W. B. Meyer). Cambridge University Press, Cambridge, pp. 393–408.

Houghton R. A., Hobbie J. E., Melillo J. M., Moore B., Peterson B. J., Shaver G. R., and Woodwell G. M. (1983) Changes in the carbon content of terrestrial biota and soils between 1860 and 1980: a net release of CO_2 to the atmosphere. *Ecol. Monogr.* **53**, 235–262.

Houghton R. A., Davidson E. A., and Woodwell G. M. (1998) Missing sinks, feedbacks, and understanding the role of terrestrial ecosystems in the global carbon balance. *Global Biogeochem. Cycles* **12**, 25–34.

Houghton R. A., Hackler J. L., and Lawrence K. T. (1999) The US carbon budget: contributions from land-use change. *Science* **285**, 574–578.

Houghton R. A., Skole D. L., Nobre C. A., Hackler J. L., Lawrence K. T., and Chomentowski W. H. (2000) Annual fluxes of carbon from deforestation and regrowth in the Brazilian Amazon. *Nature* **403**, 301–304.

House J. I., Prentice I. C., Ramankutty N., Houghton R. A., and Heimann M. (2003) Reconciling apparent inconsistencies in estimates of terrestrial CO_2 sources and sinks. *Tellus* **(55B)**, 345–363.

Hu S., Chapin F. S., Firestone M. K., Field C. B., and Chiariello N. R. (2001) Nitrogen limitation of microbial decomposition in a grassland under elevated CO_2. *Nature* **409**, 188–191.

Hurtt G. C., Pacala S. W., Moorcroft P. R., Caspersen J., Shevliakova E., Houghton R. A., and Moore B., III. (2002) Projecting the future of the US carbon sink. *Proc. Natl. Acad. Sci.* **99**, 1389–1394.

Jarvis P. and Linder S. (2000) Constraints to growth of boreal forests. *Nature* **405**, 904–905.

Jobbággy E. G. and Jackson R. B. (2000) The vertical distribution of soil organic carbon and its relation to climate and vegetation. *Ecol. Appl.* **10**, 423–436.

Joos F., Plattner G.-K., Stocker T. F., Marchal O., and Schmittner A. (1999a) Global warming and marine carbon cycle feedbacks on future atmospheric CO_2. *Science* **284**, 464–467.

Joos F., Meyer R., Bruno M., and Leuenberger M. (1999b) The variability in the carbon sinks as reconstructed for the last 1000 years. *Geophys. Res. Lett.* **26**, 1437–1440.

Joos F., Prentice I. C., and House J. I. (2002) Growth enhancement due to global atmospheric change as predicted by terrestrial ecosystem models: consistent with US forest inventory data. *Global Change Biol.* **8**, 299–303.

Keeling C. D. (1973) Industrial production of carbon dioxide from fossil fuels and limestone. *Tellus* **25**, 174–198.

Keeling C. D., Bacastow R. B., Carter A. F., Piper S. C., Whorf T. P., Heimann M., Mook W. G., and Roeloffzen H. (1989) A three-dimensional model of atmospheric CO_2 transport based on observed winds: 1. Analysis of observational data. In *Aspects of Climate Variability in the Pacific and the Western Americas.* Geophysical Monograph 55 (ed. D. H. Peterson). American Geophysical Union, Washington, DC, pp. 165–236.

Keeling C. D., Chin J. F. S., and Whorf T. P. (1996a) Increased activity of northern vegetation inferred from atmospheric CO_2 observations. *Nature* **382**, 146–149.

Keeling C. D., Piper S. C., Bacastow R. B., Wahlen M., Whorf T. P., Heimann M., and Meijer H. A. (2001) *Exchanges of Atmospheric CO_2 and $^{13}CO_2$ with the Terrestrial Biosphere and Oceans from 1978 to 2000: I. Global Aspects.* Scripps Institution of Oceanography, Technical Report SIO Reference Series, No. 01-06 (Revised from SIO Reference Series, No. 00-21), San Diego.

Keeling R. F. and Garcia H. (2002) The change in oceanic O_2 inventory associated with recent global warming. *Proc. US Natl. Acad. Sci.* **99**, 7848–7853.

Keeling R. F. and Shertz S. R. (1992) Seasonal and interannual variations in atmospheric oxygen and implications for the global carbon cycle. *Nature* **358**, 723–727.

Keeling R. F., Piper S. C., and Heimann M. (1996b) Global and hemispheric CO_2 sinks deduced from changes in atmospheric O_2 concentration. *Nature* **381**, 218–221.

Kempe S. (1979) Carbon in the rock cycle. In *The Global Carbon Cycle* (eds. B. Bolin, E. T. Degens, S. Kempe, and P. Ketner). Wiley, New York, pp. 343–377.

Knorr W. and Heimann M. (2001) Uncertainties in global terrestrial biosphere modeling: 1. A comprehensive sensitivity analysis with a new photosynthesis and energy balance scheme. *Global Biogeochem. Cycles* **15**, 207–225.

Koch G. W. and Mooney H. A. (1996) Response of terrestrial ecosystems to elevated CO_2: a synthesis and summary. In *Carbon Dioxide and Terrestrial Ecosystems* (eds. G. W. Koch and H. A. Mooney). Academic Press, San Diego, pp. 415–429.

Köhl M. and Päivinen R. (1997) *Study on European Forestry Information and Communication System.* Office for Official Publications of the European Communities, Luxembourg, Volumes 1 and 2, 1328 pp.

Kohlmaier G. H., Weber M., and Houghton R. A. (eds.) (1998) *Carbon Dioxide Mitigation in Forestry and Wood Industry.* Springer, Berlin.

Körner C. (2000) Biosphere responses to CO_2-enrichment. *Ecol. Appl.* **10**, 1590–1619.

Kurz W. A. and Apps M. J. (1999) A 70-year retrospective analysis of carbon fluxes in the Canadian forest sector. *Ecol. Appl.* **9**, 526–547.

Laws E. A., Falkowski P. G., Smith W. O., Ducklow H., and McCarthy J. J. (2000) Temperature effects on export production in the open ocean. *Global Biogeochem. Cycles* **14**, 1231–1246.

Lemon E. (1977) The land's response to more carbon dioxide. In *The Fate of Fossil Fuel CO_2 in the Oceans* (eds. N. R. Andersen and A. Malahoff). Plenum Press, New York, pp. 97–130.

Levitus S., Antonov J. I., Boyer T. P., and Stephens C. (2000) Warming of the world ocean. *Science* **287**, 2225–2229.

Lloyd A. H. and Fastie C. L. (2002) Spatial and temporal variability in the growth and climate response of treeline trees in Alaska. *Climatic Change* **52**, 481–509.

Longhurst A., Sathyendranath S., Platt T., and Caverhill C. (1995) An estimate of global primary production in the ocean from satellite radiometer data. *J. Plankton Res.* **17**, 1245–1271.

Luo Y. Q., Reynolds J., and Wang Y. P. (1999) A search for predictive understanding of plant responses to elevated $[CO_2]$. *Global Change Biol.* **5**, 143–156.

Luo Y., Wan S., Hui D., and Wallace L. L. (2001) Acclimatization of soil respiration to warming in a tall grass prairie. *Nature* **413**, 622–625.

Magill A. H., Aber J. D., Hendricks J. J., Bowden R. D., Melillo J. M., and Steudler P. A. (1997) Biogeochemical response of forest ecosystems to simulated chronic nitrogen deposition. *Ecol. Appl.* **7**, 402–415.

Magill A., Aber J., Berntson G., McDowell W., Nadelhoffer K., Melillo J., and Steudler P. (2000) Long-term nitrogen additions and nitrogen saturation in two temperate forests. *Ecosystems* **3**, 238–253.

Makipaa R. (1995) Effect of nitrogen input on carbon accumulation of boreal forest soils and ground vegetation. *Forest Ecol. Manage.* **79**, 217–226.

Malhi Y., Nobre A. D., Grace J., Kruijt B., Pereira M. G. P., Culf A., and Scott S. (1998) Carbon dioxide transfer over a central Amazonian rain forest. *J. Geophys. Res.* **103**, 31593–31612.

Malhi Y., Phillips O., Kruijt B., and Grace J. (2001) The magnitude of the carbon sink in intact tropical forests: results from recent field studies. In *6th International Carbon Dioxide Conference, Extended Abstracts*. Tohoku University, Sendai, Japan, pp. 360–363.

Marland G., Andres R. J., Boden T. A., and Johnston C. (1998) *Global, Regional and National CO₂ Emission Estimates from Fossil Fuel Burning, Cement Production, and Gas Flaring: 1751–1995* (revised January 1998). ORNL/CDIAC NDP-030/R8, http://cdiac.esd.ornl.gov/ndps/ndp030.html

Martin J. H. (1990) Glacial–interglacial CO₂ change: the iron hypothesis. *Paleoceanography* **5**, 1–13.

Masarie K. A. and Tans P. P. (1995) Extension and integration of atmospheric carbon dioxide data into a globally consistent measurement record. *J. Geophys. Res.* **100**, 11593–11610.

Matear R. J. and Hirst A. C. (1999) Climate change feedback on the future oceanic CO₂ uptake. *Tellus* **51B**, 722–733.

McGuire A. D., Sitch S., Clein J. S., Dargaville R., Esser G., Foley J., Heimann M., Joos F., Kaplan J., Kicklighter D. W., Meier R. A., Melillo J. M., Moore B., Prentice I. C., Ramankutty N., Reichenau T., Schloss A., Tian H., Williams L. J., and Wittenberg U. (2001) Carbon balance of the terrestrial biosphere in the twentieth century: Analyses of CO₂, climate and land use effects with four process-based ecosystem models. *Global Biogeochem. Cycles* **15**, 183–206.

Mclaughlin S. and Percy K. (2000) Forest health in North America: some perspectives on actual and potential roles of climate and air pollution. *Water Air Soil Pollut.* **116**, 151–197.

McNeil B. I., Matear R. J., Key R. M., Bullister J. L., and Sarmiento J. L. (2003) Anthropogenic CO₂ uptake by the ocean based on the global chlorofluorocarbon data set. *Science* **299**, 235–239.

McNulty S. G., Aber J. D., and Newman S. D. (1996) Nitrogen saturation in a high elevation spruce-fir stand. *Forest Ecol. Manage.* **84**, 109–121.

Miller S. D., Goulden M. L., Menton M. C., da Rocha H. R., Freitas H. C., Figueira A. M., and Sousa C. A. D. Tower-based and biometry-based measurements of tropical forest carbon balance. *Ecol. Appl.* (in press).

Mooney H. A., Canadell J., Chapin F. S., Ehleringer J., Körner C., McMurtrie R., Parton W. J., Pitelka L., and Schulze E.-D. (1999) Ecosystem physiology responses to global change. In *Implications of Global Change for Natural and Managed Ecosystems: A Synthesis of GCTE and Related Research* (ed. B. H. Walker, W. L. Steffen, J. Canadel, and J. S. I. Ingram). Cambridge University Press, Cambridge, pp. 141–189.

Myneni R. B., Los S. O., and Asrar G. (1995) Potential gross primary productivity of terrestrial vegetation from 1982–1990. *Geophys. Res. Lett.* **22**, 2617–2620.

Myneni R. B., Keeling C. D., Tucker C. J., Asrar G., and Nemani R. R. (1997) Increased plant growth in the northern high latitudes from 1981 to 1991. *Nature* **386**, 698–702.

Nadelhoffer K. J., Emmett B. A., Gundersen P., Kjønaas O. J., Koopmans C. J., Schleppi P., Teitema A., and Wright R. F. (1999) Nitrogen deposition makes a minor contribution to carbon sequestration in temperate forests. *Nature* **398**, 145–148.

National Research Council (NRC) (1983) *Changing Climate*. National Academy Press, Washington, DC.

Neftel A., Moor E., Oeschger H., and Stauffer B. (1985) Evidence from polar ice cores for the increase in atmospheric CO₂ in the past two centuries. *Nature* **315**, 45–47.

Nepstad D. C., Verissimo A., Alencar A., Nobre C., Lima E., Lefebvre P., Schlesinger P., Potter C., Moutinho P.,

Mendoza E., Cochrane M., and Brooks V. (1999) Large-scale impoverishment of Amazonian forests by logging and fire. *Nature* **398**, 505–508.

Oechel W. C., Hastings S. J., Vourlitis G., Jenkins M., Riechers G., and Grulke N. (1993) Recent change of arctic tundra ecosystems from a net carbon dioxide sink to a source. *Nature* **361**, 520–523.

Oechel W. C., Cowles S., Grulke N., Hastings S. J., Lawrence B., Prudhomme T., Riechers G., Strain B., Tissue D., and Vourlitis G. (1994) Transient nature of CO₂ fertilization in Arctic tundra. *Nature* **371**, 500–503.

Oren R., Ellsworth D. S., Johnsen K. H., Phillips N., Ewers B. E., Maier C., Schäfer K. V. R., McCarthy H., Hendrey G., McNulty S. G., and Katul G. G. (2001) Soil fertility limits carbon sequestration by forest ecosystems in a CO₂-enriched atmosphere. *Nature* **411**, 469–472.

Orr J. C. (2000) OCMIP carbon analysis gets underway. *Research GAIM* **3**(2), 4–5.

Orr J. C. and Dutay J.-C. (1999) OCMIP mid-project workshop. *Res. GAIM Newslett.* **3**, 4–5.

Orr J., Maier-Reimer E., Mikolajewicz U., Monfray P., Sarmiento J. L., Toggweiler J. R., Taylor N. K., Palmer J., Gruber N., Sabine C. L., Le Quéré C., Key R. M., and Boutin J. (2001) Estimates of anthropogenic carbon uptake from four 3-D global ocean models. *Global Biogeochem. Cycles* **15**, 43–60.

Pacala S. W., Hurtt G. C., Baker D., Peylin P., Houghton R. A., Birdsey R. A., Heath L., Sundquist E. T., Stallard R. F., Ciais P., Moorcroft P., Caspersen J. P., Shevliakova E., Moore B., Kohlmaier G., Holland E., Gloor M., Harmon M. E., Fan S.-M., Sarmiento J. L., Goodale C. L., Schimel D., and Field C. B. (2001) Consistent land- and atmosphere-based US carbon sink estimates. *Science* **292**, 2316–2320.

Page S. E., Siegert F., Rieley L. O., Boehm H.-D. V., Jaya A., and Limin S. (2002) The amount of carbon released from peat and forest fires in Indonesia during 1997. *Nature* **420**, 61–65.

Peterson B. J. and Melillo J. M. (1985) The potential storage of carbon by eutrophication of the biosphere. *Tellus* **37B**, 117–127.

Phillips D. L., Brown S. L., Schroeder P. E., and Birdsey R. A. (2000) Toward error analysis of large-scale forest carbon budgets. *Global Ecol. Biogeogr.* **9**, 305–313.

Phillips O. L., Malhi Y., Higuchi N., Laurance W. F., Núñez P. V., Vásquez R. M., Laurance S. G., Ferreira L. V., Stern M., Brown S., and Grace J. (1998) Changes in the carbon balance of tropical forests: evidence from land-term plots. *Science* **282**, 439–442.

Phillips O. L., Malhi Y., Vinceti B., Baker T., Lewis S. L., Higuchi N., Laurance W. F., Vargas P. N., Martinez R. V., Laurance S., Ferreira L. V., Stern M., Brown S., and Grace J. (2002) Changes in growth of tropical forests: evaluating potential biases. *Ecol. Appl.* **12**, 576–587.

Plattner G.-K., Joos F., and Stocker T. F. (2002) Revision of the global carbon budget due to changing air-sea oxygen fluxes. *Global Biogeochem. Cycles* **16**(4), 1096, doi:10.1029/2001GB001746.

Post W. M., Emanuel W. R., Zinke P. J., and Stangenberger A. G. (1982) Soil carbon pools and world life zones. *Nature* **298**, 156–159.

Powell D. S., Faulkner J. L., Darr D. R., Zhu Z., and MacCleery D. W. (1993) *Forest resources of the US, 1992*. General Technical Report RM-234. USDA Forest Service, Rocky Mountain Forest and Range Experiment Station, Fort Collins, CO.

Prather M. and Ehhalt D. (2001) Atmospheric chemistry and greenhouse gases. In *Climate Change 2001: The Scientific Basis. Contribution of Working Group I to the 3rd Assessment Report of the Intergovernmental Panel on Climate Change* (eds. J. T. Houghton, Y. Ding, D. J. Griggs, M. Noguer, P. J. van der Linden, X. Dai, K. Maskell, and C. A. Johnson). Cambridge University Press, Cambridge, UK and New York, pp. 239–287.

Prentice I. C., Farquhar G. D., Fasham M. J. R., Goulden M. L., Heimann M., Jaramillo V. J., Kheshgi H. S., Le Quéré C., Scholes R. J., and Wallace D. W. R. (2001) The carbon cycle and atmospheric carbon dioxide. In *Climate Change 2001: The Scientific Basis. Contribution of Working Group I to the 3rd Assessment Report of the Intergovernmental Panel on Climate Change* (eds. J. T. Houghton, Y. Ding, D. J. Griggs, M. Noguer, P. J. van der Linden, X. Dai, K. Maskell, and C. A. Johnson). Cambridge University Press, Cambridge, UK and New York, pp. 183–237.

Raich J. W. and Schlesinger W. H. (1992) The global carbon dioxide flux in soil respiration and its relationship to vegetation and climate. *Tellus* **44B**, 81–99.

Randerson J. T., Thompson M. V., Conway T. J., Fung I. Y., and Field C. B. (1997) The contribution of terrestrial sources and sinks to trends in the seasonal cycle of atmospheric carbon dioxide. *Global Biogeochem. Cycles* **11**, 535–560.

Randerson J. T., Field C. B., Fung I. Y., and Tans P. P. (1999) Increases in early season ecosystem uptake explain recent changes in the seasonal cycle of atmospheric CO_2 at high northern latitudes. *Geophys. Res. Lett.* **26**, 2765–2768.

Raynaud D. and Barnola J. M. (1985) An Antarctic ice core reveals atmospheric CO_2 variations over the past few centuries. *Nature* **315**, 309–311.

Revelle R. and Suess H. E. (1957) Carbon dioxide exchange between atmosphere and ocean and the question of an increase of atmospheric CO_2 during the past decades. *Tellus* **9**, 18–27.

Rice A. H., Pyle E. H., Saleska S. R., Hutyra L., de Camargo P. B., Portilho K., Marques D. F., and Wofsy S. C. Carbon balance and vegetation dynamics in an old-growth Amazonian forest. *Ecol. Appl.* (in press).

Richey J. E., Melack J. M., Aufdenkampe A. K., Ballester V. M., and Hess L. L. (2002) Outgassing from Amazonian rivers and wetlands as a large tropical source of atmospheric CO_2. *Nature* **416**, 617–620.

Saleska S. R., Miller S. D., Matross D. M., Goulden M. L., Wofsy S. C., da Rocha H., de Camargo P. B., Crill P. M., Daube B. C., Freitas C., Hutyra L., Keller M., Kirchhoff V., Menton M., Munger J. W., Pyle E. H., Rice A. H., and Silva H. Carbon fluxes in old-growth Amazonian rainforests: unexpected seasonality and disturbance-induced net carbon loss (in press).

Sampson R. N. and Scholes R. J. (2000) Additional human-induced activities—article 3.4. In *Land Use, Land-use Change, and Forestry. A Special Report of the IPCC* (eds. R. T. Watson, I. R. Noble, B. Bolin, N. H. Ravindranath, D. J. Verardo, and D. J. Dokken). Cambridge University Press, New York, pp. 181–281.

Sarmiento J. L. (1993) Ocean carbon cycle. *Chem. Eng. News* **71**, 30–43.

Sarmiento J. L. and Sundquist E. T. (1992) Revised budget for the oceanic uptake of anthropogenic carbon dioxide. *Nature* **356**, 589–593.

Sarmiento J. L., Hughes T. M. C., Stouffer R. J., and Manabe S. (1998) Simulated response of the ocean carbon cycle to anthropogenic climate warming. *Nature* **393**, 245–249.

Saugier B., Roy J., and Mooney H. A. (2001) Estimations of global terrestrial productivity: converging toward a single number? In *Terrestrial Global Productivity* (eds. J. Roy, B. Saugier, and H. A. Mooney). Academic Press, San Diego, California, pp. 543–557.

SCEP (Study of Critical Environmental Problems) (1970) *Man's Impact on the Global Environment*. The MIT Press, Cambridge, Massachusetts.

Schimel D. S., Alves D., Enting I., Heimann M., Joos F., Raynaud D., and Wigley T. (1996) CO_2 and the carbon cycle. In *Climate Change 1995* (eds. J. T. Houghton, L. G. M. Filho, B. A. Callendar, N. Harris, A. Kattenberg, and K. Maskell). Cambridge University Press, Cambridge, pp. 76–86.

Schimel D., Melillo J., Tian H., McGuire A. D., Kicklighter D., Kittel T., Rosenbloom N., Running S., Thornton P.,

Ojima D., Parton W., Kelly R., Sykes M., Neilson R., and Rizzo B. (2000) Contribution of increasing CO_2 and climate to carbon storage by ecosystems in the United States. *Science* **287**, 2004–2006.

Schimel D. S., House J. I., Hibbard K. A., Bousquet P., Ciais P., Peylin P., Braswell B. H., Apps M. J., Baker D., Bondeau A., Canadell J., Churkina G., Cramer W., Denning A. S., Field C. B., Friedlingstein P., Goodale C., Heimann M., Houghton R. A., Melillo J. M., Moore B., III, Murdiyarso D., Noble I., Pacala S. W., Prentice I. C., Raupach M. R., Rayner P. J., Scholes R. J., Steffen W. L., and Wirth C. (2001) Recent patterns and mechanisms of carbon exchange by terrestrial ecosystems. *Nature* **414**, 169–172.

Schlesinger W. H. and Lichter J. (2001) Limited carbon storage in soil and litter of experimental forest plots under increased atmospheric CO_2. *Nature* **411**, 466–469.

Shaw M. R., Zavaleta E. S., Chiariello N. R., Cleland E. E., Mooney H. A., and Field C. B. (2003) Grassland responses to global environmetal changes suppressed by elevated CO_2. *Science* **298**, 1987–1990.

Shvidenko A. Z. and Nilsson S. (1997) Are the Russian forests disappearing? *Unasylva* **48**, 57–64.

Smith H. J., Fischer H., Wahlen M., Mastroianni D., and Deck B. (1999) *Nature* **400**, 248–250.

Spiecker H., Mielikainen K., Kohl M., and Skovsgaard J. (eds.) (1996). *Growth Trends in European Forest—Studies from 12 Countries*. Springer, Berlin.

Stephens B. B., Keeling R. F., Heimann M., Six K. D., Murnane R., and Caldeira K. (1998) Testing global ocean carbon cycle models using measurements of atmospheric O_2 and CO_2 concentration. *Global Biogeochem. Cycles* **12**, 213–230.

Sundquist E. T. (1986) Geologic analogs: their value and limitation in carbon dioxide research. In *The Changing Carbon Cycle. A Global Analysis* (eds. J. R. Trabalka and D. E. Reichle). Springer, New York, pp. 371–402.

Sundquist E. T. and Broecker W. S. (eds.) (1985). *The Carbon Cycle and Atmospheric CO_2: Natural Variations Archean to Present*, Geophysical Monograph 32. American Geophysical Union, Washington, DC.

Tamm C. O., Aronsson A., and Popovic B. (1995) Nitrogen saturation in a long-term forest experiment with annual additions of nitrogen. *Water Air Soil Pollut.* **85**, 1683–1688.

Tans P. P., Fung I. Y., and Takahashi T. (1990) Observational constraints on the global atmospheric CO_2 budget. *Science* **247**, 1431–1438.

Tans P. P., Fung I. Y., and Enting I. G. (1995) Storage versus flux budgets: the terrestrial uptake of CO_2 during the 1980s. In *Biotic Feedbacks in the Global Climatic System. Will the Warming Feed the Warming* (eds. G. M. Woodwell and F. T. Mackenzie). Oxford University Press, New York, pp. 351–366.

Tans P. P., Bakwin P. S., Bruhwiler L., Conway T. J., Dlugokencky E. J., Guenther D. W., Kitzis D. R., Lang P. M., Masarie K. A., Miller J. B., Novelli P. C., Thoning K. W., Vaughn B. H., White J. W. C., and Zhao C. (2001) Carbon cycle. In *Climate Monitoring and Diagnostics Laboratory Summary Report No. 25 1998–1999* (eds. R. C. Schnell, D. B. King, and R. M. Rosson). NOAA, Boulder, CO, pp. 24–46.

Taylor J. A. and Orr J. C. (2000) The natural latitudinal distribution of atmospheric CO_2. *Global Planet. Change* **26**, 375–386.

Tian H., Melillo J. M., Kicklighter D. W., McGuire A. D., Helfrich J. V. K., Moore B., and Vorosmarty C. J. (1998) Effect of interannual climate variability on carbon storage in Amazonian ecosystems. *Nature* **396**, 664–667.

Tissue D. T. and Oechel W. C. (1987) Response of *Eriophorum vaginatum* to elevated CO_2 and temperature in the Alaskan tussock tundra. *Ecology* **68**, 401–410.

Trabalka J. R. (ed.) (1985) *Atmospheric Carbon Dioxide and the Global Carbon Cycle*. DOE/ER-0239, US Department of Energy, Washington, DC.

UNFCCC (1992) *Text of the United Nations Framework Convention on Climate Change.* (UNEP/WMO Information Unit on Climate Change.), Geneva, Switzerland, 29pp.

Valentini R., Matteucci G., Dolman A. J., Schulze E.-D., Rebmann C., Moors E. J., Granier A., Gross P., Jensen N. O., Pilegaard K., Lindroth A., Grelle A., Bernhofer C., Grünwald T., Aubinet M., Ceulemans R., Kowalski A. S., Vesala T., Rannik Ü., Berbigier P., Loustau D., Gudmundsson J., Thorgeirsson H., Ibrom A., Morgenstern K., Clement R., Moncrieff J., Montagnani L., Minerbi S., and Jarvis P. G. (2000) Respiration as the main determinant of European forests carbon balance. *Nature* **404**, 861–865.

Wanninkhof R. (1992) Relationship between wind-speed and gas-exchange over the ocean. *J. Geophys. Res.* **97**, 7373–7382.

Watson R. T., Rodhe H., Oeschger H., and Siegenthaler U. (1990) Greenhouse gases and aerosols. *Climate Change, The IPCC Scientific Assessment* (eds. J. T. Houghton, G. J. Jenkins, and J. J. Ephraums). Cambridge University Press, Cambridge, pp. 1–40.

Watson R. T., Noble I. R., Bolin B., Ravindranath N. H., Verardo D. J., and Dokken D. J. (eds.) (2000) *Land Use, Land-Use Change, and Forestry*. A Special Report of the IPCC, Cambridge University Press, New York.

White A., Cannell M. G. R., and Friend A. D. (1999) Climate change impacts on ecosystems and the terrestrial carbon sink: a new assessment. *Global Environ. Change* **9**, S21–S30.

Woodwell G. M. (1983) Biotic effects on the concentration of atmospheric carbon dioxide: a review and projection. In *Changing Climate*. National Academy Press, Washington, DC, pp. 216–241.

Woodwell G. M. and Mackenzie F. T. (eds.) (1995) *Biotic Feedbacks in the Global Climatic System. Will the Warming Feed the Warming?* Oxford University Press, New York.

Woodwell G. M. and Pecan E. V. (eds.) (1973) *Carbon and the Biosphere*, US Atomic Energy Commission, Symposium Series 30. National Technical Information Service, Springfield, Virginia.

Geochemistry of Earth Surface Systems
ISBN: 978-0-08-096706-6

pp. 87–128

4

The Global Sulfur Cycle

P. Brimblecombe

University of East Anglia, Norwich, UK

4.1	ELEMENTARY ISSUES	130
	4.1.1 History	130
	4.1.2 Isotopes	130
	4.1.3 Allotropes	131
	4.1.4 Vapor Pressure	131
	4.1.5 Chemistry	132
4.2	ABUNDANCE OF SULFUR AND EARLY HISTORY	134
	4.2.1 Sulfur in the Cosmos	134
	4.2.2 Condensation, Accretion, and Evolution	135
	4.2.3 Sulfur on the Early Earth	135
4.3	OCCURRENCE OF SULFUR	137
	4.3.1 Elemental Sulfur	137
	4.3.2 Sulfides	137
	4.3.3 Evaporites	137
	4.3.4 The Geological History of Sulfur	137
	4.3.5 Utilization and Extraction of Sulfur Minerals	138
4.4	CHEMISTRY OF VOLCANOGENIC SULFUR	139
	4.4.1 Deep-sea Vents	139
	4.4.2 Aerial Emissions	139
	4.4.3 Fumaroles	140
	4.4.4 Crater Lakes	140
	4.4.5 Impacts of Emissions on Local Environments	141
4.5	BIOCHEMISTRY OF SULFUR	141
	4.5.1 Origin of Life	141
	4.5.2 Sulfur Biomolecules	141
	4.5.3 Uptake of Sulfur	142
4.6	SULFUR IN SEAWATER	143
	4.6.1 Sulfate	143
	4.6.2 Hydrogen Sulfide	143
	4.6.3 OCS and Carbon Disulfide	144
	4.6.4 Organosulfides	144
	4.6.5 Coastal Marshes	146
4.7	SURFACE AND GROUNDWATERS	146
4.8	MARINE SEDIMENTS	147
4.9	SOILS AND VEGETATION	147
4.10	TROPOSPHERE	149
	4.10.1 Atmospheric Budget of Sulfur Compounds	149
	4.10.2 Hydrogen Sulfide	150
	4.10.3 Carbonyl Sulfide	150
	4.10.4 Carbon Disulfide	150
	4.10.5 Dimethyl Sulfide	150

4.10.6	*Dimethylsulfoxide and Methanesulfonic Acid*	151
4.10.7	*Sulfur*	152
4.10.8	*Sulfur Dioxide*	152
4.10.9	*Aerosol Sulfates and Climate*	154
4.10.10	*Deposition*	155
4.11	ANTHROPOGENIC IMPACTS ON THE SULFUR CYCLE	155
4.11.1	*Combustion Emissions*	155
4.11.2	*Organosulfur Gases*	156
4.11.3	*Acid Rain*	156
4.11.4	*Water and Soil Pollutants*	157
4.11.5	*Coastal Pollution*	158
4.12	SULFUR IN UPPER ATMOSPHERES	158
4.12.1	*Radiation Balance and Sulfate Particles*	158
4.12.2	*Ozone*	159
4.12.3	*Aircraft*	159
4.13	PLANETS AND MOONS	160
4.13.1	*Venus*	160
4.13.2	*Jupiter*	160
4.13.3	*Io*	160
4.13.4	*Europa*	160
4.14	CONCLUSIONS	161
	REFERENCES	162

4.1 ELEMENTARY ISSUES

4.1.1 History

Sulfur is one of the elements from among the small group of elements known from ancient times. Homer described it as a disinfectant, and in *In Fasti* (IV, 739–740) Ovid wrote: "caerulei fiant puro de sulpure fumi, tactaque fumanti sulpure balet ovis...," which explains how the blue smoke from burning pure sulfur made the sheep bleat. Less poetical Roman writers gave more detail and distinguished among the many forms of elemental sulfur, which could be mined from volcanic regions. It was used in trade and craft activities such as cleaning of wool. In addition to the native form, sulfur is also widely found as sulfate and sulfide minerals. It was important to alchemists, because they were able to produce sulfuric acid by heating the mineral, green vitriol, $FeSO_4 \cdot 7H_2O$ and then condensing the acid from the vapor:

$$FeSO_4 \cdot 7H_2O_{(s)} \rightarrow H_2SO_{4(l)} + FeO_{(s)} + 6H_2O_{(g)}$$

Sulfuric acid rose to become such an important industrial chemical that the demand for sulfur in the production of the acid has sometimes been considered an indicator of national wealth.

Early geochemists, such as Victor Moritz Goldschmidt (1888–1947), recognized the difficulties associated with the geochemistry of a highly mobile and biologically active element such as sulfur. Fortunately, the composition of various reservoirs was known by the beginning of the twentieth century and carefully arranged analyses in terms of reservoirs: the crust, waters, and the atmosphere. The collected data were compiled and clearly arranged by Frank W. Clarke in his *US Geological Survey (USGS) Bulletins: The Data of Geochemistry* from 1908. Soon after Waldemar Lindgren, who had also been with the USGS and became head of the Massachusetts Institute of Technology's Department of Geology, gave the "story of sulfur" in an address on the concentration and circulation of the elements (Lindgren, 1923). The story is essentially a sulfur cycle with the notion of the volcanic release of reduced sulfur and oxidation and the transport as soluble sulfates to the sea. He recognized that sulfate in the oceans would rapidly have dominated over chloride were it not for biologically mediated reduction. Later Conway (1942) speculated on the importance of oceanic hydrogen sulfide as a source to the sulfur cycle. The idea of elemental cycles formed a central part of Rankama and Sahama's (1950) *Geochemistry*. There were many early sulfur cycles drawn up including those of Robinson and Robins, Granat *et al.*, Kellogg *et al.*, and Friend *et al.* that formed the basis of more recent cycles (e.g., Brimblecombe *et al.*, 1989; Rodhe, 1999) (Figure 1).

4.1.2 Isotopes

There are four stable isotopes of sulfur as listed in Table 1. The isotopic abundances vary slightly and this is frequently used to distinguish the source of the element. Because measurement of absolute isotope abundance is difficult, relative isotopic ratios are measured by comparison with

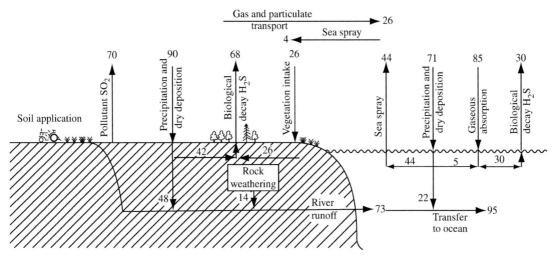

Figure 1 A generalized geochemical cycle for sulfur of the early 1970s. Note the large emissions of hydrogen sulfide from the land and oceans and that volcanic sulfur emissions are neglected (units: Tg (s) a^{-1}).

Table 1 Stable isotopes of sulfur.

Isotope	Mass	Abundance (%)
S-32	31.97207	94.93
S-33	32.971456	0.76
S-34	33.967886	4.29
S-36	35.96708	0.02

Table 2 Radioactive isotopes of sulfur.

Isotope	Half-life	Principal radiation (MeV)
S-29	0.19 s	p
S-30	1.4 s	β^+ 5.09, β^+ 4.2, γ 0.687
S-31	2.7 s	β^+ 4.42, γ 1.27
S-35	88 d	β^- 0.167
S-37	5.06 min	β^- 4.7, β^- 1.6, γ 3.09
S-38	2.87 h	β^- 3.0, β^- 1.1, γ 1.27

the abundance of the natural isotopes in a standard sample. The Canyon Diablo meteorite has been used as a standard for sulfur isotopes.

There are nine known radioactive isotopes and six are listed in Table 2. Sulfur-35 has the longest half-life and is produced by cosmogenic synthesis in the upper atmosphere: cosmogenic S-35 (Tanaka and Turekian, 1991) is sufficiently long lived to be useful in determining overall removal and transformation rates of SO$_2$ from the atmosphere and an estimated dry deposition flux to total flux ratio is $\sim$0.20 in the eastern US (Turekian and Tanaka, 1992).

4.1.3 Allotropes

In 1772 Lavoisier proved that the sulfur is an elementary substance, which was not necessarily obvious given that it is characterized by a number of allotropes (i.e., different forms). Sulfur has

more allotropes than any element because of the readiness to form S—S bonds (Table 3). These bonds can be varied both in terms of length and angle, so open and cyclic allotropes of S$_n$ are known where n ranges between 2 and 20. The familiar form of yellow orthorhombic sulfur (α-sulfur) is a cyclic crown S$_8$ ring. Two other S$_8$ ring forms are known the β-orthorhombic and the γ-orthorhombic found at higher temperatures. Engel in 1891 prepared a rhombohedral form ε-sulfur, which was ultimately shown to be an S$_6$ ring. The S$_6$ and S$_8$ rings have equally spaced sulfur atoms, which are essentially equivalent. This is not true of the S$_7$ form, which is found in four crystalline modifications. Here the interatomic distances vary between 199.3 pm and 218.1 pm. This latter distance is exceptionally large compared to the 2.037–2.066 pm typical of other sulfur rings.

4.1.4 Vapor Pressure

Sulfur vapor is a relatively volatile element and the vapor contains polyatomic species over the range S$_2$–S$_{10}$, with S$_7$ the main form at high temperatures (see Figure 2). The strong sulfur–sulfur double bond in S$_2$ (422 kJ mol^{-1}) means that monatomic sulfur is found only at very high temperatures ($>$2,200 °C).

Molten sulfur is known from volcanic lakes (Oppenheimer and Stevenson, 1989). The elemental liquid is a complex material. Elemental sulfur melts at $\sim$160 °C giving a yellow liquid, which becomes brown and increasingly viscous as the temperature rises in the range 160–195 °C, which is interpreted as a product of polymerization into forms that can contain more than 2×10^5 sulfur atoms. As temperature increases above this, the chain length (and thus viscosity) decreases to

Table 3 Sulfur allotropes.

Allotrope	Color	Density (g cm^{-3})	m.p. (°C)	Interatomic (distance/pm)
S$_2$ (gas)	Blue-violet		s	188.9
S$_3$ (gas)	Cherry red		s	
S$_6$	Orange red	2.209	d > 50	205.7
S$_7$	Yellow	2.183(@−110°)	d ∼ 39	199.3–218.1
αS$_8$	Yellow	2.069	112.8	203.7
βS$_8$	Yellow	1.94–2.01	119.6	204.5
γS$_8$	Light yellow	2.19	106.8	
S$_9$	Intense yellow		s < rt	
S$_{10}$	Yellow green	2.103(@−110°)	d > 0	205.6
S$_{12}$	Pale yellow	2.036	148	205.3
S$_{18}$	Lemon yellow	2.090	128	205.9
S$_{20}$	Pale yellow	2.016	124	204.7

Source: Greenwood and Earnshaw (1997). d decomposition, s stable at high temperature, rt room temperature.

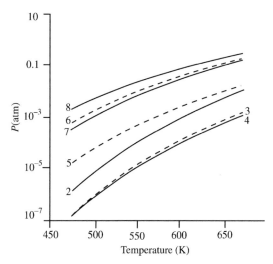

Figure 2 Partial pressure of sulfur polymers as a function of temperature. The numbers refer to the number of atoms in the polymer.

∼100 by 600 °C. If molten sulfur is cooled rapidly by pouring into water, it condenses into *plastic sulfur* that can be stretched as long fibers, which appear to be helical chains of sulfur atoms with ∼3.5 atoms for each turn of the helix (Cotton *et al.*, 1999).

4.1.5 Chemistry

Sulfur is very reactive and will combine directly with many elements. It will equilibrate with hydrogen to form hydrogen sulfide at elevated temperatures and will burn in oxygen and fluorine to form sulfur dioxide and sulfur hexafluoride. Reactions with chlorine and bromine are also known along with reactions with many metals even at low temperatures where it can be observed tarnishing copper, silver, and mercury.

Hydrogen sulfide is the only stable sulfane and is a poisonous gas with the smell of rotten eggs.

Although the smell is noticeable at 0.02 ppm, it can rapidly anesthetize the nose; so higher toxic concentrations can sometimes pass unnoticed. Much less is known about prolonged exposure to low concentrations, but there are associations with *persistent neurobehavioral dysfunction* (Kilburn and Warshaw, 1995). The gas occurs widely in volcanic regions and as the product of protein degradation. It is slightly soluble in water (K_H, 0.1 mol L^{-1} atm^{-1} at 25 °C) and dissociates as a dibasic weak acid:

$$H_2S \rightleftharpoons H^+ + HS^-$$

$$HS^- \rightleftharpoons H^+ + S^{2-}$$

The first dissociation constant takes the value $10^{-6.88 \pm 0.02}$ at 20 °C, but the second has given greater problems. Some have argued for a value close to ∼10^{-17} (Licht *et al.*, 1991; Migdisov *et al.*, 2002), while others justify the use of $10^{-14 \pm 0.2}$ at 25 °C for calculations at low ionic strength (D. J. Phillips and S. L. Phillips, 2000).

Polysulfanes with the general formula H$_2$S$_n$ ($n = 2–8$) are reactive oils in their pure state. Heating aqueous sulfide solutions yields polysulfide ions S$_n^{2-}$ most typically with $n = 3$ or 4. The presence of sulfanes has been used to explain the high concentrations of sulfur compounds in some hydrothermal fluids (Migdisov and Bychkov, 1998).

Metal sulfides often have unusual noninteger stoichiometries and may be polymorphic and alloy like. The alkali metal sulfides have antifluorite structures. The alkaline earth and less basic metals adopt an NaCl-type structure. Many of the later transition metal sulfides show an increasing tendency to covalency. Iron, cobalt, and nickel sulfides adopt a nickel arsenide structure, with each metal atom surrounded octahedrally by six sulfur atoms (Figure 3). Disulfides are also common, among which the best known is pyrite (FeS$_2$); but this is characterized by a nonstoichiometric

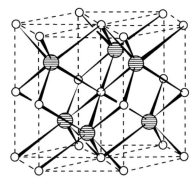

Figure 3 The nickel arsenide structure (reproduced by permission of Oxford University Press from *Structural Inorganic Chemistry*, **1945**, p. 387).

tendency, which leads to a 50–55% presence of sulfur, while retaining the nickel arsenide structure. There are more complex metal sulfides that adopt chain ring and sheet structures (Cotton *et al.*, 1999). Nonmetals often yield polymers that involve sulfur bridges.

Many oxides of sulfur are known, and although SO_2 and SO_3 are the most common, there are a range of cyclic oxides (e.g., S_nO ($n = 5$–10), S_6O_2, S_7O_2), and some thermally unstable acyclic oxides such as S_2O_2, S_2O, and SO. The oxides S_2O and SO appear to be components of terrestrial volcanic gases. More recently, there has been considerable interest in volcanogenic sulfur monoxide on the Jovian moon, Io (Russell and Kivelson, 2001; Zolotov and Fegley, 1998).

Sulfur dioxide (SO_2) is typically produced when sulfur burns in air or sulfide minerals are heated. It has been one of the most characteristic pollutants produced during the combustion of high sulfur-rich fuels. It is toxic with a characteristic choking odor. It liquefies very readily at $-10\,^\circ$C and has often been used as a nonaqueous solvent. It has been used as a bleaching agent and a preservative, and frequently found in wines. It is soluble in water, and is a weak dibasic acid, which has considerable relevance to its atmospheric chemistry.

Sulfur trioxide (SO_3) is an industrially important compound key to the production of sulfuric acid. It tends to polymeric forms both in the solid and liquid states. As a gas, the molecules have a planar triangular structure in which the sulfur atom has a high affinity for electrons. This explains its action as a strong Lewis acid towards bases that it does not oxidize. It can thus crystallize complexes with pyridine or trimethylamine. It has a very strong affinity for water and hence rapidly associates with water in the environment.

Sulfur dioxide and trioxide dissolve to give sulfurous and sulfuric acid. These acids are widely present in aqueous systems, particularly in the atmosphere. It is not possible to prepare pure sulfurous acid, but sulfuric acid can be obtained

at high concentration. Sulfurous acid undergoes a range of reactions in atmospheric droplets, which are relevant to its removal from the atmosphere. Many of these reactions oxidize it to the stronger sulfuric acid, which has been a key component of acid rain.

Sulfuric acid is a dibasic acid and although regarded as a strong acid, the second dissociation constant is rather modest ($\sim 0.01\,\mathrm{mol\,L}^{-1}$, although care must be taken to consider the nonideality of electrolyte solutions when undertaking calculations of the position of this equilibrium):

$$HSO_4^- \rightleftharpoons H^+ + SO_4^{2-}$$

The acid is a weak oxidizing agent, but has a strong affinity for water meaning that it chars carbohydrates. Although the acid is typically dilute in the environment, it may reach high concentrations (many $\mathrm{mol\,kg}^{-1}$) in atmospheric droplets. At the low temperatures found in the stratosphere, these can crystallize as a range of hydrates.

Sulfates are widely known in the environment and a few such as those of calcium, strontium, barium, and lead are insoluble. The sulfates of magnesium and the alkali metals are soluble and so are typically found in brines and salt deposits. Sulfites are rarer because of the ease of oxidation, but calcium sulfite, perhaps as the hemihydrate, can be detected on the surface of building stone in urban atmospheres containing traces of sulfur dioxide (Gobbi *et al.*, 1998).

A wide range of oxyacids is possible with more than one sulfur atom. Although the acids are not always very stable, the oxyanions are often more easily prepared. Many of these are likely intermediates in the oxidation of aqueous sulfites and occur at low concentration in water droplets in the atmosphere. Higher concentrations of sulfur oxyanions such as $S_4O_6^{2-}$, $S_5O_6^{2-}$, and $S_6O_6^{2-}$ can be found in mineralized acid-sulfate waters of volcanic crater lakes (e.g., Sriwana *et al.*, 2000).

Sulfur can form halogen compounds and nitrides. The nitrogen compounds are found as interstellar molecules and chlorides postulated on the Jovian moon, Io, but these compounds do not appear to be a significant part of biogeochemical cycles on the Earth.

Compounds of sulfur with carbon having the general formula C_xS_y are characterized through a series of linear polycarbon sulfides C_nS ($n = 2$–9), some of which can be detected in interstellar clouds. Carbon disulfide (CS_2) occurs in the atmosphere and oceans of the Earth. Carbonyl sulfide (OCS) is also well known in nature and is one of many polycarbon oxide sulfides represented by the formula OC_nS where $n < 6$.

Many organosulfur compounds occur in the environment. Divalent sulfur as thiols (e.g., CH_3SH) or thioethers (e.g., CH_3SCH_3) is analogous to alcohols and ethers. These may be present as polysulfides, e.g., the dimethylpolysulfides ($CH_3S_nCH_3$, $n = 2-4$) detected in some water bodies (Gun *et al.*, 2000). Simpler volatile compounds, most typically the methyl derivatives, are present in the atmosphere. Thiols are much more acidic than the corresponding alcohols, e.g., phenylthiol has a pK_a of 6.5. Tri- and tetra-coordinated organosulfur compounds are known in the form of sulfoxides and sulfones. Proteins often contain sulfur present in the amino acids methionine, cystine, and cysteine. Organosulfur compounds are common components of fossil fuels appearing as aliphatic or aromatic thiols or sulfides and disulfides in addition to the heterocyclic combinations as thiophenes or dibenzothiophene. Thioethers may be found in lower-grade coals (Wawrzynkiewicz and Sablik, 2002).

Oxo compounds such as alkyl sulfates and sulfonates are used as detergents in large quantities, while simple hydroxysulfonates are formed through the reactions of sulfur dioxide and aldehydes in rain and cloud water. Oxidation processes, most particularly in soils, seem to generate a range of sulfenic, sulfinic, and sulfonic acids. Sulfenic acids take on the general formula RSOH (R ≠ H) and the sulfinic acids RS(O)OH. Methyl sulfenic acid (CH_3SOH) is probably produced in the atmosphere through the oxidation of dimethyl disulfide (DMDS) (Barnes *et al.*, 1994) and the parent sulfinic acid may occur on the Jovian moon, Europa (Carlson *et al.*, 2002). The sulfonates are rather more widely known with methanesulfonic acid (CH_3SO_3H), being an important oxidation product of compounds such as dimethyl sulfide (DMS) in the Earth's atmosphere.

4.2 ABUNDANCE OF SULFUR AND EARLY HISTORY

Sulfur is the 10th most abundant element in the Sun, and has less than half the abundance of silicon. Its abundance in the Earth's crust is very much less, indicative of a mobile and reactive element as seen in Table 4, perhaps most easily seen by comparison with silicon. Note that nitrogen, whose cycle is often compared with sulfur, is also much depleted in the crust, but is unreactive and not so mineralized effectively.

4.2.1 Sulfur in the Cosmos

The universe is largely composed of hydrogen and it is in the older stars that heavier elements are found. This offers one of the central clues to the origin of the elements. The evolution of stars is interpreted with the aid of the Hertzsprung–Russell diagram. Stars that lie in the main sequence burn hydrogen into helium although the rate of this process varies with mass: heavier stars

Table 4 Average abundances of some elements in the cosmos and continental crust. This is best interpreted by comparing elements relative to silicon which take about the same values in this representation.

Element	Cosmos relative hydrogen atoms	Continental crust (%)
Hydrogen	1,000,000	
Helium	140,000	
Carbon	300	0.2
Nitrogen	91	0.006
Oxygen	680	47.2
Aluminum	1.9	7.96
Sulfur	9.5	0.07
Silicon	25	28.8
Calcium	1.7	3.85
Magnesium	29	2.2
Iron	8	4.32

Source: Wedepohl (1995).

burning up more quickly perhaps in only a few million years for the brightest. The sun will take billions of years to evolve before becoming a cooler red giant, shedding its outer layers and ending up as a white dwarf. Heavier stars evolve into super giants, which synthesize heavier elements in their interiors. They may end up exploding as supernovae, forming more elements and distributing these through the cosmos. Interstellar sulfur isotopes are synthesized by oxygen burning in massive stars. This derives from a sequence

$$^1H + {}^1H = {}^2H + e^+ + \nu$$

$$^1H + {}^2H = {}^3He + \gamma$$

$$^3He + {}^3He = {}^4He + {}^1H + {}^1H$$

A small proportion of the helium reacts by PPII and PPIII processes that yield ^7Be and ^7Li and some ^8B. In addition to the PP chains, there are CNO cycles. These are more important in heavier stars and are started by small catalytic amounts of ^{12}C, and so represent an important source of ^{13}C, ^{14}N, ^{15}N, and ^{17}O. Hydrogen burning yields helium in the stellar core, and helium burning can give rise to both ^{12}C and ^{16}O. Once helium is exhausted, carbon and oxygen burning takes place and it is at this stage the sulfur element forms

$$^{16}O + {}^{16}O = {}^{31}S + n \text{ and } {}^{16}O + {}^{16}O = {}^{32}S + \gamma$$

An equilibrium phase is possible at very high temperatures (3×10^9 K) which would lead to high abundances of elements around the same atomic number as iron, and less of lighter elements such as sulfur. However, this situation exists only in the core, and in the surrounding shells substantial amounts of other elements such as sulfur will be present.

There are a number of estimates of the isotopic abundance of sulfur in the cosmos from Galactic cosmic rays (Thayer, 1997) and interstellar clouds (Mauersberger *et al.*, 1996). Measurements from the *Ulysses High Energy Telescope* resolve the individual isotopes of sulfur (S-32, S-33, and S-34), giving a measured ratio of Galactic cosmic-ray S-34/S-32 in the heliosphere is $24.2 \pm 2.7\%$ and a ratio for S-33/S-32 in the heliosphere as $19.0 \pm 2.4\%$. The Galactic source composition for S-34/S-32 has been found to be 6.2%. The source abundance of S-33 is estimated as $2.6 \pm 2.4\%$, which is higher than, but consistent with, the solar system abundance of 0.8% (Thayer, 1997). Enrichment of S-33 in ureilite meteorites may derive from heterogeneity within the presolar nebula (Farquhar *et al.*, 2000). Molecules of CS in interstellar space suggest the abundance ratio S-34/S-36 to be $115(\pm 17)$. This is smaller than in the solar system ratio of 200 and Mauersberger *et al.* (1996) argue that it supports the idea that S-36 is, unlike the other stable sulfur isotopes, a purely secondary nucleus that is produced by s-process nucleosynthesis in massive stars.

Sulfur-containing molecules have been found in clouds in interstellar space: CS, C_2S, C_3S, SiS, SO, SO^+, H_2S, HCS^+, NS, OCS, SO_2, H_2CS, HNCS among them. Although sulfur is a very abundant and important element, it appears to be depleted in cold dense molecular clouds with embedded young stellar objects (Keller *et al.*, 2002). In dense molecular clouds solid SO_2 accounts for only 0.6–6% of the cosmic sulfur abundance (Boogert *et al.*, 1997). Keller *et al.* (2002) have determined the infrared spectra of FeS grains from primitive meteorites in the laboratory and shown a broad similarity with grains in interplanetary dust, which show an FeS peak, suggesting that the missing sulfur appears to reside as sulfides in solid grains.

4.2.2 Condensation, Accretion, and Evolution

The simplest models for the composition of the planets presume that the differences between them can be explained in terms of an equilibrium condensation. At the highest temperatures a sequence of mixed oxides of calcium, titanium, and aluminum would be found (>1,400 K). This would be followed, at lower temperatures, by metal and silicate fractions. At temperatures somewhat greater than 600 K alkali metals enter the silicate phase along with sulfur, which combines with iron at $\sim$650 K to form triolite

$$Fe + H_2S \rightleftharpoons FeS + H_2$$

Hydration reactions become possible at lower temperatures with the potential to form serpentine and talc. Although such models for the accretion of planets have significant explanatory power, the process was doubtless more complex. Many scientists have sought explanations involving the late heterogeneous accretion of infalling meteoritic materials. These would have brought in large amount of more volatile substances to planetary bodies.

After accretion there is further planetary evolution, especially through increases in temperature from the decay of long-lived radioactive elements. Planets, such as the Earth, thus become layered either through accretion or differential melting processes. The heating deep inside the planet, such as the Earth, causes outgassing of volatile materials especially water from the dehydration of tremolite:

$$Ca_2Mg_5Si_8O_{22}(OH)_2 \rightleftharpoons 2CaMgSi_2O_6 + MgSiO_3 + SiO_2 + H_2O$$

or sulfur gases:

$$FeS + 2FeO \rightleftharpoons 3Fe + SO_2$$

$$FeS + CO \rightleftharpoons Fe + OCS$$

$$FeS + H_2O \rightleftharpoons FeO + H_2S$$

along with hydrogen, nitrogen, ammonia, and traces of hydrogen fluoride and chloride. The heating was usually intense enough to cause melting. As there was insufficient oxygen available, it became associated with magnesium, silicon, and aluminum, while elements such as iron and nickel remained in an unoxidized state. Gravitational settling meant that these heavy materials moved towards the center of the Earth. Iron sulfide would also be heavy enough to settle. Thus although the Earth as a whole may be almost 2% sulfur by weight, the crust is much depleted at 700 ppm.

As dense minerals sank and formed the Earth's mantle, silicate-rich magma floated on the surface. It is this that became the crust, which represents the outermost layer of rocks forming the solid Earth. It is distinguished from the underlying mantle rocks by its composition and density. The silica-rich continental crust includes the oldest rocks as "shields" or "cratons," which in the extreme may be $\sim$4 Gyr old. Crust under the oceans is only $\sim$5 km thick compared to continental crust, which can be up to 65 km thick under mountains. The oceanic crust has a remarkably uniform composition ($49 \pm 2\%$ SiO_2) and thickness (7 ± 1 km). The ocean floor is the most dynamic part of the Earth's surface, such that no part of the oceanic crust is more than 200 Myr old. It is constantly renewed through seafloor spreading at mid-ocean ridges.

4.2.3 Sulfur on the Early Earth

Sulfur on the early Earth has been studied, because it gives insight into the oxidation state of the early atmosphere ocean system (Canfield and Raiswell, 1999). The occurrence of evaporitic sulfate from the 3.5 Ga as in the Warrawoona

Group of Western Australia suggests that sulfate must have been present in elevated concentrations at least at some sites. Although now present as barite, the original precipitation occurred as gypsum. These barites and early sulfates precipitate from the ocean. Evidence from isotope ratios in Early Archean sedimentary sulfides suggest that seawater sulfate concentrations could be as low as 1 mM (Canfield *et al.*, 2000). More recent work of Habicht *et al.* (2002) suggests even lower oceanic Archean sulfate concentrations of $<200 \,\mu M$, which is less than one-hundredth of present marine sulfate levels. They argue that such low sulfate concentrations were maintained by volcanic sulfur dioxide and were so low that they severely suppressed sulfate reduction rates allowing for a carbon cycle dominated by methanogenesis (Habicht *et al.*, 2002).

Studies of the sulfide inclusions in diamonds have given some indications of sulfur cycling on the Archean earth. These inclusions often appear to be enriched in [33]S, which has been interpreted as indicative of photolysis of volcanic SO_2 at wavelengths <193 nm. This photolysis yields elemental sulfur enriched in [33]S, while the oceanic sulfate was depleted in [33]S (see Figure 4). The elemental sulfide becomes incorporated into sulfides and can be subducted, and included in diamonds. By contrast seawater has sulfates depleted in [33]S and gave Archean barites that are also depleted in the [33]S. While such studies offer an indication of the sulfur cycle in the Archean, they can also offer insight into the nature of mantle convection through time (Farquhar *et al.*, 2002).

Some of the earliest organisms on the Earth utilized sulfur compounds particularly through anoxogenic photosynthesis. Wächtershauser and co-workers have argued that inorganic reactions involving S–S links have a reducing potential that

is strong enough to drive a reductive metabolism based on carbon dioxide fixation (e.g., Hafenbradl *et al.*, 1995):

$$FeS + H_2S \rightarrow FeS_2 + H_2$$

The sulfur-isotope signal in sedimentary record of the early Earth provides evidence for the early evolution of life. Sulfate-reducing bacteria deplete the resultant sulfides in [34]S. The isotopic record of [34]S supports this, although some evidence now suggests that seawater sulfate concentrations were so low that they severely suppressed sulfate reduction rates allowed for a carbon cycle dominated by methanogenesis (Habicht *et al.*, 2002).

Modern taxonomy based on sequences in rRNA indicates three domains of living organisms: Eucarya (includes fungi, plants, and animals), Bacteria, and Archea (prokaryotes with no nuclei or organelles that are similar to bacteria, but with distinguishing ether linked membrane lipids). Both Bacteria and Archea include organisms capable of reducing elemental sulfur and it is likely that this was one of the most primitive early metabolisms. Such sulfur may have been generated on an anoxic Earth via the reaction between SO_2 and H_2S from fumaroles (Canfield and Raiswell, 1999).

The ability to undertake anoxygenic photosynthesis is widespread among bacteria. Its presence among green nonsulfur and sulfur bacteria and the fact that this uses a single photosystem to utilize solar energy indicate that this is an ancient process, much simpler than that required in oxygenic photosynthesis. Nevertheless, much of the ability to undertake oxygenic photosynthesis is probably associated with the production of abundant organic material. During oxidation this meant the production of large amounts of reduced compounds such as H_2S and Fe^{2+}. This would have

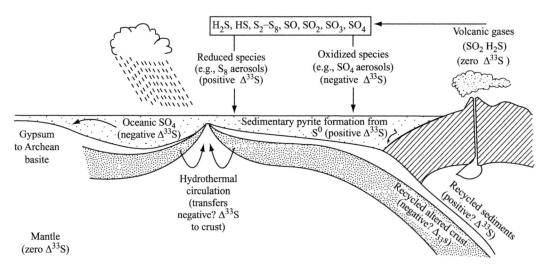

Figure 4 An Archean cycle for volcanogenic sulfur (after Farquhar *et al.*, 2002).

led to large increase in the range of environments offering potential for anoxygenic photosynthesis (Canfield and Raiswell, 1999).

An important consequence of anoxygenic photosynthesis is the production of large amounts of sulfate which can be used as an electron acceptor in the remineralization of organic carbon by sulfate-reducing bacteria. Thus, it is not surprising that sulfate-reducing bacteria are found almost as deep in the evolutionary SSU rRNA record as the earliest anoxygenic photosynthetic bacteria (Canfield and Raiswell, 1999). This is important because the photosynthesis represented an energy resource that is orders of magnitude larger than that available from oxidation–reduction reactions associated with weathering and perhaps more importantly submarine hydrothermal activity. Hydrothermal sources deliver $(0.13–1.1) \times 10^{12}$ mol yr^{-1} of S($-$II), Fe^{2+}, Mn^{2+}, H$_2$, and CH$_4$ on a global basis, sufficient to allow microorganisms capable of using hydrothermal energy to produce only limited amounts of organic carbon $(\sim(0.2–2.0) \times 10^{12}$ mol C yr$^{-1})$ (Des Marais, 2000).

4.3 OCCURRENCE OF SULFUR

4.3.1 Elemental Sulfur

Elemental sulfur is found in many parts of the world. Such native sulfur can be detected by the faint odor of sulfur dioxide gas or more often the small amount of hydrogen sulfide from reaction with traces of water. Although there are some deposits related to volcanic sublimates, the largest deposits are due to sulfate reduction. For example, deeper in the Gulf Coast of the US, salt containing gypsum as an impurity has dissolved, thus leaving the insoluble gypsum. A range of biologically mediated processes intervened. First hydrogen sulfide was produced possibly by reaction of methane with the gypsum. The hydrogen sulfide reacted with this gypsum to form free sulfur, which became trapped within limestone beds.

4.3.2 Sulfides

There are widespread deposits of sedimentary sulfides, mostly as the iron sulfides from the reduction of sulfate in seawater. Although pyrite (FeS$_2$) is an important source of sulfur in extractive mining, the deposits most utilized are often of volcanogenic origin where the deposits may consist of as much as 90% pyrite. Although most frequently found associated with iron, sulfur can also be found as calcopyrite (CuFeS$_2$), galena (PbS, lead sulfide), and cinnabar (HgS, mercury sulfide).

Extensive sulfide deposits were associated with volcanism in the earliest stages of the Earth's history. It is possible to identify at least four short epochs of massive sulfide deposition: (i) 2.69–2.72 Ga; (ii) 1.77–1.90 Ga; (iii) the Devonian–Early Carboniferous; and (iv) the Cambrian–Early Ordovician. The Devonian–Early Carboniferous is the most important period and the Cambrian–Early Ordovician less noteworthy. These processes are cyclic in nature, following tectonic cycles that include the convergence of continental masses and the formation and subsequent breakup of supercontinents. As each cycle proceeds, we see the appearance of different sequences within massive sulfide deposits (Eremin *et al.*, 2002).

4.3.3 Evaporites

Evaporites are typically found in arid regions. The salts deposited in desert basins are often sulfates, chloride, and carbonates of sodium and calcium with smaller amounts of potassium and magnesium. Saline lakes often have a composition that resembles seawater, although many have a greater abundance of carbonate. When both sodium chloride and calcium sulfate have crystallized, the remaining water may contain substantial amounts of magnesium, potassium and bromide. The dead seawater contains substantial bromide concentrations (5,900 ppm) and although it has high concentrations of calcium (1.58×10^4 ppm), sulfate concentrations are low (540 ppm) because of the low solubility of calcium sulfate.

The most important salt deposits have formed through the evaporation of seawater. As this evaporation occurs, calcium carbonate is the first mineral to precipitate, although there is insufficient carbonate to give thick deposits. Calcium sulfate deposits in substantial quantities under continued evaporation as either gypsum (CaSO$_4$·2H$_2$O) or anhydrite (CaSO$_4$). After about a half of the calcium sulfate has been deposited, anhydrite becomes the stable phase. When seawater has evaporated to 10%, its original volume halite (NaCl) starts to separate. Ultimately, polyhalite (K$_2$Ca$_2$Mg(SO$_4$)$_4$·2H$_2$O) begins to crystallize. The volume has to be reduced almost a 100-fold before salts of potassium and magnesium separate. Other sulfate minerals from such systems include magnesium sulfate (Epsom salt and kyeserite MgSO$_4$·H$_2$O), barium sulfate (barite), strontium sulfate (celestite), and sodium sulfate (glauberite).

4.3.4 The Geological History of Sulfur

The sulfur cycle over geological periods can be represented by the simplified cycle shown in Figure 5(a). A central feature of this cycle is marine sulfate, which may be crystallized as sulfate evaporites or biologically reduced to sulfidic sediments (Holser *et al.*, 1989). In the long term this sulfide becomes fixed by reaction with iron in a process that can be represented as

$$8SO_4^{2-} + 2Fe_2O_3 + 8H_2O + 15C_{(org)}$$
$$\rightarrow 4FeS_2 + 16OH^- + 15CO_2$$

The reduction is typically limited by the availability of organic carbon and often occurs in shallow waters at continental margins. Thus, global sulfide production would be dependent on the availability of biological productive areas over geological time. Sulfur-isotope data can be used to constrain simple models of the sulfur cycle over geological time and establish the size of the reservoirs as shown in Figure 5(b).

The crystallization of sulfate as evaporite deposits requires specific and relatively short-lived geographical features such as: sabkha shorelines, lagoons, and deep basins of cratonic, Mediterranean or rift valley origin, where brines can concentrate and precipitate. This suggests that sulfate removal from the oceans would be episodic over geological time, although there are problems with an imperfect preservation of deposits. Numerical models for paleozoic global evaporite deposition suggest that there were major changes in the evaporite accumulations (as in Figure 6) and rate of evaporite deposition related to alteration in the flux of sulfate to and from the ocean. These can be the result of the height distribution of the continental mass and its geographic location. Most evaporite deposition appears in the 5–45° latitude range and as seen in Figure 6 this changes significantly over geological time. The removal of large quantities of sulfate from the oceans also has implications for ocean chemistry, which would have to change, i.e., oceanic sulfate is not in steady state. Currently evaporative environments are relatively rare, as a result of a *low stand* disposition of the continents. This means that sulfate is accumulating in the oceans (Railsback, 1992).

The changing fluxes in various parts of the sulfur cycle over time imply that we have to consider the potential for significant changes in seawater composition over geological time. Although it is often assumed that in the Phanerozoic period (the last 590 Myr) seawater composition became stable, there is evidence that there were significant fluctuations in some components. Fluid inclusions in Silurian (395–430 Myr ago) suggest that the Silurian ocean had lower concentrations of Mg^{2+}, Na^+, and SO_4^{2-}, and much higher concentrations of Ca^{2+} relative to the present-day ocean composition. Concentration of sulfate was probably 11 mM then compared with 29 mM as we see nowadays. Although this could be caused by localized changes, it is also possible that global scale dolomitization of carbonate sediments or changing inputs of mid-ocean ridge brines could be responsible (Brennan and Lowenstein, 2002).

4.3.5 Utilization and Extraction of Sulfur Minerals

Elemental sulfur is widely known and has been mined since the earliest times with archeological sites in the United Arab Emirates, Yangmingshan National Park in Taiwan and from Etruscan activities in Sicily. In some parts of the world laborers still carry the sulfur out of volcanoes by hand. Sulfur in addition to being a vital ingredient of gunpowder was important to the overall industrialization of Europe. Historically the most important source was from Sicily, which has been active

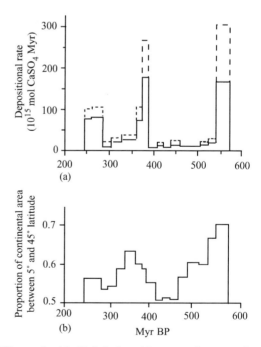

Figure 6 (a) Global deposition rates for evaporites over time. (b) The proportion of the Earth's total continental area located between paleo-latitudes 5° and 45° (after Railsback, 1992).

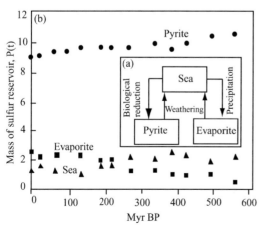

Figure 5 (a) (Inset) Simplified box model of the sulfur cycle. (b) Mass of sulfur in reservoirs in Pt (10^{21} g) (source Holser *et al.*, 1989).

for more than two thousand years. Towns such as Ravanusa became notable producers during the Spanish Bourbon rule (after 1735). Italian output peaked at 0.57 Mt in 1905 declining from that point to just a few kt by the 1970s.

The United States is now the major sulfur producer, accounting for 20% of world production. The most important sources in the US are from Louisiana and Texas, and other major producers are Japan, Canada, China, Russia, and Mexico. World sulfur production (and apparent consumption) peaked at nearly 60 Mt in 1989 and declined by almost 14% to 52.8 Mt in 1993 (Figure 7). There was a partial recovery from this time and future growth is expected.

One of the most important developments in sulfur mining was the development of the Frasch process which allowed it to be extracted from salt domes, particularly along the US Gulf Coast. Herman Frasch became involved in oil and sulfur mining and invented a process which allows liquid sulfur to be recovered with injection wells. Water at 330 °C is injected through wells into formations. The molten sulfur is extracted in a very pure form. It is an efficient method for mining sulfur, but the process can also produce saline wastewaters, increases in pH levels, and a high concentration of dissolved salts such as sodium chloride (TWRI, 1986).

Sulfur may also be recovered from H_2S in natural gas where it can be precipitated. Natural gas can contain up to 28% hydrogen sulfide gas. A range of approaches are possible, often known as Claus processes to oxidize the H_2S, often recovering more than 95% of the sulfur

$$1\frac{1}{2}O_2 + H_2S \rightarrow H_2O + SO_2$$

$$2H_2S + SO_2 \rightarrow 3S + 2H_2O$$

More than 90% of world sulfur consumption is used in the production of sulfuric acid, much of which goes to the fertilizer industry. Smaller amounts of sulfur are used in the manufacture of gunpowder, matches, phosphate, insecticides, fungicides, medicines, wood, and paper products, and in vulcanizing rubber. Despite slight uncertainties in sulfur demand in the 1990s, its use is still predicted to grow.

Sulfide ores are frequently associated with a range of important metals, most notably copper, lead, nickel, zinc, silver, and gold. This means the mining operations to extract these metals can mobilize large amounts of sulfur.

4.4 CHEMISTRY OF VOLCANOGENIC SULFUR

4.4.1 Deep-sea Vents

Deep-sea vents have only been studied in the late twentieth century, so there is still much to learn in terms of their global contribution because of their inaccessibility. However, they can affect global fluxes, and some estimates would suggest that warm ridge-flank sites may remove each year, as much as 35% of the riverine flux of sulfur to the oceans (Wheat and Mottl, 2000). The hydrothermal vents are locally important sources of sulfide-containing materials. The black smokers yield polymetal sulfides, that will oxidize to sulfur and ultimately sulfates. The reduced sulfur is also utilized by the ecological communities that develop close to the vents (Jannasch, 1989).

4.4.2 Aerial Emissions

Volcanoes represent a very large source of sulfur gases to the atmosphere. Mafic magmas such as basalts are likely to allow sulfur to be released in large quantities. However, despite the low solubility and diffusivity of sulfur in the silicic magmas typical of explosive eruptions

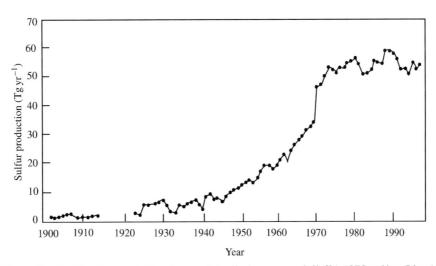

Figure 7 World sulfur production (source http://pubs.usgs.gov/of/of01-197/html/app5.htm).

large quantities of sulfur are inevitably released. The presence of fluids may account for higher than expected sulfur yields and indicates a redox control of sulfur degassing in silicic magmas. This means that different eruptions can have very different efficiencies in terms of sulfur release. Very large eruptions that involve cool silicic magmas may yield little sulfur. Krakatoa in 1883 gave $1.2 \, km^3$ of eruptive magma, with an estimated petrological yield of up to 7 Tg in reasonable agreement with ice core data, which suggests 10–18 Tg. By contrast, the Taupo eruption of 177 AD in New Zealand gave more than $35 \, km^3$ of magma, but perhaps only 0.13 Tg of sulfur (Scaillet et al., 1998).

There are numerous calculations of the multicomponent chemical equilibria in gas–solid–liquid systems within volcanoes (e.g., for Mt. St. Helens see: Symonds and Reed, 1993). These typically show the importance of SO_2 at high temperatures (see Figure 8) and a transition to a dominance of H_2S at lower temperatures in systems such as

$$SO_2 + 3H_2 \rightarrow H_2S + 2H_2O$$

where entropy changes favor fewer molecules of gas.

Much volcanism is a sporadic process and on a year-to-year basis the emissions can vary. There are also difficult problems in balancing the contributions from the more explosive processes against that from fumaroles that are more continuous. Explosive processes also place the sulfur high up in the atmosphere. There are also inputs as particulate sulfur and sulfate along with small amounts of carbon disulfide and OCS of volcanic origin, but these probably amount to no more than a $Tg(S) \, a^{-1}$ (Andres and Kasgnoc, 1998). There are lesser amounts of S_2, S_3, H_2S_2, S_2O and a range of sulfides of arsenic, lead, and antimony.

There are a variety of estimates of the total volcanic source of sulfur. A recent one considered continuous and sporadic volcanoes (Andres and

Kasgnoc, 1998). It is necessary to recognize that measurements do not include all volcanic eruptions, so assumptions have to be made about the distribution of emissions among the smaller sources. A time-averaged subaerial volcanic sulfur (S) emission rate of 10.4 Tg(S) is probably a conservative estimate of which ~65% is as SO_2 along with substantial amounts of hydrogen sulfide. The high variability of emissions is illustrated in Figure 9 as established by the TOMS emission group from satellite measurements.

4.4.3 Fumaroles

Fumaroles represent a gentler and more continuous source of sulfur. The sources can be dispersed and quite small, so the total emissions from this source are not easy to estimate. Some of them are dominated by H_2S. The sulfur gases, SO_2, H_2S, S_8, have been found in a range of fumaroles (Montegrossi et al., 2001). Although present S_8 remains a minor component several orders of magnitude below SO_2 and H_2S. The production of sulfuric acid through aerial oxidation of sulfur(IV) is the most familiar process but it can readily be produced by disproportion in fumarolic systems (Kusakabe et al., 2000):

$$3SO_2 + 3H_2O \rightarrow 2HSO_4^- + S + 2H^+$$

$$4SO_2 + 4H_2O \rightarrow 3HSO_4^- + H_2S + 3H^+$$

It is this process that can be responsible for large $\delta^{34}S_{HSO_4}$ values in crater lakes.

4.4.4 Crater Lakes

High concentrations of acids and sulfur compounds in crater lakes give rise to a complex chemistry within the waters (Sriwana et al., 2000). Reactions between hydrogen sulfide and sulfur dioxide lead to the formation of polythionates, with $S_4O_6^{2-}$, $S_5O_6^{2-}$, and $S_6O_6^{2-}$ being the most typical. The initial ratio of SO_2/H_2S controls the relative abundance of the oxyanions. The anion $S_6O_6^{2-}$ is always at the lowest concentrations, but the ratio $SO_2/H_2S > 0.07$ favors $S_4O_6^{2-} > S_5O_6^{2-}$. Polythionate concentrations vary up to a few hundred milligrams per liter and seem to be a useful, indicator of changes in the subaqueous fumarolic activity with volcanic lakes corresponding to the activity of the volcano (Takano et al., 1994). The polythionates are destroyed by increases in SO_2 input (as bisulfite):

$$(3x - 7)HSO_3^- + S_xO_6^{2-} \rightarrow (2x - 3)SO_4^{2-}$$
$$+ (2x - 4)S + (x - 1)H^+ + (x - 3)H_2O$$

Lakes of liquid sulfur at 116 °C have been found in the crater of the Poas volcano in Costa Rica. These form after evaporation of water from crater

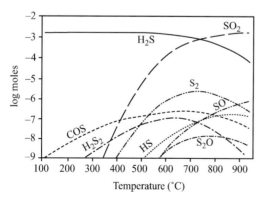

Figure 8 Multicomponent chemical equilibria in gas–solid–liquid systems within Mt. St. Helens (source Symonds and Reed, 1993).

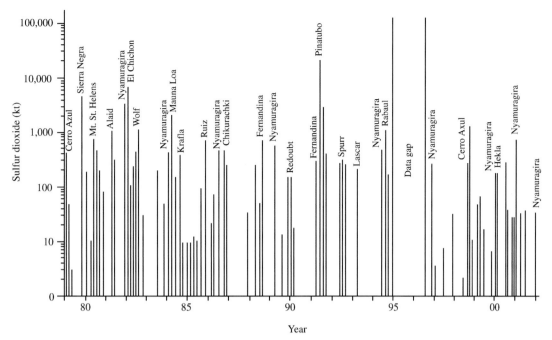

Figure 9 Volcanic sulfur emissions as established by the TOMS emission group (http://skye.gsfc.nasa.gov/).

lakes and the underlying sulfur deposits melt and are remobilized into lakes (Oppenheimer and Stevenson, 1989).

4.4.5 Impacts of Emissions on Local Environments

The volcanoes of Hawaii emit large quantities of fumarolic sulfur dioxide which causes widespread damage to human health and vegetation downwind. These emissions often take the appearance of widespread hazes and are locally known as vogs. The primarily component is of sulfuric acid and sulfate formed through oxidation of the sulfur dioxide. The vog particles also contain trace elements selenium, mercury, arsenic, and iridium (USGS, 2000). There is evidence that volcanic SO_2 and HF affect the diversity of plant communities downwind and damage crops (Delmelle *et al.*, 2002). Emissions of SO_2 and to a lesser extent HCl are responsible for substantial fluxes of acidity to soils downwind. Such acid deposition can cause extreme acid loading to local ecosystems (Delmelle *et al.*, 2001). Hydrogen sulfide concentrations can also be important in some volcanic areas (Siegel *et al.*, 1990), where the odors can be strong and increase corrosion of metals, especially those related to electrical switches and discolor lead-based paints.

4.5 BIOCHEMISTRY OF SULFUR

4.5.1 Origin of Life

An important stage in the origin of life is the abiotic synthesis of important biological molecules.

Interest has traditionally focused on the production of amino acids following the work of Stanley L. Miller, and Harold C. Urey, who created laboratory atmospheres consisting of methane, ammonia, and hydrogen that could be forced to react to give the amino acid glycine. Currently there is a view that RNA might have established what is now called the RNA world in which RNA catalyzed all the reactions to replicate proteins. Such chemistry has focused on nitrogen.

The discovery of deep-sea vents has led to speculation that they could be involved in the origin of life and here a unique sulfur chemistry could come into play (Wachtershauser, 2000). In particular, there is much interest in the potential of iron and nickel sulfides to act as catalysts in a range of reactions that can lead to keto acids such as pyruvic acids. Quite complex amino acids seem to be easily assembled in the presence of iron sulfides, e.g., the conversion of phenyl pyruvate formation of phenylalanine in a reaction where CO_2 seems to act as a catalyst (Hafenbradl *et al.*, 1995):

$$RCOCOOH + 2FeS + NH_3 + 2H^+$$
$$\rightarrow RCH(NH_2)COOH + FeS_2 + Fe^{2+} + H_2O$$

4.5.2 Sulfur Biomolecules

Although we have discussed the low oxidation state of the early atmosphere and its implications for the form of sulfur in organisms, today this chemistry is largely restricted to anoxic environments. Here sulfur is available as hydrogen sulfide and can be incorporated into amino acids such as cysteine and methionine and then into proteins. The thiolate group RS⁻ of cysteine of the thioether of methionine can act as bases or ligands for transition metals such as iron, zinc, molybdenum, and copper.

In modern organisms operating under oxic conditions this is more difficult, because sulfur arrives as sulfate and this has to be bound and reduced so that the sulfur can be utilized biologically (see Figure 10). Initially sulfate can be bound to ATP (adenosine trisphosphate) as APS (phospho-adenosine monophosphate sulfate). In the cell's Golgi apparatus it can be converted to sulfate polysaccharides. Alternatively, further reactions with ATP can lead to PAPS which can then be reduced to sulfide through sulfite using a molybdenum in the initial step and then a haem/Fe_nS_n sulfite reductase (Frausto da Silva and Williams, 2001).

Much sulfur is incorporated into proteins as the thiolate or thioether of the amino acids, cysteine, and methionine. There are also more fundamental roles for thiolate in redox reactions and in enzymes and in controls in the cytoplasm involving glutathione and thioredoxin. Glutathione is a tripeptide made up of the amino acids gamma-glutamic acid, cysteine, and glycine. The primary biological function of glutathione is to act as a nonenzymatic reducing agent to help keep cysteine thiol side chains in a reduced state on the surface of proteins. Glutathione confers protection by maintaining redox potential and communicating the redox balance between metabolic pathways at a redox potential of ~-0.1 V.

Glutathione is also involved in reductive synthesis with the enzyme thioredoxin in processes that involve redox reactions of –S–S– bridges. The –S–S– bridge is also important as a cross-linking unit in extracellular proteins. This cross-linking may have become important once the atmosphere gained its oxygen. Additionally, glutathione is utilized to prevent oxidative stress in cells by trapping free radicals that can damage DNA and RNA. There is a correlation with the speed of aging and the reduction of glutathione concentrations in intracellular fluids.

The biological importance of sulfur means that modern agriculture frequently confronts the high demand of some crops and livestock for sulfur. In Europe as the deposit of pollutant sulfur dioxide has decreased, sulfur deficiency has been more widely recognized especially in wheat, cereals, and rape-seed (Blake-Kalff *et al.*, 2001). Some animals have a particularly high demand for sulfur, and under domestication this can increase still further. It is evident in sheep where the amount of sulfur that sheep can obtain through its diet fundamentally limits wool growth. Transgenic lupin seeds have been shown to provide higher amounts of methionine and cysteine in a protein that is both rich in sulfur amino acids and stable in the sheep's rumen leading to increased wool yield (White *et al.*, 2001).

4.5.3 Uptake of Sulfur

A wide variety of forms of sulfur are used by microorganisms. Sulfur dioxide can readily deposit in wet leaf surfaces and through the

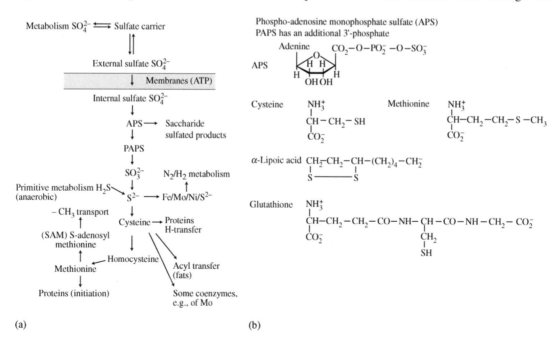

(a)   (b)

Figure 10 Scheme for the uptake and incorporation of sulfate into cells and the formulas of some biologically important sulfur-containing molecules (source Frausto da Silva and Williams, 2001).

stomata. Purple nonsulfur bacteria can utilize thiosulfate and sulfide (Sinha and Banerjee, 1997). Lichen and leafy vegetation can also act remove sulfur compounds from the atmosphere. OCS removal is the best known and probably represents a dominant sink for tropospheric OCS. The fate of the oxidation products of DMS has not always been clear. There have been some biochemical studies that suggest that they may degrade biologically. Facultatively methylotrophic species of Hyphomicrobium and Arthrobacter seem to be able to produce enzymes necessary for a reductive/oxidative pathway for dimethylsulfoxide and dimethylsulfone (although the former appears to be more generally utilizable). Methanesulfonic acid is stable to photochemical decomposition in the atmosphere, and its fate on land has been puzzling. A range of terrestrial methylotropic bacteria that appear common in soils can mineralize MSA to carbon dioxide and sulfate (Kelly and Murrell, 1999).

4.6 SULFUR IN SEAWATER

Seawater is rich in sulfur and, while containing high concentrations of inorganic sulfur, it also has an elaborate organic chemistry, driven largely by biological activities.

4.6.1 Sulfate

Most of the sulfur in seawater present as sulfate. At a salinity of 35‰ sulfate is found at $2.712 \, g \, L^{-1}$, and the ratio of sulfate chloride mass in normal seawater is 0.14. This ratio is fairly constant in the oceans. Nevertheless, it can be much lower in deep brines of the type found in the southern end of the Red Sea possibly because of clays in sediments acting as semi-permeable membranes. Even though sulfuric acid has a modest second dissociation constant, it is sufficient to ensure that the bisulfate ion is at very low concentrations in seawater.

The double charge on the sulfate ion means that it can associate with other cations, and typically seawater has been argued to have more than 30% of the sulfate associated with sodium and magnesium ion pairs. The magnesium sulfate chemistry of seawater has considerable importance in underwater acoustics. Although sound absorption coefficients are dependent on pH at frequencies below 1 kHz, they are dependent upon $MgSO_4$ over the range 10–100 kHz (Brewer *et al.*, 1995). Some models have assigned and association constant of $10^{5.09}$ to the important and powerful association between magnesium and sulfate (Millero and Hawke, 1992). However, ion association models are complex and difficult to use, so recent models of seawater and other electrolyte systems in the environment rely increasingly on the formalism of Pitzer, which treats the electrolyte in terms of ion association (Clegg and Whitfield, 1991).

Seaspray can be blown directly from the sea surface during high winds, but the particles of salt from this process tend to be large and short lived in the atmosphere so do not travel far inland. Finer particles are produced during bubble bursting. The very finest arise when the cap of the surfacing bubble shatters giving film drops that dry into salt crystals of 5–50 pg. Along with these are larger jet drops that emerge as the bubble cavity collapses and give particles of ~0.15 µg. Such processes lead to the production of around a 10–30 Pg of seasalt each year, of which some 260–770 Tg would be sulfur in the form of sulfate. High chloride and sulfate concentrations are found in coastal rainwater, but perhaps only 10% of the solutes generated through the action of wind or through bubble bursting are deposited over land, as most falls over the oceans (Warneck, 1999).

The sulfate that derives from seasalt is called seasalt sulfate. It is important to distinguish it from sulfates that are found in the marine aerosol from the oxidation of gaseous sulfur compounds, which are known as non-seasalt sulfate.

4.6.2 Hydrogen Sulfide

Hydrogen sulfide is now recognized as occurring widely throughout the oceans (Cutter *et al.*, 1999). About 12% of the total dissolved sulfide is free and the rest is largely complexed to metals in typical North Atlantic water (Cutter *et al.*, 1999). Hydrogen sulfide has long been known to occur in anoxic regions, but in surface waters of the open ocean it was thought to oxidize rapidly. Hydrogen sulfide is produced from the hydrolysis of OCS (Elliott *et al.*, 1987; Elliott *et al.*, 1989). However, studies of the North Atlantic show that the rate of total sulfide production from hydrolysis and through atmospheric input is some 20 times less than sulfide removal via oxidation and that bound to sinking particulate materials. This suggests additional sources. The similarity between the depth distribution of total sulfide and chlorophyll suggests that there is a biological involvement in the production of total sulfide, perhaps by phytoplankton in the open ocean (Radfordknoery and Cutter, 1994).

The low concentrations of hydrogen sulfide found in the marine atmosphere suggest that there might be a significant flux of the gas from the oceans, but most studies conclude that it is a rather insignificant flux (Shooter, 1999). There is also the potential for the sulfate reduction to occur on particles in the ocean (Cutter and Krahforst, 1988) and some may arise from deep-sea vents, but this source is restricted to water close to the vent (Shooter, 1999).

Hydrogen sulfide can be oxidized in less than an hour in seawater. This removal can be through oxidation by oxygen or iodate. There is a possibility of oxidation, by hydrogen peroxide, but it is probably a minor pathway (Radfordknoery and Cutter, 1994). Photo-oxidation is also possible (Pos *et al.*, 1998), along with oxidation by Fe (III) oxide particles. This latter process is dependent on the way in which the particle forms and on pH with a maximum near 6.5. The Fe(III) oxide route gives mostly elemental sulfur as a product, which may have implications for pyrite formation (Yao and Millero, 1996).

There are much greater production rates of hydrogen sulfide in some anoxic basins. The Black Sea is the largest of these where sulfide production in the water column is estimated to be $30–50\,Tg(S)\,yr^{-1}$, taking place between depths of 500 m and 2,000 m. This may amount to as much as one-third of the global sedimentary budget ($120–140\,Tg(S)\,yr^{-1}$). There is also pyrite formation in the water column of the Black Sea, through reaction with iron although this is relatively small compared with oxidation at the oxic anoxic interface (Neretin *et al.*, 2001).

4.6.3 OCS and Carbon Disulfide

Other sulfides seem readily produced by photochemical processes in seawater along with a parallel photoproduction of carbon monoxide (Pos *et al.*, 1998). Colored dissolved organic material is frequently involved in such processes. Laboratory irradiation has confirmed that cysteine and cystine are efficient precursors of CS_2 and that OH radicals are likely to be important intermediates (Xie *et al.*, 1998), but it is also likely that some of these sulfur gases are produced directly by biological processes. Ocean waters appear to be supersaturated in carbon disulfide and capable of yielding $\sim 0.1\,Tg(S)\,yr^{-1}$ (Xie and Moore, 1999).

4.6.4 Organosulfides

The discovery of substantial amounts of volatile organosulfides in the oceans was one of the major additions to the sulfur cycle in the second half of the twentieth century. The largest flux of reduced sulfur to the atmosphere from the oceans is as DMS. The importance of this compound that was largely unknown in nature until the 1970s was revealed by Lovelock *et al.* (1972) as a potential explanation for the imbalance in the sulfur cycle. Over time it has become clear that this process has important implications to the atmosphere and offers a source of sulfate to form cloud condensation nuclei.

The production of DMS is driven by the activity of microbiological organisms in the surface waters of the ocean (see Figure 11). The precursor to DMS is dimethylsulfoniopropionate (DMSP) produced within cells, but its concentration can vary over five orders of magnitude. Prymnesiophytes and some dinoflagellates are strong producers of DMSP. It is usually considered as an algal osmolyte globally distributed in the marine euphotic zone, where it represents a major form of reduced sulfur in seawater. It is typically partitioned between particulates (33%; range 6–85%;); dissolved nonvolatile degradation products (46%; range 21–74%); and in a volatile form (9%; range 2–21%) (Kiene and Linn, 2000). The natural turnover of dissolved DMSP results in the conversion of a small fraction (although large amount) to reduced sulfur gases. DMSP is released from cells during senescence, zooplankton grazing or viral attack. Once in the oceans it undergoes cleavage, which is usually microbiologically mediated or catalyzed by the enzyme DMSP lyase to DMS and acrylic acid (Kiene, 1990; Ledyard and Dacey, 1996; Malin and Kirst, 1997):

$$(CH_3)_2S^+CH_2CH_2COO^- + OH^-$$
$$\rightarrow CH_3SCH_3 + CH_2CHCOO^- + H_2O$$

The precursor DMSP is often present at concentrations an order of magnitude higher than DMS. These concentrations are so high that DMSP could support 1–13% of the bacterial carbon demand in surface waters, making it a significant substrate for bacterioplankton (Kiene *et al.*, 2000). It is likely that some 90% of the DMSP in seawater is converted to methanethiol (MeSH) via 3-methiolpropionate (Kiene and Linn, 2000). The MeSH provides sulfur for incorporation into bacterial proteins, such as methionine. If the thiol is not assimilated, it is likely to react to give dissolved nonvolatile compounds, such as sulfate and DOM–metal–MeSH complexes. High production rates of MeSH in seawater ($3–90\,nM\,d^{-1}$) offer the potential for the thiol to affect metal availability of trace metals and their chemistry in seawater (Kiene *et al.*, 2000).

Maximum concentrations of DMS in the oceans are found, at or a few meters below, the surface (as high as micrograms per liter under bloom conditions), which fall rapidly at depths where phytoplanktonic species are no longer active. In Arctic and Antarctic waters *Phaeocystis* sp. is responsible for enhanced DMS in the oceans particularly at the final stages of production of a bloom.

Questions are still asked about the role of DMS, which is sometimes seen, along with DMSP as having an antioxidant function in marine algae because the breakdown products of DMSP (i.e., DMS, acrylate, dimethylsulfoxide, and methane sulfinic acid) readily scavenge hydroxyl radicals and other reactive oxygen species (Sunda *et al.*, 2002).

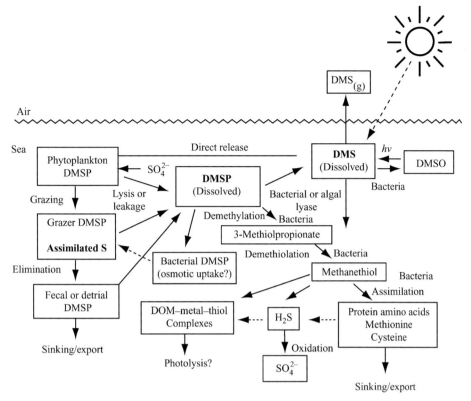

Figure 11 The biogeochemical cycles of DMSP and DMS in surface waters of the oceans (source Kiene *et al.*, 2000).

Once produced the DMS is relatively volatile and is evaded efficiently from seawater. However, the notion that the polar seas represent an underestimated source of DMS has been questioned as the concentrations of DMS in polar waters are not always high in the spring blooms (Bouillon *et al.*, 2002).

Although large quantities of DMS are transferred across the air–sea interface into the atmosphere, it can also be chemically transformed in seawater. The presence of relatively large amounts of dimethylsulfoxide (DMSO) suggested the oxidation of DMS as a likely source. Andreae (1980) proposed bacterial oxidation, while Brimblecombe and Shooter (1986) have shown that DMS is susceptible to photo-oxidation by visible light and thermal oxidation at much slower rates (Shooter and Brimblecombe, 1989). However, more recent work has indicated that DMSO is unlikely to be the only oxidation product (Hatton, 2002; Kieber *et al.*, 1996). As DMS is transparent to sunlight, photo-oxidation requires the presence of naturally occurring photosensitizers such as chlorophyll derivatives, fulvic and humic acids, or bile-type pigments. Laboratory measurements of Kieber *et al.* (1996) have indicated that the photo-oxidation of DMS is wavelength dependent and can take place in the UV and also in the range 380–460 nm. More recently, Hatton (2002) made shipboard measurements that suggested increased oxidation rates in the UV. It is not certain what fraction of DMS in the ocean is oxidized, but it may well be close to the amounts that are lost via evasion to the atmosphere.

DMS may be mixed downwards and contribute to the budget in the deep oceans. Concentrations here are low, but the water volumes are large. Transport to the deep ocean is taken to be via sinking or transport and at these depths there may be a very slow chemical oxidation to DMSO (Shooter and Brimblecombe, 1989).

Dimethylsulfoxide is found at higher concentrations than DMS in the oceans. Although oxidation of DMS has usually been seen as the main source, more recently it has become clear that phytoplankton can also biosynthesize DMSO directly (Bouillon *et al.*, 2002). Incubation experiments suggest that the biological cycling of DMSO involves both its production and consumption in seawater (Simo *et al.*, 2000). Dimethylsulfoxide is found at highest concentrations where biological activity is high, but it appears to be long lived compared with DMS. Thus, it is better mixed within the water column and found at significantly higher concentrations than DMS in deep waters (Hatton *et al.*, 1998).

These marine sources of reduced sulfur gases can be important as a source of sulfur to the continents. The gypsum accumulations of the hyper-arid Central Namib Desert seem to be mainly derived from non-seasalt sulfur, in particular oxidation products of marine DMS (Eckardt and Spiro, 1999).

4.6.5 Coastal Marshes

The salt and brackish waters found in coastal areas are also large potential sources of reduced sulfur compounds, given the high level of biological activity in such zones. The major emissions are: H_2S, OCS, DMS, and DMDS. It has been difficult to estimate the global contributions to this source because of the high degree of variability in measured fluxes from these environments, such that total flux can span four orders of magnitude. The sea–air flux of OCS from a unit area of the estuarine waters, Chesapeake Bay, are over 50 times greater than those typical of the open ocean. Nevertheless, it has not been possible to identify definite seasonal trends in the fluxes. High concentrations of OCS were found in pore waters from estuarine sediments, which were at a maximum in the summer months. Microbial sulfate reduction in sediments contributes to the estuarine budget of OCS (Zhang *et al.*, 1998). A range of factors, such as temperature and biological activity, drive the variability in emission flux, but work at the coastal margins suggests that salinity is a very important control (DeLaune *et al.*, 2002).

The types of sulfur compounds change significantly along salinity gradients in coastal marshes with emissions from saltmarsh sites mostly as DMS. Saltmarshes also showed the highest sulfur emission fluxes followed by brackish marshes, with freshwater marshes having the lowest emissions. Brackish marshes gave hydrogen sulfide, while freshwaters were predominantly a source of OCS (see Figure 12). The low emissions of hydrogen sulfide from saltmarshes were attributed to iron sulfide formation (DeLaune *et al.*, 2002).

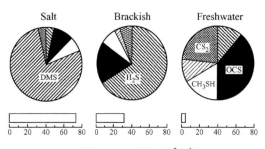

Figure 12 The relative amounts of reduced sulfur species released from various water bodies, with the total emission rate shown as histograms (source DeLaune *et al.*, 2002).

Interstitial waters of intertidal mud flats also show high concentrations (up to 1 mM) of reduced sulfur. There are wide variations in these concentrations with both season and depth indicating that annelids on tidal flats can be exposed to large concentrations of hydrogen sulfide (Thiermann *et al.*, 1996). Hydrogen sulfide emissions are much enhanced where there are significant discharges of industrial and domestic wastewaters into estuarine areas. Limited oxygen inputs and warm climate lead to optimal medium for anaerobic processes and the release of large concentrations of H_2S that affect nearby areas (Muezzinoglu *et al.*, 2000).

4.7 SURFACE AND GROUNDWATERS

Inorganic sulfur compounds are at a much lower concentration in surface waters than in the ocean. Nevertheless, rivers move a large amount of dissolved sulfate to the sea each year (Meybeck, 1987) and industrial activities and agriculture have added much to this flux. Typically estimates are perhaps a little less than a $100 \, Tg(S) \, yr^{-1}$, with additional loads of the same magnitude as those from industrial and agricultural sources. There is also some $100 \, Tg(S) \, yr^{-1}$ transported as suspended particulate matter (Brimblecombe *et al.*, 1989).

There are a number of natural sources of sulfate in rivers. In volcanic regions the sulfate in river and lake water can be derived from volcanic waters enriched in sulfur (Robinson and Bottrell, 1997). The oxidation of pyritic glaciofluvial sediments and bedrock sulfides may be an important source of sulfate in aquifers (Robinson and Bottrell, 1997; Sidle *et al.*, 2000). Chemical attack on acid-insoluble metal sulfides FeS_2, MoS_2, and WS_2 by Fe(III) hexahydrate ions generates thiosulfate. Other metal sulfides are attacked by Fe(III) ions resulting in the formation of elemental sulfur via intermediary polysulfides. Elemental sulfur and sulfate are typically the only products of FeS oxidation, whereas FeS_2 was oxidized to a variety of sulfur compounds, including intermediates such as thiosulfate, trithionate, tetrathionate, and pentathionate. Sulfur and the sulfur oxyanions are ultimately oxidized to sulfuric acid (Schippers and Jorgensen, 2001).

In extreme cases the oxidation of sulfidic floodplain sediments can cause pulses of acids in rivers during floods after long dry periods. The acids can bring large amounts of dissolved aluminum and iron into streams. Such processes can be enhanced as a result of attempts to modify drainage and mitigate floods (Sammut *et al.*, 1996).

Polysulfates, most typically thiosulfates, are found in anoxic riverine sediments. Additionally inorganic polysulfides H_2S_n ($n = 1-5$) can be present even in oxygen-rich aquatic systems.

The polysulfide ions can readily be converted to OCS or perhaps carbon disulfide under sunlight irradiation (Gun *et al.*, 2000) or via:

$$H^+ + CO + S_2^{2-} \rightarrow HS^- + OCS$$

As seen in the discussion of coastal marshes, even freshwater lakes contain a range of volatile sulfur compounds. Seasonally varying concentrations of DMS can sometimes be in the same range as in seawater, i.e., a few $nmol\,L^{-1}$. DMDS and methanthiol are also found (Gröne, 1997). DMS is the dominant organosulfur species found in lakes, along with OCS, MSH, DMDS, and CS_2. The concentrations of volatile organosulfur compounds found in lakes are dependent on their sulfate content. Hypersaline lakes with sulfate concentrations greater than 20 g SO_4^{2-} L^{-1} yield volatile organosulfur concentrations several orders of magnitude higher than typical freshwater lakes. The flux to the atmosphere from such water bodies can rival that from the ocean. However, even where the sulfate content is exceeding low there still appears to be mechanisms for the production of these compounds (Richards *et al.*, 1994). In some saline lakes dimethylpolysulfides ($CH_3S_nCH_3$, $n = 2$–4) are found probably as a result of biological methylation of polysulfides (Gun *et al.*, 2000).

4.8 MARINE SEDIMENTS

The oxidation of organic matter in anaerobic sediments can utilize a number of species as oxidants, of which sulfate is the most important. In seawater at pH values close to 8, sulfate-reducing bacteria metabolize organic matter according to the following simplified equation:

$$2CH_2O_{(s)} + SO_4^{2-}{}_{(aq)} \rightarrow 2HCO_3^{-}{}_{(aq)}$$
$$+ HS^-_{(aq)} + H^+_{(aq)}$$

This process is widespread in marine sediments but is most important at the continental margin sediments, where organic matter accumulation is fastest. Sulfate reduction can occur even meters below the sediment/water interface as long as seawater sulfate can diffuse, or be pumped by bioturbation induced through the activities of sediment-dwelling organisms. The reaction yields the bisulfide anion (HS^-), much of which diffuses upward and is reoxidized to SO_4^{2-} in oxygenated seawater closer to the sediment surface. However, ~10% of the HS^- precipitates soluble Fe(II) to yield iron monosulfide:

$$Fe^{2+}_{(aq)} + HS^-_{(aq)} \rightarrow FeS_{(s)} + H^+_{(aq)}$$

This iron monosulfides is then converted to pyrite (FeS_2) (Butler and Rickard, 2000). At Eh below $-250\,mV$ and at pH around 6, the conversion of FeS to FeS_2 occurs through the oxidation of dissolved FeS by hydrogen sulfide:

$$FeS_{(aq)} + H_2S_{(aq)} \rightarrow FeS_{2(s)} + H_{2(g)}$$

Under milder reducing conditions, the oxidation can use polysulfides, polythionates, or thiosulfate anions, e.g.,

$$FeS_{(s)} + S_2O_3^{2-}{}_{(aq)} \rightarrow FeS_{2(s)} + SO_3^{2-}{}_{(aq)}$$

The sulfite (SO_3^{2-}) anion can undergo reoxidation by oxygen closer to the sediment surface to produce polysulfur anions or even sulfate. Elemental sulfur may be found in marine sediments as a product of the oxidation of iron sulfides (Schippers and Jorgensen, 2001). The incorporation of pyrite and organosulfides, formed as a by-product of sulfate reduction in marine sediments, is a major sink for seawater SO_4^{2-}. Over geological time we have seen that this is an important process in controlling the concentrations of seawater sulfate.

An important consequence of this reduction, in terms of the human use of fossil fuels, is the incorporation of this sulfide into the formation of high sulfur coal. In addition to the sulfide as iron sulfides, organic sulfur in the coals is formed through reaction of reduced of organic materials with sulfur species. This takes place in the early stages of diagenesis by reaction of H_2S or polysulfides ($HS^-_{(x)}$) with humic substances formed by bacterial decomposition of the peat. Organic sulfur species are typically found in coals as thiols, sulfides, disulfides, and thiophenes (see Figure 13). The thiophenic fraction of organic sulfur increases with the carbon content of coals. Alkylkated thiophenes are converted to alkylated benzo(b)thiophenes and dibenzothiophenes via cyclization and subsequent aromatization reactions with coalification (Damste and Deleeuw, 1992; Wawrzynkiewicz and Sablik, 2002).

The importance of organic sulfur to the global sulfur cycle has only recently been recognized, with the awareness that organic sulfur is the second largest pool of reduced sulfur in sediments after pyrite. The incorporation of reduced inorganic sulfur into organic matter represents a significant mechanism for the preservation of functionalized organic compounds, such as thianes, thiolanes, and thiophenes but the timing of this process is not currently known (Werne *et al.*, 2000).

4.9 SOILS AND VEGETATION

Sulfur in soils is present in both inorganic and organic forms. In calcareous soils some 90% is present in the organic form. Inorganic sulfate is present in solution, adsorbed, and insoluble forms. It is this form that is typically taken up by plants. In some soils such as tundra, there are low sulfate concentrations in the water. Changes in recent years in Europe have raised interest in sulfate as a nutrient, because increasingly soils are found to be sulfur deficient. Coarse textured soils typically low in organic matter tend to lose sulfate through leaching (Havlin *et al.*, 1999). Some of the sulfate in soils comes from the atmosphere (or added as fertilizers), but it is also biologically mineralized from organic sulfur, via sulfides (see Figure 14).

Long-term studies of the nutrient balance at the Hubbard Brook Experimental Forest show that much of the SO_4^{2-} entering via atmospheric deposition passes through vegetation and microbial biomass before being released to the soil solution and stream water. Gaseous emission loss of sulfur is probably small. The residence time for S in the soil was determined to be ~9 yr (Likens *et al.*, 2002).

The sulfur in noncalcareous soils occurs mostly as organic sulfur. The typical C/N/S ratio (on a weight basis) is 120/10/1.4 and the N/S ratio

remains within a fairly narrow range 6–8.1. Organic sulfur is present in three forms: (i) HI-reducible, (ii) carbon-bonded, and (iii) residual sulfur. The HI-reducable sulfur can be reduced to H_2S with hydroiodic acid and the sulfur, is largely present as esters and ethers with C–O–S linkages. These would include arylsulfates, phenolic sulfates and sulfated polysaccharized and lipids, and represent about half the organic sulfur. Carbon bonded sulfur is found as the amino acids methionine and cystine and some 10–20% of the organic sulfur is found in this form. It also includes sulfoxides, sulfones, and sulfenic, sulfinic and sulfonic acids (Havlin *et al.*, 1999). Sulfur is additionally associated with humic and fulvic acids in a range of oxidized and reduced forms (Morra *et al.*, 1997). After conversion to inorganic sulfide, the sulfur can be oxidized quite rapidly via elemental sulfur and the thiosulfate ion through to a range of polythionates, which are ultimately converted to sulfite and thence oxidized to sulfate.

Volatilization of sulfur from soils occurs mainly as organosulfides, with DMS usually accounting for at least half (Havlin *et al.*, 1999). The emission of reduced sulfur compounds from soils and most particularly wetlands has been seen as an important source of these compounds to the global budget (Hines, 1992). Table 5 shows the biochemical origin of volatile sulfides from soils. Vegetation is also a source of trace sulfur gases and lichens notably emit H_2S and DMS (Gries *et al.*, 1994).

Sulfate reduction can take place in waterlogged soils, especially environments such as paddy fields. Soils with high sulfate content, particularly as a result of input of seawater, can over time accumulate large concentrations of sulfide through sulfate reduction. When these soils are

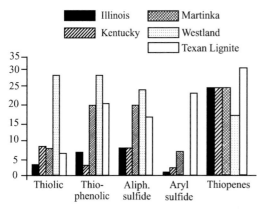

Figure 13 The relative amount of various organic forms of sulfur in some coals (source Attar, 1979).

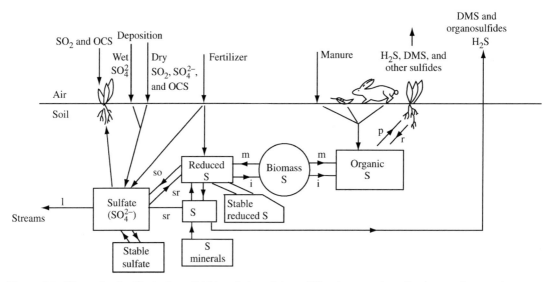

Figure 14 The cycle of sulfur in the soil (abbreviations: i, immobilization; m, mineralization; p, plant uptake; r, root exudation and turnover; so, oxidation; l, leaching; sr, reduction (after Havlin *et al.*, 1999).

exposed to air, they can become acidic through the production of sulfuric acid. This can be a serious problem to agriculture, but may also lead to the acidification of rivers.

Dusts with large amounts of sulfate, most notably as gypsum, are often driven into the air under windy conditions. In drier regions, particularly from the surfaces of dry lake beds, the fluxes can be high. The exposed areas change dramatically over geological time.

4.10 TROPOSPHERE

4.10.1 Atmospheric Budget of Sulfur Compounds

The reservoirs treated above contain most of the mass of sulfur. The amount in the atmosphere

Table 5 The biochemical origin of volatile sulfides from soils under anaerobic conditions through the microbial degradation of organic matter.

Volatile	Biochemical precursor
H_2S	Proteins, polypeptides, cystine, cysteine, glutadione
CH_3SH	Methionine, methionine sulfoxide, methionine sulfone, S-methyl cysteine
CH_3SCH_3	As for CH_3SH plus homocysteine
CH_3SSCH_3	As for CH_3SH
CS_2	Cystine, cysteine, homocysteine, lanthionine, djencolic acid
OCS	Lanthionine, djencolic acid

Adapted from Bremner and Steele by Warneck (1999).

is small compared with these reservoirs (see Figure 15). However, this does not mean that the chemistry of sulfur in the atmosphere is of proportionately small importance. The residence times of compounds in the atmosphere are frequently short. The atmosphere combines great mobility with active chemistry and the part of this biogeochemical cycles is often easy to observe because of the transparency of the gas phase. Interest in atmospheric sulfur chemistry has been further stimulated by the magnitude of anthropogenic sulfur emissions. Global anthropogenic sulfur dioxide emissions were of the same magnitude as natural emissions through the twentieth century. Although similar arguments can be made about anthropogenic contributions to the nitrogen cycle, it was the importance of emissions from high sulfur coals that have caused problems during a long period of human industrialization. On a regional scale, the oxidation of sulfur dioxide to sulfuric acid makes it the main solute in acid rain. Studies of acidification led to great improvements in the scientific understanding of the liquid phase chemistry of rain from the 1950s onwards. By comparison other aspects of liquid-phase atmospheric chemistry (such as the chemistry of water-soluble organic compounds) tended to be much neglected until the very end of the twentieth century.

The focus on oxidized forms of sulfur also meant that comparatively little was known about the more reduced sulfur gases until recently. The presence of hydrogen sulfide had been evident in volcanic regions and on coastal marshes because of its strong smell, but the detection of

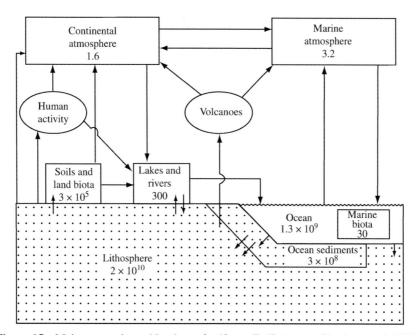

Figure 15 Major reservoirs and burdens of sulfur as Tg(S) (source Charlson *et al.*, 1992).

atmospheric organosulfides had to await the development of gas chromatographic methods and sensitive detectors.

4.10.2 Hydrogen Sulfide

The fluxes of hydrogen sulfide to the atmosphere have not been easy to establish with estimates ranging from $16\,Tg\,a^{-1}$ to $60\,Tg\,a^{-1}$ (Watts, 2000). There is a poor correlation between concentrations of hydrogen sulfide in the marine atmosphere and both biological emissions and radon concentrations, which does little to give a clearer idea of sources. The estimates made by Watts (2000) are shown in Figure 16.

In remote air hydrogen sulfide is typically in the $30{-}100\,pmol\,mol^{-1}$ range. Close to sources such as tidal mud flats it can be higher, and in volcanic regions it can be higher still. In Sulfur Bay near Rotorua New Zealand, average concentrations between $2{,}000{-}4{,}000\,ppb$ were found exposing both human and wildlife (Siegel *et al.*, 1986). The gas is oxidized in air probably through hydroxyl radical attack:

$$OH + H_2S \rightarrow HS + H_2O$$

The HS radical reacts with either oxygen or ozone, the latter giving the HSO radical. The oxidation processes ultimately yield SO_2. The typical lifetime for atmospheric H_2S is $\sim 3\,d$. Global sources and sinks of H_2S are estimated as $7.72 \pm 1.25\,Tg\,a^{-1}$ and $8.50 \pm 2.80\,Tg\,a^{-1}$, respectively, with an imbalance that was indefinable (Watts, 2000).

4.10.3 Carbonyl Sulfide

There is a range of sources for atmospheric COS and these seem fairly evenly balanced (see Figure 17). This coupled with its long residence time mean that it has a constant concentration throughout the atmosphere. It is typically found at concentrations close to 500 ppt in remote and unpolluted air. There has been some debate as to whether the oceans are a source or a sink for OCS,

but in the analysis of Watts (2000) the oceans are taken as a source. The long lifetime of $\sim 25\,yr$ is the result of a slow rate of reaction with the OH radical. This can proceed via addition to form the OCS–OH adduct, but it may decompose to the original reactants. Another possibility is that it reacts:

$$OH + OCS \rightarrow HS + CO_2$$

The long lifetime means that a significant fraction (see Figure 17) of the OCS is transferred to the stratosphere. Uptake by vegetation and deposition to soils are important loss processes, although the large uptake by oxic soils is particularly uncertain (Kjellstrom, 1998; Watts, 2000).

4.10.4 Carbon Disulfide

Carbon disulfide has a much shorter lifetime than OCS ($\sim 7\,d$), so is necessarily far more variable in concentration. At ground level, concentrations are typically in the range $15{-}30\,pmol\,mol^{-1}$, while in the unpolluted free troposphere concentrations are probably $\sim 1{-}6\,pmol\,mol^{-1}$. Oxidation of carbon disulfide proceeds:

$$OH + CS_2 + M \rightarrow HOCS_2$$

There are many ways that the adduct $HOCS_2$ can react with oxygen, one of which is an important source of OCS:

$$HOCS_2 + O_2 \rightarrow HOSO + OCS$$

or

$$HOCS_2 + O_2 \rightarrow H + SO_2 + OCS$$

Estimates of sources and sinks of CS_2 are $0.66 \pm 0.19\,Tg\,yr^{-1}$ and $1.01 \pm 0.45\,Tg\,yr^{-1}$ (Watts, 2000). The amounts that are destroyed through oxidation are probably similar to the amount that is lost to oxic soils (Watts, 2000).

4.10.5 Dimethyl Sulfide

In the early 1970s there was some concern that the sulfur cycle did not balance and that there

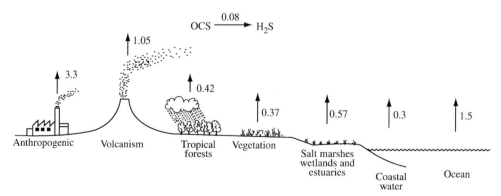

Figure 16 Sources and sinks of atmospheric hydrogen sulfide (source Watts, 2000).

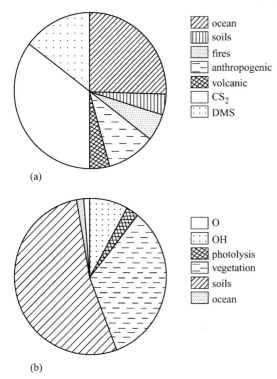

(a)

ocean
soils
fires
anthropogenic
volcanic
CS_2
DMS

(b)

O
OH
photolysis
vegetation
soils
ocean

Figure 17 Production $(1.31 \pm 0.25 \, \text{Tg yr}^{-1})$ and loss $(1.66 \pm 0.79 \, \text{Tg yr}^{-1})$ processes for OCS (source Watts, 2000).

needed to be additional global sources. When James Lovelock discovered DMS in the atmosphere of remote Ireland, this seemed a fine candidate to balance the sulfur cycle. The production of DMS in the oceans and its subsequent oxidation have been extensively investigated, particularly because of its role in producing sulfuric acid droplets that can act as an important cloud condensation nuclei in the remote marine atmosphere.

A large number of measurements of seawater DMS concentrations are available and these have been extensively used to model the flux of DMS to the atmosphere. It is not always easy to get agreement between various models and data sets. The work of Kettle *et al.* (1999) gives rather higher fluxes of DMS and shows a larger seasonal variation especially from the Southern Ocean than that of Maier-Reimer (http://www.mpimet.mpg.de/en/depts/bgc). Utilizing the data of Kettle and Andreae (2000) shows a latitudinal band of high DMS concentrations over the oceans at 60°S. Figure 18 shows global fluxes modeled by Maier-Reimer using the Kettle *et al.* data and yielding a total annual flux of 25 Tg(S) yr^{-1}. In the past estimates of marine DMS have been as high as 30–50 Tg(S) yr^{-1}, but increasingly somewhat lower values are favored and it may well be in the range 10–20 Tg(S) yr^{-1}.

Free tropospheric concentrations are low, probably less than 3 pmol mol^{-1} and its residence time is short probably ~2 d. It is much higher in surface

air close to areas of active production. Here concentrations can average 450 pmol mol^{-1} with a diurnal cycle (amplitude 85 pmol mol^{-1}) showing a decreased concentration during the day because of reactions with OH in the sunlit atmosphere (Yvon *et al.*, 1996):

$$OH + CH_3SCH_3 \rightarrow CH_3SCH_2 + H_2O$$

or via addition

$$OH + CH_3SCH_3 + M \rightarrow CH_3S(OH)CH_3$$

Some of this addition can lead onto dimethylsulfoxide, which can be detected in the atmosphere (Sciare *et al.*, 2000):

$$CH_3S(OH)CH_3 + O_2 \rightarrow CH_3SOCH_3 + HO_2$$

DMS can also react with the halogens and their oxides to yield a range of products that include DMSO.

4.10.6 Dimethylsulfoxide and Methanesulfonic Acid

Dimethylsulfoxide and methanesulfonic acid are two of the most important organic oxidation products of DMS. It is not entirely clear how methanesulfonic acid, $CH_3S(O)(O)(OH)$, forms, but methanesulfinic acid, $CH_3S(O)(OH)CH_3$ has been reported during oxidation in OH–DMS systems. Further addition of OH to methanesulfinic acid, followed by reaction with oxygen, can yield methanesulfonic acid. At lower temperatures found in the Arctic there are a wide variety of oxidation products of DMS that include the MSA, DMS, and dimethylsulfone, $CH_3S(O)(O)CH_3$.

Although DMS is only moderately soluble in rainwater, there has been some interest in the oxidation in the liquid phase, where modeling suggests that the multiphase reactions can be important. Ozone seems to be the most important oxidant, where there is a predicted lifetime for DMS of a few days in clouds. The oxidation reactions offer the potential for these heterogeneous processes to yield more soluble oxidized sulfur compounds such as DMSO and $DMSO_2$ (Betterton, 1992; Campolongo *et al.*, 1999).

The concentrations of DMSO have been measured in the southern Indian Ocean with mixing ratios range from 0.3 ppt to 5.8 ppt. Typically concentrations of DMSO in the air are ~1–2% of the DMS concentrations in air (Jourdain and Legrand, 2001). There is a seasonal cycle with a minimum in winter and a maximum in summer similar to that observed for atmospheric DMS (Ayers *et al.*, 1991). There is also a diurnal cycle for DMSO with maximum values around 09:00 and minimum ones during night, which implies OH reactions with DMS as an important source (Sciare *et al.*, 2000). Being soluble DMSO is also found in rainwater with concentrations from

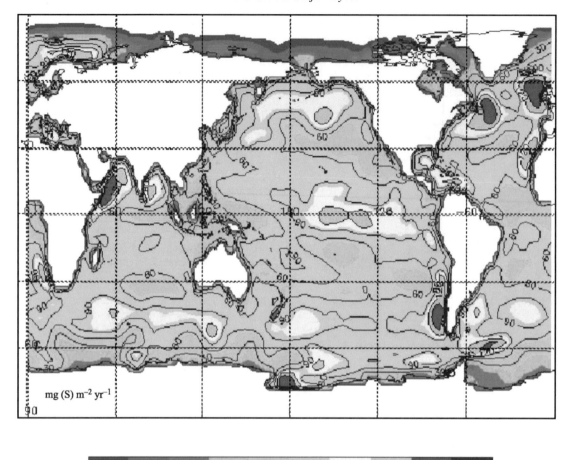

Figure 18 Global fluxes of marine DMS to the atmosphere modeled by Maier-Reimer using the Kettle *et al.* data (http://www.mpimet.mpg.de/en/depts/bgc).

7.0 nM to 369 nM and a seasonal maximum in the summer following much the same pattern as DMS in the atmosphere (Sciare *et al.*, 1998).

Dimethylsulfoxide can readily be removed onto particles and there it can undergo an efficient oxidation through to methanesulfonate. This adds a significant pathway to the gas-phase production of methanesulfonic acid, which is present largely in the submicron aerosol fraction. Peak summer concentrations are 0.6 ± 0.3 nmol m^{-3} and at times this can amount to almost a quarter of the non-seasalt sulfate in the remote marine atmosphere (Jourdain and Legrand, 2001).

Dimethylsulfoxide is also oxidized to SO_2, which can then be converted onto non-seasalt sulfate. The yield probably ranges from 50% to100% in the tropics with the potential for it to be somewhat lower, perhaps 20–40% in mid-latitudes (de Bruyn *et al.*, 2002).

4.10.7 Sulfur

Elemental sulfur has been found in marine air (e.g., Atlas, 1991) although it does not seem to

occur in continental air. The sources of this are not known, but there are hints that it is of biochemical origin and could arise from either the reduction of sulfates or the oxidation of hydrogen sulfide. Given the strong tendency of sulfur to form polyatomic molecules, it is unlikely to be present in a monatomic form.

4.10.8 Sulfur Dioxide

Volcanoes and to a small extent biomass burning represent the major natural primary sources of sulfur dioxide to the atmosphere. Further sulfur dioxide is produced through the oxidation of sulfides, in the atmosphere. In addition to the natural sources there is a very large anthropogenic source that arises from fossil fuel combustion that is comparable in magnitude to the natural sources.

Concentrations in the remote atmosphere are low. In the middle to upper troposphere SO_2 concentrations range between 20 ppt and 100 ppt. Sulfur dioxide concentrations decrease with altitude (see Figure 19), suggesting ground level

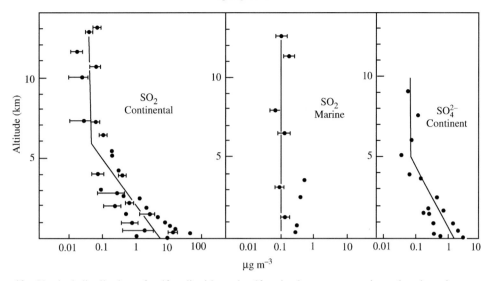

Figure 19 Vertical distribution of sulfur dioxide and sulfate in the remote continental and marine atmosphere (Atlantic) (source Warneck, 1999).

sources and oxidation to sulfate within the atmosphere. This means that the sulfate concentrations decline more slowly with altitude than those of SO_2. Over the oceans there is little change in SO_2 concentrations with height and these are typically $\sim$40 ppt. The most obvious source for marine SO_2 is the oxidation of reduced sulfur compounds such as DMS, although long-range transport from the continents must also contribute, most particularly in the upper troposphere (Warneck, 1999).

Sulfur dioxide can undergo homogeneous oxidation through attack by the OH radical:

$$SO_2 + OH \rightarrow HOSO_2$$

$$HOSO_2 \rightarrow SO_3 + HO_2$$

The sulfur trioxide reacts very rapidly with water to form a $H_2O \cdot SO_3$ complex that reacts with a further water to form the aquated sulfuric acid molecule. It is also likely that SO_3 reacts with water in aerosols. Although there are possibilities that SO_3 could react with ammonia to form sulfamic acid, most of the trioxide will be converted to sulfuric acid (Findlayson-Pitts and Pitts, 2000).

The heterogeneous oxidation can be significant, but in the presence of liquid water there is, as we will see, an important heterogeneous process for the production of sulfate from SO_2. Sulfur dioxide is not especially soluble in water, but subsequent equilibria increase the partitioning into cloud water:

$$SO_{2(g)} + H_2O \rightleftharpoons H_2SO_{3(aq)}$$

$$H_2SO_{3(aq)} \rightleftharpoons H^+_{(aq)} + HSO^-_{3(aq)}$$

$$HSO^-_{3(aq)} \rightleftharpoons H^+_{(aq)} + SO^{2-}_{3(aq)}$$

$K_H = mH_2SO_3/pSO_2, 5.4\,mol\,L^{-1}\,atm^{-1}$ at $15\,°C$

$K' = mH^+ mHSO^-_3/mH_2SO_3, 0.027\,mol\,L^{-1}$ at $15\,°C$

$K'' = mH^+ SO^{2-}_{3(aq)}/mH_2SO_3 \sim 10^{-7}\,mol\,L^{-1}$ at $15\,°C$

Strictly, SO_2 dissolves in water as $(SO_2)_{aq}$ with little forming sulfurous acid, $H_2SO_{3(aq)}$, but it is usual to neglect the distinctions between these two species. The ionization equilibria are typically fast and in the case of the hydration of aqueous SO_2 the hydration reaction proceeds with rate a constant of $3.4 \times 10^6\,s^{-1}$ which allows the formation of the bisulfite anion to be exceedingly rapid. Although $H_2SO_{3(aq)}$ is a dibasic acid, the second dissociation constant is so small that the bisulfite anion (HSO^-_3) dominates as the subsequent dissociation to the sulfite ion $SO^{2-}_{3(aq)}$ would not be important except in the most alkaline of solutions. At around pH 5.4 in a typical cloud with a gram of liquid water in each cubic meter, SO_2 will partition equally into both phases, because of the hydrolysis reactions.

The dissolution of sulfur dioxide in water at the Earth's surface, most particularly seawater, which is alkaline, represents an important sink. Land surfaces, especially those covered by vegetation, also represent a removal process as SO_2 is dry deposited to these surfaces. Although we use the term dry deposition, the vegetated surfaces act as if they were wet and gas exchange takes place with effectively wet surfaces on or inside leaves, via access through the stomata. Such dry deposition is most significant where SO_2 concentrations are high (polluted regions).

The other removal process, wet deposition, removes sulfur from the atmosphere as sulfates in rain. This would be the fate of sulfuric acid produced via the homogeneous oxidation of SO_2, but oxidation also proceeds within droplets. Aqueous sulfur dioxide is oxidized only slowly by dissolved oxygen, but the production of sulfuric acid, which is much stronger, leads to acidification

$$\frac{1}{2}O_2 + HSO_3^- \rightarrow H^+ + SO_4^{2-}$$

This process can be catalyzed by iron, manganese, and other transition metals in the atmosphere, e.g.,

$$M(III)(OH)_n + HSO_3^- \rightarrow M(II)(OH)_{n-1} + SO_3^- + H_2O$$

This initiates a radical chain that leads to sulfate production:

$$SO_3^- + O_2 \rightarrow SO_5^-$$

$$SO_5^- + SO_3^{2-} \rightarrow SO_4^- + SO_4^{2-}$$

$$SO_4^- + SO_3^{2-} \rightarrow SO_3^- + SO_4^{2-}$$

In cloud droplets in remote regions the metal concentrations are likely to be low. Here more typically the reaction proceeds with oxidants such as dissolved hydrogen peroxide (or other atmospheric peroxides) and ozone. Hydrogen peroxide is an especially important droplet phase oxidant, because the gas is very soluble in water so can dissolve from the atmosphere. Additionally, it is readily produced within droplets in the atmosphere via photochemical processes. Oxidation by hydrogen peroxide is also significant, because the reaction is faster in acidic solutions, which means that the oxidation process does not become much slower as droplets become more acidic with the production of sulfuric acid. This oxidation can be represented as

$$ROOH + HSO_3^- \rightleftharpoons ROOSO_2^- + H_2O$$

$$ROOSO_2^- \rightarrow ROSO_3^-$$

$$ROSO_3^- + H_2O \rightarrow ROH + SO_4^{2-} + H^+$$

Ozone although abundant has relatively low solubility in water and the oxidation reaction does not increase in rate at low pH values. However, the oxidation does appear to increase with ionic strength. The overall reaction can be written as

$$HSO_3^- + O_3 \rightarrow H^+ + SO_4^{2-} + O_2$$

and may proceed via the generation of aqueous OH radicals which can react with HSO_3^-:

$$OH + HSO_3^- \rightarrow SO_3^- + H_2O$$

The SO_3 radical can enter into the radical chains previously described.

A range of other oxidants dissolve or are produced in cloud droplets, such as is peroxynitric acid, which contributes to the production of sulfate:

$$HOONO_2 + HSO_3^- \rightarrow 2H^+ + NO_3^- + SO_4^{2-}$$

The relative importance of the various heterogeneous oxidation pathways depends on pH. At pH values below ~4.5 the hydrogen peroxide pathway typically dominates. In urban areas hydrogen peroxide may not be abundant enough to be the most important oxidant. Here transition metal catalysts can enhance the rates considerably, especially if there are alkaline materials from fly ash or ammonia to neutralize the growing acidity of droplet phases, which otherwise limits SO_2 solubility.

When we consider of the heterogeneous oxidation of SO_2, it has to include not only the oxidation of the SO_2 within the droplet phase, but the transfer of further SO_2 into the droplet and the overall depletion of the gas in the air mass as a whole. In general the overall oxidation in the remote atmosphere is rather slow and takes 2–4 d, but under some conditions it can be much faster. The depletion rates of sulfur dioxide in volcanic plumes can sometimes be very fast with residence times of as little as 15 min in moist plumes, where catalytic mechanisms similar to urban air masses probably occur.

There are a number of other reactions of SO_2 in solution most notably the formation of complexes with soluble aldehydes that also partition effectively into droplets. Formaldehyde, acetaldehyde, glyoxal, and methylglyoxal are the most common of these reactive aldehydes:

$$HCHO + SO_2 = CH_2(OH)SO_3^-$$

The formation of such complexes can result in an order of magnitude increase in the overall solubility of SO_2 in droplets, but some of the aldehyde reactions with dissolved SO_2 can be relatively slow so they may not compete with oxidation to sulfate (Findlayson-Pitts and Pitts, 2000).

4.10.9 Aerosol Sulfates and Climate

Sulfate is the ultimate product of the oxidation of SO_2. We might expect small concentrations of other species, such as dithionates that are intermediates in the oxidation process, although these have not been detected. Concentrations of sulfate particles in the remote atmosphere are typically at a few nanomoles per cubic meter. In the marine atmosphere the sulfate is found both in coarse particles where it derives from seasalt and in fine particles around a micron in diameter as sulfuric acid. This non-seasalt sulfate in the smaller particle sizes is an important cloud condensation nucleus (CCN) in the atmosphere.

Over the oceans much of the non-seasalt sulfate derived from the biological production of DMS in seawater (Ayers and Gillett, 2000), which has suggested a biological coupling between climate and living organisms. This coupling has been seen as support (Charlson *et al.*, 1987) for James

Lovelock's *Gaia Hypothesis*. This is a popular concept that goes well beyond accepting that there are biological impacts on the geochemical cycling. It sees the atmosphere as an extension of the biosphere, such that it becomes regulated in a similar way to homeostasis within cells. The regulation of the surface temperature of the Earth has been very important for life. It is possible to argue that enhanced DMS from marine sources would lead to sulfate particles and thus to more CCN. This would be expected to increase the number of cloud droplets and cloud albedo so more light would be reflected into space leading to a cooler Earth. If the Earth cooled there could be a gradual reduction in the biological activity of the oceans and thence less clouds and lower albedo. Such a feedback offers the potential for temperature regulation. The notion has clear attractions, but is difficult to assess and raises, in addition to scientific questions, a range of much more philosophical ones. These include some about the nature of hypotheses through to concerns that we might be assigning altruism to ecosystems.

The role that DMS plays in producing more clouds has offered the possibility for humans to control climate. In the Southern Ocean where iron is a limiting nutrient it would be possible to add large quantities of iron to fertilize the ocean and thus increase DMS emissions and hence cloudiness thus inducing lower temperatures (Watson and Liss, 1998). Experiments show that iron can indeed fertilize the ocean and increase phytoplanktonic activity, but it would be risky to undertake such an experiment with our only planet.

In continental air sulfate tends to be associated with finer particles, and as ammonia is more likely to be present in the air this can neutralize the sulfuric acid with the formation of ammonium sulfate- or bisulfate-containing particles over land. Sulfuric acid can displace chloride from seasalt aerosols and represent a source of hydrogen chloride:

$$H_2SO_4 + Cl^- \rightarrow HCl_{(g)} + HSO_4^-$$

4.10.10 Deposition

We have already seen that sulfur dioxide can be removed from the atmosphere in a number of ways. It is soluble in water, especially at the higher pH values associated with seawater. The rapid absorption of the gas into water means that it is readily transferred into the oceans or other water bodied. Vegetation also acts as an important sink for sulfur dioxide. This can be particularly efficient when the vegetation is wet, such as when it is covered by dew or rainwater. However, even when the vegetation is not wet, sulfur dioxide can enter the leaves through the stomata and enhance deposition. This process is called dry deposition because although it involves water, the water is not in the atmosphere.

The alternative process, wet deposition, deposits the sulfur in rain or other forms of precipitation. Here it is largely as sulfate, which has been incorporated from aerosols or through oxidation of dissolved sulfur dioxide. Sulfate particles can also be dry deposited to the Earth's surface.

The magnitude of the various processes varies with locality. Over vegetated continental areas typically a fifth of the sulfur is dry-deposited as SO_2. In areas of high SO_2 concentrations much higher amounts can be deposited. Where sulfate concentrations in particulate material are high, dry deposition rates can be greater. The balance of wet and dry deposition of sulfate particles over the ocean is uncertain, while some authors suggest dry deposition dominates others favor wet deposition (Warneck, 1999).

4.11 ANTHROPOGENIC IMPACTS ON THE SULFUR CYCLE

Human activities have vastly affected the sulfur cycle (Brimblecombe *et al.*, 1989). The sulfur released from combustion of fossil fuels for example, exceeds the average natural releases into the atmosphere. Thus sulfur has long been seen as a pollutant central to the acid rain debate of the 1980s. However human progress has had other impacts on the cycle.

4.11.1 Combustion Emissions

Sulfur emissions from the combustion of high sulfur coals has been a problem from the thirteenth century when the fossil fuel began to be used in London after the depletion of nearby wood supplies. The intensity of coal use increased reaching its peak within the early twentieth century in Europe and North America. Although the use of coal has declined in these areas, the late twentieth century saw a profound increase in coal use in developing countries, most notably in Asia (see Figure 20). Here emissions have continued to grow with the enormous pressures for industrial development, although changing patterns of fuel use here may lead to decreased emissions in the twenty-first century.

The combustion of fuels leads to release of SO_2 in a simple, but effective oxidation:

$$S + O_{2(g)} \rightarrow SO_{2(g)}$$

Many refining and extractive processes release large amounts of air pollution containing SO_2. For example, sulfide ores have been roasted in the past with uncontrolled emissions

$$Ni_2S_3 + 4O_2 \rightarrow 2NiO + 3SO_2$$

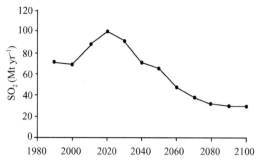

Figure 20 Predicted sulfur emissions from anthropogenic sources (http://sres.ciesin.org).

This sulfur dioxide often destroyed large tracts of vegetation downwind from smelters, such as those at Sudbury in Canada. However, changes in the processes and the construction of a very tall chimney stack have lessened the problems.

The decline in sulfur emissions in Europe and North America has come as part of a shift away from coal as a fuel in all but extremely large industrial plants. Sulfur in coal is about half as pyrites, which is relatively easy to remove, but the rest is organically bound which makes it difficult to remove at an economic rate. Improved controls on stack emissions increasingly rely on the treatment of exhaust gases. In the past this was sometimes by scrubbing the exhaust gas with water to dissolve the SO_2, but the late twentieth century saw a range of well-developed methods. The use of lime (calcium hydroxide) or limestone (calcium carbonate) slurries to absorb sulfur dioxide is widely adopted. The main product, calcium sulfate, is notionally not seen as an environmental problem by-product, although it can be contaminated with trace metals. The process is also hampered by the large amounts of lime that can be required. Regenerative desulfurization processes such as the Wellman-Lord procedure absorb SO_2 into sodium sulfite solutions converting them to sodium bisulfite. The SO_2 is later degassed and can be used as a feedstock for the production of sulfuric acid, for example.

Sulfur is also found in petroleum in organic forms. It can occur at high concentration in some residual oils. This sulfur can be removed by catalytic hydrodesulfurization, but it leads to fuels that tend to become waxy at low temperature. In vehicles catalytic converters have been used to remove nitrogen oxides, carbon monoxide, and hydrocarbons from exhaust streams. However under fuel rich driving cycles (i.e., lots of accelerating and decelerating), hydrogen gas is produced in the exhaust. Three-way catalysts containing cerium dioxide store sulfur from the exhaust stream, under driving conditions, as cerium sulfate. This can at other times be reduced by hydrogen gas to form hydrogen sulfide, which creates a noticeable odor, where traffic is heavy (Watts and Roberts, 1999).

4.11.2 Organosulfur Gases

Although the natural sources of organosulfur compounds are best known, there are a number of anthropogenic sources of this class of air pollutant. These are often released from sulfur-rich wastewaters and sewage sludges where DMDS is of particular concern because of its odor problems. MSH and DMS can cause similar problems along with hydrogen sulfide, OCS and carbon disulfide.

OCS is found as a major sulfur compound in anodic gases of commercial aluminum smelters. Studies suggest a specific OCS emission of $1-7\,kg\,t^{-1}$ (Al). In 1993 aluminum production was responsible for between $0.02\,Tg(S)\,yr^{-1}$ and $0.14\,Tg(S)\,yr^{-1}$ of OCS emissions, which is only a small fraction of the annual global budget (Harnisch *et al.*, 1995). Other sources of OCS are coal combustion $0.019\,Tg(S)\,yr^{-1}$, industry $0.001\,Tg(S)\,yr^{-1}$, the wear on tyres $0.04\,Tg(S)\,yr^{-1}$ and the combustion products from automobiles $0.002\,Tg(S)\,yr^{-1}$ (Kjellstrom, 1998). Carbon disulfide also has a range of anthropogenic sources such as chemical production $0.26\,Tg(S)\,yr^{-1}$, industry $0.002\,Tg(S)\,yr^{-1}$, aluminum production $0.004\,Tg(S)\,yr^{-1}$, and combustion products from automobiles $0.0002\,Tg(S)\,yr^{-1}$ (Kjellstrom, 1998).

There are more complex organosulfur compounds associated with the aerosols. Even relatively volatile compounds such as thiophene can bind to particulate materials (Huggins *et al.*, 2000). The emissions from sulfur-containing fuels may be characterized by the presence of dibenzothiophene and benzonaphthothiophene. Coal combustion (notably lignite in an atmospheric fluidized bed) yields a range of heterocyclic compounds containing oxygen, nitrogen, or sulfur, often with three or four rings. The types of sulfur compounds found include: dibenzothiophene, methyldibenzothiophene, dibenzothieno (3,2-b) (1)benzothiophene, benzo(b)napthothiophene, and benzo(2,3)phenanthro-(4,5-bcd) (Stefanova *et al.*, 2002).

Water-soluble organic compounds in urban atmospheric particles can also contain organosulfur compounds. Methanesulfonic acid and hydroxymethanesulfonic acid have been found as the major organosulfur compounds in urban aerosols, most particularly in particles with the diameter range of $0.43-1.1\,\mu m$. Monomethyl hydrogen sulfate has also been detected on urban particles from localities where no oil or coal power plant exist (Suzuki *et al.*, 2001).

4.11.3 Acid Rain

Rain is naturally acidic because of the weakly acidic carbon dioxide, but the oxidation of sulfur dioxide leads to the much stronger sulfuric acid. The large-scale use of coal caused this to become a problem in nineteenth-century Europe. Remote observations of polluted black rain in Scotland and Scandinavia were made in the late nineteenth century. Some of the most significant early work came from Norway with work of the geologist Waldemar Brøgger and later Amund Helland, who reported on the loss of fish stocks from the 1890s, possibly because of acidification. Modern studies by Knut Dahl and Haakon Torgersen's in the 1920s and 1930s confirmed these observations and the importance of adding lime to reduce the acidity of streams. Rainfall composition was determined by agricultural networks set up to monitor the flux of nutrients to crops in the late nineteenth century. In parallel this also established the impact of combustion processes in the rural environment. The monitoring work recommenced after World War II and gave the information that established the modern picture of acidification of precipitation by the late 1960s. Despite this acid rain did not emerge as a global environmental issue until the 1980s, with concerns over the impact of acidification to the ecosystems across Europe and North America.

The hemispheric changes in precipitation chemistry have been reflected in the records left in high latitude snow, particularly in Greenland. Here the pre-industrial snow shows about equal input of sulfur from marine sources and industrial emissions. From the late nineteenth century, there is evidence of increased anthropogenic impact (Patris *et al.*, 2002), with much of the sulfate arising from North America (Goto-Azuma and Koerner, 2001). The anthropogenic sulfate is sufficient to displace hydrogen chloride from seasalt aerosols such that increasing inputs of anthropogenic HCl can be found in alpine and Arctic ice core records of the twentieth century (Legrand *et al.*, 2002).

The declines in sulfur emissions from the North Atlantic sector have made it is easy for politicians to believe that the acid rain problem has gone away. It is true, emissions here are lower, with a shift away from high sulfur fuels, most particularly coal. In parallel, the amount of sulfur deposited in rain Europe and North America has declined. Indeed, the decline in some parts of the UK and Germany has been so large that crops such as oats and oil seed rape can suffer from sulfur deficiency.

However, the decreases in deposited sulfur are not always matched by equivalent improvements in the amount of acid brought down in rain. Declining sulfur emissions have not always been accompanied by declines in the emission of the nitrogen oxides, which give rise to atmospheric nitric acid in precipitation. Differences in the mode of oxidation that forms nitric acid (a homogeneous oxidation) mean that nitric acid rain has a different geographical distribution to that of

sulfuric acid rain. This distribution of acidity can easily change. Furthermore, calcium was once more abundant in dusts and available to neutralize some of the acidity. In recent decades, the amount of alkaline particulate material has declined, perhaps because there is less dust from unsealed roads and less grit from industry and power generation.

However, not all that has been learnt about acid rain in temperate regions is easily applicable in the Asian context. Research and regulation needs to face a different acid rain problem here. Entirely new ecosystem will be confronted by acidic deposition. We have to recognize some of the novel factors along the Asia-Pacific rim have been present for many centuries. Kosa dust and forest fires are a feature of the region, although we know only a little of their history. Alkaline dust offers the potential to buffer acids in rainfall. Forest fires can produce acids, but can also generate large amounts of alkaline ash that disperses along with the acids. However, such neutralization processes are not always well understood. The greater prevalence of acid rain into the tropics, where soils are often deeply weathered, makes available new routes for mobilizing toxic metals within ecosystems.

4.11.4 Water and Soil Pollutants

Agriculture and industrial activities have been responsible for doubling the load of sulfate in rivers and the concentrations can be much enhanced in rivers that pass through regions with high anthropogenic activity. In addition, industrial releases sometimes have catastrophic impacts. Examples of such a water pollution incident have occurred when toxic water and tailings from pyrite mining are accidentally released into river basins. In April 1998, releases from a pyrite mine in Aznalcollar, southern Spain, spilled into the Agrio and Guadiamar River Basin affecting some $40 \, km^2$. Rapid oxidation meant an increase in concentration and the solubilization of the zinc, cadmium and copper (Simon *et al.*, 2001).

Linear alkylbenzene sulfonates (from detergents) are found in the range $0–5.0 \, mg \, kg^{-1}$ in both freshwater and marine sediments. Higher levels occasionally found in sediments are associated with untreated sewage effluent (Cavalli *et al.*, 2000).

Benzothiazoles from vulcanization of rubber tires are found so enriched in some situations that they are used as a marker for street run-off water. Thiocyanatomethylthiobenzothiazole used as a fungicide in wood protection and antifouling paints may degrade into benzothiazoles and methylthiobenzothiazole is used in vulcanization processes, so these materials find their way into rivers and ultimately estuaries. Bester *et al.* (1997) determined the presence of benzothiazole and methylthiobenzothiazole in estuarine and marine waters. Their concentrations range from $0.04\,ng\,L^{-1}$ to $1.37\,ng\,L^{-1}$. Methylthiobenzothiazole and benzothiazole vary from $0.25\,ng\,L^{-1}$ to $2.7\,ng\,L^{-1}$ in the North Sea, while $55\,ng\,L^{-1}$ methylthiobenzothiazole was found in the Elbe River.

Soils are often sulfur deficient, so sulfur can be added as a fertilizer in agricultural activities. Sulfur can be mobilized by agriculture and rivers readily polluted through the use of sulfur-rich fertilizers (Robinson and Bottrell, 1997). Although this can be added as sulfate, it is readily moved into the soil profile by rain or irrigation. It can also be applied as elemental sulfur, ammonium thiosulfate, and ammonium polysulfide. Ultimately, this fertilizer is likely to appear in runoff waters. Sulfur-containing pesticides now represent a significant input of sulfur to agricultural soils (Killham, 1994) and can pollute runoff waters as mentioned in the section above.

The widespread agricultural use of pesticides increases their concentration in river waters. Many of these contain sulfur, e.g., herbicides thiobencarb (S-4-chlorobenzyl diethylthiocarbamate) isoprothiolane (diisopropyl 1,3-dithiolan-2-ylidenemalonate), diazinon (O–O-diethyl 0-2-isopropyl-6-methylpyrimidin-4-yl) and fenitrothion (O, O-dimethyl O-4-nitro-m-toryl phosphorothioate), fungicide (isoprothiolane) and two insecticides (diazinon and fenitrothion). The concentrations of these pesticides typically show strong seasonal patterns which reflect their use (Sudo *et al.*, 2002).

Thiobencarb controls broad leaf weeds mostly in rice (so much has been used in Japan), lettuce, celery, and endive (894000lbs were used in California in 1997). Thiobencarb may be found at concentrations lower than expected from its application rate and sorption on soil particles (Sudo *et al.*, 2002) and reactions that degrade it. When thiobenzcarb is present at higher concentrations (>1 ppb), a bitter taste in drinking water can be associated with the oxidation product thiobencarb sulfoxide. Other degradation products include 2-hydroxythiobenzcarb (Fan and Alexeeff, 2000) along with 4-chlorobenzaldehyde. The pesticides thiobencarb and ethiofencarb undergo direct photolysis in aqueous solution although thiobencarb yields 4-chlorobenzaldehyde and ethiofencarb gives the expected sulfoxide (Vialaton and Richard, 2000). The sulfur-containing pesticides fenthion and the pollutant from fossil fuels dibenzothiophene can undergo photodegradation typically to the sulfones (Huang and Mabury, 2000a,b).

Windblown sulfates, typically as gypsum, are found in arid regions. Water utilization or agricultural changes can enhance the production of such dusts. The decreasing water volume in the Aral Sea has much enhanced the production of sulfate dusts over recent decades that has become associated with severe environmental problems (O'Hara *et al.*, 2000).

4.11.5 Coastal Pollution

Because seawater has high sulfate concentrations, the release of sulfate to coastal waters is not a major problem. However, when organic loads in the discharge waters are heavy, the sulfate can be reduced with a problem of bad odors. A range of larger organosulfides can be found in seawater. Dibenzothiophene is representative of a group of sulfur-containing heterocyclic organic compounds that are common as a form of organosulfur in oils. The relative stability of the dibenzothiophenes has made them useful markers in the weathering of crude oils (Barakat *et al.*, 2001). Dibenzothiophene has been found as a common constituent of coastal sediments with concentrations up to several hundred $ng\,g^{-1}$. Sediments of the South China Sea show a concentration range of 11–66 $ng\ g^{-1}$ dry sediment. Dibenzothiophene content was higher nearshore relative to offshore sediments, but clay and organic carbon contents appear as two prime factors controlling the sediment dibenzothiophene levels (Yang *et al.*, 1998). The source of this sedimentary dibenzothiophene is generally seen as the deposition of particles from combustion sources. However, it is also possible from crude oil pollution or terrestrial runoff as river sediments can be high in dibenzothiophene (West *et al.*, 1988).

4.12 SULFUR IN UPPER ATMOSPHERES

4.12.1 Radiation Balance and Sulfate Particles

Sulfuric acid is an important component of upper planetary atmospheres. The background sulfur found in the Earth's stratosphere derives from OCS that is resistant to attack by the OH radical in the troposphere. This means that there is a flux of OCS to the stratosphere of 0.03–0.06 Tg $(S)\,yr^{-1}$. Although resistant to oxidation in the lower atmosphere, there is the potential for photochemistry or chemistry involving atomic oxygen in the stratosphere (Warneck, 1999):

$$OCS + h\nu \rightarrow S + CO$$
$$O + OCS \rightarrow SO + CO$$

Subsequently,

$$S + O_2 \rightarrow SO + O$$
$$SO + O_2 \rightarrow O + SO_2$$

The sulfur dioxide will ultimately be converted to sulfuric acid through addition of OH much as was shown for the troposphere. Oxidative processes lead to sulfuric acid, which has such a strong affinity for water that even under conditions where its abundance is low solution droplets form, mostly through the condensation onto small nuclei in equatorial regions. The stratospheric sulfate aerosol has a size range 0.1–0.3 µm and is found between the tropopause and ~30 km. This is often called the Junge layer after Christian Junge, who discovered it in the 1960s.

During periods of intense volcanic activity large quantities of SO_2 can be injected into the stratosphere, increasing the concentration of sulfate aerosols. In 1991 Mt. Pinatubo, in the Philippines, injected some 20 Tg of SO_2 into the stratosphere. Under normal conditions aerosol sulfate concentrations are 1–10 particles cm^{-3}, although after eruptions this can rise by as much as two orders of magnitude. Peak sulfate levels in the Junge layer can increase from around 0.1 µg m^{-3} to 40 µg m^{-3}. The sulfate layer appears to take about six months to form through the slow oxidation of SO_2 into a sulfate aerosol.

The volcanic particles, including those formed from sulfuric acid droplets, intercept incoming solar radiation. The sulfuric acid particles have a greater effect that the larger volcanic ash particles fall out more quickly. This absorption of radiation by stratospheric particles warms the stratosphere and cools the troposphere. After major eruptions this can amount to a 1° increase in the middle troposphere and changes in surface climate for a few years afterwards. Tropospheric cooling after an eruption is mitigated in the northern hemisphere winter, because tropical eruptions can induce a stronger polar vortex, with a stronger jet stream producing characteristic stationary wave pattern within the tropospheric circulation. Thus, we can find warmer northern hemisphere continents in these winters. This indirect advective effect on winter climate is stronger than the radiative cooling effect, which can dominate at lower latitudes and in the summer months. Volcanic effects play a significant role in interdecadal climate change on longer timescales (Robock, 2000).

4.12.2 Ozone

The sulfate droplets also have an important role in stratospheric chemistry. In the lower stratosphere the relative humidity is also low, so sulfuric acid concentrations can be high (65–80% H_2SO_4 mass fraction). In the higher stratosphere where water is more abundant or under conditions where the airmass cools, the sulfuric acid becomes less concentrated (around 30% H_2SO_4 mass fraction). The larger amount of water and lower temperatures allows other components of the stratosphere, most importantly nitric acid and ultimately hydrogen chloride to dissolve in the droplets. Further cooling allows various hydrates to crystallize out, although it is also possible for these solutions to remain as liquid droplets below the frost point.

These sulfate aerosols provide sites for an extensive heterogeneous chemistry and have stimulated interest in the role of polar stratospheric clouds in the depletion of ozone at high latitudes. This depletion processes have been enhanced by the increased amount of halogens introduced into the stratosphere from halogens containing compounds such as CFCs/freons. One way to view the heterogeneous chemistry in the sulfate aerosol is to consider it in terms of separating of the chlorine and the nitrogen species of the stratosphere. Chlorine and nitrogen can be found combined as chlorine nitrate, $ClONO_2$, limiting the destruction of ozone by chlorine. Heterogeneous processes within the cloud particles allow chlorine to distribute into the gas phase with nitrogen compounds as nitric acid in the solid phase, e.g.,

$$ClONO_{2(g)} + HCl_{(s)} \rightarrow Cl_{2(g)} + HNO_{3(s)}$$

This can be followed by gas-phase processes, such as photolysis:

$$Cl_{2(g)} + h\nu \rightarrow 2Cl_{(g)}$$

and a sequence of reactions:

$$2Cl_{(g)} + 2O_{3(g)} \rightarrow 2ClO_{(g)} + 2O_{2(g)}$$

$$2ClO_{(g)} + M \rightarrow Cl_2O_{2(g)} + M$$

$$Cl_2O_{2(g)} + h\nu \rightarrow ClO_{2(g)} + Cl_{(g)}$$

$$ClO_{2(g)} + M \rightarrow Cl_{(g)} + O_{2(g)} + M$$

These four equations sum

$$2O_{3(g)} \rightarrow 3O_{2(g)}$$

showing ozone destruction. One can see that the cloud chemistry of the sulfate aerosol can be important in depleting ozone. The cloud particles separate the chlorine and the nitrogen species and additionally some of the reactions proceed faster at the low temperatures found over the poles. It should also be noted that the formation of $Cl_2O_{2(g)}$ is second order so potentially sensitive to chlorine concentration.

Such heterogeneous processes are not only important in polar regions. During the Pinatubo eruption the presence of enhanced numbers of sulfate particles caused a 10–15% loss of ozone after about a year at 40° N. This gradually recovered over the following years (Warneck, 1999).

4.12.3 Aircraft

Aircraft are also an important source of sulfur in the stratosphere. Although the sulfur content of aviation fuels is rather low (typical sulfur content ~400 ppm by mass), the current subsonic fleet injects some 0.02 Tg(S) yr^{-1} compared with the OCS-derived input of 0.03–0.06 Tg(S) yr^{-1}. This is a significant fraction of the background flux, although considerably smaller than the input from large volcanic eruptions. Aircraft also emit sulfur compounds in the troposphere, but here there are other larger anthropogenic sources at these lower altitudes. Nevertheless, the aircraft emissions are very important because only a small fraction of sulfur sources at the surface of the Earth reach the upper troposphere. Such transfer to high altitudes depends on deep mid-latitude and tropical convection processes, while aircraft emissions occur at high altitude and thus do not require vertical transport. In addition, although these emissions are often in remote areas, they tend to concentrate along well-used flight paths (Fahey and Schumann, 1999).

The sulfur emitted from aircraft assists the formation of contrails, by increasing the number of ice particles and decreasing particle size. Contrails can develop into more extensive contrail cirrus clouds in air masses that are supersaturated with respect to ice and here the ice particles can grow through the uptake of water vapor. Aircraft can act as a kind of trigger for cloud formation. Currently, line-shaped contrails are more frequent over North America, the North Atlantic, and Europe. Above Europe these clouds amount to 0.5% of the daytime coverage. These contrails induce a radiative forcing that increases upper temperatures, while decreasing that at the surface. Although aviation will increase in the future, some of its effects on cirrus formation could decrease through reduced sulfur and soot emissions.

4.13 PLANETS AND MOONS

Although this chapter focuses on the Earth's biogeochemistry, it is instructive to consider what examining sulfur cycling on other bodies in the solar system reveals. The atmospheric chemistry is most studied because of the comparative effectiveness of spectroscopic methods of planetary observation.

4.13.1 Venus

Studies of Venus and most particularly the fly-by missions of early spacecraft gave an impression of a hostile environment with very high surface temperatures (735 K). Infrared observations indicated the presence of sulfur dioxide in the upper cloud layers. The sulfur dioxide (SO_2) content of atmosphere observed at the cloud tops ($\sim$50 km) varies through time and appears to have decreased by 50-fold since 1978 (Bezard *et al.*, 1993). At altitudes in the range 35–45 km region concentrations are $\sim$130 ppm and may be more stable. They are seen as a likely tracer of Venusian volcanism. The yellow color of the visual images has been attributed to suspended sulfur particles in the upper atmosphere. There has been a long debate over sulfuric acid droplets in the atmosphere of Venus, but they may be less likely than once thought. OCS can be produced in the atmosphere:

$$SO_2 + 3CO = OCS + 2CO_2$$

with nominal concentrations of 5 ppm at the surface of Venus (Hong and Fegley, 1997), but equilibrium might not be attained.

4.13.2 Jupiter

Sulfur on Jupiter is better understood since the impact of the cometary fragment *Shoemaker-Levy 9* into the Jovian atmosphere in July 1994. The impact forced a great plume of material from deep in the atmosphere up some 3,000 km. This plume was characterized by the presence of ammonia, sulfur, ammonium hydrosulfide, and helium, and led to observations of hydrogen sulfide on Jupiter. The impacts left dark spots in the atmosphere, which gave strong signatures from sulfur-bearing compounds such as diatomic sulfur (S_2), carbon disulfide (CS_2), and hydrogen sulfide (H_2S). Modeling studies have examined the post-impact sulfur photochemistry in the Jovian atmosphere (Moses *et al.*, 1995). They suggest that the sulfur polymerizes to S-8 in the first few days after an impact. Other important sulfur reservoirs are CS, whose abundance increased markedly with time. There is also the potential for sulfur to react with hydrocarbons to give thioformaldehyde or, depending on reaction rates, possibly organosulfides such as DMS. There is also the potential for NS to be significant as its loss rates may be small.

4.13.3 Io

The active volcanoes of the Jovian moon Io release large quantities of sulfur and other materials (Spencer and Schneider, 1996) that re-cover the surface at a rapid rate and maintain a tenuous atmosphere. This sulfur is largely as sulfur dioxide, which is also found as condensate that covers some three-quarters of the surface. However, sulfur is also found as elemental sulfur, with perhaps traces of hydrogen sulfide (Russell and Kivelson, 2001; Zolotov and Fegley, 1998). Low pressures in the atmosphere of Io mean that sulfur can remain in seemingly exotic forms such as sulfur monoxide (SO), which has been calculated to have an SO/SO_2 ratio of 3–10% (Zolotov and Fegley, 1998). Others suggest that OSOSO, and its cation, are likely present in the Io's atmosphere (Cacace *et al.*, 2001).

4.13.4 Europa

Recent observations from the Galileo's near infrared mapping spectrometer have suggested the presence of enhanced concentrations of sulfuric acid on the trailing side of the Jovian moon Europa. This face of Europa is struck by sulfur ions coming from Jupiter's innermost moon Io. A dark surface material, which spatially correlates with the sulfuric acid concentration, is identified

as radiolytically altered sulfur polymers. Radiolysis of the surface by magnetospheric plasma bombardment continuously cycles sulfur between three forms: sulfuric acid, sulfur dioxide, and sulfur polymers, with sulfuric acid being ~ 50 times as abundant as the other forms. Sulfur species continually undergo interconversion in radiolytic cycles as shown in Figure 21 These maintain a dynamic equilibrium between sulfur polymers S_n, sulfur dioxide SO_2, hydrogen sulfide H_2S, and $H_2SO_4 \cdot nH_2O$ (Carlson *et al.*, 2002).

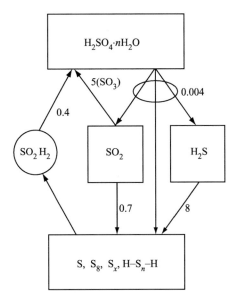

Figure 21 Radiolytic sulfur cycling on the Jovian moon Europa with reaction rates marked numerically as reaction efficiencies. The transient sulfinic acid is circled as an intermediate (source Carlson *et al.*, 2002).

4.14 CONCLUSIONS

The Earth's sulfur cycle (Figure 22) transfers enormous amounts of this biologically important element through various reservoirs each year. Sulfur has a wide range of oxidation states and shows the ability to form a large number of oxides and oxyanions, many of which are found in the environment. It also has the potential to form polymeric species with a significant number of sulfur atoms. Although the nitrogen cycle is critical to the biosphere and agricultural production, it may be fair to argue that it does not show quite the same range of polymerism. The insolubility of iron sulfides offers the potential for the burial and mineralization of large amounts of sulfur in sediments. However, sulfur shares with nitrogen a complex organic chemistry, but organosulfur compounds have not always been widely studied.

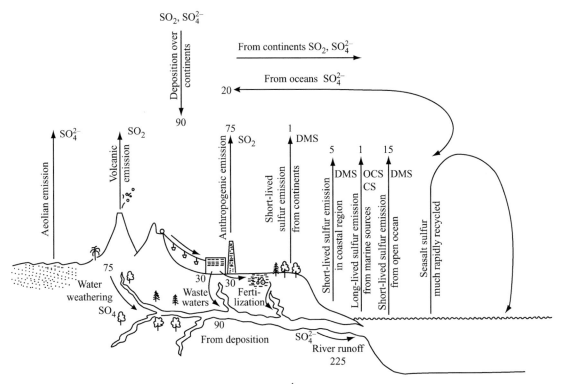

Figure 22 Global cycle showing key fluxes at Tg(S) yr^{-1}. The formula for the most significant components are marked against each flux.

Nevertheless the discovery of large fluxes of DMS to the atmosphere has revealed just how important organosulfur compounds can be.

The role that DMS plays in the formation of cloud condensation nuclei has implications for climate that has provoked a debate and interest that has gone well beyond scientists and has been seen as support for the Gaia Hypothesis.

As with the nitrogen cycle human inputs of sulfur have altered the cycles in profound ways. Human activities, although growing and dependent on economic cycles, are relatively constant compared with the high variability of some natural emissions such as explosive volcanism. Ever since the beginning of the twentieth century, global industrialization has had a remarkable effect on the sulfur cycle, perhaps most notably in the production of acid rain. There have also been increases in the release of sulfur into rivers from mining, industrialization, and particularly agriculture. Wind-blown dusts have also been enhanced usually through loss of ground cover or in some very notable incidents, the removal water from inland seas.

There will be changes in the future. Human emissions of sulfur dioxide to the atmosphere are likely to reach a maximum in the early twenty-first century. The biogeochemical cycle of sulfur seems set to undergo further change, so our retained interest is bound to unlock more of its secrets.

REFERENCES

Andres R. J. and Kasgnoc A. D. (1998) A time-averaged inventory of subaerial volcanic sulfur emissions. *J. Geophys. Res.: Atmos.* **103**(D19), 25251–25261.

Atlas E. (1991) Observation of possible elemental sulfur in the marine atmosphere and speculation on its origin. *Atmos. Environ. Part a: General Topics* **25**(12), 2701–2705.

Attar A. (1979) Evaluate sulfur in coal. *Hydrocarb. Process.* **58**, 175–179.

Ayers G. P. and Gillett R. W. (2000) DMS and its oxidation products in the remote marine atmosphere: implications for climate and atmospheric chemistry. *J. Sea Res.* **43**(3–4), 275–286.

Ayers G. P., Ivey J. P., and Gillett R. W. (1991) Coherence between seasonal cycles of dimethyl sulfide, methanesulfonate and sulfate in marine air. *Nature* **349**(6308), 404–406.

Barakat A. O., Qian Y. R., Kim M., and Kennicutt M. C. (2001) Chemical characterization of naturally weathered oil residues in arid terrestrial environment in Al-Alamein, Egypt. *Environ. Int.* **27**(4), 291–310.

Barnes I., Becker K. H., and Mihalopoulos N. (1994) An FTIR product study of the photooxidation of dimethyl disulfide. *J. Atmos. Chem.* **18**(3), 267–289.

Bester K., Huhnerfuss H., Lange W., and Theobald N. (1997) Results of non-target screening of lipophilic organic pollutants in the German Bight: I. Benzothiazoles. *Sci. Tot. Environ.* **207**(2–3), 111–118.

Betterton E. A. (1992) Oxidation of alkyl sulfides by aqueous peroxymonosulfate. *Environ. Sci. Technol.* **26**(3), 527–532.

Bezard B., Debergh C., Fegley B., Maillard J. P., Crisp D., Owen T., Pollack J. B., and Grinspoon D. (1993) The abundance of sulfur-dioxide below the clouds of Venus. *Geophys. Res. Lett.* **20**(15), 1587–1590.

Blake-Kalff M. M. A., Zhao F. J., Hawkesford M. J., and McGrath S. P. (2001) Using plant analysis to predict yield losses caused by sulphur deficiency. *Ann. Appl. Biol.* **138**(1), 123–127.

Boogert A. C. A., Schutte W. A., Helmich F. P., Tielens A. G. G. M., and Wooden D. H. (1997) Infrared observations and laboratory simulations of interstellar CH_4 and SO_2. *Astron. Astrophys.* **317**, 94–929.

Bouillon R. C., Lee P. A., de Mora S. J., Levasseur M., and Lovejoy C. (2002) Vernal distribution of dimethylsulphide, dimethylsulphoniopropionate, and dimethylsulphoxide in the North Water in 1998. *Deep-Sea Res. II: Top. Stud. Oceanogr.* **49**(22–23), 5171–5189.

Brennan S. T. and Lowenstein T. K. (2002) The major-ion composition of Silurian seawater. *Geochim. Cosmochim. Acta* **66**(15), 2683–2700.

Brewer P. G., Glover D. M., Goyet C., and Shafer D. K. (1995) The pH of the North-Atlantic Ocean—improvements to the global-model for sound-sbsorption in seawater. *J. Geophys. Res.: Oceans* **100**(C5), 8761–8776.

Brimblecombe P. and Shooter D. (1986) Photooxidation of dimethylsulfide in aqueous-solution. *Mar. Chem.* **19**(4), 343–353.

Brimblecombe P., Hammer C., Rodhe H., Ryaboshapko A., and Boutron C. F. (1989) Human influence on the sulfur cycle. In *Evolution of the Global Biogeochemical Sulphur Cycle* (eds. P. Brimblecombe and A. Y. Lein). Wiley, Chichester, vol. 39, pp. 77–121.

Butler I. B. and Rickard D. (2000) Framboidal pyrite formation via the oxidation of iron(II) monosulfide by hydrogen sulphide. *Geochim. Cosmochim. Acta* **64**(15), 2665–2672.

Cacace F., Cipollini R., de Petris G., Rosi M., and Troiani A. (2001) A new sulfur oxide, OSOSO, and its cation, likely present in the Io's atmosphere: detection and characterization by mass spectrometric and theoretical methods. *J. Am. Chem. Soc.* **123**(3), 478–484.

Campolongo F., Saltelli A., Jensen N. R., Wilson J., and Hjorth J. (1999) The role of multiphase chemistry in the oxidation of dimethylsulphide (DMS): a latitude dependent analysis. *J. Atmos. Chem.* **32**(3), 327–356.

Canfield D. E. and Raiswell R. (1999) The evolution of the sulfur cycle. *Am. J. Sci.* **299**(7–9), 697–723.

Canfield D. E., Habicht K. S., and Thamdrup B. (2000) The Archean sulfur cycle and the early history of atmospheric oxygen. *Science* **288**(5466), 658–661.

Carlson R. W., Anderson M. S., Johnson R. E., Schulman M. B., and Yavrouian A. H. (2002) Sulfuric acid production on Europa: the radiolysis of sulfur in water ice. *Icarus* **157**(2), 456–463.

Cavalli L., Cassani G., Vigano L., Pravettoni S., Nucci G., Lazzarin M., and Zatta A. (2000) Surfactants in sediments. *Tenside Surfactants Detergents* **37**(5), 282–288.

Charlson R. J., Lovelock J. E., Andreae M. O., and Warren S. G. (1987) Oceanic phytoplankton, atmospheric sulfur, cloud albedo and climate. *Nature* **326**(6114), 655–661.

Charlson R. J., Anderson T. L., and McDuff R. E. (1992) The sulfur cycle. In *Global Biogeochem. Cycles* (eds. S. S. Butcher, R. J. Charlson, G. H. Orians, and G. V. Wolfe). Academic Press, London, pp. 285–300.

Clegg S. L. and Whitfield M. (1991) Activity coefficients in natural waters. In *Activity Coefficients in Electrolyte Solutions* (ed. K. S. Pitzer). CRC Press, vol. 102A, pp. 279–434.

Conway E. J. (1942) Mean geochemical data in relation to oceanic evolution. *Proc. Roy. Irish Acad.* **48**, 119–159.

Cotton F. A., Wilkinson G., Murillo C. A., and Bochmann M. (1999) *Advanced Inorganic Chemistry*. Wiley, New York.

Cutter G. A. and Krahforst C. F. (1988) Sulfide in Surface waters of the western Atlantic-Ocean. *Geophys. Res. Lett.* **15**(12), 1393–1396.

Cutter G. A., Walsh R. S., and de Echols C. S. (1999) Production and speciation of hydrogen sulfide in surface waters of the high latitude North Atlantic Ocean. *Deep-Sea Res. II: Top. Stud. Oceanogr.* **46**(5), 991–1010.

Damste J. S. S. and Deleeuw J. W. (1992) Organically bound sulfur in coal—a molecular approach. *Fuel Process. Technol.* **30**(2), 109–178.

de Bruyn W. J., Harvey M., Cainey J. M., and Saltzman E. S. (2002) DMS and SO_2 at Baring Head, New Zealand: implications for the yield of SO_2 from DMS. *J. Atmos. Chem.* **41**(2), 189–209.

DeLaune R. D., Devai I., and Lindau C. W. (2002) Flux of reduced sulfur gases along a salinity gradient in Louisiana coastal marshes. *Estuar. Coast. Shelf Sci.* **54**(6), 1003–1011.

Delmelle P., Stix J., Bourque C. P. A., Baxter P. J., Garcia-Alvarez J., and Barquero J. (2001) Dry deposition and heavy acid loading in the vicinity of Masaya volcano, a major sulfur and chlorine source in Nicaragua. *Environ. Sci. Technol.* **35**(7), 1289–1293.

Delmelle P., Stix J., Baxter P. J., Garcia-Alvarez J., and Barquero J. (2002) Atmospheric dispersion, environmental effects and potential health hazard associated with the low-altitude gas plume of Masaya volcano, Nicaragua. *Bull. Volcanol.* **64**(6), 423–434.

Des Marais D. J. (2000) Evolution—when did photosynthesis emerge on earth? *Science* **289**(5485), 1703–1705.

Eckardt F. D. and Spiro B. (1999) The origin of sulphur in gypsum and dissolved sulphate in the Central Namib Desert, Namibia. *Sedim. Geol.* **123**(3–4), 255–273.

Elliott S., Lu E., and Rowland F. S. (1987) Carbonyl sulfide hydrolysis as a source of hydrogen-sulfide in open ocean seawater. *Geophys. Res. Lett.* **14**(2), 131–134.

Elliott S., Lu E., and Rowland F. S. (1989) Rates and mechanisms for the hydrolysis of carbonyl sulfide in natural-waters. *Environ. Sci. Technol.* **23**(4), 458–461.

Eremin N. I., Dergachev A. L., Pozdnyakova N. V., and Sergeeva N. E. (2002) Epochs of volcanic-hosted massive sulfide ore formation in the Earth's history. *Geol. Ore Dep.* **44**(4), 227–241.

Fahey D. W. and Schumann U. (1999) Aviation produced aerosol and cloudiness. In *Aviation and the Global Atmosphere* (eds. J. E. Penner, D. H. Lister, D. J. Griggs, D. J. Dokken, and M. McFarland). Cambridge University Press, Cambridge, pp. 65–120.

Fan A. M. and Alexeeff G. V. (2000) *Thiobenzcarb*. California Environmental Protection Agency, Sacramento.

Farquhar J., Jackson T. L., and Thiemens M. H. (2000) A S-33 enrichment in ureilite meteorites: evidence for a nebular sulfur component. *Geochim. Cosmochim. Acta* **64**(10), 1819–1825.

Farquhar J., Wing B. A., McKeegan K. D., Harris J. W., Cartigny P., and Thiemens M. H. (2002) Mass-independent sulfur of inclusions in diamond and sulfur recycling on early earth. *Science* **298**(5602), 2369–2372.

Findlayson-Pitts B. J. and Pitts J. N. (2000) *Chemistry of the Upper and Lower Atmosphere*. Academic Press, San Diego.

Frausto da Silva J. J. R. and Williams R. J. P. (2001) *The Biological Chemistry of the Elements*. Oxford University Press, Oxford.

Gobbi G., Zappia G., and Sabbioni C. (1998) Sulphite quantification on damaged stones and mortars. *Atmos. Environ.* **32**(4), 783–789.

Goto-Azuma K. and Koerner R. M. (2001) Ice core studies of anthropogenic sulfate and nitrate trends in the Arctic. *J. Geophys. Res.: Atmos.* **106**(D5), 4959–4969.

Greenwood N. N. and Earnshaw A. (1997) *Chemistry of the Elements*. Heinemann, Oxford.

Gries C., Nash T. H., and Kesselmeier J. (1994) Exchange of reduced sulfur gases between lichens and the atmosphere. *Biogeochemistry* **26**(1), 25–39.

Gröne T. (1997) Volatile organic sulfur species in three North Italian lakes: seasonality, possible sources and flux to the atmosphere. *Mem. Ist. ital. Idrobiol.* **56**, 77–94.

Gun J., Goifman A., Shkrob I., Kamyshny A., Ginzburg B., Hadas O., Dor I., Modestov A. D., and Lev O. (2000) Formation of polysulfides in an oxygen rich freshwater lake and their role in the production of volatile sulfur compounds in aquatic systems. *Environ. Sci. Technol.* **34**(22), 4741–4746.

Habicht K. S., Gade M., Thamdrup B., Berg P., and Canfield D. E. (2002) Calibration of sulfate levels in the Archean Ocean. *Science* **298**(5602), 2372–2374.

Hafenbradl D., Keller M., Wachtershauser G., and Stetter K. O. (1995) Primordial amino-acids by reductive amination of alpha-oxo acids in conjunction with the oxidative formation of pyrite. *Tetrahedron Lett.* **36**(29), 5179–5182.

Harnisch J., Borchers R., Fabian P., and Kourtidis K. (1995) Aluminium production as a source of atmospheric carbonyl sulfide (COS). *Environ. Sci. Pollut. Res.* **2**(3), 161–162.

Hatton A. D. (2002) Influence of photochemistry on the marine biogeochemical cycle of dimethylsulphide in the northern North Sea. *Deep-Sea Res. II.*

Hatton A. D., Turner S. M., Malin G., and Liss P. S. (1998) Dimethylsulphoxide and other biogenic sulphur compounds in the Galapagos Plume. *Deep-Sea Res. II: Top. Stud. Oceanogr.* **45**(6), 1043–1053.

Havlin J. L., D B. J., Tisdale S. L., and Nelson W. L. (1999) *Soil Fertility and Fertilizers*. Prentice-Hall, Upper Saddle River, NJ.

Hines M. E. (1992) Emissions of biogenic sulfur gases from Alaskan tundra. *J. Geophys. Res.: Atmos.* **97**(D15), 16703–16707.

Holser W. T., Maynard J. B., and Cruikshank K. M. (1989) Modelling the natural cycle of sulfur through phanerozoic time. In *Evolution of the Global Biogeochemical Sulphur Cycle* (eds. P. Brimblecombe and A. Y. Lein). Wiley, Chichester, vol. 39, pp. 21–56.

Hong Y. and Fegley B. (1997) Formation of carbonyl sulfide (OCS) from carbon monoxide and sulfur vapor and applications to Venus. *Icarus* **130**(2), 495–504.

Huang J. and Mabury S. A. (2000a) The role of carbonate radical in limiting the persistence of sulfur-containing chemicals in sunlit natural waters. *Chemosphere* **41**(11), 1775–1782.

Huang J. P. and Mabury S. A. (2000b) Hydrolysis kinetics of fenthion and its metabolites in buffered aqueous media. *J. Agri. Food Chem.* **48**(6), 2582–2588.

Huggins F. E., Shah N., Huffman G. P., and Robertson J. D. (2000) XAFS spectroscopic characterization of elements in combustion ash and fine particulate matter. *Fuel Process. Technol.* **65**, 203–218.

Jannasch H. W. (1989) Sulphur emission and transformations at deep sea hydrothermal vents. In *Evolution of the Global Biogeochemical Sulphur Cycle* (eds. P. Brimblecombe and A. Y. Lein). Wiley, Chichester, vol. 39, pp. 181–190.

Jourdain B. and Legrand M. (2001) Seasonal variations of atmospheric dimethylsulfide, dimethylsulfoxide, sulfur dioxide, methanesulfonate, and non-seasalt sulfate aerosols at Dumont d'Urville (coastal Antarctica) (December 1998 to July 1999). *J. Geophys. Res.: Atmos.* **106**(D13), 14391–14408.

Keller L. P., Hony S., Bradley J. P., Molster F. J., Waters L., Bouwman J., de Koter A., Brownlee D. E., Flynn G. J., Henning T., and Mutschke H. (2002) Identification of iron sulphide grains in protoplanetary disks. *Nature* **417**(6885), 148–150.

Kelly D. P. and Murrell J. C. (1999) Microbial metabolism of methanesulfonic acid. *Arch. Microbiol.* **172**(6), 341–348.

Kettle A. J. and Andreae M. O. (2000) Flux of dimethylsulfide from the oceans: a comparison of updated data seas and flux models. *J. Geophys. Res. Atmos.* **105**(D22), 26793–26808.

Kettle A. J., Andreae M. O., Amouroux D., Andreae T. W., Bates T. S., Berresheim H., Bingemer H., Boniforti R., Curran M. A. J., DiTullio G. R., Helas G., Jones G. B., Keller M. D., Kiene R. P., Leck C., Levasseur M., Malin G., Maspero M., Matrai P., McTaggart A. R., Mihalopoulos N., Nguyen B. C., Novo A., Putaud J. P., Rapsomanikis S., Roberts G., Schebeske G., Sharma S., Simo R., Staubes R., Turner S., and Uher G. (1999) A global database of sea surface dimethylsulfide (DMS) measurements and a procedure to predict sea surface DMS as a function of latitude, longitude, and month. *Global Biogeochem. Cycles* **13**, 399–444.

Kieber D. J., Jiao J., Kiene R. P., and Bates T. S. (1996) Impact of dimethylsulfide photochemistry on methyl sulfur cycling in the equatorial Pacific Ocean. *J. Geophys. Res.* **101**, 3715–3722.

Kiene R. P. (1990) Dimethyl sulfide production from dimethylsufoniopropionate in coastal seawater samples and bacterial cultures. *Appl. Environ. Microbiol.* **56**, 3292–3297.

Kiene R. P. and Linn L. J. (2000) The fate of dissolved dimethylsulfoniopropionate (DMSP) in seawater: tracer studies using S-35-DMSP. *Geochim. Cosmochim. Acta* **64**(16), 2797–2810.

Kiene R. P., Linn L. J., and Bruton J. A. (2000) New and important roles for DMSP in marine microbial communities. *J. Sea Res.* **43**(3–4), 209–224.

Kilburn K. H. and Warshaw R. H. (1995) Hydrogen-sulfide and reduced-sulfur gases adversely affect neurophysiological functions. *Toxicol. Ind. Health* **11**(2), 185–197.

Killham K. (1994) *Soil Ecology*. Cambridge University Press, Cambridge.

Kjellstrom E. (1998) A three-dimensional global model study of carbonyl sulfide in the troposphere and the lower stratosphere. *J. Atmos. Chem.* **29**(2), 151–177.

Kusakabe M., Komoda Y., Takano B., and Abiko T. (2000) Sulfur isotopic effects in the disproportionation reaction of sulfur dioxide in hydrothermal fluids: implications for the delta S-34 variations of dissolved bisulfate and elemental sulfur from active crater lakes. *J. Volcanol. Geotherm. Res.* **97**(1–4), 287–307.

Ledyard K. M. and Dacey J. W. H. (1996) Micronial cycling of DMSP and DMS in coastal and oligotrophic seawater. *Limonol. Oceanogr.* **41**, 33–40.

Legrand M., Preunkert S., Wagenbach D., and Fischer H. (2002) Seasonally resolved Alpine and Greenland ice core records of anthropogenic HCl emissions over the 20th century. *J. Geophys. Res.: Atmos.* **107**(D12), article no. 4139.

Licht S., Longo K., Peramunage D., and Forouzan F. (1991) Conductometric analysis of the 2nd acid dissociation-constant of H2S in highly concentrated aqueous-media. *J. Electroanal. Chem.* **318**, 111–129.

Likens G. E., Driscoll C. T., Buso D. C., Mitchell M. J., Lovett G. M., Bailey S. W., Siccama T. G., Reiners W. A., and Alewell C. (2002) The biogeochemistry of sulfur at Hubbard Brook. *Biogeochemistry* **60**(3), 235–316.

Lindgren W. (1923) Concentration and circulation of the elements from the standpoint of *Econ. Geol.* **18**, 419–442.

Lovelock J. E., Maggs R. J., and Rasmussen R. A. (1972) Atmospheric dimethyl sulfide and the natural sulfur cycle. *Nature* **237**, 452–453.

Malin G. and Kirst G. O. (1997) Algal production of dimethyl sulfide and its atmospheric role. *J. Phycol.* **33**, 889–896.

Mauersberger R., Henkel C., Langer N., and Chin Y. N. (1996) Interstellar S-36: a probe of s-process nucleosynthesis. *Astron. Astrophys.* **313**, L1–L4.

Meybeck M. (1987) Global chemical-weathering of surficial rocks estimated from river dissolved loads. *Am. J. Sci.* **287**(5), 401–428.

Migdisov A. A. and Bychkov A. Y. (1998) The behaviour of metals and sulphur during the formation of hydrothermal mercury-antimony-arsenic mineralization, Uzon caldera, Kamchatka, Russia. *J. Volcanol. Geotherm. Res.* **84**, 153–171.

Migdisov A. A., Williams-Jones A. E., Lakshtanov L. Z., and Alekhin Y. V. (2002) Estimates of the second dissociation constant of H$_2$S from the surface sulfidation of crystalline sulfur. *Geochim. Cosmochim. Acta* **66**, 1713–1725.

Millero F. J. and Hawke D. J. (1992) Ionic interactions of divalent metals in natural-waters. *Mar. Chem.* **40**(1–2), 19–48.

Montegrossi G., Tassi F., Vaselli O., Buccianti A., and Garofalo K. (2001) Sulfur species in volcanic gases. *Anal. Chem.* **73**(15), 3709–3715.

Morra M. J., Fendorf S. E., and Brown P. D. (1997) Speciation of sulfur in humic and fulvic acids using X-ray absorption near-edge structure (XANES) spectroscopy. *Geochim. Cosmochim. Acta* **61**(3), 683–688.

Moses J. I., Allen M., and Gladstone G. R. (1995) Post-Sl9 sulfur photochemistry on Jupiter. *Geophys. Res. Lett.* **22**(12), 1597–1600.

Muezzinoglu A., Sponza D., Koken I., Alparslan N., Akyarli A., and Ozture N. (2000) Hydrogen sulfide and odor control in Izmir Bay. *Water Air Soil Pollut.* **123**(1–4), 245–257.

Neretin L. N., Volkov , II, Bottcher M. E., and Grinenko V. A. (2001) A sulfur budget for the Black Sea anoxic zone. *Deep-Sea Res.: I. Oceanogr. Res. Pap.* **48**(12), 2569–2593.

O'Hara S. L., Wiggs G. F. S., Mamedov B., Davidson G., and Hubbard R. B. (2000) Exposure to airborne dust contaminated with pesticide in the Aral Sea region. *Lancet* **355**(9204), 627–628.

Oppenheimer C. and Stevenson D. (1989) Liquid sulfur lakes at Poas volcano. *Nature* **342**(6251), 790–793.

Patris N., Delmas R. J., Legrand M., De Angelis M., Ferron F. A., Stievenard M., and Jouzel J. (2002) First sulfur isotope measurements in central Greenland ice cores along the preindustrial and industrial periods. *J. Geophys. Res.: Atmos.* **107**(D11), article no. 4115.

Phillips D. J. and Phillips S. L. (2000) High temperature dissociation constants of HS and the standard thermodynamic values for S2. *J. Chem. Eng. Data* **45**, 981–987.

Pos W. H., Riemer D. D., and Zika R. G. (1998) Carbonyl sulfide (OCS) and carbon monoxide (CO) in natural waters: evidence of a coupled production pathway. *Mar. Chem.* **62**(1–2), 89–101.

Radfordknoery J. and Cutter G. A. (1994) Biogeochemistry of dissolved hydrogen-sulfide species and carbonyl sulfide in the western North-Atlantic Ocean. *Geochim. Cosmochim. Acta* **58**(24), 5421–5431.

Railsback L. B. (1992) A geological numerical-model for paleozoic global evaporite deposition. *J. Geol.* **100**(3), 261–277.

Rankama K. and Sahama T. G. (1950) *Geochemistry*. University of Chicago Press, Chicago.

Richards S. R., Rudd J. W. M., and Kelly C. A. (1994) Organic volatile sulfur in lakes ranging in sulfate and dissolved salt concentration over 5 orders of magnitude. *Limnol. Oceanogr.* **39**(3), 562–572.

Robinson B. W. and Bottrell S. H. (1997) Discrimination of sulfur sources in pristine and polluted New Zealand river catchments using stable isotopes. *Appl. Geochem.* **12**(3), 305–319.

Robock A. (2000) Volcanic eruptions and climate. *Rev. Geophys.* **38**(2), 191–219.

Rodhe H. (1999) Human impact on the atmospheric sulfur balance. *Tellus Ser. a: Dyn. Meteorol. Oceanogr.* **51**(1), 110–122.

Russell C. T. and Kivelson M. G. (2001) Evidence for sulfur dioxide, sulfur monoxide, and hydrogen sulfide in the Io exosphere. *J. Geophys. Res.: Planets* **106**, 33267–33272.

Sammut J., White I., and Melville M. D. (1996) Acidification of an estuarine tributary in eastern Australia due to drainage of acid sulfate soils. *Mar. Freshwater Res.* **47**(5), 669–684.

Scaillet B., Clemente B., Evans B. W., and Pichavant M. (1998) Redox control of sulfur degassing in silicic magmas. *J. Geophys. Res.: Solid Earth* **103**(B10), 23937–23949.

Schippers A. and Jorgensen B. B. (2001) Oxidation of pyrite and iron sulfide by manganese dioxide in marine sediments. *Geochim. Cosmochim. Acta* **65**(6), 915–922.

Sciare J., Baboukas E., Hancy R., Mihalopoulos N., and Nguyen B. C. (1998) Seasonal variation of dimethylsulfoxide in rainwater at Amsterdam island in the southern Indian Ocean: implications on the biogenic sulfur cycle. *J. Atmos. Chem.* **30**(2), 229–240.

Sciare J., Kanakidou M., and Mihalopoulos N. (2000) Diurnal and seasonal variation of atmospheric dimethylsulfoxide at Amsterdam Island in the southern Indian Ocean. *J. Geophys. Res.: Atmos.* **105**(D13), 17257–17265.

Shooter D. (1999) Sources and sinks of oceanic hydrogen sulfide—an overview. *Atmos. Environ.* **33**(21), 3467–3472.

Shooter D. and Brimblecombe P. (1989) Dimethylsulfide oxidation in the Ocean. *Deep-Sea Res. a: Oceanogr. Res. Pap.* **36**(4), 577–585.

Sidle W. C., Roose D. L., and Shanklin D. R. (2000) Isotopic evidence for naturally occurring sulfate pollution of ponds in the Kankakee River Basin, Illinois-Indiana. *J. Environ. Qual.* **29**(5), 1594–1603.

Siegel B. Z., Nachbarhapai M., and Siegel S. M. (1990) The Contribution of sulfate to rainfall pH around Kilauea volcano, Hawaii. *Water Air Soil Pollut.* **52**(3–4), 227–235.

Siegel S. M., Penny P., Siegel B. Z., and Penny D. (1986) Atmospheric hydrogen-sulfide levels at the Sulfur Bay wildlife area, Lake Rotorua, New-Zealand. *Water Air Soil Pollut.* **28**(3–4), 385–391.

Simo R., Pedros-Alio C., Malin G., and Grimalt J. O. (2000) Biological turnover of DMS, DMSP, and DMSO in contrasting open-sea waters. *Mar. Ecol.: Prog. Ser.* **203**, 1–11.

Simon M., Martin F., Ortiz I., Garcia I., Fernandez J., Fernandez E., Dorronsoro C., and Aguilar J. (2001) Soil pollution by oxidation of tailings from toxic spill of a pyrite mine. *Sci. Tot. Environ.* **279**(1–3), 63–74.

Sinha S. N. and Banerjee R. D. (1997) Ecological role of thiosulfate and sulfide utilizing purple nonsulfur bacteria of a riverine ecosystem. *FEMS Microbiol. Ecol.* **24**(3), 211–220.

Spencer J. R. and Schneider N. M. (1996) Io on the eve of the Galileo mission. *Ann. Rev. Earth Planet. Sci.* **24**, 125–190.

Sriwana T., van Bergen M. J., Varekamp J. C., Sumarti S., Takano B., van Os B. J. H., and Leng M. J. (2000) Geochemistry of the acid Kawah Putih lake, Patuha volcano, West Java, Indonesia. *J. Volcanol. Geotherm. Res.* **97**(1–4), 77–104.

Stefanova M., Marinov S. P., Mastral A. M., Callen M. S., and Garcia T. (2002) Emission of oxygen, sulphur and nitrogen containing heterocyclic polyaromatic compounds from lignite combustion. *Fuel Process. Technol.* **77**, 89–94.

Sudo M., Kunimatsu T., and Okubo T. (2002) Concentration and loading of pesticide residues in Lake Biwa Basin (Japan). *Water Res.* **36**(1), 315–329.

Sunda W., Kieber D. J., Kiene R. P., and Huntsman S. (2002) An antioxidant function for DMSP and DMS in marine algae. *Nature* **418**(6895), 317–320.

Suzuki Y., Kawakami M., and Akasaka K. (2001) H-1 NMR application for characterizing water-soluble organic compounds in urban atmospheric particles. *Environ. Sci. Technol.* **35**(13), 2656–2664.

Symonds R. B. and Reed M. H. (1993) Calculation of multicomponent chemical-equilibria in gas-solid- liquid systems- calculation methods, thermochemical data, and applications to studies of high-temperature volcanic gases with examples from Mount St-Helens. *Am. J. Sci.* **293**(8), 758–864.

Takano B., Ohsawa S., and Glover R. B. (1994) Surveillance of Ruapehu Crater Lake, New-Zealand, by aqueous polythionates. *J. Volcanol. Geotherm. Res.* **60**(1), 29–57.

Tanaka N. and Turekian K. K. (1991) Use of Cosmogenic S-35 to determine the rates of removal of atmospheric SO_2. *Nature* **352**(6332), 226–228.

Thayer M. R. (1997) An investigation into sulfur isotopes in the galactic cosmic rays. *Astrophys. J.* **482**, 792–795.

Thiermann F., Niemeyer A. S., and Giere O. (1996) Variations in the sulfide regime and the distribution of macrofauna in an intertidal flat in the North Sea. *Helgolander Meeresuntersuchungen* **50**(1), 87–104.

Turekian K. K. and Tanaka N. (1992) The use of atmospheric cosmogenic S-35 and Be-7 in determining depositional fluxes of SO_2. *Geophys. Res. Lett.* **19**(17), 1767–1770.

TWRI (1986) *Threats to Groundwater Quality.* Texas Water Resources Institute, College Station, TX.

USGS. (2000) *Volcanic Air Pollution—a Hazard in Hawaii.* US Geological Survey, Hawaii Volcanoes National Park.

Vialaton D. and Richard C. (2000) Direct photolyses of thiobencarb and ethiofencarb in aqueous phase. *J. Photochem. Photobiol. a: Chem.* **136**(3), 169–174.

Wachtershauser G. (2000) Life as we don't know it. *Science* **289**(5483), 1307–1308.

Warneck P. (1999) *Chemistry of the Natural Atmosphere.* Academic Press, San Diego.

Watson A. J. and Liss P. S. (1998) Marine biological controls on climate via the carbon and sulphur geochemical cycles. *Phil. Trans. Roy. Soc. London Ser. B: Biol. Sci.* **353**(1365), 41–51.

Watts S. F. (2000) The mass budgets of carbonyl sulfide, dimethyl sulfide, carbon disulfide and hydrogen sulfide. *Atmos. Environ.* **34**(5), 761–779.

Watts S. F. and Roberts C. N. (1999) Hydrogen sulfide from car catalytic converters. *Atmos. Environ.* **33**(1), 169–170.

Wawrzynkiewicz W. and Sablik J. (2002) Organic sulphur in the hard coal of the stratigraphic layers of the Upper Silesian coal basin. *Fuel* **81**(14), 1889–1895.

Wedepohl K. H. (1995) The composition of the continental-crust. *Geochim. Cosmochim. Acta* **59**(7), 1217–1232.

Werne J. P., Hollander D. J., Behrens A., Schaeffer P., Albrecht P., and Damste J. S. S. (2000) Timing of early diagenetic sulfurization of organic matter: a precursor-product relationship in Holocene sediments of the anoxic Cariaco Basin, Venezuela. *Geochim. Cosmochim. Acta* **64**(10), 1741–1751.

West W. R., Smith P. A., Booth G. M., and Lee M. L. (1988) Isolation and detection of genotoxic components in a Black River sediment. *Environ. Sci. Technol.* **22**(2), 224–228.

Wheat C. G. and Mottl M. J. (2000) Composition of pore and spring waters from Baby Bare: global implications of geochemical fluxes from a ridge flank hydrothermal system. *Geochim. Cosmochim. Acta* **64**(4), 629–642.

White C. L., Tabe L. M., Dove H., Hamblin J., Young P., Phillips N., Taylor R., Gulati S., Ashes J., and Higgins T. J. V. (2001) Increased efficiency of wool growth and live weight gain in Merino sheep fed transgenic lupin seed containing sunflower albumin. *J. Sci. Food Agri.* **81**(1), 147–154.

Xie H. X. and Moore R. M. (1999) Carbon disulfide in the North Atlantic and Pacific Oceans. *J. Geophys. Res.: Oceans* **104**(C3), 5393–5402.

Xie H. X., Moore R. M., and Miller W. L. (1998) Photochemical production of carbon disulphide in seawater. *J. Geophys. Res.: Oceans* **103**(C3), 5635–5644.

Yang G. P., Liu X. L., and Zhang J. W. (1998) Distribution of dibenzothiophene in the sediments of the South China Sea. *Environ. Pollut.* **101**(3), 405–414.

Yao W. S. and Millero F. J. (1996) Oxidation of hydrogen sulfide by hydrous Fe(III) oxides in seawater. *Mar. Chem.* **52**(1), 1–16.

Yvon S. A., Saltzman E. S., Cooper D. J., Bates T. S., and Thompson A. M. (1996) Atmospheric sulfur cycling in the tropical Pacific marine boundary layer (12 degrees S, 135 degrees W): a comparison of field data and model results:1. Dimethylsulfide. *J. Geophys. Res.: Atmos.* **101**(D3), 6899–6909.

Zhang L., Walsh R. S., and Cutter G. A. (1998) Estuarine cycling of carbonyl sulfide: production and sea-air flux. *Mar. Chem.* **61**(3–4), 127–142.

Zolotov M. Y. and Fegley B. (1998) Volcanic production of sulfur monoxide (SO) on Io. *Icarus* **132**, 431–434.

Geochemistry of Earth Surface Systems
ISBN: 978-0-08-096706-6

pp. 129–166

5

The History of Planetary Degassing as Recorded by Noble Gases

D. Porcelli

University of Oxford, UK

and

K. K. Turekian

Yale University, New Haven, CT, USA

5.1	INTRODUCTION	168
5.2	PRESENT-EARTH NOBLE GAS CHARACTERISTICS	168
	5.2.1 Surface Inventories	169
	5.2.2 Helium Isotopes	169
	5.2.3 Neon Isotopes	170
	5.2.4 Argon Isotopes	171
	5.2.5 Xenon Isotopes	173
	5.2.6 Noble Gas Abundance Patterns	174
	5.2.7 MORB Fluxes and Upper-mantle Concentrations	175
	5.2.8 Other Mantle Fluxes	176
	5.2.9 Subduction Fluxes	177
5.3	BULK DEGASSING OF RADIOGENIC ISOTOPES	178
	5.3.1 The ^{40}K–^{40}Ar Budget	178
	5.3.2 The ^{129}I–^{129}Xe and ^{244}Pu–^{136}Xe Budgets	178
5.4	DEGASSING OF THE MANTLE	179
	5.4.1 Early Earth Degassing	180
	5.4.2 Degassing from One Mantle Reservoir	180
	5.4.3 Multiple Mantle Reservoirs	181
	5.4.4 Interacting Reservoirs	183
	5.4.5 Open-system Models	185
	5.4.6 Boundaries within the Mantle	187
	5.4.7 Summary	187
5.5	DEGASSING OF THE CRUST	188
	5.5.1 Crustal Potassium and ^{40}Ar Budget	188
	5.5.2 Formation Time of the Crust	189
	5.5.3 Present Degassing	189
5.6	MAJOR VOLATILE CYCLES	190
	5.6.1 Carbon	190
	5.6.2 Nitrogen	192
	5.6.3 Water	193

5.7 DEGASSING OF OTHER TERRESTRIAL PLANETS   194
    5.7.1 Mars   194
    5.7.2 Venus   196
5.8 CONCLUSIONS   196
ACKNOWLEDGMENT   197
REFERENCES   197

5.1 INTRODUCTION

Noble gases provide unique clues to the structure of the Earth and the degassing of volatiles into the atmosphere. Since the noble gases are highly depleted in the Earth, their isotopic compositions are prone to substantial changes due to radiogenic additions, even from scarce parent elements and low-yield nuclear processes. Therefore, noble gas isotopic signatures of major reservoirs reflect planetary differentiation processes that generate fractionations between these volatiles and parent elements. These signatures can be used to construct planetary degassing histories that have relevance to the degassing of a variety of chemical species as well.

It has long been recognized that the atmosphere is not simply a remnant of the volatiles that surrounded the forming Earth with the composition of the early solar nebula. It was also commonly thought that the atmosphere and oceans were derived from degassing of the solid Earth over time (Brown, 1949; Suess, 1949; Rubey, 1951). Subsequent improved understanding of the processes of planet formation, however, suggests that substantial volatile inventories could also have been added directly to the atmosphere. The characteristics of the atmosphere therefore reflect the acquisition of volatiles by the solid Earth during formation (see Pepin and Porcelli, 2002), as well as the history of degassing from the mantle. The precise connection between volatiles now emanating from the Earth and the long-term evolution of the atmosphere are key subjects of modeling efforts, and are discussed below.

Major advances in understanding the behavior of terrestrial volatiles have been made based upon observations on the characteristics of noble gases that remain within the Earth. Various models have been constructed that define different components and reservoirs in the planetary interior, how materials are exchanged between them, and how the noble gases are progressively transferred to the atmosphere. While there remain many uncertainties, an overall process of planetary degassing can be discerned. The present chapter discusses the constraints provided by the noble gases and how these relate to the degassing of the volatile molecules formed from nitrogen, carbon, and hydrogen (see also Chapter 1). The evolution of particular atmospheric molecular species, such as CO_2, that

are controlled by interaction with other crustal reservoirs and which reflect surface chemical conditions, are primarily discussed elsewhere (Chapter 10).

Noble gases provide the most detailed constraints on planetary degassing. A description of the available noble gas data that must be incorporated into any Earth degassing history is provided first in Section 5.2, and the constraints on the total extent of degassing of the terrestrial interior are provided in Section 5.3. Noble gas degassing models that have been used to describe and calculate degassing histories of both the mantle (Section 5.4) and the crust (Section 5.5) are then presented. These discussions then provide the context for an evaluation of major volatile cycles in the Earth (Section 5.6), and speculations about the degassing of the other terrestrial planets (Section 5.7), Mars and Venus, that are obviously based on much more limited data. The processes controlling mantle degassing are clearly related to the structure of the mantle, as discussed in Section 5.4. An important aspect is the origin of planetary volatiles and whether initial incorporation was into the solid Earth or directly to the atmosphere. Basic noble gas elemental and isotopic characteristics are given in Ozima and Podosek (2001) and Porcelli *et al.* (2002). The major nuclear processes that produce noble gases within the solid Earth, and the half-lives of the major parental nuclides, are given in Table 1.

5.2 PRESENT-EARTH NOBLE GAS CHARACTERISTICS

There are various terrestrial reservoirs that have distinct volatile characteristics. Data from mid-ocean ridge basalts (MORBs) characterize the underlying convecting upper mantle, and are described here without any assumptions about the depth of this reservoir. Other mantle reservoirs are sampled by ocean island basalts (OIBs) and may represent a significant fraction of the mantle. Note that significant krypton isotopic variations due to radiogenic additions are neither expected nor observed, and there are no isotopic fractionation observed between any terrestrial noble gas reservoirs. Therefore, no constraints on mantle degassing can be obtained from krypton, and so krypton is not discussed further.

Table 1 Major nuclear processes producing noble-gas isotopes in the solid earth.[a]

Daughter	Nuclear process	Parent half-life	Yield (atoms/ decay)	Comments
^{3}He	^{6}Li(n, α)^{3}H(β-)^{3}He			^{3}He/^{4}He $= 1 \times 10^{-8b}$
^{4}He	α-decay of ^{238}U decay series nuclides	4.468 Ga	8[c]	
^{4}He	α-decay of ^{235}U decay series nuclides	0.7038 Ga	7[c]	^{238}U/^{235}U $= 137.88$
^{4}He	α-decay of ^{232}Th decay series nuclides	14.01 Ga	6[c]	Th/U $= 3.8$ in bulk Earth
^{21}Ne	^{18}O(α, n)^{21}Ne			^{21}Ne/^{4}He $= 4.5 \times 10^{-8b}$
^{21}Ne	^{24}Mg(n, α)^{21}Ne			^{21}Ne/^{4}He $= 1 \times 10^{-10b}$
^{40}Ar	^{40}K β^- decay	1.251 Ga	0.1048[b]	^{40}K $= 0.01167\%$ total K
^{129}Xe	^{129}I β^- decay	15.7 Ma	1	^{129}I/^{127}I $= 1.1 \times 10^{-4}$ at 4.56 Ga[d]
^{136}Xe	^{238}U spontaneous fission		4×10^{-8e}	
^{136}Xe	^{244}Pu spontaneous fission	80.0 Ma	7.00×10^{-5}	^{244}Pu/^{238}U $= 6.8 \times 10^{-3}$ at 4.56 Ga[f]

[a] From data compilations of Blum (1995), Ozima and Podosek (2001), and Pfennig *et al.* (1998).
[b] Production ratio for upper crust (Ballentine and Burnard, 2002).
[c] Per decay of series parent, assuming secular equilibrium for entire decay series.
[d] Hohenberg *et al.* (1967).
[e] Eikenberg *et al.* (1993) and Ragettli *et al.* (1994).
[f] Hudson *et al.* (1989).

5.2.1 Surface Inventories

The atmosphere is the largest accessible terrestrial noble gas reservoir, and its composition serves as a reference for measurements of other materials. The major volatile molecules of carbon, nitrogen, and hydrogen, have considerable inventories in the crust that are part of the volatile budget that has been either degassed from the mantle or initially incorporated into the atmosphere. The total surface inventory is summarized in Table 2 and includes the atmosphere, hydrosphere, and continental crust. Since helium is lost from the atmosphere, the atmospheric abundance has no significance for determining long-term evolution. Atmospheric isotopic compositions, which are generally used as standards for comparison and measurement normalization, are provided in Table 3.

5.2.2 Helium Isotopes

There are two isotopes of helium. In addition to the cosmologically produced ^{4}He and ^{3}He, ^{4}He is produced as α-particles during radioactive decay of various parent radionuclides, and the much less abundant ^{3}He is produced from ^{6}Li (Tables 1 and 2). Overall, radiogenic helium is primarily ^{4}He, with a ratio of ^{3}He/^{4}He $\sim 0.01R_A$ (Morrison and Pine, 1955), where R_A is the air value of 1.39×10^{-6}. The initial value for the Earth depends upon the origin of terrestrial noble gases, and is presumed to be that of the solar nebula of ^{3}He/^{4}He $= 120R_A$ (Mahaffy *et al.*, 1998). The solar wind value of $330R_A$ (Benkert *et al.*, 1993) was established after deuterium burning in the Sun; if terrestrial helium was captured after significant deuterium burning,

Table 2 Volatile surface inventories.

Constituent	Atmosphere (mol)	Crust (mol)
N	2.760×10^{20}	4×10^{19}
O$_2$	3.702×10^{19}	
Ar	1.651×10^{18}	
C	5.568×10^{16}	8.3×10^{21}
Ne	3.213×10^{15}	
He	9.262×10^{14}	
Kr	2.015×10^{14}	
Xe	1.537×10^{13}	

Note: Based on dry tropospheric air. Water generally accounts for $\leq 4\%$ of air. Other chemical constituents have mixing ratios less than Xe. Data from compilation by Ozima and Podosek (2001). C atmosphere data from Keeling and Whorf (2000). Crustal C from Hunt (1972) and Ronov and Yaroshevsky (1976). Crustal N from Marty and Dauphas (2003).

then this higher value would be the composition of the initial helium. The first clear evidence for the degassing of primordial volatiles still remaining within the solid Earth came from helium isotopes (Figure 1). MORB has an average of $8R_A$ (Clarke *et al.*, 1969; Mamyrin *et al.*, 1969; see Graham, 2002), and so is a mixture of radiogenic helium (that accounts for most of the ^{4}He) with initially trapped "primordial" helium (that accounts for most of the ^{3}He).

OIB has more variable ^{3}He/^{4}He ratios. Some are below those of MORB, probably due to radiogenic recycled components (e.g., Kurz *et al.*, 1982; Hanyu and Kaneoka, 1998). Due to their limited occurrence, these values are likely to represent only a small fraction of the total mantle. ^{3}He/^{4}He ratios greater than $\sim 10R_A$ provide evidence for a long-term noble gas reservoir distinct from MORB that has a time-integrated ^{3}He/(U + Th) ratio greater than that of the upper mantle (Kurz *et al.*, 1982; Allègre *et al.*, 1983) and so

Table 3 Noble-gas and major volatile isotope composition of the atmosphere.

Isotope	Relative abundances	Percent molar abundance
^{3}He	$(1.399 \pm 0.013) \times 10^{-6}$	0.000140
^{4}He	$\equiv 1$	100
^{20}Ne	9.80 ± 0.08	90.50
^{21}Ne	0.0290 ± 0.0003	0.268
^{22}Ne	$\equiv 1$	9.23
^{36}Ar	$\equiv 1$	0.3364
^{38}Ar	0.1880 ± 0.0004	0.0632
^{40}Ar	295.5 ± 0.5	99.60
^{78}Kr	0.6087 ± 0.0020	0.3469
^{80}Kr	3.9599 ± 0.0020	2.2571
^{82}Kr	20.217 ± 0.004	11.523
^{83}Kr	20.136 ± 0.021	11.477
^{84}Kr	$\equiv 100$	57.00
^{86}Kr	30.524 ± 0.025	17.398
^{124}Xe	2.337 ± 0.008	0.0951
^{126}Xe	2.180 ± 0.011	0.0887
^{128}Xe	47.15 ± 0.07	1.919
^{129}Xe	649.6 ± 0.9	26.44
^{130}Xe	$\equiv 100$	4.070
^{131}Xe	521.3 ± 0.8	21.22
^{132}Xe	660.7 ± 0.5	26.89
^{134}Xe	256.3 ± 0.4	10.430
^{136}Xe	217.6 ± 0.3	8.857
^{14}N	0.0037	0.37
^{15}N	$\equiv 1$	99.63
^{12}C	$\equiv 1$	98.63
^{13}C	0.0113	1.11
^{1}H	$\equiv 1$	99.985
^{2}H	0.00015	0.015

After Ozima and Podosek (2001) and Porcelli *et al.* (2002).

leads to the highest ^{3}He/^{4}He ratios of 32–38R_A found in Loihi Seamount, the youngest Hawaiian volcano (Kurz *et al.*, 1982; Rison and Craig, 1983; Honda *et al.*, 1993; Valbracht *et al.*, 1997), and Iceland (Hilton *et al.*, 1998b). A major issue has been determining the nature of this reservoir, the abundances of noble gases it contains, and how it degasses (see Section 5.4).

Helium isotopes have a ∼1 Myr residence time in the atmosphere prior to loss to space; therefore, their atmospheric abundances do not contain information about the integrated degassing history of the Earth. However, the large variations in helium isotope compositions in the Earth constrain mantle degassing models by (i) providing clear fingerprints of mantle volatile fluxes into the crust and atmosphere (see Sections 5.2.7 and 5.2.8); (ii) requiring several mantle noble gas reservoirs (see Section 5.4.3); (iii) indicating that the upper mantle is relatively well-mixed with respect to noble gases; and (iv) relating the sources of noble gases with heat production, since uranium and thorium are the dominant sources of both ^{4}He and heat in the mantle (see Section 5.2.7).

5.2.3 Neon Isotopes

There are three neon isotopes. The more abundant ^{20}Ne and ^{22}Ne are both essentially all primordial, as there is no significant global production of these isotopes. In contrast, ^{21}Ne is produced by nuclear reactions. In mantle-derived materials, measured ^{20}Ne/^{22}Ne ratios are greater than that of the atmosphere of 9.8, and extend toward the values of the solar wind (13.8) or implanted solar wind (12.5) (Figure 2). Since these isotopes are not produced in significant quantities in the Earth, this is unequivocal evidence for storage in the Earth of at least one nonradiogenic mantle component that is distinctive from the atmosphere and has remained trapped separately since formation of the Earth. It is likely that the ^{20}Ne/^{22}Ne ratio of the atmosphere was originally similar to the higher values now found in the mantle, and was fractionated during losses to space. This could only have occurred early in Earth history. Since the neon remaining in the mantle preserves the original isotopic composition, the difference from that of the atmosphere also limits the amount of mantle neon that can have subsequently degassed. The exact proportion depends upon how much fractionation of atmospheric neon originally occurred. While this is unconstrained, the observed fractionation is already considered quite extreme, and so it is unlikely that much lower ^{20}Ne/^{22}Ne ratios had been generated, and so only a small proportion of mantle neon is likely to have been degassed subsequently.

MORB ^{20}Ne/^{22}Ne and ^{21}Ne/^{22}Ne ratios generally are correlated (Sarda *et al.*, 1988; Moreira *et al.*, 1998), and this is likely due to mixing of variable amounts of air contamination with uniform mantle neon. The upper-mantle ^{21}Ne/^{22}Ne ratio of ∼0.074 is higher than the solar value (0.033) due to additions of nucleogenic ^{21}Ne (Table 1). OIBs with high ^{3}He/^{4}He ratios span a similar range in ^{20}Ne/^{22}Ne ratios, but with lower corresponding ^{21}Ne/^{22}Ne ratios (Sarda *et al.*, 1988; Honda *et al.*, 1991, 1993). The OIB sources therefore have higher time-integrated He/(U + Th) and Ne/(U + Th) ratios (Honda and McDougall, 1993).

The MORB helium and neon isotopic compositions can be used to calculate the ^{3}He/^{22}Ne ratio of the source region prior to any recent fractionations created during transport and eruption. Since the production ratio of ^{4}He to ^{21}Ne is fixed (Table 1), the shifts in ^{3}He/^{4}He and ^{21}Ne/^{22}Ne isotope ratios from the initial, primordial values of the Earth due to radiogenic and nucleogenic additions can be used to calculate the reservoir ^{3}He/^{22}Ne ratio. Using an uncontaminated MORB value of ^{21}Ne/^{22}Ne = 0.074 and ^{21*}Ne/^{4*}He = 4.5×10^{-8}, then ^{3}He/^{22}Ne = 11. Calculating a source value for each MORB and

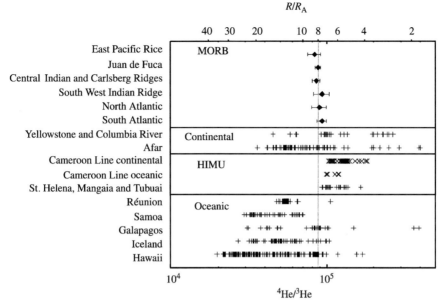

Figure 1 Helium isotope data from various mantle-derived volcanics. The upper axis is the ^{3}He/^{4}He ratio (R) normalized to the atmospheric ratio (R_A). As indicated by the data for selected segments of the MORB away from ocean islands falls almost entirely within the range of $(7-9)R_A$. While there are hotspot basalts that are characterized by high U/Pb ratios and low ^{3}He/^{4}He ratios (HIMU), many major oceanic hotspots, as well as continental hotspots, have high ^{3}He/^{4}He ratios (source Porcelli and Ballentine, 2002).

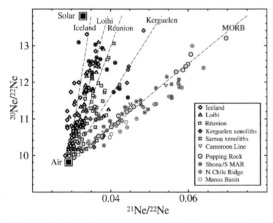

Figure 2 Neon isotope compositions of MORB and selected OIBs. The data generally fall on correlations that extend from air contamination to higher values that characterize the trapped mantle components. Islands with high ^{3}He/^{4}He ratios have lower-mantle ^{21}Ne/^{22}Ne ratios, reflecting high ^{3}He/(U + Th) and ^{22}Ne/(U + Th) ratios (source Graham, 2002).

OIB sample individually, a mantle average of 7.7 was found (Honda and McDougall, 1998). For comparison, the solar nebula value is ^{3}He/^{22}Ne = 1.9 (see Porcelli and Pepin, 2000). This value can be used to relate degassing of helium with the other noble gases (see Section 5.2.7).

Overall, neon isotopes clearly identify the noble gases presently degassing from the mantle as solar in origin. Taking this composition as the original value of the neon in the atmosphere, most of the atmospheric neon was degassed very early in Earth history and suffered substantial fractionating losses. The atmospheric noble gas inventory was therefore highly modified from that which was originally degassed from the solid Earth or added to the surface. Since noble gases generally behave similarly within the mantle and during degassing (see Section 5.2.6), the constraints on neon degassing can be applied to the other noble gases.

5.2.4 Argon Isotopes

The two minor isotopes of argon, ^{36}Ar and ^{38}Ar, are essentially all primordial, with no significant radiogenic production on a global scale. The initial ^{40}Ar/^{36}Ar ratio of the solar system was $<10^{-3}$ (Begemann *et al.*, 1976) and orders of magnitude less than any planetary values, so essentially all ^{40}Ar is radiogenic. A large range in ^{40}Ar/^{36}Ar ratios has been measured in MORBs that is likely due to mixing of variable proportions of air argon (with ^{40}Ar/^{36}Ar = 296) with a single, more radiogenic, mantle composition (Figure 3). The minimum value for this mantle composition is represented by the highest measured values of 2.8×10^4 (Staudacher *et al.*, 1989) to 4×10^4 (Burnard *et al.*, 1997). From correlations between ^{20}Ne/^{22}Ne and ^{40}Ar/^{36}Ar during step heating of a gas-rich MORB that was designed to separate contaminant air noble gases from those trapped within the glass, a maximum value of ^{40}Ar/^{36}Ar = 4.4×10^4 was obtained (Moreira *et al.*, 1998).

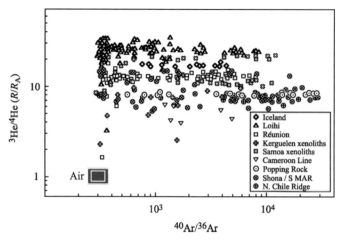

Figure 3 Helium and argon isotope compositions of MORB ("Popping Rock" from the Atlantic, Shona ridge section, and N. Chile ridge) and selected islands. $^{40}Ar/^{36}Ar$ ratios vary widely due to variable amounts of atmospheric contamination (source Graham, 2002).

Like terrestrial $^{20}Ne/^{22}Ne$ ratios, it might be expected that $^{38}Ar/^{36}Ar$ ratios vary due to different initial sources building the Earth or to early fractionation events. Measurements of MORB and OIB $^{38}Ar/^{36}Ar$ ratios typically are atmospheric within error, but have been of low precision due to the low abundance of these isotopes. While some high-precision analyses of MORB and OIB samples show $^{38}Ar/^{36}Ar$ ratios lower than that of the atmosphere and approaching solar values (see Pepin, 1998), but others do not (Kunz, 1999). While nonatmospheric ratios would limit the amount of argon transfer between the mantle and atmosphere, atmospheric ratios in the mantle could be explained either by early trapping of argon that had been fractionated or entrainment of atmospheric argon during subduction or melt formation (see Porcelli and Wasserburg, 1995b).

OIBs with high $^3He/^4He$ ratios reflecting high $^3He/(U + Th)$ ratios have been expected to have low $^{40}Ar/^{36}Ar$ ratios reflecting correspondingly high $^{36}Ar/K$ ratios. Measurements of $^{40}Ar/^{36}Ar$ in OIBs with $^3He/^4He > 10R_A$ are indeed consistently lower than MORB values. However, early values for Loihi glasses of $^{40}Ar/^{36}Ar < 10^3$ appear to reflect overwhelming contamination with air argon (Fisher, 1985; Patterson *et al.*, 1990). A study of basalts from Juan Fernandez (Farley *et al.*, 1993) found atmospheric contamination contained within phenocrysts introduced into the magma chamber, thus providing an explanation for the prevalence of air contamination of OIBs. Recent measurements of Loihi samples found higher $^{40}Ar/^{36}Ar$ values of 2,600–2,800 associated with high $^3He/^4He$ ratios (Hiyagon *et al.*, 1992; Valbracht *et al.*, 1997), while Trieloff *et al.* (2000) found values up to 8,000 on samples with $^3He/^4He = 24$ (and so with a helium composition midway between MORBs and the highest OIBs). Poreda and Farley (1992) found values of

$^{40}Ar/^{36}Ar \leq 1.2 \times 10^4$ in Samoan xenoliths that have intermediate $^3He/^4He$ ratios ($9–20R_A$). Kola Peninsula carbonatites with high $^{20}Ne/^{22}Ne$ ratios were used to calculate a mantle $^{40}Ar/^{36}Ar$ value of 5,000 (Marty *et al.*, 1998). Other attempts to remove the effects of air contamination have used associated neon isotopes and the debatable assumption that the contaminant Ne/Ar ratio is constant, and have also found $^{40}Ar/^{36}Ar$ values substantially lower than in MORBs (Sarda *et al.*, 2000). Overall, it appears that $^{40}Ar/^{36}Ar$ ratios in the high $^3He/^4He$ OIB source are >3,000 but probably <10^4, and so lower than that of the MORB source (see also Matsuda and Marty, 1995).

The unambiguous measurement of past atmospheric $^{40}Ar/^{36}Ar$ ratios would provide an important constraint on the degassing history of the atmosphere. If ^{36}Ar had largely degassed early (see below), then the past atmospheric $^{40}Ar/^{36}Ar$ ratio would reflect the past atmospheric abundance of ^{40}Ar due to subsequent ^{40}Ar degassing. Unfortunately, it has been difficult to find samples that have captured and retained atmospheric argon but do not contain significant amounts of either inherited radiogenic ^{40}Ar or potassium. Nonetheless, various studies have sought to find $^{40}Ar/^{36}Ar$ ratios that are lower than the present atmosphere, and so clearly are not dominated by either present atmosphere contamination or radiogenic ^{40}Ar. Cadogan (1977) reported a value of $^{40}Ar/^{36}Ar = 291.0 \pm 1.5$ from the 380 Myr old Rhynie chert, while Hanes *et al.* (1985) reported 25 ± 83 for an old pyroxenite sill sample from the Abitibi Greenstone Belt that contains amphibole apparently produced during deuteric alteration at 2.70 Ga. Both samples are presumed to have atmospheric argon trapped at these times. Using these data, the maximum possible changes in the atmospheric ^{40}Ar abundance is obtained by assuming that the ^{36}Ar abundance has been constant since that time, so that the differences between these $^{40}Ar/^{36}Ar$ ratios and the present atmospheric value reflect only lower ^{40}Ar abundances. In this case, the Rhynie chert data suggest that the atmospheric ^{40}Ar abundance was 1.5% lower 380 Ma ago than today. At that time, there was 1.9% less ^{40}Ar in the Earth (assuming a bulk Earth potassium concentration of 270 ppm), and it thus appears that the same fraction of terrestrial ^{40}Ar as today was in the atmosphere at that time. At 2.7 Ga ago, the calculated atmospheric ^{40}Ar abundance is 12.7% lower than at present. However, there was 66% less ^{40}Ar 2.7 Ga ago, and assuming a BSE concentration of 270 ppm K, even complete degassing of the entire Earth at that time (compared to 40% today) would not provide sufficient ^{40}Ar for the atmosphere. The alternative that the atmospheric ^{36}Ar abundance has doubled since 2.7 Ga ago is also unlikely if current arguments for early degassing of nonradiogenic nuclides are valid (see Sections 5.2.3 and 5.4.1). However, before pursuing further speculation, the possibility that the Abitibi sample contains either excess ^{40}Ar trapped during formation or subsequently produced radiogenic ^{40}Ar, and so provides an unreasonably high $^{40}Ar/^{36}Ar$ atmospheric ratio for that time, must be discounted.

Overall, argon isotope compositions indicate that the mantle is much more radiogenic than the atmosphere, and this provides an important starting point for degassing history models. It appears that there are variations in $^{40}Ar/^{36}Ar$ associated with different $^{3}He/^{4}He$ ratios in the mantle, although atmospheric contamination of samples has made this difficult to quantify. It appears that the same proportion of ^{40}Ar has been in the atmosphere over the last 380 Ma, although reliable data are required from much earlier to effectively discriminate between different degassing histories (see Section 5.4).

5.2.5 Xenon Isotopes

There are nine isotopes of xenon (see Table 3). On a global scale, there have been additions to ^{129}Xe through decay of ^{129}I ($t_{1/2} = 15.7$ Myr), which as a short-lived nuclide was only present in significant quantities early in Earth history. Additions to the heavy isotopes ^{131}Xe, ^{132}Xe, ^{134}Xe, and ^{136}Xe have also occurred by fission of ^{244}Pu ($t_{1/2} = 80$ Myr), another short-lived nuclide, and ^{238}U ($t_{1/2} = 4.5$ Gyr). The largest fission contributions are to ^{136}Xe, so this isotope is usually used as an index for fissiogenic contributions. Primordial components completely account for the other isotopes and even a dominant proportion of those with later additions.

The starting point for examining xenon isotope systematics is defining the nonradiogenic isotope composition; i.e., the proportion of primordial

xenon underlying the radiogenic and fissiogenic contributions. The light isotopes of atmospheric xenon (^{124}Xe, ^{126}Xe, ^{128}Xe, and ^{130}Xe) are related to both bulk chondritic and solar xenon by very large fractionation of $\sim 4.2\%$ per amu (Krummenacher *et al.*, 1962), which demands a strongly fractionating planetary process. However, both chondritic and solar xenon, when fractionated to match the light isotopes of the atmosphere, have proportionately more ^{134}Xe and ^{136}Xe than presently in the atmosphere and so neither can serve as the primordial terrestrial composition (see Pepin, 2000). There is no other commonly observed solar system composition that can be used to account for atmospheric xenon by simple mass fractionation and the addition of radiogenic and fissiogenic xenon. Pepin (2000) has used isotopic correlations of chondrite data to infer the presence of a mixing component, U–Xe, that has solar light isotope ratios and when highly mass-fractionated yields the light-isotope ratios of terrestrial xenon and is relatively depleted in heavy xenon isotopes. The composition of the present atmosphere then can be obtained from fractionated U–Xe by the addition of a heavy isotope component that has the composition of ^{244}Pu-derived fission xenon (Pepin, 2000). Therefore, the proportions of atmospheric xenon isotopes that are fissiogenic can be calculated. The budgets of radiogenic and fissiogenic xenon in the atmosphere are discussed in detail in Section 5.3.2 below.

MORB $^{129}Xe/^{130}Xe$ and $^{136}Xe/^{130}Xe$ ratios lie on a correlation extending from atmospheric ratios to higher values (Staudacher and Allègre, 1982; Kunz *et al.*, 1998), and likely reflect mixing of variable proportions of contaminant air xenon with an upper-mantle component having more radiogenic $^{129}Xe/^{130}Xe$ and $^{136}Xe/^{130}Xe$ ratios (Figure 4). The highest measured values thus provide lower limits for the MORB source. The MORB data demonstrate that the xenon presently in the atmosphere was in an environment with a higher Xe/I ratio than that of the xenon in the mantle, at least during the lifetime of ^{129}I. As discussed further below, this difference can be generated either by degassing processes or by early differences in mantle reservoirs related to the formation processes of the planet.

Contributions to $^{136*}Xe$ enrichments in MORBs can be from either decay of ^{238}U over Earth history or early ^{244}Pu decay, which in theory can be distinguished based on the spectrum of contributions to other xenon isotopes, although analyses have typically not been sufficiently precise to do so. More precise measurements can be obtained from the abundant xenon in some CO_2 well gases that have ^{129}Xe and ^{136}Xe enrichments similar to those found in MORBs, and are likely to be from the upper mantle (Staudacher, 1987). Precise measurements indicate that ^{244}Pu has

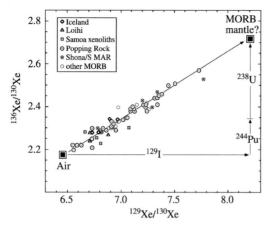

Figure 4 Xenon isotope compositions of MORB and selected ocean islands. The excesses in ^{129}Xe due to ^{129}I, and in ^{136}Xe due to ^{238}U and ^{244}Pu, are correlated due to mixing between mantle xenon and atmospheric contamination. The fraction of ^{136}Xe from ^{244}Pu calculated by Kunz *et al.* (1998) is shown for illustration (source Graham, 2002).

contributed <10–20% of the ^{136}Xe that is in excess of the atmospheric composition (Phinney *et al.*, 1978; Caffee *et al.*, 1999). An error-weighted best fit to recent precise MORB data (Kunz *et al.*, 1998) yielded a value of 32 ± 10% for the fraction of ^{136}Xe excesses relative to the atmospheric composition that are ^{244}Pu-derived, although with considerable uncertainties (Marti and Mathew, 1998). The atmosphere itself, therefore, contains ^{244}Pu-derived ^{136}Xe. Clearly, further work is warranted on the proportion of plutonium-derived heavy xenon in the mantle, although it appears that the fissiogenic xenon is dominantly derived from uranium.

It has proven to be more difficult to characterize the xenon in OIB source regions. Xenon with atmospheric isotopic ratios in high-^{3}He/^{4}He OIB samples (e.g., Allègre *et al.*, 1983) appears to be dominated by air contamination (Patterson *et al.*, 1990; Harrison *et al.*, 1999) rather than represent mantle xenon with an air composition. Although Samoan samples with intermediate (9–20R_A) helium isotope ratios have been found with xenon isotopic ratios distinct from those of the atmosphere (Poreda and Farley, 1992), the xenon in these samples may have been derived largely from the MORB source. Recently, Harrison *et al.* (1999) found slight ^{129}Xe excesses in Icelandic samples with ^{129*}Xe/^{3}He ratios that are compatible with the ratio in a gas-rich MORB, but due to the uncertainties in the data it cannot be determined whether there are indeed differences between the MORB and OIB sources. Trieloff *et al.* (2000) reported xenon isotope compositions in Loihi dunites and Icelandic glasses that were on the MORB correlation line and had values up to ^{129}Xe/^{130}Xe = 6.9. These were accompanied by

^{3}He/^{4}He ratios up to 24R_A, and so may contain noble gases from both MORB (~8R_A) and the highest ^{3}He/^{4}He ratio (37R_A) OIB source. From these data it appears that the OIB source may have xenon that is similar to that in MORB, although there may be some differences that have not been resolvable.

The relatively imprecise measured ratios of the nonradiogenic isotopes in MORB are indistinguishable from those in the atmosphere. However, more precise measurements of mantle-derived xenon in CO_2 well gases have been found to have higher $^{124–128}$Xe/^{130}Xe ratios (Phinney *et al.*, 1978; Caffee *et al.*, 1999) that can be explained by either: (i) a mixture of ~10% xenon trapped within the Earth of solar isotopic composition and ~90% atmospheric xenon (subducted or added in the crust); or (ii) a mantle xenon component that has not been isotopically fractionated relative to solar xenon to the same extent as air xenon.

In sum, nonradiogenic atmospheric xenon isotopes, like those of neon, require that early degassing of noble gases occurred when fractionating losses to space were still operating. Like argon, xenon in the upper mantle is more radiogenic than the atmosphere, and this must be a feature of any reasonable degassing model. The problems of atmospheric contamination are greatest for xenon, and so there is little definitive evidence regarding the isotopic variations in the mantle.

5.2.6 Noble Gas Abundance Patterns

Noble gas abundance patterns in MORBs and OIBs scatter greatly. This is due to sample alteration as well as fractionation during noble gas partitioning between basaltic melts and a vapor phase that may then be preferentially gained or lost by the sample. Nonetheless, MORB Ne/Ar and Xe/Ar ratios that are greater than the air values are common. An example of this pattern was found in a gas-rich, relatively uncontaminated MORB sample (Figure 5) with high ^{40}Ar/^{36}Ar and ^{129}Xe/^{130}Xe ratios and a ^{4}He/^{40}Ar ratio (~3) that is near that of production in the upper mantle (and so not fractionated) (Staudacher *et al.*, 1989; Moreira *et al.*, 1998). An upper-mantle pattern also can be calculated by assuming that the noble gases have been degassed from the mantle without substantial elemental fractionation, and the radiogenic nuclides are present in their production ratios. Using ratios of estimated upper-mantle production rates to determine the relative abundances of radiogenic ^{4}He, ^{21}Ne, ^{40}Ar, and ^{136}Xe, measured isotopic compositions can be used to determine ratios of nonradiogenic isotopes. For example, using the production ratio of ^{21}Ne/^{40}Ar and the ratio of ^{21}Ne/^{22}Ne and ^{40}Ar/^{36}Ar (taking into account the amounts of non-nucleogenic

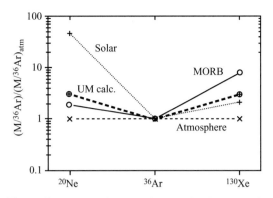

Figure 5 The relative abundances of noble gases in gas-rich MORB (Moreira *et al.*, 1998), as calculated for the upper mantle (see text), air (Tables 2 and 3), and in the solar composition (see Ozima and Podosek, 2001). The upper mantle is enriched in neon and xenon, relative to argon, compared to the air composition.

^{21}Ne) measured in basalts that are not altered by melting and transport processes, the ^{22}Ne/^{36}Ar ratio in the mantle source can be calculated. In this case, ^{22}Ne/^{36}Ar$_{MORB}$ = 0.15 × 10^{-4} and ^{130}Xe/^{36}Ar$_{MORB}$ = 3.3 × 10^{-4}, significantly higher than the corresponding atmospheric values of 0.05 × 10^{-4} and 1.1 × 10^{-4}, respectively.

It is often assumed that all the noble gases are highly incompatible during basalt genesis and so are efficiently extracted from the mantle without elemental fractionation. Experimental data for partitioning between basaltic melts and olivine are consistent with this for helium but not for the heavier noble gases, which have been found in higher concentrations in olivine than expected (Hiyagon and Ozima, 1986; Broadhurst *et al.*, 1992). However, these results may be due to experimental difficulties. Recent data (Chamorro-Perez *et al.*, 2002) indicate that the argon clinopyroxene/silicate melt partition coefficient is relatively constant and equal to ~4 × 10^{-4} at pressures up to at least 80 kbar, and more recent data indicate that neon, krypton, and xenon are similarly incompatible (Brooker *et al.*, 2003). Therefore, it appears that the noble gases are all highly incompatible in the mantle, and there is no elemental fractionation between the melt and the mantle source region. However, while noble gases transported from the mantle may not be elementally fractionated, some fractionation may occur in the highly depleted melt residue due to small differences in partition coefficients. This remains a possibility since reliable partition coefficients for most of the noble gases are unavailable.

Any fractionations that might occur during degassing of the upper mantle would be expected to be monotonic across the noble gases. Therefore, degassing alone does not explain high values for both ^{22}Ne/^{36}Ar and ^{130}Xe/^{36}Ar with respect to

atmospheric values. If the heavy noble gases were slightly more compatible (as might be plausibly assumed), then a higher ^{130}Xe/^{36}Ar value for the mantle would be generated, but accompanied by a lower ^{22}Ne/^{36}Ar ratio. It is possible that the atmospheric ^{22}Ne/^{36}Ar ratio was lowered instead during volatile losses to space. However, if the starting noble gas composition of the Earth was solar, with ^{22}Ne/^{36}Ar = 37 (Geiss *et al.*, 1972), then the upper-mantle value, which is lower than the solar value, still requires explanation. The possibilities that remain are a mantle composition that was generated during Earth formation that is different from the solar value, and subduction of atmospheric argon (see Section 5.2.9).

5.2.7 MORB Fluxes and Upper-mantle Concentrations

Noble gas concentrations vary widely in mantle-derived volcanics due to degassing during ascent and eruption, as well as from volatile redistribution in vesicles. Therefore, these data cannot be readily used to constrain the upper-mantle concentrations. An alternative approach is to compare fluxes to the atmosphere from a region with a known rate of volcanism. The largest and most clearly defined mantle volatile flux, from mid-ocean ridges (Clarke *et al.*, 1969; Craig *et al.*, 1975), is 1,060 ± 250 mol yr^{-1} ^{3}He (Lupton and Craig, 1975; Farley *et al.*, 1995), and is obtained by combining seawater ^{3}He concentrations in excess of dissolved air helium with seawater advection models. This value represents an average over the last 1,000 years, and while this is generally assumed to be the long-term average, this remains to be confirmed. Using the ^{3}He/^{4}He in vented gas, the flux of ^{4}He can be determined once the ^{3}He flux is determined. The flux of ^{4}He is approximately equal to the production rate for the upper mantle above 670 km (O'Nions and Oxburgh, 1983). This fact has been incorporated in some mantle degassing models, and is discussed further below. The fluxes of other noble gases from the mantle can be obtained from the ^{3}He fluxes and the relative abundances in MORBs (see Section 5.2.6). Assuming that radiogenic ^{4}He and ^{40}Ar are degassing together, a ^{40}Ar flux of 1.9 × 10^{31} atoms a^{-1} is obtained by using a radiogenic ratio of ^{4}He/^{40}Ar = 3, a MORB ^{3}He/^{4}He ratio of 8R_A, and noting that ~90% of mantle ^{4}He is radiogenic.

The concentration of ^{3}He in the mantle can be determined by dividing the flux of ^{3}He into the oceans by the rate of production of melt that is responsible for carrying this ^{3}He from the mantle, which is equivalent to the rate of ocean crust production of 20 km^{3} yr^{-1} (Parsons, 1981). MORBs that degas quantitatively to produce a ^{3}He flux of 1,060 mol yr^{-1} must have an average ^{3}He content of 1.96×10^{-14} mol g^{-1} or 4.4×10^{-10} cm^3 ^{3}He (STP) g^{-1}. This ^{3}He concentration is within a factor of 2 of that obtained for the most gas-rich basalt glass of $\sim 10.0 \times 10^{-10}$ cm^3 ^{3}He(STP) g^{-1} (Sarda *et al.*, 1988; Moreira *et al.*, 1998). Assuming that MORB is generated by an average of 10% partial melting, the source region contains 2×10^{-15} mol g^{-1} $(1.2 \times 10^9$ atoms ^{3}He g$^{-1})$. The mantle concentration of ^{3}He and ^{4}He cannot be usefully compared with the ^{3}He and ^{4}He inventories of the atmosphere since ^{3}He and ^{4}He are lost to space with a residence time of a few million years. A comparison, however, can be made with argon. Using the MORB ^{3}He/^{36}Ar ratio of 1.7, then the upper mantle has $(0.8–2.9) \times 10^9$ atoms ^{36}Ar g^{-1}. This is highly depleted compared to the benchmark value of 3×10^{12} atoms ^{36}Ar g^{-1} obtained by dividing the atmospheric inventory by the mass of the upper mantle. The abundances of other noble gas isotopes can be obtained similarly.

The flux of ^{4}He can be compared with that of heat, since ^{4}He is produced along with heat during radioactive decay. The Earth's global heat loss amounts to 44 TW (Pollack *et al.*, 1993). Subtracting the heat production from the continental crust (4.8–9.6 TW), and the core (3–7 TW; Buffett *et al.*, 1996) leaves 9.6–14.4 TW to be accounted for by present-day radiogenic heating and 17.8–21.8 TW as a result of secular cooling (largely from earlier heat production). O'Nions and Oxburgh (1983) pointed out that the present-day ^{4}He mantle flux was produced along with only 2.4 TW of heat, an order of magnitude less than that produced along with all the radiogenic heat presently reaching the surface, and suggested that this could be achieved by a boundary layer in the mantle through which heat could pass, but behind which helium was trapped. It has been suggested that heat and helium are transported to the surface at hotspots, where the bulk of the helium is lost, while the heat is lost subsequently at ridges (Morgan, 1998). However, evidence for such a hotspot helium flux at present or in the past, and formulation of a mantle noble gas model incorporating this suggestion, are unavailable. The possibility that the separation is due to the different mechanisms that extract heat and helium from the mantle was investigated with a secular cooling model of the Earth (van Keken *et al.*, 2001). It was found that the ratio of the surface fluxes of ^{4}He and heat were substantially higher than presently observed except for rare excursions. However, the ratio of surface fluxes would more closely match the model results if the surface helium flux were three times greater. It is possible that since the observed helium flux represents an average over only 1,000 yr, it does not represent the flux over somewhat longer timescales (Ballentine *et al.*, 2002). However, this is difficult to assess. The alternative that remains is that some boundary is

needed for separating uranium and thorium, from the upper-mantle reservoir, and allowing heat to pass more efficiently than helium, although the configuration of this boundary is unconstrained by geochemistry. The implications of this are considered further in discussions of mantle models. This significance for mantle degassing is that there are considerable amounts of radiogenic noble gases maintained behind such a boundary. Bercovici and Karato (2003) have proposed a model of extraction of incompatible elements at depth in the upper mantle by melting and entrainment in the down going slab.

5.2.8 Other Mantle Fluxes

The flux of ^{3}He from intraplate volcanic systems is dominantly subaerial and so it is not possible to obtain directly time-integrated flux values for even a short geological period. While the Loihi hotspot in the Pacific is submarine, calculation of ^{3}He fluxes into the ocean using ocean circulation models have not required a large flux from this location that is comparable to the ^{3}He plumes seen over ridges (Gamo *et al.*, 1987; Farley *et al.*, 1995), although recent data has seen an extensive ^{3}He plume from Loihi (Lupton, 1996).

Helium fluxes from OIB could be calculated if the rates of magmatism are known, along with the helium concentrations of the source regions or concentrations of undegassed magmas (see Porcelli and Ballentine, 2002). Estimates of the rate of intraplate magma production vary from 1% to 12% of the MORB production rate (Reymer and Schubert, 1984; Schilling *et al.*, 1978; Batiza, 1982; Crisp, 1984). The total noble gas flux from ocean islands is related to the mantle source concentrations and whether some source domains are more gas-rich than the MORB source mantle. OIBs are typically more extensively degassed than MORBs. This is readily explained for those that are erupted at shallower depths or subaerially. Also, OIBs may be more volatile-rich than MORBs and, therefore, may degas more effectively (Dixon and Stolper, 1995). High water contents of basalts lower CO_2 solubility, and appear to have lowered helium contents in lavas with high ^{3}He/^{4}He ratios along the Reykjanes Ridge (Hilton *et al.*, 2000). Estimates for the mantle source with high ^{3}He/^{4}He ratios have ranged from much lower than that of the MORB source, perhaps due to prior melting (Hilton *et al.*, 1997), to up to 15 times higher (Moreira and Sarda, 2000; Hilton *et al.*, 2000). It should be emphasized that in many locations, the high ^{3}He/^{4}He source component makes up a small fraction of the sample source and, if gas-rich, may substantially affect the helium composition but not substantially increase the helium flux. Therefore, in many cases, source concentrations may be closer to that of MORBs. In this case, the overall

^{3}He flux relative to that from MORBs is proportional to the relative melt production rate, or 1–12% that of MORBs. It is not clear to what extent more gas-rich OIBs augment this flux.

Continental settings provide a small but significant flux from the mantle. Regional groundwater systems provide time-integrated records of this flux over large areas, but are based on short time-scales and dependent on the hydrogeological model used. In the Pannonian basin (4,000 km^2 in Hungary), for example, the flux of ^{3}He has been estimated to be from 0.8–5 × 10^4 ^{3}He atoms m^{-2} s^{-1} (Stute *et al.*, 1992) to 8 × 10^4 ^{3}He atoms m^{-2} s^{-1} (Martel *et al.*, 1989). Taking an area of 2 × 10^{14} m^2 for continents and assuming 10% is under extension, this yields a total ^{3}He flux of 8.4–84 mol ^{3}He yr^{-1} (Porcelli and Ballentine, 2002). This value compares with a mantle flux of <3 mol ^{3}He yr^{-1} through the stable continental crust (O'Nions and Oxburgh, 1988).

The flux of mantle ^{3}He to the atmosphere at subduction zones from the upper mantle can be estimated from the volume of convergent zone volcanics. The estimate of Reymer and Schubert (1984) is 5% that at mid-ocean ridges. If this is largely generated by similar degrees of melting of a similar source as MORB, then the ^{3}He flux is only 5% that at the mid-ocean ridges, or 50 mol ^{3}He yr^{-1} (see also Hilton *et al.*, 2002). An estimated CO$_2$ flux at convergent margins of 3 × 10^{11} mol yr^{-1} (Sano and Williams, 1996) can also be used to estimate the ^{3}He flux. Using a mid-ocean ridge ratio of CO$_2$/^{3}He = 2 × 10^9 (Marty and Jambon, 1987) gives 150 mol ^{3}He yr^{-1}, although the convergent margin CO$_2$/^{3}He ratio may be much greater due to the presence of recycled carbon.

In summary, the dominant noble gas flux from the mantle is from mid-ocean ridges. Other fluxes may be negligible or augment the ridge flux by up to an additional 25%.

5.2.9 Subduction Fluxes

It is possible that atmospheric noble gases that are incorporated in oceanic crust materials, including altered MORBs, hydrothermally altered crust, and sediments are subducted into the mantle and thus are mixed with primordial noble gas constituents. Available data indicate that holo-crystalline MORBs and oceanic sediments contain atmospheric noble gases that are greatly enriched in the heavier noble gases. However, because concentrations vary by several orders of magnitude, an accurate average value cannot be easily determined. Measurements of pelagic sediments, with 3 × 10^{15} g yr^{-1} subducted (Von Huene and Scholl, 1991), have (0.05–7) × 10^{10} atoms ^{130}Xe g^{-1} (Podosek *et al.*, 1980; Matsuda and Nagao, 1986; Staudacher and Allègre, 1988) with a mean of 6 × 10^9 atoms g^{-1}. Measurements of holocrystalline basalts found (4–42) × 10^7 atoms ^{130}Xe g^{-1} (Dymond and Hogan, 1973; Staudacher and Allègre, 1988), with a mean of 2 × 10^8 atoms ^{130}Xe g^{-1}. While the depth over which addition of atmospheric gases are added by alteration is unknown, it might be assumed that this occurs over the same depth as low temperature enrichment of alkaline elements of ~600 m (Hart and Staudigel, 1982). In this case, an estimated 7 × 10^{15} g yr^{-1} of this material is subducted. In total, these numbers result in 2 × 10^{25} atoms ^{130}Xe yr^{-1} (33 mol yr^{-1}) reaching subduction zones in sediments and altered basalt (Porcelli and Wasserburg, 1995a). This can be compared to the estimate of Staudacher and Allègre (1988) determined by assuming that 40–80% of the subducting flux of oceanic crust (6.3 × 10^{16} g yr^{-1}) is altered and contains atmosphere-derived noble gases, and that 18% of this mass is ocean sediment. In this case, a similar flux of 13.8–39 mol ^{130}Xe yr^{-1} was obtained, along with 1.9–2.5 × 10^6 mol ^{4}He yr^{-1}, 1.9–5.9 × 10^3 mol ^{20}Ne yr^{-1}, 5.8–21.9 × 10^3 mol ^{36}Ar yr^{-1}, and 3.0–12.3 × 10^3 mol ^{84}Kr yr^{-1}. Subduction of noble gases at these rates over 10^9 yr would have resulted in 1%, 90%, 110%, and 170% of the respective inventories of ^{20}Ne, ^{36}Ar, ^{84}Kr, and ^{130}Xe to be in the upper mantle (Allègre *et al.*, 1986); however, the fraction stripped from subducting materials during deformation and arc volcanism before entering the mantle is unknown but it is likely to be substantial.

Subduction zone processing and volcanism may return much of the noble gases in the slab to the atmosphere. However, it is possible that the total amounts of noble gases reaching subduction zones are sufficiently high that subduction into the deeper mantle (i.e., beyond the zone of magma generation) of only a small fraction may have a considerable impact upon the composition of argon and xenon in the upper mantle (Porcelli and Wasserburg, 1995a,b). Staudacher and Allègre (1988) argued that subducting argon and xenon must be almost completely lost to the atmosphere during subduction zone magmatism, or the high ^{129}Xe/^{130}Xe and ^{136}Xe/^{130}Xe in the upper mantle would not have been preserved throughout Earth history. However, this conclusion is dependent upon a model of unidirectional degassing to generate the upper-mantle xenon isotope composition (see Section 5.4.3), and as discussed in Section 5.4.5 the contrary view that subducted noble gases are mixed with nonrecycled, mantle-derived xenon to produce the upper-mantle composition (Porcelli and Wasserburg, 1995a,b) is compatible with the mantle data. Note that, as discussed above, the ^{128}Xe/^{130}Xe ratio measured in mantle-derived xenon trapped in CO$_2$ well gases may be interpreted as a mixture of ~90% subducted xenon with 10% trapped solar xenon. Further, direct input of the subducted slab into a gas-rich deeper reservoir that has $^{40}Ar/^{36}Ar$ values that are significantly lower than in the MORB source mantle (e.g., Trieloff *et al.*, 2000) is also possible.

5.3 BULK DEGASSING OF RADIOGENIC ISOTOPES

The total fraction of a species that has been degassed to the atmosphere can only be calculated for radiogenic nuclides, which have total planetary abundances that are constrained by the parent element abundances. Most attention has focused on ^{40}Ar, although as discussed below ^{136}Xe also provides valuable constraints. Similar calculations cannot be done for ^{4}He, which does not accumulate in the atmosphere, nor for neon isotopes, since the production rate of ^{21}Ne, as well as the amount of nonradiogenic ^{21}Ne in the atmosphere, are too uncertain. Also, the amount of radiogenic ^{129}Xe in the bulk Earth is not well constrained due to early losses from the planet.

5.3.1 The ^{40}K–^{40}Ar Budget

Potassium is a moderately volatile element and is depleted by a factor of ~8 in the bulk silicate Earth compared to CI chondrites, but a precise and unambiguous concentration is difficult to obtain. Estimates have been made by comparison with uranium, which like potassium is highly incompatible during melting and so is not readily fractionated between MORB and the upper mantle. There is little debate regarding the concentration of uranium, which is obtained from concentration in carbonaceous chondrites and, by assuming that refractory elements (e.g., calcium, uranium, thorium) are unfractionated from solar values in the bulk Earth (e.g., O'Nions *et al.*, 1981). In this case, a bulk silicate Earth concentration of 21 ppb uranium is obtained (Rocholl and Jochum, 1993). If it is assumed that the MORB source value of $K/U = 1.27 \times 10^{4}$ (Jochum *et al.*, 1983) is the same as that of the bulk silicate Earth, then there is 270 ppm potassium that has produced a total of 2.4×10^{42} atoms ^{40}Ar. This is a widely accepted value. Note that the core is not a significant repository of either potassium or ^{40}Ar (Chabot and Drake, 1999).

The ^{40}Ar in the atmosphere (9.94×10^{41} atoms) is essentially entirely radiogenic. Assuming the crust has 0.91–2.0 wt.% K (Taylor and McLennan, 1985; Rudnick and Fountain, 1995; Wedepohl, 1995), and noting that the continents have a mean K–Ar age of 1×10^{9} a (Hurley and Rand, 1969), equivalent to a ratio of $^{40}Ar/K = 9.1 \times 10^{-6}$, yields an amount of crustal ^{40}Ar that is only 3.1–6.8% of that in the atmosphere. Therefore, in total, 41% of the ^{40}Ar that has been produced is now in the atmosphere (Allègre *et al.*, 1986, 1996; Turcotte and

Schubert, 1988). Thus, a significant reservoir of ^{40}Ar remains in the Earth. How this relates to degassing of nonradiogenic noble gas isotopes is the subject of degassing models, but identifying the fraction of radiogenic isotopes in the atmosphere provides an important overall constraint. It has sometimes been assumed that the mantle reservoir that is rich in ^{40}Ar is the same as that with high $^{3}He/^{4}He$ and so is also rich in ^{3}He, although this is not the case of all mantle models (see Porcelli and Ballentine, 2002).

Uncertainty arises in the above calculation when considering that the depleted MORB-source mantle may not have a bulk silicate Earth K/U ratio, since a significant fraction of either element may have been preferentially added into the upper mantle by subduction. The bulk of the potassium and uranium originally in the upper mantle is now in the continental crust, which, therefore, may be expected to have a bulk silicate Earth K/U ratio. But this is not sufficiently well constrained due to wide variations in the potassium and uranium contents of the more differentiated continental crustal rocks. It has been suggested that the bulk Earth K/U ratio is much lower than the MORB value (Albarède, 1998; Davies, 1999). In this case, the potassium content of the Earth is much lower, and so a greater fraction of the total ^{40}Ar has degassed. This would make the ^{40}Ar budget compatible with geophysical models that have convection and mixing throughout the mantle, and so would imply that the depleted mantle that serves as the MORB source represents the bulk of the whole mantle. In such a case, there is no need to isolate and maintain a gas-rich mantle reservoir making up 60% of the mantle. However, it has been argued that the relative proportions of moderately volatile elements in the Earth lie on compositional trends defined by chondritic meteorite classes (Allègre *et al.*, 1995; Halliday and Porcelli, 2001), and trends in meteoritic Rb/Sr versus K/U are compatible with a terrestrial value of $K/U = 1.27 \times 10^{4}$. Only a modest reduction in the terrestrial potassium content would still be compatible with these relationships.

5.3.2 The ^{129}I–^{129}Xe and ^{244}Pu–^{136}Xe Budgets

The amount of short-lived isotope ^{129}I ($t_{1/2} = 15.7$ Myr) that was in the Earth or Earth-forming materials and produced $^{129*}Xe$ can be estimated from the present bulk silicate Earth concentration of stable ^{127}I. Iodine is a volatile element and is highly depleted in the Earth relative to chondrites. Wänke *et al.* (1984) estimated a bulk silicate Earth stable ^{127}I concentration of 13 ppb based on the analysis of a fertile xenolith judged to represent the undepleted mantle, and this is the generally accepted estimate. McDonough and Sun (1995) arrived at a similar value of 11 ppb. Alternatively, using a bulk crust abundance of 8.6×10^{18} g I (Muramatsu and Wedepohl, 1998) and an upper-mantle concentration of 0.8 ppb (Déruelle *et al.*, 1992), and assuming that the crust was derived from 25% of the mantle, a bulk silicate Earth value of 9 ppb is obtained (Porcelli and Ballentine, 2002). However, based on unpublished data, Déruelle *et al.* (1992) quote a concentration for the crust that is 4.3 times higher, raising the possibility that the silicate Earth concentration of iodine is substantially higher than typically estimated, although such high values have not been adopted. At 4.57 Ga ago, $(^{129}I/^{127}I)_0 = 1.1 \times 10^{-4}$ based on meteorite data (Hohenberg *et al.*, 1967; Brazzle *et al.*, 1999). For a silicate Earth value of 13 ppb ^{127}I, 2.7×10^{37} atoms of $^{129*}Xe$ were produced in the Earth or Earth-forming materials since formation of the solar system.

Plutonium is a highly refractory element represented naturally by a single isotope, ^{244}Pu ($t_{1/2} = 80$ Myr), which produces heavy xenon isotopes by fission. It is assumed that plutonium was incorporated into Earth-forming materials unfractionated relative to other refractory elements such as uranium. Meteorite data suggests that at 4.56 Ga, $(^{244}Pu/^{238}U)_0 = 6.8 \times 10^{-3}$ (Hudson *et al.*, 1989) and this is the commonly accepted value. In this case the silicate Earth, or Earth-forming materials, with a present-day bulk silicate Earth value of 21 ppb U (O'Nions *et al.*, 1981), initially had 0.29 ppb Pu. This produced 2.0×10^{35} atoms of $^{136*}Xe$ since the start of the solar system. Other work on meteorites (Hagee *et al.*, 1990) calculated values of $(4-7) \times 10^{-3}$, with the higher number considered more likely to represent the solar value, although a lower value remains a possibility. The amount produced by ^{238}U in the bulk silicate Earth (7.5×10^{33} atoms ^{136}Xe) is much less, and so bulk silicate Earth (and atmospheric) $^{136*}Xe$ is dominantly plutonium-derived; even if xenon was lost over the first 10^8 a of Earth history so that half of the plutonium-derived xenon was lost, plutonium-derived ^{136}Xe is still 13 times more abundant in the Earth.

The greatest difficulty in constraining the global xenon budget has been in calculating the abundances of radiogenic xenon in the atmosphere. The composition for nonradiogenic atmospheric xenon (Section 5.2.5) provides ratios of $^{129}Xe/^{130}Xe = 6.053$ and $^{136}Xe/^{130}Xe = 2.075$ as the present best estimates of the isotopic composition of nonradiogenic terrestrial xenon (Pepin, 2000). Therefore, $6.8 \pm 0.30\%$ of atmospheric ^{129}Xe ($^{129*}Xe_{atm} = 1.7 \times 10^{35}$ atoms) and $4.65 \pm 0.5\%$ of atmospheric ^{136}Xe ($^{136*}Xe_{atm} = 3.81 \times 10^{34}$ atoms) are radiogenic. The $^{136*}Xe$ in the atmosphere is 20% of the total ^{136}Xe produced by ^{244}Pu in the bulk silicate Earth. However, the $^{129*}Xe_{atm}$ is only 0.8% of the total ^{129}Xe produced

since 4.57 Ga; such a low value cannot be accounted for by incomplete degassing of the mantle nor from any uncertainties in the estimated amount of $^{129*}Xe$, and requires losses to space over an early period that is short relative to the longer time constant of ^{136}Xe production.

The depletion of radiogenic xenon in the atmosphere due to losses from the Earth to space must have occurred during early Earth history, when such heavy species could have been lost either from protoplanetary materials or from the growing Earth. Wetherill (1975) proposed that a "closure age" of the Earth could be calculated by assuming a two-stage history that involved essentially complete loss of $^{129*}Xe$ and $^{136*}Xe$ initially, followed by complete closure against further loss. The "closure age" also can be calculated by combining the ^{129}I–^{129}Xe and ^{244}Pu–^{136}Xe systems (Pepin and Phinney, 1976) to obtain a closure age of 82 Myr. If radiogenic ^{136}Xe was lost from the entire planet over about one half-life of ^{244}Pu (80 Myr), then ~40% of the ^{136}Xe remaining in the Earth is in the atmosphere, compatible with the fraction of ^{40}Ar in the atmosphere (using a silicate Earth value of 270 ppm K). Note that in the first 100 Ma, only 6% of the ^{40}Ar now present was produced, and losses over this time would not have significantly changed the ^{40}Ar budget (Davies, 1999).

The coincidence between the ^{40}K–^{40}Ar and ^{244}Pu–^{136}Xe budgets is remarkable considering that these values are the result of a series of independent, albeit somewhat uncertain, estimates. As mentioned above, in order to adjust these numbers to accommodate a greater fraction of degassing, it has been suggested that there is greater depletion of the moderately volatile potassium in the Earth (Albarède, 1998); however, the coincidence with the Pu–^{136}Xe budget requires a similarly lowered estimate of the amount of short-lived, refractory ^{244}Pu in the solar nebula. Clearly, these two factors are unrelated. It might be assumed that the ^{40}Ar budget might reflect processing of a greater fraction of the mantle, but with subduction returning a considerable fraction of the potassium over geological time, thereby creating domains in the mantle that have been degassed early but now contain a considerable budget of ^{40}Ar by subsequent production. However, this is not possible for the ^{136}Xe budget; all plutonium-derived ^{136}Xe was produced early, and the present budget reflects the total processing of the mantle. Therefore, it appears that the noble gas budget requires that a considerable fraction of the mantle has not been degassed to the atmosphere.

5.4 DEGASSING OF THE MANTLE

The present value for the mid-ocean ridge flux of ^3He to the atmosphere, if constant over 4.5 Ga, would result in a total of 4.5×10^{12} mol ^{3}He degassed. Using the mantle value of ^{3}He/^{22}Ne = 11 (see Section 5.2.3), this corresponds to a total of 5×10^{13} mol ^{22}Ne. This is only 2% of the ^{22}Ne presently in the atmosphere (Table 2). Therefore, present fluxes of noble gases from the mantle, applied over the history of the Earth, are insufficient to provide the inventories in the atmosphere. A degassing history that involves stronger degassing in the past is required, and a consideration of degassing models is needed to address the issue of the time dependence of the fluxes. An important factor in degassing models is the extent of noble gas recycling into the mantle by subduction. While subduction is unlikely to have a significant effect on the atmospheric inventory, it may impact the characteristics of noble gases in depleted mantle reservoirs may be significant. Therefore, models either assume that subduction of noble gases does not occur, or explicitly incorporates the effects of atmospheric inputs to the mantle.

5.4.1 Early Earth Degassing

Models for the evolution of terrestrial noble gases must necessarily consider appropriate starting conditions. Early degassing is likely to be very vigorous due to high accretional impact energies during Earth formation over 10^8 yr. Loss of major volatiles from impacting materials, and noble gases, may occur after only 10% of the Earth has accreted (Ahrens et al., 1989). In this case, much of the volatiles are added directly to the atmosphere rather than being degassed from the solid planet. Volatiles may also have been captured directly from the solar nebula (e.g., Porcelli et al., 2001), followed by modifications due to losses to space. Various models for the origin of relatively volatile elements in the Earth have accounted for terrestrial volatiles by late infall of volatile-rich material (e.g., Turekian and Clark, 1969, 1975; Dreibus and Wänke, 1989; Owen et al., 1992). The relative uniformity of lead and strontium isotopes in the mantle suggests that relatively volatile elements such as rubidium and lead that would also be supplied by late-accreting volatile-rich material were subsequently mixed into the deep mantle (Gast, 1960). However, loss of noble gases from impacting materials directly into the atmosphere likely inhibited their incorporation into the growing solid Earth. Therefore, noble gases supplied to the Earth in this way were unlikely to have been initially uniformly distributed in the solid Earth. It is also clear that very strong degassing of the Earth occurred during the extended period of planetary formation.

Atmospheric noble gases were also likely to have been lost to space during accretion by atmospheric erosion (Ahrens, 1993), when large impactors impart sufficient energy to the atmosphere for the constituents to reach escape velocity. In this way, each large impact during later accretion can drive away a substantial portion of the previously degassed atmosphere. Therefore, the present atmospheric abundances do not necessarily reflect the total amounts of nonradiogenic and early-produced nuclides that were degassed from accreting materials. Also, it has been suggested that the strong fractionation of neon and xenon isotopes in the atmosphere is due to hydrodynamic escape (Hunten et al., 1987; Sasaki and Nakazawa, 1988; Pepin, 1991), where loss of hydrogen from the atmosphere entrains heavier species, leaving behind a fractionated residue. Such losses would not have affected gases within the Earth, and so would have generated isotopic contrasts between the atmosphere and internal terrestrial reservoirs. Such loss processes can account for the losses of xenon isotopes produced by short-lived ^{129}I and ^{244}Pu. In sum, degassing histories must include strong early degassing of the Earth as it accretes, and consider that the abundances degassed were not fully retained in the atmosphere.

5.4.2 Degassing from One Mantle Reservoir

The simplest case for atmosphere formation is unidirectional degassing from a single solid Earth reservoir, which is represented by the MORB source region. Early models focused on argon isotopes. Generally, the key assumption is that the rate of degassing at any time is directly proportional to the total amount of argon present in the mantle at that time. Also, there is no return flux from the atmosphere by subduction. Then

$$\frac{d^{36}Ar_m}{dt} = -\alpha(t)^{36}Ar_m \qquad (1)$$

and

$$\frac{d^{40}Ar_m}{dt} = -\alpha(t)^{40}Ar_m + \lambda_{40}y^{40}K_m \qquad (2)$$

where $\alpha(t)$ is the time-dependent degassing proportionality constant, $\lambda_{40} = 5.543 \times 10^{-10}$ a^{-1} is the total decay rate of ^{40}K, $y = 0.1048$ is the fraction of decays of ^{40}K that yield ^{40}Ar (Table 1), and $^{36}Ar_m$, $^{40}Ar_m$, and $^{40}K_m$ are the total Earth abundances. In the simplest case, $\alpha(t)$ is a constant (Turekian, 1959; Ozima and Kudo, 1972; Fisher, 1978). This is reasonable if the mantle is well mixed and has been melted and degassed at a constant rate at mid-ocean ridges. Assuming there was no argon initially in the atmosphere, then the only free variable is α; in this case

$$\left(\frac{^{40}\text{Ar}}{^{36}\text{Ar}}\right)_{\text{atm}} = \left(\frac{\alpha}{\alpha - \lambda}(1 - e^{-\lambda t}) - \frac{\lambda}{\alpha - \lambda} \times (1 - e^{-\alpha t})\right)$$
$$\left(\frac{y^{40}\text{K}_{\text{m}}e^{\lambda t}}{^{36}\text{Ar}_{\text{atm}}}\right)$$

(3)

Using a BSE value of 270 ppm K, a value of $\alpha = 1.82 \times 10^{-10}$ is obtained. From Equation (1), $^{36}\text{Ar}_{\text{atm}} = {}^{36}\text{Ar}_{\text{m0}} (1 - e^{-\alpha t})$, so that the fraction of nonradiogenic ^{36}Ar that has degassed, $^{36}\text{Ar}_{\text{atm}}/{}^{36}\text{Ar}_{\text{m0}}$, is 0.62. The $^{40}\text{Ar}/^{36}\text{Ar}$ ratio of the mantle is then

$$\left(\frac{^{40}\text{Ar}}{^{36}\text{Ar}}\right)_{\text{m}} = \frac{\lambda_{40}}{\alpha - \lambda_{40}}(e^{(\alpha - \lambda_{40})t} - 1)\frac{y^{40}\text{K}_{\text{m}}e^{\lambda_{40}t}}{^{36}\text{Ar}_{\text{m0}}}$$

(4)

A value of $(^{40}\text{Ar}/^{36}\text{Ar})_{\text{mn}} = 520$ is calculated from Equation (4) for the mantle. Once higher values were measured in MORB samples, such a simple formulation no longer appeared valid. Higher ratios can be obtained if an early catastrophic degassing event occurred, removing a fraction f of the ^{36}Ar from the mantle into the atmosphere (Ozima, 1973). In this case, the term $(1-f)^{36}\text{Ar}_{\text{m0}}$ can be substituted for $^{36}\text{Ar}_{\text{m0}}$ (see Ozima and Podosek, 1983). Then for a mantle with $^{40}\text{Ar}/^{36}\text{Ar} = 4 \times 10^4$ (see Section 5.2.4), 98.6% of ^{36}Ar was degassed initially. Alternatively, a more complicated degassing function that is steeply diminishing with time (such as $\alpha(t) = \alpha e^{-\beta t}$) can be used to match the present isotope compositions (Sarda *et al.*, 1985; Turekian, 1990), and so also involves early degassing of the bulk of the atmospheric ^{36}Ar. Regardless of the formulation used, such early degassing is required by the high measured $^{40}\text{Ar}/^{36}\text{Ar}$ ratios and may reflect extensive devolatilization of impacting material during accretion or a greater rate of mantle melting very early in Earth history due to higher heat flow.

Another aspect of mantle degassing that can be represented in model calculations is the transfer of potassium from the upper mantle into the continental crust so that the mantle potassium content becomes time dependent. This requires including the crust as an additional model reservoir. The continents may be modeled as either attaining their complete mass very early or more gradually using some growth function (see, e.g., Ozima, 1975; Hamano and Ozima, 1978; Sarda *et al.*, 1985). However, all model formulations qualitatively agree that ^{36}Ar degassing dominantly occurred very early in Earth history.

A different perspective on ^{40}Ar degassing has been provided by Schwartzman (1973), who argued that potassium is likely to be transferred "coherently" out of the mantle with ^{40}Ar; i.e., any ^{40}Ar that has been produced by potassium will be degassed when the potassium is transferred to the crust. While the K/^{40}Ar ratio of magmas leaving the mantle and that of the mantle source region are thus assumed to be approximately equal, the highly depleted residue could still be fractionated, so that very radiogenic-derived $^{40}\text{Ar}/^{36}\text{Ar}$ ratios could develop. It has been pointed out that, using the budgets of potassium and ^{40}Ar, this implies that the potassium in the crust should fully account for the ^{40}Ar in the atmosphere (Coltice *et al.*, 2000). This is only the case if the continental crust contains 2.0% K, which is higher than most estimates (see Section 5.5.1), and so additional ^{40}Ar was provided by additional potassium that has been recycled and which corresponds with up to ~30% of the potassium that is now in the continents (Coltice *et al.*, 2000).

Similar considerations used in the ^{40}Ar modeling have been applied to xenon (see Thomsen, 1980; Staudacher and Allègre, 1982; Turner, 1989). In these studies, there is greater resolution of the timing of early degassing due to the short half-lives of the parent nuclides ^{129}I and ^{244}Pu. The higher $^{129}\text{Xe}/^{130}\text{Xe}$ ratio of the upper mantle is interpreted as resulting from an increase in the I/Xe ratio during the lifetime of ^{129}I due to degassing of xenon to the atmosphere. Regardless of the exact degassing history used, this requires that strong degassing occur very early in Earth history, compatible with the results of the argon studies. Models of degassing from a single mantle reservoir have also been applied to helium (Turekian, 1959; Tolstikhin, 1975), although there are greater degrees of freedom since the atmosphere does not preserve a record of the total abundances of helium isotopes that have been degassed.

The strong early degassing inferred in these models can be identified with the degassing that likely occurred during extended accretion of the Earth. These models have not been modified to include possible substantial losses to space. This would not affect the ^{40}Ar budget, and would require stronger early degassing of ^{36}Ar and modification of the total amount degassed.

5.4.3 Multiple Mantle Reservoirs

The major shortcoming of the single reservoir degassing models is that mantle heterogeneities, especially regarding $^3\text{He}/^4\text{He}$ ratios, cannot be explained. The second generation of mantle degassing models developed with growing evidence that, in addition to the MORB-source reservoir, another helium source was required to account for the high $^3\text{He}/^4\text{He}$ ratios found in ocean island basalts (Hart *et al.*, 1979; Kurz *et al.*, 1982; Allègre *et al.*, 1983, 1986). These models also incorporate the degassing of a single mantle reservoir to the atmosphere. However, to explain the high OIB ^{3}He/^{4}He ratios, there is an additional underlying gas-rich reservoir that is isolated from the degassing upper mantle. Therefore, these layered mantle models can be considered to incorporate two separate systems; the upper-mantle atmosphere that evolves according to the systematics described in Section 5.4.2, and the lower mantle (Figure 6). There is no interaction between these two systems, and the lower mantle is completely isolated, except for a minor flux to the surfaced at ocean islands that marks its existence. In order to characterize the lower-mantle reservoir, it is further assumed that the mantle was initially uniform in noble gas and parent-isotope concentrations, so that both systems had the same starting conditions.

The deep reservoir with high ^{3}He/^{4}He ratios is assumed to have evolved approximately as a closed system for noble gases and has bulk silicate Earth parent-nuclide concentrations. Assigning the highest OIB ^{3}He/^{4}He ratios to this reservoir, a comparison between the total production of ^{4*}He

and the shift in ^{3}He/^{4}He from the initial terrestrial value to the present value provides an estimate of the ^{3}He concentration in this reservoir:

$$\left(\frac{^4\text{He}}{^3\text{He}}\right)_{present} - \left(\frac{^4\text{He}}{^3\text{He}}\right)_{initial} = \frac{^{4*}\text{He}}{^3\text{He}} \quad (5)$$

For a bulk silicate Earth concentration of 21 ppb U (O'Nions *et al.*, 1981) and Th/U = 3.8 (e.g., Doe and Zartman, 1979), a total of 1.02×10^{15} atoms ^{4*}He g^{-1} is produced over 4.5 Ga. Assigning this reservoir an Iceland value of ^{3}He/^{4}He = 37R_A and an initial value of ^{3}He/^{4}He = 120R_A (see Section 5.2.2), then the reservoir has 7.6×10^{10} atoms ^{3}He g^{-1}. The concentration of another noble gas is required for comparison of lower-mantle noble-gas abundances with the atmosphere. Using ^{3}He/^{22}Ne = 11 (see Section 5.2.3), a concentration in a closed system lower mantle of 7×10^9 atoms ^{22}Ne g^{-1} is obtained. A benchmark for comparison is the atmospheric ^{22}Ne abundance divided by the mass of the upper mantle $(1 \times 10^{27}\text{g})$ of

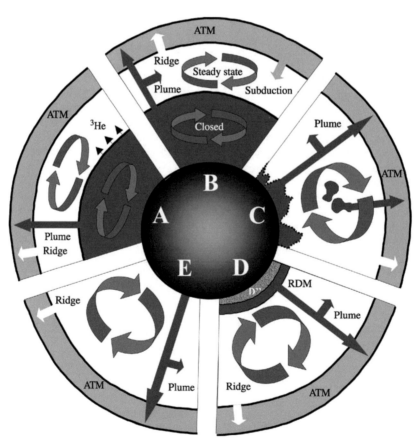

Figure 6 A range of mantle models for the distribution and fluxes of noble gases in the Earth. Layered mantle models with the atmosphere derived from the upper mantle involve either progressive unidirectional depletion of the upper mantle (A) or an upper mantle subject to inputs from subduction and the deeper mantle, and has steady state concentrations (B). Whole mantle convection models involve degassing of the entire mantle, with helium with high ^{3}He/^{4}He ratios found in OIB stored in either a deep variable-thickness layer (C), a layer of subducted material at the core–mantle boundary (D), or the core (E). (source Porcelli and Ballentine, 2002).

1.8×10^{11} atoms ^{22}Ne g^{-1}, which is much higher, and might be taken to indicate that the atmosphere source reservoir was more gas-rich than any deep isolated reservoir. However, such an undegassed reservoir has a ^{3}He concentration that is still ~40 times greater than that of the MORB source (see Section 5.2.7), and so is still relatively gas-rich.

Nonradiogenic argon and xenon isotope concentrations of such a lower-mantle reservoir cannot be directly calculated without assuming either specific lower-mantle Ar/Ne and Xe/Ne ratios or argon and xenon isotopic compositions. Since modification of the noble gases in the atmosphere has occurred, the relative abundances of the interior of the Earth are unlikely to match those now observed in the atmosphere. For example, a closed-system lower mantle with ^{40}Ar/^{36}Ar $\geq 3,000$ (see Section 5.2.4) and 270 ppm K has ^{40}Ar $= 5.7 \times 10^{14}$ atoms g^{-1} and so ^{36}Ar $\leq 1.9 \times 10^{11}$ atoms g^{-1}. The reservoir then has a ratio of ^{22}Ne/^{36}Ar ≥ 0.9, which is much greater than the air value of 0.05. Note that some calculations have assumed that the lower-mantle concentration is equal to the atmospheric inventory divided by the mass of the upper mantle (e.g., Hart et al., 1979), which is based on the idea of an initially uniform distribution of ^{36}Ar. This is compatible with a lower-mantle ratio of ^{40}Ar/^{36}Ar $= 300$. Evidence for higher lower-mantle ratios might be taken to reflect partial degassing of the deep mantle. However, it is possible that there are high ^{40}Ar/^{36}Ar ratios in the lower mantle that are due not to degassing; but rather to a lower initial trapped ^{36}Ar concentration (Porcelli and Wasserburg, 1995b). As discussed in Sections 5.2.4 and 5.2.5, the isotopic compositions of argon and xenon that accompany the high ^{3}He/^{4}He ratios in the source region of ocean island basalts are still too poorly constrained to make firmer conclusions regarding the concentrations of the heavier noble gases.

Note that these closed-system considerations do not require assumptions regarding the size of the reservoir. While models have assumed that it constitutes the entire mantle below 670 km (e.g., Allègre et al., 1986), it can involve a portion of a stratified mantle or material that is distributed as heterogeneities within another mantle reservoir. However, a large deep-mantle reservoir is compatible with the K–^{40}Ar budget and heat-^{4}He balance.

It should be emphasized that while an undepleted, undegassed mantle reservoir is a component in many models, there is no direct evidence that such a reservoir does indeed exist. An implication of these arguments is that the deep Earth reservoir also has not suffered removal of lithophile elements, and so contains bulk silicate Earth isotopic signatures for strontium, neodymium, hafnium, and lead. Evidence has not been found

for such a component in OIBs. However, it is possible that, due to high ratios of noble gases to lithophile elements, the involvement of small amounts of material in OIB source regions imparts deep-mantle noble gas signatures but not those of other elements. Models that have attempted to describe the mantle domains with high ^{3}He/^{4}He ratios as being gas-poor (Graham et al., 1990) have not been incorporated into a convincing model for global mantle noble gas evolution (see Porcelli and Ballentine, 2002).

5.4.4 Interacting Reservoirs

An important constraint that leads to a complete re-evaluation of the degassing models described above comes from consideration of fissiogenic ^{136}Xe and ^{129}Xe together (Ozima et al., 1985). Coupled shifts in ^{136*Pu}Xe/^{130}Xe and ^{129}Xe/^{130}Xe are proportional to the ^{244}Pu/^{129}I ratio of the source reservoir. On a ^{129}Xe/^{130}Xe versus ^{136}Xe/^{130}Xe plot (Figure 4), the slope of the line from the values of nonradiogenic atmospheric xenon and through the composition of the present atmosphere (which has fissiogenic ^{136}Xe largely from ^{244}Pu, and radiogenic ^{129}Xe from ^{129}I) provides the ^{136*Pu}Xe/^{129*}Xe ratio of the atmosphere, and so the ^{244}Pu/^{129}I ratio of the source of the atmospheric xenon. If degassing occurred before these parent elements became extinct, then the source would have an increased ^{129}I/^{130}Xe and ^{244}Pu/^{130}Xe ratio, and so would be expected to have higher ^{136*Pu}Xe/^{130}Xe and ^{129}Xe/^{130}Xe ratios now than the atmosphere. Evidence from MORBs indicates that the upper mantle indeed has such higher ratios. Another feature of any xenon that has remained within this source reservoir is that it will have a higher ^{136*Pu}Xe/^{129*}Xe ratio than that of air xenon; i.e., it will lie above the line that defining this ratio in Figure 4 (Ozima et al., 1985). This can be most easily envisaged by considering that degassing occurred early and in a single event. At the time of xenon loss from the mantle to the atmosphere, the proportion of undecayed ^{244}Pu will be greater than that of ^{129}I (which has a much shorter half-life). Therefore, the remaining ^{244}Pu and ^{129}I will produce xenon with a higher ^{136*Pu}Xe/^{129*}Xe ratio than is in the atmosphere. This is true regardless of whether or not degassing actually occurred as a single event. Note that mantle xenon will also have fissiogenic ^{136}Xe from decay of ^{238}U produced during later mantle evolution (see Section 5.2.5), and so to compare the present upper-mantle ^{136*}Xe/^{129*}Xe ratio generated from short-lived parent nuclides to that of the atmosphere, the proportion of mantle ^{136*}Xe that is plutonium-derived must be determined. As seen in Figure 7, MORB data do fall above the line through the atmosphere. However, any correction for

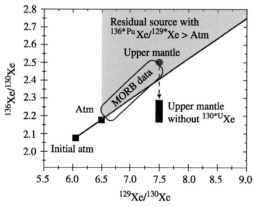

Figure 7 The relationship between the xenon isotope compositions of the atmosphere, initial atmosphere, and the upper mantle as sampled by MORB. The line connecting the initial atmosphere (fractionated U–Xe) and the present atmosphere has a slope equal to the ratio of plutonium-derived ^{136}Xe to radiogenic ^{129}Xe (^{136*}Xe/^{129*}Xe) in the atmosphere. Any xenon that remained in the solid reservoir from where this was degassed must have a greater value for this ratio and so lie in the shaded region (Ozima *et al.*, 1985). While measured MORBs do so, upper-mantle compositions that have been corrected for ^{238}U-derived ^{136}Xe (based on MORB data of Kunz *et al.*, 1998 and on CO_2 well gas data of Phinney *et al.*, 1978) do not, indicating that upper-mantle xenon is not the residual left from atmosphere degassing.

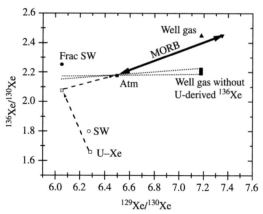

Figure 8 The xenon isotope compositions of U–Xe and solar wind xenon, both nonfractionated and fractionated to match the light xenon isotopes, are compared to the value of the atmosphere. It is clear that fractionated solar wind xenon cannot serve as the nonradiogenic composition of the atmosphere since it has a higher ^{136}Xe/^{130}Xe ratio. In order for the upper mantle to have the same ^{136*}Xe/^{129*}Xe ratio as the atmosphere, the nonradiogenic composition of the atmosphere must lie on the dotted lines, implying that there is very little plutonium-derived ^{136}Xe. This is contrary to the inferred ^{244}Pu budget of the Earth.

additions from ^{238}U will lower the ^{136}Xe/^{130}Xe ratio, and unless essentially all of ^{136*}Xe in MORB is ^{244}Pu-derived, the corrected value for upper-mantle xenon will fall below the atmosphere line and so cannot be the residue left after atmospheric xenon degassing. The data for CO_2 well gases and MORB indicate that a large fraction of the ^{136*}Xe is in fact from uranium (see Section 5.2.5), which is consistent with grow-in in the upper mantle due to the presently inferred ^{238}U/^{130}Xe ratio (see Porcelli and Wasserburg, 1995a). Therefore, it appears inescapable that the xenon presently found in the upper mantle is not the residue from degassing of the atmosphere. Since the daughter xenon isotopes how found in the atmosphere must have been derived from the ^{244}Pu and ^{129}I that were in the upper mantle (see Section 5.3.2), the xenon now found there must have been introduced from a deeper reservoir after the atmosphere was removed.

The main uncertainty in the evaluation of xenon isotopes is the composition of terrestrial nonradiogenic xenon (see Section 5.2.5). However, the arguments presented above appear to be robust when considering the possible compositions. Solar wind xenon or U–Xe do not match the terrestrial light xenon isotope composition without extensive fractionation. Only U–Xe, when fractionated, provides a plausible precursor

for the atmosphere (Figure 8). While the errors on this composition are small (Pepin, 2000), it is worth considering whether another composition is plausible. Due to the magnitude of the excess of ^{129}Xe in the atmosphere, the amount of radiogenic ^{129}Xe present has not been debated, so that the ^{129}Xe/^{130}Xe ratio of nonradiogenic terrestrial xenon is relatively well fixed. As shown in Figure 8, the CO_2 well gas data can be residual of atmospheric degassing only if there is almost no plutonium-derived ^{136}Xe in the atmosphere (i.e., the ^{136}Xe/^{130}Xe ratios of the initial and present atmospheres are very similar). However, this would require a substantial downward revision of the amount of ^{244}Pu that was present in Earth-forming materials that has been estimated completely independently and is consistent with the ^{40}Ar budget of the Earth (Section 5.3.2). There appears to be no reason to make such a re-evaluation, and, as discussed in Section 5.4.5 below, there are mantle models that can satisfy these constraints.

Another constraint on the origin of the xenon now found in the upper mantle is obtained from the composition of fissiogenic ^{136}Xe. In a closed-system reservoir, plutonium-derived ^{136*}Xe will dominate over uranium-derived ^{136*}Xe (see Section 5.3.2). In a system that has been closed throughout solar system history and starting with a chondritic ^{244}Pu/^{238}U ratio of 0.0068, fission of ^{244}Pu will be 27 times that produced by ^{238}U (Porcelli and Wasserburg, 1995a). If xenon was lost over the first 100 Myr (about one half-life of ^{244}Pu), plutogenic ^{136}Xe will still dominate. Therefore, xenon from such a reservoir that has ^{129}Xe excesses will have accompanying ^{136*}Xe that is dominantly plutonium-derived. MORB and CO_2 well gas data indicate that uranium-derived ^{136*}Xe dominates upper-mantle ^{136*}Xe. This requires that xenon in this mantle reservoir has been in an environment with a much higher ^{238}U/^{130}Xe ratio after extinction of ^{244}Pu, either due to a large decrease in Xe/U after the decay of ^{244}Pu, or has received xenon from another reservoir. As discussed above, based on xenon isotope systematics the xenon in the upper mantle has clearly been introduced from the deep mantle, and therefore this suggests that the ^{136*Pu}Xe and ^{129*}Xe was generated in the deep mantle, and the upper mantle has served as the locale for significant late ^{136*U}Xe additions.

Another constraint on the relationship between atmospheric and mantle xenon is obtained by considering the nonradiogenic xenon isotopes. The nonradiogenic xenon isotopes presently in the atmosphere are highly fractionated with respect to the composition of xenon that is likely to have been initially trapped by the Earth. The fractionation most likely occurred during subsequent losses to space under conditions that were only present very early in Earth history. In this case, the degassing of nonradiogenic isotopes occurred early while such processes were still possible (see Pepin, 1991). In contrast, as discussed above, nonradiogenic xenon isotopes in the mantle are fractionated with respect to atmospheric xenon, with e.g., ^{128}Xe/^{130}Xe ratios that are intermediate between the solar and atmospheric ratio. This requires retention in the mantle of a component that was either only somewhat fractionated relative to solar xenon (due to fractionation processes before accretion by presently unknown processes), and therefore is less fractionated than atmospheric xenon, or more likely that is unfractionated solar xenon but has been mixed with subducted xenon. In the latter case, the xenon in the upper mantle is a mixture between xenon that has been stored in the deeper mantle since initial trapping with subducted xenon. Models of xenon evolution in the mantle therefore should have a return flux from the atmosphere, rather than simple unidirectional solid Earth degassing.

In sum, the xenon isotopes point to a more complicated mantle degassing history than simple unidirectional degassing into the atmosphere. It appears that first nonradiogenic noble gases were degassed to the atmosphere early. The solid reservoir volume from which these nonradiogenic gases were lost is unconstrained. This atmospheric xenon was fractionated by losses to space over 10^8 yr, so that the atmosphere became enriched in the heavy isotopes. Radiogenic xenon was then degassed quantitatively from at least 40% of the

mantle (based on the ^{136}Xe–Pu budget). This occurred either by degassing of xenon directly from the mantle, or by transport of parent plutonium and iodine to the crust first, followed by crustal degassing. Xenon contained in another, deep-mantle reservoir and which is not residual from atmosphere removal was subsequently added to the upper mantle. The U/Xe ratio in the upper mantle was greater than that of the deeper reservoir, so that radiogenic ^{136*}Xe in the upper mantle became dominated by uranium-derived xenon.

An important implication of these considerations is that since the noble gases in the upper mantle are not the residual from atmosphere degassing, they provide no information regarding the degassing history of the atmosphere. Rather, the present composition of the upper mantle reflects interactions between different noble gas reservoirs. The disposition of these reservoirs, and the transfer of noble gases toward the surface, controls the overall degassing of the Earth, and these factors are considered further in the degassing models below. It should be emphasized that this is a crucial conclusion, since it renders invalid all calculations of degassing of the solid Earth based upon matching upper-mantle isotope compositions by simple degassing into the atmosphere. Rather, the relationship between mantle noble gas compositions and the degassing of the atmosphere is more complex.

5.4.5 Open-system Models

Another set of models still includes two mantle reservoirs, a MORB-source upper mantle and a deeper, gas-rich reservoir. However, these steady-state box models (Figure 6) are distinguished from the limited interaction box models discussed in Section 5.4.2 in being based upon the open interaction between the upper mantle and both the lower mantle (O'Nions and Oxburgh, 1983) and the atmosphere (Porcelli and Wasserburg, 1995a). Therefore, the constraints from xenon isotopes discussed above (see Section 5.4.4) are naturally accommodated. The steady-state model has been applied to helium isotope and heat fluxes (O'Nions and Oxburgh, 1983; Kellogg and Wasserburg, 1990), the U–Pb system (Galer and O'Nions, 1985), and the other noble gases (O'Nions and Tolstikhin, 1994; Porcelli and Wasserburg, 1995a,b). The central focus of the model is not degassing of the upper mantle to form the atmosphere, but rather mixing in the upper mantle. There are upper-mantle noble gas inputs by radiogenic production from decay of uranium, thorium, and potassium. In addition, atmospheric argon and xenon are subducted into the upper mantle, and lower-mantle noble gases are transported into the upper mantle within mass fluxes of upwelling material that carries all the deep-mantle noble gases together. The lower mantle is assumed to be an approximately closed system, in common with other mantle models (see Section 5.4.3). Noble gases from these sources comprise the outflows at mid-ocean ridges. The isotopic systematics of the different noble gases are linked by the assumption that transfer of noble gases from the upper mantle to the atmosphere by volcanism, as well as the transfer from the lower into the upper mantle by bulk mass flow, occurs without elemental fractionation. All model calculations assume that the upper mantle extends down to the 670 km discontinuity and so comprises 25% of the mantle. Plumes carrying lower-mantle material arise from the 670 km discontinuity.

It has been assumed that upper-mantle concentrations and isotopic compositions are in steady state, so that the inflows and outflows are equal. Since the main outflow is at mid-ocean ridges, this flux can be used to determine the upper-mantle residence times. The upper mantle is degassed to the atmosphere according to a rate constant determined by the rate of melting at ridges, and this then fixes upper-mantle residence times. The calculated values are short ($\sim$1.4 Ga) and imply that nonradiogenic primordial noble gases that cannot be supplied by subduction (including solar helium and neon) are provided from the lower mantle. An important implication of this conclusion is that the composition of the upper mantle does not carry any information regarding earlier mantle volatile history, nor how steady state conditions were reached. In this case, an extended degassing history of the mantle that includes the much earlier rigorous degassing that generated most of the atmosphere, cannot be inferred from noble gases presently found in the upper mantle.

In the open-system models, primordial nonradiogenic noble gases presently seen in the upper mantle originate in the lower mantle, and their fluxes to the atmosphere are moderated by the rates of transfer from the source region and through the upper mantle. As discussed above, the lower-mantle source region is assumed to be a closed system, and so has bulk Earth concentrations of radiogenic isotopes. However, since the lower-mantle argon and xenon isotope compositions, as represented by OIB noble gases, cannot be well constrained by available data, the nonradiogenic argon and xenon isotope concentrations are also not known. However, these can be calculated in the model from the balance of fluxes into the upper mantle. The MORB ^{3}He/^{4}He ratio is a result of mixing between lower-mantle helium and production of ^{4}He in the upper mantle (O'Nions and Oxburgh, 1983). Since the flux of the latter is fixed, the rate of helium transfer from the lower mantle can be calculated (Kellogg and Wasserburg, 1990), and this defines the degassing rate of the lower mantle. Neon, which follows helium,

is transferred similarly, with decreases in ^{3}He/^{4}He ratios accompanied by increases in ^{21}Ne/^{22}Ne ratios due to coupled production of ^{4}He and ^{21}Ne. The MORB ^{40}Ar/^{36}Ar ratio (when corrected for atmospheric contamination of basalts) is a mixture of lower-mantle argon (obtained from the coupling with helium fluxes), radiogenic ^{40*}Ar produced in the upper mantle (calculated from the upper-mantle potassium concentration), and subducted air argon. When there is no subduction, the lower mantle is calculated to have (^{40}Ar/^{36}Ar) = 9,700. The MORB ^{136}Xe/^{130}Xe ratio (when corrected for air contamination) is the result of mixing between lower-mantle xenon, fissiogenic ^{136*}Xe produced in the upper mantle, and subducted air xenon (Porcelli and Wasserburg, 1995a). The lower-mantle ratios are established early in Earth history by decay of ^{129}I and ^{244}Pu, and the ^{129}Xe/^{130}Xe ratio is constrained to be at least as great as that in the upper mantle. Once lower-mantle xenon is transported into the upper mantle, it is augmented by U-derived ^{136*}Xe. Subduction of atmospheric xenon is also possible, lowering the ^{136}Xe/^{130}Xe and ^{129}Xe/^{130}Xe ratios. Therefore, the model is consistent with the isotopic evidence that upper-mantle xenon does not have a simple direct relationship to atmospheric xenon.

The lower mantle has a ^{40}Ar/^{36}Ar ratio that is much higher than the atmospheric value. This can be compared to the reservoirs that supplied the atmosphere, which includes all late-accreting sources and the uppermost fraction of the mantle that has degassed ^{36}Ar early and ^{40}Ar subsequently. The higher ^{40}Ar/^{36}Ar ratio of the lower mantle implies the ^{36}Ar/K ratio of the lower mantle is much lower than the bulk ratio of the reservoirs that supplied the atmosphere. Assuming potassium was initially uniformly distributed throughout the BSE, this implies very heterogeneous incorporation of ^{36}Ar. This contrasts with the typical *a priori* assumption of an initially uniform distribution of ^{36}Ar. The ^{244}Pu-derived ^{136}Xe and radiogenic ^{129}Xe in the upper mantle are derived from the lower mantle. This lower-mantle compositions, obtained by subtracting from MORB xenon the ^{238}U-derived ^{136}Xe produced in the upper mantle, corresponds to closure times that are similar to that of the atmosphere, indicating that early losses occurred from the deep mantle as well (Porcelli *et al.*, 2001). These losses must have been prior to the assumed closed-system evolution.

One feature of the model is that the atmosphere generally has no *a priori* connection with other reservoirs. It simply serves as a source for subducted gases, and no assumptions are made about its origin. Since no assumption is made about the initial distribution of noble gases in the Earth, the atmospheric abundances cannot be used to derive lower-mantle concentrations. The daughter isotopes that are presently found in the atmosphere clearly were originally degassed from the upper mantle, possibly along with nonradiogenic nuclides. However, this occurred before the present character of the upper mantle was established. Therefore, the upper mantle no longer contains information regarding atmosphere formation.

5.4.6 Boundaries within the Mantle

The principal objection has been based on geophysical arguments advanced for greater mass exchange with the lower mantle below 670 km, and so the difficulty of maintaining a distinctive deep-mantle reservoir. The clearest indication of the scale of mantle convection comes from seismic tomography imaging of subducting slabs and mantle plumes that cross the 670 km discontinuity (e.g., Creager and Jordan, 1986; Grand, 1987, 1994; van der Hilst *et al.*, 1997). Plumes may arise from boundary layer instabilities within the mantle or at the core–mantle boundary. There is growing evidence that plumes such as those that form the Hawaiian and Icelandic hotspots have their origin at the core–mantle boundary rather than at a depth of 670 km. Ultrahigh seismic velocity zones at the boundary beneath these islands have been described as plume-induced (Helmberger *et al.*, 1998; Russell *et al.*, 1998). Recent images of the Iceland plume are compatible with a deep origin for this plume (Shen *et al.*, 1998), and a mantle structure extending to the core (Bijwaard and Spakman, 1999). A plume extending to a depth of at least 2,000 km also has been imaged beneath central Europe (Goes *et al.*, 1999), and hotspots in Hawaii, Iceland, South Pacific, and East Africa have been shown to be located above slow anomalies in the lower mantle that extend to the core (Zhao, 2001).

Although geophysical observations point to the present-day mantle convecting as a single layer, it has been argued (Allègre, 1997) that models requiring long-term mantle layering can be reconciled with geophysical observations for present-day whole mantle convection if the mode of mantle convection changed less than 1 Ga ago from layered to whole mantle convection. In this case, the mode of volatile transfer from the deep mantle, and so the rate of deep-mantle degassing, has changed dramatically. However, this idea is at odds with the thermal history of the mantle.

Kellogg *et al.* (1999) developed a model in which mantle below ~1,700 km has a composition, and so density, that is sufficiently different from that of the shallower mantle to largely avoid being entrained and homogenized in the overlying convecting mantle. The boundary with the overlying mantle has a variable depth, and is much deeper where there is geophysical evidence for deeper slab penetration or plumes arising from

the core–mantle boundary. This model preserves a region in the mantle behind which the radioelements and primitive noble gases can be preserved, while accommodating many geophysical observations. There is no geophysical evidence for such a boundary, although whether it could elude seismic detection is debated (Kellogg *et al.*, 1999; Vidale *et al.*, 2001). Also, if the overlying mantle has the composition of the MORB source, then the abyssal layer must contain a large proportion of the heat-producing elements, and it is not yet clear what this effect would have on the thermal stability of the layer or temperature contrast with the overlying mantle. Coltice and Ricard (1999) suggested an alternative model in which helium with high ^{3}He/^{4}He ratios are stored in the narrow zone around the core that is composed of subducted material. This would also be consistent with the geophysical evidence for whole mantle convection, and provide a source for all primordial noble gases. However, it does not provide a reservoir for the undegassed radionuclides (Section 5.3), and it is unclear how such a reservoir would have high initial ^{3}He concentrations (see Porcelli and Ballentine, 2002). Other reservoir configurations have also been suggested for reservoirs containing high ^{3}He/^{4}He ratios, such as in small domains embedded throughout the mantle, but have not been incorporated into comprehensive models that explain all of the noble gas and geophysical observations (see Porcelli and Ballentine, 2002).

The core has also been considered as a source of noble gases. During formation of the Earth, the core may have incorporated sufficient quantities of solar helium to now supply ocean islands with high ^{3}He/^{4}He ratios (Porcelli and Halliday, 2001). In this case, such ratios trace the interaction between the core and mantle. However, the core cannot be used to explain the radiogenic isotope budgets (see Section 5.3) nor the heat–^{4}He budget, and so must be coupled with a mantle model for these features.

5.4.7 Summary

Early degassing models have described the formation of the atmosphere by progressive degassing of the upper mantle, leaving the presently observed noble gases behind to generate strongly radiogenic compositions. However, it is now clear that upper-mantle noble gases are not residual from atmosphere degassing, and have been introduced into the upper mantle from a deeper reservoir after quantitative removal of atmospheric constituents. Nonetheless, the conclusions regarding strong early degassing are compatible with the events of early Earth history. A complete model of noble gas evolution in the mantle and atmosphere remains to be constructed, although elements of earlier models may survive. It might be possible to reformulate the open-system models for a larger upper mantle, or greater mass fluxes between reservoirs, although in some cases non-steady-state upper-mantle concentrations may be required. Unfortunately, the particular configuration of the reservoirs will remain speculative until the geophysical constraints are fully clarified.

A general outline of the history of mantle degassing can be constructed from the observations and constraints discussed above (Porcelli *et al.*, 2003):

(i) Noble gases are acquired throughout the mantle during formation of the Earth, and a substantial inventory is retained within the growing planet, although concentrations in the deeper mantle are lower.

(ii) Radiogenic nuclides produced in the first 10^8 a are degassed from throughout the entire mantle, while isotopic differences are generated due to lower ^{130}Xe concentrations in the deep mantle.

(iii) Strong fractionation of xenon and neon occurred during losses to space. The most plausible mechanism yet formulated involves hydrodynamic escape. Xenon is depleted by at least 10^2. This sets the currently observed fractionation of nonradiogenic xenon in the atmosphere. This may have overlapped with the degassing of the mantle. Nonfractionating losses by atmospheric erosion may have also occurred. All daughter noble gas isotopes that had been degassed to that time were lost to space.

(iv) After losses to space has terminated at ~100 Ma, the radiogenic and fissiogenic xenon produced by the ^{244}Pu and ^{129}I remaining in the upper 40% of the Earth is degassed into the atmosphere, either directly from the mantle, or by transfer of the parent elements into any early crust followed by crustal degassing. This was not accompanied by a significant amount of nonradiogenic (and nonfractionated) xenon.

(v) 40% of the mantle loses ^{40}Ar to the atmosphere and potassium to the crust, from where any further ^{40}Ar produced is largely degassed.

(vi) Gases in 60% of the mantle are largely isolated from the atmosphere. However, a small amount of noble gases has leaked into the upper mantle, where it has mixed with radiogenic nuclides produced there, to generate the presently observed MORB compositions.

The composition of the upper mantle clearly does not provide information on the degassing history of the mantle, but rather reflects the small fluxes into this reservoir from the deep mantle, from production within the upper mantle, and possibly from subduction. The transfer of ^{40}Ar

from the crust throughout much of Earth history has been dictated by the degassing of the crust, where the potassium has been stored.

5.5 DEGASSING OF THE CRUST

The processes involved in the formation of the crust are likely to release any volatiles into the atmosphere that were present in the mantle source region, and so the crust is not expected to be a significant reservoir of primordial volatiles. However, gaseous radiogenic nuclides that are produced within the crust do not readily escape. The most extensively studied has been ^{40}Ar.

5.5.1 Crustal Potassium and ^{40}Ar Budget

There have been a number of studies examining the total potassium content of the continental crust, which is a key parameter for the average petrologic composition of the crust. Estimates of K_2O in the crust, which has a total mass of 2.05×10^{25} g (Cogley, 1984) range from 0.96% to 2.4% (e.g., Weaver and Tarney, 1984; Taylor and McLennan, 1985; Rudnick and Fountain, 1995; Wedepohl, 1995). Taylor and McLennan (1985) and McLennan and Taylor (1996) have emphasized that the extensive data for heat flow can be used to constrain the total abundances of the heat-producing elements. For an average crustal heat flux away from recently disturbed regions of ~48 mW m^{-2} (Nyblade and Pollack, 1993) and a reduced heat flux (i.e., heat conducted through the crust from the mantle) of 27 mW m^{-2} (Morgan, 1984), and assuming that the heat-producing elements are in the ratios Th/U = 3.8×10^4 and K/U = 1.27×10^4, a value of 0.96% K_2O was obtained (McLennan and Taylor, 1996). The corresponding total crustal budget is therefore 1.6×10^{23} g K, and this produces 1.7×10^{31} atoms ^{40}Ar a^{-1}. Using the values of 21 ppb U and K/U = 1.27×10^4 for the bulk silicate Earth, 15% of the total terrestrial potassium is in the crust. Further consideration of the mantle contribution to crustal heat flow has suggested that models with K_2O up to ~2% may be possible (Rudnick *et al.*, 1998), so that ~30% of the terrestrial potassium may be in the crust, and producing twice as much ^{40}Ar.

As discussed above, the K/U ratio was obtained for the MORB source, and is often taken to represent the bulk Earth. This can be applied to the crust only by assuming that there has been no fractionation of potassium and uranium during crust formation, and no preferential recycling of either element. It should be noted that using this ratio, ~85% of the heat currently produced in the crust is generated by uranium and thorium (see, e.g., Turcotte and Schubert, 1982), and so heat flow is a good constraint for the total crustal uranium and thorium budgets, but a poor constraint for the potassium budget, unless the K/U ratio is known very accurately. Therefore, it is possible that the K/U ratio of the crust is somewhat different from the MORB ratio largely due to a different potassium content. In this case, the composition of the deeper mantle (not represented by MORB) may be different.

The amount of ^{40}Ar that is in the crust can be calculated from the mean K–^{40}Ar age of the continental crust of 960 Ga (Hurley and Rand, 1969), which corresponds to a $^{40}Ar/K$ ratio of 1.6×10^{17} atoms $^{40}Ar \, g^{-1}$ K. This is 3–5% of the amount of ^{40}Ar in the atmosphere.

The composition of the crust provides some interesting constraints on the composition of the mantle. If the upper mantle that supplies MORB has ~50 ppm K (e.g., Hofmann, 1988), and extends down to a depth of 670 km (about one-quarter of the mantle), then it has only a small fraction of the total terrestrial potassium. Therefore, the 70–85% of the terrestrial potassium (for a bulk silicate Earth K/U = 1.27×10^4) is in the deeper mantle. Since ~40% of the terrestrial ^{40}Ar is in the atmosphere, some of this deep-mantle material that is not depleted in potassium must have degassed. It has been argued that the upper mantle has a higher potassium concentration of ~100 ppm (Korenaga and Kelemen, 2000), so that ~40% of the terrestrial potassium is in the crust and upper mantle and can supply the atmosphere. In this case, there has been net transfer of potassium to the upper 25% of the Earth. Clearly, how much of the Earth has been degassed and depleted requires further constraints on the potassium budget of both the upper mantle and crust. Coltice *et al.* (2000) pointed out that more potassium is needed to supply the atmosphere than is in the crust, and since potassium and argon are both highly incompatible and so extracted together from the mantle, the additional potassium needed has been subducted.

5.5.2 Formation Time of the Crust

There has been considerable discussion about the age of the crust. Various growth curves have been proposed, from slow steady growth based on K–Ar ages (Hurley and Rand, 1969) to rapid early growth (see Armstrong, 1991), and various intermediate histories (see discussion in Taylor and McLennan, 1985). The current rate of crustal formation is ~1 $km^3 \, yr^{-1}$, and so is insufficient to generate the full mass of the crust over 4.5 Ga. Crustal formation was clearly much more rapid in the past, and the general consensus is that the average crustal formation age is substantially greater than the K–^{40}Ar age. Therefore, it appears that ^{40}Ar has been progressively degassed from the crust.

Considering that potassium now in the crust accounts for ~30–70% of the ^{40}Ar in atmosphere, crustal degassing may be an important component of the solid Earth degassing history (see Hamano and Ozima, 1978). If the crust formed very early, then release of ~30–70% of the ^{40}Ar in the atmosphere has been controlled by crustal processing. For slower crustal growth histories, the proportion degassed directly from the crust would depend upon the time constant for growth compared to the 1.4×10^9 a half-life of ^{40}K.

5.5.3 Present Degassing

The present rate of continental degassing of ^{40}Ar into the atmosphere cannot be readily measured, and so must be inferred by considering the mechanisms of degassing. Studies of radiogenic ^4He in groundwaters from the Great Artesian Basin in Australia (Torgersen and Clarke, 1985; Torgersen and Ivey, 1985) concluded that the higher ^4He concentrations in older waters could be explained by a steady influx from production in the underlying crust. The required calculated flux was found to equal the entire production rate in the crust, and so suggested that the continental crust degasses by continuous release of all radiogenic ^4He from uranium- and thorium-bearing host rocks. A similar conclusion was drawn from data from another basin as well (Heaton, 1984). Additional work on ^{40}Ar in the Great Artesian Basin extended this conclusion to ^{40}Ar (Torgersen *et al.*, 1989), implying that the continental flux to the atmosphere is $(1.7–3.3) \times 10^{31}$ atoms $^{40}Ar \, a^{-1}$. However, there have been doubts that helium and argon can be so readily released from minerals and transported effectively across the crust. These doubts have generated discussions regarding the interpretations of groundwater flow rates and so noble gas accumulation rates (e.g., Mazor, 1995; Bethke *et al.*, 1999).

An important factor is how radiogenic ^{40}Ar is released from host rocks. The release of ^{40}Ar from potassium-bearing minerals by diffusion has been extensively examined as part of thermal evolution studies, and the blocking temperatures of major minerals such as feldspars and biotites correspond to quantitative retention within the top ~10 km of the crust (McDougall and Harrison, 1999). The upper continental crust, down to ~10 km, is estimated to have 3.4% K_2O and 2.8 ppm U (Taylor and McLennan, 1985). This is consistent with the observation that the concentration of heat-producing elements decreases with depth (e.g., Jaupart *et al.*, 1981), although potassium may not decrease with depth as strongly as uranium and thorium (e.g., Rudnick and Fountain, 1995). Nonetheless, a dominant fraction of the potassium in the crust is below the blocking temperature of ^{40}Ar. Further, the ^{40}Ar produced in the deeper crust is more likely to be transported episodically to the surface during mobilization of fluids that may also be responsible for metamorphic processes (e.g., Etheridge *et al.*, 1984).

A process that clearly leads to widespread release of ^{40}Ar is weathering at the surface. In a comprehensive survey of river chemistry, Martin and Meybeck (1979) reported an average concentration of 1.4 ppm K in a total water discharge of 3.74×10^{16} L yr^{-1}, leading to a total of 5.2×10^{13} g K. This potassium is likely to be released largely from the weathering of feldspars (e.g., Wollast and Mackenzie, 1983). Assuming that this K is derived from weathering of crystalline rocks with a mean age of 960 Ma (Hurley and Rand, 1969) with 1.6×10^{17} atoms ^{40}Ar g^{-1} of potassium, this leads to release of 0.83×10^{31} atoms a^{-1}. Mechanical weathering of the continental crust leads to discharge to the oceans of 1.6×10^{16} g a^{-1} (Milliman and Meade, 1983), which contains an average concentration of 2% K (Martin and Meybeck, 1979) or 3×10^{14} g K. If this material is derived from cannibalization of earlier sediments (see, e.g., Veizer and Jansen, 1985) with an average age of $\sim$500 Ma (with 0.6×10^{17} atoms ^{40}Ar g^{-1} of potassium), then this material contained 1.8×10^{31} atoms ^{40}Ar when mobilized, which is twice that released during dissolution of potassium. Of course, the average K–Ar ages of materials that provide either detrital or dissolved constituents remains highly uncertain. The long-term rate of mechanical weathering may differ considerably from this because of recent changes in discharge due to human activities and due to the deposition within drainage basins. However, a much greater uncertainty lies in estimating how much of this material has lost ^{40}Ar. Limited studies have found that the K–Ar ages of surface sediments often reflect those of the source rock, even where significant clay mineral formation has occurred (e.g., Hurley *et al.*, 1961). Hurley (1966) summarized available data and suggested that detrital ages are gradually wiped out during burial. Later studies in the Gulf of Mexico confirmed that K–Ar ages decreased with depth at rates that depend upon mineral size fractions (Weaver and Wampler, 1970; Aronson and Hower, 1976), and likely reflect the redistribution of potassium, and loss of ^{40}Ar, during diagenetic illite formation. Overall, it is clear that some poorly constrained proportion of the ^{40}Ar contained in sediments is released. Perhaps as much ^{40}Ar is released by diagenesis of sediments as derived from primary chemical weathering.

Another mechanism for degassing ^{40}Ar is thermal processing due to tectonic thickening of the crust in orogenic belts (see, e.g., England and Thompson, 1984). Veizer and Jansen (1979, 1985)

used the distribution of basement K–Ar ages to model the thermal cycling rate of the crust, and obtained a rate of $\sim$2 $\times$ 10^{16} g a^{-1}. It is possible that during such events, ^{40}Ar in the lower crust is also transported to the surface, and so degassing of the entire crustal thickness may occur together. For an average crustal composition with 1% K and 1.6×10^{17} atoms ^{40}Ar g^{-1} of potassium (assuming an average K/Ar age of the crust of 960 Ma), 3.2×10^{31} atoms ^{40}Ar a^{-1} are released. Given the range of possible crustal potassium contents discussed above, this value may be as much as 2 times greater.

In summary, chemical weathering and orogenic processing of the crust generate a flux of $\sim$(3.8–7) $\times$ 10^{31} atoms ^{40}Ar a^{-1} to the atmosphere, and this may be augmented by a presently poorly constrained flux associated with potassium in riverine sediments. Nonetheless, thermal processing of the continental crust appears to dominate the ^{40}Ar flux. The total flux is more than the present-day crustal production rate of 1.7–3.4×10^{31} atoms ^{40}Ar a^{-1}. If the flux does indeed exceed the present production rate, then the ^{40}Ar reservoir size of the continents is diminishing.

The present flux of ^{40}Ar from mid-ocean ridges is 1.9×10^{31} atoms a^{-1} (Section 5.2.7), and so substantially lower than the continental flux, indicating that at present continental processing is more important than mantle processing for ^{40}Ar release to the atmosphere.

5.6   MAJOR VOLATILE CYCLES

Many of the important characteristics of the atmosphere are related to the concentrations of particular molecular species, such as O_2. Here the absolute abundances of carbon, nitrogen, and hydrogen (as H_2O) as supplied from the mantle are considered in the context of models of noble gas degassing from the solid Earth. A convenient reference for comparing surface volatile inventories to mantle reservoirs is obtained by dividing the surface volatiles into the mass of the upper mantle, the minimum size of the source reservoir. However, this should not be taken to imply a particular model of degassing.

5.6.1   Carbon

Crustal carbon. In contrast to the noble gases, carbon near the planetary surface is concentrated in crustal rocks, and is largely divided between carbonates, with an isotopic composition of $\delta^{13}C = 0$‰ and sedimentary organic carbon with $\delta^{13}C = -25$‰. While Hoefs (1969) estimated a total of 2.6×10^{22} g of C, with 66% in sedimentary rocks, Hunt (1972) reassessed the budget using sedimentary rock abundance data from Ronov and Yaroshevsky (1969) and carbon concentration data compilations to obtain a higher value of 9×10^{22} g. Ronov and Yaroshevsky (1976) updated their earlier work with new additional data to obtain a total budget of 1.2×10^{23} g. A similar value (6×10^{22} g C) was obtained by Li (1972) when requiring that the crust have an isotopic composition similar to that of MORB (see below) of approximately $\delta^{13}C = -5$; conversely, this agreement supports the notion that the crust and mantle are isotopically similar. Therefore, considering the uncertainties involved, a value of 1.0×10^{23} g C, or 8.3×10^{21} mol, with perhaps a 20% uncertainty, appears to be reasonable (Table 3).

Carbon in MORB and the upper mantle. MORB entering the ocean crust are generally oversaturated in carbon as CO_2, and exhibit exsolved CO_2-rich vesicles. Carbon contents of MORB vary widely due to degassing, but a ratio of $C/^3He = 2 \times 10^9$ appears to characterize the undegassed magmas (Marty and Jambon, 1987). Combined with the initial concentration of ^3He obtained from the total ^3He ridge flux (see Section 5.2.7), this provides a value of 510 ppm carbon in undegassed MORB, within the range of other estimates of undegassed MORB of 900–1,800 ppm carbon (Holloway, 1998) and 400 ppm (Pineau and Javoy, 1994). Assuming carbon is incompatible, this corresponds to an upper mantle (to a depth of 670 km) concentration of 50 ppm. Therefore, the upper mantle of 1×10^{27} g contains about half the carbon that is present in the crust. If this value applies to the entire mantle, then only about a third of the terrestrial carbon is at the surface. The carbon isotope composition of the upper mantle is $\delta^{13}C = -5\permil$ based on MORB data (e.g., Javoy and Pineau, 1991), and is similar to that of the surface. The carbon brought to the surface by MORB is quantitatively degassed during eruption and subsequent alteration, and results in a flux of 2×10^{12} mol C yr^{-1}.

There may also be a significant amount of carbon in the core, with values of up to 4% possible, although this depends upon the amount available in the early Earth (Wood, 1993; Halliday and Porcelli, 2001). The amount of carbon that is supplied across the core–mantle boundary into the mantle is not known.

Carbon at hotspots. Ocean islands with high $^3He/^4He$ ratios clearly contain distinctive volatile components. However, there are also some data regarding the carbon flux of hotspot sources. The ratio of hotspot volcanics is an important parameter in further defining the carbon cycle. It has been argued that the $C/^3He$ ratio is equal to that of the upper mantle (Trull *et al.*, 1993). Alternatively, Poreda *et al.* (1992) argued that the ratio is higher

and is 6×10^9 in Iceland, while Kingsley and Schilling (1995) argued from mixing relationships along the Mid-Atlantic Ridge that the Iceland source was $\sim 10^{10}$. Hilton *et al.* (1998a) found values of $(2-5) \times 10^9$ for Loihi, but also argued that degassing highly modified the ratio, and concluded that the source ratio was unknown. It should be noted that plumes appear to contain contributions not only from a high $^3He/^4He$ source region, but also subducted components (e.g., Sobolev *et al.*, 2000) that may contribute carbon that increases the $C/^3He$ ratio and mask the carbon accompanying helium with high $^3He/^4He$ ratios. Measured $\delta^{13}C$ values scatter around upper-mantle values (Trull *et al.*, 1993), and a consistent composition distinctive from that of MORB has not been unambiguously documented. The flux of carbon from ocean islands depends on the accepted $C/^3He$ as well as the flux of ^3He relative to MORB ^3He flux (see above).

Subduction. There is a significant amount of carbon on the ocean floor that is available for subduction. The amount of carbon released by arc volcanism is $(1.6-3.1) \times 10^{12}$ mol a^{-1}. This can be compared with estimates of the amount of subducted carbon of $(0.35-2.3) \times 10^{13}$ mol yr^{-1} (Bebout, 1995; Hilton *et al.*, 2002). Therefore, while it is possible that there is no carbon return flux into the upper mantle beyond island arcs, the ranges suggest that some carbon does survive in the downgoing slabs, and this flux may even be greater than the degassing flux observed at mid-ocean ridges of 0.2×10^{13} mol a^{-1}. Volcanic arcs have a range of $\sim\delta^{13}C = -2\permil$ to $-12\permil$ (Bebout, 1995), reflecting the variable proportions of subducted components.

The global carbon cycle. The flux of carbon brought to the surface by MORB of 2×10^{12} molC yr^{-1} can supply the entire crustal carbon inventory in 4.2 Ga; if there was stronger earlier degassing, there must have been a carbon return flux to the mantle. Much of the surface inventory indeed may have been degassed early along with noble gases (Section 5.4.1). It is possible that the surface inventory was initially greater than at present due to the difficulties of subducting carbon in hotter, younger slabs (Des Marais, 1985; McCulloch, 1993; Zhang and Zindler, 1993), although this effect may have been countered by increased incorporation of carbon into ocean crust by hydrothermal alteration (Sleep and Zahnle, 2001). If the carbon in the upper mantle is largely recycled, it may now be present in a steady-state abundance. This would naturally explain the high upper-mantle carbon concentration. However, since surface carbon is divided at the surface between carbonate carbon (with $\delta^{13}C = 0\permil$) and organic carbon (with $\delta^{13}C \sim -25\permil$), the subducted carbon at any location need not have the average crustal $\delta^{13}C$. Therefore, in order to reconcile the isotopic similarity between the mantle and surface, carbonate and organic carbon must be subducted in the same proportions as at the surface. Otherwise, shifts in the composition away from that of the bulk Earth will occur in both the crust and mantle. There are no obvious controls that would demand this condition. The isotopic compositions of carbon in back arc basin basalts scatter, and can be ~5‰ lighter than that of MORB carbon (Mattey *et al.*, 1984), reflecting mixing of organic carbon with MORB carbon.

An alternative view that naturally explains the isotopic similarity between the crust and upper mantle is that the carbon degassing from the mantle is primordial carbon that was trapped during solid Earth formation, representing the continuing unidirectional upward transfer of carbon. In this case, any carbon that is subducted does not constitute a dominant fraction of carbon in the upper mantle. However, carbon is not depleted in the upper mantle to the same extent as other highly incompatible elements such as the noble gases. An indication of this is the mantle $C/^{36}Ar$ ratio, which is 10^2 times greater than that of the surface (Marty and Jambon, 1987). This observation has been explained by having less net degassing of carbon early in Earth history, either due to higher recycling rates or more compatible behavior under more reducing conditions (Marty and Jambon, 1987). As discussed above (Section 5.4), the nonradiogenic noble gases may be supplied from a deeper mantle reservoir to the MORB source reservoir; if carbon is similarly supplied, then using the calculated ^3He content of a gas-rich reservoir (see Section 5.4.3) and MORB $^{36}Ar/^3He$ (see Section 5.2.6) and $^{36}Ar/C$ ratios as reflecting the deeper mantle, a carbon concentration of 3,000 ppm is obtained. This value is an order of magnitude less than values for carbonaceous chondrites (Kerridge, 1985), and would be even lower if some of the upper-mantle ^{36}Ar is from subduction.

5.6.2 Nitrogen

Nitrogen at the surface. Nitrogen is highly atmophile, although there is a significant fraction of surface nitrogen in the continental crust (see Table 3). Crustal nitrogen is isotopically heavier than atmospheric nitrogen, and the total nitrogen at the surface has $\delta^{15}N = 2‰$ relative to the atmospheric value of 0‰ (Marty and Humbert, 1997). The total budget of 4.5×10^{21} g, when divided into the upper mantle, yields 4.5 ppm for the bulk silicate Earth.

Nitrogen in MORB and the upper mantle. The concentration of N_2 in undegassed MORB can be determined from the value of $N_2/^{40}Ar = 120 \pm 20$ obtained from MORB data (Marty and Humbert, 1997; Marty and Zimmermann, 1999) to be

1.63 ppm. This corresponds to an upper-mantle concentration of 0.16 ppm, assuming nitrogen is incompatible during melting (Marty and Humbert, 1997). The amount of nitrogen in the upper mantle is therefore 1.6×10^{20} g, which is only ~3% that in the crust and atmosphere. The MORB flux is equivalent to a flux of 5.0×10^{10} mol yr^{-1}, or 9% of the surface N_2 over 4.5 Ga. Note that Javoy (1998) argued that nitrogen is relatively compatible, with an upper-mantle concentration of up to 40 ppm (Cartigny *et al.*, 2001). However, this is from mass balance calculations based upon model assumptions that volatiles in the upper mantle and on the surface are a mixture of enstatite chondrites and a late veneer of CI chondrites, and such a model has not been widely adopted.

The nitrogen isotopic composition of the upper mantle has been estimated from MORB (Marty and Humbert, 1997; Marty and Zimmermann, 1999) and diamond (Javoy *et al.*, 1984; Cartigney *et al.*, 1998; Boyd and Pillinger, 1994) data to be approximately $\delta^{15}N = -4‰$. Values as low as $\delta^{15}N = -25‰$ have been found in diamonds, indicating that there is another reservoir in the mantle.

Nitrogen at hotspots. $\delta^{15}N$ values from hotspots has been found to be largely positive and as high as $\delta^{15}N = +8‰$ as (Dauphas and Marty, 1999; Marty and Dauphas, 2003). It has been argued that these values are due to the presence of subducted components in mantle plumes (Marty and Dauphas, 2003), and is supported by the lack of correlation with helium isotope compositions.

Subduction. There are some data for the subduction of nitrogen. Metasedimentary rocks and organics in pelagic sediments are isotopically heavy, with organic nitrogen having $\delta^{15}N = +2‰$ to +10‰ in marine sediments (Peters *et al.*, 1978) and $\delta^{15}N = +2‰$ to +15‰ in metasediments (Haendel *et al.*, 1986; Bebout and Fogel, 1992). Metamorphosed complexes are also isotoipcally heavy (Bebout, 1995). Arc volcanics have an average of $\delta^{15}N = +7‰$ (Sano *et al.*, 1998). The subduction flux has been estimated from an average sediment subduction rate of 3.5×10^{15} g yr^{-1} (Ito *et al.*, 1983) and an average concentration of 125–600 ppm from the Catalina Schist to be $(0.44–2.1) \times 10^{12}$ g yr^{-1} (Bebout, 1995), although N in altered seafloor basalts may also carry nitrogen (Hall, 1989). The flux to the atmosphere due to arc volcanism is 4.5×10^{11} g a^{-1} (Hilton *et al.*, 2002). Therefore, it is possible that no nitrogen is returned to the mantle, although a flux equal to that at mid-ocean ridges (and so supporting steady state mantle and surface abundances), or higher (supporting arguments for the net inflow of nitrogen (Javoy, 1998) is also possible.

The global nitrogen cycle. A salient feature of the nitrogen cycle is that mantle nitrogen is isotopically lighter than both the surface and subducted nitrogen. The atmosphere is either derived from another component than the mantle or has suffered preferential loss of ^{14}N. The atmospheric composition therefore is inherited from the processes of Earth formation. Subsequently, the rates of mantle–crust exchange are constrained by these isotopic differences. There are limited possibilities for the source of isotopically light nitrogen in the mantle. Data for carbonaceous and ordinary chondrites typically show $\delta^{15}N > 0‰$ (Kerridge, 1985). However, E chondrites have been found to have consistently lower values, with an average of about $\delta^{15}N < -35‰$, and with concentrations on the order of 500 ppm (Grady *et al.*, 1986). Therefore, it has been proposed that E chondrites are the source of mantle nitrogen (Javoy, 1998). Since the meteorite concentrations are 4×10^3 times greater than presently found in the mantle, if a substantial portion of the Earth was derived from E chondrites, substantial nitrogen losses must have occurred during accretion (Javoy, 1998); conversely, only a small fraction of material needed to have been incorporated in an early protoplanet prior to sufficiently large impacts cause volatile loss (Ahrens, 1993). Alternatively, solar nitrogen may also be isotopically very light. Recent data for lunar samples indicate that solar nitrogen may be as light as $\delta^{15}N = -380‰$ (Hashizume *et al.*, 2000; Owen *et al.*, 2001). The consequences of incorporating solar nitrogen in the Earth has not been fully explored, although due to the high solar Ne/N ratio, incorporation of a significant amount of solar nitrogen will require strong elemental fractionation.

If nitrogen is highly incompatible along with the noble gases, then it likely is introduced into the upper mantle from another reservoir as well. Since this nitrogen may be isotopically very light, it appears that there is also a subducted, isotopically heavy, nitrogen component in the upper mantle, and this provides the dominant fraction of nitrogen. The specific proportions depend upon the isotopic compositions of each. In the case of solar nitrogen, only 3% of MORB nitrogen is solar (3 ppb), with the remainder provided by subduction. A much larger fraction of nitrogen derived from enstatite chondrites is required. If this nitrogen was introduced with ^3He from a deep-mantle source that also supplies plumes, then OIBs would be expected to have very light $\delta^{15}N$ values. This is not the case; rather, nitrogen appears to be dominated by subducted components. However, it is clear that both deep-mantle and subducted components are present in OIB source regions, and so primordial nitrogen may be masked. Alternatively, the composition of nitrogen subducted into the upper mantle prior to the Archean was isotopically light, and has persisted in the upper mantle over a long residence time (Marty and Dauphas, 2003).

5.6.3 Water

Crustal H_2O. The terrestrial oceans have $\delta D = 0‰$ with an inventory of 1.4×10^{24} g H_2O, equivalent to 120 ppm H when divided into the mass of the upper mantle. There is an additional 20% in other surface reservoirs and crustal rocks, and the total composition is $\sim\delta D = -10‰$ to $-18‰$ (Taylor and Sheppard, 1986; Lécuyer *et al.*, 1998).

H_2O in MORB and the upper mantle. Water is not so readily lost by vesiculation, with measurements of unaltered MORB glasses of 0.12% to 0.33 wt.% H_2O (Pineau and Javoy, 1994; Jambon, 1994; Sobolev and Chaussidon, 1996). Water behaves incompatibly during melting (see Jambon and Zimmermann, 1990), and so a concentration of 120–330 ppm is obtained for the mantle source region (see also Thompson, 1992). The upper mantle, therefore, contains at least as much water as the hydrosphere. The flux at ridges totals $(1–3) \times 10^{14}$ g a^{-1}. The hydrogen isotopic composition is distinctive from that of the surface, with $\delta D = -71‰$ to $-91‰$ (Craig and Lupton, 1976; Kyser and O'Neil, 1984; Poreda *et al.*, 1986).

H_2O at hotspots. There are some hydrogen isotope data available for hotspots. Poreda *et al.* (1986) found a correlation between helium and hydrogen isotopes along the Reykjanes Ridge, with values of up to $\delta D = -50‰$, significantly higher than in MORB. These samples also had an increase in water content toward Iceland, with values of up to $\sim 0.35\%$. A higher water content was also found in samples from the Azores Platform, with 0.52 wt.% H_2O. Rison and Craig (1983) found glasses from Loihi had $\delta D = -69‰$ to $-74‰$, within the MORB range. Lighter compositions, down to $-125‰$, have been found in Hawaii and elsewhere (Deloule *et al.*, 1991; Hauri, 2002), which suggests that isotopically light, juvenile hydrogen remains in the mantle.

Subduction. The flux of H_2O entering subduction zones of 8.8×10^{14} g yr^{-1} or greater (Ito *et al.*, 1983; Bebout, 1995) is substantially greater than the ridge flux, and has been interpreted as a net inflow of water (Ito *et al.*, 1983). However, a substantial fraction is returned during arc metamorphism, dewatering, and volcanism. Data for boron isotopes and B/H_2O relationships suggest that only $\sim 20\%$ of the water is recycled (Chaussidon and Jambon, 1994), compatible with a steady-state mantle concentration. Due to the difficulties of obtaining total long-term values for subduction and return of water, concrete constraints on the balance of H_2O will probably be impossible.

The global H_2O cycle. The main feature of the solid Earth hydrogen cycle is the large contrast between the isotopic composition of the surface of $\delta D = -10‰$ to $-20‰$ and the upper mantle of $\delta D = -80‰$. There have been two contrasting ideas. It has been argued that upper-mantle water is derived entirely from subduction. The low δD ratio of MORB could be controlled by those of dewatered metasedimetary and metamorphosed mafic subducted rocks that have compositions that are close to upper-mantle values, with $\delta D = -50‰$ to $-80‰$ (Magaritz and Taylor, 1976; Bebout, 1995). The average of these rocks may be somewhat heavier than MORB values, although Bell and Ihinger (2000) suggested that somewhat lighter hydrogen might be in nominally anhydrous minerals that may preferentially retain hydrogen during subduction. Alternatively, this isotopic signature has been ascribed to a primordial component established and isolated early in Earth history (e.g., Craig and Lupton, 1976; Lécuyer et al., 1998; Dauphas et al., 2000). It is possible that this is stored with ^3He in the deep mantle. If the H_2O and ^3He are both derived from the same reservoir, then using the MORB ratio of H_2O to ^3He, and an undegassed mantle ^3He concentration (Section 5.4.3), this reservoir contains 900–2,300 ppm H, which is much higher than the surface reservoir divided into the upper mantle, although somewhat below CI chondrites (Kerridge, 1985). Isotopically, CI chondrites are too heavy, and enstatite chondrites, with approximately $-460‰$, provide a better source, but contain only 50 ppm (Javoy, 1998). Overall, it appears likely that water in the upper mantle is dominated by subducted components, although hydrogen from the lower mantle may accompany ^3He upwards, and may also account for some isotopically light mantle values.

5.7   DEGASSING OF OTHER TERRESTRIAL PLANETS

Understanding of the degassing of Mars and Venus is incomplete due to the lack of data regarding interior volatile reservoirs, and uncertainties in whether differences between the surface inventories with those on Earth are due to different degassing histories or to initial differences generated during planet formation. There are manifestations of volcanic activity that can be used to deduce at least a generalized history of mantle melting, but of course the precise timing and relationships to mantle structure are not well constrained. In this review, some of the information regarding the extent of planetary degassing relative to that of the Earth is considered.

5.7.1   Mars

Mars is characterized by low noble gas abundances in the atmosphere, with equivalent total planet concentrations that are 10 times less than on Earth. The amount of martian carbon is also not well known; while the atmosphere has only

7×10^{18} g C, equivalent to 0.01 ppm for the bulk planet (Owen et al., 1977), a large fraction may be stored in the polar regolith.

While the amount of water that has been observed on Mars is low, there is evidence that there was substantially more in the past. Measurements of atmospheric water vapor on Mars have found D/H values ~ 5 times that of the Earth and has been fractionated due to Jeans escape of hydrogen to space (Owen et al., 1988). Models of hydrogen atmospheric losses (Donahue, 1995) and morphological data of features generated by surface water (Carr, 1986) both suggest that originally there may have been the equivalent of up to 500 m of water, or $\sim 7 \times 10^{22}$ g H_2O. The total mass of Mars is 0.11 that of the Earth, and so both planets originally may have had similar bulk water concentrations. This would also require degassing of water on both planets to similar extents.

The atmosphere of Mars has a high value of $\delta^{15}N = (+620 \pm 160)‰$ (Nier and McElroy, 1977) that may be due to fractionating losses, and $(4.7 \pm 1.2) \times 10^{17}$ g N, which is equivalent to 0.7 ppb when divided into the mass of the entire planet (Owen et al., 1977). Therefore, Mars appears to have 10^{-4} times the nitrogen on the Earth. Mathew et al. (1998) reported evidence for a component in a martian meteorite with $\delta^{15}N < -22‰$, suggesting that, like the Earth, the solid planet may contain nitrogen that is isotopically lighter than the atmosphere, and consistent with the modification of atmospheric nitrogen isotopes by losses.

Radiogenic ^{40}Ar. The atmospheric $^{40}Ar/^{36}Ar$ ratio has been measured by Viking to be $3,000 \pm 400$ (Pepin and Carr, 1992), although a lower value of $\sim 1,800$ has been deduced from meteorite data (Pepin and Carr, 1992; Bogard, 1997; Bogard et al., 2001). Based on the Viking data, there are $(7.0 \pm 1.4) \times 10^{39}$ atoms ^{40}Ar in the atmosphere. The Mars mantle has been estimated to have 305 ppm K (Wänke and Dreibus, 1988), so that 3.3×10^{41} atoms ^{40}Ar have been produced in Mars. Therefore, only 2% of martian ^{40}Ar has degassed to the atmosphere, and most of the planet interior has retained the ^{40}Ar produced throughout its history. The history of degassing of ^{40}Ar from the interior has been discussed in several studies (Volkov and Frenkel, 1993; Sasaki and Tajika, 1995; Hutchins and Jakosky, 1996; Tajika and Sasaki, 1996). There are considerable uncertainties in the history of martian volcanism and the amounts of volatiles that have been lost to space that must be resolved before more definitive degassing histories can be constructed. Tajika and Sasaki (1996) argue that much of the ^{40}Ar has degassed from relatively recent volcanic regions, in contrast to early degassing of other volatiles, while Hutchins and Jakosky (1996) conclude that in order to account for volatiles that have been lost to space, much degassing must have occurred by processes other than volcanic outgassing.

The radiogenic ^{129}Xe budget. Martian atmospheric xenon clearly contains a considerable fraction of radiogenic ^{129}Xe. It has been estimated that the silicate portion of Mars contains 32 ppb I (Wänke and Dreibus, 1988). Assuming that ^{129}I/^{127}I $= 1.1 \times 10^{-4}$ at 4.57 Ga (Hohenberg *et al.*, 1967), then 8.44×10^{36} atoms ^{129}Xe have been produced in Mars or precursor materials. Using fractionated CI chondrite xenon or solar xenon for the nonradiogenic light xenon isotope composition, the atmosphere is calculated to contain only 0.092% of what has been produced. Assuming there is none remaining in the planet, this corresponds to a closure age of 160 Myr. However, if only 2% has degassed (like ^{40}Ar) to the atmosphere, then a closure age of 70 Myr is obtained. This value is similar to that of the Earth; it suggests that there may also have been losses of volatiles from Mars over the same extended period of accretion.

The fissiogenic ^{136}Xe budget. The amount of ^{136}Xe produced in Mars or accreting materials, assuming that the silicate portion of Mars has 16 ppb of ^{238}U at present (Wänke and Dreibus, 1988) and initially had ^{244}Pu/^{238}U $= 0.0068$ (Hudson *et al.*, 1989), is 1.9×10^{34} atoms ^{136}Xe from ^{244}Pu and 7.2×10^{32} atoms ^{136}Xe from ^{238}U. In contrast, there is a total of 2.8×10^{33} atoms ^{136}Xe in the atmosphere. Up to $\sim$5% of the atmospheric ^{136}Xe may be plutonium-derived; if so, and the closure age for Mars is 70 Ma, then 1–2% of the ^{244}Pu produced in the solid planet has degassed. This is consistent with the ^{129}Xe and ^{40}Ar budgets. It has been argued that plutogenic ^{136*}Xe could make up much less that 5% of the total atmospheric inventory, requiring even less planetary degassing and greater very early isolation of interior volatiles from the atmosphere. Reports of significant abundances of ^{244}Pu fission xenon in several SNC meteorites do in fact point strongly to its efficient retention in the martian crust (Marty and Marti, 2002; Mathew and Marti, 2002). Further discussion of the abundances of daughter xenon isotopes in the atmosphere is provided by Swindle and Jones (1997).

Martian mantle noble gases. Martian meteorites contain components other than those derived directly from the atmosphere (see detailed discussion by Swindle (2002)). Information on the relative abundances of the heavier noble gases in the mantle (Ott, 1988; Mathew and Marti, 2001) suggests that the ^{84}Kr/^{132}Xe ratio is at least 10 times lower than both the martian atmosphere and the solar composition. If this is truly a source feature, it indicates that heavy noble gases trapped within the planet suffered substantially different elemental fractionation than the atmosphere and have not subsequently formed a dominant fraction of the atmosphere. However, it is not possible at present to conclusively determine whether the measured elemental abundance ratios reflect an interior reservoir that was initially different from atmospheric noble gases, rather than due either to planetary processing or transport and incorporation into the samples.

The noble gas isotopic composition of the martian interior is only available for xenon. Data for the martian meteorite Chassigny found xenon with little scope for radiogenic additions (Ott, 1988; Mathew and Marti, 2001), indicating that this reservoir had a high Xe/Pu ratio, at least during the lifetime of ^{244}Pu. Data from other meteorites indicate that there are other interior martian reservoirs that contain solar xenon but with resolvable fissiogenic contributions (see Mathew and Marti, 2002; Marty and Marti, 2002), and so have had lower Xe/Pu ratios. Data for ^{129}Xe and ^{136}Xe have been used to argue that there were substantial losses of xenon from the mantle within the first 35 Myr (Marty and Marti, 2002). The martian mantle appears to have nonfractionated solar xenon (Ott, 1988). This contrasts with the fractionated character of atmospheric xenon, and is consistent with fractionation of the atmosphere after strong early degassing and minimal recycling of xenon into the mantle.

Martian degassing history. The budget for radiogenic argon and xenon indicate that only a small fraction of the planet has degassed since very early in planetary history. A consequence of this is that noble gases were retained within the planet during formation from a very early stage, as indicated by the ^{129}Xe budget. This is supported by evidence for xenon-rich mantle domains that for a body of at least this size, accretion does not necessarily lead to strong degassing. The low atmospheric abundances of nonradiogenic noble gases may be partly due to retention in the largely undegassed planet, but the much higher ^{40}Ar/^{36}Ar ratio relative to the terrestrial value indicates that the initial inventory of nonradiogenic species after planet formation was likely lower. The composition of nitrogen and water at the surface of Mars has been strongly affected by the history of losses to space, with no evidence for significant fluxes back into the planet.

The observation that the martian atmosphere has a higher ^{129}Xe/^{130}Xe ratio than the mantle, in contrast to the situation observed on Earth, has led to speculations about the mechanisms fractionating iodine from xenon. Both of these elements are highly incompatible, and so melting and crust formation are expected to transport them together out of the mantle. Available data suggest that iodine is more incompatible, so that the residue is expected to have a higher I/Xe ratio. In order to generate a higher atmospheric ^{129}Xe/^{130}Xe ratio, Musselwhite *et al.* (1991) suggested that the iodine was sequestered in the crust by hydrothermal alteration while xenon was lost to space. Alternatively, Musselwhite and Drake (2000) suggest that the iodine was preferentially retained within a magma ocean. These models are based upon the assumption that the mantle noble gases are the residue that is complementary to the atmosphere; this is not the case on the Earth and it is not clear whether this is true for Mars.

5.7.2 Venus

Venus is similar in size to the Earth and might be expected to have differentiated to a similar extent. However, while the early accretion history might have been similar (with the exception of the absence of a moon-forming event), silicate differentiation did not proceed according to the familiar plate tectonic mechanisms. There is, of course, no data on the interior of Venus, and so planetary degassing characteristics must be deduced from limited atmospheric data and observations of volcanic activity at the surface.

The measured $^{40}Ar/^{36}Ar$ ratio is 1.11 ± 0.02, substantially less radiogenic than the terrestrial atmosphere. For a mixing ratio of 21–48 ppm (Donahue and Pollack, 1983), the atmosphere contains $(1.8–4.4) \times 10^{41}$ atoms ^{40}Ar. Divided by the mass of the planet, this corresponds to $(3.6–9.0) \times 10^{13}$ atoms $^{40}Ar\ g^{-1}$. This is 0.2–0.5 times the value for the Earth. However, Venus appears to be deficient in potassium. Data for the K/U ratio of the surface indicate that K/U $=$ $7,220 \pm 1,220$ (Kaula, 1999), or 0.57 ± 0.10 times that of the value of 1.27×10^4 commonly taken for the Earth. Assuming that Venus has the same uranium concentration as the total Earth (including the core) of 14 ppb, then $12 - 28\%$ of the ^{40}Ar produced in Venus is now in the atmosphere (see Kaula, 1999). This indicates that a substantial inventory of ^{40}Ar remains within the planet, possibly also accompanied by up to an equivalent fraction of nonradiogenic noble gases. In contrast, at least 40% of terrestrial ^{40}Ar is in the atmosphere.

Venus has about twice as much carbon (like nitrogen) at the surface than the Earth, equivalent to 26 ppm when divided into the bulk planet (von Zahn et al., 1983); whether Venus is more rich in carbon therefore depends upon what volume of the mantle has degassed and how much remains in the mantle (Lécuyer et al., 2000). From the radionuclide budget discussed above, it appears that Venus may have degassed to a similar extent as the Earth, but is unlikely to have been substantially more degassed. Therefore, it appears that the high abundances of carbon and nitrogen reflect greater total planetary abundances. The Venus atmosphere has ~200 ppm H_2O (Hoffman et al., 1980) and a D/H ratio of $(1.6 \pm 0.2) \times 10^{-2}$,

i.e., $\sim 10^2$ times that of the Earth (Donahue et al., 1982). It has been suggested that Venus originally had the same D/H value as the Earth, but has lost at least one terrestrial ocean volume of water by hydrodynamic escape, thereby generating an enrichment in deuterium (Donahue et al., 1982). The ratio of water to carbon and nitrogen therefore may have been similar to that of the Earth.

Venus is also rich in nonradiogenic noble gases, with the absolute abundance of ^{36}Ar on Venus exceeding that on Earth by a factor >70. This is clearly due to the amount of noble gases initially supplied and retained by the planet.

Venus is similar to the Earth in mass and composition, and so might be expected to evolve similarly. However, Venus has a hot, insulating atmosphere and does not have the features of plate tectonics. These two features appear to be related. Venus and Earth may have started with similar atmospheres, but the Earth suffered the consequences of a late moon-forming impact. The consequence is that the atmosphere of Venus remained sufficiently insulating to maintain temperatures that prevented the formation of liquid water at the surface, and so oceans. This may have led to less effective crustal recycling, and ultimately the inhibition of plate tectonics (see Kaula, 1990). However, the styles of early Venus evolution, and the transitions between different tectonic styles, are debated. Various models have been presented for the degassing of ^{40}Ar from the mantle and crust and into the Venus atmosphere, but these are dependent upon the history of tectonic activity and heat loss on the planet (Sasaki and Tajika, 1995; Turcotte and Schubert, 1988; Namiki and Solomon, 1998; Kaula, 1999). Indeed, the amount of ^{40}Ar in the atmosphere provides a constraint on the total amount of mantle melting and transfer of potassium to the crust (Kaula, 1999). The fraction of ^{40}Ar that has degassed, $\sim 12–28\%$, is substantially lower than that of the Earth. Kaula (1999) has discussed various mechanisms that may account for decreased degassing of Venus. The possibility that this may reflect layering on Venus is difficult to assess, considering that relating the limited apparent degassing of the Earth to mantle structure has proven so controversial.

5.8 CONCLUSIONS

The degassing of the Earth is an integral part of the formation and thermal evolution of the planet. Degassing histories have often naturally been based on noble gas abundances and isotopic compositions. Radiogenic isotopes provide the strongest constraints on the total volume of the silicate Earth that has degassed to the atmosphere, and indicate that a large portion of the Earth remains undegassed. Nonradiogenic noble gases, incorporated during Earth formation, strongly degassed during accretion and the extreme thermal conditions on the forming Earth. The upper mantle, down to some as yet unresolved depth, is now highly degassed. Silicate differentiation producing the continental crust also likely promoted degassing, and continual processing of the continental crust has left it relatively degassed of even radiogenic daughters produced within the crust. Other volatiles have also been effectively removed from the upper mantle, although elements such as carbon may have substantial subduction fluxes that are the dominant inputs.

An outstanding question is how much of the mantle still maintains high volatile concentrations. This involves resolution of the nature of the high ^{3}He/^{4}He OIB-source region. Most models equate this with undepleted, undegassed mantle, although some models invoke depletion mechanisms. However, none of these has matched the end-member components seen in OIB lithophile isotope correlations. It remains to be demonstrated that a primitive component is present and so can dominate the helium and neon isotope signatures in OIB. The heavy-noble-gas characteristics in OIB must still be documented. It is not known to what extent major volatiles are stored in the deep Earth and associated with these noble gas components.

As understanding of terrestrial noble gas geochemistry has evolved, various earlier conclusions now appear to be incorrect. As discussed above, the very radiogenic argon and xenon isotope ratios of the upper mantle are not the result of early degassing of this mantle reservoir, since this is a model-dependent conclusion based on the assumption that upper-mantle noble gases are residual from atmosphere degassing. However, it is now clear that xenon isotope systematics precludes such a relationship (Ozima *et al.*, 1985). While a complete description of mantle noble gases remains to be formulated, it is clear that there are other mechanisms that can account for the observed xenon isotope variations. Nonetheless, early transfer of volatiles to the atmosphere probably did occur and was caused by impact degassing.

The interactions between the atmosphere and the mantle now appear to be more complex. The subduction of heavy noble gases may have a marked impact on mantle isotope compositions. The earlier conclusion that this was not possible was based on models of the isolation of the upper mantle or arguments about preservation of non-atmospheric ^{129}Xe/^{130}Xe ratios in the mantle. In fact, upper-mantle nonradiogenic xenon isotopes could be dominated by subducted xenon and admixed with very radiogenic xenon, and some models explicitly incorporate subducted xenon fluxes. Until more is conclusively known about argon and xenon isotopic variations in the

mantle, subduction must be considered a potentially important process.

In devising atmosphere formation histories, it is also now clear that the present is not the key to the past. Although primordial noble gases continue to degas, their isotopic compositions do not match those of the atmosphere and limit their contribution to a small fraction of the present atmospheric inventory. Volatile species continue to be added to the atmosphere, but the dominant inputs occurred during very early Earth history.

There are, of course, many questions regarding terrestrial noble gases that remain to be explored. Some of the issues that are critical to making advances in global models of noble gas behavior include the partitioning of noble gases into the core and between mantle minerals and silicate melts. While the common assumptions regarding general behavior may very well be correct, the effects of fractionation between noble gases cannot be clearly assessed. Further refinement of planetary degassing models will come from greater resolution of the nature of the reservoirs that remain undegassed within the planet.

ACKNOWLEDGMENT

A very useful review by N. Daup that was provided under very short notice is much appreciated.

REFERENCES

Ahrens T. J. (1993) Impact erosion of terrestrial planetary atmospheres. *Ann. Rev. Earth Planet. Sci.* **21**, 525–555.

Ahrens T. J., O'Keefe J. D., and Lange M. A. (1989) Formation of atmospheres during accretion of the terrestrial planets. In *Origin and Evolution of Planetary and Satellite Atmospheres* (eds. S. K. Atreya, J. B. Pollack, and M. S. Matthews). University of Arizona Press, Tucson, pp. 328–385.

Albarède F. (1998) Time-dependent models of U–Th–He and K–Ar evolution and the layering of mantle convection. *Chem. Geol.* **145**, 413–429.

Allègre C. J. (1997) Limitation on the mass exchange between the upper and lower mantle: the evolving convection regime of the Earth. *Earth Planet. Sci. Lett.* **150**, 1–6.

Allègre C. J., Staudacher T., Sarda P., and Kurz M. (1983) Constraints on evolution of Earth's mantle from rare gas systematics. *Nature* **303**, 762–766.

Allègre C. J., Staudacher T., and Sarda P. (1986) Rare gas systematics: *formation* of the atmosphere, evolution and structure of the Earth's mantle. *Earth Planet. Sci. Lett.* **87**, 127–150.

Allègre C. J., Poirier J.-P., Humler E., and Hofmann A. W. (1995) The chemical composition of the Earth. *Earth Planet. Sci. Lett.* **134**, 515–526.

Allègre C. J., Hofmann A. W., and O'Nions R. K. (1996) The argon constraints on mantle structures. *Geophys. Res. Lett.* **23**, 3555–3557.

Armstrong R. L. (1991) The persistent myth of crustal growth. *Austral. J. Earth Sci.* **38**, 613–630.

Aronson J. L. and Hower J. (1976) Mechanism of burial metmorphism of argillaceous sediment: 2. Radiogenic argon evidence. *Geol. Soc. Am. Bull.* **87**, 738–744.

Ballentine C. J. and Burnard P. G. (2002) Production, release, and transport of noble gases in the continental crust. *Rev. Mineral. Geochem.* **47**, 481–538.

Ballentine C. J., van Keken P. E., Porcelli D., and Hauri E. K. (2002) Numerical models, geochemistry and the zero-paradox noble-gas mantle. *Phil. Trans. Roy. Soc. London A* **360**, 2611–2631.

Batiza R. (1982) Abundances, distribution and sizes of volcanoes in the Pacific Ocean and implications for the origin of non-hotspot volcanoes. *Earth Planet. Sci. Lett.* **60**, 195–206.

Bebout G. E. (1995) The impact of subduction-zone metamorphism on mantle-ocean chemical cycling. *Chem. Geol.* **126**, 191–218.

Bebout G. E. and Fogel M. (1992) Nitrogen isotope composition of metasedimentary rocks in the Catalina Schist, California: implications for metorphic devolitization histor. *Geochim. Cosmochim. Acta* **56**, 2839–2849.

Begemann R., Weber H. W., and Hintenberger H. (1976) On the primordial abundance of argon-40. *Astrophys. J.* **203**, L155–L157.

Bell D. R. and Ihinger P. D. (2000) The isotopic composition of hydrogen in nominally anhydrous mantle minerals. *Geochim. Cosmochim. Acta* **64**, 2109–2118.

Benkert J.-P., Baur H., Signer P., and Wieler R. (1993) He, Ne, and Ar from solar wind and solar energetic particles in lunar ilmenites and pyroxenes. *J. Geophys. Res.* **98**, 13147–13162.

Bercovici D. and Karato S.-I. (2003) Whole-mantle convection and the transition-zone water filter. *Nature* **425**, 39–44.

Bethke C. M., Zhao X., and Torgersen T. (1999) Groundwater flow and the ^{4}He distribution in the Great Artesian Basin of Australia. *J. Geophys. Res.* **104**, 12999–13011.

Bijwaard H. and Spakman W. (1999) Tomographic evidence for a narrow whole mantle plume below Iceland. *Earth Planet. Sci. Lett.* **166**, 121–126.

Blum J. D. (1995) Isotopic decay data. In *Global Earth Physics: A Handbook of Physical Constants* (ed. T. J. Ahrens). American Geophysical Union, Washington, DC, pp. 271–282.

Bogard D. D. (1997) A reappraisal of the Martian Ar-36/Ar-38 ratio. *J. Geophys. Res.* **102**, 1653–1661.

Bogard D. D., Clayton R. N., Marti K., Owen T., and Turner G. (2001) Martian volatiles: isotopic composition, origin, and evolution. *Space Sci. Rev.* **96**, 425–458.

Boyd S. R. and Pillinger C. T. (1994) A preliminary study of ^{15}N/^{14}N in octahedral growth form diamonds. *Chem. Geol.* **116**, 43–59.

Brazzle R. H., Pravdivtseva O. V., Meshik A. P., and Hohenberg C. M. (1999) Verification and interpretation of the I–Xe chronometer. *Geochim. Cosmochim. Acta* **63**, 739–760.

Broadhurst C. L., Drake M. J., Hagee B. E., and Bernatowicz T. J. (1992) Solubility and partitioning of Ne, Ar, Kr, and Xe in minerals and synthetic basaltic melts. *Geochim. Cosmochim. Acta* **56**, 709–723.

Brooker R. A., Du Z., Blundy J. D., Kelley S. P., Allan N. L., Wood B. J., Chamorro E. M., Wartho J.-A., and Purton J. A. (2003) The "zero charge" partitioning behaviour of noble gases during mantle melting. *Nature* **423**, 738–741.

Brown H. (1949) Rare gases and the formation of the Earth's atmosphere. In *The Atmospheres of the Earth and Planets* (ed. G. P. Kuiper). University of Chicago Press, Chicago, pp. 258–266.

Buffett B. A., Huppert H. E., Lister J. R., and Woods A. W. (1996) On the thermal evolution of the Earth's core. *J. Geophys. Res.* **101**, 7989–8006.

Burnard P. G., Graham D., and Turner G. (1997) Vesicle specific noble gas analyses of Popping Rock: implications for primordial noble gases in Earth. *Science* **276**, 568–571.

Cadogan P. H. (1977) Paleoatmospheric argon in Rhynie chert. *Nature* **268**, 38–41.

Caffee M. W., Hudson G. U., Velsko C., Huss G. R., Alexander E. C. Jr., and Chivas A. R. (1999) Primordial noble cases from Earth's mantle: identification of a primitive volatile component. *Science* **285**, 2115–2118.

Carr M. H. (1986) Mars: a water-rich planet? *Icarus* **68**, 187–216.

Cartigny P., Harris J. W., and Javoy M. (1998) Eclogitic diamond formation at Jwaneng: no room for a recycled component. *Science* **280**, 1421–1424.

Cartigny P., Jendrzejewski N., Pineau F., Petit E., and Javoy M. (2001) Volatile (C, N, Ar) variability in MORB and the respective roles of mantle source heterogeneity and degassing: the case of the Southwest Indian Ridge. *Earth Planet. Sci. Lett.* **194**, 241–257.

Chabot N. L. and Drake M. J. (1999) Potassium solubility in metal: the effects of composition at 15 kbar and 1900 degrees C on partitioning between iron alloys and silicate melts. *Earth Planet. Sci. Lett.* **172**, 323–335.

Chamorro-Perez E. M., Brooker R. A., Wartho J.-A., Wood B. J., Kelley S. P., and Blundy J. D. (2002) Ar and K partitioning between clinopyroxene and silicate melt to 8 GPa. *Geochim. Cosmochim. Acta* **66**, 507–519.

Chaussidon M. and Jambon A. (1994) Boron content and isotopic composition of oceanic basalts: geochemical and cosmochemical implications. *Earth Planet. Sci. Lett.* **121**, 277–291.

Clarke W. B., Beg M. A., and Craig H. (1969) Excess ^{3}He in the sea: evidence for terrestrial primordial helium. *Earth Planet. Sci. Lett.* **6**, 213–220.

Cogley J. C. (1984) Continental margins and the extent and number of the continents. *Rev. Geophys. Space Phys.* **22**, 101–122.

Coltice N. and Ricard Y. (1999) Geochemical observations and one layer mantle convection. *Earth Planet. Sci. Lett.* **174**, 125–137.

Coltice N., Albarède F., and Gillet P. (2000) ^{40}K–^{40}Ar constraints on recycling continental crust into the mantle. *Science* **288**, 845–847.

Craig H. and Lupton J. E. (1976) Primordial neon, helium, and hydrogen in oceanic basalts. *Earth Planet. Sci. Lett.* **31**, 369–385.

Craig H., Clarke W. B., and Beg M. A. (1975) Excess ^{3}He in deep water on the East Pacific Rise. *Earth Planet. Sci. Lett.* **2**, 125–132.

Creager K. C. and Jordan T. H. (1986) Slab penetration into the lower mantle beneath the Marianas and other island arcs of the northwest Pacific. *J. Geophys. Res.* **91**, 3573–3589.

Crisp J. A. (1984) Rates of magma emplacement and volcanic output. *J. Volcanol. Geotherm. Res.* **89**, 3031–3049.

Dauphas N. and Marty B. (1999) Heavy nitrogen in carbonatities of the Kola Peninsula: a possible signature of the deep mantle. *Science* **286**, 2488–2490.

Dauphas N., Robert F., and Marty B. (2000) The late asteroidal and cometary bombardment of Earth as recorded in water deuterium to protium ratio. *Icarus* **148**, 508–512.

Davies G. F. (1999) Geophysically constrained mantle mass flows and the Ar-40 budget: a degassed lower mantle? *Earth Planet. Sci. Lett.* **166**, 149–162.

Deloule E., Albarède F., and Sheppard S. M. F. (1991) Hydrogen isotope heterogeneities in the mantle from ion probe analysis of amphiboles from ultramafic rocks. *Earth Planet. Sci. Lett.* **105**, 543–553.

Déruelle B., Dreibus G., and Jambon A. (1992) Iodine abundances in oceanic basalts: implications for Earth dynamics. *Earth Planet. Sci. Lett.* **108**, 217–227.

Des Marais D. J. (1985) Carbon exchange between the mantle and crust and its effect upon the atmosphere: today compared to Archean time. In *The Carbon Cycle and Atmospheric CO$_2$: Natural Variations Archean to Present* (eds. E. T. Sundquist and W. S. Broecker). American Geophysical Union, Washington, DC, pp. 602–611.

Dixon J. E. and Stolper E. M. (1995) An experimental study of water and carbon dioxide solubilities in mid-ocean ridge basaltic liquids: 2. Applications to degassing. *J. Petrol.* **36**, 1633–1646.

Doe B. R. and Zartman R. E. (1979) Plumbotectonics: I. The Phanerozoic. In *Geochemistry of Hydrothermal Ore Deposits* (ed. H. L. Barnes). Wiley, New York, pp. 22–70.

Donahue T. M. (1995) Evolution of water reservoirs on Mars from D/H ratios in the atmosphere and crust. *Nature* 374, 432–434.

Donahue T. M. and Pollack J. B. (1983) Origin and evolution of the atmosphere of Venus. In *Venus* (eds. D. Hunten, L. Colin, T. Donahue, and V. Moroz). University of Arizona Press, Tucson, pp. 1003–1036.

Donahue T. M., Hoffman J. H., Hodges R. R. Jr., and Watson A. J. (1982) Venus was wet: a measurement of the ratio of deuterium to hydrogen. *Science* 216, 630–633.

Dreibus G. and Wänke H. (1989) Supply and loss of volatile constituents during accretion of terrestrial planets. In *Origin and Evolution of Planetary and Satellite Atmospheres* (eds. S. K. Atreya, J. B. Pollack, and M. S. Matthews). University of Arizona Press, Tucson, pp. 268–288.

Dymond J. and Hogan L. (1973) Noble gas abundance patterns in deep sea basalts—Primordial gases from the mantle. *Earth Planet. Sci. Lett.* 20, 131–139.

Eikenberg J., Signer P., and Wieler R. (1993) U–Xe, U–Kr, and U–Pb systematics for dating uranium minerals and investigations of the production of nucleogenic neon and argon. *Geochim. Cosmochim. Acta* 57, 1053–1069.

England P. C. and Thompson A. B. (1984) Pressure–temperature–time paths of regional metamorphism: I. Heat transfer during evolution of regions of thickened continental crust. *J. Petrol.* 25, 894–928.

Etheridge M. A., Wall V. J., Cox S. F., and Vernon R. H. (1984) High fluid pressures during regional metamorphism and deformation: implications for mass transport and deformation mechanisms. *J. Geophys. Res.* 89, 4344–4358.

Farley K. A., Basu A. R., and Craig H. (1993) He, Sr, and Nd isotopic variations in lavas from the Juan Fernandez archipelago. *Contrib. Mineral. Petrol.* 115, 75–87.

Farley K. A., Maier-Reimer E., Schlosser P., and Broecker W. S. (1995) Constraints on mantle He-3 fluxes and deep-sea circulation from an oceanic general circulation model. *J. Geophys. Res.* 100, 3829–3839.

Fisher D. E. (1978) Terrestrial potassium abundances as limits to models of atmospheric evolution. In *Terrestrial Rare Gases* (ed. M. Ozima). Japan Scientific Societies Press, Tokyo, pp. 173–183.

Fisher D. E. (1985) Noble gases from oceanic island basalts do not require an undepleted mantle source. *Nature* 316, 716–718.

Galer S. J. G. and O'Nions R. K. (1985) Residence time of thorium, uranium, and lead in the mantle with implications for mantle convection. *Nature* 316, 778–782.

Gamo T., Ishibashi J.-I., Sakai H., and Tilbrook B. (1987) Methane anomalies in seawater above the Loihi seamount summit area, Hawaii. *Geochim. Cosmochim. Acta* 51, 2857–2864.

Gast P. W. (1960) Limitations on the composition of the upper mantle. *J. Geophys. Res.* 65, 1287–1297.

Geiss J., Buehler F., Cerutti H., Eberhardt P., and Filleaux C. H. (1972) Solar wind composition experiments. *Apollo 15 Preliminary Sci. Report*, chap. 15.

Goes S., Spakman W., and Bijwaard H. (1999) A lower mantle source for central European volcanism. *Science* 286, 1928–1931.

Grady M. M., Wright I. P., Carr L. P., and Pillinger C. T. (1986) Compositional differences in estatite chondrites based on carbon and nitrogen stable isotope measurements. *Geochim. Cosmochim. Acta* 50, 2799–2813.

Graham D. W. (2002) Noble gases in MORB and OIB: observational constraints for the characterization of mantle source reservoirs. *Rev. Mineral. Geochem.* 47, 247–318.

Graham D. W., Lupton F., Albarède F., and Condomines M. (1990) Extreme temporal homogeneity of helium isotopes at Piton de la Fournaise, Réunion Island. *Nature* 347, 545–548.

Grand S. P. (1987) Tomographic inversion for shear velocity beneath the North American plate. *J. Geophys. Res.* 92, 14065–14090.

Grand S. P. (1994) Mantle shear structure beneath the Americas and surrounding oceans. *J. Geophys. Res.* 99, 66–78.

Haendel D., Mühle K., Nitzsche H., Stiehl G., and Wand U. (1986) Isotopic variations of the fixed nitrogen in metamorphic rocks. *Geochim. Cosmochim. Acta* 50, 749–758.

Hagee B., Bernatowicz T. J., Podosek F. A., Johnson M. L., Burnett D. S., and Tatsumoto M. (1990) Actinide abundances in ordinary chondrites. *Geochim. Cosmochim. Acta* 54, 2847–2858.

Hall A. (1989) Ammonium in spilitized basalts of southwest England and its implications for the recycling of nitrogen *Geochem. J.* 23, 19–23.

Halliday A. N. and Porcelli D. (2001) In search of lost planets—the paleocosmochemistry of the inner solar system. *Earth Planet. Sci. Lett.* 192, 545–559.

Hamano Y. and Ozima M. (1978) Earth-atmosphere evolution model based on Ar isotopic data. In *Terrestrial Rare Gases* (ed. M. Ozima). Japan Scientific Societies Press, Tokyo, pp. 155–171.

Hanes J. A., York D., and Hall C. M. (1985) An ^{40}Ar/^{39}Ar geochronological and electron microprobe investigation of an Archaean pyroxenite and its bearing on ancient atmospheric compositions. *Can. J. Earth Sci.* 22, 947–958.

Hanyu T. and Kaneoka I. (1998) Open system behavior of helium in case of the HIMU source area. *Geophys. Res. Lett.* 25, 687–690.

Harrison D., Burnard P., and Turner G. (1999) Noble gas behaviour and composition in the mantle: constraints from the Iceland plume. *Earth Planet. Sci. Lett.* 171, 199–207.

Hart R., Dymond J., and Hogan L. (1979) Preferential formation of the atmosphere-sialic crust system from the upper mantle. *Nature* 278, 156–159.

Hart S. R. and Staudigel H. (1982) The control of alkalies and uranium in seawater by ocean crust alteration. *Earth Planet. Sci. Lett.* 58, 202–212.

Hashizume K., Chaussidon M., Marty B., and Robert F. (2000) Solar wind record on the moon: deciphering presolar from planetary nitrogen. *Science* 290, 1142–1145.

Hauri E. (2002) SIMS analysis of volatiles in silicate glasses: 2. Isotopes and abundances in Hawaiian melt inclusions. *Chem. Geol.* 183, 115–141.

Heaton T. H. E. (1984) Rates and sources of 4He accumulation in groundwater. *Hydrol. Sci. J.* 29, 29–47.

Helmberger D. V., Wen L., and Ding X. (1998) Seismic evidence that the source of the Iceland hotspot lies at the core–mantle boundary. *Nature* 396, 251–255.

Hilton D. R., McMurty G. M., and Kreulen R. (1997) Evidence for extensive degassing of the Hawaiian mantle plume from helium–carbon relationships at Kilauea volcano. *Geophys. Res. Lett.* 24, 3065–3068.

Hilton D. R., McMurtry G. M., and Goff F. (1998a) Large variations in vent fluid $CO_2/^3He$ ratios signal rapid changes in magma chemistry at Loihi seamount, Hawaii. *Nature* 396, 359–362.

Hilton D. R., Grönvold K., Sveinbjornsdottir A. E., and Hammerschmidt K. (1998b) Helium isotope evidence for off-axis degassing of the Icelandic hotspot. *Chem. Geol.* 149, 173–187.

Hilton D. R., Thirlwall M. F., Taylor R. N., Murton B. J., and Nichols A. (2000) Controls on magmatic degassing along the Reykjanes Ridge with implications for the helium paradox. *Earth Planet. Sci. Lett.* 183, 43–50.

Hilton D. R., Fischer T. P., and Marty B. (2002) Noble gases in subduction zones and volatile recycling. *Rev. Mineral. Geochem.* 47, 319–370.

Hiyagon H. and Ozima M. (1986) Partition of gases between olivine and basalt melt. *Geochim. Cosmochim. Acta* 50, 2045–2057.

Hiyagon H., Ozima M., Marty B., Zashu S., and Sakai H. (1992) Noble gases in submarine glasses from mid-oceanic ridges and Loihi Seamount—Constraints on the early history of the Earth. *Geochim. Cosmochim. Acta* 56, 1301–1316.

Hoefs J. (1969) Carbon. In *Handbook of Geochemistry* (ed. K. H. Wedepohl). Springer, Berlin.

Hofmann A. W. (1988) Chemical differentiation of the Earth: the relationship between mantle, continental crust, and oceanic crust. *Earth Planet. Sci. Lett.* **90**, 297–314.

Hoffman J. H., Hodges R. R., Donahue T. M., and McElroy M. B. (1980) Composition of the Venus lower atmosphere from the Pioneer Venus mass spectrometer. *J. Geophys. Res.* **85**, 7882–7890.

Hohenberg C. M., Podosek F. A., and Reynolds J. H. (1967) Xenon–iodine dating: sharp isochronism in chondrites. *Science* **156**, 233–236.

Holloway J. R. (1998) Graphite-melt equilibria during mantle melting: constraints on CO_2 in MORB magmas and the carbon content of the mantle. *Chem. Geol.* **147**, 89–97.

Honda M. and McDougall I. (1993) Solar noble gases in the Earth—the systematics of helium–neon isotopes in mantle-derived samples. *Lithos* **30**, 257–265.

Honda M. and McDougall I. (1998) Primordial helium and neon in the Earth—a speculation on early degassing. *Geophys. Res. Lett.* **25**, 1951–1954.

Honda M., McDougall I., Patterson D. B., Doulgeris A., and Clague D. A. (1991) Possible solar noble-gas component in Hawaiian basalts. *Nature* **349**, 149–151.

Honda M., McDougall I., Patterson D. B., Doulgeris A., and Clague D. A. (1993) Noble gases in submarine pillow basalt glasses from Loihi and Kilauea, Hawaii—a solar component in the Earth. *Geochim. Cosmochim. Acta* **57**, 859–874.

Hudson G. B., Kennedy B. M., Podosek F. A., and Hohenberg C. M. (1989) The early solar system abundance of ^{244}Pu as inferred from the St. Severin chondrite. *Proc. 19th Lunar Planet. Sci. Conf.* 547–557.

Hunt J. M. (1972) Distribution of carbon in crust of Earth. *Bull. Am. Assoc. Petrol. Geol.* **56**, 2273–2277.

Hunten D. M., Pepin R. O., and Walker J. C. G. (1987) Mass fractionation in hydrodynamic escape. *Icarus* **69**, 532–549.

Hurley P. M. (1966) K–Ar dating of sediments. In *Potassium Argon Dating* (eds. O. A. Schaeffer and J. Zähringer). Springer, New York, pp. 134–151.

Hurley P. M. and Rand J. R. (1969) Pre-drift continental nuclei. *Science* **164**, 1229–1242.

Hurley P. M., Brookins D. G., Pinson W. H., Hart S. R., and Fairbairn H. W. (1961) K–Ar age studies of Mississippi and other river sediments. *Geol. Soc. Am. Bull.* **72**, 1807–1816.

Hutchins K. S. and Jakosky B. M. (1996) Evolution of martian atmospheric argon: implications for sources of volatiles. *J. Geophys. Res.* **101**, 14933–14949.

Ito E., Harris D. M., and Anderson A. T., Jr. (1983) Alteration of oceanic crust and geologic cycling of chlorine and water. *Geochim. Cosmochim. Acta* **47**, 1613–1624.

Jambon A. (1994) Earth degassing and large-scale geochemical cycling of volatile elements. *Rev. Mineral.* **30**, 479–517.

Jambon A. and Zimmermann J. L. (1990) Water in oceanic basalts—evidence for dehydration of recycled crust. *Earth Planet. Sci. Lett.* **101**, 323–331.

Jaupart C., Sclater J. G., and Simmons G. (1981) Heat flow studies: constraints on the distribution of uranium, thorium, and potassium in the continental crust. *Earth Planet. Sci. Lett.* **52**, 328–344.

Javoy M. (1998) The birth of the Earth's atmosphere: the behavior and fate of its major elements. *Chem. Geol.* **147**, 11–25.

Javoy M. and Pineau F. (1991) The volatiles record of a "Popping Rock" from the Mid-Atlantic Ridge at 14 ° N: chemical and isotopic composition of gas trapped in the vesicles. *Earth Planet. Sci. Lett.* **107**, 598–611.

Javoy M., Pineau F., and Demaiffe Dl. (1984) Nitrogen and carbon isotopic composition in the diamonds of Mbuji Mayi (Zaire). *Earth Planet. Sci. Lett.* **68**, 399–412.

Jochum K. P., Hofmann A. W., Ito E., Seufert H. M., and White W. M. (1983) K, U, and Th in mid-ocean ridge basalt glasses and heat production, K/U and K/Rb in the mantle. *Nature* **306**, 431–436.

Kaula W. M. (1990) Venus: a contrast in evolution to Earth. *Science* **247**, 1191–1196.

Kaula W. M. (1999) Constraints on Venus evolution from radiogenic argon. *Icarus* **139**, 32–39.

Keeling C. D. and Whorf T. P. (2000) Atmospheric CO_2 records from sites in the SIO air sampling network. *In Trends: A Compendium of Data on Global Change*. Carbon Dioxide Information Analysis Center, Oak Ridge National Laboratory, Oak Ridge, TN.

Kellogg L. H. and Wasserburg G. J. (1990) The role of plumes in mantle helium fluxes. *Earth Planet. Sci. Lett.* **99**, 276–289.

Kellogg L. H., Hager B. H., and van der Hilst R. D. (1999) Compositional stratification in the deep mantle. *Science* **283**, 1881–1884.

Kerridge J. F. (1985) Carbon, hydrogen, and nitrogen in carbonaceous chondrites: abundances and isotopic compositions in bulk samples. *Geochim. Cosmochim Acta* **49**, 1707–1714.

Kingsley R. H. and Schilling J.-G. (1995) Carbon in Mid-Atlantic Ridge basalt glasses from 28-degrees-N to 63-degrees-N—evidence for a carbon-enriched azores mantle plume. *Earth Planet. Sci. Lett.* **129**, 31–53.

Korenaga J. and Kelemen P. B. (2000) Major element heterogeneity in the mantle source of the North Atlantic igneous province. *Earth Planet. Sci. Lett.* **184**, 251–268.

Krummenacher D., Merrihue C. M., Pepin R. O., and Reynolds J. H. (1962) Meteoritic krypton and barium versus the general isotopic anomalies in meteoritic xenon. *Geochim. Cosmochim. Acta* **26**, 231–249.

Kunz J. (1999) Is there solar argon in the Earth's mantle? *Nature* **399**, 649–650.

Kunz J., Staudacher T., and Allègre C. J. (1998) Plutonium-fission xenon found in Earth's mantle. *Science* **280**, 877–880.

Kurz M. D., Jenkins W. J., and Hart S. R. (1982) Helium isotopic systematics of oceanic islands and mantle heterogeneity. *Nature* **297**, 43–46.

Kyser T. K. and O'Neil J. R. (1984) Hydrogen isotope systematics of submarine basalts. *Geochim. Cosmochim. Acta* **48**, 2123–2133.

Lécuyer C., Gillet P., and Robert F. (1998) The hydrogen isotope composition of seawater and the global water cycle. *Chem. Geol.* **45**, 249–261.

Lécuyer C., Simon L., and Guyot F. (2000) Comparison of carbon, nitrogen, and water budgets on Venus and the Earth. *Earth Planet. Sci. Lett.* **181**, 33–40.

Li Y.-H. (1972) Geochemical mass balance among lithosphere, hydrosphere, and atmosphere. *Am. J. Sci.* **272**, 119–137.

Lupton J. E. (1996) A far-field hydrothermal plume from Loihi Seamount. *Science* **272**, 976–979.

Lupton J. E. and Craig H. (1975) Excess ^{3}He in oceanic basalts, evidence for terrestrial primordial helium. *Earth Planet. Sci. Lett.* **26**, 133–139.

Magaritz M. and Taylor H. P., Jr. (1976) Oxygen, hydrogen, and carbon isotope studies of the Franciscan Formation Coast Ranges, California. *Geochim. Cosmochim. Acta* **40**, 215–234.

Mahaffy P. R., Donahue T. M., Atreya S. K., Owen T. C., and Niemann H. B. (1998) Galileo probe measurements of D/H and ^{3}He/^{4}He in Jupiter's atmosphere. *Space Sci. Rev.* **84**, 251–263.

Mamyrin B. A., Tolstikhin I. N., Anufriev G. S., and Kamensky I. L. (1969) Anomalous isotopic composition of helium in volcanic gases. *Dokl. Akad. Nauk SSSR* **184**, 1197–1199 (in Russian).

Marti K. and Mathew K. J. (1998) Noble-gas components in planetary atmospheres and interiors in relation to solar wind and meteorites. *Proc. Indian Acad. Sci. (Earth Planet. Sci.)* **107**, 425–431.

Martel D. J., Deak J., Dovenyi P., Horvath F., O'Nions R. K., Oxburgh E. R., Stegna L., and Stute M. (1989) Leakage of helium from the Pannonian Basin. *Nature* **432**, 908–912.

Martin J.-M. and Meybeck M. (1979) Elemental mass-balance of material carried by major world rivers. *Mar. Chem.* **7**, 173–206.

Marty B. and Dauphas N. (2003) The nitrogen record of crust–mantle interaction and mantle convection from Archean to present. *Earth Planet. Sci. Lett.* **206**, 397–410.

Marty B. and Humbert F. (1997) Nitrogen and argon isotopes in oceanic basalts. *Earth Planet. Sci. Lett.* **152**, 101–112.

Marty B. and Jambon A. (1987) C/^{3}He in volatile fluxes from the solid Earth—implications for carbon geodynamics. *Earth Planet. Sci. Lett.* **83**, 16–26.

Marty B. and Marti K. (2002) Signatures of early differentiation on Mars. *Earth Planet. Sci. Lett.* **196**, 251–263.

Marty B. and Zimmermann L. (1999) Volatiles (He, C, N, Ar) in mid-ocean ridge basalts: assessment of shallow-level fractionation and characterization of source composition. *Geochim. Cosmochim. Acta* **63**, 3619–3633.

Marty B., Tolstikhin I., Kamensky I. L., Nivin V., Balaganskaya E., and Zimmermann J.-L. (1998) Plume-derived rare gases in 380 Ma carbonatites from the Kola region (Russia) and the argon isotopic composition of the deep mantle. *Earth Planet. Sci. Lett.* **164**, 179–192.

Mathew K. J. and Marti K. (2001) Early evolution of martian volatiles: nitrogen and noble gas components in ALH84001 and Chassigny. *J. Geophys. Res.* **106**, 1401–1422.

Mathew K. J. and Marti K. (2002) Martian atmospheric and interior volatiles in the meteorite Nakhla. *Earth Planet. Sci. Lett.* **199**, 7–20.

Mathew K. J., Kim J. S., and Marti K. (1998) martian atmospheric and indigenous components of xenon and nitrogen in the Shergotty, Nakhla, and Chassigny group meteorites. *Meteorit. Planet. Sci.* **33**, 655–664.

Matsuda J.-I. and Marty B. (1995) The ^{40}Ar/^{36}Ar ratio of the undepleted mantle: a re-evaluation. *Geophys. Res. Lett.* **22**, 1937–1940.

Matsuda J.-I. and Nagao K. (1986) Noble gas abundances in a deep-sea core from eastern equatorial Pacific. *Geochem. J.* **20**, 71–80.

Mattey D. P., Carr R. H., Wright I. P., and Pillinger C. T. (1984) Carbon isotopes in submarine basalts. *Earth Planet. Sci. Lett.* **70**, 196–206.

Mazor E. (1995) Stagnant aquifer concept: 1. Large scale artesian systems—Great Artesian Basin, Australia. *J. Hydrol.* **173**, 219–240.

McCulloch M. T. (1993) The role of subducted slabs in an evolving Earth. *Earth Planet. Sci. Lett.* **115**, 89–100.

McDonough W. F. and Sun S. S. (1995) The composition of the Earth. *Chem. Geol.* **120**, 223–253.

McDougall I. and Harrison T. M. (1999) *Geochronology and Thermochronology by the ^{40}Ar/^{39}Ar Method*. Oxford University Press, Oxford.

McLennan S. M. and Taylor S. R. (1996) Heat flow and the chemical composition of continental crust. *J. Geol.* **104**, 369–377.

Milliman J. D. and Meade R. H. (1983) World-wide delivery of river sediment to the oceans. *J. Geol.* **91**, 1–21.

Moreira M. and Sarda P. (2000) Noble gas constraints on degassing processes. *Earth Planet. Sci. Lett.* **176**, 375–386.

Moreira M., Kunz J., and Allègre C. J. (1998) Rare gas systematics in Popping Rock: isotopic and elemental compositions in the upper mantle. *Science* **279**, 1178–1181.

Morgan P. (1984) The thermal structure and thermal evolution of the continental lithosphere. *Phys. Chem. Earth* **15**, 107–193.

Morgan J. P. (1998) Thermal and rare gas evolution of the mantle. *Chem. Geol.* **145**, 431–445.

Morrison P. and Pine J. (1955) Radiogenic origin of the helium isotopes in rock. *Ann. NY Acad. Sci.* **62**, 69–92.

Muramatsu Y. W. and Wedepohl K. H. (1998) The distribution of iodine in the Earth's crust. *Chem. Geol.* **147**, 201–216.

Musselwhite D. S. and Drake M. J. (2000) Early outgassing of Mars: implications from experimentally determined solubility of iodine in silicate magmas. *Icarus* **148**, 160–175.

Musselwhite D. S., Drake M. J., and Swindle T. D. (1991) Early outgassing of Mars supported by differential water solubility of iodine and xenon. *Nature* **352**, 697–699.

Namiki N. and Solomon S. C. (1998) Volcanic degassing of argon and helium and the history of crustal production on Venus. *J. Geophys. Res.* **103**, 3655–3677.

Nier A. O. and McElroy M. B. (1977) Composition and structure of Mars' upper atmosphere: results from the neutral mass spectrometers on Viking 1 and 2. *J. Geophys. Res.* **82**, 4341–4349.

Nyblade A. A. and Pollack H. N. (1993) A global analysis of heat flow from Precambrian terrains: implications for the thermal structure of Archean and Proterozoic lithosphere. *J. Geophys. Res.* **98**, 12207–12218.

O'Nions R. K. and Oxburgh E. R. (1983) Heat and helium in the Earth. *Nature* **306**, 429–431.

O'Nions R. K. and Oxburgh E. R. (1988) Helium, volatile fluxes and the development of continental crust. *Earth Planet. Sci. Lett.* **90**, 331–347.

O'Nions R. K. and Tolstikhin I. N. (1994) Behaviour and residence times of lithophile and rare gas tracers in the upper mantle. *Earth Planet. Sci. Lett.* **124**, 131–138.

O'Nions R. K., Carter S. R., Evensen N. M., and Hamilton P. J. (1981) Upper mantle geochemistry. In *The Sea* (ed. C. Emiliani). Wiley, , vol. 7, pp. 49–71.

Ott U. (1988) Noble gases in SNC meteorites: Shergotty, Nakhla, Chassigny. *Geochim. Cosmochim. Acta* **52**, 1937–1948.

Owen T., Biemann K., Rushneck D. R., Biller J. E., Howarth D. W., and Lafleur A. L. (1977) The composition of the atmosphere at the surface of Mars. *J. Geophys. Res.* **82**, 4635–4639.

Owen T., Maillard J. P., Debergh C., and Lutz B. (1988) Deuterium on Mars: the abundance of HDO and the value of D/H. *Science* **240**, 1767–1770.

Owen T., Bar Nun A., and Kleinfeld I. (1992) Possible cometary origin of heavy noble gases in the atmospheres of Venus, Earth, and Mars. *Nature* **358**, 43–46.

Owen T., Mahaffy P. R., Niemann H. B., Atreya S., and Wong M. (2001) Protosolar nitrogen. *Astrophys. J.* **553**, L77–L79.

Ozima M. (1973) Was the evolution of the atmosphere continuous or catastrophic? *Nature Phys. Sci.* **246**, 41–42.

Ozima M. (1975) Ar isotopes and Earth-atmosphere evolution models. *Geochim. Cosmochim. Acta* **39**, 1127–1134.

Ozima M. and Kudo K. (1972) Excess argon in submarine basalts and an Earth-atmosphere evolution model. *Nature Phys. Sci.* **239**, 23–24.

Ozima M. and Podosek F. A. (1983) *Noble Gas Geochemistry*. Cambridge University Press, Cambridge.

Ozima M. and Podosek F. A. (2001) *Noble Gas Geochemistry* 2nd edn. Cambridge University Press, Cambridge.

Ozima M., Podozek F. A., and Igarashi G. (1985) Terrestrial xenon isotope constraints on the early history of the Earth. *Nature* **315**, 471–474.

Parsons B. (1981) The rates of plate creation and consumption. *Geophys. J. Roy. Astron. Soc.* **67**, 437–448.

Patterson D. B., Honda M., and McDougall I. (1990) Atmospheric contamination: a possible source for heavy noble gases basalts from Loihi Seamount, Hawaii. *Geophys. Res. Lett.* **17**, 705–708.

Pepin R. O. (1991) On the origin and early evolution of terrestrial planet atmospheres and meteoritic volatiles. *Icarus* **92**, 1–79.

Pepin R. O. (1998) Isotopic evidence for a solar argon component in the Earths mantle. *Nature* **394**, 664–667.

Pepin R. O. (2000) On the isotopic composition of primordial xenon in terrestrial planet atmospheres. *Space Sci. Rev.* **92**, 371–395.

Pepin R. O. and Carr M. (1992) Major issues and outstanding questions. In *Mars* (eds. H. H. Kieffer, B. M. Jakosky, C. W. Snyder, and M. S. Matthews). University of Arizona Press, Tucson, pp. 120–143.

Pepin R. O. and Phinney D. (1976) The formation interval of the Earth. *Lunar Sci.* **VII**, 682–684.

Pepin R. O. and Porcelli D. (2002) Origin of noble gases in the terrestrial planets. *Rev. Mineral. Geochem.* **47**, 191–246.

Peters K. E., Sweeney R. E., and Kaplan I. R. (1978) Correlation of carbon and nitrogen stable isotope ratios in sedimentary organic matter. *Limnol. Oceanogr.* **23**, 598–604.

Pfennig G., Klewe-Nebenius H., and Seelann-Eggebert W. (1998) *Karlsruhe Chart of the Nuclide*, 6th edn. (revised reprint). Institut für Instrumentelle Analytik, Karlsruhe.

Phinney D., Tennyson J., and Frick U. (1978) Xenon in CO_2 well gas revisited. *J. Geophys. Res.* **83**, 2313–2319.

Pineau F. and Javoy M. (1994) Strong degassing at ridge crests: the behaviour of dissolved carbon and water in basalt glasses at 14 °N, Mid-Atlantic Ridge. *Earth Planet. Sci. Lett.* **123**, 179–198.

Podosek F. A., Honda M., and Ozima M. (1980) Sedimentary noble gases. *Geochim. Cosmochim. Acta* **44**, 1875–1884.

Pollack H. N., Hurter S. J., and Johnson J. R. (1993) Heat flow from the Earth's interior: analysis of the global data set. *Rev. Geophys.* **31**, 267–280.

Porcelli D. and Ballentine B. J. (2002) Models for the distribution of terrestrial noble gases and evolution of the atmosphere. *Rev. Mineral. Geochem.* **47**, 411–480.

Porcelli D. and Halliday A. N. (2001) The core as a possible source of mantle helium. *Earth Planet. Sci. Lett.* **192**, 45–56.

Porcelli D. and Pepin R. O. (2000) Rare gas constraints on early Earth history. In *Origin of the Earth and Moon* (eds. R. M. Canup and K. Righter). University of Arizona Press, Tucson, pp. 435–458.

Porcelli D. and Wasserburg G. J. (1995a) Mass transfer of xenon through a steady-state upper mantle. *Geochim. Cosmochim. Acta* **59**, 1991–2007.

Porcelli D. and Wasserburg G. J. (1995b) Mass transfer of helium, neon, argon, and xenon through a steady-state upper mantle. *Geochim. Cosmochim. Acta* **59**, 4921–4937.

Porcelli D., Woolum D., and Cassen P. (2001) Deep Earth rare gases: initial inventories, capture from the solar nebula, and losses during Moon formation. *Earth Planet. Sci. Lett.* **193**, 237–251.

Porcelli D., Ballentine C. J., and Wieler R. (2002) An introduction to noble gas geochemistry and cosmochemistry. *Rev. Mineral. Geochem.* **47**, 1–18.

Porcelli D., Pepin R. O., Ballentine C. J., and Halliday A. (2003) Xe and degassing of the Earth (in preparation).

Poreda R. J. and Farley K. A. (1992) Rare gases in Samoan xenoliths. *Earth Planet. Sci. Lett.* **113**, 129–144.

Poreda R., Schilling J. G., and Craig H. (1986) Helium and hydrogen isotopes in ocean ridge basalts north and south of Iceland. *Earth Planet. Sci. Lett.* **78**, 1–17.

Poreda R., Craig H., Arnorsson S., and Welhan J. A. (1992) Helium isotopes in Icelandic geothermal systems: 1. ^{3}He, gas chemistry, and ^{13}C relations. *Geochim. Cosmochim. Acta* **56**, 4221–4228.

Ragettli R. A., Hebeda E. H., Signer P., and Wieler R. (1994) Uranium–xenon chronology: precise determiantion of $\alpha_{sf} *^{136}Y_{sf}$ for spontaneous fission of ^{238}U. *Earth Planet. Sci. Lett.* **128**, 653–670.

Reymer A. and Schubert G. (1984) Phanerozoic addition rates to the continental crust. *Tectonics* **3**, 63–77.

Rison W. and Craig H. (1983) Helium isotopes and mantle volatiles in Loihi Seamount and Hawaiian Island basalts and xenoliths. *Earth Planet. Sci. Lett.* **66**, 407–426.

Rocholl A. and Jochum K. P. (1993) Th, U, and other trace elements in carbonaceous chondrites—implications for the terrestrial and solar system Th/U ratios. *Earth Planet. Sci. Lett.* **117**, 265–278.

Ronov A. B. and Yaroshevsky A. A. (1969) Chemical composition of the Earth's crust. *Am. Geophys. Union Geophys. Mon. Ser.* **13**, 37–57.

Ronov A. B. and Yaroshevsky A. A. (1976) A new model for the chemical structure of the Earth's crust. *Geochem. Int.* **13**, 89–121.

Rubey W. W. (1951) Geological history of seawater. *Bull. Geol. Soc. Am.* **62**, 1111–1148.

Rudnick R. and Fountain D. M. (1995) Nature and composition of the continental crust: a lower crustal perspective. *Rev. Geophys.* **33**, 267–309.

Rudnick R. L., McDonough W. F., and O'Connell R. J. (1998) Thermal structure, thickness, and composition of continental lithosphere. *Chem. Geol.* **145**, 395–412.

Russell S. A., Lay T., and Garnero E. J. (1998) Seismic evidence for small-scale dynamics in the lowermost mantle at the root of the Hawaiian hotspot. *Nature* **369**, 225–258.

Sano Y. and Williams S. (1996) Fluxes of mantle and subducted carbon along convergent plate boundaries. *Geophys. Res. Lett.* **23**, 2746–2752.

Sano Y., Takahata N., Nishio Y., and Marty B. (1998) Nitrogen recycling in subduction zones. *Geophys. Res. Lett.* **25**, 2289–2292.

Sarda P., Staudacher T., and Allègre C. J. (1985) $^{40}Ar/^{36}Ar$ in MORB glasses: constraints on atmosphere and mantle evolution. *Earth Planet. Sci. Lett.* **72**, 357–375.

Sarda P., Staudacher T., and Allègre C. J. (1988) Neon isotopes in submarine basalts. *Earth Planet. Sci. Lett.* **91**, 73–88.

Sarda P., Moreira M., Staudacher T., Schilling J. G., and Allègre C. J. (2000) Rare gas systematics on the southernmost Mid-Atlantic Ridge: constraints on the lower mantle and the Dupal source. *J. Geophys. Res.* **105**, 5973–5996.

Sasaki S. and Nakazawa K. (1988) Origin and isotopic fractionation of terrestrial Xe: hydrodynamic fractionation during escape of the primordial H_2–He atmosphere. *Earth Planet. Sci. Lett.* **89**, 323–334.

Sasaki S. and Tajika E. (1995) Degassing history and evolution of volcanic activities of terrestrial planets based on radiogenic noble gas degassing models. In *Volatiles in the Earth and Solar System*, AIP Conf. Proc. 341 (ed. K. A. Farley). AIP Press, New York, pp. 186–199.

Schilling J.-G., Unni C. K., and Bender M. L. (1978) Origin of chlorine and bromine in the oceans. *Nature* **273**, 631–636.

Schwartzman D. W. (1973) Argon degassing and the origin of the sialic crust. *Geochim. Cosmochim. Acta* **37**, 2479–2495.

Shen Y., Solomon S. C., Bjarnason I. T., and Wolfe C. J. (1998) Seismic evidence for a lower-mantle origin of the Iceland plume. *Nature* **395**, 62–65.

Sleep N. H. and Zahnle K. (2001) Carbon dioxide cycling and implications for climate on ancient Earth. *J. Geophys. Res.* **106**, 1373–1399.

Sobolev A. V. and Chaussidon M. (1996) H_2O concentrations in primary melts from supra-subduction zones and mid-ocean ridges: implications for H_2O storage and recycling in the mantle. *Earth Planet. Sci. Lett.* **137**, 45–55.

Sobolev A. V., Hofmann A. W., and Nikogosian I. K. (2000) Recycled oceanic crust observed in "ghost plagioclase" within the source of Mauna Loa lavas. *Nature* **404**, 986–990.

Staudacher T. (1987) Upper mantle origin for Harding County well gases. *Nature* **325**, 605–607.

Staudacher T. and Allègre C. J. (1982) Terrestrial xenology. *Earth Planet. Sci. Lett.* **60**, 389–406.

Staudacher T. and Allègre C. J. (1988) Recycling of oceanic crust and sediments: the noble gas subduction barrier. *Earth Planet. Sci. Lett.* **89**, 173–183.

Staudacher T., Sarda P., and Allègre C. J. (1989) Noble gases in basalt glasses from a Mid-Atlantic Ridge topographic high at 14° N: geodynamic consequences. *Earth Planet. Sci. Lett.* **96**, 119–133.

Stute M., Sonntag C., Deak J., and Schlosser P. (1992) Helium in deep circulating groundwater in the Great Hungarian Plain—Glow dynamics and crustal and mantle helium fluxes. *Geochim. Cosmochim. Acta* **56**, 2051–2067.

Suess H. E. (1949) The abundance of noble gases in the Earth and the cosmos. *J. Geol.* **57**, 600–607 (in German).

Swindle T. D. (2002) Martian noble gases. *Rev. Mineral. Geochem.* **47**, 171–190.

Swindle T. D. and Jones J. H. (1997) The xenon isotopic composition of the primordial martian atmosphere:

contributions from solar and fission components. *J. Geophys. Res.* **102**, 1671–1678.

Tajika E. and Sasaki S. (1996) Magma generation on Mars constrained from an ^{40}Ar degassing model. *J. Geophys. Res.* **101**, 7543–7554.

Taylor S. R. and McLennan S. M. (1985) *The Continental Crust: Its Composition and Evolution.* Blackwell, Oxford.

Taylor H. P., Jr and Sheppard S. M. F. (1986) Igneous rocks: I. Processes of isotopic fractionation and isotope systematics. *Rev. Mineral.* **16**, 227–271.

Thompson A. B. (1992) Water in the Earth's upper mantle. *Nature* **358**, 295–302.

Thomsen L. (1980) ^{129}Xe on the outgassing of the atmosphere. *J. Geophys. Res.* **85**, 4374–4378.

Tolstikhin I. N. (1975) Helium isotopes in the Earth's interior and in the atmosphere: a degassing model of the Earth. *Earth Planet. Sci. Lett.* **26**, 88–96.

Torgersen T. and Clarke W. B. (1985) Helium accumulation in groundwater: I. An evaluation of sources and the continental flux of crustal ^{4}He in the Great Artesian Basin, Australia. *Geochim. Cosmochim. Acta* **49**, 1211–1218.

Torgersen T. and Ivey G. N. (1985) Helium accumulation in groundwater: II. A model for the accumulation of crustal ^{4}He degassing flux. *Geochim. Cosmochim. Acta* **49**, 2445–2452.

Torgersen T., Kennedy B. M., Hiyagon H., Chiou K. Y., Reynolds J. H., and Clarke W. B. (1989) Argon accumulation and the crustal degassing flux of ^{40}Ar in the Great Artesian Basin, Australia. *Earth Planet. Sci. Lett.* **92**, 43–56.

Trieloff M., Kunz J., Clague D. A., Harrison D., and Allègre C. J. (2000) The nature of pristine noble gases in mantle plumes. *Science* **288**, 1036–1038.

Trull T., Nadeau S., Pineau F., Polvé M., and Javoy M. (1993) C–He systematics in hotspot xenoliths: implications for mantle carbon contents and carbon recycling. *Earth Planet. Sci. Lett.* **118**, 43–64.

Turcotte D. L. and Schubert G. (1982) *Geodynamics.* Wiley, New York.

Turcotte D. L. and Schubert G. (1988) Tectonic implications of radiogenic noble gases in planetary atmospheres. *Icarus* **74**, 36–46.

Turekian K. K. (1959) The terrestrial economy of helium and argon. *Geochim. Cosmochim. Acta* **17**, 37–43.

Turekian K. K. (1990) The parameters controlling planetary degassing based on ^{40}Ar systematics. In *From Mantle to Meteorites* (eds. K. Gopolan, V. K. Gaur, B. L. K. Somayajulu, and J. D. MacDougall). Indian Academy of Sciences, Bangalore, pp. 147–152.

Turekian K. K. and Clark S. P., Jr. (1969) Inhomogeneous accumulation of the Earth from the primitive solar nebula. *Earth Planet. Sci. Lett.* **6**, 346–348.

Turekian K. K. and Clark S. P., Jr. (1975) The non-homogeneous accumulation model for terrestrial planet formation and the consequences for the atmosphere of Venus. *J. Atmos. Sci.* **32**, 1257–1261.

Turner G. (1989) The outgassing history of the Earth's atmosphere. *J. Geol. Soc. London* **146**, 147–154.

Valbracht P. J., Staudacher T., Malahoff A., and Allègre C. J. (1997) Noble gas systematics of deep rift zone glasses from Loihi Seamount, Hawaii. *Earth Planet. Sci. Lett.* **150**, 399–411.

van der Hilst R. D., Widiyantoro S., and Engdahl E. R. (1997) Evidence for deep mantle circulation from global tomography. *Nature* **386**, 578–584.

van Keken P. E., Ballentine C. J., and Porcelli D. (2001) A dynamical investigation of the heat and helium imbalance. *Earth Planet. Sci. Lett.* **188**, 421–443.

Veizer J. and Jansen S. L. (1979) Basement and sedimentary recycling and continental evolution. *Am. J. Sci.* **87**, 341–370.

Veizer J. and Jansen S. L. (1985) Basement and sedimentary recycling: 2. Time dimension to global tectonics. *J. Geol.* **93**, 625–643.

Vidale J. E., Schubert G., and Earle P. S. (2001) Unsuccessful initial search for a mid-mantle chemical boundary with seismic arrays. *Geophys. Res. Lett.* **28**, 859–862.

Volkov V. P. and Frenkel M. Y. (1993) The modelling of Venus degassing in terms of K–Ar system. *Earth Moon Planets* **62**, 117–129.

Von Huene R. and Scholl D. W. (1991) Observations at convergent margins concerning sediment subduction, subduction erosion, and the growth of continental crust. *Rev. Geophys.* **29**, 279–316.

von Zahn U., Kumar S., Niemann H., and Prinn R. (1983) Composition of the Venus atmosphere. In *Venus* (eds. D. Hunten, L. Colin, T. Donahue, and V. Moroz). University of Arizona Press, Tucson, pp. 299–430.

Wänke H. and Dreibus G. (1988) Chemical composition and accretion history of terrestrial planets. *Phil. Trans. Roy. Soc. London* **A325**, 545–557.

Wänke H., Dreibus G., and Jagoutz E. (1984) Mantle chemistry and accretion history of the Earth. *Archaean Geochemistry* (eds. A. Kröner, G. N. Hanson, and A. M. Goodwin). Springer, Berlin, pp. 1–24.

Weaver B. L. and Tarney J. (1984) Empirical approach to estimating the composition of the continental crust. *Nature* **310**, 575–577.

Weaver C. E. and Wampler J. M. (1970) K, Ar, Illite burial. *Geol. Soc. Am. Bull.* **81**, 3423–3430.

Wedepohl K. H. (1995) The composition of the continental crust. *Geochim. Cosmochim. Acta* **59**, 1217–1232.

Wetherill G. (1975) Radiometric chronology of the early solar system. *Ann. Rev. Nuclear Sci.* **25**, 283–328.

Wollast R. and Mackenzie F. T. (1983) The global cycle of silica. In *Silicon Geochemistry and Biogeochemistry* (ed. S. R. Aston). Academic Press, New York, pp. 39–76.

Wood B. J. (1993) Carbon in the core. *Earth Planet. Sci. Lett.* **117**, 593–607.

Zhao D. (2001) Seismic structure and origin of hotspots and mantle plumes. *Earth Planet. Sci. Lett.* **192**, 251–265.

Zhang Y. and Zindler A. (1993) Distribution and evolution of carbon and nitrogen in Earth. *Earth Planet. Sci. Lett.* **117**, 331–345.

Geochemistry of Earth Surface Systems
ISBN: 978-0-08-096706-6

pp. 167–204

6

Natural Weathering Rates of Silicate Minerals

A. F. White
US Geological Survey, Menlo Park, CA, USA

NOMENCLATURE		206
6.1	INTRODUCTION	206
6.2	DEFINING NATURAL WEATHERING RATES	207
6.3	MASS CHANGES RELATED TO CHEMICAL WEATHERING	207
	6.3.1 Bulk Compositional Changes in Regoliths	209
	6.3.2 Small-scale Changes in Mineral and Rock Compositions	210
	6.3.3 Changes Based on Solute Compositions	211
	6.3.3.1 Characterization of fluid transport	214
	6.3.3.2 Weathering based on solutes in soils	215
	6.3.3.3 Weathering based on solutes in groundwater	216
	6.3.3.4 Weathering based on surface-water solutes	218
6.4	TIME AS A FACTOR IN NATURAL WEATHERING	220
	6.4.1 Comparison of Contemporary and Geologic Rates	220
	6.4.2 Utilization of Soil Chronosequences	220
6.5	NORMALIZATION OF WEATHERING TO SURFACE AREA	221
	6.5.1 Definitions of Natural Surface Areas	221
	6.5.2 Measurements of Specific Surface Areas	221
	6.5.3 Surface Roughness	222
6.6	TABULATIONS OF WEATHERING RATES OF SOME COMMON SILICATE MINERALS	222
	6.6.1 Elemental Fluxes	222
	6.6.2 Mineral Fluxes	222
	6.6.3 Specific Mineral Rates	223
	6.6.4 Normalizing Rate Data	223
6.7	FACTORS INFLUENCING NATURAL WEATHERING RATES	224
	6.7.1 Mineral Weatherability	224
	6.7.2 Solute Chemistry and Saturation States	226
	6.7.3 Coupling the Effect of Hydrology and Chemical Weathering	227
	6.7.3.1 Initial stages of weathering	227
	6.7.3.2 Late-stage weathering	228
	6.7.4 Role of Climate on Chemical Weathering	229
	6.7.4.1 Temperature	229
	6.7.4.2 Precipitation and recharge	230
	6.7.4.3 Coupling climate effects	231
	6.7.5 Role of Physical Weathering	231
	6.7.5.1 Transport versus chemical weathering regimes	231
	6.7.5.2 Physical development of regoliths	232
6.8	WEATHERING RATES IN GEOCHEMICAL MODELS	233
6.9	SUMMARY	234
REFERENCES		235

NOMENCLATURE

b_{solid}	solid weathering gradient (m kg mol^{-1})
b_{solute}	solute weathering gradient (m L mol^{-1})
c	concentration of weatherable component in solute (mol)
$C_{j,w}$	final concentration of weatherable component j in regolith (mol m^{-3})
$C_{j,p}$	initial concentration of weatherable component j in protolith (mol m^{-3})
$C_{i,w}$	final concentration of inert component i in regolith (mol m^{-3})
$C_{i,p}$	initial concentration of inert component i in protolith (mol m^{-3})
d	mineral grain diameter (m)
D_j	diffusion coefficient (m^2 s^{-1})
E_a	activation energy (kJ mol^{-1})
G	pit growth rate (m s^{-1})
ΔG	excess free energy of reaction (kJ mol^{-1})
h_g	gravitational head (m m^{-1})
h_p	pressure head (m m^{-1})
∇H	hydraulic gradient (m m^{-1})
IAP	ionic activity product
k	intrinsic rate constant
K_m	hydraulic conductivity (m s^{-1})
K_s	mineral solubility product
m_j	atomic weight of component j (g)
ΔM	change in mass due to weathering (mol)
n	number of etch pits
P	annual precipitation (m)
q_h	flux density of water (m s^{-1})
R_{solid}	reaction rate based on solid concentrations (mol m^{-2} s^{-1})
R_{solute}	reaction rate based on solute concentrations (mol m^{-2} s^{-1})
R'	gas constant (J mol^{-1} K^{-1})
S	surface area (m^2)
T	temperature (K)
t	duration of weathering (s)
v_x	half wide opening rate (m s^{-1})
V_m	molar volume (m^3)
V_s	volume of solute (m^3)
W	etch diameter (m)
z	distance (m)
ϵ	volumetric strain
ϕ	porosity (m^3 m^{-3})
λ	surface roughness (m^2 m^{-2})
υ	solute weathering velocity (m s^{-1})
θ	pit wall slope (m m^{-1})
ρ	specific mineral density (m^3 g^{-1})
ρ_p	density of regolith (cm^3 g^{-1})
ρ_w	density of protolith (cm^3 g^{-1})
τ_j	mass transfer coefficient
τ_s	sheer modulus
ω	solid-state weathering velocity (m s^{-1})
Ω	thermodynamic saturation state

6.1 INTRODUCTION

Silicates constitute more than 90% of the rocks exposed at Earth's land surface (Garrels and Mackenzie, 1971). Most primary minerals comprising these rocks are thermodynamically unstable at surface pressure/temperature conditions and are therefore susceptible to chemical weathering. Such weathering has long been of interest in the natural sciences. Hartt (1853) correctly attributed chemical weathering to "the efficacy of water containing carbonic acid in promoting the decomposition of igneous rocks." Antecedent to the recent interest in the role of vegetation on chemical weathering, Belt (1874) observed that the most intense weathering of rocks in tropical Nicaragua was confined to forested regions. He attributed this effect to "the percolation through rocks of rain water charged with a little acid from decomposing vegetation." Chamberlin (1899) proposed that the enhanced rates of chemical weathering associated with major mountain building episodes in Earth's history resulted in a drawdown of atmospheric CO_2 that led to periods of global cooling. Many of the major characteristics of chemical weathering had been described when Merrill (1906) published the groundbreaking volume *Rocks, Rock Weathering, and Soils*.

The major advances since that time, particularly during the last several decades, have centered on understanding the fundamental chemical, hydrologic, and biologic processes that control weathering and in establishing quantitative weathering rates. This research has been driven by the importance of chemical weathering to a number environmentally and economically important issues. Undoubtedly, the most significant aspect of chemical weathering is the breakdown of rocks to form soils, a process that makes life possible on the surface of the Earth. The availability of many soil macronutrients such as magnesium, calcium, potassium, and PO_4 is directly related to the rate at which primary minerals weather. Often such nutrient balances are upset by anthropogenic activities. For example, Huntington *et al.* (2000) show that extensive timber harvesting in the southeastern forests of the United States, which are underlain by intensely weathered saprolites, produces net calcium exports that exceed inputs from weathering, thus creating a long-term regional problem in forest management.

The role of chemical weathering has long been recognized in economic geology. Tropical bauxites, which account for most of world's aluminum ores, are typical examples of residual concentration of silicate rocks by chemical weathering over long time periods (Samma, 1986). Weathering of ultramafic silicates such as peridotites forms residual lateritic deposits that contain significant deposits of nickel and cobalt. Ores generated by chemical mobilization include uranium deposits that are produced by weathering of granitic rocks under oxic conditions and subsequent concentration by sorption and precipitation (Misra, 2000).

Over the last several decades, estimating rates of silicate weathering has become important in addressing new environmental issues. Acidification of soils, rivers, and lakes has become a major concern in many parts of North America and Europe. Areas at particular risk are uplands where silicate bedrock, resistant to chemical weathering, is overlain by thin organic-rich soils (Driscoll *et al.*, 1989). Although atmospheric deposition is the most important factor in watershed acidification, land use practices, such as conifer reforestation, also create acidification problems (Farley and Werritty, 1989). In such environments, silicate hydrolysis reactions are the principal buffer against acidification. As pointed out by Drever and Clow (1995), a reasonable environmental objective is to decrease the inputs of acidity such that they are equal to or less than the rate of neutralization by weathering in sensitive watersheds.

The intensive interest in past and present global climate change has renewed efforts to understand quantitatively feedback mechanisms between climate and chemical weathering. On timescales longer than a million years, atmospheric CO_2 levels have been primarily controlled by the balance between the rate of volcanic inputs from the Earth's interior and the rate of uptake through chemical weathering of silicates at the Earth's surface (Ruddiman, 1997). Weathering is proposed as the principal moderator in controlling large increases and decreases in global temperature and precipitation through the greenhouse effects of CO_2 over geologic time (R. A. Berner and E. K. Berner, 1997). Weathering processes observed in paleosols, discussed elsewhere in this volume, have also been proposed as indicating changes in Archean atmospheric CO_2 and O_2 levels (Ohmoto, 1996; Rye and Holland, 1998).

6.2 DEFINING NATURAL WEATHERING RATES

Chemical weathering is characterized in either qualitative or quantitative terms. Qualitative approaches entail comparing the relative weathering of different minerals or the weathering of the same mineral in different environments. Early observations by Goldich (1938) showed that the weathering sequence of igneous rocks in the field was the reverse of Bowen's reaction series that ranked minerals in the order of crystallization from magmas. Thermodynamic and kinetic considerations have added new dimensions to characterizing the relative weatherability of various minerals and rocks (Nesbitt and Markovics, 1997). Silicate

weathering has long been recognized as exhibiting consistent trends with time and climate (Jenny, 1941; Oillier, 1984). Such studies provide valuable insights into weathering processes and into the influence of environmental conditions such as climate, vegetation, and geomorphology. A number of these issues are discussed elsewhere in this volume (see Chapter 7).

The primary focus of this chapter is on the development of rates that quantitatively describe silicate mineral and rock weathering. The advantages of this approach are that such rates are related to reaction mechanisms and can be used as predictive tools in estimating how weathering will behave under various environmental conditions.

The weathering rate R (mol m^{-2} s^{-1}) of a primary silicate mineral is commonly defined by the relationship

$$R = \frac{\Delta M}{S \cdot t} \qquad (1)$$

where ΔM (mol) is the mass change due to weathering, S (m^2) is the total surface area involved and t (s) is the duration of the reaction. The weathering rate is therefore equivalent to a chemical flux. Much of the following discussion will center on how the terms on the right side of Equation (1) are determined and quantified. This discussion is applicable in varying degrees to other types of minerals, such as carbonates, sulfates, and sulfides that commonly weather at significantly faster rates.

An important feature of silicate weathering is the extremely large spatial and temporal scales over which the parameters describing weathering are measured (Equation (1)). Natural weathering can be characterized by changes in microscopic surface morphologies of mineral grains, by solid and solute changes in soil profiles, from solute fluxes in small watersheds and large river basins and by continental and global-scale element cycles. Clearly, the degree to which individual weathering parameters are characterized depends strongly on the magnitude of these scales. Several other chapters in this volume deal in greater detail with some of these specific weathering environments, i.e., soils (see Chapter 7), glacial environments, watersheds (see Chapter 8), and river systems (see Chapter 9).

6.3 MASS CHANGES RELATED TO CHEMICAL WEATHERING

Silicate weathering involves hydrolysis reactions that consume reactant species, i.e., primary minerals and protons and form weathering products, i.e., solute species and secondary minerals. A typical hydrolysis reaction is the weathering of albite feldspar to form kaolinite

$$2NaAlSi_3O_8 + 2H^+ + H_2O$$
$$\rightarrow Al_2Si_2O_5(OH)_4 + 2Na^+ + 4SiO_2 \quad (2)$$

which consumes hydrogen ions and water and produces kaolinite and solute sodium and SiO_2. For a more detailed discussion of the reaction mechanisms of silicate weathering, the reader is referred to White and Brantley (1995 and references therein), Lasaga (1998), and Brantley.

The rate of weathering is reflected by the mass change ΔM with time of any of the individual components in a reaction if the overall stoichiometry is known. The consumption of hydrogen ions can, in principle, determine the rate of plagioclase reaction (Equation (2)), However, such measurements are difficult due to the many sources and sinks of hydrogen ion in the weathering environment, e.g., CO_2, organic acids and ion exchange reactions. Use of secondary minerals to define weathering rates is also difficult due to compositional heterogeneities and nonconservative behavior. In Equation (2), aluminum is completely immobilized in kaolinite during plagioclase weathering. However, in low-pH and/or organic-rich weathering environments, aluminum exhibits significant mobility and the quantity of kaolinite produced no longer directly reflects the amount of plagioclase that has reacted.

Most weathering rate studies have focused on measuring decreases in the solid-state reactant and/or increases in one or more of the mobile solute products. Examples of such changes due to weathering are shown in Figure 1 for solid-state and pore-water sodium distributions with depth in a 160 kyr old marine terrace in California. Mineralogical analyses indicate that significant sodium is contained only in plagioclase. Therefore, both the decrease in solid-state sodium with decreasing depth and the increase in solute sodium with increasing depth can be used to define the rate of plagioclase weathering (Equation (2)).

The above composition trends (Figure 1) can be conceptualized as solid and solute weathering gradients describing mineral weathering (White, 2002). Under simple steady-state conditions, as shown in Figure 2, gradients, b_{solid} and b_{solute}, are defined in terms the respective weathering rates R_{solid} and R_{solute} and weathering velocities ω and υ (m s^{-1}). The state-state weathering velocity is the rate at which the weathering front propagates through the regolith and the solute weathering velocity is the rate of solute transport or fluid flux.

Increasing the weathering rate produces a shallower weathering gradient in which a measured solid or solid component, such as sodium, increases more rapidly with depth in the regolith. Increasing weathering velocity, due to the rate at which weathering front or the pore water moves through the regolith steepens the weathering gradient. The slope of these gradients, measured under field conditions (Figure 1), can be converted

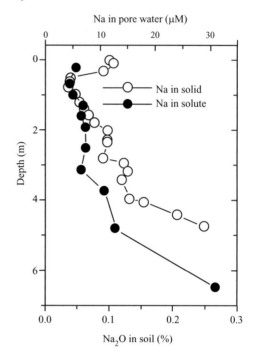

Figure 1 Distributions of sodium in the solids and pore waters of a Santa Cruz marine terrace, California, USA.

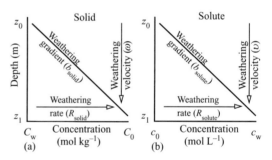

Figure 2 Schematic showing (a) solid and (b) solute distributions of a mobile element in a weathering regolith. C_w is the concentration in the most weathered material at shallow depth z_0 and C_p is the initial protolith concentration at depth z_1. c_w defines the corresponding solute weathered concentration and c_o is the initial solute concentration. The weathering gradients b_{solid} and b_{solute}, which describe the solid and solute elemental distributions, are defined in terms of respective weathering rates (R_{solid} and R_{solute}) and velocities (ω and υ) (after White, 2002).

to weathering rates in combination with additional physical and chemical parameters (White, 2002).

Defining the stoichiometry of the weathering reaction is often more difficult than suggested by plagioclase weathering (Equation (2)). The first stage of biotite weathering, for example, involves the oxidation of ferrous atoms and the concurrent release of potassium to form oxybiotite (Amonette *et al.*, 1988). Depending on the weathering environment, biotite subsequently undergoes direct transformation to kaolinite with congruent release of magnesium (Murphy *et al.*, 1998) or it can weather to vermiculite with magnesium partially retained in the solid (Fordham, 1990). Therefore, knowledge of the specific reaction pathway is required in order to define the rate of mineral weathering.

6.3.1 Bulk Compositional Changes in Regoliths

Determining weathering rates from solid-state mass changes (ΔM in Equation (1)) involves comparing concentration differences of a mineral, element or isotope in a weathered regolith with the corresponding concentration in the initial protolith (Merrill, 1906; Barth, 1961). For a component j, such as sodium in Figure 1, regolith concentrations are defined respectively as $C_{j,w}$ and $C_{j,p}$. A comparison of these concentrations is relatively straightforward for regoliths such as saprolites or laterites that are developed *in situ* on crystalline bedrock. The determinations of the parent concentrations are more difficult for sedimentary deposits, due to heterogeneities produced during deposition, e.g., the Santa Cruz marine terraces (Figure 1). In such cases, the protolith is commonly represented by the deepest, least weathered portion of the profile (Harden, 1987).

The measured ratio $C_{j,w}/C_{j,p}$ is dependent not only the weatherability of component j but also on gains and losses of other components in the regolith, as well as external factors such as compaction or dilation of the soil or regolith. The most common method for overcoming such effects is to compare the ratio of the mobile component j to the ratio of an additional component i, which is chemically inert during weathering. Such a comparison is commonly defined in terms of the mass transfer coefficient τ_j such that (Brimhall and Dietrich, 1987)

$$\tau_j = \frac{C_{j,w}/C_{j,p}}{C_{i,w}/C_{i,p}} - 1 \qquad (3)$$

When $\tau_j = 1$, no mobilization of j occurs, if $\tau_j = 0$, j has been completely mobilized. The concurrent volume change due to compaction or dilation by weathering and bioturbation is defined by the volumetric strain ϵ such that

$$\varepsilon = \frac{\rho_p C_{i,p}}{\rho_w C_{i,w}} - 1 \qquad (4)$$

where ρ_p and ρ_w are respective densities of the protolith and regolith. Positive values of ϵ denote regolith expansion and negative values indicate collapse. A value of $\varepsilon = 0$ denotes isovolumetric weathering. The reader is referred to Amundson (see Chapter 7) for a more detailed discussion of mass transfer coefficients and volumetric strain.

Conservative elements (C_i) include zirconium (Harden, 1987), titanium (Johnsson *et al.*, 1993), rare earth elements, and niobium (Brimhall and Dietrich, 1987). Considerable disagreement occurs in the literature as to the relative mobility of these elements under differing weathering conditions (Hodson, 2002). Also, these elements are concentrated in the heavy mineral fractions and are often not suitable for describing weathering ratios in depositional environments where selective concentration and winnowing occurs. For such conditions, relatively inert minerals such as quartz can be considered (Sverdrup, 1990; White, 1995).

The applicability of the above approach (Equations (3) and (4)) in describing elemental mobilities is shown in Figure 3 for a 350 kyr old regolith at Panola, Georgia, USA (White *et al.*, 2001; White *et al.*, 2002). This profile consists of a dense unstructured kaolinitic soil, overlying a porous saprolite resting on granodiorite bedrock. Saprolites, common in many subtropical to tropical weathering environments, are clay-rich regoliths that retain the original bedrock texture. Volume changes in the Panola saprolite and bedrock, calculated from zirconium, titanium, and niobium concentrations, center close to zero, i.e., $\varepsilon = 0$ (Equation (4)), indicating that weathering is isovolumetric (Figure 3(a)). In contrast, the shallowest soils have undergone volume increases due to bioturbation and the introduction of organic matter.

Values for τ_j (Equation (3)) cluster near zero in the Panola bedrock below 10.5 m indicating no elemental mobility (Figure 3(b)). Calcium and sodium are almost completely mobilized at depths of between 9.5 m and 4.5 m, where silicon is moderately depleted. Bedrock weathering is highly selective, with other elements remaining relatively immobile. Potassium and magnesium are mobilized in the overlying saprolite and iron and aluminum are strongly depleted in the upper soil and enriched in the lower soil horizons, which is indicative of downward iron and aluminum mobilization and subsequent re-precipitation.

The total elemental mass loss or gain occurring in the Panola regolith ($\Delta M_{j,\text{solid}}$) is determined by integrating the mass transfer coefficient τ_j over the regolith depth z (m) and unit surface area (1 m^2) such that (Chadwick *et al.*, 1990; White *et al.*, 1996)

$$\Delta M_{j,\text{solid}} = \left(\rho_p \frac{C_{j,p}}{m_j} 10^4 \right) \int_0^z -(\tau_j) \mathrm{d}z \qquad (5)$$

where m_j is the atomic weight of element j. The elemental losses in each horizon of the Panola profile are calculated from Equation (5) and tabulated in Table 1.

Element mobilities in the Panola regolith can be converted to changes in primary and secondary mineral abundances (Figure 4) using a series of linear equations describing mineral stochiometries. Table 1 indicates that initial bedrock weathering

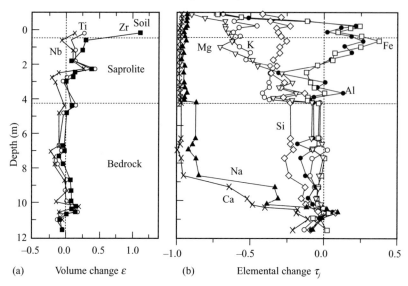

Figure 3 (a) Volume and (b) elemental changes in the Panola regolith, Georgia, USA. Volume changes are calculated separately assuming conservancy of Zr, Nb, and Ti. Positive values of ϵ (Equation (4)) indicate dilation and negative values indicate compaction. Positive values of τ_j (Equation (3)) indicate elemental enrichment and negative values denote elemental mobilization and loss (after White *et al.*, 2001).

Table 1 Elemental mobility (kmol m^3) and total % change in the Panola regolith. Negative values indicate element increases.

	Bedrock	*Saprolite*	*Soil*	*%loss*
Ca	4,550	40	−10	77
Na	8,190	670	−90	67
Si	15,030	5,520	−160	16
K	−33	3,360	10	18
Mg	−229	630	−20	16
Fe	−550	−130	540	−5
Al	1,520	−2,030	1,780	0

removes most of the plagioclase. K-feldspar weathering predominates within the overlying saprolite and biotite weathering occurs both in the saprolite and shallower soil. Secondary kaolinite is produced within the bedrock from plagioclase weathering and in the saprolite and deeper soils from K-feldspar and biotite weathering. The loss of kaolinite in the shallowest soils implies either dissolution or physical translocation to deeper soil horizons. Iron oxides are formed by oxidation of iron released from biotite weathering in the saprolite (Figure 4). The abundance of iron-oxides, like that of kaolinite, decreases in the shallow soils.

Mass balance calculations produce important insights into element and mineral mobilities in regoliths in different weathering environments, in addition to providing quantitative estimates of mass changes required in calculating chemical weathering rates (ΔM, Equation (1)). Rates of plagioclase and biotite weathering in the Panola regolith, based on this approach, are listed in Table 2 (nos. 6 and 38). Additional examples of

weathering mass balances are contained in Merritts *et al.* (1992), Stewart *et al.* (2001) and Amundson (see Chapter 7).

6.3.2 Small-scale Changes in Mineral and Rock Compositions

In addition to bulk changes in regolith composition, chemical weathering is also characterized by smaller-scale physical and chemical changes in individual mineral grains and rock surfaces. The formation of weathering and hydration rinds is commonly observed on volcanic rocks. Rind thicknesses are measured using optical and scanning electron microscopy (SEM). The most pervasive use of hydration rinds has been in the study of natural glasses in which thicknesses are used to date cultural artifacts (Friedman *et al.*, 1994). Weathering rinds developed on volcanic clasts have also been evaluated in terms of weathering duration and used as a tool for dating glacial deposits and other geomorphologic features (Colman, 1981). The weathering reactions producing rinds are considered to be transport-controlled and limited by diffusion of reactants and products through the altered surfaces of parent material.

Changes in mineral surface morphologies have also been extensively correlated with silicate weathering. Dorn (1995) used the digital processing of backscatter electron imagery to determine void spaces produced by weathering of plagioclase in Hawaiian basalts and to determine climatic effects on weathering rates.

Of all the morphological changes associated with weathering, the development of etch pits has received the most attention. Examples of

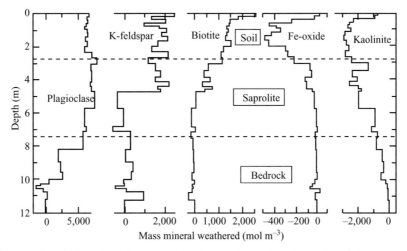

Figure 4 Primary minerals lost (positive values) and secondary minerals produced (negative values) during weathering of the Panola regolith. Horizontal lines denote regolith horizons (after White *et al.*, 2001).

etch pits developed on microcline and hornblende during soil weathering are shown by SEM images in Figure 5. Recently atomic force microscopy and interferometry have contributed higher resolution and greater depth sensitivities to etch pit imaging (Luttge and Lasaga, 1999; Maurice *et al.*, 2002).

Increases in etch pit formation have been translated into weathering rates by MacInnis and Brantley (1993) using a pit-size-distribution model (PSD) that considers the number of pits *n* as an inverse exponential function of the pit width *W* (cm) divided by the growth rate G (m s^{-1}) such that

$$n = n^0 \exp\left(\frac{-W}{\tau_s G}\right) \quad (6)$$

where τ_s is the shear modulus and n^0 is the initial pit nucleation density. Based on this equation, a plot of log *n* versus *W* will produce a straight line with a slope of $-1/\tau_s G$. This relationship was derived for several soil profiles (Figure 6) using hornblende and K-feldspar data from Cremeens *et al.* (1992).

The contribution of etch pits to the bulk weathering rate of a mineral depends on the number, density and rate of dissolution of each pit. By integrating over the PSD and assuming a pyramidal geometry for each pit, MacInnis and Brantley (1993) derived a simple expression for the weathering rate

$$R_{\text{solid}} = \frac{2n^0 v_x \tan\theta}{V_m} (\tau_s G)^3 \quad (7)$$

where V_m is the molar volume (m^3 mol^{-1}), θ is the average slope of the wall of the etch pits and v_x is the half-width opening rate ($= 0.5G$). Etch pit distributions produced quantitative weathering rates for hornblende and K-feldspar that are comparable to natural rates estimated by other techniques (Table 2, nos. 22 and 34).

In addition to characterizing mass losses, textural features can be used to discern the mechanisms and chemical environment associated with silicate dissolution. It is widely accepted that pitted surfaces, such as that shown in Figure 5, indicate surface reaction whereas smooth rounded surfaces result from diffusion-controlled dissolution (Lasaga, 1998).

6.3.3 Changes Based on Solute Compositions

The alternative method for calculating rates is based on the mass of solutes produced during weathering (Equation (1)). This approach commonly involves comparing changes between initial and final solute concentrations Δc_j (M) in a known volume of water V_s (L) such that

$$\Delta M_{j,\text{solute}} = \Delta c_j V_s \quad (8)$$

The above equation is analogous to Equation (5), which describes mass changes in the solid state. The calculation of mineral weathering rates based on solute concentrations is complicated by the fact that individual solute species are commonly produced by more than one weathering reaction. In granite weathering, for example, aqueous silicon is produced not only from plagioclase weathering (Equation (2)), but also by K-feldspar, biotite, hornblende and quartz dissolution. In addition, secondary minerals such as kaolinite take up aqueous silicon (Equation (2)).

The problem of multi-mineral sources for solutes was initially addressed using a spread sheet approach popularized by Garrels and Mackenzie (1967) in a study of weathering contributions from springs in the Sierra Nevada. This approach has since been incorporated into geochemical computer codes which simultaneously calculate mineral masses. Depending on the

Table 2 Specific weathering rates of common silicates (rates in parentheses are BET surface area-normalized rates).

	Weathering environment	log rate (mol m^{-2} s^{-1})	Long time (yr)	pH	Surface measurement	Surface area S (m^2 g^{-1})	Grain size (μm)	Roughness (λ)	References
Plagioclase									
1	Davis Run saprolite, VA, USA	−16.4	5.93	ND	BET	1.0	ND	ND	White *et al.* (2001)
2	China Hat Soil, Merced CA, USA	−16.3	6.48	6.0–7.0	BET	1.5	500	331	White *et al.* (1996)
3	Turlock Lake Soil, Merced, CA, USA	−16.0	5.78	6.0–7.0	BET	1.48	730	477	White *et al.* (1996)
4	Modesto Soil, Merced, CA, USA	−15.8	4.00	6.0–7.0	BET	0.26	320	37	White *et al.* (1996)
5	Riverbank Soil, Merced, CA, USA	−15.7	5.40	6.0–7.0	BET	0.46	650	132	White *et al.* (1996)
6	Panola bedrock, GA, USA	−15.7	5.70	6.5	BET	1.0*	500	221	White *et al.* (2001)
7	Modesto Soil, Merced, CA, USA	−15.1	4.00	6.0–7.0	BET	0.26	320	37	White *et al.* (1996)
8	Filson Ck. Watershed, MN, USA	−14.7 (−15.3)	4.00	5.0	GEO	0.50	63–125	21	Siegal and Pfannkuch (1984)
9	Bear Brook Watershed, ME, USA	−14.5 (15.2)	4.00	5.8	GEO	0.50	63–125	15	Schnoor (1990)
10	Crystal Lake Aquifer, WI, USA	−13.7 (15.3)	4.00	6.0–7.5	GEO	0.12	1–100	2	Kenoyer and Bowser (1992)
11	Bear Brook watershed soil, ME, USA	−13.3 (−15.2)	4.00	3.0–4.5	GEO	0.020	1–4,000	1	Swoboda-Colberg and Drever (1992)
12	Gardsjon watershed soil, Sweden	−13.1	4.00	6.0	BET	1.53	NA	ND	Sverdrup (1990)
13	Loch Vale Nano-Catchment, CO, USA	−12.8	4.00	7.0	BET	0.21	53–208	9	Clow and Drever (1996)
14	Coweeta Watershed, NC, USA	−12.5 (−14.9)	5.50	6.8	GEO	0.003	1,000	1	Velbel (1985)
15	Trnavka River watershed, Czech.	−12.1 (−14.0)	4.00	5.0	GEO	0.057	ND	1	Paces (1986)
16	Plastic Lake, Ontario, Canada	−11.8 (−13.7)	4.00	ND	GEO	0.030	100	1	Kirkwood and Nesbitt (1991)
K-feldspar									
17	Davis Run Saprolite, VA, USA	−16.8	5.93	ND	BET	1.0	ND	ND	White *et al.* (2001)
18	China Hat Soil, Merced CA, USA	−16.6	6.48	6.0–7.0	BET	0.81	500	179	White *et al.* (1996)
19	Riverbank Soil, Merced CA, USA	−16.4	5.52	6.0–7.0	BET	0.94	650	270	White *et al.* (1996)
20	Turlock Lake Soil, Merced, CA, USA	−16.3	5.78	6.0–7.0	BET	0.81	730	261	White *et al.* (1996)
21	Upper Modesto Soil, Merced, CA, USA	−15.3	4.00	6.0–7.0	BET	0.26	170	20	White *et al.* (1996)
22	Adams County, IL	−14.7 (−16.6)	4.10	6.0–7.0	GEO	0.02	53–100	1	Brantley *et al.* (1993)
23	Loch Vale Nano-Catchment, CO, USA	−13.8	4.00	7.0	BET	NA	NA	ND	Clow and Drever (1996)
24	Bear Brook watershed soil, ME, USA	−13.3 (−15.2)	4.00	4.5	GEO	0.020	1–4,000	1	Swoboda-Colberg and Drever (1992)

25	Gardsjon watershed soil, Sweden	−13.3	4.08	5.6–6.1	BET	1.53		ND	ND	Sverdrup (1990)
26	Surface exposures of Shap Granite	−12.9 (−14.8)	3.70	ND	GEO	ND		NA	1	Lee *et al.* (1998)
27	Plastic Lake, Ontario, Canada	−11.8 (−13.6)	4.00	ND	GEO	0.03		100	1	Kirkwood and Nesbitt (1991)
	Hornblende									
28	China Hat Soil, Merced CA, USA	−17.0	6.48	6.0–7.0	BET	0.67		500	179	White *et al.* (1996)
29	Turlock Lake Soil, Merced, CA, USA	−16.4	5.78	6.0–7.0	BET	0.67		730	261	White *et al.* (1996)
30	Riverbank Soil, Merced, CA, USA	−16.0	5.40	6.0–7.0	BET	0.72		650	250	White *et al.* (1996)
31	Riverbank Soil, Merced, CA, USA	−15.9	5.11	6.0–7.0	BET	0.72		650	250	White *et al.* (1996)
32	Lower Modesto Soil, Merced, CA, USA	−15.7	4.60	6.0–7.0	BET	0.34		320	58	White *et al.* (1996)
33	Bear Brook watershed soil ME, USA	−14.5 (−16.4)	4.00	4.5	GEO	0.020		1–4,000	1	Swoboda-Colberg and Drever (1993)
34	Adams County, IL	−14.1 (−16.0)	4.10	ND	GEO	0.02		53–100	1	Brantley *et al.* (1993)
35	Lake Gardsjon	−13.6	4.00	4.0	BET	1.53		NA	ND	Sverdrup (1990)
36	Plastic Lake, Ontario, Canada	−12.3 (−14.2)	4.00	ND	GEO	0.03		100	1	Kirkwood and Nesbitt (1991)
	Biotite									
38	Panola, GA, USA	−16.5	5.70	5.0–5.5	BET		5	100–500	775	White (2002)
39	Crystal Lake Aquifer, WI, USA	−15.5 (−17.4)	4.00	6.0–7.5	GEO		12	1–100	1	Kenoyer and Bowser (1992)
40	Rio Icacos, PR	−15.4	5.48	4.5–5.5	BET	8.1		200–1,200	2,900	White (2002)
41	Rio Icacos, PR	−15.0	5.48	4.5–5.5	BET	8.1		200–1,200	2,900	Murphy *et al.* (1998)
42	Loch Vale Nano-Catchment, CO, USA	−14.1	4.00	7.0	BET	3.2		NA	ND	Clow and Drever (1996)
43	Bear Brook watershed soil, ME, USA	−14.0 (−15.9)	4.00	4.5	GEO	0.0800		1–4,000	1	Swoboda-Colberg and Drever (1992)
44	Coweeta, NC, USA	−12.9 (−14.8)	5.48	6.0	GEO	0.006		1	1	Velbel (1985)
	Quartz									
45	Rio Icacos, PR	−15.1	5.48	4.5–5.5	BET	0.3		450		Schulz and White (1999)

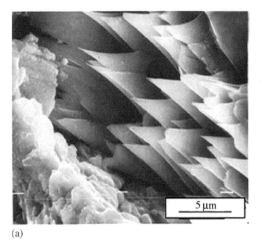 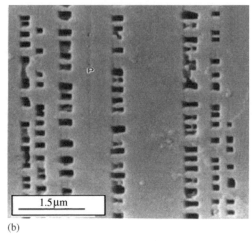

(a)   (b)

Figure 5 Etch pit morphologies: (a) denticulated margins on weathered hornblende formed by side-by-side coalescence of lenticular etch pits, Blue Ridge soil, North Carolina, USA (source Velbel, 1989) and (b) etch pits along a microperthite lamellae, weathered Shap Granite, England (source Lee and Parsons, 1995).

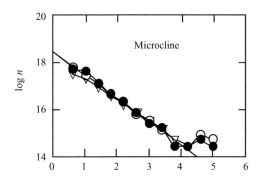

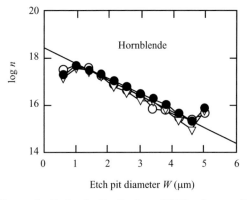

Etch pit diameter W (μm)

Figure 6 Etch pit distributions (PSD) of naturally weathered microcline and hornblende in soils from Illinois, USA (after Brantley *et al.*, 1993).

6.3.3.1 Characterization of fluid transport

Water is the medium in which weathering reactants and products are transported to and from the minerals undergoing reaction. Clearly, a detailed discussion of hydrologic principles is beyond the scope of this chapter, but several aspects critical to understanding natural weathering will be briefly mentioned.

In the case of diffusive transport, bulk water is immobile and movement of solutes occurs by Brownian motion. Under such conditions, the flux Q_j of a solute reactant or product is determined by the product of the diffusion coefficient D ($m^2 s^{-1}$) and the concentration gradient as given by Fick's First Law (Allen and Cherry, 1979)

$$Q_j = -D_j \frac{\partial c_j}{\partial z} \qquad (9)$$

The diffusion coefficients of solutes are commonly on the order of $10^{-9} m^2 s^{-1}$ making diffusive transport effective only at or near the mineral surface. On a nanometer-scale, solutes diffuse along and across water layers structurally influenced by the silicate surface. Microscopic diffusion occurs through stagnant water contained within internal pore spaces and etch pits or along grain boundaries.

Solutes also moved through the weathering environment along with the advective flow of water. In this case, the solute flux is the product of the solute concentration and the flux density of water q_h, i.e.,

$$Q_j = c_j q_h \qquad (10)$$

where q_h ($m s^{-1}$) is defined as the product of the hydraulic conductivity K_m ($m s^{-1}$), and the hydraulic gradient ∇H ($m\, m^{-1}$) (Hillel, 1982)

$$q_h = -K_m \nabla H \qquad (11)$$

number of minerals and solutes considered, non-unique results are commonly generated that require independent confirmation of the actual weathering reactions (Parkhurst and Plummer, 1993). A detailed discussion of the various aspects of mass balance calculations is presented by Bricker *et al.*

The above relationships (Equations (9)–(11)) define the basic processes by which solutes are transported during chemical weathering.

6.3.3.2 Weathering based on solutes in soils

The number of studies characterizing weathering rates based on soil/regolith pore-water solutes is relatively limited, e.g., Sverdrup (1990) and White *et al.* (2002). This deficiency is somewhat surprising considering the general accessibility of the soil/regolith environment and the relative ease with which pore waters can be sampled using lysimeters and suction water devices. An example of solute distributions generated by such sampling is shown in Figure 7 for a deep weathering profile in the tropical Luquillo Mountains of Puerto Rico (White *et al.*, 1998).

The upper 0.5 m of this regolith is a dense kaolinitic clay overlying ~10 m of saprolite resting on a quartz diorite bedrock consisting predominately of quartz, plagioclase, hornblende, and biotite. As indicated, significant variations in solute potassium, magnesium, and silicon occur in the upper clay zone due to the effects of evapotranspiration and biological cycling of mineral nutrients. At greater depths (>2 m), this variability is damped with solute species exhibiting consistent increases with depth. Such distributions reflect both the rate at which reaction products are contributed from chemical weathering and the rate of pore-water movement through the soil (Figure 2). Most soil pore-water studies assume one-dimensional vertical advective transport in which progressive increases in solute concentrations with depth reflect increasing reaction times along a single flow path. Flow paths in soils with significant topographic relief or macropore flow are more complex.

Another complexity is that pore-water flow in soils commonly occurs under unsaturated or vadose conditions. This is shown in Figure 8(a) for the Luquillo regolith in which fluid saturation varies between 65% and 95% over a depth of 7 m (White *et al.*, 1998). In such a case, the hydraulic conductivity K_m (Equation (11)) is strongly dependent on the soil moisture content (Stonestrom *et al.*, 1998). Experimental conductivities produced for cores taken from the Luquillo regolith (Figure 8 (c)) decrease between 2 and 4 orders of magnitude with less than a 30% decrease in moisture saturation. This is caused by the entrapment of air within the pore spaces, which physically obstructs the movement of water.

The hydraulic gradient ∇H in Equation (11) also becomes dependent on water content during unsaturated flow such that (Hillel, 1982)

$$\frac{dH}{dz} = \frac{dh_g}{dz} + \frac{dh_p}{dz} \qquad (12)$$

where dh_g/dz is the gravitational potential and dh_p/dz is the matric potential describing the capillary tension on soil mineral surfaces. The inverse relationship between soil-water saturation and the matric or capillary tension is shown in Figures 8(a) and (b) for the Luquillo regolith. Pore waters close to saturation at shallow depths have pressure heads close to zero. The most undersaturated pore waters at intermediate depths exhibit the greatest negative pressure heads.

Based on the solute and hydrologic information outlined above, Murphy *et al.* (1998) derived a rate equation for biotite weathering in the Luquillo regolith

$$R_{biotite} = \left[\frac{q_h}{\beta V_{biotite} \rho (1 - \phi) S} \right] \frac{dc}{dz} \qquad (13)$$

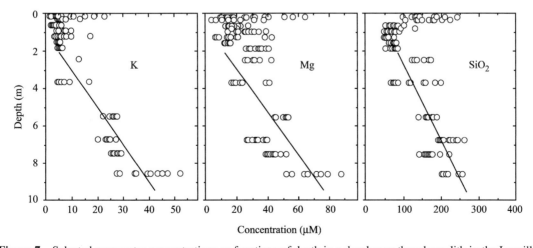

Figure 7 Selected pore-water concentrations as functions of depth in a deeply weathered regolith in the Luquillo Mountains of Puerto Rico. Diagonal solid lines are linear fits to the weathering gradients used by Murphy *et al.* (1998) to calculate biotite weathering rates (Equation (13)) (after White *et al.*, 1998).

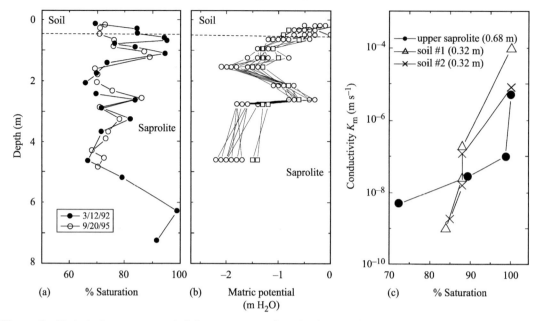

Figure 8 Hydrologic parameters defining pore-water flow in the Luquillo regolith: (a) moisture saturation, (b) matric potential describing capillary tension, and (c) experimental hydrologic conductivities as functions of moisture saturation at several regolith depths (after White *et al.*, 1998).

in which the weathering rate is the product of the water flux density q_h (Equation (11)) and the solute gradient describing the linear change in solute magnesium or potassium with depth in the regolith dc/dz. $V_{biotite}$ is the volume fraction of biotite in the regolith, β is the stiochometric coefficient describing the magnesium and potassium content of the biotite. ρ, ϕ, and S are the respective density, porosity, and BET surface area terms. The resulting biotite weathering rate in the Luquillo regolith, based on magnesium and potassium pore-water solute distributions, is reported in Table 2 (no. 41).

6.3.3.3 *Weathering based on solutes in groundwater*

Several characteristics distinguish chemical weathering in aquifers from soils. Groundwater weathering occurs under phreatic conditions in which fluid flow is governed by Darcy's law and is proportional to the product of the saturated hydraulic conductivity and the gravitational gradient (Equation (11)). Groundwater movement commonly occurs as matrix flow through unconsolidated sediments and bedrock or as fracture flow. The saturated hydraulic conductivities are dependent on porosity, tortuosity, and other factors intrinsic to the geologic material and vary over a wide range from 10^{-4} m s^{-1} for clean sand to 10^{-12} m s^{-1} for crystalline rocks (Allen and Cherry, 1979).

The physical isolation of groundwater aquifers relative to that of soils is advantageous for characterizing weathering rates because biologic and short-term climatic perturbations are diminished.

However, because of the greater isolation, the characterization of mineral distributions, aquifer heterogeneities and mineral surface areas becomes more difficult. The earliest and simplest characterization of chemical weathering in groundwater was based on spring or seep discharge (Garrels and Mackenzie, 1967). In such cases, the changes in solute concentration due to weathering are assumed to be the difference between the discharge compositions and atmospheric contributions in precipitation (Equation (8)). The fluid flux is the measured spring or seep discharge. The disadvantage of such an approach is that no information on the spatial extent of the weathering environment is known, making normalization of the weathering rate to a surface area impossible.

A more rigorous and intrusive approach to groundwater weathering is to characterize changes in solute concentrations measured in an array of wells in an aquifer. Determination of chemical weathering rates based on this approach requires that changes in solute concentrations occur along defined flow paths and that the rate of groundwater flow is known. This approach is commonly referred to as an inverse problem in which weathering reactions are computationally fitted to account for the observed chemical changes in the groundwater (Parkhurst and Plummer, 1993). Such studies tie the rates of chemical weathering closely to a detailed understanding of fluid flow in an aquifer, which is commonly described by groundwater flow models.

The above approach is illustrated in the determination of chemical weathering rates in groundwaters in glacial sediments in northern Wisconsin, USA (Kenoyer and Bowser, 1992; Bullen *et al.*, 1996). The flow path is defined by subaerial recharge from Crystal Lake moving downgradient across a narrow isthmus and discharging into Big Muskellunge Lake (cross-sections in Figure 9). Water levels and *in situ* slug tests, measured in an array of piezometers, determined the hydraulic conductivities and gradients plotted in Figures 9(a) and (b). Groundwater velocities were determined by application of Darcy's law to these measurements and were checked using tracer tests. The resulting fluid residences in groundwater at various positions in the aquifer were then calculated based on the distance along the flow path (Figure 9(c)).

Groundwater solutes, sampled from the piezometers, exhibited significant trends as shown by silicon and calcium distributions plotted in Figure 10. The core of the groundwater at the upgradient end of the cross-section is recharged directly from Crystal Lake. Significant changes

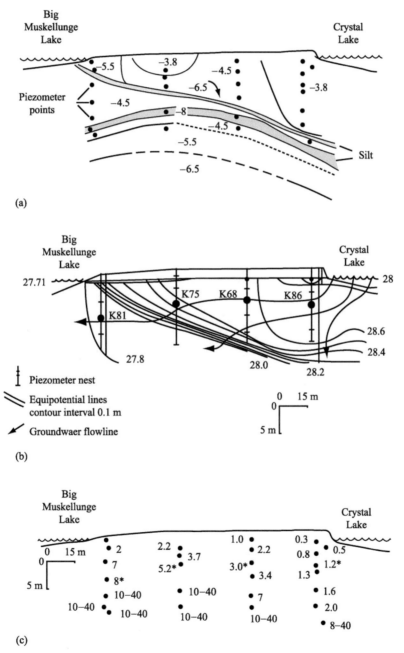

(a)

(b)

(c)

Figure 9 Hydrologic properties of a sandy silicate aquifer in northern Wisconsin: (a) hydrologic conductivities (log K_m, ms^{-1}), (b) groundwater potentials (m), and (c) groundwater residence times (yr) (reproduced by permission of American Geophysical Union from *Water Resources Research*, 1992, 28, 579–589).

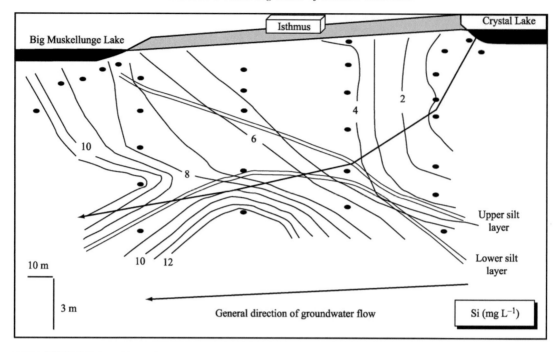

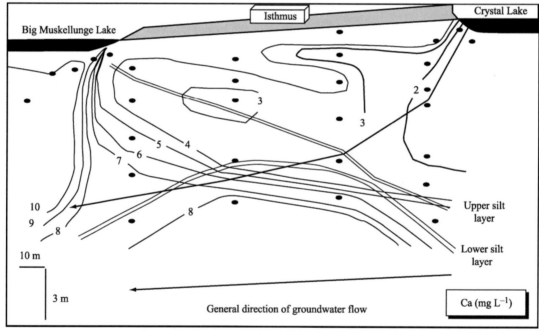

Figure 10 Cross-sections showing the vertical distribution of Si and Ca in groundwater in the North Wisconsin aquifer (after Bullen *et al.*, 1996).

in the chemistry of this plume take place along the flow path due to the dissolution of well-characterized silicate minerals that compose the aquifer. Various proportions of minerals phases were computationally dissolved using the PHREEQE code (Parkhurst and Plummer, 1993) until the best fit to the solute trends along the flow path was obtained. The resulting mineral masses were than divided by the fluid residence time and estimates of mineral surface area to produce the

plagioclase and biotite weathering rates tabulated in Table 2 (nos. 10 and 39).

6.3.3.4 Weathering based on surface-water solutes

The composition and concentrations of solutes in surface waters in catchments and rivers are used extensively to characterize silicate weathering rates. As in the case of soils and aquifers, the determination of such rates in watersheds is closely associated with an understanding of the hydrology. Commonly, solute compositions and fluid fluxes are determined at a gauged position on a stream or river. Fluid flow, expressed in terms of stage, is continuously measured, and water samples are periodically taken. Since a strong inverse correlation commonly exists between stage and solute concentration, a number of approaches were developed to produce a continuous estimate of discharge-weighted solute compositions (Likens *et al.*, 1977; Zeman and Slaymaker, 1978). Integrating both the discharge and solute concentrations over a given time interval, commonly on an annual basis, and dividing this output by the geographical area of the watershed produces a solute flux $Q_{watershed}$ (Equation (14)), commonly expressed as mol ha^{-1} yr^{-1}.

The weathering components associated with watershed discharge must be corrected for other potential sources and sinks within the watershed including atmospheric and anthropogenic inputs, biological cycling, and changes in ion exchange processes in the soil. Such a relationship can be represented by the expression (Paces, 1986)

$$Q_{weathering} = Q_{watershed\ output} - Q_{precipitation}$$
$$- Q_{anthropedgenic} \pm Q_{biology}$$
$$\pm Q_{exchange} \qquad (14)$$

Usually, for pristine watersheds, biologic and exchange reactions are assumed to be at steady state and the weathering flux is the difference between the watershed output and the precipitation input. In perturbed watersheds, the calculations become more complex due to agricultural inputs or changes in the exchangeable ions due to watershed acidification.

The measurement of solute fluxes in surface-water discharge is an indirect approach to estimating chemical weathering rates. Due to low mineral-to-fluid ratios and short residence times, minimal silicate weathering occurs in the streambed and the hyporheic environment. Rather, surface-water solutes represent discharges from other weathering environments that are spatially and temporally integrated by the watershed flux.

The presence of multiple solute sources was demonstrated for variations in discharge from the Panola watershed in central Georgia, USA (Hooper *et al.*, 1990). As shown in Figure 11, variations in solute silicon and magnesium, are explained as a mixture of three end-member sources; a groundwater component contained in fractured granite, a hill-slope component representing waters from soils and saprolites and an organic component consisting of near-surface runoff in the shallow soils. To further complicate the interpretation of the discharge flux, the relative proportions of these components vary seasonally.

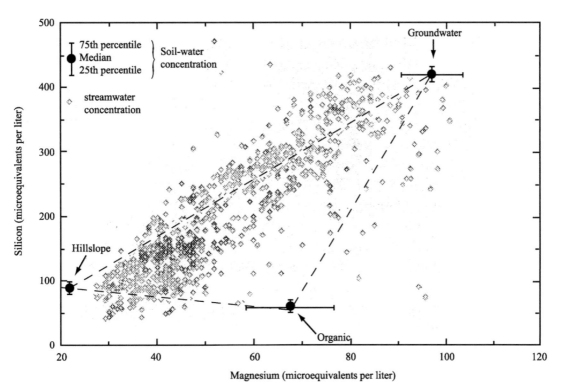

Figure 11 Dissolved silicon and magnesium in discharge from the Panola catchment, Georgia, USA. Dashed lines encompass the hillslope, groundwater, and shallow soil (organic) inputs (after Hooper *et al.*, 1990).

Hill-slope waters dominate during the wet season (January to May), groundwater dominates at low flow during seasonally dry periods, and the organic component is present only during intense storm events. Different weathering processes and rates are associated with each of these environments.

The above example serves as a cautionary note to the complexities involved in interpreting solute discharge fluxes and calculating weathering rates even in an intensely studied watershed. However, significant efforts, such as those contained in Trudgill (1995), are directed to resolving the various hydrologic and geochemical sources of watershed discharge using a number of computational approaches. The compilation of watershed solute fluxes is effective in establishing the impact of rock type, precipitation, temperature, and other parameters on chemical weathering rates, e.g., Dethier (1986), Bluth and Kump (1994), and White and Blum (1995). Such fluxes are also used to calculate silicate weathering rates (Table 2, nos. 9, 10, and 14).

6.4 TIME AS A FACTOR IN NATURAL WEATHERING

The weathering rate is inversely proportional to duration of chemical weathering (Equation (1)), i.e., the longer the time required to produce a given change in mass, the slower is the reaction rate. The duration of chemical weathering in natural systems involves very different timescales depending on whether changes in solid-state or solute concentrations are measured. Solid-state mass differences commonly reflect weathering over geologic timescales of thousands to millions of years (t_{solid}). The duration of chemical weathering reflected in solute compositions (t_{solute}) is equivalent to the residence time of the water in the regolith, commonly the time elapsed since initial recharge as precipitation. For most weathering environments such as soils, watersheds, and shallow groundwaters, solute residence times range from days to decades.

Recent advances have significantly increased our ability to quantitatively establish the duration of weathering associated with both solid-state and solute mass changes. Of particular importance is the use of cosmogenic isotopes such as ^{36}Cl, ^{26}Al, and ^{10}Be in age dating geomorphic surfaces and the use of ^{36}Cl, ^{3}He/^{4}He, and chloroflurocarbons to establish fluid residence times over time spans of years to tens of thousands of years.

6.4.1 Comparison of Contemporary and Geologic Rates

Rates based on solute compositions are measurements of present-day weathering that correlate with contemporary environmental conditions including hydrology, climate, and vegetation. Weathering rates based on solid-state composition changes reflect average rates integrated over geologic time during which conditions influencing weathering can only be indirectly estimated. However, under steady-state conditions, assuming constant surface area, the respective weathering rates R_{solid} and R_{solute} must be equal and proportional to the respective mass changes with time, i.e.,

$$R_{solid} = R_{solute} = \frac{\Delta M_{solid}}{t_{solid}} = \frac{\Delta M_{solute}}{t_{solute}} \quad (15)$$

Some authors have documented that present-day base cation fluxes measured in North America and Northern Europe watersheds are significantly faster than long-term past weathering rates due to the impacts of acidic precipitation (April *et al.*, 1986; Kirkwood and Nesbitt, 1991; Land *et al.*, 1999). Likewise, Cleaves (1993) found watershed solute discharge in the Piedmont of the eastern USA, to be 2–5 times faster than for past periglacial periods represented by long-term saprolite weathering. He attributed these weathering rate differences to past periods of lower precipitation, colder temperatures and lower soil gas CO_2.

In contrast, Pavich (1986) concluded that long-term weathering rates of saprolites in the Virginia Piedmont are comparable to current-day weathering rates based on stream solute fluxes. Rates of saprolite formation are also found to be similar to current weathering rates in the Luquillo Mountains of Puerto Rico (White *et al.*, 1998) and in the Panola watershed in northern Georgia, USA (White *et al.*, 2002). As the literature indicates, considerable debate exists regarding the extent to which contemporary fluxes correspond to increased weathering or result from other contributions such as enhanced cation exchange rates, deforestation and other parameters contained in Equation (14).

6.4.2 Utilization of Soil Chronosequences

A particularly valuable approach to investigating the effect of time on chemical weathering has been the use of soil chronosequences which are defined as a group of soils that differ in age and, therefore, in duration of weathering. These soils have similar parent materials and have formed under similar climatic and geomorphic conditions (Jenny, 1941). Individual soils, therefore, provide "snapshots" of the progressive nature of chemical weathering with time.

Chronosequences include marine and alluvial deposits and lava flows. Studies of chronosequences (e.g., Bockheim, 1980; Birkland, 1984; Harden, 1987; Jacobson *et al.*, 2002) have quantified systematic changes in weathering properties of soils with ages ranging from 1 kyr to 4,000 kyr. Mass balance studies have characterized mineralogical changes (Mokma *et al.*, 1973; Mahaney and

Halvorson, 1986), chemical changes (Brimhall and Dietrich, 1987; Merritts *et al.*, 1992), and development of etch pitting (Hall and Horn, 1993) as functions of soil age. An important conclusion of these chronosequence studies is that the rates of primary silicate weathering generally decrease with duration of weathering.

6.5 NORMALIZATION OF WEATHERING TO SURFACE AREA

The mass losses or gains resulting from weathering are normalized to the surface area in the rate equation (Equation (1)). On a fundamental level, the rate of reaction is directly proportional to the reactive surface area, which defines the density of reactive sites on a silicate surface at which hydrolysis reactions occur. While such site distributions are characterized in terms of ligand exchange sites and dislocation densities under laboratory conditions (Blum and Lasaga, 1991), natural weathering studies generally equate reactive surface area with the physical surface area of the weathering environment.

6.5.1 Definitions of Natural Surface Areas

Surface areas in natural weathering studies are geographic, volumetric, or mineral specific. The weathering rate based on geographic surface area is defined as the mass of silicate mineral weathered per unit area of Earth's surface. Generally, this surface area is not corrected for aspect or slope of the terrain. Geographic surface areas are used extensively to describe weathering rates based on solute discharge in watersheds, which are normalized to basin areas ranging from small experimental catchments to the world's rivers (see Chapters 9 and 8).

Rates normalized to geographic surface area contain no information on the vertical dimension of the weathering environment, i.e., there is no distinction between weathering in a soil that is one meter or ten meters thick. One approach in overcoming this problem is to employ a volumetric surface area, which is commonly done for groundwater systems. Weathering rates are defined either in terms of the volume of the aquifer or of the groundwater. Commonly, surface areas contained in this volume are estimated based on the distributions of fracture surfaces or on sizes and distributions of pore spaces (Paces, 1973; Gislason and Eugster, 1987).

The more fundamental approach to addressing the physical dimensions involved in weathering is to characterize the surface areas of the individual minerals, i.e., the specific mineral surface area S ($m^2 g^{-1}$). The extent to which this specific surface area scales directly with the reactive surface areas in natural environments is a matter of

considerable debate, particularly in regard to the accessibility of water. For unsaturated environments, such as those in most soils, the wetted surface area may be considerably less than the physical surface area of contained mineral grains (Drever and Clow, 1995). In addition, surface areas of microscopic features such as external pits and internal pores may be associated with stagnant water that is thermodynamically saturated and not actively involved in weathering reactions (Oelkers, 2001).

6.5.2 Measurements of Specific Surface Areas

The scale of the measurement technique operationally defines the specific surface area of a mineral. Geometric estimates of surface area S_{Geo} involve grain size analyses with physical dimensions commonly on the scale of millimeters to centimeters. Brunauer, Emmett, and Teller (BET) surface measurements S_{BET} utilize isotherms describing low temperature sorption of N_2 or argon on mineral surfaces. The scale of BET measurements, defined by atomic dimensions, is on the order of nanometers. As expected, specific surface areas determined by geometric techniques produce significantly smaller values of physical surface area than do BET measurements because the former technique does not consider microscopic irregularities on the external surface nor the internal porosity of mineral grains that have undergone significant weathering. Recently developed techniques such as atomic force microscopy (Maurice *et al.*, 2002) and interferometry (Luttge *et al.*, 1999) have the potential to span the gap between the scales of geometric and BET surface area measurements.

BET surface areas are almost universally measured in experimental weathering studies using freshly ground and prepared minerals. BET surface areas are also commonly measured on samples of natural unconsolidated soils, saprolites, and sediments. Alteration and destruction of the physical fabric of such materials by disaggregation is required for such measurements but produces uncertainties in the reported data. In addition, secondary clay and oxyhydroxides commonly coat natural silicate surfaces and contribute to the formation of mineral-organic aggregates (Tisdall, 1982; Brantley *et al.*, 1999; Nugent *et al.*, 1998). When not removed, these coatings can produce erroneously high surface areas. Alternatively, removing these phases, using mechanical and chemical methods, may expose additional silicate surfaces that are not present in the natural weathering environment (White *et al.*, 1996).

In spite of the above issues, consistent trends of increasing surface area with increasing intensity of natural weathering are observed (Brantley *et al.*, 1999). An example of this increase is shown in Figure 12(a) in which BET surfaces of

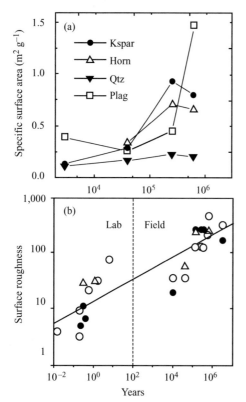

Figure 12 Surface area parameters: (a) surface area of primary silicate minerals contained in soils of the Merced chronosequence (after White *et al.*, 1996). (b) Surface roughness of silicate minerals as functions of the duration of experimental and natural weathering (source White and Brantley, 2003).

primary minerals increase with increasing age of soils in the Merced chronosequence (White *et al.*, 1996). The extent of this increase depends on the specific mineral phase. The more readily weathered aluminosilicates exhibit greater surface area increases than quartz. Application of BET measurements to characterize fractures and porosity in consolidated rocks has remained generally untested; these surface area estimates have been confined to geometric approaches (Paces, 1973; Gislason and Eugster, 1987).

6.5.3 Surface Roughness

Surface roughness λ is defined as the ratio of the geometric (S_{Geo}) to the BET (S_{BET}) surface areas (Helgeson, 1971). For a perfectly smooth surface, without internal porosity, the two surface area measurements should be the same, i.e., $\lambda = 1$. For nonideal surfaces of geometric spheres, the roughness can be related directly to the particle diameter d and the mineral density ρ such that

$$\lambda = \frac{S_{BET}}{S_{GEO}} = \frac{\rho \, d s_{BET}}{6} \qquad (16)$$

White and Peterson (1990) calculated an average roughness factor of $\lambda = 7$ for a wide size range of fresh silicate surfaces. Surface roughness generally increases during weathering as shown in Figure 12(b) for a compilation of surface roughnesses measured for experimentally and naturally reacted silicate minerals (after White and Brantley, 2003).

6.6 TABULATIONS OF WEATHERING RATES OF SOME COMMON SILICATE MINERALS

The diverse approaches described above have produced a large number of published weathering rates that are not always comparable in terms of units and dimensions. In general, weathering rates can be classified as being element or mineral specific. In the following sections, weathering rates based on geographic and volumetric surface areas are referred to as weathering fluxes, and weathering based on specific surface areas are referred to as mineral specific weathering rates.

6.6.1 Elemental Fluxes

The most prominent examples of element-based rates are those reported for annual stream or river discharge and normalized to the watershed area. The relatively straightforward nature of these measurements, coupled with a significant amount of watershed research related to land management issues, has produced a large database describing such fluxes. A tabulation of these data is beyond the scope of the present paper and the reader is referred to previous compilations (Dethier, 1986; Meybeck, 1979; Bluth and Kump, 1994; White and Blum, 1995).

Watershed fluxes are equivalent to chemical denudation commonly defined as the sum of rock-derived base cations solubilized per unit area of geographic surface (Bluth and Kump, 1994; Drever and Clow, 1995). These rates serve as a means to evaluate the importance of a number of environmental controls on chemical weathering including precipitation, temperature, vegetation, and rock type. These topics are addressed briefly in later sections of this chapter as well as elsewhere in this volume (see Chapter 8).

6.6.2 Mineral Fluxes

The alternative approach is to report silicate weathering fluxes on a mineral specific basis. As previously discussed, this requires distributing elemental fluxes among mineral phases using mass balance approaches described in this chapter and elsewhere in this volume.

Mineral flux data in the literature are less common than element-based data. Table 3 contains a tabulation of weathering rates for the common

Table 3 Weathering rates for selected silicate minerals based on average annual watershed discharge.

Watershed	Rate[a]	An[b]	References
Plagioclase			
Emerald Lake, CA, USA	44	0.24	Williams *et al.* (1993)
Loch Vale Watershed, CO, USA	86	0.27	Mast *et al.* (1990)
Gigndal Watershed, Norway	100	0.30	Frogner (1990)
Silver Creek, ID, USA	102	0.19	Clayton (1986)
Lapptrasket Basin, Sweden	133	0.00	Andersson-Calles and Ericksson (1979)
Pond Branch, MD, USA	148	0.22	Cleaves *et al.* (1970)
Velen Basin, Sweden	148	0.00	Andersson-Calles and Ericksson (1979)
Rio Parana, Brazil	174	0.49	Benedetti *et al.* (1994)
Kassjoan Basin, Sweden	181	0.00	Andersson-Calles and Ericksson (1979)
Plastic Lake, Canada	185	0.34	Kirkwood and Nesbitt (1991)
Filson Creek, MN, USA	235	0.64	Siegal and Pfannkuch (1984)
Catoctin Mtn., MI, USA	316	0.00	Katz (1989)
K-feldspar			
Pond Branch, MD, USA	11	na	Cleaves *et al.* (1970)
Silver Creek, ID, USA	44	na	Clayton (1986)
Plastic Lake, Canada	69	na	Kirkwood and Nesbitt (1991)
Biotite			
Lapptrasket Basin, Sweden	3	na	Andersson-Calles and Ericksson (1979)
Loch Vale Watershed, CO, USA	26	na	Mast *et al.* (1990)
Hornblende			
Lapptrasket Basin, Sweden	2	na	Andersson-Calles and Ericksson (1979)
Plastic Lake, Canada	8	na	Kirkwood and Nesbitt (1991)
Emerald Lake, CA, USA	24	na	Williams *et al.* (1993)

[a] $mol\ ha^{-1}yr^{-1}$.
[b] Mole fraction anorthite.

silicate minerals based on average annual watershed fluxes and geographic areas. Significantly more data are available for plagioclase feldspar than for other mineral phases, which partly reflects its common occurrence in crystalline rocks. In addition, the rate of plagioclase weathering is the easiest to calculate, because of the conventional assumption that it scales directly with the sodium discharge flux after correction for precipitation inputs. Weathering rates of other silicates are more difficult to determine from elemental fluxes because their major components are commonly derived from more than one mineral.

The reported weathering rates in Table 3 are normalized to the geographic surface area of the watershed but not to the relative proportions or surface areas of the specific mineral present. The presence of relatively large amounts of plagioclase in most igneous rocks accounts for the fact that the rates of plagioclase weathering, calculated on a watershed basis, are more rapid than those of other silicates such as hornblende, which is counter to Goldichs's order of weatherability (Goldich, 1938).

6.6.3 Specific Mineral Rates

Natural weathering rates, normalized to specific mineral surface areas, are reported in Table 2 for a number of common silicate minerals (after White and Brantley, 2003). Also included in the Table 2 are the approximate pH ranges, ages of the weathering environment and parameters defining the physical surface area of the minerals.

Large variations are apparent in the reported weathering rates for each of the minerals. For example, reported plagioclase rates vary from $4 \times 10^{-17}\ mol\ m^{-2}\ s^{-1}$ in the Davis Run saprolite in Virginia (White *et al.*, 2001) to $1.5 \times 10^{-12}\ mol\ m^{-2}\ s^{-1}$ in the Plastic Lake watershed in Ontario, Canada (Kirkwood and Nesbitt, 1991). Rates based on geometric surface areas are more rapid than those based on BET measurements. This correlation, in part, explains age trends in the rate data. Rates for younger weathering environments are based primarily on watershed fluxes and geometric surface area estimates. In nearly all cases, these watersheds are developed on glaciated topographies developed over the last 12 kyr. In contrast, weathering rates in older soils, varying in ages from 10 kyr to 3,000 kyr, are based on BET surface measurement techniques.

6.6.4 Normalizing Rate Data

One approach for overcoming the effect of different methods of surface area measurement is to normalize all the weathering rates by a surface roughness factor λ (Equation (16)), which converts geometric-based rates to their BET equivalent. This is done by statistically describing the relationship between increasing surface roughness and time for data presented in Figure 12(b) such that (White and Brantley, 2003)

$$\lambda = 1.13t^{0.182} \qquad (17)$$

The resulting roughness-normalized weathering rates are in parentheses to the right of reported rates based on geometric surface areas (Table 2).

This approach is an obvious simplification, because surface morphologies and roughnesses developed during weathering are mineral specific (Figure 5). In addition, controversy exists whether weathering rates are directly dependent on surface roughness. However, considering the large discrepancies between rates calculated by geometric versus BET derived surface areas, normalizing to an estimated surface roughness factor is warranted. It is required in any comparison with experimental weathering rates, which are almost exclusively normalized to BET surface areas.

After correction for surface roughness, a significant trend persists between decreasing weathering rates and increasing weathering duration for the rate data in Table 2. This is shown for the plagioclase and K-feldspar data plotted on log–log scales in Figure 13. Also included on the plots are experimental dissolution rates of fresh feldspars summarized by White and Brantley (2003). Experimental rates are orders of magnitude faster than natural rates, an observation that has been frequently made in the weathering literature (Schnoor, 1990; Brantley, 1992; Velbel, 1993; Drever and Clow, 1995).

Regression fits to the plagioclase and K-feldspar data (dashed lines in Figure 13) produce the respective power rate laws

$$\begin{aligned} R_{\text{plag}} &= 12.4t^{-0.54}, & r &= 0.01 \\ R_{K-\text{spar}} &= 12.1t^{-0.54}, & r &= 0.83 \end{aligned} \qquad (18)$$

that describe the progressive decrease in plagioclase and K-feldspar weathering with time. Equation (18) is similar to a relationship developed by Taylor and Blum (1995) to describe the decrease in chemical fluxes in a suite of soils with time and is comparable to parabolic kinetics describing transport control in the hydration of obsidian and the development of rock weathering rinds (Coleman, 1981; Friedman *et al.*, 1994).

6.7 FACTORS INFLUENCING NATURAL WEATHERING RATES

Weathering rates are dependent on a number of factors that can be classified as either intrinsic or extrinsic to a specific mineral (White and Brantley, 2003). Intrinsic properties are physical or chemical characteristics such as mineral composition, surface area, and defect densities. If intrinsic properties dominate weathering, such characteristics should be transferable between environments, e.g., laboratory and field rates of the same mineral

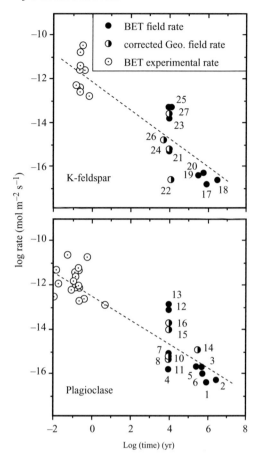

Figure 13 Relationship between weathering rate and reaction time for plagioclase and alkali feldspars. Labels for field rates correspond to those in Table 2. Experimental rates are tabulated in White and Brantley (2003).

should be comparable. Extrinsic features reflect environmental conditions external to the silicate phase that impact chemical weathering such as solution composition, climate, and biological activity. These processes are dependent on external environmental conditions that are difficult to recreate fully under laboratory simulations.

6.7.1 Mineral Weatherability

Mineral composition and structure are the primary intrinsic factors controlling weathering rates. Based on early weathering studies, Goldich (1938) observed that the weathering sequence for common igneous rocks in the field was the reverse of Bowen's reaction series that ranked minerals in the order of crystallization from magma. Amphiboles and pyroxenes are expected to weather faster than feldspars which weather faster than quartz. In addition, field data suggest that volcanic glasses weather an order of magnitude faster than crystalline minerals of comparable composition (Gislason *et al.*, 1996).

Although no fundamental method exists for predicting the weathering order of all silicate minerals, certain trends in rates between minerals are apparent. Silicate weathering is commonly viewed as a ligand exchange process with the metal ions bonded in the mineral structure. The rates are dependent on the relative strengths of coordinated metal ions within the mineral structure relative to the strength of the metal ligand bond. Casey and Westrich (1992) showed that for simple compositional series, such as olivines, the dissolution rate decreased with the increase in the metal valence state. Evidence also indicates that the relative weathering rates of multi-oxides such as feldspars approximate the relative dissolution rates of the single oxide components (Oelkers, 2001). A more detailed discussion of the relationship between weathering mechanisms and rates is presented elsewhere in this volume.

For structurally complex minerals undergoing incongruent or stepwise weathering in the natural environment, the relative rates become very dependent on specific reaction pathways. In the case of micas, for example, biotite and muscovite are structurally similar but weather at very different rates as shown by residual biotite and muscovite compositions in the saprolite developed on the Panola Granite (White *et al.*, 2002). Biotite exhibits a progressive loss of potassium and increase in the Al/Si ratio that correlates with increasing weathering intensity and decreasing depth in the profile (Figure 14(a)). These changes are driven by the relatively rapid oxidation of ferrous iron in biotite (Amonette *et al.*, 1988). As shown by the electron backscatter image in Figure 14(b), this reaction produces epitaxial replacement of biotite by kaolinite. In contrast, the muscovite, which is structurally similar, but contains low concentrations of iron, weathers very slowly, exhibiting no potassium loss (Figure 14(a)). The dramatic differences in the behavior of mica minerals emphasize the importance of mineral-specific reactions in the weathering environment.

Mineral surfaces are heterogeneous substrates possessing compositional and structural features that may differ from those of the bulk mineral phase (Hochella *et al.*, 1991). Compositional differences within single mineral grains due to zonation and exsolution can influence reaction rates. As predicted by the Bowen reaction sequence, the crystallization of plagioclase phenocrysts often produces calcic cores surrounded by more sodic rims. During the reverse process of chemical weathering, the plagioclase cores weather more rapidly, producing preferential release of solute calcium relative to sodium (Clayton, 1986). Such incongruent dissolution is a process that has been invoked to explain the commonly observed excess solute calcium to sodium ratios in watershed fluxes relative to that predicted from the bulk plagioclase stoichiometry (Stauffer, 1990).

Incongruent weathering also occurs at exsolution features. Detailed SEM studies of perthitic textures of naturally weathered feldspars by Lee and Parsons (1995) showed that the etch density was strongly dependent on the composition of these exsolution features, i.e., the sodium component dissolved more rapidly than the potassium component (Figure 5(b)). The highest densities of defect structures are also created along the strained lamellae. Weathering rates of specific

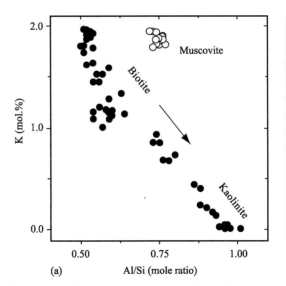

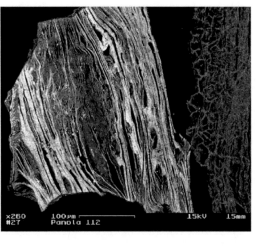

(a)   (b)

Figure 14 Mica weathering in a saprolite profile from central Georgia, USA. (a) changes in K and Al/Si mole ratios in coexisting biotite and muscovite and (b) SEM backscatter photomicrograph of residual biotite grains at 2.8 m depth. Bright layers in the left grain are unaltered biotite and gray areas are kaolinite replacement. Right grain is completely kaolinized biotite with surficial boxwork structure (after White *et al.*, 2002).

minerals may therefore be dependent on the crystallization history of igneous rocks as well as the effects of tectonics and deformation on defect formation.

Weathering acts on mineral surfaces by decreasing the overall surface free energy by selectively dissolving more soluble components and attacking structural defects and dislocations. With continued weathering, the overall surface energy and the corresponding weathering rate may decrease, as observed for the feldspar data plotted in Figure 13. The progressive occlusion of the reactive mineral substrate by secondary clays and iron and aluminium oxides, the formation of depleted leached layers and the adsorption of organic compounds, have also been proposed as mechanisms to decrease surface reactivity with time (Banfield and Barker, 1994; Nugent *et al.*, 1998). Whether these secondary coatings represent effective barriers to the transport of reactants and products from the reactive silicate surface is still actively debated.

6.7.2 Solute Chemistry and Saturation States

Of the extrinsic factors influencing natural weathering rates, solute compositions have the most direct impact. Chemical weathering is ultimately dependent on the concentration of reactants that complex with and detach the oxygen-bonded metal species in the silicate structure (Casey and Ludwig, 1995). As documented by numerous experimental studies, the principal species involved in this reaction are hydrogen ions, although other complexing agents such as organic anions can participate in this process. In contrast, some solute species, such as aluminum and sodium ions, inhibit experimental weathering rates by interfering with and competing with the ligand exchange processes (Oelkers and Schott, 1995; Stillings *et al.*, 1996).

The direct effect of aqueous species on natural weathering is difficult to demonstrate. For example, a summary of experimental findings indicates that feldspar weathering rates decrease about two orders of magnitude over a pH range of 2–6 (Blum and Stillings, 1995). However, when this effect is looked for in natural systems, the evidence is ambiguous. Over a 25 yr period, the pH of solute discharge from Hubbard Brook has increased from pH 4.1 to 4.4 due to a gradual recovery from acid deposition (Driscoll *et al.*, 1989). However, solute silicon concentrations, a good indicator of silicate weathering, have remained essentially unchanged (Figure 15). Although, these results do not disprove the effect of hydrogen ion on weathering rates, as demonstrated experimentally, it does serve to show that other processes sometimes overwhelm expected results for natural systems.

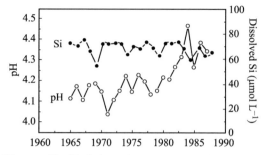

Figure 15 Variations in mean annual pH and dissolved silica concentration in Hubbard Brook, New Hampshire, USA (after Driscoll *et al.*, 1989).

Experimental rates of silicate dissolution decrease as solutions approach thermodynamic equilibrium (Burch *et al.*, 1993; Taylor *et al.*, 2000). The saturation state Ω is defined as the product of the solute activities (IAP) divided by the saturation constant of the specific mineral K_s and is related to the net free energy of reaction ΔG (kJ mol^{-1}) by the relationship

$$\Omega = \exp\left(\frac{\Delta G}{R'T}\right) \qquad (19)$$

where T is the temperature and R' is the gas constant. When the system is undersaturated, i.e., Ω is <1, ΔG is negative, leading to dissolution. At a $\Omega = 1$, saturation is achieved and $\Delta G = 0$. If $\Omega > 1$, the system is supersaturated, ΔG is positive and precipitation may occur.

Based on transition-state theory, the relationship between the reaction rate and the solute saturation state is represented as (Nagy *et al.*, 1991)

$$R = k[1 - \Omega^m]^n \qquad (20)$$

where k is the intrinsic rate constant and m and n are orders of reaction. In strongly under-saturated solutions (large negative values of ΔG), net detachment reactions dominate over attachment reactions, the dissolution rate is independent of saturation state and R is directly proportional to the intrinsic rate constant. At near saturation (small values of ΔG), Equation (20) reduces to a form in which R becomes linearly dependent on ΔG. At equilibrium ($\Delta G = 0$) the weathering rate is zero. Although the relationship between decreasing reaction rates and approach to equilibrium (Equation (20)) has been verified for a few minerals at low pH and elevated temperature, no comparable studies have been carried out under conditions more representative of natural weathering.

Calculations involving mineral saturation state (Ω in Equations (19) and (20)) are dependent on accurate characterization of the thermodynamic states of the reactants and products and are commonly calculated using speciation codes. Although the equilibrium constants for most simple primary silicates have been determined, thermodynamic data do not exist for many complex silicates or for solid solutions.

White (1995) found that the apparent thermodynamic supersaturation of silicate minerals in most soil pore waters resulted from excessive values for total dissolved aluminum. In reality, much of this aluminum is complexed with dissolved organics in shallow soils and does not contribute to the thermodynamic saturation state of silicate minerals. Solubility calculations involving low dissolved organic concentrations in deeper soil horizons and in groundwater appear to produce much clearer equilibrium relationships (Paces, 1972; Stefansson and Arnorsson, 2000; Stefansson, 2001).

Natural weathering involves much longer times and smaller solute/solid ratios than occur in experimental studies. Such weathering may therefore occur much closer to thermodynamic equilibrium. This difference in saturation states may explain two common discrepancies observed when comparing experimental and natural weathering rates. As has been frequently noted by others and documented in detail in the Figure 13, natural weathering rates are commonly several orders magnitude slower than experimental rates. Further, the relative weatherability of minerals in the natural environment often differs from that predicted on the basis of experimental dissolution rates. For example, in a review of experimental feldspar dissolution studies, Blum and Stillings (1995) concluded that under neutral to acidic pH conditions, the rates of sodic plagioclase and K-feldspar dissolution were essentially the same. However, K-feldspar is commonly much more resistant to weathering than is plagioclase during natural weathering (Nesbitt *et al.*, 1997).

Weathering of the Panola Granite is an extreme example of the relative weathering rates of feldspars (White *et al.*, 2001). As shown by the mass balance calculations (Figures 3 and 4), kaolinization of plagioclase occurs to depths of 10 m in the granite regolith (Figure 4). This is evident in SEM images (Figure 16(a)) in which, except for residual rims, plagioclase grains are completely replaced by kaolinite. In contrast, K-feldspar in the immediate proximity of plagioclase grains remains pristine and unaffected by weathering.

The significant difference in feldspar weathering rates at depth in the Panola was explained by White *et al.* (2001) by the difference in their respective thermodynamic saturation states (Figures 16(b) and (c)). Soil pore waters are generally undersaturated with respect to both plagioclase and K-feldspar (Figures 16(b) and (c)). In contrast, groundwater in the underlying granite bedrock is saturated with K-feldspar but undersaturated with more soluble plagioclase. Equation (20) predicts that plagioclase will weather in contact with groundwater while K-feldspar will remain stable.

6.7.3 Coupling the Effect of Hydrology and Chemical Weathering

The thermodynamic saturation state is dependent on solute concentrations, which in turn, are controlled by the volume and residence times of fluids moving through the weathering environment. The roles of hydrology and chemical weathering are commonly coupled in a weathering regolith and can be viewed as co-evolving as the intensity of the weathering increases. This effect is evident both in the initial phases of chemical weathering in bedrock environments and development of argillic horizons in older soils.

6.7.3.1 Initial stages of weathering

The kaolinization of plagioclase at depth in the Panola granite is clearly an example of the initial stage of chemical weathering in which other primary mineral phases remain unreacted (Figure 16(a)). This process was modeled by White *et al.* (2001) as a coupled relationship between weathering rates and the development of secondary permeability as shown in Figure 17. At the initiation of weathering, the permeability of the fresh granite is extremely low, placing severe constraints on the fluid flux q_p and the mass of feldspar that can dissolve before becoming thermodynamically saturated. Under such conditions, weathering is limited by the availability of water and not by the kinetic rate of feldspar weathering.

Over long times, slow rates of transport-limited weathering occur, resulting in mass loss from the granite (Figure 3(b)). Based on density estimates, the conversion of plagioclase to kaolinite (Equation (2)) produces a porosity increase of $\sim$50%. This change in porosity slowly increases the flux density of water (Equation (11)), plotted as the ratio of q_s/q_p in Figure 17. Increased pore-water flow accelerates saturation-limited weathering and produces greater porosity and even higher fluid fluxes. This coupled feedback accelerates plagioclase weathering, which gradually shifts from a transport limited to a kinetic limited reaction (Figure 17). At a q_s/q_p ratios >150, plagioclase weathering becomes completely controlled by kinetics and no longer reflects additional increases in conductivity.

The rate of K-feldspar weathering shows a comparable transition from transport to kinetic control but at significant higher flux ratios due to its lower solubility and not to slightly slower reaction kinetics (Figure 17). Concurrent plagioclase dissolution enhances this effect by producing solutes, principally silicon, which further suppresses K-feldspar dissolution by increasing the saturation state. Nahon and Merino (1997) and Soler and Lasaga (1996) present additional discussions of the role of solute transport and

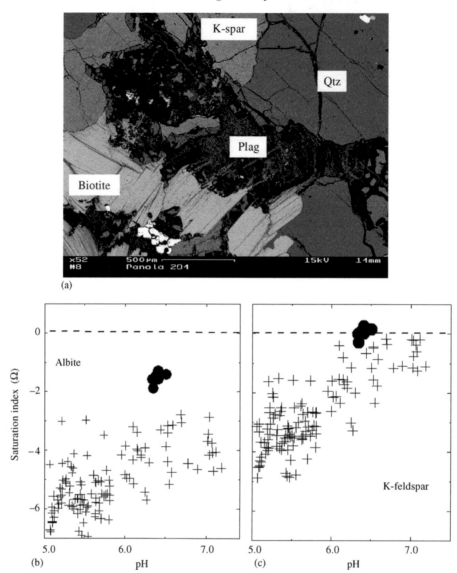

Figure 16 Primary silicate weathering in the Panola regolith: (a) backscatter SEM micrograph of weathered granite at a depth 7.5 m. The center grain is a plagioclase phenocryst almost completely replaced by kaolinite (dark region). Minor residual plagioclase remains on outside rims of grain. Saturation states (log Ω) of soil pore water (+) and groundwater (•) relative to (b) plagioclase (albite) and (c) K-feldspar (after White *et al.*, 2001).

solubility on textures and mineral distributions in saprolites and other regoliths.

6.7.3.2 Late-stage weathering

Weathering in older soils produces decreasing permeabilities due to *in situ* secondary mineral formation and the development of hard pans and argillic horizons. Such zones of secondary clay and iron-oxides are clearly evident in the increased aluminum and iron at a depth of a meter in the Panola regolith (Figure 3). Low permeabilities in such features are commonly related to the absence of continuous pores due to the formation of thick cutans of clay (O'Brien and Buol, 1984). This process is enhanced by physical

translocation and collapse of saprolite structures (Torrent and Nettleton, 1978).

Argillic horizons often correlate with the maximum depth of effective evapotranspiration, commonly 1–2 m. Water loss initiates the precipitation of secondary clays and oxides from solutes. Such precipitation may also be related to the loss of dissolved organic species and the decomplexation of soluble aluminum. The resulting low permeabilities further retard the downward percolation of pore water, commonly creating transient perched water tables directly above the hardpans. Periodic drying in this zone focuses addition secondary mineral precipitation in the vicinity of the hardpan, which then leads to a lower permeability and more clay formation.

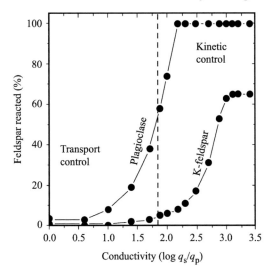

Figure 17 The percent of plagioclase and K-feldspar reacted as a function of the ratio of secondary to primary flux densities of water (q_s/q_p) in the Panola Granite. The vertical dashed line is the approximate flux ratio at which plagioclase is completly weathered and the K-feldspar weathering has not yet commenced. This condition evident in thin section in Figure 16 (after White *et al.*, 2001).

The progressive development of such argillic features with increasing weathering intensity is commonly documented in soil chronosequence studies (Harden, 1990). An example of the development of argillic horizons is shown for the Merced chronosequenence in Figure 18 (White *et al.*, 2003). Over a time span of 40–600 kyr, the maximum clay content in these horizons increases from ~5% to ~40%. As expected, the development of these intense argillic horizons has a profound effect on the water content within and below these features.

The Merced site has a Mediterranean climate with pronounced semiannual wet and dry precipitation cycles. This seasonal variability is delineated in the soil moisture contents by the open and closed symbols in Figure 18. In the youngest soil, with a poorly developed argillic horizon, the climate signal is propagated to the base of the profile. In the older soils, with increasingly more developed argillic zones, this seasonal signal becomes progressively lost, reflecting decreases in the infiltration rate of pore waters through soil. In these older soils, solubility calculations indicate that relatively long pore-water residence times result in solutes that are thermodynamically saturated with respect to the primary feldspars, a situation leading to progressively slower weathering as a function of soil age (Table 2).

The preceding examples indicate that regolith development is coupled in terms of weathering and hydrology. Ultimately, the impact of chemical weathering on hydrology is related to changes in secondary porosity and permeability, which, in turn, increase or decrease the degree of thermodynamic saturation and the weathering rate.

6.7.4 Role of Climate on Chemical Weathering

The term "weathering" implies that chemical weathering is related to climate. This relationship is important both from the standpoint of the potential long-term feedback during much of Earth's history as well as the current impacts on changes in nutrient cycling produced by present-day CO_2 inputs to the atmosphere. The importance of climate relative to other controls on weathering, in particular, topography and tectonics remains controversial (Ruddiman, 1997, and references therein). Examples of contrasting results comparing weathering rates in small-scale catchments are contained in White and Blum (1995), who found that climate was the most important control, whereas Riebe *et al.* (2001) determined that topography and physical erosion were the dominant factors.

A definitive study would involve a weathering environment that has undergone consistent sustained climate change. However, the long-term data required for such a study are not available and surrogate weathering studies comparing spatially separated climatic regimes are used. The utility of such comparisons depends on the ability to isolate the effects of climate from other variables influencing weathering including lithology, geomorphology, and vegetation. Some of these parameters often correlate with climate, making the isolation of individual variables difficult. Physical erosion generally increases with precipitation, exposing fresher mineral surfaces and increasing weathering rates. Likewise, higher rainfall may produce greater plant productivity, increased soil CO_2 and higher dissolved organic concentration, all of which tend to vary systematically with climate.

The ability to isolate climate effects decreases as the scale of the weathering process increases. For example, a number of studies comparing weathering rates in soils and small catchments have found a significant climate effect (Velbel, 1993; White and Blum, 1995; Dessert *et al.*, 2001). In contrast, comparison of solute concentrations and fluxes originating from large scale river systems commonly fail to detect a climate signature (Edmond *et al.*, 1995; Huh *et al.*, 1998).

6.7.4.1 Temperature

The effect of temperature on weathering is the easiest climate parameter to predict on a fundamental basis. The rates of most chemical reactions, including silicate hydrolysis, increase exponentially with temperature according to the Arrhenius expression. This relationship can be

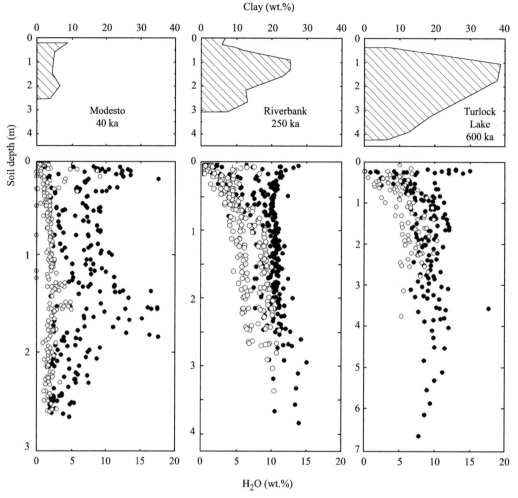

Figure 18 Correlation between clay and water contents of soils of different ages in the Merced chronosequence, California. The closed circles are soil moistures measured during the wet season (December to May) and open circles are data from the dry season (June to November) (after White *et al.*, 2003).

represented as the ratio of reaction rates R/R_0 at different temperatures T and T_0 (K) (Brady and Carroll, 1994):

$$\frac{R}{R_0} = \exp\left[\frac{E_a}{R'}\left(\frac{1}{T_0} - \frac{1}{T}\right)\right] \qquad (21)$$

The activation energy E_a (kJ mol^{-1}K^{-1}) of most common silicate minerals ranges between 50 kJ and 120 kJ (Brady and Carroll, 1994; Casey and Sposito, 1992; White *et al.*, 1999). Equation (21) predicts that rates should increase by about an order of magnitude between 0 °C and 25 °C, the temperature range encountered in most natural weathering. If temperature were the only variable in weathering, the effect should be readily observable when comparing weathering environments at substantially different temperatures.

Temperature effects on the rate of chemical weathering are observed in well-characterized environments. Velbel (1993) estimated elevation-dependent temperature differences in the Coweeta watershed in North Carolina, USA (10.6–11.7 °C) and calculated an activation energy of 77 kJ mol^{-1} for plagioclase weathering. Dorn and Brady (1995) used plagioclase porosity formed by etch pitting in Hawaiian basalt flows at different elevations and temperatures (12.5–23.3 °C) to calculate an activation energy of 109 kJ mol^{-1} (Figure 19(a)). Dessert *et al.* (2001) calculated an E_a of 42 kJ mol^{-1} for basalt weathering (2–27 °C) from a compilation of river fluxes (Figure 19(b)). Studies of larger river systems have tended to discount the importance of temperature as a significant control on weathering (Edmond *et al.*, 1995).

6.7.4.2 Precipitation and recharge

A number of studies have observed a linear correlation between precipitation or runoff and solute fluxes or chemical denudation rates (Dunne, 1978; Dethier, 1986; Stewart *et al.*, 2001). The relationship between solute silicon fluxes and

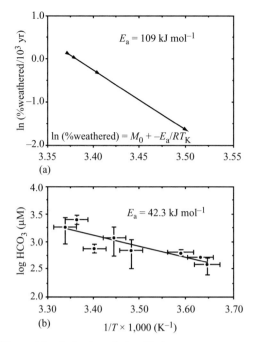

Figure 19 Arrhenius relationship between temperature and rates for basalt weathering: (a) rates based on porosity increases in plagioclase from Hawaiian basalt flows (Dorn and Brady, 1995) and (b) rates based on solute bicarbonate fluxes from a global distribution of watersheds underlain by basalt (Dessert *et al.*, 2001).

runoff of the large watershed data set plotted in Figure 20 shows comparable relationships for granitic and basaltic rock types (Bluth and Kump, 1994).

As pointed out by Drever and Clow (1995), mineral surfaces undergo weathering only in the presence of liquid water. Increases in precipitation increase regolith recharge and the wetted surface areas of minerals, thereby promoting increased weathering. In addition, increased water flow increases the rates at which reactants, such as dissolved CO_2 and organic acids, are flushed through the system. Finally, increased precipitation dilutes the solute concentrations of pore waters in the regolith, decreasing thermodynamic saturation.

6.7.4.3 *Coupling climate effects*

In order to explain anomalously rapid rates of chemical weathering in upland topical watersheds such as in Puerto Rico and Malaysia, White and Blum (1995) and White *et al.* (1999) proposed that a coupled climate effect in which the solute fluxes were proportional to the product of a linear precipitation function and an exponential temperature function such that

$$Q = (aP)\exp\left[-\frac{E_a}{R}\left(\frac{1}{T} - \frac{1}{T_0}\right)\right] \quad (22)$$

Equation (22) was fitted to data for well-characterized upland catchments underlain by granitoid rocks. A three-dimensional plot generated for silicon fluxes is shown in Figure 21. The net predicted effect is to reinforce weathering in watersheds with both high temperatures and precipitation and to decrease weathering fluxes in watersheds with low temperature and precipitation. This conclusion is in agreement with Meybeck (1994), who has shown that maximum concentrations of dissolved species from chemical weathering occur in rivers draining humid mountainous terrains and that minimum fluxes occur in arid and arctic rivers.

6.7.5 Role of Physical Weathering

Chemical weathering, in combination with physical erosion, are the processes that produce global element cycling. R. A. Berner and E. K. Berner (1997) estimated that the combined denudation rate of the continents is 252 km yr^{-1} while the average chemical denudation rate is $\sim$20% of that value. As expected, the absolute rates and relative ratios of physical to chemical weathering are strongly related to differences in topography. For example, the chemical denudation rate is less than 10% of the physical denudation rate for watersheds of high relief such as the Ganges River, which drains the Himalayas, whereas it approaches 45% of the physical denudation rate for lowland rivers such as the Congo.

6.7.5.1 *Transport versus chemical weathering regimes*

A important relationship between physical and chemical weathering was proposed by Carson and Kirby (1972) and Stallard and Edmond (1983), who differentiated mineral selectivity in terms of weathering-limited and transported-limited weathering regimes. In the weathering-limited case, mechanical erosion is faster than chemical weathering. Therefore the most reactive phases will be available for weathering. An example is the oxidation of ferrous-containing biotite to oxybiotite with the release of interlayer potassium. This rapid weathering reaction is documented by high potassium fluxes in present-day glacial watersheds in which large amounts of fresh rock are exposed by physical denudation (Andersson *et al.*, 1997; Blum and Erel, 1997). Such excess potassium is not observed in watersheds that are geomorphically older.

Under transport-limited weathering, the amount of fresh rock available to weathering is limited. Chemical weathering is faster then physical weathering and available minerals ultimately contribute to the solute load in proportion to their abundance in the bedrock. Such is the case for

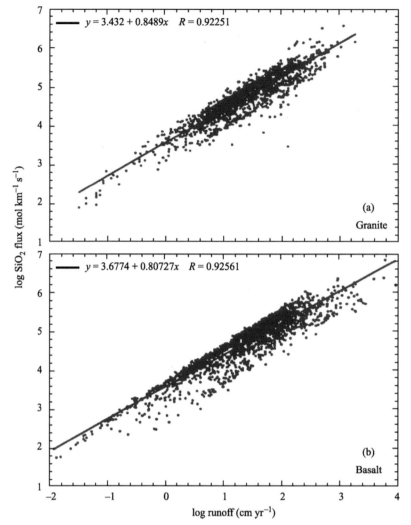

Figure 20 Relationship between Si fluxes and runoff from watersheds draining: (a) granitic and (b) basaltic watershed (after Bluth and Kump, 1994).

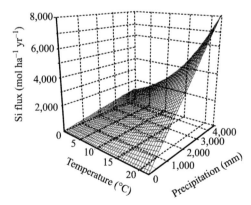

Figure 21 Three-dimensional surface representing the optimized fit of watershed Si fluxes to Equation (22) as functions of precipitation and temperature (after White and Blum, 1995).

the nearly complete destruction of aluminosilicates from old laterites and saprolites in which base cations are all effectively removed from the regolith.

6.7.5.2 *Physical development of regoliths*

In addition to controlling the relative proportions of minerals available for weathering, physical erosion also has a profound effect on the physical development of weathering environments. Regoliths can be viewed as aggrading, steady-state, or degrading systems depending on whether their weathering thicknesses increase, remain constant or decrease with time due to physical erosion. These conditions control the position of the weathering gradient shown in Figure 2.

An example of an aggrading regolith is one for which the present topographic surface represents the original depositional surface. A lack of physical erosion on flat alluvial or marine terraces can be verified by the concentrations of cosmogenic isotopes (Pavich *et al.*, 1986; Perg *et al.*, 2001) and is an important basis for using soil chronosequences in weathering studies. In terms of the definitions of Stallard and Edmond (1983) such as system would shift progressively from a chemically-limited to a transport-limited regime as primary minerals become more isolated from the surficial weathering environment.

A degrading weathering environment is one in which rates of physical erosion exceed the rate of chemical weathering and the regolith thickness decreases. Such decreases may result from natural catastrophic events such as landslides in tropical terrains or as a result of human influences such as deforestation, which result in increased rates of erosion (Larsen and Concepicon, 1998). Under such conditions, the weathering environment will shift from weathering limited to transport limited.

As discussed by Stallard (1995), given enough time, a dynamic equilibrium may be reached between the chemical and physical weathering rates. This situation has been documented for thick saprolite sequences, as developed at Panola and elsewhere in the southeastern USA. In these cases, the physical and chemical denudation rates are comparable (Pavich, 1986). Under such a "conveyor belt" scenario, the regolith depth will remain constant as additional rock is chemically weathered at depth and soils are physically removed at the surface. The position of the weathering gradient, defining the change in chemistry and mineralogy, will remain constant with depth (Figure 2).

6.8 WEATHERING RATES IN GEOCHEMICAL MODELS

As discussed by Warfvinge (1995), the incorporation of weathering kinetics into geochemical codes can be discussed in terms of the degree to which rates are coupled to other geochemical and hydrologic processes. A significant number of models have been developed to explain and predict the impact of acid precipitation on watershed solutes. The reaction rates in such models are commonly coupled to hydrogen ion concentration by the relationship

$$R = k \cdot [H^+]^n \tag{23}$$

where the rate constant k is a fitted parameter or else is determined from empirical or experimental data. With the exponential term $n = 0$, the weathering rate is independent of pH, as is assumed in the MAGIC code (Cosby *et al.*, 1985). At other powers of n, the rate becomes exponentially dependent on pH, as in the SMART code (De Vries and Kros, 1989) where $n = 0.5$. Such models have been applied in a number of detailed studies of watershed fluxes and have served as a basis for environmental regulation of acid emissions.

Another class of models relevant to chemical weathering is based on the reaction path approach originally developed by Helgeson *et al.* (1970). EQ3/EQ6 (Wolery *et al.* (1990), PHREEQC (Parkhurst and Plummer, 1993), and PATHARC.94 (Gunter *et al.*, 2000) are some codes currently used

to describe the progressive reaction of primary silicates and the precipitation of secondary phases as a function of time and mass. These codes are discussed in Nordstrom. They commonly permit the introduction of user-defined silicate reaction rates. Such models also commonly consider solubility controls on reaction kinetics as defined by Equation (20). Such models generally assume that reactions occur across a constant surface area and in a given volume of water.

A final group of more computationally complex codes consider additional factors pertinent to rates of chemical weathering. The PROFILE and SAFE codes (Sverdrup and Warfvinge, 1995) describe the evolution of solute compositions during weathering in soils and catchments and require detailed site-specific data including the number and thickness of soil horizons, bulk density, moisture content, and surface areas. These codes use an internal database describing the kinetic weathering rates of common silicate minerals. The rate coefficients are dependent on base cation and aluminum concentrations and the degree of saturation based on transition state theory. Fluid transport is one dimensional with flow dependent on relative inputs and outputs from precipitation and evapotranspiration unless otherwise specified by independent means. Surface areas vary with the degree of surface wetting depending on moisture saturation. A critical evaluation of the PROFILE code is presented by Hodson *et al.* (1997).

Other models directly couple chemical reaction with mass transport and fluid flow. The UNSATCHEM model (Suarez and Simunek, 1996) describes the chemical evolution of solutes in soils and includes kinetic expressions for a limited number of silicate phases. The model mathematically combines one- and two-dimensional chemical transport with saturated and unsaturated pore-water flow based on optimization of water retention, pressure head, and saturated conductivity. Heat transport is also considered in the model. The IDREAT and GIMRT codes (Steefel and Lasaga, 1994) and Geochemist's Workbench (Bethke, 2001) also contain coupled chemical reaction and fluid transport with input parameters including diffusion, advection, and dispersivity. These models also consider the coupled effects of chemical reaction and changes in porosity and permeability due to mass transport.

An example of the results generated by the last type of coupled code is that of Soler and Lasaga (1996), who used the IDREAT to model bauxite formation from a granite protolith. As mentioned in the introduction, bauxite formation is an example of an important economic resource produced by long-term chemical weathering. The main objective of the study was to reproduce the mineral succession, and especially the thickness of the transition zone containing gibbsite and kaolinite

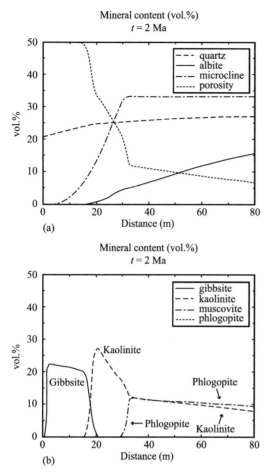

Figure 22 A 1DREACT computer simulation of: (a) residual primary silicate distributions and secondary porosity and (b) secondary mineral distributions with depth for Los Pijiguaos bauxite deposit, Venezuela after 2 Ma years of weathering (after Soler and Lasaga, 1998).

by making use of kinetic reaction rates for the primary phases albite, microcline quartz, and phlogopite. The calculations also made use of laboratory-based pH and saturation-dependent rate laws as well as field-based estimates of dispersion, diffusion, and hydraulic conductivity.

A sample of the calculation (Figure 22) shows the resulting primary and secondary mineral distributions in the Los Pijiguaos bauxite deposit, Venezuela after 2 Myr of weathering. The simulation was run with a nucleation barrier for muscovite precipitation. Due to solubility constraints, albite is selectively lost and microcline preserved in the profile, which is consistent with results observed by White *et al.* (2001) at Panola (Figures 16 and 17). The succession and relative thickness of the gibbsite and kaolinite were shown to be dependent on the kinetic rates at which the various mineral phases dissolve and precipitate. The increase in regolith porosity, as the granite is converted to saprolite, is also demonstrated.

6.9 SUMMARY

This paper reviews the chemical, physical, and hydrologic processes that control silicate mineral weathering rates at the Earth's surface. Quantitative rates of weathering are important in understanding reaction mechanisms and in addressing a number of economic and environment issues. Silicate weathering rates are fluxes that describe mass transport across an interface over a given time interval.

Mass change, defined in terms of elements, isotopes, or mineral abundances is determined from either solid-state or solute compositions. Solid-state mass changes in a bulk regolith must be corrected for the mobility of all components as well as volume and density changes due to compaction and bioturbation. This correction is commonly done by normalizing the concentrations of reactive species against an element or mineral considered to be resistant to chemical weathering. Solid-state weathering is also characterized on microscopic levels based on weathering rind thickness and mass losses due to porosity and etch pit development.

The calculation of rates based on changes in solute species concentrations in soils, aquifers, and watersheds requires partitioning the reactant between sources produced by primary mineral dissolution and sinks created by secondary mineral precipitation. Calculation of weathering rates based on solute transport requires knowing the nature and rate of fluid flow through soils, aquifers, and watersheds.

The time-period over which weathering is being evaluated is very different depending upon whether changes in solid-state or solute concentrations are measured. Solid-state mass differences reflect weathering over geologic timescales, whereas solute compositions reflect the residence time of the water. These mass losses or gains are normalized to surface area. Rates of reaction are directly proportional to the density of reactive sites on a silicate surface. Natural rates equate this density with the physical surface area of the weathering environment, generally defined on a geographic, volumetric, or mineral-specific basis. Specific surface areas are either geometric estimates based on particle size dimensions or are based on BET gas sorption isotherms, which include surface areas associated with microscopic roughness and internal porosity. The difference in the scales of these measurements produces major differences in surface-normalized mineral weathering rates.

The majority of rate studies are based on watershed solute fluxes normalized to catchment area and are equivalent to rates of chemical denudation. Previous efforts have tabulated these rates, which have proved to be valuable in evaluating the importance of a number of environmental controls on chemical weathering including precipitation, temperature, vegetation, and rock type. The present chapter summarizes literature data for the rates of weathering of several common silicate minerals based on geographic and specific surface areas. In the latter case, differences between geometric and BET-based surface areas are removed by normalizing the rates against an estimated time-dependent surface area roughness factor. Results confirm that a significant correlation exists between decreases in weathering rates and increases in weathering duration.

Factors affecting weathering rates can be categorized as either intrinsic or extrinsic to a specific silicate mineral. The primary intrinsic factors controlling weathering rates are mineral composition and structure. Reaction rates are also affected by compositional differences within single-mineral grains, such as zonation and exsolution features, and by secondary surface coatings and leached layers.

Of the extrinsic factors influencing natural weathering rates, solute composition has the most direct impact. Solutions provide reactants that complex with and detach the oxygen-bonded metal species in the silicate structure. In addition, reaction rates depend on the thermodynamic saturation state of the dissolving phase. Unlike most experimental studies, natural weathering commonly involves long times and small solute/solid ratios, producing solutes close to thermodynamic saturation. This difference in saturation states explains, in part, why natural weathering rates are commonly orders magnitude slower than experimental rates and why the relative weatherability of minerals in the natural environment is often different from that predicted by experimental dissolution rates.

The origin of the term "weathering" implies that chemical weathering is related to climate. However, the exact nature of this relationship remains an area of significant controversy. Studies comparing weathering rates in soils and small catchments often find a significant climate effect, whereas solute concentrations and fluxes originating from large river systems often fail to detect a climate signature. This discrepancy is related to the difficulty of isolating the effects of climate from other often co-dependent parameters, a difficulty that increases as the scale of the weathering processes increases. An exponential relationship between reaction rate and temperature is observed in some weathering environments. Several studies have also documented linear increases in weathering with precipitation due to increases in wetted surface areas and thermodynamic reaction affinities.

As expected, the ratio of physical to chemical weathering is strongly related to differences in topography. An important relationship between physical weathering and chemical weathering is summarized in terms of weathering-limited and transport-limited weathering regimes. In addition to controlling the relative proportions of minerals available for weathering, physical erosion also has a profound effect on the physical development of the weathering environment. Regoliths can be viewed as aggrading, steady-state, or degrading systems, depending on whether their thickness increases, remains constant, or decreases with time due to physical erosion.

The incorporation of weathering kinetics into geochemical codes is classified in terms of the level at which the rates are coupled to other geochemical and hydrologic processes. A significant number of models explain and predict the impact of acid precipitation on watershed solutes. Another class of models is based on a reaction-path approach, which models the progressive loss of primary silicates and the precipitation of secondary phases as functions of time and mass. A final group of more computationally complex codes considers additional factors pertinent to rates of chemical weathering, including fluid and heat transport and the coupling of chemical reaction to changes in porosity and permeability.

REFERENCES

Allen R. and Cherry J. A. (1979) *Groundwater*. Prentice Hall, Englewood Cliffs, NJ.

Amonette J., Ismail F. T., and Scott A. D. (1988) Oxidation of biotite by different oxidizing solutions at room temperature. *Soil Sci. Am. J.* **49**, 772–777.

Andersson-Calles U. M. and Ericksson E. (1979) Mass balance of dissolved inorganic substances in three representative basins in Sweden. *Nordic Hydrol.* **28**, 99–114.

Andersson S. P., Drever J. I., and Humphrey N. F. (1997) Chemical weathering in glacial environments. *Geology* **25**, 399–402.

April R., Newton R., and Coles L. T. (1986) Chemical weathering in two Adirondack watersheds: past and present-day rates. *Geol. Soc. Am. Bull.* **97**, 1232–1238.

Banfield J. F. and Barker W. W. (1994) Direct observation of reactant–product interfaces formed in natural weathering of exsolved, defective amphibole to smectite: evidence for episodic, isovolumetric reactions involving structural inheritance. *Geochim. Cosmochim. Acta* **58**, 1419–1429.

Barth T. F. (1961) Abundance of the elements, aerial averages and geochemical cycles. *Geochim. Cosmochim. Acta* **23**, 1–8.

Belt T. (1874) *The Naturalist in Nicaragua*. Chicago: University of Chicago press.

Benedetti M. F., Menard O., Noack Y., Caralho A., and Nahon D. (1994) Water–rock interactions in tropical catchments: field rates of weathering and biomass impact. *Chem. Geol.* **118**, 203–220.

Berner R. A. and Berner E. K. (1997) Silicate weathering and climate. In *Tectonic Uplift and Climate Change* (ed. W. F. Ruddiman). Plenum, NY, pp. 353–364.

Bethke C. M. (2001) The GEOCHEMIST WORKBENCH: a users guide to Rxn, ACT2, Tact, React and Gtplot. University of Illinois, Urbana.

Birkland P. W. (1984) Holocene soil chronofunctions, Southern Alps, New Zealand. *Geoderma* **34**, 115–134.

Blum A. E. and Lasaga A. C. (1991) The role of surface speciation in the dissolution of albite. *Geochim. Cosmochim. Acta* **55**, 2193–2201.

Blum A. E. and Stillings L. L. (1995) Feldspar dissolution kinetics. In *Chemical Weathering Rates of Silicate Minerals*, Reviews in Mineralogy, 31 (eds. A. F. White and S. L. Brantley). Mineralogical Society of America, Washington, DC, pp. 176–195.

Blum J. D. and Erel Y. (1997) Rb–Sr isotope systematics of a granitic soil chronosequence: the importance of biotite weathering. *Geochim. Cosmochim. Acta* **61**, 3193–3204.

Bluth G. S. and Kump L. R. (1994) Lithologic and climatic controls of river chemistry. *Geochim. Cosmochim. Acta* **58**, 2341–2359.

Bockheim J. G. (1980) Solution and use of chronofunctions in studying soil development. *Geoderma* **24**, 71–85.

Brady P. V. and Carroll S. A. (1994) Direct effects of CO_2 and temperature on silicate weathering: possible implications for climate control. *Geochim. Cosmochim. Acta* **58**, 1853–1863.

Brantley S. L. (1992) Kinetics of dissolution and precipitation—experimental and field results. In *Water–Rock Interaction WRI7* (eds. Y. K. Kharaka and A. S. Maest). A. A. Balkema, Rotterdam, pp. 3–6.

Brantley S. L., Blai A. C., Cremens D. L., MacInnis I., and Darmody R. G. (1993) Natural etching rates of feldspar and hornblende. *Aquat. Sci.* **55**, 262–272.

Brantley S. L., White A. F., and Hodson M. E. (1999) Surface areas of primary silicate minerals. In *Growth, Dissolution, and Pattern Formation in Geosystems* (eds. B. Jamtveit and P. Meakin). Kluwer Academic, Dordrecht, pp. 291–326.

Brimhall G. H. and Dietrich W. E. (1987) Constitutive mass balance relations between chemical composition, volume, density, porosity, and strain in metasomatic hydrochemical systems: results on weathering and pedogenesis. *Geochim. Cosmochim. Acta* **51**, 567–587.

Bullen T. D., Krabbenhoft D. P., and Kendal C. (1996) Kinetic and mineralogic controls on the evolution of groundwater chemistry and $^{87}Sr/^{86}Sr$ in a sandy silicate aquifer, northern Wisconsin, USA. *Geochim. Cosmochim. Acta* **60**, 1807–1821.

Burch T. E., Nagy K. L., and Lasaga A. C. (1993) Free energy dependence of albite dissolution kinetics at 80°C, and pH 8.8. *Chem. Geol.* **105**, 137–162.

Carson M. A. and Kirby M. J. (1972) *Hillslope, Form, and Process.* Cambridge: Cambridge University Press.

Casey W. H. and Ludwig C. (1995) Silicate mineral dissolution as a ligand exchange reaction. In *Chemical Weathering Rates of Silicate Minerals*, Reviews in Mineralogy, 31 (eds. A. F. White and S. L. Brantley). Mineralogical Society of America, Washington, DC, pp. 241–294.

Casey W. H. and Sposito G. (1992) On the temperature dependence of mineral dissolution rates. *Geochim. Cosmochim. Acta* **56**, 3825–3830.

Casey W. H. and Westrich H. R. (1992) Control of dissolution rates of orthosilicate minerals by divalent metal-oxygen bonds. *Nature* **355**, 157–159.

Chadwick O. A., Brimhall G. H., and Hendricks D. M. (1990) From black box to a grey box: a mass balance interpretation of pedogenesis. *Geomorphology* **3**, 369–390.

Chamberlin T. C. (1899) An attempt to frame a working hypothesis of the cause of glacial periods on an atmospheric basis. *J. Geol.* **7**, 545–584.

Clayton J. L. (1986) An estimate of plagioclase weathering rate in the Idaho batholith based upon geochemical transport rates. In *Rates of Chemical Weathering of Rocks and Minerals* (eds. S. Coleman and D. Dethier). Academic Press, Orlando, pp. 453–466.

Cleaves A. T., Godfrey A. E., and Bricker O. P. (1970) Geochemical balance of a small watershed and its geomorphic implications. *Geol. Soc. Am. Bull.* **81**, 3015–3032.

Cleaves E. T. (1993) Climatic impact on isovolumetric weathering of coarse-grained schist in the northern Piedmont province of the central Atlantic states. *Geomophology* **8**, 191–198.

Clow D. W. and Drever J. I. (1996) Weathering rates as a function of flow through an alpine soil. *Chem. Geol.* **132**, 131–141.

Coleman S. M. (1981) Rock-weathering rates as functions of time. *Quat. Res.* **15**, 250–264.

Cosby B. J., Wright R. F., Hornberger G. M., and Galloway J. N. (1985) Modelling the effects of acid deposition: assessment of a lumped parameter model for soil water and stream chemistry. *Water Resour. Res.* **21**, 51–63.

Cremeens D. L., Darmody R. G., and Norton L. D. (1992) Etch-pit size and shape distribution on orthoclase and pyriboles in a loess catena. *Geochim. Cosmochim. Acta* **56**, 3423–3434.

De Vries W. and Kros J. (1989) Simulation of the long term soil response to acid deposition in various buffer ranges. *Water Air Soil Pollut.* **48**, 349–390.

Dessent C., Dupre B., Francois L., Schott J., Gaillard J., Chakrapani G., and Bajpai S. (2001) Erosion of Deccan traps determined by river geochemistry: impact on global climate and the $^{87}Sr/^{86}Sr$ ratio of seawater. *Earth Planet. Sci. Lett.* **188**, 459–474.

Dethier D. P. (1986) Weathering rates and the chemical flux from catchment in the Pacific northwest USA. In *Rates of Chemical Weathering of Rocks and Minerals*, (eds. S. Coleman and D. Dethier). Academic Press, Orlando, pp. 503–530.

Dorn R. (1995) Digital processing of back-scatter electron imagery: a microscopic approach to quantifying chemical weathering. *Geol. Soc. Am. Bull.* **107**, 725–741.

Dorn R. I. and Brady P. V. (1995) Rock-based measurement of temperature-dependent plagioclase weathering. *Geochim. Cosmochim. Acta* **59**, 2847–2852.

Drever J. I. and Clow D. W. (1995) Weathering rates in catchments. In *Chemical Weathering Rates of Silicate Minerals*, Reviews in Mineralogy (eds. A. F. White and S. L. Brantley). Mineralogical Society of America, Washington, DC, pp. 463–481.

Driscoll C. T., Likens G. E., Hedlin L. O., Eaton J. S., and Bormann F. H. (1989) Changes in the chemistry of surface waters. *Environ. Sci. Technol.* **23**, 137–142.

Dunne T. (1978) Rates of chemical denudation of silicate rocks in tropical catchments. *Nature* **274**, 244–246.

Edmond J. M., Palmer M. R., Measures C. I., Grant B., and Stallard R. F. (1995) The fluvial geochemistry and denudation rate of the Guayana shield in Venezuela, Colombia, and Brazil. *Geochim. Cosmochim. Acta* **59**, 3301–3325.

Farley D. A. and Werritty A. (1989) Hydrochemical budgets for the Loch Dee experimental catchments, southwest Scotland (1981–1985). *J. Hydrol.* **109**, 351–368.

Fordham A. W. (1990) Weathering of biotite into dioctahedral clay minerals. *Clay Min.* **25**, 51–63.

Friedman I., Trembour F. W., Smith F. L., and Smith G. I. (1994) Is obsidian hydration dating affected by relative humidity? *Quat. Res.* 185–190.

Frogner T. (1990) The effect of acid deposition on cation fluxes in artificially acidified catchments in western Norway. *Geochim. Cosmochim. Acta* **54**, 769–780.

Garrels R. M. and Mackenzie F. T. (1967) Origin of the chemical composition of some springs and lakes. In *Equilibrium Concepts in Natural Water Systems*, Adv. Chem. Series 67 (ed. W. Stumm). American Chemical Society, Washington, DC, pp. 222–242.

Garrels R. M. and Mackenzie F. T. (1971) *Evolution of Sedimentary Rocks.* NY: W. W. Norton.

Gislason S. R. and Eugster H. P. (1987) Meteoric water-basalt interactions: II. A field study in NE Iceland. *Geochim. Cosmochim. Acta* **51**, 2841–2855.

Gislason S. R., Arnorsson S., and Armannsson H. (1996) Chemical weathering of basalt in southwest Iceland: effects of runoff, age of rocks and vegetative/glacial cover. *Am. J. Sci.* **296**, 837–907.

Goldich S. S. (1938) A study of rock weathering. *J. Geol.* **46**, 17–58.

Gunter W. D., Perkins E. H., and Hutcheon I. D. (2000) Aquifer disposal of acid gases: modeling of water–rock reactions for trapping of acid wastes. *Appl. Geochem.* **15**, 1085–1095.

Hall R. D. and Horn L. L. (1993) Rates of hornblende etching in soils in glacial deposits of the northern Rocky mountains (Wyoming-montana, USA): influence of climate and characteristics of the parent material. *Chem. Geol.* **105**, 17–19.

Harden J. W. (1987) Soils developed in granitic aavium near Merced, California. *US Geol. Surv. Bull.* **1590-A**, 121p.

Harden J. W. (1990) Soil development on stable landforms and implications for landscape studies. *Geomorphology* **3**, 391–398.

Hartt C. F. (1853) *Geologia e geografia fidsca do Brasil.* Companhia ditoria Nacional.

Helgeson H. C. (1971) Kinetics of mass transfer among silicates and aqueous solutions. *Geochim. Cosmochim. Acta* **35**, 421–469.

Helgeson H. C., Brown T. H., Nigrini A., and Jones T. A. (1970) Calculation of mass transfer in geochemical processes involving aqueous solutions. *Geochim. Cosmochim. Acta* **24**, 569–592.

Hillel D. (1982) *Introduction to Soil Physics.* Orlando: Academic Press.

Hochella M. F., Jr., Eggleston C. M., Johnsson P. A., Stipp S. L., Tingle T. N., Blum A. E., White A. F., and Fawcett J. J. C. (1991) Examples of mineral surface structure, composition, and reactivity observed at the molecular and atomic levels. In *Program with Abstracts—Geological Association of Canada, Mineralogical Association of Canada, Canadian Geophysical Union, Joint Annual Meeting* **16**, 56, Ontario.

Hodson M. E. (2002) Experimental evidence for the mobility of Zr and other trace elements in soils. *Geochim. Cosmochim. Acta* **66**, 819–828.

Hodson M. E., Langan S. J., and Wilson M. J. (1997) A critical evaluation of the use of the PROFILE model in calculating mineral weathering rates. *Water Air Soil Pollut.* **98**, 79–104.

Hooper R. P., Christophersen N., and Peters N. E. (1990) Modelling stream water chemistry as a mixture of soil water end-members-an application to the Panola mountain catchment, Georgia, USA. *J. Hydrol.* **116**, 321–343.

Huntington T. G., Hooper R. P., Johnson C. E., Aulenbach B. T., Cappellato R., and Blum A. E. (2000) Calcium depletion in forest ecosystems of the southeastern United States. *Soil Sci. Soc. Am. J.* **64**, 1845–1858.

Huh Y., Panteleyev G., Babich D., Zaitsev A., and Edmond J. (1998) The fluvial geochemistry of the rivers of eastern Siberia: II. Tributaries of the Lena, Omloy, Yana, Indigirka, Kolyma, and Anadyr draining the collisional/accretionary zone of the Verkhoyansk and Cherskiy ranges. *Geochim. Cosmochim. Acta* **62**, 2053–2075.

Jacobson A. D., Blum J. D., Chanberlain C. P., Poage M. A., and Sloan V. F. (2002) Ca/Sr and Sr isotope systematics of a Himalayan glacial chronosequence: carbonate versus silicate weathering rates as a function of landscape surface age. *Geochim. Cosmochim. Acta* **66**, 13–27.

Jenny H. (1941) *Factors of Soil Formation.* NY: McGraw-Hill.

Johnsson M. J., Ellen S. D., and McKittrick M. A. (1993) Intensity and duration of chemical weathering: an example from soil clays of the southeastern Koolau mountains, Oahu, Hawaii. *Geol. Soc. Am. Spec. Publ.* **284**, 147–170.

Katz B. G. (1989) Influence of mineral weathering reactions on the chemical composition of soil water, springs, and groundwater, Catoctin mountains, Maryland. *Hydrol. Process.* **3**, 185–202.

Kenoyer G. J. and Bowser C. J. (1992) Groundwater evolution in a sandy silicate aquifer in northern Wisconsin: 1. Patterns and rates of change. *Water Resour. Res.* **28**, 579–589.

Kirkwood S. E. and Nesbitt H. W. (1991) Formation and evolution of soils from an acidified watershed: Plastic Lake, Ontario, Canada. *Geochim. Cosmochim. Acta* **55**, 1295–1308.

Land M., Ingri J., and Ohlander B. (1999) Past and present weathering rates in northern Sweden. *Appl. Geochem.* **14**, 761–774.

Larsen M. C. and Concepicon I. M. (1998) Water budgets of forested and agriculturally developed watersheds in Puerto Rico: proceedings, tropical hydrology and Caribbean water resources. In *Proc. American Water Resources Association Annual Meeting San Juan Puerto Rico* (ed. R. I. Segarra-Garcia), pp. 199–204.

Lasaga A. C. (1998) *Kinetic Theory in the Earth Sciences.* Princeton: Princeton University Press.

Lee M. R. and Parsons I. (1995) Microtextural controls of weathering of perthitic alkali feldspars. *Geochim. Cosmochim. Acta* **59**, 4465–4492.

Lee M. R., Hodson M. E., and Parsons I. (1998) The role of intragranular microtextures and microstructures in chemical and mechanical weathering: direct comparisons of experimentally and naturally weathered alkali feldspars. *Geochim. Cosmochim. Acta* **62**, 2771–2788.

Likens G. E., Bormann F. H., Pierce R. S., Eaton J. S., and Johnson N. M. (1977) *Biogeochemistry of a Forested Ecosystem.* Berlin: Springer-Verlag.

Luttge A. B., Bolten E. W., and Lasaga A. C. (1999) An interferometric study of the dissolution kinetics of anorthite: the role of reactive surface area. *Am. J. Sci.* **299**, 652–678.

MacInnis I. N. and Brantley S. L. (1993) Development of etch pit size distributions on dissolving minerals. *Chem. Geol.* **105**, 31–49.

Mahaney W. C. and Halvorson D. L. (1986) Rates of mineral weathering in the wind river mountains, western Wyoming. In *Rates of Chemical Weathering of Rocks and Minerals* (eds. S. Coleman and D. Dethier). Academic Press, Orlando, pp. 147–168.

Mast M. A., Drever J. I., and Barron J. (1990) Chemical weathering in the Loch Vale watershed, Rocky Mountain National Park, Colorado. *Water Resour. Res.* **26**, 2971–2978.

Maurice P. A., McKnight D. M., Leff L., Fulghum J. E., and Gooseff M. (2002) Direct observations of aluminosilicate weathering in the hyporheic zone of an Antarctic dry valley stream. *Geochim. Cosmochim. Acta* **66**, 1335–1347.

Merino E., Nahon D., and Wang Y. (1993) Kinetic and mass transfer of pseudomorphic replacement: application to replacement of parent minerals and kaolinite by Al, Fe, and Mn oxides during weathering. *Am. J. Sci.* **293**, 135–155.

Merrill G. P. (1906) *A Treatise on Rocks, Rock Weathering and Soils.* NY: McMillian.

Merritts D. J., Chadwick O. A., Hendricks D. M., Brimhall G. H., and Lewis C. J. (1992) The mass balance of soil evolution on late quaternary marine terraces, northern California. *Geol. Soc. Am. Bull.* **104**, 1456–1470.

Meybeck M. (1979) Concentrations des eaux fluviales en elements majeurs et apports en solution aux oceans. *Revue De Geologie Dynamique et De Geographie Physique* **21**, 215–246.

Meyback M. (1994) *Material Fluxes on the Surface of the Earth.* Washington: National Academy Press.

Misra K. C. (2000) *Understanding Mineral Deposits.* Rotterdam: Kluwer Academic.

Mokma D. L., Jackson M. L., Syers J. K., and Steens P. R. (1973) Mineralogy of a chronosequence of soils from greywacke and mica-schist alluvium. *NZ J. Sci.* **16**, 769–797.

Murphy S. F., Brantley S. L., Blum A. E., White A. F., and Dong H. (1998) Chemical weathering in a tropical watershed, Luquillo mountains, Puerto Rico: II. Rate and mechanism of biotite weathering. *Geochim. Cosmochim. Acta* **62**, 227–243.

Nagy K. L., Blum A. E., and Lasaga A. C. (1991) Dissolution and precipitation kinetics of kaolinite at 80°C and pH 3: the dependence on the saturation state. *Am. J. Sci.* **291**, 649–686.

Nahon D. and Merino E. (1997) Pseudomorphic replacement in tropical weathering: evidence, geochemical consequences, and kinetic-rheological origin. *Am. J. Sci.* **297**, 393–417.

Nesbitt H. W. and Markovics G. (1997) Weathering of granodioritic crust, long-term storage of elements in weathering

Natural Weathering Rates of Silicate Minerals

profiles, and petrogenesis of siliciclastic sediments. *Geochim. Cosmochim. Acta* **61**, 1653–1670.

Nesbitt H. W., Fedo C. M., and Young G. M. (1997) Quartz and feldspar stability, steady and non steady state weathering and petrogenesis of siliciclastic sands and muds. *J. Geol.* **105**, 173–191.

Nugent M. A., Brantley S. L., Pantano C. G., and Maurice P. A. (1998) The influence of natural mineral coatings on feldspar weathering. *Nature* **396**, 527–622.

O'Brien E. L. and Buol S. W. (1984) Physical transformations in a vertical soil-saprolite sequence. *Soil Sci. Am. J.* **48**, 354–357.

Oelkers E. H. (2001) General kinetic description of multioxide silicate mineral and glass dissolution. *Geochim. Cosmochim. Acta* **65**, 3703–3719.

Oelkers E. H. and Schott J. (1995) Experimental study of anorthite dissolution and the relative mechanism of feldspar hydrolysis. *Geochim. Cosmochim. Acta* **59**, 5039–5053.

Ohmoto H. (1996) Evidence in pre-2.2 Ga paleosols for the early evolution of atmospheric oxygen and terrestrial biota. *Geology* **24**, 1135–1138.

Oillier C. (1984) *Weathering*. London: Longman.

Paces T. (1972) Chemical characteristics and equilibrium between groundwater and granitic rocks. *Geochim. Cosmochim. Acta* **36**, 217–240.

Paces T. (1973) Steady-state kinetics and equilibrium between ground water and granitic rock. *Geochim. Cosmochim. Acta* **37**, 2641–2663.

Paces T. (1986) Rates of weathering and erosin derived from mass balance in small drainage basins. In *Rates of Chemical Weathering of Rocks and Minerals* (eds. S. Coleman and D. L. Dethier). Academic Press, Orlando, pp. 531–550.

Parkhurst D. L. and Plummer L. N. (1993) Geochemical models. In *Regional Ground–Water Quality* (ed. W. M. Alley). Van Nostrand Reinhold, , pp. 199–225.

Pavich M. J. (1986) Processes and rates of saprolite production and erosion on a foliated granitic rock of the Virgina piedmont. In *Rates of Chemical Weathering of Rocks and Minerals* (eds. S. Coleman and D. Dethier). Academic Press, Orlando, pp. 551–590.

Pavich M. J., Brown L., Harden J., Klein J., and Middleton R. (1986) [10]Be distribution in soils from Merced River terraces, California. *Geochim. Cosmochim. Acta* **50**, 1727–1735.

Perg L. A., Andersson R. S., and Finkel R. C. (2001) Use of a new [10]Be and [26]Al inventory method to date marine terraces, Santa Cruz, California, USA. *Geology* **29**, 879–882.

Riebe C. S., Kirchner J. W., Granger D. E., and Finkel R. C. (2001) Strong tectonic and weak climatic control of long-term chemical weathering rates. *Geology* **29**, 511–514.

Ruddiman W. F. (1997) *Tectonic Uplift and Climate Change*. NY: Plenum.

Rye R. and Holland H. D. (1998) Palosols and the evolution of atmospheric oxygen: a critical review. *Am. J. Sci.* **298**, 621–672.

Samma J. C. (1986) *Ore Fields and Continental Weathering*. NY: Van Nostrand.

Schnoor J. L. (1990) Kinetics of chemical weathering: a comparison of laboratory and field rates. In *Aquatic Chemical Kinetics* (ed. W. Stumm). Wiley, NY, pp. 475–504.

Schulz M. S. and White A. F. (1999) Chemical weathering in a tropical watershed, Luquillo mountains, Puerto Rico: III. Quartz dissolution rates. *Geochim. Cosmochim. Acta* **63**, 337–350.

Siegal D. and Pfannkuch H. O. (1984) Silicate mineral dissolution at pH 4 and near standard temperature and pressure. *Geochim. Cosmochim. Acta* **48**, 197–201.

Soler J. M. and Lasaga A. C. (1996) A mass transfer model of bauxite formation. *Geochim. Cosmochim. Acta* **60**, 4913–4931.

Stallard R. F. (1995) Tectonic, environmental, and human aspects of weathering and erosion: a global review using a steady-state perspective. *Ann. Rev. Earth Planet. Sci.* **23**, 11–39.

Stallard R. F. and Edmond J. M. (1983) Geochemistry of the Amazon: 2. The influence of geology and weathering environment on dissolved load. *J. Geophys. Res.* **88**, 9671–9688.

Stauffer R. E. (1990) Granite weathering and the sensitivity of Alpine Lakes to acid depositon. *Limnol. Oceangr.* **35**, 1112–1134.

Stefansson A. (2001) Dissolution of primary minerals of basalt in natural waters: I. Calculation of mineral solubilities from 0°C to 350°C. *Chem. Geol.* **172**, 225–250.

Stefansson A. and Arnorsson S. (2000) Feldspar saturation state in natural waters. *Geochim. Cosmochim. Acta* **64**, 2567–2584.

Steefel C. I. and Lasaga A. C. (1994) A coupled model for the transport of multiple chemical species and kinetic precipitation/dissolution reactions with application to reactive flow in a single phase hydrothermal system. *Am. J. Sci.* **294**, 529–592.

Stewart B. W., Capo R. C., and Chadwick O. A. (2001) Effects of rainfall on weathering rate, base cation provenance, and Sr isotope composition of Hawaiian soils. *Geochim. Cosmochim. Acta* **65**, 1087–1099.

Stillings L. L., Drever J. I., Brantley S., Sun Y., and Oxburgh R. (1996) Rates of feldspar dissolution at pH 3–7 with 0–8 M oxalic acid. *Chem. Geol.* **132**, 79–89.

Stonestrom D. A., White A. F., and Akstin K. C. (1998) Determining rates of chemical weathering in soils-solute transport versus profile evolution. *J. Hydrol.* **209**, 331–345.

Suarez D. L. and Simunek J. (1996) Solute transport modelling under variably saturated water flow conditions. In *Reactive Transport in Porous Media* (eds. P. C. Lichtner, C. I. Steefel, and E. H. Oelkers). Mineralogical Society of America, Washington DC, pp. 230–268.

Sverdrup H. U. (1990) *The Kinetics of Base Cation Release due to Chemical Weathering*. Lund: Lund University.

Sverdrup K. and Warfvinge P. (1995) Estimating field weathering rates using laboratory kinetics. In *Chemical Weathering Rates of Silicate Minerals*, Reviews in Mineralogy 31 (eds. A. F. White and S. L. Brantley). Mineralogical Society of America, Washington DC, pp. 485–541.

Swoboda-Colberg N. G. and Drever J. I. (1992) Mineral dissolution rates: a comparison of laboratory and field studies. In *Water–Rock Interaction WRI 7* (eds. Y. K. Kharaka and A. S. Maest). Balkema, Rotterdam, pp. 115–117.

Taylor A. and Blum J. D. (1995) Relation between soil age and silicate weathering rates determined from the chemical evolution of a glacial chronosequence. *Geology* **23**, 979–982.

Taylor A. S., Blum J. D., and Lasaga A. C. (2000) The dependence of labradorite dissolution and Sr isotope release rates on solution saturation state. *Geochim. Cosmochim. Acta* **64** (14): 2389–2400.

Tisdall J. M. (1982) Organic matter and water-stable aggregates in soils. *J. Soil Sci.* **33**, 141–163.

Torrent J. and Nettleton W. D. (1978) Feedback processes in soil genesis. *Geoderma* **20**, 281–287.

Trudgill S. T. (1995) *Solute Modelling in Catchment System*. NY: Wiley.

Velbel M. A. (1985) Geochemical mass balances and weathering rates in forested watersheds of the southern Blue ridge. *Am. J. Sci.* **285**, 904–930.

Velbel M. A. (1989) Weathering of hornblende to ferruginous products by dissolution-reprecipitation mechanism: petrography and stoichiometry. *Clays Clay Min.* **37**, 515–524.

Velbel M. C. (1993) Temperature dependence of silicate weathering in nature: how strong a feedback on long-term accumulation of atmospheric CO_2 and global greenhouse warming. *Geology* **21**, 1059–1062.

Warfvinge P. (1995) Basic principals of frequently used models. In *Solute Modeling in Catchment Systems* (ed. S. T. Trudgill). Wiley, NY, pp. 57–71.

White A. F. (1995) Chemical weathering rates in soils. In *Chemical Weathering Rates of Silicate Minerals* (eds. A. F. White and S. L. Brantley). Mineralogical Society of America, Washington, DC, pp. 407–458.

White A. F. (2002) Determining mineral weathering rates based on solid and solute weathering gradients and velocities: application to biotite weathering in saprolites. *Chem. Geol.* **190**, 69–89.

White A. F. and Blum A. E. (1995) Effects of climate on chemical weathering rates in watersheds. *Geochim. Cosmochim. Acta* **59**, 1729–1747.

White A. F. and Brantley S. L. (1995) Chemical weathering rates of silicate minerals. In *Reviews in Mineralogy*. Mineralogical Society of America, Washington DC, vol. 31, 584pp.

White A. F. and Brantley S. L. (2003) The effect of time on the weathering of silicate minerals: why to weathering rates differ in the laboratory and field? *Chem. Geol.* (in press).

White A. F. and Peterson M. L. (1990) Role of reactive surface area characterization in geochemical models. In *Chemical Modelling of Aqueous Systems II*. Am. Chem. Soc. Symp. Ser. 416 (eds. D. C. Melchior and R. L. Bassett, Am. Chem. Soc., Washington DC, pp. 461–475.

White A. F., Blum A. E., Schulz M. S., Bullen T. D., Harden J. W., and Peterson M. L. (1996) Chemical weathering of a soil chronosequence on granitic alluvium: 1. Reaction rates based on changes in soil mineralogy. *Geochim. Cosmochim. Acta* **60**, 2533–2550.

White A. F., Blum A. E., Schulz M. S., Vivit D. V., Larsen M., and Murphy S. F. (1998) Chemical weathering in a tropical watershed, Luquillo mountains, Puerto Rico: I. Long-term versus short-term chemical fluxes. *Geochim. Cosmochim. Acta* **62**, 209–226.

White A. F., Blum A. E., Bullen T. D., Vivit D. V., Schulz M., and Fitzpatrick J. (1999) The effect of temperature on experimental and natural weathering rates of granitoid rocks. *Geochim. Cosmochim. Acta* **63**, 3277–3291.

White A. F., Blum A. E., Stonestrom D. A., Bullen T. D., Schulz M. S., Huntington T. G., and Peters N. E. (2001) Differential rates of feldspar weathering in granitic regoliths. *Geochim. Cosmochim. Acta* **65**, 847–869.

White A. F., Blum A. E., Schulz M. S., Huntington T. G., Peters N. E., and Stonestrom D. A. (2002) Chemical weathering of the Panola granite: solute and regolith elemental fluxes and the dissolution rate of biotite. In *Water–Rock Interaction, Ore Deposits, and Environmental Geochemisty: a Tribute to David A. Crerar* Spec. Publ. 7 (eds. R. Hellmann and S. A. Wood). The Geochemical Society, St Louis, pp. 37–59.

(in press)White A. F., Schulz M. J., Davison V., Blum A. E., Stonestrom D. A., and Harden J. W. (2003) Chemical weathering rates of a soil chronosequence on granitic alluvium: III. The effects of hydrology, biology, climate on pore water compositions and weathering fluxes. *Geochim. Cosmochim. Acta* (in press).

Williams M. W., Brown A. D., and Melack J. M. (1993) Geochemical and hydrologic controls on the composition of surface water in a high-elevation basin, Sierra Nevada, California. *Limnol. Oceanogr.* **38**, 775–797.

Wolery T. J., Jackson K. J., Bourcier W. L., Bruton C. J., and Viani B. E. (1990) Current status of the EQ3/6 software package for geochemical modelling. In *Chemical Modelling of Aqueous Systems II* (eds. D. C. Melchior and R. L. Bassett). American Chemical Society, Washington, DC, pp. 104–116.

Zeman L. J. and Slaymaker O. (1978) Mass balance model for calculation of ionic input loads in atmospheric fallout and discharge from a mountainous basin. *Hydrol. Sci.* **23**, 103–117.

Published by Elsevier Ltd.

Geochemistry of Earth Surface Systems
ISBN: 978-0-08-096706-6

7

Soil Formation

R. Amundson

University of California, Berkeley, CA, USA

7.1	INTRODUCTION	241
7.2	FACTORS OF SOIL FORMATION	242
7.3	SOIL MORPHOLOGY	242
7.4	MASS BALANCE MODELS OF SOIL FORMATION	243
	7.4.1 Mass Balance Evaluation of the Biogeochemistry of Soil Formation	246
	7.4.2 Mass Balance of Soil Formation versus Time	250
	7.4.2.1 Temperate climate	250
	7.4.2.2 Cool tropical climate	251
	7.4.2.3 Role of atmospheric inputs on chemically depleted landscapes	253
	7.4.3 Mass Balance Evaluation of Soil Formation versus Climate	253
7.5	PROCESSES OF MATTER AND ENERGY TRANSFER IN SOILS	255
	7.5.1 Mechanistic Modeling of the Organic and Inorganic Carbon Cycle in Soils	257
	7.5.1.1 Modeling carbon movement into soils	260
	7.5.1.2 Modeling carbon movement out of soils	261
	7.5.1.3 Processes and isotope composition of pedogenic carbonate formation	262
	7.5.2 Lateral Transport of Soil Material by Erosion	267
7.6	SOIL DATA COMPILATIONS	271
7.7	CONCLUDING COMMENTS	271
	REFERENCES	272

...that the Earth has not always been here—that it came into being at a finite point in the past and that everything here, from the birds and fishes to the loamy soil underfoot, was once part of a star. I found this amazing, and still do.

Timothy Ferris (1998)

7.1 INTRODUCTION

Soil is the biogeochemically altered material that lies at the interface between the lithosphere (Volume 3) and the atmosphere (Volume 4). *Pedology* is the branch of the natural sciences that concerns itself, in part, with the biogeochemical processes (Volume 8) that form and distribute soil across the globe. Pedology originated during the scientific renaissance of the nineteenth century as a result of conceptual breakthroughs by the Russian scientist Vassali Dochuchaev (Krupenikov, 1992; Vil'yams, 1967) and conceptual and administrative efforts by the American scientist Eugene Hilgard (Jenny, 1961; Amundson and Yaalon, 1995).

Soil is the object of study in pedology, and while the science of pedology has a definition that commands some general agreement, there is no precise definition for soil, nor is there likely ever to be one. The reason for this paradox is that soil is a part of a continuum of materials at the Earth's surface (Jenny, 1941). At the soil's base, the exact line of demarcation between "soil" and "nonsoil" will forever elude general agreement, and horizontal changes in soil properties may occur so gradually that similar problems exist in delineating the boundary between one soil "type" and another. The scientific path out of this conundrum is to divide the soil continuum, albeit arbitrarily, into *systems* that suit the need of the investigator. Soil systems are necessarily open to their surroundings, and through them pass matter and energy which measurably alter the properties of the system over timescales from seconds to millennia. It was the recognition by Dokuchaev (1880), and later the American scientist Hans Jenny (1941), that the properties of the soil system are controlled by *state factors* that ultimately formed the framework of the fundamental paradigm of pedology.

The purpose of this chapter is to present an abridged overview of the factors and processes that control soil formation, and to provide, where possible, some general statements of soil formation processes that apply broadly and commonly.

7.2   FACTORS OF SOIL FORMATION

Jenny (1941) applied principles from the physical sciences to the study of soil formation. Briefly, Jenny recognized that soil systems (or if the aboveground flora and fauna are considered, *ecosystems*) exchange mass and energy with their surroundings and that their properties can be defined by a limited set of *independent variables*. From comparisons with other sciences, Jenny's *state factor model* of soil formation states that

$$\underbrace{\text{Soils/ecosystems}}_{\text{dependent variables}} = f \underbrace{\left(\begin{array}{l} \text{initial state of system,} \\ \text{surrounding environment,} \\ \text{elapsed time} \end{array} \right)}_{\text{independent variables}} \quad (1)$$

From field observations and the conceptual work of Dokuchaev, a set of more specific environmental factors have been identified which encompass the controls listed above:

$$\underbrace{\text{Soils/ecosystems}}$$
$$= f(\underbrace{\text{climate, organisms,}}_{\text{surrounding environment}}$$
$$\underbrace{\text{topography, parent material,}}_{\text{initial state of system}} \text{time,} \dots) \quad (2)$$

These so-called "state factors of soil formation" have the following important characteristics: (i) they are independent of the system being studied and (ii) in many parts of the Earth, the state factors vary independently of each other (though, of

course, not always). As a result, through judicious site (system) selection, the influence of a single factor can be observed and quantified in nature.

Table 1 provides a brief definition of the state factors of soil formation. A field study designed to observe the influence of one state factor on soil properties or processes is referred to as a *sequence*, e.g., a series of sites which have similar state factor values except climate is referred to as a *climosequence*. Similar sequences can, and have been, established to examine the effect of other state factors on soils. An excellent review of soil state factor studies is presented by Birkeland (1999). An informative set of papers discussing the impact of Jenny's state factor model on advances in pedology, geology, ecology, and related sciences is presented in Amundson *et al.* (1994a,b).

The state factor approach to studying soil formation has been, and continues to be, a powerful quantitative means of linking soil properties to important variables (Amundson and Jenny, 1997). As an example, possibly the best characterized soil versus factor relationship is the relationship of soil organic carbon and nitrogen storage to climate (mean annual temperature and precipitation) (Figure 1). The pattern—increasing carbon storage with decreasing temperature and increasing precipitation—illustrated in Figure 1 is the result of nearly six decades of work, and is based on thousands of soil observations (Miller *et al.*, 2002). This climatic relationship is important in global change research and in predicting the response of soil carbon storage to climate change (Schlesinger and Andrews, 2000). However, the relationship, no matter how valid, provides no insight into the rates at which soils achieve their carbon storage, nor the mechanisms involved in the accumulation. Thus, other approaches, again amenable to systems studies, have been applied in pedology to quantify soil formation. These are discussed in later sections.

7.3   SOIL MORPHOLOGY

A trend in present-day pedology is to incorporate ever more sophisticated chemical and mathematical tools into our understanding of soil and

Table 1   The major state factors of soil and ecosystem formation, and a brief outline of their characteristics.

State factor	Definition and characteristics
Climate	Regional climate commonly characterized by mean annual temperature and precipitation
Organisms	Potential biotic flux into system (as opposed to what is present at any time)
Topography	Slope, aspect, and landscape configuration at time $t = 0$
Parent material	Chemical and physical characteristics of soil system at $t = 0$
Time	Elapsed time since system was formed or rejuvenated
Humans	A special biotic factor due to magnitude of human alteration of Earth's and humans' possession of variable cultural practices and attitudes that alter landscapes

Sources: Jenny (1941) and Amundson and Jenny (1997).

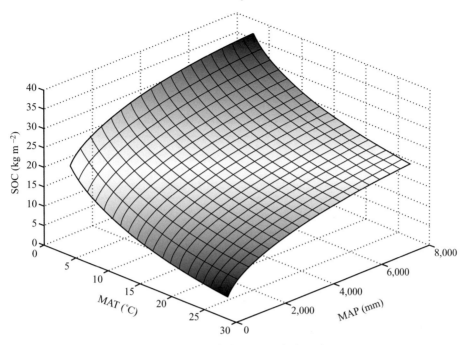

Figure 1 The distribution of global soil C in relation to variations in mean average temperature (MAT) and precipitaiton (MAP). The curve is derived from a multiple regression model of published soil C data versus climate (Amundson, 2001) (reproduced by permission of Annual Reviews from *Ann. Rev. Earth Planet. Sci.* **2001**, *29*, 535–562).

their formation. Yet, an examination of soils *in situ* is required, in order to develop the appropriate models or to even logically collect samples for study.

Soil *profiles* are two-dimensional, vertical exposures of the layering of soils. The net result of the transport of matter and energy is a vertical differentiation of visible, distinctive layers called soil *horizons*. Soil horizons reflect the fact that soil formation is a depth-dependent process. They are layers that are readily identified by visual and tactile procedures (field based) that have been developed over the years (Soil Survey Staff, 1993). A nomenclature has developed over the past century, first started by the pioneering Russian scientists in the nineteenth century, that involves the "naming"of soil horizons on the basis of *how they differ from the starting parent material*. Therefore, horizon naming requires data acquisition and hypothesis development. Soil horizon names are commonly assigned from field-based data, and may ultimately be modified as a result of subsequent laboratory investigations.

The present US horizon nomenclature has two components: (i) an upper case, "master horizon" symbol and (ii) a lower case "modifier" that provides more information on the horizon characteristics or the processes that formed it. Tables 2(a) and (b) provide definitions of both the common master and modifier symbols. The detailed rules

for their use can be found in the *Soil Survey Manual* (Soil Survey Staff, 1993).

Most soil process models are (roughly) continuous with depth. However, during the observation of many soil profiles, it is apparent that horizons do not always, or even commonly, grade gradually into one another. Sharp or abrupt horizon boundaries are common in soils around the world. This indicates that our concepts and models of soil formation capture only a part of the long-term trajectory of soil development. Some processes are not continuous with depth (the formation of carbonate horizons for example), while some may be continuous for some time period and then, due to feedbacks, change their character (the formation of clay-rich horizons which, if they reach a critical clay content, restrict further water and clay transport. This causes an abrupt buildup of additional clay at the top of the horizon). In the following sections, the author examines various approaches to understanding soil formation, and examines some of their attributes and limitations.

7.4 MASS BALANCE MODELS OF SOIL FORMATION

Detailed chemical analyses of soils and the interpretation of that data relative to the composition of the parent material have been performed since

Table 2(a)   A listing, and brief definitions, of the nomenclature used to identify master soil horizons.

Master horizons	Definition and examples of lower case modifiers
O	Layers dominated by organic matter. State of decomposition determines type: highly (Oa), moderately (Oe), or slightly (Oi)[a] decomposed
A	Mineral horizons that have formed at the surface of the mineral portion of the soil or below an O horizon. Show one of the following: (i) an accumulation of humified organic matter closely mixed with minerals or (ii) properties resulting from cultivation, pasturing, or other human-caused disturbance (Ap)
E	Mineral horizons in which the main feature is loss of silicate clay, iron, aluminum, or some combination of these, leaving a concentration of sand and silt particles
B	Horizons formed below A, E, or O horizons. Show one or more of the following: (i) illuvial[b] concentration of silicate clay (Bt), iron (Bs), humus (Bh), carbonates (Bk), gypsum (By), or silica (Bq) alone or in combination; (ii) removal of carbonates (Bw); (iii) residual concentration of oxides (Bo); (iv) coatings of sesquioxides[c] that make horizon higher in chroma or redder in hue (Bw); (v) brittleness (Bx); or (vi) gleying[d] (Bg).
C	Horizons little affected by pedogenic processes. May include soft sedimentary material (C) or partially weathered bedrock (Cr)
R	Strongly indurated[e] bedrock
W	Water layers within or underlying soil

Source: Soil Survey Staff (1999).
[a] The symbols in parentheses illustrate the appropriate lower case modifiers used to describe specific features of master horizons.
[b] The term illuvial refers to material transported into a horizon from layers above it.
[c] The term sesquioxide refers to accumulations of secondary iron and/or aluminum oxides.
[d] Gleying is a process of reduction (caused by prolonged high water content and low oxygen concentrations) that results in soil colors characterized by low chromas and gray or blueish hues.
[e] The term indurated means strongly consolidated and impenetrable to plant roots.

Table 2(b)   Definitions used to identify the subordinate characteristics of soil horizons.

Lower case modifiers of master horizons	Definitions (relative to soil parent material)
a	Highly decomposed organic matter (O horizon)
b	Buried soil horizon
c	Concretions or nodules of iron, aluminum, manganese, or titanium
d	Noncemented, root restricting natural or human-made (plow layers, etc.) root restrictive layers
e	Intermediate decomposition of organic matter (O horizon)
f	Indication of presence of permafrost
g	Strong gleying present in the form of reduction or loss of Fe and resulting color changes
h	Accumulation of illuvial complexes of organic matter which coat sand and silt particles
i	Slightly decomposed organic matter (O horizon)
j	Presence of jarosite (iron sulfate mineral) due to oxidation of pyrite in previously reduced soils
k	Accumulation of calcium carbonate due to pedogenic processes
m	Nearly continuously cemented horizons (by various pedogenic minerals)
n	Accumulation of exchangeable sodium
o	Residual accumulation of oxides due to long-term chemical weathering
p	Horizon altered by human-related activities
q	Accumulation of silica (as opal)
r	Partially weathered bedrock
s	Illuvial accumulation of sesquioxides
ss	Presence of features (called slickensides) caused by expansion and contraction of high clay soils
t	Accumulation of silicate clay by weathering and/or illuviation
v	Presence of plinthite (iron rich, reddish soil material)
w	Indicates initial development of oxidized (or other) colors and/or soil structure
x	Indicates horizon of high firmness and brittleness
y	Accumulation of gypsum
z	Accumulation of salts more soluble than gypsum (e.g., Na_2CO_3)

Source: Soil Survey Staff (1999).

nearly the origins of pedology (Hilgard, 1860). Yet, quantitative estimates of total chemical denudation, and associated physical changes that occur during soil formation, were not rigorously performed until the late 1980s when Brimhall and co-workers (Brimhall and Dietrich, 1987; Brimhall *et al.*, 1991) began applying a mass balance model originally derived for ore body studies to the soil environment. Here the author presents the key components of this model, and reports the results of its application to two issues: (i) the behavior of many of the chemical elements in soil formation and (ii) general trends of soil physical and chemical behavior as a function of time during soil formation.

A representation of a soil system during soil formation is shown in Figure 2. While the figure illustrates a loss of volume during weathering, volumetric increases can also occur, as will be shown later. The basic expression, describing mass gains or losses of a given chemical element (j), in the transition from parent material (p) to soil (s) in terms of volume (V), bulk density (ρ), and chemical composition (C) is

$$m_{j,\,\text{flux}} = m_{j,\,s} - m_{j,\,p} \qquad (3)$$

where m is the mass of element j added/lost (flux) in the soil (s) or parent material (p). Incorporating

volume, density, and concentration (in percent) into the model gives

$$\underbrace{m_{j,\,flux}}_{\substack{\text{mass of element } (j)\text{ into/out} \\ \text{of parent material volume}}} = \underbrace{\frac{V_s\rho_s C_{j,s}}{100}}_{\substack{\text{mass of element } (j) \text{ in} \\ \text{soil volume of interest}}}$$

$$- \underbrace{\frac{V_p\rho_p C_{j,p}}{100}}_{\substack{\text{mass of element } (j) \text{ in} \\ \text{parent material volume}}}$$

(4)

Definitions of all terms used in these mass balance equations are given in Table 3. The 100 in the denominator is needed only if concentrations are in percent.

During soil development, volumetric collapse (ΔV, Figure 2) may occur through weathering losses while expansion may occur through biological or physical processes. Volumetric change is defined in terms of strain (ε):

$$\varepsilon_{i,\,s} = \frac{\Delta V}{V_p} = \left(\frac{V_s}{V_p} - 1\right) = \left(\frac{\rho_p C_{i,\,p}}{\rho_s C_{i,\,s}} - 1\right) \quad (5)$$

where the subscript i refers to an immobile, index element. Commonly zirconium, titanium, or other members of the titanium or rare earth groups of the periodic table are used as index elements.

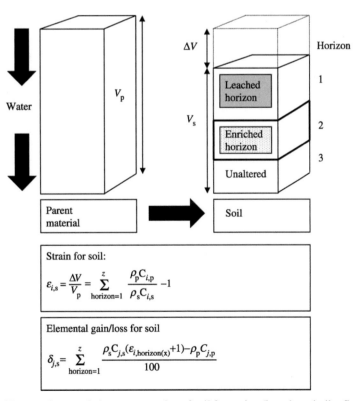

Strain for soil:

$$\varepsilon_{i,s} = \frac{\Delta V}{V_p} = \sum_{\text{horizon}=1}^{z} \frac{\rho_p C_{i,p}}{\rho_s C_{i,s}} - 1$$

Elemental gain/loss for soil

$$\delta_{j,s} = \sum_{\text{horizon}=1}^{z} \frac{\rho_s C_{j,s}(\varepsilon_{i,\text{horizon}(x)} + 1) - \rho_p C_{j,p}}{100}$$

Figure 2 Diagram illustrated a mass balance perspective of soil formation (based on similar figures in Brimhall and Dietrich, 1987).

Table 3 Definition of parameters used in mass balance model.

Parameter	Definition
V_p (cm^3)	Volume of parent material
V_s (cm^3)	Volume of soil
ρ_p (g cm^{-3})	Parent material bulk density
ρ_s (g cm^{-3})	Soil bulk density
$C_{j,p}$ (%,ppm)	Concentration of mobile element j in parent material
$C_{j,s}$ (%,ppm)	Concentration of mobile element j in soil
$C_{i,p}$ (%,ppm)	Concentration of immobile element i in parent material
$C_{i,s}$ (%,ppm)	Concentration of immobile element i in soil
$m_{j,\text{flux}}$ (g cm^{-3})	Mass of element j added or removed via soil formation
$\varepsilon_{i,s}$	Net strain determined using element i
τ	Fractional mass gain or loss of element j relative to immobile element i
$\delta_{j,s}$ (g cm^{-3})	Mass gain or loss per unit volume of element j relative to immobile element i.

Sources: Brimhall and Dietrich (1987) and Brimhall *et al.* (1992).

The fractional mass gain or loss of an element j relative to the mass in the parent material (τ) is defined by combining Equations (3)–(5):

$$\tau = \frac{m_{j,\text{flux}}}{m_{j,p}} = \left(\frac{\rho_s C_{j,s}}{\rho_p C_{j,p}} (\varepsilon_{i,s} + 1) - 1 \right) \quad (6)$$

Through substitution, Equation (6) reduces to

$$\tau = \frac{R_s}{R_p} - 1 \quad (7)$$

where $R_s = C_{j,s}/C_{i,s}$ and $R_p = C_{j,p}/C_{i,p}$. Thus, τ can be calculated readily from commonly available chemical data and does not require bulk density data. Absolute gains or losses of an element in mass per unit volume of the parent material ($\delta_{j,s}$) can be expressed as

$$\delta_{j,s} = \frac{m_{j,\text{flux}}}{V_p} = \frac{\rho_s C_{j,s}(\varepsilon_{i,s} + 1) - \rho_p C_{j,p}}{100}$$
$$= \frac{\tau C_{j,p} \rho_p}{100} \quad (8)$$

In applying the mass balance expressions, analyses are commonly performed by soil horizon, and total gains or losses (or collapse or expansion) can be plotted by depth or integrated for the whole soil profile (Figure 2).

7.4.1 Mass Balance Evaluation of the Biogeochemistry of Soil Formation

The chemical composition of soils is the result of a series of processes that ultimately link the soil to the history of the universe (Volume 1), with the principal processes of chemical differentiation being: (i) chemical evolution of universe/solar system; (ii) chemical differentiation of Earth from the solar system components; and (iii) the biogeochemical effects of soil formation on crustal chemistry.

The chemical composition of the solar system (Figure 3) has been widely discussed (Greenwood and Earnshaw, 1997; Chiappini, 2001). Today,

99% of the universe is comprised of hydrogen and helium, which were formed during the first few minutes following the big bang. The production of elements of greater atomic number requires a series of nuclear processes that occur during star formation and destruction. Thus, the relative elemental abundance versus atomic number is a function of the age of the universe and the number of cycles of star formation/termination that have occurred (e.g., Allègre, 1992).

The chemical composition of average crustal rock (Taylor and McLennan, 1985; Bowen, 1979) relative to the solar system reveals systematic differences (Brimhall, 1987) (Figure 4) that result from elemental fractionation during: (i) accretion of the Earth (and the interior planets) (Allègre, 1992) and (ii) differentiation of the core, mantle, and crust (Brimhall, 1987) and possibly unique starting materials (Drake and Righter, 2002). In general, the crust is depleted in the noble gases (group VIIIA) and hydrogen, carbon, and nitrogen, while it is enriched in many of the remaining elements. For the remaining elements, there is a trend toward decreasing enrichment with increasing atomic number within a given period, due to increasing volatility with increasing atomic number (Brimhall, 1987). The depletion of the siderophile elements in the crust relative to the solar system has been attributed to their concentration within the core (Brimhall, 1987), though the crust composition may also reflect late-stage accretionary processes (Delsemme, 2001).

The result of these various processes is that the Earth's crust, the parent material for soils, is dominated (in mass) by eight elements (oxygen, silicon, aluminum, iron, calcium, sodium, magnesium, and potassium). These elements, with the exception of oxygen, are not the dominant elements of the solar system. Thus, soils on Earth form in a matrix dominated by oxygen and silicon, the elements which form the backbone of the silicate minerals that dominate both the primary and secondary minerals found in soils.

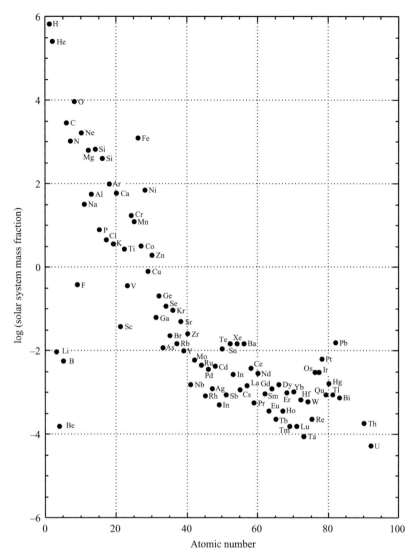

Figure 3 The log of the mass fraction (mg kg^{-1}) of elements in the solar system arranged by atomic number (source Anders and Grevesses, 1989, table 1).

There are a variety of compilations of the concentrations of many of the chemical elements for both crustal rocks (see above and Volume 3) and for soils (Bowen, 1979; Shacklette and Boerngen, 1984). In the case of soils, the samples analyzed are usually from a standard surface sampling depth, or from the uppermost horizon. Thus, these samples give a somewhat skewed view of the overall process of soil formation because, as will be discussed, soil formation is a depth-dependent process. Nonetheless, the data do provide a general overview of soil biogeochemistry that is applicable across broad geographical gradients.

When analyzing large chemical data sets, it is common to evaluate the behavior of elements in soils, and how they change during soil formation, by dividing the mass concentration of the elements in soils by that in crustal rocks, with the resulting ratio being termed the *enrichment factor*—values less than 1 indicating loss, more than 1 indicating gains. A disconcerting artifact of this analysis is that some mobile elements, particularly silicon, commonly show enrichment factors greater than 1. Silicon is one of the major elements lost via chemical weathering, having an annual flux to the ocean of 6.1×10^{12} mol Si yr^{-1} (Tréguer *et al.*, 1995), so that there is a large net loss of the element from landscapes. The reason for the apparent enrichment is that although silicon is lost via weathering, the concentration of chemically resistant silicates (e.g., quartz) leads to a relative retention of the element. These discrepancies can be avoided by relating soil and parent material concentrations to immobile index elements such as zirconium.

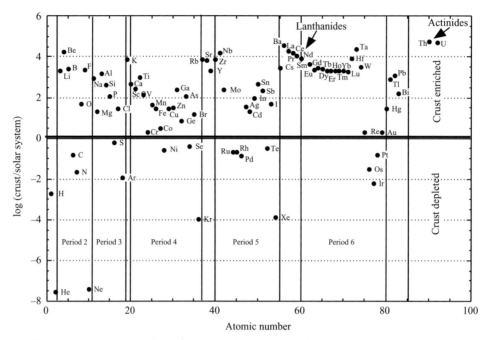

Figure 4 The log of the ratio of the average chemical concentration of elements in the Earth's upper crust (mg kg^{-1}) to that of the solar system (mg kg^{-1}). Data on the geochemistry of the upper crust from Taylor and McLennan (1985, table 2.15) with supplemental data from Bowen (1979, table 3.3). The chemistry of the solar system from Anders and Grevesse (1989).

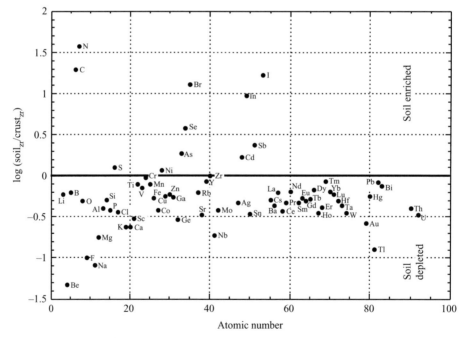

Figure 5 The log of the ratio of the mass fraction of an element in soil (relative to Zr) to that in the crust (relative to Zr). Soil data from Bowen (1979, table 4.4) and the crust data from Taylor and McLennan (1985, table 2.25) with supplemental data from Bowen (1979, table 3.3).

The present analysis uses τ, the fractional elemental enrichment factor relative to the parent material (Equation (7)). The normalization of elemental concentrations to immobile elements provides an accurate assessment of biogeochemical behavior during soil formation. Figure 5 illustrates the relative chemical composition of soil surface samples versus that of crustal rock

$(\log(\tau+1))$, where positive values indicate soil enrichment and negative values indicate soil depletion, relative to the crust. Elemental losses are due to chemical weathering, and the ultimate removal of weathering products to oceans. Elemental gains are due primarily to biological processes—the addition of elements to soils, primarily by land plants.

The comparison (soils to average continental crust) indicates that soils are: (i) particularly depleted, due to aqueous weathering losses, in the alkali metals and alkaline earths (particularly magnesium, sodium, calcium, potassium, and beryllium) and some of the halides; (ii) depleted, to a lesser degree, in silicon, iron, and aluminum; and (iii) enriched in carbon, nitrogen, and sulfur. The losses are clearly due to chemical weathering, as the chemical composition of surface waters illustrates an enrichment of these same elements relative to that of the crust (Figure 6). Plants directly assimilate elements from soil water (though they exhibit elemental selectivity across the root interface (Clarkson, 1974)), and are therefore enriched, relative to the crust, in elements derived from chemical weathering.

The key elemental addition to soils by plants is carbon, because photosynthesis greatly increases plant carbon content relative to the crust. Globally, *net primary production* (NPP) (gross photosynthetic carbon fixation–plant respiration) is ~ 60 Gt C yr^{-1}, an enormous carbon flux rate that nearly equals ocean/atmosphere carbon exchange (Sundquist, 1993). In addition to enriching the soil

in carbon, the variety of organic molecules produced during the cycling of this organic material, coupled with the CO_2 generated in the soil by the decomposition of the organic compounds by heterotrophic microorganisms, greatly accelerate rates of chemical weathering (see Chapter 6). As a result, plants are responsible not only for enrichments of soil carbon, but also for enhanced rates of chemical weathering.

Second only to carbon inputs, nitrogen fixation by both symbiotic and nonsymbiotic organisms comprises an enormous biologically driven elemental influx to soils. Biological nitrogen fixation occurs via the following reaction (Allen *et al.*, 1994):

$$N_{2(atmosphere)} + 10H^+ + nMgATP + 8e^-$$
$$= 2NH_4^+ + H_2 + nMgADP + nP_i(n \geq 16)$$

where P_i is inorganic P. The breakage of the triple bonds in N_2 is a highly energy demanding process (thus the consumption of ATP), and in nature microorganisms have developed symbiotic relations with certain host plants (particularily legumes), deriving carbon sources from the host plant, and in turn enzymatically reducing atmospheric N_2 to NH_4^+, a form which is plant available and becomes part of plant proteins.

Globally, it is estimated that, prior to extensive human activity, biological nitrogen fixation was $\sim(90-140)\times 10^{12}$ g N yr^{-1} (Vitousek *et al.*, 1997a). This rate is increasing because of the agriculturally induced nitrogen fixation. The ability to

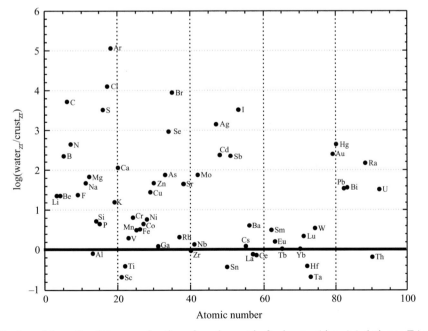

Figure 6 The log of the ratio of the mass fraction of an element in freshwater (rivers) (relative to Zr) to that of the crust (relative to Zr) as a function of atomic number. Water data from Bowen (1979, table 2.3) and crust data from Taylor and McLennan (1985, table 2.25) with supplemental data from Bowen (1979, table 3.3).

fix nitrogen is one of the most fundamental biological developments on Earth (Navarro-Gonzalez *et al.*, 2001), since nitrogen availability is one of the key limiting elements to plant growth, and hence to virtually all biogeochemical processes.

The summary of this brief discussion is that weathering losses plus plant additions characterize soil formation. This model, while capturing some important themes, neglects one of the key characteristics of soils—the distinctive and widely varying ways in which their properties vary with depth. Mass balance analyses have been applied to complete soil profiles along gradients of landform age, giving us a general perspective on the rates and directions of physical and chemical changes of soils with time. This is discussed in the next section.

7.4.2 Mass Balance of Soil Formation versus Time

7.4.2.1 *Temperate climate*

The main conclusions that can be summarized by mass balance analyses of soil formation over time in nonarid environments are that: in early phases of soil formation, the soil experiences volumetric dilation due to physical and biological processes; the later stages of soil formation are characterized by volumetric collapse caused by large chemical losses of the major elements that, given sufficient time, result in nutrient impoverishment of the landscape. The key studies that contribute to this understanding are summarized below.

On a time series of Quaternary marine terraces in northern California, Brimhall *et al.* (1992) conducted the first mass balance analysis of soil formation over geologic time spans. This analysis provided quantitative data on well-known qualitative observations of soil formation: (i) the earliest stages of soil formation (on timescales of 10^1–10^3 yr) are visually characterized by loss of sedimentary/rock structure, the accumulation of roots and organic matter, and the reduction of bulk density; and (ii) the later stages of soil development ($>10^3$ yr) are characterized by the accumulation of weathering products (iron oxides, silicate clays, and carbonates) and the loss of many products of weathering.

Figure 7 shows the trend in ε, volumetric strain (Equation (5)), over $\sim 2.40 \times 10^5$ yr. The data show the following physical changes: (i) large volumetric expansion ($\varepsilon > 0$) occurred in the young soil (Figure 7(a)); (ii) integrated expansion for the whole soil declined with age (Figure 7(b)); and (iii) the cross-over point between expansion and collapse ($\varepsilon < 0$) moved progressively toward the soil surface with increasing age (Figure 7(a)).

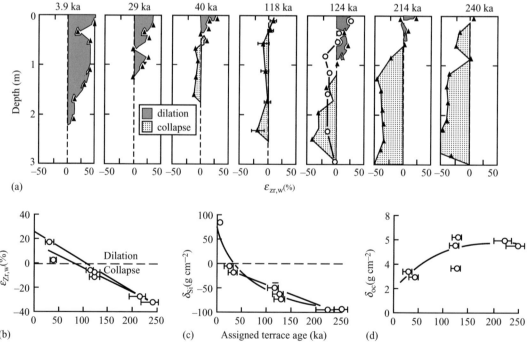

Figure 7 (a) Volumetric strain ($\varepsilon_{Zr,w}$) plotted against depth for soils on a marine terrace chronosequence on the Mendocino Coast of northern California; (b) average strain for entire profiles versus time (integrated strain to sampling depth divided by sampling depth); (c) integrated flux of Si (δ_{Si}) for entire profiles versus time; and (d) integrated flux of organic C versus time (Brimhall *et al.*, 1992) (reproduced by permission of the American Association from the Advancement of Science from *Science* **1992**, *255*, 695).

Biological processes, along with abiotic mixing mechanisms, drive the distinctive first phases of soil formation. The large positive strain (expansion) measured in the young soil on the California coast was due to an influx of silicon-rich beach sand (Figure 7(c)) and the accumulation of organic matter from plants (Figure 7(d)). In many cases, there is a positive relationship between the mass influx of carbon to soil (δ_{oc}) and strain; Jersak *et al.* (1995)). Second, in addition to adding carbon mass relative to the parent material, the plants roots (and other subterranean organisms) expand the soil, create porosity, and generally assist in both mixing and expansion. Pressures created by growing roots can reach 15 bar (Russell, 1977), providing adequate forces to expand soil material. Brimhall *et al.* (1992) conducted an elegant lab experiment showing the rapid manner in which roots can effectively mix soil, and incorporate material derived from external sources. Over several hundred "root growth cycles" using an expandable/collapsible tube in a sand mixture (Figure 8(a)), they demonstrated considerable expansion and depth of mixing (Figure 8(b)), with an almost linear relation between expansion and depth of translocation of externally added materials (Figures 8(c) and (d)).

The rate of physical mixing and volumetric expansion caused by carbon additions declines quickly with time. Soil carbon accumulation with time (Figure 7(d)) can be described by the following first-order decay model (Jenny *et al.*, 1949):

$$\frac{dC}{dt} = I - kC \qquad (9)$$

where I is plant carbon inputs (kg m^{-2} yr^{-1}), C the soil carbon storage (kg m^{-2}), and k the decay constant (yr^{-1}). Measured and modeled values of k for soil organic carbon (Jenkinson *et al.*, 1991; Raich and Schlesinger, 1992) indicate that steady state should be reached for most soils within ~10^2–10^3 yr. Thus, as rates of volumetric expansion decline, the integrated effects of mineral weathering and the leaching of silicon (Figure 7(c)), calcium, magnesium, sodium, potassium, and other elements begin to become measurable, and over time tend to eliminate the measured expansion not only near the surface (Figure 7(a)), but also for the whole profile (Figure 7(b)).

7.4.2.2 Cool tropical climate

The integrated mass losses of elements over time are affected by parent material mineralogy,

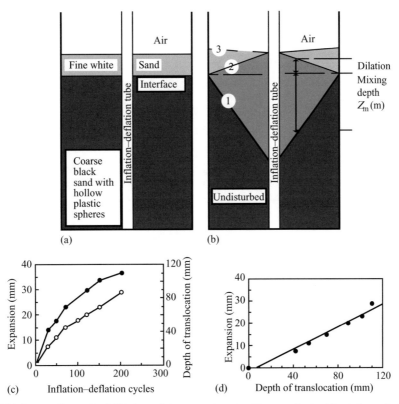

Figure 8 (a) Initial state of a cyclical dilation mixing experiment, with a surgical rubber tube embedded in a sandy matrix; (b) features after mixing: line 1 is depth of mixing after mixing, line 2 is the dilated surface, and line 3 is the top of the overlying fine sand lense; (c) expansion (o) and depth of mixing (●) as a function of mixing cycles; and (d) relationship of soil expansion to mixing depth (Brimhall *et al.*, 1992) (reproduced by permission of the American Association from the Advancement of Science from *Science* **1992**, 255, 695).

climate, topography, etc. The mass balance analysis of soil formation of the temperate California coast (Brimhall *et al.*, 1991; Chadwick *et al.*, 1990; Merritts *et al.*, 1992) is complemented by an even longer time frame on the Hawaiian Islands (Vitousek *et al.*, 1997b; Chadwick *et al.*, 1999). The Hawaiian chronosequence encompasses ~4 Myr in a relatively cool, but wet, tropical setting. Because of both steady and cyclic processes of erosion and deposition, few geomorphic surfaces in temperate settings on Earth are older than Pleistocene age. Yet the exceptions to this rule: the Hawaiian Island chronosequence, river terrace/glacial outwash sequence in

the San Joaquin Valley of California (e.g., Harden, 1987; White *et al.*, 1996), and possibly others provide glimpses into the chemical fate of the Earth's surface in the absence of geological rejuvenation.

The work by Vitousek and co-workers on Hawaii demonstrates that uninterrupted soil development on million-year timescales in those humid conditions depletes the soil in elements essential to vegetation and, ultimately, the ecosystem becomes dependent on atmospheric sources of nutrients (Chadwick *et al.*, 1999). Figures 9(a)–(g) illustrate: (i) silicon and alkali and alkaline earth metals are progressively depleted, and nearly

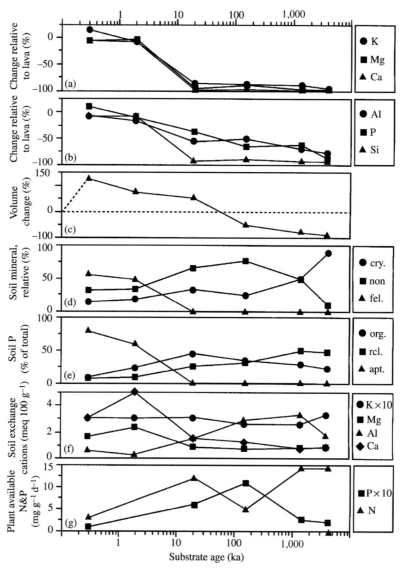

Figure 9 Weathering, mineralogical changes, and variations in plant-available elements in soils (to 1 m in depth) as a function of time in Hawaii: (a) total soil K, Mg, and Ca; (b) total soil Al, P, and Si; (c) volumetric change; (d) soil feldspar, crystalline and noncrystalline secondary minerals; (e) soil P pools (organic, recalcitrant, apatite); (f) exchangeable K, Mg, Al, and Ca; and (g) changes in resin-extractable (biologically available) inorganic N and P (Vitousek *et al.*, 1997b) (reproduced by permission of the Geological Society of America from *GSA Today*, **1997b**, *7*, 1–8).

removed from the upper 1 m; (ii) soil mineralogy shifts from primary minerals to secondary iron and aluminum oxides with time; (iii) phosphorus in primary minerals is rapidly depleted in the early stages of weathering, and the remaining phosphorus is sequestered into organic forms (available to plants through biocycling) and relatively inert oxides and hydroxides. As a result, in later stages of soil formation, the soils become phosphorus limited to plants (Vitousek *et al.*, 1997b). In contrast, in very early phases of soil formation, soils have adequate phosphorus in mineral forms, but generally lack nitrogen (Figure 9(g)) due to an inadequate time for its accumulation through a combination of nitrogen fixation (which is relatively a minor process on Hawaii; Vitousek *et al.* (1997b) and atmospheric deposition of NO_3^-, NH_4^+, and organic N (at rates of $\sim$5–50 kg N yr^{-1}; Heath and Huebert (1999)). The rate of nitrogen accumulation in soil and the model describing it generally parallel the case of organic carbon, because carbon storage hinges on the availability of nitrogen (e.g., Parton *et al.*, 1987).

In summary, soils in the early stages of their development contain most of the essential elements for plant growth with the exception of nitrogen. Soil nitrogen, like carbon, reaches maximum steady-state values in periods on the order of thousands of years and, as a result, NPP of the ecosystems reach maximum values at this stage of soil development (Figure 10). Due to progressive removal of phosphorus and other plant essential elements (particularly calcium), plant productivity declines (Figure 10), carbon inputs to the soil decline, and both soil carbon and nitrogen storage begin a slow decline (Figure 9). This trend, because of erosive rejuvenating processes, is rarely observed in climatically and tectonically active parts of the Earth. Alternatively, low latitude, tectonically stable continental regions may reflect these long-term processes. Brimhall and Dietrich (1987) and Brimhall *et al.* (1991, 1992) discuss the pervasive weathering and elemental losses from cratonal regions such as Australia and West Africa. These regions, characterized by an absence of tectonic activity and glaciation, and by warm (and sometimes humid) climates, have extensive landscapes subjected to weathering on timescales of millions of years. However, over such immense timescales, known and unknown changes in climate and other factors complicate interpretation of the soil formation processes. An emerging perspective is that these and other areas of the Earth experience atmospheric inputs of dust and dissolved components that wholly or partially compensate chemical weathering losses, ultimately creating complex soil profiles and ecosystems which subsist on the steady but slow flux of atmospherically derived elements (Kennedy *et al.*, 1998; Chadwick *et al.*, 1999).

7.4.2.3 Role of atmospheric inputs on chemically depleted landscapes

The importance of dust deposition, and its impact on soils, is not entirely a recent observation (Griffin *et al.*, 2002). Darwin complained of Saharan dust while aboard the Beagle (Darwin, 1846), and presented some discussion of its composition and the research on the phenomenon at the time. In the pedological realm, researchers in both arid regions (Peterson, 1980; Chadwick and Davis, 1990) and in humid climates recognized the impact of aerosol imputs on soil properties. With respect to the Hawaiian Islands—a remarkably isolated volcanic archipelago—Jackson *et al.* (1971) recognized the presence of prodigous quantities of quartz in the basaltic soils of Hawaii. Oxygen-isotope analyses of these quartz grains showed that the quartz was derived from continental dust from the northern hemisphere, a source now well constrained by atmospheric observations (Nakai *et al.*, 1993) and the analysis of Pacific Ocean sediment cores (Rea, 1994).

The work by Chadwick *et al.* (1999) has demonstrated that the calcium and phosphorus nutrition of the older Hawaiian Island ecosystems depends almost entirely on atmospheric sources. With respect to calcium, strontium-isotope analyses of soils and plants indicates that atmospherically derived calcium (from marine sources) increases from less than 20% to more than 80% of total plant calcium with increasing soil age. With respect to phosphorus, the use of rare earth elements and isotopes of neodymium all indicated that from $\sim$0.5 to more than 1.0 mg P m^{-2} yr^{-1} is delivered in the form of dust each year and, in the old soils, the atmospheric inputs approach 100% of the total available phosphorus at the sites. Brimhall *et al.* (1988, 1991) demonstrated that much of the zirconium in the upper part of the soils in Australia and presumably other chemical constituents are derived from atmospheric inputs. In summary, it is clear that the ultimate fate of soils in the absence of geological rejuvenation, *in humid climates*, is a subsistence on atmospheric elemental sources.

7.4.3 Mass Balance Evaluation of Soil Formation versus Climate

Hyperarid regions offer a unique view of the importance of atmospheric elemental inputs to soils, because in these areas, the inputs are not removed by leaching. The western coasts of southern Africa and America lie in hyperarid climates that have likely persisted since the Tertiary (Alpers and Brimhall, 1988). These regions, particularly the Atacama desert of Chile, are known for their commercial-grade deposits of sulfates, iodates, bromates, and particularly nitrates (Ericksen, 1981;

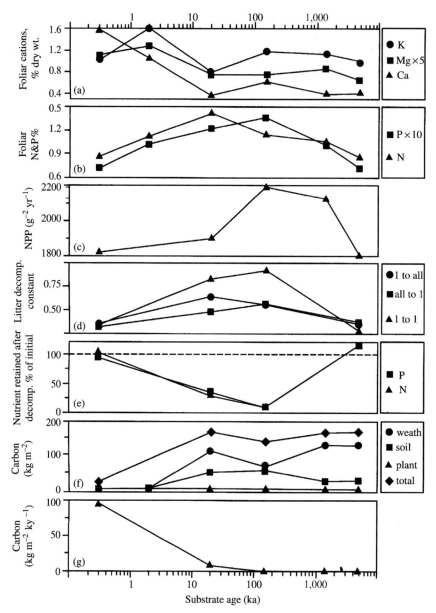

Figure 10 Plant nutrients, production and decomposition, and carbon sinks during soil development in Hawaii: (a) K, Mg, and Ca in canopy leaves of dominant tree (*Metrosideros*); (b) N and P in leaves of dominant tree; (c) changes in NPP of forests; (d) decomposition rate constant (*k*) of *metrosideros* leaves decomposed in the site collected (1-to-1), collected at each site and transferred to common site (all-to-1), and collected from one site and transferred to all sites (1-to-all); (e) fraction of N and P in original leaves remaining after 2 yr of decomposition; (f) carbon storage in the ecosystems in the form of plant biomass, soil organic matter, and as CO_2 consumed during silicate weathering; and (g) instantaneous rate of total ecosystem carbon storage (Vitousek *et al.*, 1997b) (reproduced by permission of the Geological Society of America from *GSA Today*, **1997b**, *7*, 1–8).

Böhlke *et al.*, 1997). These deposits may form due to several processes and salt sources, including deflation around playas, spring deposits (Rech *et al.*, 2003), and other settings. However, many of the deposits are simply heavily concentrated soil horizons, formed by the long accumulation of soluble constituents over long time spans (Erickson, 1981). The ultimate origin of these salts (and their

elements) is in some cases obscure, but it is clear that they arrive at the soils via the atmosphere. In Chile, sources of elements may be from fog, marine spray, reworking of playa crusts, and more general long-distance atmospheric sources (Ericksen, 1981; Böhlke *et al.*, 1997).

Recent novel research by Theimens and co-workers has convincingly demonstrated the atmospheric origin (as opposed to sea spray, playa reworking, etc.) of sulfates in Africa and Antarctica (Bao *et al.*, 2000a,b, 2001). Briefly, these researchers have shown that as sulfur undergoes chemical reactions in the stratosphere and troposphere, a mass-independent fractionation of oxygen isotopes in the sulfur oxides occurs. The mechanism for these fractionations is obscure (Thiemens, 1999), but the presence of mass-independent ratios of ^{17}O and ^{18}O in soil sulfate accumulations is a positive indicator of an atmospheric origin. In both the Nambian desert (Bao *et al.*, 2001) and Antarctica (Bao *et al.*, 2000a), there was an increase in the observed ^{17}O isotopic anomaly with distance from the ocean, possibly due to a decrease in the ratio of sea salt-derived sulfate in atmospheric sulfate.

There have been few, if any, systematic pedological studies of the soils of these hyperarid regions. Here, we present a preliminary mass balance analysis of soil formation along a precipitation gradient in the presently hyperarid region of the Atacama desert, northern Chile (Sutter *et al.*, 2002). Along a south-to-north gradient, precipitation decreases from $\sim$15 mm yr^{-1} to $\sim$2 mm yr^{-1} (http://www.worldclimate.com/worldclimate/index. htm). For the study, three sites were chosen $\sim$50 km inland on the oldest (probably Mid- to Early Pleistocene) observable fluvial landform (stream terrace or alluvial fan) in the region. Depths of observation were restricted by the presence of salt-cemented soil horizons.

Using the mass balance equations presented earlier and titanium as an immobile index element, the volumetric and major element changes with precipiation were calculated (Figure 11). The calculations for ε show progressive increases in volumetric expansion with decreasing precipitation (approaching 400% in some horizons at Yungay, the driest site) (Figure 11(a)). These measured expansions are due primarily to the accumulation and retention of NaCl and CaSO$_4$ minerals. For example, Yungay has large expansions near the surface and below 120 cm. Figure 11(c) shows that sulfur (in the form of CaSO$_4$) is responsible for the upper expansion, while chlorine (in the form of NaCl) is responsible for the lower horizon expansion. As the chemical data indicate, the type and depth of salt movement is climatically related: the mass of salt changes from (Cl, S) to (S, CaCO$_3$) to (CaCO$_3$) with increasing rainfall (north to south). Additionally, the depth of S and CaCO$_3$ accumulations increase with rainfall.

These data suggest that for the driest end-member of the transect, the virtual absence of chemical weathering (for possibly millions of years), and the pervasive input and retention of atmospherically derived chemical constituents (due to a lack of leaching), drives the long-term trajectory of these soils toward continued volumetric expansion (due to inputs) and the accretion, as opposed to the loss, of plant-essential elements. Therefore, it might be hypothesized that there is a critical water balance for soil formation (precipitation–evapotranspiration) at which the long-term accumulation of atmospherically derived elements exceeds weathering losses, and landscapes undergo continual dilation as opposed to collapse. The critical climatic cutoff point is likely to be quite arid. In the Atacama desert, the crossover point between the accretion versus the loss of soluble atmospheric inputs such as nitrate and sulfate is somewhere between 5 mm and 20 mm of precipitation per year. These Pleistocene (or older) landscapes have likely experienced changes in climate (Betancourt *et al.*, 2000; Latorre *et al.*, 2002), so the true climatic barrier to salt accumulations is unknown. Nonetheless, it is clear that the effect of climate drives strongly contrasting fates of soil formation (collapse and nutrient impoverishment versus dilation and nutrient accumulation) over geological time spans.

7.5 PROCESSES OF MATTER AND ENERGY TRANSFER IN SOILS

Chadwick *et al.* (1990) wrote that depth-oriented mass balance analyses change the study of soil formation from a "black to gray box." The "grayness" of the mass balance approach is due to the fact that it does not directly provide insight into mechanisms of mass transfer, and it does not address the transport of heat, water, and gases. The mechanistic modeling and quantification of these fluxes in field settings is truly in its infancy, but some general principles along with some notable success stories have emerged on the more mechanistic front of soil formation. In this section, the author discusses the general models that describe mass transfer in soils, and examines in some detail how these models have been successfully used to describe observed patterns in soil gas and organic matter concentrations, and their isotopic composition.

The movement of most constituents in soils can ultimately be described as variants of diffusive (or in certain cases, advective) processes. The study of these processes, and their modeling, has long been the domain of *soil physics*, an experimental branch of the soil sciences. There are several good textbooks in soil physics that provide introductions to these processes (Jury *et al.*, 1991; Hillel, 1998, 1980a,b). However, it is fair to say that the application of this work to natural soil processes, and to natural soils, has been minimal given the focus on laboratory or highly controlled field experiments. However, notable exceptions to this trend exist, exceptions initiated by biogeochemists who adapted or modified these principles to illuminate the soil "black box."

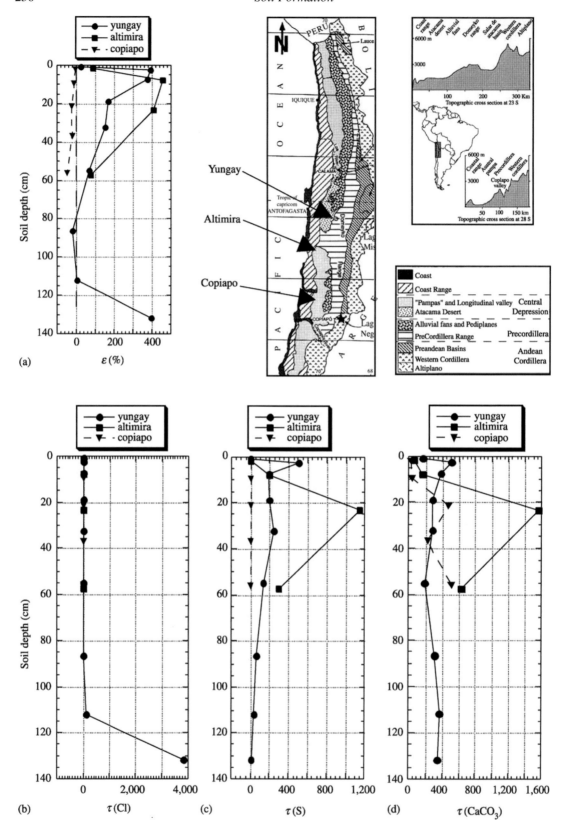

Figure 11 (a) Volumetric change along climate gradient in Atacama desert, northern Chile; (b) fractional mass gains of Cl; (c) fractional mass gains of S; and (d) fractional mass gains of $CaCO_3$ for three sites sampled along a precipitation gradient (Sutter *et al.*, 2002).

There are few, if any, cases where these various models have been fully coupled to provide an integrated view of soil formation. However, one group of soil processes that has been extensively studied and modeled comprise the soil carbon cycle. We review the mechanisms of this cycle and the modeling approaches that have received reasonably wide acceptance in describing the processes.

7.5.1 Mechanistic Modeling of the Organic and Inorganic Carbon Cycle in Soils

The processing of carbon in soils has long received attention due to its importance to agriculture (in the form of organic matter) and the marked visual impact it imparts to soil profiles. A schematic perspective of the flow of carbon through soils is given in Figure 12. Carbon is fixed from atmospheric CO_2 by plants, enters soil in organic forms, undergoes decomposition, and is cycled back to the atmosphere as CO_2. In semi-arid to arid regions, a fraction of the CO_2 may ultimately become locked in pedogenic $CaCO_3$, whereas in humid regions a portion may be leached out as dissolved organic and inorganic carbon with groundwater. On hillslopes, a portion of the organic carbon may be removed by erosion (Stallard, 1998). Most studies of the organic part of the soil carbon assume the latter three mechanisms to be of minor importance (to be discussed more fully below) and consider the respiratory loss of CO_2 as the main avenue of soil carbon loss.

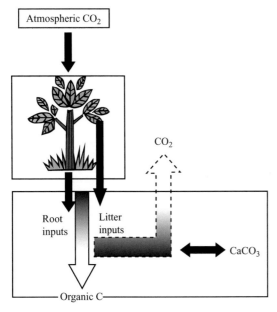

Figure 12 Schematic diagram illustrating C flow through terrestrial ecosystems.

Jenny *et al.* (1949) were among the first to apply a mathematical framework to the soil carbon cycle. For the organic layer at the surface of forested soils, Jenny applied, solved, and evaluated the mass balance model given by Equation (9) discussed earlier. This approach can be applied to the soil organic carbon pool as a whole. This has proved useful in evaluating the response of soil carbon to climate and environmental change (Jenkinson *et al.*, 1991).

The deficiency of the above model of soil carbon is that it ignores the interesting and important variations in soil carbon content with soil depth. The depth variations of organic carbon in soils vary widely, suggesting a complex set of processes that vary from one environment to another. Figure 13 illustrates just three commonly observed soil carbon (and nitrogen) trends with depth: (i) exponentially declining carbon with depth (common in grassland soils or Mollisols); (ii) randomly varying carbon with depth (common on young fluvial deposits or, as here, in highly stratified desert soils or Aridisols); and (iii) a subsurface accumulation of carbon below a thick plant litter layer on the soil surface (sandy, northern forest soils or Spodosols). The mechanisms controlling these distributions will be discussed, in reverse order, culminating with a discussion of transport models used to describe the distribution of carbon in grasslands.

Spodosols are one of the 12 soil orders in the USDA Soil Taxonomy. The soil orders, and their key properties, are listed in Table 4. The distribution of the soil orders in the world is illustrated in Figure 14. With respect to Spodosols, which are common to the NE USA, Canada, Scandinavia, and Russia, the key characteristics that lead to their formation are: sandy or coarse sediment (commonly glacial outwash or till), deciduous or coniferous forest cover, and humid, cool to cold, climates. Organic matter added by leaf and branch litter at the surface accumulates to a steady-state thickness of organic horizons (Figure 13). During the decomposition of this surface material, soluble organic molecules are released which move downward with water. The organic molecules have reactive functional groups which complex with iron and aluminum, stripping the upper layers of the mineral soils that contain these elements and leaving a white, bleached layer devoid, or depleted in iron, aluminum, and organic matter (De Coninck, 1980; Ugolini and Dahlgren, 1987). As the organics move downward, they become saturated with respect to their metal constituents, and precipitate from solution, forming an organo/metal-rich subsurface set of horizons. In many of these forests, tree roots are primarily concentrated in the organic layers above the mineral soil, so that addition of carbon from roots is a minor input of carbon to these systems.

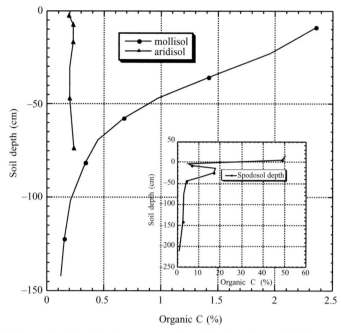

Figure 13   Soil organic C depth in three contrasting soil orders (source Soil Survey Staff, 1975).

Table 4   The soil orders of the US Soil Taxonomy and a brief definition of their characteristics.

Order	Characteristics
Alfisols	Soils possessing Bt horizons with >35% base saturation; commonly Pleistocene aged
Andisols	Soils possessing properties derived from weathered volcanic ash, such as low bulk density, high extractable Al, and high P retention
Aridisols	Soils of arid climates that possess some degree of subsurface horizon development; commonly Pleistocene aged
Entisols	Soils lacking subsurface horizon development due to young age, resistant parent materials, or high rates of erosion
Gelisols	Soils possessing permafrost and/or evidence of cryoturbation
Histosols	Soils dominated by organic matter in 50% or more of profile
Inceptisols	Soils exhibiting "incipient" stages of subsurface horizon development due to age, topographic position, etc.
Mollisols	Soils possessing a relatively dark and high C and base saturation surface horizon; commonly occur in grasslands
Oxisols	Soils possessing highly weathered profiles characterized by low-cation-exchange-capacity clays (kaolinite, gibbsite, etc.), few remaining weatherable minerals, and high clay; most common on stable, tropical landforms
Spodosols	Soils of northern temperate forests characterized by intense (but commonly shallow) biogeochemical downward transport of humic compounds, Fe, and Al; commonly Holocene aged
Ultisols	Soils possessing Bt horizons with <35% base saturation; commonly Pleistocene aged
Vertisols	Soils composed of >35% expandable clay, possessing evidence of shrink/swell in form of cracks, structure, or surface topography

After Soil Survey Staff (1999).

In sharp contrast to the northern forests, stratified, gravelly desert soils sometimes exhibit almost random variations in carbon with depth. In these environments, surface accumulation of plant litter is negligible (except directly under shrubs) due to low plant production, wind, and high temperatures which accelerate decomposition when water is available. In these soils, where plant roots are extensively distributed both horizontally and vertically to capture water, it appears that an important process controlling soil carbon distribution is the direct input of carbon from decaying plant roots or root exudates. In addition, the general lack of water movement through the soils

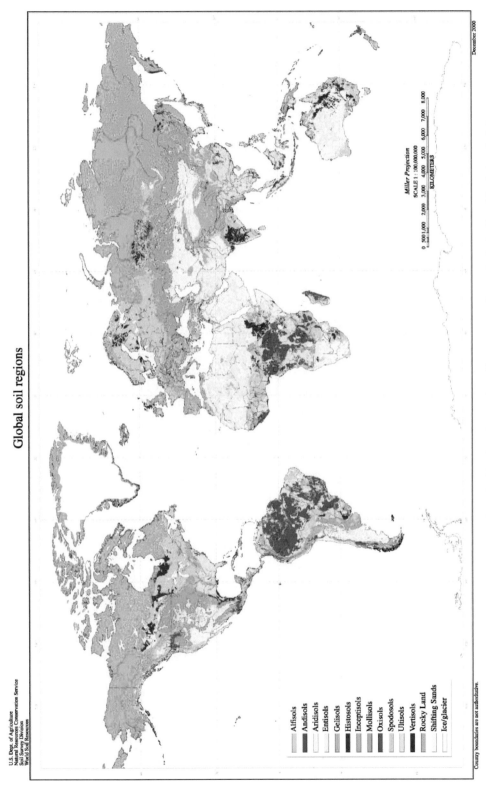

Figure 14 The global distribution of soil orders (source http://www.nrcs.usda.gov/technical/worldsoils/mapindx/).

inhibits vertical transport of carbon, and root-derived organic matter is expected to remain near the sites of emplacement.

Much work has been devoted to examining the role of carbon movement in the formation of Spodosols, but the modeling of the organic carbon flux, with some exceptions (e.g., Hoosbeek and Bryant, 1995), in these or other soils is arguably not as developed as it is for soils showing a steadily declining carbon content with depth that is found in the grassland soils of the world. In grassland soils, the common occurrence of relatively unstratified Holocene sediments, and continuous grassland cover, provides the setting for soil carbon fluxes dependent strongly on both root inputs and subsequent organic matter transport. Possibly the first study to attempt to model these processes to match both total carbon distribution and its ^{14}C content was by O'Brien and Stout (1978). Variations and substantial extensions of this work have been developed by others (Elzein and Balesdent, 1995). To illustrate the approach, the author follows the work of Baisden *et al.* (2002).

7.5.1.1　Modeling carbon movement into soils

The soil carbon mass balance is hypothesized to be, for grassland soils, a function of plant inputs (both surface and root), transport, and decomposition:

$$\frac{dC}{dt} = \underbrace{-v\frac{dC}{dz}}_{\substack{\text{downward}\\\text{advective}\\\text{transport}}} - \underbrace{kC}_{\text{decomposition}} + \underbrace{\frac{F}{L}e^{-z/L}}_{\substack{\text{plant inputs}\\\text{distributed}\\\text{exponentially}}} \quad (10)$$

where $-v$ is the advection rate (cm yr^{-1}), z the soil depth (cm), F the total plant carbon inputs (g cm^{-2}yr^{-1}), and L the e-folding depth (cm).

For the boundary conditions that $C = 0$ at $z = \infty$ and $-v(dC/dz) = F_A$ at $z = 0$ (where F_A are aboveground and F_B the belowground plant carbon inputs), the steady-state solution is

$$C(z) = \underbrace{\frac{F_A}{v}e^{-kz/v}}_{\substack{\text{above--ground}\\\text{input/transport}}} + \underbrace{\frac{F_B}{kL-v}e^{-kz/v}(e^{z(kL-v)/vL}-1}_{\text{root input/transport}}$$

$$(11)$$

This model forms the framework for examining soil carbon distribution with depth. It contains numerous simplifications of soil processes such as steady state, constant advection and decomposition rates versus depth, and the assumption of one soil carbon pool. Recent research on soil carbon cycling, particularly using ^{14}C, has revealed that soil carbon consists of multiple pools of differing residence times (Trumbore, 2000). Therefore, in modeling grassland soils in California, Baisden *et al.* (2002) modified the soil carbon model above by developing linked mass balance models for three carbon pools of increasing residence time. Estimates of carbon input parameters came from direct surface and root production measurements. Estimates of transport velocities came from ^{14}C measurements of soil carbon versus depth, and other parameters were estimated by iterative processes. The result of this effort for a ~2×10^5 yr old soil (granitic alluvium) in the San Joaquin Valley of California is illustrated in Figure 15. The goodness of fit suggests that the model captures at least the key processes distributing carbon in this soil. Model fitting to observed data became more difficult in older soils with dense or cemented soil horizons, presumably due to changes in transport velocities versus depth (Baisden *et al.*, 2002).

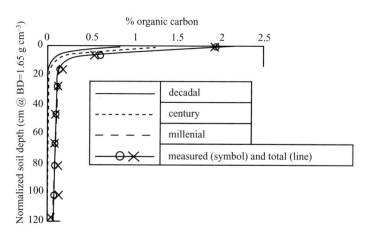

Figure 15 Measured total organic C versus depth and modeled amounts of three fractions of differing residence time (~10^0 yr, 10^2 yr, and 10^3 yr) for a ~600 ka soil formed on granitic alluvium in the San Jaoquin Valley of California (source Baisden, 2000).

7.5.1.2 *Modeling carbon movement out of soils*

The example above illustrates that long-observed soil characteristics are amenable to analytical or numerical modeling, and it illustrates the importance of transport in the vertical distribution of soil properties, in this case organic carbon. As the model and observations indicate, the primary pathway for carbon loss from soil is the production of CO_2 from the decomposition of the organic carbon. It has long been recognized that CO_2 leaves the soil via diffusion, and various forms of Fick's law have been applied to describing the transport of CO_2 and other gases in soils (e.g., Jury *et al.*, 1991; Hillel, 1980b). However, the application of these models to natural processes and to the issue of stable isotopes in the soil gases is largely attributable to the work of Thorstenson *et al.* (1983) and, in particular, of Cerling (1984). Cerling's (1984) main interest was in describing the carbon-isotope composition of soil CO_2, which he recognized ultimately controls the isotope composition of pedogenic carbonate. We begin with the model describing total CO_2 diffusion, then follow that with the extension of the model to soil carbon isotopes.

Measurements of soil CO_2 concentrations versus depth commonly reveal an increase in CO_2 content with depth. The profiles and the maximum CO_2 levels found at a given depth are climatically controlled (Amundson and Davidson, 1990) due to rates of C inputs from plants, decomposition rates, etc. Given that most plant roots and soil C are concentrated near the surface, the production rates of CO_2 would be expected to decline with depth. Cerling developed a production/diffusion model to describe steady-state soil CO_2 concentrations:

$$\varepsilon\frac{dCO_2}{dt} = 0 = \underbrace{D\frac{d^2CO_2}{dz^2}}_{\text{net diffusion}} + \underbrace{\phi}_{\substack{\text{biological}\\\text{production}}} \quad (12)$$

where ε is free air porosity in soil, CO_2 the concentration of CO_2 (mol cm^{-3}), and D the effective diffusion coefficient of CO_2 in soil (cm^2 s^{-1}), z the soil depth (cm), and $\phi = CO_2$ production (mol cm^{-3} s^{-1}). Using reported soil respiration rates, and reasonable parameter values for Equation (11), the CO_2 concentration profiles for three strongly contrasting ecosystems are illustrated in Figure 16.

For the boundary conditions of an impermeable lower layer and $CO_2(0) = CO_2(\text{atm})$, the solution to the model (with exponentially decreasing CO_2 production with depth) is

$$CO_2(z) = \frac{\phi_z = 0 z_0^2}{D}(1 - e^{-z/z_0}) + C_{\text{atm}} \quad (13)$$

where z_0 is the depth at which the production is $\phi_{z=0}/e$. At steady state, the flux of CO_2 from the

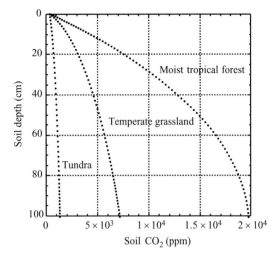

Figure 16 Calculated soil CO_2 concentrations versus depth for three contrasting ecosystems using Equation (13) and the following data: total soil respiration rates (tundra = 60 g C m^{-2}yr^{-1}, grassland = 442 g C m^{-2} yr^{-1}, and tropical forest = 1,260 g C m^{-2}yr^{-1}), $D_s = 0.021$ cm^{2}s^{-1}, atmospheric $CO_2 = 350$ ppm, and $L = 100$ cm (respiration data from Raich and Schlesinger, 1992).

soil to the atmosphere is simply the first derivative of Equation (13) evaluated at $z = 0$ (e.g., Amundson *et al.*, 1998). The production, and transport of CO_2, is accompanied by the consumption and downward transport of O_2, which is driven and described by analogous processes and models (e.g., Severinghaus *et al.*, 1996).

Cerling's primary objective was the identification of the processes controlling the carbon-isotope composition of soil CO_2, and a quantitative means of describing the process. In terms of notation, carbon isotopes in compounds are evaluated as the ratio (R) of the rare to common stable isotope of carbon (^{13}C/^{12}C) and are reported in delta notation: $\delta^{13}\text{C}(‰) = (R_s/R_{\text{std}} - 1)1,000$, where R_s and R_{std} refer to the carbon-isotope ratios of the sample and the international standard, respectively (Friedman and O'Neil, 1977).

In terms of the controls on the isotopic composition of soil CO_2, the ultimate source of the carbon is atmospheric CO_2, which has a relatively steady δ^{13}C value of about $-7‰$ (a value which has been drifting recently toward more negative values due to the addition of fossil fuel CO_2 (e.g., Mook *et al.*, 1983)). The isotopic composition of atmospheric CO_2 is also subject to relatively large temporal changes due to other changes in the carbon cycle, such as methane hydrate releases, etc. (Jahren *et al.*, 2001; Koch *et al.*, 1995). Regardless of the isotopic value, atmospheric CO_2 is utilized by plants through photosynthesis. As a result of evolutionary processes, three photosynthetic pathways have evolved in land plants: (i) C$_3$: δ^{13}C = $\sim -27‰$, (ii) C$_4$: $\sim -12‰$, and

(iii) CAM: isotopically intermediate between C_3 and C_4, though it is commonly close to C_3. It is believed that C_3 photosynthesis is an ancient mechanism, whereas C_4 photosynthesis (mainly restricted to tropical grasses) is a Cenozoic adaptation to decreasing CO_2 levels (Cerling *et al.*, 1997) or to strong seasonality and water stress (e.g., Farquhar *et al.*, 1988). Once atmospheric carbon is fixed by photosynthesis, it may eventually be added to soil as dead organic matter through surface litter, root detritus, or as soluble organics secreted by living roots. This material is then subjected to microbial decomposition (see Volume 8), and partially converted to CO_2 which then diffuses back to the overlying atmosphere. An additional source of CO_2 production is the direct respiration of living roots, which is believed to account for ~50% of the total CO_2 flux out of soils (i.e., soil respiration; Hanson *et al.* (2000)). During decomposition of organic matter, there is a small (~2‰ or more) discrimination of carbon isotopes, whereby ^{12}C is preferentially lost as CO_2 and ^{13}C remains as humic substances (Nadelhoffer and Fry, 1988). Thus, soil organic matter commonly shows an enrichment (which increases with depth due to transport processes) of ^{13}C relative to the source plants (see Amundson and Baisden (2000) for an expanded discussion of soil organic carbon and nitrogen isotopes and modeling.

Cerling recognized that $^{12}CO_2$ and $^{13}CO_2$ can be described in terms of their own production and tranport models, and that the isotope ratio of CO_2 at any soil depth is described simply by the ratio of the ^{13}C and the ^{12}C models. For the purposes of illustration here, if we assume that the concentration of ^{12}C can be adequately described by that of total CO_2 and that CO_2 is produced at a constant rate over a given depth L, then the model describing the steady-state isotopic ratio of CO_2 at depth z is (see Cerling and Quade (1993) for the solution where the above assumptions are not applied)

$$R_s^{13} = \frac{\left(\phi R_p^{13}/D_s^{13}\right)(Lz - z^2/2) + C_{atm}R_{atm}^{13}}{(\phi/D_s)(Lz - z^2/2) + C_{atm}} \quad (14)$$

where R_s, R_p, and R_{atm} refer to the isotopic ratios of soil CO_2, plant carbon, and atmospheric CO_2, respectively, and D^{13} is the diffusion coefficient of $^{13}CO_2$ which is $D_s/1.0044$. The $\delta^{13}C$ value of the CO_2 can be calculated by inserting R_s into the equation given above.

Quade *et al.* (1989a) examined the $\delta^{13}C$ value of soil CO_2 and pedogenic carbonate along elevation (climate) gradients in the Mojave Desert/ Great Basin and utilized the models described above to analyze the data. Quade *et al.* found that there were systematic trends in soil CO_2 concentrations, and $\delta^{13}C$ values, due to changes in CO_2 production rates with increasing elevation, and

soil depth (Figure 17). The primary achievement of this work was that the observations were fully explainable using the mechanistic model represented here by Equation (13), a result that opened the door for the use of pedogenic carbonates in paleoenvironmental (e.g., Quade *et al.*, 1989b) and atmospheric p_{CO_2} (Cerling, 1991) studies.

7.5.1.3 *Processes and isotope composition of pedogenic carbonate formation*

In arid and semi-arid regions of the world, where precipitation is exceeded by potential evapotranspiration, soils are incompletely leached and $CaCO_3$ accumulates in significant quantities. Figure 18 illustrates the global distribution of carbonate in the upper meter of soils. As the figure illustrates, there is a sharp boundary between calcareous and noncalcareous soil in the USA at about the 100th meridian. This long-recognized boundary reflects the soil water balance. Jenny and Leonard (1939) examined the depth to the top of the carbonate-bearing layer in soils by establishing a climosequence (precipitation gradient) along an east to west transect of the Great Plains (Figure 19). They observed that at constant mean average temperature (MAT) below 100 cm of mean average precipitation (MAP), carbonate appeared in the soils, and the depth to the top of the carbonate layer decreased with decreasing precipitation. An analysis has been made of the depth to carbonate versus precipitation relation for the entire USA (Royer, 1999). She found that, in general, the relation exists broadly but as the control on other variables between sites (temperature, soil texture, etc.) is relaxed, the strength of the relationship declines greatly.

In addition to the depth versus climate trend, there is a predictable and repeatable trend of carbonate amount and morphology with time (Gile *et al.* (1966); Figure 20) due to the progressive accumulation of carbonate over time, and the ultimate infilling of soil porosity with carbonate cement, which restricts further downward movement of water and carbonate.

The controls underlying the depth and amount of soil carbonate hinge on the water balance, Ca^{+2} availability, soil CO_2 partial pressures, etc. Arkley (1963) was the first to characterize these processes mathematically. His work has been greatly expanded by McFadden and Tinsley (1985), Marian *et al.* (1985), and others to include numerical models. Figure 21 illustrates the general concepts of McFadden and Tinsley's numerical model, and Figure 22 illustrates the results of model predictions for a hot, semi-arid soil (see the figure heading for model parameter values). These predictions generally mimic observations of carbonate distribution in desert soils, indicating that many of the key processes have been identified.

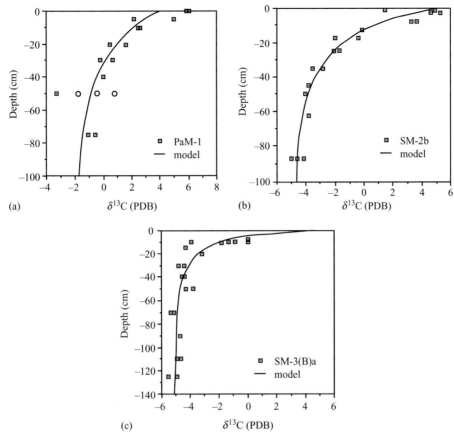

Figure 17 Measured (points) and modeled (curves) soil carbonate $\delta^{13}C$ values for three soils along an elevation (climate) gradient in the Mojave desert/Great Basin of California and Nevada. Modeled carbonate values based on Equation (14) plus the fractionation between CO_2 and carbonate ($\sim 10‰$). The elevations (and modeled soil respiration rates that drive the curve fit) are: (a) 330 m (0.18 mmol CO_2 m^{-2} h^{-1}); (b) 1,550 m (0.4 mmol CO_2 m^{-2} h^{-1}); and (c) 1,900 m (1.3 mmol CO_2 m^{-2} h^{-1}). The $\delta^{13}C$ value of soil organic matter (CO_2 source) was about $-21‰$ at all sites. Note that depth of atmospheric CO_2 isotopic signal decreases with increasing elevation and biological CO_2 production (Quade *et al.*, 1989a) (reproduced by permission of Geological Society of America from *Geol. Soc. Am. Bull.* **1989**, *101*, 464–475).

The general equation describing the formation of carbonate in soils is illustrated by the reaction

$$CO_2 + H_2O + Ca^{+2} = CaCO_3 + 2H^+$$

From an isotopic perspective, in unsaturated soils, soil CO_2 represents an infinite reservoir of carbon and soil water an infinite reservoir of oxygen, and the $\delta^{13}C$ and $\delta^{18}O$ values of the pedogenic carbonate (regardless of whether its calcium is derived from silicate weathering, atmospheric sources, or limestone) are entirely set by the isotopic composition of soil CO_2 and H_2O. Here we focus mainly on the carbon isotopes. However, briefly for completeness, we outline the oxygen-isotope processes in soils. The source of soil H_2O is precipitation, whose oxygen-isotope composition is controlled by a complex set of physical processes (Hendricks *et al.*, 2000), but which commonly shows a positive correlation with MAT

(Rozanski *et al.*, 1993). Once this water enters the soil, it is subject to transpirational (largely unfractionating) and evaporative (highly fractionating) losses. Barnes and Allison (1983) presented an evaporative soil water model that consists of processes for an: (i) upper, vapor transport zone and (ii) a liquid water zone with an upper evaporating front. The model describes the complex variations observed in soil water versus depth following periods of extensive evaporation (Barnes and Allison, 1983; Stern *et al.*, 1999). Pedogenic carbonate that forms in soils generally mirrors these soil water patterns (e.g., Cerling and Quade, 1993). In general, in all but hyperarid, poorly vegetated sites (where the evaporation/transpiration ratio is high), soil $CaCO_3$ $\delta^{18}O$ values roughly reflect those of precipitation (Amundson *et al.*, 1996).

The carbon-isotope model and its variants have, it is fair to say, revolutionized the use of soils and

Soil inorganic carbon

U.S. Department of Agriculture
Natural Resources Conservation Service
Soil Survey Division
World Soil Resources

Washington D.C. 2000

SIC kg m^{-2}

0
0 – 4
4 – 8
8 – 16
16 – 32
32 – 64
64 – 128
> 128

Miller Projection
SCALE 1 : 100,000,000

0   500 1,000   2,000   3,000   4,000   5,000   6,000   7,000   8,000
KILOMETERS

Figure 18   Global distribution of pedogenic carbonate (source http://www.nrcs.usda.gov/technical/worldsoils/mapindx/).

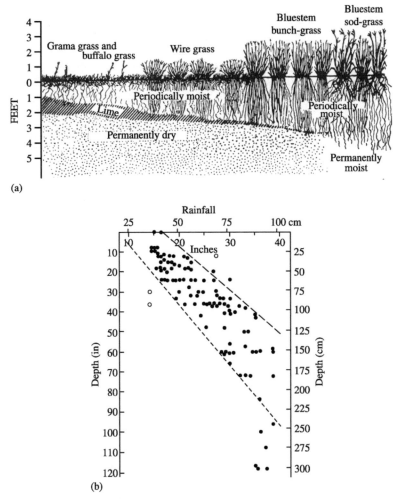

(a)

(b)

Figure 19 (a) Schematic view of plant and soil profile changes on an east (right) to west (left) transect of the Great Plains and (b) measured depth to top of pedogenic carbonate along the same gradient (source Jenny, 1941).

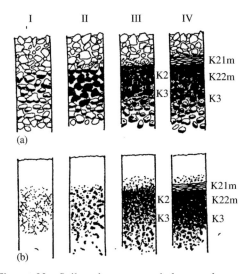

(a)

(b)

Figure 20 Soil carbonate morphology and amount versus time for: (a) gravelly and (b) fine-grained soils (Gile *et al.*, 1966) (reproduced by permission of Williams and Wilkins from *Soil Sci.* **1966**, *101*, 347–360).

paleosols in paleobotany and climatology. Some of the major achievements and uses of the model include the following.

- The model adequately describes the observed increases in the $\delta^{13}C$ value of both soil CO_2 and $CaCO_3$ with depth (Figure 17).
- The model clearly provides a mechanistic understanding of why soil CO_2 is enriched in ^{13}C relative to plant inputs (steady-state diffusional enrichment of ^{13}C).
- The model indicates that for reasonable rates of CO_2 production in soils, the $\delta^{13}C$ value of soil CO_2 should, at a depth within 100 cm of the surface, represent the $\delta^{13}C$ value of the standing biomass plus 4.4‰. The $\delta^{13}C$ of $CaCO_3$ will also reflect this value, plus an equilibrium fractionation of ~10‰ (depending on temperature). Therefore, if paleosols are sampled below the "atmospheric mixing zone," whose thickness depends on CO_2 production rates

Atmosphere

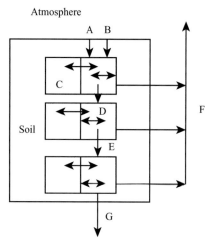

Figure 21 Diagram of compartment model for the numerical simulation of calcic soil development. Compartments on left represent solid phases and compartments on right represent aqueous phases. Line A represents precipitation, line B represents dust influx, line C represents the transfer of components due to dissolution and precipitation, line D represents transfer between aqueous phases, line E represents downward movement of solutes due to gravitational flow of soil water, line F represents evapotranspirational water loss, and line G represents leaching losses of solutes (McFadden *et al.*, 1991) (reproduced by permission of Soil Science Society of America from *Occurrence, Characteristics, and Genesis of Carbonate, Gypsum and Silica Accumulations in Soils*, **1991**).

(Figure 17), the $\delta^{13}C$ value of the carbonate will provide a guide to the past vegetation (Cerling *et al.*, 1989).

Cerling (1991) recognized that Equation (14), if rearranged and solved for C_{atm}, could provide a means of utilizing paleosol carbonates for reconstructing past atmospheric CO_2 partial pressures. To do so, the values of the other variables (including the $\delta^{13}C$ value of the atmospheric CO_2—see Jahren *et al.* (2001)) must be known, which for soils of the distant past is not necessarily a trivial problem. Nonetheless, an active research field has developed using this method, and a compilation of calculated atmospheric CO_2 levels is emerging (Ekart *et al.*,1999; Mora *et al.*, 1996), with estimates that correlate well with model calculations by Berner (1992).

Cerling's (1984) approach to modeling stable carbon isotopes in soil CO_2 has been expanded and adapted to other isotopes in soil CO_2: (i) $^{14}CO_2$—for pedogenic carbonate dating (Wang *et al.*, 1994; Amundson *et al.*, 1994a,b) and soil carbon turnover studies (Wang *et al.*, 2000) and (ii) $C^{18}O^{16}O$—for hydrological tracer applications and, more importantly, as a means to constrain the controls on global atmospheric CO_2–^{18}O budgets (Hesterberg and Siegenthaler, 1991; Amundson *et al.*, 1998; Stern *et al.*, 1999, 2001; Tans, 1998). The processes controlling the isotopes, and the complexity of the

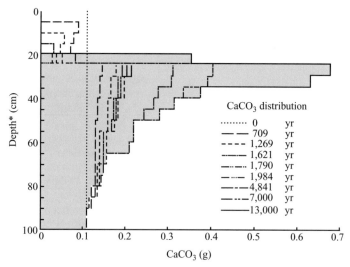

Figure 22 Predicted pattern of Holocene pedogenic carbonate accumulation in a cm² column in a semi-arid, thermic climate (leaching index = 3.5 cm). External carbonate flux rate = 1.5×10^{-4} g cm⁻² yr⁻¹, p_{CO_2} = $1.5 \times 10^{-3.3}$ atm in compartment 1 increasing to $10^{-2.5}$ atm in compartment 5 (20–25 cm). Below compartment 5, the p_{CO_2} decreases to a minimum value of 10^{-4} atm in compartment 20 (95–100 cm). Dotted line shows carbonate distribution at $t = 0$. Gray area indicates final simulated distribution. Depth* = absolute infiltration depth in <2 mm fraction (McFadden *et al.*, 1991) (reproduced by permission of Soil Science Society of America from *Occurrence, Characteristics, and Genesis of Carbonate, Gypsum and Silica Accumulations in Soils*, **1991**).

models, greatly increase from ^{14}C to ^{18}O (see Amundson *et al.* (1998) for a detailed account of all soil CO_2 isotopic models).

7.5.2 Lateral Transport of Soil Material by Erosion

The previous section, devoted to the soil carbon cycle, emphasized the conceptual and mathematical focus on the vertical transport of materials. Models emphasizing vertical transport dominate present pedological modeling efforts. However, for soils on hillslopes or basins and floodplains, the lateral transport of soil material via erosive processes exerts an overwhelming control on a variety of soil properties. The study of the mechanisms by which soil is physically moved—even on level terrain—is an evolving aspect of pedology. Here we focus on new developments in the linkage between geomorphology and pedology in quantifying the effect of sediment transport on soil formation. The focus here is on natural, undisturbed landscapes as opposed to agricultural landscapes, where different erosion models may be equally or more important.

This section begins by considering soils on divergent, or convex, portions of the landscape. On divergent hillslopes, slope increases with distance downslope such that (for a two-dimensional view of a hillslope)

$$\frac{\Delta \text{slope}}{\Delta \text{distance}} = \left(\frac{\partial}{\partial x}\right)\left(\frac{\partial z}{\partial x}\right)$$
$$= \text{negative quantity} \qquad (15)$$

where z and x are distances in a vertical and horizontal direction, respectively. On convex slopes, there is an ongoing removal of soil material due to transport processes. In portions of the landscape where the derivative of slope versus distance is positive (i.e., convergent landscapes), there is a net accumulation of sediment, which will be discussed next.

In many divergent landscapes (those not subject to overland flow or landslides), the ongoing soil movement may be almost imperceptible on human timescales. Research by geomorphologists (and some early naturalists such as Darwin) has shed light on both the rate and mechanisms of this process. In general, a combination of physical and biological processes, aided by gravity, can drive the downslope movement of soil material. These processes are sometimes viewed as roughly diffusive in that soil is randomly moved in many directions, but if a slope gradient is present, the net transport is down the slope. Kirkby (1971), in laboratory experiments, showed that wetting and drying cycles caused soil sensors to move in all directions,

but in a net downslope direction. Black and Montgomery (1991) examined pocket gopher activity on sloping landscapes in California, discussing their underground burrowing and transport mechanisms, and the rate at which upthrown material then moves downslope. Tree throw also contributes to diffusive-like movement (Johnson *et al.*, 1987). In a truly far-sighted study, Darwin (1881) quantified both the mass and the volume of soil thrown up by earthworms, and the net amount moved downslope by wind, water, and gravity.

The general transport model for diffusive-like soil transport is (Kirkby, 1971)

$$Q = -\rho_s K \frac{dz}{dx} \qquad (16)$$

where Q is soil flux per unit length of a contour line (ρ_s) (g cm^{-1}yr^{-1}) and K is transport coefficient (cm^2 yr^{-1}). The concepts behind this expression are generally attributed to Davis (1892) and Gilbert (1909). However, Darwin (1881) clearly recognized this principle in relation to sediment transport by earthworms on slopes. For several sites in England (based on a few short-term measurements), Darwin reported Q and slope, allowing us to calculate a K based solely on earthworms. The values of ~0.02 cm^2 yr^{-1}, while significant, are about two orders of magnitude lower than overall K values calculated for parts of the western USA that capture the integrated biological and abiotic mixing/transport mechanisms (McKean *et al.*, 1993; Heimsath *et al.*, 1999).

At this point, it is worth commenting further on the effect and role of bioturbation in soil formation processes whether it results in net downslope movement or not. Beginning with Darwin, there has been a continued, but largely unappreciated, series of studies concerning the effects of biological mixing on soil profile features (Johnson *et al.*, 1987; Johnson, 1990). These studies concluded that most soils (those maintaining at least some plant biomass to support a food chain) have a "biomantle" (a highly mixed and sorted zone) that encompasses the upper portions of the soil profile (Johnson, 1990). The agents are varied, but include the commonly distributed earthworms, gophers, ground squirrels, ants, termites, wombats, etc. In many cases, the biomantle is equivalent to the soil's A horizons. The rapidity at which these agents can mix and sort the biomantle is impressive. The invasion of earthworms into the Canadian prairie resulted in complete homogenization of the upper 10 cm in 3 yr (Langmaid, 1964). Johnson (1990) reports that Borst (1968) estimated that the upper 75 cm of soils in the southern San Joaquin Valley of California are mixed in 360 yr by ground squirrels. Darwin reported that in England, ~1.05×10^4 kg of soil per hectare are passed through earthworms yearly, resulting in a turnover time for the upper 50 cm of the soil of just over 700 yr. The deep burial of Roman structures and artifacts in England by earthworm casts supports this estimated cycling rate.

Bioturbation on level ground may involve a large gross flux of materials (as indicated above), but no net movement in any direction due to the absence of a slope gradient. Nonetheless, the mixing has important physical and chemical implications for soil development: (i) rapid mixing of soil and loss of stratification at initial stages of soil formation; (ii) rapid incorporation of organic matter (and the subsequent slowed decomposition of this material below ground); (iii) periodic cycling of soil structure which prevents soil horizonation within the biomantle; and (iv) in certain locations, striking surficial expressions of biosediment movement caused by nonrandom sediment transport. The famed "mima mounds" of the Plio-Pleistocene fluvial terraces and fans of California's Great Valley (Figure 23(a)) are a series of well-drained sandy loam to clay loam materials that overlie a relatively level, impermeable layer (either a dense, clay-rich horizon or a silica cemented hardpan) (Figure 23(b)). The seasonally water-logged conditions that develop over the impermeable layers have probably caused gophers to move soil preferentially into better drained landscape segments, thereby producing this unusual landscape. These landform features are now relatively rare due to the expansion of agriculture in the state, but the original extent of this surface feature is believed to have been more than 3×10^5 ha (Holland, 1996).

The conceptual model for diffusive soil transport down a hillslope is shown in Figure 24 (Heimsath *et al.*, 1997, 1999). In any given section of the landscape, the mass of soil present is the balance of transport in, transport out, and soil production (the conversion of rock or sediment to soil). If it is assumed that the processes have been operating for a sufficiently long period of time, then the soil thickness is at steady state. The model describing this condition is

$$\rho_s \frac{\partial h}{\partial t} = 0 = \underbrace{\rho_r \phi}_{\text{soil production}} + \underbrace{\left(K\rho_s \frac{\partial^2 z}{\partial x^2} \right)}_{\substack{\text{balance between} \\ \text{diffusive inputs/losses}}} \quad (17)$$

where h is the soil thickness (cm), ρ_s and ρ_r the soil and rock bulk density (g cm^{-3}), respectively, and ϕ the soil production rate (cm yr^{-1}).

Soil production is a function of soil depth (Heimsath *et al.*, 1997), parent material, and environmental conditions (Heimsath *et al.*, 1999). As soil thickens, the rate of the conversion of the underlying rock or sediment to soil decreases.

This has been shown using field observations of the relation between soil thickness and the abundance of cosmogenic nuclides (^{10}Be and ^{26}Al) in the quartz grains at the rock–soil interface (Figure 24). From this work, soil production can be described by

$$\phi = \rho_r \phi_0 e^{-\alpha h} \quad (18)$$

where ϕ and ϕ_0 are soil production for a given soil thickness and no soil cover, respectively; α the constant (cm^{-1}), and h the soil thickness (cm). The soil production model described by Equation (18) is applicable where physical disruption of the bedrock is the major soil production mechanism. By inserting Equation (18) into Equation (17), and rearranging, one can solve for soil thickness at any position on a hillslope:

$$h = \frac{1}{\alpha}\left(-\ln\left(-\frac{K\rho_s \partial^2 z}{\phi_0 \rho_r \partial x^2} \right) \right) \quad (19)$$

As this equation indicates, the key variables controlling soil thickness on hillslopes are slope curvature, the transport coefficient, and the soil production rate. The importance of curvature is intuitive in that changes in slope gradient drive the diffusive process. The value of the transport coefficient K varies greatly from one location to another (Fernandes and Dietrich, 1997; Heimsath *et al.*, 1999), and seems to increase with increasing humidity and decreasing rock competence (Table 5). Production, like transport, is likely dependent on climate and rock composition (Heimsath *et al.*, 1999).

The production/transport model of sediment transport provides a quantitative and mechanistic insight into soil profile thickness on hillslopes, an area of research that has received limited, but in some cases insightful, attention in the pedological literature. The study of the distribution of soil along hillslope gradients are called *toposequences* or *catenas*. In many areas, there is a well-known relationship between soil properties and hillslope position (e.g., Nettleton *et al.* (1968) for toposequences in southern California, for example), but the processes governing the relations were rather poorly known—at least on a quantitative level.

Beyond the effect of slope position (and environmental conditions) on soil thickness, a key factor is the amount of time that a soil on a given hillslope has to form. Time is a key variable that determines the amount of weathering and horizon formation that has occurred (Equation (2)). Yet, constraining "soil age" on erosional slopes has historically been a challenging problem. Here the author takes a simple approach and views soil age on slopes in terms of residence time (τ), where

$$\tau = \frac{\text{soilmass/area}}{\rho_r \phi_0 e^{-\alpha h}} = \frac{\text{soilmass/area}}{\rho_s K(\partial^2 z/\partial x^2)} \quad (20)$$

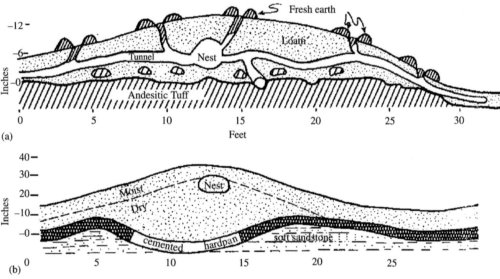

Figure 23 (a) Mima mounds (grass-covered areas) and vernal pools (gravel veneered low areas) on a Plio-Pleistocene aged fluvial terrace of the Merced river, California. Observations indicate that the landscape is underlain by a dense, clay pan (formed by long intervals of soil Bt horizon formation) capped by a gravel lense (the vernal pool gravels extend laterally under the Mima mounds). (b) Schematic diagram of "pocket gopher theory" of mima mound formation, illustrating preferential nesting and soil movement in/toward the well-drained mound areas and away from the seasonally wet vernal pools (Arkley and Brown, 1954) (reproduced by permission of Soil Science Society of America from *Soil Sci. Soc. Am. Proc.* **1954**, *18*, 195–199).

This is a pedologically meaningful measure that defines the rate at which soil is physically moved through a soil "box." As the expression indicates, soil residence time increases with decreasing curvature (and from Equation (19) above, with increasing soil thickness). From measured soil production rates in three contrasting environments, soil residence times on convex hillslopes may vary from a few hundred to 10^5 yr (Figure 25). These large time differences clearly help to explain many of the differences in soil profile development long observed on hillslope gradients.

On depositional landforms, sloping areas with concave slopes (convergent curvature) or level areas on floodplains, the concept of residence time can also be applied and quantitative models of soil formation can be derived. In these settings, residence time can be viewed as the amount of time required to fill a predetermined volume (or thickness) of soil with incoming sediment.

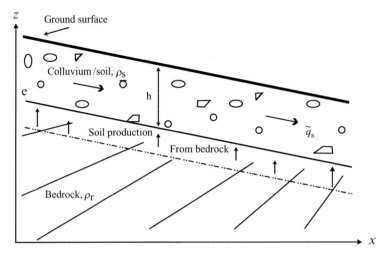

Figure 24 Schematic diagram of soil production and downslope transport on diffusion-dominated hillslopes (Heimsath *et al.*, 1997) (reproduced by permission of Nature Publishing Group from *Nature* 1997, *388*, 358–361).

Table 5 Soil production (P_0), erosion, and diffusivity values for three watersheds in contrasting climates and geology.

Location	Tennessee Valley, CA	Nunnock River, Australia	Black Diamond, CA
Bedrock	Sandstone	Granitic	Shale
Vegetation	Grass and coastal chaparral	Schlerophyll forest	Grass
Precipitation (mm yr^{-1})	760	910	450
Major transport mechanism	Gophers	Wombats, tree throw	Soil creep
Erosion rate (g m^{-2} yr^{-1})	50: backslope 130: shoulder	20: backslope 60: shoulder	625 average
P_0 (mm kyr^{-1})	77 ± 9	53 ± 3	2,078
K (cm^2 yr^{-1})	25	40	360

Sources: Heimsath (1999) and McKean *et al.* (1993).

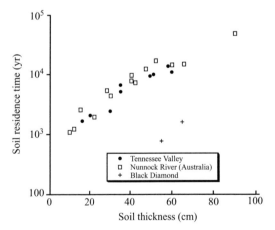

Figure 25 Calculated soil residence time versus soil thickness in three contrasting watersheds using Equation (20) and data from Table 5.

On active floodplains of major rivers, the soil residence time can be in most cases no longer than a few hundred years. Soil profiles here exhibit stratification, buried, weakly developed horizons, and little if any measurable chemical weathering.

In a landmark study of soils on depositional settings, Jenny (1962) examined soils on the floodplain of the Nile River, Eygpt. Available data suggested that the sedimentation rate prior to the construction of the Aswan Dam was 1 mm yr^{-1}, giving a residence time of 1,500 yr for a soil 150 cm thick. Jenny observed that in these soils, used for agriculture for millennia, nitrogen (the most crop-limiting element in agriculture) decreased with depth. This trend was not due to plant/atmospheric nitrogen inputs and crop cycling which creates the standard decrease in carbon and nitrogen with depth (see previous section on soil organic carbon). Instead, the continual deposition of nitrogen-rich sediment derived from soils in the Ethopian highlands was the likely source.

Jenny demonstrated that the observed nitrogen decrease with depth was a result of microbial degradation of the nitrogen-bearing organic matter in the sediment. The mineralized nitrogen was used by the crops and was then removed from the site. Stated mathematically

$$N(z) = N(0)e^{-kz} \qquad (21)$$

where $N(z)$ and $N(0)$ are the soil nitrogen at depth z and 0, respectively, and k the decomposition rate constant. Jenny's analysis showed that on the Nile floodplain, in the absence of nitrogen fertilization, long-term high rates of crop production could only be maintained by continued flood deposition of Nile sediments.

More generally, on concave (or convergent) slopes, where sedimentation is caused by long-term diffusional soil movement, one can calculate rates of sedimentation (or, of course, measure them directly) using a version of the geomorphic models discussed above. In deposition settings, soil production from bedrock or sediment can be viewed as being zero, and the change in soil height with time is (Fernandes and Dietrich, 1997)

$$\frac{\partial h}{\partial t} = K \frac{\partial^2 z}{\partial x^2} \qquad (22)$$

Soils in these depositional settings are sometimes referred to as "cumulic" soils in the US Soil Taxonomy, and commonly exhibit high concentrations of organic carbon and nitrogen to depths of several meters. This organic matter is partially due to the burial of organic matter produced *in situ*, but also reflects the influx of organic-matter-rich soil from upslope positions. This process of organic carbon burial has attracted the attention of the global carbon cycle community, because geomorphic removal and burial of organic carbon may represent a large, neglected, global carbon flux (Stallard, 1998; Rosenblum *et al.*, 2001). Beyond the carbon burial, the relatively short residence time of any soil profile thickness inhibits the accumulation of chemical weathering products, vertical transport, and horizon development.

In summary, an under-appreciated fact is that all soils—regardless of landscape position—are subject to slow but measurable physical and (especially) biological turbation. On level terrain, the soil/sediment flux is relatively random. It results in extensive soil surface mixing and, only in certain circumstances, striking directional movement. The addition of a slope gradient, in combination with this ongoing soil turbation, results in a net downslope movement dependent on slope curvature and the characteristics of the geological and climatic settings. The thickness of a soil, and its residence time, is directly related to the slope curvature. These conditions also determine the expression of chemical weathering and soil horizonation that can occur on a given landscape position.

7.6 SOIL DATA COMPILATIONS

Soils, and their properties, are now being used in many regional to global biogeochemical analyses. The scientific and education community is fortunate to have a wealth of valuable, high-quality, data on soil properties available on the worldwide web and in other electronic avenues. Here the author identifies some of these data sources.

During soil survey operations, numerous soil profile descriptions, and large amounts of laboratory data, are generated for the soils that are being mapped. Generally, the soils being described, sampled, and analyzed are representatives of *soil series*, the most detailed (and restrictive) classification of soils in the USA. There are $\sim 2.1 \times 10^4$ soil series that have been identified and mapped in the USA. The locations, soil descriptions, and lab data for many of these series are now available on the worldwide web via the USDA-NRCS (Natural Resource Conservation Service) National Soil Survey Center web site (http://www.statlab.iastate.edu/soils/ssl/natch_data.html). The complete data set is archived at the National Data Center, and is available to researchers.

The "characterization" data sheet is likely to be of interest to most investigators. In general, these data sheets are somewhat regionally oriented, reporting different chemical analyses for different climatic regions. For example, salt chemistry is reported for arid regions, whereas a variety of iron and aluminum oxide analyses are reported for humid, northern regions. Some major analyses of each soil horizon that are commonly reported include: particle size data, organic carbon and nitrogen, chemically extractable metals, cation exchange capacity and base saturation, bulk density, water retention, extractable bases and acidity, calcium carbonate, pH, chemistry of water extracted from saturated pastes, clay mineralogy, and sand and silt mineralogy (to list some of the major analyses). A complete and thorough discussion of the types of soil analyses, and the methods used, are presented by the Soil Survey Staff (1995), and on the web.

7.7 CONCLUDING COMMENTS

The scientific study of soils is just entering its second century. An impressive understanding of soil geography and, to a lesser degree, soil processes has evolved. There is growing interest in soils among scientists outside the agricultural sector, particularly geochemists and ecologists. Soils are central in the present attention to the global carbon cycle and its management, and as the human impact on the planet continues to increase, soils and their properties and services are being considered from a geobiodiversity perspective (Amundson, 1998; Amundson *et al.*, 2003). In many ways, the study of soils is at a critical and exciting point, and interdisciplinary cross-fertilization of the field is sure to lead to exciting new advances that bring pedology back to its multidisciplinary origins.

REFERENCES

Allègre C. (1992) *From Stone to Star: A View of Modern Geology*. Harvard University Press, Cambridge.

Allen R. M., Chatterjee R., Madden M. S., Ludden P. W., and Shar V. K. (1994) Biosynthesis of the iron–molybdenum co-factor of nitrogenase. *Crit. Rev. Biotech.* **14**, 225–249.

Alpers C. N. and Brimhall G. H. (1988) Middle Miocene climatic change in the Atacama desert, northern Chile: evidence for supergene mineralization at La Escondida. *Geol. Soc. Am. Bull.* **100**, 1640–1656.

Amundson R. (1998) Do soils need our protection? *Geotimes* March, 16–20.

Amundson R. (2001) The carbon budget in soils. *Ann. Rev. Earth Planet. Sci.* **29**, 535–562.

Amundson R. and Baisden W. T. (2000) Stable isotope tracers and mathematical models in soil organic matter studies. In *Methods in Ecosystem Science* (eds. O. E. Sala, R. B. Jackson, H. A. Mooney, and R. W. Howarth). Springer, New York, pp. 117–137.

Amundson R. and Davidson E. (1990) Carbon dioxide and nitrogenous gases in the soil atmosphere. *J. Geochem. Explor.* **38**, 13–41.

Amundson R. and Jenny H. (1997) On a state factor model of ecosystems. *Bioscience* **47**, 536–543.

Amundson R. and Yaalon D. (1995) E. W. Hilgard and John Wesley Powell: efforts for a joint agricultural and geological survey. *Soil Sci. Soc. Am. J.* **59**, 4–13.

Amundson R., Harden J., and Singer M. (1994a) *Factors of Soil Formation: A Fiftieth Anniversary Retrospective*, SSSA Special Publication No. 33. Soil Science Society of America, Madison, WI.

Amundson R., Wang Y., Chadwick O., Trumbore S., McFadden L., McDonald E., Wells S., and DeNiro M. (1994b) Factors and processes governing the ^{14}C content of carbonate in desert soils. *Earth Planet Sci. Lett.* **125**, 385–405.

Amundson R., Chadwick O. A., Kendall C., Wang Y., and DeNiro M. (1996) Isotopic evidence for shifts in atmospheric circulation patterns during the late Quaternary in mid-North America. *Geology* **24**, 23–26.

Amundson R., Stern L., Baisden T., and Wang Y. (1998) The isotopic composition of soil and soil-respired CO_2. *Geoderma* **82**, 83–114.

Amundson R., Guo Y., and Gong P. (2003) Soil diversity and landuse in the United States. *Ecosystems*, doi: 10.1007/s10021-002-0160-2.

Anders E. and Grevesse N. (1989) Abundances of the elements: meteoric and solar. *Geochim. Cosmochim. Acta* **53**, 197–214.

Arkley R. J. (1963) Calculations of carbonate and water movement in soil from climatic data. *Soil Sci.* **96**, 239–248.

Arkley R. J. and Brown H. C. (1954) The origin of mima mound (hog wallow) microrelief in the far western states. *Soil Sci. Am. Proc.* **18**, 195–199.

Baisden W. T. (2000) Soil organic matter turnover and storage in a California annual grassland chronosequence. PhD Dissertation, University of California, Berkeley.

Baisden W. T., Amundson R., Brenner D. L., Cook A. C., Kendall C., and Harden J. W. (2002) A multi-isotope C and N modeling analysis of soil organic matter turnover and transport as a function of soil depth in a California annual grassland chronosequence. *Glob. Biogeochem. Cycles*, **16**(4), 1135, doi: 10.1029/2001/GB001823, 2002.

Bao H., Campbell D. A., Bockheim J. G., and Theimens M. H. (2000a) Origins of sulphate in Antarctic dry-valley soils as deduced fro anomalous ^{17}O compositions. *Nature* **407**, 499–502.

Bao H., Thiemens M. H., Farquhar J., Campbell D. A., Lee C. C.-W., Heine K., and Loope D. B. (2000b) Anomalous ^{17}O compositions in massive sulphate deposits on the Earth. *Nature* **406**, 176–178.

Bao H., Thiemens M. H., and Heine K. (2001) Oxygen-17 excesses of the Central Namib gypcretes: spatial distribution. *Earth Planet. Sci. Lett.* **192**, 125–135.

Barnes C. J. and Allison G. B. (1983) The distribution of deuterium and ^{18}O in dry soils: I. Theory. *J. Hydrol.* **60**, 141–156.

Betancourt J. L., Latorre C., Rech J. A., Quade J., and Rlander K. A. (2000) A 22,000-year record of monsoonal precipitation from Chile's Atacama desert. *Science* **289**, 1542–1546.

Black T. A. and Montgomery D. R. (1991) Sediment transport by burrowing animals, Marin county, California. *Earth Surf. Proc. Landforms* **16**, 163–172.

Berner R. A. (1992) Palaeo-CO_2 and climate. *Nature* **358**, 114.

Birkeland P. W. (1999) *Soils and Geomorphology*, 3rd edn. Oxford University Press, New York.

Böhlke J. K., Erickson G. E., and Revesz K. (1997) Stable isotope evidence for an atmospheric origin of desert nitrate deposits in northern Chile and southern California, USA. *Chem. Geol.* **136**, 135–152.

Borst G. (1968) The occurrence of crotovinas in some southern California soils. *Trans. 9th Int. Congr. Soil Sci.*, Adelaide **2**, 19–22.

Bowen H. J. M. (1979) *Environmental Chemistry of the Elements*. Academic Press, London.

Brimhall G. (1987) Preliminary fractionation patterns of ore metals through Earth history. *Chem. Geol.* **64**, 1–16.

Brimhall G. H. and Dietrich W. E. (1987) Constitutive mass balance relations between chemical composition, volume, density, porosity, and strain in metasomatic hydrochemical systems: results on weathering and pedogenesis. *Geochim. Cosmochim. Acta* **51**, 567–587.

Brimhall G. H., Lewis C. J., Ague J. J., Dietrich W. E., Hampel J., Teague T., and Rix P. (1988) Metal enrichment in bauxite by deposition of chemically-mature eolian dust. *Nature* **333**, 819–824.

Brimhall G. H., Lewis C. J., Ford C., Bratt J., Taylor G., and Warren O. (1991) Quantitative geochemical approach to pedogenesis: importance of parent material reduction, volumetric expansion, and eolian influx in laterization. *Geoderma* **51**, 51–91.

Brimhall G., Chadwick O. A., Lewis C. J., Compston W., Williams I. S., Danti K. J., Dietrich W. E., Power M. E., Hendricks D., and Bratt J. (1992) Deformational mass balance transport and invasive processes in soil evolution. *Science* **255**, 695–702.

Cerling T. E. (1984) The stable isotopic composition of modern soil carbonate and its relationship to climate. *Earth Planet. Sci. Lett.* **71**, 229–240.

Cerling T. E. (1991) Carbon dioxide in the atmosphere: evidence from Cenozoic and Mesozoic paleosols. *Am. J. Sci.* **291**, 377–400.

Cerling T. E. and Quade J. (1993) Stable carbon and oxygen isotopes in soil carbonates. In *Climate Change in Continental Isotopic Records* (eds. P. K. Swart, K. C. Lohmann, J. McKenzie, and S. Sabin). Geophysical Monograph 78. American Geophysical Union, Washington, DC, pp. 217–231.

Cerling T. E., Quade J., Wang Y., and Bowman J. R. (1989) Carbon isotopes in soils and palaeosols as paleoecologic indicators. *Nature* **341**, 138–139.

Cerling T. E., Harris J. M., MacFadden B. J., Leakey M. G., Quade J., Eisenmann V., and Ehleringer J. R. (1997) Global vegetation change through the Miocene/Pliocene boundary. *Nature* **389**, 153–158.

Chadwick O. A. and Davis J. O. (1990) Soil-forming intervals caused by eoloian sediment pulses in the Lahontan basin, northwestern Nevada. *Geology* **18**, 243–246.

Chadwick O. A., Brimhall G. H., and Hendricks D. M. (1990) From a black to a gray box—a mass balance interpretation of pedogenesis. *Geomorphology* **3**, 369–390.

Chadwick O. A., Derry L. A., Vitousek P. M., Huebert B. J., and Hedin L. O. (1999) Changing sources of nutrients during four million years of ecosystem development. *Nature* **397**, 491–497.

Chiappini C. (2001) The formation and evolution of the Milky Way. *Am. Sci.* **89**, 506–515.

Clarkson D. T. (1974) *Ion Transport and Cell Structure in Plants*. McGraw-Hill, New York, 350pp.

Darwin C. (1846) An account of the fine dust which often falls on vessels in the Atlantic ocean. *Quat. J. Geol. Soc. London* **2**, 26–30.

Darwin C. (1881) *The Formation of Vegetable Mould through the Action of Worms*. Appleton, New York.

Davis W. M. (1892) The convex profile of badland divides. *Science* **20**, 245.

De Coninck F. (1980) Major mechanisms in formation of spodic horizons. *Geoderma* **24**, 101–128.

Delsemme A. H. (2001) An argument for the cometary origin of the biosphere. *Am. Sci.* **89**, 432–442.

Dokuchaev V. V. (1880) Protocol of the meeting of the branch of geology and mineralogy of the St. Petersburg Society of Naturalists. *Trans. St. Petersburg Soc. Nat.* **XII**, 65–97 (Translated by the Department of Soils and Plant Nutrition, University of California, Berkeley).

Drake M. J. and Righter K. (2002) Determining the composition of the earth. *Nature* **416**, 39–51.

Ekart D. D., Cerling T. E., Montanañez I. P., and Tabor N. J. (1999) A 400 million year carbon-isotope record of pedogenic carbonate: implications for paleoatmospheric carbon dioxide. *Am. J. Sci.* **299**, 805–827.

Elzein A. and Balesdent J. (1995) Mechanistic simulation of vertical distribution of carbon concentrations and residence times in soils. *Soil Sci. Soc. Am. J.* **59**, 1328–1335.

Ericksen G. E. (1981) Geology and origin of the Chilean nitrate deposits. Geological Survey Professional Paper 1188, US Geological Survey, 37pp.

Farquhar G. D., Hubick K. T., Condon A. G., and Richards R. A. (1988) Carbon-isotope fractionation and plant water-use efficiency. In *Stable Isotopes in Ecological Research* (eds. P. W. Rundell, J. R. Ehleringer, and K. A. Nagy). Springer, New York, pp. 21–40.

Fernandes N. F. and Dietrich W. E. (1997) Hillslope evolution by diffusive processes: the timescale for equilibrium adjustments. *Water Resour. Res.* **33**, 1307–1318.

Ferris T. (1998) Seeing in the dark. *The New Yorker* August 10, 55–61.

Friedman I. and O'Neil J. R. (1977) Compilation of stable isotope fractionation factors of geochemical interest. In *Data of Geochemistry*, 6th edn., US Geological Survey Professional Paper 440-KK (ed. M. Fleischer). chap. KK.

Gilbert G. K. (1909) The convexity of hilltops. *J. Geol.* **17**, 344–350.

Gile L. H., Peterson F. F., and Grossman R. B. (1966) Morphological and genetic sequences of carbonate accumulation in desert soils. *Soil Sci.* **101**, 347–360.

Greenwood N. N. and Earnshaw A. (1997) *Chemistry of the Elements*, 2nd edn. Heinemann, Oxford.

Griffin D. W., Kellogg C. A., Garrison V. H., and Shinn E. A. (2002) The global transport of dust. *Am. Sci.* **90**, 228–235.

Hanson P. J., Edwards N. T., Garten C. T., and Andrews J. A. (2000) Separating root and soil microbial contributions to soil respiration: a review of methods and observations. *Biogeochemistry* **48**, 115–146.

Harden J. (1987) Soils developed in granitic alluvium near Merced, California. *US Geol. Surv. Bull.* **1590-A**, pp. A1–A65.

Heath J. A. and Huebert B. J. (1999) Cloudwater deposition as a source of fixed nitrogen in a Hawaiian montane forest. *Biogeochemistry* **44**, 119–134.

Hendricks M. B., DePaolo D. J., and Cohen R. C. (2000) Space and time variation of delta O-18 and delta D in precipitation: can paleotemperature be estimated from ice cores? *Global Biogeochem. Cycles* **14**, 851–861.

Heimsath A. M., Dietrich W. E., Nishiizumi K., and Finkel R. C. (1997) The soil production function and landscape equilibrium. *Nature* **388**, 358–361.

Heimsath A. M., Dietrich W. E., Nishiizumi K., and Finkel R. C. (1999) Cosmogenic nuclides, topography, and the spatial variation of soil depth. *Geomorphology* **27**, 151–172.

Hesterberg R. and Siegenthaler U. (1991) Production and stable isotopic composition of CO_2 in a soil near Bern, Switzerland. *Tellus, Ser. B* **43**, 197–205.

Hilgard E. W. (1860) Report on the geology and agriculture of the state of Mississippi, Jackson, MS.

Hillel D. (1980a) *Fundamentals of Soil Physics*. Academic Press, New York.

Hillel D. (1980b) *Applications of Soil Physics*. Academic Press, New York.

Hillel D. (1998) *Environmental Soil Physics*. Academic Press, San Diego.

Holland R. F. (1996) Great Valley vernal pool distribution, Photorevised 1996. In *Ecology, Conservation, and Management of Vernal Pool Ecosystems*, Proceedings of a 1996 Conference, California Native Plant Society, Sacramento, CA (eds. C. W. Witham, E. T. Bauder, D. Belk, W. R. Ferren, Jr., and R. Ornduff), pp. 71–75.

Hoosbeek M. R. and Bryant R. B. (1995) Modeling the dynamics of organic carbon in a Typic Haplorthod. In *Soils and Global Change* (eds. R. Lal, J. M. Kimble, E. R. Levine, and B. A. Stuart). CRC Lewis Publishers, Chelsea, MI, pp. 415–431.

Jackson M. L., Levelt T. W. M., Syers J. K., Rex R. W., Clayton R. N., Sherman G. D., and Uehara G. (1971) Geomorphological relationships of tropospherically derived quartz in the soils of the Hawaiian Islands. *Soil Sci. Soc. Am. Proc.* **35**, 515–525.

Jahren A. H., Arens N. C., Sarmiento G., Guerrero J., and Amundson R. (2001) Terrestrial record of methane hydrate dissociation in the early Cretaceous. *Geology* **29**, 159–162.

Jenkinson D. S., Adams D. E., and Wild A. (1991) Model estimates of CO_2 emissions from soil in response to global warming. *Nature* **351**, 304–306.

Jenny H. (1941) *Factors of Soil Formation: A System of Quantitative Pedology*. McGraw-Hill, New York.

Jenny H. (1961) *E. W. Hilgard and the Birth of Modern Soil Science*. Collan della Rivista "Agrochimica", Pisa, Italy.

Jenny H. (1962) Model of a rising nitrogen profile in Nile Valley alluvium, and its agronomic and pedogenic implications. *Soil Sci. Soc. Am. Proc.* **27**, 273–277.

Jenny H. and Leonard C. D. (1939) Functional relationships between soil properties and rainfall. *Soil Sci.* **38**, 363–381.

Jenny H., Gessel S. P., and Bingham F. T. (1949) Comparative study of decomposition rates of organic matter in temperate and tropical regions. *Soil Sci.* **68**, 419–432.

Jersak J., Amundson R., and Brimhall G. (1995) A mass balance analysis of podzolization: examples from the northeastern United States. *Geoderma* **66**, 15–42.

Johnson D. L. (1990) Biomantle evolution and the redistribution of earth materials and artifacts. *Soil Sci.* **149**, 84–102.

Johnson D. L., Watson-Stegner D., Johnson D. N., and Schaetzl R. J. (1987) Proisotopic and proanisotropic processes of pedoturbation. *Soil Sci.* **143**, 278–292.

Jury W. A., Gardner W. R., and Gardner W. H. (1991) *Soil Physics*, 5th edn. Wiley, New York.

Kirkby M. J. (1971) Hillslope process-response models based on the continuity equation. *Inst. Prof. Geogr. Spec. Publ.* **3**, 15–30.

Koch P. L., Zachos J. C., and Dettman D. L. (1995) Stable isotope stratigraphy and paleoclimatology of the Paleogene Bighorn Basin. *Palaeogeogr. Palaeoclimat. Palaeoecol.* **115**, 61–89.

Kennedy M. J., Chadwick O. A., Vitousek P. M., Derry L. A., and Hendricks D. M. (1998) Changing sources of base cations during ecosystem development, Hawaiian Islands. *Geology* **26**, 1015–1018.

Krupenikov I. A. (1992) *History of Soil Science: From its Inception to the Present.* Amerind Publishing, New Delhi.

Langmaid K. K. (1964) Some effects of earthworm inversion in virgin podzols. *Can. J. Soil Sci.* **44**, 34–37.

Latorre C., Betancourt J. L., Rylander K. A., and Quade J. (2002) Vegetation invasions into absolute desert: a 45,000 yr rodent midden record from the Calama-Salar de Atacama basins, northern Chile (lat. 22°–24° S). *Geol. Soc. Am. Bull.* **114**, 349–366.

Marian G. M., Schlesinger W. H., and Fonteyn P. J. (1985) Caldep: a regional model for soil $CaCO_3$ (calcite) deposition in southwestern deserts. *Soil Sci.* **139**, 468–481.

McFadden L. D. and Tinsley J. C. (1985) Rate and depth of pedogenic-carbonate accumulation in soils: formation and testing of a compartment model. In *Quaternary Soils and Geomorphology of the Southwestern United States.* (ed. D. L. Weide). Geological Society of America Special Paper 203. Geological Society of America, Boulder, pp. 23–42.

McFadden L. D., Amundson R. G., and Chadwick O. A. (1991) Numerical modeling, chemical, and isotopic studies of carbonate accumulation in soils of arid regions. In *Occurrence, Characteristics, and Genesis of Carboante, Gypsum, and Silica Accumulations in Soils* (ed. W. D. Nettleon). Soil Science Society of America, Madison, WI, pp. 17–35.

McKean J. A., Dietrich W. E., Finkel R. C., Southon J. R., and Caffee M. W. (1993) Quantification of soil production and downslope creep rates from cosmogenic ^{10}Be accumulations on a hillslope profile. *Geology* **21**, 343–346.

Merritts D. J., Chadwick O. A., Hendricks D. M., Brimhall G. H., and Lewis C. J. (1992) The mass balance of soil evolution on late Quaternary marine terraces, northern California. *Geol. Soc. Am. Bull.* **104**, 1456–1470.

Miller A. J., Amundson R., Burke I. C., and Yonker C. (2003) The effect of climate and cultivation on soil organic C and N. *Biogeochemistry* (in press).

Mook W. G., Koopmans M., Carter A. F., and Keeling C. D. (1983) Seasonal, latitudinal, and secular variations in the abundance and isotopic ratios of atmospheric carbon dioxide: 1. Results from land stations. *J. Geophys. Res.* **88**(C15), 10915–10933.

Mora C. I., Driese S. G., and Colarusso L. A. (1996) Middle to late Paleozoic atmospheric CO_2 levels from soil carbonate and organic matter. *Science* **271**, 1105–1107.

Nadelhoffer K. and Fry B. (1988) Controls on the nitrogen-15 and carbon-13 abundances in forest soil organic matter. *Soil Sci. Soc. Am. J.* **52**, 1633–1640.

Nakai S., Halliday A. N., and Rea D. K. (1993) Provenance of dust in the Pacific Ocean. *Earth Planet. Sci. Lett.* **119**, 143–157.

Navarro-Gonzalez R., McKay C. P., and Mvondo D. N. (2001) A possible nitrogen crisis for Archaean life due to reduced nitrogen fixation by lightning. *Nature* **412**, 61–64.

Nettleton W. D., Flach K. W., and Borst G. (1968) *A Toposequence of Soils on Grus in the Southern California Peninsular Range.* US Dept. Agric. Soil Cons. Serv., Soil Surv. Invest. Rep. No. 21, 41p.

O'Brien J. B. and Stout J. D. (1978) Movement and turnover of soil organic matter as indicated by carbon isotope measurements. *Soil Biol. Biochem.* **10**, 309–317.

Parton W. J., Schimel D. S., Cole C. V., and Ojima D. S. (1987) Analysis of factors controlling soil organic matter levels in Great Plains grasslands. *Soil Sci. Soc. Am. J.* **5**, 1173–1179.

Peterson F. D. (1980) Holocene desert soil formation under sodium salt influence in a playa-margin environment. *Quat. Res.* **13**, 172–186.

Quade J., Cerling T. E., and Bowman J. R. (1989a) Systematic variations in the carbon and oxygen isotopic composition of pedogenic carbonate along elevation transects in the southern Great Basin, United States. *Geol. Soc. Am. Bull.* **101**, 464–475.

Quade J., Cerling T. E., and Bowman J. R. (1989b) Development of the Asian monsoon revealed by marked ecological shift during the latest Miocene in northern Pakistan. *Nature* **342**, 163–166.

Raich J. W. and Schlesinger W. H. (1992) The global carbon dioxide flux in soil respiration and its relationship to vegetation and climate. *Tellus* **B44**, 48–51.

Rea D. K. (1994) The paleoclimatic record provided by eolian deposition in the deep sea: the geological history of wind. *Rev. Geophys.* **5**, 193–259.

Rech J. A., Quade J., and Hart B. (2002) Isotopic evidence for the source of Ca and S in soil gypsum, anhyrite and calcite in the Atacama desert, Chile. *Geochim. Cosmochim. Acta* **67** (4), 575–586.

Rosenblum N. A., Doney S. C., and Schimel D. S. (2001) Geomorphic evolution of soil texture and organic matter in eroding landscapes. *Global Biogeochem. Cycles* **15**, 365–381.

Royer D. L. (1999) Depth to pedogenic carbonate horizon as a paleoprecipitation indicator? *Geology* **27**, 1123–1126.

Rozanski K., Araguás- Araguás L., and Gonfiantini R. (1993) Isotopic patterns in modern global precipitation. In *Climate Change in Continental Isotopic Records.* American Geophysical Union Monograph 78 (ed. P. K. Swart, *et al.*). American Geophysical Union, Washington, DC, pp. 1–36.

Russell R. S. (1977) *Plant Root Systems: Their Function and Interaction with the Soil.* McGraw-Hill, London.

Schlesinger W. H. and Andrews J. A. (2000) Soil respiration and the global carbon cycle. *Biogeochemistry* **48**, 7–20.

Severinghaus J. P., Bender M. L., Keeling R. F., and Broecker W. S. (1996) Fractionation of soil gases by diffusion of water vapor, gravitational settling, and thermal diffusion. *Geochim. Cosmochim. Acta* **60**, 1005–1018.

Shacklette H. T. and Boerngen J. G. (1984) Element concentrations in soils and other surficial materials of the conterminous United States. US Geological Survey Professional Paper 1270.

Soil Survey Staff (1975) *Soil Taxonomy: A Basic System of Soil Classification for Making and Interpreting Soil Surveys,* Agr. Handbook 426. US Government Printing Office, Washington, DC.

Soil Survey Staff (1993) *Soil Survey Manual.* USDA Handbook No. 18. US Government Printing Office, Washington, DC.

Soil Survey Staff (1995) *Soil Survey Laboratory Information Manual.* Soil Survey Investigations Report No. 45, Version 1.0, US Government Printing Office, Washington, DC.

Soil Survey Staff (1999) *Keys to Soil Taxonomy,* 8th edn. Pocahontas Press, Blacksburg, VA.

Stallard R. F. (1998) Terrestrial sedimentation and the carbon cycle: coupling weathering and erosion to the carbon cycle. *Global Biogeochem. Cycles* **12**, 231–257.

Stern L. A., Baisden W. T., and Amundson R. (1999) Processes controlling the oxygen-isotope ratio of soil CO_2: analytic and numerical modeling. *Geochim. Cosmochim. Acta* **63**, 799–814.

Stern L. A., Amundson R., and Baisden W. T. (2001) Influence of soils on oxygen-isotope ratio of atmospheric CO_2. *Global Biogeochem. Cycles* **15**, 753–759.

Sundquist E. T. (1993) The global carbon dioxide budget. *Science* **259**, 934–941.

Sutter B., Amundson R., Ewing S., Rhodes K. W., and McKay C. W. (2002) The chemistry and mineralogy of Atacama Desert soils: a possible analog for Mars soils. American Geophysical Union Fall Meeting, San Francisco, pp. 71A–0443.

Tans P. P. (1998) Oxygen isotopic equilibrium between carbon dioxide and water in soils. *Tellus* **B50**, 163–178.

Taylor S. R. and McLennan S. M. (1985) *The Continental Crust: Its Composition and Evolution.* Blackwell, Oxford.

Thiemens M. H. (1999) Mass-independent isotope effects in planetary atmospheres and the early solar system. *Science* **283**, 341–345.

Thorstenson D. C., Weeks E. P., Haas H., and Fisher D. W. (1983) Distribution of gaseous $^{12}CO_2$, $^{13}CO_2$, and $^{14}CO_2$ in sub-soil unsaturated zone of the western US Great Plains. *Radiocarbon* **25**, 315–346.

Tréguer P., Nelson D. M., Van Bennekom A. J., DeMaster D. J., Leynaert A., and Quéguiner S. E. (1995) The silica balance in the world ocean: a reestimate. *Science* **268**, 375–379.

Trumbore S. E. (2000) Age of soil organic matter and soil respiration: radiocarbon constraints on belowground dynamics. *Ecol. Appl.* **10**, 399–411.

Ugolini F. C. and Dahlgren R. (1987) The mechanism of podzolization as revealed by soil solution studies. In *Podzols et Podolization* (eds. D. Fighiand and A. Chavell). Assoc. Fr. Estude Sol, Plassier, France, pp. 195–203.

Vil'yams V. R. (1967) V. V. Dokuchaev's role in the development of soil science. In *Russian Chernozem*. (translated by Israel Program for Scientific Translations, Jerusalem).

Vitousek P. M., Aber J. D., Howarth R. W., Likens G. E., Matson P. A., Schindler D. W., Schlesinger W. H., and Tilman D. G. (1997a) Human alterations of the global nitrogen cycle: sources and consequences. *Ecol. Appl.* **7**, 737–750.

Vitousek P. M., Chadwick O. A., Crews T. E., Fownes J. H., Hendricks D. M., and Herbert D. (1997b) Soil and ecosystem development across the Hawaiian Islands. *GSA Today* **7**, 1–8.

Wang Y., Amundson R., and Trumbore S. (1994) A model for soil $^{14}CO_2$ and its implications for using ^{14}C to date pedogenic carbonate. *Geochim. Cosmochim. Acta* **58**, 393–399.

Wang Y., Amundson R., and Niu X.-F. (2000) Seasonal and altitudinal variation in decomposition of soil organic matter inferred from radiocarbon measurements of soil CO_2 flux. *Global Biogeochem. Cycles* **14**, 199–211.

White A. F., Blum A. E., Schulz M. S., Bullen T. D., Harden J. W., and Peterson M. L. (1996) Chemical weathering rates of a soil chronosequence on granitic alluvium: I. Quantification of mineralogical and surface area changes and calculation of primary silicate reaction rates. *Geochim. Cosmochim. Acta* **60**, 2533–2550.

8

Global Occurrence of Major Elements in Rivers

M. Meybeck

University of Paris VI, CNRS, Paris, France

8.1	INTRODUCTION	277
8.2	SOURCES OF DATA	278
8.3	GLOBAL RANGE OF PRISTINE RIVER CHEMISTRY	278
8.4	SOURCES, SINKS, AND CONTROLS OF RIVER-DISSOLVED MATERIAL	279
	8.4.1 Influence of Lithology on River Chemistry	281
	8.4.2 Carbon Species Carried by Rivers	281
	8.4.3 Influence of Climate on River Chemistry	283
8.5	IDEALIZED MODEL OF RIVER CHEMISTRY	284
8.6	DISTRIBUTION OF WEATHERING INTENSITIES AT THE GLOBAL SCALE	286
8.7	GLOBAL BUDGET OF RIVERINE FLUXES	287
8.8	HUMAN ALTERATION OF RIVER CHEMISTRY	290
8.9	CONCLUSIONS	291
	REFERENCES	291

8.1 INTRODUCTION

Major dissolved ions (Ca^{2+}, Mg^{2+}, Na^+, K^+, Cl^-, SO_4^{2-}, HCO_3^-, and CO_3^{2-}) and dissolved silica (SiO_2) in rivers have been studied for more than a hundred years for multiple reasons: (i) geochemists focus on the origins of elements and control processes, and on the partitioning between dissolved and particulate forms; (ii) physical geographers use river chemistry to determine chemical denudation rates and their spatial distribution; (iii) biogeochemists are concerned with the use of carbon, nitrogen, phosphorus, silica species, and other nutrients by terrestrial and aquatic biota; (iv) oceanographers need to know the dissolved inputs to the coastal zones, for which rivers play the dominant role; (v) hydrobiologists and ecologists are interested in the temporal and spatial distribution of ions, nutrients, organic carbon, and pH in various water bodies; (vi) water users need to know if waters comply with their standards for potable water, irrigation, and industrial uses.

The concentrations of the major ions are commonly expressed in $mg\,L^{-1}$; they are also reported in $meq\,L^{-1}$ or $\mu eq\,L^{-1}$, which permits a check of the ionic balance of an analysis: the sum of cations (Σ^+ in $eq\,L^{-1}$) should equal the sum of anions (Σ^- in $eq\,L^{-1}$). Dissolved silica is generally not ionized at pH values commonly found in rivers; its concentration is usually expressed in $mg\,L^{-1}$ or in $\mu mol\,L^{-1}$. Ionic contents can also be expressed as percent of Σ^+ or Σ^- ($\%C_i$), which simplifies the determination of ionic types. Ionic ratios (C_i/C_j) in $eq\,eq^{-1}$ are also often tabulated (Na^+/Cl^-, Ca^{2+}/Mg^{2+}, Cl^-/SO_4^{2-}, etc.). As a significant fraction of sodium can be derived from atmospheric sea salt and from sedimentary halite, a chloride-corrected sodium concentration is commonly reported ($Na^\# = Na^+ - Cl^-$ (in $meq\,L^{-1}$)). The export rate of ions and silica, or the yield ($Y_{Ca^{2+}}, Y_{SiO_2}$) at a given station is the average mass transported per year divided by the drainage area: it is expressed in units of $t\,km^{-2}\,yr^{-1}$ (equal to $g\,m^{-2}\,y^{-1}$) or in $eq\,m^{-2}\,yr^{-1}$.

This chapter covers the distribution of riverine major ions, carbon species, both organic and inorganic, and silica over the continents, including internal regions such as Central Asia, and also the major factors such as lithology and climate that control their distribution and yields.

Based on an unpublished compilation of water analyses in 1,200 pristine and subpristine basins, I am presenting here an idealized model of global river chemistry. It is somewhat different from the model proposed by Gibbs (1970), in that it includes a dozen major ionic types. I also illustrate the enormous range of the chemical composition of rivers—over three orders of magnitude for concentrations and yields—and provide two global average river compositions: the median composition and the discharge-weighted composition for both internally and externally draining regions of the world.

A final section draws attention to the human alteration of river chemistry during the past hundred years, particularly for Na^+, K^+, Cl^-, and SO_4^{2-}; it is important to differentiate anthropogenic from natural inputs. Trace element occurrence is covered by Gaillardet (see Chapter 9).

8.2 SOURCES OF DATA

Natural controls of riverine chemistry at the global scale have been studied by geochemists since Clarke (1924) and the Russian geochemists Alekin and Brazhnikova (1964). Regional studies performed prior to industrialization and/or in remote areas with very limited human impacts are rare (Kobayashi, 1959, 1960) and were collected by Livingstone (1963). Even some of his river water data are affected by mining, industries, and the effluents from large cities. These impacts are obvious when the evolution of rivers at different periods is compared; Meybeck and Ragu (1996, 1997) attempted this comparison for all the rivers flowing to the oceans from basins with areas exceeding $10^4 \, km^2$. More recently interest in the carbon cycle and its riverine component has been the impetus for a new set of field studies (Stallard and Edmond, 1981, 1983; Degens *et al.*, 1991; Gaillardet *et al.*, 1997, 1999; Huh *et al.*, 1998/1999; Millot *et al.*, 2002; Guyot, 1993) that build on observations made since the 1970s (Reeder *et al.*, 1972; Stallard and Edmond, 1981, 1983). These data were used to construct the first global budgets of river dissolved loads (Meybeck, 1979) and their controls (Holland, 1978; Meybeck, 1987, 1994; Bluth and Kump, 1994; Berner, 1995; Stallard, 1995a,b).

Most of the annual means derived from these data have been collected into a global set of pristine or subpristine rivers and tributaries (PRISRI, $n = 1,200$) encompassing all the continents, including exorheic and endorheic (internal drainage) runoff. The largest basins, such as the Amazon, Mackenzie, Lena, Yenisei, and Mekong, have been subdivided into several smaller subbasins. In some regions (e.g., Indonesia, Japan) the size of

the river basins included in these summaries may be less than $1,000 \, km^2$. In the northern temperate regions only the most reliable historic analyses prior to 1920 have been included. Examples are analyses performed by the US Geological Survey (1909–1965) in the western and southwestern United States and in Alaska.

In order to study the influence of climate, total cationic contents (Σ^+) of PRISRI rivers have been split into classes based on annual runoff (q in mm yr^{-1}) and Σ^+ ($meq \, L^{-1}$). A medium-sized subset has also been used (basin area from $3,200 \, km^2$ to $200,000 \, km^2$, $n = 700$). The PRISRI data base covers the whole globe but has poor to very poor coverage of western Europe, Australia, South Africa, China, and India due to the lack of data for pre-impact river chemistry in these regions. PRISRI includes rivers that flow permanently ($q > 30 \, mm$ yr^{-1}) and seasonally to occasionally ($3 < q < 30$ $mm \, yr^{-1}$). Lake outlets may be included in PRISRI.

8.3 GLOBAL RANGE OF PRISTINE RIVER CHEMISTRY

The most striking observation in global river chemistry is the enormous range in concentrations (C_i), ionic ratios (C_i/C_j), and the proportions of ions in cation and anion sums ($\%C_i$) as illustrated by the 1% and 99% quantiles (Q_1 and Q_{99}) of their distribution (Table 1). The Q_{99}/Q_1 ratio of solute concentrations is the lowest for potassium and silica, about two orders of magnitude, and it is very high for chloride and sulfate, exceeding three orders of magnitude. The concept of "global average river chemistry," calculated from the total input of rivers to oceans divided by the total volume of water, can only be applied for global ocean chemistry and elemental cycling; it is not a useful reference in either weathering studies, river ecology, or water quality.

Ionic ratios and ionic proportion distributions also show clearly that all major ions, except potassium, can dominate in multiple combinations: ionic ratios also range over two to three orders of magnitude, and they can be greater or less than unity for all except the Na^+/K^+ ratio, in which sodium generally dominates. River compositions found in less than 1% of analyses can be termed *rare*; for analyses from Q_1 to Q_{10} and Q_{90} to Q_{99}, I propose the term *uncommon*, from Q_{10} to Q_{25} and Q_{75} to Q_{90}, *common*, and between Q_{25} and Q_{75}, *very common*. An example of this terminology is shown in the next section (Figure 2) for dissolved inorganic carbon (DIC).

When plotted on a log-probability scale (Henry's law diagram) the distributions of elemental concentrations show four patterns (Figure 1): (i) lognormal distribution as for potassium in Thailand (a, D), that can be interpreted as a single source of the element and limited control on its concentration; (ii) retention at lower concentrations

Table 1 Global range of pristine river chemistry (medium-sized basins).

	Ca^{2+}	Mg^{2+}	Na^+	K^+	Cl^-	SO_4^{2-}	HCO_3^-	SiO_2	Σ^+
Ionic contents (C_i)									
Q_{99}	9,300	5,900	14,500	505	17,000	14,500	5,950	680	32,000
Q_1	32	10	18	3.9	3.7	5	47	3.3	128
Q_{99}/Q_1	290	590	805	129	4,600	2,900	126	206	250
Ionic proportions ($\%C_i$)									
Q_{99}	84	48	72	19.5	69	67	96		
Q_1	11	0.1	1	0.1	0.1	0.1	9		

	$\dfrac{Ca^{2+}}{Mg^{2+}}$	$\dfrac{Ca^{2+}}{Na^+}$	$\dfrac{Mg^{2+}}{Na^+}$	$\dfrac{Na^+}{K^+}$	$\dfrac{Na^+}{Cl^-}$	$\dfrac{Ca^{2+}}{SO_4^{2-}}$	$\dfrac{SO_4^{2-}}{HCO_3^-}$	$\dfrac{Cl^-}{SO_4^{2-}}$	$\dfrac{SiO_2}{\Sigma^+}$
Ionic ratios (C_i/C_j)									
Q_{99}	20.3	56	20	164	29	51	3.5	8.5	1.3
Q_1	0.01	0.14	0.01	0.95	0.33	0.19	0.01	0.01	0.0

C_i: ionic contents ($\mu eq\,L^{-1}$ and $\mu mol\,L^{-1}$ for silica); $\%C_i$: proportion of ions in the sum of total cations or anions; C_i/C_j: ionic ratios; and Q_1 and Q_{99}: lowest and highest percentiles of distribution.

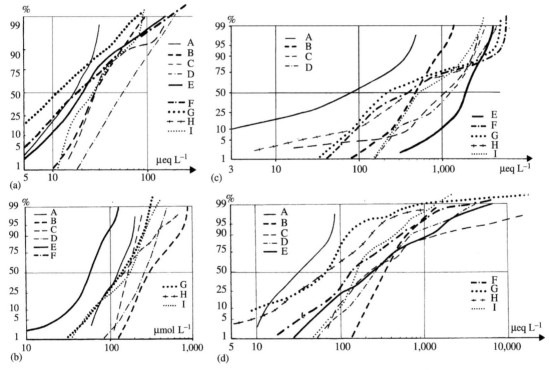

Figure 1 Cumulative distributions of major dissolved elements in pristine regions: (a) K^+, (b) SiO_2, (c) HCO_3^-, and (d) Na^+. A: Central and lower Amazon, B: Japan, C: Andean Amazon Basin, D: Thailand, E: Mackenzie Basin, F: French streams, G: temperate stream model, H: monolithologic miscellaneous French streams, I: major world rivers (source Meybeck, 1994).

(there is a significant break in the distribution as for silica in the Mackenzie Basin (b, E)); (iii) retention at higher concentrations (there is a significant break in the distribution as for bicarbonate (DIC) (c, C, E, F, G, I)); and (iv) additional source at higher concentrations (the break in the distribution is in the opposite direction, as for chloride, suggesting another source of material than the one observed between Q_{10} and Q_{90} (d, all except central Amazon)).

8.4 SOURCES, SINKS, AND CONTROLS OF RIVER-DISSOLVED MATERIAL

The observed distribution patterns have been interpreted by many authors (Likens *et al.*, 1977; Holland, 1978; Drever, 1988; Hem, 1989; E. K. Berner and R. A. Berner, 1996) in terms of multiple sources, sinks, and controls on riverine chemistry (Table 2). The sources of ions may be multiple: rainfall inputs, generally of oceanic

Table 2 Dominant sources, sinks, and controls of major ions in present day rivers.

	Major ions								Controls						
	SiO_2	K^+	Na^+	Cl^-	SO_4^{2-}	Mg^{2+}	Ca^{2+}	DIC	$T°$	τ	q	d	QH	V	T
Natural sources															
Atmosphere		◄---	---	---	---	---►		↔			−				+
Silicate weathering	◄—	—	—►		◄—	—	—►		+		+	×		+	+
Pyrite	◄—	—	—►		◄—	—	—	—►			+				
Carbonate						◄—	—	—►			+			+	
Gypsum					◄—	—	—►				+				
Halite			◄—	—►							+				
Deep waters	◄—	—	—	—	—	—	—	—►							+
Natural sinks															
Terrestrial vegetation		◄---	---	---	---	---	---	—►	+		+			+	−
Soils	◄—	—	—	—	—	—	—	—►			−		×		−
Lakes	↔									+			×		
									prod	pop	fert	treat	irrig	nb	
Anthropogenic sources															
Mines		◄---	---	---	---	---	---►		+				+		
Industries		◄---	---	---	---	---	---►		+				+		
Cities			◄—	—	—►					+					
Agriculture	◄—	—	—	—	—►						+				
Anthropogenic skins															
Reservoirs	↔					◄—	—	—►						+	
Irrigated soils	◄—	—	—	—	—	—	—	—►					+		

+, −: increase and decrease with related control; ×: complex relation; DIC: dissolved inorganic carbon.
$T°$: temperature; q: runoff; d: distance to ocean; QH: quaternary history; V: terrestrial biomass; τ: water residence time; T: volcanism, tectonic uplift and rifting.
Prod: production; pop: urban population; fert: fertilization rate; treat: waste water treatment and recycling; irrig: water loss through irrigation; nb: volume of reservoirs and eutrophication.

origin, (rich in NaCl and also in $MgSO_4$), differential weathering of silicate minerals and carbonate minerals, dissolution of evaporitic minerals contained in some sedimentary rocks (gypsum and anhydrite, halite) or leached during rainstorms from surficial soils of semi-arid regions (Garrels and Mackenzie, 1971; Drever, 1988; Stallard, 1995a,b). Sinks are also multiple: silica may be retained in lakes and wetlands due to uptake by aquatic biota. Carbonate minerals precipitate in some eutrophic lakes and also precipitate when the total dissolved solids increases, generally above $\Sigma^+ = 6\,meq\,L^{-1}$, due to evaporation.

In semi-arid regions ($30 < q < 140$ mm yr^{-1}) and arid regions (chosen here as $3 < q < 30$ mm yr^{-1}) surface waters gradually evaporate and become concentrated, reaching saturation levels of calcium and magnesium carbonates first, and then of calcium sulfate (Holland, 1978): these minerals are deposited in soils and in river beds during their drying stage, a process termed evaporation/crystallization by Gibbs (1970).

There is also growing evidence of active recycling of most major elements, including silica, in terrestrial vegetation, particularly in forested areas (Likens et al., 1977). Each element may, therefore, have multiple natural sources and sinks. The main controlling factors for each element are also multiple (Table 2). They can be climatic: higher temperature generally increases mineral

dissolution; water runoff increases all weathering rates. Conversely, soil retention increases at lower runoff. The weathering of silicate and carbonate minerals is also facilitated by organic acids generated by terrestrial vegetation. Higher lake residence times ($\tau > 6$ months) favor biogenic silica retention and the precipitation of calcite (e.g., 30 g $SiO_2\,m^{-2}\,yr^{-1}$ and 300 g $CaCO_3\,m^{-2}$ yr^{-1} in Lake Geneva). The distance to the oceans is the key factor controlling the inputs of sea salts.

Tectonics also exercises important regional controls: active tectonics in the form of volcanism and uplift is associated with the occurrence at the Earth's surface of fresh rock that is more easily weathered than surficial rocks that have been exposed to weathering for many millions of years on a stable craton (Stallard, 1995b): there, the most soluble minerals have been dissolved and replaced by the least soluble ones such as quartz, aluminum and iron oxides, and clays. Active tectonics also limits the retention of elements by terrestrial vegetation, and retention in arid soils due to high mechanical erosion rates. Rifting is generally associated with inputs of saline deep waters that can have a major influence on surface water chemistry at the local scale. In formerly glaciated shields as in Canada and Scandinavia, glacial abrasion slows the development of a weathered soil layer. This limits silicate mineral weathering even under high runoff. These low relief glaciated areas are also characterized by a very high lake density, an order of magnitude higher than in most other regions of the world; lakes are sinks for silica and nutrients. The effects of Quaternary history in semi-arid and arid regions can be considerable, due to the inheritance of minerals precipitated under past climatic conditions. Due to the effect of local lithology, distance to the ocean, regional climate and tectonics, and the occurrence of lakes, river chemistry in a given area may be very variable.

8.4.1 Influence of Lithology on River Chemistry

Lithology is an essential factor in determining river chemistry (Garrels and Mackenzie, 1971; Drever, 1988, 1994), especially at the local scale (Strahler stream orders 1–3) (Miller, 1961; Meybeck, 1986). On more regional scales, there is generally a mixture of rock types, although some large river basins (area $> 0.1\,\text{Mkm}^2$) may contain one major rock type such as a granitic shield or a sedimentary platform. When selecting nonimpacted monolithologic river basins in a given region such as France (Table 3, A–E), the influence of climate, tectonics, and distance from the ocean can be minimized, revealing the dominant control of lithology, which in turn depends on (i) the relative abundance of specific minerals and (ii) the sensitivity of each mineral to weathering. The weathering scale most commonly adopted is that of Stallard (1995b) for mineral stability in tropical soils: quartz $\gg$ K-feldspar, micas $\gg$ Na-feldspar $>$ Ca-feldspar, amphiboles $>$ pyroxenes $>$ dolomite $>$ calcite $\gg$ pyrite, gypsum, anhydrite $\gg$ halite (least stable).

Table 3F provides examples of stream and river chemistry under peculiar conditions: highly weathered quartz sand (Rio Negro), peridotite (Dumbea), hydrothermal inputs (Semliki and Tokaanu), evaporated (Saoura), black shales (Powder and Redwater), sedimentary salt deposits (Salt). They illustrate the enormous range of natural river chemistry (outlets of acidic volcanic lakes are omitted here).

Table 3 lists examples of more than a dozen different chemical types of river water. Although Ca^{2+} and HCO_3^- are generally dominant, Mg^{2+} dominance over Ca^{2+} can be found in rivers draining various lithologies such as basalt, peridotite, serpentinite, dolomite, coal, or where hydrothermal influence is important (Semliki). Sodium may dominate in sandstone basins, in black shales (Powder, Redwater in Montana), in evaporitic sedimentary basins (Salt), in evaporated basins (Saoura), and where hydrothermal and volcanic influence is important (Semliki, Tokaanu). K^+ rarely exceeds 4% of cations, except in some clayey sands, mica schists, and trachyandesite; it exceeds 15% in extremely dilute waters of Central Amazonia and in highly mineralized waters of rift lake outlets (Semliki, Ruzizi).

The occurrence of highly reactive minerals, such as evaporitic minerals, pyrite and even calcite, in low proportions—a percent or less—in a given rock, e.g., calcareous sandstone, pyritic shale, marl with traces of anhydrite, granite with traces of calcite, may determine the chemical character of stream water (Miller, 1961; Drever, 1988). In a study of 200 streams from monolithologic catchments underlain by various rock types under similar climatic conditions in France, the relative weathering rate based on the cation sum (Meybeck, 1986) ranges from 1 for quartz sandstone to 160 for gypsiferous marl.

When the lithology in a given region is fairly uniform, the distribution of major-element concentrations is relatively homogeneous with quantile ratios Q_{90}/Q_{10} well under 10 as, for example, in Japan (mostly volcanic), the Central Amazon Basin (detrital sand and shield) (Figure 1, distributions B and A); when a region is highly heterogeneous with regards to lithology, as in the Mackenzie Basin and in France, the river chemistry is much more heterogeneous (Figure 1, distributions E and F) and quantile ratios Q_{90}/Q_{10}, Q_{99}/Q_1 may reach those observed at the global scale (Figure 1, distributions G) (Meybeck, 1994). In the first set, regional geochemical background compositions can be easily defined, but not in the second set.

8.4.2 Carbon Species Carried by Rivers

The carbon cycle and its long-term influence on climate through the weathering of fresh silicate rocks (Berner *et al.*, 1983) has created a new interest in the river transfer of carbon. Bicarbonate (HCO_3^-) is the dominant form of DIC in the pH range of most world rivers ($6 < \text{pH} < 8.2$); carbonate (CO_3^{2-}) is significant only at higher pH, which occurs in a few eutrophic rivers such as the Loire River, where pH exceeds 9.2 during summer algal blooms, and in waters that have undergone evaporation. Undissociated dissolved CO_2 is significant only in very acidic waters rich in humic substances such as the Rio Negro (Amazonia), but this is unimportant at the global scale. In terms of fluxes, bicarbonate DIC dominates (Table 4). It has two different sources: (i) carbonic acid weathering of noncarbonate minerals, particularly of silicates such as feldspars, micas, and olivine, (ii) dissolution of carbonate minerals such as calcite and dolomite, in which half of the resulting DIC originates from soil and/or atmospheric CO_2, and half from the weathered rock. Other forms of riverine carbon include particulate inorganic carbon (PIC) due to mechanical erosion in carbonate terrains, and dissolved and particulate organic carbon (DOC and POC) that are largely due to soil leaching and erosion; fossil POC in loess and shale may also contribute to river POC; organic pollution and algal growth in eutrophic lakes and rivers contribute minor fluxes (Meybeck, 1993).

Table 3 A–E: composition of pristine waters draining single rock types in France (medians of analyses corrected for atmospheric inputs) (from 3 to 26 analyses in each class, except for estimates that are based on one analysis only). F: other river chemistry from various origins uncorrected for atmospheric inputs (Meybeck, 1986). Cation and anion proportions in percent of their respective sums (Σ^+ and Σ^-).

	SiO_2 ($\mu mol^{-1}L^{-1}$)	Σ^+ ($\mu eq^{-1}L^{-1}$)	Ca^{2+} (%)	Mg^{2+} (%)	Na^+ (%)	K^+ (%)	Cl^- (%)	SO_4^{2-} (%)	HCO_3^- (%)
A. *Noncarbonate detrital rocks*									
Quartz sand and sandstones	170	170	30	20	45	5	0	(40)	(60)
Clayey sands	135	300	53	17	13	17	0	(30)	(70)
Arkosic sands	200	400	48	35	12	5	0	(20)	(80)
Graywacke	90	350	58	22	18	2	0	(20)	(80)
Coal-bearing formations	150	5,000	30	55	14	1	0	20–90	80–10
B. *Carbonate-containing detrital rocks*									
Shales	90	500	60	30	8	1.5	0	25	75
Permian shales	175	2,200	53	35	8	4	(0–20)	10	90–70
Molasse	280	2,500	80	17	3	0.3	0	2	98
Flysch	50	2,200	79	19	1.5	0.5	0	2	98
Marl	90	3,000	83	14	2.5	0.5	0	2	98
C. *Limestones*									
Limestones	60	4,500	95	3	0.6	0.4	0	2.5	97.5
Dolomitic limestones	60	4,500	72	26	0.6	0.4	0	3.5	96.5
Chalk[a]	200	4,500	95	2.5	2.2	0.5	0	2.5	97.5
Dolomite[a]	67	5,900	54	46	0.1	0.1	0	8.5	91.5
D. *Evaporites*									
Gypsum marl[a]	160	22,000	77	22	0.2	0.2	0	83	17
Salt and gypsum marl[a]	133	27,500	34	32	33	(0.8)	36	44	20
E. *Plutonic, metamorphic, and volcanic rocks*									
Alkaline granite, gneiss, mica schists	140	130	15	15	65	5	0	(30)	(70)
Calc-alkaline granite, gneiss, mica schists	100	300	54	25	17	4	0	(15)	(85)
Serpentinite	225	1,500	38	60	7	1	0	(7)	93
Peridotite	180	600	5	93	1	1	0	(15)	(85)
Amphibolite	65	1,600	85	17	2	1	0	(7)	(93)
Marble[a]	150	3,400	86	11	1.8	0.6	0	(12)	88
Basalt	200	500	42	38	17	2.5	0	(2)	98
Trachyandesite	190	220	32	25	35	8	0	(2)	98
Rhyolite	190	550	53	15	25	2	0	(2)	98
Anorthosite	260	400	45	25	26	4	0	(2)	98
F. *Miscellaneous river waters (not rain corrected)*									
Rio Negro tributaries[b]	75	18.1	10.5	16.5	51.9	20.9			
Dumbea (New Caledonia)	232	1,175	3.9	84.7	10.9	0.3	14.2	7.1	78.5
Cusson (Landes, France)[b]	251	1,463	13.8	21.2	61.5	3.4	69	12.3	18.7
Semliki (Uganda)[c]	213	8,736	6.4	36.2	39.6	17.5	14.6	23.3	62.0
Powder (Montana)	148	20,200	27.2	20.3	51.5	0.9	15.0	62.8	22.2
Saoura (Marocco)[d]		26,150	23.3	16.8	59.2	0.7	62.8	27.7	9.6
Redwater (Montana)	116	40,700	10.7	24.5	64.1	0.6	1.0	72.2	26.7
Tokaanu (New Zeal.)[c]	4,760	41,600	14.4	3.0	79.5	3.1	90.6	5.8	3.5
Salt (NWT, Canada)	20	312,000	9.7	1.8	88.4	0.1	89.7	9.3	1.0

[a] Estimates.
[b] Rain dominated.
[c] Hydrothermal inputs.
[d] Evaporated.

The "ages" of these carbon species, i.e., the time since their original carbon fixation fall into two categories: (i) those from 0 to ~1,000 yr, representing the fast cycling external part of the cycle are termed total atmospheric carbon (TAC); (ii) those from 50 kyr (Chinese Loess Plateau) to 100 Myr representing "old" carbon from sedimentary rocks. The sensitivity of these transfers to global change is complex (Table 4).

River carbon transfers are also very variable at the global scale: DIC concentrations range from 0.06 mg DIC L^{-1} (Q_1) to 71 mg DIC L^{-1}

Table 4 Riverine carbon transfer and global change.

	Sources	Age (yr)	Flux[a] $(10^{12} \text{ g C yr}^{-1})$	Sensitivity to global change						
				A	B	C	D	E	F	
PIC	Geologic	10^4–10^8	170	●					●	
DIC	Geologic	10^4–10^8	140		●	●			●	
	Atmospheric	0–10^2	245		●	●			●	TAC
DOC	Soils	10^0–10^3	200				●		●	
	Pollution	10^{-2}–10^{-1}	(15) ?					●		
CO_2	Atmospheric	0	(20–80)		●	●	●			
POC	Soils	10^0–10^3	(100)	●					●	
	Algal	10^{-2}	(<10)					●	●	
	Pollution	10^{-2}–10^0	(15) ?					●		
	Geologic	10^4–10^8	(80)	●					●	

A = land erosion, B = chemical weathering, C = global warming and UV changes, D = eutrophication, E = organic pollution, F = basin management, TAC = Total atmospheric carbon.
[a] Present global flux to oceans mostly based on Meybeck (1993), 10^{12} g C yr^{-1}.

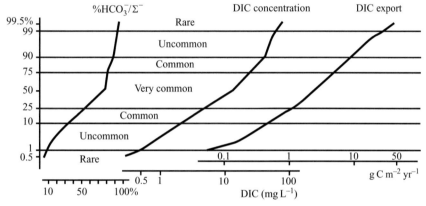

Figure 2 Global distribution of DIC concentration, ratio and export rate (yield) in medium-sized basins (3,500–200,000 km^2).

(Q_{99}); DIC export, or yield, ranges from 0.16 g C m^{-2} yr^{-1} (Q_1) to 33.8 g C m^{-2} yr^{-1} (Q_{99}). In 50% of river basins bicarbonate makes up more than 80% of the anionic charge. The distribution quantiles for these variables are shown in Figure 2.

Organic acids, particularly those present in wetland-rich basins, also play an important role in weathering (Drever, 1994; Viers *et al.*, 1997).

8.4.3 Influence of Climate on River Chemistry

The influence of air temperature is nearly impossible to measure for large basins due to its spatial heterogeneity and its absence in most databases. Some studies of mountain streams have considered the effect of air temperature (Drever and Zobrist, 1992). It is generally agreed that silicate weathering is more rapid in wet tropical regions (other than areas of very low relief) due to the combined effect of high temperature, vegetation impact and, most of all, high runoff.

The influence of the water balance can be studied easily; annual river runoff, which is generally documented, integrates entire river basins (Meybeck, 1994; White and Blum, 1995).

The set of PRISRI rivers has been subdivided into 12 classes of runoff for which the median elemental concentrations have been determined (Figure 3). The Q_{25} and Q_{75} quantiles and, to some extend the Q_{10} to Q_{90} quantiles follow the same patterns for all elements. On the basis of Na$^+$, K$^+$, Cl$^-$, and SO$_4^{2-}$ two different clusters can be distinguished: (i) above 140 mm yr^{-1} the influence of runoff is not significant, (ii) below 140 mm yr^{-1} there is a gradual increase in elemental concentrations as runoff decreases. This pattern is interpreted as due to evaporation. For other elements there is a gradual increase of Ca^{2+}, Mg^{2+}, and HCO$_3^-$ from the highest runoff to 85 mm yr^{-1} or 50 mm yr^{-1}, then a stabilization of these concentrations at lower runoff values ($q < 50$ mm yr^{-1}). The second part of this pattern is most probably due to the precipitation of carbonate minerals as a result of evaporative

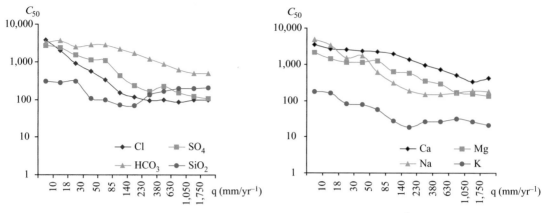

Figure 3 Median concentrations C_{50} in 12 classes of river runoff (q_1: <10 mm yr^{-1}, then 10–18, 18–30, 30–50, 50–85, 85–140, 140–230, 230–380, 380–630, 630–1,050, 1,050–1,750 and q_{12}: >1,750 mm yr^{-1}) (ions in μeq L^{-1}, silica in μmol L^{-1}) (total sample, $n = 1,091$).

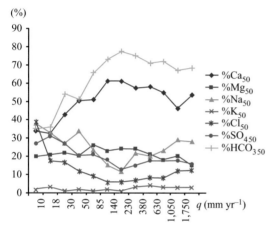

Figure 4 Median proportions (in percent of cations or anions sum) in 12 classes of river runoff (q_1: <10 mm yr^{-1}, then 10–18, 18–30, 30–50, 50–85, 85–140, 140–230, 230–380, 380–630, 630–1,050, 1,050–1,750 and q_{12}: >1,750 mm yr^{-1}, sample $n = 1,091$).

concentration. The decrease in Ca^{2+}, Mg^{2+}, and HCO_3^- concentrations at very high runoff, while the silicate-weathering related Na^+ and K^+ are stable, could be due to a lower occurrence of carbonate rocks in wetter regions. This does not imply a climate control of these rock types so much as the dominance of the Amazon and Congo basins, which are underlain by silicate rocks, and volcanic islands in the population of high-runoff rivers. The silica pattern is even more complex: the increase in silica concentration with runoff from 140 mm yr^{-1} to >1,750 mm yr^{-1} could be linked with a greater occurrence of crystalline rocks, particularly of volcanic rocks, in wetter regions as in volcanic islands (Iceland, Japan, Indonesia) or coasts (Washington, Oregon), which are particularly abundant in the PRISRI database. These interactions between climate and lithology are possible but need to be verified. The lower silica value from 30 mm yr^{-1} to 140 mm yr^{-1}

could also be attributed to the importance of the Canadian and Siberian rivers in these classes, where SiO_2 retention in lakes is important.

The relative median proportions of ions in the runoff classes of the PRISRI rivers confirm this pattern (Figure 4): below 140 mm yr^{-1} the $\%HCO_3^-$ and $\%Ca^{2+}$ drop from 75% and 60%, respectively, to ~33%, while those of Cl^- and SO_4^{2-} increase from 6% to 39% and from 13% to 27%, respectively: the proportion of K^+ does not vary significantly. There is a significant increase in the median proportions of Na^+ and Cl^- above 1,050 mm yr^{-1} that could correspond with the increased marine influence in these categories. The median proportion of Mg^{2+} is the most stable of all ions: from 15% to 26%.

8.5 IDEALIZED MODEL OF RIVER CHEMISTRY

Gibbs (1970) defined three main categories of surface continental waters: rain dominated, with Na^+ and Cl^- as the major ions, weathering dominated, with Ca^{2+} and HCO_3^- as the major ions, and evaporation/crystallization dominated, with Na^+ and Cl^- as the major ions. This typology can still be used, but it is too simplified. According to the PRISRI database there are many more types of water and controlling mechanisms, although the ones described by Gibbs may account for ~80% of the observed chemical types. The sum of cations (Σ^+), which is a good indicator of weathering, ranges from 50 μeq L^{-1} to 50,000 μeq L^{-1}, and the corresponding types of water, characterized by the dominant ions may reach as many as a dozen (Table 3). The true rain dominance type is found near the edges of continents facing oceanic aerosol inputs, as in many small islands and in Western Europe: it is effectively of the Na^+–Cl^- type and Σ^+ may be as high as 1 meq L^{-1}, for example, the Cusson R. (Landes, France) draining arkosic sands

(Table 3, F). In continental interiors, in some rain forests, this water type occurs with much lower ionic concentrations ($\Sigma^+ < 0.1\,meq\,L^{-1}$) in areas with highly weathered soils and very low mechanical erosion rates, such as the Central Amazon Basin (Rio Negro tributaries, Table 3F) or Cameroon: the very limited cationic inputs in rainfall are actively utilized and recycled by the forest and may even be stored (Likens *et al.*, 1977; Viers *et al.*, 1997). If the DOC level is high enough ($\sim 10\,mg\,L^{-1}$) the pH is often so low, that HCO_3^- is insignificant: the dominant anions are SO_4^{2-} and organic anions or Cl^-. This water chemistry is actually controlled by the terrestrial vegetation and can have a variety of ionic assemblages. The silica generated by chemical weathering is exported as dissolved SiO_2 and also as particulate biogenic SiO_2 from phytoliths or sponge spicules, as in the Rio Negro Basin.

Numerous water types reflect weathering control; these include Mg^{2+}–HCO_3^-, Ca^{2+}–HCO_3^-, Na^+–HCO_3^-, $Mg^{2+} - SO_4^{2-}$, $Ca^{2+} - SO_4^{2-}$, $Na^+ - SO_4^{2-}$, Na^+–Cl^- types (Table 3). The evaporation–crystallization control found in semiarid and arid regions such as Central Asia (Alekin and Brazhnikova, 1964) also gives rise to multiple water types: $Mg^{2+} - SO_4^{2-}$, $Ca^{2+} - SO_4^{2-}$, $Na^+ - SO_4^{2-}$, Na^+–Cl^-, Mg^{2+}–HCO_3^-, and even Mg^{2+}–Cl^-. It is difficult to document the precipitation–redissolution processes (either during rare rain storms for occasional streams or in allochtonous rivers flowing from wetter headwaters such as the Saoura in Marocco, Table 3F), which are likely to have $\Sigma^+ > 6\,meq\,L^{-1}$ and/or runoff below 140–85 mm yr^{-1}. In some Canadian and Siberian basins, very low runoff (<85 mm yr^{-1}) is not associated with very high evaporation but with very low precipitation: weathering processes still dominate.

In rift and/or volcanic regions and in recent mountain ranges such as the Caucasus, hydrothermal inputs may add significant quantities of dissolved material (Na^+, K^+, Cl^-, SO_4^{2-}, SiO_2) to surface waters. The Semliki River, outlet of Lake Edward, is particularly enriched in K^+ (Table 3, F); the Tokaanu River (New Zealand) drains a hydrothermal field with record values of silica, Na^+ and Cl^- (Table 3, F).

Overall, on the basis of ionic proportions and total concentrations, only 8.2% of the rivers (in number) in the PRISRI database can be described as evaporation controlled, 2.6% as rain dominated and vegetation controlled, and 89.2% as weathering dominated, including rivers affected by large water inputs.

A tentative reclassification of the major ionic types is presented on Figure 5 showing the occurrence of major ion sources for different classes of Σ^+. The rainfall dominance types also include rivers with vegetation control and correspond to multiple ionic types ($Ca^{2+} - SO_4^{2-}$, Na^+–Cl^-, Ca^{2+}–Cl^-, Na^+–HCO_3^-). Other water types

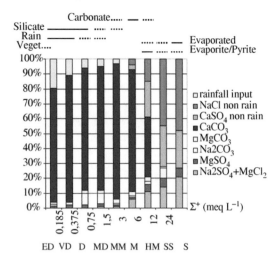

Figure 5 Idealized occurrence of water types (dominant ions and dominant control factors) per classes of increasing cationic content (Σ^+ in meq L^{-1}). ED: extremely dilute waters, VD: very dilute, D: dilute, MD: medium dilute, MM: medium mineralized, HM: highly mineralized, SS: subsaline, S: saline.

correspond to various rock weathering modes, including Mg^{2+}–HCO_3^- waters observed in many volcanic regions. The most common type, Ca^{2+}–HCO_3^-, dominates from $\Sigma^+ < 0.185\,meq\,L^{-1}$ to $6\,meq\,L^{-1}$: in the most dilute waters it originates from silicate rock weathering, above $1.5\,meq\,L^{-1}$ from weathering of carbonate minerals. This water type virtually disappears above $12\,meq\,L^{-1}$. "Non-rain" $NaCl$ and $CaSO_4$ water types appear gradually above $6\,meq\,L^{-1}$. However, with these water types it is impossible to differenciate between evaporite rock dissolution, which can be observed even in the wet tropics (Stallard and Edmond, 1981) or in the Mackenzie (Salt R., Table 3F), from the leaching of salinized soils very common in Central Asia (Alekin and Brazhnikova, 1964). $Mg^{2+} - SO_4^{2-}$, $Na^+ - SO_4^{2-}$, and Mg^{2+}–Cl^- types are occasionally found in streams with Σ^+ values below $3\,meq\,L^{-1}$; they are commonly the result of pyrite weathering that can lead to very high Σ^+ as in the Powder and Redwater Rivers (Table 3, F).

The global proportions of the different water types depend on the global representativeness of the database. The exact occurrence of water types can only be estimated indirectly through modelling on the basis of lithologic maps, water balance, and oceanic fallout. Moreover, it will depend on spatial resolution: a very fine scale gives more importance to the smallest, rain-dominated coastal basins. In PRISRI, 70% of the basins have areas exceeding 3,200 km^2, thus limiting the appearance of oceanic influence.

A river salinity scale based on Σ^+ is also proposed (Figure 5). The least mineralized waters ($\Sigma^+ < 0.185\,meq\,L^{-1}$), termed here "extremely dilute" correspond to a concentration of total dissolved solids of $\sim 10 \text{mg L}^{-1}$ in NaCl equivalent. The most mineralized waters ($\Sigma^+ > 24 \text{ meq L}^{-1}$) are here termed "saline" up to 1.4 g L^{-1} NaCl equivalent, a value slightly less than the conventional limit of 3 g L^{-1} NaCl adopted for "saline" lakes.

8.6 DISTRIBUTION OF WEATHERING INTENSITIES AT THE GLOBAL SCALE

Present-day weathering intensities can theoretically be assessed by the export of dissolved material by rivers. Yet many assumptions and corrections have to be made: (i) human impacts (additional sources and sinks) should be negligible, (ii) atmospheric inputs should be subtracted, (iii) products of chemical weathering should not be carried as particulates (e.g., phytoliths), nor (iv) accumulated within river basins in lakes or soils. In addition, it must be remembered that 100% of the HCO_3^- may originate from the atmosphere in noncarbonate river basins (calcite is only found in trace amounts in granites) and $\sim 50\%$ in carbonate terrains. In basins of mixed lithologies the proportions range between 50% and 100%. The total cation export or yield Y^+ (which excludes silica) is used to express the weathering intensity. Y^+, expressed in eq m^{-2} yr^{-1}, the product of annual runoff (m yr^{-1}) and Σ^+ (meq L^{-1}), is extremely variable at the Earth's surface since it combines both runoff variability and river chemistry variability (Table 5).

The concentration of elements that are derived from rock weathering (Ca^{2+}, Mg^{2+}, Na^+, K^+), are less variable than runoff even in the driest conditions. The opposite is observed for Cl^- and SO_4^{2-}, which are characterized by very low Q_1 quantiles. The retention of silica, particularly under the driest conditions, makes its yield the most variable. The lowest yearly average runoff in this data set (3.1 mm yr^{-1} for Q_1) actually corresponds to the conventional limit for occasional river flow (3 mm yr^{-1}). Under such extremely arid conditions, flow may occur only few times per hundred years, as for some tributaries of Lake Eyre in Central Australia. The other runoff quantile Q_{99} corresponds to the wettest regions of the planet bordering the coastal zone.

Table 5 Global distribution of ionic (eq m^{-2} yr^{-1}) and silica (mol m^{-2} yr^{-1}) yields and annual runoff (q in mm yr^{-1}) in medium-sized basins (3,200–200,000 km^2, $n = 685$).

Y_i	Ca^{2+}	Mg^{2+}	Na^+	K^+	Cl^-	SO_4^{2-}	HCO_3^-	Σ^+	SiO_2	q
Q_{99}	2.77	0.95	1.3	0.115	1.05	1.25	2.82	3.0	0.82	3,040
Q_1	0.0045	0.002	0.002	0.0002	0.0003	0.0004	0.0046	0.0005	0.0001	3.1
Q_{99}/Q_1	615	475	650	575	3500	3100	613	600	8200	980

Q_1 and Q_{99}: lowest and highest percentiles.

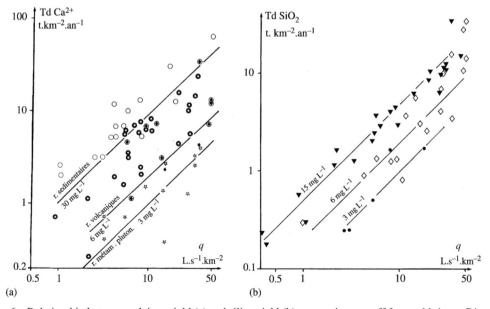

Figure 6 Relationship between calcium yield (a) and silica yield (b) versus river runoff for world rivers. Diagonals represent lines of equal concentrations. (a) Calcium yield for carbonated basins (average 30 mg Ca^{2+}L^{-1}), volcanic basins (6 mg L^{-1}), shield and plutonic (3 mg L^{-1}). (b) Silica yield for tropical regions (average 15 mg SiO$_2$ L^{-1}), temperate regions (6 mg L^{-1}), and cold regions (3 mg L^{-1}) (source Meybeck, 1994).

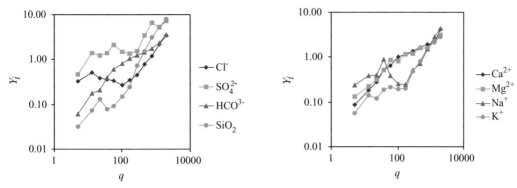

Figure 7 Median yields of ions and silica (Y_i in meq m^{-2} yr^{-1}) for 12 classes of river runoff ($n = 1{,}091$, all basins). Yields $= C_{i50} \times q_{50}$ in each class.

Since the variability of runoff, represented by the percentile ratio Q_{99}/Q_1, is generally much greater than the variability of concentration (Tables 1 and 5), ionic yields primarily depend on runoff. With a given rock type, there is a strong correlation between yield and runoff, that corresponds to clusters of similar concentrations as for Ca^{2+} and silica (Figure 6). This influence was used by Meybeck (1994), Bluth and Kump (1994), and Ludwig *et al.* (1998) to model ionic yields and inputs of carbon species to the ocean. If all PRISRI basins are considered, this runoff control on chemical yields is still observed, although the relationship is complex. The data set is subdivided into 12 classes of runoff from $q < 10$ mm yr^{-1} to $q > 1{,}750$ mm yr^{-1} (see Figure 3), for which the median yields of major ions and silica have been determined (Figure 7). In this log–log diagram, domains of equal concentration are parallel to the diagonal (1:1) line. Several types of evolution can be observed:

(i) sulfate and chloride yields are fairly constant below 140 mm yr^{-1}, suggesting that lower runoff is compensated by a concentration increase through evaporation (except for SO_4^{2-} below 10 mm yr^{-1});

(ii) above 140 mm yr^{-1} all ion and silica yields are primarily linked to runoff;

(iii) above 140 mm yr^{-1} the median concentrations of Ca^{2+}, Mg^{2+}, and HCO_3^- decrease gradually, whereas silica and potassium concentrations increase gradually, suggesting a greater influence of silicate weathering relative to carbonate weathering;

(iv) below 140 mm yr^{-1} median Ca^{2+} and Mg^{2+} concentrations are constant and median HCO_3^- concentration decreases slightly, which suggest a regulation mechanism through precipitation of carbonate minerals in semi-arid and arid regions;

(v) potassium yield is stable between 30 mm yr^{-1} and 140 mm yr^{-1}; below 30 mm yr^{-1} it decreases, suggesting retention; and

(vi) median silica yield is the most complex: there is certainly a retention in the arid regions. These observed trends need to be confirmed particularly below 30 mm runoff, where there are fewer silica analyses in the PRISRI base than analyses of major ions.

Detailed weathering controls, particularly for silicate rocks, can be found in Holland (1978), Drever and Clow (1995), Drever and Zobrist (1992), Gallard *et al.* (1999), Stallard (1995a,b), and White and Blum (1995).

8.7 GLOBAL BUDGET OF RIVERINE FLUXES

The unglaciated land surface of the present Earth amounts to $\sim$133 Mkm^2. Excluding land below a 3 mm yr^{-1} runoff threshold (at a 0.5° resolution), $\sim$50 Mkm^2 can be considered as nonexposed to surface water weathering (arheic, where $q < 3$ mm yr^{-1}), whether draining to the ocean (exorheic) or draining internally such as the Caspian Sea basin (endorheic) (Table 6). The land area effectively exposed to weathering by meteoric water is estimated to be $\sim$82.8 Mkm^2 (rheic regions); this, in turn, has to be divided into exorheic regions (76.1 Mkm^2) and endorheic regions (6.7 Mkm^2). Since the weathering intensity in the rheic regions is highly variable (see above), three main groupings have been made on the basis of weathering intensity or runoff: the least active or oligorheic regions (36.4 Mkm^2), the regions of medium activity or mesorheic (42.3 Mkm^2), and the most active or hyperrheic regions (4.1 Mkm^2). The corresponding ionic fluxes can be computed on the basis of the PRISRI database. The hyperrheic regions, which are exclusively found in land that drains externally, represent 37% of all DIC fluxes for only 2.75% of the land area; the mesorheic regions (28.3% of land area) contribute 55% of the DIC fluxes, and the oligorheic regions

Table 6 Distribution of global land area (Mkm²) exposed to chemical weathering and to river transfer of soluble material. A: percent of land area (nonglaciated area also contains alpine glaciers). B: percent of weathering generated fluxes (e.g., DIC flux). C: percent of river fluxes to oceans.

				A % land area	B % weath. flux	C % flux to ocean
Total land 149 Mkm²	Glaciated 16 Mkm²			10.7	0.1?	0.1?
	Nonglaciated 133 Mkm²	Arheic 50.2 Mkm²	Endorheic[a]	33.7	0	0
			Exorheic[a]			
		Rheic 82.8 Mkm²	Oligorheic { Endorheic	3.4	1	0
			Exorheic	21	7	7
			Mesorheic { Endorheic	1.1	2	0
			Exorheic	27.3	53	55
			Hyperrheic Exorheic	2.75	37	38

[a] Potentially.

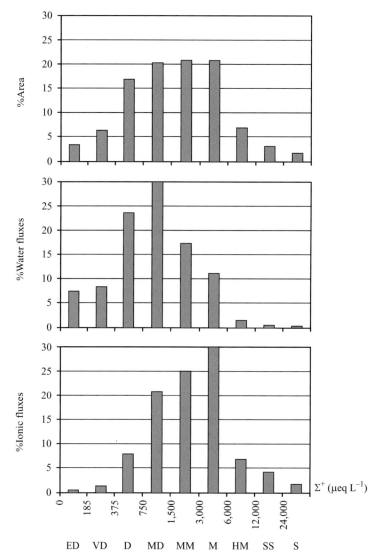

Figure 8 Distribution of global river attributes from the PRISRI database for classes of increasing total cations (Σ^+) based on PRISRI rivers (rheic regions). ED: extremely dilute waters, VD: very dilute, D: dilute, MD: medium dilute, MM: medium mineralized, M: mineralized, HM: highly mineralized, SS: subsaline, S: saline.

Table 7 WWA ionic concentrations (C_i^* in μeq L^{-1}, and μmol L^{-1} for silica) and yields (Y_d^* in g m^{-2} yr^{-1}, Y_i^* in meq m^{-2} yr^{-1}, and mmol m^{-2} yr^{-1} for silica), ionic ratios (C_i/C_j^* in eq eq^{-1}) and relative ionic proportion (%C_i^*). Same variables for the WSM determined for mesobasins (3,000–200,000 km^2) Na$^{\#}$ = Na–Cl; q = global average runoff (mm yr^{-1}).

	SiO_2	Ca^{2+}	Mg^{2+}	Na^+	K^+	Cl^-	SO_4^{2-}	HCO_3^-	Σ^+	Σ^-	q
World river inputs to oceans											
Clarke (1924) C_i	161	843	231	208	45	132	208	969	1,327	1,309	(320)
Livingstone (1963) C_i	217	748	337	274	59	220	233	958	1,418	1,411	314
Meybeck (1979)											
pristine C_i	174	669	276	224	33	162	172	853	1,202	1,187	374
1970 C_i	174	733	300	313	36	233	239	869	1,382	1,341	374
World weighted average (WWA) (endorheic + exorheic) (this work)											340
C_i^*	145	594	245	240	44	167	175	798	1,125	1,139	
Y_d^*	2.98	4.04	1.01	1.88	0.60	2.0	2.87	16.54			
Y_i^*		202	83.3	81.8	15.3	56.8	59.7	271.3	382	388	
$\%C_i^*$		52.2	21.5	23	3.3	14.7	15.4	69.9	100	100	
World spatial median (WSM) (endorheic + exorheic) (this work)											222
C_{i50}	134	1,000	375	148	25.5	96	219	1,256	1,548	1,571	
Y_{d50}	1.82	3.78	0.91	1.31	0.29	0.95	2.66	13.3			
Y_{i50}		188	75	57	7.4	22	55	278	327	300	
$\%C_{i50}$		64.6	24.1	9.6	1.7	6.1	13.9	80			

	$\dfrac{Ca^{2+}}{Mg^{2+}}$	$\dfrac{Ca^{2+}}{Na^+}$	$\dfrac{Na^+}{K^+}$	$\dfrac{Na^+}{Cl^-}$	$\dfrac{Ca^{2+}}{SO_4^{2-}}$	$\dfrac{Mg^{2+}}{Na^+}$	$\dfrac{Cl^-}{SO_4^{2-}}$	$\dfrac{SO_4^{2-}}{HCO_3^-}$	$\dfrac{Na^{\#}}{K^+}$	$\dfrac{Ca^{2+}}{Na^{\#}}$	$\dfrac{Mg^{2+}}{Na^{\#}}$	$\dfrac{SiO_2}{\Sigma^+}$
WWA												
$(C_i/C_j)^*$	2.42	2.47	5.5	1.43	3.39	1.02	0.95	0.22	1.65	8.1	3.35	0.13
WSM												
$(C_i/C_j)_{50}$	2.32	2.54	6.55	2.27	2.97	1.14	0.48	0.26	3.0	4.3	1.9	0.12

(24.4% of land area) contribute only 8% of these fluxes (Table 6). If a finer spatial distribution of these fluxes is extracted from the PRISRI data, 1% of the land surface (carbonate rocks in very wet regions) contributes 10% of the river DIC fluxes. Similar figures are found for all other ions.

These budgets can also be broken down into classes on the basis of ionic contents (Figure 8), assuming the PRISRI database to be fully representative at the global scale. The extremely dilute waters correspond in PRISRI to ∼3% of the land area, but they are areas of very high runoff, therefore, their weight in terms of water volume is ∼7.5%. However, their influence on the global ionic fluxes is very low, less than 1%. At the other end of the salinity scale the subsaline and saline waters, mostly found in semi-arid and arid areas, correspond to ∼5% of the PRISRI data set; they contribute much less than 1% of runoff but ∼7% of the ionic fluxes.

Although we have drawn attention to the extreme variability of ionic and silica concentrations, ionic proportions, ionic ratios, ionic and silica yields throughout this chapter, two sets of global averages are proposed here (Table 7). The world spatial median values (WSM) of the medium-sized PRISRI data set (3,200–200,000 km^2 endorheic and exorheic basins) correspond

to the river water chemistry most commonly found on continents at this resolution. The world weighted average (WWA) has been computed by summing the individual ionic fluxes of the largest 680 basins, including endorheic basins (Aral, Caspian, Titicaca, Great Basin, Chad basins), in the PRISRI set. The runoff values in both averages are different, although very close considering the global range. The WSM lists higher concentrations than the WWA because the dry and very dry regions are more common in the database than the very humid regions. Ionic ratios are more similar between the two averages. The composition of river inputs to the oceans is not stable: (i) it has varied since the Late Glacial Maximum and during the geological past (Kump and Arthur, 1997); (ii) present-day river chemistry is much altered by human activities. Four estimates of average exorheic rivers are presented in Table 7: considering the natural variations of river chemistry, these averages are relatively close to each other and to WWA and WSM, the differences are due to the nature of the data sets and to the inclusion or exclusion of presently altered rivers. A re-estimation of the pristine inputs (under present-day climatic conditions) should now be done with the new data, although human impacts are difficult to quantify.

8.8 HUMAN ALTERATION OF RIVER CHEMISTRY

River chemistry is very sensitive to alteration by many human activities, particularly mining and the chemical industries, but also to urbanization as urban wastewaters are much more concentrated than rural streams, and to agriculture through the use of fertilizers (Table 2) (Meybeck *et al.*, 1989; Meybeck, 1996; Flintrop *et al.*, 1996). New sinks are also created such as reservoirs (calcite and silica trapping; enhanced evaporation) and irrigated soils, which may retain soluble elements if they are poorly drained. New controls correspond to these anthropogenic influences, such as mining and industrial production, rate of urbanization and population density, fertilization rate, irrigation rate and practices, and construction and operation of reservoirs. Wastewater treatment and/or recycling can be effective as a control on major ions originating from mines (petroleum and gas exploitation, coal and lignite, pyritic ores, potash and salt mines) and industries, yet their effects are

seldom documented. Urban wastewater treatment does not generally affect the major ions. Human impacts on silica are still poorly studied apart from retention in reservoirs. At the pH values common in surface waters, silica concentration is limited and evidence of marked excess silica in rivers due to urban or industrial wastes has not been observed by this author. The gradual alteration of river chemistry was noted very early, for example, for SO_4^{2-} (Berner, 1971) and Cl^- (Weiler and Chawla, 1969) in the Mississippi and the Saint Lawrence systems. Regular surveys made since the 1960's and comparisons with river water analyses performed a hundred years ago reveal a worldwide increase in Na^+, Cl^-, and SO_4^{2-} concentrations, whereas Ca^{2+}, Mg^{2+}, and HCO_3^- concentrations are more stable (Meybeck *et al.*, 1989; Kimstach *et al.*, 1998). Some rivers affected by mining (Rhine, Weser, Vistula, Don) may be much more altered than rivers affected by urbanization and industrialization only (Mississippi, Volga, Seine) (Figure 9). When river water is diverted and used for irrigation, there is a gradual

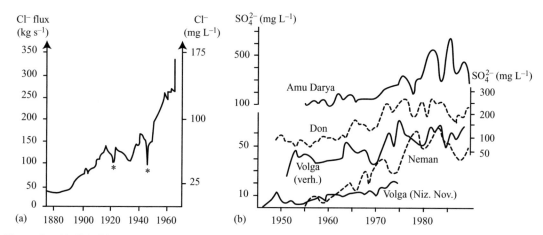

Figure 9 (a) Chloride evolution (fluxes, concentrations) in the Rhine River at mouth from 1,875 to 1,970 (ICPR, 1975). (b) Sulfate evolution in selected rivers of the former USSR, Volga at Nizhny Novgorod and Neman at Sovetsk (lower scale), Don at Aksai, and Volga at Verhnelebyazhye (medium scale), Amu Darya at Kzyl Djar (upper scale) (Tsikurnov in Kimstach *et al.*, 1998). [*] reduced mining activity in the Rhine Basin due to crisis and conflicts.

Table 8 Excess loads of major elements normalized to basin population in industrialized regions ($kg\,cap^{-1}\,yr^{-1}$).

	SiO_2	Ca^{2+}	Mg^{2+}	Na^+	K^+	Cl^-	SO_4^{2-}	HCO_3^-	Period
Per capita loads in residential urban sewage ($kg\,cap^{-1}yr^{-1}$)									
Montreal	ND	3.2	0.65	6.6	1.0	8.2	13.5	24	1970s
US sewer	2.4	3	1.5	14	2	15	6	20	1960s
Brussels	1.2	2.6	ND	9.5	1.6	8.4	5.8	14.7	1980s
Paris	0.2	1.2	0.7	6.4	1.5	6.3	11.0	14.5	1990s
Evolution of rivers[a] ($kg\,cap^{-1}yr^{-1}$)									
	ND	51	17	85	6	100	136	ND	1900–1970
Estimated anthropogenic input to ocean ($10^6\,t\,yr^{-1}$)									
	ND	47	10.5	78	5	93	124	100	1970

Sources: Meybeck (1979) and Meybeck *et al.* (1998).
[a]Mississippi, St. Lawrence, Seine, Rhine, Odra, Wisla.

increase of ionic content, particularly Na^+, K^+, Cl^-, and SO_4^{2-} as for the Colorado, Murray, Amu Darya (Figure 9). Salinization may also result from agriculture as for the Neman River (Figure 9). Salinization is discussed in detail by Vengosh.

It is difficult to assess the different anthropogenic sources of major ions which depend on the factors mentioned above. These human factors also vary in time for a given society and reflect the different environmental concerns of these societies, resulting in multiple types of river–society relationships (Meybeck, 2002). In the developed regions of the northern temperate zone it is now difficult to find a medium-sized basin that is not significantly impacted by human activities. In industrialized countries, each person generates dissolved salt loadings that eventually reach river systems (Table 8).

These anthropogenic loads are higher for Na^+, Cl^-, and SO_4^{2-} relative to natural loads. This partially explains the higher sensitivity of river chemistry to human development. At a certain population density in impacted river basins the anthropogenic loads equal (Na^+, K^+, Cl^-, SO_4^{2-}) or greatly exceed the natural ones (NO_3^-, PO_4^{3-}), defining a new era, the Anthropocene (Meybeck, 2002) when humans control geochemical cycles.

The silica trend in world rivers has recently attracted attention: dissolved silica is decreasing in impounded and/or eutrophied rivers, at the same time as nitrate is increasing in agricultural basins. As a result the Si : N ratio, which was generally well above 10 a hundred years ago, has dropped below 1.0 g g^{-1} in rivers such as the Mississippi, resulting in a major shift in the coastal algal assemblages (Rabalais and Turner, 2001), leading to dystrophic coastal areas. This trend, which is also observed for other river systems (Seine, Danube), could expand in the next decades due to increased fertilizer use and to increased reservoir construction.

8.9 CONCLUSIONS

Riverine chemistry is naturally highly variable at the global scale, which confirms the observation made by (the former) Soviet geochemists on a 20 Mkm^2 subset of land area (Alekin and Brazhnikova, 1964) and the conclusions of Clarke (1924), Livingstone (1963), Holland (1978), and Drever (1988). Ionic concentrations and yields (weathering rates) commonly vary over two to three orders of magnitude: only a fraction of the surface of the continents is actively exposed to weathering by meteoric water. The related river water types are multiple. There are at least a dozen types depending on surficial rock exposed to weathering, the water balance, and atmospheric inputs. Gibbs' (1970) global scheme for water chemistry holds for ~80% of river waters but is oversimplified for the remaining 20%. In very

dilute waters (cation sum $<0.185\,\mu eq\,L^{-1}$) water chemistry is probably controlled by vegetation. In these areas water chemistry is likely to be very sensitive to climate change and to forest cutting, although the related ionic fluxes are small. Each major ion and silica should be considered individually, since its sources, sinks and controls, both natural and anthropogenic, are different.

In the northern temperate regions of the world from North America to East Asia, the human impacts should be carefully filtered from the natural influence: the study of pristine river geochemistry will be more and more limited to the subarctic regions, to some remaining undeveloped tropical regions, and to small temperate areas of the southern hemisphere such as South Chile and New Zealand. There is a global-scale increase of riverine Na^+, K^+, Cl^-, and SO_4^{2-}. HCO_3^- is still very stable. Impacts of riverine changes on the Earth System (Li, 1981) should be further addressed.

REFERENCES

Alekin O. A. and Brazhnikova L. V. (1964) *Runoff of Dissolved Substances from the USSR Territory* (in Russian). Nauka, Moscow.

Berner E. K. and Berner R. A. (1996) *Global Environment, Water, Air, and Geochemical Cycles*. Prentice Hall, Englewoods Cliff, NJ.

Berner R. A. (1971) Worldwide sulfur pollution of rivers. *J. Geophys. Res.* **76**, 6597–6600.

Berner R. A. (1996) Chemical weathering and its effect on atmospheric CO_2 and climate. In *Chemical Weathering Rates of Silicate Minerals, Reviews in Mineralogy* (eds. A. F. White and S. L. Brantley). Mineralogical Society of America, Washington, DC, vol. 31, pp. 565–583.

Berner R. A., Lasaga A. C., and Garrels R. M. (1983) The carbonate–silicate geochemical cycle and its effect on atmospheric carbon dioxide over the past 100 millions years. *Am. J. Sci.* **301**, 182–204.

Bluth G. J. S. and Kump L. R. (1994) Lithologic and climatologic controls of river chemistry. *Geochim. Cosmochim. Acta* **58**, 2341–2359.

Clarke F. W. (1924) *The Data of Geochemistry*, 5th edn. US Geol. Surv. Bull. vol. 770, USGS, Reston, VA.

Degens E. T., Kempe S., and Richey J. E. (eds.) (1991) *Bio-Geochemistry of Major World Rivers*. Wiley, New York.

Drever J. I. (1988) *The Geochemistry of Natural Waters*, 2nd edn. Prentice Hall, Englewood Cliff, NJ, 437pp.

Drever J. I. (1994) The effect of land plants on weathering rates of silicate minerals. *Geochim. Cosmochim. Acta* **58**, 2325–2332.

Drever J. I. and Clow D. W. (1995) Weathering rates in catchments. In *Chemical Weathering Rates of Silicate Minerals, Reviews in Mineralogy* (eds. A. F. White and S. L. Brantley). Mineralogical Society of America, Washington, DC, vol. 31, pp. 463–483.

Drever J. I. and Zobrist J. (1992) Chemical weathering of silicate rocks as a function of elevation in the southern Swiss Alps. *Geochim. Cosmochim. Acta* **56**, 3209–3216.

Flintrop C., Hohlmann B., Jasper T., Korte C., Podlaha O. G., Sheele S., and Veizer J. (1996) Anatomy of pollution: rivers of North Rhine-Westphalia, Germany. *Am. J. Sci.* **296**, 58–98.

Gaillardet J., Dupré B., Allègre C. J., and Négrel P. (1997) Chemical and physical denudation in the Amazon River Basin. *Chem. Geol.* **142**, 141–173.

Gaillardet J., Dupré B., Louvat P., and Allègre C. J. (1999) Global silicate weathering and CO_2 consumption rates deduced from the chemistry of large rivers. *Chem. Geol.* **159**, 3–30.

Garrels R. M. and Mackenzie F. T. (1971) *The Evolution of Sedimentary Rocks.* W. W. Norton, New York.

Gibbs R. J. (1970) Mechanism controlling world water chemistry. *Science* **170**, 1088–1090.

Guyot J. L. (1993) Hydrogéochimie des fleuves de l'Amazonie bolivienne. Etudes et Thèses, ORSTOM, Paris, 261pp.

Hem J. D. (1989) *Study and Interpretation of the Chemical Characteristics of Natural Water.* US Geol. Surv., Water Supply Pap. 2254, USGS, Reston, VA.

Holland H. D. (1978) *The Chemistry of the Atmosphere and Oceans.* Wiley-Interscience, New York.

Huh Y. (1998/1999) The fluvial geochemistry of the rivers of Eastern Siberia: I to III. *Geochim. Cosmochim. Acta* **62**, 1657–1676; **62**, 2053–2075; **63**, 967–987.

ICPR (International Commission for the Protection of the Rhine) (1975) ICPR/IKSR Bundesh f. Gewässerkunde, Koblenz (Germany).

Kimstach V., Meybeck M., and Baroudy E. (eds.) (1998) *A Water Quality Assessment of the Former Soviet Union.* E and FN Spon, London.

Kobayashi J. (1959) Chemical investigation on river waters of southeastern Asiatic countries (report I). The quality of waters of Thailand. Bericht. Ohara Inst. für Landwirtschaft Biologie, vol. 11(2), pp. 167–233.

Kobayashi J. (1960) A chemical study of the average quality and characteristics of river waters of Japan. Bericht. Ohara Inst. für Landwirtschaft Biologie, vol. 11(3), pp. 313–357.

Kump L. R. and Arthur M. A. (1997) Global chemical erosion during the Cenozoic: weatherability balances the budget. In *Tectonic and Climate Change* (ed. W. F. Ruddiman). Plenum, New York, pp. 399–426.

Li Y. H. (1981) Geochemical cycles of elements and human perturbations. *Geochim. Cosmochim. Acta* **45**, 2073–2084.

Likens G. E., Bormann F. H., Pierce R. S., Eaton J. S., and Johnson N. M. (1977) *Biogeochemistry of a Forested Ecosystem.* Springer, Berlin.

Livingstone D.A. (1963) *Chemical Composition of Rivers and Lakes.* Data of Geochemistry. US Geol. Surv. Prof. Pap. 440 G, chap. G, G1–G64.

Ludwig W., Amiotte-Suchet P., Munhoven G., and Probst J. L. (1998) Atmospheric CO_2 consumption by continental erosion: present-day controls and implications for the last glacial maximum. *Global Planet. Change* **16–17**, 107–120.

Meybeck M. (1979) Concentration des eaux fluviales en éléments majeurs et apports en solution aux océans. *Rev. Géol. Dyn. Géogr. Phys.* **21**(3), 215–246.

Meybeck M. (1986) Composition chimique naturelle des ruisseaux non pollués en France. *Sci. Geol. Bull.* **39**, 3–77.

Meybeck M. (1987) Global chemical weathering of surficial rocks estimated from river dissolved loads. *Am. J. Sci.* **287**, 401–428.

Meybeck M. (1993) Riverine transport of atmospheric carbon: sources, global typology and budget. *Water Air Soil Pollut.* **70**, 443–464.

Meybeck M. (1994) Origin and variable composition of present day riverborne material. In *Material Fluxes on the Surface of the Earth. Studies in Geophysics* (ed. Board on Earth Sciences and Resources–National Research Council). National Academic Press, Washington, DC, pp. 61–73.

Meybeck M. (1996) River water quality, global ranges time and space variabilities. *Vehr. Int. Verein. Limnol.* **26**, 81–96.

Meybeck M. (2002) Riverine quality at the Anthropocene: propositions for global space and time analysis, illustrated by the Seine River. *Aquat. Sci.* **64**, 376–393.

Meybeck M. and Ragu A. (1996) River Discharges to the Oceans. An assessment of suspended solids, major ions, and nutrients. Environment Information and Assessment Report. UNEP, Nairobi, 250p.

Meybeck M. and Ragu A. (1997) Presenting Gems Glori, a compendium of world river discharge to the oceans. *Int. Assoc. Hydrol. Sci. Publ.* **243**, 3–14.

Meybeck M., Chapman D., and Helmer R. (eds.) (1989) *Global Fresh Water Quality: A First Assessment.* Basil Blackwell, Oxford, 307pp.

Meybeck M., de Marsily G., and Fustec E. (eds.) (1998) *La Seine en Son Bassin.* Elsevier, Paris.

Miller J. P. (1961) *Solutes in Small Streams Draining Single Rock Types, Sangre de Cristo Range, New Mexico.* US Geol. Surv., Water Supply Pap., 1535-F.

Millot R., Gaillardet J., Dupré B., and Allègre C. J. (2002) The global control of silicate weathering rates and the coupling with physical erosion: new insights from rivers of the Canadian Shield. *Earth Planet. Sci. Lett.* **196**, 83–98.

Rabalais N. N. and Turner R. E. (eds.) (2001) *Coastal Hypoxia.* Coastal Estuar. Studies, **52**, American Geophysical Union.

Reeder S. W., Hitchon B., and Levinston A. A. (1972) Hydrogeochemistry of the surface waters of the Mackenzie river drainage basin, Canada: I. Factors controlling inorganic composition. *Geochim. Cosmochim. Acta* **36**, 825–865.

Stallard R. F. (1995a) Relating chemical and physical erosion. In *Chemical Weathering Rates of Silicate Minerals,* Reviews in Mineralogy (eds. A. F. White and S. L. Brantley). Mineralogical Society of America, Washington, DC, pp. 543–562.

Stallard R. F. (1995b) Tectonic, environmental and human aspects of weathering and erosion: a global review using a steady-state perspective. *Ann. Rev. Earth Planet. Sci.* **23**, 11–39.

Stallard R. F. and Edmond J. M. (1981) Geochemistry of the Amazon: 1. Precipitation chemistry and the marine contribution to the dissolved load at the time of peak discharge. *J. Geophys. Res.* **86**, 9844–9858.

Stallard R. F. and Edmond J. M. (1983) Geochemistry of the Amazon: 2. The influence of geology and weathering environment on the dissolved load at the time of peak discharge. *J. Geophys. Res.* **86**, 9844–9858.

USGS (1909–1965) *The Quality of Surface Waters of the United States.* US Geol. Surv., Water Supply Pap., 236, 237, 239, 339, 363, 364, 839, 970, 1953, Reston, VA.

Viers J., Dupré B., Polvé M., Schott J., Dandurand J. L., and Braun J. J. (1997) Chemical weathering in the drainage basin of a tropical watershed (Nsimi-Zoetele site, Cameroon): comparison between organic poor and organic rich waters. *Chem. Geol.* **140**, 181–206.

Weiler R. R. and Chawla V. K. (1969) Dissolved mineral quality of the Great Lakes waters. *Proc. 12th Conf. Great Lakes Res.: Int. Assoc. Great Lakes Res.,* 801–818.

White A. T. and Blum A. E. (1995) Effects of climate on chemical weathering in watersheds. *Geochim. Cosmochim. Acta* **59**, 1729–1747.

9

Trace Elements in River Waters

J. Gaillardet

Institut de Physique du Globe de Paris, France

and

J. Viers and B. Dupré

Laboratoire des Mécanismes et Transferts en Géologie, Toulouse, France

9.1	INTRODUCTION	294
9.2	NATURAL ABUNDANCES OF TRACE ELEMENTS IN RIVER WATER	295
	9.2.1 *Range of Concentrations of Trace Elements in River Waters*	306
	9.2.2 *Crustal Concentrations versus Dissolved Concentrations in Rivers*	307
	9.2.3 *Correlations between Elements*	308
	9.2.4 *Temporal Variability*	310
	9.2.5 *Conservative Behavior of Trace Elements in River Systems*	311
	9.2.6 *Transport of Elements*	311
9.3	SOURCES OF TRACE ELEMENTS IN AQUATIC SYSTEMS	312
	9.3.1 *Rock Weathering*	312
	9.3.2 *Atmosphere*	314
	9.3.3 *Other Anthropogenic Contributions*	314
9.4	AQUEOUS SPECIATION	315
9.5	THE "COLLOIDAL WORLD"	318
	9.5.1 *Nature of the Colloids*	318
	9.5.2 *Ultrafiltration of Colloids and Speciation of Trace Elements in Organic-rich Rivers*	319
	9.5.3 *The Nonorganic Colloidal Pool*	321
	9.5.4 *Fractionation of REEs in Rivers*	322
	9.5.5 *Colloid Dynamics*	324
9.6	INTERACTION OF TRACE ELEMENTS WITH SOLID PHASES	325
	9.6.1 *Equilibrium Solubility of Trace Elements*	325
	9.6.2 *Reactions on Surfaces*	327
	9.6.3 *Experimental Adsorption Studies*	327
	9.6.4 *Adsorption on Hydrous Oxides in River Systems*	328
	9.6.5 *The Sorption of REEs: Competition between Aqueous and Surface Complexation*	330
	9.6.6 *Importance of Adsorption Processes in Large River Systems*	330
	9.6.7 *Anion Adsorption in Aquatic Systems*	331
	9.6.8 *Adsorption and Organic Matter*	332
	9.6.9 *Particle Dynamics*	333
9.7	CONCLUSION	333
	ACKNOWLEDGMENTS	335
	REFERENCES	335

9.1 INTRODUCTION

Trace elements are characterized by concentrations lower than $1\,mg\,L^{-1}$ in natural waters. This means that trace elements are not considered when "total dissolved solids" are calculated in rivers, lakes, or groundwaters, because their combined mass is not significant compared to the sum of Na^+, K^+, Ca^{2+}, Mg^{2+}, H_4SiO_4, HCO_3^-, CO_3^{2-}, SO_4^{2-}, Cl^-, and NO_3^-. Therefore, most of the elements, except about ten of them, occur at trace levels in natural waters. Being trace elements in natural waters does not necessarily qualify them as trace elements in rocks. For example, aluminum, iron, and titanium are major elements in rocks, but they occur as trace elements in waters, due to their low mobility at the Earth's surface. Conversely, trace elements in rocks such as chlorine and carbon are major elements in waters.

The geochemistry of trace elements in river waters, like that of groundwater and seawater, is receiving increasing attention. This growing interest is clearly triggered by the technical advances made in the determination of concentrations at lower levels in water. In particular, the development of inductively coupled plasma mass spectrometry (ICP-MS) has considerably improved our knowledge of trace-element levels in waters since the early 1990s. ICP-MS provides the capability of determining trace elements having isotopes of interest for geochemical dating or tracing, even where their dissolved concentrations are extremely low.

The determination of trace elements in natural waters is motivated by a number of issues. Although rare, trace elements in natural systems can play a major role in hydrosystems. This is particularly evident for toxic elements such as aluminum, whose concentrations are related to the abundance of fish in rivers. Many trace elements have been exploited from natural accumulation sites and used over thousands of years by human activities. Trace elements are therefore highly sensitive indexes of human impact from local to global scale. Pollution impact studies require knowledge of the natural background concentrations and knowledge of pollutant behavior. For example, it is generally accepted that rare earth elements (REEs) in waters behave as good analogues for the actinides, whose natural levels are quite low and rarely measured. Water quality investigations have clearly been a stimulus for measurement of toxic heavy metals in order to understand their behavior in natural systems.

From a more fundamental point of view, it is crucial to understand the behavior of trace elements in geological processes, in particular during chemical weathering and transport by waters. Trace elements are much more fractionated by weathering and transport processes than major elements, and these fractionations give clues for understanding the nature and intensity of the weathering + transport processes. This has not only applications for weathering studies or for the past mobilization and transport of elements to the ocean (potentially recorded in the sediments), but also for the possibility of better utilization of trace elements in the aqueous environment as an exploration tool.

In this chapter, we have tried to review the recent literature on trace elements in rivers, in particular by incorporating the results derived from recent ICP-MS measurements. We have favored a "field approach" by focusing on studies of natural hydrosystems. The basic questions which we want to address are the following: What are the trace element levels in river waters? What controls their abundance in rivers and fractionation in the weathering + transport system? Are trace elements, like major elements in rivers, essentially controlled by source-rock abundances? What do we know about the chemical speciation of trace elements in water? To what extent do colloids and interaction with solids regulate processes of trace elements in river waters? Can we relate the geochemistry of trace elements in aquatic systems to the periodic table? And finally, are we able to satisfactorily model and predict the behavior of most of the trace elements in hydrosystems?

An impressive literature has dealt with experimental works on aqueous complexation, uptake of trace elements by surface complexation (inorganic and organic), uptake by living organisms (bioaccumulation) that we have not reported here, except when the results of such studies directly explain natural data. As continental waters encompass a greater range of physical and chemical conditions, we focus on river waters and do not discuss trace elements in groundwaters, lakes, and the ocean. In lakes and in the ocean, the great importance of life processes in regulating trace elements is probably the major difference from rivers.

Section 9.2 of this chapter reports data. We will review the present-day literature on trace elements in rivers to show that our knowledge is still poor. By comparing with the continental abundances, a global mobility index is calculated for each trace element. The spatial and temporal variability of trace-element concentrations in rivers will be shown to be important. In Section 9.3, sources of trace elements in river waters are indicated. We will point out the great diversity of sources and the importance of global anthropogenic contamination for a number of elements. The question of inorganic and organic speciation of trace elements in river water will then be addressed in Section 9.4, considering some general relationships between speciation and placement in the periodic table. In Section 9.5, we will show that studies on organic-rich rivers have led to an exploration of the "colloidal world" in rivers. Colloids are small particles, passing through the conventional filters used to separate dissolved and suspended loads in rivers. They appear as major carriers of trace elements in rivers and considerably complicate aqueous-speciation calculation. Finally, in Section 9.6, the significance of interactions between solutes and solid surfaces in river waters will be reviewed. Regulation by surfaces is of major importance for a great range of elements. Although for both colloids and surface interactions, some progress has been made, we are still far from a unified model that can accurately predict trace-element concentrations in natural water systems. This is mainly due to our poor physical description of natural colloids, surface site complexation, and their interaction with solutes.

9.2 NATURAL ABUNDANCES OF TRACE ELEMENTS IN RIVER WATER

Trace-element abundances in the dissolved load of a number of worldwide rivers, classified by continents, are compiled in Table 1, along with the corresponding references. These data have been measured after filtration of the river sample using either 0.2 μm or 0.45 μm filters. We have tried to report, as far as possible, "natural" river systems by discarding data from highly polluted systems. For the Amazon and Orinoco rivers, we have reported data for the dissolved load filtered using filters or membranes with smaller pore size. A major feature of trace-element abundances in river waters is that they are strongly dependent upon filtration characteristics. For simplicity, elements have been ranked in alphabetical order. Most often, data for large rivers are a compendium of data derived from different bibliographical sources and therefore have not been measured on the same river sample. However, in the case of the Amazon basin, two different analyses, corresponding to two different sampling dates are reported for the major tributaries. These comparisons show the high degree of variability of trace-element concentrations in river waters through time. When time series are available, a mean value is given in Table 1. It is apparent from these data that the Americas and Africa are the places where most of our information on natural levels of trace elements in waters has been obtained. By contrast, rivers of Asia have received relatively little attention, except for a number of key elements of geochemical importance such as osmium, strontium, and uranium.

North America. A substantial literature exists on the trace-element levels in North American rivers. We have focused in Table 1 on the largest systems. Smaller and pristine watersheds from northern Canada have also been reported for

comparison. The St. Lawrence system is represented by the St. Lawrence itself and by the Ottawa River, which, due to its status as an international standard of river water (SLRS 4), appears as the most often measured river. The Mistassini River is a black (high humic acid content) river discharging into St. Jean Lake and represents the North American end-member in terms of trace-element levels. Other large river systems include the Mackenzie and Peel rivers, the Fraser, the Columbia, Mississippi, Connecticut and Hudson rivers. Small rivers from the Mackenzie Basin have been reported, because they drain relatively pristine regions of North America with well-characterized geological substratum. The Indin River drains the Slave Province, dominated by old granites. The Beatton River is a typical black river and suspended sediment-rich river draining the interior sedimentary platform. Other rivers drain the western Cordillera, composed of volcanic and volcaniclastic rocks (Upper Yukon and Skeena rivers) or sedimentary rocks (Peel river).

Europe. Situation is far less favorable in Europe. Fewer data are available in the standard scientific literature and most of the rivers are strongly impacted by pollution. Only the rivers from the northern part of Europe (Kola peninsula) reported by Pokrovski and Schott (2002) are probably close to the natural background. The enhancement of dissolved concentrations for a number of elements due to human activities is apparent from the data on the Seine River. Data for REEs in the Rhine River and streams from the Vosges mountains in Alsace and central Europe have been reported for comparison. All these locations are known to have suffered from acid deposition in the very recent past.

Africa. Recently published studies have considerably improved our knowledge of trace-element concentrations in the rivers of central Africa. We have reported in Table 1 data for the Congo–Zaire River (the second largest river in the world in terms of discharge) and its main tributaries (Ubangui, Kasai, and Zaire), data from the upper Niger and Douna (the main tributary of the Niger river) and from smaller rivers in Cameroon. The Mengong River is a small stream that has been shown to have extreme concentrations of certain trace elements. All the rivers reported for Africa drain the humid and forested part of the continent.

South America. Due to its importance in terms of water discharge and to generally high trace-element levels, the Amazon river system has been well documented for a number of elements, including their seasonal variations. In Table 1, we have reported different analyses of the Amazon River and its major tributaries to show their temporal variability at a given location. Results of ultrafiltration experiments for the Amazon and Orinoco rivers demonstrate that the concentration

Table 1 Database of trace element concentration in the dissolved load ($<0.2\ \mu m$) of rivers. All concentrations in ppb (μgL^{-1}) except for Ra (fgL^{-1}) and Os (pgL^{-1}). DOC, TSS, TDS are Dissolved Organic Carbon ($mg\ L^{-1}$), total suspended solid ($mg\ L^{-1}$), and total dissolved solutes ($mg\ L^{-1}$). Water discharge and surface area are in $m^3\ s^{-1}$ and $10^3\ Km^2$, respectively.

Element	References	pH	DOC	TSS	TDS	Discharge	Surface area	Ag	Al	As
Africa										
Oubangui	1	6.81		30	36.1	3,500	475		12	
Zaire	1	5.98		31	48.68	17,000	1,660		46	
Kasai	1	6.35		17		11,000	900		51	
Congo at Brazzaville	1, 2, 3	6.4		21		39,100	3,500		76	
Niger	2, 4	7.00	1.5	43	44	907	141		76	
Douna	2, 4									
Nyong	5	5.60	23	5	22	340	29		215	
Sanaga	6	7.43	3.82				133		29	0.17
Nyong	6	5.88	14.1			340	29		159	0.11
Mengong	6	4.62	24						480	0.11
Europe										
Seine at Paris, Fr.	7			40	400	260	44		16	2.71
Garonne River Fr.	8, 9			128	400	540	55			
Rhine in Alsace, Fr.	10	8.01	1.95						<50	
Vosges Stream, Fr.	10	7.10	2.03						76	
Harz Mountains, Ger.	11	5.10							1,080	0.37
Kalix River, Sweden, 1991	12, 45	6.95	3.4			296	24		27	
Kalix River, 1977 May	12					296	24			
Idel river	13	6.85	12.3			2	1		44	0.21
N. America										
St. Lawrence	2, 14, 15, 16, 46	8.00		11	183	10,700	1,020		15	0.91
SLRS 4	17								53	0.70
Ottawa	14								67	0.45
Mistassini, Can.	14	5.50	26	5		195	10		174	0.12
Mackenzie	2, 14, 18, 19	8.10	5.8	300	226	9,000	1,680		18	0.50
Peel, Can.	14	8.10	12.9	250	167	690	71		22	0.35
Indin River, Can.	14	6.70			12	8	2		28	0.14
Beatton, Can.	14	7.20		810	68				91	0.80
Upper Yukon, Can.	2, 14	7.70							25	0.62
Skeena, Can.	14	7.50	4.48	60	53	962	42		33	0.21
Fraser River, Can.	19, 20	7.34	3.39	175	93	3630	238		19	0.52
Columbia River	8, 19	7.85		64	115	5941	670			
Californian Streams	21									
Connecticut	22	7.00	3.16		70	540	25			
Hudson River	22, 23, 46				126	621	35	0.004		
Upper Mississippi	24	7.70		462						

Missouri	24, 25	7.70		2,332				
Ohio	25	8.20		177				
Illinois	25			102				
Mississippi at Mouth	3, 15, 19, 22, 25, 46	7.80		860	280	18,400	2,980	
S. America								
Amazon, mean value	**3, 15, 26, 27, 28, 29, 46, 48**	**6.89**	**5.05**	**182**	**44**	**205,000**	**6,100**	**9.4**
Amazon < 0.2 µm	**19, 30**	**7.10**	**5.05**	**182**	**44**	**205,000**	**6,100**	**6.2**
Amazon < 100 kDa	30		2.96					0.8
Amazon < 5kDa	30		1.11					0.5
Negro < 0.2 µm	26	*4.85*						*113.9*
Negro < 0.2 µm	30	5.87	7.17					97.0
Solimoes < 0.2 µm	26	*7.10*						*171.4*
Solimoes < 0.2 µm	30	7.66	2.76					5.8
Madeira < 0.2 µm	26	*6.73*						*2.6*
Madeira < 0.2 µm	30	7.54	11.1					4.3
Trompetas < 0.2 µm	26	*6.10*						*39.2*
Trompetas < 0.2 µm	30	6.54	1.85					7.9
Tapajos < 0.2 µm	26	*6.68*						*15.2*
Tapajos < 0.2 µm	30	7.45	1.48					4.0
Rio Beni at Riberalta	31					8,262	243	
Mamore	31					8,392	599	
Rio Beni at Rurrenabaque	31					2,025	68	
Orinoco < 0.2 µm	19, 27, 29, 30, 46, 48	6.51	5.42	132	25	35,900	1100	61.8
Orinoco < 10kDa	27, 30		4.3					5.6
Caroni at Cuidad Bolivar	19, 27, 29, 30, 32	5.58	3.71					16.1
Asia								
Ob	33		7.4–9.95	40	126	13,500	2,990	
Yenisei	33		4	23	112	19,800	2,500	
Lena	2, 19, 34	7.60		30	112	19,100	1,200	0.15
Changjiang	2, 3, 9, 19, 27, 35, 36, 37, 3	7.80		520	221	29,400	1,808	0.83
Huanghe	2, 15, 38, 39, 40, 47	8.30		27,000	460	1300	752	2.00
Xijiang	2, 15, 47	7.70		190	161	11,500	437	
Ganges	19, 29, 41, 42, 43, 46, 47, 4	7.70		1,100	182	15,600	1,050	
Mekong	2, 15, 47	7.80		321	263	14,800	795	
Brahmaputra	9, 19, 41, 42, 46, 47, 48	7.40		1,060	101	16,100	580	
Indus	9, 41, 44, 47	7.80		2,780	302	2880	960	
Shinano, Jpn.	44, 47	7.11				475	11	
World average							*1,200*	***32***
Riverine flux (kt yr⁻¹)							23	***0.62***

(continued)

Table 1 (continued).

Element	B	Be	Ba	Cd	Ce	Co	Cr	Cs	Cu	Dy	Er	Eu	Fe	Ga
Africa														
Oubangui			17		0.4920	0.077	0.533	0.008		0.043	0.029	0.0160	60	
Zaire			24.8		0.6610	0.075	0.386	0.0026		0.053	0.033	0.0220	202	
Kasai			20		0.4520	0.0580	0.4	0.009		0.038	0.025	0.0140	108	
Congo at Brazzaville	3.1		30		0.6890	0.0594	0.501	0.016	0.630	0.006	0.033	0.0170	179	
Niger	3.2				0.1710	0.0400	0.450	0.0100					105	
Douna														
Nyong			18		1.3200	0.3637			0.952	0.119	0.071	0.0255	241	
Sanaga			27		0.1780	0.0590			2.030	0.015	0.016		31	
Nyong			19		0.8060	0.2530			1.397	0.086	0.055		174	0.0220
Mengong			24		0.8274	0.4307				0.059	0.027		614	0.1087
Europe														
Seine at Paris, Fr.	25.0		32	0.0600	0.0600	0.1800	11.460		3.530				302	
Garonne River Fr.					0.0810					0.002	0.004	0.0016		
Rhine in Alsace, Fr.					0.0096						0.001	0.0003		
Vosges Stream, Fr.					0.0930					0.020	0.012	0.0087		
Harz Mountains, Ger.			13	0.4200	0.2170	0.2600	<0.85		0.820				525	
Kalix River, Sweden, 1991		0.61												
Kalix River, 1977 May														
Idel river	39.0		5	0.0200	0.2430	0.0840	1.07	0.0112	0.456	0.013	0.008	0.0039	666	0.0088
N. America														
St. Lawrence	24.8		23	0.0114	0.0600	0.0632		0.0052	0.936	0.005	0.004	0.0030	111	
SLRS 4	6.0	0.008	13	0.0140	0.3600	0.0480	0.37	0.0090	1.930	0.024	0.013	0.0080	108	0.0119
Ottawa	3.3		15	0.0207	0.5599	0.0746		0.0061	1.144	0.040	0.023	0.0112	112	
Mistassini, Can.			8	0.0873	1.1771	0.1221		0.0070	1.578	0.036	0.022	0.0120	170	
Mackenzie	11.8		56	0.1838	0.0266	0.0682	0.375	0.0066	1.609	0.003		0.0023	119	
Peel, Can.			62	0.0347	0.0310	0.1426	0.294	0.0037	1.043	0.001	0.003	0.0019	152	
Indin River, Can.			3		0.1305	0.0158			0.841				50	
Beatton, Can.			47	0.1206	0.4750	0.1515	0.741	0.0028	2.594	0.072	0.056	0.0290	739	
Upper Yukon, Can.			50	0.0906	0.0311	0.0616	0.241	0.0049	1.306	0.009		0.0010	102	
Skeena, Can.			13	0.0194	0.0637	0.0794		0.0006	1.077				52	0.0083
Fraser River, Can.			15		0.0600	0.0800	2.1		1.040	0.009		0.0030	47	
Columbia River					0.0583							0.0016		
Californian Streams														0.001–0.006
Connecticut					0.0258					0.004	0.003	0.0009	10	
Hudson River					0.0621					0.013	0.008	0.0026		
Upper Mississippi			73						1.850					

Missouri			80						2.010					
Ohio			32						1.741					
Illinois			59						1.984					
Mississippi at Mouth	37.8		62			0.0074			1.60–2.24	0.004	0.004	0.0007		
S. America														
Amazon, mean value	**6.1**	**0.0095**	**21**	**0.1781**	**0.1766**	**0.2180**	**0.717**		**1.463**	**0.033**	**0.018**	**0.0104**	**0.0174**	
Amazon < 0.2 μm			**28**			**0.0680**				**0.011**	**0.006**	**0.0027**	**0.0217**	**43**
Amazon < 100 kDa			26			0.0200				0.004	0.002	0.0007		23
Amazon < 5 kDa			17			0.0067				0.000	0.000			13
Negro < 0.2 μm			*6*		*0.1241*	*0.4150*			*0.399*	*0.023*	*0.015*	*0.0088*	*0.0050*	*117*
Negro < 0.2 μm			7			0.5853				0.030	0.018	0.0078		
Solimoes < 0.2 μm	*3.8*		*28*		*0.1643*	*0.3630*			*1.542*	*0.044*	*0.028*	*0.0150*	*0.0390*	*351*
Solimoes < 0.2 μm			29			0.0528				0.007	0.004	0.0019		53
Madeira < 0.2 μm	*3.4*		*18*		*0.0176*	*0.1380*			*0.863*	*0.024*	*0.013*	*0.0083*	*0.0025*	*18*
Madeira < 0.2 μm			32			0.0074				0.001	0.001		0.0169	26
Trompetas < 0.2 μm	*1.5*		*14*		*0.1274*	*0.9080*			*0.269*	*0.044*	*0.028*	*0.0105*	*0.0059*	*87*
Trompetas < 0.2 μm			15			0.1300				0.007	0.005	0.0017		30
Tapajos < 0.2 μm			*21*		*0.0195*	*0.1150*			*0.227*	*0.012*	*0.008*	*0.0033*	*0.0033*	
Tapajos < 0.2 μm			18			0.0277				0.003	0.002	0.0007	0.0178	11
Rio Beni at Riberalta			30	0.0081					1.517					
Mamore			4	0.0091					1.997					
Rio Beni at Rurrenabaque			23	0.0011					0.710					
Orinoco < 0.2 μm		0.009	8			0.5207		0.007–0.013		0.056	0.031	0.0140	0.1176	142
Orinoco < 10 kDa			8			0.1703				0.020	0.012	0.0047	0.1143	15
Caroni at Cuidad Bolivar	2.4	0.0135	7			0.1443		0.006		0.012	0.006	0.0032	0.1027	16
Asia														
Ob				0.0006–0.0008					1.8–2.4					24–36
Yenisei	4.7			0.0012–0.0018					1.39–1.91					14–17.8
Lena	12.5			0.0089					0.755					24.3
Changjiang	150.0			0.0033		0.1150			1.66			0.0050		31
Huanghe	6.0			0.0011–0.0055		0.0059–0.0295			0.96–1.6					1.4–25
Xijiang	17.8	0.00056												
Ganges	15.0													
Mekong	20.9													
Brahmaputra														
Indus						0.0024				0.001	0.001	0.0002		
Shinano, Jpn.						0.0834				0.012	0.007	0.003		
World average	*10.2*	*0.0089*	*23*	*0.08*	*0.148*	*0.2620*	*0.7*	*0.011*	*1.48*	*0.03*	*0.02*	*0.0098*	*0.03*	*66*
Riverine flux (kt yr⁻¹)	*380*	*0.33*	*860.2*	*3*	*5.5*	*9.800*	*26*	*0.4*	*55*	*1.1*	*0.75*	*0.37*	*1.1*	*2470*

(continued)

Table 1 (continued).

Element	Gd	Ge	Hf	Ho	La	Li	Lu	Mn	Mo	Nb	Nd	Ni	Os (pg L^{-1})	P
Africa														
Oubangui	0.0510		0.0042	0.0090	0.249		0.0040				0.277	1.15		
Zaire	0.0630		0.0057	0.0110	0.349		0.0040				0.360	1.02		
Kasai	0.0470		0.0038	0.0080	0.189		0.0030				0.241	0.41		
Congo at Brazzaville	0.0660	0.0066	0.0067	0.0120	0.319		0.0045				0.350	0.934	6.7	
Niger			0.0030		0.091						0.085	0.29	5.3	
Douna								0.50						
Nyong	0.1343			0.0199	0.538		0.0080	29.72			0.690	0.70		
Sanaga	0.0240				0.09			0.44			0.084			
Nyong	0.094	0.0065		0.0165	0.349		0.0105	22.61			0.505	1.18		
Mengong	0.0551	0.0799		0.0116	0.348		0.0053	20.02			0.416	5.04		
Europe														
Seine at Paris, Fr.	0.0088			0.0016	0.030		0.0006	3.76			0.030	5.06	41.8	
Garonne River Fr.	0.0025			0.0004	0.047		0.0004				0.038			
Rhine in Alsace, Fr.	0.0037			0.0041	0.005		0.0014				0.005			
Vosges Stream, Fr.					0.153						0.245			
Harz Mountains, Ger.					0.480	2.00		48.00				0.92		
Kalix River, Sweden, 1991					0.155			9.40						
Kalix River, 1977 May														
Idel river	0.0190	0.0082		0.0027	0.151	0.80	0.0015	22.80	0.112		0.141	0.35		2.67
N. America														
St. Lawrence	0.0059	0.0031	0.0031	0.0013	0.029		0.0006	6.28	1.292	0.0021	0.038	1.33	22.8	
SLRS 4	0.0342	0.0100		0.0047	0.287	0.54	1.9000	3.37	0.210		0.269	0.82		
Ottawa	0.0593	0.0086	0.0034	0.0078	0.411		0.0037	14.86	0.199	0.0045	0.411	0.83		
Mistassini, Can.	0.0618	0.0058	0.0070	0.0071	0.635		0.0025	11.31	0.039	0.0107	0.547	0.47		
Mackenzie	0.0019			0.0005	0.002	4.60		1.28	1.067	0.0012	0.019	1.83	25.5	
Peel, Can.	0.0116			0.0020	0.002		0.0024	4.54	1.078	0.0019	0.004	2.68		
Indin River, Can.	0.1599	0.0049		0.0101	0.099	0.91		1.89			0.091	0.64		1.82
Beaton, Can.	0.0136	0.0196	0.1106	0.0010	0.090			2.98	0.301	0.0069	0.042	5.14		
Upper Yukon, Can.		0.0014			0.001	0.64	0.0006	2.29	1.055	0.0019	0.007	10.39	24.7	
Skeena, Can.					0.051	0.35		5.37	0.418		0.081	0.91		
Fraser River, Can.	0.0110				<0.05	1.05		5.40	1.330		0.044	1.86		
Columbia River	0.0065	0.0138		0.0009	0.030	1.46	0.0007				0.023			
Californian Streams														
Connecticut	0.0047				0.021		0.0008				0.020			
Hudson River	0.0190						0.0007				0.060			

S. America and Asia sections are in italic in the original; italic sample rows are noted.

Upper Mississippi	0.0042						0.41	1.114				1.66	
Missouri	0.0219						0.44	1.613				1.53	
Ohio							0.46	1.258				1.12	
Illinois							0.70	2.314				2.92	
Mississippi at Mouth	0.0042				0.008	10.00	0.0006	0.66–1.82	1.63–2.69		0.011	1.12–1.77	14.45
S. America													
Amazon, mean value	**0.0356**	**0.0048**		**0.0064**	**0.106**	**0.91**	**0.0020**	**50.73**	**0.175**		**0.136**	**0.74**	**4.6**
Amazon < 0.2 μm	**0.0123**	**0.0074**		**0.0021**	**0.032**	**2.46**	**0.0009**	**3.31**			**0.042**		
Amazon < 100 kDa	0.0043	0.0076		0.0007	0.010		0.0003	2.90			0.013		
Amazon < 5 kDa	0.0007	0.0061		0.0001	0.006			2.00			0.003		
Negro < 0.2 μm	*0.0350*			*0.0050*	*0.151*		*0.0016*	*8.24*			*0.172*	*0.21*	
Negro < 0.2 μm	0.0432	0.0046		0.0061	0.208		0.0023	7.35			0.211		
Solimoes < 0.2 μm	*0.0490*			*0.0093*	*0.166*	*1.02*	*0.0037*	*14.56*			*0.226*	*0.92*	
Solimoes < 0.2 μm	0.0089	0.0093		0.0017	0.050		0.0006	6.54			0.032		
Madeira < 0.2 μm	*0.0260*			*0.0053*	*0.054*	*1.18*	*0.0014*				*0.100*	*0.57*	
Madeira < 0.2 μm	0.0018	0.0041		0.0003	0.005		0.0001	3.29			0.005		
Trompetas < 0.2 μm	*0.0485*			*0.0093*	*0.266*	*0.41*	*0.0037*	*8.62*			*0.309*	*0.12*	*2.57*
Trompetas < 0.2 μm	0.0102	0.0049		0.0016	0.044		0.0008	1.36			0.053		
Tapajos < 0.2 μm	*0.0114*			*0.0020*	*0.228*		*0.0009*	*1.34*			*0.072*	*0.22*	
Tapajos < 0.2 μm	0.0037	0.0053		0.0007	0.016		0.0004	0.46			0.018		
Rio Beni at Riberalta								4.13	0.380			0.91	
Mamore								113.52	0.240			1.11	
Rio Beni at Rurrenabaque								2.37	0.218			0.79	
Orinoco < 0.2 μm	0.0737			0.0107	0.177	0.32	0.0043	6.82			0.289		11.09
Orinoco < 10 kDa	0.0256			0.0040	0.049		0.0018	5.24			0.094		6.1
Caroni at Cuidad Boliver	0.0147			0.0021	0.067	0.16	0.0009	5.57			0.078	3.3	6.61
Asia													
Ob												1.24–1.42	
Yenisei												0.52–0.55	
Lena		0.0122				1.33						0.38	8.2
Changjiang					0.005	3.44		1.00				0.15	13.9
Huanghe							0.0020	0.55–2.2			0.070	0.30–0.59	42.1
Xijiang													8.3
Ganges						3.47							32.0
Mekong													17.2
Brahmaputra						2.61							9.9
Indus					0.003		0.0002				0.003		11.2
Shinano, Jpn.					0.037		0.0016				0.050		
World average	*0.04*	*0.0068*	*0.0059*	*0.0071*	*0.120*	*1.84*	*0.0024*	*34*	*0.420*	*0.0017*	*0.152*	*0.801*	*9.0*
Riverine flux (kt yr⁻¹)	*1.5*	*0.25*	*0.22*	*0.27*	*4.5*	*69*	*0.09*	*1270*	*16*	*0.063*	*5.7*	*30*	*0.33.10-3*

(continued)

Table 1 (continued).

Element	Pb	Pd	Pr	Ra (fg L^{-1})	Re	Rb	Sb	Sc	Se	Sm	Sr	Ta	Tb	Th	Ti
Africa															
Oubangui			0.069			2.7		0.055		0.0600	15.0		0.0070	0.042	
Zaire			0.093			3.9		0.067		0.0820	21.0		0.0100	0.056	
Kasai			0.052			2.7		0.062		0.0470	10.5		0.0060	0.023	
Congo at Brazzaville			0.089			3.1		0.087		0.0620	11.5		0.0097	0.065	
Niger	0.039					3.86					26.4			0.013	
Douna															
Nyong			0.179			4.18				0.1362	9.7		0.0178	0.121	
Sanaga			0.024			6.16					30.3			0.012	0.231
Nyong			0.114			3.68				0.1210	12.4			0.111	0.199
Mengong			0.096			0.73				0.0780	17.9			0.137	5.808
Europe															
Seine at Paris, Fr.	0.220					1.40		1.340			227.0			0.010	
Garonne River Fr.			0.005							0.0082			0.0012		
Rhine in Alsace, Fr.			0.001							0.0012			0.0003		
Vosges Stream, Fr.			0.049							0.0500			0.0046		
Harz Mountains, Ger.	3.800					5.90	0.190				20.0				
Kalix River, Sweden, 1991															
Kalix River, 1977 May															
Idel river	0.119		0.037			0.96	0.027			0.0240	16.8		0.0025	0.022	1.070
N. America															
St. Lawrence	0.233		0.009	2		1.04	0.205			0.0067	177.2		0.0010	0.004	0.509
SLRS 4	0.084	0.021	0.069			1.53	0.270		0.230	0.0574	28.2		4.3000		1.460
Ottawa	0.105		0.104			1.55	0.057			0.0705	50.7		0.0073	0.027	1.854
Mistassini, Can.	0.113		0.149	48		1.14	0.023			0.0783	11.4		0.0085	0.041	2.278
Mackenzie	0.771		0.012			0.66	0.121			0.0055	237.8	0.0009		0.634	0.423
Peel, Can.	1.129		0.012			0.36	0.120			0.0149	154.0	0.0029	0.0025	0.588	0.574
Indin River, Can.		0.001	0.023			1.77	0.005			0.0152	10.5				0.112
Beatton, Can.	0.269		0.132			0.30	0.102			0.1849	62.7	0.1484	0.0289	1.054	1.200
Upper Yukon, Can.	0.818		0.006			0.90	0.150			0.0056	162.2		0.0010	0.988	0.768
Skeena, Can.		0.028	0.015			0.20	0.044			0.0230	78.4				0.372
Fraser River, Can.	0.078		0.011			0.91	0.053	0.141		0.0110	108.0				0.680
Columbia River			0.010							0.0435			0.0012		
Californian Streams															
Connecticut										0.0042					
Hudson River										0.0119					
Upper Mississippi	0.008			4–31		1.24									

	(1)	(2)	(3)	(4)	(5)	(6)	(7)	(8)	(9)	(10)	(11)	(12)	(13)	(14)	(15)
Missouri	0.006														
Ohio	0.007														
Illinois	0.035														
Mississippi at Mouth	0.011–0.016			5–30						0.0030					
S. America															
Amazon, mean value	**0.064**		**0.031**	**9–31**	**0.00020**	**1.49**	**0.061**	**1.540**	**0.051**	**0.0349**	**25.8**		**0.0043**		**0.006**
Amazon < 0.2 μm			**0.009**			**1.89**		**1.580**		**0.0100**	**51.2**		**0.0017**		
Amazon < 100 kDa			0.003			1.79		1.550		0.0041	47.2		0.0007		
Amazon < 5 kDa			0.001			1.29					31.3				
Negro < 0.2 μm	*0.170*		*0.047*			*1.13*				*0.0380*	*3.6*		*0.0040*		
Negro < 0.2 μm			0.054			1.73		0.900		0.0390	4.2		0.0056		0.053
Solimoes < 0.2 μm	*0.151*		*0.052*			*1.59*				*0.0520*	*45.7*		*0.0067*		*0.010*
Solimoes < 0.2 μm			0.007			1.69		1.770		0.0082	61.5		0.0014		0.002
Madeira < 0.2 μm	*0.005*		*0.022*			*1.34*				*0.0311*	*19.2*		*0.0048*		
Madeira < 0.2 μm			0.001			1.94		1.510		0.0014	55.5		0.0002		0.001
Trompetas < 0.2 μm	*0.052*		*0.080*			*2.95*				*0.0596*	*6.7*		*0.0061*		*0.121*
Trompetas < 0.2 μm			0.013			4.04		1.260		0.0094	9.6		0.0012		0.016
Tapajos < 0.2 μm	*0.061*		*0.017*			*2.75*				*0.0181*	*9.9*		*0.0015*		
Tapajos < 0.2 μm			0.004			2.08		1.410		0.0040	6.5		0.0005		0.002
Rio Beni at Riberalta						1.01					42.9				
Mamore						1.44					31.4				
Rio Beni at Rurrenabaque						0.90					48.3				
Orinoco < 0.2 μm	0.062			12–17	0.00083	1.50	0.032–0.050	0.560		0.0682	8.0		0.0098		0.073
Orinoco < 10 kDa	0.020					1.43		0.620		0.0234	7.5		0.0035		0.026
Caroni at Cuidad Bolivar	0.019					1.13	0.019–0.020	0.530		0.0153	2.9		0.0020		0.017
Asia															
Ob	0.011–0.017														
Yenisei	0.005–0.006														
Lena	0.019			50											
Changjiang	0.054								0.22–0.23	0.0150	210				
Huanghe	0.010–4.1										1140				
Xijiang											110				
Ganges				45–90	0.00170						90				
Mekong											298				
Brahmaputra											59				
Indus				31	0.00011					0.0007	324				
Shinano, Jpn.										0.0110					
World average	*0.079*	*0.028*	*0.04*	*24*	*0.0004*	*1.63*	*0.07*	*1.2*	*0.07*	*0.036*	*60.0*	*0.0011*	*0.0055*	*0.041*	*0.489*
Riverine flux (kt yr⁻¹)	*3*	*1.05*	*1.5*	*0.9.10-6*	*0.015*	*60.962*	*2.6*	*45*	*2.6*	*1.3*	*2240*	*0.04*	*0.2*	*1.5*	*18*

(continued)

Table 1 (continued).

Element	Tl	Tm	U	V	W	Y	Yb	Zn	Zr
Africa									
Oubangui		0.0040	0.055				0.0240		
Zaire		0.0050	0.071				0.0270		
Kasai		0.0030	0.027				0.0190		
Congo at Brazzaville		0.0035	0.049				0.0290		
Niger			0.020	0.590				0.89	0.120
Douna									0.395
Nyong		0.0085	0.029	0.645			0.0597		
Sanaga			0.028			0.0870		1.02	0.038
Nyong		0.0085	0.022			0.4610	0.0530	1.81	0.355
Mengong		0.0051	0.022			0.2821	0.0311	3.12	0.592
Europe									
Seine at Paris, Fr.			0.820	2.850		0.0500		4.98	
Garonne River Fr.		0.0006	0.750				0.0036		
Rhine in Alsace, Fr.							0.0018		
Vosges Stream, Fr.							0.0120		
Harz Mountains, Ger.	0.0400		0.060	0.400					
Kalix River, Sweden, 1991			0.090			1.4000		27.00	
Kalix River, 1977 May									
Idel river		0.0015	0.038	0.442		0.0920	0.0079	6.30	0.130
N. America									
St. Lawrence	0.0076	0.0006	0.373	0.439		0.0320	0.0029	2.58	0.022
SLRS 4		0.0002	0.050	0.350		0.1460	0.0120	1.24	0.120
Ottawa		0.0035	0.072	0.341		0.2173	0.0201	3.53	0.086
Mistassini, Can.		0.0025	0.022	0.324		0.2033	0.0191	3.79	0.047
Mackenzie		0.0016	0.730	0.253		0.0313	0.0073	0.50	0.054
Peel, Can.		0.0011		0.236		0.0574		0.88	0.038
Indin River, Can.				0.009		0.0533		1.52	0.037
Beatton, Can.		0.0087		0.398		0.8936		1.34	0.710
Upper Yukon, Can.				0.347		0.0283	0.0040	2.29	0.041
Skeena, Can.				0.106		0.1412			0.048
Fraser River, Can.			0.330	0.390		0.0690			
Columbia River							0.0045		
Californian Streams					0.1–180				
Connecticut							0.0047		

Hudson River				0.0091		
Upper Mississippi		1.285	2.055		0.21	
Missouri		1.142	0.638		0.12	
Ohio		0.333	0.581		0.17	
Illinois		1.404	1.770		0.98	
Mississippi at Mouth		0.62–1.3	0.82–1.84	0.0044	0.18–0.35	
S. America						
Amazon, mean value	**0.0033**	**0.052**	**0.703**	**0.0159**	**0.45**	
Amazon < 0.2 μm	**0.0009**	**0.055**		**0.0051**	**0.76**	**0.027**
Amazon < 100 kDa	0.0003	0.022		0.0016		0.004
Amazon < 5 kDa		0.004			0.80	
Negro < 0.2 μm	*0.0024*	*0.019*		*0.0100*	*1.80*	
Negro < 0.2 μm	0.0025	0.034		0.0169	1.21	0.068
Solimoes < 0.2 μm	*0.0045*	*0.040*		*0.0214*	*2.35*	
Solimoes < 0.2 μm	0.0006	0.050		0.0037	3.01	0.008
Madeira < 0.2 μm	*0.0025*	*0.023*		*0.0092*	*0.67*	
Madeira < 0.2 μm	0.0001	0.026		0.0007	0.67	0.001
Trompetas < 0.2 μm	*0.0041*	*0.044*		*0.0264*	*1.15*	
Trompetas < 0.2 μm	0.0006	0.024		0.0043	1.16	0.026
Tapajos < 0.2 μm	*0.0013*	*0.019*		*0.0055*	*1.02*	
Tapajos < 0.2 μm	0.0003	0.015		0.0019	0.75	0.003
Rio Beni at Riberalta		0.033			0.46	
Mamore		0.042			0.27	
Rio Beni at Rurrenabaque		0.060			0.40	
Orinoco < 0.2 μm	0.0043	0.049			1.75	0.105
Orinoco < 10 kDa	0.0018	0.023			2.42	0.029
Caroni at Cuidad Bolivar	0.0009	0.012			1.53	0.070
Asia						
Ob						
Yenisei						
Lena					0.36	
Changjiang		1.100		0.0080	0.039–0.078	
Huanghe		7.500			0.065–0.32	
Xijiang						

(continued)

Table 1 (continued).

Element	Tl	Tm	U	V	W	Y	Yb	Zn	Zr
Ganges			2.000						
Mekong									
Brahmaputra			1.000						
Indus			4.940			0.0009			
Shinano, Jpn.						0.0071			
World average		*0.0033*	*0.372*	*0.71*	*0.1*	*0.0400*	*0.0170*	*0.60*	*0.039*
Riverine flux (kt yr^{-1})		*0.12*	*14*	*27*	*3.7*	*1.5*	*0.6*	*23*	*1.5*

(1) Dupré et al. (1996), (2) Levasseur et al. (1999), (3) Froelich et al. (1985), (4) Picouet et al. (2001), (5) Viers et al. (2000), (6) Viers et al. (2000), (7) Roy (1996), (8) Keasler and Loveland (1982), (9) Chabaux et al. (2001), (10) Tricca et al. (1999), (11) Frei et al. (1998), (12) Ingri et al. (2000), (13) Pokrovski and Schot (2002), (14) Gaillardet et al. (2003), (15) Lemarchand et al. (2000), (16) Andrae and Froelich (1985), (17) Yeghicheyan et al. (2000), (18) Vigier et al. (2001), (19) Huh et al. (1998), (20) Cameron et al. (1995), (21) Johannesson et al. (1999), (22) Sholkovitz (1995), (23) Benoit (1995), (24) Shiller and Mao (2000), (26) Gaillardet et al. (1997), (27) Yee et al. (1987), (28) Seyler and Boaventura (2002), (29) Brown et al. (1992a), (30) Deberdt et al. (2002), (31) Elbaz-Poulichet et al. (1999), (32) Edmond et al. (1995), (33) Dai and Martin (1995), (34) Martin et al. (1993), (35) Zhang et al. (1998), (36) Shiller and Boyle (1985), (37) Edmond et al. (1985), (38) Huang et al. (1988), (39) Zhang (1994), (40) Zhang et al. (1993), (41) Sharma et al. (1999), (42) Sarin et al. (1990), (43) Dalai et al. (2001), (44) Goldstein and Jacobsen (1988), (45) Porcelli et al. (1997), (46) Chabaux et al. (2003), (47) Gaillardet et al. (1999a), (48) Colodner et al. (1993).

of a number of elements in waters depends on filtration pore size. Finally, data for some Andean tributaries of the Madeira River have been reported. They show remarkably similar levels to those of the Amazon. Trace-element data for rubidium, caesium, barium, and uranium, have been measured for the Guyana shield.

Asia. Large rivers of Asia are clearly the less well documented in terms of trace-element concentrations. This is mainly due to their low abundances of trace elements, probably related to their high pH character. A couple of studies have focused on the riverine input of metals to the Arctic and Pacific oceans. Himalayan rivers have not been documented for REEs (except the Indus river), but have been analyzed for particular elements such as strontium, uranium, osmium, and radium. There is clearly a need for data on trace elements in the rivers of Asia, particularly in the highly turbid peri-Himalayan rivers.

9.2.1 Range of Concentrations of Trace Elements in River Waters

Based on the compilation of Table 1, an attempt has been made to compute the mean value of trace-element input to the estuaries and ocean by selecting the largest rivers (Congo, Niger, Sanaga, Seine, Garonne, Kalix, Idel, St. Lawrence, Mackenzie, Peel, Fraser, Columbia, Connecticut, Hudson, Mississippi, Amazon, Orinoco, Ob, Yenisei, Lena, Changjiang, Huanghe, Xijiang, Ganges, Mekong, Brahmaputra, and Indus). This calculation is based on the assumption that our database is representative of most of the rivers and a total water discharge of $3.74 \times 10^4 \, km^{-3} \, yr^{-1}$ (E. K. Berner and R. A. Berner, 1996).

This mean value should be considered as a first-order approximation, because for the majority of elements, the number of analyses is small. Because our best information on trace concentrations is from humid tropical areas, the mean values proposed here are dominated by rivers such as the Amazon, Orinoco, and Congo, and may therefore be overestimates for a number of elements (such as REEs). For trace metals, the role of global pollution is an unresolved issue that may also contribute to enhancing the mean value proposed in Table 1 (especially for trace metals). These values compare relatively well with the previous estimates of Martin and Meybeck (1979) for selected trace elements, with the exception of lower concentrations for some metals in the present study.

The results of Table 1 are summarized in Figure 1, in which elements are ranked according to the order of magnitude of their concentration in river water. Trace-element concentrations in river waters span 10 orders of magnitude, from elements present in extremely low concentrations such as radium (on the order of $fg \, L^{-1}$) to

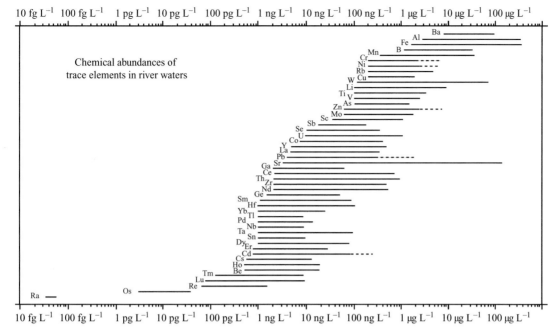

Figure 1 Graphical representation of the order of magnitude of natural trace-element concentrations in the river dissolved load. World average values derived from Table 1.

elements present in relatively high concentrations such as iron, aluminum, and barium whose concentration range exceeds $10\,\mu g\,L^{-1}$. In between, the vast majority of trace elements have dissolved concentrations between $1\,ng\,L^{-1}$ and $1\,\mu g\,L^{-1}$ (heavy REEs (HREEs), transition metals). The platinum group elements (PGEs) are represented by osmium and show very low concentrations, $\sim 10\,pg\,L^{-1}$. Very few data exist for PGEs in river waters.

9.2.2 Crustal Concentrations versus Dissolved Concentrations in Rivers

The abundances of trace elements in rivers depends both on their abundances in the continental crust and their mobility during weathering and transport. In order to depict a global "solubility" trend of trace elements, dissolved concentrations (Cw) are normalized to those of the upper continental crust (Cc) (Figure 2). Data from the continental crust are from Li (2000). In this figure, major elements in river waters are also shown and all normalized concentrations are compared to the value for sodium. It is important to note that the Cw/Cc ratio is a global mobility index rather than a solubility index because, as will be shown below, a number of very different processes contribute to the occurrence of trace elements in river dissolved load. In addition, for a number of rarely measured elements, the concentrations in the upper continental crust may well not be correct. The graph of Figure 2 is therefore a first-order approach.

A rough classification of trace-element mobility in river waters can be drawn. The first group comprises the highly mobile elements, having mobility close to or greater than that of sodium. It consists of chlorine, carbon, sulfur, rhenium, cadmium, boron, selenium, arsenic, antimony, molybdenum, calcium, magnesium, and strontium. The case of palladium is uncertain because few data are available, and its enrichment in Figure 2 could well be a result of an incorrect continental crust concentration. The following group of moderately mobile elements includes uranium, osmium, silicon, lithium, tungsten, potassium, manganese, barium, copper, radium, rubidium, cobalt, and nickel. Their mobility is ~ 10 times less than that of sodium. The third group of elements contains the REEs, zinc, chromium, yttrium, vanadium, germanium, thorium, lead, caesium, beryllium, gallium, iron, and hafnium. Their mobility is 10–100 times less than that of sodium. We will call them the nonmobile elements. Finally, the last category, the most immobile elements, includes niobium, titanium, zirconium, aluminum, and tantalum, with mobility indexes more than 100 times lower than that of sodium. Depending on weathering, soil and river conditions, it is clear that some elements of these groups can pass to another group (e.g., for highly variable elements such as aluminum, iron, rubidium, lithium, and manganese), but the picture described here should be considered as an average global trend. At present, the database of trace elements in river waters is too incomplete, but there is probably much information on the processes that control the distribution of trace elements in river

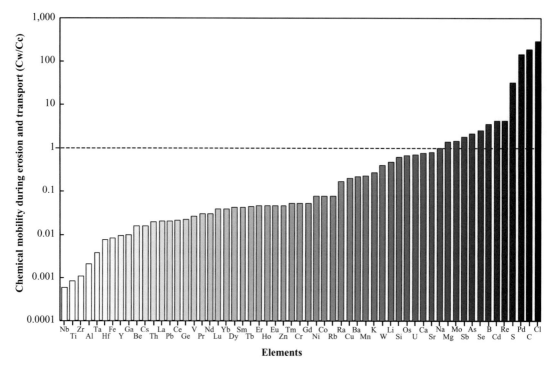

Figure 2 Normalization diagram of dissolved concentrations in rivers (Cw, Table 1) to continental abundances (Cc, Li, 2000). All values are compared to that of Na. This graph shows the increasing mobility of the elements in the weathering + transport processes, from the left to the right.

waters to be gained by addressing the variability of trace-element concentrations in rivers. Part of this variability corresponds to seasonal variability (see below).

9.2.3 Correlations between Elements

Generally, trace-element variations are not independent, and a number of authors have reported good correlations among trace elements or between trace elements and major elements. These correlations allow us to isolate groups of elements that present similar behavior during weathering and transport.

Interrelationships between trace elements and major elements or within the group of trace elements give information not only on their origin, but also about the mechanisms controlling the transport of the elements in rivers.

A clear distinction exists between trace elements whose abundance follows the abundance of major elements or TDS, and those whose concentrations are decoupled from the variations of major ions. This is illustrated by the dendrogram (Figure 3) deduced from a cluster analysis of trace-element concentration in the river waters of the Congo Basin (from Dupré *et al.*, 1996). In the Congo waters, three categories are revealed by a cluster analysis. In the first set of elements, uranium, rubidium, barium, and strontium are closely related to variations of major elements.

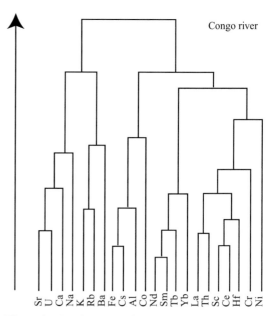

Figure 3 Dendrograms of cluster analysis of the trace-element concentrations in the dissolved load of the Congo rivers (Dupré *et al.*, 1996) showing affinity groupings of elements.

By contrast, REE variations are not correlated with those of the previous set but rather co-vary with thorium, scandium, and chromium (elements of the second set). A third set of elements showing good intercorrelation consists of caesium, cobalt, iron, and aluminum. The behavior of the latter elements is decoupled from that of the major elements but close to that of the REE group. Elements from both second and third sets show decreasing concentrations with increasing pH, in contrast to elements of the first set.

On a global scale, alkali and alkaline earth trace elements are strongly correlated to either sodium, calcium, or TDS in rivers (Edmond *et al.*, 1995, 1996). Beryllium is an alkaline-earth element with a peculiar behavior. As shown by Brown *et al.* (1992a,b), it is easily adsorbed by particle surfaces. Like beryllium, caesium has a behavior that contrasts with that of the other alkali elements. In the Upper Beni river (Bolivia) good correlations between molybdenum, zinc, cadmium, rubidium, strontium, barium and major solutes have been reported, indicating that under those conditions, trace metals can also co-vary with major elements (Elbaz-Poulichet *et al.*, 1999). From the largest tributaries of the Amazon basin, Seyler and Boaventura (2001) reported that vanadium, copper, arsenic, barium, and uranium concentrations are strongly positively correlated with major elements and pH. A correlation of vanadium with HCO_3 has also been reported by Shiller and Mao (2000) in Californian streams. Other trace metals, in nonorganic contexts, have been shown to follow major elements, e.g., the correlation between copper and silica concentrations reported by Zhang and Huang (1992) for the Huanghe River.

Selenium concentrations correlate with those of calcium and sulfate in the Orinoco river system (Yee *et al.*, 1987). Similarly, co-variations of tin or antimony with sulfate are documented by Cameron *et al.* (1995) in the Fraser river system. Little information is available on the behavior of the platinum group elements. Osmium and rhenium appear to correlate with the variations of major elements (Levasseur *et al.*, 1999; Dalai *et al.*, 2001). A good correlation between rhenium concentration and the sum of major cations is reported by Dalai *et al.* (2001) for the rivers of Yamuna and Ganga river systems in India. More generally, elements known to be present as oxyanions in oxidized waters also exhibit some coherence with major solute variations. A good correlation between tungsten or boron concentrations and chlorine has been reported by Johannesson *et al.* (1999) in rivers from California and Nevada, suggesting an essentially conservative behavior for these elements in the watershed and a strong contribution from evaporite weathering. Correlation between boron and major solutes is also reported by Lemarchand *et al.* (2000) at a global scale. In California and Nevada, the correlation between molybdenum or vanadium with chlorine is poor compared to tungsten or boron, suggesting that secondary processes operate for the regulation of these elements. From the 56 rivers analyzed for

germanium by Froelich *et al.* (1985), a relatively good correlation between germanium and silicon on a global scale is observed, provided contaminated rivers are disregarded.

The behavior of uranium varies considerably across samples studied. Palmer and Edmond (1993) have reported increasing trends of uranium concentration with river alkalinity in the Orinoco, Amazon and Ganga river basins, showing the importance of limestone and black shale dissolution for the control of uranium concentrations in river waters. The association of uranium and major soluble elements is also reported by Elbaz-Poulichet *et al.* (1999) for the upper Amazonian basins of Bolivia. However, under the organic-rich conditions of Scandinavian rivers (Porcelli *et al.*, 1997) or African rivers (Viers *et al.*, 1997), uranium concentrations can be decoupled from those of major elements, due to the existence of a colloidal fraction of uranium.

The best examples of decoupling between major solutes and trace elements in river waters come from the REEs. In most cases, dissolved REE concentrations are insensitive to major solute concentrations or even increase when solute concentrations are low. Parameters such as dissolved organic carbon (DOC) and pH appear to control the REE concentrations, and those of a number of associated elements, in river waters. Figure 4 is from Deberdt *et al.* (2002) and shows the extensive database of neodymium concentrations in river water ($<0.2\,\mu m$ or $0.45\,\mu m$) as a function of pH. Given the very good correlation coefficients between the different REE elements, this correlation indicates that the lowest concentrations of REEs in river waters are found in the rivers having the highest pH. This graph explains

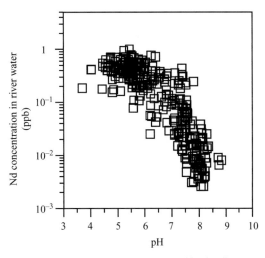

Figure 4 Nd concentrations measured in the dissolved phase ($<0.2\,\mu m$) of a great variety of rivers draining different regions of the world. This graph is based on the compilation of Deberdt *et al.* (2002) and shows the global control of dissolved REE concentrations by pH.

why the majority of dissolved REE concentrations published so far are for African or Northern rivers of low pH. The rivers of Asia, being much influenced by carbonate dissolution, have high pH and alkalinity values and therefore present very low REE concentrations.

Correlated with REE concentrations in river waters are elements such as thorium, yttrium, and, to a lesser extent, iron, aluminum, gallium, zirconium, manganese, and zinc (Laxen *et al.*, 1984; Shiller and Boyle, 1985; Dupré *et al.*, 1996; Viers *et al.*, 1997; Pokrovski and Schott, 2002). The close association of these elements in river waters with DOC has been shown by several authors (Perdue *et al.*, 1976; Sholkovitz *et al.*, 1978; Sholkovitz, 1978; Elderfield *et al.*, 1990; Viers *et al.*, 1997; Ingri *et al.*, 2000; Tricca *et al.*, 1999; Deberdt *et al.*, 2002).

9.2.4 Temporal Variability

The temporal variability of trace-element concentrations has been poorly addressed in the literature. The variability of trace-element concentration may be important even on a diel basis. Brick and Moore (1996) have reported diel variations, for example, in the Upper Clark Fork river, Montana. This river is a high-gradient river rich in metals because of mining. Correlating with pH variations between day and night, dissolved manganese and zinc concentrations increase two- and three-fold at night, while major solutes and water discharge show no evidence of variations.

Tenfold daily to weekly variations of manganese have been described in the Kalix River (Ponter *et al.*, 1992) between winter (minimum concentrations) and snow-melt in May. These variations are attributed to the input of manganese-rich mire waters (due to rapid oxidation–reduction processes) and are shown to be associated with significant variations of dissolved REEs, aluminum, and iron. A sevenfold increase of lanthanum is observed and is associated with the increase of water discharge and DOC concentrations and the decrease of pH occurring at spring flood (Figure 18; Ingri *et al.*, 2000). It is important to note that even the REE patterns are affected, changing from a relatively flat to an LREE-enriched pattern when water discharge increases. In the Kalix River, these variations in the dissolved load are associated with variations in particulate concentrations. A similar cycling of manganese and iron concentrations has been reported for the St. Lawrence system by Cossa *et al.* (1990).

A similar seasonal variability is observed in the organic-rich rivers of Africa. Viers *et al.* (2000) have classified the trace elements as a function of their variability (standard deviation/mean value) during the period 1994–1997. While major elements show the lowest variability (except potassium), trace elements show high variability, ranging from 80% for rubidium and manganese to 40–50% for DOC, REEs, thorium, aluminum, and iron. The temporal evolution of elements such as thorium or REEs strictly mimics that of DOC and shows the highest concentrations during the high-water stage period. Major elements follow the reverse tendency.

The variability of very large river systems has been addressed for the Mississippi (Shiller and Boyle, 1987; Shiller, 1997, 2002) and the Amazon (Seyler and Boaventura, 2002). In the Mississippi, the largest amplitude of dissolved concentrations is observed for manganese (50 times increase with water discharge increase) and iron (eightfold variation). All other trace elements analyzed show lower concentration variability (from 1.5-fold for barium and rubidium to fourfold for lead and molybdenum). Certain trace elements vary in phase with manganese and iron, being maximum at high-water stages, e.g., zinc and lead, some others show an opposite seasonal behavior, such as vanadium, molybdenum, uranium and, to a lesser extent, copper, nickel, and cadmium. REEs (Shiller, 2002) co-vary with manganese or zinc, the LREEs showing much more variability than the HREEs. These seasonal trends for the lower Mississippi are remarkably similar to those observed for the Kalix River. In the Amazon system, a time series of trace-element concentrations at Obidos (800 km upstream from the mouth of the Amazon) shows (Seyler and Boaventura, 2002) that several patterns of variation exist. As in the Mississippi river, manganese shows the largest variations (10-fold) and exhibits maximum concentrations at high-water stage (but one month after the water peak). Elements such as cadmium ($\times 8$) and cobalt ($\times 4$) show similar patterns. The opposite trend is observed for antimony, molybdenum, copper, strontium, barium, vanadium, and major elements, with maximum concentrations at low-water stage and less variability. The elements uranium, rubidium, nickel, and chromium do not show significant or systematic variations during the hydrological cycle. So far, variations of REEs, aluminum, and iron have not been documented in the Amazon.

The conclusion of this rapid review of the literature on temporal variations of trace-element abundances in river waters is that the range of variation can be very large. Temporal series are thus necessary not only to compute riverine budgets to the oceans but also to understand the processes controlling trace-element concentrations. Manganese and iron appear to be the most variable elements in rivers, as well as a number of elements whose chemistries seem to be associated with those of manganese and iron. As will be discussed later, manganese and iron variations are consistent with redox changes within the river system. We would stress that, in order to understand controls on trace-element concentrations in watersheds and to refine load budgets of elements to the ocean, future studies will have to take into account temporal variations.

9.2.5 Conservative Behavior of Trace Elements in River Systems

The question of seasonal variability of trace elements in river systems raises the question of their conservative or nonconservative behavior during the mixing of tributaries. Major elements and trace elements of high mobility indexes generally have conservative behavior during the mixing of tributaries. This is apparent for the Amazon River (Seyler and Boaventura, 2002). Conversely, if redox cycling and associated exchange with solid phases, related to temperature or biological activity within the river system, control the temporal variations of manganese, iron, and associated elements, it is highly probable that their behavior will not be conservative. In the Mississippi River, seasonal changes in trace-element concentrations are not explained by the hydrological mixing of tributaries, except for barium and uranium (Shiller and Boyle, 1987; Shiller, 1997). The same conclusion is reached from times-series comparison in the Amazon basin by Seyler and Boaventura (2002). The downstream evolution of major and trace elements along the Solimoes and Amazon rivers (from 3,000 km upstream to the river mouth) illustrates conservative and nonconservative behavior in the Amazon system. The patterns of manganese, vanadium, and nickel are shown on Figure 5. While vanadium, like the major elements, shows a downstream decrease of concentration, consistent with the inputs of the dilute lowland rivers, manganese and nickel show increasing and constant concentrations, respectively. In the case of manganese and nickel, a substantial source is necessary that cannot be represented by the minor input of lowland tributaries, but rather by processes releasing these elements into the dissolved load.

9.2.6 Transport of Elements

Rivers transport material both in dissolved form and as solid load (suspended matter and bottom sands). The dominant form of transport of trace elements per liter of river water depends on both the mobility of the element in the weathering + transport process and on the amount of solids transported annually by the river. Two contrasted examples of big rivers from the Amazon basin are shown on Figure 6. The Solimoes River is characterized by relatively high concentrations of suspended sediments ($230\,mg\,L^{-1}$), whereas the Rio Negro is characterized by very low suspended sediment concentrations (less than $10\,mg\,L^{-1}$). This discrepancy is related to contrasting weathering regimes in the two basins. In the Solimoes, even

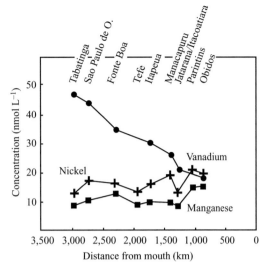

Figure 5 Downstream evolution of vanadium, nickel, and manganese concentrations measured in the dissolved load of the Amazon river showing different possible behaviors of trace elements in river systems. Major solutes and TDS follow the similar pattern to that of vanadium and correspond to a dilution of Andean waters by waters from the Amazonian lowlands (source Seyler and Boaventura, 2002).

the most mobile elements, such as sodium, are transported dominantly in a solid form. This also applies to the highly turbid rivers of Asia. Conversely, rivers from humid tropical regions such as the Rio Negro can show a significant fraction of their trace elements transported in the dissolved load, both because the suspended matter concentration is low (less than $10\,mg\,L^{-1}$) and because their dissolved load is generally enriched in the so-called "nonmobile and highly insoluble" elements of Figure 2. In the Rio Negro, half of the REEs, thorium, and aluminum are transported in the dissolved load.

If TSS is total suspended sediment concentration in the river, K_d, the ratio of concentration in the suspended sediments over dissolved concentration, then the proportion D of a given element in the dissolved phase can be expressed as

$$D = \frac{1}{1 + K_d \cdot \text{TSS}}$$

The typical values of K_d can be estimated using Table 1 and average values of concentrations in suspended sediments (Gaillardet *et al.*, 1999b). Except for the most mobile elements, K_d values are not very different from the continental crust normalized concentrations shown on Figure 2. We have plotted in Figure 7 the proportion D as a function of TSS for three types of elements: thorium (immobile element), nickel (intermediate), and boron (mobile element), corresponding to the K_d values of 1,000, 100, and 10, respectively. This graph shows that at the world average value of TSS ($350\,mg\,L^{-1}$), the dominant form of

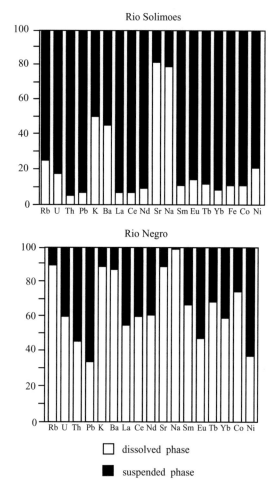

Figure 6 Proportion of elements transported in dissolved (<0.2 μm) and particulate forms by contrasted rivers of the Amazon Basin: the Solimoes (a turbid river, mostly influenced by the Andes) and the Rio Negro (a typical lowland, black river with very low suspended sediments yields).

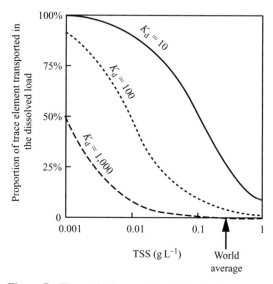

Figure 7 Theoretical proportion of dissolved transport as a function of the concentration of riverine suspended sediments for three values of the ratio of concentration in the suspended load over concentration in the dissolved load. $K_d = 10$, 100, and 1,000 correspond to the most mobile elements (Na, B, Re, Se, As, Sb), the intermediate elements (Cu, Ni, Cr, Ra) and highly immobile elements (Th, Al, Ti, Zr) respectively (see Figure 2).

transport is the solid form. Only the most mobile elements can be significantly transported in a dissolved form.

9.3 SOURCES OF TRACE ELEMENTS IN AQUATIC SYSTEMS

Continental crust is the ultimate source of trace elements in hydrologic systems. Trace elements are introduced in the river basin by rock weathering, atmospheric dry and wet deposition and by anthropogenic activities. The sketch diagram of Figure 8 summarizes the natural and anthropogenic sources of elements in aquatic environments (modified after Foster and Charlesworth, 1996).

9.3.1 Rock Weathering

The chemical weathering of rocks results in the release of the most soluble elements and, in

the case of silicate rocks, leaves a residue, which, conversely, is enriched in the insoluble elements. The normalization of soil materials or river suspended sediments, taken as a natural average of the finest soil materials, allows a first-order examination of the behavior of trace elements during chemical weathering. For example, the suspended sediments of some large unpolluted rivers have been normalized to the upper-crust concentrations (Gaillardet *et al.*, 1999b) in Figure 9. Elements are classified according to their magmatic compatibility. Most of the trace elements analyzed on those patterns have continental-like abundances. Only a few elements are depleted. They include soluble elements (rubidium, uranium, barium, and major elements) and elements concentrated in heavy minerals (zirconium, hafnium). The complementary reservoirs are, respectively, the dissolved phase and bottom sands. For elements that are not depleted by chemical weathering and transport processes, their ratio in the particles is expected to be similar to that of the source rock from which they are derived. For example, the Sm/Nd ratio of suspended sediments is a good integrated signature of the rocks that have undergone weathering throughout the drainage basin. The Sm/Nd isotopic system is therefore insensitive to the weathering history of sediment, and it records episodes of crustal formation rather than weathering + deposition cycles. This property is the basis of the use of neodymium isotopes as tracers of crustal growth.

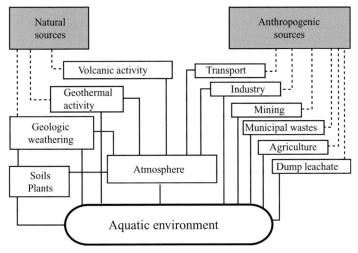

Figure 8 Pathways of trace elements to the aquatic system.

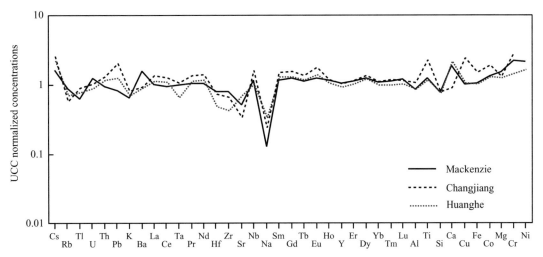

Figure 9 Concentrations measured in the suspended sediments of large river systems normalized to the upper continental crust UCC (Li, 2000). The order of element on the *x*-axis is an order of magmatic compatibily (source Gaillardet *et al.*, 1999b).

In the dissolved load, the preferential dissolution of the different types of lithology or mineral can cause large variations in trace-element abundances in rivers. A classical tracer of rock dissolution is strontium. The isotopic ratio of strontium in river waters is, to a first approximation, explained by mixing between different sources. Rain, evaporite dissolution, and carbonate and silicate weathering are the predominant sources of strontium in rivers. Examples of the use of Ca/Sr, Na/Sr, and Sr isotopic ratios to quantify the proportion of strontium derived from carbonate weathering in large basins can be found in Négrel *et al.* (1993) and Gaillardet *et al.* (1999a). Levasseur (1999) have used isotopic ratios of osmium in river waters to determine the origin of osmium in the largest rivers. As a global average, osmium in river waters is derived from silicates (14%),

carbonates (55%), and shales (31%). Uranium in river basins (Palmer and Edmond, 1993) is associated with carbonate dissolution. A number of trace elements also appear to be good tracers of black shale weathering (Peucker-Ehrenbrink and Hannigan, 2002; Dalai *et al.*, 2001, Pierson-Wickman *et al.* (2000) Chabaux *et al.*, 2001). These rocks are enriched in elements that have a strong affinity for organic matter and are stable in reducing conditions (uranium, metals, PGE); their dissolution can lead to substantial enrichments. The correlation between tungsten and chlorine in California streams reflects the importance of evaporite dissolution to the riverine budget of tungsten (Johannesson *et al.*, 2000).

Trace-element abundance in rivers depends not only on their abundance in source rocks but also on the weathering style. Some authors have proposed that trace-element ratios involving chemically similar elements might be fractionated during the weathering of silicates and could serve as a proxy for weathering intensity. It has, for example, been shown that significant variations in the Ge/Si or Ga/Al ratios occur in river waters and that these variations are related to the degree of weathering (Murnane and Stallard, 1990; Shiller and Frilot, 1995). A weathering-limited regime would produce high Ga/Al and low Ge/Si in the river dissolved load because of slight differences in solubility between elements. In transport-limited regimes, these ratios should approach those of the source rocks. However, as we will see later, the chemistry of the river, especially organic-matter complexation, can significantly obscure the weathering signal. Shiller and Mao (2000) have shown this for vanadium. Although vanadium in rivers is essentially derived from silicate weathering, variations in the V/Si ratios are observed due to minor inputs from sulfide dissolution, increased solubilization of vanadium by dissolved organic matter, and variations of the vanadium content of source rocks. In spite of these complications, research is needed to evaluate the use of trace element in rivers as tracers of specific types of mineral/rock weathering or as indexes of weathering regimes. A vast majority of dissolved trace elements in natural waters are not provided by rock weathering alone; other sources will be discussed below.

9.3.2 Atmosphere

Trace elements are transported in the atmosphere in the form of wind-blown soil particles, fine volcanic products, sea salts, ashes from forest fires, and biogenic aerosols. The atmospheric input of trace elements can be significant, depending on their abundance in rain and aerosol solubility. Due to the ability of aerosols to travel over long distances in the atmosphere, it is difficult, based on present-day sampling, to determine the natural background concentrations of elements in rain water and aerosols. In remote sites of Niger and the Ivory Coast, Freydier et al. (1998) have shown that the emission of terrigeneous particles of crustal composition is the main source of trace elements in atmospheric deposition of this region. Where the amount of crustal particles in the atmosphere is low, the contribution of other sources becomes evident (ocean, vegetation, human activities). In both sites, a 50-fold enrichment of zinc with respect to the continental crust is observed in rainwater, possibly derived from the vegetation or from remote industrial sources. In the Amazon basin (Artaxo et al., 1990), biogenic aerosols are enriched in potassium, phosphorus, sulfur, and zinc, which can explain the enrichment of rain waters. The impact of human activities on rain-water chemistry has been clearly demonstrated in large cities and even in regions far from industrial centers (e.g., Berg et al., 1994). In Western Europe, rain waters in Paris have concentrations more than 100 times those in the mean continental crust for the elements cadmium, zinc, lead, copper, bismuth, nickel, chromium, cobalt, and vanadium (Roy, 1996). In those samples, arguments based on lead isotopes have shown that 100% of the excess of metals can be attributed to anthropogenic sources. These elements are introduced into the atmosphere by coal-burning or metal smelting that produces fine particles. The atmosphere is therefore extremely sensitive to anthropogenic contamination. Nriagu (1979) has calculated that global atmospheric emissions increased exponentially over the twentieth century.

9.3.3 Other Anthropogenic Contributions

Apart from the atmosphere, there are a number of potential point and non-point sources of contamination in hydrological systems. The industrial revolution, since the beginning of the twentieth century, has caused a drastic increase in the exploitation and processing of metals, resulting in their release into the environment and the release of associated elements with no economic value (e.g., arsenic). In southwestern France, Schäfer and Blanc (2002) have shown that the geochemistry of suspended sediments of the largest rivers can be related to the occurrence of ore deposits in the upper basins of the Massif Central and Pyrenees. The release of metals is also associated with the use of metals and other trace elements in paints, cements, pharmaceutical products, wood treatment, water treatment, plastics and electronic items, and fertilizers. These activities can release trace metals directly into the river catchment (Foster and Charlesworth, 1996; Luck and Ben Othman, 1998; De Caritat et al., 1996). For example, positive gadolinium anomalies have been reported in the dissolved load of European rivers and attributed to the use of gadopentetic acid in magnetic resonance imagining (Bau and Dulski, 1996). In the Seine River (Roy, 1996), lead isotope studies have shown that almost half of the lead transported by the river (adsorbed and dissolved) is of anthropogenic origin. In Mediterranean catchments, it has been shown by Luck and Ben Othman (1998) that lead and other metals brought in by rain waters are stored in soils and remobilized during floods as particles. In the Rhine River, the concentration of metals associated with suspended sediments decreases by a factor of 3 between low water discharge and high water discharge (Foster and Charlesworth, 1996). The same effect is observed in the rivers of southwestern France (Schäfer and Blanc, 2002) and is interpreted as a "dilution" of pollution by natural erosion of uncontaminated soil particles at high water discharges. Over longer timescales, there are a few sites where metal pollution has been monitored in rivers. Records from the river Rhine are available in the Netherlands (Foster and Charlesworth, 1996). A steady (between 2 and 20 times) increase in the concentrations of zinc, lead, chromium, copper, arsenic, nickel, cobalt, cadmium, and mercury is observed over the twentieth century and a slight decline after 1975, clearly showing the sensitivity of aquatic systems to human activities. For many trace metals, anthropogenic contributions from all sources far exceeds natural levels. Nriagu and Pacyna (1988) estimated that the man-induced mobilization of arsenic, cadmium, copper, mercury, molybdenum, nickel, lead, antimony, selenium, vanadium, and zinc far exceeds the natural fluxes. For arsenic, cadmium, lead, selenium, and mercury they report enrichment factors due to human activities of 3, 7.6, 24.1, 2.8, and 11.3, respectively, with respect to the natural levels.

Atmospheric acid deposition influences the mobility of trace elements, depending on the source rocks. Central Europe has been strongly affected by acid deposition during the twentieth century. Frei *et al.* (1998) have shown that the concentrations of trace elements in the Upper Ecker drainage and in the Northern Harz Mountains (Germany) are high and are related to the neutralizing capacity of the source rocks. Acidic waters have the highest trace-element concentrations.

As a conclusion, the origin of trace elements in river waters can be very diverse. Elucidating the sources and quantifying the proportion of elements derived from each of these sources is not an easy task, and it has been addressed by the use of isotopic ratios (strontium, neodymium, lead) and enrichment factors of trace elements with respect to well-characterized reservoirs (upper continental crust, ocean). However, we are far from being able to trace the origin of all elements in natural waters, especially when their biogeochemical cycle is complex and involves biomass or an atmospheric subcycle. In addition, human activities have led to a generalized perturbation of element abundances in the atmosphere, soils, and waters; it is very difficult, if not impossible, to assess the natural levels of trace elements in river waters and particles. It is, for a number of elements such as metals, difficult to estimate their natural rate of transport in rivers. The largest rivers with high water discharge and active erosion are probably the best-suited rivers for assessing the natural background of elements in rivers.

The behavior of trace elements in aquatic systems not only depends on the sources but it is also strongly controlled by the soil and in-stream processes, particularly through aqueous (organic and inorganic) complexation and reactions with solids.

9.4 AQUEOUS SPECIATION

The speciation of a given element is the molecular form under which it is transported in hydrosystems. Like some of the major elements, trace elements are transported in surface waters in complexed forms. Aqueous complexes correspond to the association of a cation and an anion or neutral molecule, called the ligand. Ligands can be inorganic and organic. It is rarely possible to measure the concentration of an individual complex (Pereiro and Carro Díaz, 2002; Ammann, 2002). Most of the techniques used to measure trace-element concentrations give the total concentration, whatever the speciation. The proportion of elements corresponding to the different chemical forms has therefore to be calculated, depending on the total concentration, pH, Eh conditions, the major element chemical composition of the water, and the complexation constants of the assumed complexes. Knowing the concentrations of the individual species of a given element in waters is of crucial importance in order to predict its toxicity and bioavailability. For example, the toxicity of metals depends more on their chemical form in waters rather than their total concentration (Morel and Hering, 1993), in particular, because complexation may enhance or inhibit adsorption on surfaces.

In river waters, inorganic ligands are essentially H_2O, OH^-, HCO_3^-, CO_3^{2-} and, to a lesser extent, Cl^-, SO_4^{2-}, F^-, and NO_3^-. Organic ligands are small organic weak bases such as oxalate or acetate and low-molecular-weight humic acids containing phenolate and carboxylate groups.

Speciation calculations are performed according to mass balance equations and mass law equations corresponding to the different complexation reactions. As an example, we show in Figure 10 the calculated speciation of thorium in pure water (Langmuir and Herman, 1980). The situation

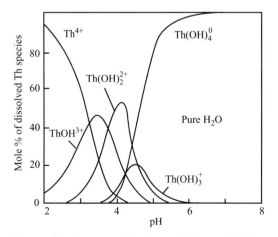

Figure 10 An example of speciation calculation. Distribution of thorium complexes versus pH at 25 °C with $\Sigma Th = 0.01\ \mu g\,L^{-1}$ in pure water (after Langmuir and Herman, 1980).

would be more complex in natural waters, depending on the abundance and nature of the inorganic and organic ligands, but no free thorium is predicted and $Th(OH)_4^0$ is still the dominant complex at common river water pH.

Complexation constants of trace elements in aqueous solution are determined using a great variety of techniques (including conductimetry, potentiometry, spectrophotometry, solvent extraction, calorimetry, ion exchange) and are reported in numerous publications. Critical studies reporting complexation constants can be found in Baes and Mesmer (1976), Martell and Smith (1977), or Martel and Hancock (1996). The values for complexation reactions are still a matter of debate, given the high possible number of complexes between trace elements and natural ligands, especially poorly known organic ligands. A number of computer codes have been developed in order to calculate the speciation of trace elements in waters. These computer codes basically use the mass action laws of the different complexation reactions, activity coefficients and conservation equations (e.g., WATEQF, Plummer *et al.*, 1984).

A number of authors have investigated the interesting issue of relating the aqueous speciation rules of trace elements to the periodic table (Turner *et al.*, 1981; Langmuir, 1997). The classical "Goldschmidt plot" plots the charge of the cation (in the complex) versus the cationic radius. Three groups of elements can be distinguished: those forming oxyanions in solution (carbon, arsenic, boron, nitrogen,...), elements forming hydroxycations and hydoxyanions (thorium, aluminum, iron, titanium, lead) and elements forming free cations or aquocations (sodium, potassium, silver). More precisely, following the Born theoretical treatment of ion–water interaction (see Turner *et al.*, 1981), the ability of cations to attract electrons can be approximated by the polarizing power, Z^2/r, where Z is the ion charge and r the ionic radius. Figure 11 is a graph of the intensity of hydrolysis

$$M^{n+} + H_2O = (MOH)^{(n-1)+} + H^+$$

as a function of the polarizing parameter and allows a classification of trace elements in solution.

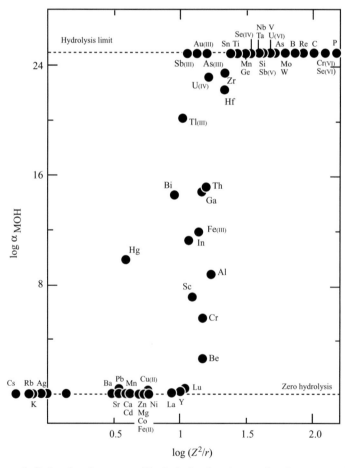

Figure 11 The "Born plot" showing the extent of hydrolysis of a given cation (α_{MOH} quantifies the affinity with OH^-) as a function of the polarizing power (Z^2/r) where r and Z are, respectively, the radius and charge of the cation (after Turner *et al.*, 1981).

A high polarizing power (>1.3) means that the element easily attracts electrons and is fully hydrolyzed. The speciation of such elements is in the form $MO_p(OH)_q$ and depends on pH, via the dissociation constant. Examples of such elements and their associated acid form are arsenic, boron, chromium, germanium, molybdenum, niobium, tantalum, rhenium, antimony, silicon, selenium, uranium, and vanadium. They exist, for example, in the form of $As(OH)_3$, $B(OH)_3$, $Ge(OH)_4$, $Sn(OH)_4$, CrO_4^{2-}, MoO_4^{2-}, WO_4^{2-}, ReO_4^{2-},..., and behave as strong or weak acids in solution. These elements are known as fully hydrolyzed elements and are belong to groups III—VII in the periodic table.

Elements with intermediate and low Z^2/r are capable of producing free cationic forms in solution. Their speciation is a question of competition among the major ligand species (Cl^-, SO_4^{2-}, OH^-, CO_3^{2-}). These elements are the alkalis, alkaline earths, REEs, and metals. They can be classified according to their tendency to form covalent bonds. The concept of hard and soft ions was introduced by Pearson and is equivalent to the concept of (a)- and (b)-type cations introduced by Ahrland (Langmuir, 1997). The cations are Lewis acids (electron acceptor) and the ligands, Lewis bases (electron donor). "Soft" means that the electrons of the species can easily be deformed whereas "hard" means that the species are rigid. Soft species easily form covalent bonds. Hard species prefer electrostatic bonding. Soft cations form strong covalent bonds with soft ligands, and hard acids form strong ionic bonds with hard ligands (Langmuir, 1997). Complexes between hard (soft) acids and soft (hard) bases are

weak and rare. Alkaline earths, La, Ti^{4+}, Y^{3+}, Sc^{3+}, Ga^{3+}, In^{3+}, Sn^{4+}, Cr^{3+}, Mn^{3+}, Fe^{3+}, Co^{3+}, U^{4+}, Th^{4+}, and actinides are classified as hard acids and form strong complexes with the major ligands found in natural freshwaters: F^-, OH^-, SO_4^{2-}, CO_3^{2-}, or HCO_3^- (hard bases). Complexes with Cl^- (hard–soft ligands) are weaker. Soft acids comprise Cu^+, Ag^+, Au^+, Au^{3+}, Pd^{2+}, Cd^{2+},..., while Zn^{2+}, Fe^{2+}, Co^{2+}, Ni^{2+}, Cu^{2+}, Pb^{2+} are classified as borderline acids due to their intermediate behavior. Turner *et al.* (1981) have reported correlations between the log of complex stability constants for all major freshwater ligands and $Z^2/(r+0.85)$. Significant linear correlations between the strength of the complex and the electrostatic parameter are observed. Hard ligands (such as F^-, SO_4^{2-}), exhibit a rapid increase of stability constant with polarizing power. Soft ligands, such as Cl^-, show only a slow increase of the affinity constants with $Z^2/(r+0.85)$. Intermediate ligands (OH^-, CO_3^{2-}) show a clear increase of association constant with the polarizing parameter. For example, Figure 12 represents the case of carbonate complexes. Such relations are remarkable, as they relate the complexation constants of trace elements in solution to relatively simple properties of the periodic table.

The results of cation speciation calculations for freshwaters lead to clear conclusions (Turner *et al.*, 1981). The cations with low polarizing numbers are very weakly complexed in freshwaters and exist dominantly as free ions (lithium, strontium, rubidium, caesium, barium). Conversely, elements of high polarizing power (hafnium, thorium, aluminum, scandium,...) have their speciation dominated by hydrolyzed species and correspond to the

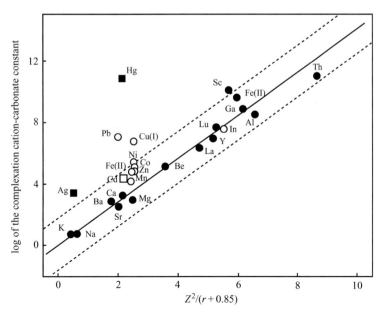

Figure 12 Linear relation between the complexation constant of cations and carbonate in solution as a function of $Z^2/(r+0.85)$, where 0.85 is an empirical correction of the Born function (after Turner *et al.*, 1981).

elements of the transition zone in the Born graph (Figure 11). A third group of elements corresponds to soft cations (Ag^+, Au(I), Tl(I)) in association with soft ligands (essentially Cl^-). The last group of elements consists of the REEs, transition metals, cadmium, which exist both as free ions and carbonate or hydroxyl complexes, depending on the pH. The speciation of REEs in natural waters has been modeled by several authors (Cantrell and Byrne, 1987; Wood, 1990; Millero, 1992; Lee and Byrne, 1993; Johannesson et al., 1996, and references therein, Liu and Byrne, 1997). The dicarbonato ($Ln(CO_3)_2^-$) and carbonato ($LnCO_3^+$) REE carbonate complexes are shown to be the dominant forms of REEs in groundwaters from Nevada and California, having pH values between 7 and 9. Stability constants show that the formation of REE carbonate complexes increases with increasing atomic number across the REE suite. The formation of other complexes such as sulfate, phosphate, or hydroxide is less important but also tends to favor the HREEs. Fractionation of REEs between source rocks and associated groundwaters is not observed in groundwaters from Nevada and California (Johannesson et al., 1996), but HREE enrichments occur in some cases, reflecting the dominance of HREE carbonate complexes, more stable in solution than the LREE carbonate complexes.

Organic speciation has been poorly documented, compared to inorganic speciation. However, as shown by several authors, a good correlation is observed between the first hydrolysis constant and the association of the cation with small organic molecules such as oxalate (Langmuir, 1997). Because humic molecules are large, poorly characterized molecules, they considerably complicate speciation calculations in freshwaters. Most often, these molecules can no longer be considered as dissolved ligands, but rather as "colloids".

Finally, it is interesting to note that the relation between the global mobility of trace elements in the aquatic system (Figure 2) and the speciation patterns summarized above is not direct. Although most of the fully hydrolyzed species appear as mobile elements (phosphorus, carbon, rhenium, boron, arsenic, molybdenum), some elements in this family are amongst the least mobile elements (titanium, germanium, niobium, tantalum). Conversely, the elements with zero hydrolysis (Figure 11) are generally mobile elements but exceptions exist such as caesium, rubidium, lead, and lanthanum. This absence of correlation clearly indicates that aqueous complexation is not the only process to control trace-element abundances in aquatic systems.

9.5 THE "COLLOIDAL WORLD"

It is now well known that trace-element concentrations in continental waters depend on the size of the pore filters used to separate the particulate from the dissolved fraction. This is apparent in Table 1, where results from the Amazon and Orinoco are reported using two filtration sizes: the conventional 0.2 μm filtration and filtration with membranes of smaller cutoff size (ultrafiltration). These results suggest the presence in solution of very small (submicrometric) particles that pass through filters during filtration. The view that trace elements can be separated into "particulate" and "dissolved" fractions can thus no longer be held; this has led authors to operationally define a colloidal fraction (0.20 μm or 0.45 μm to 1 nm) and a truly dissolved fraction (<1 nm) (e.g., Buffle and Van Leeuwen, 1992; Stumm, 1993). The existence of a colloidal phase has a major influence on the speciation calculation schemes presented above (based only on aqueous complexation), as the apparent solubility of trace elements will be enhanced by the presence of colloids. The dynamics of colloids also completely change the way reactive solute transport is modeled and are an issue for water quality standards since the toxicity of trace elements depends not only on their abundance and speciation but also on their bioavailability.

Although, strictly speaking, colloids consist of small particles and could be treated as a particular case of reactive surfaces, we summarize below the recent literature on the nature and reactivity of these phases.

9.5.1 Nature of the Colloids

The separation and characterization of submicron-sized particles in water is difficult, in particular because of artifacts from sampling and concentration techniques. Lead et al. (1997) have presented a critical review of the different techniques for separation and analysis of colloids (filtration, dialysis, centrifugation, but also voltametry, gels (DET/DGT), field-flow fractionation, SPLITT). Ultrafiltration membranes have been developed with nominal cutoff sizes ranging from the thousands of daltons (Da) to hundreds of thousands of daltons, which have been used to separate the colloidal pool into several fractions.

Figure 13 (after Buffle and Van Leeuwen, 1992) shows an example of the distribution of mineral and organic colloids as a function of size from ångstrom to micrometer scales. This diagram shows that there is no clear boundary between dissolved and colloidal substances or between colloids and particulates.

Colloids can be organic or inorganic. Even if they are not separated from the dissolved load by classical filtration, colloids have the physicochemical properties of a solid. Colloids are finely divided amorphous substances or solids with very high specific surface areas and strong adsorption capacities. It is shown by Perret et al. (1994) for

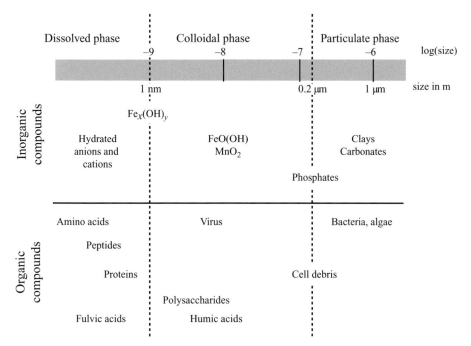

Figure 13 Distribution of mineral and organic colloids as a function of size in aquatic systems (after Buffle and Van Leeuwen, 1992).

the Rhine River that the colloids contribute less than 2% of the total particle volume and mass, but represent a dominant proportion of the available surface area for adsorption of pollutants. The abundance of colloids, their fate, through coagulation and sedimentation processes in natural waters therefore control the abundance of a number of elements.

The most common mineral colloids are metallic oxyhydroxides (e.g., mainly iron-, aluminum- and manganese-oxyhydroxides), clays, and siliceous phases. The reaction between trace elements in water and surface hydroxyl groups (S–OH) is, to a first order, analogous to the formation of aqueous complexes. Organic colloids are mainly humic and fulvic acid (humic substances) and derive from an incomplete degradation of soil organic matter. Humic substances represent 70–90% of DOC in wetland areas (Thurman, 1985) and are usually responsible for the brown color of water. Despite their heterogeneity and complexity, humic substances are characterized by similar functional groups (carboxyls, quinones, phenolic OH groups) and the presence of aliphatic and aromatic components (Stevenson, 1994). Metal binding in natural environments is, however, mainly related to carboxylic and phenolic groups (Perdue *et al.*, 1984).

Inorganic and organic colloids are most often intimately associated. Association of organic colloids, with clays (Koshinen and Harper, 1990) and with iron oxyhydroxides (Tipping, 1981; Gu *et al.*, 1995; Herrera Ramos and McBride, 1996) are commonly described in natural waters and have been the subject of laboratory experiments.

For example, Perret *et al.* (1994) have sampled, fractionated, and characterized submicron particles from the Rhine and showed that these colloids consist of a tight association between organic and inorganic material.

9.5.2 Ultrafiltration of Colloids and Speciation of Trace Elements in Organic-rich Rivers

Several authors have recently performed ultrafiltration of natural waters (river, groundwater, estuaries) in order to determine the role of colloids in the transport of major and trace elements. As noted by some authors (Sholkovitz, 1995; Horowitz *et al.*, 1996; Hoffmann *et al.*, 2000), these experiments require caution; mass balance calculations must be performed to check the ultrafiltration procedure, as some truly dissolved elements can be retained inside the filtration system.

The main advantage of ultrafiltration techniques is that they permit a direct determination of the association constants between colloids and trace elements in natural systems. This is illustrated by a series of papers dedicated to organic-rich tropical rivers of the Nyong River Basin (Cameroon, Africa) (Viers *et al.*, 1997, 2000; Dupré *et al.*, 1999). The Nyong basin river waters exhibit low major cation concentrations (i.e., sodium, potassium, calcium, magnesium) but high concentrations of some trace elements (aluminum, iron, thorium, zirconium, REEs), silica, and DOC. Figure 14 shows the behavior of some trace elements (samarium, aluminum, strontium), silicon, and organic carbon

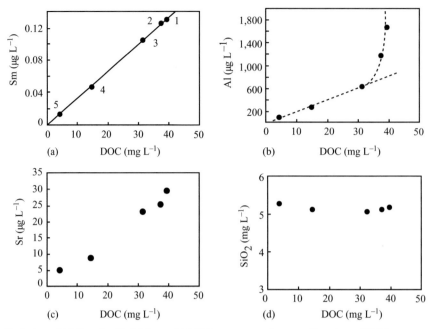

Figure 14 Variation of (a) Sm, (b) Al, (c) Sr, and (d) SiO$_2$ concentrations as a function of DOC concentrations in the different filtrates of the Awout River. 1: <0.20 μm (FF), 2: <0.20 μm (TF), 3: <300 kDa (TF), 4: <5 kDa (TF), 5: <1 kDa (TF). FF and TF mean frontal filtration and tangential filtration, respectively (source Dupré *et al.*, 1999).

during successive filtrations through decreasing pore size membranes (0.2 μm, 300 kDa, 5 kDa, and 1 kDa) of the Awout River (Dupré *et al.*, 1999). Concentrations decrease when samples are filtered through progressively finer pore-size membranes. Three patterns of element concentration as a function of DOC (taken as an index of colloid abundance) are observed. Silica shows no concentration variation as a function of DOC. The second type of behavior is that of yttrium and REEs. The relationship between these elements and DOC defines a straight line going through zero, which means that they are progressively and completely removed from the solution jointly with DOC. A group of elements composed of rubidium, strontium, copper, cobalt, manganese, chromium, vanadium, and major cations presents the same type of relation but a certain amount remains into the solution when all DOC is removed. These linear relationships suggest that the ability of these elements to form complexes with organic colloids remains constant over the whole molecular size range of the colloid materials examined. For some other elements, plots of (aluminum, iron, gallium, thorium, uranium) concentration versus DOC exhibit a nonlinear relationship. This suggests that, for this set of elements, the binding capacities (i.e., the quantity of element bound to organic material per unit mass of carbon) in the organic colloids are a function of the size of colloids. Several authors, with contradictory conclusions, have discussed this hypothesis. Lakshman *et al.* (1993) separated, by ultrafiltration, three fractions of a fulvic acid extracted from a soil

(<500 Da; 500–1,000 Da; 1,000–10,000 Da) and demonstrated that the aluminum binding constant of organic material increases with the molecular weight of the fractions. Conversely, Burba *et al.* (1995), working on humic substance fractions (<1,000 Da; 1,000–5,000 Da; 5,000–10,000 Da), showed that the lower-molecular-weight fractions exhibit the highest complexing capacity for copper. Eyrolle *et al.* (1993, 1996) obtained similar results.

One of the major results of ultrafiltration techniques conducted on the Nyong basin rivers is the decrease of major cation (strontium, rubidium, barium) concentrations with decreasing filter pore size. This result is unexpected, as these cations are known to have a very low affinity for organic matter. Eyrolle *et al.* (1996) observed the same intriguing concentration decrease of cations (sodium, magnesium, potassium, calcium) during ultrafiltration of organic-rich river waters from South America and proposed that these cations were incorporated in plant debris. Dupré *et al.* (1999), alternatively, suggested that there are artifacts associated with the ultrafiltration of organic-rich waters. In the pH range of most natural waters (4–8), the surfaces of organic colloids are negatively charged. During the ultrafiltration process, the coating of the membrane by organic colloids disturbs the charge equilibrium of the solution. As a consequence, retention of free cations occurs at the membrane surface in order to maintain charge balance. This view is supported by the results of isotopic tracing (Dupré *et al.*, 1999). The natural isotopic ratio of strontium ($^{87}Sr/^{86}Sr$)

and the isotopic ratios of strontium ($^{86}Sr/^{84}Sr$) and barium ($^{138}Ba/^{135}Ba$) in spiked and filtered samples are constant. This shows that there is one pool of strontium and that strontium, barium and by extension, the other cations are present in an exchangeable form at the surface of colloids. In order to avoid the artifact, that would lead to an incorrect determination of complexation constants, Dupré et al. (1999) performed ultrafiltrations at low pH or by adding to the solution a strongly complexing cation such as lanthanum. In both cases, colloids are neutralized and no filtration artifact occurs, allowing Dupré et al. (1999) to calculate metal-humate stability constants by comparing the abundances of elements in the different ultrafiltration fractions with a speciation model (BALANCE, Akinfiev, 1976). Stability constants are determined in order to fit the results from the ultrafiltration. The $\log K$ values calculated in this way for the metal-humate complexes decrease in the order: Al, Ga, Fe, Th, U, Y, REEs (more than 7) $\gg$ Cr (5.5) $\gg$ Co (3) $>$ Rb, Ba, Sr, Mn, Mg ($\cong$2). As shown in Figure 15, a remarkable relationship is observed between the values calculated for element-humate stability constant and the first hydrolysis constant of the element. The same kind of linear relationship was used by Martell and Hancock (1996), to predict the formation constant for metal ion complexes with unidentate organic ligands. Figure 15 shows, however, that the elements yttrium and REEs appear to lie outside the general trend, having a higher stability constant than metals that have similar first hydrolysis constants. This result is not in agreement with the literature for the complexation of REEs with simple organic acids,

which shows that REEs have stability constants comparable to those of copper or zinc (Martell and Hancock, 1996). The filtration experiments performed at low pH and with high lanthanum concentration show that yttrium and the REEs exhibit a very distinct behavior as a function of concentration. This result is in good agreement with a study by Hummel et al. (1995) on the binding of radionuclides (americium, curium, and europium) by humic substances. These authors showed that "complexation at trace-metal concentration on humic acids is rather different than complexation at high metal loading." They explained this discrepancy by the presence of two different sites having different intrinsic constants. In particular, there is a small amount of strong complexing sites, which raises the overall complexation constant at low concentration levels. The same explanation holds for the experiments of Dupré et al. (1999), who proposed two types of site for REEs: a first site with a large stability constant ($\log K > 7$) and a second site with a small stability constant, comparable to that of copper ($\log K \cong 4.5$). The first kind of site, present in low concentration, plays a role when there is a low REE concentration, whereas the second site, present in higher concentration, operates at high REE content. The first site has not been characterized, but the second may be related to the carboxylic functional group.

This example illustrates the complexity of colloid–solute interactions in river waters.

9.5.3 The Nonorganic Colloidal Pool

In the tropical organic-rich waters of Cameroon, inorganic colloids also explain the behavior of some elements. The relationship between aluminum concentration and DOC, for example, (Figure 14) reflects the combination of two linear relationships with different slopes. The part with the larger slope ([Al] $> 700\,\mu g\,L^{-1}$) can be assigned to the retention of kaolinitic particles by the 300 kDa membrane. Using X-ray diffraction (XRD), transmission electron microscopy (TEM), Fourier-transform infrared spectroscopy (FTIR), electron paramagnetic resonance spectroscopy (EPR), and visible diffuse reflectance spectroscopy (DRS) techniques performed on the same samples, Olivié-Lauquet et al. (1999, 2000) showed that iron and manganese in the colloidal fraction were both in the form of hydroxides and organic complexes. They also documented the presence of euhedral particles of kaolinite in the colored waters ($<0.22\,\mu m$). This result suggests that only the first part of the curve ([Al] $<700\,\mu g\,L^{-1}$) corresponds to aluminum bound to humic substances. Even if organic matter plays a key role in controlling trace-element levels in Cameroon, these results emphasize the difficulty of deciphering the roles of organic and inorganic

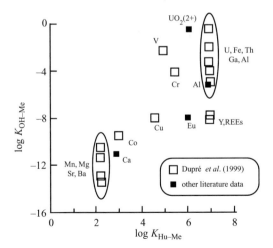

Figure 15 Correlation between metal-humate stability constants ($\log K_{Hu-Me}$) and the first hydrolysis constant ($\log K_{OH-Me}$) of the corresponding metal (Dupré et al., 1999). Squares correspond to humic-metal association constants determined by ultrafiltration of organic-rich waters. K_{Hu-Me} corresponds to the conditional constant of the reaction $Me^{n+} + Hu^- \rightarrow MeHu^{(n-1)+}$.

colloids since metallic oxyhydroxide and organic matter are intimately associated in the solution and in the colloidal phase.

On Earth, organic-rich rivers occur in boreal regions as well as in wet tropical areas. The role of colloids in boreal organic-rich rivers has been addressed by Pokrovski and Schott (2002) and Ingri *et al.* (2000). For rivers of Karelia (northern Russia) Pokrovski and Schott (2002) reported that, conversely to what has been observed in the tropics (Viers *et al.*, 1997), 90% of the organic matter is concentrated in the smallest filtrate (1–10 kDa) in the form of fulvic acids. Pokrovski and Schott (2002) showed the presence of two types of colloids: one type composed of iron oxyhydroxides and another composed of organic matter. The elements aluminum and iron do not show any significant relation with organic matter, which indicates their control by inorganic colloids (aluminosilicates and/or metallic oxyhydroxides). Three groups of elements are deduced by the authors: (i) those not affected by the ultrafiltration procedure and which are present in the form of true ions (i.e., calcium, magnesium, lithium, sodium, potassium, strontium, barium, rubidium, caesium, silicon, boron, arsenic, antimony, molybdenum); (ii) those present in the fraction smaller than 1–10 kDa under the form of inorganic or organic complexes (manganese, cobalt, nickel, copper, zinc, cadmium and for some rivers lead, chromium, yttrium, HREEs, uranium); and (iii) those strongly associated with large iron colloids (phosphorus, aluminum, gallium, REEs, lead, vanadium, chromium, tungsten, titanium, germanium, zirconium, thorium, uranium). No relation between uranium and iron or DOC was reported by Pokrovski and Schott (2002) for the rivers of Karelia, in contrast to the observations of Porcelli *et al.* (1997) for the Kalix river in Sweden, where between 30% and 90% of uranium is associated with organic colloids of >10 kDa molecular weight.

These preliminary investigations on the role of colloids in high-latitude rivers clearly shows differences compared to rivers from the tropics, even if the total dissolved organic content is similar. This observation shows that there is a potential interesting climatic control on the nature and dynamics of colloids.

9.5.4 Fractionation of REEs in Rivers

The geochemistry of REEs has proven to be extremely powerful for the study of the genesis of igneous rocks (Henderson, 1984; Taylor and McLennan, 1985). Several attempts have been made over the last decades to generalize their use to surface water geochemistry (Martin *et al.*, 1976; Goldstein and Jacobsen, 1988; Elderfield *et al.*, 1990; Sholkovitz, 1992, 1995; Gaillardet

et al., 1997). These papers have clearly shown that REEs are among the less mobile elements in weathering and transport by waters. They explained variations in concentrations of REEs in river waters by the existence of colloids that enhance their apparent solubility. REEs are therefore excellent tracers for the dynamics of the colloidal pool in rivers. We have already mentioned that the concentrations of neodymium measured in the <0.2 μm fraction of rivers decrease with pH. This decrease in pH is associated with a fractionation of REEs. High-pH river waters have the most enriched HREE patterns, close to that of seawater (Gaillardet *et al.*, 1997). The REE patterns of seawater have been modeled (see Sholkovitz, 1995 for references); it has been concluded that fractionation for lanthanum to lutetium is a result of gradual differences in affinity of REEs for adsorption to particles and for complexation with ligands in solution. LREEs are preferentially adsorbed on surfaces while HRREs are preferentially complexed with carbonate in solution. The behavior of REEs in river water is far less constrained, due to the diversity of river chemistries and the complexity of solution–colloid interactions. Qualitatively at least, REE concentrations and fractionation in river waters are the result of pH-dependent reactions in solution and at the interface with colloids. Again, ultrafiltration techniques have proved to be extremely useful.

A recent compilation on REE data in river waters has been made by Deberdt *et al.* (2002). Figure 16 presents the REE concentrations in the smaller-pore-size ultrafiltrates (from 3 kDa to 100 kDa) normalized to those determined in the corresponding solution fraction (<0.2 μm) for different types of rivers. For all these rivers, the concentrations of REEs measured in the solution fraction are strongly to moderately reduced by ultrafiltration. This confirms that REEs are strongly controlled by colloidal matter whatever the pH conditions.

The comparison of REE concentrations for the <0.2 μm fraction with the lower-filter-size fraction shows that there is no unique pattern of colloidal material when rivers of different pH and different environments are compared. Ultrafiltration experiments conducted by Deberdt *et al.* (2002) on rivers from the Amazon and Orinoco basins as well as on Cameroon Rivers show slightly depleted LREE patterns to flat REE patterns when the colloidal fraction is normalized to the bulk solution. The results obtained by Sholkovitz (1995) and Ingri *et al.* (2000) for rivers of boreal environments, the Mississippi, Hudson, and Connecticut rivers, and for the Murray river by Douglas *et al.* (1995) show that the smaller ultrafiltration fractions (e.g., <5 kDa or 3 kDa) are more clearly enriched in HREEs compared with the solution fraction (<0.20 μm). In these

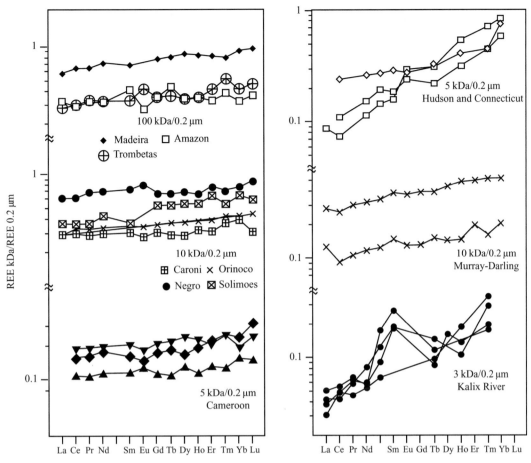

Figure 16 REE concentrations measured in the lower pore size ultrafiltrates (from 3 kDa to 100 kDa) normalized to those found in the solution ($<0.2\,\mu m$) fraction in a panel of world rivers (source Deberdt *et al.*, 2002).

rivers, colloids are therefore LREE enriched and the enrichment of HREEs in the truly dissolved fractions is explained by the authors as the direct consequence of aqueous complexation with common inorganic ions such as carbonate, phosphate, fluoride, nitrate, sulfate, and chloride (Sholkovitz, 1995), which tends to enhance their solubility. Therefore, there is no systematic relationship between LREE depletion and pH or DOC, although more work is necessary to establish the precise relationship between filtered and ultrafiltered REE patterns and river chemistry on a global scale.

Using the REE concentration obtained in the $<5\,kDa$ fraction and assuming it to be the free REE concentration (REE^{3+}), Deberdt *et al.* (2002) deduced the speciation of REE in the whole $<0.20\,\mu m$ fraction using the complexation constants with the most common inorganic ligands in solution and taking into account organic colloidal complexation (carboxylic and phenolic sites). The difference between calculated and observed concentration in the ultrafiltrates is attributed to mineral colloidal material.

The sum of dissolved REE in solution can be described as follows:

$$[REE]_{colinorg}$$
$$= [REE]_{<0.3\mu m} - [REE]_{free}$$
$$\times \left(1 + \sum \frac{K_L a_L^n \gamma_{REE}}{\gamma_{REE(L)}a} + \sum \frac{K_U a_U^n \gamma_{REE}}{\gamma_{REE(U)}a}\right)$$

where the subscript *colinorg* refer to inorganic colloids, L and U to solution ligands and humic substances. K are the complexation constants, a and γ are, respectively, the activity and activity coefficients. These calculations demonstrated the general and unexpected predominance of the mineral colloidal fraction in natural river waters (Figure 17). Mineral colloids account for more than 60% of the total neodymium content whatever the pH range (moderately acid to basic) in the rivers studied (Caroni, Orinoco, Amazon, Madeira, Solimoes). Colloidal neodymium can be adsorbed on submicron colloids such as iron oxyhydroxides or clay minerals (e.g., kaolinite), or can occur as submicron colloids such as REE phosphate or carbonate minerals, which contain

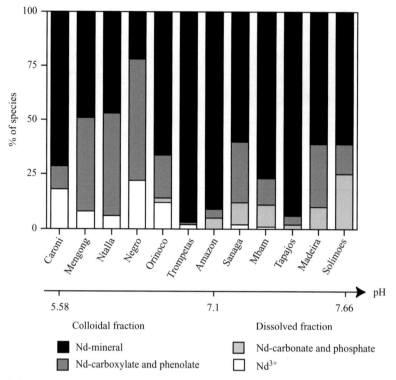

Figure 17 Speciation of Nd in a number of world rivers calculated in the solution fraction (<0.2 μm) by considering inorganic complexes (carbonate and phosphate), mineral and organic colloidal Nd species (dicarboxylic and diphenolic acids) (source Deberdt *et al.*, 2002).

structural REEs. Note that the variable origin of neodymium in waters seems be supported by the nonhomogeneous neodymium isotopic composition measured in the different filtrates of the Mengong (Viers and Wasserburg, 2002). As the physical and chemical properties (e.g., complexation constants) of the hypothetical phases cited above remain poorly documented, it is not possible so far to model the abundances of mineral colloids-related REEs.

This attempt to distinguish between the organic and mineral colloidal pools is highly approximate, but it does show that understanding the abundance of REEs and, by extension, other "insoluble" elements requires a better physical characterization of colloidal phases and association constants with individual elements. Experimental work aimed at determining the association constants of elements with surface sites coupled with field data are necessary. The diversity of colloids, the diversity of associations between elements and colloids, and the existence of filtration artifacts are intrinsic impediments toward a better description and prediction of trace-element concentration in river waters.

9.5.5 Colloid Dynamics

As colloids are a major carrier of trace elements in river waters, their behavior (coagulation, adsoprtion, and oxidation) as function of pH,

increasing solid load, and ionic strength is crucial to predict and model the behavior of trace elements. Ingri *et al.* (2000) have clearly shown that the strong seasonality of lanthanum abundances in the Kalix River is totally controlled by the dynamics of organic and inorganic colloids (Figure 18). McKnight and Bencala (1989, 1990) and McKnight *et al.* (1989) coupled the use of lithium as a conservative tracer and iron, aluminum, and DOC concentrations in acidic mountain streams in the Colorado Rocky Mountains to show that colloid processes (e.g., precipitation of aluminum at stream confluences, sorption of dissolved organic material by hydrous iron and aluminum oxides in a stream confluence) were controlling element concentrations along the river channel. Finally, the importance of colloid dynamics is revealed in estuaries. Siberian rivers (Ob and Yenisei) have low suspended sediment load and the behavior of metals (cadmium, lead, copper, nickel, iron) in the estuary is dominated by the coagulation/sedimentation processes of riverine colloids. Clearly, the input of rivers to the ocean cannot be predicted without paying attention to the behavior of riverine colloids. For example, Dai and Martin (1995) report that copper (or nickel), although associated with colloids, behaves conservatively in the estuarine mixing zone, whereas organic carbon does not. They interpret this unexpected behavior by suggesting that

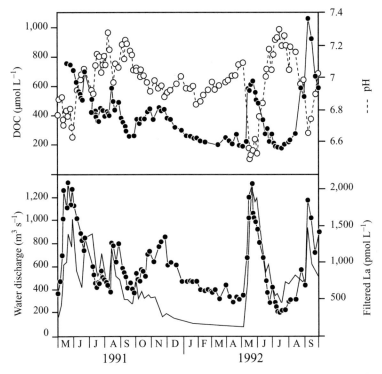

Figure 18 Dissolved organic carbon (DOC; <0.7 μm), filtered La (<0.45 μm), pH, and water discharge in the Kalix River (after Ingri *et al.*, 2000).

colloidal organic matter is composed of at least two fractions: a refractory fraction and a labile fraction, which is removed during mixing with seawater. It is possible that the refractory fraction may be the degradation products of diatoms, given the good co-variance between silicon and copper in the estuary.

It is interesting to come back to the figure of continental crust-normalized abundances in river waters (Figure 2) in the light of the speciation considerations developed in this section. It seems that the affinity for OH^- and organic-complexation sites decreases from the highly immobile elements (e.g. aluminum, gallium), to the nonmobile elements (REEs, zinc,...), moderately mobile elements and mobile elements. The fact that the elements having the highest affinity for complexation by sites on colloids are the least mobile on a global average indicates that processes other than aqueous complexation control their abundance in aquatic systems.

9.6 INTERACTION OF TRACE ELEMENTS WITH SOLID PHASES

9.6.1 Equilibrium Solubility of Trace Elements

When a solution equilibrates with a mineral containing a trace element as a major component, the concentration in the solution can be predicted by thermodynamics. In nature, these phases are mostly aluminum, iron, and manganese oxides. A good case study describing solubility of aluminum and iron in the mineral phase is available from the Nsimi-Zoetele watershed in Cameroon (Viers *et al.*, 1997). Aluminum concentrations measured in organic-poor waters have been plotted on a solubility diagram on Figure 19. Ultrafiltered (5000Da) organic-rich waters similarly plot on the solubility curves of pure well-crystallized kaolinite (observed in the soil profiles). The data suggests that both aluminum and iron concentrations in the waters of the Nsimi watershed are controlled by the solubility of mineral phases produced by rock weathering. Another example is given by Fox (1988), in which dissolved iron concentrations in the Delaware river were determined by dialysis followed by atomic absorption spectroscopy. The concentrations of iron were near saturation with colloidal ferric hydroxides. This is not, however, the case for other rivers such as the Amazon, Zaire, Negro, and Mullica, in which filterable iron concentrations are much higher than values corresponding to equilibrium with ferric hydroxides. The presence of colloids that are not removed by the filtration explains this feature.

For solubility equilibrium to predict the aqueous concentration of a trace element, thermodynamic equilibrium is required and the solid phases must also be identified. In the case of oxyhydroxides and carbonates, it is reasonable to assume a

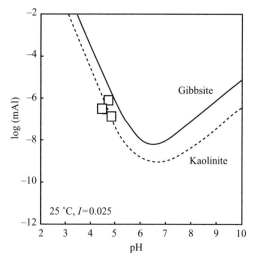

Figure 19 Saturation state of organic-poor waters from the Nsimi-Zoétélé watershed with respect to secondary aluminous phases (i.e., gibbsite, kaolinite). log(mAl) represents the logarithm of the Al concentration measured in the 5 kDa filtrate (source Viers *et al.*, 1997).

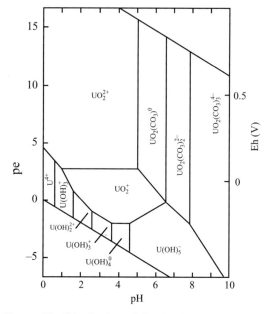

Figure 20 Distribution of dissolved uranium species in the system U–O–H_2O–CO_2 at 25 °C, assuming a P_{CO_2} of 0.01 atm (after Drever, 1997).

close approach to equilibrium because the characteristic reaction times of dissolution of these minerals are in the range of a few days to a few hundred years (Bruno *et al.*, 2002). For silicate phases the assumption of thermodynamic equilibrium is more problematic due to the low reaction rates compared to the residence time of waters in hydrosystems. Examples of codes and database used by modelers to calculate the speciation and the solubility of a number of trace elements can be found in Bruno *et al.* (2002).

Iron, like manganese, uranium, and other metals, can exist in nature in more than one oxidation state. The solubility of such elements in aquatic systems depends on the concentration of oxygen, or more generally on the redox state of the fluvial system. Redox processes have been shown to control the behavior of a number of elements in lakes and in the pore waters of sediments where pe–pH diagrams are classically used to calculate the domains of stability of the solids that precipitate (e.g., MnO_2, $Fe(OH)_3$, FeS_2). The kinetics of redox reactions in natural systems is generally slow, which makes the use of pe less straightforward than pH (Michard, 1989). Even if pe–pH diagrams are theoretical, they offer a reasonable model for understanding the behavior of a number of redox-sensitive elements such as iron, manganese or uranium. The pe–pH diagram of uranium is shown in Figure 20 for a CO_2-rich water. The solubility of uranium is sensitive to the presence of vanadium and the diagram of Figure 20 is no longer valid for water with vanadium concentrations above 0.1 mg L^{-1}.

Most often, running waters are oxygenated, but anoxic conditions can prevail locally in

watersheds (e.g., in lakes, in mires or anoxic soils) and affect the whole river system. Redox reactions can be abiotic or biologically mediated but, as stated by Morel and Hering (1993), "life is by nature a redox process." The ultimate reductant in nature is organic matter made by photosynthetic organisms, while the ultimate oxidant is molecular oxygen (Morel and Hering, 1993). Light and temperature have a clear impact on the rates of redox reactions: examples of the importance of redox reactions in river systems on the regulation of dissolved trace elements and temporal variations are given below.

Nonphotosynthetic iron photoreduction and oxidation have been shown to control the diel cyclicity of dissolved ferrous ion concentrations in a small acidic mountain stream (McKnight *et al.*, 1989). The cycling of manganese and iron concentrations in the Kalix watershed has been especially documented by Ponter *et al.* (1992), Ingri and Widerlund (1994), Ingri *et al.* (2000), and Porcelli *et al.* (1997). The Kalix River is an organic-rich Scandinavian river dominated by the coniferous forest. Mires cover 20% of the drainage area. The concentration of suspended matter is very low with peak concentrations of ~10 mg L^{-1}. Tenfold seasonal variations of manganese concentrations in the dissolved load are essentially due to the input of manganese-rich lake waters after breakup of the ice in June. The manganese concentrations are then regulated in the river during summer by precipitation of manganese oxyhydroxide phases in the suspended sediments. This precipitation of manganese-rich oxides occurs when temperature gets higher than 15 °C and is probably biologically mediated. The precipitation of iron-rich particles is also reported in the Kalix River during summer and this leads to a depletion of iron in solution.

Similar variations of manganese and iron concentrations are observed in larger river systems and are attributed to similar mechanisms. According to Shiller (1997), the strong seasonality of iron and manganese concentrations in the Mississippi traces a redox cyclicity. Within the drainage basin of the Mississippi, as in the Kalix river system, a number of environments are likely to have variable redox conditions through time, including mires and bogs, stagnant streams, bottom waters of stratified lakes and reservoirs, and river-bed sediments.

9.6.2 Reactions on Surfaces

Rather than a clear distinction between colloids and suspended material, separated by a $0.2\,\mu m$ filtration, a continuum of particle radii exists between the smallest and the largest particles in water. Generally, the particle size distribution follows a power-law function in the form

$$N(r) = Ar^{-p}$$

with p close to 4 on the range $1–100\,\mu m$ in aquatic systems (Morel and Herring, 1993). The chemical reactivity of these particles is inversely correlated to their size, so that the smallest particles play the major role. The amount of suspended sediments in rivers fluctuates widely from a few milligrams per liter (lowland rivers) up to several grams per liter, for example, for the circum-Himalayan rivers. In rivers, particles encompass a wide range of chemical and mineralogical composition (biological debris, organic substances, oxyhydroxides, clays, rock fragments). These sediments are either produced *in situ* or most generally are derived from the erosion of soils.

Like colloidal material, surfaces have complexing sites for trace elements and the same formalism as that described for colloids can be used. Understanding the partitioning of metals and more generally trace elements between water and solids is crucial for fundamental studies on transport, bioavailability, and fate of trace elements in river systems. For example, the spatial and temporal trends of metal or radionuclide partitioning between dissolved and suspended solids is a major issue for understanding and predicting the pathways of pollutants in the environment. As a consequence, an impressive literature focuses on experimental studies of trace-element adsorption/desorption on synthetic surfaces (mostly hydrous oxides). However, field-based studies aimed at assessing the importance of adsorption mechanisms as a regulating process of trace-element concentrations in natural waters are sparse. We will briefly summarize some

of the major ideas that emerged from experimental studies before reviewing the few "natural" studies.

9.6.3 Experimental Adsorption Studies

The uptake of trace elements by surfaces is due to the presence of complexing sites similar to those complexing ions in solution (e.g., OH, COOH groups). The difficulty of studying the trace-element–surface interaction is the geometry of the sites at the solid surface. The different processes possibly occurring at the surface and influencing the ion complexation are described in Morel and Herring (1993). In particular, the physical proximity of sites and the long-range nature of electrostatic interactions are major differences between reactions in solution and on solid surfaces. There are several types of solids capable of interacting with solutes: oxides surfaces, carbonate and sulfide surfaces, organic surfaces, and clay mineral surfaces. Most of the literature on the experimental determination of adsorption properties is based on metal oxides (iron, manganese, aluminum, silicon). Oxide surfaces are covered with surface hydroxyl groups represented by (S–OH). Following Schindler and Stumm (1987), the adsorption of metals involves one or two surface hydroxyls:

$$S - OH + M^{z+} = S - OM^{(z-1)+} + H^+$$

The adsorption of anions is generally by ligand exchange involving one or two hydroxyls:

$$S - OH + L^- = S - L^+ + OH^-$$

These reactions can be treated in the same way as equilibria in solution. Equilibrium adsorption constants can thus be defined, which quantify the affinity of the cation/anion for the surface.

For example, in the case of cation sorption, the adsorption constant will be

$$K_{ads} = \left[S - OM^{(z-1)+} \right] \cdot [H^+] / [S - OH] \cdot [M^{z+}]$$

where the "constant" K_{ads} takes into account electrostatic interactions at the mineral surface (Drever, 1997). It follows that adsorption of cations and anions is strongly dependent upon pH. Cation adsorption increases with increasing pH. The range of pH at which adsorption starts depends on both the acid–base properties of the surface and the metal adsorption constant. The percentage of adsorption on FeOOH for selected cations as a function of pH is shown in Figure 21(a) (Sigg *et al.*, 2000). Conversely, the adsorption of anions decreases with increasing pH. Very often, the exchange of ligands is associated with acid–base reactions in solution. Figure 21(b) shows some selected adsorption curves for anions on FeOOH. The association of a ligand in solution, a trace metal in solution and a

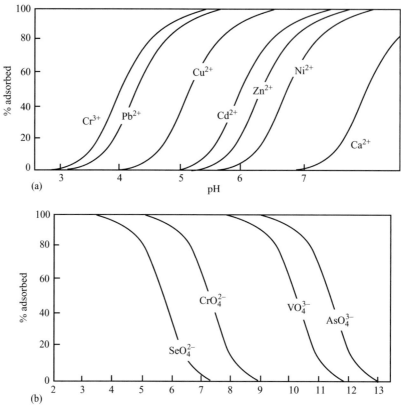

Figure 21 Proportion of cations and anions adsorbed at the surface of hydrous oxides, at high ratio of adsorption sites to adsorbing cations. Adsorption constants are from Dzomback and Morel (1990) (after Sigg *et al.*, 2000).

surface can lead to the formation of ternary surface complexes. In this case, a metal–ligand association is adsorbed on the surface.

Finally, finely divided hydrous oxides of iron, aluminum, manganese, and silicon are the dominant sorbents in nature because they are common in soils and rivers, where they tend to coat other particles. This is the reason why numerous laboratory researchers have been studying the uptake of trace elements by adsorption on hydrous oxides (Dzomback and Morel, 1990). Partition coefficients (concentration in solid/concentration in the solution) for a number of trace elements and a great variety of surfaces have been determined. The comparison of these experimental K_d with natural K_d values should give information on the nature of the material on which trace elements adsorb in natural systems and allow quantitative modeling.

The comparison of experimental adsorption coefficients and measured partition coefficients in natural systems remains difficult because partition coefficients are often defined with respect to bulk solid concentration and not the desorbable element concentration. Except in a few case studies of river systems having very low detrital suspended sediments concentration (e.g., iron in the Kalix River, Ingri *et al.* (2000)), there is no

agreement between adsorption and experimental partition coefficients. Another difficulty has been pointed out by Benoit (1995) and Benoit and Rozan (1998) and is known as the "particle concentration effect" (PCE). The PCE is a decline in partition coefficient as suspended particulate matter concentration increases in the river, although thermodynamics predicts no variation of K_d with solute or particle concentration. PCE occurs owing to both the increasing contribution of colloidal material as water discharge and particle discharge increase (Morel and Gschwend, 1987) and the contribution of coarser particles (with lower specific surface area and complexing site density) in periods of high suspended sediment concentrations and high water discharge. The PCE illustrates quite well the inherent difficulties in comparing experimental and field approaches.

9.6.4 Adsorption on Hydrous Oxides in River Systems

A series of studies have shown that adsorption onto hydrous oxides is a major mechanism regulating the concentration of trace elements in river waters. Several authors have described the formation of freshwater Fe–Mn coatings on gravels in well-oxygenated rivers. The use of the radioactive tracers ^{60}Co, ^{90}Sr, and [137]Cs (Cerling and Turner, 1982) clearly demonstrated the rapid and nonreversible (in oxic conditions) adsorption of cobalt and caesium onto river Fe–Mn coatings. Reversible adsorption of strontium was also observed.

A study of iron, cadmium and lead mobility in remote mountain streams of California by Erel *et al.* (1990) showed that the excess of atmospheric pollution-derived lead and cadmium is rapidly removed downstream. The comparison of truly dissolved, colloidal, and surface particle concentrations measured in the stream with the results of a model of equilibrium adsorption indicates that the mechanism of removal in this organic-poor environment is essentially by uptake onto hydrous iron oxides. The experimentally determined partition coefficients (Dzomback and Morel, 1990) explain the behavior of lead; however, they fail to explain the cadmium removal. It is proposed by the authors that cadmium is taken up by surfaces other than hydrous iron oxides.

Another attempt at modeling trace metals (copper, nickel, lead, and zinc) in river water is presented by Mouvet and Bourg (1983) for the Meuse River. In this study, K_d partition coefficients (between adsorbed and dissolved species) are calculated, based on field measurements and compared to model calculations in which the stability constants were estimated from the adsorption of trace metals by sediments of the Meuse River in laboratory experiments. The adsorption curves of copper, zinc, and cadmium by the Meuse River bottom sediments are shown in Figure 22. Under the natural conditions, most of the copper and lead are adsorbed onto the river suspended sediments. Nickel is the least solid-bound element and zinc has an intermediate behavior. The results of the model show a general agreement between predicted and observed partition coefficients (Figure 22). Especially for the periods of high suspended sediment concentrations (TSS > 30 mg L^{-1}), adsorption onto

suspended solids explains relatively well the measured river concentrations. When the suspended sediment concentrations are low, the calculated percentages of adsorbed elements are underestimated, suggesting the importance of the association of organic material and suspended material in the river. The poor affinity of nickel in the Meuse River for suspended sediments is due to its dominant aqueous speciation as the $NiCO_3$ complex.

The adsorption of thorium has been documented experimentally by several studies (Langmuir and Herman, 1980, and references therein). As with other cationic trace elements, sorption onto surfaces is maximum at high pH and corresponds, in the absence of dissolved organic ligands, to the scavenging of hydroxy complexes. Interestingly, it has been shown for thorium that the presence of organic ligands in solution, such as fulvic acids and EDTA, inhibits adsorption and favors desorption, indicating that the affinity of thorium for organic sites is greater than for surface sites. However, our current knowledge of adsorption coefficients for the natural materials is not sufficient to allow a quantitative modeling of the role of thorium adsorption on surfaces. Because the stability of manganese and iron oxides is redox sensitive, the behavior of trace elements controlled by adsorption on these surfaces is expected to vary with redox state. Two examples of this are given below.

In the Clark Fork river, Montana, Brick and Moore (1996) showed that the diel cyclicity of dissolved manganese and zinc and acid-soluble particulate aluminum, iron, manganese, copper, and zinc are related to diel changes in biogeochemical processes, such sorption or redox changes, related to pH and dissolved oxygen cycles caused by photosynthesis and respiration in the river. The highest pH values, associated with the lowest dissolved concentrations, are consistent with this scenario. It is also possible that the evapotranspiration by stream-bank vegetation could change the contribution of the hyporheic zone (reduced waters sampled from piezometers beneath the river) as a function of time and induce the observed variations in trace-element concentrations. The uptake of alkali and alkaline-earth elements on suspended iron and manganese oxides in the Kalix River has been documented by Ingri and Widerlund (1994). In this river, a significant part of sodium, potassium, calcium, magnesium, and also trace elements such as strontium and barium are scavenged by manganese and iron oxides that constitute most of the suspended sediments of the river. The calculated Kd for strontium is, however, two orders of magnitude larger than the Kd values determined experimentally by Dzomback and Morel (1990), which suggested to the authors that DOC might play an additional role

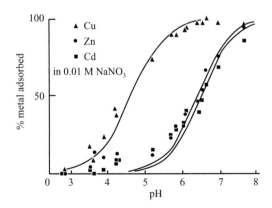

Figure 22 Adsorption of trace metals by the Meuse River bottom sediments: measured data in NaNO$_3$ 0.01 M and interpretation (curves) in terms of adsorption.

in the scavenging of alkali and alkaline-earth cations. The Kalix River is, however, not representative of a "world average" river, as the composition of its suspended load is dominated by minerals that are usually minor phases in the suspended sediments of large rivers. At a global scale, the adsorption of alkali and alkaline-earths does not affect the conservative behavior of alkali and alkaline-earths in river systems.

The strong seasonal variability of trace concentrations in the Mississippi River water has been ascribed to the local dynamic balance of redox conditions (Shiller and Boyle, 1997). The variations in redox conditions not only affects the dissolved concentrations of iron and manganese but also the concentrations of elements sorbed onto iron and manganese oxides such as zinc and lead. Reducing conditions will destabilize manganese and iron oxides and release zinc and lead in waters. Conversely, the reduced forms of molybdenum, vanadium, and uranium are more readily adsorbed on sediment surfaces (Emerson and Huested, 1991); these elements will tend to be scavenged under reducing conditions. If the variations in the redox state of rivers are bacterially mediated, then the seasonal variations of elements such as zinc, lead, molybdenum, vanadium, or uranium in rivers are of biological origin and should be affected by nutrient supply and pollution.

9.6.5 The Sorption of REEs: Competition between Aqueous and Surface Complexation

The particle/solution interactions of REEs have attracted the attention of a number of workers trying to model the REE pattern of seawater or groundwaters (e.g., Turner *et al.*, 1981; Erel and Stolper, 1992; Byrne and Kim, 1990). Freshwater systems are more complex and, as of early 2000s, no model taking into account complexation by colloids, surface adsorption and complexation by inorganic ligands has been attempted. The question of the adsorption of REEs onto suspended solids in freshwaters has been addressed by Elderfield *et al.* (1990) and Sholkovitz (1995).

Suspended sediments from the Connecticut and Mississippi rivers were leached with acetic acid or seawater in order to remove the more labile (adsorbed) fraction of REEs on sediments. The so-mobilized REEs have very low concentrations. Although significant differences exist in the leached REE patterns, the desorption of REEs from suspended sediments preferentially releases LREEs. The enrichment of dissolved HREEs in estuarine waters also confirms the preferential scavenging of LREEs on surfaces. This allowed Sholkovitz (1995) to conclude that the composition of REEs in the dissolved form of rivers is mainly controlled by surface reactions. The trend of an overall LREE

enrichment in the adsorbed component relative to the dissolved component is in agreement with quantitative thermodynamic models of REE adsorption (e.g., Byrne and Kim, 1990; Erel and Morgan, 1991; Erel and Stolper, 1992) in which the competitive complexation of REEs between the solution and the oxygen-donor groups (e.g., carbonate, hydroxide, phosphate) on particle surfaces is postulated. Erel and Stolper (1992) reported a linear relationship between adsorption constants and the first hydroxide binding constants, which is supported by the data of Dzomback and Morel (1990). A strong cerium anomaly in the Mississippi River dissolved load is attributed by Sholkovitz (1995) to the oxidation of dissolved Ce(III) to particulate Ce(IV) oxides in the suspended sediments of the river. These reactions would produce a cerium anomaly only in rivers of high pH and with abundant surface areas on suspended particles. Bau (1999) has experimentally observed oxidative scavenging of cerium during the sorption of REEs onto iron oxyhydroxides.

9.6.6 Importance of Adsorption Processes in Large River Systems

The particles transported by large rivers are a complex mixing of primary minerals, carbonates, clays, oxides and biogenic remains. The assessment of adsorption processes in controlling the levels of trace elements in large rivers has been documented by a couple of studies that will be described below.

Various chemical extraction techniques have been introduced in order to selectively remove metals from the different adsorption or complexation sites of natural sediments (e.g., Tessier *et al.*, 1979; Erel *et al.*, 1990; Leleyter *et al.*, 1999). It is, for example, shown by Leleyter *et al.* (1999) that between 20% and 60% of REE in various suspended river sediments are removed by successive extractions by water, by $Mg(NO_3)_2$ (exchangeable fraction), sodium acetate (acid-soluble fraction), $NH_2OH + HCl$ (manganese oxide dissolution); ammonium oxalate (iron oxide dissolution) and a mixture of $H_2O_2 + HNO_3$ (oxidizable fraction). The complexity of the extraction procedure and the absence of consensus on the specificity of the reagents for particular mineral phases make the results of sequential extraction procedures difficult to interpret.

The dissolved concentrations of zinc in the Yangtze, Amazon, and Orinoco rivers have been shown to strongly decrease with pH, between 5 and 8.5 (Shiller and Boyle, 1985). This decrease is similar to that obtained in an experiment in which the pH of unfiltered Mississippi River water was adjusted to various pH values, and to isotherms of zinc adsorption on various natural and synthetic metal oxides surfaces (Figure 23). According to

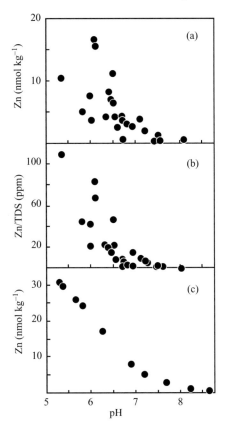

Figure 23 (a) Dissolved Zn concentrations and (b) dissolved Zn concentrations normalized to TDS as a function of pH in large rivers from the Changjiang, Amazon and Orinoco river basins. (c) Zn in pH-adjusted aliquots of Mississippi River (after Shiller and Boyle, 1985).

Shiller and Boyle (1985), alkaline rivers have less dissolved zinc because they are more suspended sediment-rich that acidic rivers. K_d values between adsorbed (acid-leachable) zinc and dissolved zinc approach 10^5 for the Mississippi, which indicates that the Mississippi transports 10 times more adsorbed zinc than dissolved zinc.

Zhang *et al.* (1993) have examined trace-element data from *in situ* measurements and laboratory experiments. The analysis of both dissolved concentrations and bulk particulate concentrations allowed them to calculate "partition coefficients." They showed, in the Huanghe River, significant variations in K_d for the different elements. The lowest K_d values are reported for copper and the highest for iron. Parameters such as pH, temperature and total suspended matter concentrations affected trace metal partitioning coefficients in the Huanghe, although the dominant effect was due to pH. As expected, adsorption of metals was favored at high pH. These experiments showed a great affinity of the Huanghe suspended solids for trace metals such as cadmium, copper and lead and showed that the kinetics of

adsorption are rapid. Thus, river particulates may have the potential of regulating trace metal inputs to aquatic systems from pollution.

Beryllium is an alkaline-earth elements whose behavior drastically differs from that of the other alkaline-earth elements. Its low mobility in natural waters is attributed to its affinity for surfaces. Laboratory experiments have been performed to examine the partitioning of ^{7}Be between sediments from natural systems and water (You *et al.*, 1989). The partition coefficient depends strongly on pH in the range 2–7. The curve of K_d as a function of pH (Figure 24) can be explained by a thermodynamic model by taking into account beryllium speciation in freshwater and the partition coefficient of each species. Data on ^{10}Be and ^{9}Be measured in the Orinoco system (Brown *et al.*, 1992b) also fit well the model curve and experimental data. Note that the K_d values given here are the partition coefficients between soluble beryllium and exchangeable beryllium (leached from sediments with hydroxylamine). The high K_d value of beryllium clearly shows that beryllium is transported mainly in adsorbed form. The K_d values measured for beryllium between surfaces and dissolved load show that for the sediment yields present in the Orinoco, 90% of beryllium is transported in a suspended form. The low mobility of beryllium in river waters is therefore due to its attraction for surfaces. Similar to beryllium is the case of caesium, an alkali metal whose behavior in freshwaters differs from that of sodium because of its adsorption properties.

9.6.7 Anion Adsorption in Aquatic Systems

Boron is present in freshwater as nonionized boric acid and negatively charged borate ion $(B(OH)_4^-)$. The ability of boron to adsorb on surfaces is a well-known characteristic of its geochemical cycle at the surface of the Earth. Spivack *et al.* (1987) determined partition coefficients of boron in the marine sediments off the Mississippi and showed that K_d ($B_{adsorbed}/B_{aqueous}$) is close to 1.5. The two isotopes of boron do not behave similarly during adsorption on surfaces: the light isotope is preferentially attached to surfaces. This adsorption of boron onto particles has a significant effect on the boron isotopic composition of seawater, but is probably not significant in freshwaters due to the relatively low amount of dissolved boron in freshwaters.

Among the other trace anions, the concentration of arsenic in natural waters is probably controlled by solid–solution interaction (Smedley and Kinniburgh, 2002). In sediments, the element that most frequently correlates with arsenic is iron and numerous studies have been reported on the sorption of arsenic (as arsenate ion, $HAsO_4^{2-}$) onto iron oxides, manganese oxides or aluminum

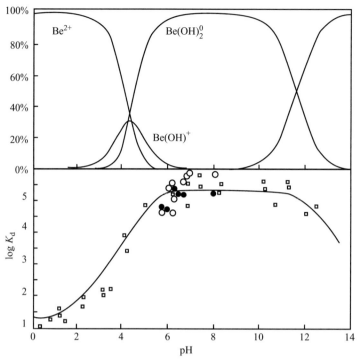

Figure 24 Bulk partitioning coefficients between dissolved and adsorbed Be on particles measured in the Orinoco for ^{9}Be (filled circles) and ^{10}Be (open circles). Open squares are experimental distribution coefficients determined by You *et al.* (1989) by adsorption of Be onto riverine particles. The line is deduced from a speciation-dependent model of adsorption (after Brown *et al.*, 1992b).

oxides (see references in Smedley and Kinniburgh (2002)). K_d values up to 10^6 (L kg^{-1}) are found for the experimental adsorption of arsenic onto hydrous ferric oxides. The desorption of arsenic, like the desorption of other anion-forming elements (vanadium, boron, molybdenum, selenium, and uranium), especially from iron oxides, is one of the key processes invoked to explain high levels of arsenic in natural waters and associated toxicity problems. This desorption is favored at pH values above 8 and when iron oxides undergo reductive dissolution.

9.6.8 Adsorption and Organic Matter

The extrapolation of experimental studies on trace-metal adsorption to natural waters is difficult: a particular problem is the formation of both ternary surface complexes involving dissolved organic matter and the aqueous complexation of trace elements by dissolved organic matter. The number of studies trying to shed light on the complexity of these interactions by combined studies of organic matter in the suspended phase and the dissolved phase is incredibly low.

Sholkovitz and Copland (1980) used the organic-rich water of the Luce River and a tributary of the Luce River, Scotland, to investigate the relations between pH-related adsorption and organic complexation in solution. They concluded

that, under their experimental conditions, the control of organic matter is much more important for determining the levels of dissolved trace elements in the river than adsorption onto the suspended sediments (composed either of clays plus organics or iron oxides plus organics). These results are not in agreement with, and are even opposite to, those inferred from artificial solutions and chemical models (see Sholkovitz and Copland (1980) for references).

Shafer *et al.* (1997) addressed the influence of dissolved organic carbon, suspended particulates and hydrology on the concentration and partitioning of trace metals in two contrasting (one agricultural, one forested) Wisconsin (USA) watersheds. As expected, the forested watershed had higher DOC levels (mean value of 7 mg L^{-1} versus 4 mg L^{-1}), but lower suspended particulate matter (SPM) (mean value of 2 mg L^{-1} versus 18 mg L^{-1}). The high concentrations of DOC and the low concentrations of potential inorganic ligands in the two rivers suggested that the dominant speciation of aluminum, cadmium, copper, lead, and zinc in solution was organic. This study shows the partitioning preferences of selected metals in these two DOC-rich and SPM-poor watersheds. Lead appears as the most strongly adsorbed element with an intermediate affinity for dissolved organic matter. Kd values for lead are uniform in low-DOC and high-SPM waters, but somewhat variable in high-DOC and low clay environments. Copper exhibits the opposite behavior. It shows a stronger affinity for DOC than for surfaces. As a consequence, copper partitioning is rather insensitive to variations in DOC concentration. The intermediate affinity of zinc for SPM and DOC makes zinc partitioning highly sensitive to changes in either SPM or DOC. Accordingly, Kd values are ranked in the order Pb > Zn > Cd > Cu. Note that the trends observed in these two Wisconsin watersheds are probably not representative of larger systems, which generally have higher suspended sediment concentrations.

As shown by Laxen (1985), the adsorption of cadmium, copper, and nickel onto hydrous ferric oxides in the presence of dissolved humic substances is significantly modified. The adsorption of cadmium and nickel is enhanced by the presence of humics. The model favored by the authors is the complexation of metals with adsorbed humics, which is stronger than binding with the dissolved humics. The adsorption of copper is enhanced but more sensitive to the competitive formation of soluble copper–humic complexes. An important implication of these experiments is that metals will be more strongly adsorbed at low pH values in the presence of humics than in their absence. Radiotracer experiments have shown that, in the ocean, interactions of organic-rich colloids and particles (through sorption and coagulation) occur in estuaries. This mechanism is termed "colloidal pumping" and plays a crucial role in the fate of many trace elements in estuarine waters (Wen *et al.*, 1997). The extent of this mechanism in rivers has not been explored.

Finally, one should not end this section without mentioning the uptake of trace elements by living biomass. This process can be of quantitative importance in river systems of low velocity or draining lakes or stagnant water bodies. For example, a study in the St. Lawrence river (Quémerais and Lum, 1997) has shown that an uptake of cadmium by phytoplankton occurs in the summer. In lakes, the uptake of trace elements by living organisms has been demonstrated for a number of trace elements with known biological activity (copper, zinc, iron, molybdenum, manganese), or unknown biological function (cadmium, arsenic) (cf. Sigg *et al.*, 2000).

9.6.9 Particle Dynamics

Regulation of concentration of trace elements by surfaces plays a key role in river water systems. This applies to both colloids and larger particles. Over the past decades, the physicochemical understanding of sorption processes has considerably improved. Mass transfer codes such as PHREEQE (Parkhurst *et al.*, 1990), PHREEQC (Parkhurst, 1995), SOLMINEQ.88 (Kharaka

et al., 1988), MINTEQA2 (Allison *et al.*, 1991), MINTEQ(4.00) (Eary and Jenne, 1992), MINEQL + (Schecher and McAvoy, 1991), and EQ3/6 (Wollery, 1992) perform specialized calculations and predict changes in solution chemistry caused by different processes such as dissolution/precipitation, ion exchange/adsorption, or mixing of water masses. However, so far, these models, which explain the results from the laboratory experiments well, are not able, in general, to predict the pathways of trace elements in the natural system. This is due to the complexity of the natural systems characterized by the diversity of substrates, of surface sites, of ligands in solution, and the abundance of organic material, the presence of colloids that pass through filters and preclude the strict separation between aqueous and solid phase concentrations, and finally the dynamics of colloids and particles within the river system. The transfer of experimental data to natural systems is not straightforward because natural surfaces are far more complex than the experimentally synthesized ones. Studies of the kinetics of adsorption/desorption of trace elements on natural sediments should be encouraged, even if severe experimental artifacts exist. More data on concentrations of dissolved, adsorbed and colloidal material in natural systems are necessary, particularly on a worldwide scale, in order to improve our knowledge of trace-element behavior in natural waters and to improve the quality of modeling and the accuracy of prediction. Because trace metals are carried on the finer fraction of sediments in river systems (Horovitz, 1991), due to a surface area effect, the sedimentary dynamics and the postdepositional stability of sediments (residence times) in watersheds are important factors to take into account for the transport of trace elements. For example, the preferential deposition of fine-grained material in floodplains, lakes or reservoirs will strongly influence the pathways of metals through catchments. An integrated approach of chemistry, geochemistry, and geomorphology is therefore in order.

9.7 CONCLUSION

This review focuses on the concentration of trace elements occurring in river waters, based on the data obtained from natural sources. The determination of concentrations of many trace elements in river waters is facilitated in particular by the technical advances of ICP-MS allowing rapid measurement of a large number of elements. Trace-element measurements are not only motivated by the requirements of pollution and toxicology studies, but also by the need for better understanding and modeling of the behavior of elements during continental weathering and transport, which constitute the two major aspects of the geological cycle.

So far, the global systematics of trace elements in river waters are still poorly understood and limited either to industrial countries, where the natural levels are overwhelmed by anthropogenic inputs, or to regions of the world where rivers are organic-rich, namely, the tropics and the high-latitude areas. Continents such as Asia, Australia and the numerous Pacific peninsulas have been very poorly investigated, except for a few elements of geochemical importance (osmium, uranium, thorium, strontium).

Trace-element concentrations in river waters span over 10 orders of magnitude, similar to the range of crustal abundances. However, the normalization of trace-element levels in river waters to mean continental abundances shows that the abundance of trace elements in river waters depends not only on their continental abundances, but also on their mobility in weathering and transport processes. Normalized abundances of elements in river water vary over five orders of magnitude from highly immobile elements such as titanium, niobium, zirconium, aluminum, and tantalum to highly mobile elements such as selenium, rhenium, boron, arsenic, molybdenum, and sodium. An important feature of trace-element concentrations in river water is a large variability in space and time, even over very short periods of time (days). In addition, the abundance of elements such as REEs is a function of river chemistry as exemplified by the worldwide correlation observed between dissolved neodymium and pH. It is a major characteristic of trace-element concentrations in river waters that they not only depend on their abundance in the source rocks but that they are also strongly dependent on the chemical conditions prevailing in the river. There are a few exceptions to this rule, notably the most mobile elements that correlate fairly well with the major elements.

The mobility of trace elements in river waters results from a complex combination of several factors: their solubility in water, the input to the system of nonweathering sources such as atmospheric and/or anthropogenic sources, the ability of the elements to be complexed by fine ($<0.2\,\mu m$) colloidal material, and the affinity of the elements for solids (adsorption, co-precipitation, solubility equilibrium). The "chemical" solubility of elements depends on the solubility of the minerals that contain the elements and is approximately related to the atomic and molecular properties of the elements. Elements of small atomic radii tend to form polycharged oxyanions that make them soluble (e.g., boron, arsenic, tungsten, vanadium). Elements of higher atomic radii, and of $+1$ or $+2$ valency tend to be soluble as free cations (sodium, potassium, barium, lithium, strontium). Elements such as aluminum, REEs, thorium, and iron fall in between and form poorly soluble

oxyanions or hydroxyanions. Nonweathering inputs comprise atmospheric inputs, anthropogenic inputs or plant inputs. Owing to the global present-day contamination of the atmosphere by anthropogenic emissions (e.g., combustion dusts, detrital particles, industrial aerosols, gases etc.), it is difficult to estimate the natural input of trace elements to hydrosystems, but even in the most remote areas, rain can be a significant source of elements transported by the rivers. In the industrialized countries and to a lesser extent, at a global scale, clear evidence of direct contamination of rivers by anthropogenic sources exists, especially for heavy metals and other metals.

Although it controls the mobility of trace elements in water, the aqueous speciation of many elements in river waters is not well known, principally when organic complexes are involved. However, one of the most striking characteristics of the behavior of trace elements for which chemical aqueous species are known, is that their concentrations are usually not predictable by classical speciation calculations involving the major inorganic or organic ligands in solution. This is due to two competitive mechanisms. The first is the existence in rivers, especially in rivers of low pH, of a colloidal phase made of intimately coupled organic and inorganic material, which can pass through the pores of filtration membranes (conventionally $0.2\,\mu m$; and therefore analyzed and considered as dissolved material). The more the colloidal material "contaminates" the dissolved phase, the higher the concentrations of chemically insoluble elements will be. Although colloidal material is poorly known, and its affinity for trace elements still debated, the affinity of organic colloid for trace metals increases with the first hydrolysis constant of the metal. The competing mechanism that tends to lower the levels of "dissolved" trace elements is the affinity of the elements for particles larger than $0.2\,\mu m$, which leads to the removal of these elements from the dissolved phase during filtration. In rivers, a great variety of surfaces can exist and uptake mechanisms can be very diverse. Adsorption or co-precipitation on hydrous oxides, organic particles or clays, and uptake by living organisms are possible mechanisms, and they are not easy to distinguish. Although colloids are small particles and should obey the same formalism as surface complexation on larger particles, our understanding and modeling of colloidal and surface uptake in natural hydrosystems is far from being clear and complete. Surface sorption, like aqueous complexation, increases with the first hydrolysis constant. Although, an impressive literature exists on the interaction between solutes and hydrous oxides, the use of the experimentally derived affinity constants adapting to the natural systems is a difficult task. The principal obstacle in this attempt is the complexity of naturally occurring material, the poor physical description we have of colloids, the potential experimental artifacts associated with the isolation of colloids (fractionation, coagulation), and the extremely variable chemistry of rivers from one continent to another. The pH-sensitivity of stability of colloids in the aquatic system, as well as the varying extent of adsorption of ions as a function of pH, can explain the dependence of concentration of the trace elements in river water on pH and other chemical variables.

Much work is clearly needed to gain an understanding of the control on concentrations of trace elements in river waters and aquatic systems. It seems clear that further studies should focus on the physical characterization of the colloid pool and comprehension of its dynamics. Estuarine studies have shown that the behavior of elements in the mixing zones between rivers and ocean water is controlled by the flocculation, coagulation, or degradation of colloidal material. The technological advances in filtration membranes should allow a better size fractionation and isolation of colloids. Nanofiltration and dialysis are other promising techniques that need to be developed and should lead to a better characterization of the "true" dissolved pool and of the colloidal-size material. So far, trace-element concentrations have been measured in a restricted number of environments, which may not be typical. Extended investigations on rivers of high pH, high suspended material yields, and mountainous rivers, or rivers of variable climatic or lithologic settings are necessary to get a better world-scale overview of trace-element levels and controlling parameters. From a more theoretical point of view, the relationships between the periodic table properties and the geochemical behavior of trace element in natural hydrosystems, through organic and inorganic complexation, solid uptake and affinity for life in river systems need to be explored.

Finally, the recent technical progress in the new generation of multicollector ICP–MS instruments that enable the measurement of the isotopic ratios of a number of trace metals, whose isotopes are fractionated by complexation processes, is a new challenge for aquatic sciences, which, no doubt, will improve our knowledge and prediction of trace-element dynamics in hydrosystems.

ACKNOWLEDGMENTS

J. I. Drever and M. Meybeck provided the first impulse for this chapter. We thank O. Pokrovski, J. Schott, J. L. Dandurand, S. Deberdt, R. Millot, P. Seyler, E. Lemarchand, and B. Bourdon for their help and discussions. This chapter benefited from a detailed review by J. I. Drever. This is IPGP contribution no. 1947.

REFERENCES

Allison J. D., Brown D. S., and Novo-Gradac K. J. (1991) *MINTEQA2: A Geochemical Assessment Data Base and Test Cases for Environmental Systems*, vers. 3.0 user's manual. Report EPA/600/3-91/-21. Athens, GA, US EPA.

Ammann A. A. (2002) Speciation of heavy metals in environmental water by ion chromatography coupled to ICP-MS. *Anal. Bioanal. Chem.* **372**, 448–452.

Akinfiev N. N. (1976) "Balance": IBM computer code for calculating mineral aqueous solution-gas equilibria. *Geochemistry* **6**, 882–890 (in Russian).

Artaxo P., Maenhaut W., Storms H., and Van Grieken R. (1990) Aerosol characteristics and sources for the Amazon basin during the wet season. *J. Geophys. Res.* **95**, 16971–16985.

Baes C. F. and Mesmer R. E. (1976) *The Hydrolysis of Cations*. Wiley, New York, 489pp.

Bau M. (1999) Scavenging of dissolved yttrium and rare earths by precipitating iron hydroxide: experimental evidence for Ce oxidation, Y–Ho fractionation and lanthanide tetrad effect. *Geochim. Cosmochim. Acta*, 67–77.

Bau M. and Dulski P. (1996) Anthropogenic origin of positive gadolinium anomalies in river waters. *Earth. Sci. Planet. Lett.* **143**, 345–355.

Benoit G. (1995) Evidence of the particle concentration effect for lead and other metals in fresh waters based on ultrafiltration technique analyses. *Geochim. Cosmochim. Acta* **59**(13), 2677–2687.

Benoit G. and Rozan T. F. (1998) The influence of size distribution on the particle concentration effect and trace metal partitioning in rivers. *Geochim. Cosmochim. Acta* **63**(1), 113–127.

Berg T., Royset O., and Steinnes E. (1994) Trace element in atmospheric precipitation at Norwegian background stations (1989–1990) measured by ICP-MS. *Atmos. Environ.* **28**, 3519–3536.

Berner E. K. and Berner R. A. (1996) In *Global environment: water, air, and geochemical cycles*. Prentice-Hall, Englewood Cliffs, NJ, 376p.

Brick C. M. and Moore J. N. (1996) Diel variation of trace metals in the upper Clark Fork river, Montana. *Environ. Sci. Technol.* **30**, 1953–1960.

Brown E. T., Measures C. I., Edmond J. M., Bourlès D. L., Raibeck G. M., and Yiou F. (1992a) Continental inputs of beryllium to the oceans. *Earth Planet. Sci. Lett.* **114**, 101–111.

Brown E., Edmond J. M., Raibeck G. M., Bourlès D. L., Yiou F., and Measures C. I. (1992b) Beryllium isotope geochemistry in tropical river basins. *Geochim. Cosmochim. Acta* **56**, 1607–1624.

Bruno J., Duro L., and Grivé M. (2002) The applicability and limitations of thermodynamic geochemical models to stimulate trace element behavior in natural waters: lessons from natural analogue studies. *Chem. Geol.* **190**, 371–393.

Buffle J. and Van Leeuwen H. P. (1992) Environmental particles: 1. In *Environmental Analytical and Physical Chemistry Series*. Lewis Publishers, London, 554p.

Burba P., Shkinev V., and Spivakov B. Y. (1995) On-line fractionation and characterisation of aquatic humic substances by means of sequential-stage ultrafiltration. *J. Anal. Chem.* **351**, 74–82.

Byrne R. H. and Kim K. H. (1990) Rare earth element scavenging in seawater. *Geochim. Cosmochim. Acta* **54**, 2645–2656.

Cameron E. M., Hall G. E. M., Veizer J., and Roy Krouse H. (1995) Isotopic and elemental hydrogeochemistry of a major river system: Fraser river, British Columbia, Canada. *Chem. Geol.* **122**, 149–169.

Cantrell K. J. and Byrne R. H. (1987) Rare earth element complexation by carbonate and oxalate ions. *Geochim. Cosmochim. Acta* **51**, 597–605.

Cerling T. E. and Turner R. R. (1982) Formation of freshwater Fe–Mn coatings on gravel and the behavior of ^{60}Co, ^{90}Sr, and ^{137}Cs in a small watershed. *Geochim. Cosmochim. Acta* **46**, 1333–1343.

Chabaux F., Riotte J., Clauer N., and France-Lanord Ch. (2001) Isotopic tracing of the dissolved U fluxes in Himalayan rivers: implications for the U oceanic budget. *Geochim. Cosmochim. Acta* **65**, 3201–3217.

Chabaux F., Riotte J., and Dequincey O. (2003) U–Th–Ra fractionation during weathering and river transport. In *U-series Geochemistry and Applications*, Rev. Mineral. Geochem. (eds. B. Bourdon, S. Turner, G. Henderson, and C. C. Lundstrom). Min. Soc. Am. and Geochem. Soc. (in press).

Colodner D., Sachs J., Ravizza G., Turekian K., Edmond J., and Boyle E. (1993) The geochemical cycle of rhenium: a reconnaissance. *Earth Planet. Sci. Lett.* **117**, 205–221.

Cossa D., Tremblay G. H., and Gobeil C. (1990) Seasonality in iron and manganese concentrations in the St. Lawrence river. *Sci. Tot. Environ.* **97/98**, 185–190.

Dai M. and Martin J.-M. (1995) First data on trace metal level and behaviour in two major Arctic river-estuarine systems (Ob and Yenisey) and in the adjacent Kara Sea, Russia. *Earth Planet. Sci. Lett.* **131**, 127–141.

Dalai T. K., Singh S. K., Triverdi J. R., and Krishnaswami S. (2001) Dissolved rhenium in the Yamuna river system and the Ganga in the Himalaya: role of black shale weathering on the budgets of Re, Os, and U in rivers and CO_2 in the atmosphere. *Geochim. Cosmochim. Acta* **66**(1), 29–43.

Deberdt S., Viers J., and Dupre B. (2002) New insights about the rare earth elements (REE) mobility in river waters. *Bull. Soc. Geol. France* **173**(n°2), 147–160.

De Caritat P., Reimann C., Ayras M., Niskavaara H., Chekushin V. A., and Pavlov V. A. (1996) Stream water geochemistry from selected catchments on the Kola peninsula (NW Russia) and the neighbouring areas of Finland and Norway: 1. Element levels and sources. *Aquat. Geochem.* **2**, 149–168.

Douglas G. B., Gray C. M., Hart B. T., and Beckett R. (1995) A strontium isotopic investigation of the origin of suspended particulate matter (SPM) in the Murray-Darling river system, Australia. *Geochim. Cosmochim. Acta* **59**, 3799–3815.

Drever J. I. (1997) *Geochemistry of Natural Waters,* 3rd edn. Prentice-Hall, Englewood Cliffs, NJ.

Dupré B., Gaillardet J., and Allègre C. J. (1996) Major and trace elements of river-borne material: the Congo Basin. *Geochim. Cosmochim. Acta* **60**, 1301–1321.

Dupré B., Viers J., Dandurand J. L., Polvé M., Bénézeth P., Vervier P., and Braun J. J. (1999) Major and trace elements associated with colloids in organic-rich river waters: ultrafiltration of natural and spiked solutions. *Chem. Geol.* **160**, 63–80.

Dzomback D. A. and Morel F. M. M. (1990) *Surface Complexation Modelling.* Wiley, New York.

Eary L. E. and Jenne E. A. (1992) *Version 4.00 of the MINTEQ Code.* Report PNL-8190/UC-204. Richland, WA, Pacific Northwest Laboratory.

Edmond J. M., Spivack A., Grant B. C., Hu Ming-Hui, Chen Zexiam, Chen Sung, and Zeng Xiushau (1985) Chemical dynamics of the Changjiang estuary. *Cont. Shelf. Res.* **4** (1/2) 17–36.

Edmond J. M., Palmer M. R., Measures C. I., Grant B., and Stallard R. F. (1995) The fluvial geochemistry and denudation rate of the Guyana shield in Venezuela, Columbia and Brazil. *Geochim. Cosmochim. Acta* **59**, 3301–3325.

Edmond J. M., Palmer M. R., Measures C. I., Brown E. T., and Huh Y. (1996) Fluvial geochemistry of the eastern slope of the northeastern Andes and its foredeep in the drainage of the Orinoco in Colombia and Venezuela. *Geochim. Cosmochim. Acta* **60**, 2949–2975.

Elbaz-Poulichet F., Seyler P., Maurice-Bourgoin L., Guyot J. L., and Dupuy C. (1999) Trace element geochemistry in the upper Amazon drainage basin (Bolivia). *Chem. Geol.* **157**, 319–334.

Elderfield H., Upstill-Goddard R., and Sholkovitz E. R. (1990) The rare earth element in rivers, estuaries, and coastal seas and their significance to the composition of ocean waters. *Geochim. Cosmochim. Acta* **54**, 971–991.

Emerson S. R. and Huested S. S. (1991) Ocean anoxia and the concentrations of molybdenum and vanadium in seawater. *Mar. Chem.* **34**, 177–196.

Erel Y. and Morgan J. J. (1991) The effect of surface reactions on the relative abundances of trace metals in deep sea waste. *Geochim. Coschim. Acta* **55**, 1807–1813.

Erel Y. and Stolper E. M. (1992) Modelling of rare-earth element partitioning between particles and solution in aquatic environments. *Geochim. Cosmochim. Acta* **57**, 513–518.

Erel Y., Morgan J. J., and Patterson C. C. (1990) Natural levels of lead and cadmium in remote mountain stream. *Geochim. Cosmochim. Acta* **55**, 707–719.

Eyrolle F., Fevrier D., and Benaim J. U.-Y. (1993) Etude par D. P. A. S. V. de l'aptitude de la matière organique colloïdale a fixer et a transporter les métaux: exemples de bassins versants en zone tropicale. *Environ. Technol.* **14**, 701–717.

Eyrolle F., Benedetti M. F., Benaim J.-Y., and Fevrier D. (1996) The distributions of colloidal and dissolved organic carbon, major elements and trace elements in small tropical catchments. *Geochim. Cosmochim. Acta* **60**, 3643–3656.

Foster I. D. L. and Charlesworth S. M. (1996) Heavy metals in the hydrological cycle: trends and explanation. *Hydrol. Process.* **10**, 227–261.

Fox L. E. (1988) The solubility of colloidal ferric hydroxide and its relevance to iron concentrations in river water. *Geochim. Cosmochim. Acta* **52**, 771–777.

Frei M., Bielert U., and Heinrichs H. (1998) Effects of pH, alkalinity and bedrock chemistry on metal concentrations of springs in an acidified catchment (Ecker dam, Harz Mountains, FRG). *Chem. Geol.* **170**, 221–242.

Freydier R., Dupré B., and Lacaux J. P. (1998) Precipitation chemistry in intertropical Africa. *Atmos. Environ.* **32**, 749–765.

Froelich P. N., Hambrick G. A., Andreae M. O., Mortlock R. A., and Edmond J. M. (1985) The geochemistry of inorganic germanium in natural waters. *J. Geophys. Res.* **90**(C1), 1133–1141.

Gaillardet J., Dupré B., Allègre C. J., and Négrel P. (1997) Chemical and physical denudation in the Amazon river basin. *Chem. Geol.* **142**, 141–173.

Gaillardet J., Dupré B., Louvat P., and Allègre C. J. (1999a) Global silicate weathering of silicates estimated from large river geochemistry. *Chem. Geol. (Spec. Issue Carbon Cycle 7)* **159**, 3–30.

Gaillardet J., Dupré B., and Allegre C. J. (1999b) Geochemistry of large river suspended sediments: silicate weathering or crustal recycling? *Geochim. Cosmochim. Acta* **63**(23/24), 4037–4051.

Gaillardet J., Millot R., and Dupré B. (2003) Trace elements in the Mackenzie river basin (in preparation).

Goldstein S. J. and Jacobsen S. B. (1988) Rare earth elements in river waters. *Earth Planet. Sci. Lett.* **89**, 35–47.

Gu B., Schmitt J., Chen Z., Liang L., and McCarthy J. F. (1995) Adsorption and desorption of different organic matter fractions on iron oxide. *Geochim. Cosmochim. Acta.* **59**, 219–229.

Henderson P. (1984) General geochemical properties and abundances of the rare earth elements. In *Developments in Geochemistry: 2. Rare Earth Element Geochemistry* (ed. P. Henderson). Elsevier, Amsterdam, pp. 1–32.

Herrera Ramos A. C. and McBride M. B. (1996) Goethite dispersibility in solutions of variable ionic strength and soluble organic matter content. *Clay Clays Min.* **44**, 286–296.

Hoffmann S. R., Shafer M. M., Babiarz C. L., and Armstrong D. E. (2000) A critical evaluation of tangential-flow ultrafiltration for trace metals studies in freshwater systems: 1. Organic carbon. *Environ. Sci. Technol.* **34**, 3420–3427.

Horowitz A. J., Lum K. R., Garbarino J. R., Gwendy E. M. H., Lemieux C., and Demas C. R. (1996) The effect of membrane filtration on dissolved trace element concentrations. *Water Air Soil Pollut.* **90**, 281–294.

Huang W. W., Martin J. M., Seyler , Zhang J., and Zhong X. M. (1988) Distribution and behavior of arsenic in the Huanghe (Yellow river) estuary and Bohai Sea. *Mar. Chem.* **25**, 75–91.

Huh Y., Chan L. H., Zhang L., and Edmond J. M. (1998) Lithium and its isotopes in major world rivers: implications for weathering and the oceanic budget. *Geochim. Cosmochim. Acta* **62**, 2039–2051.

Hummel W., Glaus M., Van Loon L. R. (1995) Binding of radionuclides by humic substances: the "Conservative Roof" approach. In *Proceedings of an NEA Workshop, September 14–16, 1994, Bad Zurzach, Switzerland*. OECD documents, ISBN no. 92-94, 251–262.

Ingri J. and Widerlund A. (1994) Uptake of alkali and alkaline-earth elements on suspended iron and manganese in the Kalix River, northern Sweden. *Geochim. Cosmochim. Acta* **58**, 5433–5442.

Ingri J., Widerlund A., Land M., Gustafsson Ö., Andersson P., and Öhlander B. (2000) Temporal variations in the fractionation of the rare earth elements in a boreal river: the role of colloidal particles. *Chem. Geol.* **166**, 23–45.

Johannesson K., Stetzenbach K., Hodge V. F., and Lyons W. B. (1996) Rare earth element complexation behavior in circum-neutral pH groundwaters: assessing the role of carbonate and phosphate ions. *Earth Planet. Sci. Lett.* **139**, 305–319.

Johannesson K. H., Lyons W. B., Graham E. Y., and Welch C. A. (2000) Oxyanion concentrations in eastern Sierra Nevada rivers: 3. Boron, Molybdenum, Vanadium, and Tungsten. *Aquat. Geochem.* **6**, 19–46.

Kharaka Y. K., Gunter W. D., Aggarwal P. K., Perkins E. H., and Debraal J. D. (1988) SOLMINEQ: 88. A computer program for geochemical modelling of water rock interactions. In *US Geological Survey Water Resources Inv. 88-4227, Meno Park, CA: US Geological Survey*.

Keasler K. M. and Loveland W. D. (1982) Rare earth elemental concentrations in some Pacific Northwest rivers. *Earth Planet. Sci. Lett.* **61**, 68–72.

Koshinen W. C. and Harper S. S. (1990) The retention process: mechanisms. In *Pesticides in the Soil Environment: Processes, Impacts, and Modelling*. Book Ser. 2 (ed. H. H. Cheng). Soil Science Society of America, Madison, WI, pp. 51–77.

Lakshman S., Mills R., Patterson H., and Cronan C. (1993) Apparent differences in binding site distributions and aluminum(III) complexation for three molecular weight fractions of a coniferous soil fulvic acid. *Anal. Chim. Acta* **282**, 101–108.

Langmuir D. (1997) *Aqueous Environmental Geochemistry*. Prentice-Hall, New Jersey, 600p.

Langmuir D. and Herman J. S. (1980) The mobility of thorium in natural waters at low temperatures. *Geochim. Cosmochim. Acta* **44**, 1753–1766.

Laxen D. P. H. (1985) Trace metals adsorption/co-precipitation on hydrous ferric oxide under realistic conditions. *Water Res.* **19**, 1229–1236.

Laxen D. P. H., Davison X., and Wook C. (1984) Manganese chemistry in rivers and streams. *Geochim. Cosmochim. Acta* **48**, 2107–2111.

Lead J. R., Davison W., Hamilton-Taylor J., and Buffle J. (1997) Characterizing colloidal material in natural waters. *Aquat. Geochem.* **3**, 213–232.

Lee J. H. and Byrne R. H. (1993) Complexation of trivalent rare earth elements by carbonate ions. *Geochim. Cosmochim. Acta* **57**, 295–302.

Leleyter L., Probst J. L., Depetris P., Haida S., Mortatti J., Rouault R., and Samuel J. (1999) Distribution des terres rares dans les sediments fluviaux: fractionnement entre les phases labiles et résiduelles. *C. R. Acad. Sci. Paris* **329**, 45–52.

Lemarchand D., Gaillardet J., Lewin E., and Allègre C. J. (2000) Boron isotopes river fluxes: limitation for seawater pH reconstruction over the last 100 Myr. *Nature* **408**, 951.

Levasseur S. (1999) Contribution à l'étude du cycle externe de l'osmium. PhD Thesis, University of Paris 7.

Levasseur S., Birck J. L., and Allègre C. J. (1999) The osmium riverine flux and the oceanic mass balance of osmium. *Earth Planet. Sci. Lett.* **174**, 7–23.

Li Y. (2000) A compendium of geochemistry. In *From Solar Nebula to the Human Brain*. Princeton University Press.

Liu X. and Byrne R. H. (1997) Rare earth and yttrium phosphate solubilities in aqueous solution. *Geochim. Cosmochim. Acta* **61**, 1625–1633.

Luck J. M. and Ben Othman D. (1998) Geochemistry and water dynamics: II. Trace metals ad Pb–Sr isotopes as tracers of water movements and erosion processes. *Chem. Geol.* **150**, 263–282.

Martin J. M. and Meybeck M. (1979) Elemental mass balance of materiel carried by world major rivers. *Mar. Chem.* **7**(2), 173–206.

Martin J. M., Hogdahl O., and Philippot J. C. (1976) Rare earth element supply to the ocean. *J. Geophys. Res.* **81**(18), 3119–3124.

Martin J. M., Guan D. N., Elbaz-Poulichet F., Thomas A. J., and Gordeev V. V. (1993) Preliminary assessment of the distributions of some trace elements (As, Cd, Cu, Fe, Ni, Pb, Zn) in a pristine aquatic environment: the Lean River estuary (Russia). *Mar. Chem.* **43**, 185–199.

Martel A. E. and Hancock R. D. (1996) Metal complexes in Aqueous solutions. Plenum, New York, 253pp.

Martell A. E. and Smith R. M. (1977) Critical stability constants. In *Other Organic Ligands*. Volume 3, Plenum, New York.

McKnight D. M. and Bencala K. E. (1989) Reactive iron transport in an acidic mountain stream in Summit County, Colorado: a hydrologic perspective. *Geochim. Cosmochim. Acta* **53**, 2225–2234.

McKnight D. M. and Bencala K. E. (1990) The chemistry of iron, aluminium, and dissolved organic material in three acidic, metal-enriched, mountains streams, as controlled by watershed and in-stream processes. *Water Resour. Res.* **26** (n°12), 3087–3100.

McKnight D. M., Kimball B. A., and Bencala K. E. (1989) Iron photoreduction and oxydation in an acidic mountain stream. *Science* **240**, 637–640.

Michard G. (1989) Equilibres chimiques dans les eaux naturelles. Edition Publisud, Paris.

Millero F. J. (1992) Stability constants for the formation of rare earth inorganic complexes as a function of ionic strength. *Geochim. Cosmochim. Acta* **56**, 3123–3132.

Morel F. M. M. and Gschwend P. M. (1987) The role of colloids in the partitioning of solutes in natural waters. In *Aquatic Surface Chemistry* (ed. W. Stumm). Wiley, New York, pp. 405–422.

Morel F. M. M. and Hering J. G. (1993) *Principles and Applications of Aquatic Chemistry*. Wiley, New York.

Mouvet C. and Bourg A. C. M. (1983) Speciation (including adsorbed species) of copper, lead, nickel, and zinc in the Meuse River: observed results compared to values calculated with a chemical equilibrium computer program. *Water Res.* **6**, 641–649.

Murnane R. J. and Stallard R. J. (1990) Germanium and silicon in rivers of the Orinoco drainage basin. *Nature* **344**, 749–752.

Négrel P., Allègre C. J., Dupré B., and Lewin E. (1993) Erosion sources determined by inversion of major and trace element ratios in river water: the Congo Basin case. *Earth Planet. Sci. Lett.* **120**, 59–76.

Niriagu J. O. (1979) Global inventory of natural and anthropogenic emissions of trace metals to the atmosphere. *Nature* **279**, 409–411.

Niriagu J. O. and Pacyna J. M. (1988) Quantitative assessment of worldwide contamination of air, water, and soils by trace metals. *Nature* **333**, 134–139.

Olivié-Lauquet G., Allard T., Benedetti M., and Muller J. P. (1999) Chemical distribution of trivalent iron in riverine material from a tropical ecosystem: a quantitative EPR study. *Water Res.* **33**, 2276–2734.

Olivié-Lauquet G., Allard T., Bertaux J., and Muller J. P. (2000) Crystal chemistry of suspended matter in a tropical hydrosystem, Nyong basin (Cameroon, Africa). *Chem. Geol.* **170**, 113–131.

Palmer M. R. and Edmond J. M. (1993) Uranium in river water. *Geochim. Cosmochim. Acta* **57**, 4947–4955.

Parkhurst D. L. (1995) *Users Guide to PHREEQC: A Computer Program for Speciation, Reaction-path, Advective-transport, and Inverse Geochemical Calculations.* US Geological Survey Water Resources Inv. Report, 95-4227.

Parkhurst D. L., Thorstenson D. C., and Plummer N. L. (1990) *PHREEQE: A Computer Program for Geochemical Calculations.* Rev. US Geological Survey Water Resources Inv. Report, 80-96.

Perdue E. M., Beck K. C., and Reuter J. H. (1976) Organic complexes of iron and aluminum in natural waters. *Nature* **260**, 418–420.

Perdue E. M., Reuter J. H., and Parrish R. S. (1984) A statistical model of proton binding by humus. *Geochim. Cosmochim. Acta* **48**, 1257–1263.

Pereiro R. and Carro Díaz A. (2002) Speciation of mercury, tin, and lead compounds by gas chromatography with microwave-induced plasma and atomic-emission detection (GC–MIP–AED). *Anal. Bioanal. Chem.* **372**, 74–90.

Perret D., Newman M. E., Negre J. C., Chen Y., and Buffle J. (1994) Submicron particles in the Rhine river: I. Physico-chemical characterization. *Water Res.* **28**, 91–106.

Peucker-Ehrenbrink and Hannigan (2000) Effects of black shale weathering on the mobility of rhenium and platinum group elements. *Geology* **28**, 475–478.

Picouet C., Dupré B., Orange D., and Valladon M. (2001) Major and trace element geochemistry of the upper Niger (Mali): physical and chemical weathering rates and CO_2 consumption. *Chem. Geol.* 93–124.

Pierson-Wickman A. C., Reisberg L., and France-Lanord C. (2000) The Os isotopic composition of Himalayan river bedloads and bedrocks: importance of black shales. *Earth Planet. Sci. Lett.* **176**, 201–216.

Plummer L. E., Jones B. F., and Truesdell A. H. (1984) *WATEQF—A FORTRAN IV Version of WATEQ: A Computer Program for Calculating Chemical Equilibria of Natural Waters.* Rev. US Geological Survey Water Resources Inv. Report 76-13. US Geological Survey, Reston, VA.

Pokrovski O. S. and Schott J. (2002) Iron colloids/organic matter associated transport of major and trace elements in small boreal rivers and their estuaries (NW Russia). *Chem. Geol.* **190**, 141–181.

Ponter C., Ingri J., and Boström K. (1992) Geochemistry of manganese in the Kalic river, northern Sweden. *Geochim. Cosmochim. Acta* **56**, 1485–1494.

Porcelli D., Andersson P. S., Wasserburg G. J., Ingri J., and Baskaran M. (1997) The importance of colloids and mires for the transport of uranium isotopes through the Kalix River watershed and Baltic Sea. *Geochim. Cosmochim. Acta* **61**, 4095–4113.

Quémerais B. and Lum K. R. (1997) Distribution and temporal variation of Cd in the St. Lawrence river basin. *Aquat. Sci.* **59**, 243–259.

Roy S. (1996) Utilisation des isotopes du plomb et du strontium comme traceurs des apports atnthropiques et naturels dans les précipitations et rivières du bassin de Paris. PhD Thesis, Université Paris 7.

Sarin M. M., Krishnaswami S., Somayajulu B. L. K., and Moore W. S. (1990) Chemistry of U, Th, and Ra isotopes in the Ganga–Brahmaputra river system: weathering processes and fluxes to the Bay of Bengal. *Geochim. Cosmochim. Acta* **54**, 1387–1396.

Schäfer J. and Blanc G. (2002) Relationship between ore deposits in river catchments and geochemistry of suspended particulate matter from six rivers in southwest France. *Sci. Tot. Environ.* **298**(1–3), 103–118.

Schecher W. D. and McAvoy D. C. (1991) *MINEQL+: A Chemical Equilibrium Program for Personal Computers,* user's manual ver. 2.1. Edgewater, MD, Environ. Res. Software.

Schindler P. W. and Stumm W. (1987) The surface chemistry of oxides, hydroxides and oxide minerals. In *Aquatic Surface Chemistry* (ed. W. Stumm). Wiley, New York, pp. 83–110.

Seyler P. and Boaventura G. (2001) Trace metals in the main-stem river. In *The Biogeochemistry of the Amazon Basin and its Role in a Changing World* (eds. M. McClain, R. L. Victoria, and J. E. Richey). Oxford University Press, Oxford, pp. 307–327.

Seyler P. and Boaventura G. (2002) Distribution and partition of trace elements in the Amazon basin. In Hydrological Processes, Special Issue of International symposium on Hydrological and Geochemical Processes in Large Scale River Basins, Nov. 15–19, 1999, Manaus, Brésil.

Shafer M. M., Overdier J. T., Hurley J. P., Armstrong J., and Webb D. (1997) The influence of dissolved organic carbon, suspended particulates, and hydrology on the concentration, partitioning and variability of trace metals in two contrasting Wisconsin watersheds (USA). *Chem. Geol.* **136**, 71–97.

Sharma M., Wasserburg G. J., Hofmann A. W., and Chakrapani G. J. (1999) Himalayan uplift and osmium isotopes in oceans and rivers. *Geochim. Cosmochim. Acta* **63**, 4005–4012.

Shiller A. M. (1997) Dissolved trace elements in the Mississippi River: seasonal, interannual, and decadal variability. *Geochim. Cosmochim. Acta* **51**(20), 4321–4330.

Shiller A. M. (2002) Seasonality of dissolved rare earth elements in the lower Mississippi River. *Geochem. Geophys. Geosys.* **3**(11), 1068.

Shiller A. M. and Boyle E. (1985) Dissolved zinc in rivers. *Nature* **317**, 49–52.

Shiller A. M. and Boyle E. (1987) Variability of dissolved trace metals in the Mississippi River. *Geochim. Cosmochim. Acta* **51**, 3273–3277.

Shiller A. M. and Boyle E. (1997) Trace elements in the Mississippi River delta outflow region: behavior at high discharge. *Geochim. Cosmochim. Acta* **55**, 3241–3251.

Shiller A. M. and Mao L. (2000) Dissolved vanadium in rivers: effects of silicate weathering. *Chem. Geol.* **165**, 13–22.

Shiller A. M. and Frilot D. M. (1995) The geochemistry of gallium relative to aluminium in Californian streams. *Geochim. Cosmochim. Acta* **60**, 1323–1328.

Sholkovitz E. R. (1978) The flocculation of dissolved Fe, Mn, Al, Cu, Ni, Co, and Cd during estuarine mixing. *Earth Planet. Sci. Lett.* **41**, 77–86.

Sholkovitz E. R. (1992) Chemical evolution of REE: fractionation between colloidal and solution phases of filtered river water. *Earth Planet. Sci. Lett.* **114**, 77–84.

Sholkovitz E. R. (1995) The aquatic chemistry of rare earth elements in rivers and estuaries. *Aquat. Chem.* **1**, 1–34.

Sholkovitz E. R. and Copland D. (1980) The coagulation, solubility and adsorption properties of Fe, Mn, Cu, Ni, Cd, Co, and humic acids in a river water. *Acta* **45**, 181–189.

Sholkovitz E. R., Boyle E. R., and Price N. B. (1978) Removal of dissolved humic acid and iron during estuarine mixing. *Earth Planet. Sci. Lett.* **40**, 130–136.

Sigg L., Behra P., and Stumm W. (2000) *Chimie des Milieux Aquatiques,* Dunod, Parris, 3rd edn.

Smedley P. L. and Kinniburg D. G. (2002) A review of the source, behavior and distribution of arsenic in natural waters. *Appl. Geochem.* **17**, 517–568.

Spivack A. J., Palmer M. R., and Edmond J. M. (1987) The sedimentary cycle of the boron isotopes. *Geochim. Cosmochim. Acta* **51**, 1939–1949.

Stevenson F. J. (1994) *Humus Chemistry: Genesis, Composition, Reactions,* 2nd edn. Wiley, New York.

Stumm W. (1993) Aquatic colloids as chemical reactants: surface structure and reactivity. *Coll. Surf.* **A73**, 1–18.

Taylor S. R. and McLennan S. M. (1985) *The Continental Crust: Its Composition and Evolution*. Blackwell, Oxford, 312p.

Tessier A., Campbell P. G. C., and Bisson M. (1979) Sequential extraction procedure for the speciation of particulate trace metals. *Anal. Chem.* **51**, 844–851.

Thurman E. M. (1985) *Organic Geochemistry of Natural Waters*. Nijhoff and Junk publishers, Dordrecht.

Tipping E. (1981) The adsorption of humic substances by iron oxides. *Geochim. Cosmochim. Acta* **45**, 191–199.

Tricca A., Stille P., Steinmann M., Kiefel B., Samuel J., and Eikenberg J. (1999) Rare earth elements and Sr and Nd isotopic compositions of dissolved and suspended loads from small river systems in the Vosges mountains (France), the river Rhien and groundwater. *Chem. Geol.* **160**, 139–158.

Turner D. R., Whitfield M., and Dickson A. G. (1981) The equilibrium speciation of dissolved components in freshwater and seawater at 25 °C and 1 atm pressure. *Geochim. Cosmochim. Acta* **45**, 855–881.

Viers J. and Wasserburg G. J. (2002) Behavior of Sm and Nd in a lateritic soil profile. *Geochim. Cosmochim. Acta* (in press).

Viers J., Dupré B., Polvé M., Schott J., Dandurand J. L., and Braun J. J. (1997) Chemical weathering in the drainage basin of a tropical watershed (Nsimi-Zoetele site, Cameroon) comparison between organic-poor and organic-rich waters. *Chem. Geol.* **140**, 181–206.

Viers J., Dupré B., Deberdt S., Braun J. J., Angeletti B., Ndam Ngoupayou J., and Michard A. (2000) Major and traces elements abundances, and strontium isotopes in the Nyong basin rivers (Cameroon) constraints on chemical weathering processes and elements transport mechanisms in humid tropical environments. *Chem. Geol.* **169**, 211–241.

Vigier N., Bourdon B., Turner S., and Allègre C. J. (2001) Erosion timescales derived from U-decay series measurements in rivers. *Earth Planet. Sci. Lett.* **193**, 549–563.

Wollery T. J. (1992) EQ3/6, A software package for geochemical modelling of aqueous systems: package overview and installation guide (ver. 7.0). UCRL-MA-110662: Part I. Lawrence Livermore Natl. Lab.

Wood S. A. (1990) The aqueous geochemistry of the rare earth elements and yttrium. *Chem. Geol.* **82**, 159–186.

Wen L. S., Santschi P. H., and Tang D. (1997) Interactions between radioactively labelled colloids and natural particles: evidence for colloidal pumping. *Geochim. Cosmochim. Acta* **61**, 2867–2878.

Yee H. S., Measures C. I., and Edmond J. M. (1987) Selenium in the tributaries of the Orinoco in Venezuela. *Nature* **326**, 686–689.

Yeghicheyan D., Carignan J., Valladon M., Bouhnik Le Coz M., Le Cornec F., Castrec Rouelle M., Robert M., Aquilina L., Aubry E., Churlaud C., Dia A., Deberdt S., Dupré B., Freydier R., Gruau G., Hénin O., de Kersabiec A. M., Macé J., Marin L., Morin N., Petitjean P., and Serrat E. (2001) A compilation of silicon and thirty one trace elements measured in the natural river water reference SLRS-4 (NRC-CNRS). *Geostand. Newslett.* **35**(2–3), 465–474.

You C. F., Lee T., and Li Y. H. (1989) The partition of Be between soil and water. *Chem. Geol.* **77**, 105–118.

Zhang J. (1994) Geochemistry of trace metals from Chinese river/estuary systems: an overview. *Estuar. Coast. Shelf Sci.* **41**, 631–658.

Zhang J. and Huang W. W. (1992) Dissolved trace metals in the Huanghe, the most turbid river large river in the world. *Water Res.* **27**(1), 1–8.

Zhang J., Huang W. W., and Wang J. H. (1993) Trace element chemistry of the Huanghe (Yellow river), China: examination of the data from *in situ* measurements and laboratory approach. *Chem. Geol.* **114**, 83–94.

Zhang C., Wang L., Zhang, S., and Li X. (1998) Geochemistry of rare earth elements in the mainstream of the Yangtze river, China. *Appl. Geochem.* **13**, 451–462.

Geochemistry of Earth Surface Systems
ISBN: 978-0-08-096706-6

pp. 293–340

10

The Geologic History of the Carbon Cycle

E. T. Sundquist and K. Visser
US Geological Survey, Woods Hole, MA, USA

10.1	INTRODUCTION	341
10.2	MODES OF CARBON-CYCLE CHANGE	342
	10.2.1 *The Carbon Cycle over Geologic Timescales*	342
	10.2.1.1 *The "carbon dioxide" carbon cycle*	342
	10.2.1.2 *The "methane" carbon subcycle*	348
	10.2.2 *Timescales of Carbon-cycle Change*	350
10.3	THE QUATERNARY RECORD OF CARBON-CYCLE CHANGE	351
	10.3.1 *Analysis of CO_2 and CH_4 in Ice Cores*	352
	10.3.2 *Holocene Carbon-cycle Variations*	355
	10.3.3 *Glacial/interglacial Carbon-cycle Variations*	359
	10.3.3.1 *Carbon-cycle influences on glacial/interglacial climate*	361
	10.3.3.2 *Climate influences on glacial/interglacial carbon cycling*	362
	10.3.3.3 *Carbon/climate interactions at glacial terminations*	365
10.4	THE PHANEROZOIC RECORD OF CARBON-CYCLE CHANGE	367
	10.4.1 *Mechanisms of Gradual Geologic Carbon-cycle Change*	367
	10.4.1.1 *The carbonate weathering-sedimentation cycle*	367
	10.4.1.2 *The silicate-carbonate weathering-decarbonation cycle*	368
	10.4.1.3 *The organic carbon production-consumption-oxidation cycle*	369
	10.4.2 *Model Simulations of Gradual Geologic Carbon-cycle Change*	371
	10.4.3 *Geologic Evidence for Phanerozoic Atmospheric CO_2 Concentrations*	373
	10.4.4 *Abrupt Carbon-cycle Change*	374
10.5	THE PRECAMBRIAN RECORD OF CARBON-CYCLE CHANGE	375
10.6	CONCLUSIONS	376
	ACKNOWLEDGMENTS	376
	REFERENCES	376

10.1 INTRODUCTION

Geologists, like other scientists, tend to view the global carbon cycle through the lens of their particular training and experience. The study of Earth's history requires a view both humbled by the knowledge of past global transformations and emboldened by the imagination of details not seen in the fragments of the rock record. In studying the past behavior of the carbon cycle, geologists are both amazed by unexpected discoveries and reassured by the extent to which "the present is the key to the past." Understanding the present-day carbon cycle has become a matter of societal urgency because of concerns about the effects of human activities on atmospheric chemistry and global climate. This public limelight has had far-reaching consequences for research on the geologic history of the carbon cycle as well as for studies of its present and future. The burgeoning new "interdiscipline" of biogeochemistry claims among its adherents many geologists as well as biologists, chemists, and other scientists. The pace of discovery demands that studies of the geologic history of the carbon cycle cannot be isolated from the context of present and future events.

This chapter describes the behavior of the carbon cycle prior to human influence. It describes events and processes that extend back through geologic time and include the exchange of carbon between the Earth's surface and the long-term reservoirs in the lithosphere. Chapter 3 emphasizes carbon exchanges that are important over years to decades, with a focus on relatively recent human influences and prospects for change during the coming century. presents an overview of the biogeochemistry of methane, again with emphasis on relatively recent events. In these chapters as well as in the present chapter, relationships between the carbon cycle and global climate are a central concern. Together, these chapters provide an overview of how our knowledge of the present-day carbon cycle can be applied both to contemporary issues and to the record of the past. Similarly, these chapters collectively reflect the collaborative efforts of biogeochemists to utilize information about past variations in the carbon cycle to understand both Earth's history and modern changes.

This chapter begins with an overview of the carbon exchanges and processes that control the variations observed in the geologic record of the carbon cycle. Then examples of past carbon-cycle change are described, beginning with the most recent variations seen in cores drilled from glaciers and the sea floor, and concluding with the distant transformations inferred from the rock record of the Precambrian. Throughout this treatment, three themes are prominent. One is that different processes control carbon cycling over different timescales. A second theme is that relatively "abrupt" changes have played a central role in the evolution of the carbon cycle throughout Earth's history. The third theme is that the geologic cycling of carbon over all timescales passes through the atmosphere and the hydrosphere, and "it is this common course that unites the entire carbon cycle and allows even its most remote constituents to influence our environment and biosphere" (Des Marais, 2001).

The description of geologic events in this chapter includes examples from a broad span of the geologic record, but does not distribute attention in proportion to the distribution of geologic time in Earth's history. Readers will note, in particular, a disproportional emphasis on the Quaternary period, the most recent but briefest of geologic periods. The reason for this emphasis is twofold. First, the quality of the available Quaternary record of carbon-cycle change is far better than that available for earlier geologic periods. Second, the Quaternary record reveals a particularly illuminating array of details about interactions

among the atmosphere, the biosphere, and the hydrosphere—the subset of the carbon cycle that must be understood in order to comprehend carbon cycling over nearly all times and timescales.

10.2 MODES OF CARBON-CYCLE CHANGE

10.2.1 The Carbon Cycle over Geologic Timescales

Throughout Earth's history, the principal forms of carbon in the atmosphere have been carbon dioxide (CO_2) and methane (CH_4). These gases have played crucial, yet distinct, roles in the development of life forms and the alteration of Earth's surface environment. Carbon dioxide is a principal medium of photosynthesis, metabolism, and organic decomposition. Through its transformation in weathering and carbonate precipitation, it supplies a major portion of the Earth's sedimentary rocks and contributes to the cycling of volatiles through the lithosphere. Methane represents the anaerobic side of carbon cycling, from its importance in microbial metabolism to its release from organic matter trapped in rocks and sediments. While the major transfers of mass in the carbon cycle are usually associated with the cycling of carbon dioxide through the atmosphere, methane may have played a more important role in the past, and may be a more sensitive indicator of changes in Earth's processes. Together, carbon dioxide and methane are the primary compounds through which carbon cycling over all timescales has influenced the Earth's surface.

Figures 1 and 2 depict the carbon reservoirs and fluxes that affect atmospheric CO_2 and CH_4, respectively, over geologic timescales. Estimates are shown for both the most recent glacial period and the Late Holocene Epoch prior to human influence. Numerical values in these figures are shown in the units most commonly used in the literature: petagrams carbon (Pg C, 10^{15} g carbon) for the transfer and storage of carbon and carbon dioxide, and teragrams methane (Tg CH_4, 10^{12} g methane) for the transfer and storage of methane. (For direct comparison between CO_2 and CH_4, Table 1(c) includes estimates of CH_4 fluxes and reservoirs in the units for carbon transfer and storage (Pg C).) In Figures 1 and 2, the vertical alignment of each reservoir represents the approximate timescale over which it may significantly influence the atmosphere. The array of processes that can affect atmospheric CO_2 and CH_4, and the spectrum of timescales over which these processes act, comprise a principal topic of this chapter.

10.2.1.1 The "carbon dioxide" carbon cycle

Estimates of many Late Holocene carbon fluxes shown in Figure 1 are derived from values

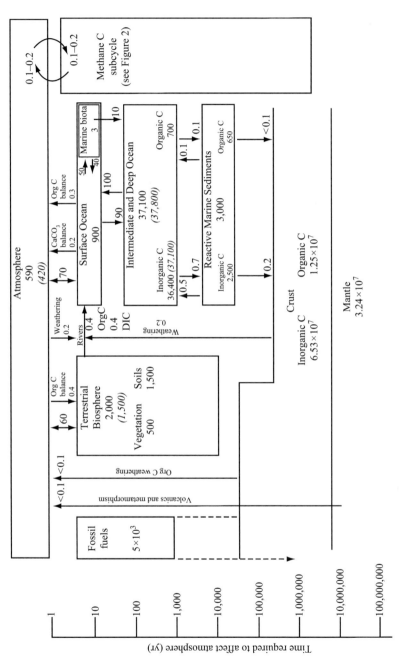

Figure 1 Reservoirs (Pg C; boxes) and fluxes (Pg C yr^{-1}; arrows) of the pre-industrial global carbon cycle. Values for glacial periods, where available, are shown in parentheses. Values for the glacial and pre-industrial (1750 AD) atmospheric carbon reservoir were calculated using CO$_2$ concentrations of 200 ppmv and 278 ppmv, respectively, and the equation M_a (g) $= (P_a \times 10^{-6}) \times (12.01/28.97) \times (5.12 \times 10^{21})$, where P_a is the atmospheric CO$_2$ concentration in ppmv, 12.01 is the atomic weight of carbon, 28.97 is the effective atmospheric molecular weight, and 5.12×10^{21} g is the mass of the (dry) atmosphere. See Section 10.2.1.1 for further derivation of estimates for glacial reservoirs. The vertical bar on the left shows the approximate time (in years) necessary for the different reservoirs to affect the atmosphere. The data for the figure are primarily from Chapter 3, Li (2000), Sarmiento and Gruber (2002), Sundquist (1985), and Sundquist (1993) (see also Table 1(a)). Atmospheric "balance" fluxes are shown to indicate the small net atmospheric exchange required to maintain a steady state with respect to sedimentation of organic carbon and calcium carbonate. The terrestrial biosphere and oceanic reservoir values are rounded to the nearest 100 Pg C, total reactive marine sediments to the nearest 1,000 Pg C, and all other reservoirs to the number of significant figures shown, according to the references cited in Table 1(a). All fluxes are rounded to one significant figure.

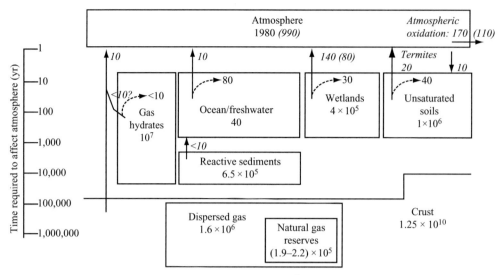

Figure 2 Reservoirs (Tg CH$_4$ or Tg C) and fluxes (Tg CH$_4$ yr^{-1}) of the natural, pre-industrial methane carbon subcycle. Values for glacial periods, where available, are shown in parentheses. Values for reactive sediments, wetlands, unsaturated soils, and crustal reservoirs represent organic carbon that might be converted to methane, and are all given as Tg C. Values for the glacial and pre-industrial (1750 AD) atmospheric reservoir were calculated using concentrations of 350 ppbv and 700 ppbv, respectively, and the equation M_a (g) $= (P_a \times 10^{-9}) \times (16.04/28.97) \times (5.12 \times 10^{21})$, where P_a is the atmospheric CH$_4$ concentration in ppbv, 16.04 is the molecular weight of methane, 28.97 is the effective atmospheric molecular weight, and 5.12×10^{21} g is the mass of the (dry) atmosphere. See Section 10.2.1.2 for further derivation of glacial flux estimates. The vertical bar on the left shows the approximate time (in years) necessary for the different reservoirs to affect the atmosphere. Estimates of methane consumption within reservoirs, from are shown as dashed arrows (see Section 10.2.1.2). Gross production of methane within a reservoir can be calculated by adding the flux to the atmosphere and the consumption value. The flux and consumption values are rounded to the nearest 10 Tg CH$_4$ or Tg C. See Table 1(c) for the data sources.

given by Houghton (see Chapter 3) and by Sarmiento and Gruber (2002) for the contemporary carbon cycle, adjusted for the estimates of human influence since ~1750 AD. The adjustments for human influence are based on relatively well-known estimates of the historical cumulative increase in atmospheric CO$_2$ (190 Pg C), and the corresponding uptake of CO$_2$ by the oceans (155 Pg C). These estimates are calculated from the estimates in (see Chapter 3) of the Treatise for the period 1850–2000 AD, corrected to the period 1750–2000 AD by adding 15 Pg C to the cumulative atmospheric increase (assuming a global mean atmospheric CO$_2$ concentration of 285 ppmv in 1850) and 15 Pg C to the cumulative oceanic increase. As described in Chapter 3, these estimates can be used with the estimated cumulative CO$_2$ production from burning fossil fuels (275 Pg C) to infer the net cumulative change in the amount of carbon stored in land plants and soils. This change is not applied to the estimates shown in Figure 1 because it is too small (<50 Pg C) to be considered significant relative to probable uncertainties in the total size of the reservoirs. The adjustments for human influence can be compared to uncertainties suggested by the range of recent estimates in reservoir sizes and fluxes in Table 1(a). Table 1(b) shows details of recent estimates of carbon in fossil fuel resources.

Figure 1 and Table 1(a) include estimates of the "geologic" components of the carbon cycle (weathering, river transport, sedimentation, volcanic/metamorphic emissions, and the rock and sediment reservoirs), so called because they are generally more important over longer timescales. Some sediment processes (the burial and remineralization of organic matter and calcium carbonate) may affect the atmosphere over timescales as short as a few thousand years. The sediment carbon reservoirs associated with these timescales are depicted in Figure 1 and Table 1(a) as "reactive sediment" reservoirs (Sundquist, 1985). Although the estimated magnitude of these reservoirs is somewhat arbitrary and process-dependent (see Broecker and Takahashi, 1977; Sundquist, 1985), they are large enough to affect ocean–atmosphere chemistry significantly through such processes as carbonate dissolution and organic carbon diagenesis.

Another large and potentially reactive sediment carbon reservoir is the pool of methane hydrates, shown as part of the "methane carbon subcycle" in Figure 2. Methane is produced most commonly by anoxic bacterial metabolism of CO$_2$ or of organic substrates derived ultimately from photosynthesis of CO$_2$. In the presence of oxygen or other electron acceptors containing oxygen, methane is oxidized to CO$_2$. Thus the cycling of methane is

Table 1(a) Representative range of recent estimates for natural, pre-industrial reservoirs and geologic fluxes of carbon (see Sundquist (1985) for a summary of older published estimates).

Reservoir	Size (Pg C)	References
Atmosphere	590	See Figure 1 caption
Oceans	$(3.71–3.90) \times 10^4$	Sundquist, 1985, 1993; see Chapter 3
Surface layer—inorganic	700–900	Sundquist, 1985; Sarmiento and Gruber, 2002; see Chapter 3
Deep layer—inorganic	$(3.56–3.80) \times 10^4$	Sundquist, 1993; Sarmiento and Gruber, 2002; see Chapter 3
Total organic	685–700	Sharp, 1997; Hansell and Carlson, 1998 (depths >1,000 m); Doval and Hansell, 2000
Aquatic biosphere	1–3	Falkowski *et al.*, 2000; Sarmiento and Gruber, 2002
Terrestrial biosphere and soils	$2.0–2.3 \times 10^3$	Sundquist, 1993; Sarmiento and Gruber, 2002; see Chapter 3
Vegetation	500–600	Sundquist, 1993; see Chapter 3
Soil	$(1.5–1.7) \times 10^3$	See Chapter 3
Reactive marine sediments	3,000	Sundquist, 1985
Inorganic	2,500	
Organic	650	
Crust	$(7.78–9.0) \times 10^7$	Holland, 1978; Li, 2000
Sedimentary carbonates	6.53×10^7	Li, 2000
Organic carbon	1.25×10^7	Li, 2000
Mantle	3.24×10^8	Des Marais, 2001
Fossil fuel reserves and resources	$(4.22–5.68) \times 10^3$	See Table 1(b) for references
Oil (conv. and unconv.)	636–842	
Natural gas (conv. and unconv.)	483–564	
Coal	$(3.10–4.27) \times 10^3$	

Flux	Size (Pg C yr^{-1})	References
Carbonate burial	0.13–0.38	Berner *et al.*, 1983; Berner and Berner, 1987; Meybeck, 1987; Drever *et al.*, 1988; Milliman, 1993; Wollast, 1994
Organic carbon burial	0.05–0.13	Lein, 1984; Berner, 1982; Berner and Raiswell, 1983; Dobrovolsky, 1994; Schlesinger, 1997
Rivers (dissolved inorganic carbon)	0.39–0.44	Berner *et al.*, 1983; Berner and Berner, 1987; Meybeck, 1987; Drever *et al.*, 1988
Rivers (total organic carbon)	0.30–0.41	Schlesinger and Melack, 1981; Meybeck, 1981; Meybeck, 1988; Degens *et al.*, 1991
Rivers-dissolved organic carbon	0.21–0.22	Meybeck, 1981; Meybeck, 1988; Spitzy and Leenheer, 1991
Rivers-particular organic carbon	0.17–0.30	Meybeck, 1981; Milliman *et al.*, 1984; Ittekot, 1988; Meybeck, 1988
Volcanism	0.04–0.10	Gerlach, 1991; Kerrick *et al.*, 1995; Arthur, 2000; Kerrick, 2001; Morner and Etiope, 2002
Mantle exchange	0.022–0.07	Sano and Williams, 1996; Marty and Tolstikhin, 1998; Des Marais, 2001

depicted in Figure 1 as a subcycle of the "carbon dioxide" carbon cycle. The methane carbon subcycle is further described in Section 10.2.1.2. It is important to note that, during the early history of the Earth, the processes now associated with the cycling of CO_2 may have emerged initially as a subcycle of the more prevalent cycling of methane (see Section 10.5).

The amount of CO_2 in the glacial atmosphere can be calculated accurately from the concentration of CO_2 in bubbles of air trapped in ice that formed during the most recent glacial period.

Table 1(b)   Global fossil fuel energy reserves and resources, in exajoules (EJ), and the equivalent carbon content in petagrams carbon (Pg C). Fossil fuel reserves are identified deposits that can be recovered using current technology under existing economic conditions. Resources are defined as "the occurrences of material in recognizable form" (WEC, 2000) but not extractable under current economic or technological conditions. Resources are essentially the amount of fossil fuels of "foreseeable economic interest." The differences between reserve and resource estimates illustrate the wide range of inherent uncertainties (geological, economic, and technological) in the resource estimates. "Additional occurrences" refers to fossil fuels that are not believed to be potentially recoverable (Moomaw, 2001). Unconventional occurrences differ from conventional occurrences by either "the nature of existence (being solid rather than liquid for oil) or the geological location (coal bed methane or clathrates)." Unconventional occurrences include oil shale, tar sands, coalbed methane, clathrates (Moomaw, 2001).

	EJ^c	$Pg\ C^d$
Reserves		
Oil—conventional[1,2,3,4,5,6]	5,908–6,759	111–127
Oil—unconventional[1,2]	6,624–8,639	124–162
Natural gas—conventional[1,2,3,4,5,6]	5,058–6,311	73–91
Natural gas—unconventional[1,2]	8,102–8,594	117–125
Coal[1,2,4,6]	28,825–41,985	499–1,094
Resources		
Oil—conventional[1,2,3,5,a]	6,490–14,055	122–264
Oil—unconventional[1,2,a]	14,860–15,397	279–289
Natural gas—conventional[1,2,3,5,b]	9,355–12,488	136–181
Natural gas—unconventional[1,2,b]	10,787–11,458	157–167
Coal[1,2]	100,352–125,059	2,605–3,177
Additional occurrences		
Oil—unconventional[1,2]	61,008–85,004	1,145–1,596
Natural gas—unconventional[1,2]	15,979–17,904	232–260
Coal[1,2]	120,986–134,280	3,122–3,411

References: [1]Nakicenovic *et al.*, 1998; [2]Moomaw 2001 (IPCC); [3]USGS World Energy Assessment, 2000; [4]WEC, 2000; [5]Masters *et al.*, 1994; [6]Oil and Gas Journal, 2001.
[a] For oil resources, the USGS World Energy Assessment (2000) estimated that there is a 5% probability that 16,466 EJ is recoverable, a 50% probability of 8,872 EJ and a 95% probability of 4,283 EJ. The mean values given in the assessment are used in the ranges shown above. [b]For natural gas resources, the USGS World Energy Assessment (2000) estimated that there is a 5% probability that 15,207 EJ is recoverable, a 50% probability of 8,057 EJ and a 95% probability of 3,946 EJ. The mean values given in the assessment are used in the ranges shown above. [c]The CDIAC "Common Energy Unit Conversion Factors Table 4" (O'Hara, 1990; http://cdiac.esd.ornl.gov/pns/convert.html) was used to convert all energy units to Exajoules. [d]The energy units were converted to equivalent carbon content (Pg C) using the methods detailed in Sundquist (1985) as well as the conversion factor for the average carbon content of coal [1 TJ=25.4 mt carbon (ORNL Bioenergy Feedstock Development Programs, 2003—http://bioenergy.ornl.gov/papers/misc/energy_conv.html)] in cases where the breakdown of coal rank was not available.

(These measurements are further described in Section 10.3.1.) The glacial atmospheric CO_2 concentration also provides a way to estimate the dissolved inorganic carbon (DIC) content of the glacial ocean surface mixed layer. DIC exists primarily in four forms: dissolved CO_2, carbonic acid (H_2CO_3), bicarbonate ions (HCO_3^-), and carbonate ions (CO_3^{2-}). The ionic forms may combine with other ions to form complex ions in solutions such as seawater. Chemical reaction among the dissolved inorganic carbon species is rapid, and they occur in proportions that can be calculated from well-known thermodynamic equilibrium relationships. CO_2 exchange between the atmosphere and the mixed layer is so rapid that, for global estimates like those shown in Figure 1, the ocean surface layer can be considered to be near chemical equilibrium with the atmosphere (Sundquist and Plummer, 1981). The glacial-to-interglacial increase in ocean surface DIC, corresponding to the atmospheric CO_2 increase

(170 Pg C) shown in Figure 1, is estimated to have been ~30 Pg C (Sundquist, 1993).

Estimates of other carbon reservoirs and fluxes during the most recent glacial period depend largely on estimates of the amount of carbon stored in the terrestrial biosphere. Because the transfer of carbon over timescales of thousands of years is limited largely to translocation among the atmosphere, the biosphere, and the oceans, a loss or gain of carbon in one of these reservoirs can be used as a good approximation for a corresponding gain or loss in the others. (see Section 10.2.2 for discussion of this "rule" and its possible exceptions.) Estimates of the glacial terrestrial biosphere are referenced to the Late Holocene or present-day terrestrial biosphere, either by reconstruction of ecological changes or by inference from the shift in the isotopic composition of oceanic dissolved carbon. There is a broad consensus that the amount of carbon stored in vegetation and soils was smaller during the most recent glacial period than during the Holocene Epoch (Shackleton, 1977; Crowley, 1995). Paleoecological reconstructions, based on data from soils and sediments (see, e.g., Adams *et al.*, 1990; Adams and Faure, 1998) or on models of glacial climate/ vegetation relationships (see, e.g., Prentice *et al.*, 1993; Otto *et al.*, 2002), yield estimates of glacial carbon storage ranging from a few hundred to more than one thousand Pg C less than carbon storage in Holocene plants and soils. This broad range of estimates reflects diverse sources of uncertainty, including limitations in paleoclimatic and paleoecological data, climate models, and assumptions about effects of atmospheric CO_2, interactions between climate and vegetation, and amounts of carbon stored in various biomes and coastal environments (Prentice and Fung, 1990; Van Campo *et al.*, 1993; Francois *et al.*, 1999).

The carbon isotope record of marine sediments provides an important constraint on glacial/interglacial changes in carbon storage by the terrestrial biosphere. During photosynthesis, plants preferentially assimilate $^{12}CO_2$, leaving the atmosphere relatively enriched in $^{13}CO_2$. Carbon assimilated during photosynthesis by land plants may be depleted in ^{13}C by $\sim$3–25‰ relative to the inorganic carbon source. (Carbon isotope compositions are denoted by per mil (‰) differences in the ratio of ^{13}C to ^{12}C relative to a standard.) The extent of photosynthetic ^{13}C depletion varies primarily because plants may use several different biochemical pathways to assimilate carbon. The ribulose-1,5-bisphosphate carboxylation, or "C_3," pathway (also known as the Calvin cycle) is the most common and reduces the $^{13}C : ^{12}C$ ratio by $\sim$15 to 25‰ (O'Leary, 1988). The phosphenolpyruvate carboxylation, or "C_4," pathway is observed most commonly in corn and other grasses. Although the C_4 pathway operates in close association with ribulose-1,5-bisphosphate carboxylation, the resulting reduction in the $^{13}C : ^{12}C$ ratio is only $\sim$3–8‰ (Deleens *et al.*, 1983). Because the C_3 pathway is the most common among land plants, this pathway probably determined the average $^{13}C : ^{12}C$ ratio of the atmospheric CO_2 transferred to and from vegetation and soils during glacial/interglacial transitions.

Exchange of atmospheric CO_2 with the oceans extends the isotopic fractionation effect of land plants to the ocean DIC reservoir. Over timescales of ocean mixing, the carbon isotope effect of land plants is diluted but nevertheless measurable in the larger oceanic DIC reservoir. A significant global change in the amount of photosynthesized carbon stored in plants and soils would be expected to change the isotopic composition of carbon in atmospheric CO_2 and oceanic DIC. The carbon isotope compositions of both atmospheric CO_2 and oceanic DIC are also affected by

other factors, most notably sea-surface gas exchange, chemical reactions among DIC species, and photosynthesis by marine organisms. For example, photosynthesis by marine organisms reduces the $^{13}C : ^{12}C$ ratio by $\sim$10–30‰ (Deines, 1980). Marine photosynthesis is widely believed to occur primarily via the C_3 biochemical pathway, but effects of C_4 photosynthesis by diatoms may account for part of the relatively wide range of carbon isotope ratios observed in planktonic ecosystems (Reinfelder *et al.*, 2000). Glacial/ interglacial changes in the factors affecting oceanic isotope fractionation must be quantified in order to discriminate the carbon isotopic signature of changes in the terrestrial biosphere (Hofmann *et al.*, 1999).

The isotopic fractionation effects of marine photosynthesis, gas exchange, and chemical reactions are most pronounced in atmospheric CO_2 and ocean surface DIC. Deep-ocean DIC is relatively unaffected by these factors. Thus, the carbon isotope effects of changes in the terrestrial biosphere are best seen in the fossils of organisms that form their shells in the deep ocean. Glacial/ interglacial isotopic changes in oceanic DIC are recorded in the calcareous shells of organisms deposited in sediments, and isotopic changes in atmospheric CO_2 are recorded in ice cores. These records generally indicate a glacial-to-interglacial increase in the $^{13}CO_2 : ^{12}CO_2$ ratio of oceanic DIC by $\sim$0.3–0.5‰ (Sarnthein *et al.*, 1988; Curry *et al.*, 1988; Duplessy *et al.*, 1988; Ku and Luo, 1992; Crowley, 1995), and a corresponding increase in atmospheric CO_2 by $\sim$0.1–0.5‰ (Leuenberger *et al.*, 1992; Smith *et al.*, 1999). Mass balance calculations based on these isotopic shifts imply a glacial-to-interglacial expansion of $\sim$400–800 Pg C in the amount of organic carbon stored in plants and soils (Sundquist, 1993; Bird *et al.*, 1996). Sundquist (1993) suggested a range of 450–750 Pg C, while Bird *et al.* (1996) suggested a range of 300–700 Pg C based on somewhat different assumptions. Here we derive an estimated range of 400–800 Pg C from the mass balance equations of Bird *et al.* (1996) applied to the range of atmospheric and oceanic carbon isotope shifts given in the text, and to a range of −22‰ to −25‰ (relative to the common PDB standard) for the ^{13}C content of the carbon taken up by the biosphere during deglaciation. These calculations do not take into account the possible effect of increased methane production and oxidation during the period of deglaciation (Maslin and Thomas, 2003); see Section 10.3.3.3.

The procedures described above provide a reasonable basis for estimating the redistribution of carbon storage among the atmosphere, terrestrial biosphere, and oceans during the most recent glacial period. The mass balance of glacial carbon losses and gains, relative to Holocene values, can be tallied as follows:

Glacial carbon losses:

Organic carbon in plants and soils	400–800 Pg C
Atmosphere	170
Ocean surface DIC	30
Total losses	600–1,000 Pg C

Glacial carbon gains:

Deep-ocean DIC	600–1,000 Pg C

Although these calculations are used to derive the estimates of glacial carbon reservoirs shown in Figure 1, glacial carbon fluxes are much less certain and are not shown.

10.2.1.2 The "methane" carbon subcycle

Estimates of Late Holocene CH_4 fluxes and reservoirs are shown in Figure 2. Many of the flux estimates in this figure are derived from the values for natural sources presented by Reeburgh and Lelieveld *et al.* (1998). Estimates of the carbon reservoirs that contribute to methane cycling have been added. The wetland soil and sediment organic carbon reservoirs estimated in Figure 2 are also represented in the soil and sediment reservoirs depicted in Figure 1, because these reservoirs may yield both CO_2 and CH_4 depending on environmental conditions. Uncertainties in the values shown in Figure 2 can be inferred from the ranges of estimates listed in Table 1(c).

Global fluxes of CH_4 to and from the atmosphere are generally smaller than those of CO_2 (Tables 1(a) and 1(c) and Figures 1 and 2). Although CH_4 may be less significant in exchange of carbon mass, atmospheric CH_4 can be an important indicator of changes in carbon cycling that do not entail large transfers of carbon. Moreover, atmospheric CH_4 has direct importance for the carbon cycle due to its disproportionate influence on climate.

A molecule of CH_4 absorbs more than 20 times as much long-wave radiation as a molecule of CO_2 (Ramaswamy *et al.*, 2001). In considering the climatic effects of different gases, their radiative properties are often compared using calculated indices such as the greenhouse warming potential (GWP; Shine *et al.*, 1990), which is defined as the time-integrated radiative effect following an instantaneous injection of a particular gas relative to the effect of an injection of an equal mass of CO_2. The GWP is referenced to specific integration times in order to adjust for the different time trajectories of instantaneous trace gas injections. The GWP of CH_4 is 7 times that of CO_2 over a 500 yr period (Ramaswamy *et al.*, 2001). The increase in

atmospheric CH_4 that occurred during the time between the most recent glacial period and the Late Holocene Epoch (Figure 2) had a direct radiative effect that was approximately one-fifth that of the corresponding increase in atmospheric CO_2 (Figure 1) (Petit *et al.*, 1999). The relative impact of methane's radiative properties on the global greenhouse effect is diminished by its low concentration and short lifetime in the atmosphere, but CH_4 clearly plays an important role in the complex interactions between the carbon cycle and the climate system.

As depicted in Figure 2 and detailed in the rate of CH_4 release to the oceans and atmosphere is only a fraction of the rate at which CH_4 is produced in anoxic environments. In wetlands and marine sediments, zones of CH_4 production are frequently overlain by or interspersed with zones where aerobic or anaerobic methanotrophic microbes are active consumers of CH_4 (Reeburgh, 1976; Scranton and Brewer, 1978; Yavitt *et al.*, 1988; Whalen and Reeburgh, 1990; Moosavi *et al.*, 1996; R. S. Hanson and T. E. Hanson, 1996; Orphan *et al.*, 2001; Hinrichs and Boetius, 2002). Although some CH_4 may bypass oxidation in these zones by transport via bubbles or vascular plants (Dacey and Klug, 1979; Bartlett *et al.*, 1988), significant quantities of CH_4 are consumed in these layers and therefore are not released to the atmosphere and oceans (Galchenko *et al.*, 1989; Oremland and Culbertson, 1992; Reeburgh *et al.*, 1993). This process is difficult to quantify, but it is probably a significant component of the "methane" carbon subcycle (King, 1992). Several studies suggest that additional CH_4 may be produced in anoxic microsites within oxic soils and ocean environments, which are on the whole net consumers of atmospheric CH_4 (Yavitt *et al.*, 1995; Holmes *et al.*, 2000; von Fischer and Hedin, 2002). Thus the extent and magnitude of the methane carbon subcycle may be greater than suggested by the fluxes and reservoirs shown in Figure 2.

Surprisingly, CH_4 has an atmospheric lifetime longer than that of CO_2. "Atmospheric lifetime" is calculated here as the ratio of abundance in the atmosphere to the sum of annual sources or removals, assuming a steady state. Using the fluxes represented in Figure 1, CO_2 would have an atmospheric lifetime of $\sim$4–5 yr. However, this calculation does not reflect the full extent of cycling of CO_2 through plants because it considers only net primary production and heterotrophic respiration. When the cycling of CO_2 via gross primary productivity and autotrophic respiration is taken into account, the exchange of CO_2 between the atmosphere and the terrestrial biosphere is approximately doubled, and the atmospheric lifetime of CO_2 is reduced to $\sim$3 yr. The atmospheric lifetime of CH_4 is calculated from the

Table 1(c)   Representative range of recent estimates for natural (pre-industrial) fluxes and reservoirs of methane.

Reservoir	Size (Tg CH$_4$ or Tg C)	Size (Pg C)	References
Atmosphere	1980 Tg CH$_4$	1.5	See Figure 2 caption
Oceans	22–65 Tg CH$_4$	0.016–0.049	Holmes et al., 2000; Kelley and Jeffrey, 2002
Wetlands	$(2.2–4.9)\times10^5$ Tg C	224–489	Gorham, 1991; Botch et al., 1995; Lappalainen, 1996; Schlesinger, 1997, after Schlesinger, 1977; Clymo et al., 1998
Reactive marine sediments (org. C)	6.5×10^5 Tg C	650	Sundquist, 1985
Non-wetland soils	9.7×10^5 Tg C	970	Schlesinger, 1997, after Schlesinger, 1977
Geological sources			
Crust	1.25×10^{10} Tg C	1.25×10^7	Li, 2000
Hydrates	$5\times10^5 – 2.4\times10^7$ Tg CH$_4$	$4\times10^2 – 1.8\times10^4$	Kvenvolden and Lorenson, 2001
Dispersed gas in sedimentary basins	1.6×10^6 Tg CH$_4$	1.2×10^3	Hunt, 1996
Natural gas reserves (part of dispersed gas in sed. basins)	$(2.5–2.9)\times10^3$ Tg CH$_4$	190–216	See Table 1(b) for references

Flux	Size (Tg CH$_4$ yr^{-1})	Size[a] (Pg C yr^{-1})	References
Oceans	0.4–20	0–0.01	Ehhalt, 1974; Watson et al., 1990; Fung et al., 1991; Chappellaz et al., 1993b; Lambert and Schmidt, 1993; Bange et al., 1994; Bates et al., 1996; Lelieveld et al., 1998; Holmes et al., 2000; Kelley and Jeffrey, 2002
Marine sediments	0.4–12.2	0–0.01	Judd, 2000
Wetlands	92–260	0.07–0.19	Fung et al., 1991; Bartlett and Harriss, 1993; Chappellaz et al., 1993b; Cao et al., 1996; Hein et al., 1997; Matthews, 2000; Walter et al., 2001
Termites	2–22	0–0.02	Cicerone and Oremland, 1988; Chappellaz et al., 1993b; Sanderson, 1996; Sugimoto et al., 1998
Wild fires	2	0	Levine et al., 2000
Geological sources (Hydrates, volcanoes, natural gas seeps, geothermal)	5–65	0–0.05	Chappellaz et al., 1993b; Hovland, 1993; Lacroix, 1993; Hornafius et al., 1999; Judd, 2000, after Lacroix, 1993; Etiope and Klusman, 2002; Judd et al. 2002
Total source	159–290	0.12–0.22	McElroy, 1989; Chappellaz et al., 1993b; Thompson et al., 1993; Martinerie et al., 1995; Etheridge et al., 1998; Brook et al., 2000

[a] Fluxes less than 0.01 Pg C yr^{-1} are rounded to 0.

fluxes depicted in Figure 2 to be ~11 yr. For the glacial values shown in Figure 2, the atmospheric lifetime of CH_4 is decreased to ~8 yr. Because the atmospheric oxidation of CH_4 is coupled to the abundance of OH and other reactants, its atmospheric lifetime may vary (Prinn *et al.*, 1983; Lelieveld *et al.*, 1998; Breas *et al.*, 2002).

CO_2 is removed from the atmosphere by exchange with the biosphere, the oceans, and the lithosphere. The primary mechanism for removal of atmospheric CH_4 is chemical oxidation in the atmosphere without further exchange. Thus, while atmospheric CO_2 is controlled by exchange with other components of the carbon cycle, the concentration of CH_4 in the atmosphere reflects a balance between its rate of supply to the atmosphere and its oxidation in the atmosphere. This fundamental difference in control mechanisms affects the time dependence of responses of atmospheric CO_2 and CH_4 to relatively abrupt perturbations. The response of CH_4 will occur rapidly through effects on its rate of oxidation in the atmosphere, whereas the response of CO_2 will be mediated by the more complex array of processes that govern its rate of exchange with other large carbon reservoirs and the rates of response of those reservoirs. Thus, even though CH_4 has a longer atmospheric lifetime than CO_2, the response of atmospheric CH_4 to perturbations will tend to be more rapid.

Methane is oxidized primarily in the troposphere by reactions involving the hydroxyl radical (OH). Methane is the most abundant hydrocarbon species in the atmosphere, and its oxidation affects atmospheric levels of other important reactive species, including formaldehyde (CH_2O), carbon monoxide (CO), and ozone (O_3) (Wuebbles and Hayhoe, 2002). The chemistry of these reactions is well known, and the rate of atmospheric CH_4 oxidation can be calculated from the temperature and concentrations of the reactants, primarily CH_4 and OH (Prinn *et al.*, 1987). Tropospheric OH concentrations are difficult to measure directly, but they are reasonably well constrained by observations of other reactive trace gases (Thompson, 1992; Martinerie *et al.*, 1995; Prinn *et al.*, 1995; Prinn *et al.*, 2001). Thus, rates of tropospheric CH_4 oxidation can be estimated from knowledge of atmospheric CH_4 concentrations. And because tropospheric oxidation is the primary process by which CH_4 is removed from the atmosphere, the estimated rate of CH_4 oxidation provides a basis for approximating the total rate of supply of CH_4 to the atmosphere from all sources at steady state (see Section 10.2.2) (Cicerone and Oremland, 1988).

Atmospheric CH_4 concentrations during the Late Holocene Epoch and the most recent glacial period are known from analyses of air bubbles in ice cores (see Section 10.3). These concentrations provide the basis for calculating the rates of

atmospheric CH_4 oxidation and total rates of atmospheric CH_4 supply shown in Figure 2 (Chappellaz *et al.*, 1993b; Martinerie *et al.*, 1995; Lelieveld *et al.*, 1998). Although these overall budgets are reasonably well determined, the relative contributions of individual sources contributing to the atmospheric CH_4 supply are less well known (Cicerone and Oremland, 1988). Likewise, the sizes of the reservoirs other than atmospheric CH_4 can only be approximated.

Methane hydrates comprise the largest CH_4 reservoir in the methane carbon subcycle. Methane hydrates are formed when abundant dissolved methane accumulates under specific conditions of cold temperature and high pressure. These conditions commonly occur in marine sediments below water depths of a few hundred meters, and in continental sediments at high latitudes (Figure 3). When environmental conditions in these locations change, methane hydrates may become unstable and yield large quantities of dissolved and gaseous methane. The sensitivity of methane hydrates to changing environmental conditions implies that they may have played a role in past carbon-cycle changes (see Sections 10.3.3 and 10.4.4).

10.2.2 Timescales of Carbon-cycle Change

In studying the many factors that contribute to carbon-cycle change over geologic timescales, geochemists commonly observe that the relative importance of various processes and reservoirs depends on the timescale under consideration. The time axes shown in Figures 1 and 2 portray approximate timescales over which various fluxes and reservoirs may influence the atmosphere. With important exceptions (see below), these timescales can be used to categorize components of the carbon cycle into time-related frames of reference (Sundquist, 1986). Over relatively short timescales (up to ~10^3 yr), the most common variations in atmospheric CO_2 and CH_4 involve exchange with only the terrestrial biosphere and the oceans. Over somewhat longer timescales (~10^3–10^5 yr), the frame of reference must be expanded to include "reactive" carbon in the uppermost layers of marine sediments. Over timescales of millions of years and longer, the frame of reference must include carbon in the Earth's crust.

These frames of reference define fundamentally different modes of carbon-cycle change. Over the timescales of Quaternary glacial/interglacial cycles, changes in atmospheric CO_2 and CH_4 reflect primarily the redistribution of carbon among the atmosphere, biosphere, and oceans, with important contributions from reactive sediments. Over timescales of millions of years and longer, atmospheric CO_2 and CH_4 are controlled not only by carbon

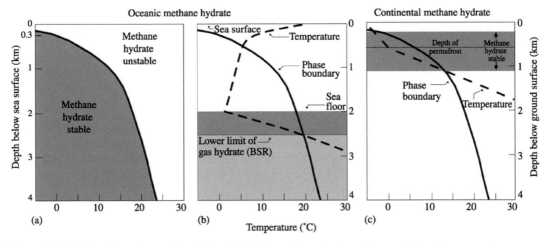

Figure 3 (a) Phase boundary and stability fields for methane hydrate. The vertical axis represents pressure, increasing downward and plotted as equivalent depth (km) assuming hydrostatic pressure. (b) The same phase boundary as shown in (a) in a marine environment, with seafloor inserted at 2 km, and a typical marine temperature profile through the water column and seafloor (dashed line). The dark gray area denotes the vertical extent of methane hydrate under the specified conditions (after Dillon, 2001, figure 2). (c) The phase boundary of methane hydrate in a high-latitude continental environment, plotted with a typical temperature (geothermal) gradient (dashed line) and showing depths assuming a hydrostatic pressure gradient. The dark gray area denotes the vertical extent of methane hydrate stability, under the specified conditions (after Kvenvolden and Lorenson, 2001, figure 1).

cycling at the Earth's surface, but also by the balance between long-term releases from the Earth's crust and carbon burial in sediments. With respect to the ocean–atmosphere–biosphere system, these relatively short-term and long-term carbon-cycle modes can be characterized respectively by internal redistributions and external exchange. Both modes of change operate simultaneously, and sorting out their effects is one of the most challenging problems in understanding the geologic evidence for carbon-cycle change (see, e.g., Sundquist (1991)).

The dynamic behavior of the carbon cycle and other complex systems may tend toward conditions of no change or "steady state" when exchanges are balanced by feedback loops. For example, model simulations of historical and projected effects of anthropogenic CO_2 and CH_4 emissions are usually based on an assumed carbon-cycle steady state before the onset of human influence. It is important to understand that the concept of steady state refers to an approximate condition within the context of a particular time-dependent frame of reference. Sundquist (1985) examined this problem rigorously using eigenanalysis of a hierarchy of carbon-cycle box models in which boxes were mathematically added and "lumped" to span a broad range of timescales. This analysis demonstrated that the nature and accuracy of any geochemical steady-state approximation depends on the timescale of interest and the defined frame of reference for that timescale. For example, the particular steady-state approximation used in projecting atmospheric CO_2 and CH_4 concentrations for the next century might not be appropriate for

projections over timescales of hundreds to thousands of years (Sundquist, 1990b).

Recent studies of the geologic history of the carbon cycle have revealed many surprising examples of relatively abrupt change that cannot be characterized by the time-scale relationships depicted in Figures 1 and 2. While time-dependent frames of reference and steady-state approximations have provided a basis for many advances in understanding the carbon cycle, it is clear that these concepts are not adequate to guide studies of events such as bolide impacts or methane hydrate outbursts. A rigorous theoretical treatment of abrupt events is beyond the scope of this chapter. We offer two examples and some general comments in Section 10.4.4.

10.3 THE QUATERNARY RECORD OF CARBON-CYCLE CHANGE

The study of the Quaternary history of the global carbon cycle is powerfully constrained by the analysis of air trapped in ice from the continental ice sheets of Greenland and Antarctica. Cores taken from glacial ice and firn have yielded samples of air ranging in age from a few decades to more than 400,000 yr. The ice cores provide direct evidence of past variations in atmospheric concentrations of CO_2, CH_4, other greenhouse gases, aerosols, and dust. These variations in atmospheric chemistry reflect variations in global carbon cycling. The ice cores also offer a record of climate change that can be closely correlated with the record of carbon-cycle variations. The ice-core record from ice sheets provides compelling evidence for the intimate association between climate and the carbon cycle over a broad range of timescales.

10.3.1 Analysis of CO_2 and CH_4 in Ice Cores

The analysis of ice-core gas chemistry presents a very challenging technical problem. The gas samples are extremely small (a 10 g sample of ice might yield less than 1 cm^3 of trapped air (Schwander and Stauffer, 1984)), and they are very susceptible to alteration during extraction and analysis. Because CO_2 might dissolve in melted ice or condensed water vapor, the extraction of samples of this gas must be conducted by crushing the ice in cold and completely dry conditions (Delmas et al., 1980). Methane is generally not extracted with CO_2 because friction during the dry crushing process may produce CH_4 (Stauffer et al., 1985). Fortunately, CH_4 is less soluble in water than CO_2, so the extraction of samples for CH_4 can be accomplished by melting the sample (see, e.g., Rasmussen and Khalil (1984) or by special grinding techniques (Etheridge et al., 1988)). Other sampling and analytical difficulties include the effects of breakdown of clathrates, which are the principal form of *in situ* gas storage in the deepest and coldest ice (Neftel et al., 1983). Techniques to address these problems have been pioneered by the ice-core laboratories of the University of Bern in Switzerland, the Laboratoire de Glaciologie et de Géophysique de l'Environment in Grenoble, France, and the Division of Atmospheric Research at CSIRO in Australia (see the many references to work from these laboratories in Table 2).

Ice cores can be dated by counting annual layers, modeling ice accumulation and flow, and correlating recognizable events with other datable records (see, e.g., Dansgaard et al., 1969; Johnsen et al., 1972; Hammer et al., 1978). However, the entrapped air is significantly younger than the surrounding ice, because the air is not completely isolated from the atmosphere until burial to depths that may exceed 100 m (Schwander and Stauffer, 1984). An additional problem arises because the process of complete bubble enclosure (coinciding with the conversion of granular firn to solid ice) occurs gradually during ice accumulation. Thus, the bubbles of air trapped at any particular depth are not only younger than the surrounding ice, but they also represent a range of ages reflecting the range of depths and times of bubble enclosure. Ice-core gas samples are generally too small, and many are too old, for radiocarbon dating of the enclosed carbon compounds. Thus the dating of ice-core gas records is very difficult.

Ice-core gas samples are typically dated by applying an age-difference correction relative to the age of the enclosing ice, based on the offset inferred from an estimate of the depth of bubble enclosure. An age range is also assigned based on the estimated rate of bubble enclosure during ice accumulation. The dating of gas samples therefore depends on assumptions about the physical properties of firn and ice during periods of past accumulation. Fortunately, these properties can be estimated from models confirmed by observations in modern firn layers (Schwander and Stauffer, 1984) and extrapolated to past climatic conditions inferred from the ice record itself (Schwander et al., 1997). Corrections and uncertainties are larger for ice that accumulated more slowly, and thus they vary according to changes in accumulation conditions from location to location and through time (Schwander et al., 1997). For example, the age difference between the ice and the air entrapped during the most recent 200 years at the Siple Station (in Ellsworth Land, West Antarctica) is estimated to be 80–85 yr (Neftel et al., 1985), whereas the corresponding age difference for the more slowly accumulating ice at the Dome C site (on the polar plateau in East Antarctica) approaches 5,500 yr for samples from the last glacial maximum (LGM) ($\sim$18–24 ka) (Monnin et al., 2001). The range of gas ages at any given depth likewise depends on the accumulation rate of the ice. Where accumulation rates are relatively low, the range is generally 5–10% of the age difference between the ice and the enclosed air. The relative range may be higher in ice that accumulates more rapidly. For ice cores that provide records of relatively recent trends, estimates of the gas age difference correction and age range can be confirmed by comparison with historical atmospheric measurements (see, e.g., Levchenko et al., 1996). However, these estimates are less certain for longer ice-core records. Uncertainties in the dating of ice-core gases are an important concern in efforts to correlate the ice record of carbon-cycle changes with other variations recorded in the ice itself.

Some ice-core gas samples do not represent old air because they have undergone postdepositional alteration. In glaciated regions where seasonal melting occurs, the frozen and buried melt layers are enriched in CO_2 (Neftel et al., 1983). High CO_2 concentrations may also be caused by chemical reactions among other constituents buried in the firn and ice. For example, many features of the CO_2 record in ice cores from Greenland are probably not reliable because of reactions between carbonate dust and trace acidic compounds in the ice (Delmas, 1993; Anklin et al., 1995). Oxidation of organic compounds may likewise produce CO_2 that contaminates the CO_2 in entrapped air (Haan and Raynaud, 1998; Tschumi and Stauffer, 2000). *In situ* bacterial activity is another possible source of contamination (Sowers, 2001).

Other postdepositional changes may be caused by physical fractionation of gases in the thick and

Table 2 Carbon dioxide data from polar ice cores. Carbon dioxide concentrations and carbon isotope values (in per mil relative to the PDB standard) are listed for the LGM (~18–24 ka), the Holocene Epoch (1–10 ka), and the most recent pre-industrial part of the Holocene Epoch (1000–1800 AD). Key references for the CO_2 data from each core site are listed in the right-hand column. The superscript letters in the data table indicate the references from which the indicated values were derived. Values for Vostok (LGM and Holocene) were calculated for this table from the numerical data cited.

Site	Location	Data type	LGM	Holocene	Pre-industrial Holocene	References
Vostok	78° 28′ S, 106° 48′ E	CO_2	187.7 ± 2–3 ppmv[a]	266.2 ± 2–3 ppmv[a]		Lorius et al., 1985; Barnola et al., 1987; Barnola et al., 1991; Jouzel et al., 1993; Fischer et al., 1999; Petit et al., 1999[a]
Law Dome	66° 46′ S, 112° 48′ E	CO_2			275–284 (± 1.2) ppmv (1000–~1800) with lower levels during 1550–1800 AD[a], 279.2 ppmv (1006 AD)[b]	Pearman et al., 1986; Etheridge et al., 1988; Etheridge et al., 1996[a], Francey et al., 1999[b]; Joos et al., 1999; Gillett et al., 2000
		$\delta^{13}CO_2$			-6.44 ± 0.013‰ (1006 AD)	Francey et al., 1999
Taylor Dome	77° 48′ S, 158° 43′ E	CO_2	186–190 ppmv[a,d]	~275 ppmv ± 7.5 (avg Hol.)[b]; 268 ± 1 ppmv (10.5 ka); 260 ± 1 ppmv (8.2 ka); slow increase to 285 ppmv (1 ka)[c]		Smith et al., 1999[a]; Fischer et al., 1999[b]; Indermuhle et al., 1999[c]; Indermuhle et al., 2000[d]
		$\delta^{13}CO_2$	-7.5 to -7.0‰ (18–16.5 ka)[a]	-6.4 ± 1.6‰ (avg)[a], -6.6‰ (11 ka), -6.3‰ (8 ka), slow decrease to -6.1‰ (1 ka)[b] (uncertainty: ± 0.085‰)	-6.5‰[b]	Smith et al., 1999[a], Indermuhle et al., 1999[b]
Siple Dome	75° 55′ S, 83° 55′ W	CO_2			280 ± 5 ppmv (1750); 279 ± 3 ppmv (1734–1756 & 1754–1776)[a]; 276.8 ppmv (1744)[b]	Neftel et al., 1985[a]; Friedli et al., 1986[b]
		$\delta^{13}CO_2$			-6.48 ± 0.15‰ (1744); -6.41‰ (avg of 3 samples from before 1800	Friedli et al., 1986
Byrd	80° 01′ S, 119° 31′ W	CO_2	~200 ppmv[a]	unreliable Holocene values, no results from 8.7–2.2 ka (brittle zone)[b]; ~280 ppmv (with 40 ppmv fluctuations-early Hol)[a]		Berner et al., 1980; Neftel et al., 1988[a]; Stauffer et al., 1998; Indermuhle et al., 1999[b]
		$\delta^{13}CO_2$	-6.84 ± 0.12‰	early Holocene: -6.65 ± 13‰; avg: -6.78‰ (excluding samples from brittle zone)	-6.52 ± 0.12‰	Leuenberger et al., 1992

Continued

Table 2 Continued

Site	Location	Data type	LGM	Holocene	Pre-industrial Holocene	References
South Pole	90° S, 0°	CO_2			283 ± 5 ppmv (avg 430–770 yr BP)[b]; 278 ± 3 ppmv (1660–1880)[a]; 281 ± 3 ppmv (1450–1670)[a]; 279 ± 4 ppmv (950–1170)[a]	Neftel et al., 1985[a]; Friedli et al., 1986[b]; Siegenthaler et al., 1988
		$\delta^{13}CO_2$			−6.69 ± 0.8‰ (avg. 430–770 yr BP) (uncertainty: 0.22‰)[a]; −6.34 ± 0.3‰ (500–1000 yr BP) (uncertainty: 0.22‰)[b]	Friedli et al., 1984; Stauffer and Oeschger, 1985[b]; Siegenthaler et al., 1988
Dome C	75° 06′ S, 123° 24′ E	CO_2	188 ppmv[a]	265 ppmv (Early Holocene): variation during Hol: 260–280 ppmv[a]		Lorius et al., 1979; Delmas et al., 1980; Monnin et al., 2001[a]; Fluckiger et al., 2002
D47	67° 23′ S, 154° 03′ E	CO_2		240 ppmv (6.0 ka) to 270 ppmv (6.8 ka): slow increase to 280 ppmv	273.2 ppmv (914 AD)	Barnola et al., 1995
D57	68° 11′ S, 137° 33′ E	CO_2			284.8 ppmv (1310 AD)	Barnola et al., 1995; Raynaud and Barnola, 1985
GISP2	72° 36′ N, 38° 30′ W	CO_2	~195 ppmv[b]	275 ± 9 ppmv (3.1–1.9 ka)[b]	280 ± 5 ppmv[a]	Wahlen et al., 1991[a]; Smith et al., 1997[b]
GRIP	72° 35′ N, 37° 38′ W	CO_2	205 ± 19 ppmv[b]		280–287.4 ppmv (946 AD)[a]	Barnola et al., 1995[a]; Anklin et al., 1995; Anklin et al., 1997[b]
Camp Century	77° 12′ N, 61° 06′ W	CO_2	~200 ppmv			Berner et al., 1980

porous firn layer. Gravitational fractionation causes gases with higher molecular weights to be enriched at the depths of bubble enclosure. Although this process has a relatively minor effect (less than 1%) on concentrations of gases like CO_2 and CH_4 in air, it can significantly alter the more subtle concentration differences among species that differ only in their isotopic compositions (Craig *et al.*, 1988; Schwander, 1989). Similarly, thermal fractionation among different isotopic species may occur under conditions when a change in temperature at the ice surface is large and persistent enough to cause a temperature gradient in the firn (Severinghaus *et al.*, 1998). Additional fractionation may occur due to molecular size exclusion during bubble enclosure (Craig *et al.*, 1988). These effects must be taken into account in interpreting the stable isotopic compositions of ice-core gas samples.

Because of the possibility of postdepositional alteration, and the other problems relating to extraction, analysis, and dating described above, ice-core records are most compelling when they can be shown to agree across different locations, accumulation conditions, and analytical procedures. These conditions have been demonstrated for many ice-core gas records. For example, Table 2 shows the agreement among records of CO_2 and CH_4 concentrations during the LGM ($\sim$18–24 ka) and the Holocene Epoch. The agreement among some ice-core gas records is so substantial that more detailed time-dependent features have proven to be an important tool in the correlation of variations in other properties of the ice cores. Millennial variations in atmospheric methane concentrations, recorded in ice cores from both Antarctica and Greenland, provide a basis for detailed correlation between ice-core records from these locations (Steig *et al.*, 1998; Blunier *et al.*, 1998; Blunier *et al.*, 1999; Blunier and Brook, 2001). More gradual variations in the isotopic composition of atmospheric oxygen are likewise reflected in gases extracted from ice cores in both Greenland and Antarctica (Sowers and Bender, 1995). Because variations in the isotopic composition of atmospheric oxygen are controlled mainly by the isotopic composition of oxygen in seawater, the isotopic record of oxygen in ice-core air provides not only a tool for correlation among ice-core gas records, but also a robust means of correlation with the widely used marine oxygen isotope record measured in marine sediments (Shackleton, 2000).

10.3.2 Holocene Carbon-cycle Variations

Recent anthropogenic influences on greenhouse gases are clearly documented in ice-core records. Several ice cores from Antarctica provide strong evidence that the atmospheric concentration of CO_2 before the onset of human influence was

280 ± 5 ppmv, and ice cores from both Antarctica and Greenland document CH_4 levels of 650–730 ppbv before human influence (see Table 2 and references therein). Modern (2000 AD) atmospheric concentrations of CO_2 and CH_4 are near 370 ppmv and 1,800 ppmv, respectively (Blasing and Jones, 2002), reflecting the rapid increases due to anthropogenic sources during the last two centuries. The ice-core record also shows that the carbon isotope ratio of atmospheric CO_2, expressed as $\delta^{13}CO_2$, was $\sim$−6.4‰ before human influence (Friedli *et al.*, 1986; Francey *et al.*, 1999), compared to its present value near −7.9‰ (Francey *et al.*, 1999), reflecting the influence of ^{13}C-depleted anthropogenic CO_2 added to the atmosphere. ($\delta^{13}CO_2$ denotes the $^{13}CO_2 : {}^{12}CO_2$ ratio difference in per mil relative to the PDB standard. See Section 10.2.1.1.) The effects of human activities on atmospheric CO_2 and CH_4 are discussed in other chapters of this treatise (see Chapter 3). These recent trends provide an important test of the geologic ice-core record, because the ice-core gas measurements can be closely matched for recent decades with records from firn air and atmospheric measurements (Figures 4(a) and (b)). Measurements of carbon dioxide and its carbon isotopes in firn and in the uppermost sections of ice cores taken at the Siple Station and Law Dome sites in Antarctica show close agreement with the record of direct atmospheric measurements (Neftel *et al.*, 1985; Pearman *et al.*, 1986; Friedli *et al.*, 1986; Etheridge *et al.*, 1996; Francey *et al.*, 1999). Extrapolated trends in ice-core methane measurements at these sites (Stauffer *et al.*, 1985; Etheridge *et al.*, 1992) and at the Dye 3 site in Greenland (Craig and Chou, 1982) are likewise consistent with recent atmospheric measurements. These observations provide powerful support for the validity of ice-core gas records extending back in time.

Models of the modern global carbon cycle typically assume a pre-disturbance steady state in calculations of historical and future human influences on atmospheric greenhouse gases. The ice-core gas record provides a basis for testing this steady-state assumption. The record shows that atmospheric levels of both CO_2 and CH_4 varied during the most recent 10,000 years prior to human influence. The documented Holocene variations, while small and slow compared to the recent anthropogenic changes in atmospheric chemistry, provide important information about interactions among greenhouse gases, climate, and the global carbon cycle. Implications for the assumption of steady state in models are discussed below and in Section 10.2.2.

Subtle variations in atmospheric CO_2 during the most recent millennium, first hinted in some of the earliest Antarctic ice-core measurements (Raynaud and Barnola, 1985; Stauffer *et al.*, 1988; Etheridge *et al.*, 1988), have been

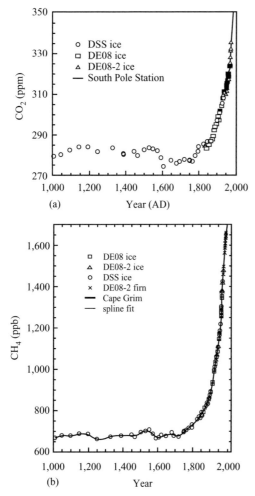

and Rasmussen, 1989). CH_4 measurements in ice from the Law Dome site in Antarctica show small variations that parallel the CO_2 trends over the most recent millennium, including a decrease of ~ 40 ppbv coinciding with the CO_2 decrease in the sixteenth and seventeenth centuries (Etheridge *et al.*, 1998). The decrease in atmospheric CO_2 and CH_4 during the sixteenth and seventeenth centuries coincides with a maximum in $\delta^{13}CO_2$ at $\sim -6.3‰$ (Francey *et al.*, 1999). Because variations in terrestrial sources and sinks are the most likely cause of the CH_4 and carbon isotope trends, they are probably associated with changes in terrestrial carbon cycling, although simultaneous oceanic changes cannot be ruled out (Trudinger *et al.*, 1999; Joos *et al.*, 1999).

Times of slightly increased and decreased CO_2 and CH_4 during the most recent millennium appear to coincide roughly with the times of slightly warmer and cooler climate known as the "Medieval Warm Period" and the "Little Ice Age." The extent and significance of these climatic variations are somewhat unclear (see, e.g., Bradley (1999) and references therein), and the implied changes in climate forcing by the greenhouse gases are very small (Etheridge *et al.*, 1998). Nevertheless, these coincidences of greenhouse gas and climate trends are consistent with an understanding of close coupling between climate and carbon cycling that emerges from observations over a very broad range of timescales.

The likelihood of a "Little Ice Age" carbon-cycle perturbation poses a particular concern for the assumption of pre-anthropogenic steady state in models of the modern carbon cycle. Part of the increase in concentrations of CO_2 and CH_4 during recent centuries may be due to trends associated with the natural climate anomaly. This association may seriously affect estimates of eighteenth- and early nineteenth-century human impacts on the terrestrial biosphere based on greenhouse gas trends. More importantly, it may imply the need for climate and carbon cycling to be coupled in models of the modern carbon cycle (Enting, 1992). The maximum natural rates of change in atmospheric CO_2 and CH_4 during the most recent millennium were at least an order of magnitude smaller than current anthropogenic rates of change, and the natural variations were much less persistent than the increasing trends of the most recent two centuries. Thus the primary concern raised by natural carbon-cycle variations during the most recent millennium is not that they may be significant relative to anthropogenic changes, but that they suggest a degree of global carbon/climate coupling that is not well represented in current models of present-day and potential future carbon cycling.

Coupling between variations in carbon cycling and climate is also a prominent theme

Figure 4 (a) Carbon dioxide concentrations over the last 1,000 yrs derived from three ice cores from Law Dome, Antarctica, and from air samples collected at South Pole Station since the late 1950s (Etheridge *et al.* (1996), reproduced by permission of the American Geophysical Union from *J. Geophys. Res.*, **1996**, *101*, 4115–4128 (Figure 4); the air sample data are from Keeling, 1991). (b) Methane concentrations over the last 1,000 yr derived from three Antarctic ice cores, the Antarctic firn, and archived air samples collected since 1978 at Cape Grim, Tasmania, Australia (Etheridge *et al.* (1998); reproduced by permission of the American Geophysical Union from *J. Geophys. Res.*, **1998**, *103*, 15979–15993 (figure 2); the air sample data are from Langenfelds *et al.*, 1996).

confirmed by more detailed and precise analyses (Figure 5). CO_2 concentrations appear to have risen to near 285 ppmv in the thirteenth century AD, then decreased to ~ 275 ppmv during the sixteenth century AD, and then increased again during the eighteenth century AD (Barnola *et al.*, 1995; Etheridge *et al.*, 1996; Indermuhle *et al.*, 1999). Some of the earliest ice-core CH_4 measurements also suggested a period of decreased concentrations during the sixteenth and seventeenth centuries (Rasmussen and Khalil, 1984; Khalil

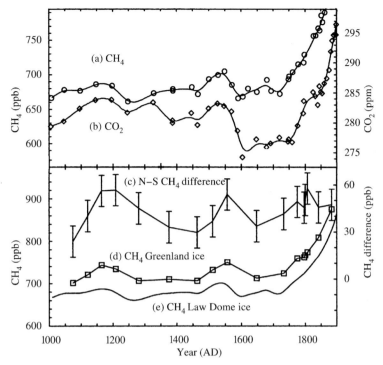

Figure 5 Variability of (a) methane (using the same dataset as that of Figure 4(b)) and (b) carbon dioxide (using the same dataset as that of Figure 4(a)) concentrations over the past 1,000 yr. (c) The interpolar difference in methane, which is the difference between the Greenland ice-core record (d) and the Law Dome methane record (e, same as curve a) for corresponding years. Estimated 1σ uncertainty for the interpolar difference is 10 ppb (Etheridge *et al.* (1998); reproduced by permission of the American Geophysical Union from *J. Geophys. Res.*, **1998**, *103*, 15979–15993 (figure 3)).

in efforts to understand the ice-core gas records extending back through the most recent 10,000 years. The records of CO_2, $\delta^{13}CO_2$, and CH_4 for this period show variations that are larger and more complex than those of the most recent millennium. Measurements from the Taylor Dome and Dome C sites provide evidence that atmospheric CO_2 concentrations decreased from ~270 ppmv at 10.5 ka to 260 ppmv at 8 ka and then gradually increased to values near 285 ppmv at 1 ka (Figure 6; (Indermuhle *et al.*, 1999; Fluckiger *et al.*, 2002)). The Holocene $\delta^{13}CO_2$ record from Taylor Dome is much less detailed and precise, and cannot be confirmed by overlap with the $\delta^{13}CO_2$ record of the most recent millennium (Francey *et al.*, 1999), but it appears to show a maximum near −6.3‰ between 7 ka and 8 ka and a minimum near −6.6‰ between 2 ka and 3 ka (Indermuhle *et al.*, 1999). Measurements in ice cores from both Greenland and Antarctica show that atmospheric methane concentrations also varied significantly during the Holocene Epoch (Figure 6). Concentrations decreased from levels near 700 ppbv at 10 ka to levels less than 600 ppbv at 5 ka, followed by a gradual return to values near 700 ppbv at 1 ka (Blunier *et al.*, 1995; Chappellaz *et al.*, 1997; Fluckiger *et al.*, 2002).

Carbon-cycle responses to Holocene climate change have been hypothesized to explain the atmospheric CO_2 and CH_4 trends observed in the ice-core record of this period. The early Holocene expansion of terrestrial vegetation and soils in areas that were previously glaciated has been suggested as the cause of the early Holocene decrease in CO_2 (Indermuhle *et al.*, 1999). Likewise, the increase in atmospheric CO_2 between 8 ka and 1 ka has been attributed to a release of biospheric carbon caused by a global trend during that period toward cooler and drier conditions (Indermuhle *et al.*, 1999). In contrast, one explanation for the Late Holocene increase in atmospheric CH_4 is an expansion of boreal wetland source areas (Blunier *et al.*, 1995; Chappellaz *et al.*, 1997; Velichko *et al.*, 1998). (Some investigators have suggested that the Late Holocene increase in atmospheric CH_4 may be partly due to emissions from the onset of early human rice cultivation (Subak, 1994; Chappellaz *et al.*, 1997; Ruddiman and Thomson, 2001)). Oceanic responses to climate change have also been suggested to explain the Holocene CO_2 and $\delta^{13}CO_2$ trends (Indermuhle *et al.*, 1999; Broecker *et al.*, 2001). This variety of hypotheses about Holocene carbon cycling is not surprising, because the Holocene variations in CO_2, $\delta^{13}CO_2$, and CH_4 are clearly

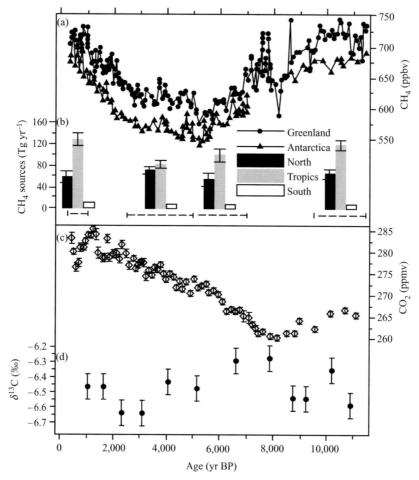

Figure 6   The Holocene ice-core record of the variability of methane, carbon dioxide, and $\delta^{13}CO_2$. (a) The methane record from ice cores in Greenland and Antarctica, showing significant variability including variability in the interpolar gradient (Chappellaz *et al.*, 1997). (b) The results of a three-box model used to infer the latitudinal distribution sources of methane at four different time spans during the Holocene Epoch (Chappellaz *et al.*, 1997). (c, d) The carbon dioxide concentration and its isotopic composition from Taylor Dome, Antarctica (Indermuhle *et al.*, 1999). Note the dissimilarities among the trends in this figure (Raynaud *et al.* (2000); reproduced by permission of Elsevier from *Quat. Sci. Rev.*, **2000**, *19*, 9–17 (figure 1)).

dissimilar (Figure 6). Moreover, as discussed in Section 10.2.1.2, significant changes in atmospheric CH_4 can be caused by changes in specific sources and sinks that do not necessarily require the larger translocations of carbon needed to change atmospheric CO_2 and $\delta^{13}CO_2$. It seems likely that a variety of processes, reflecting the regional and global complexity of both the carbon cycle and the climate system, contributed to the trends observed in the ice-core record of Holocene CO_2 and CH_4.

Important clues to these processes come from analysis of details available in the Holocene ice-core CH_4 record. The nature of CH_4 cycling through the atmosphere (see Section 10.2.1.2) assures that any significant change in CH_4 sources will quickly affect atmospheric CH_4 concentrations. Although the atmospheric lifetime of CO_2 is shorter than that of CH_4, a change in CO_2 sources must be

much larger (in terms of carbon transfer) to affect the larger mass of atmospheric CO_2, and the response of atmospheric CO_2 will be attenuated and prolonged by its exchange with the ocean surface and the terrestrial biosphere (see Section 10.2.1.1). Thus, atmospheric CH_4 concentrations are more susceptible than CO_2 concentrations to rapid variations resulting from changes in sources. Abrupt changes and "spikes" are more common in the ice-core CH_4 record than in the ice-core CO_2 record. For example, the Greenland GRIP and GISP2 ice cores (see references in Table 2) record a sharp CH_4 decrease at 8.2 ka (see Figure 6). This feature corresponds to a widespread climatic event linked to outburst flooding from the melting of the Laurentide ice sheet (Barber *et al.*, 1999). The ice-core record of abrupt changes in atmospheric CH_4 is an important source of information about links between climate and carbon-cycle changes.

Whereas the oxidation of CH_4 occurs in the atmosphere worldwide (see Section 10.2.1.2), the sources of CH_4 are not distributed uniformly. The dominant natural sources of CH_4 are wetlands, which occur today mainly in the tropics and the northern hemisphere. The CH_4 produced by northern wetland sources is oxidized during its transit southward in the atmosphere, resulting in a north-to-south decreasing atmospheric CH_4 concentration gradient, well documented in ice-core CH_4 measurements of air from Greenland and Antarctica (Rasmussen and Khalil, 1984; Nakazawa *et al.*, 1993). The gradient appears to have varied during the most recent millennium, with atmospheric CH_4 concentrations over Greenland exceeding those over Antarctica by 24–58 ± 10 ppbv (Etheridge *et al.*, 1998) (Figure 5). During the most recent 10,000 years, the difference between Greenland and Antarctica CH_4 concentrations ranged from 33 ± 7 ppbv during the middle Holocene Epoch (~7–5 ka) to values as high as 50 ± 3 ppbv during the Late Holocene Epoch (~5–2.5 ka) and 44 ± 4 ppbv during the Early Holocene Epoch (11.5–9.5 ka) (Chappellaz *et al.*, 1997; Figure 6). The ice core record of the atmospheric CH_4 gradient implies that northern wetlands (as well as tropical wetlands) have been an important CH_4 source throughout the Epoch. The variations in the interhemispheric CH_4 gradient, together with the trends in CH_4 concentrations described above and shown in Figure 6, have been used to infer changes in the magnitude and relative importance of tropical versus northern wetland CH_4 sources (Chappellaz *et al.*, 1997; Etheridge *et al.*, 1998). Recent anthropogenic sources of CH_4, which occur predominantly in the northern hemisphere, have caused the modern north-to-south CH_4 concentration gradient to be greater than the Holocene gradient by a factor of about three (Chappellaz *et al.*, 1997). Anthropogenic sources of CO_2 have similarly caused a present-day north-to-south atmospheric CO_2 gradient of several ppmv (Pearman *et al.*, 1983; Fung *et al.*, 1983; Enting and Mansbridge, 1991) (see Chapter 3). A small south-to-north decreasing CO_2 gradient has been hypothesized for the Late Holocene atmosphere prior to human influence (Keeling *et al.*, 1989; Taylor and Orr, 2000), but the proposed gradient is difficult to verify in ice-core records because it requires correction for postdepositional artifacts in Greenland ice, and its magnitude is close to the measurement precision for CO_2 in ice-core samples (but see Barnola (1999)).

10.3.3 Glacial/interglacial Carbon-cycle Variations

The ice coring effort at Vostok Station in East Antarctica required several decades and a remarkable combination of international (Russian, French, and American) technological and analytical expertise. The resulting record (Figure 7; Jouzel *et al.*, 1987; Barnola *et al.*, 1987; Chappellaz *et al.*, 1990; Jouzel *et al.*, 1993; Petit *et al.*, 1999) extends continuously to depths exceeding 3 km and provides detailed evidence for the close relationship between climate and carbon cycling throughout the four most recent glacial/interglacial cycles. The Vostok ice-core data reveal the range of natural variations in atmospheric CO_2 and CH_4 during this period, and demonstrate that present-day levels of CO_2 and CH_4 are much higher than they have been at any time during the most recent 420,000 years (Petit *et al.*, 1999). The Vostok record is a preeminent constraint on analyses and hypotheses concerning variations in climate and the carbon cycle over glacial/interglacial timescales.

The most conspicuous feature of the Vostok data (Figure 7) is the close similarity among trends in CO_2, CH_4, and local temperature (calculated from the ratio of deuterium to hydrogen in the ice). These trends all show a pattern of relatively rapid onsets of warm (interglacial) conditions, followed by more gradual transitions to cool (glacial) conditions, repeated in cycles ~100 kyr long. This "sawtoothed" pattern resembles characteristics long observed in paleoclimate records from marine sediments worldwide (see, e.g., Emiliani, 1966). The low CO_2 and CH_4 concentrations observed in ice that formed during glacial periods confirmed earlier measurements in ice from the most recent glacial period collected at other sites in both Greenland and Antarctica (Berner *et al.*, 1980; Delmas *et al.*, 1980; Stauffer *et al.*, 1988). At times of minimum glacial temperatures, CO_2 concentrations dropped to values near 180 ppmv, and then rapidly rose to values of 280–300 ppmv during interglacials. Similarly, glacial periods coincided with minimum CH_4 concentrations of 320–350 ppbv, and CH_4 concentrations rose to maximum values of 650–770 ppbv during interglacials. The Vostok record revealed that maximum and minimum temperature, CO_2, and CH_4 levels were similar during each of the four most recent glacial cycles, including the Holocene Epoch prior to human influence. Other detailed features of the CO_2, CH_4, and isotopic temperature variations appear to be related: statistical correlation of the Vostok data yields overall r^2 values exceeding 0.7 for both CO_2 and CH_4 with respect to the ice isotopic composition, which is a proxy for local temperatures (Petit *et al.*, 1999).

The correlation among these records is so high that local or regional causes—rather than global conditions—might be considered to explain such a close correspondence. However, there is excellent agreement among glacial/interglacial ice-core CH_4 records from Antarctica and Greenland (Brook *et al.*, 1996; Brook *et al.*, 2000), and

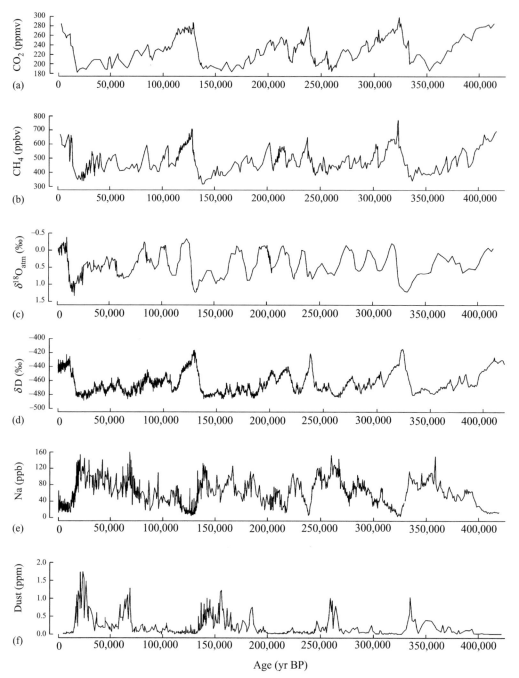

Figure 7 The Vostok ice-core record (Petit *et al.*, 1999). (a)–(c) Carbon dioxide, methane, and oxygen isotope ratios (expressed as $\delta^{18}O$) of oxygen in air extracted from the ice. (d) Hydrogen isotope ratios (expressed as δ^2H, or δD) in the ice. This record is a proxy for local temperature. (e) Sodium in the ice, a measure of sea salt aerosol deposition. (f) Dust in the ice, a measure of continental aerosol deposition. (a)–(c) are plotted against the estimated age of the ice air; (d)–(f) are plotted against the estimated age of the ice.

among glacial/interglacial CO_2 records from diverse locations in Antarctica (see, e.g., Fischer *et al.*, 1999). Unfortunately, due to higher levels of reactive impurities in Greenland ice (see Section 10.3.1), there is no ice-core CO_2 record from Greenland that can be compared in detail to the Antarctic CO_2 measurements. But the Antarctic CO_2 measurements are much less susceptible to

effects of impurities (Anklin *et al.*, 1997). The agreement among CH_4 measurements in the same ice cores suggests strongly that the Antarctic ice-core CO_2 record, like the CH_4 record, can be considered a record of global glacial/interglacial changes in atmospheric composition. Likewise, the Vostok deuterium temperature record and other Antarctic and Greenland records of stable isotopes in ice can be linked to well-known global climate variations by direct matching of obvious similarities to marine sediment isotope records (Shackleton *et al.*, 2000; Shackleton, 2001) and by correlations between oxygen isotope variations in the ice-core air and in marine sediments (Bender *et al.*, 1999; Shackleton, 2000).

The Southern Ocean might be an important regional influence on both temperature and CO_2 in Antarctica. But if this influence contributes to the close correlation of temperature and CO_2 in the Vostok record, then the Southern Ocean is likely part of a complex interaction among processes at high southern latitudes that plays an important role in both global climate and the global carbon cycle (Petit *et al.*, 1999). There is no known local or regional CH_4 source near Antarctica; thus, the close correlation of temperature and CH_4 at Vostok seems to imply an even broader scope of interconnected global processes. The Vostok ice-core record provides compelling evidence that climate and carbon-cycle variations were closely interactive throughout the Late Pleistocene period.

This understanding of Pleistocene climate/carbon coupling is illustrated by the fact that many paleoclimatologists conclude that the variable greenhouse effect of atmospheric CO_2 and CH_4 has contributed significantly to glacial/interglacial climate change, while many geochemists consider climate change to have been an essential driver for glacial/interglacial variations in the carbon-cycle processes that control atmospheric CO_2 and CH_4.

10.3.3.1 Carbon-cycle influences on glacial/ interglacial climate

Paleoclimatologists have analyzed the timing of Pleistocene climate cycles and confirmed statistical similarities to the periodicities of cyclic changes in orbital parameters that control the seasonal and latitudinal distribution of solar radiation (insolation) reaching the Earth's surface (Broecker and van Donk, 1970; Shackleton and Opdyke, 1976; Imbrie *et al.*, 1992; Imbrie *et al.*, 1993). The influence of these orbital parameters on global climate was hypothesized and calculated by Milutin Milankovitch (1879–1958), for whom the identified periodicities have come to be called "Milankovitch cycles." Although these periodicities (predominantly at 100,000 yr, 41,000 yr, 23,000 yr, and 19,000 yr) have been identified in many paleoclimate records, there is still no consensus concerning the exact mechanisms by which variations in the Earth's orbital configuration affect climate. Mathematical models and other analyses of global climate suggest strongly that the insolation changes directly attributed to orbital causes are not sufficient to explain the magnitude and the relative importance of the periodicities observed in differences between glacial and interglacial climates (Berger, 1979, 1988; Rind *et al.*, 1989; Imbrie *et al.*, 1993). Thus, the study of Pleistocene climate requires a focus on the processes that might have been capable of amplifying the subtle effects of orbital changes.

Paleoclimatologists have identified several possible amplifying feedback in the differences observed between glacial and interglacial climatic conditions. One important feedback is the "ice–albedo" feedback, which is caused by changes in the reflection of solar radiation back into space by expanding and retreating glaciers. The changing greenhouse effect is another likely amplifying feedback, through variations in the retention of heat in the atmosphere caused by rising and falling concentrations of greenhouse gases, including CO_2 and CH_4 (Broecker, 1982; Manabe and Broccoli, 1985).

No single feedback mechanism appears to be sufficient to explain the full magnitude or character of the climatic amplification between Late Pleistocene glacial and interglacial conditions. Variations in heat absorption due to greenhouse gases might have accounted for about half of the total glacial–interglacial difference in radiative forcing (Lorius *et al.*, 1990). The glacial–interglacial difference in the greenhouse effect is attributed primarily to CO_2, which contributed ~5 times as much as CH_4 and N_2O combined to the change in radiative forcing (Petit *et al.*, 1999). Thus, the variable greenhouse effect is viewed as an important contributing feedback mechanism in the glacial–interglacial climate system, but the influence of greenhouse gases is viewed as operating in a very complex association with other important feedback (Hewitt and Mitchell, 1997; Felzer *et al.*, 1998; Berger *et al.*, 1998; Petit *et al.*, 1999; Yoshimori *et al.*, 2001).

In addition to contributing to the global temperature differences between glacial and interglacial conditions, the greenhouse gas feedback might interact in complex ways with other aspects of Pleistocene climate variability. For example, one of the major mysteries of Pleistocene climate is the unexplained appearance, beginning ~700 ka, of the pronounced 100 kyr cycles that dominate paleoclimate records such as the Vostok data (Shackleton and Opdyke, 1976). Because the magnitude of the orbital insolation effects with 100 kyr periodicity is relatively weak in comparison to the effects with other periodicities, the 100 kyr cycles would be expected to be less conspicuous than those at the other orbital periodicities (as evident in paleoclimate records prior to 700 ka). Long-term feedback processes have been suggested as a way to explain the enhancement of the longer cycles. Several studies have suggested that the variable greenhouse effect of CO_2 or CH_4 might provide the needed long-term feedback because of the long response times associated with some components of the carbon cycle (see Section 10.2) (Pisias, 1984; Saltzman, 1987; Shackleton, 2000; Kennett *et al.*, 2003). Alternatively, others have suggested that the changing character of Late Pleistocene climate variability, including the unexplained appearance of pronounced 100 kyr periodicity, might be a response of shifting climate thresholds associated with a long-term decrease in atmospheric CO_2 since the Pliocene Epoch (Berger *et al.*, 1999).

10.3.3.2   *Climate influences on glacial/interglacial carbon cycling*

Geochemists have intensely debated the processes that caused atmospheric CO_2 and CH_4 concentrations to change between glacial and interglacial periods. The broad scope of this debate reflects the many ways that climate change can influence the global carbon cycle.

The influence of glacial/interglacial climate on the terrestrial carbon cycle is seen most dramatically in the ice-core CH_4 record. Because rates of atmospheric CH_4 oxidation during glacial periods were probably not drastically different from those during interglacials (see Section 10.2.1.2), differences between glacial and interglacial atmospheric CH_4 levels are attributed primarily to variations in the principal natural CH_4 source, wetlands. Wetlands occur primarily in tropical regions, where they are sensitive to variations in monsoonal rainfall, and in northern midlatitude and boreal regions, where they are sensitive to the changing extent of glacial ice and associated conditions of temperature and water balance (Matthews and Fung, 1987; Bubier and Moore, 1994). It is very likely that the distribution and extent of these wetland CH_4 sources were affected by differences between glacial and interglacial climate (Prell and Kutzbach, 1987; Khalil and Rasmussen, 1989; Chappellaz *et al.*, 1990; Petit-Maire *et al.*, 1991; Crowley, 1991; Chappellaz *et al.*, 1993a,b; Velichko *et al.*, 1998). The ice-core record of atmospheric CH_4 can be used to infer changes not only in the magnitude of wetland CH_4 sources, but also in their geographic distribution (see Section 10.3.2). The CH_4 concentration gradient between Greenland and Antarctica during the most recent glacial period ranged from ~40 ppbv during relatively warm periods to near zero during the coldest part of the last glacial period (Brook *et al.*, 2000; Dallenbach *et al.*, 2000). Thus, northern CH_4 sources appear to have been significant during all but the coldest glacial climates, and northern wetlands may have persisted throughout most of the last glacial period.

Climatic influences are apparent not only in the record of differences between glacial and interglacial atmospheric CH_4 concentrations and budgets, but also in more frequent variations that occurred over millennial timescales extending throughout the most recent glacial period (Figure 8). These variations correlate very closely with the abrupt interstadial warming events

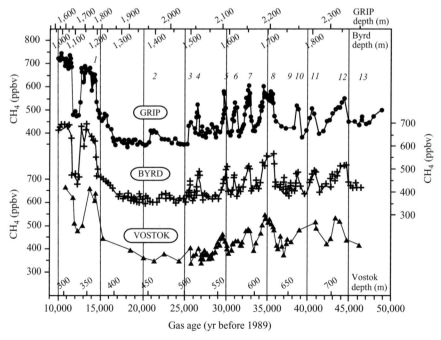

Figure 8   Methane variations during the most recent glacial period as recorded in ice cores from Greenland (GRIP) and Antarctica (Byrd, Vostok) (Blunier *et al.*, 1998; Stauffer *et al.*, 1998). The records were correlated by statistical matching of the variations shown (Blunier *et al.* (1999); reproduced by permission of Kluwer Academic/Plenum Publishers from *Reconstructing Ocean History: A Window into the Future*, **1999**, pp. 121–138 (figure 1)).

known as Dansgaard–Oeschger events, documented in the stable isotope record in Greenland ice cores (Oeschger *et al.*, 1984; Dansgaard *et al.*, 1984; Chappellaz *et al.*, 1993a; Brook *et al.*, 1996). The millennial-scale CH_4 variations, which range in magnitude from ~50 ppbv to ~300 ppbv, can be readily correlated in ice cores from both Greenland and Antarctica. In addition to providing a global record of carbon-cycle response to climate change, they have been used as a correlation tool to demonstrate synchronies and asynchronies between the Dansgaard–Oeschger events and corresponding variations in Antarctic climate (Steig *et al.*, 1998; Blunier *et al.*, 1998; Blunier and Brook, 2001).

The correlation of millennial climate fluctuations in Greenland and Antarctica is very important in efforts to understand the source of atmospheric CO_2 variations over millennial timescales during the most recent glacial period (Figure 9). These variations, on the order of 20 ppmv, are not as conspicuous as the millennial-scale CH_4 variations, and they appear to have occurred only in association with the largest and longest Dansgaard–Oeschger events (Stauffer *et al.*, 1998). Larger glacial CO_2 variations, in excess of 50 ppmv, were observed to correlate with Dansgaard–Oeschger events in ice cores from Greenland (Stauffer *et al.*, 1984). However, these features are now definitively interpreted as artifacts of postdepositional chemical reactions within the ice (Delmas, 1993; Tschumi and Stauffer, 2000). The smaller glacial CO_2 variations seen in Antarctic ice cores do not appear to be affected by postdepositional reactions (Stauffer *et al.*, 1998). Detailed correlation, based on the more clearly

defined millennial-scale CH_4 variations, suggests that the glacial variations in atmospheric CO_2 more closely paralleled the less abrupt and out-of-phase temperature variations recorded in Antarctic ice cores (Blunier *et al.*, 1999; Indermuhle *et al.*, 2000). Given these subtle differences in sensitivity and phasing, atmospheric CO_2 and CH_4 appear to have been associated with different components of glacial climate variability at millennial timescales. Similar differences are observed in the record of CO_2 and CH_4 variations during the Holocene Epoch (see Section 10.3.2).

These differences do not obscure the obvious similarities between the larger CO_2 and CH_4 trends that characterize the longer glacial/interglacial cycles (Figure 7). Because the differences between glacial and interglacial CH_4 levels are attributed to changes in terrestrial sources, it seems likely that the parallel changes in atmospheric CO_2 might also reflect changes in the terrestrial biosphere. Unfortunately, the changes that most likely controlled glacial/interglacial CH_4 sources would be expected to affect CO_2 sources in the opposite manner. Wetlands—particularly peatlands—are known to store far more carbon per unit area than the soils that form under drier climatic conditions (Schlesinger, 1997), and there is some evidence of a direct relationship between the intensity of CH_4 emissions and the rate of carbon production and storage among individual wetland ecosystems (Whiting and Chanton, 1993). Thus, an expansion of wetlands and their CH_4 emissions during interglacial times would be associated with a decrease in soil CO_2 emissions. Indeed, as we have seen in Section 10.2.1.1, most estimates of climatic effects on terrestrial carbon storage (by wetlands and other ecosystems) suggest that the global land surface stored significantly less carbon during the most recent glacial period than during the Holocene Epoch. While these terrestrial trends might help to explain glacial/interglacial differences in atmospheric CH_4, they require that we look elsewhere in order to understand the glacial/interglacial CO_2 budget.

As discussed in Section 10.2.1.1, the deep ocean is the largest and most probable reservoir capable of large transfers of carbon over the timescales of glacial/interglacial transitions. All of the processes that account for deep-ocean carbon storage are susceptible to effects of climate change. These processes can be summarized in three general categories that can be described as the "solubility pump," the "soft-tissue pump," and the "carbonate pump" (Volk and Hoffert, 1985).

The oceanic solubility pump stores CO_2 in the deep ocean by simple dissolution of atmosphere CO_2. As deep-ocean water forms by sinking from the surface, it carries with it the CO_2 dissolved by gas exchange with the atmosphere. Because deep

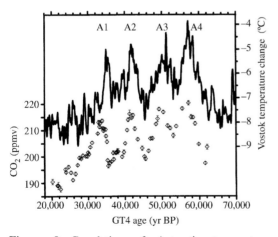

Figure 9 Correlation of Antarctic temperature variations (solid line, calculated from hydrogen isotope ratios in the ice at Vostok) and CO_2 variations in air extracted from the Taylor Dome ice core. The temperature curve is a running mean of the data from Petit *et al.* (1999) (Indermuhle *et al.* (2000); reproduced by permission of the American Geophysical Union from *Geophys. Res. Lett.*, **2000**, *27*, 735–738 (figure 2)).

water forms where the ocean surface is coldest, and because the solubility of CO_2 in seawater is highest at low temperatures, the amount of CO_2 dissolved in sinking deep water tends to be greater than the average amount of CO_2 dissolved in global ocean surface waters. The solubility pump is sensitive to the conditions that affect local gas exchange at the sites of deep-water formation. Climate affects all of these conditions, including temperature, salinity, wind, sea state, sea-ice cover, planktonic metabolism, and prevailing currents and vertical mixing patterns. The effects of climate change on the solubility pump have long comprised a central topic in studies of glacial/ interglacial CO_2 change (Newell *et al.*, 1978), and the importance of these effects continues to be a topic of intense debate (see, e.g., Bacastow (1996) and Toggweiler *et al.* (2003)).

The oceanic soft-tissue pump stores carbon in the deep ocean by the sinking of organic debris produced by microorganisms at the ocean surface. Much of the sinking organic material is oxidized in the deep ocean, yielding CO_2 and smaller quantities of nitrate, phosphate, and other nutrients. Like the solubility pump, the soft-tissue pump is sensitive to climate change. The assimilation of organic matter by photosynthesis at the ocean surface depends on temperature, salinity, and the availability of light and nutrients. Nitrate and phosphate are supplied to the ocean surface primarily by the upward mixing of deeper waters, where their concentration ratio is nearly the same as in the sinking organic material from which they are derived. Thus, the soft-tissue pump is sensitive not only to direct effects of climate such as light and temperature at the ocean surface, but also to more complex effects such as the impact of climate on ocean mixing and the cycling of nutrients. Changes in oceanic phosphate levels may have occurred as a result of the glacial/interglacial exposure and flooding of continental shelves (and their "reactive sediment" reservoir; see Figure 1) during changes in sea level (Broecker, 1982). Changes in nitrate concentrations may have resulted from glacial/interglacial variations in ocean mixing and other conditions that affect rates of oceanic nitrogen fixation and denitrification (McElroy, 1983; Altabet *et al.*, 1999). Changes in the availability of iron and other trace nutrients may have been caused by glacial/ interglacial alterations in the windborne transport of continental dust to ocean areas (Martin *et al.*, 1990; Martin, 1990). The oceanic soft-tissue pump encompasses a wide array of potential climate sensitivities because it links deep-ocean carbon storage to the oceanic cycles of nitrogen, phosphorus, and other nutrients.

The oceanic carbonate pump removes carbon from the ocean surface by the production of calcium carbonate ($CaCO_3$) in the shells, tests, and skeletal frameworks of various marine organisms. Some of this $CaCO_3$ settles as debris into the deep ocean, where much is dissolved and added to the deep-ocean DIC reservoir. The $CaCO_3$ that does not dissolve is buried in sediments, where it may persist to comprise a major component of the carbon cycle over timescales of millions of years. The carbonate pump differs from the solubility and soft-tissue pumps in three fundamental ways. First, unlike the solubility and soft-tissue pumps, the carbonate pump transfers significant quantities of alkalinity as well as carbon from the ocean surface to the deep sea. The result of this "alkalinity pump" is to partially offset the effect of deep-ocean carbon storage on atmospheric CO_2. This offset results from the effect of alkalinity removal on the chemical equilibrium between dissolved CO_2 and the other dissolved inorganic carbon species in ocean surface waters. For every mole of $CaCO_3$ removed from the ocean surface, ~0.6 mol of CO_2 are released to the atmosphere (Sundquist, 1993). Because many of the organisms that produce $CaCO_3$ in the ocean surface are restricted to warmer waters and/or shallow shelf substrates, the alkalinity pump is sensitive to climate and sea-level change.

Second, unlike the solubility and soft-tissue pumps, the carbonate pump supplies a large reservoir of "reactive sediments" (Figure 1) that almost certainly affected glacial/interglacial ocean carbon storage by contributing large net removals or additions through changes in $CaCO_3$ production and dissolution. The glacial/interglacial exposure and flooding of warm shallow-water shelf areas probably affected rates of $CaCO_3$ production and dissolution in coral reefs and other carbonate-rich sedimentary environments (Berger, 1982; Berger and Keir, 1984). Glacial/interglacial fluctuations in the extent of deep-sea $CaCO_3$ dissolution are one of the most widespread and conspicuous features in the Quaternary marine sediment record (Broecker, 1971; Peterson and Prell, 1985; Farrell and Prell, 1989). The influence of these fluctuations on atmospheric CO_2 probably occurred primarily through their effect on oceanic alkalinity. For example, a decrease in deep-sea $CaCO_3$ dissolution tends to decrease deep-ocean alkalinity as well as DIC. When these changes are mixed upward to the ocean surface, atmospheric CO_2 will increase in a manner identical to the chemical equilibrium effect of removing $CaCO_3$ from the ocean surface. Conversely, an increase in deep-sea dissolution will tend to increase deep-ocean alkalinity and DIC, and every mole of dissolved $CaCO_3$ that consequently reaches the ocean surface will cause ~0.6 mol of CO_2 to be absorbed from the atmosphere (Sundquist, 1993).

The third characteristic that distinguishes the carbonate pump is the fact that it is metered by the chemical equilibrium relationship between seawater and the carbonate minerals (primarily calcite) in the reactive sediment pool. Changes in the DIC and/or alkalinity of deep-ocean waters may cause changes in carbonate dissolution through their effect on the state of saturation of the water with respect to the carbonate minerals calcite and aragonite. Although carbonate dissolution is also affected by the diagenesis of organic matter in sediment pore waters (Archer and Maier-Reimer, 1994), chemical equilibrium appears to be the primary control on the distribution of carbonate dissolution in the deep sea (Broecker and Takahashi, 1978; Plummer and Sundquist, 1982). The response time of this process is on the order of 5–10 kyr (Sundquist, 1990a; Archer *et al.*, 1997), providing a mechanism for maintaining the overall balance between the oceanic supply and removals of both alkalinity and DIC over longer timescales (see Section 10.4.1.1). Thus, the carbonate pump is an important mechanism not only in the glacial/interglacial storage of oceanic carbon, but also in the carbon cycle over much longer timescales (Garrels *et al.*, 1976; Sundquist, 1991; Archer *et al.*, 1997).

The oceanic carbon pumps (solubility, soft-tissue, and carbonate) are inherently linked. The amount of CO_2 dissolved in ocean surface waters is controlled not only by its equilibrium solubility, but also by the relative rates of gas exchange, ocean mixing, and soft-tissue and carbonate production. Nutrients that are necessary for soft-tissue assimilation arrive at the ocean surface with excess dissolved CO_2 from deeper waters, and these nutrients nourish carbonate as well as soft-tissue production. Carbonate production may be affected by dissolved CO_2 concentrations (Riebesell *et al.*, 2000) and by competition from production by non-carbonate-producing organisms (Nozaki and Yamamoto, 2001). The "rain ratio" of $CaCO_3$ to organic carbon debris reaching the sea floor is a very important factor controlling the magnitude and distribution of deep-sea $CaCO_3$ dissolution (Archer, 1991; Archer and Maier-Reimer, 1994). Any change in deep-ocean carbon storage will be accompanied by a $CaCO_3$ dissolution response within 5–10 kyr (see above). Efforts to explain the role of the oceans in controlling glacial/interglacial CO_2 levels must address all of these interconnections in a manner consistent with the evidence available in ice and marine sediment cores. Progress in these important endeavors is detailed elsewhere in this treatise (see Chapters 13 and 14).

10.3.3.3 *Carbon/climate interactions at glacial terminations*

Given the many ways in which glacial/interglacial climate change have probably affected the carbon cycle, and given the likely importance of changes in atmospheric CO_2 and CH_4 as amplifiers of climate change, it is very difficult to separate cause from effect in the evolution of Quaternary climate and carbon cycling. Some of the most important evidence for mechanisms of change comes from detailed analysis of the sequence of events that occurred during the terminations of glacial periods. These events are the most rapid and dramatic variations in the pattern that characterizes Late Quaternary climate and carbon-cycle records. They are not only one of the most important aspects of glacial/interglacial change, but also among the best opportunities for detailed global correlation using the methods described in Section 10.3.1.

Analysis of glacial terminations in ice-core data has suggested that the onset of Antarctic warming may have preceded the onset of rising CO_2 levels by a few hundred years (Fischer *et al.*, 1999). However, there is a general consensus that this difference is too small to be distinguished from errors due to sampling and dating the ice cores (see Section 10.3.1), and that the initial increases in CO_2, CH_4 and Antarctic temperatures occurred simultaneously within the available dating resolution (Petit *et al.*, 1999; Monnin *et al.*, 2001). More significant timing differences have provided a basis for describing a general sequence of events that characterize at least the two most recent termination episodes (Broecker and Henderson, 1998; Petit *et al.*, 1999). First, the initiation of the termination is characterized by a dramatic decrease in the delivery of continental dust to Antarctica and nearby oceans. Second, atmospheric CO_2 and CH_4 rise in synchrony with Antarctic temperatures. Finally, several thousand years later, the rising temperatures and greenhouse gas concentrations are joined by the oxygen isotope trends in air and seawater that indicate melting of the continental ice sheets primarily in the northern hemisphere.

These events occurred within the context of very complex and still controversial relationships between northern and southern hemisphere climate trends and the influence of Earth's variable orbital configuration (see, e.g., Steig *et al.*, 1998; Bender *et al.*, 1999; Henderson and Slowey, 2000; Alley *et al.*, 2002). Asynchronies between northern and southern climate are particularly (perhaps uniquely) apparent in the record of the most recent glacial termination. As shown in Figure 10, these asynchronies affected atmospheric CO_2 and CH_4 (Monnin *et al.*, 2001). The Antarctic warming trend was interrupted by the Antarctic Cold Reversal, which appears to have coincided with a temporary 2 kyr cessation in the rise of atmospheric CO_2 concentrations. Atmospheric CH_4 levels, on the other hand, increased dramatically during this period, then fell precipitously to near-glacial values for ~1.5 kyr, and then rose to

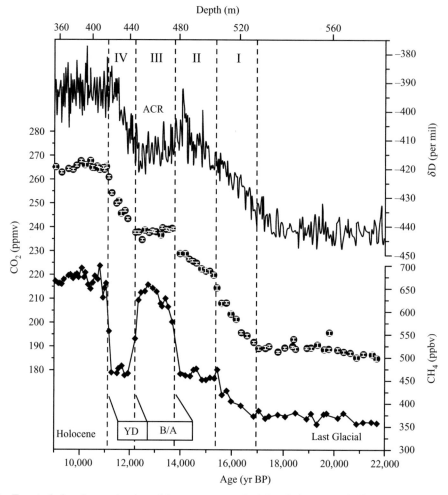

Figure 10 Events during the termination of the most recent glacial period, as recorded in the Dome C ice core. The top curve (δD) is a proxy for local temperatures, showing the Antarctic Cold Reversal (ACR). The solid circles represent CO_2 (means and error bars of six samples each). The diamonds represent CH_4 (uncertainty is 10 ppb). The depth scale at the top refers only to the CO_2 and CH_4 records. Bars indicate the Younger Dryas (YD) and Bolling/ Allerod (B/A) events, which are recorded in the isotope ratios of Greenland ice cores (Monnin *et al.* (2001); reproduced by permission of the American Association for the Advancement of Science from *Science*, **2001**, *291*, 112–114 (figure 1)).

Holocene values just as suddenly. These CH_4 trends, observed in ice-core records from both Antarctica and Greenland, correlate closely with the warm Bølling/Allerød event followed by the cold Younger Dryas event in the Greenland ice-core isotope record (Blunier *et al.*, 1997). Thus, the CH_4 record during the most recent glacial termination seems to reflect a strong influence by climate variations that are most conspicuous in the northern hemisphere (Dallenbach *et al.*, 2000; Brook *et al.*, 2000). The CO_2 record of this period more closely resembles the record of Antarctic climate, supporting suggestions that the Southern Ocean played a key role in the oceanic variations needed to account for changes in atmospheric CO_2 and terrestrial carbon storage (Francois *et al.*, 1997; Blunier *et al.*, 1997; Broecker and Henderson, 1998; Toggweiler, 1999; Petit *et al.*,

1999; Stephens and Keeling, 2000). Although more specific mechanisms continue to elude explanation, the importance of CO_2 and CH_4 as climate amplifiers seems to be reinforced by the observation that the melting of continental ice sheets—and hence the ice–albedo amplifying feedback—was not widespread until relatively late in the sequence of glacial termination events.

One of the most remarkable features of the carbon cycle during the Quaternary period is the rapidity of many global atmospheric CH_4 changes observed in the ice-core record. Using an elegant analysis of the ice-core evidence for transient thermal diffusion in nitrogen isotopes, Severinghaus *et al.* (1998) were able to delineate the relative timing of the CH_4 and temperature increases at the end of the Younger Dryas event without the ambiguities inherent in the age difference between the ice and the enclosed air (see Section 10.3.1). They concluded that atmospheric CH_4 began to rise within 0–30 yr following the sudden temperature increase. The speed of these changes has been cited in support of the hypothesis that climatically important changes in atmospheric CH_4 throughout the Quaternary period have been caused by sudden release from the very large volume of methane hydrates stored in marine and continental sediments (Nisbet, 1990, 1992; Kennett *et al.*, 2003) (see Section 10.2.1.2). Others have argued for rapid changes in CH_4 release from wetlands and a less significant role for CH_4 as a driver of climate change (Brook *et al.*, 1999; Severinghaus and Brook, 1999; Brook *et al.*, 2000). The disparity of these views, and the intensity of the debate, serves to remind us that our understanding of carbon-cycle behavior during the Quaternary period is still quite limited, despite the wealth of detailed information available in the record of ice and sediment cores.

10.4 THE PHANEROZOIC RECORD OF CARBON-CYCLE CHANGE

We have emphasized in Section 10.3 that the carbon-cycle variations observed on glacial/interglacial time scales during the Quaternary period can be understood in terms of the redistribution of carbon among the atmosphere, oceans, biosphere, and reactive sediments. Over timescales of millions of years and longer, many carbon-cycle trends cannot be explained by similar redistributions among Earth-surface reservoirs. In the pre-Quaternary past, although Earth-surface reallocations likely caused some degree of carbon-cycle variability, the geologic record reveals more substantial changes that appear to have required the influence of imbalances in exchange between the Earth's surface carbon reservoirs and the rocks of the Earth's crust. In this section we describe the types of crustal exchange processes that have been identified as causes of gradual geologic carbon-cycle change.

10.4.1 Mechanisms of Gradual Geologic Carbon-cycle Change

As shown in Figures 1 and 2, the amount of carbon in the Earth's crust vastly exceeds the amount stored in the atmosphere, biosphere, and oceans combined. A persistent imbalance in the exchange of crustal carbon could, in principle, cause a drastic depletion or buildup of carbon at the Earth's surface (Holland, 1978). Therefore, the study of gradual geologic carbon-cycle change involves seeking both the potential causes of change and the feedback mechanisms that might limit the extent of change (Berner and Caldeira,

1997; Berner, 1999). Feedback mechanisms have been identified in the balances between carbonate weathering and sedimentation, between silicate weathering and metamorphic decarbonation, and between organic carbon production and oxidation.

10.4.1.1 The carbonate weathering-sedimentation cycle

The most abundant anion delivered by rivers to the oceans is bicarbonate ion (HCO_3^-), and most of the bicarbonate alkalinity in rivers comes from the weathering of carbonate rocks (Meybeck, 1987). The chemical weathering of limestones and dolostones by dissolved CO_2 can be represented by the reactions for dissolution of calcite and dolomite:

$$CaCO_3 + CO_2 + H_2O \rightarrow Ca^{2+} + 2HCO_3^- \quad (1a)$$

$$CaMg(CO_3)_2 + 2CO_2 + 2H_2O$$
$$\rightarrow Ca^{2+} + Mg^{2+} + 4HCO_3^- \quad (1b)$$

In the oceans, reef and planktonic organisms precipitate calcium carbonate (both calcite and aragonite), which comprises a major component of marine sediments. The precipitation of calcium carbonate is essentially the reverse of reaction (1a) above:

$$Ca^{2+} + 2HCO_3^- \rightarrow CaCO_3 + CO_2 + H_2O \quad (2a)$$

Although dolomite is abundant in Proterozoic and Paleozoic rocks, its relative contribution to more recent carbonate sediments is significantly less (Daly, 1909), and it comprises only $\sim$10% of modern carbonate sediments (Holland and Zimmerman, 2000). Thus, the precipitation of dolomite,

$$Ca^{2+} + Mg^{2+} + 4HCO_3^- \rightarrow CaMg(CO_3)_2$$
$$+ 2CO_2 + 2H_2O \quad (2b)$$

is not as significant today as it has been during some of the geologic past. The weathering of abundant dolostones and the lack of comparable Mg-carbonate sedimentation during the last 40 Myr may have contributed to a non-steady-state increase in the oceanic concentration of Mg^{2+} (Zimmerman, 2000; Horita *et al.*, 2002). But as described in Section 10.3.3.2, the oceanic carbonate pump should have responded rapidly to any imbalance in oceanic bicarbonate associated with the Mg^{2+} imbalance. The mechanism of this response would have been an enhancement of $CaCO_3$ removal brought about effectively by diminished carbonate dissolution. The rate of reaction (2a) changes effectively to match any change in the sum of the rates of reactions (1a) and (1b). Simple stoichiometry requires that this condition returns as much CO_2 to the ocean/atmosphere system via reaction (2a) as the amount of CO_2 consumed in reactions (1a) and (1b). Changes in atmospheric CO_2 might occur in association with the response of the carbonate pump to an imbalance between carbonate weathering and sedimentation, but these changes would involve the reactive sediments of the Earth-surface system (see Figure 1 and Table 1(a)) rather than changes in the exchange of crustal carbon (Sundquist, 1991).

Interestingly, the Late Tertiary increase in oceanic Mg^{2+} concentrations appears to have been accompanied by a decrease in oceanic Ca^{2+} concentrations (Horita *et al.*, 2002), and there is strong evidence that the rate of present-day $CaCO_3$ sedimentation significantly exceeds the rate of Ca^{2+} input to the oceans (Milliman, 1993). However, interpretation of these trends must consider not only the balance between carbonate weathering and sedimentation, but also the importance of changes in the silicate–carbonate weathering–decarbonation cycle (see Section 10.4.1.2), variations in hydrothermal and cation exchange reactions involving mid-ocean ridge basalts and sediments, and fluctuations in the relative magnitudes of shallow-water and deep-water carbonate sedimentation (Spencer and Hardie, 1990; Hardie, 1996; Holland and Zimmerman, 2000). A full discussion of dolomite weathering and sedimentation is beyond the scope of this chapter. The Tertiary trends described above may be related to the proliferation of planktonic calcareous organisms and the consequent shift of $CaCO_3$ deposition to the deep sea (Holland and Zimmerman, 2000); thus, they may not be analogous to earlier Phanerozoic trends in carbonate mineralogy and oceanic cation concentrations inferred from the record of nonskeletal carbonates and fluid inclusions (Sandberg, 1983, 1985; Horita *et al.*, 2002). The relationships described above demonstrate that steady-state feedback in the long-term cycling of carbon must be viewed in the context of significant non-steady-state trends in the chemistry of major dissolved oceanic cations. The carbonate weathering-sedimentation cycle is sometimes ignored or trivialized in treatments of the geological carbon cycle, because reactions (1) and (2) do not appear to have an impact on long-term atmospheric CO_2 trends. It is important to remember that the weathering and sedimentation of carbonate minerals are, like other carbon-cycle feedback systems, coupled to the global cycling of other materials.

10.4.1.2 The silicate-carbonate weathering-decarbonation cycle

Although the chemical weathering of silicate minerals contributes less than the weathering of carbonates to the cycling of materials from the land surface to ocean waters and sediments, silicate weathering presents a much more complex challenge to geochemists seeking to discern the feedback mechanisms that assure a balance in crustal carbon exchange. This balance can be illustrated by the following cyclic reactions:

Weathering:

$$3H_2O + 2CO_2 + CaSiO_3$$
$$\rightarrow Ca^{2+} + 2HCO_3^- + Si(OH)_4 \quad (3a)$$

$$3H_2O + 2CO_2 + MgSiO_3$$
$$\rightarrow Mg^{2+} + 2HCO_3^- + Si(OH)_4 \quad (3b)$$

Sedimentation:

$$Ca^{2+} + 2HCO_3^- + Si(OH)_4$$
$$\rightarrow CaCO_3 + SiO_2 + 3H_2O + CO_2 \quad (4a)$$

$$Mg^{2+} + 2HCO_3^- + Si(OH)_4$$
$$\rightarrow MgCO_3 + SiO_2 + 3H_2O + CO_2 \quad (4b)$$

Decarbonation:

$$CaCO_3 + SiO_2 + 3H_2O + CO_2$$
$$\rightarrow 3H_2O + 2CO_2 + CaSiO_3 \quad (5a)$$

$$MgCO_3 + SiO_2 + 3H_2O + CO_2$$
$$\rightarrow 3H_2O + 2CO_2 + MgSiO_3 \quad (5b)$$

Reactions (3)–(5) are an oversimplification of the wide variety of contributing minerals and reactions, but they can be used to illustrate the most important aspects of the silicate–carbonate weathering–decarbonation cycle. Like carbonate weathering, silicate weathering consumes CO_2 and yields cations and bicarbonate ions that are delivered by rivers to the oceans. However, the sedimentation of carbonate minerals releases only some of the CO_2 consumed during weathering. The remaining CO_2 is released by decarbonation reactions that occur during burial, subduction, and other tectonic processes.

Arrhenius (1896) and Chamberlin (1898) proposed feedback linkages among tectonic activity, atmospheric CO_2, chemical weathering, and carbonate deposition. Budyko and Ronov (1979) were the first to use silicate–carbonate feedback linkages to estimate past atmospheric CO_2 concentrations. They observed a correlation between volcanic and carbonate rock abundances in the sediments of the Russian platform, and they calculated Phanerozoic atmospheric CO_2 concentrations by assuming a simple proportionality to rates of carbonate deposition. Walker *et al.* (1981) contributed a very important extension to this idea, proposing that the feedback mechanism is mediated by the climatic effect of atmospheric

CO_2. According to this hypothesis, higher atmospheric CO_2 would increase the greenhouse effect, raising global temperatures and silicate weathering reaction rates, and providing a negative feedback to the CO_2 increase. Conversely, an increase in silicate weathering would lead to a decrease in atmospheric CO_2 and global temperatures, providing a negative feedback to the weathering increase. The CO_2–climate feedback was adopted as a central concept in the time-dependent model of Berner *et al.* (1983), popularly known as the BLAG model (see Section 10.4.2).

Much debate has been devoted to the question of which of the reactions (3)–(5) "control" atmospheric CO_2. The BLAG model (Berner *et al.*, 1983; Berner, 1991) suggested that rates of silicate weathering (reaction (3)) have adjusted essentially instantaneously to maintain a steady state with respect to rates of CO_2 production from decarbonation (reaction (5)). In this view, the rate-limiting step in the cycle is, therefore, the tectonically induced production of CO_2. Others have emphasized the importance of continental uplift as a factor that could affect atmospheric CO_2 through changes in weathering without necessarily changing the rate of decarbonation CO_2 supply (Raymo *et al.*, 1988; Edmond *et al.*, 1995). The two views are not incompatible because steady-state weathering feedback can incorporate effects of uplift (Sundquist, 1991; Berner, 1991). There is now a broad consensus that both weathering and tectonism are critical to the relationship between atmospheric CO_2 and the silicate–carbonate cycle (see, e.g., Berner, 1994; Bickle, 1996; Berner and Kothavala, 2001; Ruddiman, 1997; Wallmann, 2001a).

It is very difficult to infer present-day values for the carbon fluxes and relationships relevant to the silicate–carbonate weathering–decarbonation cycle (see Table 1(a)), and it is even more daunting to estimate how these interactions might have varied in the geologic past (Kump *et al.*, 2000; Boucot and Gray, 2001). The difficulty is further complicated by the suggestion that low-temperature hydrothermal reactions may precipitate significant quantities of $CaCO_3$ from seawater circulating through the upper ocean crust (Staudigel *et al.*, 1990; Caldeira, 1995; Alt and Teagle, 1999). The overall stoichiometry of these "sea-floor weathering" reactions is very complex, but they appear to be a process of global if enigmatic significance to the long-term oceanic carbon and alkalinity balance (Wallmann, 2001a). Thus reactions (3)–(5) must be considered illustrative not only in the sense that they are idealized representations of more complicated reactions, but also in the sense that they may not represent all of the basic silicate–carbonate reaction types that contribute exchange of carbon between the Earth's crust and surface reservoirs.

10.4.1.3 The organic carbon production-consumption-oxidation cycle

Since the Precambrian, oxygenic photosynthesis and aerobic respiration have been the dominant reactions in the cycling of organic carbon (Des Marais *et al.*, 1992) (see Section 10.5). The production of organic matter by oxygenic photosynthesis can be illustrated by the production of glucose:

$$6CO_2 + 6H_2O \rightarrow C_6H_{12}O_6 + 6O_2 \qquad (6)$$

The transformation of glucose and other carbohydrates into the vast array of compounds that are buried and further transformed in sediments is beyond the scope of this chapter. The fundamental relationship in reaction (6) is the production of oxygen that accompanies the production of organic carbon. The aerobic cycling of organic carbon is completed by oxidative consumption, which can be illustrated by the reverse of reaction (6):

$$C_6H_{12}O_6 + 6O_2 \rightarrow 6CO_2 + 6H_2O \qquad (7)$$

Reactions (6) and (7) demonstrate that the cycling of organic carbon is inherently associated with the cycling of oxygen. This relationship has two very important consequences. First, a net excess of organic matter production relative to oxidation is accompanied by net production of oxygen. Conversely, a net excess of oxidation is accompanied by net consumption of oxygen. Thus, as long as reactions (6) and (7) have dominated the cycling of organic carbon, net burial of organic matter has been associated with production of atmospheric oxygen over timescales of millions of years and longer (Garrels and Lerman, 1981; Shackleton, 1987; Kump *et al.*, 1991; Des Marais *et al.*, 1992). Second, through its linkage to the cycling of oxygen, the geological carbon cycle has also been intimately connected to the biogeochemical cycles of sulfur (see below), phosphorus, iron, nitrogen, and other elements that are cycled in forms that are sensitive to the oxidation state of their environment (Ingall and Van Cappellen, 1990; Van Cappellen and Ingall, 1996; Petsch and Berner, 1998; Berner, 1999).

In keeping with our perspective that the modern cycling of methane can be viewed as a subcycle of the CO_2 carbon cycle, we represent microbial methane production by the fermentation of glucose:

$$C_6H_{12}O_6 \rightarrow 3CH_4 + 3CO_2 \qquad (8)$$

and we represent methane oxidation by the stoichiometrically consistent reaction:

$$3CH_4 + 6O_2 \rightarrow 6H_2O + 3CO_2 \qquad (9)$$

The sum of reactions (8) and (9) is identical to reaction (7), demonstrating the treatment of methane production and oxidation as a subcycle.

Under anaerobic conditions, microbial consumption of methane may occur by reactions such as:

$$3CH_4 + 3SO_4^{2-} \rightarrow 3HCO_3^- + 3HS^- + 3H_2O \tag{10}$$

Reaction (10) illustrates an important connection between the cycling of carbon and sulfur. The bisulfide produced by this kind of reaction is often precipitated and buried in marine sediments as iron sulfide (Berner, 1982). Over geological timescales, the removal of iron sulfide from the Earth-surface sulfur cycle affects the oxygen cycle in a manner analogous to the net removal (burial) of organic carbon from the Earth-surface carbon cycle. The burial of reduced forms of carbon and sulfur allows oxygen to accumulate in the atmosphere (Garrels and Lerman, 1984; Berner, 1987; Berner, 2001). Thus, the anaerobic consumption of methane may have played an important part in the geologic history of atmospheric oxygen.

Global trends in the organic carbon production-consumption-oxidation cycle can be discerned from the carbon isotope record of marine carbonates. Because the formation of organic matter selectively assimilates ^{12}C (see Section 10.2.1.1), changes in the relative proportions of net removals or additions of organic and inorganic carbon are reflected in the ratio of ^{13}C to ^{12}C in the carbon of the oceans and atmosphere. Although the carbon isotope ratios of marine carbonate shells are affected by environmental and biological variables as well as by the isotopic composition of oceanic carbon, the record from fossil shells shows significant trends through the Phanerozoic Eon that most likely represent global changes in the ratio of ^{13}C to ^{12}C in oceanic DIC (Figure 11). More generally, the marine carbonate record is characterized by $^{13}C : {}^{12}C$ ratios that are consistently greater than the corresponding ratio of the carbon released to the Earth's surface environment from the mantle (Schidlowski, 2001). This relationship implies that the burial of carbonates has been accompanied by burial of ^{13}C-depleted organic matter. If mantle carbon reaches the Earth's surface with a $\delta^{13}C$ value of $-5\permil$ (Deines, 1992), and if carbonate and organic carbon are buried with $\delta^{13}C$ values of $0\permil$ and $-25\permil$, respectively, then the steady-state ratio of organic carbon burial to carbonate carbon burial can be calculated by isotopic mass balance to be 1 : 4 (Schidlowski, 2001). Variations such as those depicted in Figure 11 probably reflect transient changes in the relative rates of organic carbon and carbonate burial (Schidlowski and Junge, 1981; Berner and Raiswell, 1983; Garrels and Lerman, 1984; Berner, 1987), as well as possible changes in the degree of fractionation of carbon isotopes due to changes in ocean/atmosphere chemistry and other environmental factors (Hayes *et al.*, 1999; Berner *et al.*, 2000; Berner, 2001). This record is an important constraint on models of the geologic carbon cycle (see Section 10.4.2).

The fractionation of carbon isotopes between ^{13}C-depleted organic matter and ^{13}C-enriched

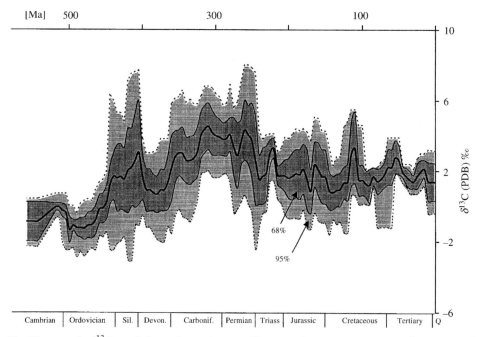

Figure 11 Phanerozoic $\delta^{13}C$ trends in marine carbonates. The central curve represents running means calculated using 20 Ma window and 5 Ma forward step. The shaded areas show relative uncertainties (Veizer *et al.* (1999); reproduced by permission of Elsevier from *Chem. Geol.*, **1999**, *161*, 59–88 (figure 10)).

carbonates is also observed in Precambrian sediments extending back to the Archean Eon (Schidlowski *et al.*, 1983; Schidlowski, 2001). Prior to the early Proterozoic Eon, the cycling of organic carbon probably occurred through reactions and life forms that did not require significant levels of atmospheric oxygen (Holland, 1994). These early modes of organic carbon cycling and their relationships to ocean/atmosphere chemistry are discussed briefly in Section 10.5 and more extensively in Holland (2003).

10.4.2 Model Simulations of Gradual Geologic Carbon-cycle Change

The mechanisms of gradual geologic carbon-cycle change are complex and poorly understood. They span a broad range of timescales, and they include many poorly understood relationships and feedback. They present challenges to our understanding of the cycling of many other elements in addition to carbon. Mathematical modeling is an essential tool in the quest for this understanding. Models provide a crucial device for quantifying relationships, formulating and testing hypotheses, and comparing simulated results to the evidence of the geologic record. Biogeochemists use models of the global carbon cycle as an iterative self-teaching tool, reconfiguring and reformulating as new ideas and evidence emerge.

The most widely cited and applied example of this iterative modeling approach is the work of Berner and coworkers on the development of the GEOCARB models (Berner *et al.*, 1983; Lasaga *et al.*, 1985; Berner, 1990, 1991, 1994, Berner and Kothavala, 2001). This effort, which began with the publication of the BLAG model twenty years before the writing of this chapter, is distinguished by its prolonged capacity to test innovative ideas and to incorporate evolving insights (including many initiated by the criticisms of colleagues). The core of the GEOCARB model is a simple steady-state ocean-atmosphere carbon mass balance expression:

$$F_{wc} + F_{mc} + F_{wg} + F_{mg} = F_{bc} + F_{bg} \quad (11)$$

where F_{wc} is the rate of weathering of carbonates (reaction (1)), F_{mc} is the production of CO_2 by decarbonation (reaction (5)), F_{wg} is the rate of production of CO_2 by oxidative weathering of sedimentary organic matter (reaction (7)), F_{mg} is the rate of production of CO_2 by the metamorphism of sedimentary organic matter (reactions (7)–(9)), F_{bc} is the rate of burial of sedimentary carbonates (reactions (2) and (4)), and F_{bg} is the rate of burial of sedimentary organic material (reaction (6)). Each of these terms is defined in the model as a function of variables that represent the estimated effects of subduction and tectonics (for F_{mc} and F_{mg}) and of land area, relief, climate,

vegetation, and atmospheric CO_2 (for the weathering fluxes). The rate of CO_2 consumption during weathering of silicate minerals is expressed by the equation:

$$F_{ws} = F_{bc} - F_{wc} \quad (12)$$

which simply represents the steady-state approximation that the rate of carbonate burial will be equal to the sum of the rates of bicarbonate delivered to the oceans from weathering of silicate and carbonate minerals. Using a model configured to represent the time-dependent behavior of ocean chemistry and reactive sediment interactions, Sundquist (1991) confirmed that carbonate sedimentation does indeed respond very quickly (within a few thousand years) to balance changes in weathering, and determined that the overall response time of weathering to changes in the terms in Equations (11) and (12) was on the order of a few hundred thousand years. Thus the steady-state approximation of these equations appears to be reasonable for the timescales to which GEOCARB has been applied.

The calculation of past CO_2 levels using GEOCARB begins with the iterative solution of equations (11) and (12) and a parallel mass balance equation for carbon isotopes to yield fluxes that are consistent with the carbonate carbon isotope record (Figure 11) and a variety of other sources of information for the factors that determine relative variations of the individual fluxes through time. Further iteration is then applied to derive atmospheric CO_2 concentrations consistent with the steady-state weathering fluxes. GEOCARB results have shown the same general pattern through more than ten years of modification and refinement (Figure 12). The calculated CO_2 trend shows very high levels in the Early Paleozoic Era, a significant decline during the Devonian and Carboniferous periods, high values during the Early Mesozoic Era, and a decline beginning at ~170 Ma toward the relatively low values of the Cenozoic Era.

Like all geochemical simulations requiring broad assumptions and extrapolations, the GEOCARB results are not immune to criticism. Some have argued that the model does not adequately represent the effects of pre-Devonian weathering (Boucot and Gray, 2001), and others have taken issue with the inference that atmospheric CO_2 plays a significant role in the climatic history of the Phanerozoic Eon (see, e.g., Cowling, 1999; Veizer *et al.*, 2000; Wallmann (2001b). While a number of independent modeling efforts have yielded results similar to those of GEOCARB (see, e.g., Tajika, 1998; Wallmann, 2001a), others have not (see, e.g., Rothman, 2002). There is general agreement that all models of Phanerozoic carbon cycling are poorly constrained by the incomplete and circumstantial nature of the

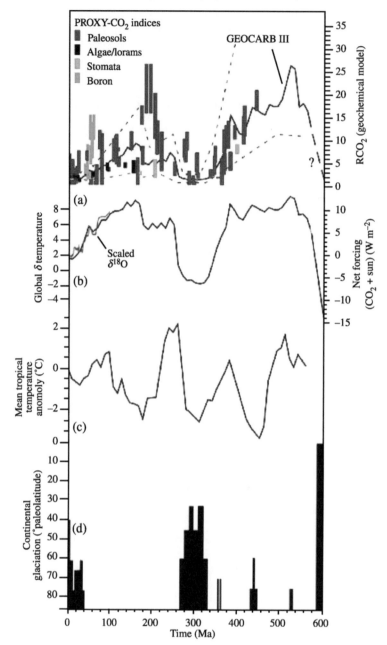

Figure 12 Phanerozoic model results (Berner and Kothavala, 2001) compared to climate indicators. Crowley and Berner (2001) describe this figure as follows: "(a) Comparison of CO_2 concentrations from the GEOCARB III model (Berner and Kothavala, 2001) with a compilation of proxy-CO_2 evidence (vertical bars; Royer *et al.*, 2001). Dashed lines: estimates of uncertainty in the geochemical model values (Berner and Kothavala, 2001). Solid line: conjectured extension to the Late Neoproterozoic (~590–600 Ma). R_{CO_2}, ratio of CO_2 levels with respect to the present (300 ppm). Other carbon-cycle models (Tajika, 1998; Wallmann, 2001b) for the past 150 Myr are in general agreement with the results from this model. (b) Radiative forcing for CO_2 calculated from (Kiehl and Dickinson, 1987) and corrected for changing luminosity (Crowley and Baum, 1993) after adjusting for an assumed 30% planetary albedo. Deep-sea oxygen isotope data over the past 100 Ma (Douglas and Woodruff, 1981; Miller *et al.*, 1987) have been scaled to global temperature variations according to Crowley (2000). (c) Oxygen isotope-based low-latitude paleotemperatures (Veizer *et al.*, 2000). (d) Glaciological data for continental-scale ice sheets modified from Crowley (2000) and Crowell (1999) based on many sources. The duration of the Late Neoproterozoic glaciation is a subject of considerable debate" (Crowley and Berner (2001); reproduced by permission of the American Association for the Advancement of Science from *Science*, **2001**, *292*, 870–872 (figure 1)).

geologic evidence. The ongoing assessment and evolution of the GEOCARB model exemplifies the effective use of what information is available.

10.4.3 Geologic Evidence for Phanerozoic Atmospheric CO₂ Concentrations

Figure 12 shows some of the geologic evidence to which the GEOCARB model results are compared. The general Phanerozoic pattern of atmospheric CO_2 (Figure 12, panel a) and derived radiative forcing (panel b) appears to correspond to the pattern of continental glaciation episodes (panel d) in a manner consistent with a major effect of CO_2 on global climate. However, this relationship does not appear to be robust in comparing the CO_2 and derived radiative forcing to trends in low-latitude temperatures derived from the oxygen isotope ratios of marine calcite fossils (panel c). These discrepancies remain unresolved. It seems likely that the record of Phanerozoic climate and carbon cycling, like the more detailed record of Quaternary climate and carbon cycling described in Section 10.3, reflects the influence on climate of many complex factors including—but not limited to—atmospheric CO_2.

Figure 12 also displays some of the proxy data to which model simulations of atmospheric CO_2 can be more directly compared. These proxy data have recently been reviewed by Royer *et al.* (2001) and Beerling and Royer (2002). All of the proxy measures are limited by significant uncertainties.

The carbon isotope ratios of pedogenic carbonates have been used to infer atmospheric CO_2 concentrations from calculations based on a model for steady-state diffusive mixing with soil-respired CO_2 in the soil profile (Cerling, 1991; Cerling, 1992; Ekart *et al.*, 1999; Ghosh *et al.*, 2001). Significant sources of uncertainty in this approach include the dependence of soil CO_2 diffusion on temperature and moisture, the contribution of C_3 versus C_4 plants to respired CO_2, and the somewhat arbitrary choice of values for the mole fraction of respired CO_2 at depth in the soil.

The extent of carbon isotope fractionation during photosynthesis is sensitive to atmospheric CO_2 concentrations. This sensitivity has been used to estimate atmospheric CO_2 concentrations from fossil marine and terrestrial plant residues (Rau *et al.*, 1989; Popp *et al.*, 1989; Marino and McElroy, 1991; Freeman and Hayes, 1992; Pagani *et al.*, 1999a,b; Grocke, 2002; Beerling and Royer, 2002; Pagani, 2002). Uncertainties in this method include (for marine plants) variations in the dissolved CO_2 concentration of ocean surface waters and effects of temperature, salinity, and cell size and growth rate; and (for land plants) variations in canopy CO_2 concentrations and

effects of temperature and leaf-to-air vapor pressure. Analysis of alkenones extracted from marine plankton appears to show some promise of narrowing uncertainties but yields disparate estimates of Miocene CO_2 levels (Freeman and Hayes, 1992; Pagani, 2002). Analysis of fossil land plants has demonstrated some encouraging correlations with carbon isotope anomalies recorded in marine carbonates (see, e.g., Koch *et al.*, 1992), but efforts to reconstruct past atmospheric CO_2 levels are much more tentative (Beerling and Royer, 2002).

Land plants take up CO_2 through leaf pores called stomates. C_3 plants appear to regulate the number of stomates per unit of leaf area (stomatal density) to maximize CO_2 uptake while minimizing water loss through the stomates. The stomatal density of many modern plants has been shown to be inversely related to ambient CO_2 concentrations (Woodward, 1987; Beerling and Royer, 2002), and this relationship has been applied to analysis of stomatal densities of fossil plant leaves (Beerling, 1993; McElwain and Chaloner, 1995; Rundgren and Beerling, 1999; Retallack, 2001, 2002; Beerling and Royer, 2002). Although there is impressive support for this approach in its validation for subtle Holocene CO_2 trends (Rundgren and Beerling, 1999), its application to a larger range of CO_2 variations and plant species remains uncertain.

The ratio of the boron isotopes ^{11}B and ^{10}B is known to depend on ambient pH in the boron incorporated in the carbonate shells of marine foraminifera (Sanyal *et al.*, 1996; Sanyal *et al.*, 2001). The use of boron isotopes as an indicator for seawater paleo-pH has been extended to the calculation of past CO_2 concentrations in ocean surface waters and in the atmosphere (Spivack, 1993; Sanyal *et al.*, 1997; Pearson and Palmer, 1999, 2000; Sanyal and Bijma, 1999; Palmer and Pearson, 2003). Sources of uncertainty in these estimates include the fractionation of boron isotopes during incorporation in carbonate shells, effects of diagenesis, the assumptions needed to calculate CO_2 concentrations from pH, and the influence of changing boron isotope ratios in ambient seawater (Lemarchand *et al.*, 2000; Lemarchand *et al.*, 2002). The latter problem is especially serious for estimates based on samples older than the 15 Myr residence time of boron in the oceans.

Given the discrepancies and uncertainties enumerated above, it is clear that the study of Phanerozoic carbon-cycle change does not enjoy the luxury of a paleo-CO_2 "gold standard" analogous to the ice-core records of the Late Quaternary period. Measurements that reflect the global mass balance of carbon, such as the carbon isotope record of marine carbonates (Figure 11), remain the most powerful kind of evidence for analysis of gradual geologic carbon-cycle change. This approach requires geochemical models to quantify and test hypothesized relationships among global fluxes and reservoirs.

10.4.4   Abrupt Carbon-cycle Change

Like the concept of steady state described in Section 10.2.2, the concept of "abrupt" change requires the context of a particular timescale and frame of reference. In Section 10.3, we described examples of abrupt change in the context of the Late Quaternary ice-core record. Abrupt changes in the context of the Phanerozoic rock record might be viewed as leisurely transitions from the perspective of Quaternary timescales. A formal treatment of abrupt change is beyond the scope of this chapter. However, we describe two examples that seem to qualify as "abrupt" changes in the carbon cycle by any definition.

About 55 Myr ago, a brief period known as the Late Paleocene thermal maximum (Zachos *et al.*, 1993) is defined in marine and terrestrial records by a sudden increase in proxy temperature estimates and a very large decrease in the ratio of ^{13}C to ^{12}C (Figure 13). Most of the carbon isotope shift appears to have occurred within a period of a few thousand years (Norris and Rohl, 1999), suggesting that the cause might have been a redistribution of carbon from the terrestrial biosphere to the oceans and atmosphere. But the magnitude of the shift, $\sim-2.5\permil$ to $-3\permil$, would have required the transfer of an amount of terrestrial organic carbon equivalent at least to virtually the entire modern terrestrial biosphere, including soils (Dickens *et al.*, 1995). Instead, the most likely cause is a

sudden transfer of methane hydrate to the ocean–atmosphere system (Dickens *et al.*, 1995, 1997; Dickens, 2000). The extremely depleted ratio of ^{13}C to ^{12}C in methane means that only nearly one-third as much methane carbon must be transferred to account for the carbon isotope shift. Even so, the amount estimated to have caused the Late Paleocene isotope shift is on the order of 1,000–2,000 Pg C. The precise mechanism of this methane release is not clear, but the hypothesized release appears to be consistent with the evidence for abrupt global warming and (assuming that the methane was rapidly oxidized to CO_2) for $CaCO_3$ dissolution (Thomas and Shackleton, 1996). The carbon isotope anomaly seems to have decayed somewhat exponentially over ~150 kyr (Figure 13), which is comparable to the modern residence time of oceanic carbon with respect to sediment burial (see Figure 1). From evidence for an episode of widespread biogenic barium coincident with the carbon isotope anomaly, Bains *et al.* (2000) suggested that oceanic productivity increased in response to a warming-induced enhancement of continental weathering and delivery of nutrients to the oceans. The Late Paleocene thermal maximum seems to provide not only an example of an abrupt event, but also an opportunity to document consequent feedback mechanisms ranging from climate change to carbonate dissolution to continental weathering and oceanic productivity.

The Cretaceous–Tertiary boundary, ~65 Myr ago, is defined by a bolide impact event putatively associated with the Chicxulub Crater in Mexico (Alvarez *et al.*, 1980). Among the cataclysmic environmental effects of this event, a large but

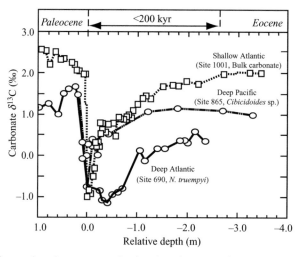

Figure 13 High-resolution carbon isotope records showing the anomaly corresponding to the Late Paleocene Thermal Maximum. The data are from three widely separated Ocean Drilling Program cores (Kennett and Stott, 1991; Bralower *et al.*, 1995; Bralower *et al.*, 1997) plotted on a common depth scale with the $\delta^{13}C$ minimum at 0.0 m (Dickens (2001; reproduced by permission of the American Geophysical Union from *Natural Gas Hydrates: Occurrence, Distribution, and Detection*, **2001**, pp. 19–38 (figure 1)).

uncertain quantity of CO_2 was likely injected into the ocean–atmosphere system from the volatilization of $CaCO_3$ rocks at the impact site (O'Keefe and Ahrens, 1989; Pope *et al.*, 1997; Pierazzo *et al.*, 1998). Not surprisingly, a large simultaneous release of methane hydrate has also been hypothesized (Max *et al.*, 1999). Although the effects of these carbon transfers are greatly complicated by the wide range of biological and geochemical consequences of the impact, they provide another opportunity to document and analyze carbon-cycle feedback mechanisms.

An understanding of geologically abrupt change requires not only evidence of the abrupt event itself, but also analysis of the sequence of subsequent responses. These responses may "cascade" through a wide range of timescales, reflecting and perhaps shedding new light on many of the relationships between processes and timescales that characterize more gradual change. Further geological evidence is needed to determine the extent to which carbon-cycle change has occurred by abrupt rather than gradual transitions. If abrupt change turns out to be a significant mode of long-term carbon-cycle evolution, new theoretical understandings may be needed to accommodate more pronounced effects of non-steady-state relationships (see, e.g., Sundquist, 1991).

10.5 THE PRECAMBRIAN RECORD OF CARBON-CYCLE CHANGE

Many aspects of the Precambrian history of the carbon cycle are summarized in Holland (2003). Although the carbon isotope record suggests the common existence of ^{13}C-depleted organic carbon of biological origin by 3 Ga, the atmosphere contained very little oxygen and the carbon cycle did not begin to resemble its modern form until the Late Archean Eon and the transition to the Proterozoic Eon (Des Marais *et al.*, 1992; Kasting, 1993; Des Marais, 2001). The earliest forms of life appear to have been based on fundamentally different modes of carbon cycling in which organic synthesis and metabolism occurred without the production and consumption of oxygen. Anoxygenic photosynthesis is observed in several forms of bacteria today, and the antiquity of these forms is suggested by analyses based on genomic sequencing and the molecular structure and function of key protein complexes (Mathis, 1990; Schubert *et al.*, 1998; Rhee *et al.*, 1998; Xiong *et al.*, 2000; Hedges *et al.*, 2001). Early forms of life may have cycled carbon through redox reactions involving hydrogen, sulfur, iron, or nonbiotic organic compounds (see, e.g., Walker, 1987; Widdel *et al.*, 1993; MacLeod *et al.*, 1994; Russell and Hall, 1997; Pace, 1997; Rasmussen, 2000; Kelley *et al.*, 2001).

The origin of oxygenic photosynthesis is attributed to the cyanobacteria, which appear to have evolved hundreds of millions of years before the rise in atmospheric oxygen (Buick, 1992; Brocks *et al.*, 1999; Summons *et al.*, 1999; Des Marais, 2000). During the period of transition to oxygenated conditions, the oxygen produced by photosynthesis was probably consumed by reaction with reduced compounds except in particular oxygenated environments (Cloud Jr., 1968; Walker *et al.*, 1983; Kasting, 1987, 1993). The quantitative pace of the rise in atmospheric oxygen is a topic of lively debate (Holland, 1994, 2003; Ohmoto, 1996; Holland and Rye, 1997; Lasaga and Ohmoto, 2002). The expansion of oxygenated conditions may have been affected by the evolving differentiation of the Earth's mantle and crust, changes in the oxidation state of volcanic gases, episodes of tectonic activity, hydrogen escape to space from the upper atmosphere, and the increasing availability of stable cratonic platforms for more productive life forms and more active subaerial weathering (Walker *et al.*, 1983; Des Marais *et al.*, 1992; Kasting *et al.*, 1993; Kasting, 1993, 2001; Hoehler *et al.*, 2001; Catling *et al.*, 2001; Holland, 2002). The variety of processes affecting atmospheric oxygen must also have affected the cycling of carbon, and the atmosphere may have contained both CO_2 and CH_4 at this time (Kasting, 1993; Hayes, 1994; Rye *et al.*, 1995), but see Kasting *et al.* (1983). The cycling of CH_4 by microbial production and consumption may have been so important that this period has been dubbed the "Age of Methanotrophs" (Hayes, 1994). High atmospheric concentrations of CO_2 and/or CH_4 were probably necessary in order to offset the lowered luminosity of the Sun during this period (Sagan and Mullen, 1972; Owen *et al.*, 1979; Kuhn and Kasting, 1983; Kasting *et al.*, 1988; Rye *et al.*, 1995).

As the oxidation state of the Earth's surface gradually changed, the anaerobic oxidation of methane via reaction (10) may have been an important transitional pathway (Hinrichs and Boetius, 2002). The first significant increase in atmospheric oxygen (the so-called "Great Oxidation Event") has been inferred from a positive anomaly in the carbon isotope ratios of marine carbonates deposited between 2.22 Ga and 2.06 Ga (Karhu and Holland, 1996). This shift is interpreted as an effect of increased organic carbon burial, leading to more atmospheric oxygen as discussed in Section 10.4.1.3. The timing of this carbon and oxygen event is generally supported in the sediment record by the earliest appearance of redbeds (oxidized subaerial sediments) and by the disappearance of banded iron formations and detrital pyrite and uraninite (Cloud Jr., 1968; Walker *et al.*, 1983; Holland, 1994). Further support has been documented in the disappearance of sediment sulfur isotope signatures that could only have originated through mass-independent fractionation in a low-oxygen atmosphere (Farquhar *et al.*, 2000).

A second positive carbon isotope excursion is apparent in marine carbonates formed during the Neoproterozoic. This feature is punctuated by large oscillations associated with glaciogenic sediments that have been attributed to so-called "snowball Earth" conditions (Hoffman *et al.*, 1998; Hoffman and Schrag, 2002). Declining concentrations of both CO_2 and CH_4 have been hypothesized as causes of the onset of intense global glaciation during this time, and increasing CO_2 has been hypothesized as a cause of subsequent deglaciation (Hoffman and Schrag, 2002; Pavlov *et al.*, 2003).

10.6 CONCLUSIONS

It is clear that the carbon cycle and global climate are linked in many ways throughout the history of the Earth. But it is equally apparent that the complex interactions evident in the geologic record defy simple attribution of cause and effect. Collection of more data and further analysis and modeling will continue to improve our understanding of these interactions, which are now so important to the near-term relationship between human activities and the global environment.

Until recently, the paradigm for analysis of the geologic carbon cycle has been to focus on particular processes and carbon stores that are thought to be relevant to chosen timescales of interest (see, e.g., Sundquist, 1985; Sundquist, 1986). This paradigm has provided a platform for development of a range of models, each appropriate to particular timescales. But the geologic record is revealing a growing array of "abrupt" events that cannot be analyzed in this way. New models are needed with capabilities to span a broad range of timescales, from the abruptness of the events to the prolonged cascade of subsequent effects. The need for these models is urgent because the effects of current human activities must be viewed as a geologically abrupt event by any standard.

ACKNOWLEDGMENTS

Work on this chapter was supported by the National Research Program of the US Geological Survey. We are grateful to Dick Holland, Skee Houghton, Clive Oppenheimer, Bill Reeburgh, and Roberta Rudnick for providing preprints of their chapters for this treatise, enabling us to benefit from and complement their work. We also thank Richie S. Williams and Dick Holland for thoughtful and constructive comments. Finally, in this twentieth anniversary year of the BLAG publication, it is fitting for a paper on the geologic history of the carbon cycle to take special note of Bob Berner's prolonged and insightful contributions. In that spirit, we add that this chapter benefited from the collaborative musical inspiration of Miles Davis, Cannonball Adderly, Paul Chambers, James Cobb, John Coltrane, Bill Evans, and Wynton Kelly.

REFERENCES

Adams J. M. and Faure H. (1998) A new estimate of changing carbon storage on land since the last glacial maximum, based on global land ecosystem reconstruction. *Global Planet. Change* **16–17**, 3–24.

Adams J. M., Faure H., Faure-Denard L., McGlade J. M., and Woodward F. I. (1990) Increases in terrestrial carbon storage from the last glacial maximum to the present. *Nature* **348**, 711–714.

Alley R. B., Brook E. J., and Anandakrishnan S. (2002) A northern lead in the orbital band: north-south phasing of ice-age events. *Quat. Sci. Rev.* **21**, 431–441.

Alt J. C. and Teagle D. A. H. (1999) The uptake of carbon during alteration of oceanic crust. *Geochim. Cosmochim. Acta* **63**, 1527–1536.

Altabet M. A., Murray D. W., and Prell W. L. (1999) Climatically linked oscillations in Arabian Sea denitrification over the past 1 my: implications for the marine N cycle. *Paleoceanography* **14**, 732–743.

Alvarez L. A., Alvarez W., Asaro F., and Michel H. V. (1980) Extraterrestrial cause for the Cretaceous–Tertiary extinction. *Science* **208**, 1095–1108.

Anklin M., Barnola J.-M., Schwander J., Stauffer B., and Raynaud D. (1995) Processes affecting the CO_2 concentrations measured in Greenland ice. *Tellus* **47B**, 461–470.

Anklin M., Schwander J., Stauffer B., Tschumi J., Fuchs A., Barnola J. M., and Raynaud D. (1997) CO_2 record between 40 and 8 kyr BP from the Greenland Ice Core Project ice core. *J. Geophys. Res.-Oceans* **102**, 26539–26545.

Archer D. (1991) Modeling the calcite lysocline. *J. Geophys. Res.-Oceans* **96**, 17037–17050.

Archer D. and Maier-Reimer E. (1994) Effect of deep-sea sedimentary calcite preservation on atmospheric CO_2 concentration. *Nature* **367**, 260–263.

Archer D., Kheshgi H., and Maier-Reimer E. (1997) Multiple time-scales for neutralization of fossil fuel CO_2. *Geophys. Res. Lett.* **24**, 405–408.

Arrhenius S. (1896) On the influence of carbonic acid in the air upon the temperature of the ground. *Phil. Mag. J. Sci.* **41**, 237–276.

Arthur M. A. (2000) Volcanic contributions to the carbon and sulfur geochemical cycles and global change. In *Encyclopedia of Volcanoes* (eds. H. Sigurdsson, B. F. Houghton, S. R. McNutt, H. Rymer, and J. Stix). Academic Press, San Diego, pp. 1045–1056.

Bacastow R. B. (1996) The effect of temperature change of the warm surface waters of the oceans on atmospheric CO_2. *Global Biogeochem. Cycles* **10**, 319–333.

Bains S., Norris R. D., Corfield R. M., and Faul K. L. (2000) Termination of global warmth at the Palaeocene/Eocene boundary through productivity feedback. *Nature* **407**, 171–174.

Bange H. W., Bartell U. H., Rapsomanikis S., and Andreae M. O. (1994) Methane in the Baltic and North Seas and a reassessment of the marine emissions of methane. *Global Biogeochem. Cycles* **8**, 465–480.

Barber D. C., Dyke A., Hillaire-Marcel C., Jennings A. E., Andrews J. T., Kerwin M. W., Bilodeau G., McNeely R., Southon J., Morehead M. D., and Gagnon J.-M. (1999) Forcing of the cold event of 8,200 years ago by catastrophic drainage of Laurentide lakes. *Nature* **400**, 344–348.

Barnola J. M. (1999) Status of the atmospheric CO_2 reconstruction from ice cores analyses. *Tellus* **51B**, 151–155.

Barnola J. M., Raynaud D., Korotkevich Y. S., and Lorius C. (1987) Vostok ice core provides 160,000-year record of atmospheric CO_2. *Nature* **329**, 408–414.

Barnola J. M., Pimienta P., Raynaud D., and Korotkevich Y. S. (1991) CO_2–climate relationship as deduced from the Vostok ice core: a reexamination based on new measurements and on a re-evaluation of the air dating. *Tellus* **43B**, 83–90.

Barnola J. M., Anklin M., Porcheron J., Raynaud D., Schwander J., and Stauffer B. (1995) CO_2 evolution during the last millennium as recorded by Antarctic and Greenland ice. *Tellus* **47B**, 264–272.

Bartlett K. B. and Harriss R. C. (1993) Review and assessment of methane emissions from wetlands. *Chemosphere* **26**, 261–320.

Bartlett K. B., Crill P. M., Sebacher D. I., Harriss R. C., Wilson J. O., and Melack J. M. (1988) Methane flux from the central Amazonian floodplain. *J. Geophys. Res.-Atmos.* **93**, 1571–1582.

Bates T. S., Kelley K. C., Johnson J. E., and Gammon R. H. (1996) A reevaluation of the open ocean source of methane to the atmosphere. *J. Geophys. Res.-Atmos.* **101**, 6953–6961.

Beerling D. J. (1993) Changes in the stomatal density of *Betula nana* leaves in response to increases in atmospheric carbon dioxide concentration since the late-glacial. *Spec. Pap. Palaeontol.* **49**, 181–187.

Beerling D. J. and Royer D. L. (2002) Fossil plants as indicators of the Phanerozoic global carbon cycle. *Ann. Rev. Earth. Planet. Sci.* **30**, 527–556.

Bender M., Malaize B., Orchardo J., Sowers T., and Jouzel J. (1999) High precision correlations of Greenland and Antarctic ice core records over the last 100 kyr. In *Mechanisms of Global Climate Change at Millennial Time Scales: Geophysical Monograph 112* (eds. P. U. Clark, R. S. Webb, and L. D. Keigwin). American Geophysical Union, Washington, DC, pp. 149–164.

Berger A. (1988) Milankovitch theory and climate. *Rev. Geophys.* **26**, 624–657.

Berger A., Loutre M. F., and Gallee H. (1998) Sensitivity of the LLN climate model to the astronomical and CO_2 forcings over the last 200 ky. *Clim. Dyn.* **14**, 615–629.

Berger A., Li X. S., and Loutre M. F. (1999) Modeling northern hemisphere ice volume over the last 3 Ma. *Quat. Sci. Rev.* **18**, 1–11.

Berger A. L. (1979) Insolation signatures of Quaternary climatic change. *Il Nuovo Cimento* **2C**, 63–87.

Berger W. H. (1982) Increase of carbon dioxide in the atmosphere during deglaciation: the coral reef hypothesis. *Naturwissenschaften* **69**, 87–88.

Berger W. H. and Keir R. S. (1984) Glacial-Holocene changes in atmospheric CO_2 and the deep-sea record. In *Climate Processes and Climate Sensitivity: Geophysical Monograph 29* (eds. J. E. Hansen and T. Takahashi). American Geophysical Union, Washington, DC, pp. 337–351.

Berner E. K. and Berner R. A. (1987) *The Global Water Cycle.* Prentice-Hall, Englewood Cliffs.

Berner R. A. (1982) Burial of organic carbon and pyrite sulfur in the modern ocean: its geochemical and environmental significance. *Am. J. Sci.* **282**, 451–473.

Berner R. A. (1987) Models for carbon and sulfur cycles and atmospheric oxygen: application to Paleozoic geologic history. *Am. J. Sci.* **287**, 177–196.

Berner R. A. (1990) Atmospheric carbon dioxide levels over Phanerozoic time. *Science* **249**, 1382–1386.

Berner R. A. (1991) A model for atmospheric CO_2 over Phanerozoic time. *Am. J. Sci.* **291**, 339–376.

Berner R. A. (1994) Geocarb II: a revised model of atmospheric CO_2 over Phanerozoic time. *Am. J. Sci.* **294**, 56–91.

Berner R. A. (1999) A new look at the long-term carbon cycle. *GSA Today* **9**, 1–6.

Berner R. A. (2001) Modeling atmospheric O_2 over Phanerozoic time. *Geochim. Cosmochim. Acta* **65**, 685–694.

Berner R. A. and Caldeira K. (1997) The need for mass balance and feedback in the geochemical carbon cycle. *Geology* **25**, 955–956.

Berner R. A. and Kothavala Z. (2001) Geocarb III: a revised model of atmospheric CO_2 over Phanerozoic time. *Am. J. Sci.* **301**, 182–204.

Berner R. A. and Raiswell R. (1983) Burial of organic carbon and pyrite sulfur in sediments over Phanerozoic time: a new theory. *Geochim. Cosmochim. Acta* **47**, 855–862.

Berner R. A., Lasaga A. C., and Garrels R. M. (1983) The carbonate-silicate geochemical cycle and its effect on atmospheric carbon dioxide over the past 100 million years. *Am. J. Sci.* **283**, 641–683.

Berner R. A., Petsch S. T., Lake J. A., Beerling D. J., Popp B. N., Lane R. S., Laws E. A., Westley M. B., Cassar N., Woodward F. I., and Quick W. P. (2000) Isotope fractionation and atmospheric oxygen: implications for Phanerozoic O_2 evolution. *Science* **287**, 1630–1633.

Berner W., Oeschger H., and Stauffer B. (1980) Information on the CO_2 cycle from ice core studies. *Radiocarbon* **22**, 227–235.

Bickle M. J. (1996) Metamorphic decarbonation, silicate weathering and the long-term carbon cycle. *Terra Nova* **8**, 270–276.

Bird M. I., Lloyd J., and Farquhar G. D. (1996) Terrestrial carbon-storage from the last glacial maximum to the present. *Chemosphere* **33**, 1675–1685.

Blasing T. J. and Jones S. (2002) *Current Greenhouse Gas Concentrations Trends: A Compendium of Data on Global Change.* Carbon Dioxide Information Analysis Center, US Department of Energy, Oak Ridge, TN. http://cdiac.esd.ornl.gov/pns/current_ghg.html.

Blunier T. and Brook E. J. (2001) Timing of millennial-scale climate change in Antarctica and Greenland during the last glacial period. *Science* **291**, 109–112.

Blunier T., Chappellaz J. A., Schwander J., Barnola J.-M., Desperts T., Stauffer B., and Raynaud D. (1993) Atmospheric methane, record from a Greenland ice core over the last 1000 years. *Geophys. Res. Lett.* **20**, 2219–2222.

Blunier T., Chappellaz J., Schwander J., Stauffer B., and Raynaud D. (1995) Variations in the atmospheric methane concentration during the Holocene epoch. *Nature* **374**, 46–49.

Blunier T., Schwander J., Stauffer B., Stocker T., Dallenbach A., Indermuhle A., Tschumi J., Chappellaz J., Raynaud D., and Barnola J.-M. (1997) Timing of Antarctic Cold Reversal and the atmospheric CO_2 increase with respect to the Younger Dryas event. *Geophys. Res. Lett.* **24**, 2683–2686.

Blunier T., Chappellaz J., Schwander J., Dallenbach A., Stauffer B., Stocker T. F., Raynaud D., Jouzel J., Clausen H. B., Hammer C. U., and Johnsen S. J. (1998) Asynchrony of Antarctic and Greenland climate change during the last glacial period. *Nature* **394**, 739–743.

Blunier T., Stocker T. F., Chappellaz J., and Raynaud D. (1999) Phase lag of Antarctic and Greenland temperature in the last glacial and link between CO_2 variations and Heinrich events. In *Reconstructing ocean history: A window into the future (Proceedings of the Sixth International Conference on Paleoceanography)* (eds. F. Abrantes and A. Mix). Kluwer Academic/Plenum, New York, pp. 121–138.

Botch M. S., Kobak K. I., Vinson T. S., and Kolchugina T. P. (1995) Carbon pools and accumulation in peatlands of the former Soviet Union. *Global Biogeochem. Cycles* **9**, 37–46.

Boucot A. J. and Gray J. (2001) A critique of Phanerozoic climatic models involving changes in the CO_2 content of the atmosphere. *Earth Sci. Rev.* **56**, 1–159.

Bradley R. S. (1999) *Paleoclimatology: Reconstructing Climates of the Quaternary.* Harcourt Academic Press, New York.

Bralower T. J., Zachos J. C., Thomas E., Parrow M., Paull C. K., Kelly D. C., Premoli Silva I., Sliter W. V., and

Lohmann K. C. (1995) Late Paleocene to Eocene paleoceanography of the equatorial Pacific Ocean: stable isotopes recorded at Ocean Drilling Program Site 865, Allison Guyot. *Paleoceanography* **10**, 841–865.

Bralower T. J., Thomas D. J., Zachos J. C., Hirschmann M. M., Rohl U., Sigurdsson H., Thomas E., and Whitney D. L. (1997) High-resolution records of the late Palaeocene thermal maximum and circum-Caribbean volcanism: is there a causal link? *Geology* **25**, 963–966.

Breas O., Guillou C., Reniero F., and Wada E. (2002) The global methane cycle: isotopes and mixing ratios, sources and sinks. *Isotopes Environ. Health. Stud.* **37**, 257–379.

Brocks J. J., Logan G. A., Buick R., and Summons R. E. (1999) Archean molecular fossils and the early rise of eukaryotes. *Science* **285**, 1033–1036.

Broecker W. S. (1971) Calcite accumulation rates and glacial to interglacial changes in ocean mixing. In *The Late Cenozoic Glacial Ages* (ed. K. K. Turekian). Yale University Press, New Haven, CT, pp. 239–265.

Broecker W. S. (1982) Ocean chemistry during glacial time. *Geochim. Cosmochim. Acta* **46**, 1689–1705.

Broecker W. S. and Henderson G. M. (1998) The sequence of events surrounding Termination II and their implications for the cause of glacial-interglacial CO_2 changes. *Paleoceanography* **13**, 352–364.

Broecker W. S. and Takahashi T. (1977) Neutralization of fossil fuel CO_2 by marine calcium carbonate. In *The Fate of Fossil Fuel CO_2 in the oceans* (eds. N. R. Andersen and A. Malahoff). Plenum, New York, pp. 213–242.

Broecker W. S. and Takahashi T. (1978) The relationship between lysocline depth and *in situ* carbonate ion concentration. *Deep Sea Res.* **25**, 65–95.

Broecker W. S. and van Donk J. (1970) Insolation changes, ice volumes, and the O^{18} record in deep-sea cores. *Rev. Geophys. Space Phys.* **8**, 169–198.

Broecker W. S., Lynch-Stieglitz J., Clark E., Hajdas I., and Bonani G. (2001) What caused the atmosphere's CO_2 content to rise during the last 8000 years? *Geochem. Geophys. Geosys.* **2**, 2001GC000177.

Brook E. J., Sowers T., and Orchardo J. (1996) Rapid variations in atmospheric methane concentration during the past 110,000 years. *Science* **273**, 1087–1091.

Brook E. J., Severinghaus J., Harder S., and Bender M. (1999) Atmospheric methane and millennial scale climate change. In *Mechanisms of Global Climate Change at Millennial Time Scales: Geophysical Monograph 112* (eds. P. U. Clark, R. S. Webb, and L. D. Keigwin). American Geophysical Union, Washington, DC, pp. 165–175.

Brook E. J., Harder S., Severinghaus J., Steig E. J., and Sucher C. M. (2000) On the origin and timing of rapid changes in atmospheric methane during the last glacial period. *Global Biogeochem. Cycles* **14**, 559–572.

Bubier J. L. and Moore T. R. (1994) An ecological perspective on methane emissions from northern wetlands. *Trends Ecol. Evol.* **9**, 460–464.

Budyko M. I. and Ronov A. B. (1979) Chemical evolution of the atmosphere in the Phanerozoic. *Geochem. Int.* **16**, 1–9.

Buick R. (1992) The antiquity of oxygenic photosynthesis: evidence from stromatolites in sulphate-deficient Archaean lakes. *Science* **255**, 74–77.

Caldeira K. (1995) Long-term control of atmospheric carbon dioxide: low-temperature seafloor alteration or terrestrial silicate-rock weathering? *Am. J. Sci.* **295**, 1077–1114.

Cao M., Marshall S., and Gregson K. (1996) Global carbon exchange and methane emissions from natural wetlands: application of a process-based model. *J. Geophys. Res.-Atmos.* **101**, 14399–14414.

Catling D. C., Zahnle K. J., and McKay C. P. (2001) Biogenic methane, hydrogen escape, and the irreversible oxidation of early Earth. *Science* **293**, 839–843.

Cerling T. E. (1992) Use of carbon isotopes in paleosols as an indicator of the pCO_2 of the paleoatmosphere. *Global Biogeochem. Cycles* **6**, 307–314.

Cerling T. H. (1991) Carbon dioxide in the atmosphere: evidence from Cenozoic and Mesozoic paleosols. *Am. J. Sci.* **291**, 377–400.

Chamberlin T. C. (1898) The influence of great epochs of limestone formation upon the constitution of the atmosphere. *J. Geol.* **6**, 609–621.

Chappellaz J., Barnola J. M., Raynaud D., Korotkevich Y. S., and Lorius C. (1990) Ice-core record of atmospheric methane over the past 160,000 years. *Nature* **345**, 127–131.

Chappellaz J., Blunier T., Raynaud D., Barnola J.-M., Schwander J., and Stauffer B. (1993a) Synchronous changes in atmospheric CH_4 and Greenland climate between 40 and 8 kyr BP. *Nature* **366**, 443–445.

Chappellaz J. A., Fung I. Y., and Thompson A. M. (1993b) The atmospheric CH_4 increase since the Last Glacial Maximum: 1. Source estimates. *Tellus* **45B**, 228–241.

Chappellaz J., Blunier T., Kints S., Dallenbach A., Barnola J.-M., Schwander J., Raynaud D., and Stauffer B. (1997) Changes in the atmospheric CH_4 gradient between Greenland and Antarctica during the Holocene. *J. Geophys. Res.-Atmos.* **102**, 15987–15997.

Cicerone R. J. and Oremland R. S. (1988) Biogeochemical aspects of atmospheric methane. *Global Biogeochem. Cycles* **2**, 299–327.

Cloud P. E., Jr. (1968) Atmospheric and hydrospheric evolution on the primitive Earth. *Science* **160**, 729–736.

Clymo R. S., Turunen J., and Tolonen K. (1998) Carbon accumulation in peatland. *Oikos* **81**, 368–388.

Cowling S. A. (1999) Plants and temperature-CO_2 uncoupling. *Science* **285**, 1500–1501.

Craig H. and Chou C. C. (1982) Methane: the record in polar ice cores. *Geophys. Res. Lett.* **9**, 1221–1224.

Craig H., Horibe Y., and Sowers T. (1988) Gravitational separation of gases and isotopes in polar ice caps. *Science* **242**, 1675–1678.

Crowell J. C. (1999) *Pre-Mesozoic ice Ages: Their Bearing on understanding the Climate System: GSA Memoir 192*. Geological Society of America, Boulder.

Crowley T. J. (1991) Ice-age methane variations. *Nature* **353**, 122–123.

Crowley T. J. (1995) Ice age terrestrial carbon changes revisited. *Global Biogeochem. Cycles* **9**, 377–389.

Crowley T. J. (2000) Carbon dioxide and Phanerozoic climate: an overview. In *Warm Climates in Earth History* (eds. B. T. Huber, K. G. MacLeod, and S. L. Wing). Cambridge University Press, Cambridge, pp. 425–444.

Crowley T. J. and Baum S. K. (1993) Effect of decreased solar luminosity on late Precambrian ice extent. *J. Geophys. Res.-Atmos.* **98**, 16723–16732.

Crowley T. J. and Berner R. A. (2001) CO_2 and climate change. *Science* **292**, 870–872.

Curry W. B., Duplessy J. C., Labeyrie L. D., and Shackleton N. J. (1988) Changes in the distribution of $\delta^{13}C$ of deep water $\sum CO_2$ between the last glaciation and the Holocene. *Paleoceanography* **3**, 317–341.

Dacey J. W. H. and Klug M. J. (1979) Methane efflux from lake sediments through water lilies. *Science* **203**, 1253–1255.

Dallenbach A., Blunier T., Fluckiger J., Stauffer B., Chappellaz J., and Raynaud D. (2000) Changes in the atmospheric CH_4 gradient between Greenland and Antarctica during the last glacial and the transition to the Holocene. *Geophys. Res. Lett.* **27**, 1005–1008.

Daly R. A. (1909) First calcareous fossils and the evolution of the limestones. *Geol. Soc. Am. Bull.* **20**, 155–170.

Dansgaard W., Johnsen S. J., Moller J., and Langway C. C., Jr. (1969) One thousand centuries of climatic record from Camp Century on the Greenland ice sheet. *Science* **166**, 377–381.

Dansgaard W., Johnsen S. J., Clausen H. B., Dahl-Jensen D., Gundestrup N., Hammer C. U., and Oeschger H. (1984) North Atlantic climatic oscillations revealed by deep Greenland ice cores. In *Climate Processes and Climate Sensitivity: Geophysical Monograph 29* (eds. J. E. Hansen and T. Takahashi). American Geophysical Union, Washington, DC, pp. 288–298.

Degens E. T., Kempe S., and Richey J. E. (1991) Summary: biogeochemistry of major world rivers. In *SCOPE 42: Biogeochemistry of Major World Rivers* (eds. E. T. Degens, S. Kempe, and J. E. Richey). Wiley, New York, pp. 323–347.

Deines P. (1980) The isotopic composition of reduced organic carbon. In *Handbook of Environmental Isotope Geochemistry* (eds. P. Fritz and J. Ch. Fontes). Elsevier, New York, vol. 1, pp. 329–406.

Deines P. (1992) Mantle carbon: concentration, mode of occurrence, and isotopic composition. In *Early Organic Evolution: Implications for Mineral and Energy Resources* (eds. M. Schidlowski, S. Golubic, M. M. Kimberley, D. M. McKirdy, and P. A. Trudinger). Springer, Berlin, pp. 133–146.

Deleens E., Ferhi A., and Queiroz O. (1983) Carbon isotope fractionation by plants using the C4 pathway. *Physiol. Veg.* 21, 897–905.

Delmas R. J. (1993) A natural artefact in Greenland ice-core CO_2 measurements. *Tellus* 45B, 391–396.

Delmas R., Ascencio J.-M., and Legrand P. (1980) Polar ice evidence that atmospheric CO_2 20,000 yr BP was 50% of present. *Nature* 284, 155–157.

Des Marais D. J. (2000) Evolution: when did photosynthesis emerge on Earth? *Science* 289, 1703–1705.

Des Marais D. J. (2001) Isotopic evolution of the biogeochemical carbon cycle during the Precambrian. *Rev. Mineral. Geochem.* 43, 555–578.

Des Marais D. J., Strauss H., Summons R. E., and Hayes J. M. (1992) Carbon isotope evidence for the stepwise oxidation of the Proterozoic environment. *Nature* 359, 605–609.

Dibb J. E., Rasmussen R. A., Mayewski P. A., and Holdsworth G. (1993) Northern Hemisphere concentrations of methane and nitrous oxide since 1800: results from the Mt. Logan and 20D ice cores. *Chemosphere* 27, 2413–2423.

Dickens G. R. (2000) Methane oxidation during the Late Palaeocene Thermal Maximum. *Bull. Soc. Geol. France* 171, 37–49.

Dickens G. R. (2001) Modeling the global carbon cycle with a gas hydrate capacitor: significance for the Latest Paleocene Thermal Maximum. In *Natural Gas Hydrates: Occurrence, Distribution, and Detection: Geophysical Monograph 124* (eds. C. K. Paull and W. P. Dillon). American Geophysical Union, Washington, DC, pp. 19–38.

Dickens G. R., Oneil J. R., Rea D. K., and Owen R. M. (1995) Dissociation of oceanic methane hydrate as a cause of the carbon isotope excursion at the end of the Paleocene. *Paleoceanography* 10, 965–971.

Dickens G. R., Castillo M. M., and Walker J. C. G. (1997) A blast of gas in the latest Paleocene: simulating first-order effects of massive dissociation of oceanic methane hydrate. *Geology* 25, 259–262.

Dillon W. P. (2001) Gas hydrate in the ocean environment. In *Encyclopedia of Physical Science and Technology* (ed. R. Meyers). Academic Press, San Diego, vol. 6, pp. 473–486.

Dobrovolsky V. V. (1994) *Biogeochemistry of the World's Land*. CRC Press, Boca Raton, Florida.

Douglas R. G. and Woodruff F. (1981) Deep sea benthic foraminifera. In *The Sea* (ed. C. Emiliani). Wiley, New York, vol. 7, pp. 1233–1327.

Doval M. D. and Hansell D. A. (2000) Organic carbon and apparent oxygen utilization in the western South Pacific and the central Indian Oceans. *Mar. Chem.* 68, 249–264.

Drever J. I., Li Y.-H., and Maynard J. B. (1988) Geochemical cycles: the continental crust and the oceans. In *Chemical Cycles in the Evolution of the Earth* (eds. C. B. Gregor, R. M. Garrels, F. T. Mackenzie, and J. B. Maynard). Wiley, New York, pp. 17–53.

Duplessy J. C., Shackleton N. J., Fairbanks R. G., Labeyrie L., Oppo D., and Kallel N. (1988) Deepwater source variations during the last climatic cycle and their impact on the global deepwater circulation. *Paleoceanography* 3, 343–360.

Edmond J. M., Palmer M. R., Measures C. I., Grant B., and Stallard R. F. (1995) The fluvial geochemistry and denudation rate of the Guayana Shield in Venezuela, Colombia, and Brazil. *Geochim. Cosmochim. Acta* 59, 3301–3325.

Ehhalt D. H. (1974) The atmospheric cycle of methane. *Tellus* 26, 58–70.

Ekart D. D., Cerling T. E., Montanez I. P., and Tabor N. J. (1999) A 400 million year carbon isotope record of pedogenic carbonate: implications for paleoatmospheric carbon dioxide. *Am. J. Sci.* 299, 805–827.

Emiliani C. (1966) Paleotemperature analysis of Caribbean cores P6304-8 and P6304-9 and generalized temperature curve for the past 425,000 years. *J. Geol.* 74, 109–126.

Enting I. G. (1992) The incompatibility of ice-core CO_2 data with reconstructions of biotic CO_2 sources (II): the influence of CO_2–fertilised growth. *Tellus* 44B, 23–32.

Enting I. G. and Mansbridge J. V. (1991) Latitudinal distribution of sources and sinks of CO_2: results of an inversion study. *Tellus* 43B, 156–170.

Etheridge D. M., Pearman G. I., and de Silva F. (1988) Atmospheric trace-gas variations as revealed by air trapped in an ice core from Law Dome, Antarctica. *Ann. Glaciol.* 10, 28–33.

Etheridge D. M., Pearman G. I., and Fraser P. J. (1992) Changes in tropospheric methane between 1841 and 1978 from a high accumulation-rate Antarctic ice core. *Tellus* 44B, 282–294.

Etheridge D. M., Steele L. P., Langenfelds R. L., Francey R. J., Barnola J.-M., and Morgan V. I. (1996) Natural and anthropogenic changes in atmospheric CO_2 over the last 1000 years from air in Antarctic ice and firn. *J. Geophys. Res.-Atmos.* 101, 4115–4128.

Etheridge D. M., Steele L. P., Francey R. J., and Langenfelds R. L. (1998) Atmospheric methane between 1000 AD and present: evidence of anthropogenic emissions and climatic variability. *J. Geophys. Res.-Atmos.* 103, 15979–15993.

Etiope G. and Klusman R. W. (2002) Geologic emissions of methane to the atmosphere. *Chemosphere* 49, 777–789.

Falkowski P., Scholes R. J., Boyle E., Canadell J., Canfield D., Elser J., Gruber N., Hibbard K., Hogberg P., Linder S., Mackenzie F. T., Moore I. B., Pedersen T., Rosenthal Y., Seitzinger S., Smetacek V., and Steffen W. (2000) The global carbon cycle: a test of our knowledge of Earth as a system. *Science* 290, 291–296.

Farquhar J., Bao H. M., and Thiemens M. (2000) Atmospheric influence of Earth's earliest sulfur cycle. *Science* 289, 756–758.

Farrell J. W. and Prell W. L. (1989) Climatic change and $CaCO_3$ preservation: an 800,000 year bathymetric reconstruction from the central equatorial Pacific Ocean. *Paleoceanography* 4, 447–466.

Felzer B., Webb T., and Oglesby R. J. (1998) The impact of ice sheets, CO_2, and orbital insolation on late Quaternary climates: sensitivity experiments with a general circulation model. *Quat. Sci. Rev.* 17, 507–534.

Fischer H., Wahlen M., Smith J., Mastroianni D., and Deck B. (1999) Ice core records of atmospheric CO_2 around the last three glacial terminations. *Science* 283, 1712–1714.

Fluckiger J., Monnin E., Stauffer B., Schwander J., Stocker T. F., Chappellaz J., Raynaud D., and Barnola J.-M. (2002) High-resolution Holocene N_2O ice core record and its relationship with CH_4 and CO_2. *Global Biogeochem. Cycles* 16, 10.29/2001GB001417.

Francey R. J., Allison C. E., Etheridge D. M., Trudinger C. M., Enting I. G., Leuenberger M., Langenfelds R. L., Michel E., and Steele L. P. (1999) A 1000-year high precision record of $\delta^{13}C$ in atmospheric CO_2. *Tellus* 51B, 170–193.

Francois L. M., Godderis Y., Warnant P., Ramstein G., de Noblet N., and Lorenz S. (1999) Carbon stocks and isotopic budgets of the terrestrial biosphere at mid-Holocene and last glacial maximum times. *Chem. Geol.* 159, 163–189.

Francois R., Altabet M. A., Yu E.-F., Sigman D. M., Bacon M. P., Frank M., Bohrmann G., Bareille G., and Labeyrie L. D. (1997) Contribution of Southern Ocean surface-water stratification to low atmospheric CO_2 concentrations during the last glacial period. *Nature* **389**, 929–935.

Freeman K. H. and Hayes J. M. (1992) Fractionation of carbon isotopes by phytoplankton and estimates of ancient CO_2 levels. *Global Biogeochem. Cycles* **6**, 185–198.

Friedli H., Moor E., Oeschger H., Siegenthaler U., and Stauffer B. (1984) $^{13}C/^{12}C$ ratios in CO_2 extracted from Antarctic ice. *Geophys. Res. Lett.* **11**, 1145–1148.

Friedli H., Lotscher H., Oeschger H., Siegenthaler U., and Stauffer B. (1986) Ice core record of the $^{13}C/^{12}C$ ratio of atmospheric CO_2 in the past two centuries. *Nature* **324**, 237–238.

Fung I., Prentice K., Matthews E., Lerner J., and Russell G. (1983) Three-dimensional tracer model of atmospheric CO_2: response to seasonal exchanges with the terrestrial biosphere. *J. Geophys. Res.-Oceans* **88**, 1281–1294.

Fung I., John J., Lerner J., Matthews E., Prather M., Steele L. P., and Fraser P. J. (1991) Three-dimensional model synthesis of the global methane cycle. *J. Geophys. Res.-Atmos.* **96**, 13033–13065.

Galchenko V. F., Lein A., and Ivanov M. (1989) Biological sinks of methane. In *Exchange of Trace Gases between Terrestrial Ecosystems and the Atmosphere* (eds. M. O. Andreae and D. S. Schimel). Wiley, New York, pp. 59–71.

Garrels R. M. and Lerman A. (1981) Phanerozoic cycles of sedimentary carbon and sulfur. *Proc. Natl. Acad. Sci.* **78**, 4652–4656.

Garrels R. M. and Lerman A. (1984) Coupling of the sedimentary sulfur and carbon cycles-an improved model. *Am. J. Sci.* **284**, 989–1007.

Garrels R. M., Lerman A., and Mackenzie F. T. (1976) Controls of atmospheric O_2 and CO_2: past, present, and future. *Am. Sci.* **64**, 306–315.

Gerlach T. M. (1991) Present-day CO_2 emissions from volcanoes. *EOS, Trans., AGU* **72**, 249, 254–255.

Ghosh P., Ghosh P., and Bhattacharya S. K. (2001) CO_2 levels in the Late Palaeozoic and Mesozoic atmosphere from soil carbonate and organic matter, Satpura basin, Central India. *Palaeogeogr. Palaeoclimatol. Palaeoecol.* **170**, 219–236.

Gillett R. W., van Ommen T. D., Jackson A. V., and Ayers G. P. (2000) Formaldehyde and peroxide concentrations in Law Dome (Antarctica) firn and ice cores. *J. Glaciol.* **46**, 15–19.

Gorham E. (1991) Northern peatlands: role in the carbon cycle and probable responses to climatic warming. *Ecol. Appl.* **1**, 182–195.

Grocke D. R. (2002) The carbon isotope composition of ancient CO_2 based on higher-plant organic matter. *Phil. Trans. Roy. Soc. London Ser. A: Math. Phys. Eng. Sci.* **360**, 633–658.

Haan D. and Raynaud D. (1998) Ice core record of CO variations during the two last millennia: atmospheric implications and chemical interactions within the Greenland ice. *Tellus* **50B**, 253–262.

Hammer C. U., Clausen H. B., Dansgaard W., Gundestrup N., Johnsen S. J., and Reeh N. (1978) Dating of Greenland ice cores by flow models, isotopes, volcanic debris, and continental dust. *J. Glaciol.* **20**, 3–26.

Hansell D. A. and Carlson C. A. (1998) Deep-ocean gradients in the concentration of dissolved organic carbon. *Nature* **395**, 263–266.

Hanson R. S. and Hanson T. E. (1996) Methanotrophic bacteria. *Microbiol. Rev.* **60**, 439–471.

Hardie L. A. (1996) Secular variation in seawater chemistry: an explanation for the coupled secular variation in the mineralogies of marine limestones and potash evaporites over the past 600 m. y. *Geology* **24**, 279–283.

Hayes J. M. (1994) Global methanotrophy at the Archean-Proterozoic transition. In *Early Life on Earth: Nobel Symposium No. 84* (ed. S. Bengtson). Columbia University Press, New York, pp. 220–236.

Hayes J. M., Strauss H., and Kaufman A. J. (1999) The abundance of ^{13}C in marine organic matter and isotopic fractionation in the global biogeochemical cycle of carbon during the past 800 Ma. *Chem. Geol.* **161**, 103–125.

Hedges S. B., Chen H., Kumar S., Wang D. Y. C., Thompson A. S., and Watanabe H. (2001) A genomic timescale for the origin of eukaryotes. *BMC Evol. Biol.* **1**, article 4: (1471-2148/1/4).

Hein R., Crutzen P. J., and Heimann M. (1997) An inverse modeling approach to investigate the global atmospheric methane cycle. *Global Biogeochem. Cycles* **11**, 43–76.

Henderson G. M. and Slowey N. C. (2000) Evidence from U–Th dating against Northern Hemisphere forcing of the penultimate deglaciation. *Nature* **404**, 61–66.

Hewitt C. D. and Mitchell J. F. B. (1997) Radiative forcing and response of a GCM to ice age boundary conditions: cloud feedback and climate sensitivity. *Clim. Dyn.* **13**, 821–834.

Hinrichs K.-U. and Boetius A. (2002) The anaerobic oxidation of methane: new insights in microbial ecology and biogeochemistry. In *Ocean Margin Systems* (eds. G. Wefer, D. Billett, D. Hebbeln, B. B. Jorgensen, M. Schluter, and T. van Weering). Springer, New York, pp. 457–477.

Hoehler T. M., Bebout B. M., and Des Marais D. J. (2001) The role of microbial mats in the production of reduced gases on the early Earth. *Nature* **412**, 324–327.

Hoffman P. F. and Schrag D. P. (2002) The snowball Earth hypothesis: testing the limits of global change. *Terra Nova* **14**, 129–155.

Hoffman P. F., Kaufman A. J., Halverson G. P., and Schrag D. P. (1998) A Neoproterozoic snowball earth. *Science* **281**, 1342–1346.

Hofmann M., Broecker W. S., and Lynch-Stieglitz J. (1999) Influence of a $[CO_{2(aq)}]$ dependent biological C-isotope fractionation on glacial C-13/C-12 ratios in the ocean. *Global Biogeochem. Cycles* **13**, 873–883.

Holland H. D. (1978) *Chemistry of the Atmosphere and Oceans.* Wiley, New York.

Holland H. D. (1994) Early Proterozoic atmospheric change. In *Early Life on Earth: Nobel Symposium No. 84* (ed. S. Bengtson). Columbia University Press, New York, pp. 237–244.

Holland H. D. (2002) Volcanic gases, black smokers, and the Great Oxidation Event. *Geochim. Cosmochim. Acta* **66**, 3811–3826.

Holland H. D. (2003) Discussion of the article by A. C. Lasaga and H. Omoto on "The oxygen geochemical cycle: dynamics and stability," Geochim. Cosmochim. Acta **66**, 361–381, 2002. *Geochim. Cosmochim. Acta* **67**, 787–789.

Holland H. D. and Rye R. (1997) Evidence in pre-2.2 Ga paleosols for the early evolution of atmospheric oxygen and terrestrial biota: comment and reply. *Geology* **25**, 857–859.

Holland H. D. and Zimmerman H. (2000) The dolomite problem revisited. *Int. Geol. Rev.* **42**, 481–490.

Holmes M. E., Sansone F. J., Rust T. M., and Popp B. N. (2000) Methane production, consumption, and air-sea exchange in the open ocean: an evaluation based on carbon isotopic ratios. *Global Biogeochem. Cycles* **14**, 1–10.

Horita J., Zimmermann H., and Holland H. D. (2002) Chemical evolution of seawater during the Phanerozoic: implications from the record of marine evaporites. *Geochim. Cosmochim. Acta* **66**, 3733–3756.

Hornafius J. S., Quigley D., and Luyendyk B. P. (1999) The world's most spectacular marine hydrocarbon seeps (Coal Oil Point, Santa Barbara Channel, California): quantification of emissions. *J. Geophys. Res.-Oceans* **104**, 20703–20711.

Hovland M., Judd A. G., and Burke J. R. A. (1993) The global flux of methane from shallow submarine sediments. *Chemosphere* **26**, 559–578.

Hunt J. M. (1996) *Petroleum Geochemistry and Geology.* W. H. Freeman, New York.

Imbrie J., Boyle E. A., Clemens S. C., Duffy A., Howard W. R., Kukla G., Kutzbach J., Martinson D. G., McIntyre A., Mix A. C., Molfino B., Morley J. J., Peterson L. C., Pisias N. G., Prell W. G., Raymo M. E., Shackleton N. J., and Toggweiler J. R. (1992) On the structure and origin of major glaciation cycles: 1. Linear responses to Milankovitch forcing. *Paleoceanography* 7, 701–738.

Imbrie J., Berger A., Boyle E. A., Clemens S. C., Duffy A., Howard W. R., Kukla G., Kutzbach J., Martinson D. G., McIntyre A., Mix A. C., Molfino B., Morley J. J., Peterson L. C., Pisias N. G., Prell W. G., Raymo M. E., Shackleton N. J., and Toggweiler J. R. (1993) On the structure and origin of major glaciation cycles: 2. The 100,000-year cycle. *Paleoceanography* 8, 699–735.

Indermuhle A., Stocker T. F., Joos F., Fischer H., Smith H. J., Wahlen M., Deck B., Mastroianni D., Tschumi J., Blunier T., Meyer R., and Stauffer B. (1999) Holocene carbon-cycle dynamics based on CO_2 trapped in ice at Taylor Dome, Antarctica. *Nature* 398, 121–126.

Indermuhle A., Monnin E., Stauffer B., Stocker T. F., and Wahlen M. (2000) Atmospheric CO_2 concentration from 60 to 20 kyr BP from the Taylor Dome ice core, Antarctica. *Geophys. Res. Lett.* 27, 735–738.

Ingall E. D. and Van Cappellen P. (1990) Relation between sedimentation rate and burial of organic phosphorus and organic carbon in marine sediments. *Geochim. Cosmochim. Acta* 54, 373–386.

Ittekot V. (1988) Global trends in the nature of organic matter in river suspensions. *Nature* 332, 436–438.

Johnsen S. J., Dansgaard W., Clausen H. B., and Langway C. C., Jr. (1972) Oxygen isotope profiles through the Antarctic and Greenland ice sheets. *Nature* 235, 429–434.

Joos F., Meyer R., Bruno M., and Leuenberger M. (1999) The variability in the carbon sinks as reconstructed for the last 1000 years. *Geophys. Res. Lett.* 26, 1437–1440.

Jouzel J., Lorius C., Petit J. R., Genthon C., Barkov N. I., Kotlyakov V. M., and Petrov V. M. (1987) Vostok ice core: a continuous isotope temperature record over the last climatic cycle (160,000 years). *Nature* 329, 403–408.

Jouzel J., Barkov N. I., Barnola J.-M., Bender M., Chappellaz J., Genthon C., Kotlyakov V. M., Lipenkov V., Lorius C., Petit J.-R., Raynaud D., Raisbeck G., Ritz C., Sowers T., Sievenard M., Yiou F., and Yiou P. (1993) Extending the Vostok ice-core record of paleoclimate to the penultimate glacial period. *Nature* 364, 407–412.

Judd A. G. (2000) Geological sources of methane. In *Atmospheric Methane: Its Role in the Global Environment* (ed. M. A. K. Khalil). Springer, New York, pp. 280–303.

Judd A. G., Hovland M., Dimitrov L. I., Garcia Gil S. G., and Jukes V. (2002) The geological methane budget at continental margins and its influence on climate change. *Geofluids* 2, 109–126.

Karhu J. A. and Holland H. D. (1996) Carbon isotopes and the rise of atmospheric oxygen. *Geology* 24, 867–870.

Kasting J. F. (1987) Theoretical constraints on oxygen and carbon dioxide concentrations in the Precambrian atmosphere. *Precamb. Res.* 34, 205–228.

Kasting J. F. (1993) Earth's early atmosphere. *Science* 259, 920–926.

Kasting J. F. (2001) The rise of atmospheric oxygen. *Science* 293, 819–820.

Kasting J. F., Zahnle K. J., and Walker J. C. G. (1983) Photochemistry of methane in the Earth's early atmosphere. *Precamb. Res.* 20, 121–148.

Kasting J. F., Toon O. B., and Pollack J. B. (1988) How climate evolved on the terrestrial planets. *Sci. Am.* 256, 90–97.

Kasting J. F., Eggler D. H., and Raeburn S. P. (1993) Mantle redox evolution and the case for a reduced Archean atmosphere. *J. Geol.* 101, 245–257.

Keeling C. D. (1991) *Atmospheric CO_2—modern record, South Pole. Trends:* A Compendium of Data on Global Change. Carbon Dioxide Information Analysis Center, US Department of Energy, Oak Ridge, TN.

Keeling C. D., Bacastow R. B., Carter A. F., Piper S. C., Whorf T. P., Heimann M., Mook W. G., and Roeloffzen H. (1989) A three-dimensional model of atmospheric CO_2 transport based on observed winds: 1. Analysis of observational data. In *Aspects of Climate Variability in the Pacific and the Western Americas: Geophysical Monograph 55* (ed. D. H. Peterson). American Geophysical Union, Washington, DC, pp. 165–236.

Kelley C. A. and Jeffrey W. H. (2002) Dissolved methane concentration profiles and air-sea fluxes from 41°S to 27°N. *Global Biogeochem. Cycles* 16, 10.1029/2001GB001809.

Kelley D. S., Karson J. A., Blackman D. K., Fruh-Green G. L., Butterfield D. A., Lilley M. D., Olson E. J., Schrenk M. O., Roe K. K., Lebon G. T., Rivizzigno P. and the AT3-60 Shipboard Party. (2001) An off-axis hydrothermal vent field near the Mid-Atlantic Ridge at 30N. *Nature* 412, 145–149.

Kennett J. P. and Stott L. D. (1991) Abrupt deep sea warming, paleoceanographic changes and benthic extinctions at the end of the Palaeocene. *Nature* 353, 319–322.

Kennett J. P., Cannariato K. G., Hendy I. L., and Behl R. J. (2003) *Methane Hydrates in Quaternary Climate Change: The Clathrate Gun Hypothesis.* American Geophysical Union, Washington, DC.

Kerrick D. M. (2001) Present and past nonanthropogenic CO_2 degassing from the solid earth. *Rev. Geophys.* 39, 565–585.

Kerrick D. M., McKibben M. A., Seward T. M., and Caldeira K. (1995) Convective hydrothermal CO_2 emission from high heat flow regions. *Chem. Geol.* 121, 285–293.

Khalil M. A. K. and Rasmussen R. A. (1989) Climate-induced feedback for the global cycles of methane and nitrous oxide. *Tellus* 41B, 554–559.

Kiehl J. T. and Dickinson R. E. (1987) A study of the radiative effects of enhanced atmospheric CO_2 and CH_4 on early Earth surface temperatures. *J. Geophys. Res.-Atmos.* 92, 2991–2998.

King G. M. (1992) Ecological aspects of methane oxidation, a key determinant of global methane dynamics In *Advances in Microbial Ecology* (ed. K. C. Marshall). Plenum, New York, vol. 12, pp. 431–468.

Koch P. L., Zachos J. C., and Gingerich P. D. (1992) Correlation between isotope records in marine and continental carbon reservoirs near the Paleocene/Eocene boundary. *Nature* 358, 319–322.

Ku T.-L. and Luo S. (1992) Carbon isotopic variations on glacial-to-interglacial time scales in the ocean: modeling and implications. *Paleoceanography* 7, 543–562.

Kuhn W. R. and Kasting J. F. (1983) Effects of increased CO_2 concentrations on surface temperature of the early Earth. *Nature* 301, 53–55.

Kump L. R., Kasting J. F., and Robinson J. M. (1991) Atmospheric oxygen variation through geologic time–introduction. *Palaeogeogr. Palaeoclimatol. Palaeoecol.* 97, 1–3.

Kump L. R., Brantley S. L., and Arthur M. A. (2000) Chemical weathering, atmospheric CO_2, and climate. *Ann. Rev. Earth. Planet. Sci.* 28, 611–667.

Kvenvolden K. A. and Lorenson T. D. (2001) The global occurrence of natural gas hydrate. In *Natural Gas Hydrates: Occurrence, Distribution, and Detection: Geophysical Monograph 124* (eds. C. K. Paull and W. P. Dillon). American Geophysical Union, Washington, DC, pp. 3–18.

Lacroix A. V. (1993) Unaccounted-for sources and isotopically enriched methane and their contribution to the emissions inventory: a review synthesis. *Chemosphere* 26, 507–557.

Lambert G. and Schmidt S. (1993) Reevaluation of the oceanic flux of methane: uncertainties and long term variations. *Chemosphere* 26, 579–589.

Langenfelds R. L., Fraser P. J., Francey R. J., Steele L. P., Porter L. W., and Allison C. E. (1996) The Cape Grim air archive: the first seventeen years, 1978–1995. In *Baseline Atmospheric Program Australia* (ed. R. J. Francey et al.). Bureau of Meteorology and CSIRO Division of Atmospheric Research, Melbourne, Australia, pp. 53–70.

Lappalainen E. (1996) General review on world peatland and peat resources. In *Global Peat Resources* (ed. E. Lappalainen). International Peat Society, Finland, pp. 53–56.

Lasaga A. C. and Ohmoto H. (2002) The oxygen geochemical cycle: dynamics and stability. *Geochim. Cosmochim. Acta* **66**, 361–381.

Lasaga A. C., Berner R. A., and Garrels R. M. (1985) An improved geochemical model of atmospheric CO_2 fluctuations over the past 100 million years. In *The Carbon Cycle and Atmospheric CO_2: Natural Variations Archean to Present: Geophysical Monograph 32* (eds. E. T. Sundquist and W. S. Broecker). American Geophysical Union, Washington, DC, pp. 397–411.

Lein A. Y. (1984) Anaerobic consumption of organic matter in modern marine sediments. *Nature* **312**, 148–150.

Lelieveld J., Crutzen P. J., and Dentener F. J. (1998) Changing concentration, lifetime, and climate forcing of atmospheric methane. *Tellus* **50B**, 128–150.

Lemarchand D., Gaillardet J., Lewin E., and Allegre C. J. (2000) The influence of rivers on marine boron isotopes and implications for reconstructing past ocean pH. *Nature* **408**, 951–954.

Lemarchand D., Gaillardet J., Lewin E., and Allegre C. J. (2002) Boron isotope systematics in large rivers: implications for the marine boron budget and paleo-pH reconstruction over the Cenozoic. *Chem. Geol.* **190**, 123–140.

Leuenberger M., Siegenthaler U., and Langway C. C. (1992) Carbon isotope composition of atmospheric CO_2 during the last ice age from an Antarctic ice core. *Nature* **357**, 450–488.

Levchenko V. A., Francey R. J., Etheridge D. M., Tuniz C., Head J., Morgan V. I., Lawson E., and Jacobsen G. (1996) The ^{14}C "bomb spike" determines the age spread and age of CO_2 in Law Dome firn and ice. *Geophys. Res. Lett.* **23**, 3345–3348.

Levine J. S., Cofer W. R., and Pinto J. P. (2000) Biomass burning. In *Atmospheric Methane: Its Role in the Global Environment* (ed. M. A. K. Khalil). Springer, New York, pp. 190–201.

Li Y.-H. (2000) *A compendium of geochemistry: from solar nebula to the human brain.* Princeton University Press, Princeton.

Lorius C., Merlivat L., Jouzel J., and Pourchet M. (1979) A 30,000-yr isotope climatic record from Antarctic ice. *Nature* **280**, 644–648.

Lorius C., Jouzel J., Ritz C., Merlivat L., Barkov N. I., Korotkevich Y. S., and Kotlyakov V. M. (1985) A 150,000-year climatic record from Antarctic ice. *Nature* **316**, 591–596.

Lorius C., Jouzel J., Raynaud D., Hansen J., and Le Treut H. (1990) The ice-core record: climate sensitivity and future greenhouse warming. *Nature* **347**, 139–145.

MacLeod G., McKeown C., Hall H. J., and Russell M. J. (1994) Hydrothermal and oceanic pH conditions of possible relevance to the origin of life. *Orig. Life Evol. Biosph.* **23**, 19–41.

Manabe S. and Broccoli A. J. (1985) A comparison of climate model sensitivity with data from the last glacial maximum. *J. Atmos. Sci.* **42**, 2643–2651.

Marino B. D. and McElroy M. B. (1991) Isotopic composition of atmospheric CO_2 inferred from carbon in C4 plant cellulose. *Nature* **349**, 127–131.

Martin J. H. (1990) Glacial-interglacial CO_2 change: the iron hypothesis. *Paleoceanography* **5**, 1–13.

Martin J. H., Fitzwater S. E., and Gordon R. M. (1990) Iron deficiency limits phytoplankton growth in Antarctic waters. *Global Biogeochem. Cycles* **4**, 5–12.

Martinerie P., Brasseur G. P., and Granier C. (1995) The chemical composition of ancient atmospheres: a model study constrained by ice core data. *J. Geophys. Res.-Atmos.* **100**, 14291–14304.

Marty B. and Tolstikhin I. N. (1998) CO_2 fluxes from mid-ocean ridges, arcs, and plumes. *Chem. Geol.* **145**, 233–248.

Maslin M. A. and Thomas E. (2003) Balancing the deglacial global carbon budget: the hydrate factor. *Quat. Sci. Rev.* **22**, 1729–1736.

Masters C. D., Attanasi E. D., and Root D. H. (1994) USGS World Petroleum Assessment and Analysis. In *Proceedings of the 14th World Petroleum Congress.* Wiley, New York.

Mathis P. (1990) Compared structure of plant and bacterial photosynthetic reaction centers. Evolutionary implications. *Biochim. Biophys. Acta (BBA): Bioenergetics* **1018**(2–3): 163–167.

Matthews E. (2000) Wetlands. In *Atmospheric Methane: Its Role in the Global Environment* (ed. M. A. K. Khalil). Springer, New York, pp. 202–233.

Matthews E. and Fung I. (1987) Methane emission from natural wetlands: global distribution, area, and environmental characteristics of sources. *Global Biogeochem. Cycles* **1**, 61–86.

Max M. D., Dillon W. P., Nishimura C., and Hurdle B. G. (1999) Sea-floor methane blow-out and global firestorm at the K–T boundary. *Geo. Mar. Lett.* **18**, 285–291.

McElroy M. B. (1983) Marine biological controls on atmospheric CO_2 and climate. *Nature* **302**, 328–329.

McElroy M. B. (1989) Studies of polar ice: insights for atmospheric chemistry. In *The Environmental Record in Glaciers and Ice Sheets* (eds. H. Oeschgers and Langway, Jr.). Wiley, New York, pp. 363–377.

McElwain J. C. and Chaloner W. G. (1995) Stomatal density and index of fossil plants track atmospheric carbon dioxide in the Paleozoic. *Ann. Bot.* **76**, 389–395.

Meybeck M. (1981) River transport of organic carbon to the ocean. In *Flux of Organic Carbon by Rivers to the Oceans.* Carbon Dioxide Effects Research and Assessment Program(eds. G. E. Likens and F. T. Mackenzie). US DOE, Washington, DC, pp. 219–269.

Meybeck M. (1987) Global chemical weathering of surficial rocks estimated from river dissolved loads. *Am. J. Sci.* **288**, 401–428.

Meybeck M. (1988) How to establish and use world budgets of riverine materials. In *Physical and Chemical Weathering in Geochemical Cycles* (eds. A. Lerman and M. Meybeck). Kluwer Academic, Dordrecht, pp. 247–272.

Miller K. G., Fairbanks R. G., and Mountain G. S. (1987) Tertiary oxygen isotope synthesis, sea level history, and continental margin erosion history. *Paleoceanography* **2**, 1–19.

Milliman J. D. (1993) Production and accumulation of calcium carbonate in the ocean: budget of a nonsteady state. *Global Biogeochem. Cycles* **7**, 927–957.

Milliman J. D., Quinchun X., and Zuosheng Y. (1984) Transfer of particulate organic carbon and nitrogen from the Changjiang River to the ocean. *Am. J. Sci.* **284**, 824–834.

Monnin E., Indermuhle A., Dallenbach A., Fluckiger J., Stauffer B., Stocker T. F., Raynaud D., and Barnola J. M. (2001) Atmospheric CO_2 concentrations over the last glacial termination. *Science* **291**, 112–114.

Moomaw W. R., Moreira J. R., Blok K., Greene D. L., Gregory K., Jaszay T., Kashiwagi T., Levine M., McFarland M., Prasad N. S., Price L., Rogner H.-H., Sims R., Zhou F., and Zhou P. (2001) Technological and economic potential of greenhouse gas emissions reduction. In *Climate Change 2001: Mitigation. Contribution of Working Group III to the Third Assessment Report of the Intergovernmental Panel on Climate Change* (eds. B. Metz, O. Davidson, R. Swart, and J. Pan). Cambridge University Press, Cambridge, pp. 171–299.

Moosavi S. C., Crill P. M., Pullman E. R., Funk D. W., and Peterson K. M. (1996) Controls on CH_4 flux from an Alaskan boreal wetland. *Global Biogeochem. Cycles* **10**, 287–296.

Morner N.-A. and Etiope G. (2002) Carbon degassing from the lithosphere. *Global Planet. Change* **33**, 185–203.

Nakazawa T., Machida T., Tanaka M., Fujii Y., Aoki S., and Watanabe O. (1993) Differences of the atmospheric CH_4 concentration between the Arctic and Antarctic regions in pre-Industrial/pre-Agricultural era. *Geophys. Res. Lett.* **20**, 943–946.

Nakicenovic N., Grubler A., and McDona A. (1998) *Global Energy Perspectives.* Cambridge University Press, New York.

Neftel A., Oeschger H., Schwander J., and Stauffer B. (1983) Carbon dioxide concentration in bubbles of natural cold ice. *J. Phys. Chem.* **87**, 4116–4120.

Neftel A., Moor E., Oeschger H., and Stauffer B. (1985) Evidence from polar ice cores for the increase in atmospheric CO_2 in the past two centuries. *Nature* **315**, 45–47.

Neftel A., Oeschger H., Staffelbach T., and Stauffer B. (1988) CO_2 record in the Byrd ice core 50,000–5,000 years BP. *Nature* **331**, 609–611.

Newell R. E., Navato A. R., and Hsiung J. (1978) Long-term global sea surface temperature fluctuations and their possible influence on atmospheric CO_2 concentrations. *Pure Appl. Geophys.* **116**, 351–371.

Nisbet E. G. (1990) The end of the ice age. *Can. J. Earth Sci.* **27**, 148–157.

Nisbet E. G. (1992) Sources of atmospheric CH_4 in early postglacial time. *J. Geophys. Res.-Atmos.* **97**, 12859–12867.

Norris R. D. and Rohl U. (1999) Carbon cycling and chronology of climate warming during the Palaeocene/Eocene transition. *Nature* **401**, 775–778.

Nozaki Y. and Yamamoto Y. (2001) Radium 228 based nitrate fluxes in the eastern Indian Ocean and the South China Sea and a silicon-induced "alkalinity pump" hypothesis. *Global Biogeochem. Cycles* **15**, 555–567.

O'Hara F. Jr. (1990) *CDIAC Glossary: Carbon Dioxide and Climate.* ORNL/CDIAC-39. Carbon Dioxide Information and Analysis Center, Oak Ridge National Laboratory, Oak Ridge, TN. http://cdiac.esd.ornl.gov/pns/convert.html.

O'Keefe J. D. and Ahrens T. J. (1989) Impact production of CO_2 by the Cretaceous/Tertiary extinction bolide and the resultant heating of the Earth. *Nature* **338**, 247–249.

O'Leary M. H. (1988) Carbon isotopes in photosynthesis. *BioScience* **38**, 328–335.

Oeschger H., Beer J., Siegenthaler U., and Stauffer B. (1984) Late glacial climate history from ice cores. In *Climate Processes and Climate Sensitivity: Geophysical Monograph 29* (eds. J. E. Hansen and T. Takahashi). American Geophysical Union, Washington, DC, pp. 299–306.

Ohmoto H. (1996) Evidence in pre-2.2 Ga paleosols for the early evolution of atmospheric oxygen and terrestrial biota. *Geology* **24**, 1135–1138.

Oil and Gas Journal (2001) Worldwide look at reserves and production. *Oil Gas J.* 126–127.

Oremland R. S. and Culbertson C. W. (1992) Importance of methane-oxidizing bacteria in the methane budget as revealed by the use of a specific inhibitor. *Nature* **356**, 421–423.

ORNL Bioenergy Feedstock Development Programs (2003) *Bioenergy conversion factors.* http://bioenergy.ornl.gov/papers/misc/energy_conv.html.

Orphan V. J., House C. H., Hinrichs K.-U., McKeegan K. D., and DeLong E. F. (2001) Methane-consuming Archaea revealed by directly coupled isotopic and phylogenetic analysis. *Science* **293**, 484–487.

Otto D., Rasse D., Kaplan J., Warnant P., and Francois L. (2002) Biospheric carbon stocks reconstructed at the last glacial maximum: comparison between general circulation models using prescribed and computed sea surface temperatures. *Global Planet. Change* **33**, 117–138.

Owen T., Cess R. D., and Ramanathan V. (1979) Enhanced CO_2 greenhouse to compensate for reduced solar luminosity on early Earth. *Nature* **277**, 640–642.

Pace N. R. (1997) A molecular view of microbial diversity and the biosphere. *Science* **276**, 734–774.

Pagani M. (2002) The alkenone-CO_2 proxy and ancient atmospheric carbon dioxide. *Phil. Trans. Roy. Soc. London Ser. A: Math. Phys. Eng. Sci.* **360**, 609–632.

Pagani M., Arthur M. A., and Freeman K. H. (1999a) Miocene evolution of atmospheric carbon dioxide. *Paleoceanography* **14**, 273–292.

Pagani M., Freeman K. H., and Arthur M. A. (1999b) Late Miocene atmospheric CO_2 concentrations and the expansion of C-4 grasses. *Science* **285**, 876–879.

Palmer M. R. and Pearson P. N. (2003) A 23,000-year record of surface water pH and pCO_2 in the western equatorial Pacific Ocean. *Science* **300**.

Pavlov A. A., Hurtgen M. T., Kasting J. F., and Arthur M. A. (2003) Methane-rich Proterozoic atmosphere? *Geology* **31**, 87–90.

Pearman G. I., Hyson P., and Fraser P. J. (1983) The global distribution of atmospheric carbon dioxide: 1. Aspects of observations and modeling. *J. Geophys. Res.-Oceans* **88**, 3581–3590.

Pearman G. I., Etheridge D., de Silva F., and Fraser P. J. (1986) Evidence of changing concentrations of atmospheric CO_2, N_2O and CH_4 from air bubbles in Antarctic ice. *Nature* **320**, 248–250.

Pearson P. N. and Palmer M. R. (1999) Middle Eocene seawater pH and atmospheric carbon dioxide concentrations. *Science* **284**, 1824–1826.

Pearson P. N. and Palmer M. R. (2000) Atmospheric carbon dioxide concentrations over the past 60 million years. *Nature* **406**, 695–699.

Peterson L. C. and Prell W. L. (1985) Carbonate dissolution in recent sediments of the eastern equatorial Indian Ocean: preservation patterns and carbonate loss above the lysocline. *Mar. Geol.* **64**, 259–290.

Petit J. R., Jouzel J., Raynaud D., Barkov N. I., Barnola J. M., Basile I., Bender M., Chappellaz J., Davis M., Delaygue G., Delmotte M., Kotlyakov V. M., Legrand M., Lipenkov V. Y., Lorius C., Pepin L., Ritz C., Saltzman E., and Stievenard M. (1999) Climate and atmospheric history of the past 420,000 years from the Vostok ice core, Antarctica. *Nature* **399**, 429–436.

Petit-Maire N., Fontugne M., and Rouland C. (1991) Atmospheric methane ratio and environmental changes in the Sahara and Sahel during the last 130 kyrs. *Palaeogeogr. Palaeoclimatol. Palaeoecol.* **86**, 197–204.

Petsch S. T. and Berner R. A. (1998) Coupling the geochemical cycles of C, P, Fe, and S: the effect on atmospheric O_2 and the isotopic records of carbon and sulfur. *Am. J. Sci.* **298**, 246–262.

Pierazzo E., Kring D. A., and Melosh H. J. (1998) Hydrocode simulation of the Chicxulub impact event and the production of climatically active gases. *J. Geophys. Res.-Planet* **103**, 28607–28625.

Pisias N. G. S. N. J. (1984) Modeling the global climate response to orbital forcing and atmospheric carbon dioxide changes. *Nature* **310**, 757–759.

Plummer L. N. and Sundquist E. T. (1982) Total individual ion activity coefficients of calcium and carbonate in seawater at 25°C and 35‰ salinity, and implications to the agreement between apparent and thermodynamic constants of calcite and aragonite. *Geochim. Cosmochim. Acta* **46**, 247–258.

Pope K. O., Baines K. H., Ocampo A. C., and Ivanov B. A. (1997) Energy, volatile production, and climatic effects of the Chicxulub Cretaceous/Tertiary impact. *J. Geophys. Res.-Planet* **102**, 21645–21664.

Popp B. N., Takigiku R., Hayes J. M., Louda J. W., and Baker E. W. (1989) The post-Paleozoic chronology and mechanism of ^{13}C depletion in primary marine organic matter. *Am. J. Sci.* **289**, 436–454.

Prell W. L. and Kutzbach J. E. (1987) Monsoon variability over the past 150,000 years. *J. Geophys. Res.-Atmos.* **92**, 8411–8425.

Prentice I., Sykes M., Lautenschlager M., Harrison S., Denissenko O., and Bartlein P. (1993) Modelling global vegetation patterns and terrestrial carbon storage at the last glacial maximum. *Global Ecol. Biogeogr. Lett.* **3**, 67–76.

Prentice K. C. and Fung I. Y. (1990) The sensitivity of terrestrial carbon storage to climate change. *Nature* **346**, 48–51.

Prinn R., Cunnold D., Rasmussen R., Simmonds P., Alyea F., Crawford A., Fraser P., and Rosen R. (1987) Atmospheric trends in methylchloroform and the global average for the hydroxyl radical. *Science* **238**, 945–950.

Prinn R. G., Simmonds P. G., Rasmussen R. A., Rosen R. D., Alyea F. N., Cardelino C. A., Crawford A. J., Cunnold D. M., Fraser P. J., and Lovelock J. E. (1983) The atmospheric lifetime experiment: 1. Introduction, instrumentation, and overview. *J. Geophys. Res.-Oceans* **88**, 8353–8367.

Prinn R. G., Weiss R. F., Miller B. R., Huang J., Alyea F. N., Cunnold D. M., Fraser P. J., Hartley D. E., and Simmonds P. G. (1995) Atmospheric trends and lifetime of CH_3CCl_3 and global OH concentrations. *Science* **269**, 187–192.

Prinn R. G., Huang J., Weiss R. F., Cunnold D. M., Fraser P. J., Simmonds P. G., McCulloch A., Harth C., Salameh P., O'Doherty S., Wang R. H. J., Porter L., and Miller B. R. (2001) Evidence for substantial variations of atmospheric hydroxyl radicals in the past two decades. *Science* **292**, 1882–1888.

Ramaswamy V., Boucher O., Haigh J., Hauglustaine D., Haywood J., Myhre G., Nakajima T., Shi G. Y., Solomon S., Betts R., Charlson R., Chuang C., Daniel J. S., Del Genio A., van Dorland R., Feichter J., Fuglestvedt J., de. F. Forster P. M., Ghan S. J., Jones A., Kiehl J. T., Koch D., Land C., Lean J., Lohmann U., Minschwaner K., Penner J. E., Roberts D. L., Rodhe H., Roelofs G. J., Rotstayn L. D., Schneider T. L., Schumann U., Schwartz S. E., Schwarzkopf M. D., Shine K. P., Smith S., Stevenson D. S., Stordal F., Tegen I., Zhang Y., and Jones A. (2001) Radiative forcing of climate change. In *Climate Change 2001: The Scientific Basis. Contribution of Working Group I to the Third Assessment Report of the Intergovernmental Panel on Climate Change* (eds. J. T. Houghton, Y. Ding, D. J. Griggs, M. Noguer, P. J. van der Linden, X. Dai, K. Maskell, and C. A. Johnson). Cambridge University Press, Cambridge, pp. 350–416.

Rasmussen B. (2000) Filamentous microfossils in a 3,235-million-year-old volcanogenic massive sulphide deposit. *Nature* **405**, 676–679.

Rasmussen R. A. and Khalil M. A. K. (1984) Atmospheric methane in the recent and ancient atmospheres: concentrations, trends, and interhemispheric gradient. *J. Geophys. Res.-Atmos.* **89**, 11599–11605.

Rau G. H., Takahashi T., and Des Marais D. J. (1989) Latitudinal variations in plankton $\delta^{13}C$: implications for CO_2 and productivity in past oceans. *Nature* **341**, 516–518.

Raymo M. E., Ruddiman W. F., and Froelich P. N. (1988) Influence of late Cenozoic mountain building on ocean geochemical cycles. *Geology* **16**, 649–653.

Raynaud D. and Barnola J. M. (1985) An Antarctic ice core reveals atmospheric CO_2 variations over the past few centuries. *Nature* **315**, 309–311.

Raynaud D., Chappellaz J., Barnola J. M., Korotkevich Y. S., and Lorius C. (1988) Climatic and CH_4 cycle implications of glacial-interglacial CH_4 change in the Vostok ice core. *Nature* **333**, 655–657.

Raynaud D., Barnola J. M., Chappellaz J., Blunier T., Indermuhle A., and Stauffer B. (2000) The ice record of greenhouse gases: a view in the context of future changes. *Quat. Sci. Rev.* **19**, 9–17.

Reeburgh W. S. (1976) Methane consumption in Cariaco Trench waters and sediments. *Earth Planet. Sci. Lett.* **28**, 337–344.

Reeburgh W. S., Whalen S. C., and Alperin M. J. (1993) The role of methylotrophy in the global methane budget. In *Microbial Growth on C1 Compounds* (eds. J. C. Murrell and D. P. Kelley). Intercept Press, Andover, UK, pp. 1–14.

Reinfelder J. R., Kraepiel A. M., and Morel F. M. M. (2000) Unicellular C_4 photosynthesis in a marine diatom. *Nature* **407**, 996–999.

Retallack G. J. (2001) A 300-million-year record of atmospheric carbon dioxide from fossil plant cuticles. *Nature* **411**, 287–290.

Retallack G. J. (2002) Carbon dioxide and climate over the past 300 Myr. *Phil. Trans. Roy. Soc. London Ser. A: Math. Phys. Eng. Sci.* **360**, 659–673.

Rhee K.-H., Morris E. P., Barber J., and Kuhlbrandt W. (1998) Three-dimensional structure of the plant photosystem II reaction centre at 8A resolution. *Nature* **396**, 283–286.

Riebesell U., Zondervan I., Rost B., Tortell P. D., Aeebe R. E., and Morel F. M. M. (2000) Reduced calcification of marine plankton in response to increased atmospheric CO_2. *Nature* **407**, 364–367.

Rind D., Peteet D., and Kukla G. (1989) Can Milankovitch orbital variations initiate the growth of ice sheets in a general circulation model? *J. Geophys. Res.-Atmos.* **94**, 12851–12871.

Rothman D. H. (2002) Atmospheric carbon dioxide levels for the last 500 million years. *Proc. Natl. Acad. Sci.* **99**, 4167–4171.

Royer D. L., Berner R. A., and Beerling D. J. (2001) Phanerozoic atmospheric CO_2 change: evaluating geochemical and paleobiological approaches. *Earth Sci. Rev.* **54**, 349–392.

Ruddiman W. F. (1997) *Tectonic Uplift and Climate Change*. Plenum, New York.

Ruddiman W. F. and Thomson J. S. (2001) The case for human causes of increased atmospheric CH_4. *Quat. Sci. Rev.* **20**, 1769–1777.

Rundgren M. and Beerling D. (1999) A Holocene CO_2 record from the stomatal index of subfossil *Salix herbacea L.* leaves from northern Sweden. *Holocene* **9**, 509–513.

Russell M. J. and Hall A. J. (1997) The emergence of life from iron monosulphide bubbles at a submarine hydrothermal redox and pH front. *J. Geol. Soc.* **154**, 377–402.

Rye R., Kuo P. H., and Holland H. D. (1995) Atmospheric carbon dioxide concentrations before 2.2 billion years ago. *Nature* **378**, 603–605.

Sagan C. and Mullen G. (1972) Earth and Mars: evolution of atmospheres and surface temperatures. *Science* **177**, 52–56.

Saltzman B. (1987) Carbon dioxide and the $\delta^{18}O$ record of late-Quaternary climatic change: a global model. *Clim. Dyn.* **1**, 77–85.

Sandberg P. A. (1983) An oscillating trend in Phanerozoic nonskeletal carbonate mineralogy. *Nature* **305**, 19–22.

Sandberg P. A. (1985) Nonskeletal aragonite and pCO_2 in the Phanerozoic and Proterozoic. In *The Carbon Cycle and Atmospheric CO_2: Natural Variations Archean to Present: Geophysical Monograph 32* (eds. E. T. Sundquist and W. S. Broecker). American Geophysical Union, Washington, DC, pp. 585–594.

Sanderson M. G. (1996) Biomass of termites and their emissions of methane and carbon dioxide: a global database. *Global Biogeochem. Cycles* **10**, 543–557.

Sano Y. and Williams S. N. (1996) Fluxes of mantle and subducted carbon along convergent plate boundaries. *Geophys. Res. Lett.* **23**, 2749–2752.

Sanyal A. and Bijma J. (1999) A comparative study of the northwest Africa and eastern equatorial Pacific upwelling zones as sources of CO during glacial periods based on boron isotope paleo-pH estimation. *Paleoceanography* **14**, 753–759.

Sanyal A., Hemming N. G., Broecker W. S., Lea D. W., Spero H. J., and Hanson G. N. (1996) Oceanic pH control on the boron isotopic composition of foraminifera: evidence from culture experiments. *Paleoceanography* **11**, 513–517.

Sanyal A., Hemming N. G., Broecker W. S., and Hanson G. N. (1997) Changes in pH in the eastern equatorial Pacific across stage 5-6 boundary based on boron isotopes in foraminifera. *Global Biogeochem. Cycles* **11**, 125–133.

Sanyal A., Bijma J., Spero H., and Lea D. W. (2001) Empirical relationship between pH and the boron isotope composition of *Globigerinoides sacculifer*: implications for the boron isotope paleo-pH proxy. *Paleoceanography* **16**, 515–519.

Sarmiento J. L. and Gruber N. (2002) Sinks for anthropogenic carbon. *Phys. Today* **5**, 30–36.

Sarnthein M., Winn K., Duplessy J. C., and Fontugne M. R. (1988) Global variations of surface ocean productivity in low and mid latitudes: influence on CO_2 reservoirs of the deep ocean and atmosphere during the last 21,000 years. *Paleoceanography* **3**, 361–399.

Schidlowski M. (2001) Carbon isotopes as biogeochemical recorders of life over 3.8 Ga of Earth history: evolution of a concept. *Precamb. Res.* **106**, 117–134.

Schidlowski M. and Junge C. E. (1981) Coupling among the terrestrial sulfur, carbon and oxygen cycles: numerical modeling based on revised Phanerozoic carbon isotope record. *Geochim. Cosmochim. Acta* **45**, 589–594.

Schidlowski M., Hayes J. M., and Kaplan I. R. (1983) Isotopic inferences of ancient biochemistries: carbon, sulfur, hydrogen, nitrogen. In *Earth's Earliest Biosphere: Its Origin and Evolution* (ed. J. W. Schopf). Princeton University Press, Princeton, NJ, pp. 149–186.

Schlesinger W. H. (1977) Carbon balance in terrestrial detritus. *Ann. Rev. Ecol. Sys.* **8**, 51–81.

Schlesinger W. H. (1997) *Biogeochemistry: An Analysis of Global Change.* Academic Press, New York.

Schlesinger W. S. and Melack J. M. (1981) Transport of organic carbon in the world's rivers. *Tellus* **33B**, 172–187.

Schubert W.-D., Klukas O., Saenger W., Witt H. T., Fromme P., and Kraub N. (1998) A common ancestor for oxygenic and anoxygenic photosynthetic systems: a comparison based on the structural model of photosystem I. *J. Mol. Biol.* **280**, 297–314.

Schwander J. (1989) The transformation of snow to ice and the occlusion of gases. In *The Environmental Record in Glaciers and Ice Sheets* (eds. H. Oeschger and C. C. Langway). Wiley, New York, pp. 53–67.

Schwander J. and Stauffer B. (1984) Age difference between polar ice and the air trapped in its bubbles. *Nature* **311**, 45–47.

Schwander J., Sowers T., Barnola J. M., Blunier T., Fuchs A., and Malaize B. (1997) Age scale of the air in the Summit ice—implication for glacial-interglacial temperature change. *J. Geophys. Res.-Atmos.* **102**, 19483–19493.

Scranton M. I. and Brewer P. G. (1978) Consumption of dissolved methane in the deep ocean. *Limnol. Oceanogr.* **23**, 1207–1213.

Severinghaus J. P. and Brook E. J. (1999) Abrupt climate change at the end of the last glacial period inferred from trapped air in polar ice. *Science* **286**, 930–934.

Severinghaus J. P., Sowers T., Brook E. J., Alley R. B., and Bender M. L. (1998) Timing of abrupt climate change at the end of the Younger Dryas interval from thermally fractionated gases in polar ice. *Nature* **391**, 141–146.

Shackleton N. J. (1977) Carbon-13 in *Uvigerina*: tropical rainforest history and the equatorial Pacific carbonate dissolution cycles. In *The Fate of Fossil Fuel CO_2 in the Oceans* (eds. N. R. Andersen and A. Malahoff). Plenum, New York, pp. 401–427.

Shackleton N. J. (1987) The carbon isotope record of the Cenozoic: history of organic carbon burial and of oxygen in the ocean and atmosphere. In *Marine Petroleum Source Rocks: Geological Society Special Publication 26* (eds. J. Brooks and A. J. Fleet). Blackwell, Boston, pp. 423–434.

Shackleton N. J. (2000) The 100,000-year ice-age cycle identified and found to lag temperature, carbon dioxide, and orbital eccentricity. *Science* **289**, 1897–1902.

Shackleton N. J. (2001) Climate change across the hemispheres. *Science* **291**, 58–59.

Shackleton N. J. and Opdyke N. D. (1976) Oxygen isotope and paleomagnetic stratigraphy of Pacific core V28-239 late Pliocene to latest Pleistocene. In *Investigation of Late Quaternary Paleoceanography and Paleoclimatology:*

Geological Society of America Memoir 145 (eds. R. M. Cline and J. D. Hays). Geological Society of America, Boulder, CO, pp. 449–464.

Shackleton N. J., Hall M. A., and Vincent E. (2000) Phase relationships between millennial-scale events 64,000–24,000 years ago. *Paleoceanography* **15**, 565–569.

Sharp J. H. (1997) Marine dissolved organic carbon: are the older values correct? *Mar. Chem.* **56**, 265–277.

Shine K. P., Derwent R. G., Wuebbles D. J., and Morcrette J.-J. (1990) Radiative forcing of climate. In *Climate Change: The IPCC Scientific Assessment* (eds. J. T. Houghton, G. J. Jenkins, and J. J. Ephraums). Cambridge University Press, New York, pp. 45–68.

Siegenthaler U., Friedli H., Loetscher H., Moor E., Neftel A., Oeschger H., and Stauffer B. (1988) Stable-isotope ratios and concentration of CO_2 in air from polar ice cores. *Ann. Glaciol.* **10**, 151–156.

Smith H. J., Wahlen M., Mastroianni D., and Taylor K. C. (1997) The CO_2 concentration of air trapped in GISP2 ice from the last glacial maximum-Holocene transition. *Geophys. Res. Lett.* **24**, 1–4.

Smith H. J., Fischer H., Wahlen M., Mastroianni D., and Deck B. (1999) Dual modes of the carbon cycle since the last glacial maximum. *Nature* **400**, 248–250.

Sowers T. (2001) N_2O record spanning the penultimate deglaciation from the Vostok ice core. *J. Geophys. Res.-Atmos.* **106**, 31903–31914.

Sowers T. and Bender M. (1995) Climate records covering the last deglaciation. *Science* **269**, 210–214.

Sowers T., Brook E., Etheridge D., Blunier T., Fuchs A., Leuenberger M., Chappellaz J., Barnola J. M., Wahlen M., Deck B., and Weyhenmeyer C. (1997) An interlaboratory comparison of techniques for extracting and analyzing trapped gases in ice cores. *J. Geophys. Res.-Oceans* **102**, 26527–26538.

Spencer R. J. and Hardie L. A. (1990) Control of seawater composition by mixing of river waters and mid-ocean ridge hydrothermal brines. In *Fluid-mineral interactions: a tribute to H. P. Eugster: Geochemical Society Special Publication 2* (eds. R. J. Spencer and I.-M. Chou). Geochemical Society, San Antonio, pp. 409–419.

Spitzy A. and Leenheer J. (1991) Dissolved organic carbon in rivers. In *SCOPE 42: Biogeochemistry of Major World Rivers* (eds. E. T. Degens, S. Kempe, and J. E. Richey). Wiley, New York, pp. 213–232.

Spivack A. J. (1993) Foraminiferal boron isotope ratios as a proxy for surface ocean pH over the past 21 Myr. *Nature* **363**, 149–151.

Staudigel H. R., Hart S. R., Schmincke H. U., and Smith B. M. (1990) Cretaceous ocean crust at DSDP Sites 417 and 418: carbon uptake from weathering versus loss by magmatic outgassing. *Geochim. Cosmochim. Acta* **53**, 3091–3094.

Stauffer B. and Oeschger H. (1985) Gaseous components in the atmosphere and the historic record revealed by ice cores. *Ann. Glaciol.* **7**, 54–60.

Stauffer B., Hofer H., Oeschger H., Schwander J., and Siegenthaler U. (1984) Atmospheric CO_2 concentration during the last glaciation. *Ann. Glaciol.* **5**, 160–164.

Stauffer B., Fischer G., Neftel A., and Oeschger H. (1985) Increase of atmospheric methane recorded in Antarctic ice core. *Science* **229**, 1386–1388.

Stauffer B., Lochbronner E., Oeschger H., and Schwander J. (1988) Methane concentration in the glacial atmosphere was only half that of the pre-industrial Holocene. *Nature* **332**, 812–814.

Stauffer B., Blunier T., Dallenbach A., Indermuhle A., Schwander J., Stocker T. F., Tschumi J., Chappellaz J., Raynaud D., Hammer C. U., and Clausen H. B. (1998) Atmospheric CO_2 concentration and millennial-scale climate change during the last glacial period. *Nature* **392**, 59–62.

Steig E. J., Brook E. J., White J. W. C., Sucher C. M., Bender M. L., Lehman S. J., Morse D. L.,

Waddington E. D., and Clow G. D. (1998) Synchronous climate changes in Antarctica and the North Atlantic. *Science* **282**, 92–95.

Stephens B. B. and Keeling R. F. (2000) The influence of Antarctic sea ice on glacial-interglacial CO_2 variations. *Nature* **404**, 171–174.

Subak S. (1994) Methane from the House of Tudor and Ming Dynasty: anthropogenic emissions in the sixteenth century. *Chemosphere* **29**, 843–854.

Sugimoto A., Inoue T., Kirtibutr N., and Abe T. (1998) Methane oxidation by termite mounds estimated by the carbon isotopic composition of methane. *Global Biogeochem. Cycles* **12**, 595–605.

Summons R. E., Jahnke L. L., Hope J. M., and Logan G. A. (1999) 2-Methylhopanoids as biomarkers for cyanobacterial oxygenic photosynthesis. *Nature* **400**, 554–557.

Sundquist E. T. (1985) Geological perspectives on carbon dioxide and the carbon cycle. In *The Carbon Cycle and Atmospheric CO_2: Natural Variations Archean to Present: Geophysical Monograph 32* (eds. E. T. Sundquist and W. S. Broecker). American Geophysical Union, Washington, DC, pp. 5–60.

Sundquist E. T. (1986) Geologic analogs: their value and limitations in carbon dioxide research. In *The Changing Carbon Cycle, A Global Analysis* (eds. J. R. Trabalka and D. E. Reichle). Springer, New York, pp. 371–402.

Sundquist E. T. (1990a) Influence of deep-sea benthic processes on atmospheric CO_2. *Phil. Trans. Roy. Soc. London Ser. A: Math. Phys. Eng. Sci.* **331**, 155–165.

Sundquist E. T. (1990b) Long-term aspects of future atmospheric CO_2 and sea-level changes. *Sea Level Change: Natl. Res. Council,* 193–207.

Sundquist E. T. (1991) Steady- and non-steady-state carbonate-silicate controls on atmospheric CO_2. *Quat. Sci. Rev.* **10**, 283–296.

Sundquist E. T. (1993) The global carbon dioxide budget. *Science* **259**, 934–941.

Sundquist E. T. and Plummer L. N. (1981) Carbon dioxide in the ocean surface layer: some modeling considerations. In *Carbon Cycle Modelling* (ed. B. Bolin). Wiley, New York, pp. 259–270.

Tajika E. (1998) Climate change during the last 150 million years: reconstruction from a carbon cycle model. *Earth Planet. Sci. Lett.* **160**, 695–707.

Taylor J. A. and Orr J. C. (2000) The natural latitudinal distribution of atmospheric CO_2. *Global Planet. Change* **26**, 375–386.

Thomas E. and Shackleton N. J. (1996) The Paleocene-Eocene benthic foraminiferal extinction and stable isotope anomalies. In *Correlations of the early Paleogene in Northwest Europe: Geological Society Special Publication 101* (eds. R. O. Knox, R. M. Corfield, and R. E. Dunay). Geological Society of London, London, pp. 401–411.

Thompson A. M. (1992) The oxidizing capacity of the Earth's atmosphere: probable past and future changes. *Science* **256**, 1157–1165.

Thompson A. M., Chappellaz J. A., Fung I. Y., and Kucsera T. L. (1993) The atmospheric CH4 increase since the Last Glacial Maximum: 2. Interactions with oxidants. *Tellus* **45B**, 242–257.

Toggweiler J. R. (1999) Variation of atmospheric CO_2 by ventilation of the ocean's deepest water. *Paleoceanography* **14**, 571–588.

Toggweiler J. R., Gnanadesikan A., Carson S., Murnane R., and Sarmiento J. L. (2003) Representation of the carbon cycle in box models and GCMs: 1. Solubility pump. *Global Biogeochem. Cycles* **17**, 1026, doi:10.1029/2001GB001401.

Trudinger C. M., Enting I. G., Francey R. J., Etheridge D. M., and Rayner P. J. (1999) Long-term variability in the global carbon cycle inferred from a high-precision CO_2 and delta C-13 ice-core record. *Tellus* **51B**, 233–248.

Tschumi J. and Stauffer B. (2000) Reconstructing past atmospheric CO_2 concentration based on ice-core analyses: open questions due to *in situ* production of CO_2 in the ice. *J. Glaciol.* **46**, 45–53.

US Geological Survey World Energy Assessment Team (2000) *World Petroleum Assessment, 2000*. USGS Digital Data Series 60.

Van Campo E., Guiot J., and Peng C. (1993) A data-based reappraisal of the terrestrial carbon budget at the last glacial maximum. *Global Planet. Change* **8**, 189–201.

Van Cappellen P. and Ingall E. D. (1996) Redox stabilization of the atmosphere and oceans by phosphorus-limited marine productivity. *Science* **271**, 493–496.

Veizer J., Ala D., Azmy K., Bruckschen P., Buhl D., Bruhn F., Carden G. A. F., Diener A., Ebneth S., and Godderis Y. (1999) $^{87}Sr/^{86}Sr$, $\delta^{13}C$ and $\delta^{18}O$ evolution of Phanerozoic seawater. *Chem. Geol.* **161**, 59–88.

Veizer J., Godderis Y., and Francois L. M. (2000) Evidence for decoupling of atmospheric CO_2 and global climate during the Phanerozoic eon. *Nature* **408**, 698–701.

Velichko A. A., Kremenetski C. V., Borisova O. K., Zelikson E. M., Nechaev V. P., and Faure H. (1998) Estimates of methane emission during the last 125,000 years in Northern Eurasia. *Global Planet. Change* **16–17**, 159–180.

Volk T. and Hoffert M. I. (1985) Ocean carbon pumps: analysis of relative strengths and efficiencies in ocean-driven atmospheric CO_2 changes. In *The Carbon Cycle and Atmospheric CO_2: Natural Variations Archean to Present: Geophysical Monograph 32* (eds. E. T. Sundquist and W. S. Broecker). American Geophysical Union, Washington, DC, pp. 99–110.

von Fischer J. C. and Hedin L. O. (2002) Separating methane production and consumption with a field-based isotope pool dilution technique. *Global Biogeochem. Cycles* **16**, 10.1029/2001GB001448.

Wahlen M. D., Allen D., and Deck B. (1991) Initial measurements of CO_2 concentrations (1530-1940 AD) in air occluded in the GISP2 ice core from central Greenland. *Geophys. Res. Lett.* **18**, 1457–1460.

Walker J. C. G. (1987) Was the Archean biosphere upside down? *Nature* **329**, 710–712.

Walker J. C. G., Hays P. B., and Kasting J. F. (1981) A negative feedback mechanism for the long-term stabilization of Earth's surface temperature. *J. Geophys. Res.-Oceans* **86**, 9776–9782.

Walker J. C. G., Klein C., Schidlowski M., Schopf J. W., Stevenson D. J., and Walter M. R. (1983) Environmental evolution of the Archean-Early Proterozoic Earth. In *Earth's Earliest Biosphere: Its Origin and Evolution* (ed. J. W. Schopf). Princeton University Press, Princeton, NJ, pp. 260–290.

Wallmann K. (2001a) Controls on the Cretaceous and Cenozoic evolution of seawater composition, atmospheric CO_2 and climate. *Geochim. Cosmochim. Acta* **65**, 3005–3025.

Wallmann K. (2001b) The geological water cycle and the evolution of marine delta O-18 values. *Geochim. Cosmochim. Acta* **65**, 2469–2485.

Walter B. P., Heimann M., and Matthews E. (2001) Modeling modern methane emissions from natural wetlands: 1. Model description and results. *J. Geophys. Res.-Atmos.* **106**, 34189–34206.

Watson R. T., Rodhe H., Oeschger H., and Siegenthaler U. (1990) Greenhouse gases and aerosols. In *Climate Change: The IPCC Scientific Assessment* (eds. J. T. Houghton, G. J. Jenkins, and J. J. Ephraums). Cambridge University Press, Cambridge, pp. 1–40.

Whalen S. C. and Reeburgh W. S. (1990) Consumption of atmospheric methane by tundra soils. *Nature* **346**, 160–162.

Whiting G. J. and Chanton J. P. (1993) Primary production control of methane emission from wetlands. *Nature* **364**, 794–795.

Widdel F., Schnell S., Heising S., Ehrenreich A., Assmus B., and Schink B. (1993) Ferrous iron oxidation by anoxygenic phototrophic bacteria. *Nature* **362**, 834–836.

Wollast R. (1994) The relative importance of biomineralization and dissolution of $CaCO_3$ in the global carbon cycle. In *Past and Present Biomineralization Processes* (ed. F. Doumenge). Musee Oceanographie, Monaco, pp. 13–35.

Woodward F. I. (1987) Stomatal numbers are sensitive to increases in CO_2 from pre-industrial levels. *Nature* **327**, 617–618.

World Energy Council (WEC) (2000) *Survey of Energy Resources*. World Energy Council, London.

Wuebbles D. J. and Hayhoe K. (2002) Atmospheric methane and global change. *Earth Sci. Rev.* **57**, 177–210.

Xiong J., Fischer W. M., Inoue K., Nakahara M., and Bauer C. E. (2000) Molecular evidence for the early evolution of photosynthesis. *Science* **289**, 1724–1730.

Yavitt J. B., Lang G. E., and Downey D. M. (1988) Potential methane production and methane oxidation rates in peatland ecosystems of the Appalachian Mountains, United States. *Global Biogeochem. Cycles* **2**, 253–268.

Yavitt J. B., Fahey T. J., and Simmons J. A. (1995) Methane and carbon dioxide dynamics in a northern hardwood ecosystem. *Soil. Sci. Soc. Am. J.* **59**, 796–804.

Yoshimori M., Weaver A. J., Marshall S. J., and Clarke G. K. C. (2001) Glacial termination: sensitivity to orbital and CO_2 forcing in a coupled climate system model. *Clim. Dyn.* **17**, 571–588.

Zachos J. C., Lohmann K. C., Walker J. C. G., and Wise S. W. (1993) Abrupt climate change and transient climates during the Paleogene: a marine perspective. *J. Geol.* **101**, 191–213.

Zimmerman H. (2000) Tertiary seawater chemistry–implications from fluid inclusions in primary marine halite. *Am. J. Sci.* **300**, 723–767.

Published by Elsevier Ltd.

Geochemistry of Earth Surface Systems
ISBN: 978-0-08-096706-6

11
Organic Matter in the Contemporary Ocean

T. I. Eglinton and D. J. Repeta

Woods Hole Oceanographic Institution, MA, USA

11.1	INTRODUCTION	389
11.2	RESERVOIRS AND FLUXES	390
	11.2.1 Reservoirs	390
	11.2.2 Fluxes	392
	11.2.2.1 Terrigenous organic matter fluxes to the oceans	392
	11.2.2.2 Water column fluxes and the burial of organic carbon in sediments	392
11.3	THE NATURE AND FATE OF TERRIGENOUS ORGANIC CARBON DELIVERED TO THE OCEANS	393
	11.3.1 Background	393
	11.3.2 Terrestrial Organic Matter in River Systems	394
	11.3.3 Quantitative Importance of Terrigenous Organic Carbon in Marine Sediments	396
	11.3.3.1 Black carbon	397
11.4	A BIOPOLYMERIC ORIGIN FOR OCEANIC DISSOLVED ORGANIC MATTER	398
	11.4.1 Background	398
	11.4.2 High Molecular Weight Dissolved Organic Matter: Biopolymers or Geopolymers?	399
	11.4.2.1 Acylpolysaccharides in high molecular weight dissolved organic matter	400
	11.4.2.2 Proteins in high molecular weight dissolved organic matter	402
	11.4.3 Gel Polymers and the Cycling of High Molecular Weight Dissolved Organic Matter	404
11.5	EMERGING PERSPECTIVES ON ORGANIC MATTER PRESERVATION	406
	11.5.1 Background	406
	11.5.2 Compositional Transformations Associated with Sedimentation and Burial of Organic Matter	406
	11.5.3 Controls on Organic Matter Preservation	409
	11.5.3.1 Physical protection	410
	11.5.3.2 Role of anoxia	411
	11.5.3.3 Chemical protection	412
11.6	MICROBIAL ORGANIC MATTER PRODUCTION AND PROCESSING: NEW INSIGHTS	414
	11.6.1 Background	414
	11.6.2 Planktonic Archea	415
	11.6.3 Anaerobic Methane Oxidation	415
11.7	SUMMARY AND FUTURE RESEARCH DIRECTIONS	417
	ACKNOWLEDGMENTS	418
	REFERENCES	418

11.1 INTRODUCTION

This chapter summarizes selected aspects of our current understanding of the organic carbon (OC) cycle as it pertains to the modern ocean, including underlying surficial sediments. We briefly review present estimates of the size of OC reservoirs and the fluxes between them. We then proceed to highlight advances in our understanding that have occurred since the late 1980s, especially those which have altered our perspective of the ways organic matter is cycled in the oceans. We have focused on specific areas where substantial progress has been made, although in most cases our understanding remains far from complete. These are the fate of terrigenous OC inputs in the ocean, the composition of oceanic dissolved organic matter (DOM), the mechanisms of OC preservation, and new insights into microbial inputs and processes. In each case, we discuss prevailing hypotheses concerning the composition and fate of organic matter derived from the different inputs, the reactivity and relationships between different organic matter pools, and highlight current gaps in our knowledge.

The advances in our understanding of organic matter cycling and composition has stemmed largely from refinements in existing methodologies and the emergence of new analytical capabilities. Molecular-level stable carbon and nitrogen isotopic measurements have shed new light on a range of biogeochemical processes. Natural abundance of radiocarbon data has also been increasingly applied as both a tracer and source indicator in studies of organic matter cycling. As for ^{13}C, bulk ^{14}C measurements are now complemented by measurements at the molecular level, and the combination of these different isotopic approaches has proven highly informative. The application of multinuclear solid- and liquid-state nuclear magnetic resonance (NMR) spectroscopy has provided a more holistic means to examine the complex array of macromolecules that appears to comprise both dissolved and particulate forms of organic matter. New liquid chromatography/mass spectrometry techniques provide structural information on polar macromolecules that have previously been beyond the scope of established methods. In addition to technological advances, large multidisciplinary field programs have provided important frameworks and contexts within which to interpret organic geochemical data, while novel sampling techniques have been developed that allow for the collection of more representative samples and their detailed analytical manipulation. Two particular analytical approaches are highlighted in this chapter—NMR spectroscopy as a powerful tool for structural characterization of complex macromolecules, and compound specific carbon isotope (^{13}C and ^{14}C) analysis as probes for the cycling of organic matter in the ocean through space and time.

Finally, we outline new as well as unresolved questions which provide future challenges for marine organic biogeochemists, and discuss emerging analytical approaches that may shed new light on organic matter cycling in the oceans. For example, (i) the source of "old" dissolved organic carbon (DOC) in the deep sea has yet to be resolved; (ii) the molecular-level composition of the majority of organic matter buried in marine sediments evades elucidation; and (iii) while planktonic archea have been found to be amongst the most abundant organisms in the ocean, their role in biogeochemical cycles and their legacy in the sedimentary record are only beginning to be considered. Such fundamental observations and questions continue to challenge us, and limit our understanding of the processes underpinning organic matter cycling in the oceans.

The discipline of marine organic geochemistry has expanded and evolved greatly in recent years. Hence, we have had to be selective in our coverage of new developments. This review, therefore, is by no means comprehensive and there are many important and exciting aspects of marine organic geochemistry that we have not covered. Comprehensive discussions of some of these aspects are to be found in the following review papers and chapters: soil OC (Hedges and Oades, 1997), terrestrial OC inputs to the oceans (Hedges et al., 1997; Schlunz and Schneider, 2000), organic matter preservation (Tegelaar et al., 1989; de Leeuw and Largeau, 1993; Hedges and Keil, 1995), lipid biomarkers (Volkman et al., 1998), bacterial contributions (Sinninghe Damsté and Schouten, 1997), deep biosphere (Parkes et al., 2000), eolian inputs (Prospero et al., 2003), black carbon (BC) (Schmidt and Noack, 2000), gas hydrates (Kvenvolden, 1995), water column particulate organic matter (POM) (Wakeham and Lee, 1993), carbon isotopic systematics (Hayes, 1993), and use of ^{14}C and ^{13}C as tracers of OC input (Raymond and Bauer, 2001b).

In this chapter two pools of organic matter (OM) are discussed in detail. Particulate organic matter is manifestly heterogeneous, composed of all sorts of particles resulting from a wide range of inputs and a multitude of processes acting on them. In effect, sedimentary POM is chemically and spatially heterogeneous and much effort needs to be focused on sampling, fractionation, and bulk characterization rather than on detailed molecular-level studies. This situation contrasts sharply with the study of DOM, which, despite its largely macromolecular nature, appears to be remarkably uniform in composition throughout the oceans. Here, the prime need is for studies of the colloid processes involved and detailed molecular-level analysis of the composition and conformation of the refractory DOM in order to provide a basis for explaining its apparent lack of bioavailability, and to answer the question: why does DOM persist for years, even millennia, in the deep ocean?

11.2 RESERVOIRS AND FLUXES

11.2.1 Reservoirs

Figure 1 depicts the major components of the OC cycle on and in the Earth's crust. Greater than

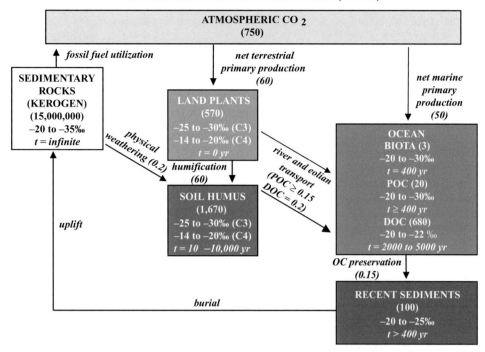

THE GLOBAL ORGANIC CARBON CYCLE (ca. 1950)

Figure 1 The global organic carbon cycle. Numbers in parentheses are approximate reservoir sizes (10^{15} g C = Gt) and italicized are approximate fluxes (10^{15} g C yr^{-1}). Nonitalicized numbers are approximate ranges for stable carbon isotopic compositions (δ^{13}C, per mil) and italicized numbers are approximate radiocarbon ages (yr BP) (after Hedges, 1992).

99.9% of all carbon in the Earth's crust is stored in sedimentary rocks (Berner, 1989). About 20% of this total ($\sim 1.5 \times 10^{7}$ Gt) is organic, and the majority (>90%) of the OC in these consolidated sediments is "kerogen," operationally defined as macromolecular material that is insoluble in common organic solvents and nonoxidizing acids (Durand, 1980). Most kerogen is finely disseminated in sedimentary rocks (shales and limestones) which, on average, contain $\sim 1\%$ organic matter. Organic-rich deposits that include the World's fossil fuel reserves (coal, oil shales, and petroleum) account for less than 0.1% of total sedimentary OC (Hunt, 1996). Of the small fraction of organic matter that is not in the form of relict carbon sequestered in ancient sedimentary rocks, almost two-thirds resides on the continents. Approximately 25% of this "terrestrial" OC (~ 570 Gt) is in the form of standing biomass (plant tissue), with a further 70 Gt of plant litter on the soil surface (Post, 1993), and almost 1,600 Gt residing within the upper 1 m of soils and peat deposits (Eswaran et al., 1993).

Marine biota comprise only ~ 3 Gt of OC, and sinking and suspended particulate OC account for a further 10–20 Gt. The majority of OC in the oceans is in the form of DOC (680 Gt) and organic matter sequestered in the upper meter of marine sediments (~ 100 Gt). Concentrations of marine DOC are highest in the upper ocean, and in the coastal zone. Typical open ocean DOC concentrations in surface seawater range from 60 μM to 80 μM. In the coastal zone, concentrations may climb to in excess of 200 μM, although concentrations rapidly decrease within a few kilometers of shore (Vlahos et al., 2002). The inventory of marine OC is fixed by the concentration of DOC in the deep ocean, which is relatively constant at 42 μM, although small variations of a few μM C have been reported (Hansell and Carlson, 1998). These variations are intriguing, in that they suggest active cycling of DOC in the deep sea. North Atlantic Deep Water, Antarctic Bottom Water, and other deep-water masses all carry the same burden of DOC, even though they are formed at different latitudinal extremes and under very different forcing conditions. Why the concentration of DOC is so constant in the deep sea is a mystery, but the narrow range of DOC values measured in the global ocean implies a very tightly controlled feedback between production and degradation.

Approximately 100 Gt of organic matter is sequestered in the upper meter of marine sediments. In the modern ocean, $\sim 90\%$ of OC burial occurs under oxygenated bottom waters along continental margins (e.g., Hedges and Keil, 1995). Up to 45% of OC burial occurs in deltaic sediments, which, in spite of nature as loci of major riverine inputs, appear to include a major fraction of marine-derived organic matter (Keil *et al.*, 1997). An equivalent amount of OC is buried in nondeltaic sediments that are also proximal to land masses (Hedges and Keil, 1995), primarily on continental shelves and upper slopes (Premuzic *et al.*, 1982). Of the remainder, ~7% of OC is buried beneath highly productive regions with associated oxygen minimum zones (OMZs) and in anoxic basins. The balance (~5%) is buried pelagic abyssal sediments.

Much of the discussion in this chapter concerning sedimentary OC is centered on continental margin sediments, because this is the major locus of OC burial. In contrast, most DOM is held in the deep ocean basins, and hence discussion of the latter is largely in the context of the pelagic water column.

11.2.2 Fluxes

11.2.2.1 *Terrigenous organic matter fluxes to the oceans*

OC fluxes from sedimentary rock weathering on land are not well constrained but on geological timescales are believed to match OC burial in sediments (Berner, 1989). Superimposed on this background of relict OC from sedimentary rock weathering are fluxes associated with terrestrial primary production. The global rate of net terrestrial photosynthesis is estimated to be in the range of $60\,Gt\,yr^{-1}$ (Post, 1993). Approximately two-thirds of the resulting total plant litter is oxidized rapidly to CO_2 (Post, 1993), while the remainder enters the soil cycle and is subject to further oxidation. Organic matter pools within soils exhibit different reactivities and turnover times that range from decades to millenia (Torn *et al.*, 1997). Over geologic timescales, however, the pervasive and continuous oxidative degradation and leaching and erosion processes on the continents result in little long-term storage of organic matter on the continents (Hedges *et al.*, 1997). However, some fraction of this terrestrial (vascular plant-derived) OC and sedimentary rock-derived (relict) OC escapes oxidation and is delivered to the oceans. The delivery of terrigenous OC to the oceans is primarily *via* riverine or atmospheric (eolian) processes.

Riverine fluxes. Approximately 0.2 Gt each of dissolved and particulate OC are carried from land to sea annually by rivers (Ludwig *et al.*, 1996). Much of this riverine organic matter appears to be soil derived based on its chemical characteristics (Meybeck, 1982; Hedges *et al.*, 1994), although autochthonous sources may be important for the dissolved fraction (Repeta *et al.*, 2002). It is now recognized that, on a global basis, riverine discharge is dominated by low-latitude tropical rivers. This not only includes major systems such as the Amazon, and Congo, but also includes the numerous smaller rivers draining mountainous tropical regions (Nittrouer *et al.*, 1995), most notably in Papua New Guinea and other parts of Oceania, which are estimated to account for nearly 50% of the global flux of river sediment to the oceans (Milliman and Syvitski, 1992). During the present-day high sea-level stand, much of the particulate OC associated with riverine discharge is trapped and buried on continental shelves (Berner, 1982; Hedges, 1992). However, some rivers discharge much of their terrestrial OC load beyond the shelf due either to turbidity flows down submarine canyons (e.g., Congo, Ganges, Brahmaputra), to the presence of a narrow shelf (e.g., on the eastern flank of Papua New Guinea), or the influence of ice-rafting as an additional mode of sediment entrainment and export on polar margins (e.g., Macdonald *et al.*, 1998).

Eolian fluxes. Eolian fluxes of organic matter from land to sea are much less well constrained than riverine inputs. They have been estimated to be $<0.1\,Gt\,yr^{-1}$ (Romankevich, 1984). While these flux estimates imply lesser importance of eolian inputs compared to riverine OC contributions, this mode of delivery may be significant in a regional context. In particular, marine locations downwind from major dust sources (principally in eastern Asia and western Africa) are influenced profoundly by eolian inputs of OC and other detrital components. In addition, eolian transport can deliver terrigenous materials to remote locations of the oceans, far from the influence of rivers. For such regions (e.g., central equatorial Pacific Ocean) eolian OC fluxes may be important both in terms of POM in the water column and underlying sediments (Gagosian and Peltzer, 1986; Zafiriou *et al.*, 1985; Prospero *et al.*, 2003; Eglinton *et al.*, 2002).

11.2.2.2 *Water column fluxes and the burial of organic carbon in sediments*

The turnover time and fluxes of DOC into the ocean are obtained by comparing the reservoir size and radiocarbon age. The ocean inventory of DOC is ~680 Gt, and nearly all of this carbon resides in the deep sea, where concentration profiles and radiocarbon values are constant with depth. DOC ages by ~1,000 yr as deep seawater moves from the Atlantic to the Pacific Basin, but even in the Atlantic, DOC radiocarbon values are significantly depleted relative to dissolved inorganic carbon (DIC) (Druffel *et al.*, 1992). DOC persists in seawater through several ocean mixing cycles. The average age of DOC in the deep ocean is ~5,000 yr. Assuming a steady-state ocean, and that all this carbon is synthesized with a radiocarbon age equivalent to atmospheric carbon dioxide via marine production, then the annual flux of

DOC into and out of the deep-sea reservoir is $\sim 0.1\,Gt\,C\,yr^{-1}$. This flux is comparable to the delivery of terrestrial DOC to the ocean by rivers, but very small compared to annual marine production $(60-75\,Gt\,C\,yr^{-1})$. Only 0.1–0.2% of annual marine production needs to be fixed into the permanent reservoir to maintain it, making the processes that sequester and remove carbon nearly impossible to track.

Marine photosynthesis by unicellular phytoplankton produces OC at a comparable rate to land plants (Hedges, 1992). Only $\sim 10\%$ of the net primary production escapes the upper 100 m of the water column. This vertical export occurs in the form of sinking fecal pellets produced by zooplankton that graze upon the phytoplankton, and as aggregates of cellular debris ("marine snow") (Alldredge et al., 1993; Alldredge and Silver, 1998). The rain of particulate organic carbon (POC) out of the surface ocean attenuates exponentially through the water column, and only $\sim 10\%$ of the OC sinking out of the euphotic zone reaches an average seafloor depth of 4,000 m (Suess, 1980). Subsequent to losses in organic matter through the water column, a further 90% or more of that deposited on the seafloor is degraded, leaving $\sim 0.1\%$ of organic material originally synthesized in the surface ocean to be ultimately preserved in sediments underlying most of the open ocean (Wakeham et al., 1997). Global burial efficiencies exceed 0.1%, because a significant amount of organic matter is deposited on continental margins and in oxygen-deficient regions where burial efficiencies are considerably higher (Berner, 1989). Estimates for the global rate of OC burial in marine sediments range from $0.1\,Gt\,yr^{-1}$ to $0.6\,Gt\,yr^{-1}$ (Berner, 1989).

In addition to vertical transport, export of OC from the margins to the ocean interior is being increasingly recognized as of significance (Bauer and Druffel, 1998; Thomsen and Van Weering, 1998; Ransom et al., 1998a). Some regions of the coastal ocean produce more OC than they respire (Smith and Hollibaugh, 1993), suggesting that a fraction of this nonrespired, unburied OC is available for export from margins to the deep ocean (Wollast, 1991). Lateral transport of organic matter from margins to pelagic and abyssal environments has also been invoked to help explain carbon and oxygen imbalances in the deep ocean (Smith et al., 1994; Jahnke et al., 1990). Radiocarbon studies also provide evidence for basin-ward export of OM from the ocean margins. For example, Bauer and co-workers (Bauer and Druffel, 1998; Bauer et al., 2001) observed suspended POM (SPOM) (and DOM) in Mid-Atlantic Bight slope and rise waters that are concurrently older and higher in concentration than in the adjacent North Atlantic gyres. While there are several potential origin(s) for this old,

^{14}C-depleted carbon, sediment resuspension and advection from the shelf and upper slope (Anderson et al., 1994; Churchill et al., 1994) is a likely explanation.

In addition, the chemical nature of advected particulate matter may favor its preservation over biogenic debris directly produced in surface waters. Even when vertical transport of recently produced surface ocean-derived material is rapid (e.g., seasonal thermocline breakdown, rapidly sinking POM), this fresher material may be more susceptible to degradation relative to older, margin-derived material. Thus, the ^{14}C age and concentration of suspended POC in the deep ocean may be maintained by greater relative inputs from the margins than from recent surface production, and may be partly responsible for the apparent old age of POC observed in deep North Atlantic and Pacific central gyres. It remains uncertain whether or not the presence of "old" POC and DOC in slope and rise waters (Bauer and Druffel, 1998) reflects pre-aged terrigenous organic matter (from continental soils or sedimentary rocks) rather than organic matter of marine origin produced on the margin and temporarily sequestered in shelf and upper slope sediments.

11.3 THE NATURE AND FATE OF TERRIGENOUS ORGANIC CARBON DELIVERED TO THE OCEANS

11.3.1 Background

A long-standing paradox evident from global OC flux estimates (Figure 1) is that while the combined global discharge of particulate and dissolved OC from rivers is twice the OC burial rate in marine sediments, OM in both the water column and underlying sediments is apparently dominated by autochthonous inputs (e.g., Hedges and Mann, 1979; Gough et al., 1993; Aluwihare et al., 1997). This implies that most terrigenous OM delivered to the oceans must be efficiently mineralized, and that the ocean is operating as a net heterotrophic system, accumulating less sedimentary OC than it receives via riverine discharge alone (Smith and MacKenzie, 1987). Potential explanations for this paradox are emerging as a result of recent studies, and our improved understanding of the composition of OM in the ocean.

There are several unresolved issues underlying present estimates and assumptions on the abundance and fate of terrigenous organic matter in the oceans. The first is the composition and proportion of terrigenous OC that is buried in margin sediments proximal to the continental source, a second is the proportion of terrestrial OC that enters the deep ocean, and a third is the variability in input composition and flux, mode of delivery, and geographic distribution.

11.3.2 Terrestrial Organic Matter in River Systems

One particularly intriguing question is whether a significant component of deep-sea DOC is of terrestrial origin. The annual flux of DOC through rivers is of the same magnitude as the annual flux of DOC out of and into the deep ocean reservoir, and measurements of $DO^{14}C$ in rivers show the carbon to have largely modern radiocarbon values (see Figure 2). An annual flux of 0.1 Gt C with a modern radiocarbon value would support the deep-sea DOC reservoir. Lignin oxidation products, which are good biomarkers for terrestrial OC, have been measured at low concentrations in open ocean DOM, providing molecular-level confirmation of a terrestrial origin for at least some fraction of this carbon (Meyers-Schulte and Hedges, 1986; Opsahl and Benner, 1997). Terrestrial carbon could also enter the DOC reservoir by desorption from POM. As discussed later in this chapter, POM enters the ocean with a coating of terrestrial OC that is rapidly replaced by marine OC. The load of organic matter introduced by POM could be injected into seawater as DOM. Radiocarbon values for riverine POM are highly variable, but are often depleted relative to modern carbon. If new DOC is pre-aged in this manner, then the annual flux of DOC would be in excess of $0.1\,\text{Gt C yr}^{-1}$.

Marine DOC has stable carbon isotope values between −21‰ and −22‰ (Druffel *et al.*, 1992) consistent with a largely marine source. While these data seem to exclude a significant contribution from C3 terrestrial plants, there is increasing evidence for an important contribution from C4 plants to persistent POM in marine sediments on the continental shelf and slope (see below). Desorption of C4 plant carbon and incorporation into oceanic DOC would be difficult to detect by isotopic or molecular biomarker analyses.

The sources, abundances, and compositions of SPOM carried by rivers vary significantly, depending on the characteristics of the drainage basin (Onstad *et al.*, 2000; Raymond and Bauer, 2001a). For example, Raymond and Bauer (2001b) demonstrate significant spatial and temporal variation in $\Delta^{14}C$ and $\delta^{13}C$ values ($\Delta^{14}C$ is the measured ^{14}C concentration normalized to pre-industrial atmospheric values following Stuiver and Pollach (1977). $\Delta^{14}C = (\%\text{modern} \times e^{\lambda t} - 1) \times 1{,}000$, reported in per mil (‰), where $\lambda = {}^{14}C$ decay constant and $t =$ calendar age) of riverine dissolved and particulate OM, reflecting varying source and ages of terrestrial and aquatic productivity. As an example of the heterogeneity in SPOM signatures, Figure 2 shows the radiocarbon contents (expressed as $\Delta^{14}C$) of suspended OC reported for a range of river systems. Some systems transport predominantly "fresh" carbon as indicated by the presence of "bomb" ^{14}C ($\Delta^{14}C$ values greater than 0‰ reflect the atmospheric signature from aboveground nuclear weapons testing during the 1950s and early 1960s), while others carry carbon that is dominated by material older than 5×10^4 yr ($\Delta^{14}C$, −1,000‰).

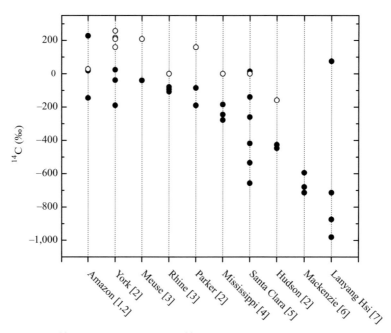

Figure 2 Variations in the ^{14}C content, expressed as $\Delta^{14}C$ (‰) of DOC (open symbols) and suspended POC (closed symbols) from a range of river systems. Data from: [1] Hedges *et al.* (1986); [2] Raymond and Bauer (2001b); [3] Megens *et al.* (2001); [4] Goni *et al.* (unpublished); [5] Masiello and Druffel (2001); [6] Eglinton *et al.* (unpublished); [7] Kao and Lui (1996).

For some river systems, the ^{14}C contents of SPOM vary temporally, presumably reflecting variations in sediment provenance, mode of erosion or other characteristics of the drainage basin. Thus, contrary to earlier notions that rivers exclusively export ^{14}C-enriched OM to the ocean (Hedges *et al.*, 1986), many rivers export a significant fraction of old, ^{14}C-depleted DOC and POC to the oceans (Raymond and Bauer, 2001a). This old SPOM could reflect pre-aged, vascular plant-derived OC stored in an intermediate reservoir (e.g., soils), and/or contributions of relict OC from sedimentary rock weathering, and/or contributions from aquatic production utilizing old DIC (the "hardwater" effect).

The age variations highlighted above are undoubtedly coupled to differences in the chemical composition and reactivity of SPOM. Onstad *et al.* (2000) examined elemental, stable carbon isotope and lignin phenol characteristics of SPOM from rivers draining the south central US. Variations in $\delta^{13}C$ values, ranging from $-18.5\permil$ to $-26.4\permil$, were attributed to the contributions from C3 and C4 plants in the catchment area, and hence to temperature and hydrologic patterns in the drainage basin. Lignin–phenol compositions reflect degraded, angiosperm-rich vegetation. Results from this and other studies indicate that highly degraded soil OM is a major component of fine-grained POM transported by rivers, and that most riverine OM residing in the particulate fraction is associated with mineral phases.

In large fluvial systems, estuaries and deltas serve as the interface between the rivers and the ocean, and are sites of intensive organic matter reworking and production (Hedges and Keil, 1999). Recent work has revealed that extensive removal of terrestrial OC from suspended particles occurs at these locations. In one approach to quantify terrestrial OC losses, Keil *et al.* (1997) argued that detrital mineral surface area can serve as a conservative tracer for riverine discharged POM. Accordingly, changes in OC to specific mineral surface area (OC : SA) ratios should indicate net OC exchange between upstream and downstream locations. When applied in conjunction with $\delta^{13}C$ as a source indicator of marine and terrestrial OC, this can yield estimates of the fraction of terrestrial OC entering the ocean as riverine SPOM that is deposited in deltaic systems (Figure 3). Using this approach, Keil *et al.* (1997) calculate average OC loadings within the Amazon River ($0.67\,\mathrm{mg\,C\,m^{-2}}$) that are approximately twice those of the Amazon Delta ($0.35\,\mathrm{mg\,C\,m^{-2}}$), while $\delta^{13}C$ measurements suggest that approximately two-thirds of the TOC in deltaic sediment is terrestrial. Together, these data imply that >70% of the Amazon fluvial POM evades sequestration in deltaic sediments. Extrapolating losses of riverine SPOM for a range of river/delta systems (Columbia, USA; Fly, New Guinea; and Huange-He, China), Keil *et al.* (1997) calculate a global loss of fluvial POM in delta regions of $\sim 0.1\,\mathrm{Gt}$.

The magnitude of this loss is thus substantial, and comparable to flux estimates for the delivery of terrigenous OC to the oceans (Figure 1). However, while the above studies imply low burial efficiencies for fluvial POM in deltaic environments, it is uncertain whether the apparent losses of riverine POM reflect its complete mineralization or export to the ocean interior either in dissolved or particulate form (Edmond *et al.*, 1981).

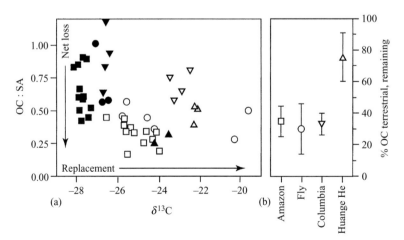

Figure 3 Loss of terrestrial OC in deltaic systems. (a) Organic carbon to mineral surface area ratio (OC : SA) plotted against bulk stable carbon isotopic compositions for riverine suspended sediments (closed symbols) and deltaic surface sediments (open symbols). A shift to lower OC : SA values indicates net loss of organic matter, and a shift to heavier (i.e., ^{13}C-enriched) isotopic compositions indicates increasing contributions from marine organic matter. (b) The average (± 1 SD) total amount of terrestrial OC persisting in deltaic sediments, based on the changes in OC : SA and $\delta^{13}C$ composition between river suspended sediments and deltaic sediments for four river systems (after Keil *et al.*, 1997).

Moreover, the extent of terrestrial OC export and burial from river systems that do not form deltaic deposits is less well constrained.

11.3.3 Quantitative Importance of Terrigenous Organic Carbon in Marine Sediments

There is still much debate about the abundance and composition of terrestrial OC in both margin and pelagic sediments. Terrigenous organic matter contributions may be substantial for several reasons.

- As pointed out above, the flux of particulate OC emanating from the continents carried by rivers alone is sufficient to account for the burial of OC in marine sediments.
- Most OC burial in surface sediments occurs along continental margins, proximal to the supply of terrestrial OC.
- Much of the terrestrial OC delivered to the oceans will have been subject to extensive degradative processes in soils and rivers, implying that the fraction that survives these processes might be refractory with high potential for preservation.

Despite these considerations, terrestrial OC contributions to the marine sedimentary OM

pool have generally been considered to be low. These conclusions are based on compositional studies, including bulk stable carbon isotopic ($\delta^{13}C$) and elemental (atomic $C_{organic}/N_{total}$) and biomarker compositions. Loss of terrestrial OC in deltaic systems has already been discussed (Keil *et al.*, 1997). Hedges and Parker (1976) also observed sharp reductions in the abundance of lignin-derived phenols (as determined by yields after CuO oxidation) in Gulf of Mexico sediments with increasing distance from riverine sources (Figure 4). Compositional information stemming from the same measurements indicates that the lignin present in these samples originates from nonwoody angiosperm sources (e.g., grasses), and is highly modified. Both the terrestrial biomarker abundances and bulk stable carbon isotopic compositions (Hedges and Parker, 1976) are in accord with the replacement of terrestrial C3 vegetation with marine OC for sediments deposited progressively further offshore (Figure 4). Thus, despite proximity of a major river system (Mississippi/Atchafalaya), an important source of terrigenous sediment to the shelf, bulk sedimentary OM composition fails to indicate the presence of significant terrestrial input. Moreover, Gough *et al.* (1993) observed only trace quantities of lignin phenols in abyssal North Atlantic Ocean sediments.

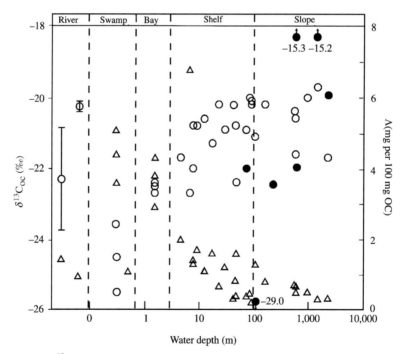

Figure 4 Variations in $\delta^{13}C$ of total OC (left axis: ‰, open circles) and the abundance of lignin-derived syringyl and vanillyl phenols (right axis: Λ, mg per 100 mg OC, triangles) in surface sediments from various water depths of the Gulf of Mexico in the vicinity of the outflow from the Mississippi River system (after Goni *et al.*, 1997). Also shown are average values for the same parameters determined on suspended particulate matter from the upper and lower reaches of the Mississippi River (extreme left column, error bars = 1 SD; data from Onstad *et al.*, 2000). In addition, $\delta^{13}C$ values are indicated for one lignin phenol (syringic acid filled circles), for selected shelf and slope samples (after Goni *et al.*, 1997).

However, some studies indicate that data of the type reported above may lead to underestimates of the proportion of terrigenous organic matter in marine sediments. For example, Onstad *et al.* (2000) showed that the Mississippi currently discharges isotopically "heavy" ($\delta^{13}C \sim -20\text{‰}$), lignin-poor POM that is difficult to distinguish from marine plankton remains in sediments from the Gulf of Mexico (Figure 4). Moreover, Goni *et al.* (1997, 1998) analyzed surface sediments from two offshore transects in the northwestern Gulf of Mexico using a range of techniques, including compound-specific $\delta^{13}C$ analysis of lignin-derived phenols. Bulk OC radiocarbon analyses of core top sediments yield depleted $\Delta^{14}C$ values, indicating that a significant fraction of the sedimentary carbon is "pre-aged," and most likely of allochthonous origin. Lignin phenol $\delta^{13}C$ values for inner shelf sediments are relatively depleted (average, -26.3‰), consistent with C3 vascular plant inputs, but are markedly enriched in ^{13}C at the slope sites (average, -17.5‰ for the two deepest samples) (Figure 4). Goni *et al.* (1997) interpret these molecular and isotopic compositions to indicate that a significant fraction ($>50\%$) of the lignin, and by inference land-derived OC, in slope sediments ultimately originated from C4 plants. Consistent with Onstad *et al.* (2000) the source of this material is likely to be soil organic matter eroded from the extensive grasslands of the Mississippi River drainage basin. The mixed C3 and C4 land plant sources, the highly degraded state of this material, and the differential transport effects (i.e., winnowing, resuspension, and lateral advection through nepheloid transport) hampers its recognition and quantification in shelf and slope sediments. These findings bear upon other river-dominated margins where drainage basins may include a significant proportion of C4 vegetation, and highlight the difficulty in quantitatively assessing terrigenous OC inputs.

Prahl *et al.* (1994) also concluded that terrestrial OC contributes significantly to Washington Margin sediments. These authors determined bulk elemental and stable carbon isotopic compositions and concentrations of a range of vascular plant biomarkers (epicuticular wax-derived *n*-alkanes, lignin-derived phenols and cutin-derived hydroxyalkanoic acids) for sediments from the Columbia River basin and adjacent margin. Using end-member values determined empirically by two independent means, Prahl *et al.* (1994) estimated terrestrial OC contributions of $\sim 60\%$, 30%, and $<15\%$ for sediments on the shelf and slope of the Washington margin, and the adjacent Cascadia Basin, respectively.

As noted earlier, prior studies have focused on only a few major river systems (e.g., Mississippi, Amazon, Columbia) that may not be representative of fluvial inputs worldwide. While our perspective on terrigenous OC inputs has been biased towards large river systems, there is a growing body of evidence that the majority of terrestrial OC delivery to the oceans occurs via numerous small rivers draining mountain regions in the tropics (Milliman and Syvitski, 1992). The contrasting modes of OC delivery between large deltas (passive, hydrodynamic control) and small mountainous rivers (episodic, high-energy, poorly sorted particles) could have a significant influence on the proportion and location of terrestrial OC that is deposited in marine sediments (Leithold and Blair, 2001; Masiello and Druffel, 2001). In addition to terrestrial OC inputs at low latitudes, our knowledge of terrestrial OC inputs to the high-latitude oceans is limited. Here, ice-rafting serves as another mechanism for export of terrestrial OC to the ocean basins. While some data indicates that terrestrial OC may indeed comprise a substantial fraction of the OM buried in Arctic sediments (Schubert and Stein, 1996; Macdonald *et al.*, 1998), traditional tools for source apportionment are not well suited to address this problem. For example, bulk stable carbon isotopes are of limited use as source indicators because of the isotopically depleted nature of phytoplankton in the polar oceans (Rau *et al.*, 1989; Goericke and Fry, 1994). Similarly, while eolian OC inputs to the oceans are generally considered to be minor, major dust fluxes emanating from West Africa carry a substantial C4 plant signature (Huang *et al.*, 2000; Eglinton *et al.*, 2002), confounding estimates of marine and terrestrial OC in underlying sediments based on bulk stable carbon isotopic measurements.

11.3.3.1 Black carbon

One group of exclusively terrestrially derived organic components found in marine sediments which has been the subject of renewed interest in recent years are the carbonaceous particles from incomplete combustion processes. These products, collectively termed "black carbon" (Goldberg, 1985), are ubiquitous in the environment, and may comprise a significant fraction of OC in contemporary marine sediments (Middelburg *et al.*, 2000; Masiello and Druffel, 1998; Gustafsson and Gschwend, 1998). Schmidt and Noack (2000) reviewed the state of knowledge of BC in soils and sediments. BC can be produced via condensation of volatiles to highly graphitized particles ("soot-BC") or by formation of solid residues through direct carbonization of particulate plant material ("char-BC"). Both forms of BC are relatively inert and are distributed globally by water (fluvial) and wind (atmospheric) transport. Although BC is directly emitted and transported during biomass burning, available radiocarbon data suggest that some BC is sequestered in soils and aquatic sediments for millennia prior to subsequent export to the oceans (Masiello and Druffel, 1998; Eglinton *et al.*, 2002).

Limited progress has been made on standardizing techniques for the quantitation and characterization of BC. These techniques and definitions have primarily been developed with regard to BC in atmospheric particles. The problem of defining and quantifying BC in sediments adds a further layer of complexity, and hence reservoir sizes and fluxes are subject to considerable uncertainty. Current estimates of global BC production (0.05–$0.27\,\mathrm{Gt\,yr^{-1}}$; Kuhlbusch and Crutzen, 1995) are of the same order as those of riverine input of POC to the ocean and OC burial in marine sediments. If these flux estimates are correct, BC may represent a significant sink in the global OC cycle (Kuhlbusch, 1998), and constitute a significant fraction of the carbon buried in soils (Skjemstad *et al.*, 1996; Schmidt *et al.*, 1999) and marine sediments (Middelburg *et al.*, 2000; Masiello and Druffel, 1998).

Gustafsson and Gschwend (1998) determined concentrations and fluxes of BC in modern ocean margin sediments off northeastern USA using a thermal oxidation method (Gustafsson *et al.*, 1997). Core-top BC concentrations indicate that BC comprises a significant component of OC budgets for coastal sediments. Down-core trends were consistent with anthropogenic fossil-fuel combustion dominating BC input, while the fractional abundance of BC increased in deeper sediment sections, implying that it is resistant to degradation and may be selectively preserved (Wolbach and Anders, 1989). Suman *et al.* (1997) calculated the pre-industrial global burial of BC to be $\sim\!10\,\mathrm{Tg\,yr^{-1}}$, which corresponds to $\sim\!11\%$ of the estimated global marine sediment burial of TOC.

Middelburg *et al.* (1999) compared BC determined using thermal and chemical (hot nitric acid) oxidative pre-treatments in a range of marine sediments. They found that the latter significantly overestimates combustion-derived phases. Nevertheless, the lower BC estimates obtained from the thermal oxidation method account for between 15% and 30% of total OC. Examination of BC concentrations across a relict oxidation front in a Madeira Abyssal Plain turbidite provided evidence for significant BC degradation under prolonged exposure to oxygenated bottom waters.

Masiello and Druffel (1998) measured the abundance and radiocarbon content of BC (isolated by wet chemical oxidation) in sediment cores from two deep Pacific Ocean sites. They found BC comprises 12–31% of the total sedimentary OC, and was between 2,400 ^{14}C yr and 13,000 ^{14}C yr older than non-BC sedimentary OC (Figure 5). For sediment intervals deposited prior to the Industrial era (i.e., free of BC inputs from

fossil fuel utilization), the authors argue that the older ages for BC must be due to storage in an intermediate reservoir before deposition. Possible intermediate pools are oceanic DOC and terrestrial soils. They conclude that if DOC is the intermediate reservoir, then BC comprises 4–22% of the DOC pool. If soils are the intermediate reservoir, then the importance of riverine OC has been underestimated.

11.4 A BIOPOLYMERIC ORIGIN FOR OCEANIC DISSOLVED ORGANIC MATTER

11.4.1 Background

As deep water upwells to the surface, some 30–$40\,\mu\mathrm{M}$ carbon is added to the dissolved phase (Peltzer and Hayward, 1995). Although the mechanisms that fix carbon into DOC are not known, it has long been recognized that this new carbon represents water-soluble by-products of algal photosynthesis (Duursma, 1963; Menzel, 1974). Phytoplankton in laboratory culture and in seawater have been shown to exude OC (Hellebust, 1965; Iturriaga and Zsolnay, 1983), and the release of dissolved OC through exudation, grazing, cell lysis, or other processes fuels much of the microbial growth and respiration in the marine environment (Hellebust, 1965; Nagata and Kirchman, 1992). Stable isotope measurements of carbon standing stocks throughout the entire water column confirm an autochthonous origin for marine DOC, with values identical to marine particulate matter ($-21‰$ to $-22‰$) (Druffel *et al.*, 1992). As DOC cycles through the water column, new carbon is somehow altered and sequestered into the more permanent, deep ocean carbon reservoir.

Although DOC production is a function of primary production, DOC accumulation is not. There are a few reports of transient, localized increases in DOC inventory following algal blooms (Holmes *et al.*, 1967), and annual cycles of DOC inventory in the upper ocean share features in common with annual cycles of primary production (Duursma, 1963; Holmes *et al.*, 1967; Carlson *et al.*, 1994). However, DOC concentrations are not well correlated with either phytoplankton biomass or primary productivity (Carlson *et al.*, 1994; Chen *et al.*, 1995). The accumulation of DOC in surface seawater results from a decoupling of production and removal processes over annual to decadal timescales, and it is convenient to subdivide DOC into at least three distinct reservoirs of differing reactivity (Carlson and Ducklow, 1995). The largest reservoir in terms of *production* is very reactive, and supports most secondary production in the ocean. This fraction of DOC consists of a few $\mu\mathrm{M}$ C of soluble biochemicals

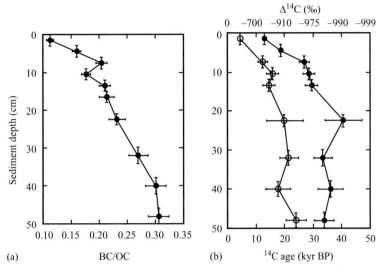

Figure 5 Abundance of radiocarbon age of black carbon in slowly accumulating (~2.5 cm kyr⁻¹) deep-sea sediments from the Southern Ocean (54°S 176° 40′W): (a) a plot of the ratio of black carbon to total organic carbon (BC/OC) with sediment depth and (b) $\Delta^{14}C$ (per mil) and ^{14}C age (kyr BP) of BC (solid symbols) and non-BC sedimentary OC (open symbols) as a function of depth (after Masiello and Druffel, 1998).

(proteins, carbohydrates, lipids, etc.), which has a turnover time of hours to days, and may equal 10–30% of total primary production (Vaccaro *et al.*, 1968; Mague *et al.*, 1980; Jørgensen *et al.*, 1993). The largest fraction of DOC in terms of total inventory is the nonreactive fraction. Nearly all the DOC in the deep ocean, and half the DOC in the surface ocean, is considered to be non-reactive based on radiocarbon measurements, which show this fraction to be 4,000–6,000 yr old (Williams *et al.*, 1969; Druffel *et al.*, 1992). This nonreactive component of DOC persists through several ocean-mixing cycles before remineralization or sequestration as particulate OC. Finally, the 30–40 μM DOC equal to the difference between deep and surface seawater DOC concentration values is considered to be reactive DOC, and cycles on seasonal to decadal timescales. Reactive DOC is produced in the surface ocean, and largely consumed as surface water is subducted into the deep ocean. The composition, cycling, and fate of reactive DOC has been the focus of much of the research on DOC completed since the 1990s.

11.4.2 High Molecular Weight Dissolved Organic Matter: Biopolymers or Geopolymers?

The very old radiocarbon age of DOC was originally attributed to the structural complexity of the organic constituents that make up this fraction of marine organic matter. Simple biochemicals produced by marine bacteria, algae, and animals were thought to react through abiotic or geochemically mediated reactions to form more complex humic-like substances that are metabolically unavailable

to marine microorganisms (Gagosian and Stuermer, 1977). The concept of organic matter humification in seawater follows earlier models of humification reactions in soils, and early studies on marine DOC sought to characterize and compare marine humic substances with humic substances extracted from terrestrial and freshwater environments in an attempt to better understand the reactions in each environment (Hatcher *et al.*, 1980; François, 1990) These studies show marine DOM to have a low average molecular weight (<1,000 Da), to be relatively rich in nitrogen and aliphatic carbon, and poor in aromatic carbon (Stuermer and Harvey, 1974; Stuermer and Payne, 1976). Chemical differences between marine and terrestrially derived humic substances could be attributed directly to the nature of biochemicals prevalent in each environment, and the differences in reaction conditions (e.g., light, presence of catalytic surfaces) (François, 1990; Malcolm, 1990). Marine and terrestrial humic substances represent two, chemically distinct, carbon reservoirs.

Marine humic substances are isolated by adsorption onto hydrophilic resins and contribute only a small fraction (10% or less) of total marine DOC. The difficulty in sampling DOC in the presence of much more abundant inorganic salts significantly slowed progress in marine DOM research for over a decade. Application of ultrafiltration, and especially large volume ultrafiltration made a much larger fraction (up to 40%) of DOC available for study (Buesseler *et al.*, 1996). Ultrafiltration does not select strongly for the chemical characteristics of the organic matter as does resin adsorption, but selects on the basis of molecular size instead, and therefore preferentially isolates the high molecular weight (HMW) fraction of DOM. Initial studies of HMW DOM composition showed this fraction to be chemically distinct from humic substances (Benner *et al.*, 1992).

Our understanding of DOM composition and cycling has undergone a rapid change since the 1990s. Chemical studies of HMW DOM now show a composition that is rich in specific poly-saccharides and proteins and remarkably uniform across diverse environments. These discoveries led Aluwihare *et al.* (1997) to propose that a major fraction of HMW DOM arises directly from biosynthesis. The concept that marine DOM has a large component of metabolically resistant biploymers is a sharp departure from earlier ideas that described DOM as a mixture of simple biomolecules that had experienced abiotic transformation (geopolymerization) into HMW substances (fulvic and humic substances). Support for the directly formed biopolymer hypothesis comes from the chemical composition of HMW DOM itself.

11.4.2.1 Acylpolysaccharides in high molecular weight dissolved organic matter

NMR spectroscopy has proven to be the most effective technique for characterizing carbon functional groups in HMW DOM. The ^{13}C-NMR spectrum of HMW DOM collected at 15 m in the North Pacific Ocean surface is given in Figure 6(a). Nearly identical spectra have been collected from the Atlantic Ocean, as well as from some lakes and rivers (McKnight *et al.*, 1997; Repeta *et al.*, 2002). All ^{13}C-NMR spectra display a rather simple pattern of broad resonance from carboxyl (CO–(OH or NH); 175 ppm), alkene/aromatic (C=C;140 ppm), anomeric (O–C–O; 100 ppm), alcoholic (H–C–OH; 70 ppm), and alkyl (CH$_x$;10–40 ppm) carbon (Figure 6(a)). From these spectra we can quantify the relative amounts of carbon associated with each class of functional group, and by inference, identify the major biochemical units in DOM. ^{13}C-NMR shows HMW DOM to be especially rich in carbohydrate (100 ppm (O–C–O), and 70 ppm (H–C–OH)), which accounts for 76% of the carbon in the spectrum. The ratio of alcohol to anomeric carbon (70–100 ppm) is 4.5, within the expected range for most sugars (4–5). The importance of carbohydrate is confirmed through molecular-level analyses of monosaccharides which show neutral sugars to be abundant in HMW DOM (Sakugawa and Handa, 1985; McCarthy *et al.*, 1994; Aluwihare *et al.*, 1997; Biersmith and Benner, 1998; Borch and Kirchman, 1998). However, the amount of carbohydrate determined by NMR is much higher than that measured by molecular-level techniques (76% versus 20%, respectively for Figure 6(a)). Reconciling NMR

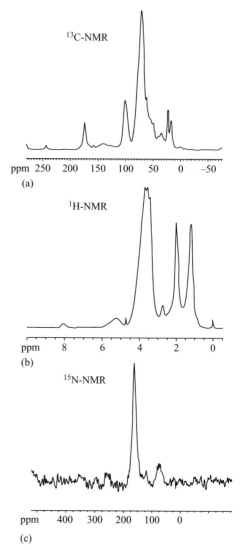

Figure 6 NMR spectra of HMW DOM from surface seawater. (a) ^{13}C-NMR spectra can be used to quantitatively determine the functional groups, and by inference, the relative importance of different biochemical classes in HMW DOM. The spectra highlight the importance of carbohydrates (100 ppm and 70 ppm), carboxylic acids (175 ppm), and alkyl carbon (10–30 ppm). (b) ^{1}H-NMR also show the importance of carbohydrates (4 ppm) and alkyl carbon (1 ppm), but additionally show that acetyl groups most likely bound to carbohydrate are an important components. ^{15}N-NMR show that 80–90% of HMW DON is amide, while 10–20% is free amine. Quantitative analyses for acetate and nitrogen suggest that most amide in surface seawater is bound as *N*-acetyl amino sugars and protein residues. In the deep ocean HMW DOM however, most amide is nonhydrolyzable, and is of unknown molecular environment.

data and molecular-level techniques is currently a major challenge in understanding the composition of HMW DOM.

Sugar distributions are dominated by six neutral sugars—fructose, galactose, glucose, mannose, rhamnose, and xylose—which occur in approximately equimolar amounts. A seventh neutral sugar (arabinose), and two amino sugars (*N*-acetyl glucosamine and *N*-acetyl galactosamine) are also abundant, but occur at only half to one quarter the concentrations of other sugars. The relative amounts of these nine sugars, and their contribution to surface water HMW DOC vary remarkably little in fresh and marine waters (Figure 7). The uniformity in molecular and spectroscopic properties of many HMW DOM samples suggests that the fraction of HMW DOM enriched in carbohydrate, acetate, and lipids is a well-defined family of biopolymers synthesized by marine algae (Aluwihare *et al.*, 1997; Aluwihare and Repeta, 1999). These polymers, referred to as acylpolysaccharide (APS) are ubiquitous in natural waters (Repeta *et al.*, 2002).

The ^{13}C-NMR spectrum of HMW DOM also includes contributions from carboxyl (CO–(OH or NH), 5% of total carbon), and alkyl (CH$_x$, 14% total carbon) functional groups, which may derive from proteins, lipids, or carbohydrates (deoxy- and methyl sugars). Hydrolysis of HMW DOM followed by extraction with organic solvent yields

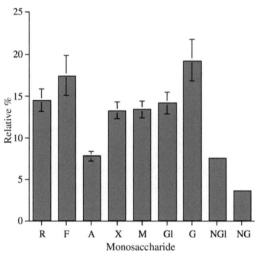

Figure 7 The distribution of neutral and basic monosaccharides in HMW DOM. Hydrolysis of HMW DOM followed by monosaccharide analysis show a remarkably uniform distribution of sugars in samples from surface and deep ocean waters in all ocean basins. Important neutral sugars include rhamnose (R), fucose (F), arabinose (A), xylose (X), mannose (M), glucose (Gi), and galactose (G). Error bars show 1 σ in relative abundance of 12 HMW DOM samples collected from the Atlantic Ocean and Pacific Ocean between 1994 and 2001. Little change in the relative abundance of sugars is observed between ocean basins. Important amino sugars include *N*-acetyl glucosamine (NGl) and *N*-acetyl galactosamine (NG), which are less abundant than the neutral sugars. If all acetate is bound as *N*-acetyl amino sugars, then monosaccharide analyses significantly under-recover the amount of amino sugars in HMW DOM.

4–8% of the total carbon in HMW DOM as acetic acid. Acetyl is easily recognized in the ^{1}H-NMR spectra of HMW DOM, where it appears as a broad singlet centered at 2 ppm (Figure 6(b)). Free acetic acid and its derivatives are not retained by ultrafiltration, and the acetyl in HMW DOM must be covalently bound to macromolecular material, most likely as an *N*-acetyl amino sugar. Acetyl contributes up to half the carboxyl carbon in the ^{13}C-NMR spectrum.

Elemental C/N and C/P ratios of HMW DOM lie between 13–16 and 300–800, respectively (McCarthy *et al.*, 1996; Karl and Bjorkman, 2002). Although C/P ratios are so high that they prohibit a major contribution from organophosphorous compounds to HMW DOC, C/N ratios are sufficiently low that N-containing biochemicals must contribute an important fraction of DOM carbon. Natural abundance ^{15}N-NMR measurements show that 80–90% of HMW dissolved organic nitrogen (DON) is amide (R-CON; Figure 6(c)) (McCarthy *et al.*, 1997). Free amine groups contribute the remaining 10–20% of HMW DON that can be observed by NMR. Small amounts of other components, such as aromatic N-containing compounds, may also be present but are below the limit of detection by this technique. The amount of nitrogen in HMW DON is approximately equal to the amount of carboxylic carbon. If nearly all *N* is amide, then conversely nearly all carboxyl carbon must be amide as well, and in agreement with molecular-level analyses which show only trace amounts of biomarker lipids (<0.1% total carbon), little carboxyl carbon can be components of other organic acids (fatty acids, etc.; Mannino and Harvey, 2000).

Several important biochemicals are amides, of which proteins and *N*-acetyl amino sugars are the most abundant in marine organisms, where proteins contribute up to 80%, and amino sugars up to 10% or the total nitrogen. Molecular-level analyses show both these biochemical classes to be present in HMW DON, but only at low concentrations. For most samples, amino acids account for <15–20% of HMW DON, and aminosugars <1% of the HMW DON (McCarthy *et al.*, 1996; Kaiser and Benner, 2000; Mannino and Harvey, 2000). Either HMW DON amide is derived from some other component, or the analytical protocols used in molecular-level analyses are not appropriate for measuring these biochemicals in HMW DON.

The discrepancy between molecular-level and NMR measurements of amide nitrogen can best be resolved by using the two techniques interactively. Proteins and *N*-acetyl amino sugars both yield amino compounds (free amino acids and amino sugars, respectively) on treatment with acid (Figure 8). Therefore, acid hydrolysis of HMW DON should be accompanied by a change in the ^{15}N-NMR chemical shift from amide to

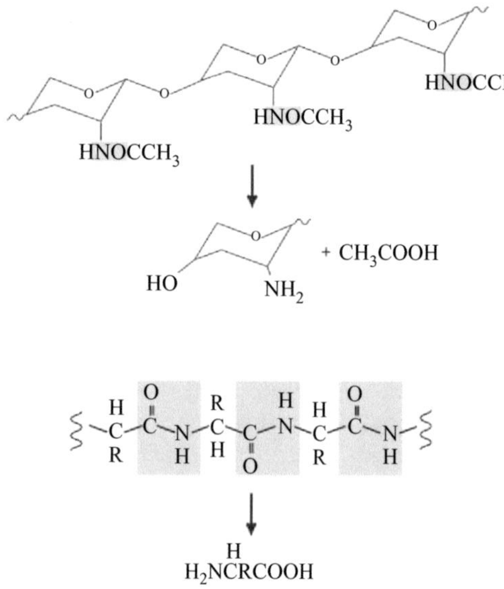

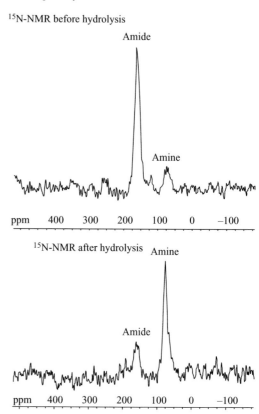

Figure 8 The effect of mild acid hydrolysis on amides in HMW DOM. Two potentially important classes of biochemicals that likely contribute to HMW DOM are (poly)-*N*-acetyl amino sugars (top) and proteins (bottom). Mild acid hydrolysis of (poly)-*N*-acetyl amino sugars will yield free acetic acid, but will not depolymerize the polysaccharide. The generation of acetic acid will be accompanied by a shift in the ^{15}N-NMR from amide to amine. In contrast, mild acid hydrolysis of proteins does not yield acetic acid, but may depolymerize the protein macromolecular segments to yield free amino acids. Free amino acids can be quantified by chromatographic techniques and compared to the shift from amide (protein) to amine (free amino acid) in ^{15}N-NMR.

Figure 9 The effect of mild acid hydrolysis on ^{15}N-NMR of HMW DOM. Nitrogen in HMW DOM is primarily amide (180 ppm), with smaller amounts of free amine (90 ppm). Treatment of HMW DOM with dilute hydrochloric acid increases the amount of amine and decreases the amount of amide. The decrease in amide equals the amount of acetic acid and amino acids released by hydrolysis of poly-*N* acetyl amino sugars and proteins. The relative amount of protein and amino sugar can be determined by the ratio of acetic acid to amino acids in the hydrolysis product.

amine (Figure 9). *N*-acetyl amino sugars will also yield free acetic acid with hydrolysis, and the high concentrations of acetate in HMW DOM along with pyrolysis MS measurements of acetamide suggest that much of HMW DON is *N*-acetyl amino sugars (Boon *et al.*, 1998). The amount of acetate bound as amide can be quantified by comparing the amount of free acetic acid released by hydrolysis with the hydrolytic yield of free amine observed by ^{15}N-NMR. Treatment of HMW DOM with mild acid converts ∼60–70% of amide to free amine. This amount equals the molar sum of free amino acids (10–20% total N, from protein hydrolysis) and acetic acid (90–50% total N, from *N*-acetyl amino sugars) measured by molecular-level techniques after HMW DON hydrolysis. The agreement between NMR and molecular-level analyses indicates that 40–50% of HMW DON is derived from *N*-acetyl amino sugars and that molecular-level analyses underestimate the amount of amino sugars in HMW DON by at least an order of magnitude.

Our knowledge of APS structure and composition is still evolving; however, Aluwihare and Repeta (1999) suggest that APS is a biopolymer that is largely, perhaps exclusively, composed of carbohydrate units. This biopolymer is approximately half neutral sugar and half *N*-acetyl amino sugar (Figure 10). Together these two components contribute 40–50% of the total carbon, 60–70% of the total carbohydrate, and an equal amount of the total new (or modern) carbon in HWM DOC (Druffel *et al.*, 1992; Guo *et al.*, 1994). Further work is needed to establish, on a molecular level, the abundance of *N*-acetyl amino sugars in APS. Are the neutral sugars and amino sugars covalently linked into a common macromolecule, and if so, what is the full structure of this polysaccharide?

11.4.2.2 *Proteins in high molecular weight dissolved organic matter*

A small fraction of the carbon, and a larger fraction of the nitrogen in HMW DOM, is protein.

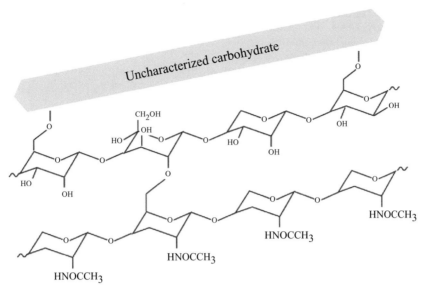

Figure 10 Hypothetical structure of APS in HMW DOM. Spectroscopic and molecular-level analyses suggest that approximately one-third of APS is neutral sugars, one-third amino sugars, and one-third unidentified carbohydrate. The consistency in monosaccharide distribution, and the fixed relative amount of acetate to total carbohydrate suggests these portions of HMW DOM are coupled into the same macromolecular structure. Further work needs to be done to establish the coupling of neutral and amino sugars directly, and to identify the unknown component of APS to bring molecular-level and spectroscopic measurements into better agreement.

Tanoue *et al.* (1995) analyzed HMW DOM by gel electrophoresis and found a complex mixture of proteins with masses between 14 kDa and 66 kDa. Two major components were noted at 48 kDa and ~34–39 kDa. The 48 kDa protein was purified for *N*-terminal sequencing and shown to have significant homology with porin-P from the gram negative bacterium *Pseudomonas aeruginosa*. Tanoue's data are the most direct evidence to date that resistant biopolymers selectively survive degradation and accumulate as oceanic DOM.

Additional evidence for a bacterial contribution to HMW DOM proteins comes from molecular-level analyses of dissolved amino acids. Hydrolysis of HMW DON releases 11–29% of the nitrogen as amino acids (McCarthy *et al.*, 1996). Specific amino acids include common protein amino acids, as well as β-alanine and γ-aminobutyric acid which are nonprotein amino acid degradation products. The distribution of amino acids is similar to that of fresh plankton cells, suspended particulate matter, and total dissolved amino acids. However, stereochemical analyses show HMW DOM amino acids to be elevated in the D-enantiomer, with D/L ratios for alanine, aspartic acid, glutamic acids, and serine ranging from 0.1 to 0.5 (McCarthy *et al.*, 1998). Racemization of phytoplankton-derived L-amino acids is too slow at ocean temperatures to yield such high D/L ratios, but bacteria can synthesize D-amino acids, and it is likely that the D-amino acids in HMW DOM result from bacterial bioplymers rich in these particular amino acids. The high D/L ratios of some amino

acids and the abundance of amide nitrogen in HMW DOM ¹⁵N-NMR spectra led McCarthy *et al.* (1998) to postulate that peptidoglycan may be one such biopolymer that is significantly enriched in HMW DOM. Most gram-negative bacteria produce peptidoglycan as part of their cell membrane and, like porin-P discussed above, are therefore a potential source for HMW DON in the ocean.

Further chemical characterization of HMW DOM is needed to verify this hypothesis. Peptidoglycan is a polymer of *N*-acetyl glucosamine, muramic acid, and amino acids. *N*-acetyl glucosamine has already been identified by molecular-level techniques in HMW DOM. The concentration of *N*-acetyl glucosamine is low, but consistent with the amounts of D-amino acids in samples, assuming all D-amino acids are part of a peptidoglycan biopolymer. However, muramic acid has not been detected in HMW DOM samples, and the absence of muramic acid suggests pepidoglycan is not present (Repeta, unpublished). Lactic acid, a component of muramic acid, is recovered from HMW DOM in amounts of ~0.5% of total carbon, or again at levels consistent with peptidoglycan. The analytical protocols used to release lactic acid from HMW DOM are nonspecific, and it is not known whether lactic acid comes from muramic acid, or from the degradation of another polymer. Molecular-level techniques used to quantify muramic acid may suffer from the same uncertainties as those discussed previously for *N*-acetyl amino sugars, for example, current techniques may be inappropriate for the degradation resistant biopolymers in HMW DOM, and therefore they may not as yet provide accurate data on the distributions of molecular constitutents in HMW DOM. Assuming all alanine, serine, aspartic, and glutamic acid in HMW DOM is bound as peptidoglycan, then this bacterial polymer could contribute up to a maximum of 9% of the total HMW DOC. However, if these amino acids are assumed to occur only in porin-P or other bacterial proteins, then the contribution of bacterial carbon to HMW DOC would be much lower.

11.4.3 Gel Polymers and the Cycling of High Molecular Weight Dissolved Organic Matter

HMW biopolymers are part of the colloid-sized fraction of marine organic matter that can be visualized and enumerated with transmission electron microscopy (TEM) (Wells and Goldberg, 1991, 1993). Colloids range in size from a few nanometers up to $\sim$1 μm in size. Very small colloids ($<$30 nm) are irregular in shape, while larger colloids ($\sim$30–60 nm) are more spherical assemblies of 2–5 nm sized subparticles. Concentrations of small colloids ($<$200 nm) range from non-detectable ($<10^4$ colloids ml^{-1}) to $>9 \times 10^9$ ml^{-1}, making them the most abundant particles in seawater. Vertical profiles of colloids show a highly variable distribution throughout the water column, characteristic of a dynamic reservoir that is rapidly cycling (Wells and Goldberg, 1994). Colloids have low electron opacity by TEM, consistent with a largely organic composition, and slight enrichments of some trace metals (iron, cobalt, chromium, nickel, and vanadium) and other elements (silicon, aluminum, and calcium). Santschi *et al.* (1998) examined colloids using atomic force microscopy (AFM), which gives comparable results to TEM but has a greater resolving power for small colloids. Visualization of colloids by AFM shows fibrils, elongated particles $\sim$1 nm in cross-section by 100–200 nm long, to be ubiquitous and a major component of colloidal material. Chemical analyses of samples artificially enriched with fibrils (through laboratory manipulation) show a parallel enrichment in carbohydrate, which may be the biochemical component of fibril material.

Colloids have been described as classic polymer gels, and their dynamics in seawater may be understood in terms of polymer gel theory (Chin *et al.*, 1998). Polymer gels are stable, three-dimensional networks of polymers and seawater (Figure 11). Gels assemble spontaneously from DOM to form nanometer to micrometer sized particles. The kinetics of polymer gel assembly has been monitored by flow cytometry

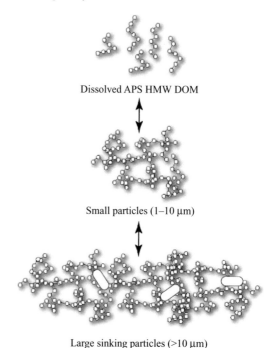

Dissolved APS HMW DOM

Small particles (1–10 μm)

Large sinking particles (>10 μm)

Figure 11 Formation of gel polymers by APS. The APS fraction of HMW DOM may entangle and spontaneously assemble into small particles. Further assembly will increase particle size and be accompanied by colonization by bacteria or incorporation into large, rapidly sinking particles.

and dynamic laser scattering, which show a rapid formation of particles in filtered seawater. In less than 30 min DOM can assemble into micrometer-sized particles that continue to grow as long as free polymers remain in solution (Chin *et al.*, 1998). Gels are stabilized by the presence of divalent cations, particularly Ca^{+2} and Mg^{+2}, and will disassemble in the presence of EDTA which out competes natural polymers for these cations. The composition of the gels is unknown, but the polymers that form the gels are presumed to be organic. Gels stain positively for a number of biochemicals, including carbohydrates, proteins, and lipids. The picture of colloid formation that emerges from these experiments is consistent with observations of natural colloids in seawater. Dissolved organic polymers, 1–10 nm in size, spontaneously assemble in seawater due to the relatively high concentrations of calcium ions present. Assembly is rapid and reversible, and the gel particles will grow to at least several microns in size. Particles may continue to grow until they sink out of the water column, or are removed by other processes. Polymer gel assembly couples the dissolved and particulate reservoirs of OC through the rapid exchange of colloids (Figure 12).

The polymers active in gel polymer assembly may be aged polysaccharides. Santschi *et al.*

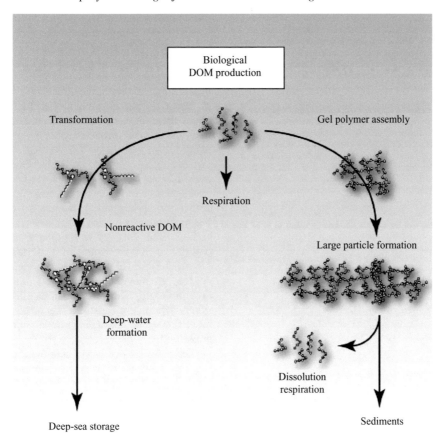

Figure 12 DOM cycling in the ocean. DOM is initially produced as a by-product of marine production in the mixed layer. Carbon stable isotope data suggests that some fraction of this organic matter becomes incorporated into the longer-term reservoir of rather refractory DOM to be subducted into the deep ocean. Newly produced DOM may also assemble into particles to enter the POM cycle. POM is oxidized, or transported into the deep ocean on large, rapidly sinking particles where biological activity may cause further oxidation, or drive re-dissolution to HMW DOM.

(1998) noted that fibril enriched samples of HMW DOM were rich in carbohydrate and radiocarbon. APSs are also rich in carbohydrate, and isotopic analyses of monosaccharides show this fraction to be enriched in radiocarbon relative to total DOC. APSs have a size distribution consistent with polymers that assemble into polymer gels, and are recovered from seawater as isolates rich in calcium. NMR spectra (Figure 13) and molecular-level analyses of 3 kDa, 10 kDa, and 100 kDa HMW DOM show the same major resonances and the same distribution of neutral sugars as 1 kDa HMW DOM (Figure 7). It has not been established if these size fractions represent true APS polymers of increasing molecular weight, or assemblies of smaller (1–3 kDa) sized APS.

One promising approach that may help to establish APS as the fraction of HMW DOM that assembles into particulate matter is to compare the chemical composition of polymer gel assemblies with HMW DOM. In one such experiment, HMW DOM recovered by ultrafiltration of seawater or spent culture media was redissolved in seawater and agitated by bubbling to produce particles which collect at the top of a bubble tower (Gogou and Repeta, unpublished). Particles formed by bubbling have the same neutral sugars in approximately the same proportions as APS. NMR data likewise show particles to be rich in carbohydrate and have the same major resonances as APS, although the relative amount of major biochemicals differs between the two samples. This and other similar approaches further support the hypothesis that APS is the reactive fraction of HMW DOM that undergoes spontaneous assembly into polymer gels and larger particles.

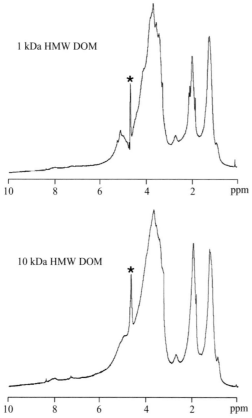

1 kDa HMW DOM

10 kDa HMW DOM

Figure 13 NMR spectra showing the similar spectral characteristics of 1 kDa and 10 kDa HMW DOM. Spectroscopic and chemical analyses of different-sized fractions of HWM DOM show each fraction to have essentially the same spectroscopic and elemental properties, and the same distribution of neutral sugars. The very HMW DOM (>10 kDa) may be made up of either assemblies of 1–10 kDa APS, or true polymers of increasing molecular weight that spontaneously assemble in seawater. Further work is needed to establish if APS is the component of HMW DOM that forms gel polymer assemblies.

11.5 EMERGING PERSPECTIVES ON ORGANIC MATTER PRESERVATION

11.5.1 Background

The vast majority of POM that sinks from the surface ocean is recycled during passage through the water column and burial in the upper sediment layers. The portion that escapes recycling to be sequestered in the underlying sediments serves to modulate atmospheric carbon dioxide concentrations over geological timescales, and provides a valuable archive of past ocean and climate conditions. The mechanisms by which OM survives degradation in the water column and in sediments, and the composition of the residual material are questions that have challenged marine organic chemists for many years. These processes have

remained unclear, partly due to the difficulty of resolving the composition of the residual organic matter at depth using traditional wet chemical procedures and chromatographic and/or mass spectrometric techniques. In general, even for organic matter residing in the surface ocean, and in material that has recently exited the photic zone, there is a significant fraction of the OM that is no longer recognizable as biochemicals using traditional assays (Wakeham *et al.*, 1997). For example, a recent molecular-level survey of over 100 amino acids, sugars, and lipids in the water column of the central Pacific Ocean (Wakeham *et al.*, 1997) failed to account for ~15% of the molecules composing "plankton" and missed greater than 75% of the organic molecules in particulate debris raining in a matter of days to the ocean floor (Figure 14). Our assessment of the degree of diagenetic alteration undergone by organic matter using the abundance and compositional parameters based on the minor fraction of identifiable biochemicals is thus subject to major uncertainty.

The processes of signal attenuation and modification continue to varying extents on material after deposition and vary, depending on burial conditions. Moreover, our ability to structurally identify organic matter decreases as it becomes further removed from its biological source(s). Thus, we can account for less of the sedimentary organic debris leaving the photic zone derived from primary production compared to that in the original phytoplankton biomass, and organic matter in the underlying sediments contains less recognizable biological constituents than that in sinking particles. Indeed, often >80% of organic matter in surficial marine sediments remains unaccounted for in terms of readily distinguishable organic molecules (Hedges *et al.*, 2000). Broad structural features of this "molecularly uncharacterized component" have been gleaned from bulk elemental and spectral analyses. However, without detailed molecular-level information the origins, reactions, and fates of this fraction are likely to remain obscure. To quote Hedges *et al.* (2000), "biogeochemists of today are playing with an extremely incomplete deck of surviving molecules, among which most of the trump cards that molecular knowledge would supply remain masked."

11.5.2 Compositional Transformations Associated with Sedimentation and Burial of Organic Matter

By virtue of where, when, and how the various organic matter inputs were formed and transported to the underlying sediments, it is possible to exploit specific chemical and isotopic characteristics to make inferences about the sources and

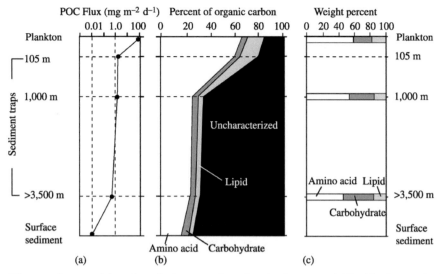

Figure 14 Fluxes and composition of particulate organic carbon in the equatorial Pacific Ocean. (a) POC fluxes $(mg\,m^{-2}\,d^{-1})$. (b) Corresponding fractions of amino acid, carbohydrate, lipid, and molecularly uncharacterized carbon (biochemical class-carbon as a percentage of total OC) in plankton, sediment traps (105 m, ~1,000m, >3,500 m) and surface sediment samples. The fraction of molecularly uncharacterized organic carbon (calculated as the difference between total OC and the sum of amino acid + carbohydrate + lipid) increases with more extensive degradation to become the major constituent in deeper POC samples (after Wakeham *et al.*, 1997). (c) Calculated weight percentages of amino acid, carbohydrate, and lipid in plankton and in sinking (sediment trap) particles in the upper and lower water column as determined by solid-state ^{13}C-NMR spectroscopy (source Hedges *et al.*, 2001).

composition of sedimentary organic matter. Much of this information is inaccessible at the bulk level. For example, bulk elemental compositions and stable carbon isotopic compositions are often insufficiently unique to distinguish and quantify sedimentary inputs. Abundances and distributions of source-specific organic compounds ("biomarkers") can help to identify specific inputs. However, this molecular marker approach suffers from the fact that the source diagnostic marker compounds are typically present as trace components, and extrapolation of abundances to infer overall organic matter contributions is therefore subject to considerable uncertainty. Isotopic measurements at the molecular level have provided a means to bridge the information gap between bulk and biomarker composition.

New insights into composition and transformation in sinking and SPOM have been gleaned from isotopic analyses performed in conjunction with compositional studies (e.g., Wang *et al.*, 1996, 1998; Megens *et al.*, 2001). For example, Druffel *et al.* (1992) measured ^{14}C, ^{13}C, bulk carbon, and biochemical constituents in dissolved and particulate carbon pools from the North Central Pacific Ocean and subtropical North Atlantic (Sargasso Sea). The decrease in Δ^{14}C values of suspended and sinking POC with depth has been interpreted in terms of incorporation of low-reactivity OM into the POC pool, possibly via DOC sorption or heterotrophy (Druffel *et al.*, 1992). ^{14}C and ^{13}C measurements on different classes of biochemical revealed that lipids were much "older" than

corresponding amino acid and carbohydrate fractions in detrital aggregates, sediment floc, and sediments. These data indicate differences in decomposition and chemical behavior for different classes of biochemical in the deep ocean (Figure 15; Wang *et al.*, 1998).

Similar trends in organic matter reactivity and isotopic composition are evident in sediments. For example, Wang and co-workers (Wang *et al.*, 1996; Wang and Druffel, 2001) measured ^{14}C and ^{13}C compositions of total hydrolysable amino acids (THAAs), total carbohydrates (TCHO), and total lipids in deep-sea sediment profiles (Figure 15). Based on sedimentary concentration profiles, and using a "multi-G" model considering both labile and refractory organic fractions, Wang *et al.* (1998) calculated that degradation rate constants were in the order THAA ≈ TCHO > TOC ≈ TN > Total Lipid, indicating their relative reactivities in the sediment during early diagenesis. This is in good agreement with the order of average Δ^{14}C values in the sediment (THAA, −275‰; TCHO, −262‰; TOC, −371‰; lipid, −506‰), indicating that differential decomposition of organic matter may be a major process controlling the observed Δ^{14}C signatures. Alternatively, these results may indicate sorption and/or biological incorporation of "old" DOC into POC pool.

^{14}C age differences observed between biochemical classes are both expressed and magnified at the level of individual organic compounds (e.g., Eglinton *et al.*, 1997; Pearson *et al.*, 2001).

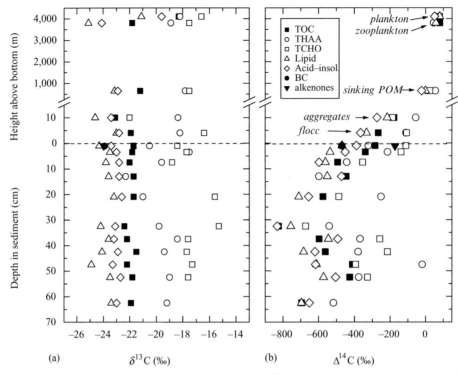

Figure 15 Variations in: (a) stable carbon isotopic composition ($\delta^{13}C$, ‰) and (b) radiocarbon content ($\Delta^{14}C$, ‰) of water column and sedimentary organic matter in the northeastern Pacific Ocean (Stn M, 34° 50'N, 123° 00'W; 4,100 m water depth). Samples: phytoplankton (1 m), zooplankton (100 m), sinking particulate material collected in a sediment trap 650 m above the seafloor, detrital aggregates, and surface flocc material isolated from the sediment surface and sediment core samples. Fractions measured: Total organic carbon (solid squares), total hydrolyzable amino acids (THAA, open circles), total carbohydrates (TCHO, open squares), lipids (open triangles), acid insoluble organic matter (open diamond), BC (solid circles),and alkenones (solid triangles). With the exception of black carbon (Masiello and Druffel, 1998) and alkenone (Ohkouchi and Eglinton, unpublished) data, all measurements are from Wang *et al.* (1996, 1998), or Wang and Druffel (2001).

This is because compound class measurements integrate molecular species derived from potentially diverse sources, whereas the full range of isotopic variability is retained in source-specific molecules. Intermolecular isotopic variability is most evident in the lipids, a biochemical class common to all organisms, but with specific compounds that may be unique to a subgroup of organisms, or even individual species.

The diversity in stable carbon isotopic compositions evident at the molecular level became apparent with the advent of gas chromatography coupled to isotope ratio mass spectrometry (IRM-GC-MS; e.g., Hayes *et al.*, 1990). More recently, ^{14}C age variations among different biomarkers have been investigated (e.g., Eglinton *et al.*, 1997; Pearson *et al.*, 2001). Substantial variability has been observed where inputs from diverse sources are encountered, even within depth horizons representing short periods of sediment accumulation. For example, surficial sediments from the Bermuda Rise in the northern Sargasso Sea deposited over a time span of less than 300 yr yield over a 3×10^4 yr spread in ^{14}C ages for different organic compounds (Figure 16).

These different organic compounds that can be confidently assigned to marine photoautotrophs (alkenones), vascular plant waxes (long-chain fatty acids), and fossil (mainly thermogenic) organic matter (even carbon-numbered long chain *n*-alkanes). Their ^{14}C contents can provide valuable information on the relative importance and mode of delivery of these different inputs. In this particular example, substantial age variations are evident within a single compound class (e.g., fatty acids), highlighting the isotopic heterogeneity evident at the molecular level. Distinct ^{14}C age differences are also revealed between phytoplankton markers (alkenones) and planktonic forams—nominally both tracers of surface ocean DIC (Pearson *et al.*, 2000). These variations are potentially interpretable in terms of the provenance, modes, and timescales of delivery of different sedimentary constituents. However, even at this molecular level, it is important to recognize that ^{14}C data for a specific compound still reflect a population of otherwise identical molecules which will likely have different origins and have experienced diverse histories. For example, in the case of the alkenones from the Bermuda Rise

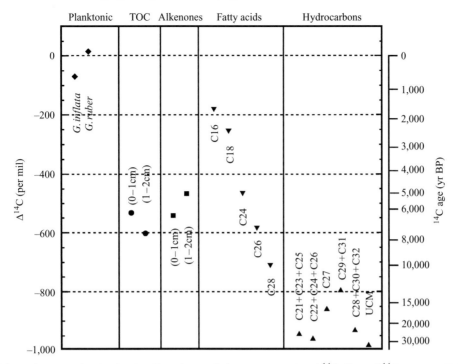

Figure 16 Bulk and molecular-level radiocarbon variations (expressed as $\Delta^{14}C$, ‰, and ^{14}C age, yr BP) in surface (0–3 cm) sediment from the northeastern Bermuda Rise in the subtropical North Atlantic (33° 41′ N, 57° 36′ W, ~4,500 m): planktonic foraminiferal calcite (diamonds), total OC (circles), C_{37}–C_{39} alkenones (squares), fatty acids (down triangles), and hydrocarbons (up triangles). Carbon numbers of fatty acids and hydrocarbons are adjacent to the symbols (UCM = unresolved complex mixture of hydrocarbons). The sedimentation rate at this site during the Holocene averages 10–20 cm kyr^{-1} (source Ohkouchi and Eglinton, unpublished).

(Figure 16), the ^{14}C age of the C_{37}–C_{39} compounds isolated from a single sediment sample undoubtedly reflects two major populations: alkenone molecules input directly from the overlying water column (and therefore presumably of similar age to the planktonic forams) and alkenone molecules which had been biosynthesized several millennia ago, stored on the margins, and subsequently transported to this location *via* lateral advection in deep currents or turbidity layers (Ohkouchi *et al.*, 2002). Although such studies are revealing the extraordinary complexity of organic matter in marine sediments, there is the exciting potential for these molecular-level age variations to provide invaluable novel insights into oceanic processes.

11.5.3 Controls on Organic Matter Preservation

Compositional and isotopic studies have shown that a wide range of organic materials are sequestered in marine sediments. These include direct or indirect products of marine photoautotrophy, vascular plant debris, and relict OC derived from sedimentary rock weathering.

The survival of any organic matter in the sedimentary record seems remarkable, given the efficiency of OM recycling in the water column and surface sediments. The means by which organic matter escapes degradation has been the subject of much debate and research. While the processes and mechanisms have remained elusive, it is evident that there are several principal factors that contribute to OM preservation in sediments, including organic matter source/provenance, the time interval over which sedimenting materials are exposed to oxic degradation, and the availability of (and physical associations with) detrital mineral phases. It seems likely that all of these factors play a role to some extent, and that their relative importance will depend on the specific depositional circumstances. Distinguishing between these factors is difficult, because they often vary in concert, and we have limited ability to recognize compositional features resulting from a given input or depositional characteristic. For example, higher burial efficiencies are encountered in many continental margin sediments yet it is unclear whether this result reflects greater inputs of recalcitrant terrigenous carbon, higher OC fluxes stemming from high primary production in coastal waters, or shorter oxygen exposure times due to the elevated fluxes of OC and other materials. We summarize prevailing concepts and supporting evidence in the following sections.

11.5.3.1 Physical protection

One major determinant in organic matter preservation appears to be the interaction of organic molecules with inorganic materials (Mayer, 1994, 1999; Keil *et al.*, 1994a,b; Hedges and Keil, 1995). These interactions provide a means of physically protecting labile biochemicals from degradation in the water column and underlying sediments.

Recently, Hedges *et al.* (2001) analyzed organic matter in both surface plankton and sinking particulate matter using solid-state ^{13}C-NMR spectroscopy. They observed that, despite extensive signal attenuation, minimal changes in bulk organic composition occurred, with apparently labile biochemicals such as amino-acid-like material and carbohydrates dominating organic matter content throughout the entire water column (Figure 14). These NMR measurements do not exclude the possibility of subtle modifications to these biopolymers. Nevertheless, the compositional similarity between phytoplankton biomass and the small remnant of organic matter reaching the ocean interior does imply that the formation of "unusual" macromolecules, either by chemical recombination (e.g., melanoidin formation (Tissot and Welte, 1984)) or microbial biosynthesis, is not a major process controlling the preservation of POM in pelagic waters. Instead, Hedges *et al.* (2001) suggest that OM might be shielded from degradation through association with the inorganic matrix (e.g., opal, coccoliths, and detrital aluminosilicates). This mineral matter makes up most (>80%) of the mass of sinking particles and, in addition to their protective role, also serves as ballast, expediting OM transport through the water column. Moreover, most sedimentary organic matter cannot be physically separated from its mineral matrix, indicating that the majority of OM is in intimate association with mineral phases.

Observations of organic carbon-to-mineral surface area (OC:SA) ratios by Mayer (1994) and others have revealed that sediments accumulating along continental shelves and upper slopes (excluding deltas) characteristically exhibit surface area loadings approximately equivalent to a single molecular covering on detrital mineral grains, the so-called "monolayer-equivalent" coating (Figure 17). Experiments have shown that a fraction of this organic matter is bound reversibly, and is intrinsically labile (Keil *et al.*, 1994a), but apparently escapes mineralization through association with minerals. This appears to hold true for situations where the OM passes relatively rapidly through oxygenated bottom and pore waters prior to sequestration in the underlying anoxic sediments.

In summary, the above observations and relationships indicate that the supply and availability

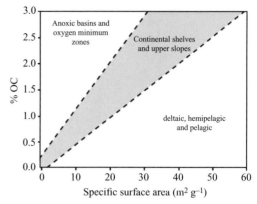

Figure 17 Generalized relationship between weight percent OC (% OC) and specific mineral surface area (SA) for marine sediments. The shaded area represents the boundaries (OC:SA ratio of 0.5–1.1 mg C m^{-2}), within which most continental shelf and upper slope sediments (outside the direct influence of rivers) fall. Sediments underlying anoxic basins and OMZs associated with high productivity (upwelling) margins tend to exhibit OC:SA ratios greater than 1.1 mg C m^{-2} whereas deltaic and abyssal sediments exhibit OC:SA ratios of less than 0.5 mg C m^{-2} (after Mayer, 1994; Hedges and Keil, 1995).

of mineral surface area may be a primary control on OM preservation, particularly in open continental shelf and upper slope sediments where ~45% of all carbon burial takes place in the contemporary ocean. However, while a clear first-order relationship between mineral surface area and OC loading has been apparent for some time, the exact mode of this association remained unclear. Specifically, it was uncertain whether adsorbed organic matter is dispersed over all mineral surfaces or is more localized in occurrence. Theoretical considerations and empirical relationships have suggested the former to be the case (Keil *et al.*, 1994b), but recent work has not supported this paradigm. Transmission electron microscopic examination of sedimentary mineral grains (e.g., Ransom *et al.*, 1997, 1998b) has revealed that OM is not uniformly distributed across all mineral surfaces, but instead is preferentially associated with minerals of high surface area, particularly smectite-rich clays. Ransom *et al.* (1998a) have concluded that the association is a function of differences in the site density and chemistry of the clays, as well as differences in their flocculation behavior. Similarly, studies of gas adsorption in model systems and marine aluminosilicate sediments (Mayer, 1999) have shown that the sediments with low to moderate loading of OM (<3 mg OC m^{-2}) have generally less than 15% of their surface coated. These data imply that the abundance of nonspherical, high surface area-to-volume particles—such as clays, oxyhydroxides, and inorganic biogenic debris (e.g., diatom frustules)—controls specific surface area in most continental margin sediments. Thus, although OM is associated with mineral grains (Mayer *et al.*, 1993; Mayer, 1994; Keil *et al.*, 1994b), it is not adsorbed as a monolayer (Mayer, 1999), but must instead be locally concentrated on certain surfaces. Mayer (1999) argued that the term "monolayer equivalent" should, therefore, be removed from usage, and there appears good justification for so doing. Hence, although there appears to be a clear relationship between OC : SA and organic matter preservation, the mechanistic basis for this relationship still remains to be determined. Mayer (1999) notes that most surface area appears to be within mesopores of <10 nm diameter (Mayer, 1994), and these pores may be too small to allow attack by extracellular enzymes, hence offering protection from degradation (Mayer, 1994).

In addition to physical protection via association with detrital mineral phases, preservation of labile organic matter through association with other matrices is also possible. In particular, "encapsulation" of labile organic matter within refractory organic polymers (Knicker *et al.*, 1996; Zang *et al.*, 2001) or within biogenic mineral lattices is a possible means of stymieing organic matter degradation during burial.

11.5.3.2 Role of anoxia

While a protective role of minerals is inferred from the consistency of OC : SA values in open continental margin sediments, there are depositional environments that yield OC : SA values well above those predicted from sorptive controls, and where other modes of preservation must be invoked. In particular, sediments underlying anoxic or low oxygen bottom waters, or where molecular oxygen does not penetrate to a significant extent into the sediment, are characteristically rich in organic matter and tend to exhibit elevated levels of OC with OC : SA values in excess of those characteristic of margin sediments overlain by oxygenated waters (Bergamaschi *et al.*, 1997; Keil and Cowie, 1999; Figure 17). While such depositional settings account for only ~5% of total OC burial, they are of importance in relation to understanding the formation of organic-rich sedimentary rocks that are responsible for reserves of petroleum. These organic-rich sediments also represent potentially valuable, high fidelity archives of past ocean conditions, since they tend to be free from sediment mixing due to bioturbation and often accumulate relatively rapidly. They are hence conducive for preservation of labile organic compounds that carry the highest information contents. While the extent of exposure to molecular oxygen prior to entering the anoxic realm is thus frequently considered as

a master variable influencing OM preservation (e.g., Demaison, 1991), much debate persists concerning the specific factors leading to enhanced organic matter burial under oxygen deficient or anoxic conditions. Indeed, is there a causal or even correlative relationship between the presence of a minimum in bottom water oxygen (BWO) concentration and the preservation of organic matter in the underlying marine sediments (Pedersen and Calvert, 1990; Cowie and Hedges, 1992).

Various approaches have been taken towards assessing the role of oxygen in OM degradation. Laboratory incubations under controlled conditions have provided detailed information on the influence of aerobic versus anaerobic microbial degradation of phytoplankton biomass (e.g., Harvey and Macko, 1997). Investigation of natural systems includes comparison of surface sediments from depocenter and periphery of anoxic or dysoxic basins (e.g., Gong and Hollander, 1997), examination of depth transects traversing OMZs that impinge on continental margins (e.g., Keil and Cowie, 1999), and comparison of turbidites overlying pelagic sediments that were subjected to oxygen "burn-down" (Cowie *et al.*, 1995; Prahl *et al.*, 1997; Hoefs *et al.*, 1998, 2002; Middelburg *et al.*, 1999). Keil and Cowie (1999) examined OC : SA relationships in relation to BWO concentrations in sediments from the NE Arabian Sea. Sediments deposited under the oxygen minimum had OC : SA ratios in excess of 1.1 mg OC m^{-2}, while samples shallower or deeper than the oxygen-depleted water mass (BWO > 35 μM) exhibited OC : SA ratios that fall within the typically observed range (0.5–1.1 mg OC m^{-2}) (Figure 18(a)). These data indicate that organic matter preservation is enhanced within the general locale of the BWO minimum in NE Arabian Sea sediments. While OM loadings are 2–5 times the monolayer equivalent in anoxic sediments, there often remains a direct relationship between OM content and mineral surface area. Hedges and Keil (1995) speculate that organic materials sorbed in excess of a monolayer may be partially protected as result of equilibration with DOM-rich porewaters, and brief oxygen exposure times.

Thus, Hedges and Keil (1995) argue that organic matter preservation throughout much of the ocean may be controlled largely by competition between sorption on mineral surfaces at different protective thresholds and oxic degradation. The presence of reducing conditions slows OC degradation, leading to elevated OC contents. As an extension of this line of thinking, Hedges *et al.* (1999) hypothesize that organic matter preservation in continental margin sediments is controlled by the average residential period experienced by the accumulating organic particles in the water column and in the oxygenated pore waters

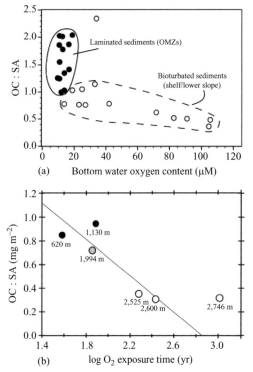

(a)   Bottom water oxygen content (μM)

(b)   log O_2 exposure time (yr)

Figure 18 Oxygen controls on the degradation of OC. (a) Organic carbon to mineral surface area ratio (OC : SA) for core top sediments from the northeastern Arabian Sea plotted as a function of BWO content. Open symbols denote bioturbated sediments lying under relatively oxygenated waters on the shelf and lower slope. Solid symbols denote laminated sediments lie beneath water with the lowest BWO values (i.e., OMZ) (after Keil and Cowie, 1999). (b) Organic carbon to surface area (OC : SA) ratios for surface sediments from the Washington margin (northeastern Pacific Ocean) as a function of OET. Samples with short OETs derive from the shelf and upper slope (solid and shaded symbols). Samples with longer OETs are from the lower slope and Cascadia basin (open symbols). Water depths, in meters, are indicated (after Hedges *et al.*, 1999).

immediately beneath the sediment–water interface. Trends in oxygen penetration depth, organic elemental composition, and mineral surface area for surface sediments collected along an offshore transect across the Washington continental shelf, slope, and adjacent Cascadia basin support this notion. Sediment accumulation rates decrease and dissolved molecular oxygen penetration depths increase offshore, resulting in a seaward increase in oxygen exposure times from decades (mid-shelf and upper slope) to millennia (outer Cascadia basin). Organic contents and compositions were essentially constant at each site, but varied between sites. In particular, OC : SA ratios decreased and indicators for the level of degradation increased with increasing oxygen exposure time (Figure 18(b)), indicating that sedimentary

organic matter experiences a clear "oxic effect" (Hedges *et al.*, 1999). The oxygen exposure time concept helps to integrate contributing factors such as sediment accumulation rate and BWO levels that have been invoked in the past to explain OM preservation in accumulating continental margin sediments. However, we are still lacking a detailed mechanistic basis for this observed control on degradation exerted by BWO conditions.

11.5.3.3 Chemical protection

Intrinsically refractory biomolecules. While evidence exists for the physical protection of organic compounds from degradation (e.g., via association with mineral surfaces), and for enhanced preservation due to limited exposure to oxygenated conditions, organic matter persists in many oceanic sediments where conditions are not conducive for OM preservation based on either of the above criteria. For example, deltaic sediments often exhibit OC : SA coatings that are significantly lower than the "monolayer equivalent" (Keil *et al.*, 1997). Submonolayer equivalent OM coatings are also observed in deep-sea turbidite sediments where organic-rich margin sediments, originally sequestered under anoxic conditions on the margins, have been redeposited in abyssal locations and are exposed to oxygenated bottom waters. Submonolayer organic coatings are also observed in continental rise and abyssal plain sediments where slower sediment accumulation rates and deeper O_2 penetration depths result in increased oxygen exposure times, and little OM preservation.

The residual organic matter in such sediments is often inferred to be highly refractory, and it is assumed that the structural attributes of the organic matter may be pivotal in dictating its survival. The concept of "selective preservation" of one type of natural product over another was introduced by Tegelaar *et al.* (1989) and others. Evidence in support of selective preservation stems from analyses of insoluble macromolecules in sediments and in precursor organisms (Gelin *et al.*, 1999). Solid-state NMR spectra and pyrolytic degradation of OM in many recent and ancient sediments indicate the presence of a highly aliphatic component(s) (Van de Meent *et al.*, 1980; Eglinton, 1994). The nature and origin of this type of aliphatic component had been the subject of considerable debate until it was found that several types of organisms synthesize highly aliphatic biopolymers (Largeau *et al.*, 1984; Nip *et al.*, 1986; Goth *et al.*, 1988; Zelibor *et al.*, 1988, Gelin, 1996, Gelin *et al.*, 1997, 1999). These natural products were then proposed as the source of this sedimentary component. The recalcitrance of these aliphatic macromolecules is indicated by their relative enrichment in oxidized layers of pelagic turbidite sequences (Hoefs *et al.*, 1998).

The concept of chemical recalcitrance is not restricted to aliphatic biopolymers and one guideline for assessing reactivity and conformational relationships pertinent to the preservation of other types of biomacromolecules are structural comparisons between proteins of the same generic type occurring in hyperthermophiles and their low-temperature counterparts. X-ray-based structural studies have revealed that very minor differences evident in molecular structure and conformation brought about by a few α-amino acid substitutions can have dramatic effects on the reactivities and thermal stabilities of such molecules (Danson and Hough, 1998). The corollary for stability of marine organic matter is that similar minor changes in molecular content and architecture induced by mineralization and diagenetic processes may lead to sharp contrasts in biochemical reactivity and hence preservation (Eglinton, 1998; Hedges *et al.*, 2000).

Formation of organosulfur compounds. One factor that distinguishes organic matter accumulation under anoxic or suboxic conditions from their oxic counterparts is the presence of sulfide from microbial sulfate reduction in bottom waters and sediment pore waters. It has been established that certain organic compounds can readily react with reduced sulfur species (H_2S, polysulfides) under ambient conditions (Vairavamurthy and Mopper, 1987), providing a potential means of sequestering labile, extremely oxygen-sensitive organic molecules, such as functionalized or polyunsaturated lipids. Evidence for the reaction of organic matter with sulfides stems from laboratory studies (e.g., Krein and Aizenshtat, 1994; Schouten *et al.*, 1994), together with down-core profiles of the content and isotopic composition of organically bound sulfur (Francois, 1987; Mossmann *et al.*, 1991; Eglinton *et al.*, 1994; Putschew *et al.*, 1996; Hartgers *et al.*, 1997).

Numerous studies have investigated potential mechanisms of OM sulfurization during early diagenesis. Sulfur is considered to react in both an intramolecular and intermolecular fashion (Figure 19). Intramolecular incorporation leads to the formation of cyclic OSC (e.g., thianes, thiolanes, and thiophenes). In intermolecular reactions, the formation of sulfide or polysulfide bridges between molecules generates a wide variety of sulfur-cross-linked macromolecules (Sinninghe Damsté *et al.*, 1989; Kohnen *et al.*, 1991).

Formation of organically bound sulfur is promoted by the availability of reactive organic matter (i.e., bearing the appropriate type, position, and number of functional groups), an excess of reduced sulfur species (e.g. HS^-, polysulfides), and a limited supply of reactive iron, which would otherwise outcompete OM for sulfides (Canfield, 1989; Hartgers *et al.*, 1997). These conditions are characteristic of anoxic basins and OMZs underlying productive upwelling systems on the continental margins. The high flux of labile OM to the sediment provides an abundant source of reactive OM that both possesses the requisite functional groups, and fuels bacterial sulfate reduction, providing reduced sulfides and polysulfides. Shelf and slope sediments distal from sources of terrigenous sediment receive only limited amounts of reactive iron, except in regions of major eolian dust input. In this context, organic matter preservation through formation of organically bound sulfur is unlikely to be important on a global basis.

Most attention has been focused on the sulfurization of functionalized lipids and its influence on lipid preservation and resulting biomarker

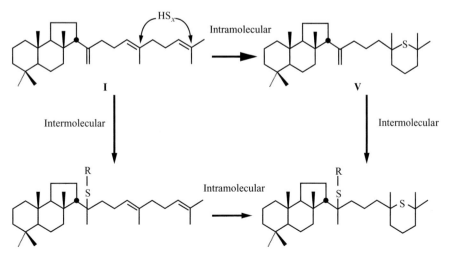

Figure 19 Model for the intramolecular and intermolecular reaction of reduced sulfur species with functionalized lipids. Proposed reaction scheme for sulfur incorporation into (17*E*)-13β(*H*)-malabarica-14(27),17,21-triene (I) (after Werne *et al.*, 2000).

fingerprints (e.g., Kohnen *et al.*, 1991). However, lipids comprise a relatively minor fraction of the input organic matter, and hence the significance of this process as an organic matter preservation mechanism has been uncertain. Recently, it has been shown that carbohydrates can be preserved in a similar manner (Sinninghe Damsté *et al.*, 1998). Carbohydrates comprise a large fraction of the carbon fixed by photoautotrophs, especially certain diatoms which are often the dominant primary producers in upwelling systems. These biopolymers are generally considered to be highly labile and therefore poorly preserved in sediments. However, based on the premise that carbohydrates tend to be isotopically enriched relative to lipids, Sinninghe Damsté *et al.* (1998) interpret strong relationships between OC content, organic sulfur content, and $\delta^{13}C$ of bulk OC in Jurassic age sediments as a consequence of variable preservation of ^{13}C enriched carbohydrates through sulfurization reactions.

It remains unclear whether sulfurization of organic matter results in a net increase in OC preservation, for example, whether sulfurization acts to transform organic matter, but not preserve it. A key issue with respect to this question is the timing of these reactions in relation to competing diagenetic reactions that remove the organic substrates. Circumstantial evidence suggests that these reactions take place quite rapidly (Eglinton *et al.*, 1994); however, the identification of both precursor and product (Figure 19) in age-dated sediments from the anoxic Cariaco Basin provides a direct estimate of reaction rates (Werne *et al.*, 2000; Figure 20). While these results apply only for the precursor in question, they imply that sulfur incorporation is far from instantaneous.

11.6 MICROBIAL ORGANIC MATTER PRODUCTION AND PROCESSING: NEW INSIGHTS

11.6.1 Background

Compared to OM inputs from sedimentation of biogenic particles resulting from primary production in the overlying water column and transported terrigenous materials, the role of prokaryotic organisms in the production and processing of organic matter is a poorly understood component of the oceanic carbon cycle. Prokaryotes, which include the bacteria and archea, can be divided into several classes: cyanobacteria, anoxygenic photosynthetic bacteria, heterotrophic bacteria, methane oxidizing bacteria, and methanogenic archea (Balows *et al.*, 1992). All of these types of prokaryotes have a major impact on either the production or the mineralization of OC in the

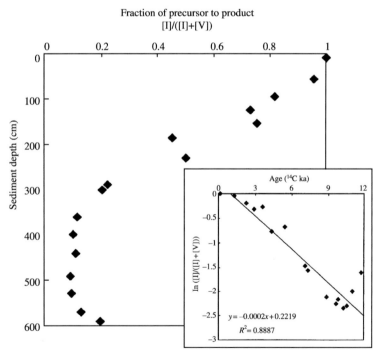

Figure 20 Progress of intramolecular sulfurization of malabaricatriene (I) with depth in anoxic Cariaco basin sediments. Plot of the ratio of the concentrations of I to the sum of the concentrations of I + V (see Figure 19 for structures) as a function of sediment depth. The progress of transformation of I–V with increasing sediment depth is indicated by the steady decrease in the relative abundance of the precursor lipid (I). Inset: plot of ln ([I]/([I] + [V])) versus sediment age (= time), used to empirically determine the first-order rate constant for sulfur incorporation (after Werne *et al.*, 2000).

water column and in sediments. Indeed, viable microbial communities have now been shown to extend hundreds of meters into the sediment column (e.g., Parkes *et al.*, 1994, 2000), indicating that fresh microbial biomass may also constitute some fraction of the OM hitherto considered as "preserved" OM remnants from the overlying water column. Thus, the extent to which OM buried in sediments represents organic remnants directly inherited from the overlying surface ocean versus bacterial debris is a subject of some debate. The fact that benthic microorganisms are the last "filter" that OM passes through during burial suggests that their biomass and products might become sequestered irrespective of their reactivity (i.e., there is no one else to "eat" them). It is difficult to distinguish between these inputs, both because of the modified nature of sedimentary OM, and the fact that heterotrophic organisms utilizing multicarbon substrates will inherit the isotopic characteristics of the OM they act upon (Hayes, 1993).

De novo biosynthesis and associated reworking of organic matter by prokaryotes does not imply that all of this microbially processed organic material is transferred to the sedimentary record. Evidence of contributions from prokaryotic organisms is abundant in the form of molecular fossils (e.g., hopanoids) that are virtually ubiquitous in the sedimentary record. However, while the presence of specific biomarkers may be diagnostic of specific prokaryotic inputs, their abundance is not necessarily in proportion to these inputs (the same is, of course, true for all biomarkers and incidentally POM in general). Sinninghe Damsté and Schouten (1997) reviewed evidence for prokaryotic biomass inputs to sedimentary organic matter. They argue that several lines of evidence point to limited contributions or poor preservation of bacterial biomass in the sedimentary record. Their conclusions are based on, among other things: (i) isotopic mass balance in various sedimentary settings; (ii) the effects of bacterial oxidation in organic matter-rich turbidites; and (iii) the absence of apparently recalcitrant biomacromolecules, such as aliphatic biopolymers, in prokaryotes.

In addition to the prokaryotes whose identity, physiology, and ecological role has been established, relatively recent studies using culture-independent r-RNA analyses of environmental samples have revealed genetic diversity within natural microbial communities that greatly exceeds estimates based on classical microbiological techniques. These findings underline our presently limited view of microbial activity and its biogeochemical consequences. Below, two examples are provided that highlight the potential influences of microbial processes on sedimentary organic matter composition.

11.6.2 Planktonic Archea

Recent culture-independent, r-RNA gene surveys have indicated the ubiquity and importance of planktonic archea in the ocean, particularly in subsurface waters. For example, Karner *et al.* (2001) found pelagic crenarcheota comprised a large fraction of total marine picoplankton, equivalent in cell numbers to bacteria at depths greater than 1,000 m (Figure 21). The fraction of crenarchaeota increased with depth, reaching 39% of the total DNA-containing picoplankton detected. Moreover, the high proportion of cells containing significant amounts of r-RNA suggests that most pelagic deep-sea microorganisms are metabolically active. The oceans are estimated to harbor $\sim 1.3 \times 10^{28}$ archeal cells and 3.1×10^{28} bacterial cells, suggesting that pelagic crenarchaeota represent one of the ocean's single most abundant cell types.

The physiologies, ecological niches, and biogeochemical roles of these organisms are yet to be fully determined. Their imprint on the sedimentary record is only now being appreciated. Sinninghe Damsté *et al.* (2002) quantified intact tetraether lipids of marine planktonic crenarchaeota in SPOM from the NE Arabian Sea. In contrast to eukaryotic and bacterial lipids (sterols and fatty acids, respectively), which were highest in surface waters, maximum concentrations of crenarcheol (Figure 22), generally occurred at 500 m, near the top of the OMZ (Figure 21). This indicates that these crenarcheota are not restricted to the photic zone of the ocean (consistent with molecular biological studies). Sinninghe Damsté *et al.* (2002) suggest that the coincidence of maximum abundances of crenarcheotal membrane lipids with the core of the OMZ indicates these organisms are probably facultative anaerobes. Moreover, calculations of cell numbers (based on membrane lipid concentrations) support other recent estimates for their significance in the World's oceans. Schouten *et al.* (1998) investigated acyclic and cyclic biphytane carbon skeletons derived from planktonic archea in a number of lacustine and marine sediments. They found these compounds to be amongst the most abundant lipids in sediments, and sometimes were present in greater proportion to those synthesized by eukaryotes and bacteria, indicating that these organisms may be an important source of sedimentary organic matter.

11.6.3 Anaerobic Methane Oxidation

A second newly recognized group of prokaryotes are the methane oxidizing archea. Nearly 90% of the methane produced in anoxic marine sediments is recycled through anaerobic microbial oxidation processes (Cicerone and Oremland,

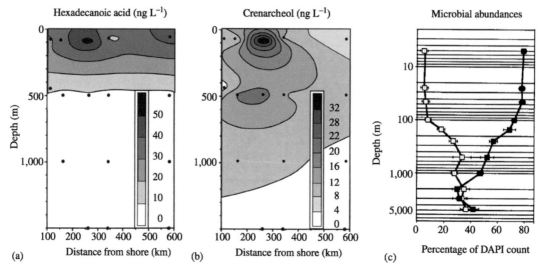

Figure 21 Comparison of vertical distribution of biomarker and microbial abundances in oceanic water columns. (a) Contour plots of concentration (ngL^{-1}) of hexadecanoic acid. (b) Crenarcheol at various depths in the water column and distances from shore on a northwest-to-southeast transect off Oman in the Arabian Sea (after Sinninghe Damsté *et al.*, 2002). Hexadecanoic acid serves as a biomarker proxy for eukaryotic and bacterial biomass and clearly shows the expected surface maximum, with concentrations dropping off steeply with increasing water depth. In contrast, crenarcheol, a molecular biomarker for planktonic crenarcheota, shows two maxima with one near 50 m and the other ~500 m. (c) Vertical distributions of microbial concentrations in the North Pacific subtropical gyre: bacteria (solid squares) and planktonic crenarcheota (open squares). Effectively, there are two microbial domains which were determined using a DAPI nucleic acid stain (after Karner *et al.*, 2001). These data show the increasing proportion of planktonic archea in deep waters, with the result that at depths greater than 2,000 m, the crenarcheota are as abundant as the bacteria.

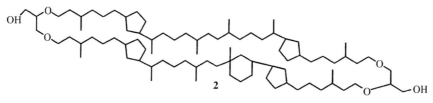

Figure 22 Molecular structure of crenarcheol (after Sinninghe Damsté *et al.*, 2002).

1988; Reeburgh *et al.*, 1991). However, the organisms and biochemical processes responsible for the anaerobic oxidation of methane (AMO) have largely evaded elucidation until recently. Convergent lines of molecular, carbon-isotopic, and phylogenetic evidence have now implicated archea in consortia with sulfate reducing bacteria as the primary participants in this process (Hoehler *et al.*, 1994; Boetius *et al.*, 2000; Pancost *et al.*, 2000; Hinrichs *et al.*, 1999).

Lipids of these organisms are distinguished by their ether-linked nature and highly ^{13}C-depleted values. Hinrichs *et al.* (2000) investigated microbial lipids associated with AMO in gas hydrate-bearing sediments from the Eel River Basin, offshore Northern California, as well as sediments from a methane seep in the Santa Barbara Basin. In addition to archeal markers (*sn*-2-hydroxyarchaeol), these lipids are accompanied by additional ^{13}C-depleted glycerol ethers and fatty acids. Hinrichs *et al.* speculate that these ^{13}C-depleted

lipids are produced by (unknown) sulfate reducing bacteria growing syntrophically with the methane utilizing archea. Interestingly, these authors note that at all of the methane seep sites examined, preservation of aquatic products is enhanced because enhanced consumption of sulfate by the methane oxidizing consortium depletes the sulfate pool that would otherwise have been available for remineralization of materials from the water column.

The identification of ^{13}C-depleted (δ^{13}C values as low as -58 per mil) archeal cyclic biphytanes in particulate matter from the Black Sea, where more than 98% of the methane released from sediments is apparently oxidized anaerobically (Reeburgh *et al.*, 1991), provides evidence for AMO in euxinic waters (Schouten *et al.*, 2001). However, the same isotopically depleted compounds were not detected in the underlying sediments, suggesting that the responsible organisms are in low abundance and/or leave no characteristic molecular fingerprint in the sedimentary record.

11.7 SUMMARY AND FUTURE RESEARCH DIRECTIONS

Our understanding of the composition and cycling of organic matter in the oceans has changed tremendously since the 1990s. Through the use of new analytical tools, we have a greater appreciation of the complexity of dissolved and POM while, at the same time, we can view its biomolecular building blocks with greater clarity. Despite these advances, many important unanswered questions provide impetus for continued research.

Questions of the origin and composition of marine macromolecular organic matter continue to challenge us. A key point that has emerged, especially from NMR studies, is that the formation of randomly cross-linked macromolecules ("humic substances") is no longer considered a primary pathway in the formation of either HMW DOM or sedimentary OM. Rather, these materials seem to be comprised of biochemicals that have been directly inherited from their biological precursors (primarily marine algae and bacteria) with minimal alteration. Three questions immediately arise: (i) Why are traditional wet chemical approaches unable to detect much of these apparently intact biopolymers? (ii) Given this close similarity to natural products, why is it that this material not readily biodegraded in the water column and sediments? (iii) What are the processes that render normally labile macromolecules unavailable to heterotrophic microorganisms and their hydrolytic enzymes? These issues are particularly perplexing in the case of deep-water DOM, where physical mechanisms of preservation such as sorption on mineral grains or physical encapsulation cannot be invoked, where a large fraction of the macromolecular pool is polysaccharide in nature and, paradoxically, exhibits radiocarbon ages of several millennia! It is evident that our knowledge of the fundamental factors that control organic matter degradation and transformation is far from complete. However, subtle modifications of the molecular structure of a biochemical (e.g., changes in protein conformation) are known to influence susceptibility to attack by degrading chemicals and enzymes. New analytical methods that can detect small changes in biopolymer structure are urgently required, together with new experimental approaches that can assess the impact of such molecular rearrangements on organic matter preservation at a mechanistic level.

Our limited understanding of OM composition is not restricted to the macromolecular pool. Our knowledge of the initial biological signatures carried in particles and DOM also remains highly limited. Surveys of marine microbes for their molecular signatures have largely been dictated by what can be grown axenically in culture. However, many important organisms are not available through laboratory culture, including the most abundant photoautotroph in the sea, Prochlorococus, and some of the most abundant autotrophic microbes, the planktonic crenarcheota. Clearly, given the recently recognized prevalence of these microbes in the contemporary ocean, it will be important to further study their roles in carbon and other biogeochemical cycles. The molecular signature that these organisms carry is only now being studied and much work is required to accurately interpret their corresponding molecular stratigraphic records.

A second key question for future research concerns the fate of terrestrial organic matter in the oceans. Two lines of evidence that have long argued against a significant terrestrial input are the stable carbon isotopic (δ^{13}C) and elemental C_{org}/N composition of oceanic dissolved and particulate organic matter. Recent findings of extensive terrestrial organic matter degradation and exchange in deltaic systems have reinforced this notion. However, interpretation of bulk elemental and isotopic data is not necessarily straightforward. Organic matter–mineral associations, while a key determinant in OM preservation, are highly dynamic, and fluvial and eolian contributions of ^{13}C-enriched, nitrogen-replete soil OM derived from C4 vascular plant debris blur traditional distinctions based on these parameters, undermining their diagnostic capability. Several other new lines of evidence, including the recognition that terrestrial OC inputs to the oceans are largely dominated by numerous small, low latitude rivers draining mountainous regions, the recognition that black carbon in marine sediments could account for up to 30% of total OC in some continental margin deposits, and finally that relict OC eroding from sedimentary rocks on the continent represents an additional terrigenous component, all suggest that terrestrial OC may be an underestimated component of the oceanic carbon reservoir.

Multiple lines of further investigation are clearly warranted in order to address this important and complex issue. Small, low-latitude river systems have been undersampled, and our understanding of the composition and mode of terrestrial OM export from the continental shelf and slope is not well developed. At a mechanistic level, the manner in which different forms of organic are associated with particles, and move between dissolved and solid phases both during transit from the land to the oceans, and within sediment pore waters, requires detailed investigation. The timescales over which organic matter exchange and degradation take place in relation to the timescales for terrestrial OC delivery to the oceans is an important consideration in this context. Our ability to recognize and quantify terrigenous OC in the marine environment must also be improved. With respect to the preservation of POM in sediments, reactivity depends on both the intrinsic lability of the molecular structures present and on the associated material matrix. While a strong causal relationship has been established between the properties of the matrix (e.g., mineral type and surface area) and efficiency of organic matter preservation, we are lacking a mechanistic basis for it. Future research strategies need to address the nature of the associations of OM with mineral surfaces and their effects on the reactivities toward chemical and biological agents. Similarly, while recent studies have demonstrated the link between bottom and pore-water oxygen concentration and the extent of sedimentary OM degradation, we do not yet know whether the key reactant is molecular oxygen itself.

One striking aspect of marine organic geochemistry is that its material basis extends from the smallest molecules (methane, α-amino acids, short-chain fatty acids, etc.) through the ubiquitous but complex, HMW DOM to a highly diverse range of organic particles (BC, marine snow, necromass, etc.). No single set of tools is adequate to address the composition of all these materials and the ways in which they function. A variety of different analytical approaches is required. Bulk chemical, physical, and isotopic characterization, microscopy (TEM, SEM, etc.), spectroscopic techniques (NMR, etc.), fractionation and chromatographic techniques, and chemical degradation, have all been used to derive useful information. As we look ahead to the future, it is evident that many exciting opportunities lie at the interface between molecular biology and molecular organic geochemistry. For example, a better knowledge of the genetic machinery that directs the synthesis and degradation of biopolymers may help to unravel questions of OC preservation as DOC in the deep ocean and as POC in marine sediments. Such studies may also help to close the gap between organic matter characterization at the bulk and molecular level.

Finally, it is clear that human activity has and will continue to have an impact on the oceanic carbon cycle and associated biogeochemical processes. By the beginning of the 1990s, atmospheric CO_2 comprised $\sim 750\,GtC$, and this has continued to rise at a rate of $\sim 3.4\,Gt\,yr^{-1}$. The oceans represent a major sink for anthropogenic CO_2 and it is imperative that we develop a better understanding of how organic matter cycling is impacted in order to assess the long-term consequences and health of the planet.

ACKNOWLEDGMENTS

The authors wish to dedicate this contribution to the memory of John I. Hedges. John was a source of inspiration; his contributions to marine organic geochemistry are immeasurable, but clearly evident from the extensive citations of his work throughout this chapter. The field of marine organic geochemistry has suffered a tremendous setback with his untimely departure.

The authors wish to thank Lucinda Gathercole for assistance with the preparation of this manuscript and Thomas Wagner for constructive comments on this paper. They also wish to acknowledge support from NSF Grant OCE-9907129 (T.I.E.) and the Department of Energy. This is Woods Hole Oceanographic Institution Contribution No. 10993.

REFERENCES

Alldredge A. L. and Silver M. W. (1998) Characteristics, dynamics and significance of marine snow. *Prog. Oceanogr.* **20**, 41–82.

Alldredge A. L., Passow U., and Logan B. E. (1993) The abundance and significance of a class of large, transparent organic particles in the ocean. *Deep-Sea Res. I* **40**, 1131–1140.

Aluwihare L. I. and Repeta D. J. (1999) A comparison of the chemical composition of oceanic DOM and extracellular DOM produced by marine algae. *Mar. Ecol. Prog. Ser.* **186**, 105–117.

Aluwihare L. I., Repeta D. J., and Chen R. F. (1997) A major biopolymeric component to dissolved organic carbon in seawater. *Nature* **387**, 166–167.

Anderson R., Rowe G., Kemp P., Trumbore S., and Biscaye P. (1994) Carbon budget for the mid-slope depocenter of the middle Atlantic bight. *Deep-Sea Res. Part II* **41**, 669–703.

Balows A., Trüper H. G., Dworkin M., Harder W., and Schleifer K.-H. (1992) *The Prokaryotes*. Springer, Berlin.

Barlow R. G. (1982) Phytoplankton ecology in the southern Benguela current: III. Dynamics of a bloom. *J. Exp. Mar. Bio. Ecol.* **63**, 239–248.

Bauer J. E. and Druffel E. R. M. (1998) Ocean margins as a significant source of organic matter to the deep open ocean. *Nature* **392**, 482–485.

Bauer J. E., Druffel E. R. M., Wolgast D. M., and Griffin S. (2001) Sources and cycling of dissolved and particulate organic radiocarbon in the northwest Atlantic continental margin. *Global Biogeochem. Cycles* **15**(3), 615–636.

Benner R., Pakulski J. D., McCarthy M., Hedges J. I., and Hatcher P. G. (1992) Bulk chemical characteristics of dissolved organic matter in the ocean. *Science* **255**(20 March 1992), 1561–1564.

Bergamaschi B. A., Tsamakis E., Keil R. G., Eglinton T. I., Montlucon D. B., and Hedges J. I. (1997) The effect of grain size and surface area on organic matter, lignin and carbohydrate concentrations and molecular compositions in Peru margin sediments. *Geochim. Cosmochim. Acta* **61**, 1247–1260.

Berner R. A. (1982) Burial of organic matter and pyrite in the modern ocean: its geochemical and environmental significance. *Am. J. Sci.* **282**, 451–475.

Berner R. A. (1989) Biogeochemical cycles of carbon and sulfur and their effect on atmospheric oxygen over phanerozoic time. *Palaeogeogr. Palaeoclimatol. Palaeoecol.* **75**, 97–122.

Biersmith A. and Benner R. (1998) Carbohydrates in phytoplankton and freshly produced dissolved organic matter. *Mar. Chem.* **63**, 131–144.

Boetius A., Ravenschlag K., Schubert C. J., Rickert D., Widdel F., Gieseke A., Amann R., Jorgensen B. B., Witte U. and Pfannkuche O. (2000) A marine microbial consortium apparently mediating anaerobic oxidation of methane. *Nature* **407**(6804) 623–626.

Boon J. J., Klap V. A., and Eglinton T. I. (1998) Molecular characterization of microgram amounts of oceanic colloidal organic matter by direct temperature resolved ammonia chemical ionization mass spectrometry. *Org. Geochem.* **29**, 1051–1061.

Borch N. H. and Kirchman D. L. (1998) Concentration and composition of dissolved combined neutral sugars (polysaccharides) in seawater determined by HPLC-PAD. *Mar. Chem.* **57**, 85–95.

Buesseler K. O., Bauer J. E., Chen R. F., Eglinton T. I., Gustafsson O., Landing W., Mopper K., Moran S. B., Santschi P. H., VernonClark R., and Wells M. L. (1996) An intercomparison of cross-flow filtration techniques used for sampling marine colloids: overview and organic carbon results. *Mar. Chem.* **55**, 1–33.

Canfield D. E. (1989) Reactive iron in marine sediments. *Geochim. Cosmochim. Acta* **53**, 619–632.

Carlson C. A. and Ducklow H. W. (1995) Dissolved organic carbon in the upper ocean of the equatorial Pacific Ocean, 1992: daily and finescale vertical variations. *Deep-Sea Res. II* **42**(2–3), 639–656.

Carlson C. A., Ducklow H. W., and Michaels A. F. (1994) Annual flux of dissolved organic carbon from the euphotic zone in the northern Sargasso Sea. *Nature* **371**, 405–408.

Chen R. F., Fry B., Hopkinson C. S., Repeta D. J., and Peltzer E. T. (1995) Dissolved organic carbon on Georges Bank. *Continent. Shelf Res.* **16**, 409–420.

Chin W.-C., Orellana M. V., and Verdugo P. (1998) Spontaneous assembly of marine dissolved organic matter into polymer gels. *Nature* **391**, 568–571.

Churchill J., Wirick C., Flagg C., and Pietrafesa L. (1994) Sediments resuspension over the continental shelf east of the Delmarva Peninsula. *Deep-Sea Res. II* **41**, 341–363.

Cicerone R. J. and Oremland R. S. (1988) Biogeochemical aspects of atmospheric methane. *Global Geochem. Cycles* **2**(4), 299–327.

Cowie G. L. and Hedges J. I. (1992) The role of anoxia in organic matter preservation in coastal sediments: relative stabilities of the major biochemicals under oxic and anoxic depositional conditions. *Org. Geochem.* **19**(1–3), 229–234.

Cowie G. L., Hedges J. I., Prahl F. G., and de Lange G. J. (1995) Elemental and major biochemical changes across an oxidation front in a relict turbidite: an oxygen effect. *Geochim. Cosmochim. Acta* **59**(1), 33–46.

Danson M. J. and Hough D. W. (1998) Structure, function and stability of exoenzymes from the Archaea. *Trends Microbiol.* **6**, 307–314.

de Leeuw J. W. and Largeau C. (1993) A review of macromolecular organic compounds that comprise living organisms and their role in kerogen, coal, and petroleum formation. In *Organic Geochemistry* (eds. M. H. Engel and S. A. Macko). Plenum, New York, pp. 23–72.

Demaison G. (1991) Anoxia vs. productivity: what controls the formation of organic-carbon-rich sediments and sedimentary rocks? discussion. *Am. Assoc. Petrol. Geol. Bull.* **75** (3), 499.

Druffel E. R. M., Williams P. M., Bauer J. E., and Ertel J. R. (1992) Cycling of dissolved and particulate organic matter in the open ocean. *J. Geophys. Res.* **97**(C10), 15639–15659.

Durand B. and Nicaise G. (1980) Procedures for kerogen isolation. In *Kerogen—Insoluble Organic Matter from Sedimentary Rocks* (ed. B. Durand). Editions Technip, Paris, pp. 35–53.

Duursma E. K. (1963) The production of dissolved organic matter in the sea, as related to the primary gross production of organic matter. *Neth. J. Sea Res.* **2**, 85–94.

Edmond J. M., Boyle E. A., Grant B., and Stallard R. F. (1981) The chemical mass balance in the Amazon plume: I. The nutrients. *Deep-Sea Res.* **28A**, 1339–1374.

Eglinton G. (1998) The archaeological and geological fate of biomolecules. In *Digging for Pathogens* (ed. C. L. Greenblatt). Balakan, Rehovot, Israel, pp. 299–327.

Eglinton T. I. (1994) Carbon isotopic evidence for the origin of macromolecular aliphatic structures in kerogen. *Org. Geochem.* **21**, 721–735.

Eglinton T. I., Minor E. C., Olson R. J., Zettler E. R., Boon J. J., Noguerola A., Eijkel G., and Pureveen J. (1994) Microscale characterization of algal and related particulate organic matter by direct temperature-resolved ("in-source") mass spectrometry. *Mar. Chem.* **52**, 27–54.

Eglinton T. I., Benitez-Nelson B. C., Pearson A., McNichol A. P., Bauer J. E., and Druffel E. R. M. (1997) Variability in radiocarbon ages of individual organic compounds from marine sediments. *Science* **277**, 796–799.

Eglinton T. I., Eglinton G., Dupont L., Sholkovitz E. R., Montluçon D., and Reddy C. M. (2002) Composition, age, and provenance of organic matter in NW African dust over the Atlantic Ocean. *Geochem. Geophys. Geosys. (G3)* **3**(8).

Eswaran H., Van den Berg E., and Reich P. (1993) Organic carbon in soils of the world. *J. Soil Sci. Soc. Am.* **57**, 192–194.

Francois R. (1987) A study of sulphur enrichment in the humic fraction of marine sediments during early diagenesis. *Geochim. Cosmochim. Acta* **51**, 17–27.

François R. (1990) Marine sedimentary humic substances: structure, genesis, and properties. *Rev. Aquat. Sci.* **3**, 41–80.

Gagosian R. B. and Peltzer E. T. (1986) The importance of atmospheric input of terrestrial organic material to deep sea sediments. *Org. Geochem.* **10**, 661–669.

Gagosian R. B. and Stuermer D. H. (1977) The cycling of biogenic compounds and their diagenetically altered transformation products in seawater. *Mar. Chem.* **5**, 605–632.

Gelin F. (1996) Origin and molecular characterization of the insoluble organic matter in marine sediments. In *Annual Report Netherlands Institute for Sea Research,* pp. 43–45.

Gelin F., Boogers I., Noordeloos A. A. M., Damsté J. S. S., Riegman R., and De Leeuw J. W. (1997) Resistant biomacromolecules in marine microalgae of the classes eustigmatophyceae and chlorophyceae: geochemical implications. *Org. Geochem.* **26**, 11–12.

Gelin F., Volkman J. K., Largeau C., Derenne S., Damsté J. S. S., and De Leeuw J. W. (1999) Distribution of aliphatic, nonhydrolyzable biopolymers in marine microalgae. *Org. Geochem.* **30**(2/3), 147–159.

Goericke R. and Fry B. (1994) Variations of marine plankton $\delta^{13}C$ with latitude, temperature and dissolved CO_2 in the world oceans. *Global Biogeochem. Cycles* **8**, 85–90.

Goldberg E. D. (1985) *Black Carbon in the Environment: Properties and Distribution.* Wiley, New York.

Gong C. and Hollander D. J. (1997) Differential contribution of bacteria to sedimentary organic matter in oxic and anoxic environments, Santa Monica Basin, California. *Org. Geochem.* **26**(9/10), 545–563.

Goni M. A., Ruttenberg K. C., and Eglinton T. I. (1997) Sources and contribution of terrigenous organic carbon to surface sediments in the Gulf of Mexico. *Nature* **389**, 275–278.

Goni M. A., Ruttenberg K. C., and Eglinton T. I. (1998) A reassessment of the sources and importance of land-derived organic matter in surface sediments from the Gulf of Mexico. *Geochim. Cosmochim. Acta* **62**, 3055–3075.

Goth K., Leeuw J. d., Puttmann W., and Tegelaar E. (1988) Origin of messel oil shale kerogen. *Nature* **336**(22/29), 759–761.

Gough M. A., Fauzi R., Mantoura C., and Preston M. (1993) Terrestrial plant biopolymers in marine sediments. *Geochim. Cosmochim. Acta* **57**(5), 945–964.

Guo L., Coleman C. H., and Santschi P. H. (1994) The distribution of colloidal and dissolved organic carbon in the Gulf of Mexico. *Mar. Chem.* **45**, 105–119.

Gustafsson O. and Gschwend P. M. (1998) The flux of black carbon to surface sediments on the New England continental shelf. *Geochim. Cosmochim. Acta* **62**, 465–472.

Gustafsson O., Haghseta F., Chan C., Macfarlane J., and Gschwend P. M. (1997) Quantification of the dilute sedimentary soot phase: implications for PAH speciation and bioavailability. *Environ. Sci. Technol.* **24**, 1687–1693.

Hansell D. A. and Carlson C. A. (1998) Deep-ocean gradients in the concentration of dissolved organic carbon. *Nature* **395**, 263–266.

Hartgers W. A., Lopez J. F., Damsté J. S. S., Reiss C., Maxwell J. R., and Grimalt J. O. (1997) Sulfur-binding in recent environments: speciation of sulfur and iron and implications for the occurrence of organo-sulfur compounds. *Geochim. Cosmochim. Acta* **61**(22), 4769–4788.

Harvey H. R. and Macko S. A. (1997) Catalysts or contributors? tracking bacterial mediation of early diagenesis in the marine water column. *Org. Geochem.* **26**(9/10), 531–544.

Hatcher P. G., Rowan R., and Mattingly M. A. (1980) 1H and 13CNMR of marine humic acids. *Org. Geochem.* **2**, 77–85.

Hayes J. (1979) Sandstone diagenesis—the hole truth. *SEPM (Spec. Publ.)* **26**(March), 127–139.

Hayes J. M. (1993) Factors controlling ^{13}C contents of sedimentary organic compounds: principles and evidence. *Mar. Geol.* **113**, 111–125.

Hayes J. M., Freeman K. H., and Popp B. N. (1990) Compound-specific isotopic analyses: a novel tool for reconstruction of ancient biogeochemical processes. *Org. Geochem.* **16**(4–6), 1115–1128.

Hedges J. I. (1992) Global biogeochemical cycles: progress and problems. *Mar. Chem.* **39**, 67–93.

Hedges J. I. and Keil R. G. (1995) Sedimentary organic matter preservation: an assessment and speculative synthesis. *Mar. Chem.* **49**, 81–115.

Hedges J. I. and Keil R. G. (1999) Organic geochemical perspectives on estuarine processes: sorption reactions and consequences. *Mar. Chem.* **65**(1–2), 55–65.

Hedges J. I. and Mann D. C. (1979) The characterization of plant tissues by their lignin oxidation products **43**(11), 1803–1807.

Hedges J. I. and Oades J. M. (1997) Comparative organic geochemistries of soils and marine sediments. *Org. Geochem.* **27**, 319–361.

Hedges J. I. and Parker P. L. (1976) Land-derived organic matter in surface sediments from the Gulf of Mexico. *Geochim. Cosmochim. Acta* **40**, 1019–1029.

Hedges J. I., Ertel J. R., Quay P. D., Grootes P. M., Richey J. E., Devol A. H., Farwell G. W., Schmidt F. W., and Salati E. (1986) Organic carbon-14 in the Amazon River system. *Limnol. Oceanogr.* **231**(4742), 1129–1131.

Hedges J. I., Cowie G. L., Richey J. E., Quay P. D., Benner R., Strom M., and Forsberg B. R. (1994) Origins and processing of organic matter in the Amazon River as indicated by carbohydrates and amino acids. *Science* **39**(4), 743–761.

Hedges J. I., Keil R. G., and Benner R. (1997) What happens to terrestrial organic matter in the ocean? *Org. Geochem.* **27**(5/6), 195–212.

Hedges J. I., Hu F. S., Devol A. H., Hartnett H. E., Tsamakis E., and Keil R. G. (1999) A test for selective degradation under oxic conditions. *Am. J. Sci.* **299**, 529–555.

Hedges J. I., Eglinton G., Hatcher P. G., Kirchman D. L., Arnosti C., Derenne S., Evershed R. P., Kogel-Knabner I., Leeuw J. W. d., Littke R., Michaelis W., and Rullkotter J. (2000) The molecularly-uncharacterized component of non-living organic matter in natural environments. *Org. Geochem.* **31**, 945–958.

Hedges J. I., Baldock J. A., Gelinas Y., Lee C., Peterson M., and Wakeham S. G. (2001) Evidence for non-selective preservation of organic matter in sinking marine particles. *Nature* **409**, 801–804.

Hellebust J. A. (1965) Excretion of some organic compounds by marine phytoplankton. *Limnol. Oceanogr.* **10**, 192–206.

Hinrichs K.-U., Hayes J. M., Sylva S. P., Brewer P. G., and DeLong E. F. (1999) Methane-consuming archaebacteria in marine sediments. *Nature* **398**, 802–805.

Hinrichs K.-U., Summons R. E., Orphan V., Sylva S. P., and Hayes J. M. (2000) Molecular and isotopic analysis of anaerobic methan-oxidizing communities in marine sediments. *Org. Geochem.* **31**, 1685–1701.

Hoefs M. J. L., Damsté J. S. S., Lange G. J. D., and Leewu J. W. d. (1998) Changes in kerogen composition across an oxidation front in Madeira abyssal plain turbidites as revealed by pyrolysis GC–MS. In *Proceedings of the Ocean Drilling Program, Scientific Results* (eds. P. P. E. Weaver, H.-U. Schmincke, J. V. Firth, and W. Duffield). vol. 157, pp. 591–607.

Hoefs M. J. L., Rijpstra W. I. C., and Damsté J. S. S. (2002) The influence of toxic degradation on the sedimentary biomarker record: I. Evidence from Madeira Abyssal plain turbidites. *Geochim. Cosmochim. Acta* **66**, 18.

Hoehler T. M., Alperin M. J., Albert D. B., and Martens C. S. (1994) Field and laboratory studies of methane oxidation in an anoxic marine sediment: evidence for a methanogen-sulfate reducer consortium. *Global Biogeochem. Cycles* **8**, 451–463.

Holmes R. W., Williams P. M., and Epply R. W. (1967) Red water in La Jolla Bay. *Limnol. Oceanogr.* **12**, 503–512.

Huang Y., Dupont L., Sarnthein M., Hayes J. M., and Eglinton G. (2000) Mapping of C4 Plant input from north west Africa into north east Atlantic sediments. *Geochim. Cosmochim. Acta* **64**, 3505–3513.

Hunt J. M. (1996) *Petroleum Geochemistry and Geology*. Freeman Press, New York.

Iturriaga R. and Zsolnay A. (1983) Heterotrophic uptake and transformation of phytoplankton extracellular products. *Botanica Marina* **26**, 375–381.

Jahnke R. A., Reimers C. E., and Craven D. B. (1990) Intensification of recycling of organic matter at the sea floor near ocean margins. *Nature* **348**(1 Novermber 1990), 50–53.

Jørgensen N. O. G., Kroer N., Coffin R. B., Yang X.-H., and Lee C. (1993) Dissolved free amino acids and DNA as sources of carbon and nitrogen to marine bacteria. *Mar. Ecol. Prog. Ser.* **98**, 135–148.

Kaiser K. and Benner R. (2000) Determination of amino sugars in environmental samples with high salt content by high performance anion-exchange chromatography and pulsed amperometric detection. *Anal. Chem.* **72**, 2566–2572.

Kao S.-J. and Liu K. K. (1996) Particulate organic carbon export from a subtropical mountainous river (Lanyang Hsi) in Taiwan. *Limnol. Oceanogr.* **41**(8), 1749–1757.

Karl D. M. and Bjorkman K. M. (2002) Dynamics of DOP. In *Biogeochemistry of Marine Dissolved Organic Matter* (ed. D. Hansell). Elsevier, London, pp. 249–366.

Karner M. B., DeLong E. F., and Karl D. M. (2001) Archaeal dominance in the mesopelagic zone of the Pacific Ocean. *Nature* **409**, 507–510.

Keil R. G. and Cowie G. L. (1999) Organic matter preservation through the oxygen-deficient zone of the NE Arabian Sea as discerned by organic carbon: mineral surface area ratios. *Mar. Geol.* **161**, 13–22.

Keil R. G., Montlucon D. B., Prahl F. G., and Hedges J. I. (1994a) Sorptive preservation of labile organic matter in marine sediments. *Nature* **370**, 549–552.

Keil R. G., Tsamakis E., Bor Fuh C., Giddings J. C., and Hedges J. I. (1994b) Mineralogical and textural controls on the organic composition of coastal marine sediments: hydrodynamic separation using SPLITT-fractionation. *Geochim. Cosmochim. Acta* **58**(2), 879–893.

Keil R. G., Mayer L. M., Quay P. D., Richey J. E., and Hedges J. I. (1997) Loss of organic matter from riverine particles in deltas. *Geochim. Cosmochim. Acta* **61**, 1507–1511.

Knicker H., Scaroni A. W., and Hatcher P. G. (1996) 13C and 15N NMR spectroscopic investigation on the formation of fossil algal residues. *Org. Geochem.* **24**, 661–669.

Kohnen M. E. L., Sinninghe Damsté J. S., and de Leeuw J. W. (1991) Biases from natural sulphurization in palaeoenvironmental reconstruction based on hydrocarbon biomarker distributions. *Nature* **349**(28 February), 775–778.

Krein E. B. and Aizenshtat Z. (1994) The formation of isoprenoid sulfur compounds during diagenesis: simulated sulfur incorporation and thermal transformation. *Org. Geochem.* **21**(10–11), 1015–1025.

Kuhlbusch T. A. J. (1998) Research: ocean chemistry: black carbon and the carbon cycle. *Science* **280**(5371), 1903–1904.

Kuhlbusch T. A. J. and Crutzen P. J. (1995) Toward a global estimate of black carbon in residues of vegetation fires representing a sink of atmospheric CO_2 and a source of O_2. *Global Biogeochem. Cycles* **9**(4), 491–501.

Kvenvolden K. A. (1995) A review of the geochemistry of methane in natural gas hydrate. *Org. Geochem.* **23**, 997–1008.

Largeau C., Casadevall E., Kadouri A., and Metzger P. (1984) Comparative study of immature torbanite and of the extant alga Botryococcus braunii. *Org. Geochem.* **6**, 327–332.

Leithold E. L. and Blair N. E. (2001) Watershed control on the carbon loading of marine sedimentary particles. *Geochim. Cosmochim. Acta* **65**(14), 2231–2240.

Ludwig W., Probst J. L., and Kempe S. (1996) Predicting the oceanic input of organic carbon by continental erosion. *Global Biogeochem. Cycles* **10**, 23–41.

Macdonald R. W., Solomon S. M., Cranston R. E., Welch H. E., Yunker M. B., and Gobeil C. (1998) A sediment and organic carbon budget for the Canadian Beaufort Shelf. *Mar. Geol.* **144**, 255–273.

Mague T. H., Friberg E., Hughers D. J., and Morris I. (1980) Extracellular release of carbon by marine phytoplankton: a physiological approach. *Limnol. Oceanogr.* **25**, 262–279.

Malcolm R. L. (1990) The uniqueness of humic substances in each of soil, stream, and marine environments. *Anal. Chim. Acta* **232**, 19–30.

Mannino A. and Harvey H. R. (2000) Biochemical composition of particles and dissolved organic matter along an estuarine gradient: sources and implications for DOM reactivity. *Limnol. Oceanogr.* **45**, 775–788.

Masiello C. A. and Druffel E. R. M. (1998) Black carbon in deep-sea sediments. *Science* **280**, 1911–1913.

Masiello C. A. and Druffel E. R. M. (2001) Carbon isotope geochemistry of the Santa Clara River. *Global Biogeochem. Cycles* **15**(2), 407–416.

Mayer L. M. (1994) Surface area control of organic carbon accumulation in continental shelf sediments. *Geochim. Cosmochim. Acta* **58**.

Mayer L. M. (1999) Extent of coverage of mineral surfaces by organic matter in marine sediments. *Geochim. Cosmochim. Acta* **63**, 207–215.

Mayer L. M., Jumars P. A., Taghon G. L., Macko S. A., and Trumbore S. (1993) Low-density particles as potential nitrogenous foods for benthos. *J. Mar. Res.* **51**(2), 373–389.

McCarthy M. D., Hedges J. I., and Benner R. (1994) The chemical composition of dissolved organic matter in seawater. *Chem. Geol.* **107**, 503–507.

McCarthy M. D., Hedges J. I., and Benner R. (1996) Major biochemical composition of dissolved high molecular weight organic matter in seawater. *Mar. Chem.* **55**, 281–297.

McCarthy M., Pratum T., Hedges J. I., and Benner R. (1997) Chemical composition of dissolved organic nitrogen in the ocean. *Nature* **390**, 150–154.

McCarthy M. D., Hedges J. I., and Benner R. (1998) Major bacterial contribution to marine dissolved organic nitrogen. *Science* **281**, 231–234.

McKnight D. M., Harnish R., Wershaw R. L., Baron J. S., and Schiff S. (1997) Chemical characteristics of particulate, colloidal, and dissolved organic material in Loch Vale watershed, Rocky Mountain National Park. *Biogeochemistry* **99**, 99–124.

Megens L., Plicht J. v.d., and Leeuw J. W. (2001) Temporal variations in ^{13}C and ^{14}C concentrations in particulate organic matter from the southern North Sea. *Geochim. Cosmochim. Acta* **65**(17), 2899–2911.

Menzel D. W. (1974) Primary productivity, dissolved and particulate organic matter and the sites of oxidation of organic matter. In *The Sea* (ed. E. D. Goldberg). Wiley, New York, London, Torondo, pp. 659–678.

Meybeck M. (1982) Carbon, nitrogen, and phosphorus transport by world rivers. *Am. J. Sci.* **282**(4), 401–450.

Meyers-Schulte K. J. and Hedges J. I. (1986) Molecular evidence for a terrestrial component of organic matter dissolved in ocean water. *Nature* **321**, 61–63.

Middelburg J. J., Nieuwenhuize J., and van Breugel P. (1999) Black carbon in marine sediments. *Mar. Chem.* **65**, 245–252.

Middelburg J. J., Barranguet C., Boschker H. T. S., Herman P. M. J., Moens T., and Heip C. H. R. (2000) The fate of intertidal microphytobenthos carbon: an *in situ* super (13)C-labeling study. *Limnol. Oceanogr.* **45**(6), 1224–1234.

Milliman J. D. and Syvitski J. P. M. (1992) Geomorphic/tectonic control of sediment discharge to the ocean: the importance of small mountainous rivers. *J. Geol.* **100**, 525–544.

Mossmann J.-R., Aplin A. C., Curtis C. D., and Coleman M. L. (1991) Geochemistry of inorganic and organic sulphur in organic-rich sediments from the Peru margin. *Geochim. Cosmochim. Acta* **55**, 3581–3595.

Nagata T. and Kirchman D. L. (1992) Release of dissolved organic matter by heterotrophic protozoa, implications for microbial food webs. *Arch. Hydrobiol.* **35**, 99–109.

Nip M., Tegelaar E. W., Leeuw J. W. D., Schenck P. A., and Holloway P. J. (1986) A new non-saponifiable highly aliphatic and resistant biopolymer in plant cuticles: evidence from pyrolysis and ^{13}C NMR analysis of present day and fossil plants. *Naturwissenschaften* **73**, 579–585.

Nittrouer C. A., Kuehl S. A., Sternberg R. W., Jr., Figueiredo A. G., Jr., and Faria L. E. C. (1995) An introduction to the geological significance of sediment transport and accumulation on the Amazon continental shelf. *Mar. Geol.* **125**, 177–192.

Ohkouchi N., Eglinton Timothy I., Keigwin Lloyd D., and Hayes John M. (2002) Spatial and temporal offsets between proxy records in a sediment drift. *Science* **298**(5596), 1224–1227.

Onstad G. D., Canfield D. E., Quay P. D., and Hedges J. I. (2000) Sources of particulate organic matter in rivers from the continental USA: lignin phenol and stable carbon isotope compositions. *Geochim. Cosmochim. Acta* **64**, 3539–3546.

Opsahl S. and Benner R. (1997) Distribution and cycling of terrigenous dissolved organic matter in the ocean. *Nature* **386**, 480–482.

Pancost R., Damsté J. S. S., de Lint S., vanderMaarel M J E C, Gottschal J. C., and Party M. S. (2000) Widespread anaerobic methane oxidation by methanogens in mediterranean sediments. *Appl. Environ. Microbiol.* **66**, 1126–1136.

Parkes R. J., Cragg B. A., Bale S. J., Getliff J. M., Goodman K., Rochelle P. A., Fry J. C., Weightman A. J., and Harvey S. M. (1994) Deep bacterial biosphere in Pacific Ocean sediments. *Nature* **371**, 410–412.

Parkes R. J., Cragg B. A., and Wellsbury P. (2000) Recent studies on bacterial populations and processes in subseafloor sediments: a review. *Hydrogeol. J.* **8**(1), 11–28.

Pearson A. and Eglinton T. I. (2000) The origin of n-alkanes in Santa Monica Basin surface sediment: a model based on compound-specific ^{14}C and ^{13}C data. *Org. Geochem.* **31**, 1103–1116.

Pearson A., Eglinton T. I., and McNichol A. P. (2000) An organic tracer for surface ocean radiocarbon. *Paleoceanography* **15**, 541–550.

Pearson A., McNichol A. P., Benitez-Nelson B. C., Hayes J. M., and Eglinton T. I. (2001) Origins of lipid biomarkers in Santa Monica Basin surface sediment: a case study using compound-specific D14C analysis. *Geochim. Cosmochim. Acta* **65**(18), 3123–3137.

Pedersen T. F. and Calvert S. E. (1990) Anoxia vs. productivity: what controls the formation of organic-carbon-rich sediments and sedimentary rocks? *Am. Assoc. Petrol. Geol. Bull.* **74**, 454–466.

Peltzer E. T. and Hayward N. A. (1995) Spatial and temporal variability of total organic carbon along 140° W in the equatorial Pacific Ocean in 1992. *Deep-Sea Res. Spec. EqPac (II)* **43**, 1155–1180.

Post W. M. (1993) Organic carbon in soil and the global carbon cycle. In *The Global Carbon Cycle* (ed. M. Heimann). Springer, New York, pp. 277–302.

Prahl F. G., Ertel J. R., Goni M. A., Sparrow M. A., and Eversmeyer B. (1994) Terrestrial OC on the Washington margin. *Geochim. Cosmochim. Acta* **58**, 3035–3048.

Prahl F. G., Lange G. J. D., Scholten S., and Cowie G. L. (1997) A case of post-depositional aerobic degradation of terrestrial organic matter in trubidite deposits from the Madeira Abyssal plain. *Org. Geochem.* **27**(3/4), 141–152.

Premuzic E. T., Benkovitz C. M., Gaffney J. S., and Walsh J. J. (1982) The nature and distribution of organic matter in the surface sediments of the world oceans and seas. *Org. Geochem.* **4**, 63–77.

Prospero J. M., Ginoux P., Torres O., Nicholson S. E., and Gill T. E. (2003) Environmental characterization of global sources of atmospheric soil dust identified with the NIMBUS-7 TOMS absorbing aerosol product. *Rev. Geophys.* (in press).

Putschew A., Scholz-Bottcher B. M., and Rullkotter J. (1996) Early diagenesis of organic matter and related sulfur incorporation in surface sediments of meromictic Lake Cadagno in the Swiss Alps. *Org. Geochem.* **25**, 379–390.

Ransom B., Bennett R. H., and Baerwald R. (1997) TEM study of in situ organic matter on continental margins: occurrence and the 'monolayer' hypothesis. *Mar. Geol.* **138**, 1–9.

Ransom B., Kim D., Kastner M., and Wainwright S. (1998a) Organic matter preservation on continental slopes: importance of mineralogy and surface area. *Geochim. Cosmochim. Acta* **62**(8), 1329–1345.

Ransom B., Shea K. F., Burkett P. J., Bennett R. H., and Baerwald R. (1998b) Comparison of pelagic and nepheloid layer marine snow: implications for carbon cycling. *Mar. Geol.* **150**, 39–50.

Rau G. H., Takahashi T., and Des Marais D. J. (1989) Latitudinal variations in plankton $\delta^{13}C$: implications for CO_2 and productivity in past oceans. *Nature* **341**(12 October 1989), 516–518.

Raymond P. A. and Bauer J. E. (2001a) Riverine export of aged terrestrial organic matter to the North Atlantic Ocean. *Nature* **409**, 497–500.

Raymond P. A. and Bauer J. E. (2001b) Use of super(14)C and super(13)C natural abundances for evaluating riverine, estuarine, and coastal DOC and POC sources and cycling: a review and synthesis. *Org. Geochem.* **32**(4), 469–485.

Reeburgh W. S., Ward B. B., Whalen S. C., Sandbeck K. A., Kilpatrick K. A., and Kerkhof L. J. (1991) Black Sea methane geochemistry. *Deep-Sea Res.* **38**(suppl. 2), S1189–S1210.

Repeta D. J., Quan T. M., Aluwihare L. I., and Accardi A. M. (2002) Chemical characterization of high molecular weight dissolved organic matter from fresh and marine waters. *Geochim. Cosmochim. Acta* **66**, 955–962.

Romankevich E. A. (1984) *Geochemistry of Organic Matter in the Ocean*, Springer, Berlin, 351pp.

Sakugawa H. and Handa N. (1985) Isolation and chemical characterization of dissolved and particulate polysaccharides in Mikawa Bay. *Geochim. Cosmochim. Acta* **49**, 1185–1193.

Santschi P. H., Balnois E., Wilkinson K. J., Zhang J., and Buffle J. (1998) *Fibrillar polysaccharides in marine macromolecular organic matter as imaged by atomic force microscopy and transmission electron microscopy*, **43**, 896–908.

Schlunz B. and Schneider R. R. (2000) Transport of terrestrial organic carbon to the oceans by rivers: re-estimating flux- and burial rates. *Int. J. Earth Sci.* **88**, 599–606.

Schmidt M. W. I. and Noack A. G. (2000) Black carbon in soils and sediments: analysis, distribution, implications, and current challenges. *Global Geochem. Cycles* **14**(3), 777–793.

Schmidt M. W. I., Skjemstad J. O., Gehrt E., and Kogel-Knabner I. (1999) Charred organic carbon in German chernozemic soils. *Euro. J. Soil. Sci.* **50**, 351–365.

Schouten S., de Graaf W., Sinninghe Damsté J. S., van Driel G. B., and de Leeuw J. W. (1994) Laboratory simulation of natural sulfurization: II. Reaction of multifunctionalized lipids with inorganic polysulfides at low temperatures. *Org. Geochem.* **22**, 825–834.

Schouten S., Hoefs M. J. L., Koopmans M. P., Bosch H.-J., and Damsté J. S. S. (1998) Structural identification, occurrence and fate of archaeal ether-bound acyclic and cyclic biphytanes and corresponding diols in sediments. *Org. Geochem.* **29**, 1305–1319.

Schouten S., Wakeham S. G., and Sinninghe Damsté J. S. (2001) Evidence for anaerobic methane oxidation by archaea in euxinic waters of the Black Sea. *Org. Geochem.* **32**, 1277–1281.

Schubert C. J. and Stein R. (1996) Deposition of organic carbon in Arctic Ocean sediments: terrigenous supply vs. marine productivity. *Org. Geochem.* **24**(4), 421–436.

Siegenthaler U. and Sarmiento J. L. (1993) Atmospheric carbon dioxide and the ocean, 365, 119–125.

Sinninghe Damsté J. S. and Schouten S. (1997) Is there evidence for a substantial contribution of prokaryotic biomass to organic carbon in phanerozoic carbonaceous sediments? *Org. Geochem.* **26**(9/10), 517–530.

Sinninghe Damsté J. W., Eglinton T. I., de Leeuw J. W., and Schenck P. A. (1989) Organic sulfur in macromolecular sedimentary organic matter: I. Structure and origin of sulfur-containing moieties in kerogen, asphaltenes and coal as revealed by flash pyrolysis. *Geochim. Cosmochim. Acta* **53**, 873–889.

Sinninghe Damsté J. S., Kok M. D., Koster J., and Schouten S. (1998) Sulfurized carbohydrates: an important sedimentary sink for organic carbon? *Earth Planet. Sci. Lett.* **164**, 7–13.

Sinninghe Damsté J. S., Rijpstra W. I. C., Hopmans E. C., Prahl F. G., Wakeham S. G., and Schouten S. (2002) Distribution of membrane lipids of planktonic Crenarchaeota in the Arabian Sea. *Appl. Environ. Microbiol.* **68**(6), 2997–3002.

Skjemstad J. O., Clarke P., Taylor J. A., Oades J. M., and McClure S. G. (1996) The chemistry and nature of protected carbon in soil. *Austral. J. Soil Res.* **34**(2), 251–271.

Smith K., Kaufman R., and Baldwin R. (1994) Coupling of near-bottom pelagic and benthic processes at abyssal depths. *Limnol. Oceanol.* **39**, 1101–1118.

Smith S. V. and Hollibaugh J. T. (1993) Coastal metabolism and the oceanic organic carbon balance. *Rev. Geophys.* **31**, 75–89.

Smith S. V. and MacKenzie F. T. (1987) The ocean as a net heterotrophic system: implications from the carbon biogeochemical cycle. *Global Biogeochem. Cycles* **1**, 187–198.

Stuermer D. and Harvey G. R. (1974) Humic substances from seawater. *Nature* **250**, 480–481.

Stuermer D. and Payne J. R. (1976) Investigations of seawater and terrestrial humic substances with carbon-13 and proton magnetic resonance. *Geochim. Cosmochim. Acta* **40**, 1109–1114.

Stuiver M. and Pollach H. A. (1977) On the reporting of ^{14}C ages. *Radiocarbon* **35**, 355–365.

Suess E. (1980) Particulate organic carbon flux in the oceans; surface productivity and oxygen utilization. *Nature (London)* **288**(5788), 260–263.

Suman D. O., Khulbusch T. A. J., and Lim B. (1997) Marine sediments: a reservoir for black carbon and their use as spatial and temporal records of combustion. In *Sedimental Records of Biomass Burning and Global Change* (ed. J. S. Clark). Springer, New York, pp. 271–293.

Tanoue E., Nishiyama S., Kamo M., and Tsugita A. (1995) Bacterial membranes: possible source of a major dissolved protein in seawater. *Geochim. Cosmochim. Acta* **59**, 2643–2648.

Tegelaar E. W., de Leeuw J. W., Derenne S., and Largeau C. (1989) A reappraisal of kerogen formation. *Geochim. Cosmochim. Acta* **53**, 3103–3106.

Thomsen L. and Van Weering T. J. (1998) Spatial and temporal variability of particulate matter in the benthic boundary layer at the NW European continental margin (Goban Spur). *Prog. Oceanogr.* **42**, 61–76.

Tissot B. P. and Welte D. H. (1984) *Petroleum Formation and Occurrence.* Springer, New York.

Torn M. S., Trumbore S. E., Chadwick O. A., Vitousek P. M., and Hendricks D. M. (1997) Mineral control on carbon storage and turnover. *Nature* 389, 170–173.

Vaccaro R. F., Hicks S. E., Jannasch H. W., and Carey F. G. (1968) The occurrence and role of glucose in seawater. *Limnol. Oceanogr.* 13, 356–360.

Vairavamurthy A. and Mopper K. (1987) Geochemical formation of organosulphur compounds (thiols) by addition of H_2S to sedimentary organic matter. *Nature* 329(6140), 623–625.

van de Meent D., Brown S. C., Philp R. P., and Simoneit B. R. T. (1980) Pyrolysis-high resolution gas chromatography and pyrolysis-gas chromatography-mass spectrometry of kerogen and kerogen precursors. *Geochim. Cosmochim. Acta* 44, 999–1013.

Vlahos P., Chen R. F., and Repeta D. J. (2002) Dissolved organic carbon in the mid-Atlantic bight. *Deep-Sea Res.* 49, 4369–4385.

Volkman J. K., Barrett S. M., Blackburn S. I., Mansour M. P., Sikes E. L., and Gelin F. (1998) Microalgal biomarkers: a review of recent research developments. *Org. Geochem.* 29, 1163–1176.

Wakeham S. G. and Lee C. (1993) Production, transport, and alteration of particulate organic matter in the marine water column. In *Organic Geochemistry* (eds. M. H. Engel and S. A. Macko). Plenum, New York, pp. 145–169.

Wakeham S. G., Lee C., Hedges J. I., Hernes P. J., and Peterson M. L. (1997) Molecular indicators of diagenetic status in marine organic matter. *Geochim. Cosmochim. Acta* 61, 5363–5369.

Wang X.-C. and Druffel E. R. M. (2001) Radiocarbon and stable carbon isotope compositions of organic compound classes in sediments from the NE Pacific and Southern Oceans. *Mar. Chem.* 73, 65–81.

Wang X.-C., Druffel E. R. M., and Lee C. (1996) Radiocarbon in organic compound classes in particulate organic matter and sediment in the deep northeast Pacific Ocean. *Geophys. Res. Lett.* 23, 3583–3586.

Wang X. C., Druffel E. R. M., Griffin S., Lee C., and Kashgarian M. (1998) Radiocarbon studies of organic compound classes in plankton and sediment of the northeastern Pacific Ocean. *Geochim. Cosmochim. Acta* 62, 1365–1378.

Wells M. L. and Goldberg E. D. (1991) Occurrence of small colloids in sea water. *Nature* 353, 342–344.

Wells M. L. and Goldberg E. D. (1993) Colloid aggregation in seawater. *Mar. Chem.* 41, 353–358.

Wells M. L. and Goldberg E. D. (1994) The distribution of colloids in the North Atlantic and Southern Oceans. *Limnol. Oceanogr.* 39, 286–302.

Werne J. P., Hollander D. J., Behrens A., Schaeffer P., Albrecht P., and Damsté J. S. S. (2000) Timing of early diagenetic sulfurization of organic matter: a precursor-product relationship in holocene sediments of the anoxic Cariaco Basin, Venezuela. *Geochim. Cosmochim. Acta* 64(10), 1741–1751.

Williams P. M., Oeschger H., and Kinney P. (1969) Natural radiocarbon activity of dissolved organic carbon in the north-east Pacific Ocean. *Nature* 224, 256–259.

Wolbach W. S. and Anders E. (1989) Elemental carbon in sediments: determination and isotopic analysis in the presence of kerogen. *Geochim. Cosmochim. Acta* 53, 1637–1647.

Wollast R. (1991) The coastal organic carbon cycle: fluxes, sources, and sinks. In *Physical, Chemical, and Earth Sciences Research Report,* vol. 9, pp. 365–381.

Zafiriou O. C., Gagosian R. B., Peltzer E. T., Alford J. B., and Loder T. (1985) Air-to-sea fluxes of lipids at Enewetak Atoll. *J. Geophys. Res.* 90(D1), 2409–2423.

Zang X., Nguyen R. T., Harvey H. R., Knicker H., and Hatcher P. G. (2001) Preservation of proteinaceous material during the degradation of the green alga Botryococcus braunii: a solid-state 2D 15N 13C NMR spectroscopy study. *Geochim. Cosmochim. Acta* 65(19), 3299–3305.

Zelibor J. L., Romankiw L., Hatcher P. G., and Colwell R. R. (1988) Comparative analysis of the chemical composition of mixed and pure cultures of green algae and their decomposed residues by ^{13}C nuclear resonance spectroscopy. *Appl. Environ. Microbiol.* 54, 1051–1060.

Geochemistry of Earth Surface Systems
ISBN: 978-0-08-096706-6

pp. 389–424

12

The Biological Pump

C. L. De La Rocha

Alfred Wegener Institute for Polar and Marine Research, Bremerhaven, Germany

NOMENCLATURE		426
12.1	INTRODUCTION	426
12.2	DESCRIPTION OF THE BIOLOGICAL PUMP	427
	12.2.1 Photosynthesis and Nutrient Uptake	428
	12.2.1.1 Levels of primary production	429
	12.2.1.2 Patterns in time and space	429
	12.2.1.3 Nutrient limitation	430
	12.2.2 Flocculation and Sinking	430
	12.2.2.1 Marine snow	430
	12.2.2.2 Aggregation and exopolymers	430
	12.2.2.3 Sinking and transport of POM to depth	431
	12.2.3 Particle Decomposition and Repackaging	431
	12.2.3.1 Zooplankton grazing	432
	12.2.3.2 Bacterial hydrolysis	432
	12.2.3.3 Geochemistry of decomposition	433
	12.2.4 Sedimentation and Burial	433
	12.2.5 Dissolved Organic Matter	433
	12.2.6 New, Export, and Regenerated Production	434
12.3	IMPACT OF THE BIOLOGICAL PUMP ON GEOCHEMICAL CYCLING	434
	12.3.1 Macronutrients	434
	12.3.1.1 Carbon	434
	12.3.1.2 Nitrogen	435
	12.3.1.3 Phosphorus	436
	12.3.1.4 Silicon	436
	12.3.2 Trace Elements	437
	12.3.2.1 Barium	437
	12.3.2.2 Zinc	438
	12.3.2.3 Cadmium	439
	12.3.2.4 Iron	440
12.4	QUANTIFYING THE BIOLOGICAL PUMP	440
	12.4.1 Measurement of New Production	440
	12.4.2 Measurement of Particle Flux	441
	12.4.2.1 Sediment traps	441
	12.4.2.2 Particle-reactive nuclides	442
	12.4.2.3 Oxygen utilization rates	442
12.5	THE EFFICIENCY OF THE BIOLOGICAL PUMP	443
	12.5.1 Altering the Efficiency of the Biological Pump	443
	12.5.1.1 In HNLC areas	443
	12.5.1.2 Through changes in community composition	444

    12.5.1.3 By varying the C:N:P ratios of sinking material   444

    12.5.1.4 By enhancing particle transport   445

12.6   THE BIOLOGICAL PUMP IN THE IMMEDIATE FUTURE   445

   12.6.1 Response to Increased CO₂   445

   12.6.2 Response to Agricultural Runoff   446

    12.6.2.1 Shift toward Si limitation   446

    12.6.2.2 Shifts in export production and deep ocean C:N:P   446

   12.6.3 Carbon Sequestration via Ocean Fertilization and the Biological Pump   446

REFERENCES   447

NOMENCLATURE

b_1 and b_2	D/W (m^{-1})
C_e	extracellular CO_2 concentration (μM)
C_{export}	export production (mmol C m^{-2} d^{-1} or mg C m^{-2} d^{-1})
C_{flux}	sinking flux of POC (mmol C m^{-2} d^{-1} or mg C m^{-2} d^{-1})
C_i	intracellular CO_2 concentration (μM)
C_{prod}	primary production (mmol C m^{-2} d^{-1} or mg C m^{-2} d^{-1})
D	organic matter decomposition rate (d^{-1})
D_{CO_2}	diffusivity of CO_2 in seawater (m^2 s^{-1})
F_{CO_2}	flux of CO_2 to cell surface (μmol s^{-1})
G	acceleration due to gravity (m s^{-2})
$H_2CO_3^*$	dissolved $CO_2 + H_2CO_3$ (μM)
k'	rate constant for $HCO_3^- \rightarrow CO_2$ (s^{-1})
R	radius (m)
W	particle sinking rate (m d^{-1})
z	depth (m)
$\Delta\rho$	density difference (g m^{-3})
μ	dynamic viscosity of seawater (Pa s)
ρ	partition coefficient
$\sum CO_2$	total CO_2 (μM)
τ	residence time (years)

12.1 INTRODUCTION

Despite having residence times (τ) that exceed the ~1,000 years mixing time of the ocean, many dissolved constituents of seawater have concentrations that vary with depth and from place to place. Silicic acid ($\tau = 15,000$ years), nitrate ($\tau = 3,000$), phosphate ($\tau = 10,000$–$50,000$ years), and dissolved inorganic carbon (DIC; $\tau = 83,000$) are generally present in low concentrations in surface waters and at much higher concentrations below the thermocline (Figure 1). Additionally, their concentrations are higher in the older deep waters of the North Pacific than they are in the younger waters of the deep North Atlantic. This is the general distribution exhibited by elements and compounds taking part in biological processes in the ocean and is generally referred to as a "nutrient-type" distribution.

Both the lateral and vertical gradients in the concentrations of nutrients result from "the biological pump" (Figure 2). Dissolved inorganic materials (e.g., CO_2, NO_3^-, PO_4^{2-}, and $Si(OH)_4$) are fixed into particulate organic matter (POM; carbohydrates, lipids, and proteins) and biominerals (silica and calcium carbonate) by phytoplankton in surface waters. These particles are subsequently transported, by sinking, into the deep. The bulk of the organic material and biominerals decomposes in the upper ocean via dissolution, zooplankton grazing, and microbial hydrolysis, but a significant supply of material survives to reach the deep sea and sediments. Thus just as biological uptake removes certain dissolved inorganic materials from surface waters, the decomposition of sinking biogenic particles provides a source of dissolved inorganic material to deeper waters. Because of this, deeper waters contain higher concentrations of biologically utilized materials than surface waters, and older deep waters contain more than deep waters more recently formed from ocean surface waters.

One side effect of the biological pump is that CO_2 is shunted from the surface ocean into the deep sea, lowering the amount present in the atmosphere. For many years it has been recognized that pre-Industrial levels of CO_2 in the atmosphere were a one-third of what they would have been, had there been no biological pump operating in the ocean (Broecker, 1982). It has also long been considered that the biological pump is not operating at its full capacity. In the so-called high-nutrient, low-chlorophyll (HNLC) areas of the ocean, a considerable portion of the nutrients supplied to the surface waters is not utilized in support of primary production, most probably due to the limitation of phytoplankton growth by an inadequate supply of trace elements (Martin and Fitzwater, 1988). The possibility that the biological pump in HNLC areas might be stimulated by massive additions of Fe, both artificially as a means of removing anthropogenic CO_2 from the atmosphere and naturally as a cause for the lower glacial atmospheric CO_2 levels (Martin, 1990), is a current focus of research and debate (Watson *et al.*, 1994; Chisholm *et al.*, 2001; de Baar *et al.*, 2005).

Although the biological pump is most popularly known for its impact on the cycling of carbon

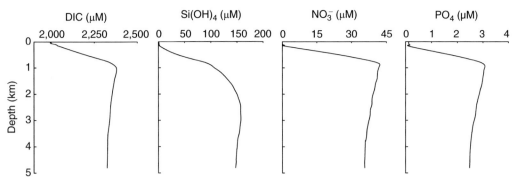

Figure 1 Idealized profiles of dissolved inorganic carbon, silicic acid, nitrate, and phosphate with depth in the ocean.

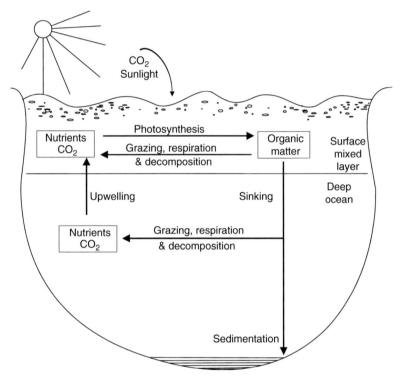

Figure 2 Simplified diagram of the biological pump.

and major nutrients, it also affects the geochemistry of other elements and compounds. The biological pump influences the cycling, concentrations, and residence times of trace elements such as Cd, Ge, Zn, Ni, Fe, As, and Se, through their incorporation into organic matter and biominerals (Bruland, 1980; Azam and Volcani, 1981; Elderfield and Rickaby, 2000). Scavenging by sinking biogenic particles and precipitation of materials in the microenvironment of organic aggregates and fecal pellets plays a major role in the marine geochemistry of elements such as Ba, Th, Pa, Be, rare earth elements (REEs), and Y (Dehairs *et al.*, 1980; Anderson *et al.*, 1990; Buesseler *et al.*, 1992; Kumar *et al.*, 1993; Zhang and Nozaki, 1996; Azetsu-Scott and Niven, 2005).

Even major elements in seawater such as Ca^{2+} and Sr^{2+} display slight surface depletions (Broecker and Peng, 1982; de Villiers, 1999) as a result of the biological pump, despite their long respective oceanic residence times of 1 and 5 Myr (Broecker and Peng, 1982; Elderfield and Schultz, 1996).

12.2 DESCRIPTION OF THE BIOLOGICAL PUMP

The biological pump can be sectioned into several major steps: the production of organic matter and biominerals in surface waters, the sinking of these particles into the deep, and the decomposition of the settling (or settled) particles. In general, phytoplankton in surface waters take up

DIC and nutrients. Carbon is fixed into organic matter via photosynthesis and, together with N, P, and trace elements, forms the carbohydrates, lipids, and proteins that comprise the bulk organic matter. Once formed, this organic matter faces the immediate prospect of decomposition back to CO_2, phosphate, ammonia, and other dissolved nutrients through consumption by herbivorous zooplankton and degradation by bacteria; most of the primary products formed will be recycled within a few hundred meters of the surface (Martin *et al.*, 1987). Some portion of the primary production will, however, be exported to deeper waters or even to the sediments before decomposition and may even escape remineralization entirely and remain in the sedimentary reservoir.

It is worth taking a closer look at the various steps in the biological pump (Figure 2). By what processes does organic matter make it from surface to deep? What factors define the proportion of surface primary production that survives transport to the seabed?

12.2.1 Photosynthesis and Nutrient Uptake

In the initial step of the biological pump, phytoplankton in sunlit surface waters converts CO_2 into organic matter via photosynthesis:

$$CO_2 + H_2O + light \rightarrow CH_2O + O_2 \qquad (1)$$

The first stable product of carbon fixation by the enzyme, ribulose bisphosphate carboxylase (Rubisco), is glyceraldehyde 3-phosphate, a 3-C sugar. This 3-C sugar is fed into biosynthetic pathways and forms the basis for all organic compounds produced by photosynthetic organisms. Fixed carbon and major trace elements such as

H, N, P, Ca, Si, Fe, Zn, Cd, Mg, I, Se, and Mo are used for the synthesis of carbohydrates, lipids, proteins, biominerals, amino acids, enzymes, DNA, and other essential compounds.

Besides C, the two main components of phytoplankton organic matter are N and P, in the roughly average molar proportion of 106:16:1 (C:N:P) (Redfield, 1934, 1958) known as the Redfield ratio. In addition, diatoms, by virtue of depositing opal (amorphous, hydrated silica) in their cell wall, have an average C/Si ratio of 8 - (Brzezinski, 1985), although this ratio may vary from at least 3–40 depending on the conditions of light, temperature, and nutrient availability (Harrison *et al.*, 1977; Brzezinski, 1985; De La Rocha *et al.*, 2000; Franck *et al.*, 2005). Coccolithophorids produce scales (coccoliths) made of $CaCO_3$ and contain 20–100 μmol of $CaCO_3$ per mol of organic C (Paasche, 1999).

POM and biominerals produced by phytoplankton contain many trace elements. The most abundant are Mg, Cd, Fe, Ca, Ba, Cu, Ni, Zn, and Al (Table 1), which are important constituents of enzymes, pigments, and structural materials. Carbonic anhydrase requires Zn or Cd (Price and Morel, 1990; Lane and Morel, 2000), nitrate reductase requires Fe (Geider and LaRoche, 1994), and chlorophyll contains Mg. Additionally, elements such as Na, Mg, P, Cl, K, and Ca are present as ions within cells and are important for osmoregulation and the maintenance of charge balance (e.g., Fagerbakke *et al.*, 1999).

A wide variety of ions may also be adsorbed onto the surfaces of biogenic particles. The removal from the water column and deposition in the sediments of particle-reactive elements such as Th (Buesseler *et al.*, 1992) and Pa (Kumar *et al.*, 1993) have

Table 1 Elemental composition of marine phytoplankton from cultures and plankton tows.

Element	Element/C ratio $(mol\ mol^{-1})$	References
N	0.15	Redfield (1958)
Si (diatoms)	0.13	Brzezinski (1985)
P	0.009	Redfield (1958)
Ca	0.03	Martin and Knauer (1973), Collier and Edmond (1984)
Fe	$2.3 \times 10^{-6} – 1.8 \times 10^{-3}$	Martin and Knauer (1973), Collier and Edmond (1984), Sunda and Huntsman (1995a)
Zn	6×10^{-5}	Martin and Knauer (1973), Collier and Edmond (1984)
Al	1×10^{-4}	Martin and Knauer (1973), Collier and Edmond (1984)
Cu	$3 \times 10^{-6} – 0.006$[a]	Martin and Knauer (1973), Collier and Edmond (1984)
Ni	$2 \times 10^{-5} – 0.006$[a]	Collier and Edmond (1984)
Cd	$5 \times 10^{-7} – 0.005$[a]	Martin and Knauer (1973), Collier and Edmond (1984)
Mn	$4 \times 10^{-6} – 0.004$[a]	Martin and Knauer (1973), Collier and Edmond (1984)
Ba	$1 \times 10^{-5} – 0.01$[a]	Martin and Knauer (1973), Collier and Edmond (1984)
Mg	0.02[a]	Martin and Knauer (1973)
Na	0.1[a]	Martin and Knauer (1973)
Sr	8×10^{-5}[a]	Martin and Knauer (1973)
Ti	1×10^{-5}[a]	Martin and Knauer (1973)
Cr	2×10^{-6}[a]	Martin and Knauer (1973)

[a] Calculated from dry weight data using an average phytoplankton C content on a dry weight basis of 50%.

been shown to correlate with the primary production of particles in the ocean. Additionally, Th has been shown to complex with colloidal, surface-reactive polysaccharides (Quigley *et al.*, 2002; Azetsu-Scott and Niven, 2005).

12.2.1.1 Levels of primary production

The amount of primary production carried out in the oceans each year has been estimated from ocean color satellite data and shipboard ^{14}C incubations to be $140\,\mathrm{gC\,m^{-2}}$, for a total of 50–60 Pg C (4–5 Pmol C) fixed in the surface ocean each year (Shushkina, 1985; Martin *et al.*, 1987; Field *et al.*, 1998). This represents roughly half of the global annual 105 Pg C fixed each year (Field *et al.*, 1998), despite the fact that marine phytoplankton comprise <1% of the total photosynthetic biomass on Earth. Extrapolation from Redfield ratios suggests the incorporation of 0.6–0.8 Pmol N and 40–50 Tmol P into biogenic particles each year in association with marine primary production. From the proportion of primary production carried out by diatoms and the average Si/C ratio of diatoms of silica, production rates of 200–280 Tmol Si yr^{-1} may be calculated (Nelson *et al.*, 1995; Tréguer *et al.*, 1995).

12.2.1.2 Patterns in time and space

Rates of primary production in upwelling regions of the ocean outpace those of nonupwelling coastal regions, which in turn are greater than rates in the oligotrophic open ocean (Figure 3; Ryther, 1969; Martin *et al.*, 1987). Open-ocean primary production levels are of the order of $130\,\mathrm{g\,C\,m^{-2}\,yr^{-1}}$, whereas in nonupwelling coastal areas and upwelling zones they are 250 and $420\,\mathrm{gCm^{-2}\,yr^{-1}}$, respectively (Martin *et al.*, 1987). However, because the open ocean constitutes 90% of the area of the ocean, the bulk (80%) of the ocean's annual carbon fixation occurs there rather than in coastal and upwelling regions.

Different types of phytoplankton dominate primary production in the different marine regimes. Diatoms perform ~75% of the primary production that occurs in upwelling and coastal regions of the ocean but <35% of that taking place in the open ocean (Nelson *et al.*, 1995). Phytoplankton biomass and primary production in the open ocean are dominated instead by prokaryotic picoplankton (Chisholm *et al.*, 1988; Liu *et al.*, 1999; Steinberg *et al.*, 2001).

Outside the tropics, levels of marine primary productivity vary systematically throughout the year (Heinrich, 1962). Standing stocks of phytoplankton and levels of primary production peak in the spring following the onset of water column stratification and the increase in available light. Depletion of nutrients in the stratified water column in summer inhibits phytoplankton growth and grazing by zooplankton reduces standing stocks. Some areas may experience a small bloom of phytoplankton in the autumn when light levels are still adequate and the onset of winter convection and overturning injects nutrients into the euphotic zone.

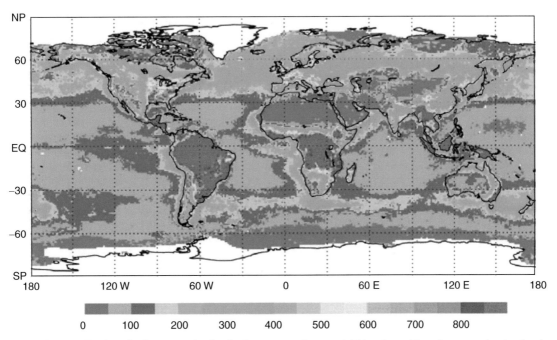

Figure 3 Distribution of primary production in the ocean and terrestrial biosphere. Net primary production levels are given in g C m^{-2} yr^{-1}. Reproduced by permission of American Association for the Advancement of Science from Field *et al.* (1998).

12.2.1.3 Nutrient limitation

The upper limit of primary production is set by the supply of nutrients (N, P, Si, and Fe) into the euphotic zone. N inputs into the surface ocean may limit the primary productivity of the whole ocean over short timescales. Over timescales approaching and exceeding the 10,000–50,000 years residence time of P (Ruttenberg, 1993; Filippelli and Delaney, 1996), P inputs limit global ocean primary production (Tyrrell, 1999).

Regionally and for different types of phytoplankton, the limitation of both the rate and overall amount of primary production is more varied. Major nutrients in HNLC areas of the ocean (such as the Southern Ocean, the Equatorial Pacific, and the North Pacific subarctic) are never completely consumed in support of primary production, because low levels of Fe limit phytoplankton growth (Martin and Fitzwater, 1988; Martin et al., 1994; Coale et al., 1996; Boyd et al., 2000). Diatoms, which require Si for growth, are often limited by low concentrations of silicic acid in surface waters (Brzezinski and Nelson, 1996; Nelson and Dortch, 1996). Growth of diazotrophic (N_2 fixing) phytoplankton such as the cyanobacterium, *Trichodesmium*, will be more susceptible to P and Fe limitation than to N limitation. Even the concentration of dissolved CO_2 in seawater (especially in the midst of a phytoplankton bloom) may limit instantaneous rates, although not ultimate levels, of primary production (Riebesell et al., 1993; Wolf-Gladrow et al., 1999).

12.2.2 Flocculation and Sinking

12.2.2.1 Marine snow

Material that reaches the deep ocean and seafloor does so not as individual phytoplankton cells slowly meandering down, but as larger, rapidly sinking particles (McCave, 1975; Suess, 1980; Billett et al., 1983; Fowler and Knauer, 1986; Alldredge and Gotschalk, 1989), which have traversed the distance between surface and deep in a space of days (Billett et al., 1983; Asper et al., 1992). These larger particles, known collectively as "marine snow," are largely aggregates (Alldredge and Silver, 1988) formed either by zooplankton, which produce mucous feeding structures and fecal pellets, or by the physical coagulation of smaller particles (McCave, 1984; Alldredge et al., 1993). Coagulation is the more important of the two formation pathways; the bulk of the organic material reaching the deep sea does so as aggregated phytoplankton that have not been ingested by zooplankton (Billett et al., 1983; Turner, 2002).

Sinking rates of marine snow are orders of magnitude greater than those of unaggregated phytoplankton cells (Smayda, 1970; Shanks and Trent, 1980; Alldredge and Gotschalk, 1989).

The shorter transit time from surface to bottom for aggregated particles results in the enhanced transport of C, N, P, Si, and other materials to the deep, despite the fact that marine snow particles are sites of elevated rates of decomposition and nutrient regeneration. Intense colonization and hydrolysis of the particles by bacteria (Smith et al., 1992; Bidle and Azam, 1999, 2001) and fragmentation and consumption of the particles by zooplankton (Steinberg, 1995; Dilling et al., 1998; Dilling and Alldredge, 2000; Goldthwait et al., 2004) reduce the vertical flux of materials to the seafloor.

12.2.2.2 Aggregation and exopolymers

Coagulation requires the success of two activities, the collision of particles and their subsequent joining to form an aggregate. In the ocean, particles collide due to processes such as shear, Brownian motion, and differential settling. The probability of particles attaching following a collision is controlled by the physical and chemical properties of the particles' surfaces (Alldredge and Jackson, 1995). The probability of sticking is greatly enhanced by exopolymers produced by phytoplankton and bacteria (Alldredge and McGillivary, 1991; Alldredge et al., 1993).

These exopolymers turn out to be important for the transport of material to the deep. In a coastal system, it was observed that transparent exopolymer particles (TEPs), which are colloidal polysaccharides exuded by phytoplankton, were required for the aggregation and sedimentation of diatoms out of the water column (Passow et al., 2001). The formation of rapidly sinking aggregates has also been shown to be controlled more by TEP abundance than by phytoplankton concentrations (Logan et al., 1995).

Little is known about the composition of the polymers responsible for particle aggregation in marine systems. They are generally composed of polysaccharides (Alldredge et al., 1993) and proteins (Long and Azam, 1996). The carbohydrate component of TEP contains glucose, mannose, arabinose, xylose, galactose, rhamnose, glucuronate, and O-methylated sugars (Janse et al., 1996; Holloway and Cowen, 1997), and is generally rich in deoxysugars (Mopper et al., 1995). Very little is known about the specific composition of TEP, and virtually nothing is known of its structural characteristics (Holloway and Cowen, 1997; Schumann and Rentsch, 1998; Engel and Passow, 2001).

Exopolymer particles may scavenge dissolved organic matter (DOM) as they form, providing a mechanism for the biological pumping of DOM into the deep sea (Engel and Passow, 2001). TEP also contains C and N in proportions exceeding Redfield ratios (Mari et al., 2001; Engel et al., 2002), providing a mechanism for pumping of carbon in excess of what would be predicted from the availability of nitrogen.

12.2.2.3 Sinking and transport of POM to depth

The flux (C_{flux}) of particulate organic carbon (POC) to depth (z) reflects the balance between the rate of decomposition (D) of the POC and the velocity at which it sinks (W) (Banse, 1990):

$$C_{flux} = C_{flux(z_0)} e^{-(D/W)z} \qquad (2)$$

The faster the particles sink into the deep, the larger the fraction of their organic matter that arrives with them. According to Stokes' law, the key players in particle sinking velocities are the size and density of the particle, with sinking velocities increasing with the square of the radius (r) of a particle or with the density difference between the particle and seawater ($\Delta\rho$):

$$W = \frac{2gr^2(\Delta\rho)}{9 \quad \mu} \qquad (3)$$

where g is the acceleration due to gravity and μ the dynamic viscosity of seawater.

Because of their small size (generally not more than a few micrometers), sinking velocities of solitary phytoplankton cells are only about a meter per day (Smayda, 1970). Particles sinking so slowly require more than a year to reach the benthos of even a relatively shallow continental shelf, and 10 years to reach the abyssal ocean floor. Given the rapid rates of microbial decomposition of organic material in the ocean and the abundance of zooplankton grazers, it is virtually impossible for such a slowly sinking particle to reach the seabed.

Sinking velocities of large (>0.5 mm) particles, in contrast, are >100 mday^{-1} (Shanks and Trent, 1980; Alldredge and Gotschalk, 1989). Transit time to the deep in this case is days to weeks, which agrees with observations of a close temporal coupling between surface production and seafloor sedimentation (e.g., Billett *et al.*, 1983; Asper *et al.*, 1992).

It might be argued that particle flux is controlled by rates of particle aggregation and sinking more than it is controlled by overall levels of primary production. Year to year variability in carbon export to deep waters correlates more strongly with the size of the dominant primary producer, for example, than with year to year variations in levels of carbon fixation (Boyd and Newton, 1995).

It has been recently suggested that minerals such as calcium carbonate, opal, and clays may drive the sedimentation of POM in the ocean by adding density, or "ballast" to organic aggregates

(Armstrong *et al.*, 2002; François *et al.*, 2002; Klaas and Archer, 2002). This hypothesis is based on several observations. Below 3,000 m, sinking particles contain a fairly constant 5 wt.% of organic carbon (Armstrong *et al.*, 2002; Passow, 2004), as if they are saturated with mineral particles. The fluxes of calcium carbonate and POC into sediment traps below 3,000 m also appear correlated (r^2 values ~0.7) (François *et al.*, 2002; Klaas and Archer, 2002). There is also a significant correlation between opal and POC fluxes (François *et al.*, 2002; Klaas and Archer, 2002) that is not as strong in large part due (Passow and De La Rocha, 2006) to high variability in the Si/C ratio of primary production in various regions of the ocean (e.g., Ragueneau *et al.*, 2000).

There are two points against the mineral "ballast" controlling the flux of POC to depth. The first is that correlation does not imply causation; from the regression data (François *et al.*, 2002; Klaas and Archer, 2002), there is no way to tell if it is the mineral fluxes that are responsible for the POC fluxes or vice versa (Passow, 2004; Passow and De La Rocha, 2006). The second is that the accumulation of minerals on organic aggregates causes aggregates to fragment into smaller particles (Passow and De La Rocha, 2006). The competing processes of density addition and size reduction mean that the relationship between sinking rates and the accumulation of minerals on aggregates is not a straightforward, linear one (Hamm, 2002). This points toward sinking of POC fluxes being important to the settling of suspended mineral fragments rather than minerals being critical to the sedimentation of POC (Passow and De La Rocha, 2006).

12.2.3 Particle Decomposition and Repackaging

Less than half of the particles produced during primary production survive zooplankton grazing and microbial attack to be exported from the euphotic zone and only about 1% endure to settle into the deep ocean and sediments (Suess, 1980; Martin *et al.*, 1987; Lutz *et al.*, 2002; Andersson *et al.*, 2004). This leaves the efficiency of the biological pump largely in the hands of processes taking place during the sinking of particles to the deep. The processes influencing the balance between particle sinking and decomposition rates are only poorly known.

Following its formation, organic matter in the ocean rapidly decomposes and there is intense recycling of elements even within the euphotic zone. The flux of POC in the ocean decreases more or less exponentially with depth below the euphotic zone (Figure 4; Suess, 1980; Martin *et al.*, 1987; Lutz *et al.*, 2002; Andersson *et al.*, 2004). The amount of organic matter exported to

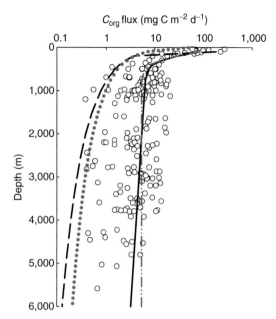

Figure 4 Sinking fluxes of organic C in the ocean as estimated from sediment traps and regional estimates of export production. Data are from tables 1 and 2 in Lutz et al., 2002. The fitted curves are equation (4), the Suess curve (Suess, 1980) (black dashed line), equation (5), the Martin curve (Martin et al., 1987) (dotted red line), equation (6), following Lutz et al. (2002) (red dashed and dotted line), and equation (7), following Andersson et al. (2004) (black solid line).

the deep ocean (C_{flux}) has been classically described as a function of primary production in the euphotic zone (C_{prod}) (Suess, 1980):

$$C_{flux}(z) = \frac{C_{prod}}{(0.0238z + 0.212)} \qquad (4)$$

where z is the depth in the water column, or as a function of export from the euphotic zone (C_{export}) (Martin et al., 1987):

$$C_{flux}(z) = C_{export}\left(\frac{z}{z_0}\right)^{-0.858} \qquad (5)$$

These curves were derived from POC flux measured via sediment traps and are widely used to describe carbon flux in global circulation models. Other curves have been proposed, based on sediment traps or sediment oxygen consumption, that take into account the increase in the refractory nature of particles with depth (Lutz et al., 2002; Andersson et al., 2004):

$$C_{flux}(z) = C_{flux_0}\left[(1-p)e^{-b_1 z} + pe^{-b_2 z}\right] \qquad (6)$$

In this approach, C_{flux_0} is partitioned (p) into a pool of fresh, labile POC degrading rapidly relative to its sinking rate and a pool too rapidly sinking or too refractory for significant decomposition to take place en route to the seabed. The two relationships for D/W are b_1 and b_2.

The simplicity of these curves belies the complexity of the processes causing the decrease in POC flux with depth. Mediating the decomposition and recycling of organic matter in the ocean are zooplankton and heterotrophic bacteria (Cho and Azam, 1988; Smith et al., 1992; Steinberg, 1995; Dilling et al., 1998; Dilling and Alldredge, 2000). Bacteria and zooplankton diminish the sinking particulate flux by consuming particles and converting them back into CO_2 and dissolved materials, and by converting large, sinking particles into smaller particles with reduced or nonexistent sinking rates.

Although most particles are broken down in the surface ocean, processes occurring in the mesopelagic and bathypelagic zones are significant enough to wipe out the regional variability in POC fluxes seen in the surface ocean (Antia et al., 2001). At 125 m in the Atlantic, for example, POC flux ranges regionally from 0.5 to 12 $gCm^{-2}yr^{-1}$. By 3,000 m, the variability has been compressed by 85% to 0.5–2.4 $gCm^{-2}yr^{-1}$.

12.2.3.1 Zooplankton grazing

Zooplankton may reduce the sinking flux of biogenic particles in the ocean in two ways. The first is by grazing upon particles which reduces the total amount of POM in the water column and shifts its occurrence from large, fast-sinking aggregates to smaller fecal pellets, which constitute only a minor portion of the sinking organic flux (Turner, 2002). In the second way the particle flux is reduced by actively breaking up aggregates into smaller particles (Dilling and Alldredge, 2001; Goldthwait et al., 2004). At stations off of southern California, for instance, the average overnight increase in the number of aggregates per liter by 15% was attributable to the fragmentation of larger particles by swimming euphausiids (Dilling and Alldredge, 2001).

The relative impact of zooplankton grazing on primary production decreases with increasing production levels; the proportion of primary production that is consumed by zooplankton decreases exponentially as productivity levels increase (Calbet, 2001). This supports the observation that the ratio of export production to total production is higher in areas of high productivity. Globally, about 12% of marine primary production, or 5.5 PgC (0.5 Pmol C), is consumed by mesozooplankton each year (Calbet, 2001).

12.2.3.2 Bacterial hydrolysis

Bacterial hydrolysis plays a major role in the decomposition of sinking and suspended matter in the ocean. Bacterial organic carbon has been observed to make up over 40% of the total POC in the water column, and the proportion of the sinking flux of carbon utilized by bacteria may be equal to 40–80% of the surface primary production in nearshore areas (Cho and Azam, 1988). Marine aggregates have been shown to contain high concentrations of hydrolytic enzymes, such as proteases, polysaccharidases, and glucosidases (Smith *et al.*, 1992). Turnover times of organic components of marine aggregates due to hydrolysis may be short, on the order of fractions of days to days (Smith *et al.*, 1992). Bacterial proteases have also been shown to enhance the dissolution rates of biogenic silica in diatom aggregates (Bidle and Azam, 2001). Much of the hydrolyzed material is not taken up by the bacteria attached to the particles but instead joins the pool of DOM present in the water (Smith *et al.*, 1992).

12.2.3.3 Geochemistry of decomposition

Ingestion of POM by zooplankton results in the respiration and excretion of a portion of that POM as CO_2, NH_4, dissolved organic nitrogen (DON), phosphate, and dissolved organic phosphorus (DOP). Assimilation efficiencies for organic matter for zooplankton grazing on phytoplankton range from 10% to 40% (Ryther, 1969; Michaels and Silver, 1988). Zooplankton do not assimilate significant quantities of silicon from diatoms consumed, leaving regeneration of silicic acid to be mediated strictly by opal dissolution rates.

12.2.4 Sedimentation and Burial

Of the net 50–60 Pg C (4–5 Pmol C) fixed into organic matter in the surface ocean each year (Shuskina, 1985; Martin *et al.*, 1987; Field *et al.*, 1998), only ~10% is exported from the euphotic zone (Buesseler, 1998). Each year roughly 0.16 Pg C (13 Tmol C) are preserved in ocean sediments (Hedges and Keil, 1995), an accumulation rate representing 0.3% of that of carbon fixation in the surface ocean.

The accumulation of sedimentary organic carbon in the ocean varies greatly from place to place, with continental margin sediments accounting for the bulk of the organic carbon buildup (Hedges and Keil, 1995; de Haas *et al.*, 2002). About 94% of the sedimentary organic carbon that is preserved in the oceans is buried on continental shelves and slopes (Hedges and Keil, 1995). This leaves only 6% of the total sedimentary organic carbon to accumulate in the open ocean. Since the open ocean, due to its vast area, plays host to 80% of the annual primary production, the accumulation of only 6% of the organic carbon suggests an overall preservation efficiency of 0.02% here. Preservation efficiency on continental shelves and slopes is, at 1.4%, much greater.

Many reasons have been given for the regional differences in the accumulation and preservation of organic carbon in the sediments. Differences in the flux of organic particles to the seafloor due to differences in primary production levels, rates of aggregation and sinking, and depth of the water column may contribute to a higher preservation efficiency. The oxygen content of bottom waters may also be important, although a correlation between burial Qefficiency and bottom water oxygen concentration is not seen (Hedges and Keil, 1995) and rates of hydrolysis of organic matter by bacteria may be high even under anoxic conditions (e.g., Arnosti *et al.*, 1994). The long-term preservation of organic material in sediments may be tied to the sorption of the organic molecules to mineral surfaces (Hedges and Keil, 1995), although the nature of the associations and the rates at which they occur have not been closely detailed.

12.2.5 Dissolved Organic Matter

DOM has not been as intensively studied as other aspects of the biological pump perhaps because DOM does not sink. However, DOM does play an active role in the biological pump in at least three ways. Much of the DOM is biologically utilizable and may directly provide P and N for primary production (Clark *et al.*, 1998; Zehr and Ward, 2002). DOM may assemble into colloidal and particulate material that can sink as well as scavenge other material to form marine snow (Alldredge *et al.*, 1993; Kepkay, 1994; Chin *et al.*, 1998). DOM is also a large reservoir of carbon in the ocean, containing at least an order of magnitude more carbon than the other marine organic carbon reservoirs (Kepkay, 1994).

Although the origins of DOM have not been fully detailed, phytoplankton are thought to serve as the dominant source of DOM to the ocean. Actively growing phytoplankton secrete DOM (Biddanda and Benner, 1997; Soendergaard *et al.*, 2000; Teira *et al.*, 2001) and the polysaccharide composition of phytoplankton exudates resembles that of the high-molecular-weight fraction of DOM (Aluwihare and Repeta, 1999). Phytoplankton DOM is also released during grazing by zooplankton (Strom *et al.*, 1997). Organic matter, such as mucus, on phytoplankton cell surfaces may also be hydrolyzed by bacteria and released as DOM (Smith *et al.*, 1995).

The exact composition of marine DOM is not known. It has so far been shown to contain neutral sugars (such as glucose, fucose, mannose, and arabinose), polysaccharides, amino acids, amides (such as chitin), phosphorus esters, and phosphonates (Benner *et al.*, 1992; McCarthy *et al.*, 1997; Clark *et al.*, 1998; Amon *et al.*, 2001). Microbial degradation is thought to play a role in setting the composition of DOM in the ocean (Amon *et al.*, 2001).

One interesting feature of DOM, at least with respect to the carbon cycle, is its enrichment in C over the Redfield ratio. The C:N:P ratios of high-molecular-weight DOM are roughly 350:20:1 (Kolowith *et al.*, 2001). C-enrichment is also observed for bulk DOM (Kaehler and Koeve, 2001). Part of this enrichment in C may be due to the enhanced remineralization of P and N from DOM (Clark *et al.*, 1998; Kolowith *et al.*, 2001), as suggested by an increase in the C/P ratio of DOM with depth (Kolowith *et al.*, 2001). DOM may also just simply be produced with high C/N and C/P ratios. TEP, for example, which forms from DOM precursors (Alldredge *et al.*, 1993; Chin *et al.*, 1998) has high C/N ratios (Engel and Passow, 2001; Mari *et al.*, 2001). The polysaccharides that eventually form TEP are exuded by phytoplankton and may represent excess photosynthate, which is carbohydrates, and lipids formed in the absence of nutrients like N, that are needed for the synthesis of compounds, like proteins (Morris, 1981).

The importance of the carbon-enriched DOM pool in the carbon cycle hinges on the size and turnover time of the reservoir. A recent estimate of dissolved organic carbon (DOC) in the ocean is 200 Pg C (Kepkay, 1994), which is more comparable to the 750 Pg of carbon present in the atmosphere than it is to the 36,000 Pg C deep sea reservoir of DIC (Sundquist, 1993). The average age of marine DOM is ~6,000 years (Williams and Druffel, 1987), although the turnover time of different fractions of the DOM pool varies from decades to nearly 20,000 years (Loh *et al.*, 2004). The bulk (~70%) of DOM is low molecular weight (Benner *et al.*, 1992) and relatively resistant to microbial degradation (Bauer *et al.*, 1992; Amon and Benner, 1994). High-molecular-weight compounds that are quickly turned over by bacterial decomposition (Amon and Benner, 1994) make up the remaining 30% of the DOM pool.

12.2.6 New, Export, and Regenerated Production

Not all of the primary production in the ocean feeds carbon into the biological pump. Most of the carbon fixed each year is converted straight back into CO_2 and dissolved nutrients by zooplankton and bacteria in the euphotic zone. These recycled nutrients may then be used to fuel further carbon fixation.

It has long been recognized (Dugdale and Goering, 1967) that a portion of the primary production (regenerated production) is supported by nutrients regenerated in the euphotic zone and another portion (new production) is supported by nutrients imported in the euphotic zone through upwelling, river inputs, nitrogen fixation, or atmospheric deposition. The ratio of new to total primary production in the ocean, known as the *f* ratio (Eppley and Peterson, 1979), is generally higher in upwelling environments than it is in oligotrophic regions of the ocean (Harrison, 1990; Laws *et al.*, 2000). On average, ~20% of the total global marine primary production is new production, ranging from 7% to 70% from region to region (Laws *et al.*, 2000).

12.3 IMPACT OF THE BIOLOGICAL PUMP ON GEOCHEMICAL CYCLING

12.3.1 Macronutrients

12.3.1.1 Carbon

The influence of the biological pump on the distribution of DIC in the ocean illustrates the influence it has on the distribution of many other elements. Low concentrations of DIC are observed in surface waters (Figure 1) due to the uptake of dissolved CO_2 (and perhaps HCO_3^-; Raven, 1997) by phytoplankton. Concentrations of dissolved CO_2 increase most rapidly just below the euphotic zone, associated with the bulk of the decomposition of POC (Martin *et al.*, 1987; Antia *et al.*, 2001). Deepwater concentrations of dissolved CO_2 are higher than surface water concentrations, and older deep waters contain more dissolved CO_2 than younger ones (Broecker and Peng, 1982).

Much of the current scientific interest in the biological pump revolves around the impact it has on levels of CO_2 in the atmosphere and, subsequently, on climate. The photosynthetic fixation of carbon into organic matter lowers the concentration of dissolved CO_2 in surface waters, allowing influx of CO_2 from the atmosphere. This fixed carbon may then be exported to deeper waters or the sediments before it decomposes back to CO_2, maintaining the observed gradient in dissolved CO_2 concentrations between waters of the surface and deep (Figure 1).

Atmospheric concentrations of CO_2 are thus lower for the given size of the oceanic DIC reservoir than they would be in the absence of this biological transport of carbon to the deep. If all life in the ocean were to die off and the ocean and atmosphere came to equilibrium with respect to CO_2, concentrations of CO_2 in the atmosphere would rise by ~140 µatm (Broecker, 1982), which is a remarkable 50% of the pre-industrial interglacial value of 280 µatm.

Details of the influence of the biological pump on the distribution and cycling of carbon in the ocean and the controlling factors are discussed later in this chapter. The next few pages are devoted to the relationship of the biological pump to the cycling of other elements. This is by no means an exhaustive overview of the biological shuffling of elements throughout the ocean, but instead a highlight of several elements of particular biogeochemical interest.

12.3.1.2 Nitrogen

Of all the elements playing an important role in the regulation of the biological pump, nitrogen is the one with the most complex biologically mediated cycling. Nitrogenous species taking part in productivity range from N_2, which may be fixed into a more universally biologically available form by nitrogen-fixing bacteria, to NO_3^-, which follows from the production of NO_2^- from NH_4^+ through nitrification (Figure 5). The denitrification pathway sequentially results in the transformation of NO_2^- to N_2O and N_2. NH_4^+ can also be directly oxidized into N_2 with NO_2^- as the oxidant, via anaerobic ammonium oxidation (anammox) (Mulder *et al.*, 1995; Thamdrup and Dalsgaard, 2002; Dalsgaard *et al.*, 2003). There are also dissolved organic forms of nitrogen, such as amides, urea, free amino acids, amines (McCarthy *et al.*, 1997), which are also biologically utilizable to varying degrees.

The average marine *f*-ratio of 0.2 (Laws *et al.*, 2000) suggests that most of the primary production in the ocean is supported by dissolved inorganic nitrogen (DIN) that has been recycled in the euphotic zone. This further suggests that the bulk of the primary production in the ocean relies on NH_4^+, since that is the predominant recycled form of nitrogen. Extrapolating from the Redfield C/N ratio of 6.6 and the average *f* ratio of 0.2 and

the overall estimate of primary production of 4–5 Pmol C in the ocean each year (Shushkina, 1985; Martin *et al.*, 1987; Field *et al.*, 1998) yields 0.7 Pmol of particulate organic nitrogen (PON) produced each year, 0.1 Pmol of which is exported out of the euphotic zone.

12.3.1.2.1 Nitrogen fixation. The two aspects of the nitrogen cycle having the greatest impact on the biological pump are nitrogen fixation and denitrification. The first provides a mechanism for drawing on the extensive atmospheric pool of N_2 gas in support of primary production. The second provides a pathway for DIN to be converted back into N_2 gas and removed from the ocean system.

Approximately 28 Tg N (2 Tmol N) are added to the marine nitrogen inventory through nitrogen fixation each year (Gruber and Sarmiento, 1997; Capone *et al.*, 2005), although the overall amount of nitrogen fixed each year is considerably higher than this. Nitrogen fixation accounts for about half of the new nitrogen used in primary production (Karl *et al.*, 1997). Only prokaryotic organisms can fix nitrogen, leaving this, at least in the ocean, in the hands of the cyanobacteria and out of the hands of eukaryotic algae such as diatoms, dinoflagellates, and coccolithophorids (unless they are hosting diazotrophic symbionts). Until recently it was believed that the filamentous, colony-forming cyanobacterium *Trichodesmium*, and cyanobacterial symbionts in diatoms were responsible for the bulk of the nitrogen fixation occurring in the ocean (Capone *et al.*, 1997). However, direct measurements of the rates of nitrogen fixation by *Trichodesmium*, coupled with knowledge of their distribution and abundance, fell significantly short of nitrogen fixation rates calculated from geochemical budgets (Gruber and Sarmiento, 1997). It has now been discovered that free-living, unicellular cyanobacteria are expressing the genes for the nitrogen-fixing enzyme, nitrogenase (Zehr *et al.*, 2001), and contribute considerably to fixed nitrogen budgets in the ocean (Montoya *et al.*, 2004).

It has been suggested that rates of nitrogen fixation are currently limited by the availability of Fe. The Fe-requirement of *Trichodesmium*, however, turns out to be much lower than previously estimated. Instead it is the availability of P that controls the upper limit of nitrogen fixation in the modern ocean (Sañudo-Wilhelmy *et al.*, 2001).

The demonstration of the P limitation of nitrogen fixation by cyanobacteria supports the notion that over geologic time phosphorus ultimately limits productivity (Tyrrell, 1999). When cyanobacteria face a shortfall of nitrogen, they fix nitrogen to meet their demands. This influx of new nitrogen to N-limited systems allows for the

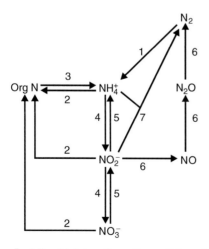

Figure 5 Microbial transformations of the nitrogen cycle. Pathways depicted are (1) N_2 fixation, (2) DIN assimilation, (3) ammonium regeneration, (4) nitrification, (5) nitrate/nitrite reduction, (6) denitrification, and (7) anaerobic ammonium oxidation (anammox).

drawdown of phosphate until the system is P-limited. Under P limitation, nitrogen fixation is curtailed (Sañudo-Wilhelmy *et al.*, 2001). Thus while nitrogen may at times limit instantaneous rates of carbon fixation, as is frequently the case in the modern ocean, over long time periods the input of phosphorus to the oceans sets the upper limit of net primary production (Tyrrell, 1999).

12.3.1.3 Phosphorus

Although there is a tendency to consider dissolved inorganic phosphorus (DIP, measured as soluble-reactive phosphate; Strickland and Parsons, 1968) as being simply PO_4^{3-}, DIP exists as a considerable number of species. At seawater pH, DIP is predominantly H_2PO_4 (87%) and only 12% PO_4^{3-} and 1% $H_2PO_4^-$ (Greenwood and Earnshaw, 1984). There are also numerous dissolved organic forms of phosphorus that are taken up by phytoplankton and used to fuel primary production in the ocean.

One interesting aspect of the phosphorus cycle is that, unlike the cases for the other major nutrient elements in the ocean, the phosphorus cycle contains sinks that are not mediated by biological activities. Dissolved phosphate may be scavenged onto iron or manganese oxyhydroxide particles associated with hydrothermal activity or reacted with basalt during the circulation of water through mid-ocean ridge hydrothermal systems (Föllmi, 1996). Although the exact values are poorly known, it is estimated that the scavenging of phosphate by the hydrothermal oxyhydroxides constitutes up to 50% of the removal flux of phosphorus from the ocean (e.g., Froelich *et al.*, 1982; Berner *et al.*, 1993). Another large inorganic sink for dissolved phosphorus is the precipitation of authigenic phosphate, which accounts for 10–40% of the removal flux. Removal of phosphorus as sedimenting POC by comparison is thought to be 20–50% of the output of phosphorus from the ocean.

Roughly, 5 Tg P (0.2 Tmol P) are removed from the ocean each year as both organic and inorganic phases. This is in reasonable balance with the roughly 5 Tg of reactive P being brought into the system each year naturally. However these natural sources constitute only half of the modern-day input of phosphorus to the ocean (Froelich *et al.*, 1982), anthropogenic inputs having doubled the annual phosphate flux. A doubling of phosphorus inputs to the ocean could have a significant impact on the biological pump, although it should be noted that the ocean is already nitrogen limited and further inputs of phosphorus will only exacerbate this. Shifts in N/P ratios of surface waters may alter the structure of the phytoplankton community. Phaeocystis, for example, grows poorly under high phosphate conditions and may see its numbers decline.

12.3.1.4 Silicon

Dissolved silicic acid is required for the growth of diatoms, which deposit opal (amorphous, hydrated silica) in their cell wall and dominate the production of opal in the modern-day ocean (Lisitzin, 1972). About 240 Tmol of silica are produced by diatoms in the surface ocean each year (Nelson *et al.*, 1995; Tréguer *et al.*, 1995). Total inputs of dissolved silicic acid to the euphotic zone are 120 Tmol Si, mostly from rivers (6 Tmol Si) and upwelling (115 Tmol Si; Tréguer *et al.*, 1995).

Half of the biogenic silica produced each year dissolves in the upper 100 m of the water column (Nelson *et al.*, 1995), and a further 47% dissolves in the deep ocean and seafloor, for a net deposition of ~6 Tmol (3% of surface production) each year (Tréguer *et al.*, 1995; DeMaster, 2002). Opal preservation efficiencies are generally highest in productive environments (e.g., 6% in the permanently open ocean zone of the Southern Ocean versus 0.4% in the oligotrophic North Atlantic). The traditional view of opal accumulation in the sediments is that it is closely linked to opal production in overlying waters (e.g., Broecker and Peng, 1982), but many factors besides opal production govern opal preservation. Opal dissolution rates more than double for every 10 °C rise in temperature (Kamitani, 1982). Aggregation has been shown to reduce rates of Si regeneration from diatoms (Bidle and Azam, 2001). Additionally, the fraction of produced silica that reaches the seabed correlates with high seasonality and high ratios of carbon export to production (Pondaven *et al.*, 2000), which also tend to be in areas where organic matter is formed in blooms and is transported quickly to the sediments as large aggregates.

12.3.1.4.1 The impact of the appearance of the diatoms on the marine silica cycle. The silica cycle is an excellent example of how much the biological pump can impact the concentration and distribution of elements in seawater. Presently, surface concentrations of silicic acid are low: the overall average silicic acid content of ocean waters is only 70 μM (Tréguer *et al.*, 1995). Prior to the appearance of the diatoms in the early Tertiary (Tappan and Loeblich, 1973), oceanic concentrations of silicic acid were ~1,000 μM (Siever, 1991). The sponges and radiolarians controlling the oceanic inventory of silicic acid at that time possessed neither the numbers nor demand for Si needed to draw down silicic acid concentrations. The increased output of biogenic silica from the ocean associated with the ascension of diatoms, with their high affinity for silicic acid and high cellular requirements for silicon, resulted in a precipitous drop in the silicic acid content of ocean waters over the late Cretaceous and Paleocene (Figure 6), stabilizing in the Eocene to the $<100\,\mu M$ values that have held ever since (Siever, 1991).

Since sponges and radiolarians are not great players in particle flux, the rise of the diatoms must have profoundly altered the partitioning of silicic acid between surface and deep. The familiar nutrient-type distribution may have only existed for the last 50–100 Myr. The approximate 14-fold drop in silicic acid concentration also suggests that the residence time of Si in the ocean dropped from 200,000 years to today's 15,000.

12.3.1.4.2 Excessive pumping of silicon. Silicic acid, by virtue of being regenerated from silica instead of from relatively labile POM, is regenerated more deeply than the other major nutrients (Figure 1; Dugdale *et al.*, 1995). This decoupling between silicic acid and the other nutrients and the fact that not all phytoplankton utilize Si means that there is no Redfield relationship between Si and C, N, or P. On average across the ocean, water upwelled into the euphotic zone, with an Si/N of roughly 0.6, contains a slight excess of nitrate over silicic acid relative to the 1:1 utilization ratio of nutrient replete diatoms (Figure 7). Below $\sim500\,m$, however, there is more silicic acid than nitrate, with deep ocean Si/N ratios being near 3 (Figure 7).

Fe limitation of diatoms increases the pumping of Si (relative to N and C) to deeper waters. Fe-limited diatoms are inhibited in their utilization of N as a result of the Fe requirement of nitrate reductase (Geider and LaRoche, 1994). However, Fe-limited diatoms continue to take up silicic acid, although at lowered rates (De La Rocha *et al.*, 2000). As a result, the Si/N ratios of Fe-limited diatoms may be as high as 2 or 3 (Takeda, 1998; Hutchins and Bruland, 1998), much higher than 1 of nutrient replete diatoms (Brzezinski, 1985).

12.3.2 Trace Elements

The biological pump influences the distribution of many elements in seawater besides C and the major nutrients. Ba, Cd, Ge, Zn, Ni, Fe, As, Se, Y, and many of the REEs show depth distributions that closely resemble profiles of the major nutrients. Additionally, Be, Sc, Ti, Cu, Zr, and Ra have profiles where concentrations increase with depth, although the correspondence of these profiles with nutrient profiles is not very tight (Nozaki, 1997).

12.3.2.1 Barium

Vertical profiles of dissolved barium (Ba^{2+}) in the ocean resemble profiles of silicic acid and alkalinity (Figure 8; Lea and Boyle, 1989; Jeandel *et al.*, 1996), suggesting that biological processes strongly influence Ba distributions throughout the ocean. However, the strict incorporation of Ba into biogenic materials is not the dominant means of Ba^{2+} removal from ocean waters. Although the similarity between the profiles of Ba^{2+} and $Si(OH)_4$ suggests a common removal phase, the amount of barium incorporated into diatom opal ($<9 \times 10^{-6}$ mol Ba mol^{-1} Si; Shemesh *et al.*, 1988) cannot account for the 2×10^{-4} mol Ba^{2+} mol^{-1} Si(OH)$_4$ slope (Jeandel *et al.*, 1996) in the ocean. Ba^{2+} appears instead to be mainly removed from seawater as barite (BaSO$_4$) formed in association with opal and decaying organic material (Dehairs *et al.*, 1980; Bishop, 1988). The exact mechanism for barite precipitation is unknown, but it is thought that it forms in the SO_4^{2-}-enriched microenvironments of decaying particles that may be thus supersaturated with respect to barite (Dehairs *et al.*, 1980; Ganeshram *et al.*, 2003).

Although the marine budget of barium is only approximately known, it appears to be both balanced and controlled by biogenic particle formation. Approximately 35 Gmol of Ba^{2+} are removed from surface waters every year (Dehairs *et al.*, 1980). Of this 35 Gmol, 60% settles as barite and the rest is incorporated into or adsorbed onto phases such as $CaCO_3$ or SiO_2 (Dehairs *et al.*, 1980; Dymond *et al.*, 1992). About 10–25 Gmol of Ba are buried on the seafloor each year (Dehairs *et al.*, 1980).

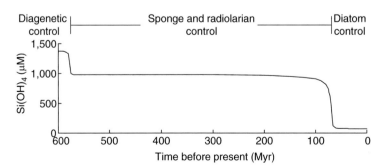

Figure 6 Estimated average marine concentrations of silicic acid over the Phanerozoic. Adapted from Siever (1991).

Because barite forms in association with organic material, there is a tight correlation ($r^2 = 0.93$) between the sedimentary fluxes of carbon and barite (Dymond *et al.*, 1992). Thus, barite

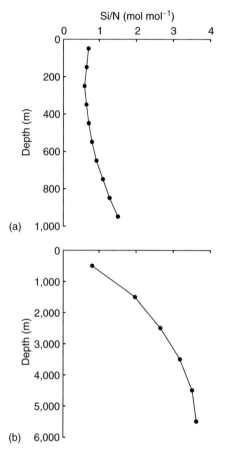

Figure 7 Ratio of silicic acid to nitrate with depth in the ocean between (a) 0 and 1000 m and (b) between 0 and 6,000 m. Profile have been constructed from volume-weighted averages of average silicic acid and nitrate concentrations in the Southern Ocean, North Atlantic, South Atlantic, North Pacific, South Pacific, and Indian Ocean from the WOCE dataset.

accumulation rates have been used to infer past levels of export production in the ocean (e.g., Dymond *et al.*, 1992; Paytan *et al.*, 1996).

12.3.2.2   Zinc

The profiles of dissolved Zn in the ocean are also similar to those of $Si(OH)_4$ (Figure 9; Bruland, 1980), but as with Ba, the main removal phase for Zn is not opal. Less than 3% of the Zn taken up by diatoms is deposited in their opaline cell wall (Ellwood and Hunter, 2000) and the Zn/Si ratio of acid-leached opal is much lower than the ratio of dissolved Zn and $Si(OH)_4$ in the water column (Bruland, 1980; Collier and Edmond, 1984). Instead most of the Zn removed from the surface ocean is bound up in POM.

Because Zn is required for the growth of phytoplankton, Zn availability affects the biological pump. Although Zn limitation of an entire phytoplankton community has never been demonstrated (e.g., Leblanc *et al.*, 2005), levels of dissolved Zn are often low enough to limit many taxa (Morel *et al.*, 1994; Sunda and Huntsman, 1995b; Timmermans *et al.*, 2001). Zn is an integral part of the enzyme, carbonic anhydrase (Morel *et al.*, 1994), which helps maintain an efficient supply of CO_2 to Rubisco (Sunda and Huntsman, 2005). Zn may also be involved in the acquisition or use of organic forms of P (Shi *et al.*, 2004).

The impact of low concentrations of Zn on phytoplankton growth varies. Some phytoplankton substitute cobalt for zinc in many enzymes (Price and Morel, 1990; Sunda and Huntsman, 1995b; Timmermans *et al.*, 2001) and can maintain maximal growth rates at low levels of Zn. Calcification aids with the acquisition of DIC, so calcareous phytoplankton such as the coccolithophorids are not as dependent on carbonic anhydrase (and therefore Zn) to maintain high rates of C fixation (Sunda and Huntsman, 1995b). Thus while low levels of Zn may not curtail overall levels of

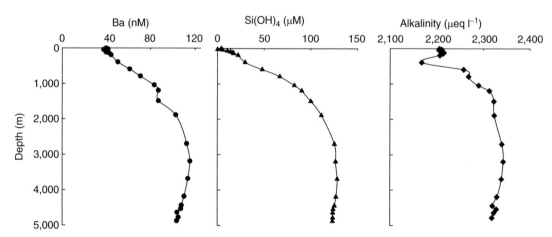

Figure 8   Profiles of Ba^{2+}, $Si(OH)_4$, and alkalinity in the Indian Ocean (06°09′ S, 50°55′ E). Data are from Station 36 of Jeandel *et al.* (1996).

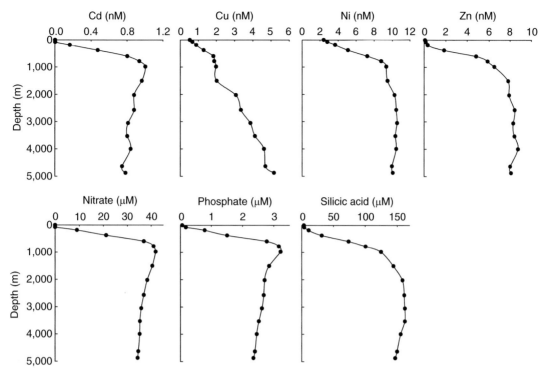

Figure 9 Profiles of Cd, Cu, Ni, Zn, nitrate, phosphate, and silicic acid in the north Pacific (32°41.0′ N, 144°59.4′ W). Plots have been redrawn from Bruland (1980).

primary production, they may shift the phyto-plankton community structure away from diatoms and toward coccolithophorids (Morel *et al.*, 1994; Sunda and Huntsman, 1995b; Timmermans *et al.*, 2001), which will shift the ratio of organic C to CaCO$_3$ of particles sinking to the deep.

12.3.2.3 Cadmium

Like Ba and Zn, Cd shows a nutrient-like distribution in the ocean (Figure 9; Bruland, 1980; Löscher *et al.*, 1998), more closely mirroring those of the labile nutrients, nitrate and phosphate than those of silicic acid and alkalinity. Cd is taken up by phytoplankton and incorporated into organic material, accounting for the similarity of its profile to that of nitrate and phosphate. Cadmium may also be adsorbed onto the surfaces of phytoplankton (Collier and Edmond, 1984).

Cd is taken up by phytoplankton slightly preferentially to phosphate (Löscher *et al.*, 1998; Elderfield and Rickaby, 2000) In general, waters with low phosphate concentrations have lower Cd/P ratios (0.1 nmol Cd (μmol P)$^{-1}$) than waters with higher phosphate concentrations where Cd/P may approach 0.4 nmol Cd (μmol P)$^{-1}$ (Elderfield and Rickaby, 2000). Additionally, surface water Cd/P ratios drop over the development of the spring bloom in the Southern Ocean (Löscher *et al.*, 1998). This may be tied to variations in the Cd/P utilization ratio of phytoplankton with dissolved CO$_2$, Mn, and Zn concentrations as well

as physiological state of the phytoplankton community (Cullen and Sherrell, 2005).

Cd also regenerates preferentially from decomposing particles (Knauer and Martin, 1981; Collier and Edmond, 1984). More labile components of decaying particles have higher Cd/P ratios than bulk decaying particles (Knauer and Martin, 1981). Box models of Cd and P cycling also require enhanced regeneration of Cd from particles for the replication of observed Cd distributions in the ocean (Collier and Edmond, 1984).

Cadmium is generally toxic to organisms and the how the marine phytoplankton utilize their cadmium is unknown. Cd may substitute for Zn in carbonic anhydrase at times when Zn is limiting (Price and Morel, 1990; Lane and Morel, 2000). It is possible that Cd may play a role in polyphosphate bodies, a form of cellular storage of P that have been shown to contain significant quantities of elements such as Ca, Zn, and Mg (Ruiz *et al.*, 2001).

Because of the similarity between Cd profiles and nutrient profiles, attempts have been made to reconstruct past phosphate concentrations from the Cd/Ca ratio of foraminifera (Boyle, 1988; Elderfield and Rickaby, 2000). Foraminifera substitute Cd^{2+} for Ca^{2+} in the lattice of their calcite tests, and the ratio of Cd/Ca incorporation varies with the Cd/Ca ratio of seawater (Boyle, 1988). The Cd/Ca ratio of foraminifera in the Southern Ocean suggests that P in surface waters was not as heavily utilized by phytoplankton during the last glacial maximum (LGM), suggesting lower levels of primary production relative to nutrient flux into the euphotic zone at that time (Elderfield and Rickaby, 2000).

12.3.2.4 Iron

Of all the trace elements whose distributions are affected by the biological pump, Fe is the one that has the most profound impact on the workings of the biological pump. Fe plays a key role in many of the crucial enzymes in biological systems, such as superoxide dismutase, ferredoxin, and nitrate reductase (Geider and LaRoche, 1994), which evolved at a time when the oceans were low in oxygen and therefore high in dissolved Fe. As a result, marine phytoplankton have a heavy demand for Fe relative to the present-day availability of dissolved Fe in the ocean. Growth of phytoplankton and rates of photosynthesis are frequently limited by the lack of Fe in surface waters (Martin and Fitzwater, 1988).

One of the major inputs of iron to the ocean comes from the dissolution of $Fe(II)$ from windborne continental dust deposited on the surface of the ocean (Zhuang et al., 1990). In oxygenic environments, such as surface ocean waters, $Fe(II)$ will quickly be oxidized to the insoluble form $Fe(III)$ and removed from seawater. $Fe(II)$ is also taken up by phytoplankton as well as complexed by ligands exuded into the water by marine organisms to prevent its precipitation as $Fe(III)$.

Profiles of dissolved Fe in seawater show the influence of both biotic and abiotic processes. At stations in the Northeast Pacific, dissolved Fe concentrations are low in surface waters reflecting biological uptake. Fe concentrations also show peak values at depth, corresponding to the oxygen minimum zone (Martin and Gordon, 1988), suggesting the abiotic reduction of $Fe(III)$ back to the soluble $Fe(II)$.

Vast tracks of the ocean, such as the Equatorial Pacific, the Northeast Pacific subarctic, the Southern Ocean, and even parts of the California upwelling zone do not have sufficient supplies of Fe to fully support phytoplankton growth (Martin and Fitzwater, 1988; Martin et al., 1994; Coale et al., 1996; Hutchins and Bruland, 1998; Boyd et al., 2000). In these areas, macronutrients such as N and P are rarely depleted. Attention has turned to these HNLC areas as sites where further primary production (and the associated drawdown of atmospheric CO_2) could occur. Increased supplies of dust stimulating the biological pump in HNLC regions may be responsible for low atmospheric CO_2 concentrations during glacial times (Martin and Gordon, 1988; Watson et al., 2000). Artificially stimulating the biological pump by seeding HNLC areas with chelated Fe has been proposed as a means of pumping into the

deep sea the 90 μatm of CO_2 in the atmosphere put there by human hands, although there is not much consensus as to the effectiveness of such an endeavor (Chisholm et al., 2001; de Baar et al., 2005).

12.4 QUANTIFYING THE BIOLOGICAL PUMP

There are many different ways to quantify the biological pump. Total levels of carbon fixation in surface waters may be estimated in bottle incubations from the uptake of $^{14}CO_2$ by phytoplankton (Steemann Nielsen, 1952) or from the deviation of oxygen isotopes from their terrestrial mass fractionation line (Luz and Barkan, 2000). At the other end of the biological pump, the rate of organic C accumulation or oxygen consumption in the sediments may be measured. And in between, the impact of the feeding and vertical migration activities of midwater organisms on the flux of particles may be investigated (e.g., Steinberg, 1995; Dilling et al., 1998). The next few pages will concentrate on the methods used to estimate export production and particle flux from the euphotic zone.

Particle flux is frequently extrapolated from the measurement of new production in surface waters (Dugdale and Goering, 1967). Export of POM out of surface waters or into the deep sea may also be estimated directly through its collection in sediment traps (e.g., Martin et al., 1987; Antia et al., 2001). Export or sedimentation of POM may also be estimated from disequilibria between two nuclides (such as ^{238}U and ^{234}Th, and ^{230}Th and ^{231}Pa) that are scavenged by particles of different degrees (Buesseler et al., 1992; Kumar et al., 1993).

Some methods for quantifying the biological pump focus on the relationship between the biological pump and CO_2. POC flux measurements or estimates of nutrient removal from surface waters may be used in conjunction with various ocean models to estimate the impact of the biological pump on atmospheric concentrations of CO_2 (Sarmiento and Toggweiler, 1984; Sarmiento and Orr, 1991). Others have focused on the importance of the ratio of POC to $CaCO_3$ to the sequestering of CO_2 in the ocean (Antia et al., 2001; Buitenhuis et al., 2001).

12.4.1 Measurement of New Production

The method in most widespread use for quantifying the biological pump hinges on the ideas that the surface ocean is at steady state on annual timescales with respect to the nitrogen budget and that nitrogen predominantly limits phytoplankton growth in the ocean. In such a system, the amount of productivity that is exported from the euphotic zone must be equal to the amount of productivity that is fueled by the input of allocthonous or "new" nitrogen to the euphotic zone (Dugdale and Goering, 1967). New production by this definition is that which is supported from dissolved nitrogen upwelled into the euphotic zone or fixed from N_2 into PON, and export production is taken to be equal to new production. For the sake of measurement, the above definition of new production has been simplified even further. Experimentally, new production is taken to be equal to the production that uses nitrate (and nitrite) as its nitrogen source, as opposed to ammonium or any of the organic forms of nitrogen.

It should be pointed out that measurement of new production is the measurement of the maximum amount of productivity that can be exported without running the system down with respect to the annual supply of nutrients to the euphotic zone. It may not be always appropriate to assume that new production and export production are equal; the fluxes will be equal only in systems that are not evolving. The validity of the assumption that new production equals export production depends on to what degree and over what timescale the nitrogen cycle in surface waters is at steady state.

At the moment the nitrogen cycle in the ocean is not at steady state on the decadal timescale. In the last few decades the N/P ratio of many oceanic waters has changed (Pahlow and Riebesell, 2000; Emerson *et al.*, 2001) and nearshore waters have shifted toward Si limitation from N limitation (Conley *et al.*, 1993) due to anthropogenic inputs of DIN to these systems and an increase in the extent of nitrogen fixation. On very long timescales, the N cycle is not at steady state either, but responds to changes in the oceanic inventory of P, which ultimately governs the rate of nitrogen fixation (Tyrrell, 1999).

Another assumption open to question is that the rate of nitrate uptake adequately represents the rate of uptake of new forms of nitrogen. This simplification has come about for two reasons. The first is that the uptake rate of nitrate by phytoplankton can be measured with reasonable ease by tracking the uptake of ^{15}N-labelled nitrate (Dugdale and Goering, 1967). The second is that nitrate is not produced in the euphotic zone to any significant degree and so its presence there can only be as a result of upwelling or atmospheric deposition.

The form of DIN released during the death, decay, and grazing of phytoplankton is ammonium which is also the most easily utilizable form of DIN to phytoplankton. Oxidized forms of DIN, such as nitrate and nitrite must be reduced to ammonium by nitrifying bacteria such as *Nitrosomonas*, *Nitrobacter*, *Nitrospira*, and *Proteobacteria* (Zehr and Ward, 2002) prior to assimilation by phytoplankton.

The classical view is that nitrification does not occur in the euphotic zone due to the inhibition of nitrifying bacteria by light (Zehr and Ward, 2002). Eukaryotic phytoplankton are also thought to outcompete nitrifying bacteria for the supplies of ammonium in surface waters. Thus, nitrate found in the euphotic zone must have had its origins outside of the euphotic zone, in deeper waters, in rivers or agricultural runoff, or from atmospheric deposition and is taken as the sole representative of new nitrogen in the euphotic zone (Dugdale and Goering, 1967).

Of course, nitrate is not likely to be the only form of allocthonous nitrogen in the euphotic zone. Concentrations of ammonium in upwelled water are not zero, although they are significantly lower than those of nitrate. DON may also serve as a significant source of new nitrogen to the euphotic zone.

There is one last word of caution concerning the use of nitrate uptake to estimate the transport of CO_2 to depth via the biological pump. Measurement of new production divulges no information concerning the depth of decomposition of the POM formed or the ratio of POC to $CaCO_3$ of the exported particles (Antia *et al.*, 2001). For instance, CO_2 from material decomposed beneath the euphotic zone, but above the maximum depth of winter mixing will be ventilated straight back out of the atmosphere. $CaCO_3$ formation, a feature that is not common to all phytoplankton, diminishes the efficiency of CO_2 drawdown with primary production (see below; Buitenhuis *et al.*, 2001).

12.4.2 Measurement of Particle Flux

Means more direct than the measurement of new production exist for the estimation of particle fluxes into the deep. Moored or free-floating traps may be used to collect sinking particles (e.g., Martin *et al.*, 1987). Alternatively, particle flux may be estimated from particle-reactive nuclides (e.g., Buesseler *et al.*, 1992). Particle flux may also be estimated from the consumption of oxygen (associated with the decomposition of sinking POM) in waters below the surface layer (e.g., Jenkins, 1982).

12.4.2.1 *Sediment traps*

The collection of particles in sediment traps, while perhaps the most direct way of measuring the sinking flux of POM, is a method not free from controversy. Sediment traps both over- and under-collect particles. Comparison of ^{234}Th accumulating in a suite of sediment traps with ^{234}Th fluxes expected from U–Th disequilibria in the upper 300 m of the ocean suggests that the particle collection efficiency of these traps ranged from 10% to 1,000% (Buesseler, 1991). Traps deployed in the deep ocean also show a considerable variability in trapping efficiencies (e.g., Scholten et al., 2001). Despite the magnitude of these biases, there is no generally applied method for correcting fluxes using particle-reactive isotopes (Antia et al., 2001).

Zooplankton actively swimming into sediment traps also serve as a source of error in flux measurements. It is impossible to differentiate these "swimmers" from zooplankton that have settled passively into the cup as part of the sinking POM flux. Swimmers may constitute as much as a quarter of the POC collected by the trap (Steinberg et al., 1998) and are generally removed from trap material prior to analysis. This introduces minimal error into the trap estimates of POC flux, as detrital zooplankton likely only comprise ~2% of the total organic matter sinking into the traps (Steinberg et al., 1998).

12.4.2.2 Particle-reactive nuclides

Radionuclides in the uranium decay series serve as useful tracers of particle flux. One type of these tracers consists of a soluble parent nuclide and a particle-reactive daughter. These soluble–particle-reactive pairs include ^{238}U–^{234}Th, ^{234}U–^{230}Th, and ^{235}U–^{231}Pa. The half-life of the parent exceeds the mixing time of the ocean and its distribution throughout the ocean is uniform. Once the soluble parent isotope decays to the particle-reactive daughter, the daughter is scavenged onto particulate material.

In systems with no particle scavenging, the activities of the parent and daughter nuclide will be in secular equilibrium. What is seen instead is that the activities of ^{234}Th, ^{230}Th, and ^{231}Pa are lower in surface waters than those of their parents (Figure 10). The difference in the activities of parent and daughter is a measure of the uptake of the daughter onto particles (Buesseler et al., 1992). With the help of a model of particle scavenging, fluxes of the particle-reactive daughter may be estimated from its vertical distribution

(e.g., Coale and Bruland, 1985; Buesseler et al., 1992). If the ratio of the particle-reactive nuclide to POC or PON is known, then the calculated flux of nuclide can be converted into an estimate of particle flux (Buesseler et al., 1992).

Relative estimates of particle flux may also be made from the ratio of two particle-reactive nuclides, such as ^{230}Th and ^{231}Pa, which are scavenged onto particles to different degrees. The half-lives of these isotopes are much larger than their residences times in the ocean (on the order of 10^4 years versus tens to hundreds of years) and thus there is no significant radioactive decay that occurs in the water column. The extent of the scavenging of ^{231}Pa, which is not as particle-reactive as ^{230}Th, is highest in areas of high particle flux (Anderson et al., 1990). Thus sediments in high flux areas exhibit $^{231}Pa/^{230}Th$ ratios in excess of the initial production ratio of 0.093, and sediments accumulating slowly exhibit ratios <0.093 (Anderson et al., 1990). $^{231}Pa/^{230}Th$ ratios have been used to infer changes in productivity and sediment accumulation rates between the present-day interglacial and the LGM, roughly 20,000 years ago (Kumar et al., 1993).

12.4.2.3 Oxygen utilization rates

The distribution of oxygen in ocean waters contains information about primary production. The amount of excess oxygen present in the seasonal thermocline in the Pacific, for example, was long ago used to suggest that ^{14}C-based estimates of primary production were severely underestimating levels of primary production in the ocean (Shulenberger and Reid, 1981). Oxygen deficiencies in deeper waters have been used to estimate levels of export production (Jenkins, 1982).

The supplies of oxygen to waters below the euphotic zone are primarily physical: advection and mixing. The removal of oxygen from these waters takes place through the oxidation of organic matter:

$$(CH_2O)_{106}(NH_3)_{16}H_3PO_4 + 138O_2$$
$$\rightarrow 106CO_2 + 16NO_3^- + HPO_4^{2-} + 122H_2O + 18H^+$$

$$(7)$$

By measuring rates of ventilation and the degree of oxygen undersaturation in deeper waters an estimate of the rates of oxygen utilization (OUR) may be made and integrated to yield the total amount of oxygen consumed beneath the euphotic zone each year (Jenkins, 1982). From this number a flux of POC and PON may be calculated. Estimates of export production based on OURs are reasonably in line with the amount of new production that could be supported from measured fluxes of NO_3^- into surface waters (Jenkins, 1988).

One advantage of using the OUR method for estimating export fluxes is that it integrates over

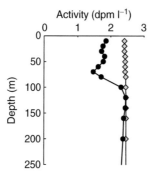

Figure 10 Profiles of ^{234}Th (black circles) and ^{238}U (gray diamonds) in the upper ocean. Redrawn from Buesseler (1991).

larger spatial and temporal scales than estimates based on sediment traps and nuclide fluxes. Also, unlike the estimates of new production based on NO_3^- uptake, OURs are directly coupled to the recycling of CO_2 via particle decomposition and are thus a more direct measure of the impact of the biological pump on atmospheric CO_2. In practice, however, the measurement of NO_3^- uptake is less technically challenging and is carried out much more frequently than are estimates from nuclides and OURs.

12.5 THE EFFICIENCY OF THE BIOLOGICAL PUMP

12.5.1 Altering the Efficiency of the Biological Pump

There is much talk concerning the "efficiency" of the biological pump. Is it pumping as much carbon to the deep sea as it could be? The general consensus is that it is not operating at its full capacity, and this is generally meant to imply that globally, the nitrate flux into the euphotic zone is not fully consumed in support of marine primary production. Broecker (1982) has estimated, for example, that if all of the nutrients supplied to present-day ocean surface waters were consumed by phytoplankton in support of primary production the atmospheric CO_2 content would drop by ∼130 ppmv.

12.5.1.1 In HNLC areas

The biological pump in HNLC areas is not operating at full efficiency on at least 2 counts. Phytoplankton growth is curtailed by the lack of availability of trace metals and so the major nutrients N and P are not completely utilized and carbon fixation does not occur to its maximum possible extent (Martin and Fitzwater, 1988). Fe limitation also heavily impacts the larger phytoplankton, like diatoms, that are important to particle flux. When a phytoplankton community is released from Fe limitation, diatom growth is stimulated more strongly than is the growth of the other phytoplankton (Cavender-Bares *et al.*, 1999; Lam *et al.*, 2001).

Given all of this, the addition of Fe to HNLC waters should result in higher levels of carbon fixation, increased growth of diatoms, local drawdown of CO_2, and enhanced export of carbon to the deep sea. Fe-addition experiments in bottles and *in situ* on the mesoscale unequivocally support the first two points. Addition of Fe to HNLC waters results in phytoplankton blooms (Martin and Fitzwater, 1988; Martin *et al.*, 1994; Coale *et al.*, 1996; Boyd *et al.*, 2000; de Baar *et al.*, 2005). Chlorophyll concentrations may quadruple and carbon-fixation rates may triple (e.g., Coale *et al.*, 1996) over the first few days after Fe addition (Figure 11).

In contrast, support for the drawdown of CO_2 following Fe addition has been mixed. In IronEx I, the first mesoscale Fe experiment, Fe addition did not result in a marked drop of CO_2 concentrations in surface waters of the Equatorial Pacific (Watson *et al.*, 1994). Some CO_2 was drawn down during IronEx II, also in the Equatorial Pacific, but the amount removed from surface waters was not enough to prevent these recently upwelled, high CO_2 waters from outgassing CO_2 to the atmosphere (Coale *et al.*, 1996). Large-scale Fe-addition experiments in the Southern Ocean have produced a significant lowering (25–30 µatm) of the CO_2 content of surface waters (Boyd *et al.*, 2000;

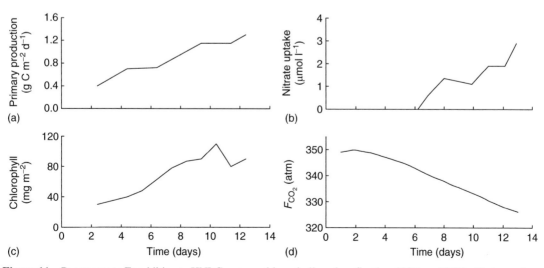

Figure 11 Responses to Fe addition to HNLC waters: chlorophyll, carbon fixation, NO_3^-, and PCO_2. Redrawn from Boyd *et al.* (2000) and Watson *et al.* (2000).

Watson *et al.*, 2000; Bozec *et al.*, 2005; Hiscock and Millero, 2005).

The one key critical point that the Fe-enrichment experiments have largely failed to show is an increase in the export of carbon to the deep sea (de Baar *et al.*, 2005). Even the Southern Ocean experiments, where a CO_2 drawdown occurred, did not result in substantial increases in export flux. Sediment traps set out at a depth of 100 m (below the fertilized patch) in the SOIREE experiment collected a similar amount of POC to traps stationed outside of the patch (Boyd *et al.*, 2000). Likewise particle flux in the SOFeX-South experiment was similar to that seen normally in the Southern Ocean (de Baar *et al.*, 2005).

There are many possible reasons for the lack of an observed increase in export flux in the meso-scale Fe-addition experiments. The additional carbon fixed might not become export flux, but instead flow through the microbial loop (microzooplankton and bacteria), which would result in the regeneration of CO_2 and nutrients in the euphotic zone. Microzooplankton biomass and bacterial activities were seen to increase in many of the mesoscale Fe-addition experiments (e.g., Boyd *et al.*, 2000), suggesting that some fraction of the carbon fixed through Fe addition was being shunted into the microbial loop instead of being pumped into deeper waters. Alternatively, the sediment traps used may simply have not been deployed for long enough or in the right location to catch the carbon sinking out as a result of the Fe-induced bloom.

12.5.1.2   Through changes in community composition

Changes in community composition should profoundly affect the efficiency of the biological pump. Shifting productivity from small cells, like photosynthetic picoplankton, which do not efficiently sink, to large cells, such as diatoms, capable of aggregating into particles that sink at hundreds of meters a day will result in the pumping of more carbon to the deep. However, there is finite amount of productivity that may be exported without running the nutrients in the system down to zero. In a steady-state system, shifting the community structure cannot increase the total amount of export production above the level of new production. But a shift in community structure may export carbon deeper into the water column before it decomposes, thus delaying the return of that carbon to the surface ocean by a significant number of years.

A shift in community composition may also be important to the biological pump if the shift is from a calcareous to a noncalcareous phytoplankton, as precipitation of $CaCO_3$ diminishes the ocean's ability to hold dissolved CO_2. DIC ($\sum CO_2$) is present in seawater as several species,

dissolved CO_2, carbonic acid, and the dissociated forms, bicarbonate ion and carbonate ion:

$$\sum CO_2 \rightleftharpoons H_2CO_3^* + HCO_3^- + CO_3^{2-} \quad (8)$$

where $H_2CO_3^*$ represents the sum of dissolved CO_2 and carbonic acid. The ratio of the three species varies with pH, with the undissociated forms dominating only at low pH and HCO_3^- favored at the typical seawater pH of 8.3 (Zeebe and Wolf-Gladrow, 2001). The saturation concentration of $H_2CO_3^*$ is controlled in large part by the amount of HCO_3^- and CO_3^{2-} present in solution. Removal of HCO_3^- or CO_3^{2-} during $CaCO_3$ formation results in a lowering of the saturation concentration of dissolved CO_2, and therefore the outgassing of CO_2 from solution to the atmosphere. Precipitation of $CaCO_3$ both produces CO_2:

$$Ca^{2+} + 2HCO_3^- \rightarrow CaCO_3 + H_2O + CO_2 \quad (9)$$

and lowers the alkalinity (approximately equal to $2[CO_3^{2-}] + [HCO_3^-]$) of surface waters (Zeebe and Wolf-Gladrow, 2001). Loss of alkalinity decreases the capacity of surface waters to play host to dissolved CO_2. In the surface ocean, under an atmosphere with a CO_2 partial pressure of 350 µatm (19 µatm less than the 2001 values; Keeling and Whorf, 2001) the precipitation of 1 mol of $CaCO_3$ results in the release of 0.6 mol of CO_2 (Ware *et al.*, 1992; Frankignoulle *et al.*, 1994).

The coccolithophorid, *Emiliania huxleyi*, contains on average 0.433 mol of $CaCO_3$ (present as scales covering the cell) for every mole of organic C (Buitenhuis *et al.*, 2001). Thus for every mole of CO_2 fixed into organic matter by coccolithophorids such as *E. huxleyi*, ~0.26 mol of CO_2 will be released due to the formation of $CaCO_3$. A shift from export production being carried out by coccolithophorids to a noncalcareous phytoplankton such as diatoms will result in an increased drawdown of CO_2 for the same amount of primary production. The fact that carbon fixation by phytoplankton that calcify is only 75% as effective at removing CO_2 as carbon fixation done by phytoplankton that do not precipitate $CaCO_3$ complicates estimates of CO_2 drawdown from the surface ocean POC export flux (Antia *et al.*, 2001).

12.5.1.3   By varying the C:N:P ratios of sinking material

The C:N:P ratios of sinking POM are not fixed; P and N are preferentially regenerated from sinking POM, which allows for the sequestering of more C in the deep ocean per unit N or P than if Redfield ratios remain unaltered during the decomposition of organic matter. It was remineralization ratios calculated from DIC and nutrient concentrations along isopycnals that initially suggested that the C/N of sinking particles might be higher than the value of 6.6 proposed by Redfield (Takahashi *et al.*, 1985). Sediment trap material also showed that the C/N and C/P ratios of sinking particles increases with depth, from near Redfield values of 6.6 and 106 at the base of the euphotic zone to ~180 and 11 by 5,000 m (Martin *et al.*, 1987).

Since these initial observations, evidence has mounted for the preferential remineralization of N and P out of POM relative to C. Ocean models with continuous vertical resolution support the preferential release of N and P from sinking POM in line with the estimates of Martin *et al.* (Shaffer, 1996). Bacteria have been shown to more rapidly degrade PON than POC (Verity *et al.*, 2000). Increases in the C/N and C/P of sinking particles have also been observed at station ALOHA near Hawaii (Christian *et al.*, 1997). And differences in the C/N ratio of suspended POC (6.4) and less buoyant POC trapped at the pycnocline (8.6) in the Kattegat, just east of the North Sea, further suggest that N is remineralized more rapidly than C at significant levels even in the euphotic zone (Olesen and Lundsgaard, 1995).

12.5.1.4 *By enhancing particle transport*

While sediment trap evidence supports the idea that the flux of organic carbon (e.g., Figure 4), the correlation between POC flux and overlying levels of primary production ($r^2 = 0.53$; Antia *et al.*, 2001) is not strong. The ratio of POC flux (at 125 m) to primary production also shows variability from region to region, ranging from 0.08 to 0.38 in the Atlantic Ocean alone (Antia *et al.*, 2001).

There are many reasons that an increased downward transport of POC out of the euphotic zone does not directly follow from an increase in productivity. Export fluxes are also controlled by the packaging of smaller more numerous particles into larger less numerous aggregates and by the rate at which the resulting aggregates sink. The depth to which POC exported from the euphotic zone is transported must be determined by the balance between the carbon content of the particle, the rate of microbial hydrolysis, and the sinking rate of the particle. Carbon fixed by marine diatoms, which are large and adept at aggregating, stands a better chance at being exported to the deep than carbon fixed by very small, nonsinking phytoplankton.

The mode of productivity is also crucial to the sinking flux of POC. Systems exporting POC in pulses (e.g., following the periodic formation of phytoplankton blooms in temperate and high-latitude areas) export a much higher proportion of their total production than do systems, such as the oligotrophic central gyres, where carbon fixation is a more static, steady process (Lampitt and Antia, 1997). This results partly from the influence of particle concentration on aggregation and partly from zooplankton population dynamics. In temperate areas, zooplankton reproduction cannot begin until after the onset of the spring phytoplankton bloom, leading to a lag in the increase in zooplankton population (and, subsequently, in the grazing pressure exerted by zooplankton) behind that of the increase in phytoplankton numbers (Heinrich, 1962) allowing for the aggregation and sinking of intact phytoplankton from the euphotic zone. In lower latitude, oligotrophic areas, standing stocks of phytoplankton and zooplankton are not as decoupled in time and there is no clear window within which a high density of phytoplankton cells may aggregate, escape predation, and sink to the deep.

12.6 THE BIOLOGICAL PUMP IN THE IMMEDIATE FUTURE

What may we expect out of the biological pump in the future? We cannot currently predict how ocean biology will respond to climate warming (Sarmiento *et al.*, 1998), but there are other questions that we may begin to ask. Will, for instance, the biological pump respond to the rise in surface ocean concentrations of DIC tied to the rise in atmospheric CO_2 fueled by anthropogenic emissions? What impact will inputs of agricultural fertilizers have on the biological pump? And what role will artificial stimulation of the biological pump through deliberate ocean fertilization play in the sequestration of excess CO_2?

12.6.1 Response to Increased CO_2

Will the biological pump respond to the increase CO_2? Increasing concentrations of CO_2 do not appear to have the significant impact on the C/N or C/P ratios of phytoplankton (Burkhardt *et al.*, 1999) that is necessary if CO_2 is to stimulate productivity in the ocean. It may be possible, however, for CO_2 levels to increase the proportion of carbon diverted into the biological pump at the expense of grazing and microbial food webs. Riebesell *et al.* (1993) suggested that carbon fixation by phytoplankton may be rate limited by the diffusive flux of CO_2 to the cell which delivers CO_2 too slowly relative to nitrate and phosphate to take up C, N, and P in Redfield proportions.

The diffusive flux of CO_2 to the surface of a phytoplankton cell (F_{CO_2}) is controlled not only by the diffusivity of CO_2 (D_{CO_2}) and the radius of the cell (r), but the concentration gradient of CO_2 between the aqueous medium and the interior of the cell ($C_e - C_i$) and the rate constant for the conversion of HCO_3^- into CO_2 (k') as the carbonate system equilibrates following CO_2 removal:

$$F_{CO_2} = -4\pi r D_{CO_2} \left(1 + \frac{r}{\sqrt{D_{CO_2}/k'}}\right)(C_e - C_i)$$

(10)

If the diffusive flux of CO_2 is controlling the delivery of CO_2 to the carbon-fixing enzyme, Rubisco, then an increase in the dissolved CO_2 content of seawater should stimulate phytoplankton growth rates. In particular, an increase in dissolved CO_2 concentrations should stimulate growth rates in the larger (i.e., aggregate forming) size classes and thus should also result in a shunting of carbon out of the mouths of zooplankton and into the biological pump (Riebesell *et al.*, 1993).

At the present time, it is not clear that the acquisition of carbon is the rate-limiting step for photosynthesis even in the larger phytoplankton cells. Phytoplankton are not limited to the passive diffusion of CO_2 to the cell surface for carbon acquisition. HCO_3^- may be taken up directly by phytoplankton and used as a source of CO_2 for photosynthesis (Korb *et al.*, 1997; Nimer *et al.*, 1997; Raven, 1997; Tortell *et al.*, 1997). The considerably greater abundance of HCO_3^- than dissolved CO_2 in seawater would imply that C-fixation rates of marine phytoplankton are not CO_2 limited and increasing concentrations of CO_2 will not have an impact on rates of primary production.

12.6.2   Response to Agricultural Runoff

Inputs of agricultural fertilizers are having more than one impact on the biological pump. A shift in the nitrate to phosphate to silicic acid ratio of natural waters is causing a shift in the phytoplankton community structure that should impact the biological cycling of carbon in aquatic systems (Conley *et al.*, 1993). Additionally, a recent shift in the C:N:P ratios of deeper waters and an increase in export production have been observed for the northern hemisphere oceans (Pahlow and Riebesell, 2000).

12.6.2.1   Shift toward Si limitation

Over the last 150 years, changes in land use patterns, human population density, and extent of the use of fertilizers have resulted in an increased flux of nitrate and phosphate to rivers, lakes, and the coastal ocean (Conley *et al.*, 1993) and even to the open ocean through atmospheric deposition, for example of anthropogenic nitrous oxides (Pahlow and Riebesell, 2000). At the same time, large freshwater systems, such as the North American Great Lakes, and coastal areas, such as the Adriatic and Baltic Seas and the Mississippi River plume, have been shifting away from N or P limitation and toward Si limitation (Conley

et al., 1993; Nelson and Dortch, 1996). The extra productivity fueled on the extra N and P has resulted in the removal of Si from these systems and a shift in the phytoplankton community structure away from diatoms. Given that rivers are significant sources of new nitrate, phosphate, and silicic acid to the global ocean, this effect is expected to spread further throughout the sea.

A shift in community structure associated with silicic acid depletion may sharply reduce the amount of carbon delivered to deep waters and sediments via the biological pump. Diatoms, which are the only major phytoplankton group requiring silicic acid, are relatively large and notable for aggregating and sinking. Diatoms feed a greater portion of the organic matter they produce into the biological pump than do most other classes of phytoplankton. Currently, diatoms perform more than 75% of the primary production that occurs in high nutrient and coastal regions of the ocean (Nelson *et al.*, 1995), the exact areas that will be impacted by this input of additional nitrate and phosphate and the exact areas where carbon tends to make it down into the sediments.

One possible further impact of the decline of the diatoms lies in the identity of their probably successor. If a $CaCO_3$-producing phytoplankter, such as coccolithophorids, step in to utilize the nitrate and phosphate the Si-starved diatoms leave behind, the CO_2 pumping efficiency of the biological pump will decline (Robertson *et al.*, 1994) even if the same amount of organic carbon continues to be removed to the deep sea and sediments each year.

12.6.2.2   Shifts in export production and deep ocean C:N:P

There is evidence for anthropogenic perturbations increasing the biological pumping of carbon into deep waters. In both the North Pacific and North Atlantic, for instance, deepwater N/P ratios have increased over the last 50 years (Pahlow and Riebesell, 2000) possibly due to atmospheric deposition of anthropogenic N and a subsequent shift toward P limitation in these areas. Concomitant with the rise in N/P is an increase in the apparent oxygen utilization (AOU) in both the North Atlantic and North Pacific, which suggests that levels of export production have also increased. This is estimated to have resulted in the increased oceanic sequestration of $0.2\,PgC\,yr^{-1}$ (Pahlow and Riebesell, 2000).

12.6.3   Carbon Sequestration via Ocean Fertilization and the Biological Pump

There have been many calls to sequester anthropogenic CO_2 in the deep ocean by stimulating primary production through the addition of literally tons of Fe to HNLC surface waters. Patents have been taken out on the idea (e.g., Markels, 2001) and, counter to the recommendations of the

American Society of Limnology and Oceanography (Ocean Fertilization Summary Report, 2001) several companies have been established to dispense carbon credits to industries willing to pay for ocean fertilization (Chisholm *et al.*, 2001). However, it remains unclear whether ocean fertilization will ever successfully transport carbon in the deep sea, how long the transported carbon might remain out of contact with the atmosphere, and what side effect large-scale fertilization will have on marine geochemistry and ecology (Fuhrman and Capone, 1991; Peng and Broecker, 1991a, b; Sarmiento and Orr, 1991; Chisholm *et al.*, 2001; Lenes *et al.*, 2001).

A perusal of the literature suggests that carbon sequestration by Fe fertilization is not a panacea for the anthropogenic carbon emissions that increased atmospheric CO_2 from 280 µatm at the start of the industrial revolution to 369 µatm by December 2000 (Keeling and Whorf, 2001). Ocean models suggest that enhanced ocean uptake of carbon with Fe fertilization of the Antarctic Ocean will at best draw down atmospheric CO_2 by 70 µatm if carried out continuously for a century (Peng and Broecker, 1991a, b) and damp the annual anthropogenic CO_2 input to the atmosphere by <30% of current annual emission levels (Joos *et al.*, 1991).

Large-scale Fe fertilization will have side effects. If Fe fertilization resulted in a complete drawdown of the nutrients available in the Southern Ocean, for example, the average O_2 content of deep waters would drop by 4–12% (Sarmiento and Orr, 1991), with areas of anoxia cropping up in the Antarctic (Peng and Broecker, 1991b) and Indian Oceans (Sarmiento and Orr, 1991). Even small-scale patches of anoxia would have a profound negative impact on the survival and distribution of metazoan fauna in the ocean and alter the balance of microbial transformations of nitrogen between reduced and oxidized phases (Fuhrman and Capone, 1991).

REFERENCES

Alldredge A. L. and Gotschalk C. C. (1989) Direct observations of the mass flocculation of diatom blooms: characteristics, settling velocities and formation of diatom aggregates. *Deep-Sea Res.* **36**, 159–171.

Alldredge A. L. and Jackson G. A. (1995) Aggregation in marine systems. *Deep-Sea Res. II* **42**, 1–7.

Alldredge A. L. and McGillivary P. (1991) The attachment probabilities of marine snow and their implications for particle coagulation in the ocean. *Deep-Sea Res.* **38**, 431–443.

Alldredge A. L., Passow U., and Logan B. E. (1993) The abundance and significance of a class of large, transparent organic particles in the ocean. *Deep-Sea Res.* **40**, 1131–1140.

Alldredge A. L. and Silver M. W. (1988) Characteristics, dynamics, and significance of marine snow. *Prog. Oceanogr.* **20**, 41–82.

Aluwihare L. I. and Repeta D. J. (1999) A comparison of the chemical characteristics of oceanic DOM and extracellular DOM produced by marine algae. *Mar. Ecol. Prog. Ser.* **186**, 105–117.

Amon R. M. W. and Benner R. (1994) Rapid cycling of high-molecular weight dissolved organic matter in the ocean. *Nature* **369**, 549–552.

Amon R. M. W., Fitznar H.-P., and Benner R. (2001) Linkages among the bioreactivity, chemical composition, and diagenetic state of marine dissolved organic matter. *Limnol. Oceanogr.* **46**, 287–297.

Anderson R. F., Lao Y., Broecker W. S., Trumbore S. E., Hofman H. J., and Wolfli W. (1990) Boundary scavenging in the Pacific Ocean: a comparison of ^{10}Be and 231 Pa. *Earth Planet. Sci. Lett.* **96**, 287–304.

Andersson J. H., Wijsman J. W. M., Herman P. J., Middelburg J. J., Soetaert K., and Heip K. (2004) Respiration patterns in the deep sea. *Geophys. Res. Lett.* **31**, doi: 10.1029/2003GL018756.

Antia A. N., Koeve W., Fischer G., Blanz T., Schulz-Bull D., Scholten J., Neuer S., Kremling K., Kuss J., Peinert R., Hebbeln D., Bathmann U., Conte M., Fehner U., and Zeitzschel B. (2001) Basin-wide particulate carbon flux in the Atlantic Ocean: regional export patterns and potential for atmospheric CO_2 sequestration. *Global Biogeochem. Cycles* **15**, 845–862.

Armstrong R. A., Lee C., Hedges J. I., Honjo S., and Wakeham S. G. (2002) A new, mechanistic model for organic carbon fluxes in the ocean based on the quantitative association of POC with ballast minerals. *Deep-Sea Res. Part II* **49**, 219–236.

Arnosti C., Repeta D. J., and Blough N. V. (1994) Rapid bacterial degradation of polysaccharides in anoxic marine systems. *Geochim. Cosmochim. Acta* **58**, 2639–2652.

Asper V. L., Deuser W. G., Knauer G. A., and Lohrenz S. E. (1992) Rapid coupling of sinking particle fluxes between surface and deep ocean waters. *Nature* **357**, 670–672.

Azam F. and Volcani B. E. (1981) Germanium–silicon interactions in biological systems. In *Silicon and Siliceous Structures in Biological Systems Silicon and Siliceous Structures in Biological Systems* (eds. T. L. Simpson and B. E. Volcani). Springer, Berlin, pp. 69–93.

Azetsu-Scott K. and Niven S. E. H. (2005) The role of transparent exopolymer particles (TEP) in the transport of Th-234 in coastal water during a spring bloom. *Cont. Shelf. Res.* **25**, 1133–1141.

Banse K. (1990) New views on the degradation and disposition of organic particles as collected by sediment traps in the open sea. *Deep-Sea Res. A* **37**, 1177–1195.

Bauer J. E., Williams P. M., and Druffel E. R. M. (1992) ^{14}C activity of dissolved organic carbon fractions in the north-central Pacific and Sargasso Sea. *Nature* **357**, 667–670.

Benner R., Pakulski J. D., McCarthy M., Hedges J. I., and Hatcher P. G. (1992) Bulk chemical characteristics of dissolved organic matter in the ocean. *Science* **255**, 1561–1564.

Berner R. A., Ruttenberg K. C., Ingall E. D., and Rao J. L. (1993) The nature of phosphorus burial in modern marine sediments. In *Interactions of C, N, P and S Biochemical Cycles and Global Change* (eds. R. Wollast, F. T. Mackenzie, and L. Chou). Springer, Berlin, pp. 365–378.

Biddanda B. and Benner R. (1997) Carbon, nitrogen, and carbohydrate fluxes during the production of particulate and dissolved organic matter by marine phytoplankton. *Limnol. Oceanogr.* **42**, 506–518.

Bidle K. D. and Azam F. (1999) Accelerated dissolution of diatom silica by natural marine bacterial assemblages. *Nature* **397**, 508–512.

Bidle K. D. and Azam F. (2001) Bacterial control of silicon regeneration from diatom detritus: significance of bacterial ectohydrolases and species identity. *Limnol. Oceanogr.* **46**, 1606–1623.

Billett D. S. M., Lampitt R. S., Rice A. L., and Mantoura R. F. C. (1983) Seasonal sedimentation of phytoplankton to the deep sea benthos. *Nature* **302**, 520–522.

Bishop J. K. B. (1988) The barite-opal-organic carbon association in oceanic particulate matter. *Nature* **332**, 341–343.

Boyd P. and Newton P. (1995) Evidence of the potential influence of planktonic community structure on the interannual variability of particulate organic carbon flux. *Deep-Sea Res. I* **42**, 619–639.

Boyd P. W., Waston A. J., Law C. S., Abraham E. R., Trull T., Murdoch R., Bakker D. C. E., Bowie A. R., Buesseler K. O., Chang H., Charette M., Croot P., Downing K., Frew R., Gall M., Hadfield M., Hall J., Harvey M., Jameson G., LaRoche J., Liddicoat M., Ling R., Maldonado M. T., McKay R. M., Nodder S., Pickmere S., Pridmore R., Rintoul S., Safi K., Sutton P., Strzepek R., Tanneberger K., Turner S., Waite A., and Zeldis J. (2000) A mesoscale phytoplankton bloom in the polar Southern Ocean stimulated by iron fertilization. *Nature* **407**, 695–702.

Boyle E. A. (1988) Cadmium: chemical tracer of deepwater paleoceanography. *Paleoceanography* **3**, 471–489.

Bozec Y., Bakker D. C. E., Hartmann C., Thomas H., Bellerby R. G. J., Nightingale P. D., Riebesell U., Watson A. J., and de Baar H. J. W. (2005) The CO_2 system in a Redfield context during an iron enrichment experiment in the Southern Ocean. *Mar. Chem.* **95**, 89–105.

Broecker W. S. (1982) Ocean chemistry during glacial time. *Geochim. Cosmochim. Acta* **46**, 1689–1705.

Broecker W. S. and Peng T.-H. (1982) *Tracers in the Sea.* Eldigio Press, Palisades, New York.

Bruland K. W. (1980) Oceanographic distributions of cadmium, zinc, nickel, and copper in the North Pacific. *Earth Planet. Sci. Lett.* **47**, 176–198.

Brzezinski M. A. (1985) The Si:C:N ratio of marine diatoms: interspecific variability and the effect of some environmental variables. *J. Phycol.* **21**, 347–357.

Brzezinski M. A. and Nelson D. M. (1996) Chronic substrate limitation of silicic acid uptake in the western Sargasso Sea. *Deep-Sea Res. II* **43**, 437–453.

Buesseler K. O. (1991) Do upper-ocean sediment traps provide an accurate record of particle flux? *Nature* **353**, 420–423.

Buesseler K. O. (1998) The decoupling of production and particulate export in the surface ocean. *Global Biogeochem. Cycles* **12**, 297–310.

Buesseler K. O., Bacon M. P., Cochran J. K., and Livingston H. D. (1992) Carbon and nitrogen export during the JGOFS North Atlantic Bloom Experiment estimated from $^{234}Th:^{238}U$ disequilibria. *Deep-Sea Res.* **39**, 1115–1137.

Buitenhuis E. T., van der Wal P., and de Baar H. J. W. (2001) Blooms of *Emiliania huxleyi* are sinks of atmospheric carbon dioxide: a field and mesocosm study derived simulation. *Global Biogeochem. Cycles* **15**, 577–587.

Burkhardt S., Zondervan I., and Riebesell U. (1999) Effect of CO_2 concentration on C:N:P ratio in marine phytoplankton: a species comparison. *Limnol. Oceanogr.* **44**, 683–690.

Calbet A. (2001) Mesozooplankton grazing effect on primary production: a global comparative analysis in marine ecosystems. *Limnol. Oceanogr.* **46**, 1824–1830.

Capone D. G., Burns J. A., Montoya J. P., Subramaniam A., Mahaffey C., Gunderson T., Michaels A. F., Carpenter E. J. (2005) Nitrogen fixation by *Trichodesmium* spp: an important source of new nitrogen to the tropical and subtropical North Atlantic Ocean. *Global Biogeochem. Cycles* **19**, doi:10.1029/2004GB002331.

Capone D. G., Zehr J. P., Paerl H. W., Bergman B., and Carpenter E. J. (1997) Trichodesmium: a globally significant marine cyanobacterium. *Science* **276**, 1221–1229.

Cavender-Bares K. K., Mann E. L., Chisholm S. W., Ondrusek M. E., and Bidigare R. R. (1999) Differential response of equatorial Pacific phytoplankton to iron fertilization. *Limnol. Oceanogr.* **44**, 237–246.

Chin W. C., Orellana M. V., and Verdugo P. (1998) Spontaneous assembly of marine dissolved organic matter into polymer gels. *Nature* **391**, 568–572.

Chisholm S. W., Falkowski P. G., and Cullen J. J. (2001) Discrediting ocean fertilization. *Science* **294**, 309–310.

Chisholm S. W., Olson R. J., Zettler E. R., Goericke R., Waterbury J. B., and Welschmeyer N. A. (1988) A novel free-living prokaryote abundant in the oceanic euphotic zone. *Nature* **334**, 340–343.

Cho B. C. and Azam F. (1988) Major role of bacteria in biogeochemical fluxes in the ocean's interior. *Nature* **332**, 441–443.

Christian J. R., Lewis M. R., and Karl D. M. (1997) Vertical fluxes of carbon, nitrogen, and phosphorus in the North Pacific subtropical gyre near Hawaii. *J. Geophys. Res.* **102**, 15667–15677.

Clark L. L., Ingall E. D., and Benner R. (1998) Marine phosphorus is selectively remineralized. *Nature* **393**, 428.

Coale K. H. and Bruland K. W. (1985) $^{234}Th:^{238}U$ disequilibria within the California current. *Limnol. Oceanogr.* **30**, 22–33.

Coale K. H., Johnson K. S., Fitzwater S. E., Gordon R. M., Tanner S., Chavez F. P., Ferioli L., Sakamoto C., Rogers P. L., Millero F., Steinberg P. L., Nightingale P. D., Cooper D., Cochlan W. P., Landry M. R., Constantinou J., Rollwagen G., Trasvina A., and Kudela R. (1996) A massive phytoplankton bloom induced by an ecosystem scale iron fertilization experiment in the equatorial Pacific Ocean. *Nature* **383**, 495–501.

Collier R. and Edmond J. (1984) The trace element geochemistry of marine biogenic particulate matter. *Prog. Oceanogr.* **13**, 113–199.

Conley D. J., Schelske C. L., and Stoermer E. F. (1993) Modification of the biogeochemical cycle of silica with eutrophication. *Mar. Ecol. Prog. Ser.* **101**, 179–192.

Cullen J. T. and Sherrell R. M. (2005) Effects of dissolved carbon dioxide, zinc, and manganese on the cadmium to phosphorus ratio in natural phytoplankton assemblages. *Limnol. Oceanogr.* **50**, 1193–1204.

Dalsgaard T., Canfield D. E., Peterson J., Thamdrup B., and Acuña-González J. (2003) Anammox is a significant pathway of N_2 production in the anoxic water column of Golfo Dulce, Costa Rica. *Nature* **422**, 606–608.

de Baar H. J. W., Boyd P. W., Coale K. H., Landry M. R., Tsuda A., Assmy P., Bakker D. C. E., Bozec Y., Barber R. T., Brzezinski M. A., Buesseler K. O., Boye M., Croot P. L., Gervais F., Gorbunov M. Y., Harrison P. J., Hiscock W. T., Laan P., Lancelot C., Law C. S., Levasseur M., Marchetti A., Millero F. J., Nishioka J., Nojiri Y., van Oijen T., Riebesell U., Rijkenberg M. J. A., Saito H., Takeda S., Timmermans K. R., Veldhuis M. J. W., Waite A. W., and Wong C. S. (2005) Synthesis of iron fertilization experiments: from the iron age in the age of enlightenment. *J. Geophys. Res. Oceans* **110**, C09S16.

Dehairs F., Chesselet R., and Jedwab J. (1980) Discrete suspended particles of barite and the barium cycle in the open ocean. *Earth Planet. Sci. Lett.* **49**, 528–550.

de Haas H., van Weering T. C. E., and de Stigter H. (2002) Organic carbon in shelf seas: sinks or sources, processes and products. *Cont. Shelf Res.* **22**, 691–717.

De La Rocha C. L., Hutchins D. A., Brzezinski M. A., and Zhang Y. (2000) Effects of iron and zinc deficiency on elemental composition and silica production by diatoms. *Mar. Ecol. Prog. Ser.* **195**, 71–79.

DeMaster D. J. (2002) The accumulation and cycling of biogenic silica in the Southern Ocean: revisiting the marine silica budget. *Deep-Sea Res. II* **49**, 3155–3167.

deVilliers S. (1999) Seawater strontium and Sr/Ca variability in the Atlantic and Pacific oceans. *Earth Planet Sci. Lett.* **171**, 623–634.

Dilling L. and Alldredge A. L. (2000) Fragmentation of marine snow by swimming macrozooplankton: a new process impacting carbon cycling in the sea. *Deep-Sea Res. I* **47**, 1227–1245.

Dilling L., Wilson J., Steinberg D., and Alldredge A. (1998) Feeding by the euphausiid *Euphausia pacifica* and the copepod *Calanus pacificus* on marine snow. *Mar. Ecol. Prog. Ser.* **170**, 189–201.

Dugdale R. C. and Goering J. J. (1967) Uptake of new and regenerated forms of nitrogen in primary productivity. *Limnol. Oceanogr.* **12**, 196–206.

Dugdale R. C., Wilkerson F. P., and Minas H. J. (1995) The role of a silicate pump in driving new production. *Deep-Sea Res. I* **42**, 697–719.

Dymond J., Suess E., and Lyle M. (1992) Barium in deep-sea sediment: a geochemical proxy for paleoproductivity. *Paleoceanography* **7**, 163–181.

Elderfield H. and Rickaby R. E. M. (2000) Oceanic Cd/P ratio and nutrient utilization in the glacial Southern Ocean. *Nature* **405**, 305–310.

Elderfield H. E. and Schultz A. (1996) Mid-ocean ridge hydrothermal fluxes and the chemical composition of the ocean. *Annu. Rev. Earth Planet. Sci.* **24**, 191–224.

Ellwood M. J. and Hunter K. A. (2000) The incorporation of zinc and iron into the frustule of the marine diatom *Thalassiosira pseudonana*. *Limnol. Oceanogr.* **45**, 1517–1524.

Emerson S., Mecking S., and Abell J. (2001) The biological pump in the subtropical North Pacific Ocean: nutrient sources, Redfield ratios, and recent changes. *Global Biogeochem. Cycles* **15**, 535–554.

Engel A., Goldthwait S., Passow U., and Alldredge A. (2002) Temporal decoupling of carbon and nitrogen dynamics in a mesocosm diatom bloom. *Limnol. Oceanogr.* **47**, 753–761.

Engel A. and Passow U. (2001) Carbon and nitrogen content of transparent exopolymer particles (TEP) in relation to their Alcian Blue absorbtion. *Mar. Ecol. Prog. Ser.* **219**, 1–10.

Eppley R. W. and Peterson B. J. (1979) Particulate organic matter flux and planktonic new production in the deep ocean. *Nature* **282**, 677–680.

Fagerbakke K. M., Norland S., and Heldal M. (1999) The inorganic ion content of native aquatic bacteria. *Can. J. Microbiol.* **45**, 304–311.

Field C. B., Behrenfeld M. J., Randerson J. T., and Falkowski P. (1998) Primary production of the biosphere: integrating terrestrial and oceanic components. *Science* **281**, 237–240.

Filippelli G. M. and Delaney M. L. (1996) Phosphorus geochemistry of equatorial Pacific sediments. *Geochim. Cosmochim. Acta* **60**, 1479–1495.

Föllmi K. B. (1996) The phosphorus cycle, phosphogenesis and marine phosphate rich deposits. *Earth Sci. Rev.* **40**, 55–124.

Fowler S. W. and Knauer G. A. (1986) Role of large particles in the transport of elements and organic compounds through the oceanic water column. *Prog. Oceanogr.* **16**, 147–194.

Franck V. M., Smith G. J., Bruland K. W., and Brzezinski M. A. (2005) Comparison of size-dependent carbon, nitrate, and silicic acid uptake rates in high- and low-iron waters. *Limnol. Oceanogr.* **50**, 825–838.

François R., Honjo S., Krishfield R., and Manganini S. (2002) Factors controlling the flux of organic carbon to the bathypelagic zone of the ocean. *Global Biogeochem. Cycles* **16**, doi:10.1029/2001GB001722, 2002.

Frankignoulle M., Canon C., and Gattuso J.-P. (1994) Marine calcification as a source of carbon dioxide: positive feedback of increasing atmospheric CO_2. *Limnol. Oceanogr.* **39**, 458–462.

Froelich P. N., Bender M. L., Luedtke N. A., Heath G. R., and De Vries T. (1982) The marine phosphorus cycle. *Am. J. Sci.* **282**, 474–511.

Fuhrman J. A. and Capone D. G. (1991) Possible biogeochemical consequences of ocean fertilization. *Limnol. Oceanogr.* **36**, 1951–1959.

Ganeshram R. S., François R., Commeau J., and Brown-Leger S. L. (2003) An experimental investigation of barite formation in seawater. *Geochim. Cosmochim. Acta* **67**, 2599–2605.

Geider R. J. and LaRoche J. L. (1994) The role of iron in phytoplankton photosynthesis, and the potential for iron limitation of primary productivity in the sea. *Photosynth. Res.* **39**, 275–301.

Goldthwait S., Yen J., Brown J., and Alldredge A. (2004) Quantification of marine snow fragmentation by swimming euphausiids. *Limnol. Oceanogr.* **49**, 940–952.

Greenwood N. N. and Earnshaw A. (1984) *Chemistry of the Elements*. Pergamon, Oxford.

Gruber N. and Sarmiento J. L. (1997) Global patterns of marine nitrogen fixation and denitrification. *Global Biogeochem. Cycles* **11**, 235–266.

Hamm C. E. (2002) Interactive aggregation and sedimentation of diatoms and clay-sized lithogenic material. *Limnol. Oceanogr.* **47**, 1790–1795.

Harrison P. J., Conway H. L., Holmes R. W., and Davis C. O. (1977) Marine diatoms grown in chemostats under silicate or ammonium limitation. III. Cellular chemical composition and morphology of *Chaetoceros debilis*, *Skeletonema costatum*, and *Thalassiosira gravida*. *Mar. Biol.* **43**, 19–31.

Harrison W. G. (1990) Nitrogen utilization in chlorophyll and primary productivity maximum layers: an analysis based on the *f*-ratio. *Mar. Ecol. Prog. Ser.* **60**, 85–90.

Hedges J. I. and Keil R. G. (1995) Sedimentary organic matter preservation: an assessment and speculative synthesis. *Mar. Chem.* **49**, 81–115.

Heinrich A. K. (1962) The life histories of planktonic animals and seasonal cycles of plankton communities in the ocean. *J. Cons. Int. Explor. Mer.* **27**, 15–24.

Hiscock W. T. and Millero F. J. (2005) Nutrient and carbon parameters during the Southern Ocean iron experiment (SOFeX). *Deep-Sea Res. I* **52**, 2086–2108.

Holloway C. F. and Cowen J. P. (1997) Development of a scanning confocal laser microscopic technique to examine the structure and composition of marine snow. *Limnol. Oceanogr.* **42**, 1340–1352.

Hutchins D. A. and Bruland K. W. (1998) Iron-limited diatom growth and Si:N uptake ratios in a coastal upwelling regime. *Nature* **393**, 561–564.

Janse I., vanRijssel M., Gottschal J. C., Lancelot C., and Gieskes W. W. C. (1996) Carbohydrates in the North Sea during spring blooms of *Phaeocystis*: a specific fingerprint. *Aquat. Microb. Ecol.* **10**, 97–103.

Jeandel C., Dupré B., Lebaron G., Monnin C., and Minster J.-F. (1996) Longitudinal distributions of dissolved barium, silica and alkalinity in the western and southern Indian Ocean. *Deep-Sea Res. I* **43**, 1–31.

Jenkins W. J. (1982) Oxygen utilization rates in North Atlantic subtropical gyre and primary production in oligotrophic systems. *Nature* **300**, 246–248.

Jenkins W. J. (1988) Nitrate flux into the euphotic zone near Bermuda. *Nature* **331**, 521–523.

Joos F. L., Sarmiento J. L., and Sigenthaler U. (1991) Estimates of the effect of Southern Ocean iron fertilization on atmospheric CO_2 concentrations. *Nature* **349**, 772–775.

Kaehler P. and Koeve W. (2001) Marine dissolved organic matter: can its C:N ratio explain carbon overconsumption? *Deep-Sea Res. I* **48**, 49–62.

Kamitani A. (1982) Dissolution rates of silica from diatoms decomposing at various temperatures. *Mar. Biol.* **68**, 91–96.

Karl D. M., Letelier R., Tupas L., Dore J., Christian J., and Hebel D. (1997) The role of nitrogen fixation in biogeochemical cycling in the subtropical North Pacific Ocean. *Nature* **388**, 533–538.

Keeling C. D. and Whorf T. P. (2005) Atmospheric CO_2 records from sites in the SIO air sampling network. In *Trends: A Compendium of Data on Global Change*. Carbon Dioxide Information Analysis Center, Oak Ridge National Laboratory, US Department of Energy, Oak Ridge, TN, USA.

Kepkay P. E. (1994) Particle aggregation and the biological reactivity of colloids. *Mar. Ecol. Prog. Ser.* **109**, 293–304.

Klaas C. and Archer D. E. (2002) Association of sinking organic matter with various types of mineral ballast in the deep sea: implications for the rain ratio. *Global Biogeochem. Cycles* **16**, doi:10.1029/2001GB001765.

Knauer G. A. and Martin J. H. (1981) Phosphorus and cadmium cycling in northeast Pacific waters. *J. Mar. Res.* **39**, 65–76.

Kolowith L. C., Ingall E. D., and Benner R. (2001) Composition and cycling of marine organic phosphorus. *Limnol. Oceanogr.* **46**, 309–321.

Korb R. A., Saville P. J., Johnston A. M., and Raven J. A. (1997) Sources of inorganic carbon for photosynthesis by three species of marine diatom. *J. Phycol.* **33**, 433–440.

Kumar N., Gwiazda R., Anderson R. F., and Froelich P. N. (1993) ^{231}Pa/^{230}Th ratios in sediments as a proxy for past changes in Southern Ocean productivity. *Nature* **362**, 45–48.

Lam P. J., Tortell P. D., and Morel F. M. M. (2001) Differential effects of iron additions on organic and inorganic carbon production by phytoplankton. *Limnol. Oceanogr.* **46**, 1199–1202.

Lampitt R. and Antia A. N. (1997) Particle flux in deep seas: regional characteristics and temporal variability. *Deep-Sea Res. I* **44**, 1377–1403.

Lane T. W. and Morel F. M. M. (2000) A biological function for cadmium in marine diatoms. *Proc. Natl. Acad. Sci. USA* **97**, 4627–4631.

Laws E. A., Falkowski P. G., Smith W. O., Jr., Ducklow H., and McCarthy J. J. (2000) Temperature effects on export production in the open ocean. *Global Biogeochem. Cycle* **14**, 1231–1246.

Lea D. W. and Boyle E. (1989) Barium content of benthic foraminifera controlled by bottom water composition. *Nature* **338**, 751–753.

Leblanc K., Hare C. E., Boyd P. W., Bruland K. W., Sohst B., Pickmere S., Lohan M. C., Buck K., Ellwood M., and Hutchins D. A. (2005) Fe and Zn effects on the Si cycle and diatom community structure in two contrasting high and low-silicate HNLC areas. *Deep-Sea Res. I* **52**, 1842–1864.

Lenes J. M., Darrow B. P., Cattrall C., Heil C. A., Callahan M., Vargo G. A., Byrne R. H., Prospero J. M., Bates D. E., Fanning K. A., and Walsh J. J. (2001) Iron fertilization and the Trichodesmium response on the West Florida shelf. *Limnol. Oceanogr.* **46**, 1261–1277.

Lisitzin A. P. (1972) Sedimentation in the World Ocean. *Soc. Econ. Paleontiol. Mineral. Spec. Publ.* **17**, 218pp.

Liu H., Landry M. R., Vaulot D., and Campbell L. (1999) Prochlorococcus growth rates in the central equatorial Pacific: an application of the f_{max} approach. *J. Geophys. Res.* **104**, 3391–3399.

Logan B. E., Passow U., Alldredge A. L., Grossart H. P., and Simon M. (1995) Rapid formation and sedimentation of large aggregates is predictable from coagulation rates (half-lives) of transparent exopolymer particles (TEP). *Deep-Sea Res. II* **42**, 203–214.

Loh A. N., Bauer J. E., and Druffel E. R. M. (2004) Variable ageing and storage of dissolved organic components in the open ocean. *Nature* **430**, 877–881.

Long R. A. and Azam F. (1996) Abundant protein-containing particles in the sea. *Aquat. Microb. Ecol.* **10**, 213–221.

Löscher B. M., de Jong J. T. M., and de Baar H. J. W. (1998) The distribution and preferential biological uptake of cadmium at 6W in the Southern Ocean. *Mar. Chem.* **62**, 259–286.

Lutz M., Dunbar R., and Caldiera K. (2002) Regional variability in the vertical flux of particulate organic carbon in the ocean interior. *Global Biogeochem. Cycles.* **16**, doi:10.1029/2000GB001383.

Luz B. and Barkan E. (2000) Assessment of oceanic productivity with the triple isotope composition of dissolved oxygen. *Science* **288**, 2028–2031.

Mari X., Beauvais S., Lemee R., and Pedrotti M. L. (2001) Non-Redfield C:N ratio of transparent exopolymeric particles in the northwestern Mediterranean Sea. *Limnol. Oceanogr.* **46**, 1831–1836.

Markels M., Jr. (2001). Method of Sequestering Carbon Dioxide with Spiral Fertilization. US Patent No. 6,200,530.

Martin J. H. (1990) Glacial–interglacial CO_2 change: the iron hypothesis. *Paleoceanography* **5**, 1–13.

Martin J. H., Coale K. H., Johnson K. S., Fitzwater S. E., Gordon R. M., Tanner S. J., Hunter C. N., Elrod V. A., Nowicki J. L., Coley T. L., Barber R. T., Lindley S., Watson A. J., Van Scoy K., and Law C. S. (1994) Testing the iron hypothesis in ecosystems of the equatorial Pacific Ocean. *Nature* **371**, 123–129.

Martin J. H. and Fitzwater S. E. (1988) Iron deficiency limits phytoplankton growth in the northeast Pacific subarctic. *Nature* **331**, 341–343.

Martin J. H. and Gordon R. M. (1988) Northeast Pacific iron distributions in relation to phytoplankton productivity. *Deep-Sea Res.* **35**, 177–196.

Martin J. H. and Knauer G. A. (1973) The elemental composition of plankton. *Geochim. Cosmochim. Acta* **37**, 1639–1653.

Martin J. H., Knauer G. A., Karl D. M., and Broenkow W. W. (1987) VERTEX: carbon cycling in the northeast Pacific. *Deep-Sea Res.* **34**, 267–285.

McCarthy M., Pratum T., Hedges J., and Benner R. (1997) Chemical composition of dissolved organic nitrogen in the ocean. *Nature* **390**, 150–154.

McCave I. N. (1975) Vertical flux of particles in the ocean. *Deep-Sea Res.* **22**, 491–502.

McCave I. N. (1984) Size spectra and aggregation of suspended particles in the deep ocean. *Deep Sea-Res.* **31**, 329–352.

Michaels A. F. and Silver M. W. (1988) Primary production, sinking fluxes and the microbial food web. *Deep-Sea Res.* **35**, 473–490.

Montoya J. P., Holl C. M., Zehr J. P., Hansen A., Villareal T. A., and Capone D. G. (2004) High rates of N_2 fixation by unicellular diazotrophs in the oligotrophic Pacific Ocean. *Nature* **430**, 1027–1031.

Mopper K., Zhou J. A., Ramana K. S., Passow U., Dam H. G., and Drapeau D. T. (1995) The role of surface-active carbohydrates in the flocculation of a diatom bloom in a mesocosm. *Deep-Sea Res. II* **42**, 47–73.

Morel F. M. M., Reinfelder J. R., Roberts S. B., Chamberlain C. P., Lee J. G., and Yee D. (1994) Zinc and carbon co-limitation of marine phytoplankton. *Nature* **369**, 740–742.

Morris I. (1981) Photosynthetic products, physiological state, and phytoplankton growth. *Can. J. Fish. Aquat. Sci.* **210**, 83–102.

Mulder A., van de Graaf A. A., Robertson L. A., and Kuenen J. G. (1995) Anaerobic ammonium oxidation discovered in a denitrifying fluidized bed reactor. *FEMS Microbiol. Ecol.* **16**, 177–184.

Nelson D. M. and Dortch Q. (1996) Silicic acid depletion and silicon limitation in the plume of the Mississippi River: evidence from kinetic studies in spring and summer. *Mar. Ecol. Prog. Ser.* **136**, 163–178.

Nelson D. M., Tréguer P., Brzezinski M. A., Leynaert A., and Quéguiner B. (1995) Production and dissolution of biogenic silica in the ocean: revised global estimates, comparison with regional data and relationship to biogenic sedimentation. *Global Biogeochem. Cycles* **9**, 359–372.

Nimer N. A., Iglesias-Rodriguez M. D., and Merrett M. J. (1997) Bicarbonate utilization by marine phytoplankton species. *J. Phycol.* **33**, 625–631.

Nozaki T. (1997) A fresh look at element distribution in the North Pacific. EOS Trans. AGU Electron. Suppl. http://www.agu.org/eos_elec/97025e.html.

Ocean Fertilization Summary Report (2001) Fertilization to transfer atmospheric CO_2 to the oceans, http://www.aslo.org/policy/docs/oceanfertsummary.pdf (accessed on August 2006).

Olesen M. and Lundsgaard C. (1995) Seasonal sedimentation of autochthonous material from the euphotic zone of a coastal system. *Estuar. Coast. Shelf Sci.* **41**, 475–490.

Paasche E. (1999) Reduced coccolith calcite production under light-limited growth: a comparative study of three clones of *Emiliania huxleyi* (Prymnesiophyceae). *Phycologia* **38**, 508–516.

Pahlow M. and Riebesell U. (2000) Temporal trends in deep ocean Redfield ratios. *Science* **287**, 831–833.

Passow U. (2004) Switching perspectives: do mineral fluxes determine particulate organic carbon fluxes or vice versa? *Geochem. Geophys. Geosyst.* **5**, doi: 10.1029/2003GC000670.

Passow U. and De La Rocha C. L. (2006) The accumulation of mineral ballast on organic aggregates. *Global Biogeochem. Cycles* **20**, doi:10.1029/2005GB002579.

Passow U., Shipe R. F., Murray A., Pak D., Brzezinski M. A., and Alldredge A. L. (2001) The origin of transparent exopolymer particles (TEP) and their role in the sedimentation of particulate matter. *Cont. Shelf Res.* **21**, 327–346.

Paytan A., Kaistner M., and Chavez F. P. (1996) Glacial to interglacial fluctuations in productivity in the Equatorial Pacific as indicated by marine barite. *Science* **274**, 1355–1357.

Peng T.-H. and Broecker W. S. (1991a) Dynamical limitations on the Antarctic iron fertilization strategy. *Nature* **349**, 227–229.

Peng T.-H. and Broecker W. S. (1991b) Factors limiting the reduction of atmospheric CO_2 by iron fertilization. *Limnol. Oceanogr.* **36**, 1919–1927.

Pondaven P., Ragueneau O., Tréguer P., Hauvespre A., Dezileau L., and Reyss J. L. (2000) Resolving the 'opal paradox' in the Southern Ocean. *Nature* **405**, 168–172.

Price N. M. and Morel F. M. M. (1990) Cadmium and cobalt substitution for zinc in a marine diatom. *Nature* **344**, 658–660.

Quigley M. S., Santschi P. H., Hung C.-C., Guo L., and Honeyman B. D. (2002) Importance of acid polysaccharides for ^{234}Th complexation to marine organic matter. *Limnol. Oceanogr.* **47**, 367–377.

Ragueneau O., Treguer P., Leynaert A., Anderson R. F., Brzezinski M. A., DeMaster D. J., Dugdale R. C., Dymond J., Fischer G., Francois R., Heinze C., Maier-Reimer E., Martin-Jezequel V., Nelson D. M., and Queguiner B. (2000) A review of the Si cycle in the modern ocean: recent progress and missing gaps in the application of biogenic opal as a paleoproductivity proxy. *Global Planet. Change* **26**, 317–365.

Raven J. A. (1997) Inorganic carbon acquisition by marine autotrophs. *Adv. Bot. Res.* **27**, 85–209.

Redfield A. C. (1934) On the proportions of organic derivatives in sea water and their relation to the composition of plankton. In *James Johnstone Memorial Volume James Johnstone Memorial Volume* (ed. R. J. Daniel). University Press of Liverpool, Liverpool, pp. 177–192.

Redfield A. C. (1958) The biological control of chemical factors in the environment. *Am. Sci.* **46**, 205–221.

Riebesell U., Wolf-Gladrow D. A., and Smetacek V. (1993) Carbon dioxide limitation of marine phytoplankton growth rates. *Nature* **361**, 249–251.

Robertson J. E., Robinson C., Turner D. M., Holligan P., Watson A. J., Boyd P., Fernandez E., and Finsh M. (1994) The impact of a coccolithophore bloom on oceanic carbon uptake in the northeast Atlantic during summer 1991. *Deep-Sea Res. I* **41**, 297–314.

Ruiz F. A., Marchesini N., Seufferheld M., and Docampo R. (2001) The polyphosphate bodies of *Chlamydomonas reinhardtii* possess a proton-pumping pyrophosphatase and are similar to acidocalcisomes. *J. Biol. Chem.* **276**, 46196–46203.

Ruttenberg K. C. (1993) Reassessment of the oceanic residence time of phosphorus. *Chem. Geol.* **107**, 405–409.

Ryther J. H. (1969) Photosynthesis and fish production in the sea. *Science* **166**, 72–76.

Sañudo-Wilhelmy S. A., Kustka A. B., Gobler C. J., Hutchins D. A., Yang M., Lwiza K., Burns J., Capone D. G., Raven J. A., and Carpenter E. J. (2001) Phosphorus limitation of nitrogen fixation by *Trichodesmium* in the central Atlantic Ocean. *Nature* **411**, 66–69.

Sarmiento J. L., Hughes T. M. C., Stouffer R. J., and Manabe S. (1998) Simulated response of the ocean carbon cycle to anthropogenic climate warming. *Nature* **393**, 245–249.

Sarmiento J. L. and Orr J. C. (1991) Three-dimensional simulations of the impact of Southern Ocean nutrient depletion on atmospheric CO_2 and ocean chemistry. *Limnol. Oceanogr.* **36**, 1928–1950.

Sarmiento J. L. and Toggweiler J. R. (1984) A new model for the role of the oceans in determining atmospheric P_{CO_2}. *Nature* **308**, 621–624.

Scholten J. C., Fietzke J., Vogler S., van der Loeff M. M. R., Mangini A., Koeve W., Waniek J., Stoffers P., Antia A., and Kuss J. (2001) Trapping efficiency of sediment traps from the deep eastern North Atlantic: ^{230}Th calibration. *Deep-Sea Res. II* **48**, 243–268.

Schumann R. and Rentsch D. (1998) Staining particulate organic matter with DTAF—a fluorescence dye for carbohydrates and protein: a new approach and application of a 2D image analysis system. *Mar. Ecol. Prog. Ser.* **163**, 77–88.

Shaffer G. (1996) Biogeochemical cycling in the global ocean. 2: New production, Redfield ratios, and remineralization in the organic pump. *J. Geophys. Res.* **101**, 3723–3745.

Shanks A. L. and Trent J. D. (1980) Marine snow: sinking rates and potential role in vertical flux. *Deep-Sea Res.* **27A**, 137–143.

Shemesh A., Mortlock R. A., Smith R. J., and Froelich P. N. (1988) Determination of Ge/Si in marine siliceous microfossils: separation, cleaning and dissolution of diatoms and radiolarian. *Mar. Chem.* **25**, 305–323.

Shi Y., Hu H., Ma R., Cong W., and Cai Z. (2004) Improved use of organic phosphate by Skeletonema costatum through regulation of Zn^{2+} concentrations. *Biotechnol. Lett.* **26**, 747–751.

Shulenberger E. and Reid J. L. (1981) The Pacific shallow oxygen maximum, deep chlorophyll maximum, and primary productivity, reconsidered. *Deep-Sea Res.* **28A**, 901–919.

Shushkina E. A. (1985) Production of principal ecological groups of plankton in the epipelagic zone of the ocean. *Oceanology* **25**, 653–658.

Siever R. (1991) Silica in the oceans: biological–geochemical interplay. In *Scientists on Gaia* (eds. S. Schneider and P. Boston). MIT Press, Cambridge, MA, pp. 287–295.

Smayda T. J. (1970) The suspension and sinking of phytoplankton in the sea. *Oceanogr. Mar. Biol. Ann. Rev.* **8**, 353–414.

Smith D. C., Simon M., Alldredge A. L., and Azam F. (1992) Intense hydrolytic enzyme activity on marine aggregates and implications for rapid particle dissolution. *Nature* **359**, 139–142.

Smith D. C., Steward G. F., Long R. A., and Azam F. (1995) Bacterial mediation of carbon fluxes during a diatom bloom in a mesocosm. *Deep-Sea Res. II* **42**, 75–97.

Soendergaard M., Williams P. J. L., Cauwet G., Riemann B., Robinson C., Terzic S., Woodward E. M. S., and Worm J. (2000) Net accumulation and flux of dissolved organic carbon and dissolved organic nitrogen in marine plankton communities. *Limnol. Oceanogr.* **45**, 1097–1111.

Steemann Neilsen E. (1952) The use of radioactive carbon (^{14}C) for measuring organic production in the sea. *J. Cons. Int. Explor. Mer.* **18**, 117–140.

Steinberg D. K. (1995) Diet of copepods (*Scolpalatum vorax*) associated with mesopelagic detritus (giant larvacean houses) in Monterey Bay, California. *Mar. Biol.* **122**, 571–584.

Steinberg D. K., Carlson C. A., Bates N. R., Johnson R. J., Michaels A. F., and Knap A. P. (2001) Overview of the US JGOFS Bermuda Atlantic Time-series Study (BATS): a decade-scale look at ocean biology and biogeochemistry. *Deep-Sea Res. II* **48**, 1405–1447.

Steinberg D. K., Pilskaln C. H., and Silver M. W. (1998) Contribution of zooplankton associated with detritus to sediment trap 'swimmer' carbon in Monterey Bay, California, USA. *Mar. Ecol. Prog. Ser.* **164**, 157–166.

Strickland J. D. H. and Parsons T. R. (1968) *A Practical Handbook of Seawater Analysis*. Fisheries Research Board of Canada, Ottawa.

Strom S. L., Benner R., Ziegler S., and Dagg M. J. (1997) Planktonic grazers are a potentially important source of marine dissolved organic carbon. *Limnol. Oceanogr.* **42**, 1364–1374.

Suess E. (1980) Particulate organic carbon flux in the oceans—surface productivity and oxygen utilization. *Nature* **288**, 260–263.

Sunda W. G. and Huntsman S. A. (1995a) Iron uptake and growth limitation in oceanic and coastal phytoplankton. *Mar. Chem.* **50**, 189–206.

Sunda W. G. and Huntsman S. A. (1995b) Cobalt and zinc interreplacement in marine phytoplankton: biological and geochemical implications. *Limnol. Oceanogr.* **40**, 1404–1417.

Sunda W. G. and Huntsman S. A. (2005) Effect of CO_2 supply and demand on zinc uptake and growth limitation in a coastal diatom. *Limnol. Oceanogr.* **50**, 1181–1192.

Sundquist E. T. (1993) The global carbon dioxide budget. *Science* **259**, 934–941.

Takahashi T., Broecker W. S., and Langer S. (1985) Redfield ratio based on chemical data from isopycnal surfaces. *J. Geophys. Res.* **90**, 6907–6924.

Takeda S. (1998) Influence of iron availability on nutrient consumption ratio of diatoms in oceanic waters. *Nature* **393**, 774–777.

Tappan H. and Loeblich A. R., Jr. (1973) Evolution of the oceanic plankton. *Earth-Sci. Rev.* **9**, 207–240.

Teira E., Pazó M. J., Serret P., and Fernández E. (2001) Dissolved organic carbon production by microbial populations in the Atlantic Ocean. *Limnol. Oceanogr.* **46**, 1370–1377.

Thamdrup B. and Dalsgaard T. (2002) Production of N_2 through anaerobic ammonium oxidation coupled to nitrate reduction in marine sediments. *Appl. Environ. Microbiol.* **68**, 1312–1318.

Timmermans K. R., Snoek J., Gerringa L. J. A., Zondervan I., and de Baar H. J. W. (2001) Not all eukaryotic algae can replace zinc with cobalt: *Chaetoceros calcitrans* (Bacillariophyceae) versus *Emiliania huxleyi* (Prymnesiophyceae). *Limnol. Oceanogr.* **46**, 699–703.

Tortell P. D., Reinfelder J. R., and Morel F. M. M. (1997) Active uptake of bicarbonate by diatoms. *Nature* **390**, 243–244.

Tréguer P., Nelson D. M., Van Bennekom A. J., DeMaster D. J., Leynaert A., and Quéguiner B. (1995) The silica balance in the world ocean: a reestimate. *Science* **268**, 375–379.

Turner J. T. (2002) Zooplankton fecal pellets, marine snow and sinking phytoplankton blooms. *Aquat. Microb. Ecol.* **27**, 57–102.

Tyrrell T. (1999) The relative influences of nitrogen and phosphorus on oceanic primary production. *Nature* **400**, 525–531.

Verity P. G., Williams S. C., and Hong Y. (2000) Formation, degradation, and mass: volume ratios of detritus derived from decaying phytoplankton. *Mar. Ecol. Prog. Ser.* **207**, 53–68.

Ware J. R., Smith S. V., and Reaka-Kudla M. L. (1992) Coral reefs: sources or sinks of atmospheric CO_2. *Coral Reefs* **11**, 127–130.

Watson A. J., Bakker D. C. E., Ridgwell A. J., Poyd P. W., and Law C. S. (2000) Effect of iron supply on Southern Ocean CO_2 uptake and implications for glacial atmospheric CO_2. *Nature* **407**, 730–733.

Watson A. J., Law C. S., Van Scoy K. A., Millero F. J., Yao W., Friedderich G. E., Liddicoat M. I., Wanninkhof R. H., Barber R. T., and Coale K. H. (1994) Minimal effect of iron fertilization on sea-surface carbon dioxide concentrations. *Nature* **371**, 143–145.

Williams P. M. and Druffel E. R. M. (1987) Radiocarbon in dissolved organic matter in the central north Pacific Ocean. *Nature* **330**, 246–248.

Wolf-Gladrow D. A., Riebesell U., Burkhardt S., and Bijma J. (1999) Direct effects of CO_2 concentration on growth and isotopic composition of marine plankton. *Tellus (B Chem. Phys. Meteorol.)* **51B**, 461–476.

Zeebe R. E. and Wolf-Gladrow D. (2001) *CO_2 in Seawater: Equilibrium, Kinetics, Isotopes.* Elsevier, Amsterdam.

Zehr J. P. and Ward B. B. (2002) Nitrogen cycling in the ocean: new perspectives on processes and paradigms. *Appl. Environ. Microbiol.* **68**, 1015–1024.

Zehr J. P., Waterbury J. B., Turner P. J., Montoya J. P., Omoregie E., Steward G. F., Hansen A., and Karl D. M. (2001) Unicellular cyanobacteria fix N_2 in the subtropical North Pacific Ocean. *Nature* **412**, 635–638.

Zhang J. and Nozaki Y. (1996) Rare earth elements and yttrium in seawater: ICP-MS determinations in the East Caroline, Coral Sea, and Southern Fiji basins of the western Pacific Ocean. *Geochim. Cosmochim. Acta* **60**, 4631–4644.

Zhuang G., Duce R. A., and Kester D. R. (1990) The dissolution of atmospheric iron in surface seawater of the open ocean. *J. Geophys. Res.* **95**, 16207–16216.

Geochemistry of Earth Surface Systems
ISBN: 978-0-08-096706-6

pp. 425–452

13

The Biological Pump in the Past

D. M. Sigman

Princeton University, NJ, USA

and

G. H. Haug

Geoforschungszentrum Potsdam, Germany

13.1	INTRODUCTION	453
13.2	CONCEPTS	458
	13.2.1 Low- and Mid-latitude Ocean	460
	13.2.2 High-latitude Ocean	465
13.3	TOOLS	469
	13.3.1 Export Production	471
	13.3.2 Nutrient Status	472
	13.3.3 Integrative Constraints on the Biological Pump	474
	13.3.3.1 Carbon isotope distribution of the ocean and atmosphere	474
	13.3.3.2 Deep-ocean oxygen content	474
	13.3.3.3 Phasing	475
13.4	OBSERVATIONS	475
	13.4.1 Low- and Mid-latitude Ocean	475
	13.4.2 High-latitude Ocean	478
13.5	SUMMARY AND CURRENT OPINION	482
	ACKNOWLEDGMENTS	483
	REFERENCES	483

13.1 INTRODUCTION

It is easy to imagine that the terrestrial biosphere sequesters atmospheric carbon dioxide; the form and quantity of the sequestered carbon, living or dead organic matter, are striking. In the ocean, there are no aggregations of biomass comparable to the forests on land. Yet biological productivity in the ocean plays a central role in the sequestration of atmospheric carbon dioxide, typically overshadowing the effects of terrestrial biospheric carbon storage on timescales longer than a few centuries. In an effort to communicate the ocean's role in the regulation of atmospheric carbon dioxide, marine scientists frequently refer to the ocean's biologically driven sequestration of carbon as the "biological pump." The original and strict definition of the biological (or "soft-tissue") pump is actually more specific: the sequestration of carbon dioxide in the ocean interior by the biogenic flux of organic matter out of surface waters and into the deep sea prior to decomposition of that organic matter back to carbon dioxide (Volk and Hoffert, 1985) (Figure 1). The biological pump extracts carbon from the "surface skin" of the ocean that interacts with the atmosphere, presenting a lower partial pressure of carbon dioxide (CO_2) to the atmosphere and thus lowering its CO_2 content.

The place of the biological pump in the global carbon cycle is illustrated in Figure 2. The atmosphere exchanges carbon with essentially three

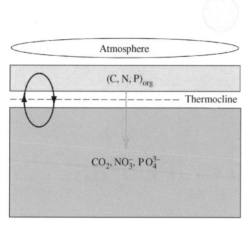

Figure 1 A schematic of the ocean's "biological pump," the sequestration of carbon and nutrients (nitrogen and phosphorus) in the ocean interior (lower dark blue box) by the growth of phytoplankton (floating unicellular algae) in the sunlit surface ocean (upper light blue box), the downward rain of organic matter out of the surface ocean and into the deep ocean (green downward arrow), and the subsequent breakdown of this organic matter back to carbon dioxide (CO_2), nitrate (NO_3^-), and phosphate (PO_4^{3-}). The nutrients and CO_2 are reintroduced to the surface ocean by mixing and upwelling (the circling arrows at left). The biological pump lowers the CO_2 content of the atmosphere by extracting it from the surface ocean (which exchanges CO_2 with the atmosphere) and sequestering it in the isolated waters of the ocean interior. In most of the low- and mid-latitude ocean, the surface is isolated from the deep sea by a temperature-driven density gradient, or "thermocline," keeping nutrient supply low and leading to essentially complete consumption of NO_3^- and PO_4^{3-} by phytoplankton at the surface. At the higher latitudes, where there is no permanent thermocline, more rapid communication with the deep sea leads to incomplete consumption of NO_3^- and PO_4^{3-}.

reservoirs: the ocean, the terrestrial biosphere, and the geosphere. The ocean holds ~50 times as much carbon as does the atmosphere, and ~20 times as much as the terrestrial biosphere. Thus, on timescales that are adequately long to allow the deep sea to communicate with the surface ocean and atmosphere ($\gtrsim 500$ yr), carbon storage in the ocean all but sets the concentration of CO_2 in the atmosphere. The effect of the biological pump is not permanent on this timescale of ocean circulation; the downward transport of carbon is balanced by the net upward flux from the CO_2-rich waters of the deep sea, which, in the absence of the biological pump, would work to homogenize the carbon chemistry of the ocean, raising atmospheric CO_2 in the process.

The dynamics of marine organic carbon can be described as a set of three nested cycles in which the biological pump is the cycle with flux and reservoir of intermediate magnitude (Figure 3).

The cycle with the shortest timescale, which operates within the surface ocean, is composed of net primary production by phytoplankton (their photosynthesis less their respiration) and heterotrophic respiration by zooplankton and bacteria that oxidize most of the net primary production back to CO_2. This cycle is by far the greatest in terms of the flux of carbon, but the reservoir of sequestered carbon that accumulates in surface waters (phytoplankton biomass, dead organic particles, and dissolved organic carbon) is small relative to the atmospheric reservoir of CO_2, and it has a short residence time (less than a year). A small imbalance in this cycle, between net primary production and heterotrophic respiration, feeds the next cycle in the form of organic carbon that sinks (or is mixed) into the deep sea. This "export production" drives the biological pump (Figure 3).

At the other extreme, a small fraction of the organic matter exported from the surface ocean escapes remineralization in the water column and sediments and is thus buried in the sediments, removing carbon from the ocean/atmosphere system (Figure 3). On the timescale of geologic processes, this carbon removal is balanced by the oxidation of the organic matter when it is exposed at the earth surface by uplift and weathering, or when it is released by metamorphism and volcanism. While the fluxes involved in this cycle are small, the reservoir is large, so that its importance increases with timescale, becoming clearly relevant on the timescale of millions of years.

The biological pump, in the strict sense, does not include the burial of organic matter in marine sediments. There are several related reasons for this exclusion. First, as described below, the variations in atmospheric carbon dioxide that are correlated with the waxing and waning of ice ages have driven much of the thinking about the role of organic carbon production on atmospheric composition. Only a very small fraction of the organic matter exported out of surface waters is preserved and buried in the underlying sediment, so that this process cannot sequester a significant amount of carbon on the timescale of millennia and thus is not a candidate process for the major carbon dioxide variations over glacial cycles. Second, on the timescale for which organic matter burial is relevant, i.e., over millions of years, it is thought to be only one of several mechanisms by which atmospheric CO_2 is regulated. Most hypotheses regarding the history of CO_2 over geological time involve weathering of rocks on land and the precipitation of carbonates in the ocean, due to both the larger fluxes and larger reservoirs involved in the geological trapping of CO_2 as carbonate as opposed to organic carbon (Figure 2). Finally, the importance of biological productivity in determining the burial rate of organic carbon is not at all clear. While some examples of

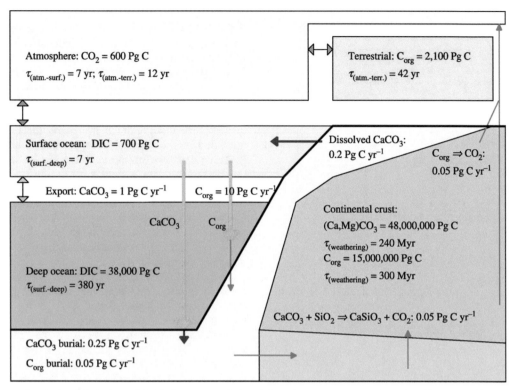

Figure 2 A simplified view of the Holocene (pre-industrial) carbon cycle (largely based on Holmen (1992)). Units of carbon are petagrams (Pg, 10^{15} g). The fluxes are colored according to the residence time of carbon in the reservoirs they involve, from shortest to longest, as follows: red, orange, purple, blue. Exchanges of the atmosphere with the surface ocean and terrestrial biosphere are relatively rapid, such that changes in the fluxes alter atmospheric CO_2 on the timescale of years and decades. Exchange between the surface and the deep ocean is such that centuries to millennia are required for a change to yield a new steady state. Because the deep-ocean carbon reservoir is large relative to the surface ocean, terrestrial biosphere, and atmosphere, its interactions with the surface ocean, given thousands of years, determine the total amount of carbon to be partitioned among those three reservoirs. In a lifeless ocean, the only cause for gradients in CO_2 between the deep ocean and the surface ocean would be temperature (i.e., the "solubility pump" (Volk and Hoffert, 1985))—deep waters are formed from especially cold surface waters, and CO_2 is more soluble in cold water. The biological pump, represented by the right downward arrow, greatly enhances the surface–deep CO_2 gradient, through the rain of organic matter ("C_{org}") out of the surface ocean. The "carbonate pump," represented by the left downward arrow, is the downward rain of calcium carbonate microfossils out of the surface ocean. Its effect is to actually raise the p_{CO_2} of the surface ocean; this involves the alkalinity of seawater and is discussed in Chapters 12 and 14. Almost all of the organic matter raining out of the surface ocean is degraded back to CO_2 and inorganic nutrients as it rains to the seafloor or once it is incorporated into the shallow sediments; only less than 1% ($\sim$0.05 Pg out of $\sim$10 Pg) is removed from the ocean/atmosphere system by burial (the downward blue arrow). This is in contrast to the calcium carbonate rain out of the surface (the downward purple arrow), $\sim$25% of which is buried. In parallel, the weathering rate of calcium carbonate on land is significant on millennial timescales (the left-pointing purple arrow), while the release of geologically sequestered C_{org} occurs more slowly, on a similar timescale as the release of CO_2 from the metamorphism of calcium carbonate to silicate minerals.

high organic carbon burial appear to be due to high oceanic productivity (Vincent and Berger, 1985), others involve diverse additional processes, such as the delivery of sediments to the ocean (FranceLanord and Derry, 1997), which influences the ease with which the organic matter rain enters the sedimentary record (Canfield, 1994). In the latter case, the distribution of oceanic productivity is of secondary importance to the effect of organic carbon burial on atmospheric CO_2.

The most direct evidence for natural variations in atmospheric CO_2 comes from the air that is trapped in Antarctic glacial ice. Records from Antarctic ice cores indicate that the concentration of CO_2 in the atmosphere has varied in step with the waxing and waning of ice ages (Barnola *et al.*, 1987; Petit *et al.*, 1999) (Figure 4). During inter-glacial times, such as the Holocene (roughly the past 10^4 years), the atmospheric partial pressure of CO_2 (p_{CO_2}) is typically near 280 ppm by volume (ppmv). During peak glacial times, such as the last glacial maximum $\sim$1.8 $\times$ 10^4 yr ago, atmospheric p_{CO_2} is 180–200 ppmv, or $\sim$80–100 ppmv lower. CO_2 is a greenhouse gas, and model calculations

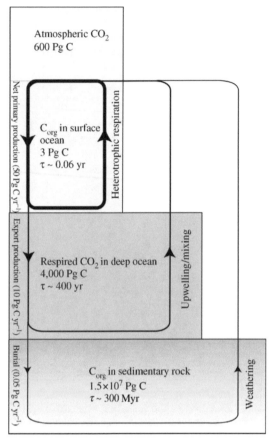

Figure 3 The nested cycles of marine organic carbon, including (i) net primary production and heterotrophic respiration in the surface ocean, (ii) export production from the surface ocean and respiration in the deep sea followed by upwelling or mixing of the respired CO_2 back to the surface ocean (the biological pump, the subject of this chapter), and (iii) burial of sedimentary organic carbon and its release to the atmosphere by weathering. The slight imbalance of each cycle fuels the longer-timescale cycles. The carbon reservoirs shown refer solely to the fraction affected by organic carbon cycling; for instance, the deep-ocean value shown is of CO_2 produced by respiration, not the total dissolved inorganic carbon reservoir.

suggest that its changes play a significant role in the energetics of glacial/interglacial climate change (Weaver *et al.*, 1998; Webb *et al.*, 1997). However, we have not yet identified the cause of these variations in CO_2. The explanation of glacial/interglacial cycles in CO_2 is a major motivation in the paleoceanographic and paleoclimatological communities, for three reasons: first, these CO_2 changes are the most certain of any in our record of the past (excluding the anthropogenic rise in CO_2); second, they show a remarkably strong connection to climate, inspiring considerations of both cause and effect; and third, they are of a timescale that is relevant to the history and future of humanity.

The ultimate pacing of glacial cycles is statistically linked to cyclic changes in the orbital parameters of the Earth, with characteristic frequencies of approximately 100 kyr, 41 kyr, and 23 kyr (Berger *et al.*, 1984; Hays *et al.*, 1976). These orbitally driven variations in the seasonal and spatial distribution of solar radiation incident on the Earth surface, known as the "Milankovitch cycles", are thought by many to be the fundamental driver of glacial/interglacial oscillations in climate. However, positive feedbacks within the Earth's climate system must amplify orbital forcing to produce the amplitude and temporal structure of glacial/interglacial climate variations. Carbon dioxide represents one such feedback. The point of greatest uncertainty in this feedback is the mechanism by which atmospheric CO_2 is driven to change.

Broecker (1982a,b) first hypothesized a glacial increase in the strength of the biological pump as the driver of lower glacial CO_2 levels. This suggestion has spawned many variants, which we will tend to phrase in terms of the "major" nutrients, nitrogen and phosphorus. These nutrients are relatively scarce in surface ocean waters but are required in large quantities to build all algal biomass. In terms of the major nutrients, the biological pump hypotheses fall into two groups: (i) changes in the low and mid-latitude surface ocean, where the major nutrients appear to limit the extraction of CO_2 by biological production, and (ii) changes in the polar and subpolar ocean regions, where the major nutrients are currently not completely consumed and not limiting. In both cases, the central biological process is "export production," the organic matter produced by phytoplankton that is exported from the surface ocean, resulting in the sequestration of its degradation products (inorganic carbon and nutrients) in the ocean interior. Here, we review the concepts, tools, and observations relevant to variability in the biological pump on the millenial timescale, with a strong focus on its potential to explain glacial/interglacial CO_2 change.

The biogenic rain to the deep sea has important mineral components: calcium carbonate, mostly from coccolithophorids (phytoplankton) and foraminifera (zooplankton), and opal, mostly from diatoms (phytoplankton) and radiolaria (zooplankton). The calcium carbonate ($CaCO_3$) component is important for atmospheric carbon dioxide in its own right (Figure 2). In contrast to organic carbon, the production, sinking, and burial of $CaCO_3$ acts to raise atmospheric carbon dioxide concentrations. This is unintuitive, in that $CaCO_3$, like organic carbon, represents a repository for inorganic carbon and a vector for the removal of this carbon from the surface ocean and atmosphere. The difference involves ocean "alkalinity." Alkalinity is the acid-titrating capacity of the ocean. As it increases, the

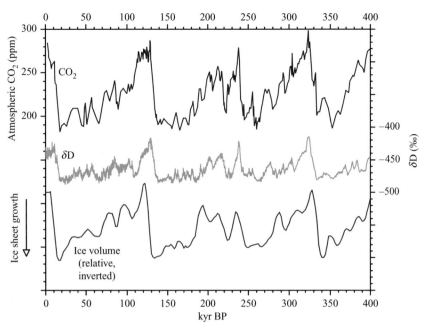

Figure 4 The history of atmospheric CO_2 back to 400 kyr as recorded by the gas content in the Vostok ice core from Antarctica (Petit *et al.*, 1999). The ratio of deuterium to hydrogen in ice (expressed as the term δD) provides a record of air temperature over Antarctica, with more negative δD values corresponding to colder conditions. The history of global ice volume based on benthic foraminiferal oxygen isotope data from deep-sea sediment cores (Bassinot *et al.*, 1994) is plotted as relative sea level, so that ice ages (peaks in continental ice volume) appear as sea-level minima, with a full glacial/interglacial amplitude for sea level change of ~130 m (Fairbanks, 1989). During peak glacial periods, atmospheric CO_2 is 80–100 ppmv lower than during peak interglacial periods, with upper and lower limits that are reproduced in each of the 100 kyr cycles. Ice cores records, including the Vostok record shown here, indicate that atmospheric CO_2 was among the first parameters to change at the termination of glacial maxima, roughly in step with southern hemisphere warming and preceding the decline in northern hemisphere ice volume.

pH of seawater rises (i.e., the concentration of H^+, or protons, decreases), and an increasing amount of CO_2 is stored in the ionic forms bicarbonate (HCO_3^-) and carbonate (CO_3^{2-}). This "storage" is achieved by the loss of H^+ from carbonic acid (H_2CO_3), which is itself formed by the combination of dissolved CO_2 with its host H_2O. Thus, an increase in ocean alkalinity lowers the p_{CO_2} of surface waters and thus the CO_2 concentration of the overlying atmosphere. The carbonate ion holds two equivalents of alkalinity for every mole of carbon: CO_3^{2-} must absorb two protons before it can leave the ocean as CO_2 gas. In the precipitation of $CaCO_3$, CO_3^{2-} is removed from the ocean, lowering the alkalinity of the ocean water and thus raising its p_{CO_2}. We can describe this in terms of a chain of reactions: the CO_3^{2-} that was lost to precipitation is replaced by the deprotonation of a HCO_3^-, generating a proton. This proton then combines with another HCO_3^- to produce H_2CO_3, which dissociates to form CO_2 and H_2O, yielding the summary reaction: $Ca^{2+} + 2HCO_3^- \rightarrow CaCO_3 + CO_2 + H_2O$. Thus, the biological precipitation of $CaCO_3$ raises the p_{CO_2} of the water in which it occurs.

The biological formation of $CaCO_3$ affects atmospheric CO_2 from two perspectives. First, the precipitation of $CaCO_3$ in surface waters and its sinking to the seafloor drives a surface-to-deep alkalinity gradient, which raises the p_{CO_2} of surface waters in a way that is analogous to the p_{CO_2} decrease due to the biological pump; this might be referred to as the "carbonate pump." Just as with the CO_2 produced in deep water by organic matter degradation, the chemical products of the deep redissolution of the $CaCO_3$ rain are eventually mixed up to the surface again, undoing the effect of their temporary (~1,000 yr) sequestration in the abyss. Second, ~25% of the $CaCO_3$ escapes dissolution and is buried, thus sequestering carbon and alkalinity in the geosphere, on a timescale of thousands to millions of years (Figure 2). An excess in the loss of alkalinity by calcium carbonate burial rate relative to the input of alkalinity by continental weathering will drive an increase in the p_{CO_2} of the whole ocean on the timescale of thousands of years. In summary, $CaCO_3$ precipitation can alter atmospheric p_{CO_2} by generating a surface-to-deep gradient in seawater alkalinity (the carbonate pump) and by changing the total amount of alkalinity in the ocean.

The $CaCO_3$ cycle is a central part of the effect of biological productivity on atmospheric CO_2.

However, it is not within the strict definition of the biological pump, which deals specifically with organic carbon. Moreover, the effect of the $CaCO_3$ rain is determined not only by the actual magnitude of the rain to the seafloor but also by its degree of preservation and burial at the seafloor, a relatively involved subject that is treated elsewhere in this volume. Thus, in our discussions below, we try as much as possible to limit ourselves to the geochemical effects of the biogenic rain of organic matter, bringing $CaCO_3$ into the discussion only when absolutely necessary and then trying to focus on its biological production and not its seafloor preservation.

13.2 CONCEPTS

At the coarsest scale, phytoplankton abundance (Figure 5(a)) and nutrient availability (Figures 5(b)–(d)) are correlated across the global surface ocean, indicating that the supply of major nutrients is a dominant control on productivity. In this context, two types of environments emerge from the global ocean distributions (Figure 5): (i) the tropical and

(a)

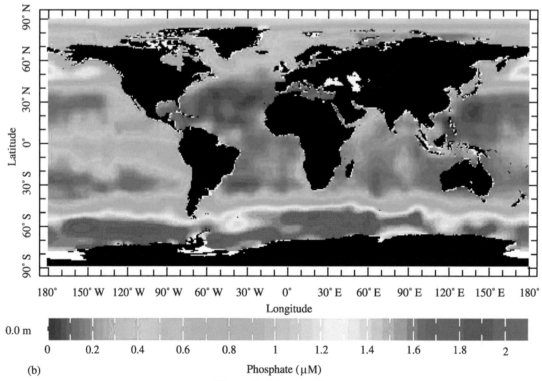

(b)

Phosphate (μM)

Figure 5 (Continued).

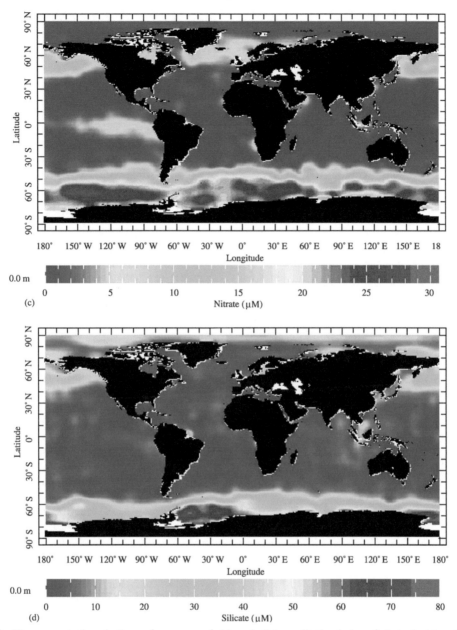

(c) Nitrate (μM)

0.0 m

(d) Silicate (μM)

0.0 m

Figure 5 The concentrations in the surface ocean of chlorophyll (a qualitative index of phytoplankton abundance and thus net primary productivity, (a)), the "major nutrients" phosphate (PO_4^{3-}, (b)) and nitrate (NO_3^-, (c)), and the nutrient silicate (SiO_4^{4-}, (d)), which is required mostly by diatoms, a group of phytoplankton that builds opal tests. The chlorophyll map in (a) is from composite data from the NASA SeaWiFs satellite project collected during 2001 (http://seawifs.gsfc.nasa.gov/SEAWIFS.html). The nutrient data in (b), (c), and (d) are from 1994 World Ocean Atlas (Conkright *et al.*, 1994), and the maps are generated from the International Research Institute for Climate Prediction Data Library (http://ingrid.ldeo.columbia.edu). Comparison of (a) with (b), (c), and (d) demonstrates that phytoplankton abundance, at the coarsest scale, is driven by the availability of these nutrients, and these nutrients are generally most available in the polar ocean and along the equatorial upwellings where nutrient-rich deep water mixes more easily to the surface. However, among the nutrient-bearing high-latitude regions, there is not a good correlation between nutrient concentration and chlorophyll, indicating that other parameters come to limit productivity in these regions. The high-latitude ocean of the southern hemisphere, which has the highest nutrient concentrations, is known as the Southern Ocean. It is composed of the more polar Antarctic zone, where the silicate concentration is high and diatom productivity is extensive, and the more equatorward Subantarctic zone, where nitrate and phosphate remain high but silicate is scarce. The strong global correlation between phosphate and nitrate, originally recognized by Redfield (Redfield, 1942, 1958; Redfield *et al.*, 1963), allows us to group the two nutrients together when considering their internal cycling within the ocean, as they are exhausted in more or less the same regions. However, their input/output budgets may change over hundreds and thousands of years, and the relationship between these two nutrients may have been different in the past.

subtropical ocean, where productivity is low and limited by the supply of nitrate and phosphate, and (ii) the subpolar and polar regions, where the supply of nitrate and phosphate is high enough so as not to limit phytoplankton growth. This distinction frames much of the conceptual discussion that follows. One important result will at this point seem counterintuitive: the global biological pump is driven from the low-latitude regions, where the biological pump is working at maximal efficiency with respect to the major nutrient supply, and weakened by the high-latitude regions, which are not maximally efficient with respect to the major nutrients and which are thus allowing deeply sequestered CO_2 to escape back to the atmosphere.

The relative importance of the low versus high latitudes in controlling atmospheric CO_2 is currently a matter of debate, with the lines drawn between the two types of models that are used to study the biological pump: geochemical box models of the ocean, the first tools available for quantifying the role of the biological pump, and the newer ocean general circulation models. Relative to box models, at least some general circulation models express a greater sensitivity of atmospheric CO_2 to the low-latitude ocean and a lesser sensitivity to the high-latitude ocean (Archer *et al.*, 2000b). This difference in sensitivity has major implications for the feasibility of different oceanic changes to explain glacial/interglacial CO_2 change. While some authors have argued that the discrepancy can be framed in terms of the amount of vertical mixing in the low versus high-latitude ocean (Broecker *et al.*, 1999), we are not convinced by this argument, as such differences should be captured by the deep-ocean temperatures in the two types of models. Rather, we favor the arguments of Toggweiler *et al.* (2003a) that the difference is mostly in the degree of air/sea equilibration of CO_2 that is allowed in the high-latitude regions of the models. If this is so, there is good reason to believe that the high latitudes are as important in the biological pump as the box models suggest (Toggweiler *et al.*, 2003a). Regardless of our view, it must be recognized that the quantitative arguments made below are based largely on our experience with box models (Hughen *et al.*, 1997; Sigman *et al.*, 1998, 1999b, 2003), and that some workers would find different areas of emphasis, although the same basic processes would probably be described (e.g., see Archer *et al.*, 2000a).

While the global correlation between nutrients and productivity is strong, it is far from perfect. For instance, the chlorophyll levels among the high-latitude regions do not correlate with their surface nutrient concentration, nor does such a correlation hold particularly well within a single high-latitude region (Figure 5). These discrepancies provide a starting point for addressing what controls polar productivity and how the different polar regions impact the biological pump. We also look to the current differences among the polar regions to understand how the high-latitude leaks in the biological pump might have changed in the past.

The general link between nutrient concentration and productivity is also broken in some low-latitude regions. For instance, coastal productivity is high even where there is no apparent elevation in major nutrients. While part of this discrepancy is probably due to nutrient inputs from the continents that are not captured in Figure 5 because they are sporadic and/or compositionally complex, the discrepancy also reflects the very rapid cycling of organic matter in the shallow ocean. We will see that the coastal regions, despite their extremely high productivity, are not central to the biological pump, illustrating the distinction between productivity and the biological pump.

While this comparison between "nutrients" and productivity is useful, the surface distributions of the nutrients are themselves not all identical. For instance, nitrate and phosphate penetrate much further north than does silicate in the Southern Ocean, the polar ocean surrounding Antarctica (Figures 5(b)–(d)). As silicate is needed by diatoms but not other types of phytoplankton, this difference has major implications for the ecology of the Southern Ocean, present and past. Moreover, since surface water from 40–50°S is fed into the thermocline that supplies the low-latitude surface ocean with nutrients, the nutrient conditions of the Southern Ocean affect the nutrient supply and ecology of lower latitudes (Brzezinski *et al.*, 2002; Matsumoto *et al.*, 2002b). With regard to nitrate and phosphate (Figures 5(b) and (c)), a more careful comparison of their concentration variations would show lower nitrate-to-phosphate ratios in the Pacific than elsewhere, indicative of the nitrate loss that is occurring in that basin (Deutsch *et al.*, 2001; Gruber and Sarmiento, 1997). The input/output budget of the marine nitrogen cycle and its affect on the ocean nitrate reservoir arises as a matter of great importance in the section that follows.

13.2.1 Low- and Mid-latitude Ocean

We can describe the global biological pump in terms of a balance of fluxes between the sunlit surface ocean and the cold, dark, voluminous deep ocean (Figure 1). Phytoplankton consume CO_2 and nutrients (in particular, the major nutrients nitrogen and phosphorus) in the surface, and subsequent processing drives a downward rain of this organic matter that is degraded after it has sunk into the ocean interior, effectively extracting the CO_2 and nutrients from the surface ocean and lowering atmospheric CO_2. Balancing this downward flux of particles is the net upward flux of dissolved inorganic carbon and nutrients into the surface, due to a combination of diffuse vertical mixing over the entire ocean and focused vertical motion (upwelling and downwelling) in specific regions.

This steady state can be described by the following expressions (Broecker and Peng, 1982), which indicate a balance between the gross upward (left-hand side) and gross downward (right-hand side) transport of phosphorus (Equation (1)) and carbon (Equation (2)):

$$[PO_4^{3-}]_{deep} \times Q = [PO_4^{3-}]_{surface} \times Q + EP_P \quad (1)$$

$$[DIC]_{deep} \times Q = [DIC]_{surface} \times Q + (EP_P \times C/P_{EP}) \quad (2)$$

where $[PO_4^{3-}]$ is the phosphate concentration and $[DIC]$ is the dissolved inorganic carbon concentration (the sum of dissolved CO_2, HCO_3^-, and CO_3^{2-}) in the deep and surface layers, Q is the exchange rate of water between the deep and surface, EP_P is the export production in terms of phosphorus, and C/P_{EP} is the carbon to phosphorus ratio of the export production. In these expressions, phosphorus is intended to represent the major nutrients in general; the distinctions between phosphorus and nitrogen will be considered later.

The biological pump is the biologically driven gradient in $[DIC]$ between the surface and the deep sea ($[DIC]_{deep} - [DIC]_{surface}$). Because the deep ocean is a very large reservoir, increasing this gradient essentially means decreasing $[DIC]_{surface}$. Because the surface ocean equilibrates with the gases of the atmosphere, an increase in the gradient lowers atmospheric CO_2 as well. Solving Equation (2) for $[DIC]_{deep} - [DIC]_{surface}$ gives

$$[DIC]_{deep} - [DIC]_{surface} = (EP_P \times C/P_{EP})Q \quad (3)$$

Solving Equation (1) for EP_P gives:

$$EP_P = \left([PO_4^{3-}]_{deep} - [PO_4^{3-}]_{surface}\right) \times Q \quad (4)$$

Substitution of Equation (4) into Equation (3) for EP_P gives the following expression for $[DIC]_{deep} - [DIC]_{surface}$:

$$\begin{aligned}&[DIC]_{deep} - [DIC]_{surface}\\&= \left([PO_4^{3-}]_{deep} - [PO_4^{3-}]_{surface}\right) \times C/P_{EP}\end{aligned} \quad (5)$$

Thus, given the assumptions above, the surface–deep gradient in inorganic carbon concentration (the "strength" of the biological pump) is determined by (i) the C/P of the organic matter that is exported from the surface ocean, and (ii) the nutrient concentration ($[PO_4^{3-}]$) gradient between the surface and deep ocean.

In the low- and mid-latitude ocean, this balance is simplified by the nearly complete extraction of $[PO_4^{3-}]$ and $[NO_3^-]$ from the surface ocean (Figures 5(b) and (c)). The warm sunlit surface layer is separated by a strong temperature-driven density gradient ("thermocline") from the cold deep ocean, preventing rapid communication between the low-latitude ocean surface and the ocean interior. As a result, nutrient supply from below occurs slowly, and phytoplankton completely strip surface waters of available nitrogen and phosphorus; i.e., low- and mid-latitude phytoplankton growth is limited by the major nutrients nitrogen and phosphorus. If the low-latitude surface ocean is so strongly limited by the major nutrients that $[PO_4^{3-}]_{surface}$ would not be raised above zero for any imaginable increase in Q (i.e., if surface ocean plant productivity remains major-nutrient limited over a broad range of surface–deep mixing rates), then changes in upwelling and vertical mixing are balanced by equivalent changes in biological export production. In this case, changes in Q could cause changes in productivity over time, but these would not change the surface–deep gradient in $[DIC]$ maintained by the biological pump.

If this is an adequate description of the real low-latitude ocean, then the strength of its biological pump (the amplitude of the surface–deep gradient in inorganic carbon concentration) is controlled solely by: (i) the carbon/nutrient ratio of the organic matter, and (ii) the nutrient content of the deep ocean (see Equation (5)). If the C/P ratio of export production increased or the $[PO_4^{3-}]$ of the deep ocean increased, then biological production at the surface would drive an increase in the downward flux of carbon that is not at first matched by any change in the upward flux of carbon dioxide associated with surface–deep mixing. A new balance between the downward and upward fluxes of carbon would eventually be reached as the $[DIC]$ gradient between the surface and the deep increases, so that a given amount of mixing transports more CO_2 back into the surface ocean. This increase in surface–deep $[DIC]$ gradient would be mostly due to a drop in dissolved inorganic carbon concentration of the surface ocean, which would lower the p_{CO_2} of the surface ocean and drive an associated decrease in atmospheric CO_2.

Biological production in the ocean tends to incorporate carbon, nitrogen, and phosphorus into biomass in the ratios of 106/16/1 (Redfield, 1942, 1958; Redfield *et al.*, 1963). These ratios determine the amount of inorganic carbon that is sequestered in the deep sea when the supply of nutrients to the surface is completely converted to export production, as is the case in the modern low- and mid-latitude ocean. It is unclear at this point how variable the C/N/P ratios of export production might be through time. When Broecker (1982a) first hypothesized a glacial increase in the strength of the biological pump as the driver of lower glacial CO_2 levels, he considered the possibility (suggested to him by Peter Weyl) that the

C/P export production changes between ice ages and interglacials, with a higher C/P ratio during the last ice age, but he could find neither a mechanism nor direct paleoceanographic evidence for such a change.

Roughly the same situation persists today. We remain exceedingly uncertain about the robustness of the "Redfield ratios" through time and space. However, we do not yet have any mechanistic understanding of their variability (Falkowski, 2000), neither have we recognized a paleoceanographic archive that would help us to assess this variability. The organic matter preserved in the sediment record represents a minute and compositionally distinct fraction of the organic matter exported from the surface ocean. We have no reason to believe that variations in the elemental ratios of this residual sedimentary organic matter will reflect that of export production, especially in the face of changing seafloor conditions.

One of the persistent views of geochemists, to the frustration of their more biologically oriented colleagues, is that the complex recycling of elements in the low-latitude surface ocean has no major significance for the low-latitude biological pump. The argument is that, given the nutrient limitation that predominates in the low-latitude surface, the amount of carbon to be exported is determined by the supply of the major nutrients from the subsurface and the C/N/P uptake ratios of phytoplankton, regardless of the upper ocean recycling of a given nutrient element before export. However, the C/N/P ratios of export production may be altered by ecological effects on the relationship between phytoplankton biomass and the organic matter actually exported from the surface ocean, even without changes in the elemental requirements of the phytoplankton themselves. As a result, the ecology of the surface ocean could drive variations in the strength of the biological pump, for instance, by changing the C/P ratio of export production independent of the C/P ratio of phytoplankton, so as to change the amount of carbon exported for a given phosphorus supply. It was with this reasoning that Weyl originally justified the idea of glacial/interglacial changes in the C/P ratio of export production (Broecker, 1982a). Terrestrial biogeochemists and limnologists are actively seeking a quantitative theory for the ecological constraints imposed by elemental cycling (Elser *et al.*, 2000); oceanographers may benefit from a similar focus on the controls on elemental ratios, in particular, in export production. In any case, while changes in the chemical composition of export production may play a role in the variability of the ocean's biological pump, we are not yet in a position to posit specific hypotheses about such changes or to recognize them in the sedimentary record.

Broecker (1982b) and McElroy (1983) initially hypothesized that the oceanic reservoirs of phosphate and nitrate, respectively, might increase during glacial times, which would allow enhanced low-latitude biological production to lower atmospheric CO_2. However, it was recognized that this basic mechanism would require large changes in nutrient reservoirs to produce the entire observed amplitude of CO_2 change. According to a box-model calculation, the immediate effect of an increase in the nutrient reservoir is to extract CO_2 from the surface ocean and atmosphere, sequestering it in the deep sea, with a 30% increase in the oceanic nutrient reservoir driving a 30–40 ppmv CO_2 decrease (Table 1, see "closed system effects") (Sigman *et al.*, 1998). The longer-term effect on CO_2 from the ocean alkalinity balance depends greatly on whether $CaCO_3$ export increases in step with C_{org} export. It is not known whether a global increase in low-latitude production would increase the $CaCO_3$ rain in step with the C_{org} rain—this would probably depend on the circumstances of the increase—so we should consider both possibilities.

If $CaCO_3$ export increases proportionately with C_{org} export (see "$CaCO_3$ production proportional to C_{org}" in Table 1), the increase in the $CaCO_3$ flux to the seafloor causes a gradual loss of alkalinity from the ocean, which works to raise atmospheric CO_2 (see "open system effects" in Table 1). As a result, only a 15–25 ppmv drop in CO_2 results from a 30% increase in the oceanic major nutrient reservoir (Table 1). However, if $CaCO_3$ production remains constant as C_{org} production increases (see "$CaCO_3$ production constant" in Table 1), the effects on the whole ocean reservoir of alkalinity are minimal (see "open system effects" in Table 1), and the increase in the oceanic nutrient reservoir is much more effective at lowering atmospheric CO_2. In this case, the box model calculation suggests that a 30% increase in the oceanic nutrient reservoir lowers atmospheric CO_2 by ~50 ppmv (Table 1), while some general circulation models would suggest declines in CO_2 of as much as 80 ppm (Archer *et al.*, 2000a). The required increase in the ocean nutrient reservoir could perhaps be lowered by a coincident decrease in $CaCO_3$ export that was modest enough not to violate observations of the

Table 1 Atmospheric CO_2 effects of a 30% increase in the ocean nutrient inventory.

CO_2 changes (ppm)	$CaCO_3$ production	
	Proportional to C_{org}	Constant
Closed system effects	−34	−43
Open system effects	+17	−3
Total CO_2 change	−17	−46

glacial-age sedimentary $CaCO_3$ distribution (Archer and Maier-Reimer, 1994; Brzezinski *et al.*, 2002; Matsumoto *et al.*, 2002b; Sigman *et al.*, 1998). Nevertheless, a 30% increase in oceanic nutrients would require a dramatic change in ocean biogeochemistry.

Given a residence time of >6 thousand years for oceanic phosphate and the nature of the input/output terms in the oceanic phosphorus budget (Froelich *et al.*, 1982; Ruttenberg, 1993), it is difficult to imagine how oceanic phosphate could vary so as to cause the observed amplitude, rate, and phasing of atmospheric CO_2 change (Ganeshram *et al.*, 2002). Entering an ice age, Broecker (1982a) considered the possibility of a large phosphorus input from the weathering of shelf sediments, as ocean water is captured in ice sheets and sea level drops. Ice-core studies dispatched with this admittedly unlikely scenario by demonstrating that, going into an interglacial period, CO_2 begins to rise before sea level rises (Sowers and Bender, 1995), so that the phosphorus inventory decrease would occur too late to explain the rise of CO_2 into interglacials.

There is growing evidence that the nitrogen cycle is adequately dynamic to allow for a large change in the oceanic nitrate reservoir on glacial/interglacial timescales (Codispoti, 1995; Gruber and Sarmiento, 1997). Denitrification, the heterotrophic reduction of nitrate to N_2 gas, is the dominant loss term in the oceanic budget of "fixed" (or bioavailable) nitrogen. Nitrogen isotope studies in currently active regions of denitrification provide compelling evidence that water column denitrification was reduced in these regions during glacial periods (Altabet *et al.*, 1995; Ganeshram *et al.*, 1995, 2002; Pride *et al.*, 1999). In addition, it has been suggested that N_2 fixation, the synthesis of new fixed nitrogen from N_2 by cyanobacterial phytoplankton, was greater during glacial periods because of increased atmospheric supply of iron to the open ocean (iron being a central requirement of the enzymes for N_2 fixation) (Falkowski, 1997). Both of these changes, a decrease in water column denitrification and an increase in N_2 fixation, would have increased the oceanic nitrate reservoir. It has been suggested that such a nitrate reservoir increase would lead to significantly higher export production in the open ocean during glacial periods, potentially explaining glacial/interglacial atmospheric CO_2 changes (Broecker and Henderson, 1998; Falkowski, 1997; McElroy, 1983).

The major conceptual questions associated with this hypothesis involve the feedback of the nitrogen cycle and the strictness with which marine organisms must adhere to the Redfield ratios, both N/P and C/N. Oceanic N_2 fixation provides a mechanism by which the phytoplankton community can add fixed nitrogen to the ocean when conditions favor it. By contrast, riverine input of phosphorus, the major source of phosphate to the ocean, is not directly controlled by marine phytoplankton. Because phytoplankton have no way to produce biologically available phosphorus when there is none available, geochemists have traditionally considered phosphorus to be the fundamentally limiting major nutrient on glacial/interglacial timescales (Broecker and Peng, 1982). For a nitrate reservoir increase alone to drive an increase in low-latitude biological production, the nutrient requirements of the upper-ocean biota must be able to deviate from Redfield stoichiometry to adjust to changes in the N/P ratio of nutrient supply, so as to fully utilize the added nitrate in the absence of added phosphate. Alternatively, if this compensatory shift in biomass composition is incomplete, the surface ocean will tend to shift toward phosphate limitation, and an increase in export production will be prevented.

It was originally suggested by Redfield (Redfield *et al.*, 1963), based on analogy with freshwater systems, that N_2 fixers in the open ocean will enjoy greater competitive success under conditions of nitrate limitation but will be discouraged in the case of phosphate limitation. Having posited this sensitivity, Redfield hypothesized that N_2 fixation acts as a negative feedback on the nitrate reservoir, varying so as to prevent large changes in the nitrate reservoir that are not associated with a coincident change in the phosphate reservoir (Redfield *et al.*, 1963). On the grand scale, this feedback must exist and must put some bounds on the degree to which the oceanic nitrate reservoir can vary independently from phosphorus. However, the quantitative constraint that this negative feedback places on the global nitrate reservoir is not yet known, with one suggested possibility being that iron is so important to N_2 fixers that changes in its supply can overpower the nitrate/phosphate sensitivity described by Redfield, leading to significant variations in the nitrate reservoir over time (Broecker and Henderson, 1998; Falkowski, 1997).

In summary, there are two components of the paradigm of phosphorus's control on the nitrate reservoir and the low-latitude biological pump, both stemming from Redfield's work: (i) oceanic N_2 fixation proceeds in phosphorus-bearing, nitrogen-poor environments, but not otherwise, and (ii) export production has a consistent C/N/P stoichiometry. The first statement would have to be inaccurate for the nitrate reservoir to vary independently of the phosphate reservoir, and the second statement would have to be inaccurate for such an independent change in the nitrate reservoir to actually cause a change in the biological pump. Partly as a reaction to studies that have considered the influence of other parameters, in particular, the supply of iron to open ocean N_2 fixation (Falkowski, 1997), some papers have restated the traditional Redfield-based paradigm (Ganeshram *et al.*, 2002; Tyrrell, 1999). However, the real challenge before us is to test and quantify the rigidity of the Redfield constraints relative to the influence of other environmental parameters. Biogeochemical studies of the modern ocean (Karl et al., 1997; Sanudo-Wilhelmy et al., 2001) and the paleoceanographic record (Haug et al., 1998) are both likely to play a role in the evaluation of these questions regarding Redfield stoichiometry and N_2 fixation, which are, in turn, critical for the hypothesis of the low latitude biological pump as the driver of glacial/interglacial CO_2 change.

The mean concentration of O_2 in the ocean interior would be lowered in the glacial ocean by an increase in low-latitude carbon export, regardless of whether it is driven by an increase in the ocean phosphorus reservoir, the ocean nitrogen reservoir, or a change in the carbon/nutrient ratio of sinking organic matter. As described below, this makes deep-ocean $[O_2]$ a possible constraint on the strength and efficiency of the biological pump in the past. Here, however, we focus on this sensitivity of deep-ocean $[O_2]$ as part of an additional negative feedback in the ocean nitrogen cycle. This feedback may restrict the variability of the oceanic nitrate reservoir, even in the case that Redfield's N_2 fixation-based feedback on the ocean nitrogen cycle is weak and ineffective.

Denitrification occurs in environments that are deficient in O_2. An increase in the ocean's nitrate content will drive higher export production (neglecting, for the moment, the possibility of phosphate limitation). When this increased export production is oxidized in the ocean interior, it will drive more extensive O_2 deficiency. This will lead to a higher global rate of denitrification, which, in turn, will lower the ocean's nitrate content. Thus, the sensitivity of denitrification to the O_2 content of the ocean interior generates a hypothetical negative feedback that may, like the N_2 fixation-based feedback, work to stabilize the nitrate content of the ocean (Toggweiler and Carson, 1995).

While it is generally true that the major nutrients nitrate and phosphate are absent in low- and mid-latitude surface ocean, there are important exceptions (Figures 5(b) and (c)). Wind-driven upwelling leads to nonzero nutrient concentrations and high biological productivity at the surface along the equator and the eastern margins of the ocean basins (e.g., off of Peru, California, and western Africa). Why have we given such short shrift to these biologically dynamic areas in our discussion of the low- and mid-latitude biological pump?

While critically important for ocean ecosystems and potentially important for interannual variations in atmospheric CO_2, the nutrient status in equatorial and coastal upwelling systems does not greatly affect atmospheric CO_2 on centennial or millennial timescales. Above, we demonstrated that greater vertical exchange between the nutrient-poor surface ocean and the deep sea causes

higher export production but does not affect the [DIC] gradient between the surface and the deep sea, so that it would not drive a change in atmospheric CO_2. Following this same reasoning, low-latitude upwellings may generate much export production, but they are not millennial-scale sinks for atmospheric CO_2 because the upwelling brings up both nutrients and respiratory CO_2 from the ocean interior. Neither do the upwellings drive a net loss of CO_2 from the ocean, because the high-nutrient surface waters of the low-latitude upwellings do not contribute appreciably to the ventilation of the ocean subsurface. Put another way, the nutrients supplied to the surface are eventually consumed as the nutrient-bearing surface water flows away from the site of upwelling, before the water has an opportunity to descend back into the ocean interior. This pattern is evident in surface ocean p_{CO_2} data (Takahashi et al., 1997). For instance, in the core of equatorial Pacific upwelling, p_{CO_2} is high both because the water is warming and because deeply sequestered metabolic CO_2 is brought to the surface and evades back to the atmosphere. Adjacent to the equatorial upwelling, the tropical and subtropical Pacific is a region of low p_{CO_2} which is associated with the consumption of the excess surface nutrients that have escaped consumption at the site of upwelling. It does not matter for atmospheric CO_2 whether the upwelled nutrients are converted to export production on-axis of the upwelling system or further off-axis. Thus, while changes in upwelling, biological production, and nutrient status are central to the history of the low-latitude ocean, they have limited importance for atmospheric CO_2 changes on glacial/interglacial timescales. Increased export production driven by higher rates of vertical mixing or upwelling would only play a role in lowering atmospheric CO_2 if it caused a change in the chemical composition of the exported organic material, such as a decrease in its $CaCO_3/C_{org}$ ratio or an increase in the mean depth at which the organic rain is metabolized in the ocean interior (e.g., Boyle, 1988b; Dymond and Lyle, 1985).

The surface chlorophyll distribution shows that coastal environments, even in regions without upwelling, are among the most highly productive in the ocean (Figure 5(a)). However, it is believed that coastal ocean productivity does not have a dominant effect on atmospheric CO_2 over centuries and millennia. As biomass does not accumulate significantly in the ocean, any long-term excess in net primary production relative to heterotrophy must lead to export of organic matter out of the surface layer, in particular, as a downward rain of biogenic particles. In the case of the coastal ocean, this export is stopped by the shallow seafloor, frequently within the depth range that mixes actively with the surface ocean. The high net primary productivity of the coastal ocean owes much to the presence of the shallow seafloor, as the nutrients released from the degradation of settled organic matter are available to the phytoplankton in the sunlit surface. Yet, just as the seafloor prevents the loss of nutrients from the coastal surface ocean, so too does it prevent the export of carbon. Almost as soon as the settled organic carbon is respired, the CO_2 product is free to diffuse back into the atmosphere. Were the same export to have occurred over the open ocean, it would have effectively sequestered the carbon in that biogenic rain within the ocean interior for roughly a thousand years. Thus, the shallow seafloor of the ocean margin leads to high primary productivity while at the same time limiting its effect on the carbon cycle over the timescale of centuries and millennia. It should be noted, however, that the ocean margins are critically important in the global carbon cycle on the timescale of millions of years, due to their role in organic carbon burial.

It has been hypothesized that a significant fraction of the organic matter raining onto the shallow margin is swept over the shelf/slope break, thus transporting organic matter into the ocean interior (Walsh *et al.*, 1981). This has been interpreted as rendering ocean margin productivity important to atmospheric CO_2 on millennial timescales, in that the organic matter being swept off the shelf would be an important route by which CO_2 is shuttled into the deep sea. However, this view should be considered carefully. The ocean margins have high net primary production because of nutrient recycling, but the recycled nutrients were produced by the decomposition of organic matter that would have otherwise been exported out of the system. Mass balance dictates that export of organic carbon from the coastal surface ocean, like export production in the open ocean surface, is set by the net supply of nutrients to the system. If a low-latitude ocean margin receives most of its net nutrient supply from the deep sea, then it is no different than the neighboring open ocean in the quantity of carbon that it can export. There is one caveat to this argument: if organic matter deposited on the shelf can have its nutrients stripped out without oxidizing the carbon back to CO_2 and this "depleted" organic carbon is then transported off the shelf and into the ocean interior, then ocean margin productivity, having been freed from the Redfield constraint on the elemental composition of its export production, could drive a greater amount of organic carbon export for a given amount of nutrient supply.

13.2.2 High-latitude Ocean

Broecker's (Broecker, 1981, 1982a,b; Broecker and Peng, 1982) initial ideas about the biological pump revolutionized the field of chemical oceanography. However, soon after their description, several groups demonstrated that his focus on the low-latitude ocean missed important aspects of the ocean carbon cycle. The thermocline outcrops at subpolar latitudes, allowing deep waters more ready access to the surface in the polar regions. In these regions, the nutrient-rich and CO_2-charged waters of the deep ocean are exposed to the atmosphere and returned to the subsurface before the available nutrients are fully utilized by phytoplankton for carbon fixation. This incomplete utilization of the major nutrients allows for the leakage of deeply sequestered CO_2 back into the atmosphere, raising the atmospheric p_{CO_2}. Work is ongoing to understand what limits phytoplankton growth in these high-latitude regions. Both light and trace metals such as iron are scarce commodities in these regions and together probably represent the dominant controls on polar productivity, with light increasing in importance toward the poles due to the combined effects of low irradiance, sea ice coverage, and deep vertical mixing (see Chapter 12). The Southern Ocean, which surrounds Antarctica, holds the largest amount of unused surface nutrients (Figures 5(b)–(d)), yet the surface chlorophyll suggests that it is perhaps the least productive of the polar oceans; iron and light probably both play a role in explaining this pattern (Martin *et al.*, 1990; Mitchell *et al.*, 1991).

The efficiency of the global biological pump with respect to the ocean's major nutrient content is determined by the nutrient status of the polar regions that ventilate the deep sea, or more specifically, the nutrient concentration of the new water that enters the deep sea from polar regions (Figure 6). Deep water is enriched in nutrients and CO_2 because of the low- and mid-latitude biological pump, which sequesters both nutrients and inorganic carbon in subsurface. In high-latitude regions such as the Antarctic, the exposure of this CO_2-rich deep water at the surface releases this sequestered CO_2 to the atmosphere. However, the net uptake of nutrients and CO_2 in the formation of phytoplankton biomass and the eventual export of organic matter subsequently lowers the p_{CO_2} of surface waters, causing the surface layer to reabsorb a portion of the CO_2 that was originally lost from the upwelled water. Thus, the ratio between export production and the ventilation of CO_2-rich subsurface water, not the absolute rate of either process alone, controls the exchange of CO_2 between the atmosphere and high-latitude surface ocean. The nutrient concentration of water at the time that it enters the subsurface is referred to as its "preformed" nutrient concentration. The preformed nutrient concentration of subsurface water provides an indicator of the cumulative nutrient utilization that it underwent while at the surface, with a higher preformed nutrient concentration indicating lower cumulative nutrient utilization at the surface and

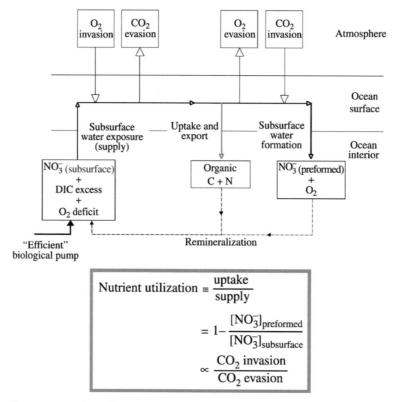

Figure 6 The effect on atmospheric CO_2 of the biological pump in a region of deep-ocean ventilation where the major nutrients are not completely consumed, such as the Antarctic zone of the Southern Ocean. In this figure, NO_3^- is intended to represent the major nutrients and thus should not be distinguished conceptually from PO_4^{3-}. The biological pump causes deep water to have dissolved inorganic carbon (DIC) in excess of its "preformed" [DIC] (the concentration of DIC in the water when it left the surface; see lower left box, which represents water in the ocean interior). The low- and mid-latitudes (and a few polar regions, such as the North Atlantic) house a biological pump that is "efficient" in that it consumes all or most of the major nutrient supply. As a result, it drives the DIC excess toward a concentration equivalent to the deep $[NO_3^-]$ multiplied by the C/N ratio of exported organic matter (left lower arrow represents this influence on deep-water chemistry). The respiration of organic matter in the ocean interior also leads deep water to have an O_2 deficit relative to saturation. In high-latitude regions such as the Antarctic, the exposure of this deep water at the surface releases this sequestered CO_2 to the atmosphere and leads to the uptake of O_2 from the atmosphere (upper left). However, the net uptake of nutrients and CO_2 in the formation of phytoplankton biomass and the eventual export of organic matter ("uptake and export") subsequently lowers the p_{CO_2} of surface waters, causing the surface layer to reabsorb a portion of the CO_2 that was originally lost from the upwelled water (upper right). The net excess in photosynthesis to respiration that drives this export out of the surface also produces O_2, some of which evades to the atmosphere, partially offsetting the initial O_2 uptake by the surface ocean. While the down-going water has lost its DIC excess by exchange with the atmosphere, a portion of the original DIC excess will be reintroduced in the subsurface when the exported organic matter is remineralized (the more complete the nutrient consumption in the surface, the greater the DIC excess that is reintroduced in the subsurface). The ratio between export and the ventilation of CO_2-rich subsurface water, which controls the net flux of CO_2 between the atmosphere and surface ocean in this region, can be related to the nutrients. Nutrient utilization, defined as the ratio of the rate of nutrient uptake (and export) to the rate of nutrient supply, expresses the difference between nutrient concentration of rising and sinking water. Nutrient utilization relates directly (through the Redfield ratios) to the ratio of CO_2 invasion (driven by organic matter export) relative to evasion (driven by deep-water exposure at the surface). The lower the nutrient utilization, the lower the ratio of invasion to evasion and thus the more this region represents a leak in the biological pump. This diagram overlooks a number of important facts. For instance, ocean/atmosphere CO_2 exchange is not instantaneous, so water can sink before it has reached CO_2 saturation with respect to the atmosphere.

thus a greater leak in the biological carbon pump. This picture overlooks a number of important facts. For instance, ocean/atmosphere CO_2 exchange is not instantaneous, so water can sink before it has reached CO_2 saturation with respect to the atmosphere. In polar regions, this typically means that deep water may leave the surface before it has lost as much CO_2 to the atmosphere as it might have (Stephens and Keeling, 2000).

The global efficiency of the biological pump can be evaluated in terms of a global mean preformed nutrient concentration for surface water that is folded into the ocean interior. In the calculation of this mean preformed nutrient concentration, each subsurface water formation term is weighted according to the volume of the ocean that it ventilates because this corresponds to the amount of CO_2 that could be sequestered in a given portion of the ocean interior. As a result, the nutrient status of surface ocean regions that directly ventilate the ocean subsurface (deep, intermediate, or thermocline waters) have special importance for atmospheric CO_2.

The high-latitude North Atlantic and the Southern Ocean are the two regions that appear to dominate the ventilation of the modern deep ocean. Through the preformed nutrient concentrations of the newly formed deep waters, their competition to fill the deep ocean largely determines the net efficiency of the global biological pump (Toggweiler *et al.*, 2003b). Of these two regions, the North Atlantic has a low preformed nutrient concentration and thus drives the ocean toward a high efficiency for the biological pump. In contrast, the preformed nutrient concentration of Southern Ocean-sourced deep water is high and thus drives the global biological pump toward a low efficiency.

Within the polar oceans involved in deep water formation, certain regions are more important than others. The Antarctic Zone, the most polar region in the Southern Ocean, is involved in the formation of both deep and intermediate-depth waters, making this region important to the atmosphere/ocean CO_2 balance. The quantitative effect of the Subantarctic Zone on atmospheric CO_2 is less certain, depending on the degree to which the nutrient status of the Subantarctic surface influences the preformed nutrient concentration of newly formed subsurface water (Antarctic Intermediate Water and Subantarctic mode water), but its significance is probably much less than that of the Antarctic.

Questions regarding the most critical regions extend to even smaller spatial scales. For instance, with regard to deep-ocean ventilation in the Antarctic, it is uncertain as to whether the entire Antarctic surface plays an important role in the CO_2 balance, or whether instead only the specific locations of active deep water formation (e.g., the Weddell Sea shelf) are relevant. This question hinges largely on whether the surface water in the region of deep water formation exchanges with the open Antarctic surface, or whether it instead comes directly from the shallow subsurface and remains isolated until sinking. If the latter is true, then the open Antarctic may be irrelevant to the chemistry of the newly formed deep water. This would be problematic for paleoceanographers, since essentially all of the reliable paleoceanographic records come from outside these rather special environments. There is, however, a new twist to this question. While deep water formation in the modern Antarctic has long been

thought of as restricted to the shelves, efforts to reproduce deep-sea nutrient chemistry and radiochemistry raise the possibility that the deep ocean may also be ventilating in a more diffuse mode throughout the open Antarctic (Broecker *et al.*, 1998; Peacock, 2001). If this is the case, the significance of the open Antarctic in the CO_2 balance is less susceptible to uncertainties about lateral exchange with the shelf sites of deep water formation. In the discussion that follows, we assume that the open Antarctic does affect the CO_2 exchanges associated with deep water formation, either by active exchange of surface water with the specific sites of deep water formation or by its actual participation in the ventilation process. In any case, we must improve our understanding of the modern physical and chemical oceanography of the polar regions if we are to determine their roles in atmospheric CO_2 change.

In their hypothesis of a glacial decrease in Southern Ocean CO_2 leak, early workers considered two causes: (i) an increase in biological export production and (ii) a decrease in the exposure rate of deep waters at the polar surface (Figure 7). One can imagine processes that would have reduced the evasion of CO_2 from the Southern Ocean by either of the two mechanisms mentioned above. For instance, an increase in the input of dust and its associated trace metals to the Southern Ocean might have driven an increase in the rate of nutrient and carbon uptake by phytoplankton (Martin, 1990). Alternatively, an increase in the salinity-driven stratification of the Antarctic and/or a decrease in wind-driven upwelling could have lowered the rate of nutrient and carbon dioxide supply to the surface (discussed further in Section 13.3) (Francois *et al.*, 1997; Keeling and Visbeck, 2001; Sigman and Boyle, 2000, 2001; Toggweiler and Samuels, 1995). Under most conditions, these two hypothesized changes do not have the same quantitative effect on atmospheric CO_2. For instance, stratification can cause a significantly larger amount of CO_2 decrease than does increased export production for a given amount of nutrient drawdown in Southern Ocean surface waters (Sigman *et al.*, 1999b).

These differences can be understood in the context of the mean preformed nutrient concentration of the ocean interior (Figure 8). There are two Southern Ocean mechanisms by which the global efficiency of the biological pump may be increased during glacial times: (i) decreasing the preformed nutrient concentration of Southern-Ocean-sourced deep water (Figures 8(c) and (d)), and (ii) decreasing the importance of Southern-Ocean-sourced deep waters in the ventilation of the deep ocean, relative to the North Atlantic or some other source of deep water with low preformed nutrients (Figures 8(b) and (d)). Increasing surface productivity would lower the preformed

468 *The Biological Pump in the Past*

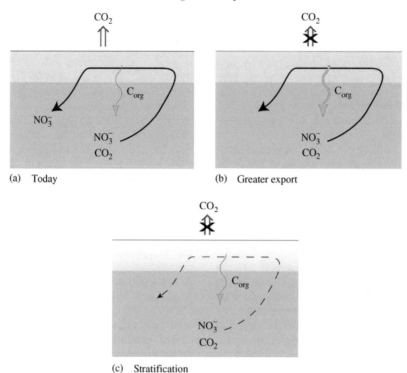

Figure 7 Schematic illustrations of the CO_2 leak from the Southern Ocean to the atmospheric that exists today and of two alternative scenarios by which this leak might have been stopped during ice ages. The incomplete consumption of nutrients in the modern Southern Ocean causes waters that come to the surface in the Southern Ocean to release CO_2 to the atmosphere because not all of the nutrient supply to the surface drives a downward rain of organic material, allowing CO_2 sequestered by the lower latitude biological pump to escape (a). During ice ages, atmospheric CO_2 could have been reduced by increasing the nutrient efficiency of the biological pump in the Southern Ocean (b and c). This could have been driven by an increase in biological productivity (b), which would have actively increased the downward flux of carbon, or by density "stratificaton" of the upper ocean, allowing less nutrient- and CO_2-rich water to the Southern Ocean surface (c). As described in Figure 8, because of the existence of other regions of deep water formation, the evasion of CO_2 is reduced upon stratification (c) even if the degree of nutrient consumption in the surface does not change.

nutrient concentration of the deep sea solely by reducing the preformed nutrient concentration of the Southern Ocean contribution to deep water (Figure 8(c)). Stratification of the Southern Ocean surface would reduce the fraction of the deep ocean ventilated by the Southern Ocean, allowing the ocean interior to migrate toward the lower preformed nutrient concentration of northern-sourced deep water (Figure 8(b)). In addition, if upon stratification phytoplankton growth rate does not decrease as much as does the supply of nutrients to the surface, then stratification would also lower the preformed nutrient concentration of southern-sourced deep waters (Figure 8(d)). In this case, stratification lowers the preformed nutrient concentration by both (i) and (ii) above, leading to a lower net preformed nutrient concentration for the ocean interior and thus making it a more potent mechanism for lowering atmospheric CO_2. Of course, there are important additional considerations. First, a decrease in Southern Ocean deep water might allow the deep-sea temperature to rise, which would push CO_2 back into

the atmosphere. This apparently did not occur during glacial times (Schrag *et al.*, 1996), suggesting that all glacial-age deep-water sources were very cold. Second, and more importantly, North Atlantic deep water formation may have been weaker during the last ice age (Boyle and Keigwin, 1982). As this deep-water source is the low preformed nutrient end-member today, its reduction would have worked to increase the preformed nutrient concentration of the deep ocean during the last ice age, potentially offsetting the effects of a decrease in Southern-sourced deep water. These questions about actual events aside, the mean preformed nutrient concentration of the deep ocean provides an important predictive index for the effect of changes in the polar ocean on atmospheric CO_2.

Some attention has focused on the prevention of CO_2 release from the Antarctic surface by sea ice cover as a potential driver of glacial/interglacial CO_2 change (Stephens and Keeling, 2000). Though the mechanism is essentially physical, it nevertheless would cause an increase in the global efficiency of the biological pump by removing its polar leak. This is an important reminder that the biological pump is not solely a biological process, but rather arises from the interaction between the biology and physics of the ocean. For prevention of CO_2 release by sea ice cover to be the sole mechanism for glacial/interglacial CO_2 change, nearly complete year-round ice coverage of the Antarctic would be required. This seems unlikely to most investigators (Crosta *et al.*, 1998). As a result, a hybrid hypothesis has been proposed for the glacial Antarctic, in which intense surface stratification and nutrient consumption during the summer was followed by wintertime prevention of gas exchange due to ice cover (Moore *et al.*, 2000). While promising, this mechanism faces a discrepancy with the evidence for lower export production in the glacial Antarctic (see below). In the modern Antarctic, nutrients are supplied to the surface largely by wintertime vertical mixing. Thus, without year-round stratification, higher annual export production would have been required to lower the surface nutrient concentration below current summertime levels, even for a brief summer period. An alternative combined gas exchange limitation/stratification mechanism that does not violate the productivity constraint is that of spatial segregation of the two mechanisms within the Antarctic (Figure 9). Near the Antarctic continental margin, the formation of thick ice cover and its associated salt rejection could drive overturning, but the near-complete ice cover in that region

could prevent CO_2 release. The ice formed along the coast would eventually drift into the open Antarctic, where it would melt, thus stratifying the water column and preventing the release of CO_2 from open Antarctic as well. In any case, ice cover provides an alternative or additional potential mechanism for preventing the release of biologically sequestered CO_2 from the Antarctic, and future paleoceanographic work must try to provide constraints on its actual importance.

13.3 TOOLS

While there are many aspects of biological productivity that one might hope to reconstruct through time, there are two parameters that are most fundamental to the biological pump: (i) export production and (ii) nutrient status. In regions where a deviation from major-nutrient limitation is highly unlikely, in particular, the tropical and subtropical ocean, export production is the major regional constraint that local paleoceanographic data can provide. For instance, if export production in the tropics was higher some time in the past, then there are at least two plausible explanations: (i) there was more rapid water exchange between the surface and the nutrient-bearing subsurface water, or (ii) there was an increase in the nutrient content of the subsurface water (and/or in N_2 fixation in the surface; see discussion above). While physically driven changes in nutrient supply (e.g., changes in upwelling intensity) have limited significance for the biological pump, changes

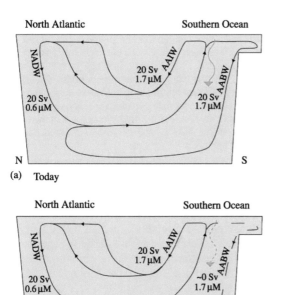

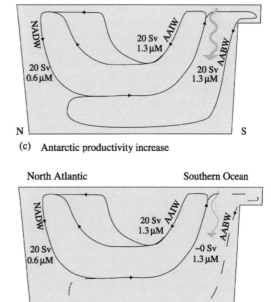

Figure 8 (Continued).

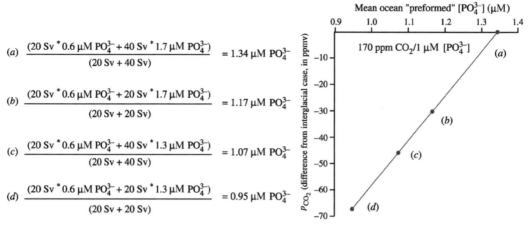

(e) Calculation of mean preformed phosphate for (a)–(d)   (f) Atmospheric CO_2 vs. mean preformed phosphate for (a)–(d)

Figure 8 Cartoons depicting the ventilation of the ocean interior by the North Atlantic and Southern Ocean and the preformed $[PO_4^{3-}]$ of the different ventilation terms, for today (a) and three hypothesized ice-age conditions ((b), (c), and (d)); calculation of the mean preformed $[PO_4^{3-}]$ of the ocean interior for each of these cases (e); and the effects of the hypothesized glacial changes (b), (c), and (d) on atmospheric CO_2 compared with the change in mean preformed $[PO_4^{3-}]$, as calculated by the CYCLOPS ocean geochemistry box model (f) (Keir, 1988; Sigman et al., 1998). In the interglacial case (a), there are three sources of subsurface water shown: North Atlantic Deep Water (NADW), Antarctic Intermediate Water (AAIW), and Antarctic Bottom Water (AABW). In this simple picture, NADW and AAIW formation are related; NADW is drawn into the Antarctic surface by the wind-driven ("Ekman") divergence of surface waters and then forms AAIW because of surface-water convergence further north (Gordon et al., 1977; Toggweiler and Samuels, 1995). The rates of formation (in Sverdrups, $10^6 \, m^3 \, s^{-1}$) and preformed $[PO_4^{3-}]$ (in μM) are from the CYCLOPS model. In (b), stratification is assumed to shut off AABW formation. Because we do not change NADW, we are forced to leave AAIW unchanged as well. In (b), export production in the Antarctic decreases proportionally with the decrease in upwelling to the surface, such that surface $[PO_4^{3-}]$ remains at 1.7 μM. This would be the expected outcome if, for instance, the Antarctic is iron limited and most of the iron comes from upwelling, during both interglacials and ice ages. In (d), the same circulation change occurs as in (b), but in this case export production does not decrease as upwelling decreases, so that surface $[PO_4^{3-}]$ drops to ~1.3 μM. This would be the expected outcome if, for instance, the Antarctic iron is supplemented by a high dust input during ice ages. In (c), circulation remains constant but Antarctic export production increases (as indicated by the increase in the downward green arrow). This would be the expected outcome if, for instance, the combined input of iron from dust supply and deep water is greater during ice ages (Martin, 1990; Watson et al., 2000). In (e), the mean preformed $[PO_4^{3-}]$ of the ocean interior is calculated as the total phosphate input to the ocean interior divided by the ventilation rate (North Atlantic terms on the left, Antarctic terms on the right). In (f), the calculated change in atmospheric p_{CO_2} (relative to the interglacial case (a)) is plotted against the mean preformed $[PO_4^{3-}]$ of each case, as calculated in (e). The strong correlation between these two parameters (a 170 ppm decrease in CO_2 for a 1 μM decrease in $[PO_4^{3-}]$) indicates that the mean preformed $[PO_4^{3-}]$ is an excellent predictor for the effect of a given change on the strength of the biological pump. In these experiments, the mean gas exchange temperature for the ocean interior was held constant, so that no change in the "solubility pump" was allowed (see text). In addition, calcium carbonate dynamics are included in these model experiments, so that a fraction of the CO_2 response (~25%) is not the result of the biological pump, in the most strict sense. Finally, none of these experiments address the evidence that NADW formation was reduced during glacial times (Boyle and Keigwin, 1982).

in deep-ocean nutrient content could drive a strengthening of the pump and thus an associated decline in atmospheric CO_2 (see discussion above). Thus, it is critical for studies of the biological pump not only to reconstruct changes in low-latitude export production but also to determine what caused the changes.

In the polar ocean, both export production and nutrient status must be known to determine the impact on atmospheric CO_2, because both of these terms are needed to determine the ratio of CO_2 supply from deep water to CO_2 sequestration by export production (Figure 7). For instance, a decrease in export production associated with an

increase in surface nutrients would imply an increased leak in the biological pump, whereas a decrease in export production associated with reduced nutrient availability would imply a smaller leak in the pump. In low-latitude regions of upwelling, nutrient status is less important from the perspective of the global biological pump. Nevertheless, since nutrient status is potentially variable in these environments, it must be constrained to develop a tractable list of explanations for an observed change in productivity.

A great many approaches have been taken to study the biological pump in the past. While there are approaches that take advantage of virtually every

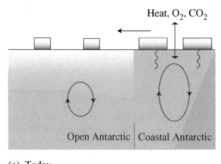

(a) Today

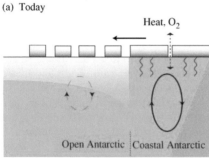

(b) Hybrid sea ice/stratification hypothesis

Figure 9 North-South (left-to-right) depth section of the shallow Antarctic, (a) under modern conditions, and (b) under glacial conditions as posited by a hybrid hypothesis involving both limitation of gas exchange by sea ice near the Antarctic continent and sea ice melt-driven stratification of the open Antarctic. During glacial times, sea ice formation was likely vigorous in the Antarctic, and the salt rejection during freezing would have worked to increase the density of the surface, perhaps driving vigorous overturning (bold loop close to the Antarctic continent) (Keeling and Stephens, 2001). While this alone would act to ventilate the deep sea and release CO_2 to the atmosphere, the ice coverage may have been adequately dense to prevent CO_2 outgassing, which occurs more slowly than the exchange of heat and O_2 (Stephens and Keeling, 2000). Here, we posit that this region of ice formation and ventilation was limited to the waters close to the Antarctic continent. The winds in the Antarctic would have drawn the ice northward, and some of it would have melted in the open Antarctic. This would have worked to strengthen the vertical stratification in the open Antarctic, which may have reduced CO_2 loss from the ocean in this region (Francois *et al.*, 1997). Depending on the predominant route of deep-ocean ventilation in the modern Antarctic (coastal or open ocean, see text), either the coastal sea ice coverage or the open Antarctic stratification may be taken as the most important limitation on CO_2 release. This hybrid hypothesis has a significant list of requirements in order to be feasible. For instance, the surface waters in the hypothesized region of convection must not mix laterally with the surface waters of the open Antarctic.

aspect of the available geologic archives, we focus here on tools to reconstruct export production and nutrient status and limit ourselves to those that are largely based on geochemical measurement. We refer the interested reader to Fischer and Wefer (1999) for additional information and references.

13.3.1 Export Production

Export production, as defined above, is the flux of organic carbon that sinks (or is mixed) out of the surface ocean and into the deep sea. While it is extremely difficult to imagine how paleoceanographic measurements can provide direct constraints on this highly specific parameter, approaches exist for the reconstruction of the flux of biogenic debris to the seafloor. To the degree that sinking biogenic debris has a predictable chemical composition and that its export out of the surface ocean is correlated with its rain rate to the seafloor, these approaches can provide insight into export production variations (Muller and Suess, 1979; Ruhlemann *et al.*, 1999; Sarnthein *et al.*, 1988) (see Chapter 12).

Only a very small fraction of the organic carbon exported out of the surface ocean accumulates in deep-sea sediments, with loss occurring both in the water column and at the seafloor. Moreover, various environmental conditions may influence the fraction of export production preserved in the sedimentary record. The mineral components of the biogenic rain (calcium carbonate, opal) do not represent a source of chemical energy to the benthos and thus may be preserved in a more predictable fashion in deep-sea sediments. Thus, paleoceanographers sometimes hope to reconstruct export production (the flux of organic carbon out of the surface ocean) from the biogenic rain of opal or carbonate to the seafloor (Ruhlemann *et al.*, 1999). We know that systematic variations in the ratio of opal or calcium carbonate to organic carbon in export production do occur, so there are many cases where changes in the mineral flux can be interpreted either as a change in the rate of export production or in its composition. Despite these limitations, we are better off with this information than without it.

The most basic strategy to reconstruct the biogenic rain to the seafloor is to measure accumulation rate in the sediments. This approach is appropriate for materials that accumulate without loss in the sediments or that are preserved to a constant or predictable degree. For the biogenic components of interest, organic carbon, opal, and calcium carbonate, the degree of preservation varies with diverse environmental variables, detracting from the usefulness of accumulation rate for the reconstruction of their rain to the seafloor. Nevertheless, because the environmental variables controlling preservation in the sediments are often linked to rain rate to the seafloor, accumulation rate can sometimes be a sound basis for at least the qualitative reconstruction of changes in some components of the biogenic rain. For instance, the fraction of the opal rain that is preserved in the sediment appears to be higher in opal-rich environments (Broecker and Peng, 1982). As a result, an observed increase in sedimentary opal accumulation rate may have been partially due to an increase in preservation. However, an increase in opal rain (and thus diatom-driven export) would have been needed to increase the sedimentary opal content in the first place.

If an age model can be developed for a deep-sea sediment core, then the accumulation histories of the various sediment components can be reconstructed. Assuming that the age model is correct, the main uncertainty in the environmental significance of the reconstruction is then the potential for lateral sediment transport. Sediments can be winnowed from or focused to a given site on the seafloor.

The geochemistry of radiogenic thorium provides a way to evaluate the effect of lateral sediment transport and related processes (Bacon, 1984; Suman and Bacon, 1989). Thorium-230 (^{230}Th) is produced at a constant rate throughout the oceanic water column by the decay of dissolved ^{234}U. As ^{230}Th is produced, it is almost completely scavenged onto particles at the site of its production. As a result, the accumulation rate of ^{230}Th in deep-sea sediments should match its integrated production in the overlying water column. If the accumulation of ^{230}Th over a time period, as defined by a sediment age model, is more or less than should have been produced in the overlying water column during that period, then sediment is being focused or winnowed, respectively.

Thorium-230 is also of great use as an independent constraint on the flux of biogenic material to the seafloor (Bacon, 1984; Suman and Bacon, 1989). Because the production rate of ^{230}Th in the water column is essentially constant over space and time, its concentration in sinking particles is diluted in high-flux environments with a large biogenic flux to the seafloor, yielding lower sedimentary concentrations of ^{230}Th in these environments. One limitation on the use of ^{230}Th is the radioactive decay of this isotope, which has a half-life of 7.5×10^4 yr. Helium-3 can apparently be used in similar ways to ^{230}Th and has the advantage that it is a stable isotope (Marcantonio *et al.*, 1996); however, it has been studied less than ^{230}Th.

The addition of other scavenged elements further enhances the utility of ^{230}Th. ^{231}Pa, like ^{230}Th, is produced throughout the ocean water column by radioactive decay of a uranium isotope (^{235}U in the case of ^{231}Pa). However, protactinium is somewhat less particle-reactive than thorium. As a result, the downward flux of ^{231}Pa to the seafloor varies across the ocean. Protactinium from low-flux regions mixes into high-flux regions to be scavenged, resulting in higher ^{231}Pa fluxes in environments with high biogenic rain rates. Thus, the ^{231}Pa/^{230}Th ratio of sediments has been studied as an index of the particle flux through the water column. This index is potentially complementary to

thorium-normalized accumulation rates in that the ^{231}Pa/^{230}Th ratio should not be affected by losses during sedimentary diagenesis. Limitations of this approach include (i) the tendency for the less easily scavenged isotope (^{231}Pa) to adsorb preferentially to certain types of particles (e.g., diatom opal), which can make it difficult to distinguish a change in the flux of particles from a compositional change, and (ii) the scavenging of protactinium on basin margins, which can vary in importance as ocean circulation changes (Chase *et al.*, 2003b; Walter *et al.*, 1999; Yu *et al.*, 2001). Beryllium-10 may be used in similar ways to ^{231}Pa but has been studied less (Anderson *et al.*, 1990).

The flux of barium to the seafloor appears to be strongly related to the rain of organic matter out of the surface ocean (Dehairs *et al.*, 1991). Apparently, the oxidation of organic sulfur produces microsites within sinking particles that become supersaturated with respect to the mineral barite ($BaSO_4$). On this basis, barium accumulation has been investigated and applied as a measure of export production in the past, representing a more durable sedimentary signal of the organic carbon sinking flux than sedimentary organic carbon itself (Dymond *et al.*, 1992; Gingele *et al.*, 1999; Paytan *et al.*, 1996). While debates continue on aspects of the biogenic barium flux, some problems with preservation are broadly recognized. In sedimentary environments with low bottom water O_2 and/or high organic carbon rain rates, active sulfate reduction in the shallow sediments can cause the barite flux to dissolve. In addition, there is some low level of barite dissolution under all conditions; if the biogenic barium flux is low, a large fraction of the barite can dissolve at the seafloor. Thus, barium accumulation studies appear to be most applicable to environments of intermediate productivity.

The rapidly growing field of organic geochemistry promises new approaches for the study of biological productivity in the past. By studying specific chemical components of the organic matter found in marine sediments, uncertainties associated with carbon source can be removed, and a richer understanding of past surface conditions can be developed (e.g., Hinrichs *et al.*, 1999; Martinez *et al.*, 1996). To some degree, organic geochemical approaches may allow us to circumvent the thorny problem of breakdown and alteration at the seafloor (Sachs and Repeta, 1999).

13.3.2 Nutrient Status

The measurable geochemical parameters currently available for addressing the nutrient status of the surface ocean include (i) the Cd/Ca ratio (Boyle, 1988a) and ^{13}C/^{12}C of planktonic (surface-dwelling) foraminiferal carbonate (Shackleton *et al.*, 1983), which are intended to record the concentration of cadmium and the $^{13}C/^{12}C$ of DIC in surface water, (ii) the nitrogen isotopic composition of bulk sedimentary organic matter (Altabet and Francois, 1994a) and microfossil-bound organic matter (Shemesh *et al.*, 1993; Sigman *et al.*, 1999b), which may record the degree of nitrate utilization by phytoplankton in surface water, (iii) the silicon isotopic composition (De La Rocha *et al.*, 1998) and Ge/Si ratio (Froelich and Andreae, 1981) of diatom microfossils, which may record the degree of silicate utilization in surface water (but see Bareille *et al.* (1998)), and (iv) the carbon isotopic composition of organic matter in sediments (Rau *et al.*, 1989) and diatom microfossils (Rosenthal *et al.*, 2000; Shemesh *et al.*, 1993), which may record the dissolved CO_2 concentration of surface water and/or the carbon uptake rate by phytoplankton.

The concentration of dissolved cadmium is strongly correlated with the major nutrients throughout the global deep ocean, and the Cd/Ca ratio of benthic (seafloor-swelling) foraminifera shells records the cadmium concentration of seawater (Boyle, 1988a). As a result, benthic foraminiferal Cd/Ca measurements allow for the reconstruction of deep-ocean nutrient concentration gradients over glacial/interglacial cycles. Planktonic foraminiferal Cd/Ca measurements in surface water provide an analogous approach to reconstruct surface ocean nutrient concentration, although this application has been less intensively used and studied.

A number of factors appear to complicate the link between planktonic foraminiferal Cd/Ca and surface nutrient concentrations in specific regions. First, temperature may have a major effect on the Cd/Ca ratio of planktonic foraminifera (Elderfield and Rickaby, 2000; Rickaby and Elderfield, 1999). Second, planktonic foraminiferal shell growth continues below the surface layer (Bauch *et al.*, 1997; Kohfeld *et al.*, 1996) and probably integrates the Cd/Ca ratio of surface and shallow subsurface waters. This is of greatest concern in polar regions such as the Antarctic, where there are very sharp vertical gradients within the depth zone in which planktonic foraminferal calcification occurs. Third, carbonate geochemistry on the deep seafloor may affect the Cd/Ca ratio of foraminifera preserved in deep-sea sediments (McCorkle *et al.*, 1995). Finally, the link between cadmium and phosphate concentration in surface waters is not as tight as it is in deeper waters (Frew and Hunter, 1992).

Because of isotopic fractionation during carbon uptake by phytoplankton, there is a strong correlation between the $^{13}C/^{12}C$ of DIC and nutrient concentration in deep waters; as a result, the $^{13}C/^{12}C$ of benthic foraminiferal fossils is a central tool in paleoceanography. The $^{13}C/^{12}C$ of planktonic foraminiferal calcite has been used as a tool to study the strength of the biological pump (Shackleton *et al.*, 1983). However, the exchange of CO_2 with the atmosphere leads to a complicated relationship between the $^{13}C/^{12}C$ of DIC and the nutrient concentration in surface waters (Broecker and Maier-Reimer, 1992), such that even a perfect reconstruction of surface water $^{13}C/^{12}C$ would not provide direct information of surface-ocean nutrient status. In addition, the $^{13}C/^{12}C$ of planktonic foraminiferal fossils found in surface sediments appears to be an imperfect recorder of the $^{13}C/^{12}C$ of DIC in modern surface waters, for a variety of reasons (Spero *et al.*, 1997; Spero and Lea, 1993; Spero and Williams, 1988). Finally, the same concerns about the calcification depth noted for Cd/Ca also apply to carbon isotopes or, for that matter, any geochemical signal in planktonic foraminiferal calcite.

During nitrate assimilation, phytoplankton preferentially consume ^{14}N-nitrate relative to ^{15}N-nitrate (Montoya and McCarthy, 1995; Waser *et al.*, 1998), leaving the surface nitrate pool enriched in ^{15}N (Sigman *et al.*, 1999a). This results in a correlation between the $^{15}N/^{14}N$ ratio of organic nitrogen and the degree of nitrate utilization by phytoplankton in surface waters (Altabet and Francois, 1994a,b). There are major uncertainties in the use of this correlation as the basis for paleoceanographic reconstruction of nutrient status, which include (i) the isotopic composition of deep-ocean nitrate through time, (ii) temporal variations in the relationship between nitrate utilization and the nitrogen isotopes in the surface ocean (i.e., the "isotope effect" of nitrate assimilation), and (iii) the survival of the isotope signal of sinking organic matter into the sedimentary record (e.g., Lourey *et al.*, in press). The nitrogen isotope analysis of microfossil-bound organic matter (Sigman *et al.*, 1999b) and of specific compound classes such as chlorophyll-degradation products (Sachs and Repeta, 1999) promises to provide tools to evaluate and circumvent the effect of diagenesis. However, it remains to be seen whether selective nitrogen pools such as microfossil-bound nitrogen are tightly linked to the nitrogen isotope ratio of the integrated sinking flux, the parameter that theoretically relates most directly to the degree of nitrate utilization in surface waters (Altabet and Francois, 1994a).

The isotopic composition of silicon in diatom opal has been investigated as a proxy for the degree of silicate utilization by diatoms, based on the fact that diatoms fractionate the silicon isotopes (^{30}Si and ^{28}Si) during uptake (De La Rocha *et al.*, 1997, 1998). This application is analogous to the use of nitrogen isotopes to study nitrate utilization, but with important differences. On the one hand, the upper ocean cycle is arguably simpler for silica than bio-available nitrogen, which bodes well for the silicon isotope system. On the other hand, there are very few regions of the surface ocean that maintain high dissolved silicate concentrations (Figure 4(d)). As a result, in regions of strong silicate gradients, the link between silicate utilization and silicon isotopic composition may be compromised by mixing processes in surface waters.

The carbon isotopic composition of sedimentary organic matter was originally developed as a paleoceanographic proxy for the aqueous CO_2 concentration of Southern Ocean surface waters (Rau et al., 1989). The aqueous CO_2 concentration is a nearly ideal constraint for understanding a region's effect on the biological pump, as it would provide an indication of its tendency to release or absorb carbon dioxide. However, it has become clear that growth rate and related parameters are as important as the concentration of aqueous CO_2 for setting the $^{13}C/^{12}C$ of phytoplankton biomass and the sinking organic matter that it yields (Popp et al., 1998). In addition, active carbon acquisition by phytoplankton is also probably important for phytoplankton $^{13}C/^{12}C$ in at least some environments (Keller and Morel, 1999). Thus, the $^{13}C/^{12}C$ of organic carbon is a useful paleoceanographic constraint (Rosenthal et al., 2000), but one that is currently difficult to interpret in isolation.

13.3.3 Integrative Constraints on the Biological Pump

Above, we have focused on approaches to reconstruct the role of a specific region of the surface ocean on the global biological pump. However, if our goal is to explain the global net effect of ocean biology on the carbon cycle, we must also search for less local, more integrative constraints on the biological pump. This is possible because the atmosphere, surface ocean and deep sea are each relatively homogenous geochemical reservoirs, while being distinct from one another. There are a number of global scale geochemical parameters that may provide important constraints on the biological pump; we describe several of these below.

13.3.3.1 Carbon isotope distribution of the ocean and atmosphere

As described above, the biological pump tends to sequester ^{12}C-rich carbon in the ocean interior. All else being equal, the stronger the global biological pump, the higher will be the $^{13}C/^{12}C$ of dissolved inorganic carbon in the surface ocean and of carbon dioxide in the atmosphere. Broecker (1982a,b) and Shackleton (see Shackleton et al., 1983) compared sediment core records of the $^{13}C/^{12}C$ of calcite precipitated by planktonic and benthic foraminifera, the goal being to reconstruct the $^{13}C/^{12}C$ difference in DIC between the surface ocean and the deep sea, a measure of the strength of the global ocean's biological pump. Indeed, this work was the first suggestion that the biological pump was stronger during ice ages, thus potentially explaining the lower CO_2 levels of glacial times. Our view of these results is now more complicated (e.g., Spero et al., 1997); however, the basic conclusion remains defensible (Hofmann et al., 1999). Reliable measurement of the $^{13}C/^{12}C$ of atmospheric CO_2 has proven challenging (Leuenberger et al., 1992; Marino and McElroy, 1991; Smith et al., 1999). Moreover, there are additional modifiers of the $^{13}C/^{12}C$ of atmospheric CO_2, such as the temperature of gas exchange. Nevertheless, these data also seem consistent with the biological pump hypothesis for glacial/interglacial CO_2 change (Smith et al., 1999).

13.3.3.2 Deep-ocean oxygen content

The atmosphere/ocean partitioning of diatomic oxygen (O_2) is a potentially important constraint on the strength of the biological pump; the stronger the pump, the more O_2 will be shuttled from the deep ocean to the surface ocean and atmosphere. The rain of organic detritus into the deep ocean drives an O_2 demand by the deep-ocean benthos as it sequesters carbon dioxide in the deep sea, while exposure of deep waters at the surface recharges them with O_2 as it allows deep waters to degas excess CO_2 to the atmosphere (Figure 6). A decrease in atmospheric CO_2 due to the biological pump should, therefore, be accompanied by a decrease in the O_2 content of the ocean subsurface.

The concentration of dissolved O_2 in the ocean interior has long been a target for paleoceanographic reconstruction (the atmospheric change in O_2 content would be minute). Sediments underlying waters with nearly no O_2 tend to lack burrowing organisms, so that sediments in these regions are undisturbed by "bioturbation" and can be laminated; this is perhaps our most reliable paleoceanographic indicator of deep-water O_2 content. Arguments have been made for surface area-normalized sedimentary organic carbon content as an index of O_2 content in some settings (Keil and Cowie, 1999). It remains to be seen whether this is complicated by the potential for changes in the rain rate of organic matter to the sediments. There are a number of redox-sensitive metals, the accumulation of which gives information on the O_2 content of the pore water in shallow sediments (Anderson et al., 1989; Crusius et al., 1996; Crusius and Thomson, 2000). Unfortunately, the O_2 content of the sediment pore waters can vary due to organic matter supply to the sediments as well as the O_2 content of the bottom water bathing the seafloor, so that these two parameters can be difficult to separate (a situation that is analogous to that for sedimentary organic carbon content). While interesting data and arguments have been put forward in support of various approaches (Hastings *et al.*, 1996; Russell *et al.*, 1996), it seems fair to argue that the paleoceanographic community still lacks a reliable set of methods for the global reconstruction of deep ocean dissolved O_2 content, especially in environments far from complete O_2 consumption.

Initial model results suggested that a biological pump mechanism for the glacial/interglacial CO_2 change would have rendered the ocean subsurface so O_2 deficient as to prevent the presence of burrowing organisms and oxic respiration over large expanses of the seafloor, which should leave some tell-tale signs in the sediment record. However, observations have changed this story significantly. For instance, the O_2 minimum, which is at intermediate depths in the modern ocean, may have migrated downward into the abyssal ocean during the last ice age (Berger and Lange, 1998; Boyle, 1988c; Herguera *et al.*, 1992; Marchitto *et al.*, 1998), so that the O_2 decrease was apparently focused in waters which are currently relatively rich in O_2, perhaps avoiding widespread anoxia at any given depth (Boyle, 1988c). For this reason, testing the biological pump hypothesis by reconstructing deep ocean O_2 will require that we do more than simply search for extensive deep-sea anoxia; rather, it will probably require a somewhat quantitative indicator of dissolved O_2 that works at intermediate O_2 concentrations.

13.3.3.3 Phasing

With adequate dating and temporal resolution in paleoceanographic and paleoclimatic records, the sequence of past events and changes can be reconstructed, providing among the most direct evidence for cause and effect. There is much information on phasing that is relevant to the biological pump and its role in glacial/interglacial CO_2 change. We limit ourselves here to one example that arises largely from ice core records: the timing of CO_2 change relative to temperature and glaciation (Broecker and Henderson, 1998 and references therein). Near the end of ice ages, atmospheric CO_2 rises well before most of the deglaciation (Monnin *et al.*, 2001; Sowers and Bender, 1995); this was referred to above as strong evidence against Broecker's shelf phosphorus hypothesis for CO_2 change (see Section 13.2.1). While the phasing of temperature is still debated, it appears that most of the warming in the high-latitude southern hemisphere preceded most of the warming in the high-latitude northern hemisphere, and that the atmospheric CO_2 rise lagged only slightly behind the southern hemisphere warming. This latter observation is roughly consistent with a variety of hypotheses for changes in the biological pump, but appears inconsistent with alternative

hypotheses that rely solely on changes in the oceanic calcium carbonate budget, which operates on a longer timescale than the biological pump (Archer and Maier-Reimer, 1994; Opdyke and Walker, 1992). With continued work, the detailed timing of CO_2 change may provide quantitative constraints on changes in the oceanic calcium carbonate budget as a partial contributor to CO_2 change; calcium carbonate plays a role in many of the biological pump hypotheses (Archer *et al.*, 2000a; Sigman *et al.*, 1998; Toggweiler, 1999).

13.4 OBSERVATIONS

To this point, a number of central observations and concepts have already been described to motivate a search for changes in the biological pump over glacial/interglacial cycles. First, there are carbon dioxide variations over glacial/interglacial cycles that are of the right magnitude and temporal structure to be caused by changes in the biological pump. Second, carbon isotope data for carbon dioxide and foraminiferal carbonate appear consistent with a biological-pump mechanism. Third, changes in the nitrogen cycle have been recognized that would tend to increase the oceanic nitrate reservoir during glacial times, although an actual increase in this reservoir has not been demonstrated; such a reservoir change might be expected to strengthen the low-latitude pump during glacial times. Finally, observations about phytoplankton in polar regions, in particular, their incomplete consumption of the major nutrients and their tendency toward iron limitation, suggest that simple changes in either the iron supply to the Antarctic surface ocean or in polar ocean circulation could lead to an increase in the efficiency of the high-latitude biological pump during glacial times. These observations, together with other ideas about the operation of the ocean carbon cycle, warrant that we consider the basic regional observations on biological productivity and nutrient status over glacial/interglacial cycles.

13.4.1 Low- and Mid-latitude Ocean

There have been many studies of the response of coastal and equatorial upwelling systems to glacial/interglacial climate change, and this overview cannot do justice to the entire body of work. The coastal upwelling regions along the western continental margins show an Atlantic/Pacific difference in their response to glacial/interglacial climate change. Studies along the western coast of Africa suggest an increase in productivity during glacial times (Summerhayes *et al.*, 1995; Wefer *et al.*, 1996). However, in the eastern Pacific, the coastal upwelling zones off of California and Mexico in the north and off of Peru in the south were less productive during the last glacial period (Dean *et al.*, 1997; Ganeshram and Pedersen, 1998; Ganeshram *et al.*, 2000; Heinze and Wefer, 1992). This response has generally been explained as the effect of continental cooling (and a large North American ice sheet in the case of the California Current) on the winds that currently drive coastal upwelling in the eastern tropical Pacific (Ganeshram and Pedersen, 1998; Herbert *et al.*, 2001).

Early paleoceanographic studies in the equatorial Pacific found that export production was greater in the equatorial Pacific during ice ages (Lyle, 1988; Pedersen, 1983; Pedersen *et al.*, 1991; Rea *et al.*, 1991), and some subsequent studies have supported this conclusion (Herguera and Berger, 1991; Murray *et al.*, 1993; Paytan *et al.*, 1996; Perks *et al.*, 2000). However, this interpretation has been put in question by reconstructions using ^{230}Th- and ^{3}He-normalized accumulation rates, ^{231}Pa/^{230}Th ratios (Marcantonio *et al.*, 2001b; Schwarz *et al.*, 1996; Stephens and Kadko *et al.*, 1997) and other evidence (Loubere, 1999). From studies along the eastern equatorial Atlantic, the consensus is for higher productivity during the last ice age than the Holocene; because of the limited extent of the Atlantic basin, it is difficult to differentiate this change from the glacial-age increase in productivity in African coastal upwelling (Lyle *et al.*, 1988; Martinez *et al.*, 1996; Moreno *et al.*, 2002; Rutsch *et al.*, 1995; Schneider *et al.*, 1996). Proxies for nutrient utilization (Altabet, 2001; Farrell *et al.*, 1995; Holmes *et al.*, 1997), pH (Sanyal *et al.*, 1997), and wind strength (Stutt *et al.*, 2002) would suggest that any export production increases that did occur in the low-latitude upwelling systems of the Pacific and Atlantic during the last glacial maximum were due to higher wind-driven upwelling, which increased the nutrient supply to the surface.

Upwelling associated with monsoonal circulation has apparently behaved predictably since the last ice age (Duplessy, 1982; Prell *et al.*, 1980). With less summertime warming of the air over Asia during the last glacial maximum, there was a weakening of the southwesterly winds of the summertime monsoon that drive coastal upwelling off the horn of Africa and Saudi Arabia, leading to a glacial decrease in export production in that region (Altabet *et al.*, 1995; Anderson and Prell, 1993; Marcantonio *et al.*, 2001a; Prell *et al.*, 1980). To the east and in the more open Indian Ocean, where the (southward) wintertime monsoonal winds drive upwelling or vertical mixing, productivity was apparently higher during the last glacial maximum, again consistent with the expected effect of a cooler Eurasian climate on the monsoon cycle (Beaufort, 2000; Duplessy, 1982; Fontugne and Duplessy, 1986). In the South China Sea, where the winter monsoon (seaward flow) is responsible for the exposure of nutrient-rich deep

waters, export production was apparently higher during the last glacial maximum, again suggesting a stronger winter monsoon associated with colder winter conditions over Eurasia (Huang *et al.*, 1997; Thunell *et al.*, 1992). The effect of the wintertime monsoon winds may also have impacted the western equatorial Pacific (Herguera, 1992; Kawahata *et al.*, 1998).

The low-productivity regions of the low- and mid-latitude ocean are an important source of information on the strength of the low-latitude biological pump, as they may be less susceptible to wind-driven changes in nutrient supply than are the upwelling regions discussed above, so that changes in productivity would be more closely tied to changes in the oceanic nutrient reservoir (and/or in N_2 fixation). As mentioned above, studies in the western equatorial Pacific indicate that the sedimentary C_{org} was higher in these regions during glacial times (Figure 10; Perks *et al.* (2000)); however, studies from the more southern western Pacific appear to suggest lower productivity (Kawahata *et al.*, 1999). Studies in the western equatorial and tropical Atlantic indicate lower $CaCO_3$ and C_{org} burial rates during glacial times (François *et al.*, 1990; Ruhlemann *et al.*, 1996). In the tropical Indian Ocean, away from the Arabian Sea, productivity was apparently higher during glacial times, but this has again been explained as the result stronger wind-driven mixing during the wintertime monsoon (Beaufort, 2000; Fontugne and Duplessy, 1986). Discrete sedimentation events of massive mat-forming diatoms are found in glacial-age sediments from the equatorial Indian; their significance is unclear (Broecker *et al.*, 2000).

As described in Section 13.2, there is substantial evidence for a change in the spatial pattern of denitrification across glacial/interglacial transitions (Altabet *et al.*, 1995, 2002; Emmer and Thunell, 2000; Ganeshram *et al.*, 1995; Pride *et al.*, 1999), and it seems likely that this led to a net global decrease in the loss rate of nitrate from the ocean during the last glacial maximum. Depending on the changes in N_2 fixation over glacial/interglacial cycles and the role of phosphorus in these changes (see Section 13.2), low-latitude productivity may or may not have responded so as to strengthen the biological pump. As discussed above, increased low-latitude productivity due to increased upwelling or vertical mixing has limited significance for the strength of the biological pump. It is a challenge to develop an approach to study past productivity changes that can distinguish between a change in the physical rate of vertical exchange and a change in the nutrient content of the subsurface. Thus, it is difficult to test the hypothesis that the ocean nutrient reservoir drove a significant glacial increase in the low- and mid-latitude biological pump, so as to

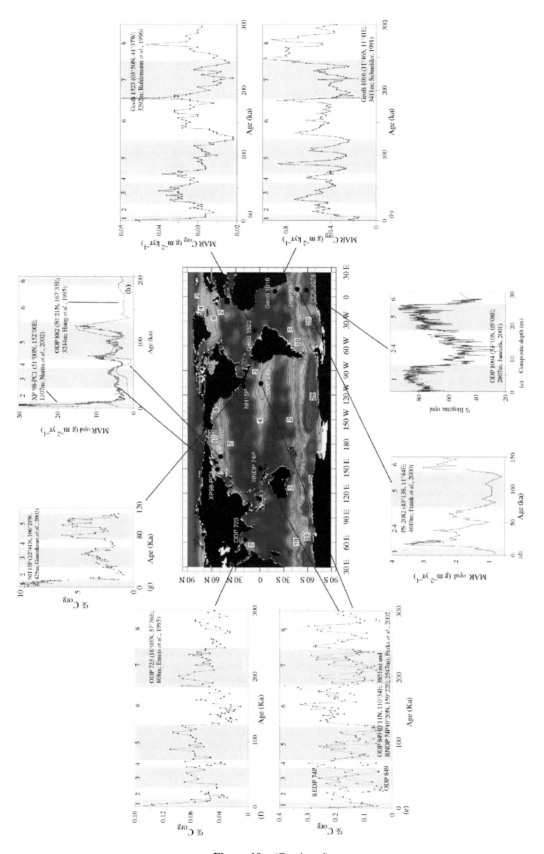

Figure 10 (Continued).

explain glacial/interglacial CO_2 changes. Nevertheless, it is notable that no direct support for the hypothesis has arisen without a focused search. Based on the overview above, the evidence for a net global increase in low-latitude productivity during the last ice age is not compelling. Moreover, the evidence that does exist for regional increases in low-latitude productivity seems easily explained in terms of changes in the wind-driven upwelling. Comparison of accumulation records for organic carbon and opal (Herguera and Berger, 1994) and of records from opposite sides of the South Atlantic basin (Ruhlemann et al., 1999) has led to the suggestion that the nutrient content of the thermocline was lower during ice ages. If such was indeed the case, it could be associated with a tendency for nutrients to shift into the deeper ocean during ice ages (Boyle, 1988b,c); nevertheless, it provides little encouragement for hypotheses of a larger oceanic nutrient reservoir during glacial times.

While it is tempting to phrase orbitally driven climate variability in terms of "ice ages" and "interglacials," this simplification breaks down frequently in the lower-latitude ocean (Clemens et al., 1991; McIntyre and Molfino, 1996; Sachs et al., 2001). The precession component of the Earth's orbital variations is by far the most directly important for the energy budget of the tropics. The Pleistocene variability of productivity in some tropical regions (Pailler et al., 2002) and coastal and equatorial upwellings (Moreno et al., 2002; Rutsch et al., 1995; Summerhayes et al., 1995) is dominated by the precession ($\sim$23 kyr) cycle, with the result that our focus on the Last Glacial Maximum/Holocene comparison can be misleading about the real timing of change. New data make this case very strongly for equatorial Pacific productivity (Perks et al., 2000). As the variations in atmospheric CO_2, glacial ice volume, and polar temperature all have less variability at the precession frequency, one would surmise that the low-latitude productivity changes dominated by precession are not major drivers of variability in the biological pump or in atmospheric CO_2. This is consistent with the conceptual point that wind-driven changes in low-latitude productivity are of limited importance to CO_2 transfer between the atmosphere and ocean.

13.4.2 High-latitude Ocean

The global high-latitude open ocean includes the Arctic, the North Atlantic, the Southern

---

Figure 10 Paleoceanographic records relevant to the history of export production from various regions of the global ocean, overlain on a map of surface nitrate concentration (in μM; Conkright et al., 1994). The records are chosen to be somewhat representative as a crude first-order paleoproductivity record for their oceanic regime; the extreme uncertainty associated with some of these reconstructions is discussed at length in Section 13.2. The records are of the sedimentary concentration or accumulation of organic carbon or biogenic opal. While the concentration records are complicated by potential changes in dilution by other sediment components, there are some circumstances where accumulation rate estimates are also problematic. The referenced sources should be consulted for information on these concerns. The biogeochemical regime of the surface ocean is roughly indicated by surface nitrate concentration (see Figure 5), which indicates whether a record is from a nutrient-poor, low-latitude region, a nutrient-bearing low-latitude upwelling region, or a nutrient-bearing polar region. Records are plotted versus time (except core ODP 1094 in the Southern Ocean) and glacial–interglacial stages are numbered, with a gray shading during interglacials. (a) The C_{org} accumulation record from core GeoB 1523 for the last 300 ka (Ruhlemann et al., 1996) conforms with the general finding that productivity was lower in the oligotrophic tropical Atlantic during the last ice age. (b) The C_{org} accumulation record from core GeoB 1016 for the last 300 ka (Schneider, 1991) is among the large body of evidence indicating that the Benguela upwelling system and the eastern equatorial Atlantic were more productive during the last ice age. However, all of the Atlantic records show a strong precessional response, indicating only a loose connection with ice volume. Ruhlemann et al. (1999) have noted an anti-correlation between eastern and western tropical Atlantic records. (c) Biogenic opal concentrations of ODP Site 1094 in the Antarctic zone of Southern Ocean (Janecek, 2001) conform with the large body of data indicating lower opal accumulation rates during ice ages. The upper 30 m of the composite sediment sequence represent approximately the last 135 ka. (d) Biogenic opal accumulation rate in core PS2062 (Frank et al., 2000) over the last 150 ka conforms with a large body of data indicating higher opal and organic carbon accumulation in the Subantarctic zone of the Southern Ocean during ice ages, opposite to the temporal pattern observed in the Antarctic. (e) The C_{org} concentration in RNDP 74P from the western equatorial Pacific and in ODP Site 849 from the eastern equatorial upwelling over the last 300 ka (Perks et al., 2000) suggests higher productivity across the entire equatorial upwelling during the last ice age but shows its dominant response to precession, not polar ice volume. (f) The C_{org} concentration in ODP Site 723 over the last 300 ka (Emeis et al., 1995) conforms with the large body of data suggesting lower productivity in the Oman upwelling system during ice ages because of a weaker summer monsoon (and thus weaker upwelling) when Eurasia is cold. (g) The C_{org} concentration in core NH15P during the last 120 ka in the coastal upwelling zone off northwestern Mexico (Ganeshram and Pedersen, 1998) conforms with a large body of data indicating lower productivity during the last ice age because the North American ice sheet caused a reduction in wind-driven coastal upwelling. (h) Biogenic opal accumulation rates in the Subarctic North Pacific (ODP Site 882; Haug et al. (1995)) and the Okhotsk Sea (Core XP98-PC1 Narita et al. (2002)) suggest lower diatom export during glacials, perhaps because of stronger salinity stratification of the upper water column and thus reduced nutrient supply. See text for more references.

Ocean, and the Subarctic North Pacific (Figure 5). We address each of these in turn.

The modern open Arctic appears to be relatively unproductive. Annually averaged solar insolation is low and sea ice cover is extensive, so that light limitation of phytoplankton growth is likely. Surface nitrate is also quite low in the Arctic surface (Figure 5(c)), at least partially due to a strong permanent "halocline," or vertical salinity gradient, that drives a strong vertical density gradient in the upper Arctic ocean, isolating fresh surface waters from saltier and thus denser nutrient-rich deep water. However, as the major nutrients are not completely absent in the surface, the major nutrients are probably not the dominant limitation for phytoplankton in the open Arctic. The few paleoproductivity studies of the open Arctic suggest that it was less productive during the last ice age (Schubert and Stein, 1996; Wollenburg et al., 2001). A range of ice age conditions may have contributed to this decrease, including greater ice cover and thus more extreme light limitation as well as sea-level-driven loss of the nutrient supply from the Pacific that currently enters from the Bering Strait. Our current understanding of the Arctic would suggest that it is not in itself a central part of ocean/atmosphere CO_2 partitioning, so little energy has been expended to understand its glacial/interglacial cycle.

The high-latitude North Atlantic has a productive spring bloom but tends towards major-nutrient limitation in the summer. This tendency toward nutrient limitation is due at least partially to the formation of deep water in the high-latitude North Atlantic. The Gulf Stream extension provides salty, warm, low-nutrient water to the high-latitude surface, and the cooling of this water eventually leads to the formation of deep waters. As a result of this steady state, neither surface waters nor underlying deep waters can supply large amounts of nutrients to the surface, with the largest nutrient source being the Antarctic Intermediate Water that penetrates from the south. The equivocal evidence for changes in North Atlantic productivity over glacial/interglacial cycles appears to suggest lower export production during the last ice age (Manighetti and McCave, 1995; Thomas et al., 1995). This is consistent with the persistent formation of a low-nutrient subsurface water in the glacial North Atlantic (Oppo and Lehman, 1993), even though this water mass was distinct from modern North Atlantic Deep Water. If, during ice ages, the North Atlantic continued to import lower-latitude surface waters and produce a subsurface water mass with low preformed nutrients, then it would have played roughly the same role in the biological pump as it does today. However, a glacial weakening of the North Atlantic source would have had implications for the global mean preformed nutrient concentration of the deep ocean (Section 13.2.2 and Figure 8).

Paleoceanographic work in the Southern Ocean indicates a strong glacial/interglacial oscillation, but with very different responses in the (poleward) Antarctic than in the (equatorward) Subantarctic. In the modern Antarctic, biological export production is dominated by diatoms, the group of phytoplankton that precipitate tests of hydrated silica ("opal"), and these microfossils are frequently the dominant component of the underlying sediments. The accumulation of diatomaceous sediments in the Antarctic was clearly lower during the last ice age (Mortlock et al., 1991), and other paleoceanographic indicators suggest that the biological export of carbon was also lower (Francois et al., 1997; Kumar et al., 1995; Rosenthal et al., 2000). However, given an apparent bias of at least some of these proxies to preferentially record the carbon export closely associated with diatom microfossils (Chase et al., 2003b; Yu et al., 2001), it remains possible that the biological export of carbon from the glacial Antarctic surface was as high or higher than during interglacial times but that it occurred in a form that was poorly preserved in the sediment record (Arrigo et al., 1999; Boyle, 1998; Brzezinski et al., 2002; DiTullio et al., 2000; Moore et al., 2000; Tortell et al., 2002).

In the Atlantic and Indian sectors of the Subantarctic, there is very strong evidence for higher export production during glacial times and an associated increase in the relative importance of diatom production (Chase et al., 2001; Francois et al., 1997; Kumar et al., 1995; Mortlock et al., 1991; Rosenthal et al., 1995). This high glacial productivity has not yet been recognized in the Pacific sector (Chase et al., 2003a). If this productivity increase was indeed absent in the Pacific sector of the Subantarctic, it provides an important constraint on its cause. For instance, this finding may be consistent with iron fertilization as the dominant driver of the glacial increase in Subantarctic productivity, as the increase in iron input may have been extremely limited in parts of the Pacific sector (Chase et al., 2003a).

While these changes in export production are critical to the workings of the ice age Southern Ocean, they do not address directly the effect of the Southern Ocean on atmospheric CO_2, which is largely determined by the competition between the upwelling of respiratory CO_2 and nutrients to the surface and export of organic matter out of the surface (Section 13.2). In the Antarctic, nitrogen isotope data suggest that the fraction of the nitrate consumed during the last ice age was higher than its current value (Francois et al., 1997; Sigman et al., 1999b). These result seem to support the long-standing hypothesis that nutrient utilization changes in Antarctic waters are a fundamental driver of glacial/interglacial changes in the atmospheric concentration of CO_2; box model calculations predict the 25–40% higher nitrate utilization (i.e., 50–65% in the glacial ocean compared to 25% during the present interglacial) could lower atmospheric CO_2 by the full glacial/interglacial amplitude under the relevant conditions (Sigman *et al.*, 1999b). Since paleoceanographic proxy data suggest that Antarctic export production was lower during the last ice age, one would infer that more complete nitrate utilization in the Antarctic was due to a lower rate of nitrate supply from the subsurface, implying that the fundamental driver of the CO_2 change was an ice age decrease in the ventilation of deep waters at the surface of the Antarctic (François *et al.*, 1997). Such a change in the Antarctic would also help to explain observations in the low-latitude ocean. For instance, the extraction of nutrients in the Antarctic should lower the nutrient content of the waters that subducted into the intermediate-depth ocean and thermocline at the equatorward margin of the Antarctic. This would work to prevent the transfer of nutrients from cold, deep ocean into the warmer, upper ocean, potentially explaining an apparent transfer of nutrients from the mid-depths to the deep ocean during glacial times (Boyle, 1988b; Herguera *et al.*, 1992; Keir, 1988; Matsumoto *et al.*, 2002a; Sigman and Boyle, 2000).

Two possible causes for Antarctic stratification have been discussed. First, the southern hemisphere westerlies winds apparently shifted northward during glacial times (Hebbeln *et al.*, 2002; McCulloch *et al.*, 2000), which should have reduced Ekman-driven upwelling in the Antarctic (Toggweiler *et al.*, 1999) ("wind-shift" mechanism). Second, the vertical gradient in density, as determined jointly by the vertical gradients in temperature and salinity, may have been stronger in the glacial Antarctic ("density" mechanism). However, both of these mechanisms are currently incomplete as explanations for the glacial reduction in atmospheric CO_2 (Keeling and Visbeck, 2001; Sigman and Boyle, 2001). As an example, we consider the "density" mechanism in more detail. In the modern Antarctic, as in most polar regions, the vertical salinity gradient tends to stratify the upper ocean, while temperature tends to destabilize it. Thus, either an increase in the relative importance of salinity or a decrease in the importance of temperature could be the fundamental trigger for the development of strong stratification in the glacial Antarctic; once stratification begins, it will be reinforced by the accumulation of freshwater in the upper layer. Given the abundance of sea ice in the glacial Antarctic (Crosta *et al.*, 1998) and the typical association of sea ice with freshwater release, it is tempting to call upon sea ice as the initial driver of stratification. However, sea ice does not represent a source of freshwater *per se*, but rather a mechanism for transporting freshwater from one region of the Antarctic to another, with net ice formation in the coastal Antarctic and net melting in the open Antarctic. As a result, stratification of the open Antarctic by the melting of sea ice might occur at the expense of the coastal Antarctic, where sea ice formation would make surface waters more saline and thus more dense, leading to overturning and the accompanying release of CO_2 to the atmosphere. Thus, we are more encouraged by hypotheses that involve changes in temperature (i.e., cooling of the deep ocean) as the trigger for stratification (Gildor and Tziperman, 2001; Gildor *et al.*, 2002), although these have been tested only in simple physical models. In any case, if the evidence for stratification could be taken as overwhelming, we would still need to determine what caused it and how much it lowered atmospheric CO_2 during glacial times.

However, the evidence for Antarctic stratification is not overwhelming; indeed, many investigators are not convinced (e.g., Anderson *et al.*, 1998; Elderfield and Rickaby, 2000). The interpretation from nitrogen isotopes of higher nitrate utilization in the Antarctic faces a number of apparent disagreements with other proxies of nutrient status, in particular, the Cd/Ca (Boyle, 1988a; Keigwin and Boyle, 1989) and $^{13}C/^{12}C$ (Charles and Fairbanks, 1988) of planktonic foraminiferal calcite. Measurements of these ratios in planktonic foraminifera indicate no clear decrease in the nutrient concentration of Antarctic surface water, while such a decrease would have been expected on the basis of the nitrogen isotope data. Each of the paleochemical proxies has significant uncertainties (see Section 13.3.2). Nevertheless, the need to define new interpretations of a variety of measurements to fit the stratification hypothesis does not inspire confidence.

Dissolved silicate is a major nutrient for diatom growth because of the silica tests that these phytoplankton precipitate. Much like nitrate and phosphate, silicate is nearly completely depleted in the low-latitude surface ocean but is found at relatively high concentrations in the modern Antarctic (Figure 5(d)). The silicon isotopic composition of diatom microfossils implies that there was a reduction in Antarctic silicate utilization during the last ice age (De La Rocha *et al.*, 1998), in contrast to the evidence for enhanced nitrate utilization. While this may indicate a disagreement among proxies, the alternative is that the difference signals a real oceanographic change. Field observations, incubations and culture studies (Franck *et al.*, 2000; Hutchins and Bruland, 1998; Takeda, 1998) indicate that iron-replete conditions (such as may result from the dustiness of the ice age atmosphere (Mahowald *et al.*, 1999)) favor a higher nitrate/silicate ratio in diatoms, and phytoplankton species composition changes may have reinforced this shift (Tortell *et al.*, 2002). Indeed, taking both the nitrogen and silicon isotope data at face value implies that the nitrate/silicate uptake ratio of Antarctic phytoplankton was higher during the last ice age, leading to lower nitrate but higher silicate concentrations in the glacial Antarctic relative to modern times (Brzezinski *et al.*, 2002). From the perspective of our efforts to reconstruct the history of export production, this possibility is problematic in that the rain of biogenic silica out of the surface ocean would have decreased independently from the rain of organic carbon.

If the change in nitrate/silicate uptake ratio was adequately great, it may have actually removed the tendency for preferential depletion of silicate relative to nitrate that is observed in the modern Antarctic, possibly increasing the silicate/nitrate ratio of Southern-source water that supplies nutrients to the low-latitude surface (Brzezinski *et al.*, 2002). A greater supply of silicate to the low-latitude ocean could have driven an increase in the importance of silica-secreting phytoplankton (diatoms) relative to $CaCO_3$-precipitating phytoplankton (coccolithophorids) (Matsumoto *et al.*, 2002b); this may explain a shift toward higher opal accumulation rates in some equatorial regions (Broecker *et al.*, 2000). If increased diatom productivity at low latitudes in the southern hemisphere occurred at the expense of coccolithophores, the high sinking rates of diatom-derived particulate matter may have facilitated the transport of organic matter through intermediate waters to the deep ocean (Boyle, 1988b), again consistent with evidence for the "nutrient deepening" during glacial times (Boyle, 1988c, 1992; Boyle *et al.*, 1995; Herguera *et al.*, 1992). In addition, a floral shift away from coccolithophores to diatoms would have also lowered the $CaCO_3$/organic carbon rain ratio, weakening the ocean carbonate pump. Both the increase in the remineralization depth of organic carbon and a decrease in the rain ratio would have worked to lower-atmospheric CO_2 (Archer and Maier-Reimer, 1994; Berger and Keir, 1984; Boyle, 1988b; Broecker and Peng, 1987; Brzezinski *et al.*, 2002; Dymond and Lyle, 1985; Keir and Berger, 1983; Matsumoto *et al.*, 2002b; Sigman *et al.*, 1998).

In the Subantarctic, planktonic foraminiferal Cd/Ca and the $^{13}C/^{12}C$ of diatom-bound organic matter are consistent with an ice age state of higher nutrient utilization (Rosenthal *et al.*, 1997, 2000). However, the consideration of proxy complications argue against such a change (Elderfield and Rickaby, 2000). The significance of the nitrogen isotope data for the glacial Subantarctic is uncertain (François *et al.*, 1997; Sigman *et al.*, 1999a). Thus, while the Subantarctic was certainly more productive during glacial times, the history of its nutrient status is unclear and deserves further

investigation. Given our current understanding of the role that the Subantarctic plays in the carbon cycle, the Subantarctic is less important in itself than in what it indicates about the nutrient status of the Antarctic zone at its (up-stream) southern end. The Subantarctic also represents the gateway by which the Antarctic affects the low-latitude ocean and is thus central in questions regarding the impact of Antarctic nutrient changes on low-latitude productivity (Brzezinski *et al.*, 2002; Matsumoto *et al.*, 2002b).

The Subarctic Pacific, like the Antarctic, is characterized by year-round nonzero concentrations of the major nutrients, nitrate and phosphate. However, the Subarctic Pacific maintains a higher degree of nutrient utilization (lower surface nutrient concentrations) than the Antarctic. In the open Subarctic Pacific, summer nitrate concentration frequently falls below 8 μM in surface water, despite nitrate concentrations of greater than 35 μM in the upwelling deep water below the permanent halocline. One important difference between the Subarctic Pacific and the Antarctic is the much stronger halocline in the former (Reid, 1969, Talley, 1993; Warren, 1983), where the extremely low salinity of the upper 400 m limits the exposure of nutrient- and CO_2-charged subsurface waters at the surface. It is roughly this salinity-driven stratification that has been proposed for the glacial Antarctic. Thus, the modern Subarctic Pacific provides something of a modern analogue for the glacial Antarctic stratification hypothesis (Haug *et al.*, 1999; Morley and Hays, 1983).

The Subarctic Pacific undergoes changes in productivity and nutrient regime over glacial/interglacial cycles, although these changes are of smaller amplitude than is observed for the Southern Ocean. Opal accumulation was apparently lower during the last glacial maximum in the western Subarctic Pacific and its marginal seas (Gorbarenko, 1996; Haug *et al.*, 1995; Narita *et al.*, 2002). This has been interpreted, in much the same way as the Antarctic changes, as indicating intensified stratification during glacial times (Narita *et al.*, 2002). In this light, the smaller glacial/interglacial signal of this region relative to the Antarctic is explained by the fact the modern Subarctic Pacific is already strongly stratified. However, the dominant feature in most sediment records from western Subarctic Pacific and Bering Sea is of a high-productivity event upon the transition from the last ice age to the Holocene (Keigwin *et al.*, 1992; Nakatsuka *et al.*, 1995), for which there is as yet no well-supported explanation. Moreover, the eastern Subarctic Pacific seems to lack a glacial/interglacial signal, instead being characterized by abrupt events of opal deposition that are not clearly linked to global climate change (McDonald *et al.*, 1999).

13.5 SUMMARY AND CURRENT OPINION

While the early focus of hypotheses for glacial/interglacial CO_2 change was on the biological pump (Broecker, 1982a,b), a number of alternative mechanisms, mostly involving the calcium carbonate budget, attracted increasing attention during the late 1980s and the 1990s (Archer and Maier-Reimer, 1994; Boyle, 1988b; Broecker and Peng, 1987; Opdyke and Walker, 1992). As problems have been recognized with these mechanisms as the sole driver of glacial/interglacial CO_2 change (Archer and Maier-Raimer, 1994; Sigman *et al.*, 1998), the biological pump is receiving attention once again. Hypotheses exist for both low- and high-latitude changes in the biological pump. Their strengths, weaknesses, and central questions are now fairly clear.

Currently, the most popular low-latitude hypotheses depend largely on whether the nitrogen cycle alone can drive a large-scale change in low-latitude export production. As of the early 2000s, while this question has not yet been directly addressed, no new observations have arisen that would overturn the traditional view that the nitrogen cycle could have had only a limited effect without cooperation from the phosphorus cycle and that the phosphorus cycle could not have been adequately dynamic over glacial/interglacial cycles (e.g., Froelich *et al.*, 1982; Haug *et al.*, 1998; Redfield, 1942, 1958; Redfield *et al.*, 1963; Ruttenberg, 1993; Sanudo-Wilhelmy *et al.*, 2001). Thus, while the low-latitude biological pump has by no means been eliminated as the driver of glacial/interglacial cycles, our attention is focused on the polar ocean, the Antarctic in particular.

We have discussed three rough categories of Antarctic change that might drive a decline in CO_2 during ice ages: (i) increased biological productivity, as in response to increased iron supply, (ii) vertical stratification of the upper ocean, and (iii) gas exchange limitation by ice cover. The productivity hypothesis is not supported by data, which suggest a smaller biogenic rain in the glacial Antarctic, although it remains possible that our proxies are telling us only about a reduction in the rain of bulk biogenic material (i.e., diatom-derived opal) and are leading us to overlook an increase in "export production," the rain of organic carbon out of the surface ocean. The gas exchange hypothesis remains possible but is difficult to test and would seem to require extreme conditions to be the sole cause for the observed amplitude of CO_2 change. The stratification hypothesis is supported by some data but conflicts with other data, or at least with the traditional interpretations of those data. While there are major concerns about each of the polar hypotheses, these hypotheses have recently been strengthened by the realization that they would have impacted the lower-latitude ocean and the $CaCO_3$ budget in ways that should have strengthened their overall effect on atmospheric CO_2 (e.g., Brzezinski *et al.*, 2002; Matsumoto *et al.*, 2002b; Toggweiler, 1999).

As is clear from the treatment above, the stratification hypothesis has particular resonance with the authors of this chapter. This is driven largely by the conceptual and observational links we observe between two nutrient-bearing polar ocean regions, the Antarctic and the Subarctic Pacific. As described above, there is a strong analogy between the Subarctic Pacific that we observe today and the glacial-age Antarctic that is posed by the stratification hypothesis. Moreover, at a major cooling event 2.7 Myr ago, it appears that the Subarctic Pacific underwent the transition to its current salinity-stratified condition (Figure 11) (Haug *et al.*, 1999), just as stratification has been hypothesized for the Antarctic on the transitions to ice age conditions. Ongoing work on other intervals of the climate record seems to support the existence of a strong generalized link between climate cooling and polar ocean stratification. The physical mechanism for this link, however, is still unresolved.

To these authors, there is no infallible recipe for the investigation of the biological pump in the past. Nevertheless, there are glaring uncertainties that require our attention, both in our concepts and our tools. The quantitative effects of the low latitude and polar ocean on atmospheric CO_2 are still a matter of debate (Broecker *et al.*, 1999; Toggweiler *et al.*, 2003a,b). Until our understanding of the modern ocean and carbon cycle has progressed to the degree that such major conceptual questions can be resolved, it will be extremely difficult to generate any consensus on the role of the biological pump in driving carbon dioxide change. After optimistic discussion of "proxy calibration" in the 1990s, we now recognize that the sedimentary and geochemical parameters we use to study the history of the biological pump are the product of multiple environmental variables, requiring that we think deeply about their significance in each case that they are applied. The quantity of work on the history of the biological pump has increased markedly over the last decade; a concerted effort to review, evaluate, and synthesize this information would greatly improve its usefulness. At the same time, we must continue to support a vanguard in search of untapped constraints. We have emphasized above that the biological pump is not strictly a biological phenomenon but rather results from the interaction of ocean biology, chemistry, and physics; as such, progress in its study is tied to our understanding of the history of the ocean in general, including its physical circulation and fundamental conditions (Adkins *et al.*, 2002).

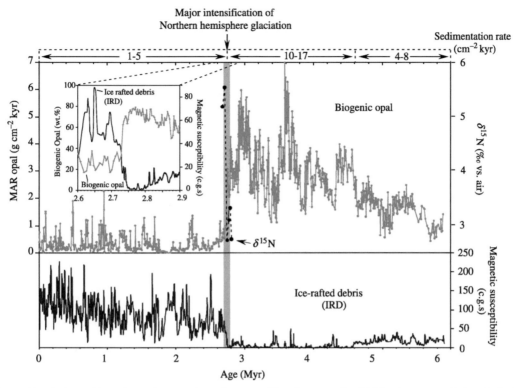

Figure 11 A 6 Myr biogenic opal record from Ocean Drilling Program Site 882 in the western Subarctic Pacific (50°21′N, 167°35′E; water depth 3,244 m; figure from Haug *et al.* (1999)), showing an approximately fourfold decrease in opal accumulation rate at 2.73 Myr ago (top panel). This abrupt drop in opal occurred at isotope stage G6 (the former isotope stage 110) synchronously with the massive onset of ice rafted debris, as indicated by the increase in magnetic susceptibility (bottom panel). Maxima in opal accumulation during the last 2.73 Myr are generally linked to interglacial times or deglaciations, as reported by Keigwin *et al.* (1992) for the last deglaciation. Since silicate is supplied by the exposure of nutrient-rich deep water, the high opal flux rates of the mid-Pliocene Subarctic Pacific require a rate of deep-water exposure of a magnitude similar to that observed in the modern Antarctic. Because silicate consumption is nearly complete in the modern Subarctic Pacific (Figure 5(d)), the sharp drop in opal flux at 2.73 Myr ago cannot be attributed simply to a decrease in the completeness of silicate consumption. Rather, it must record a decrease in the rate of exposure of silicate-rich deep water at the Subarctic Pacific surface. Sediment $\delta^{15}N$ ($(^{15}N/^{14}N_{sample}/^{15}N/^{14}N_{air}-1)\times 1,000$) increases concurrently with the sharp drop in opal accumulation at the event 2.73 Myr ago (from ~3‰ to 5‰, upper panel). This shift toward higher $\delta^{15}N$ suggests an increase in nitrate utilization (Altabet and Francois, 1994a), providing additional evidence that the decrease in opal flux resulted from a decrease in the exposure of nutrient-rich deep water, which lowered the nitrate supply and thus forced more complete nitrate consumption while decreasing the silicate available for opal production (reproduced by permission of Nature Publishing Group from *Nature 401*, 779–782).

ACKNOWLEDGMENTS

DMS is supported by the US NSF through grants OCE-9981479, OCE-0081686, DEB-0083566 (to Simon Levin), and OCE-0136449, and by British Petroleum and Ford Motor Company through the Princeton Carbon Mitigation Initiative. This chapter is dedicated to the memory of David S. Sigman, the only biochemist who would have tried to read it.

REFERENCES

Adkins J. F., McIntyre K., and Schrag D. P. (2002) The salinity, temperature, and delta O-18 of the glacial deep ocean. *Science* **298**(5599), 1769–1773.

Altabet M. A. (2001) Nitrogen isotopic evidence for micronutrient control of fractional NO_3^- utilization in the equatorial Pacific. *Limnol. Oceanogr.* **46**(2), 368–380.

Altabet M. A. and Francois R. (1994a) Sedimentary nitrogen isotopic ratio as a recorder for surface ocean nitrate utilization. *Global Biogeochem. Cycles* **8**(1), 103–116.

Altabet M. A. and Francois R. (1994b) The use of nitrogen isotopic ratio for reconstruction of past changes in surface ocean nutrient utilization. In *Carbon Cycling in the Glacial Ocean: Constraints on the Ocean's Role in Global Change* (eds. R. Zhan, M. Kaminski, L. Labeyrie, and T. F. Pederson). Springer, Berlin, Heidelberg, New York, vol. 17, pp. 281–306.

Altabet M. A., Francois R., Murray D. W., and Prell W. L. (1995) Climate-related variations in denitrification in the Arabian Sea from sediment $^{15}N/^{14}N$ ratios. *Nature* **373**, 506–509.

Altabet M. A., Higginson M. J., and Murray D. W. (2002) The effect of millennial-scale changes in Arabian Sea denitrification on atmospheric CO_2. *Nature* **415**(6868), 159–162.

Anderson D. M. and Prell W. L. (1993) A 300 kyr record of upwelling off Oman during the Late Quaternary: evidence of the Asian southwest monsoon. *Paleoceanography* **8**, 193–208.

Anderson R., Kumar N., Mortlock R., Froelich P., Kubik P., Dittrich-Hannen B., and Suter M. (1998) Late-Quaternary changes in productivity of the Southern Ocean. *J. Mar. Sys.* **17**, 497–514.

Anderson R. F., Lehuray A. P., Fleisher M. Q., and Murray J. W. (1989) Uranium deposition in Saanich Inlet sediments, Vancouver Island. *Geochim. Cosmochim. Acta* **53**(9), 2205–2213.

Anderson R. F., Lao Y., Broecker W. S., Trumbore S. E., Hofmann H. J., and Wolfi W. (1990) Boundary scavenging in the Pacific Ocean: a comparison of ^{10}Be and ^{231}Pa. *Earth Planet. Sci. Lett.* **96**, 287–304.

Archer D. and Maier-Reimer E. (1994) Effect of deep-sea sedimentary calcite preservation on atmospheric CO_2 concentration. *Nature* **367**, 260–263.

Archer D., Winguth A., Lea D., and Mahowald N. (2000a) What caused the glacial/interglacial atmospheric pCO_2 cycles? *Rev. Geophys.* **38**(2), 159–189.

Archer D. E., Eshel G., Winguth A., Broecker W. S., Pierrehumbert R., Tobis M., and Jacob R. (2000b) Atmospheric pCO_2 sensitivity to the biological pump in the ocean. *Global Biogeochem. Cycles* **14**(4), 1219–1230.

Arrigo K. R., Robinson D. H., Worthen D. L., Dunbar R. B., DiTullio G. R., VanWoert M., and Lizotte M. P. (1999) Phytoplankton community structure and the drawdown of nutrients and CO_2 in the Southern Ocean. *Science* **283**(5400), 365–367.

Bacon M. P. (1984) Glacial to interglacial changes in carbonate and clay sedimentation in the Atlantic Ocean estimated from Th-230 measurements. *Isotope Geosci.* **2**, 97–111.

Bareille G., Labracherie M., Mortlock R. A., Maier-Reimer E., and Froelich P. N. (1998) A test of (Ge/Si)$_{opal}$ as a paleorecorder of (Ge/Si)$_{seawater}$. *Geology* **26**(2), 179–182.

Barnola J. M., Raynaud D., Korotkevich Y. S., and Lorius C. (1987) Vostok ice core provides 160,000-year record of atmospheric CO_2. *Nature* **329**, 408414.

Bassinot F. C., Labeyrie L. D., Vincent E., Quidellerur X., Shackleton N. J., and Lancelot Y. (1994) The astronomical theory of climate and the age of the Brunhes-Matuyama magnetic reversal. *Earth Planet. Sci. Lett.* **126**, 91–108.

Bauch D., Carstens J., and Wefer G. (1997) Oxygen isotope composition of living *Neogloboquadrina pachyderma* (sin.) in the Arctic Ocean. *Earth Planet. Sci. Lett.* **146**(1–2), 47–58.

Beaufort L. (2000) Dynamics of the monsoon in the equatorial Indian Ocean over the last 260,000 years. *Quat Int.* **31**, 13–18.

Berger A., Imbrie J., Hays J., Kukla G., and Saltzman B. (1984) Milankovitch and Climate (vol. 1). Reidel, Boston Vol. 1, 493pp.

Berger W. H. and Keir R. S. (1984) Glacial-Holocene changes in atmospheric CO_2 and the deep-sea record. In *Climate Processes and Climate Sensitivity*. Geophyisical Monograph 29 (eds. J. E. Hansen and T. Takahashi). American Geophysical Union, Washington DC, pp. 337–351.

Berger W. H. and Lange C. B. (1998) Silica depletion in the thermocline of the glacial North Pacific: corollaries and implications. *Deep-Sea Res. II* **45**(8–9), 1885–1904.

Boyle E. A. (1988a) Cadmium: chemical tracer of deepwater paleoceanography. *Paleoceanography* **3**(4), 471–489.

Boyle E. A. (1988b) The role of vertical chemical fractionation in controlling late quaternary atmospheric carbon dioxide. *J. Geophys. Res.* **93**(C12), 15701–15714.

Boyle E. A. (1988b) Vertical oceanic nutrient fractionation and glacial/interglacial CO_2 cycles. *Nature* **331**, 55–56.

Boyle E. A. (1992) Cadmium and δ^{13}C paleochemical ocean distributions during the stage 2 glacial maximum. *Ann. Rev. Earth Planet. Sci.* **20**, 245–287.

Boyle E. A. (1998) Pumping iron makes thinner diatoms. *Nature* **393**, 733–734.

Boyle E. A. and Keigwin L. D. (1982) Deep circulation of the North Atlantic for the last 200,000 years: geochemical evidence. *Science* **218**, 784–787.

Boyle E. A., Labeyrie L., and Duplessy J.-C. (1995) Calcitic foraminiferal data confirmed by cadmium in aragonitic *Hoeglundina*: application to the last glacial maximum in the northern Indian Ocean. *Paleoceanography* **10**(5), 881–900.

Broecker W. S. (1981) Glacial to interglacial changes in ocean and atmospheric chemistry. In *Climatic Variations and Variability: Facts and Theory* (ed. A. Berger). Kluwer, Boston, pp. 111–121.

Broecker W. S. (1982a) Glacial to interglacial changes in ocean chemistry. *Progr. Oceanogr.* **2**, 151–197.

Broecker W. S. (1982b) Ocean chemistry during glacial time. *Geochim. Cosmochim. Acta* **46**, 1689–1706.

Broecker W. S. and Henderson G. M. (1998) The sequence of events surrounding Termination II and their implications for the cause of glacial–interglacial CO_2 changes. *Paleoceanography* **13**(4), 352–364.

Broecker W. S. and Maier-Reimer E. (1992) The influence of air and sea exchange on the carbon isotope distribution in the sea. *Global Biogeochem. Cycles* **6**(3), 315–320.

Broecker W. S. and Peng T.-H. (1982) *Tracers in the Sea.* Eldigio Press, Palisades, New York.

Broecker W. S. and Peng T.-H. (1987) The role of $CaCO_3$ compensation in the glacial to interglacial atmospheric CO_2 change. *Global Biogeochem. Cycles* **1**(1), 15–29.

Broecker W. S., Peacock S. L., Walker S., Weiss R., Fahrbach E., Schroeder M., Mikolajewicz U., Heinze C., Key R., Peng T. H., and Rubin S. (1998) How much deep water is formed in the Southern Ocean? *J. Geophys. Res.: Oceans* **103**(C8), 15833–15843.

Broecker W. S., Lynch-Stieglitz J., Archer D., Hofmann M., Maier-Reimer E., Marchal O., Stocker T., and Gruber N. (1999) How strong is the Harvardton-Bear constraint? *Global Biogeochem. Cycles* **13**(4), 817–820.

Broecker W. S., Clark E., Stiegglitz J. L., Beck W., Stott L. D., Hajdas I., and Bonani G. (2000) Late glacial diatom accumulation at 9° S in the Indian Ocean. *Paleoceanography* **15**, 348–352.

Brzezinski M. A., Pride C. J., Franck V. M., Sigman D. M., Sarmiento J. L., Matsumoto K., Gruber N., Rau G. H., and Coale K. H. (2002) A switch from Si(OH)$_4$ to NO$_3^-$ depletion in the glacial Southern Ocean. *Geophys. Res. Lett.* **29**(12), 10.1029/12001GL014349.

Canfield D. E. (1994) Factors influencing organic carbon preservation in marine sediments. *Deep-Sea Res.* **114**, 315–329.

Charles C. D. and Fairbanks R. G. (1988) Glacial to interglacial changes in the isotopic gradients of southern Ocean surface water. In *Geological History of the Polar Oceans: Arctic versus Antarctic* (eds. U. Bleil and J. Thiede). Kluwer Academic, , pp. 519–538.

Chase Z., Anderson R. F., and Fleisher M. Q. (2001) Evidence from authigenic uranium for increased productivity of the glacial Subantarctic Ocean. *Paleoceanography* **16**(5), 468–478.

Chase Z., Anderson R. F., Fleisher M. Q., and Kubik P. W. (2003a) Accumulation of biogenic and lithogenic material in the Pacific sector of the Southern Ocean during the past 40,000 years. *Deep-Sea Res. II* **50**(3–4), 739–768.

Chase Z., Anderson R. F., Fleisher M. Q., and Kubik P. W. (2003b) Scavenging of ^{230}Th, ^{231}Pa and ^{10}Be in the Southern Ocean (SW Pacific sector): the importance of particle flux, particle composition and advection. *Deep-Sea Res. II* **50**(3–4), 739–768.

Clemens S. C., Prell W. L., Murray D. W., Shimmield G., and Weedon G. (1991) Forcing mechanisms of the Indian Ocean monsoon. *Nature* **353**, 720–725.

Codispoti L. A. (1995) Biogeochemical cycles: Is the ocean losing nitrate? *Nature* **376**(6543), 724.

Conkright M., Levitus S., and Boyer T. (1994) *World Ocean Atlas 1994*, **vol. 1**: Nutrients. US Department of Commerce, Washington DC, pp. 16.

Crosta X., Pichon J.-J., and Burckle L. H. (1998) Application of modern analogue technique to marine Antarctic diatoms: reconstruction of maximum sea-ice extent at the Last Glacial Maximum. *Paleoceanography* **13**(3), 284–297.

Crusius J. and Thomson J. (2000) Comparative behavior of authigenic Re, U, and Mo during reoxidation and subsequent long-term burial in marine sediments. *Geochim. Cosmochim. Acta* **64**(13), 2233–2242.

Crusius J., Calvert S., Pedersen T., and Sage D. (1996) Rhenium and molybdenum enrichments in sediments as indicators of oxic, suboxic and sulfidic conditions of deposition. *Earth Planet. Sci. Lett.* **145**(1–4), 65–78.

Dean W. E., Gardner J. V., and Piper D. Z. (1997) Inorganic geochemical indicators of glacial–interglacial changes in productivity and anoxia on the California continental margin. *Geochim. Cosmochim. Acta* **61**(21), 4507–4518.

Dehairs F., Stroobants N., and Goeyens L. (1991) Suspended barite as a tracer of biological activity in the Southern Ocean. *Mar. Chem.* **35**, 399–410.

De La Rocha C. L., Brzezinski M. A., and DeNiro M. J. (1997) Fractionation of silicon isotopes by marine diatoms during biogenic silica formation. *Geochim. Cosmochim. Acta* **61**, 5051–5056.

De La Rocha C. L., Brzezinski M. A., DeNiro M. J., and Shemesh A. (1998) Silicon-isotope composition of diatoms as an indicator of past oceanic change. *Nature* **395**(6703), 680–683.

Deutsch C., Gruber N., Key R. M., Sarmiento J. L., and Ganaschaud A. (2001) Denitrification and N_2 fixation in the Pacific Ocean. *Global Biogeochem. Cycles* **15**(2), 483–506.

DiTullio G. R., Grebmeier J. M., Arrigo K. R., Lizotte M. P., Robinson D. H., Leventer A., Barry J. P., VanWoert M. L., and Dunbar R. B. (2000) Rapid and early export of *Phaeocystis antarctica* blooms in the Ross Sea, Antarctica. *Nature* **404**, 595–598.

Duplessy J. C. (1982) Glacial to interglacial contrasts in the northern Indian Ocean. *Nature* **295**, 494–498.

Dymond J. and Lyle M. (1985) Flux comparisons between sediments and sediment traps in the eastern tropical Pacific: implications for atmospheric CO_2 variations during the Pleistocene. *Limnol. Oceanogr.* **30**(4), 699–712.

Dymond J., Suess E., and Lyle M. (1992) Barium in deep-sea sediment: a geochemical proxy for paleoproductivity. *Paleoceanography* **7**(2), 163–181.

Elderfield H. and Rickaby R. E. M. (2000) Oceanic Cd/P ratio and nutrient utilization in the glacial Southern Ocean. *Nature* **405**, 305–310.

Elser J. J., Sterner R. W., Gorokhova E., Fagan W. F., Markow T. A., Cotner J. B., Harrison J. F., Hobbie S. E., Odell G. M., and Weider L. J. (2000) Biological stoichiometry from genes to ecosystems. *Ecol. Lett.* **3**, 540–550.

Emeis K.-C., Anderson D. M., Doose H., Kroon D., and Schulz-Bull D. E. (1995) Sea-surface temperatures and the history of monsoon upwelling in the NW Arabian Sea during the last 500,000 yr. *Quat. Res.* **43**, 355–361.

Emmer E. and Thunell R. C. (2000) Nitrogen isotope variations in Santa Barbara Basin sediments: implications for denitrification in the eastern tropical North Pacific during the last 50,000 years. *Paleoceanography* **15**(4), 377–387.

Fairbanks R. G. (1989) A 17,000-year glacio-eustatic sea level record: influence of glacial melting rates on the Younger Dryas event and deep-ocean circulation. *Nature* **342**, 637–642.

Falkowski P. G. (1997) Evolution of the nitrogen cycle and its influence on the biological sequestration of CO_2 in the ocean. *Nature* **387**(6630), 272–275.

Falkowski P. G. (2000) Rationalizing elemental ratios in unicellular algae. *J. Phycol.* **36**, 3–6.

Farrell J. W., Pedersen T. F., Calvert S. E., and Nielsen B. (1995) Glacial–interglacial changes in nutrient utilization in the equatorial Pacific Ocean. *Nature* **377**(6549), 514–517.

Fischer G. and Wefer G. (1999) *Use of Proxies in Paleoceanography: Examples from the South Atlantic* (eds. G. Fischer and G. Wefer). Springer, Berlin, 727pp.

Fontugne M. R. and Duplessy J.-C. (1986) Variations of the monsoon regime during the upper Quaternary: evidence from carbon isotopic record of organic matter in North Indian Ocean sediment cores. *Palaeogeogr. Paleoclimatol. Palaeoecol.* **56**, 69–88.

FranceLanord C. and Derry L. A. (1997) Organic carbon burial forcing of the carbon cycle from Himalayan erosion. *Nature* **390**(6655), 65–67.

Franck V. M., Brzezinski M. A., Coale K. H., and Nelson D. M. (2000) Iron and silicic acid availability regulate Si uptake in the Pacific Sector of the Southern Ocean. *Deep-Sea Res. II* **47**, 3315–3338.

François R., Bacon M. P., and Suman D. O. (1990) Thorium-230 profiling in deep-sea sediments: high-resolution records of flux and dissolution of carbonate in the equatorial Atlantic during the last 24,000 years. *Paleoceanography* **5**, 761–787.

François R. F., Altabet M. A., Yu E.-F., Sigman D. M., Bacon M. P., Frank M., Bohrmann G., Bareille G., and Labeyrie L. D. (1997) Water column stratification in the Southern Ocean contributed to the lowering of glacial atmospheric CO_2. *Nature* **389**, 929–935.

Frank M., Gersonde R., van der Loeff M. R., Bohrmann G., Nurnberg C. C. Kubik, P. W., Suter M., and Mangini A. (2000) Similar glacial and interglacial export bioproductivity in the Atlantic sector of the Southern Ocean: multiproxy evidence and implications for glacial atmospheric CO_2. *Paleoceanography* **15**(6), 642–658, 2000PA000497.

Frew R. D. and Hunter K. A. (1992) Influence of Southern Ocean waters on the cadmium-phosphate properties of the global ocean. *Nature* **360**(6400), 144–146.

Froelich P. N. and Andreae M. O. (1981) The marine geochemistry of germanium: ekasilicon. *Science* **213**, 205–207.

Froelich P. N., Bender M. L., Luedtke N. A., Heath G. R., and DeVries T. (1982) The marine phosphorus cycle. *Am. J. Sci.* **282**, 474–511.

Ganeshram R. S. and Pedersen T. F. (1998) Glacial–interglacial variability in upwelling and bioproductivity off NW Mexico: implications for quaternary paleoclimate. *Paleoceanography* **13**(6), 634–645.

Ganeshram R. S., Pedersen T. F., Calvert S. E., and Murray J. W. (1995) Large changes in oceanic nutrient inventories from glacial to interglacial periods. *Nature* **376**(6543), 755–758.

Ganeshram R. S., Pedersen T. F., Calvert S. E., McNeill G. W., and Fontugne M. R. (2000) Glacial–interglacial variability in denitrification in the world's oceans: causes and consequences. *Paleoceanography* **15**(4), 361–376.

Ganeshram R. S., Pedersen T. F., Calvert S. E., and Francois R. (2002) Reduced nitrogen fixation in the glacial ocean inferred from changes in marine nitrogen and phosphorus inventories. *Nature* **415**, 156–159.

Gildor H. and Tziperman E. (2001) Physical mechanisms behind biogeochemical glacial–interglacial CO_2 variations. *Geophys. Res. Lett.* **28**, 2421–2424.

Gildor H., Tziperman E., and Toggweiler J. R. (2002) Sea ice switch mechanism and glacial–interglacial CO_2 variations. *Global Biogeochem. Cycles* **16**(3), 10.1029/2001GB001446.

Gingele F. X., Zabel M., Kasten S., Bonn W. J., and Nurnberg C. C. (1999) Biogenic barium as a proxy for paleoproductivity: methods and limitations of application. In *Use of Proxies in Paleoceanography: Examples from the South Atlantic* (eds. G. Fischer and G. Wefer). Springer, Berlin, pp. 345–364.

Gorbarenko S. A. (1996) Stable isotope and lithologic evidence of late-Glacial and Holocene oceanography of the northwestern Pacific and its marginal seas. *Quat. Res.* **46**, 230–250.

Gordon A. L., Taylor H. W., and Georgi D. T. (1977) Antarctic oceanographic zonation. In *Polar Oceans* (ed. M. J. Dunbar). Arctic Institute of North America, , pp. 219–225.

Gruber N. and Sarmiento J. L. (1997) Global patterns of marine nitrogen fixation and denitrification. *Global Biogeochem. Cycles* **11**, 235–266.

Hastings D. W., Emerson S. R., and Mix A. C. (1996) Vanadium in foraminiferal calcite as a tracer for changes in the areal extent of reducing conditions. *Paleoceanography* **11**(6), 666–678.

Haug G. H., Maslin M. A., Sarnthein M., Stax R., and Tiedemann R. (1995) Evolution of northwest Pacific sedimentation patterns since 6 Ma: site 882. *Proc. ODP. Sci. Results* **145**, 293–315.

Haug G. H., Pedersen T. F., Sigman D. M., Calvert S. E., Nielsen B., and Peterson L. C. (1998) Glacial/interglacial variations in productivity and nitrogen fixation in the Cariaco Basin during the last 550 ka. *Paleoceanography* **13**(5), 427–432.

Haug G. H., Sigman D. M., Tiedemann R., Pedersen T. F., and Sarnthein M. (1999) Onset of permanent stratification in the Subarctic Pacific Ocean. *Nature* **40**, 779–782.

Hays J. D., Imbrie J., and Shackleton N. J. (1976) Variations in the Earth's orbit: pacemaker of the ice ages. *Science* **194**, 1121–1132.

Hebbeln D., Marchant M., and Wefer G. (2002) Paleoproductivity in the southern Peru-Chile Current through the last 33 000 yr. *Mar. Geol.* **186**(3–4), 487–504.

Heinze H.-M. and Wefer G. (1992) The history of coastal upwelling off Peru over the past 650,000 years. In *Upwelling Systems: Evolution Since the Early Miocene* (eds. C. P. Summerhayes, W. L. Prell, and K.-C. Emeis). Geological Society of London, , vol. 64, pp. 451–462.

Herbert T. D., Schuffert J. D., Andreasen D., Heusser L., Lyle M., Mix A., Ravelo A. C., Stott L. D., and Herguera J. C. (2001) Collapse of the California current during glacial maxima linked to climate change on land. *Science* **293**, 71–76.

Herguera J. C. (1992) Deep-sea benthic foraminifera and biogenic opal—glacial to postglacial productivity changes in the western equatorial Pacific. *Mar. Micropaleonthol.* **19**, 79–98.

Herguera J. C. and Berger W. H. (1991) Paleoproductivity from benthic foraminifera abundance—glacial to postglacial change in the west-equatorial Pacific. *Geology* **19**(12), 1173–1176.

Herguera J. C. and Berger W. H. (1994) Glacial to postglacial drop in productivity in the western equatorial Pacific—mixing rate versus nutrient concentrations. *Geology* **22**(7), 629–632.

Herguera J. C., Jansen E., and Berger W. H. (1992) Evidence for a bathyal front at 2,000 m depth in the glacial Pacific, based on a depth transect on Ontong Java Plateau. *Paleoceanography* **7**(3), 273–288.

Hinrichs K. U., Schneider R. R., Muller P. J., and Rullkotter J. (1999) A biomarker perspective on paleoproductivity variations in two Late Quaternary sediment sections from the Southeast Atlantic Ocean. *Org. Geochem.* **30**(5), 341–366.

Hofmann M., Broecker W. S., and Lynch-Stieglitz J. (1999) Influence of a [CO_2(aq)] dependent biological C-isotope fractionation on glacial C-13/C-12 ratios in the ocean. *Global Biogeochem. Cycles* **13**(4), 873–883.

Holmen K. (1992) The global carbon cycle. In *Global Biogeochemical Cycles* (eds. S. S. Butcher, R. J. Charlson, G. H. Orians, and G. V. Wolfe). Academic Press, New York, pp. 239–262.

Holmes M. E., Schneider R. R., Muller P. J., Segl M., and Wefer G. (1997) Reconstruction of past nutrient utilization in the eastern Angola Basin based on sedimentary $^{15}N/^{14}N$ ratios. *Paleoceanography* **12**(4), 604–614.

Huang C.-Y., Liew P.-M., Zhao M., Chang T.-C., Kuo C.-M., Chen M.-T., Wang C.-H., and Zheng L.-F. (1997) Deep sea and lake records of the Southeast Asian paleomonsoons for the last 25 thousand years. *Earth Planet. Sci. Lett.* **146**, 59–72.

Hutchins D. A. and Bruland K. W. (1998) Iron-limited diatom growth and Si : N uptake ratios in a coastal upwelling regime. *Nature* **393**, 561–564.

Janecek T. R. (2001) Data report: Late Pleistocene biogenic opal data for Leg 177 Sites 1093 and 1094. *Proc. Ocean Drilling Program, Sci. Results* **177**, 1–5.

Karl D., Letelier R., Tupas L., Dore J., Christian J., and Hebel D. (1997) The role of nitrogen fixation in biogeochemical cycling in the subtropical North Pacific Ocean. *Nature* **388**(6642), 533–538.

Kawahata H., Suzuki A., and Ahagon N. (1998) Biogenic sediments in the West Caroline Basin and the western equatorial Pacific during the last 330,000 years. *Mar. Geol.* **149**, 155–176.

Kawahata H., Ohkushi K. I., and Hatakeyama Y. (1999) Comparative Late Pleistocene paleoceanographic changes in the mid latitude Boreal and Austral Western Pacific. *J. Oceanogr.* **55**, 747–761.

Keeling R. F. and Stephens B. B. (2001) Antarctic sea ice and the control of Pleistocene climate instability. *Paleoceanography* **16**(1), 112–131.

Keeling R. F. and Visbeck M. (2001) Palaeoceanography: Antarctic stratification and glacial CO_2. *Nature* **412**, 605–606.

Keigwin L. D. and Boyle E. A. (1989) Late quaternary paleochemistry of high-latitude surface waters. *Paleogeogr. Palaeoclimatol. Paleoecol.* **73**, 85–106.

Keigwin L. D., Jones G. A., and Froelich P. N. (1992) A 15,000 year paleoenvironmental record from Meiji Seamount, far northwestern Pacific. *Earth Planet. Sci. Lett.* **111**, 425–440.

Keil R. G. and Cowie G. L. (1999) Organic matter preservation through the oxygen-deficient zone of the NE Arabian Sea as discerned by organic carbon: mineral surface area ratios. *Mar. Geol.* **161**(1), 13–22.

Keir R. and Berger W. (1983) Atmospheric CO_2 content in the last 120,000 years: the phosphate extraction model. *J. Geophys. Res.* **88**, 6027–6038.

Keir R. S. (1988) On the late Pleistocene ocean geochemistry and circulation. *Paleoceanography* **3**, 413–445.

Keller K. and Morel F. M. M. (1999) A model of carbon isotopic fractionation and active carbon uptake in phytoplankton. *Mar. Ecol. Prog. Ser.* **182**, 295–298.

Kohfeld K. E., Fairbanks R. G., Smith S. L., and Walsh I. D. (1996) *Neogloboquadrina pachyderma* (sinistral coiling) as paleoceanographic tracers in polar oceans: evidence from northeast water polynya plankton tows, sediment traps, and surface sediments. *Paleoceanography* **11**(6), 679–699.

Kumar N., Anderson R. F., Mortlock R. A., Froelich P. N., Kubik P., Dittrich-Hannen B., and Suter M. (1995) Increased biological productivity and export production in the glacial Southern Ocean. *Nature* **378**, 675–680.

Leuenberger M., Siegenthaler U., and Langway C. C. (1992) Carbon isotope composition of atmospheric CO_2 during the last Ice Age from an Antarctic ice core. *Nature* **357**(6378), 488–490.

Loubere P. (1999) A multiproxy reconstruction of biological productivity and oceanography in the eastern equatorial Pacific for the past 30,000 years. *Mar. Micropaleontol.* **37**, 173–198.

Lourey M. J., Trull T. W., and Sigman D. M. (2003) An unexpected decrease of δ^{15} N of surface and deep organic nitrogen accompanying Southern Ocean seasonal nitrate depletion. *Global Biogeochem. Cycles* (in press).

Lyle M. (1988) Climatically forced organic carbon burial in the equatorial Atlantic and Pacific Oceans. *Nature* **335**, 529–532.

Lyle M., Murray D., Finney B., Dymond J., Robbins J., and Brookforce K. (1988) The record of Last Pleistocene biogenic sedimentation in the eastern tropical Pacific Ocean. *Paleoceanography* **3**, 39–59.

Mahowald N., Kohfeld K., Hansson M., Balkanski Y., Harrison S. P., Prentice I. C., Schulz M., and Rodhe H. (1999) Dust sources and deposition during the last glacial maximum and current climate: a comparison of model results with paleodata from ice cores and marine sediments. *J. Geophys. Res.: Atmos.* **104**, 15895–15916.

Manighetti B. and McCave I. (1995) Depositional fluxes and paleoproductivity and ice rafting in the NE Atlantic over the past 30 ka. *Paleoceanography* **10**, 579–592.

Marcantonio F., Anderson R. F., Stute M., Kumar N., Schlosser P., and A M. (1996) Extraterrestrial He-3 as a tracer of marine sediment transport and accumulation. *Nature* **383**(6602), 705–707.

Marcantonio F., Anderson R., Higgins S., Fleisher M., Stute M., and Schlosser P. (2001a) Abrupt intensification of the SW Indian Ocean monsoon during the last delaciation: constraints from Th, Pa and He isotopes. *Earth Planet. Sci. Lett.* **184**, 505–514.

Marcantonio F., Anderson R., Higgins S., Stute M., and Schlosser P. (2001b) Sediment focussing in the central equatorial Pacific Ocean. *Paleoceanography* **16**, 260–267.

Marchitto T. M., Curry W. B., and Oppo D. W. (1998) Millennial-scale changes in North Atlantic circulation since the last glaciation. *Nature* **393**(6685), 557–561.

Marino B. D. and McElroy M. B. (1991) Isotopic composition fo atmospheric CO_2 inferred from carbon in C4 plant cellulose. *Nature* **349**, 127–131.

Martin J. H. (1990) Glacial–interglacial CO_2 change: the iron hypothesis. *Paleoceanography* **5**, 1–13.

Martin J. H., Fitzwater S. E., and Gordon R. M. (1990) Iron deficiency limits growth in Antarctic waters. *Global Biogeochem. Cycles* **4**, 5–12.

Martinez P. H., Bertrand P. H., Bouloubassi I., Bareille G., Shimmield G., Vautravers B., Grousset F., Guichard S., Ternois Y., and Sicre M.-A. (1996) An integrated view of inorganic and organic biogeochemical indicators of paleoproductivity changes in a coastal upwelling area. *Org. Geochem.* **24**, 411–420.

Matsumoto K., Oba T., Lynch-Stieglitz J., and Yamamoto H. (2002a) Interior hydrography and circulation of the glacial Pacific Ocean. *Quat. Sci. Rev.* **21**, 1693–1704.

Matsumoto K., Sarmiento J. L., and Brzezinski M. A. (2002b) Silicic acid "leakage" from the Southern Ocean as a possible mechanism for explaining glacial atmospheric $p CO_2$. *Global Biogeochem. Cycles* **16**, 10.1029/2001GB001442.

McCorkle D. C., Martin P. A., Lea D. W., and Klinkhammer G. P. (1995) Evidence of a dissolution effect on benthic foraminiferal shell chemistry: delta C-13, Cd/Ca, Ba/Ca, and Sr/Ca results from the Ontong Java Plateau. *Paleoceanography* **10**(4), 699–714.

McCulloch R. D., Bentley M. J., Purves R. S., Hulton N. R. J., Sugden D. E., and Clapperton C. M. (2000) Climate inferences from glacial and palaeoecological evidence at the last glacial termination, southern South America. *J. Quat. Sci.* **15**, 409–417.

McDonald D., Pederson T., and Crusius J. (1999) Multiple Late Quaternary episodes of exceptional diatom production in the Gulf of Alaska. *Deep-Sea Res. II* **46**, 2993–3017.

McElroy M. B. (1983) Marine biological controls on atmospheric CO_2 and climate. *Nature* **302**, 328–329.

McIntyre A. and Molfino B. (1996) Forcing of Atlantic equatorial and subpolar millennial cycles by precession. *Science* **274**(5294), 1867–1870.

Mitchell B. G., Brody E. A., Holm-Hansen O., McClain C., and Bishop J. (1991) Light limitation of phytoplankton biomass and macronutrient utilization in the Southern Ocean. *Limnol. Oceanogr.* **36**(8), 1662–1677.

Monnin E., Indermuhle A., Dallenbach A., Fluckiger J., Stauffer B., Stocker T. F., Raynaud D., and Barnola J.-M. (2001) Atmospheric CO_2 concentrations over the last glacial termination. *Science* **291**, 112–114.

Montoya J. P. and McCarthy J. J. (1995) Isotopic fractionation during nitrate uptake by marine phytoplankton

grown in continuous culture. *J. Plankton Res.* **17**(3), 439–464.

Moore J. K., Abbott M. R., Richman J. G., and Nelson D. M. (2000) The Southern Ocean at the last glacial maximum: a strong sink for atmospheric carbon dioxide. *Global Biogeochem. Cycles* **14**, 455–475.

Moreno A., Nave S., Kuhlmann H., Canals M., Targarona J., Freudenthal T., and Abrantes F. (2002) Productivity response in the North Canary Basin to climate changes during the last 250,000 yr: a multi-proxy approach. *Earth Planet. Sci. Lett.* **196**, 147–159.

Morley J. J. and Hays J. D. (1983) Oceanographic conditions associated with high abundances of the radiolarian *Cycladophora davisiana*. *Earth Planet. Sci. Lett.* **66**, 63–72.

Mortlock R. A., Charles C. D., Froelich P. N., Zibello M. A., Saltzman J., Hyas J. D., and Burckle L. H. (1991) Evidence for lower productivity in the Antarctic during the last glaciation. *Nature* **351**, 220–223.

Muller P. J. and Suess E. (1979) Productivity, sedimentation rate, and sedimentary organic matter in the oceans: I. Organic carbon preservation. *Deep-Sea Res. I* **26**, 1347–1362.

Murray R. W., Leinen M., and Isern A. R. (1993) Biogenic flux of Al to sediment in the central equatorial Pacific Ocean: evidence for increased productivity during glacial periods. *Paleoceanography* **8**(5), 651–670.

Nakatsuka T., Watanabe K., Handa N., Matsumoto E., and Wada E. (1995) Glacial to interglacial surface nutrient variation of the Bering deep basins recorded by delta-C-13 and delta-N-15 of sedimentary organic matter. *Paleoceanography* **10**(6), 1047–1061.

Narita H., Sato M., Tsunogai S., Murayama M., Nakatsuka T., Wakatsuchi M., Harada N., and Ujiié Y. (2002) Biogenic opal indicating less productive northwestern North Pacific during the glacial ages. *Geophys. Res. Lett.* **29**(15) 10.1029 22.1–22.4.

Opdyke B. N. and Walker J. C. G. (1992) Return of the coral reef hypothesis: basin to shelf potitioning of $CaCO_3$ and its effect on atmospheric CO_2. *Geology* **20**, 733–736.

Oppo D. W. and Lehman S. J. (1993) Mid-depth circulation of the subpolar North Atlantic during the last glacial maximum. *Science* **259**, 1148–1152.

Pailler D., Bard E., Rostek F., Zheng Y., Mortlock R., and van Geen A. (2002) Burial of redox-sensitive metals and organic matter in the equatorial Indian Ocean linked to precession. *Geochim. Cosmochim. Acta* **66**(5), 849–865.

Paytan A., Kastner M., Martin E. E., Macdougall J. D., and Herbert T. (1996) Glacial to interglacial fluctuations in productivity in the equatorial Pacific as indicated by marine barite. *Science* **274**(5291), 1355–1357.

Peacock S. (2001) Use of tracers to constrain time-averaged fluxes in the ocean. PhD, Columbia University.

Pedersen T. F. (1983) Increased productivity in the eastern equatorial Pacific during the last glacial maximum (19,000 to 14,000 yr B.P.). *Geology* **11**, 16–19.

Pedersen T. F., Nielsen B., and Pickering M. (1991) Timing of Late Quaternary productivity pulses in the Panama Basin and implications for atmospheric CO_2. *Paleoceanography* **6**, 657–677.

Perks H. M., Charles C. D., and Keeling R. F. (2000) Precessionally forced productivity variations across the equatorial Pacific. *Paleoceanography* **17**(3) 10.1029/2000PA000603.

Petit J. R., Jouzel J., Raynaud D., Barkov N. I., Barnola J.-M., Basile I., Bender M., Chappellaz J., Davis M., Delaygue G., Delmotte M., Kotlyakov V. M., Legrand M., Lipenkov V. Y., Lrius C., Pepin L., Ritz C., Saltzman E., and Stievenard M. (1999) Climate and atmospheric history of the past 420,000 years from the Vostok ice core, Antarctica. *Nature* **399**, 429–436.

Popp B. N., Laws E. A., Bidigare R. R., Dore J. R., Hanson K. L., and Wakeham S. G. (1998) Effect of phytoplankton cell geometry on carbon isotopic fractionation. *Geochim. Cosmochim. Acta* **62**(1), 69–77.

Prell W. L., Hutson W. H., Williams D. F., Be A. W. H., Geitzenauer K., and Molfino B. (1980) Surface circulation of the Indian Ocean during the Last Glacial maximum, approximately 18,000 yr BP. *Quat. Res.* **14**(3), 309–336.

Pride C., Thunell R., Sigman D., Keigwin L., Altabet M., and Tappa E. (1999) Nitrogen isotopic variations in the Gulf of California since the last deglaciation: response to global climate change. *Paleoceanography* **14**(3), 397–409.

Rau G. H., Tkahashi T., and Des Marais D. J. (1989) Latitudinal variations in plankton $\delta^{13}C$: implications for CO_2 and productivity in past oceans. *Nature* **341**, 516–518.

Rea D., Pisias N., and Newberry T. (1991) Late Pleistocene paleoclimatology of the central equatorial Pacific: flux patterns of biogenic sediments. *Paleoceanography* **6**, 227–244.

Redfield A. C. (1942) The processes determining the concentration of oxygen, phosphate and other organic derivatives within the depths of the Atlantic ocean. *Pap. Phys. Oceanogr. Meteorol.* **IX**.

Redfield A. C. (1958) The biological control of chemical factors in the environment. *Am. Sci.* **46**, 205–221.

Redfield A. C., Ketchum B. H., and Richards F. A. (1963) The influence of organisms on the composition of seawater. In *The Sea* (ed. M. N. Hill). Interscience, , Vol. 2, pp. 26–77.

Reid J. L. (1969) Sea surface temperature, salinity, and density of the Pacific Ocean in summer and in winter. *Deep-Sea Res. I* **16**(suppl.), 215–224.

Rickaby R. E. M. and Elderfield H. (1999) Planktonic foraminiferal Cd/Ca: Paleonutrients or paleotemperature? *Paleoceanography* **14**(3), 293–303.

Rosenthal Y., Boyle E. A., Labeyrie L., and Oppo D. (1995) Glacial enrichments of authigenic Cd and U in Subantarctic sediments—a climatic control on the elements oceanic budget. *Paleoceanography* **10**(3), 395–413.

Rosenthal Y., Boyle E. A., and Labeyrie L. (1997) Last glacial maximum paleochemistry and deepwater circulation in the Southern Ocean: evidence from foraminiferal Cadmium. *Paleoceanography* **12**(6), 787–796.

Rosenthal Y., Dahan M., and Shemesh A. (2000) Southern Ocean contributions to glacial-interglacial changes of atmospheric CO_2: evidence from carbon isotope records in diatoms. *Paleoceanography* **15**(1), 65–75.

Ruhlemann C., Muller P. J., and Schneider R. R. (1999) Organic carbon and carbonate as paleoproductivity proxies: examples from high and low productivity areas of the tropical Atlantic. In *Use of Proxies in Paleoceanography: Examples from the South Atlantic* (eds. G. Fischer and G. Wefer). Springer, Berlin, pp. 315–344.

Ruhlemann M., Frank M., Hale W., Mangini A., Multiza P., and Wefer G. (1996) Late Quaternary productivity changes in the western equatorial Atlantic: evidence from ^{230}Th-normalized carbonate and organic carbon accumulation rates. *Mar. Geol.* **135**, 127–152.

Russell A. D., Emerson S., Mix A. C., and Peterson L. C. (1996) The use of foraminiferal uranium/calcium ratios as an indicator of changes in seawater uranium content. *Paleoceanography* **11**(6), 649–663.

Rutsch H.-J., Mangini A., Bonani G., Dittrich-Hannen B., Kubik P., Suter M., and Segl M. (1995) ^{10}Be and Ba concentrations in West African sediments trace productivity in the past. *Earth Planet. Sci. Lett.* **133**, 129–143.

Ruttenberg K. C. (1993) Reassessment of the oceanic residence time of phosphorous. *Chem. Geol.* **107**, 405–409.

Sachs J. P. and Repeta D. J. (1999) Oligotrophy and nitrogen fixation during Eastern Mediterranean sapropel events. *Science* **286**, 2485–2488.

Sachs J. P., Anderson R. F., and Lehman S. J. (2001) Glacial surface temperatures of the southeast Atlantic Ocean. *Science* **293**(5537), 2077–2079.

Sanudo-Wilhelmy S., Kustka A., Gobler C., Hutchins D., Yang M., Lwiza K., Burns J., Capone D., raven J., and Carpenter E. (2001) Phosphorus limitation of nitrogen fixation by *Trichodesmium* in the central Atlantic Ocean. *Nature* **411**(6833), 66–69.

Sanyal A., Hemming N. G., Broecker W. S., and Hanson G. N. (1997) changes in *p*H in the eastern equatorial Pacific across stage 5–6 boundary based on boron isotopes in foraminifera. *Global Biogeochem. Cycles* **11**(1), 125–133.

Sarnthein M., Winn K., Duplessy J.-C., and Fontugne M. R. (1988) Global variations of surface ocean productivity in low and mid latitudes: influence on CO_2 reservoirs of the deep ocean and atmosphere during the last 21,000 years. *Paleoceanography* **3**(3), 361–399.

Schneider R. R. (1991) Spätquartäre Produktivitätsänderungen im östlichen Anfola Becken: Reaktion auf Variationen im Passat-Monsun Windsystem und der Advektion des Benguela-Küstenstroms. *Ber. Fachbereich Geowiss., Univ. Bremen* **21**, 1–198.

Schneider R. R., Muller P. J., Ruhland G., Meinecke G., Schmidt H., and Wefer G. (1996) Late Quaternary surface temperatures and productivity in the East-equatorial South Atlantic: Response to changes in Trade/Monsoon wind forcing and surface water advection. In *The South Atlantic: Present and Past Circulation* (eds. G. Wefer, W. H. Berger, G. Siedler, and D. J. Webb). Springer, Berlin, pp. 527–551.

Schrag D., Hampt G., and Murray D. (1996) Pore fluid constraints on the temperature and oxygen isotopic composition of the glacial ocean. *Science* **272**(5270), 1930–1932.

Schubert C. J. and Stein R. (1996) Deposition of organic carbon in Arctic sediments: terrigenous supply versus marine productivity. *Org. Geochem.* **24**, 421–436.

Schwarz B., Mangini A., and Segl M. (1996) Geochemistry of a piston core from Ontong Java Plateau (western equatorial Pacific): evidence for sediment redistribution and changes in paleoproductivity. *Geol. Rundsch.* **85**, 536–545.

Shackelton N. J., Hall M. A., Line J., and Cang S. (1983) Carbon isotope data in core V19-30 confirm reduced carbon dioxide concentration of the ice age atmosphere. *Nature* **306**, 319–322.

Shemesh A., Macko S. A., Charles C. D., and Rau G. H. (1993) Isotopic evidence for reduced productivity in the glacial southern ocean. *Science* **262**, 407–410.

Sigman D. M. and Boyle E. A. (2000) Glacial/interglacial variations in atmospheric carbon dioxide. *Nature* **407** (6806), 859–869.

Sigman D. M. and Boyle E. A. (2001) Palaeoceanography: Antarctic stratification and glacial CO_2. *Nature* **412**, 606.

Sigman D. M., McCorkle D. C., and Martin W. R. (1998) The calcite lysocline as a constraint on glacial/interglacial low-latitude production changes. *Global Biogeochem. Cycles* **12** (3), 409–427.

Sigman D. M., Altabet M. A., Francois R., McCorkle D. C., and Fischer G. (1999a) The $\delta^{15}N$ of nitrate in the Southern Ocean: consumption of nitrate in surface waters. *Global Biogeochem. Cycles* **13**(4), 1149–1166.

Sigman D. M., Altabet M. A., Francois R., McCorkle D. C., and Gaillard J.-F. (1999b) The isotopic compositon of diatom-bound nitrogen in Southern Ocean sediments. *Paleoceanography* **14**(2), 118–134.

Sigman D. M., Lehman S. J., and Oppo D. W. (2003) Evaluating mechanisms of nutrient depletion and ^{13}C enrichment in the intermediate-depth Atlantic during the last ice age. *Paleoceanography* **18**(3), 1072, doi: 10.1029/2002PA000818.

Smith H. J., Fischer H., Wahlen M., Mastroianni D., and Deck B. (1999) Dual modes of the carbon cycle since the Last Glacial Maximum. *Nature* **400**, 248–250.

Sowers T. and Bender M. L. (1995) Climate records covering the last deglaciation. *Science* **269**, 210–214.

Spero H. J. and Lea D. W. (1993) Intraspecific stable isotope variability in the planktic foraminifera *Globigerinoides sacculifer*: Results from laboratory experiments. *Mar. Micropaleontol.* **22**, 221–234.

Spero H. J. and Williams D. F. (1988) Extracting environmental information from planktonic foraminiferal delta 13-C. *Nature* **335**, 717–719.

Spero H. J., Bijma J., Lea D. W., and Bemis B. E. (1997) Effect of seawater carbonate concentration on foraminiferal carbon and oxygen isotopes. *Nature* **390**, 497–499.

Stephens B. B. and Keeling R. F. (2000) The influence of Antarctic sea ice on glacial–interglacial CO_2 variations. *Nature* **404**, 171–174.

Stephens M. and Kadko D. (1997) Glacial-Holocene calcium carbonate dissolution at the central equatorial Pacific seafloor. *Paleoceanography* **12**, 797–804.

Stutt J. B. W., Prins M. A., Schneider R. R., Weltje G. J., Jansen J. H. F., and Postma G. (2002) A 300 kyr record of aridity and windstrength in southwestern Africa; inferences from grain-size distributions of sediments on Walvis Ridge, SE Atlantic. *Mar. Geol.* **180**(1–4), 221–233.

Suman D. O. and Bacon M. P. (1989) Variations in Holocene sedimentation in the North-American Basin determined from Th-230 measurements. *Deep-Sea Res. A* **36**(6), 869–878.

Summerhayes C. P., Kroon D., Rosell-Mele A., Jordan R. W., Schrader H.-J., Hearn R., Villanueva J., Grimalt J. O., and Eglinton G. (1995) Variability in the Benguela Current upwelling system over the past 70,000 years. *Progr. Oceanogr.* **35**, 207–251.

Takahashi T., Feely R. A., Weiss R. F., Wanninkohf R. H., Chipman D. W., Sutherland S. C., and Takahashi T. T. (1997) Global air-sea flux of CO_2: an estimate based on measurements of sea-air pCO(2) difference. *Proc. Natl. Acad. Sci. USA* **94**(16), 8292–8299.

Takeda S. (1998) Influence of iron availability on nutrient consumption ratio of diatoms in oceanic waters. *Nature* **393**, 774–777.

Talley L. D. (1993) Distribution and formation of North Pacific intermediate water. *J. Phys. Oceanogr.* **23**, 517–537.

Thomas E., Booth L., Maslin M., and Shackleton N. J. (1995) Northeastern Atlantic benthic foraminifera during the last 45,000 years—changes in productivity seen from the bottom up. *Paleoceanography* **10**(3), 545–562.

Thunell R. C., Miao Q., Calvert S. E., and Pederson T. F. (1992) Glacial–Holocene biogenic sedimentation patters in the South China Sea: productivity variations and surface water pCO2. *Paleoceanography* **7**, 143–162.

Toggweiler J. R. (1999) Variations in atmospheric CO_2 driven by ventilation of the ocean's deepest water. *Paleoceanography* **14**(5), 571–588.

Toggweiler J. R. and Carson S. (1995) What are the upwelling systems contributing to the ocean's caron and nutrient budgets? In *Upwelling in the Ocean: Modern Processes and Ancient Records* (eds. C. P. Summerhayes, K.-C. Emeis, M. V. Angel, R. L. Smith, and B. Zeitzschel). Wiley, New York.

Toggweiler J. R. and Samuels B. (1995) Effect of Drake Passage on the global thermohaline circulation. *Deep-Sea Res. I* **42**(4), 477–500.

Toggweiler J. R., Carson S., and Bjornsson H. (1999) Response of the ACC and the Antarctic pycnocline to a meridional shift in the southern hemisphere westerlies. *EOS, Trans. AGU* **80**(49), OS286.

Toggweiler J. R., Gnanadesikan A., Carson S., Murnane R., and Sarmiento J. L. (2003a) Representation of the carbon cycle in box models and GCMs: 1. Solubility pump. *Global Biogeochem. Cycles* **17**(1) 0.1029/2001GB001401.

Toggweiler J. R., Murnane R., Carson S., Gnanadesikan A., and Sarmiento J. L. (2003b) Representation of the carbon cycle in box models and GCMs: 2. Organic pump. *Global Biogeochem. Cycles* **17**(1) 10.1029/2001GB001841.

Tortell P. D., DiTullio G. R., Sigman D. M., and Morel F. M. M. (2002) CO_2 effects on taxonomic composition and nutrient utilization in an equatorial Pacific phytoplankton assemblage. *Mar. Ecol. Progr. Ser.* **236**, 37–43.

Tyrrell T. (1999) The relative influences of nitrogen and phosphorus on oceanic primary production. *Nature* **400**, 525–531.

Vincent E. and Berger W. H. (1985) Carbon dioxide and polar cooling in the Miocene: the Monterey hypothesis. In *The Carbon Cycle and Atmospheric CO_2: Natural Variations Archean to Present* (eds. E. T. Sundquist and W. S. Broecker). American Geophysical Union, Washington DC, pp. 455–468.

Volk T. and Hoffert M. I. (1985) Ocean carbon pumps: analysis of relative strengths and efficiencies in ocean-driven atmospheric CO_2 changes. In *The Carbon Cycle and Atmospheric CO_2: Natural Variations Archean to Present* (eds. E. T. Sundquist and W. S. Broecker). American Geophysical Union, Washington DC, , pp. 99–110.

Walsh J. J., Rowe G. T., Iverson R. L., and McRoy C. P. (1981) Biological export of shelf carbon is a neglected sink of the global CO_2 cycle. *Nature* **291**, 196–201.

Walter H.-J., Rutgers van der Loeff M. M., and Francois R. (1999) Reliability of the $^{231}Pa/^{230}Th$ activity ratio as a tracer for bioproductivity of the ocean. In *Use of Proxies in Paleoceanography: Examples from the South Atlantic* (eds. G. Fischer and G. Wefer). Springer, Berlin, , pp. 393–408.

Warren B. (1983) Why is no deep water formed in the North Pacific? *J. Mar. Res.* **41**, 327–347.

Waser N. A. D., Turpin D. H., Harrison P. J., Nielsen B., and Calvert S. E. (1998) Nitrogen isotope fractionation during the uptake and assimilation of nitrate, nitrite, and urea by a marine diatom. *Limnol. Oceanogr.* **43**(2), 215–224.

Watson A. J., Bakker D. C. E., Ridgewell A. J., Boyd P. W., and Law C. S. (2000) Effect of iron supply on Southern Ocean CO_2 uptake and implications for atmospheric CO_2. *Nature* **407**(6805), 730–733.

Weaver A. J., Eby M., Fanning A. F., and Wiebe E. C. (1998) Simulated influence of carbon dioxide, orbital forcing and ice sheets on the climate of the Last Glacial Maximum. *Nature* **394**, 847–853.

Webb R. S., Lehman S. J., Rind D. H., Healy R. J., and Sigman D. M. (1997) Influence of ocean heat transport on the climate of the Last Glacial Maximum. *Nature* **385** (6618), 695–699.

Wefer G., Berger W. H., Siedler G., and Webb D. J. (1996) The South Atlantic: present and past circulation. Springer, Berlin.

Wollenburg J., Kuhnt W., and Mackensen A. (2001) Changes in Arctic Ocean paleoproductivity and hydrography during the last 145 kyr: the benthic foraminiferal record. *Paleoceanography* **16**, 65–77.

Yu E. F., Francois R., Bacon M. P., Honjo S., Fleer A. P., Manganini S. J., van der Loeff M. M. R., and Ittekot V. (2001) Trapping efficiency of bottom-tethered sediment traps estimated from the intercepted fluxes of Th-230 and Pa-231. *Deep Sea Res. I* **48**(3), 865–889.

14

The Oceanic CaCO$_3$ Cycle

W. S. Broecker

Columbia University, Palisades, NY, USA

14.1.	INTRODUCTION	491
14.2.	DEPTH OF TRANSITION ZONE	492
14.3.	DISTRIBUTION OF CO$_3^{2-}$ ION IN TODAY'S DEEP OCEAN	493
14.4.	DEPTH OF SATURATION HORIZON	495
14.5.	DISSOLUTION MECHANISMS	496
14.6.	DISSOLUTION IN THE PAST	498
14.7.	SEDIMENT-BASED PROXIES	499
14.8.	SHELL WEIGHTS	501
14.9.	THE BORON ISOTOPE PALEO pH METHOD	502
14.10.	Zn/Cd RATIOS	503
14.11.	DISSOLUTION AND PRESERVATION EVENTS	504
14.12.	GLACIAL TO INTERGLACIAL CARBONATE ION CHANGE	508
14.13.	NEUTRALIZATION OF FOSSIL FUEL CO$_2$	509
REFERENCES		509

14.1. INTRODUCTION

Along with the silicate debris carried to the sea by rivers and wind, the calcitic hard parts manufactured by marine organisms constitute the most prominent constituent of deep-sea sediments. On high-standing open-ocean ridges and plateaus, these calcitic remains dominate. Only in the deepest portions of the ocean floor, where dissolution takes its toll, are sediments calcite-free. The foraminifera shells preserved in marine sediments are the primary carriers of paleoceanographic information. Mg/Ca ratios in these shells record past surface water temperatures; temperature corrected $^{18}O/^{16}O$ ratios record the volume of continental ice; $^{13}C/^{12}C$ ratios yield information about the strength of the ocean's biological pump and the amount of carbon stored as terrestrial biomass; the cadmium and zinc concentrations serve, respectively, as proxies for the distribution of dissolved phosphate and dissolved silica in the sea. While these isotopic ratios and trace element concentrations constitute the workhorses of the field of paleoceanography, the state of preservation of the calcitic material itself has an important story to tell. It is this story with which this chapter is concerned.

In all regions of the ocean, plots of sediment composition against water depth have a characteristic shape. Sediments from mid-depth are rich in CaCO$_3$ and those from abyssal depths are devoid of CaCO$_3$. These two realms are separated by a transition zone spanning several hundreds of meters in water depth over which the CaCO$_3$ content drops toward zero from the 85–95% values which characterize mid-depth sediment. The upper bound of this transition zone has been termed the "lysocline" and signifies the depth at which dissolution impacts become noticeable. The lower bound is termed the "compensation depth" and signifies the depth at which the CaCO$_3$ content is reduced to 10%. While widely used (and misused), both of these terms suffer from ambiguities. My recommendation is that they be abandoned in favor of the term "transition zone." Where quantification is appropriate, the depth of the mid-point of CaCO$_3$ decline should be given. While the width of the zone is also of interest, its definition suffers from the same problems associated with the use of the terms "lysocline" and "compensation depth," namely, the boundaries are gradual rather than sharp.

While determinations of sediment $CaCO_3$ content as a function of water depth in today's ocean or at any specific time in the past constitute a potentially useful index of the extent of dissolution, it must be kept in mind that this relationship is highly nonlinear. Consider, for example, an area where the rain rate of calcite to the seafloor is 9 times that of noncarbonate material. In such a situation, were 50% of the calcite to be dissolved, the $CaCO_3$ content would drop only from 90% to only 82%, and were 75% dissolved away, it would drop only to 69% (see Figure 1). One might counter by saying that as the $CaCO_3$ content can be measured to an accuracy of $\pm0.5\%$ or better, one could still use $CaCO_3$ content as a dissolution index. The problem is that in order to obtain a set of sediment samples covering an appreciable range of water depth, topographic gradients dictate that the cores would have to be collected over an area covering several degrees. It is unlikely that the ratio of the rain rate of calcite to that of noncalcite would be exactly the same at all the coring sites. Hence, higher accuracy is not the answer.

14.2. DEPTH OF TRANSITION ZONE

As in most parts of today's deep ocean the concentrations of Ca^{2+} and of CO_3^{2-} are nearly constant with water depth, profiles of $CaCO_3$ content with depth reflect mainly the increase in the solubility of the mineral calcite with pressure (see Figure 2). This increase occurs because the volume occupied by the Ca^{2+} and CO_3^{2-} ions dissolved in seawater is smaller than when they are combined in the mineral calcite. Unfortunately, a sizable uncertainty exists in the magnitude of this volume difference. The mid-depth waters in the ocean are everywhere supersaturated with respect to calcite. Because of the pressure dependence of solubility, the extent of supersaturation decreases with depth until the saturation horizon is reached. Below this depth, the waters are undersaturated with respect to calcite. While it is tempting to conclude that the saturation horizon corresponds to the top of the transition zone, as we shall see, respiration CO_2 released to the pore waters complicates the situation by inducing calcite dissolution above the saturation horizon.

One might ask what controls the depth of the transition zone. The answer lies in chemical economics. In today's ocean, marine organisms manufacture calcitic hard parts at a rate several times faster than CO_2 is being added to the ocean–atmosphere system (via planetary outgassing and weathering of continental rocks) (see Figure 3). While the state of saturation in the ocean is set by the product of the Ca^{2+} and CO_3^{2-} concentrations, calcium has such a long residence (10^6 yr) that, at least on the timescale of a single glacial cycle ($\sim10^5$ yr), its concentration can be assumed to have remained unchanged. Further, its concentration in seawater is so high that $CaCO_3$ cycling within the sea does not create significant

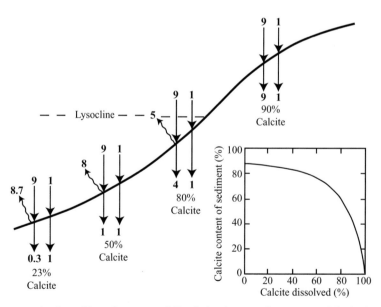

Figure 1 A diagrammatic view of how the extent of dissolution impacts the percent calcite in the sediment. In each example, the right-hand vertical arrows give the rain rate and accumulation rate of non-$CaCO_3$ debris and the left-hand vertical arrows the rain rate and accumulation rate of calcite. The wavy arrows represent the dissolution rates of calcite. As can be seen from the graph on the lower right, the percent of calcite in the sediment gives a misleading view of the fraction of the raining calcite which has dissolved, for large amounts of dissolution are required before the calcite content of the sediment drops significantly.

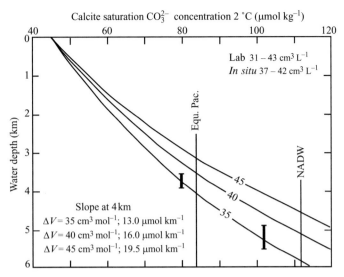

Figure 2 Saturation carbonate ion concentration for calcite at deep-water temperatures as a function of water depth (i.e., pressure). Curves are drawn for three choices of ΔV between Ca^{2+} and CO_3^{2-} ions when combined in calcite and when dissolved in seawater. The vertical black lines indicate the average CO_3^{2-} concentration in the deep equatorial Pacific and in NADW. The bold segments indicate the onset of dissolution as indicated by *in situ* experiments carried out in the deep North Pacific (Peterson, 1966) and in the deep North Atlantic (Honjo and Erez, 1978). As indicated in the upper right-hand corner, estimates of ΔV cover a wide range.

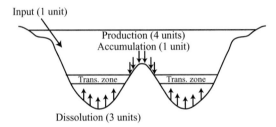

Figure 3 Marine organisms produce calcite at $\sim$4 times the rate at which the ingredients for this mineral are supplied to the sea by continental weathering and planetary outgassing. A transition zone separates the mid-depth ocean floor where calcite is largely preserved from the abyssal ocean floor where calcite is largely dissolved.

gradients. In contrast, the dissolved inorganic carbon (i.e., CO_2, HCO_3^-, and CO_3^{2-}) in the ocean is replaced on a timescale roughly equal to that of the major glacial to interglacial cycle (10^5 yr). But, since in the deep sea CO_3^{2-} ion makes up only $\sim$5% of the total dissolved inorganic carbon, its adjustment time turns out to be only about one-twentieth that for dissolved inorganic carbon or $\sim$5,000 yr. Hence, the concentration of CO_3^{2-} has gradients within the sea and likely has undergone climate-induced changes.

Hence, at least on glacial to interglacial time-scales, attention is focused on distribution of CO_3^{2-} concentration in the deep sea for it alone sets the depth of the transition zone. Thus, it is temporal changes in the concentration of carbonate ion which have captured the attention of those

paleoceanographers interested in glacial to inter-glacial changes in ocean operation. These changes involve both the carbonate ion concentration averaged over the entire deep ocean and its distribution with respect to water depth and geographic location. Of course, it is the global average carbonate ion concentration in the deep sea that adjusts in order to assure that burial of $CaCO_3$ in the sediments matches the input of CO_2 to the ocean atmosphere system (or, more precisely, the input minus the fraction destined to be buried as organic residues). For example, were some anomaly to cause the burial of $CaCO_3$ in seafloor sediments to exceed supply, then the CO_3^{2-} concentration would be drawn down. This drawdown would continue until a balance between removal and supply was restored. As already mentioned, the time constant for this adjustment is on the order of 5,000 yr.

14.3. DISTRIBUTION OF CO_3^{2-} ION IN TODAY'S DEEP OCEAN

As part of the GEOSECS, TTO, SAVE and WOCE ocean surveys, $\sum CO_2$ and alkalinity measurements were made on water samples captured at various water depths in Niskin bottles. Given the depth, temperature and salinity for these samples, it is possible to compute the *in situ* carbonate ion concentrations. LDEO's Taro Takahashi played a key role not only in the measurement programs, but also in converting the measurements to *in situ* carbonate ion concentrations. Because of his efforts and, of course, those of many others involved in these expeditions, we now have a complete picture of the distribution of CO_3^{2-} ion concentrations in the deep sea.

Below 1,500 m in the world ocean, the distribution of carbonate ion concentration is remarkably simple (see Broecker and Sutherland, 2000 for summary). For the most part, waters in the Pacific, Indian, and Southern Oceans have concentrations confined to the range $83 \pm 8\,\mu mol\,kg^{-1}$. The exception is the northern Pacific, where the values drop to as low as $60\,\mu mol\,kg^{-1}$. In contrast, much of the deep water in the Atlantic has concentrations in the range $112 \pm 5\,\mu mol\,kg^{-1}$. The principal exception is the deepest portion of the western basin where Antarctic bottom water (AABW) intrudes.

As shown by Broecker *et al.*, the deep waters of the ocean can be characterized as a mixture of two end members, i.e., deep water formed in the northern Atlantic and deep water formed in the Southern Ocean. These end members are characterized by quite different values of a quasi-conservative property, PO_4^* (i.e., $PO_4 - 1.95 + O_2/175$). Although these two deep-water sources have similar initial O_2 contents, those formed in the northern Atlantic have only roughly half the PO_4 concentration of the deep waters descending in the Southern Ocean. Thus, the northern end member is characterized by a PO_4^* value of 0.73 ± 0.03, while the southern end member is characterized by a value of $1.95 \pm 0.05\,\mu mol\,kg^{-1}$. In Figure 4 is shown a plot of carbonate ion concentration for waters deeper than 1,700 m as a function of PO_4^*.

The points are color coded according to O_2 content. As can be seen, the high O_2 waters with northern Atlantic PO_4^* values have carbonate ion concentrations of $\sim120\,\mu mol\,kg^{-1}$, while those formed in Weddell Sea and Ross Sea have values closer to $90\,\mu mol\,kg^{-1}$.

The sense of the between-ocean difference in carbonate ion concentration is consistent with the PO_4^*-based estimate that Atlantic deep water (i.e., North Atlantic deep water (NADW)) is a mixture of about 85% deep water formed in the northern Atlantic and 15% deep water formed in Southern Ocean, while the remainder of the deep ocean is flooded with a roughly 50–50 mixture of these two source waters (Broecker, 1991). The inter-ocean difference in carbonate ion concentration relates to the fact that deep water formed in the northern Atlantic has a higher CO_3^{2-} concentration than that produced in the Southern Ocean. The transition zone between NADW and the remainder of the deep ocean is centered in the western South Atlantic and extends around Africa into the Indian Ocean (fading out as NADW mixes into the ambient circumpolar deep water).

The difference in carbonate ion concentration between NADW and the rest of the deep ocean is related to the difference in PO_4 concentration. NADW has only about half the concentration of PO_4 as does, for example, deep water in equatorial Pacific. This is important because, for each mole of phosphorus released during respiration, $\sim120\,mol$ of CO_2 are also produced. This excess

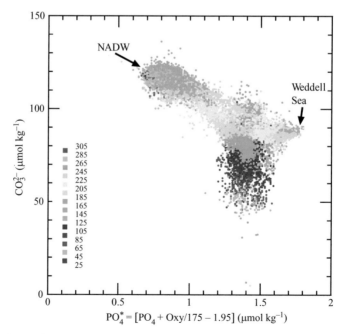

Figure 4 Plot of carbonate ion concentration as a function of PO₄* for waters deeper than 1,700 m in the world ocean. The points are coded according to O_2 content (in $\mu mol\,kg^{-1}$). This plot was provided by LDEO'S Stew Sutherland and Taro Takahashi. It is based on measurements made as part of the GEOSECS, TTO, SAVE and WOCE surveys.

CO_2 reacts with CO_3^{2-} ion to form two HCO_3^- ions. Were PO_4 content the only factor influencing the interocean difference in carbonate ion concentration, then it would be expected to be more like $90\,\mu mol\,kg^{-1}$ rather than the observed $30\,\mu mol\,kg^{-1}$. So, something else must be involved.

This something is CO_2 transfer through the atmosphere (Broecker and Peng, 1993). The high-phosphate-content waters upwelling in the Southern Ocean lose part of their excess CO_2 to the atmosphere. This results in an increase in their CO_3^{2-} ion content. In contrast, the low-PO_4-content waters reaching in the northern Atlantic have CO_2 partial pressures well below that in the atmosphere and hence they absorb CO_2. This reduces their CO_3^{2-} concentration. Hence, it is the transfer of CO_2 from surface waters in the Southern Ocean to surface waters in the northern Atlantic reduces the contrast in carbonate ion concentration between deep waters in the deep Atlantic and those in the remainder of the deep ocean.

One other factor expected to have an impact on the carbonate ion concentration in deep Pacific Ocean and Indian Ocean turns out to be less important. Much of the floor of these two oceans lies below the transition zone. Hence, most of the $CaCO_3$ falling into the deep Pacific Ocean and Indian Ocean dissolves. One would expect then that the older the water (as indicated by lower $^{14}C/C$ ratios), the higher its CO_3^{2-} ion concentration would be. While to some extent this is true, the trend is much smaller than expected. The reason is that in the South Pacific Ocean and South Indian Ocean an almost perfect chemical titration is being conducted, i.e., for each mole of

respiration CO_2 released to the deep ocean, roughly one mole of $CaCO_3$ dissolves (Broecker and Sutherland, 2000). So indeed, the older the water, the higher its $\sum CO_2$ content. But, due to $CaCO_3$ dissolution, there is a compensating increase in alkalinity such that the carbonate ion concentration remains largely unchanged. Only in the northern reaches of these oceans does the release of metabolic CO_2 overwhelm the supply of $CaCO_3$ allowing the CO_3 concentration to drop.

As in the depth range of transition zone, the solubility of $CaCO_3$ increases by $\sim14\,\mu mol\,kg^{-1}\,km^{-1}$ increase in water depth, the $30\,\mu mol\,kg^{-1}$ higher CO_3^{2-} concentration in NADW should (other things being equal) lead to a 2 km deeper transition zone in the Atlantic than in the Pacific Ocean and the Indian Ocean. In fact, this is more or less what is observed.

14.4. DEPTH OF SATURATION HORIZON

A number of attempts have been made to establish the exact depth of the calcite saturation horizon. The most direct way to do this is to suspend preweighed calcite entities at various water depths on a deep-sea mooring, then months later, recover the mooring and determine the extent of weight loss (see Figure 5). Peterson (1966) performed such an experiment at 19°N in the Pacific Ocean using polished calcite spheres and observed a pronounced depth-dependent increase in weight loss that commenced at $\sim3{,}900\,m$. Honjo and Erez (1978) performed a similar experiment at 32°N in the Atlantic and found that coccoliths, foraminifera shells and reagent calcite experienced a 25–60% weight loss at $5{,}500\,m$ but no measurable

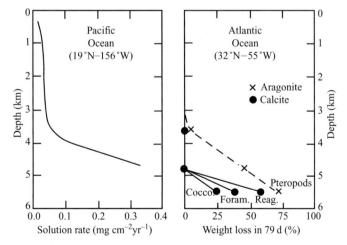

Figure 5 Results of *in situ* dissolution experiments. Peterson (1966) re-weighed polished calcite spheres after a 250 d deployment on a mooring in the North Pacific. Honjo and Erez (1978) observed the weight loss for calcitic samples (coccoliths, foraminifera and reagent calcite) and an aragonitic sample (pteropods) held at depth for a period of 79 d. While Peterson hung his spheres directly in seawater, the Honjo–Erez samples were held in containers through which water was pumped. The results suggest that the calcite saturation horizon lies at $4{,}800 \pm 200\,m$ in the North Atlantic and at about $3{,}800 \pm 200\,m$ in the North Pacific. For aragonite, which is 1.4 times more soluble than calcite, the saturation horizon in the North Atlantic is estimated to be in the range $3{,}400 \pm 200\,m$.

weight loss at 4,900 m. Thus the North Atlantic–North Pacific depth difference in the depth of the onset of dissolution is more or less consistent with expectation. Broecker and Takahashi (1978) used a combination of the depth of the onset of sedimentary $CaCO_3$ content decline and the results of a technique referred to as *in situ* saturometry (Ben-Yaakov and Kaplan, 1971) to define the depth dependence of solubility. While fraught with caveats, these results are broadly consistent with those from the mooring experiment. By measuring the composition of pore waters extracted *in situ* from sediments at various water depths, Sayles (1985) was able to calculate what he assumed to be saturation CO_3^{2-} concentrations. Finally, several investigators have performed laboratory equilibrations of calcite and seawater as a function of confining pressure. But, as each approach is subject to biases, more research is needed before the exact pressure dependence of the solubility of calcite can be pinned down.

14.5. DISSOLUTION MECHANISMS

Three possible dissolution processes come to mind. The first of these is termed water column dissolution. As foraminifera shells fall quite rapidly and as they encounter calcite undersaturated water only at great depth, it might be concluded that dissolution during fall is unimportant. But it has been suggested that organisms feeding on falling debris ingest and partially dissolve calcite entities (Milliman *et al.*, 1999). Because of their small size, coccoliths are presumed to be the most vulnerable in this regard. But little quantitive information is available to permit quantification of this mode of dissolution.

The other two processes involve dissolution of calcite after it reaches the seafloor. A distinction is made between dissolution that occurs before burial (i.e., interface dissolution) and dissolution that takes place after burial (i.e., pore-water dissolution). The former presumably occurs only at water depths greater than that of the saturation horizon. But the latter has been documented to occur above the calcite saturation horizon. It is driven by respiration CO_2 released to the pore waters.

Following the suggestion of Emerson and Bender (1981) that the release of respiration CO_2 in pore waters likely drives calcite dissolution above the saturation horizon, a number of investigators took the bait and set out to explore this possibility. David Archer, as part of his PhD thesis research with Emerson, developed pH microelectrodes that could be slowly ratcheted into the upper few centimeters of the sediment from a bottom lander. He deployed these pH microelectrodes along with the O_2 microelectrodes and was able to show that the release of respiration CO_2 (as indicated by a reduction in pore-water O_2) was accompanied by a drop in pH (and hence also of CO_3^{2-} ion concentration). Through modeling the combined results, Archer *et al.* (1989) showed that much of the CO_2 released by respiration reacted with $CaCO_3$ before it had a chance to escape (by molecular diffusion) into the overlying bottom water. As part of his PhD research, Burke Hales, a second Emerson student, improved Archer's electrode system and made measurements on the Ceara Rise in the western equatorial Atlantic (Hales and Emerson, 1997) and on the Ontong–Java Plateau in the western equatorial Pacific (Hales and Emerson, 1996) (see Figure 6). Taken together, these two studies strongly support

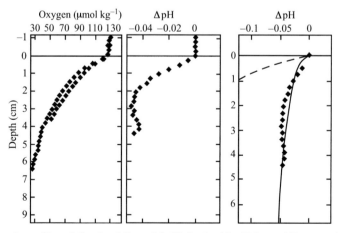

Figure 6 Microelectrode profiles of dissolved O_2 and ΔpH obtained by Hales and Emerson (1996) at 2.3 km depth on the Ontong–Java Plateau in the western equatorial Pacific. On the right are model curves showing the pH trend expected if none of the CO_2 released during the consumption of the O_2 was neutralized by reaction with sediment $CaCO_3$ (dashed curve) and a best model fit to the measured ΔpH trend (solid curve). The latter requires that much of the respiration CO_2 reacts with $CaCO_3$ before it escapes into the overlying bottom water.

the proposal that dissolution in pore waters of sediments leads to substantial dissolution of calcite. This approach has been improved upon by the addition of an LIX electrode to measure CO_2 itself and a micro-optode to measure Ca^{2+} (Wenzhöfer *et al.*, 2001).

In another study designed to confirm that most of the CO_2 released into the upper few centimeters of the sediments reacts with $CaCO_3$ before escaping to the overlying bottom water, Martin and Sayles (1996) deployed a very clever device that permitted the *in situ* collection of closely spaced pore-water samples in the upper few centimeters of the sediment column. Measurements of $\sum CO_2$ and alkalinity on these pore-water samples revealed that the gradient of $\sum CO_2$ (μmol km^{-1}) with depth is close to that of alkalinity (μequiv. kg^{-1}). This can only be the case if much of the respiration CO_2 reacts with $CaCO_3$ to form a Ca^{2+} and two HCO_3^- ions.

Dan McCorkle of Woods Hole Oceanographic Institution conceived of yet another way to confirm that pore-water respiration CO_2 was largely neutralized by reaction with $CaCO_3$. As summarized in Figure 7, he made $^{13}C/^{12}C$ ratio measurements on $\sum CO_2$ from pore-water profiles and found that the trend of $\delta^{13}C$ with excess $\sum CO_2$

is consistent with a 50–50 mixture of carbon derived from marine organic matter ($-20\permil$) and that derived from marine calcite ($+1\permil$) (Martin *et al.*, 2000). Again, these results require that a large fraction of the metabolic CO_2 reacts with $CaCO_3$.

There is, however, a fly in the ointment. Benthic flux measurements made by deploying chambers on the seafloor reveal a curious pattern (see Figure 8). R. A. Jahnke and D. B. Jahnke (2002) found that alkalinity and calcium fluxes from sediments (both high and low in $CaCO_3$ content) below the calcite saturation horizon and on low-$CaCO_3$-content sediments from above the saturation horizon yield more or less the expected fluxes. However, chambers deployed on high-$CaCO_3$-content sediments from above the saturation horizon yield no measurable alkalinity flux. Yet pore-water profiles and electrode measurements for these same sediments suggest that calcite is dissolving. Whole foraminifera shell weight and $CaCO_3$ size index measurements (see below) agree with conclusion of these authors that calcite dissolution is not taking place. R. A. Jahnke and D. B. Jahnke (2002) propose that impure $CaCO_3$ coatings formed on the surfaces of calcite grains are redissolved in

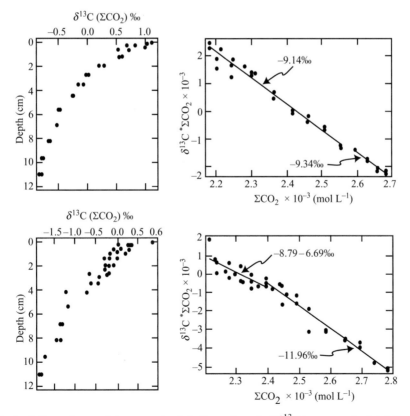

Figure 7 On the left are shown plots versus depth in the sediment of $\delta^{13}C$ in pore water total dissolved inorganic carbon (i.e., $\sum CO_2$) for two sites on the Ceara Rise (5°S in the western Atlantic) (Martin *et al.*, 2000). On the right are plots of $\delta^{13}C$ versus $\sum CO_2$. The slopes yield the isotopic composition of the excess CO_2. As can be seen, it requires that the respiration CO_2 be diluted with a comparable contribution from dissolved $CaCO_3$.

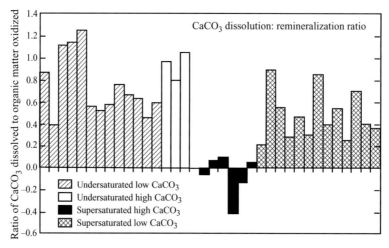

Figure 8 Summary of the ratio of CaCO₃ dissolved and organic material oxidized for bottom chamber deployments in the northeastern Pacific, Ontong–Java Plateau, Ceara Rise, Cape Verde Plateau, northwestern Atlantic continental rise and California borderland basins (R. A. Jahnke and D. B. Jahnke, 2002). The absence of measurable alkalinity fluxes from high-CaCO₃ sites bathed in supersaturated bottom water appears to be inconsistent with observations (see text).

contact with respiration CO_2-rich pore waters and that the products of this dissolution diffuse back to sediment–water interface. Based on this scenario, the reason that the these authors record no calcium or alkalinity flux is that the ingredients for upward diffusion of calcium and alkalinity are being advected downward bound to the surfaces of calcite grains. Hence, there is no net flux of either property into their benthic chamber. That such coatings form was demonstrated long ago by Weyl (1965), who showed that when exposed in the laboratory to supersaturated seawater it was the calcite crystal surfaces that achieved saturation equilibrium with seawater rather than vice versa.

Broecker and Clark (2003) fortify the mechanism proposed by R.A. Jahnke and D. B. Jahnke, providing additional evidence by proposing that it must be coatings rather than the biogenic calcite itself that dissolve. As shown in Figure 9, while on the Ontong–Java Plateau, there is a progressive decrease in shell weight and CaCO₃ size index with water depth; on the Ceara Rise, neither of these indices shows a significant decrease above a water depth of 4,100 m. This is consistent with the conclusion that no significant dissolution occurs at the depth of 3,270 m where the pore-water and chamber measurements were made.

Although Berelson *et al.* (1994) report chamber-based alkalinity fluxes from high-calcite sediment, the sites at which their studies were performed are very likely bathed in calcite-undersaturated bottom water. If so, coatings would not be expected to form.

One other observation, i.e., core-top radiocarbon ages, appears to be at odds with pore-water dissolution. The problem is as follows. To the extent that respiration CO_2-driven dissolution

occurring in the core-top bioturbated zone is homogeneous (i.e., all calcite entities lose the same fraction of their weight in a unit of time), then the core-top radiocarbon age should decrease slowly with increasing extent of dissolution. The reason is that dissolution reduces the time of residence of CaCO₃ entities in the core-top mixed layer, and hence also their apparent ^{14}C age. But, as shown by Broecker *et al.* (1999), core-top radiocarbon ages on Ontong–Java Plateau cores from a range of water depths reveal an increase rather than a decrease with water depth (see Figure 10). This increase is likely the result of dissolution that occurs on the seafloor in calcite-undersaturated bottom waters before the calcite is incorporated into the core-top mixed layer. In this case, the reduction of CaCO₃ input to the sediment leads to an increase in the average residence time of calcite in the bioturbated layer. It may be that competition between pore-water dissolution and seafloor dissolution changes with depth. As shown in Figure 3, down to about 3 km pore-water dissolution appears to have the upper hand (and hence the ^{14}C ages becomes progressively younger with water depth). Below 3 km, the situation switches and seafloor dissolution dominates (hence, the ^{14}C ages become progressively older with increasing water depth).

14.6. DISSOLUTION IN THE PAST

One of the consequences of dissolution of CaCO₃ in pore waters is that it creates an ambiguity in all of the sediment-based methods for reconstructing past carbonate ion distributions in the deep sea. By "sediment-based" methods, one means methods involving some measure of the

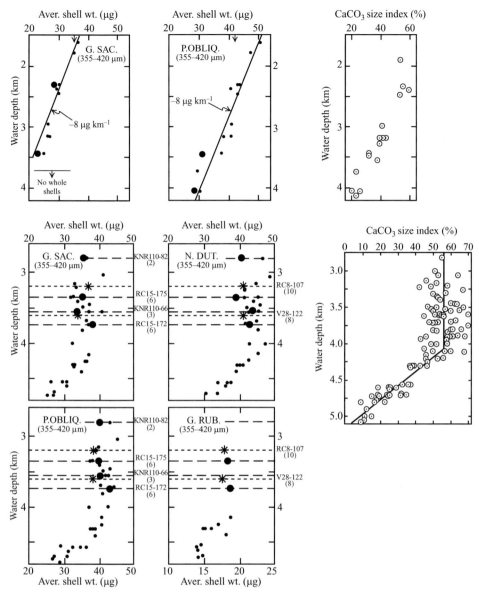

Figure 9 The upper panel shows shell weight and CaCO$_3$ size fraction results from core top covering a range of water depth on the Ontong–Java Plateau. The lower panel shows shell weight results from Ceara Rise and CaCO$_3$ size fraction results from the equatorial Atlantic.

preservation of the CaCO$_3$ contained in deep-sea sediments. The ambiguity involves the magnitude of the offset between the bottom-water and the pore-water carbonate ion concentrations. The results obtained using any such methods can be applied to time trends in bottom-water carbonate ion concentration only if the pore-water–bottom-water offset is assumed to have remained nearly constant.

Fortunately, two methods have been proposed for which this ambiguity does not exist. One involves measurements of boron isotope ratios in benthic foraminifera (Sanyal *et al.*, 1995) and the other Zn/Cd ratios in benthic foraminifera (Marchitto *et al.*, 2000). Unfortunately, as of early

2003, neither of these methods has received wide enough application to allow its utility to be proven (see below). Until this has been done, we are left with the ambiguity as to whether sediment-based methods reflect mainly changes in bottom-water CO$_3^{2-}$ or as proposed by Archer and Maier-Reimer (1994) in the pore-water-bottom-water CO$_3^{2-}$ offset.

14.7. SEDIMENT-BASED PROXIES

A number of schemes have been proposed by which changes in the carbonate ion concentration in the deep sea might be reconstructed. The most obvious of these is the record of the CaCO$_3$

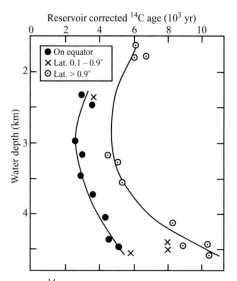

Figure 10 ¹⁴C ages (reservoir corrected by 400 yr) as a function of water depth for core-top samples from the Ontong–Java Plateau (Broecker *et al.*, 1999). As can be seen, the ages for cores taken on the equator are systematically younger than those for cores taken a degree or so off the equator. The reason is that the sedimentation rates are twice as high on, than off the equator, while the depth of bioturbation is roughly the same. The onset of the increase in core-top age occurs at a depth of ~3 km. If this onset can be assumed to represent the depth of the saturation horizon (see text), then those results suggest a value ΔV of ~45 cm³ mol⁻ for the reaction $CO_3^{2-} + Ca^{2+} \Leftrightarrow CaCO_3$ (calcite). On the other hand, this depth may represent the horizon where interface dissolution just matches pore-water dissolution.

content of the sediment. Unfortunately, as already discussed, the $CaCO_3$ content depends on the ratio of the rain rate of $CaCO_3$ to that of silicate debris as well as on the extent of dissolution of the calcite. Unless quite large, changes in the extent of dissolution cannot be reliably isolated from changes in the composition of the raining debris. Other schemes focus on the state of preservation of the calcite entities. One involves the ratio of dissolution-prone to dissolution-resistant planktonic foraminifera shells (Ruddiman and Heezen, 1967; Berger, 1970). The idea is that the lower this ratio, the greater the extent of dissolution. A variant on this approach is to measure the ratio of foraminifera fragments to whole shells (Peterson and Prell, 1985; Wu and Berger, 1989). The idea behind both approaches is that as dissolution proceeds, the foraminifera shells break into pieces. These methods suffer, however, from two important drawbacks. First, any method involving entity counting is highly labor-intensive. Second, the results depend on the initial makeup of the foraminifera population in the sediment.

Furthermore, neither of these methods has yet been calibrated against present-day pressure-normalized carbonate ion concentration nor has

either one been widely applied. At one point, the author became enamored with a simplified version of the fragment method. Instead of counting fragments (a labor-intensive task), the ratio of $CaCO_3$ in the greater than 63 μm fraction to the total $CaCO_3$ was measured, the idea being that as dissolution proceeded, calcite entities larger than 63 μm would break down to entities smaller than 63 μm. This method was calibrated by conducting measurements on core-top samples from low-latitude sediments spanning a range of water depth in all three oceans (Broecker and Clark, 1999). While these results were promising, when the method was extended to glacial sediment, it was found that the core-top calibration relationship did not apply (Broecker and Clark, 2001a). A possible reason is that the ratio of the fine (coccolith) to coarse (foraminifera) $CaCO_3$ grains in the initial material was higher during glacial time than during the Holocene.

Despite their drawbacks, these methods have led to several important findings. First, it was clearly demonstrated that during glacial time the mean depth of the transition zone did not differ greatly from today's. This finding is important because it eliminates one of the hypotheses which have been put forward to explain the lower glacial atmospheric CO_2 content, namely, the coral reef hypothesis (Berger, 1982). According to this idea, shallow-water carbonates (mainly coral and coralline algae) formed during the high-sea stands of periods of interglaciation would be eroded and subsequently dissolved during the low-sea stands of periods of glaciation, alternately reducing and increasing the sea's CO_3^{2-} concentration. But in order for this hypothesis to be viable, the transition zone would have to have been displaced downward by several kilometers during glacial time. Rather, the reconstructions suggest that the displacement was no more than a few hundred meters.

Two other findings stand out. First, as shown by Farrell and Prell (1989), at water depths in the 4 km range in the eastern equatorial Pacific, the impact of dissolution was greater during interglacials than during glacials (i.e., the transition zone was deeper during glacial time). Second, fragment-to-whole foraminifera ratios measured on a series of cores from various depths in the Caribbean Sea clearly demonstrate better preservation during glacials than interglacials (Imbrie, 1992). These findings have been confirmed by several investigators using a range of methods. Taken together, these findings gave rise to the conclusion that the difference between the depth of the transition zone in the Atlantic from that in the Pacific was somewhat smaller than now during glacial time. In addition, the existence of a pronounced dissolution event in the Atlantic Ocean at the onset of the last glacial cycle has been documented (Curry and Lohmann, 1986).

14.8. SHELL WEIGHTS

An ingenious approach to the reconstruction of the carbonate ion concentration in the deep sea was developed by WHOI's Pat Lohmann (1995). Instead of focusing on ratios of one entity to another, he developed a way to assess the extent of dissolution experienced by shells of a given species of planktonic foraminifera. He did this by carefully cleaning and sonification of the greater than 63 µm material sieved from a sediment sample. He then picked and weighed 75 whole shells of a given species isolated in a narrow size fraction range (usually 355–420 µm). In so doing, he obtained a measure of the average shell wall thickness. By obtaining shell weights for a given species from core-top samples spanning a range in water depth, Lohmann was able to show that the lower the pressure-normalized carbonate ion concentration, the smaller the whole shell weight (and hence the thinner the shell walls) (see Figure 11).

Lohmann's method seemingly has the advantage over those used previously in that no assumptions need to be made about the initial composition of the sediment. However, Barker and Elderfield (2002) make a strong case that the thickness of the foraminifera shell walls varies with growth conditions. They did so by weighing shells of temperate foraminifera from core tops from a number of locales in the North Atlantic. They found strong correlations between shell weight and both water temperature and carbonate ion concentration, the warmer the water and the higher its carbonate ion concentration, the thicker the shells. If, as Barker and Elderfield (2002) content, it is the carbonate ion concentration that drives the change in initial wall thickness, then glacial-age shells should have formed with thicker shells than do their Late Holocene counterparts. Fortunately, the ice-core-based atmospheric CO_2 record allows the carbonate ion

concentration in the glacial surface waters to be reconstructed and hence presumably also the growth weight of glacial foraminifera. At this point, however, several questions remain unanswered. For example, does the dependence of shell weight on surface water carbonate ion concentration established for temperate species apply to tropical species? Perhaps the shell weight dependence flattens as the high carbonate ion concentrations characteristic of tropical surface waters are approached. Is carbonate ion concentration the only environmental parameter on which initial shell weights depends? As discussed below, there is reason to believe that the situation is perhaps more complicated.

Lohmann's method has other drawbacks. Along with all sediment-based approaches, it suffers from an inability to distinguish changes in bottom-water carbonate ion concentration from changes in bottom-water to pore-water concentration offset. Shell thickness may also depend on growth rate and hence nutrient availability. Finally, a bias is likely introduced when dissolution becomes sufficiently intense to cause shell breakup, in which case the shells with the thickest walls are likely to be the last to break up. Nevertheless, Lohmann's method opens up a realm of new opportunities.

The sensitivity of shell-weight to pressure-normalized carbonate ion concentration (i.e., after correction for the increase in the solubility of calcite with water depth) was explored by determining the weight of Late Holocene shells from various water depths in the western equatorial Atlantic (Ceara Rise) and western equatorial Pacific (Ontong–Java Plateau). This strategy takes advantage of the contrast in carbonate ion concentration between the Atlantic and Pacific deep waters. As shown in Figure 9, shell weights for Ontong–Java Plateau samples do decrease with water depth and hence with decreasing pressure-normalized

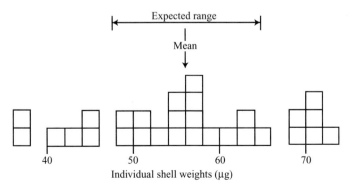

P. obliquiloculata 355–420 µm split

Figure 11 Weights of 29 individual *P. obliquiloculata* shells picked from the 355–420 µm size fraction. If all the shells had the same wall thickness, a spread in weight of 17 µg would be expected (assuming that shell weight varies with the square of size). Clearly, this indicates that shells of the same size must have a range in wall thickness. As can be seen, the observed range in weight is twice the expected range.

carbonate ion concentrations (Broecker and Clark, 2001a). However, the surprise is that there is no evidence of either weight loss or shell break at depths less than 4,200 m for Ceara Rise core-top samples. Rather, weight loss and shell breakup is evident only for samples from deeper than 4,200 m. This observation is in agreement with the benthic chamber results of R. A. Jahnke and D. B. Jahnke and hence supports the hypothesis that above the calcite saturation horizon the gradients in pore-water composition are fueled primarily by the dissolution of "Weyl" (1965) coatings rather than of the biogenic calcite itself.

The Ontong–Java results yield a weight loss of ~8 µg for each kilometer increase in water depth. In order to convert this to a dependence on pressure-normalized carbonate ion concentration, it is necessary to take into account the change in *in situ* carbonate ion concentration in the water column over the Ontong–Java Plateau water column (i.e., $CO_3^{2-} = 72 + 3(z - 2)$ µmol kg^{-1}, where z is the water depth in km) and the pressure dependence of the saturation carbonate ion concentration. The latter depends on the difference in volume between Ca^{2+} and CO_3^{2-} ions when in solution and when they are bound into calcite. The relationship is as follows:

$$\left(CO_{3\ sat}^{2-}\right)^z = \left(CO_{3\ sat}^{2-}\right)^0 e^{PV/RT}$$

where the units of z are km, of ΔV, L mol^{-1}, of R, L atm, and T, K. If ΔV is re-expressed as cm^3 mol^{-1}, the relationship becomes

$$\left(CO_{3\ sat}^{2-}\right)^z = \left(CO_{3\ sat}^{2-}\right)^0 e^{z\Delta V/225}$$

where ΔV is the volume of the ions when bound into calcite minus that when they are dissolved in seawater. $\left(CO_3^{2-}\right)^0$ is 45 mol kg^{-1} and while the exact value of ΔV remains uncertain, 40 cm^3 mol^{-1} fits most ocean observations (Peterson, 1966; Honjo and Erez, 1978; Ben-Yaakov and Kaplan, 1971; Ben-Yaakov et al., 1974). Listed in Table 1 are the saturation concentrations based on this ΔV and the slope of the solubility as a function of water depth. Also given are estimates

of the weight loss for foraminifera shells per unit decrease in carbonate ion concentration.

14.9. THE BORON ISOTOPE PALEO pH METHOD

Theoretical calculations by Kakihana *et al.* (1977) suggested that the uncharged species of dissolved borate $(B(OH)_3)$ should have a 21 per mil higher $^{11}B/^{10}B$ ratio than that for the charged species $(B(OH)_4^-)$. Hemming and Hanson (1992) demonstrated that this offset might be harnessed as a paleo pH proxy. Their reasoning was as follows. As the residence time of borate in seawater is tens of millions of years, on the timescale of glacial cycles the isotope composition of oceanic borate could not have changed. They further reasoned that it must be the charged borate species that is incorporated into marine $CaCO_3$ and hence marine calcite should have an isotope composition close to that of the charged species in seawater. This is important because as shown in Figure 12 the isotopic composition of the charged species must depend on the pH of the seawater. The higher the pH, the larger the fraction of the borate in the charged form and hence the closer its isotopic composition will be to that for bulk seawater borate. In contrast, for pH values the isotopic composition of the residual amount of charged borate must approach a value 21 per mil lower than that for bulk seawater borate. Working with a graduate student, Abhijit Sanyal, Hemming applied his method to foraminifera shells and demonstrated that indeed foraminifera shells record pH (Sanyal et al., 1995). Benthic foraminifera had the expected offset from planktonics. Glacial age *G. sacculifer*, as dictated by the lower glacial atmospheric CO_2 content, recorded a pH about 0.15 units higher than that for Holocene shells. Sanyal went on to grow planktonic foraminifera shells at a range of pH values (Sanyal et al., 1996, 2001). He also precipitated inorganic $CaCO_3$ at a range of pH values (Sanyal et al., 2000). These results yielded the expected pH dependence of boron isotope composition. However, they also revealed

Table 1 The calcite-saturation carbonate ion concentration in cold seawater and the slope of this solubility as a function of water depth based on a 1 atm solubility of 45 µmol CO_3^{2-} kg^{-1} and a ΔV of 40 cm^3 mol^{-1}.

Water depth (km)	Calcite sat. (µmol CO_3^{2-} kg^{-1})	Sol. slope (µmol kg^{-1} km^{-1})	CO_3^{2-} versus shell wt. (µmol kg^{-1} µg^{-1})
1.5	58.8	10.3	1.4
2.0	64.2	11.4	1.5
2.5	70.2	12.5	1.6
3.0	76.7	13.6	1.8
3.5	83.8	14.9	2.0
4.0	91.6	16.3	2.1
4.5	100.1	17.9	2.3

Also shown is the slope of the shell-weight loss–carbonate ion concentration relationship for various water depths. The 0.7 µmol kg^{-1} km^{-1} increase in carbonate ion concentration in the Ontong–Java Plateau deep-water column is taken into account.

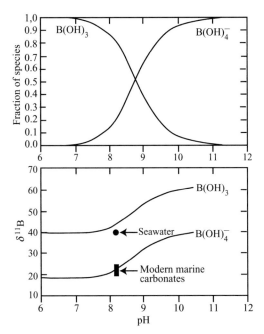

Figure 12 Speciation of borate in seawater as a function of pH (upper panel). Isotopic composition of the uncharged (B(OH)$_3$) and charged (B(OH)$_4^-$) species as a function of pH(lower panel) (Hemming and Hanson, 1992). As marine carbonates incorporate only the charged species, their isotopic composition is close to that of B(OH)$_4^-$.

sizable species-to-species offsets (as do the carbon and oxygen isotopic compositions).

The waterloo of this method came when glacial-age benthic foraminifera were analyzed. The results suggested that the pH of the glacial deep ocean was 0.3 units greater than today (Sanyal *et al.*, 1995). This corresponds to a whopping 90 µmol kg^{-1} increase in carbonate ion concentration. The result was exciting because, if correct, the lowering of the CO$_2$ content of the glacial atmosphere would be explained by a whole ocean carbonate ion concentration change. But this result was clearly at odds with reconstructions of the depth of the glacial transition zone. Such a large increase in deep-water carbonate ion concentration would require that it deepened by several kilometers. Clearly, it did not. Archer and Maier-Reimer (1994) proposed a means by which this apparent disagreement might be explained. They postulated that if during glacial time the release of metabolic CO$_2$ to sediment pore waters (relative to the input of CaCO$_3$) was larger than today's, this would have caused a shoaling of the transition zone and thereby thrown the ocean's CaCO$_3$ budget out of kilter. Far too little CaCO$_3$ would have been buried relative to the ingredient input. The result would be a steady increase in the ocean's carbonate ion inventory (see Figure 13) and a consequent progressive deepening of the transition zone. This deepening would have

continued until a balance between input and loss was once again achieved. In so doing, a several kilometer offset between the depth of the saturation horizon and the depth of the transition zone would have been created. However, this explanation raised three problems so serious that the boron isotope-based deep-water pH change has fallen into disrepute. First, it required that the change in glacial ecology responsible for the increase in the rain of organic matter be globally uniform. Otherwise, there would have been very large "wrinkles" in the depth of the glacial transition zone. No such wrinkles have been documented. Second, at the close of each glacial period when the flux of excess organic matter was shut down, there must have been a prominent global preservation event. In order to restore the saturation horizon to its interglacial position, an excess over ambient CaCO$_3$ accumulation of ~3 g cm^{-2} would have to have occurred over the entire seafloor. It would be surprising if some residue from this layer were not to be found in sediments lining the abyssal plains. It has not. These sediments have no more than 0.2% by weight CaCO$_3$. In other words, of the 3 g cm^{-2} deposited during the course of the carbonate ion drawdown, almost nothing remains. Finally, based on model simulations, Sigman *et al.* (1998) have shown that it is not possible to maintain for tens of thousands of years a several-kilometer separation between the saturation horizon and the transition zone.

This "waterloo" was unfortunate for the author considers the boron method to be basically sound and potentially extremely powerful. The answer to the benthic enigma may lie in species-to-species differences in the boron isotope "vital" effect for benthic foraminifera. The measurement method use by Sanyal *et al.* (1996) required a large number of benthic shells in order to get enough boron to analyze. This created a problem because, as benthics are rare among foraminifera shells, mixed benthics rather than a single species were analyzed. If the boron isotope pH proxy is to become an aid to deep-ocean studies, then techniques requiring smaller amounts of boron will have to be created. There also appears to be a problem associated with variable isotopic fractionation of boron during thermal ionization. As this fractionation depends on the ribbon temperature and perhaps other factors, it may introduce biases in the results for any particular sample. Hopefully, a more reproducible means of ionizing boron will be found.

14.10. Zn/Cd RATIOS

The other bottom-water CO$_3^{2-}$ ion concentration proxy is based on the Zn/Cd ratio in benthic foraminifera shells. As shown by Marchitto *et al.* (2000), the distribution coefficient of zinc

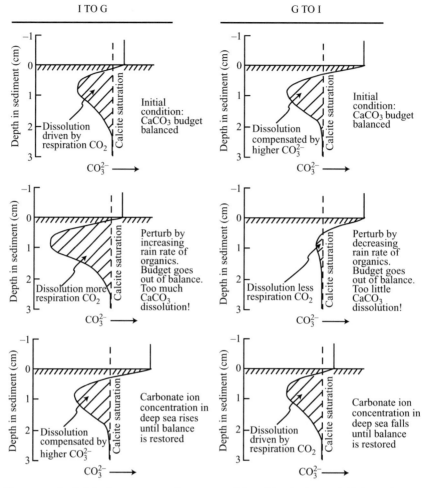

Figure 13 Shown on the left is the sequence of events envisioned by Archer and Maier-Reimer (1994) for the transition from interglacial (I) to glacial (G) conditions. An increase in respiration CO_2 release to the sediment pore waters enhances calcite dissolution, thereby unbalancing the $CaCO_3$ budget. This imbalance leads to a buildup in CO_3^{2-} ion concentration in the deep sea until it compensates for the extra respiration CO_2. On the right is the sequence of events envisioned for the transition from G to I conditions. The input of excess respiration CO_2 to the sediments ceases, thereby reducing the rate of calcite dissolution. This leads to an excess accumulation of $CaCO_3$ on the seafloor and hence to a reduction in carbonate ion concentration which continues until steady state is reestablished.

between shell and seawater depends on CO_3^{2-} ion concentration, such that the lower the carbonate ion concentration, the large the Zn/Cd ratio in the foraminifera shell. Assuming that the Zn/Cd ratio in seawater was the same during the past as it is today, the ratio of these two trace elements should serve as a paleo carbonate ion proxy. However, there are problems to be overcome. For example, in today's ocean, zinc correlates with silica and cadmium with phosphorus. As silica is 10-fold enriched in deep Pacific water relative to deep Atlantic water while phosphorus is only twofold enriched, differential redistribution of silica and phosphorus in the glacial ocean poses a potential bias. However, as at high carbonate supersaturation the distribution coefficient for zinc flattens out, it may be possible to use measurements on benthic foraminifera from sediments bathed in highly supersaturated waters to sort this out. But, as is

the case for the boron isotope proxy, much research will be required before reconstructions based on Zn/Cd ratios can be taken at face value.

14.11. DISSOLUTION AND PRESERVATION EVENTS

There are several mechanisms that might lead to carbonate ion concentration transients at the beginning and end of glacial periods. One such instigator is changes in terrestrial biomass. Shackleton (1997) was the first to suggest that the mass of carbon stored as terrestrial biomass was smaller during glacial than during interglacial periods. He reached this conclusion based on the fact that measurements on glacial-age benthic foraminifera yielded lower $\delta^{13}C$ values than those for their interglacial counterparts. Subsequent studies confirmed that this was indeed the case and when benthic foraminifera ^{13}C results were averaged over the entire deep sea, it was found that the ocean's dissolved inorganic carbon had a $^{13}C/^{12}C$ ratio 0.35 ± 0.10 per mil lower during glacial time than during the Holocene (Curry *et al.*, 1988). If this decrease is attributed to a lower inventory of wood and humus, then the magnitude of the glacial biomass decrease would have been $500 \pm 150\,Gt$ of carbon. The destruction of this amount of organic material at the onset of a glacial period would create a $20\,\mu mol\,kg^{-1}$ drop in the ocean's CO_3^{2-} concentration and hence produce a calcite dissolution event. Correspondingly, the removal of this amount of CO_2 from the ocean–atmosphere reservoir at the onset of an interglacial period would raise the carbonate ion concentration by $20\,\mu mol\,kg^{-1}$ and hence produce a calcite preservation event. This assumes that the time over which the biomass increase occurred was short compared to the CO_3^{2-} response (i.e., $\sim 5,000\,yr$). If this is not the case, then the magnitude of the carbonate ion changes would be correspondingly smaller.

Another possible instigator of such transients was proposed by Archer and Maier-Reimer (1994). Their goal was to create a scenario by which the lower CO_2 content of the glacial atmosphere might be explained. As already mentioned, it involved a higher ratio of organic carbon to $CaCO_3$ carbon in the material raining to the deep-sea floor during glacial times than during interglacial times, and hence an intensification of pore-water dissolution. As in the case for the terrestrial biomass change, such an increase would have thrown the ocean's carbon budget temporarily out of kilter. The imbalance would have been remedied by a buildup of carbonate ion concentration at the onset of glacials and a drawdown of carbonate ion concentration at the onset of interglacials (see Figure 14). Hence, it would also lead to a dissolution event at the onset of glacial episodes and a preservation event at the onset of interglacial episodes. Were the changes in organic to $CaCO_3$ rain proposed by Archer and Maier-Reimer to have explained the entire glacial to interglacial CO_2 change, then the magnitude of

the transients would have been ~ 4 times larger than that resulting from $500\,Gt\,C$ changes in terrestrial biomass.

Regardless of their origin, these dissolution events and preservation events would be short-lived. As they would disrupt the balance between burial and supply, they would be compensated by either decreased or increased burial of $CaCO_3$ and the balance would be restored with a constant time of $\sim 5,000\,yr$ (see Figure 14).

Clear evidence for the compensation for an early Holocene preservation event is seen in shell weight results form a core from $4.04\,km$ depth on the Ontong–Java Plateau in the western equatorial Pacific (see Figure 15). A drop in the weight of *P. obliquiloculata* shells of $11\,\mu g$ between about $7,500\,yr$ ago and the core-top bioturbated zone (average age $4,000\,yr$) requires a decrease in carbonate ion concentration between $7,500\,y$ ago and today (see Table 1). This Late Holocene CO_3^{2-} ion concentration drop is characterized by an up-water column decrease in magnitude becoming imperceptible at $2.31\,km$. It is interesting to note that during the peak of the preservation event, the shell weights showed only a small decrease with water depth (see Figure 15), suggesting either that the pressure effect on calcite solubility was largely compensated by an increase with depth in the *in situ* carbonate ion concentration or that the entire water column was supersaturated with respect to calcite.

In the equatorial Atlantic only in the deepest core (i.e., that from $5.20\,km$) is the Late Holocene intensification of dissolution strongly imprinted. As in this core no whole shells are preserved, the evidence for an Early Holocene preservation event is based on $CaCO_3$ size-index and $CaCO_3$ content measurements. As can be seen in Figure 15, both show a dramatic decrease starting about $7,500\,yr$ ago. As for the Pacific, the magnitude of the imprint decreases up-water column, becoming imperceptible at $3.35\,km$.

If either the biomass or the respiration CO_2 mechanisms are called upon, the magnitude of the Early Holocene CO_3^{2-} maximum must have

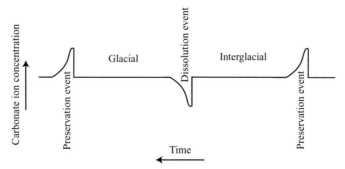

Figure 14 Idealized scenario for carbonate ion concentration changes associated with dissolution and preservation events.

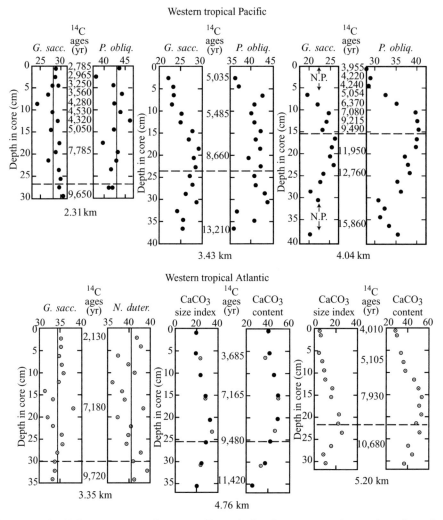

Figure 15 The Early Holocene preservation event. Records for six cores are shown: three from Ontong–Java Plateau in western tropical Pacific (BC36:0° 158°E, 2.31 km; BC51: 0° 161°E, 3.43 km and BC56: 0° 162°E, 4.04 km) and three from the western tropical Atlantic (RC15-175: 4°N, 47°W, 3.35 km; RC16-55: 10°N, 45°W, 4.76 km and RC17-30: 11°N, 41°W, 5.20 km). In the upper panel, N.P. indicates that there are no whole shells present. In the lower panel, the open circles represent measurements made on trigger weight cores and the closed circles, measurements made on piston cores. Where measurements were made on both the depth scale is that for the piston core, and the trigger weight sample depths have been multiplied by a factor of 1.5 to compensate for foreshortening (Broecker *et al.*, 1993). The shell weights are in μg, the size index is the percentage of the CaCO₃ contained in the >63 μm fraction and the calcium carbonate content is in percent. The dashed lines show the depth of the 9,500 B.P. radiocarbon-age horizon.

been uniform throughout the deep ocean. The most straightforward explanation for the up-water column reduction in the magnitude of the preservation event is that at mid-depths; the sediment pore waters are presently close to saturation with respect to calcite. Hence, the Early Holocene maximum in deep-sea CO_3 ion concentration pushed them into the realm of supersaturation. If so, there is no need to call on a depth dependence for the magnitude of the preservation event.

The post-8,000-year-ago decrease in CO_3^{2-} ion concentration of 23 μmol kg^{-1} required to explain the Late Holocene 11 μg decrease in *P. obliquilo-culata* shell weights observed in the deepest

Ontong–Java Plateau core is twice too large to be consistent with the 20 ppm increase in atmospheric CO_2 content over this time interval (Indermühle *et al.*, 1999). The significance of this remains unknown.

In the equatorial Atlantic $CaCO_3$ content, $CaCO_3$ size-index and shell-weight measurements reveal three major dissolution events, one during marine isotope stage 5d, one during 5b and one during stage 4 (see Figures 16(a) and (b)). As these events are only weakly imprinted on Pacific sediments, it appears that a major fraction of the carbonate ion reduction was the result of enhanced penetration into the deep Atlantic of low

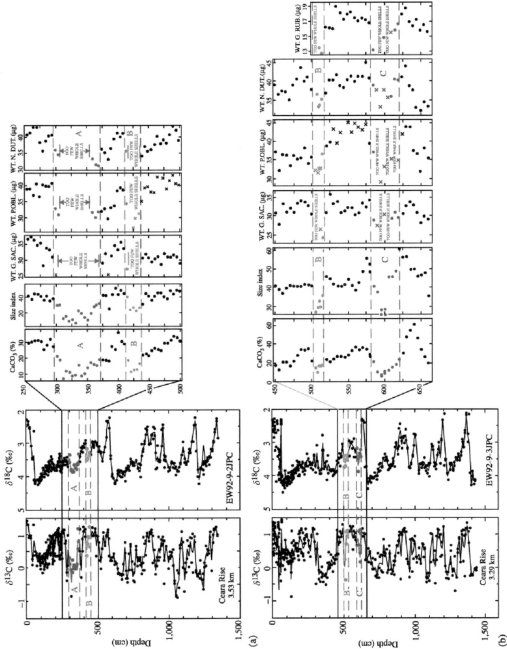

Figure 16 (A) CaCO₃, size index and shell weight results for a portion of jumbo pistoncore EW92-9-2 (a) and EW92-9-3 (b) from the northern flank of the Ceara Rise. The results in (a) document dissolution events A (stage 4) and B (stage 5b). (B) The results in (b) document dissolution events B (stage 5b) and C (stage 5d). The ¹⁸O and ¹³C records for benthic foraminifera are from Curry (1996). The xs represent samples in which only 15–30 whole shells were found.

carbonate ion concentration Southern Ocean water. If this conclusion proves to be correct, then it suggests that the balance between the density of deep waters formed in the northern Atlantic and those formed in the Southern Ocean is modulated by the strength of northern hemisphere summer insolation (i.e., by Milankovitch cycles).

14.12. GLACIAL TO INTERGLACIAL CARBONATE ION CHANGE

In addition to the preservation and dissolution event transients, there were likely carbonate ion concentration changes that persisted during the entire glacial period. These changes could be placed in two categories. One involves a change in the average CO_3^{2-} concentration of the entire deep sea necessary to compensate for a change in the ratio of calcite production by marine organisms to ingredient supply. The other involves a redistribution of carbonate ion within the deep sea due to a redistribution of phosphate (and hence also of respiration CO_2) and/or to a change in the magnitude of the flux of CO_2 through the atmosphere from the Southern Ocean to the northern Atlantic.

Based on shell-weight measurements, Broecker and Clark (2001c) attempted to reconstruct the depth distribution of carbonate ion concentration during late glacial time for the deep equatorial Atlantic Ocean and Pacific Ocean. At the time their paper was published, these authors were unaware of the dependence of initial shell weight on carbonate ion concentration in surface water

established by Barker and Elderfield (2002) for temperate species. Since during the peak glacial time the atmosphere's p_{CO_2} was ~80 ppm lower than during the Late Holocene, the carbonate ion concentration in tropical surface waters must have been 40–50 μmol kg^{-1} higher at that time. Based on the Barker and Elderfield (2002) trend of ~1 μg increase in shell weight per 9 μmol kg^{-1} increase in carbonate ion concentration, this translates to an 8 μg heavier initial shell weights during glacial time. Figure 17 shows, while this correction does not change the depth dependence or interocean concentration difference, it does greatly alter the magnitude of the change. In fact, were the correction made, it would require that the carbonate ion concentration in virtually the entire glacial deep ocean was lower during glacial time than during interglacial time. For the deep Pacific Ocean and the Indian Ocean, this flies in the face of all previous studies which conclude that dissolution was less intense during periods of glaciation than during periods of interglaciation. However, as the Broecker and Clark study concentrates on the Late Holocene while earlier studies concentrate on previous periods of interglaciation, it is possible that the full extent of the interglacial decrease in carbonate ion during the present interglacial has not yet been achieved. Of course, it is also possible that significant thickening of foraminifera shells during glacial time did not occur. Until this matter can be cleared up, reconstruction of glacial-age deep-sea carbonate ion concentrations must remain on hold.

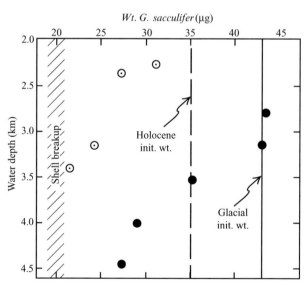

Figure 17 Average shell weights for whole *G. sacculifer* shells of glacial age as a function water depth. The open circles are for cores from the Ontong–Java Plateau, and the closed circles are for cores from the equatorial Atlantic. The vertical dashed line is the initial *G. sacculifer* weight obtained from measurements on Holocene samples from the Ceara Rise. The solid vertical line is the estimated initial weight for glacial-age *G. sacculifer* based on the assumption that the carbonate ion concentration dependence of initial shell weight established by Barker and Elderfield (2002) applies to tropical species.

14.13. NEUTRALIZATION OF FOSSIL FUEL CO$_2$

The ultimate fate of much of the CO$_2$ released to the atmosphere through the burning of coal, oil, and natural gas will be to react with the CaCO$_3$ stored in marine sediments (Broecker and Takahashi, 1977; Sundquist, 1990; Archer *et al.*, 1997). The amount of CaCO$_3$ available for dissolution at any given place on the seafloor depends on the calcite content in the sediment and the depth to which sediments are stirred by organisms. The former is now well mapped and the latter has been documented in many places by radiocarbon measurements. The amount of CaCO$_3$ available for dissolution at any given site is given by

$$\sum CaCO_3 = \frac{h\rho f_c}{1 - f_c}$$

where h is the depth of bioturbation, ρ is the water-free sediment density, and f_c is the weight-fraction calcite. The high-CaCO$_3$ sediments that drape the oceans' ridges and plateaus typically have ~90% CaCO$_3$ and a water-free density of 1 g cm^{-3}. The bioturbation depth in these sediments averages 8 cm. Hence, the upper limit on amount of CaCO$_3$ available for dissolution in such a sediment is 72 g cm^{-2}. As roughly one quarter of the seafloor is covered with calcite-rich sediments, this corresponds to ~6.3×10^{19} g CaCO$_3$ (i.e., 7,560 Gt C). This amount could neutralize 6.3×10^{17} mol of fossil fuel CO$_2$. This amount exceeds the combined oceanic inventory of dissolved CO$_3^{2-}$ (1.6×10^{17} mol) and of dissolved HBO$_3^-$ (0.8×10^{17} mol). It is comparable to the amount of recoverable fossil fuel carbon.

I say 'upper limit' because once this amount of CaCO$_3$ has been dissolved, the upper 8 cm of the sediment would consist entirely of a noncarbonate residue. As molecular diffusion through such a thick residue would be extremely slow, the rate of dissolution of CaCO$_3$ stored beneath this CaCO$_3$-free cap would be minuscule, and further neutralization would be confined to the fall to the seafloor of newly formed CaCO$_3$.

The rate of this dissolution of the CaCO$_3$ stored in the uppermost sediment will depend not only on the magnitude of the reduction of the deep ocean's CO$_3^{2-}$ content, but also on the rate at which the insoluble residue is stirred into the sediment. This bioturbation not only homogenizes the mixed layer, but is also exhumes CaCO$_3$ from beneath the mixed layer.

REFERENCES

Archer D. and Maier-Reimer E. (1994) Effect of deep-sea sedimentary calcite preservation on atmospheric CO$_2$ concentration. *Nature* **367**, 260–263.

Archer D., Emerson S., and Reimers C. (1989) Dissolution of calcite in deep-sea sediments: pH and O$_2$ microelectrode results. *Geochim. Cosmochim. Acta* **53**, 2831–2845.

Archer D., Kheshgi H., and Maier-Reimer E. (1997) Multiple timescales for neutralization of fossil fuel CO$_2$. *Geophys. Res. Lett.* **24**, 405–408.

Archer D., Kheshgi H., and Maier-Riemer E. (1998) Dynamics of fossil fuel CO$_2$ neutralization by marine CaCO$_3$. *Global Biogeochem. Cycles* **12**, 259–276.

Barker S. and Elderfield H. (2002) Response of foraminiferal calcification to glacial–interglacial changes in atmospheric carbon dioxide. *Science* **297**, 833–836.

Ben-Yaakov S. and Kaplan I. R. (1971) Deep sea *in situ* calcium carbonate saturometry. *J. Geophys. Res.* **76**, 772–781.

Ben-Yaakov S., Ruth E., and Kaplan I. R. (1974) Carbonate compensation depth: relation to carbonate solubility in ocean waters. *Science* **184**, 982–984.

Berelson W. M., Hammond D. E., and Cutter G. A. (1990) *In situ* measurements of calcium carbonate dissolution rates in deep-sea sediments. *Geochim. Cosmochim. Acta* **54**, 3013–3020.

Berelson W. M., Hammond D. E., McManus J., and Kilgore T. E. (1994) Dissolution kinetics of calcium carbonate in equatorial Pacific sediments. *Global Biogeochem. Cycles* **8**, 219–235.

Berger W. H. (1970) Planktonic foraminifera: selective solution and the lysocline. *Mar. Geol.* **8**, 111–138.

Berger W. H. (1982) Increase of carbon dioxide in the atmosphere during deglaciation: the coral reef hypothesis. *Naturwissenschaften* **69**, 87–88.

Broecker W. S. (1991) The great ocean conveyor. *Oceanography* **4**, 79–89.

Broecker W. S. and Clark E. (1999) CaCO$_3$ size distribution: a paleo carbonate ion proxy. *Paleoceanography* **14**, 596–604.

Broecker W. S. and Clark E. (2001a) Reevaluation of the CaCO$_3$ size index paleocarbonate ion proxy. *Paleoceanography* **16**, 669–771.

Broecker W. S. and Clark E. (2001b) A dramatic Atlantic dissolution event at the onset of the last glaciation. *Geochem. Geophys. Geosys.* **2**, 2001GC000185, Nov. 2.

Broecker W. S. and Clark E. (2001c) Glacial to Holocene redistribution of carbonate ion in the deep sea. *Science* **294**, 2152–2155.

Broecker W. S. and Clark E. (2001d) An evaluation of Lohmann's foraminifera-weight index. *Paleoceanography* **16**, 531–534.

Broecker W. S. and Clark E. (2002) A major dissolution event at the close of MIS 5e in the western equatorial Atlantic. *Geochem. Geophys. Geosys.* **3**(2) 10.1029/2001GC000210.

Broecker W. S. and Clark E. (2003) Pseudo-dissolution of marine calcite. *Earth Planet. Sci. Lett.* **208**, 291–296.

Broecker W. S. and Peng T.-H. (1993) Interhemispheric transport of $\sum$CO$_2$ through the ocean. In *The Global Carbon Cycle*. NATO SI Series *The Global Carbon Cycle*. NATO SI Series (ed. M. Heimann). Springer, vol. 115, pp. 551–570.

Broecker W. S. and Sutherland S. (2000) The distribution of carbonate ion in the deep ocean: support for a post-Little Ice Age change in Southern Ocean ventilation. *Geochem. Geophys. Geosys.* **1**, 2000GC000039, July 10.

Broecker W. S. and Takahashi T. (1977) Neutralization of fossil fuel CO$_2$ by marine calcium carbonate. In *The Fate of Fossil Fuel CO$_2$ in the Oceans* (eds. N. R. Andersen and A. Malahoff). Plenum, New York, pp. 213–248.

Broecker W. S. and Takahashi T. (1978) The relationship between lysocline depth and *in situ* carbonate ion concentration. *Deep-Sea Res.* **25**, 65–95.

Broecker W. S., Lao Y., Klas M., Clark E., Bonani G., Ivy S., and Chen C. (1993) A search for an early Holocene CaCO$_3$ preservation event. *Paleoceanography* **8**, 333–339.

Broecker W. S., Clark E., Hajdas I., Bonani G., and McCorkle D. (1999) Core-top ^{14}C ages as a function of water depth on the Ontong-Java Plateau. *Paleoceanography* **14**, 13–22.

Curry W. B. (1996) Late Quaternary deep circulation in the western equatorial Atlantic. In *The South Atlantic: Present and Past Circulation* (eds. G. Wefer, W. H. Berger, G. Siedler, and D. J. Webb). Springer, New York, pp. 577–598.

Curry W. B. and Lohmann G. P. (1986) Late Quaternary carbonate sedimentation at the Sierra Leone rise (eastern equatorial Atlantic Ocean). *Mar. Geol.* **70**, 223–250.

Curry W. B., Duplessy J. C., Labeyrie L. D., and Shackleton N. J. (1988) Changes in the distribution of $\delta^{13}C$ of deep water $\sum CO_2$ between the last glaciation and the Holocene. *Paleoceanography* **3**, 317–341.

Emerson S. and Bender M. (1981) Carbon fluxes at the sediment–water interface of the deep-sea: calcium carbonate preservation. *J. Mar. Res.* **39**, 139–162.

Farrell J. W. and Prell W. L. (1989) Climatic change and $CaCO_3$ preservation: an 800,000 year bathymetric reconstruction from the central equatorial Pacific Ocean. *Paleoceanography* **4**, 447–466.

Hales B. and Emerson S. (1996) Calcite dissolution in sediments of the Ontong-Java Plateau: *in situ* measurements of porewater O_2 and pH. *Global Biogeochem. Cycles* **5**, 529–543.

Hales B. and Emerson S. (1997) Calcite dissolution in sediments of the Ceara Rise: *in situ* measurements of porewater O_2, pH and $CO_{2(aq)}$. *Geochim. Cosmochim. Acta* **61**, 501–514.

Hales B., Emerson S., and Archer D. (1994) Respiration and dissolution in the sediments of the western North Atlantic: estimates from models of *in situ* microelectrode measurements of porewater oxygen and pH. *Deep-Sea Res.* **41**, 695–719.

Hemming N. G. and Hanson G. N. (1992) Boron isotopic composition and concentration in modern marine carbonates. *Geochim. Cosmochim. Acta* **56**, 537–543.

Honjo S. and Erez J. (1978) Dissolution rates of calcium carbonate in the deep ocean: an *in situ* experiment in the North Atlantic. *Earth Planet. Sci. Lett.* **40**, 226–234.

Imbrie J. (1992) On the structure and origin of major glaciation cycles: I. Linear responses to Milankovitch forcing. *Paleoceanography* **7**, 701–738.

Indermühle A., Stocker T. F., Joos F., Fischer H., Smith H. J., Wahlen M., Deck B., Mastroianni D., Techumi J., Blunier T., Meyer R., and Stauffer B. (1999) Holocene carbon-cycle dynamics based on CO_2 trapped in ice at Taylor Dome, Antarctica. *Nature* **398**, 121–126.

Jahnke R. A. and Jahnke D. B. (2002) Calcium carbonate dissolution in deep-sea sediments: implications of bottom water saturation state and sediment composition. *Geochim. Cosmochim. Acta* (submitted for publication, November, 2001).

Kakihana H., Kotaka M., Satoh S., Nomura M., and Okamoto M. (1977) Fundamental studies on the ion exchange separation of boron isotopes. *Bull. Chem. Soc. Japan* **50**, 158–163.

Lohmann G. P. (1995) A model for variation in the chemistry of planktonic foraminifera due to secondary calcification and selective dissolution. *Paleoceanography* **10**, 445–457.

Marchitto T. M., Jr., Curry W. B., and Oppo D. W. (2000) Zinc concentrations in benthic foraminifera reflect seawater chemistry. *Paleoceanography* **15**, 299–306.

Martin W. R. and Sayles F. L. (1996) $CaCO_3$ dissolution in sediments of the Ceara Rise, western equatorial Atlantic. *Geochim. Cosmochim. Acta* **60**, 243–263.

Martin W. R., McNichol A. P., and McCorkle D. C. (2000) The radiocarbon age of calcite dissolving at the sea floor:

estimates from pore water data. *Geochim. Cosmochim. Acta* **64**, 1391–1404.

Milliman J. D., Troy P. J., Balch W. M., Adams A. K., Li Y.-H., and Mackenzie F. T. (1999) Biologically mediated dissolution of calcium carbonate above the chemical lysocline? *Deep-Sea Res.* **46**, 1653–1669.

Peterson M. N. A. (1966) Calcite: rates of dissolution in a vertical profile in the central Pacific. *Science* **154**, 1542–1544.

Peterson L. C. and Prell W. L. (1985) Carbonate preservation and rates of climatic change: an 800 kyr record from the Indian Ocean. In *The Carbon Cycle and Atmospheric CO₂: Natural Variations Archean to Present*, Geophys. Monogr. Ser. 32 (eds. E. T. Sundquist and W. S. Broecker). pp. 251–270.

Reimers C. E., Jahnke R. A., and Thomsen L. (2001) In situ sampling in the benthic boundary layer. In *The Benthic Boundary Layer: Transport Processes and Biogeochemistry* (eds. B. P. Boudreau and B. B. Jørgensen). Oxford University Press, Oxford, pp. 245–268.

Ruddiman W. F. and Heezen B. C. (1967) Differential solution of planktonic foraminifera. *Deep-Sea Res.* **14**, 801–808.

Sanyal A., Hemming N. G., Hanson G. N., and Broecker W. S. (1995) Evidence for a higher pH in the glacial ocean from boron isotopes in foraminifera. *Nature* **373**, 234–236.

Sanyal A., Hemming N. G., Broecker W. S., Lea D. W., Spero H. J., and Hanson G. N. (1996) Oceanic pH control on the boron isotopic composition of foraminifera: evidence from culture experiments. *Paleoceanography* **11**, 513–517.

Sanyal A., Nugent M., Reeder R. J., and Bijma J. (2000) Seawater pH control on the boron isotopic composition of calcite: evidence from inorganic calcite precipitation experiments. *Geochim. Cosmochim. Acta* **64**, 1551–1555.

Sanyal A., Bijma J., Spero H., and Lea D. W. (2001) Empirical relationship between pH and the boron isotopic composition of *Globigerinoides sacculifer*: implications for the boron isotope paleo-pH proxy. *Paleoceanography* **16**, 515–519.

Sayles F. L. (1985) $CaCO_3$ solubility in marine sediments: evidence for equilibrium and non-equilibrium behavior. *Geochim. Cosmochim. Acta.* **49**, 877–888.

Shackleton N. J. (1977) Tropical rainforest history and the equatorial Pacific carbonate dissolution cycles. In *The Fate of Fossil Fuel CO₂ in the Oceans* (eds. N. R. Anderson and A. Malahoff). Plenum, New York, pp. 401–428.

Sigman D. M., McCorkle D. C., and Martin W. R. (1998) The calcite lysocline as a constraint on glacial/interglacial low-latitude production changes. *Global Biogeochem. Cycles* **12**, 409–427.

Sundquist E. T. (1990) Long-term aspects of future atmospheric CO_2 and sea-level changes. In *Sea-level Change*. National Research Council Studies in Geophysics (ed. R. Revelle). National Academy Press, Washington, pp. 193–207.

Wenzhöfer F., Adler M., Kohls O., Hensen C., Strotmann B., Boehme S., and Schulz H. D. (2001) Calcite dissolution driven by benthic mineralization in the deep-sea: *in situ* measurements. *Geochim. Cosmochim. Acta* **65**, 2677–2690.

Weyl P.K. (1965) The solution behavior of carbonate materials in seawater. In *Proc. Int. Conf. Tropical Oceanography, Miami Beach, Florida*, pp. 178–228.

Wu G. and Berger W. H. (1989) Planktonic foraminifera: differential dissolution and the quaternary stable isotope record in the west equatorial Pacific. *Paleoceanography* **4**, 181–198.

Geochemistry of Earth Surface Systems
ISBN: 978-0-08-096706-6

pp. 491–510

15

The Global Oxygen Cycle

S. T. Petsch

University of Massachusetts, Amherst, MA, USA

15.1	INTRODUCTION	512
15.2	DISTRIBUTION OF O_2 AMONG EARTH SURFACE RESERVOIRS	512
	15.2.1 *The Atmosphere*	512
	15.2.2 *The Oceans*	512
	15.2.3 *Freshwater Environments*	515
	15.2.4 *Soils and Groundwaters*	516
15.3	MECHANISMS OF O_2 PRODUCTION	516
	15.3.1 *Photosynthesis*	516
	15.3.2 *Photolysis of Water*	517
15.4	MECHANISMS OF O_2 CONSUMPTION	518
	15.4.1 *Aerobic Cellular Respiration*	518
	15.4.2 *Photorespiration*	518
	15.4.3 *C_1 Metabolism*	519
	15.4.4 *Inorganic Metabolism*	519
	15.4.5 *Macroscale Patterns of Aerobic Respiration*	519
	15.4.6 *Volcanic Gases*	520
	15.4.7 *Mineral Oxidation*	520
	15.4.8 *Hydrothermal Vents*	521
	15.4.9 *Iron and Sulfur Oxidation at the Oxic–Anoxic Transition*	521
	15.4.10 *Abiotic Organic Matter Oxidation*	521
15.5	GLOBAL OXYGEN BUDGETS AND THE GLOBAL OXYGEN CYCLE	522
15.6	ATMOSPHERIC O_2 THROUGHOUT EARTH'S HISTORY	522
	15.6.1 *Early Models*	522
	15.6.2 *The Archean*	524
	15.6.2.1 *Constraints on the O_2 content of the Archean atmosphere*	524
	15.6.2.2 *The evolution of oxygenic photosynthesis*	526
	15.6.2.3 *Carbon isotope effects associated with photosynthesis*	527
	15.6.2.4 *Evidence for oxygenic photosynthesis in the Archean*	528
	15.6.3 *The Proterozoic Atmosphere*	528
	15.6.3.1 *Oxygenation of the Proterozoic atmosphere*	528
	15.6.3.2 *Atmospheric O_2 during the Mesoproterozoic*	531
	15.6.3.3 *Neoproterozoic atmospheric O_2*	532
	15.6.4 *Phanerozoic Atmospheric O_2*	534
	15.6.4.1 *Constraints on Phanerozoic O_2 variation*	534
	15.6.4.2 *Evidence for variations in Phanerozoic O_2*	534
	15.6.4.3 *Numerical models of Phanerozoic oxygen concentration*	537
15.7	CONCLUSIONS	545
	REFERENCES	547

15.1 INTRODUCTION

One of the key defining features of Earth as a planet that houses an active and diverse biology is the presence of free molecular oxygen (O_2) in the atmosphere. Biological, chemical, and physical processes interacting on and beneath the Earth's surface determine the concentration of O_2 and variations in O_2 distribution, both temporal and spatial. In the present-day Earth system, the process that releases O_2 to the atmosphere (photosynthesis) and the processes that consume O_2 (aerobic respiration, sulfide mineral oxidation, oxidation of reduced volcanic gases) result in large fluxes of O_2 to and from the atmosphere. Even relatively small changes in O_2 production and consumption have the potential to generate large shifts in atmospheric O_2 concentration within geologically short periods of time. Yet all available evidence supports the conclusion that stasis in O_2 variation is a significant feature of the Earth's atmosphere over wide spans of the geologic past. Study of the oxygen cycle is therefore important because, while an equable O_2 atmosphere is central to life as we know it, our understanding of exactly why O_2 concentrations remain nearly constant over large spans of geologic time is very limited.

This chapter begins with a review of distribution of O_2 among various reservoirs on Earth's surface: air, sea, and other natural waters. The key factors that affect the concentration of O_2 in the atmosphere and surface waters are next considered, focusing on photosynthesis as the major process generating free O_2 and various biological and abiotic processes that consume O_2. The chapter ends with a synopsis of current models on the evolution of an oxygenated atmosphere through 4.5 billion years of Earth's history, including geochemical evidence constraining ancient O_2 concentrations and numerical models of atmospheric evolution.

15.2 DISTRIBUTION OF O_2 AMONG EARTH SURFACE RESERVOIRS

15.2.1 The Atmosphere

The partial pressure of oxygen in the present-day Earth's atmosphere is ~ 0.21 bar, corresponding to a total mass of $\sim 34 \times 10^{18}$ mol O_2 (0.20946 bar (force/area) multiplied by the surface area of the Earth (5.1×10^{14} m^2), divided by average gravitational acceleration g (9.8 m s^{-2}) and the formula weight for O_2 (32 g mol^{-1}) yields $\sim 34 \times 10^{18}$ mol O_2). There is a nearly uniform mixture of the main atmospheric gases (N_2, O_2, Ar) from the Earth's surface up to ~ 80 km altitude (including the troposphere, stratosphere, and mesosphere), because turbulent mixing dominates over molecular diffusion at these altitudes. Because atmospheric pressure (and thus gas

molecule density) decreases exponentially with altitude, the bulk of molecular oxygen in the atmosphere is concentrated within several kilometers of Earth's surface. Above this, in the thermosphere, gases become separated based on their densities. Molecular oxygen is photodissociated by UV radiation to form atomic oxygen (O), which is the major form of oxygen above ~ 120 km altitude.

Approximately 21% O_2 in the atmosphere represents an average composition. In spite of well-developed turbulent mixing in the lower atmosphere, seasonal latitudinal variations in O_2 concentration of ± 15 ppm have been recorded. These seasonal variations are most pronounced at high latitudes, where seasonal cycles of primary production and respiration are strongest (Keeling and Shertz, 1992). In the northern hemisphere, the seasonal variations are anticorrelated with atmospheric p_{CO_2}; summers are dominated by high O_2 (and high inferred net photosynthesis), while winters are dominated by lower O_2. In addition, there has been a measurable long-term decline in atmospheric O_2 concentration of $\sim 10^{14}$ mol yr^{-1}, attributed to oxidation of fossil fuels. This decrease has been detected in both long-term atmospheric monitoring stations (Keeling and Shertz, 1992, Figure 1(a)) and in atmospheric gases trapped in Antarctic firn ice bubbles (Figure 1(b)). The polar ice core records extend the range of direct monitoring of atmospheric composition to show that a decline in atmospheric O_2 linked to oxidation of fossil fuels has been occurring since the Industrial Revolution (Bender *et al.*, 1994b; Battle *et al.*, 1996).

15.2.2 The Oceans

Air-saturated water has a dissolved O_2 concentration dependent on temperature, the Henry's law constant k_H, and ionic strength. In pure water at 0 °C, O_2 saturation is 450 μM; at 25 °C, saturation falls to 270 μM. Other solutes reduce O_2 solubility, such that at normal seawater salinities, O_2 saturation is reduced by $\sim 25\%$. Seawater is, of course, rarely at perfect O_2 saturation. Active photosynthesis may locally increase O_2 production rates, resulting in supersaturation of O_2 and degassing to the atmosphere. Alternately, aerobic respiration below the sea surface can consume dissolved O_2 and lead to severe O_2-depletion or even anoxia.

Lateral and vertical gradients in dissolved O_2 concentration in seawater reflect balances between O_2 inputs from air–sea gas exchange, biological processes of O_2 production and consumption, and advection of water masses. In general terms, the concentration of O_2 with depth in the open ocean follows the general structures described in Figure 2. Seawater is saturated to supersaturated with O_2 in the surface mixed layer

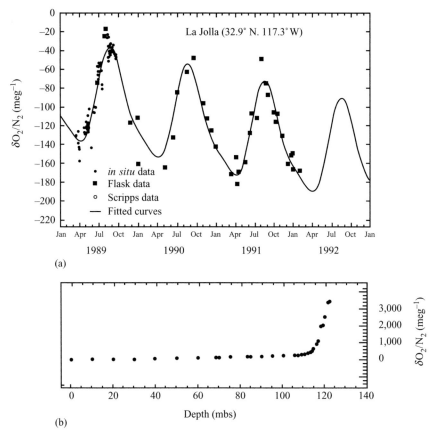

Figure 1 (a) Interannual variability of atmospheric O_2/N_2 ratio, measured at La Jolla, California. 1 ppm O_2 is equivalent to $\sim$4.8 meg^{-1} (source Keeling and Shertz, 1992). (b) Variability in atmospheric O_2/N_2 ratio measured in firn ice at the South Pole, as a function of depth in meters below surface (mbs). The gentle rise in O_2/N_2 between 40 m and 100 m reflects a loss of atmospheric O_2 during the last several centuries due to fossil fuel burning. Deeper than 100 m, selective effusion of O_2 out of closing bubbles into firn air artificially boosts O_2/N_2 ratios (source Battle *et al.*, 1996).

($\sim$0–60 m water depth). Air–sea gas exchange and trapping of bubbles ensures constant dissolution of atmospheric O_2. Because gas solubility is temperature dependent, O_2 concentrations are greater in colder high-latitude surface waters than in waters near the equator. Oxygen concentrations in surface waters also vary strongly with season, especially in high productivity waters. Supersaturation is strongest in spring and summer (time of greatest productivity and strongest water column stratification) when warming of surface layers creates a shallow density gradient that inhibits vertical mixing. Photosynthetic O_2 production exceeds consumption and exchange, and supersaturation can develop. O_2 concentrations drop below the surface mixed layer to form O_2 minimum zones (OMZs) in many ocean basins. O_2 minima form where biological consumption of O_2 exceeds resupply through advection and diffusion. The depth and thickness of O_2 minima vary among ocean basins. In the North Atlantic, the OMZ extends several hundred meters. O_2 concentrations fall from an average of $\sim$300 μM in the surface mixed layer to $\sim$160 μM at 800 m depth.

In the North Pacific, however, the O_2 minimum extends deeper, and O_2 concentrations fall to $<$100 μM. Along the edges of ocean basins, where OMZs impinge on the seafloor, aerobic respiration is restricted, sediments are anoxic at or near the sediment–water interface, and burial of organic matter in sediments may be enhanced. Below the oxygen minima zones in the open ocean, O_2 concentrations gradually increase again from 2000 m to the seafloor. This increase in O_2 results from the slow progress of global thermohaline circulation. Cold, air-saturated seawater sinks to the ocean depths at high-latitudes in the Atlantic, advecting in O_2-rich waters below the O_2 minimum there. Advection of O_2-rich deep water from the Atlantic through the Indian Ocean into the Pacific is the source of O_2 in deep Pacific waters. However, biological utilization of this deep-water O_2 occurs along the entire path from the North Atlantic to the Pacific. For this reason, O_2 concentrations in deep Atlantic water are slightly greater ($\sim$200 μM) than in the deep Pacific ($\sim$150 μM).

In some regions, dissolved O_2 concentration falls to zero. In these regions, restricted water circulation and ample organic matter supply result in biological utilization of oxygen at a rate that exceeds O_2 resupply through advection and diffusion. Many of these are temporary zones of anoxia that form in coastal regions during summer, when warming facilitates greatest water column stratification, and primary production and organic matter supply are high. Such O_2-depletion is now common in the Chesapeake Bay and Schelde estuaries, off the mouth of the Mississippi River and other coastal settings. However, there are several regions of the world oceans where stratification and anoxia

are more permanent features (Figure 3). These include narrow, deep, and silled coastal fjords, larger restricted basins (e.g., the Black Sea, Cariaco Basin, and the chain of basins along the southern California Borderlands). Lastly, several regions of the open ocean are also associated with strong O_2-depletion. These regions (the equatorial Pacific along Central and South America and the Arabian Sea) are associated with deep-water upwelling, high rates of surface water primary productivity, and high dissolved oxygen demand in intermediate waters.

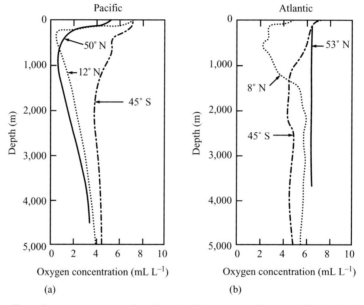

Figure 2 Depth profiles of oxygen concentration dissolved in seawater for several latitudes in the Pacific (a) and Atlantic (b). Broad trends of saturation or supersaturation at the surface, high dissolved oxygen demand at mid-depths, and replenishment of O_2 through lateral advection of recharged deep water are revealed, although regional influences of productivity and intermediate and deep-water heterotropy are also seen (source Ingmanson and Wallace, 1989).

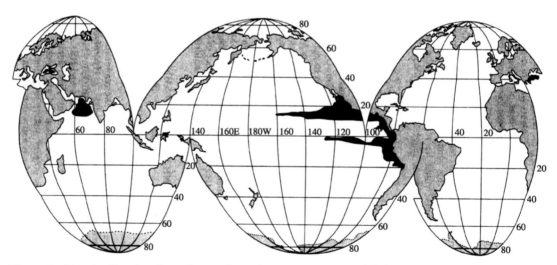

Figure 3 Map detailing locations of extensive and permanent oxygen-deficient intermediate and deep waters (Deuser, 1975) (reproduced by permission of Elsevier from *Chemical Oceanography* **1975**, p.3).

Oxygen concentrations in the pore fluids of sediments are controlled by a balance of entrainment of overlying fluids during sediment deposition, diffusive exchange between the sediment and the water column, and biological utilization. In marine sediments, there is a good correlation between the rate of organic matter supply and the depth of O_2 penetration in the sediment (Hartnett *et al.*, 1998). In coastal sediments and on the continental shelf, burial of organic matter is sufficiently rapid to deplete the sediment of oxygen within millimeters to centimeters of the sediment–water interface. In deeper abyssal sediments, where organic matter delivery is greatly reduced, O_2 may penetrate several meters into the sediment before being entirely consumed by respiration.

There is close coupling between surface water and atmospheric O_2 concentrations and air–sea gas exchange fluxes (Figure 4). High rates of marine primary productivity result in net outgassing of O_2 from the oceans to the atmosphere in spring and summer, and net ingassing of O_2 during fall and winter. These patterns of air–sea O_2 transfer relate to latitude and season: outgassing of O_2 during northern hemisphere high productivity months (April through August) are accompanied by simultaneous ingassing in southern latitudes

when and where the productivity is lowest (Najjar and Keeling, 2000). Low-latitude ocean surface waters show very little net air–sea O_2 exchange and minimal change in outgassing or ingassing over an annual cycle.

15.2.3 Freshwater Environments

Oxygen concentrations in flowing freshwater environments closely match air-saturated values, due to turbulent mixing and entrainment of air bubbles. In static water bodies, however, O_2-depletion can develop much like in the oceans. This is particularly apparent in some ice-covered lakes, where inhibited gas exchange and wintertime respiration can result in O_2-depletion and fish kills. High productivity during spring and summer in shallow turbid aquatic environments can result in extremely sharp gradients from strong O_2 supersaturation at the surface to near O_2-depletion within a few meters of the surface. The high concentration of labile dissolved and particulate organic matter in many freshwater environments leads to rapid O_2-depletion where advective resupply is limited. High rates of O_2 consumption have been measured in many temperature and tropical rivers.

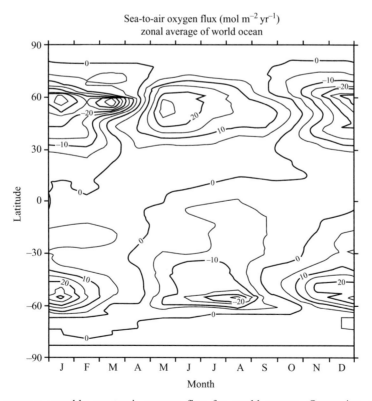

Figure 4 Zonal average monthly sea-to-air oxygen flux for world oceans. Outgassing and ingassing are concentrated at mid- to high latitudes. Outgassing of oxygen is strongest when primary production rates are greatest; ingassing is at a maximum during net respiration. These patterns oscillate during an annual cycle from northern to southern hemisphere (source Najjar and Keeling, 2000).

15.2.4 Soils and Groundwaters

In soil waters, oxygen concentrations depend on gas diffusion through soil pore spaces, infiltration and advection of rainwater and groundwater, air–gas exchange, and respiration of soil organic matter (see review by Hinkle, 1994). In organic-matter-rich temperate soils, dissolved O_2 concentrations are reduced, but not entirely depleted. Thus, many temperate shallow groundwaters contain some dissolved oxygen. Deeper groundwaters, and water-saturated soils and wetlands, generally contain little dissolved O_2. High-latitude mineral soils and groundwaters contain more dissolved O_2 (due to lower temperature and lesser amounts of soil organic matter and biological O_2 demand). Dry tropical soils are oxidized to great depths, with dissolved O_2 concentration less than air saturated, but not anoxic. Wet tropical forests, however, may experience significant O_2-depletion as rapid oxidation of leaf litter and humus occurs near the soil surface. Soil permeability also influences O_2 content, with more clay-rich soils exhibiting lower O_2 concentrations.

In certain environments, localized anomalously low concentrations of soil O_2 have been used by exploration geologists to indicate the presence of a large body of chemically reduced metal sulfides in the subsurface. Oxidation of sulfide minerals during weathering and soil formation draws down soil gas p_{O_2} below regional average. Oxidation of sulfide minerals generates solid and aqueous-phase oxidation products (i.e., sulfate anion and ferric oxyhydroxides in the case of pyrite oxidation). In some instances, the volume of gaseous O_2 consumed during mineral oxidation generates a mild negative pressure gradient, drawing air into soils above sites of sulfide mineral oxidation (Lovell, 2000).

15.3 MECHANISMS OF O_2 PRODUCTION

15.3.1 Photosynthesis

The major mechanism by which molecular oxygen is produced on Earth is through the biological process of photosynthesis. Photosynthesis occurs in higher plants, the eukaryotic protists collectively called algae, and in two groups of prokaryotes: the cyanobacteria and the prochlorophytes. In simplest terms, photosynthesis is the harnessing of light energy to chemically reduce carbon dioxide to simple organic compounds (e.g., glucose). The overall reaction (Equation (1)) for photosynthesis shows carbon dioxide and water reacting to produce oxygen and carbohydrate:

$$6CO_2 + 6H_2O \rightarrow 6O_2 + C_6H_{12}O_6 \qquad (1)$$

Photosynthesis is actually a two-stage process, with each stage broken into a cascade of chemical reactions (Figure 5). In the light reactions of photosynthesis, light energy is converted to chemical energy that is used to dissociate water to yield oxygen and hydrogen and to form the reductant NADPH from $NADP^+$.

The next stage of photosynthesis, the Calvin cycle, uses NADPH to reduce CO_2 to phosphoglyceraldehyde, the precursor for a variety of metabolic pathways, including glucose synthesis. In higher plants and algae, the Calvin cycle operates

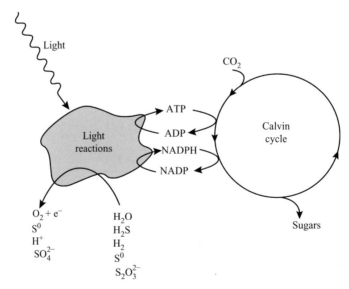

Figure 5 The two stages of photosynthesis: light reaction and the Calvin cycle. During oxygenic photosynthesis, H_2O is used as an electron source. Organisms capable of anoxygenic photosynthesis can use a variety of other electron sources ($H_2S, H_2, S^0, S_2O_3^{2-}$) during the light reactions, and do not liberate free O_2. Energy in the form of ATP and reducing power in the form of NADPH are produced by the light reactions, and subsequently used in the Calvin cycle to deliver electrons to CO_2 to produce sugars.

in special organelles called chloroplasts. However, in bacteria the Calvin cycle occurs throughout the cytosol. The enzyme ribulose-1,5-biphosphate carboxylase (rubisco) catalyzes reduction of CO_2 to phosphoglycerate, which is carried through a chain of reactions that consume ATP and NADPH and eventually yield phosphoglyceraldehyde. Most higher plants are termed C_3 plants, because the first stable intermediate formed during the carbon cycle is a three-carbon compound. Several thousand species of plant, spread among at least 17 families including the grasses, precede the carbon cycle with a CO_2-concentrating mechanism which delivers a four-carbon compound to the site of the Calvin cycle and rubisco. These are the C_4 plants. This four-carbon compound breaks down inside the chloroplasts, supplying CO_2 for rubisco and the Calvin cycle. The C_4-concentrating mechanisms is an advantage in hot and dry environments where leaf stomata are partially closed, and internal leaf CO_2 concentrations are too low for rubisco to efficiently capture CO_2. Other plants, called CAM plants, which have adapted to dry climates utilize another CO_2-concentrating mechanisms by closing stomata during the day and concentrating CO_2 at night. All higher plants, however, produce O_2 and NADPH from splitting water, and use the Calvin cycle to produce carbohydrates. Some prokaryotes use mechanisms other than the Calvin cycle to fix CO_2 (i.e., the acetyl-CoA pathway or the reductive tricarboxylic acid pathway), but none of these organisms is involved in oxygenic photosynthesis.

Global net primary production estimates have been derived from variations in the abundance and isotopic composition of atmospheric O_2. These estimates range from 23×10^{15} mol yr^{-1} (Keeling and Shertz, 1992) to 26×10^{15} mol yr^{-1}, distributed between 14×10^{15} mol yr^{-1} O_2 production from terrestrial primary production and 12×10^{15} mol yr^{-1} from marine primary production (Bender *et al.*, 1994a). It is estimated that ~50% of all photosynthetic fixation of CO_2 occurs in marine surface waters. Collectively, free-floating photosynthetic microorganisms are called phytoplankton. These include the algal eukaryotes (dinoflagellates, diatoms, and the red, green, brown, and golden algae), various species of cyanobacteria (*Synechococcus* and *Trichodesmium*), and the common prochlorophyte *Prochlorococcus*. Using the stoichiometry of the photosynthesis reaction, this equates to half of all global photosynthetic oxygen production resulting from marine primary production. Satellite-based measurements of seasonal and yearly average chlorophyll abundance (for marine systems) and vegetation greenness (for terrestrial ecosystems) can be applied to models that estimate net primary productivity, CO_2 fixation, and O_2 production. In the oceans, there are significant regional and seasonal variations in photosynthesis that result from limitations by light, nutrients, and temperature. Yearly averages of marine chlorophyll abundance show concentrated primary production at high-latitudes in the North Atlantic, North Pacific, and coastal Antarctica, in regions of seawater upwelling off of the west coasts of Africa, South America, and the Arabian Sea, and along the Southern Subtropical Convergence. At mid- and high latitudes, marine productivity is strongly seasonal, with primary production concentrated in spring and summer. At low latitudes, marine primary production is lower and varies little with season or region. On land, primary production also exhibits strong regional and seasonal patterns. Primary production rates (in g C m^{-2} yr^{-1}) are greatest year-round in the tropics. Tropical forests in South America, Africa, and Southeast Asia are the most productive ecosystems on Earth. Mid-latitude temperate forests and high-latitude boreal forests are also highly productive, with a strong seasonal cycle of greatest production in spring and summer. Deserts (concentrated at ~30°N and S) and polar regions are less productive. These features of seasonal variability in primary production on land and in the oceans are clearly seen in the seasonal variations in atmospheric O_2 (Figure 4).

Transfer of O_2 between the atmosphere and surface seawater is controlled by air–sea gas exchange. Dissolved gas concentrations trend towards thermodynamic equilibrium, but other factors may complicate dissolved O_2 concentrations. Degassing of supersaturated waters can only occur at the very surface of the water. Thus, in regions of high primary production, concentrations of O_2 can accumulate in excess of the rate of O_2 degassing. In calm seas, where the air–sea interface is a smooth surface, gas exchange is very limited. As seas become more rough, and especially during storms, gas exchange is greatly enhanced. This is in part because of entrainment of bubbles dispersed in seawater and water droplets entrained in air, which provide much more surface for dissolution or degassing. Also, gas exchange depends on diffusion across a boundary layer. According to Fick's law, the diffusive flux depends on both the concentration gradient (degree of super- or undersaturation) and the thickness of the boundary layer. Empirically it is observed that the boundary layer thickness decreases with increasing wind speed, thus enhancing diffusion and gas exchange during high winds.

15.3.2 Photolysis of Water

In the upper atmosphere today, a small amount of O_2 is produced through photolysis of water vapor. This process is the sole source of O_2 to the atmospheres on the icy moons of Jupiter (Ganymede and Europa), where trace concentrations of O_2 have been detected (Vidal *et al.*, 1997). Water vapor photolysis may also have been the source of O_2 to the early Earth before the evolution of oxygenic photosynthesis. However, the oxygen formed by photolysis would have been through reactions with methane and carbon monoxide, preventing any accumulation in the atmosphere (Kasting *et al.*, 2001).

15.4 MECHANISMS OF O_2 CONSUMPTION

15.4.1 Aerobic Cellular Respiration

In simple terms, aerobic respiration is the oxidation of organic substrates with oxygen to yield chemical energy in the form of ATP and NADH. In eukaryotic cells, the respiration pathway follows three steps (Figure 6). Glycolysis occurs throughout the cytosol, splitting glucose into pyruvic acid and yielding some ATP. Glycolysis does not directly require free O_2, and thus occurs among aerobic and anaerobic organisms. The Krebs citric acid (tricarboxylic acid) cycle and oxidative phosphorylation are localized within the mitochondria of eukaryotes, and along the cell membranes of prokaryotes. The Krebs cycle completes the oxidation of pyruvate to CO_2 from glycolysis, and together glycolysis and the Krebs cycle provide chemical energy and reductants (in the form of ATP, NADH, and $FADH_2$) for the third step—oxidative phosphorylation. Oxidative phosphorylation involves the transfer of electrons from NADH and $FADH_2$ through a cascade of electron carrying compounds to molecular oxygen. Compounds used in the electron transport chain of oxidative phosphorylation include a variety of flavoproteins, quinones, Fe–S proteins, and cytochromes. Transfer of electrons from NADH to O_2 releases considerable energy, which is used to generate a proton gradient across the mitochondrial membrane and fuel significant ATP

synthesis. In part, this gradient is created by reduction of O_2 to H_2O as the last step of oxidative phosphorylation. While eukaryotes and many prokaryotes use the Krebs cycle to oxidize pyruvate to CO_2, there are other pathways as well. For example, prokaryotes use the glyoxalate cycle to metabolize fatty acids. Several aerobic prokaryotes also can use the Entner–Doudoroff pathway in place of normal glycolysis. This reaction still produces pyruvate, but yields less energy in the form of ATP and NADH.

Glycolysis, the Krebs cycle, and oxidative phosphorylation are found in all eukaryotes (animals, plants, and fungi) and many of the aerobic prokaryotes. The purpose of these reaction pathways is to oxidize carbohydrates with O_2, yielding CO_2, H_2O, and chemical energy in the form of ATP. While macrofauna generally require a minimum of $\sim0.05–0.1$ bar ($\sim10\,\mu M$) O_2 to survive, many prokaryotic microaerophilic organisms can survive and thrive at much lower O_2 concentrations. Because most biologically mediated oxidation processes occur through the activity of aerotolerant microorganisms, it is unlikely that a strict coupling between limited atmospheric O_2 concentration and limited global respiration rates could exist.

15.4.2 Photorespiration

The active site of rubisco, the key enzyme involved in photosynthesis, can accept either CO_2 or O_2. Thus, O_2 is a competitive inhibitor of photosynthesis. This process is known as photorespiration, and involves addition of O_2 to ribulose-biphosphate. Products of this reaction enter a metabolic pathway that eventually produces CO_2. Unlike cellular respiration, photorespiration generates no ATP, but it does consume O_2. In some plants, as much as 50% of the carbon

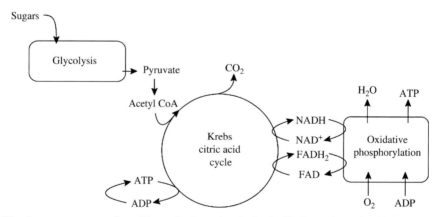

Figure 6 The three components of aerobic respiration: glycolysis, the Krebs cycle, and oxidative phosphorylation. Sugars are used to generate energy in the form of ATP during glycolysis. The product of glycolyis, pyruvate, is converted to acetyl-CoA, and enters the Krebs cycle. CO_2, stored energy as ATP, and stored reducing power as NADH and $FADH_2$ are generated in the Krebs cycle. O_2 is only directly consumed during oxidative phosphorylation to generate ATP as the final component of aerobic respiration.

fixed by the Calvin cycle is respired through photorespiration. Photorespiration is enhanced in hot, dry environments when plant cells close stomata to slow water loss, CO_2 is depleted and O_2 accumulates. Photorespiration does not occur in prokaryotes, because of the much lower relative concentration of O_2 versus CO_2 in water compared with air.

15.4.3 C₁ Metabolism

Beyond metabolism of carbohydrates, there are several other biological processes common in prokaryotes that consume oxygen. For example, methylotrophic organisms can metabolize C_1 compounds such as methane, methanol, formaldehyde, and formate, as in

$$CH_4 + NADH + H^+ + O_2 \rightarrow CH_3OH + NAD^+ + H_2O$$

$$CH_3OH + PQQ \rightarrow CH_2O + PQQH_2$$

$$CH_2O + NAD^+ + H_2O \rightarrow HCOOH + NADH + H^+$$

$$HCOOH + NAD^+ \rightarrow CO_2 + NADH + H^+$$

These compounds are common in soils and sediments as the products of anaerobic fermentation reactions. Metabolism of these compounds can directly consume O_2 (through monooxygenase enzymes) or indirectly, through formation of NADH which is shuttled into oxidative phosphorylation and the electron transport chain. Oxidative metabolism of C_1 compounds is an important microbial process in soils and sediments, consuming the methane produced by methanogenesis.

15.4.4 Inorganic Metabolism

Chemolithotrophic microorganisms are those that oxidize inorganic compounds rather than organic substrates as a source of energy and electrons. Many of the chemolithotrophs are also autotrophs, meaning they reduce CO_2 to generate cellular carbon in addition to oxidizing inorganic compounds. In these organisms, CO_2 fixation is not tied to O_2 production by the stoichiometry of photosynthesis. Hydrogen-oxidizing bacteria occur wherever both O_2 and H_2 are available. While some H_2 produced by fermentation in anoxic environments (deep soils and sediments) may escape upward into aerobic environments, most H_2 utilized by hydrogen-oxidizing bacteria derives from nitrogen fixation associated with nitrogen-fixing plants and cyanobacteria. Nitrifying bacteria are obligate autotrophs that oxidize ammonia. Ammonia is produced in many environments during fermentation of nitrogen compounds and by dissimilatory nitrate reduction. Nitrifying bacteria are common at oxic–anoxic interfaces in soils, sediments, and the water column. Other chemolithoautotrophic bacteria can oxidize

nitrite. Non-photosynthetic bacteria that can oxidize reduced sulfur compounds form a diverse group. Some are acidophiles, associated with sulfide mineral oxidation and tolerant of extremely low pH. Others are neutrophilic and occur in many marine sediments. These organisms can utilize a wide range of sulfur compounds produced in anaerobic environments, including H_2S, thiosulfate, polythionates, polysulfide, elemental sulfur, and sulfite. Many of the sulfur-oxidizing bacteria are also autotrophs. Some bacteria can live as chemoautotrophs through the oxidation of ferrous-iron. Some of these are acidophiles growing during mining and weathering of sulfide minerals. However, neutrophilic iron-oxidizing bacteria have also been detected associated with the metal sulfide plumes and precipitates at mid-ocean ridges. Other redox-sensitive metals that can provide a substrate for oxidation include manganese, copper, uranium, arsenic, and chromium. As a group, chemolithoautotrophic microorganisms represent a substantial flux of O_2 consumption and CO_2 fixation in many common marine and terrestrial environments. Because the net reaction of chemolithoautotrophy involves both O_2 and CO_2 reduction, primary production resulting from chemoautotrophs has a very different O_2/CO_2 stoichiometry than does photosynthesis. Chemolithoautotrophs use the electron flow from reduced substrates (metals, H_2, and reduced sulfur) to O_2 to generate ATP and NAD(P)H, which in turn are use for CO_2-fixation. It is believed that most aerobic chemoautotrophs utilize the Calvin cycle for CO_2-fixation.

15.4.5 Macroscale Patterns of Aerobic Respiration

On a global scale, much biological O_2 consumption is concentrated where O_2 is abundant. This includes surficial terrestrial ecosystems and marine surface waters. A large fraction of terrestrial primary production is consumed by aerobic degradation mechanisms. Although most of this is through aerobic respiration, some fraction of aerobic degradation of organic matter depends on anaerobic breakdown of larger biomolecules into smaller C_1 compounds, which, if transported into aerobic zones of soil or sediment, can be degraded by aerobic C_1 metabolizing microorganisms. Partially degraded terrestrial primary production can be incorporated into soils, which slowly are degraded and eroded. Research has shown that soil organic matter (OM) can be preserved for up to several millennia, and riverine export of aged terrestrial OM may be a significant source of dissolved and particulate organic carbon to the oceans.

Aside from select restricted basins and specialized environments, most of the marine water column is oxygenated. Thus, aerobic respiration dominates in open water settings. Sediment trap and particle flux studies have shown that substantial fractions of marine primary production are completely degraded (remineralized) prior to deposition at the sea floor, and thus by implication, aerobic respiration generates an O_2 consumption demand nearly equal to the release of oxygen associated with photosynthesis. The bulk of marine aerobic respiration occurs within the water column. This is because diffusion and mixing (and thus O_2 resupply) are much greater in open water than through sediment pore fluids, and because substrates for aerobic respiration (and thus O_2 demand) are much less concentrated in the water column than in the sediments. O_2 consumption during aerobic respiration in the water column, coupled with movement of deep-water masses from O_2-charged sites of deep-water formation, generates the vertical and lateral profiles of dissolved O_2 concentration observed in seawater.

In most marine settings, O_2 does not penetrate very far into the sediment. Under regions of high primary productivity and limited water column mixing, even if the water column is oxygenated, O_2 may penetrate 1 mm or less, limiting the amount of aerobic respiration that occurs in the sediment. In bioturbated coastal sediments, O_2 penetration is facilitated by the recharging of pore fluids through organisms that pump overlying water into burrows, reaching several centimeters into the sediment in places. However, patterns of oxic and anoxic sediment exhibit a great degree of spatial and temporal complexity as a result of spotty burrow distributions, radial diffusion of O_2 from burrows, and continual excavation and infilling through time. Conversely, in deep-sea (pelagic) sediments, where organic matter delivery is minor and waters are cold and charged with O_2 from sites of deep-water formation, O_2 may penetrate uniformly 1 m or more into the sediment.

15.4.6 Volcanic Gases

Gases emitted from active volcanoes and fumaroles are charged with reduced gases, including CO, H_2, SO_2, H_2S, and CH_4. During explosive volcanic eruptions, these gases are ejected high into the atmosphere along with H_2O, CO_2, and volcanic ash. Even the relatively gentle eruption of low-silicon, low-viscosity-shield volcanoes is associated with the release of reduced volcanic gases. Similarly, reduced gases are released dissolved in waters associated with hot springs and geysers. Oxidation of reduced volcanic gases occurs in the atmosphere, in natural waters, and on the surfaces of minerals. This is predominantly an abiotic process, although many chemolithoautotrophs have colonized the walls and channels of hot springs and fumaroles, catalyzing the oxidation of reduced gases with O_2. Much of biological diversity in hyperthermophilic environments consists of prokaryotes employing these unusual metabolic types.

Recent estimates of global average volcanic gas emissions suggest that volcanic sulfur emissions range, $0.1–1 \times 10^{12}$ mol S yr^{-1}, is nearly equally distributed between SO_2 and H_2S (Halmer *et al.*, 2002; Arthur, 2000). This range agrees well with the estimates used by both Holland (2002) and Lasaga and Ohmoto (2002) for average volcanic S emissions through geologic time. Other reduced gas emissions (CO, CH_4, and H_2) are estimated to be similar in magnitude (Mörner and Etiope, 2002; Arthur, 2000; Delmelle and Stix, 2000). All of these gases have very short residence times in the atmosphere, revealing that emission and oxidative consumption of these gases are closely coupled, and that O_2 consumption through volcanic gas oxidation is very efficient.

15.4.7 Mineral Oxidation

During uplift and erosion of the Earth's continents, rocks containing chemically reduced minerals become exposed to the oxidizing conditions of the atmosphere. Common rock-forming minerals susceptible to oxidation include olivine, Fe^{2+}-bearing pyroxenes and amphiboles, metal sulfides, and graphite. Ferrous-iron oxidation is a common feature of soil formation. Iron oxides derived from oxidation of Fe^{2+} in the parent rock accumulate in the B horizon of temperate soils, and extensive laterites consisting of iron and aluminum oxides develop in tropical soils to many meters of depth. Where erosion rates are high, iron-bearing silicate minerals may be transported short distances in rivers; however, iron oxidation is so efficient that very few sediments show deposition of clastic ferrous-iron minerals. Sulfide minerals are extremely susceptible to oxidation, often being completely weathered from near-surface rocks. Oxidation of sulfide minerals generates appreciable acidity, and in areas where mining has brought sulfide minerals in contact with the atmosphere or O_2-charged rainwater, low-pH discharge has become a serious environmental problem. Although ferrous silicates and sulfide minerals such as pyrite will oxidize under sterile conditions, a growing body of evidence suggests that in many natural environments, iron and sulfur oxidation is mediated by chemolithoautotrophic microorganisms. Prokaryotes with chemolithoautotrophic metabolic pathways have been isolated from many environments where iron and sulfur oxidation occurs. The amount of O_2 consumed annually through oxidation of Fe^{2+} and sulfur-bearing minerals is not known, but based on sulfur isotope mass balance constraints is on the order of $(0.1–1) \times 10^{12}$ mol yr^{-1} each for iron and sulfur, similar in magnitude to the flux of reduced gases from volcanism.

15.4.8 Hydrothermal Vents

The spreading of lithospheric plates along mid-ocean ridges is associated with much undersea volcanic eruptions and release of chemically reduced metal sulfides. Volcanic gases released by subaerial volcanoes are also generated by submarine eruptions, contributing dissolved reduced gases to seawater. Extrusive lava flows generate pillow basalts, which are composed in part of ferrous-iron silicates. Within concentrated zones of hydrothermal fluid flow, fracture-filling and massive sulfide minerals precipitate within the pillow lavas, large chimneys grow from the seafloor by rapid precipitation of Fe–Cu–Zn sulfides, and metal sulfide-rich plumes of high-temperature "black smoke" are released into seawater. In cooler zones, more gradual emanations of metal and sulfide rich fluid diffuse upward through the pillow basalts and slowly mix with seawater. Convective cells develop, in which cold seawater is drawn down into pillow lavas off the ridge axis to replace the water released at the hydrothermal vents. Reaction of seawater with basalt serves to alter the basalt. Some sulfide is liberated from the basalt, entrained seawater sulfate precipitates as anhydrite or is reduced to sulfide, and Fe^{2+} and other metals in basalt are replaced with seawater-derived magnesium. Oxygen dissolved in seawater is consumed during alteration of seafloor basalts. Altered basalts containing oxidized iron-mineral can extend as much as 500 m below the seafloor. Much more O_2 is consumed during oxidation of black smoker and chimney sulfides. Chimney sulfide may be only partially oxidized prior to transport off-axis through spreading and burial by sediments. However, black smoker metal-rich fluids are fairly rapidly oxidized in seawater, forming insoluble iron and manganese oxides, which slowly settle out on the seafloor, generating metalliferous sediments.

Because of the diffuse nature of reduced species in hydrothermal fluids, it is not known what role marine chemolithoautotrophic microorganisms may play in O_2 consumption associated with fluid plumes. Certainly such organisms live within the walls and rubble of cooler chimneys and basalts undergoing seafloor weathering. Because of the great length of mid-ocean ridges throughout the oceans, and the abundance of pyrite and other metal sulfides associated with these ridges, colonization and oxidation of metal sulfides by chemolithoautotrophic organisms may form an unrecognized source of primary production associated with consumption, not net release, of O_2.

15.4.9 Iron and Sulfur Oxidation at the Oxic–Anoxic Transition

In restricted marine basins underlying highly productive surface waters, where consumption of O_2 through aerobic respiration near the surface allows anoxia to at least episodically extend beyond the sediment into the water column, oxidation of reduced sulfur and iron may occur when O_2 is present. Within the Black Sea and many anoxic coastal fjords, the transition from oxic to anoxic environments occurs within the water column. At other locations, such as the modern Peru Shelf and coastal California basins, the transition occurs right at the sediment–water interface. In these environments, small concentrations of sulfide and O_2 may coexist within a very narrow band where sulfide oxidation occurs. In such environments, appreciable sulfur recycling may occur, with processes of sulfate reduction, sulfide oxidation, and sulfur disproportionation acting within millimeters of each other. These reactions are highly mediated by microorganisms, as evidenced by extensive mats of chemolithoautotrophic sulfur-oxidizing *Beggiatoa* and *Thioploca* found where the oxic–anoxic interface and seafloor coincide.

Similary, in freshwater and non-sulfidic brackish environments with strong O_2 demand, dissolved ferrous-iron may accumulate in groundwaters and anoxic bottom waters. Significant iron oxidation will occur where the water table outcrops with the land surface (i.e., groundwater outflow into a stream), or in lakes and estuaries, at the oxic–anoxic transition within the water column. Insoluble iron oxides precipitate and settle down to the sediment. Recycling of iron may occur if sufficient organic matter exists for ferric-iron reduction to ferrous-iron.

The net effect of iron and sulfur recycling on atmospheric O_2 is difficult to constrain. In most cases, oxidation of sulfur or iron consumes O_2 (there are some anaerobic chemolithoautotrophic microorganisms that can oxidize reduced substrates using nitrate or sulfate as the electron acceptor). Reduction of sulfate or ferric-iron are almost entirely biological processes, coupled with the oxidation of organic matter; sulfate and ferric-iron reduction individually have no effect on O_2. However, the major source of organic substrates for sulfate and ferric-iron reduction is ultimately biomass derived from photosynthetic organisms, which is associated with O_2 generation. The net change derived from summing the three processes (S^{2-}- or Fe^{2+}-oxidation, SO_4^{2-}- or Fe^{3+}- reduction, and photosynthesis), is net gain of organic matter with no net production or consumption of O_2.

15.4.10 Abiotic Organic Matter Oxidation

Aerobic respiration is the main means by which O_2 is consumed on the Earth. This pathway occurs throughout most Earth-surface environments: soils and aquatic systems, marine surface waters supersaturated with O_2, within the water column and the upper zones of sediments. However, reduced carbon materials are also reacted with O_2 in a variety of environments where biological activity has not been demonstrated. Among these are photo-oxidation of dissolved and particulate organic matter and fossil fuels, fires from burning vegetation and fossil fuels, and atmospheric methane oxidation. Olefins (organic compounds containing double bonds) are susceptible to oxidation in the presence of transition metapls, ozone, UV light, or gamma radiation. Low-molecular-weight oxidized organic degradation products form from oxidation reactions, which in turn may provide organic substrates for aerobic respiration or C_1 metabolism. Fires are of course high-temperature combustion of organic materials with O_2. Research on fires has shown that O_2 concentration has a strong influence on initiation and maintenance of fires (Watson, 1978; Lenton and Watson, 2000). Although the exact relationship between p_{O_2} and initiation of fire in real terrestrial forest communities is debated (see Robinson, 1989, 1991), it is generally agreed that, at low p_{O_2}, fires cannot be started even on dry wood, although smoldering fires with inefficient oxidation can be maintained. At high p_{O_2}, even wet wood can support flame, and fires are easily initiated with a spark discharge such as lightning.

Although most methane on Earth is oxidized during slow gradual transport upwards through soils and sediments, at select environments there is direct injection of methane into the atmosphere. Methane reacts with O_2 in the presence of light or metal surface catalysts. This reaction is fast enough, and the amount of atmospheric O_2 large enough, that significant concentrations of atmospheric CH_4 are unlikely to accumulate. Catastrophic calving of submarine, CH_4-rich hydrates during the geologic past may have liberated large quantities of methane to the atmosphere. However, isotopic evidence suggests that this methane was oxidized and consumed in a geologically short span of time.

15.5 GLOBAL OXYGEN BUDGETS AND THE GLOBAL OXYGEN CYCLE

One window into the global budget of oxygen is the variation in O_2 concentration normalized to N_2. O_2/N_2 ratios reflect changes in atmospheric O_2 abundance, because N_2 concentration is assumed to be invariant through time. Over an annual cycle, $\delta(O_2/N_2)$ can vary by 100 per meg or more, especially at high-latitude sites (Keeling and Shertz, 1992). These variations reflect latitudinal variations in net O_2 production and consumption related to seasonal high productivity during summer. Observations of O_2/N_2 variation have been expanded beyond direct observation (limited to the past several decades) to records of atmospheric composition as trapped in Antarctic firn

ice and recent ice cores (Battle *et al.*, 1996; Sowers *et al.*, 1989; Bender *et al.*, 1985, 1994b). These records reflect a slow gradual decrease in atmospheric O_2 abundance over historical times, attributed to the release and oxidation of fossil fuels.

As yet, details of the fluxes involved in the processes that generate and consume molecular oxygen are too poorly constrained to establish a balanced O_2 budget. A summary of the processes believed to dominate controls on atmospheric O_2, and reasonable best guesses for the magnitude of these fluxes, if available, are shown in Figure 7 (from Keeling *et al.*, 1993).

15.6 ATMOSPHERIC O_2 THROUGHOUT EARTH'S HISTORY

15.6.1 Early Models

Starting with Cloud (1976), two key geologic formations have been invoked to constrain the history of oxygenation of the atmosphere: banded iron formations (BIFs) and red beds (Figure 8). BIFs are chemical sediments containing very little detritus, and consist of silica laminae interbedded with layers of alternately high and low ratios of ferric- to ferrous-iron. As chemical sediments, BIFs imply the direct precipitation of ferrous-iron from the water column. A ferrous-iron-rich ocean requires anoxia, which by implication requires an O_2-free atmosphere and an anaerobic world. However, the ferric-iron layers in BIFs do reflect consumption of molecular O_2 (oxidation of ferrous-iron) at a rate much greater than supply of O_2 through prebiotic H_2O photolysis. Thus, BIFs may also record the evolution of oxygenic photosynthesis and at least localized elevated dissolved O_2 concentrations (Walker, 1979). Red beds are sandy sedimentary rocks rich in coatings, cements, and particles of ferric-iron. Red beds form during and after sediment deposition, and thus require that both the atmosphere and groundwater are oxidizing. The occurrence of the oldest red beds ($\sim$2.0 Ga) coincides nearly with the disappearance of BIFs (Walker, 1979), suggesting that some threshold of atmospheric O_2 concentration was reached at this time. Although the general concept of a low-O_2 atmosphere before $\sim$2.0 Ga and accumulated O_2 since that time has been agreed on for several decades, the details and texture of oxygenation of the Earth's atmosphere are still being debated.

Geochemists and cosmochemists initially looked to models of planetary formation and comparison with other terrestrial planets to understand the earliest composition of Earth's atmosphere. During planetary accretion and core formation, volatile components were liberated from a molten and slowly convecting mixture of silicates, metals,

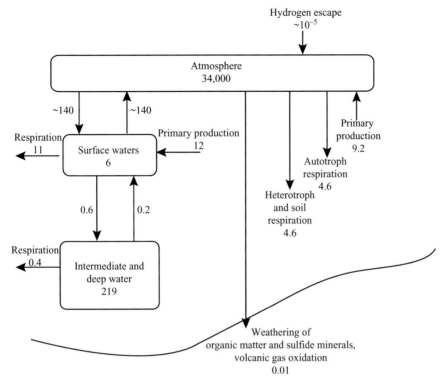

Figure 7 Global budget for molecular oxygen, including gas and dissolved O_2 reservoirs (sources Keeling *et al.*, 1993; Bender *et al.*, 1994a,b).

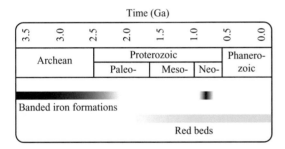

Figure 8 Archean distribution of banded iron formations, with short reoccurrence associated with widespread glaciation in the Neoproterozoic, and the Proterozoic and Phanerozoic distribution of sedimentary rocks containing ferric-iron cements (red beds). The end of banded iron formation and beginning of red bed deposition at ∼2.2 Ga has been taken as evidence for a major oxygenation event in Earth's atmosphere.

and trapped gases. The gravitational field of Earth was sufficient to retain most of the gases released from the interior. These include CH_4, H_2O, N_2, NH_3, and H_2S. Much H_2 and He released from the interior escaped Earth's gravitational field into space; only massive planets such as Jupiter, Saturn, Neptune, and Uranus have retained an H_2–He rich atmosphere. Photolysis of H_2O, NH_3, and H_2S produced free O_2, N_2, and S, respectively. O_2 was rapidly consumed by oxidation of CH_4 and H_2S to form CO_2, CO, and SO_2. High partial pressures of

CO_2 and CH_4 maintained a strong greenhouse effect and warm average Earth surface temperature (∼90 °C), in spite of much lower solar luminosity. Recognizing that the early Earth contained an atmospheric substantially richer in strong greenhouse gases compared with the modern world provided a resolution to subfreezing average Earth surface temperatures predicted for the early Earth due to reduced solar luminosity (Kasting *et al.*, 2001, 1983; Kiehl and Dickinson, 1987).

Liquid water on the early planet Earth allowed a hydrologic cycle and silicate mineral weathering to develop. Fairly quickly, much of the atmosphere's CO_2 was reacted with silicates to produce a bicarbonate-buffer ocean, while CH_4 was rapidly consumed by oxygen produced through photolysis of H_2O. Early microorganisms (and many of the most primitive organisms in existence today) used inorganic substrates to derive energy, and thrive at the high temperatures expected to be widespread during Earth's early history. These organisms include Archea that oxidize H_2 using elemental sulfur. Once photolysis and CH_4 oxidation generated sufficient p_{CO_2}, methanogenic Archea may have evolved. These organisms reduce CO_2 to CH_4 using H_2. However, sustainable life on the planet is unlikely to have developed during the first several hundred million years of Earth's history, due to large and frequent bolide impacts that would have sterilized the entire

Earth's surface prior to ~3.8 Ga (Sleep *et al.*, 1989; Sleep and Zahnle, 1998; Sleep *et al.*, 2001; Wilde *et al.*, 2001), although recent work by Valley *et al.* (2002) suggests a cool early Earth that continually supported liquid water as early as 4.4 Ga.

Much of present understanding of the earliest evolution of Earth's atmosphere can trace descent from Walker (1979) and references therein. The prebiological atmosphere (before the origin of life) was controlled principally by the composition of gases emitted from volcanoes. Emission of H_2 in volcanic gases has contributed to net oxidation of the planet through time. This is achieved through several mechanisms. Simplest is hydrodynamic escape of H_2 from Earth's gravity. Because H_2 is a strongly reducing gas, loss of H_2 from the Earth equates to loss of reducing power or net increase in whole Earth oxidation state (Walker, 1979). Today, gravitational escape of H_2 and thus increase in oxidation state is minor, because little if any H_2 manages to reach the upper levels of the atmosphere without oxidizing. Early in Earth's history, sources of H_2 included volcanic gases and water vapor photolysis. The small flux of O_2 produced by photolysis was rapidly consumed by reaction with ferrous-iron and sulfide, contributing to oxidation of the crust.

Today, volcanic gases are fairly oxidized, consisting mainly of H_2O and CO_2, with smaller amounts of H_2, CO, and CH_4. The oxidation state of volcanic gases derives in part from the oxidation state of the Earth's mantle. Mantle oxygen fugacity today is at or near the quartz-fayalite-magnesite buffer (QFM), as is the f_{O_2} of eruptive volcanic gases. Using whole-rock and spinel chromium abundance from volcanogenic rocks through time, Delano (2001) has argued that the average oxidation state of the Earth's mantle was set very early in Earth's history (~3.6–3.9 Ga) to f_{O_2} at or near the QFM buffer. Magmas with this oxidation state release volcanic gases rich in H_2O, CO_2, and SO_2, rather than more reducing gases. Thus, throughout much of Earth's history, volcanic gases contributing to the atmosphere have been fairly oxidized. More reduced magma compositions have been detected in diamond-bearing assemblages likely Hadean in age (>4.0 Ga) (Haggerty and Toft, 1985). The increase in mantle oxidation within several hundred million years of early Earth's history reveals very rapid "mantle + crust" overturn and mixing at this time, coupled with subduction and reaction of the mantle with hydrated and oxidized crustal minerals (generated from reaction with O_2 produced through H_2O photolysis).

15.6.2 The Archean

15.6.2.1 Constraints on the O_2 content of the Archean atmosphere

Several lines of geochemical evidence support low to negligible concentrations of atmospheric

O_2 during the Archean and earliest Proterozoic, when oxygenic photosynthesis may have evolved. The presence of pyrite and uraninite in detrital Archean sediments reveals that the atmosphere in the earliest Archean contained no free O_2 (Cloud, 1972). Although Archean-age detrital pyrites from South Africa may be hydrothermal in origin, Australian sediments of the Pilbara craton (3.25–2.75 Ga) contain rounded grains of pyrite, uraninite, and siderite (Rasmussen and Buick, 1999), cited as evidence for an anoxic atmosphere at this time. Although disputed (Ohmoto, 1999), it is difficult to explain detrital minerals that are extremely susceptible to dissolution and oxidation under oxidizing condition unless the atmosphere of the Archean was essentially devoid of O_2.

Archean paleosols provide other geochemical evidence suggesting formation under reducing conditions. For example, the 2.75 Ga Mount Roe #2 paleosol of Western Australia contains up to 0.10% organic carbon with isotope ratios between −33‰ and −55‰ (Rye and Holland, 2000). These isotope ratios suggest that methanogenesis and methanotrophy were important pathways of carbon cycling in these soils. For modern soils in which the bulk organic matter is strongly ^{13}C-depleted (<−40‰), the methane fueling methanotrophy must be derived from somewhere outside the soil, because reasonable rates of fermentation and methanogenesis cannot supply enough CH_4. By extension, Rye and Holland (2000) argue that these soils formed under an atmosphere rich in CH_4, with any O_2 consumed during aerobic methanotrophy having been supplied by localized limited populations of oxygenic photoautotrophs. Other paleosol studies have used lack of cerium oxidation during soil formation as an indicator of atmospheric anoxia in the Archean (Murakami *et al.*, 2001). In a broader survey of Archean and Proterozoic paleosols, Rye and Holland (1998) observe that all examined paleosols older than 2.4 Ga indicate loss of iron during weathering and soil formation. This chemical feature is consistent with soil development under an atmosphere containing $<10^{-4}$ atm O_2 (1 atm $= 1.01325 \times 10^5$ Pa), although some research has suggested that anoxic soil development in the Archean does not necessarily require an anoxic atmosphere (Ohmoto, 1996).

Other evidence for low Archean atmospheric oxygen concentrations come from studies of mass-independent sulfur isotope fractionation. Photochemical oxidation of volcanic sulfur species, in contrast with aqueous-phase oxidation and dissolution that characterizes the modern sulfur cycle, may have been the major source of sulfate to seawater in the Archean (Farquhar *et al.*, 2002; Farquhar *et al.*, 2000). Distinct shifts in $\delta^{33}S$ and $\delta^{34}S$ in sulfide and sulfate from Archean rocks occurred between 2.4–2.0 Ga, consistent with a shift from an O_2-free early atmosphere in which

SO_2 photochemistry could dominate among seawater sulfate sources to an O_2-rich later atmosphere in which oxidative weathering of sulfide minerals predominated over photochemistry as the major source of seawater sulfate. Sulfur isotope heterogeneities detected in sulfide inclusions in diamonds also are believed to derive from photochemical SO_2 oxidation in an O_2-free atmosphere at 2.9 Ga (Farquhar *et al.*, 2002). Not only do these isotope ratios require an O_2-free atmosphere, but they also imply significant contact between the mantle, crust, and atmosphere as recently as 2.9 Ga.

Nitrogen and sulfur isotope ratios in Archean sedimentary rocks also indicate limited or negligible atmospheric O_2 concentrations (Figure 9). Under an O_2-free environment, nitrogen could only exist as N_2 and reduced forms (NH_3, etc.). Any nitrate or nitrite produced by photolysis would be quickly reduced, likely with Fe^{2+}. If nitrate is not available, then denitrification (reduction of nitrate to free N_2) cannot occur. Denitrification is associated with a substantial nitrogen-isotope discrimination, generating N_2 that is substantially depleted in ^{15}N relative to

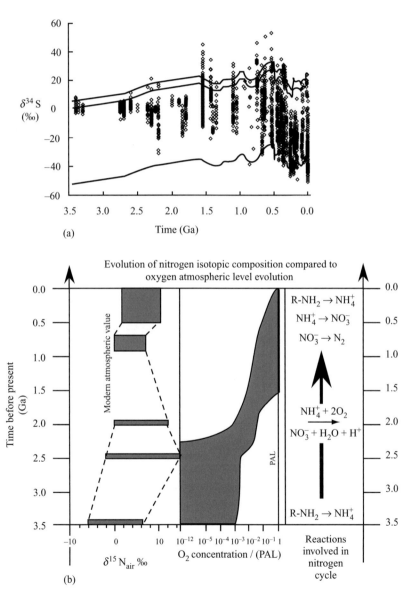

(a)

(b)

Figure 9 (a) The isotopic composition of sedimentary sulfate and sulfides through geologic time. The two upper lines show the isotopic composition of seawater sulfate (5‰ offset indicates uncertainty). The lower line indicated $\delta^{34}S$ of sulfate displaced by -55‰, to mimic average Phanerozoic maximum fractionation during bacterial sulfate reduction. Sulfide isotopic data (circles) indicated a much reduced fractionation between sulfate and sulfide in the Archean and Proterozoic (source Canfield, 1998). (b) Geologic evolution of the nitrogen isotopic composition of kerogen (left), estimated atmospheric O_2 content, and representative reactions in the biogeochemical nitrogen cycle at those estimated O_2 concentrations (sources Beaumont and Robert, 1999; Kasting, 1992).

the NO_3^- source. In the modern system, this results in seawater nitrate (and organic matter) that is ^{15}N-enriched relative to air. Nitrogen in kerogen from Archean sedimentary rocks is not enriched in [15]N (as is found in all kerogen nitrogen from Proterozoic age to the present), but instead is depleted relative to modern atmospheric N_2 by several ‰ (Beaumont and Robert, 1999). This is consistent with an Archean nitrogen cycle in which no nitrate and no free O_2 was available, and nitrogen cycling was limited to N_2-fixation, mineralization and ammonia volatilization. Bacterial sulfate reduction is associated with a significant isotope discrimination, producing sulfide that is depleted in [34]S relative to substrate sulfate. The magnitude of sulfur isotope fractionation during sulfate reduction depends in part on available sulfate concentrations. Very limited differences in sulfur isotopic ratios among Archean sedimentary sulfide and sulfate minerals (<2‰ $\delta^{34}S$) indicate only minor isotope fractionation during sulfate reduction in the Archean oceans (Canfield *et al.*, 2000; Habicht *et al.*, 2002), best explained by extremely low SO_4^{2-} concentrations ($<200\,\mu M$ in contrast with modern concentrations of $\sim 28\,mM$) (Habicht *et al.*, 2002). The limited supply of sulfate in the Archean ocean suggests that the major source of sulfate to Archean seawater was volcanic gas, because oxidative weathering of sulfide minerals could not occur under an O_2-free atmosphere. The limited sulfate concentration would have suppressed the activity of sulfate-reducing bacteria and facilitated methanogenesis. However, by $\sim 2.7\,Ga$, sedimentary sulfides that are [34]S-depleted relative to sulfate are detected, suggesting at least sulfate reduction, and by implication sources of sulfate to seawater through sulfide oxidation, may have developed (Canfield *et al.*, 2000).

15.6.2.2   The evolution of oxygenic photosynthesis

In the early Archean, methanogenesis (reaction of $H_2 + CO_2$ to yield CH_4) was likely a significant component of total primary production. O_2 concentrations in the atmosphere were suppressed due to limited O_2 production and rapid consumption with iron, sulfur, and reduced gases, while CH_4 concentrations were likely very high (Kasting *et al.*, 2001). The high methane abundance is calculated to have generated a hydrocarbon-rich smog that could screen UV light and protect early life in the absence of O_2 and ozone (Kasting *et al.*, 2001). Other means of protection from UV damage in the O_2-free Archean include biomineralization of cyanobacteria within UV-shielded iron–silica sinters (Phoenix *et al.*, 2001).

Biological evolution may have contributed to early Archean oxidation of the Earth. Catling *et al.*, (2001) have recognized that CO_2 fixation

associated with early photosynthesis may have been coupled with active fermentation and methanogenesis. Prior to any accumulation of atmospheric O_2, CH_4 may have been a large component of Earth's atmosphere. A high flux of biogenic methane is supported by coupled ecosystem–climate models of the early Earth (Pavlov *et al.*, 2000; Kasting *et al.*, 2001) as a supplement to Earth's greenhouse warming under reduced solar luminosity in the Archean. CH_4 in the upper atmosphere is consumed by UV light to yield hydrogen (favored form of hydrogen in the upper atmosphere), which escapes to space. Because H_2 escape leads to net oxidation, biological productivity and methanogenesis result in slow oxidation of the planet. Net oxidation may be in the form of direct O_2 accumulation (if CO_2 fixation is associated with oxygenic photosynthesis) or indirectly (if CO_2 fixation occurs via anoxygenic photosynthesis or anaerobic chemoautotrophy) through production of oxidized iron or sulfur minerals in the crust, which upon subduction are mixed with other crustal rocks and increase the crustal oxidation state. One system that may represent a model of early Archean biological productivity consists of microbial mats found in hypersaline coastal ponds. In these mats, cyanobacteria produce H_2 and CO that can be used as substrates by associated chemoautotrophs, and significant CH_4 fluxes out of the mats have been measured (Hoehler *et al.*, 2001). Thus, mats may represent communities of oxygenic photosynthesis, chemoautotrophy, and methanogenesis occurring in close physical proximity. Such communities would have contributed to elevated atmospheric CH_4 and the irreversible escape of hydrogen from the Archean atmosphere, contributing to oxidation of the early Earth.

The earliest photosynthetic communities may have contributed to oxidation of the early Earth through production of oxidized crustal minerals without requiring production of O_2. Crustal rocks today are more oxidized than the mantle, with compositions ranging between the QFM and hematite-magnesite f_{O_2} buffers. This is best explained as an irreversible oxidation of the crust associated with methanogenesis and hydrogen escape (Catling *et al.*, 2001). It is unlikely that O_2 produced by early photosynthesis ever directly entered the atmosphere.

Maintaining low O_2 concentrations in the atmosphere while generating oxidized crustal rocks requires oxidation mechanisms that do not involve O_2. Among these may be serpentinization of seafloor basalts (Kasting and Siefert, 2002). During seafloor weathering, ferrous-iron is released to form ferric oxyhydrides. Using H_2O or CO_2 as the ferrous-iron oxidant, H_2 and CH_4 are generated. This H_2 gas (either produced directly, or indirectly through UV decomposition of CH_4) can escape the atmosphere, resulting in net oxidation of the crust and accumulation of oxidized crustal rocks.

BIF formation in the Archean may be related to dissolved Fe^{2+} that was oxidized in shallow-water settings associated with local oxygenic photosynthesis or chemoautotrophy (Kasting and Siefert, 2002). After the biological innovation of oxygenic photosynthesis evolved, there was still a several hundred million year gap until O_2 began to accumulate in the atmosphere, because O_2 could only accumulate once the supply of reduced gases (CO and CH_4) and ferrous-iron fell below rates of photosynthetic O_2 supply.

15.6.2.3 Carbon isotope effects associated with photosynthesis

The main compound responsible for harvesting light energy to produce NADPH and splitting water to form O_2 is chlorophyll. There are several different structural variants of chlorophyll, including chlorophyll *a*, chlorophyll *b*, and bacteriochlorophylls *a–e* and *g*. Each of these shows optimum excitation at a different wavelength of light. All oxygenic photosynthetic organisms utilize chlorophyll *a* and/or chlorophyll *b*. Other non-oxygenic photoautotrophic microorganisms employ a diverse range of chlorophylls.

The earliest evidence for evolution of oxygenic photosynthesis comes from carbon isotopic signatures preserved in Archean rocks (Figure 10). CO_2 fixation through the Calvin cycle is associated with a significant carbon isotope discrimination, such that organic matter produced through CO_2 fixation is depleted in ^{13}C relative to ^{12}C by several per mil. In a closed or semi-closed system (up to and including the whole ocean–atmosphere system), isotope discrimination during fixation of CO_2 then results in a slight enrichment in the $^{13}C/^{12}C$ ratio of CO_2 not taken up during photosynthesis. Thus, a biosignature of CO_2 fixation is an enrichment in $^{13}C/^{12}C$ ratio in atmospheric CO_2 and seawater bicarbonate over a whole-Earth averaged isotope ratio. When carbonate minerals precipitate from seawater, they record the seawater isotope value. Enrichment of the $^{13}C/^{12}C$ ratio in early Archean carbonate minerals, and by extension seawater bicarbonate, is taken as early evidence for CO_2 fixation. Although anoxygenic photoautotrophs and chemoautotrophs fix CO_2 without generating O_2, these groups either do not employ the Calvin cycle (many anoxygenic photoautotrophs use the acetyl-CoA or reverse tricarboxylic acid pathways) or require O_2 (most chemolithoautotrophs). Thus, the most likely group of organisms responsible for this isotope effect are oxygenic photoautotrophs such as cyanobacteria. Kerogen and graphite that is isotopically depleted in ^{13}C is a common and continuous feature of the sedimentary record, extending as far back as ~3.8 Ga to the Isua Supracrustal Suite of Greenland (Schidlowski, 1988, 2001; Nutman *et al.*, 1997). Although some isotopically depleted graphite in the metasediments from the Isua Suite may derive from abiotic hydrothermal processes (van Zuilen *et al.*, 2002), rocks interpreted as metamorphosed turbidite deposits retain ^{13}C-depleted graphite believed to be biological in origin. Moreover, the isotopic distance between coeval carbonate and organic matter (in the form of kerogen) can be used to estimate biological productivity through time. With a few exceptions, these isotope mass balance estimates reveal that, since Archean times, global scale partitioning between inorganic and organic carbon, and thus global productivity and carbon burial, have not

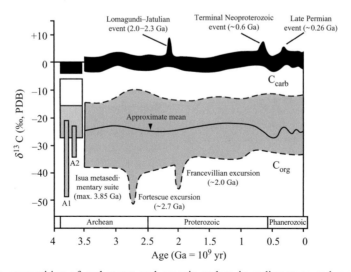

Figure 10 Isotopic composition of carbonates and organic carbon in sedimentary and metasedimentary rocks through geologic time. The negative excursions in organic matter $\delta^{13}C$ at 2.7 Ga and 2.0 Ga may relate to extensive methanogenesis as a mechanism of carbon fixation (Schidlowski, 2001) (reproduced by permission of Elsevier from *Precanb. Res.* **2001**, *106*, 117–134).

varied greatly over nearly 4Gyr of Earth's history (Schidlowski, 1988, 2001). Approximately 25% of crustal carbon burial is in the form of organic carbon, and the remainder is inorganic carbonate minerals. This estimate derives from the mass- and isotope-balance equation:

$$\delta^{13}C_{avg} = f\delta^{13}C_{organic\ matter} + (1-f) \times \delta^{13}C_{carbonate} \quad (2)$$

where $\delta^{13}C_{avg}$ is the average isotopic composition of crustal carbon entering the oceans from continental weathering and primordial carbon emitted from volcanoes, $\delta^{13}C_{organic\ matter}$ is the average isotopic composition of sedimentary organic matter, $\delta^{13}C_{carbonate}$ is the average isotopic composition of carbonate sediments, and f is the fraction of carbon buried as organic matter in sediments.

The observation that the proportion of carbon buried in sediments as organic matter versus carbonate has not varied throughout geologic time raises several intriguing issues. First, biogeochemical cycling of carbon exhibits remarkable constancy across 4Gyr of biological evolution, in spite of large-scale innovations in primary production and respiration (including anoxygenic and oxygenic photosynthesis, chemoautotrophy, sulfate reduction, methanotrophy, and aerobic respiration). Thus, with a few notable exceptions expressed in the carbonate isotope record, burial flux ratios between organic matter and carbonate have remained constant, in spite of varying dominance of different modes of carbon fixation and respiration through time, not to mention other possible controls on organic matter burial and preservation commonly invoked for Phanerozoic systems (anoxia of bottom waters, sedimentation rate, selective preservation, or cumulative oxygen exposure time). Second, the constancy of organic matter versus carbonate burial through time reveals that throughout geologic time, the relative contributions of various sources and sinks of carbon to the "ocean + atmosphere" system have remained constant. In other words, to maintain a constant carbonate isotopic composition through time, not only must the relative proportion of organic matter versus carbonate burial have remained nearly constant, but the relative intensity of organic matter versus carbonate weathering also must have remained nearly constant. In the earliest stages of Earth's history, when continents were small and sedimentary rocks were sparse, inputs of carbon from continental weathering may have been small relative to volcanic inputs. However, the several billion year sedimentary record of rocks rich in carbonate minerals and organic matter suggests that continental weathering must have formed a significant contribution to total oceanic carbon inputs fairly early in the Archean. Intriguingly, this indicates that oxidative weathering of ancient sedimentary rocks may have been active even in the

Archean, prior to accumulation of O_2 in the atmosphere. This runs counter to traditional interpretations of geochemical carbon–oxygen cycling, in which organic matter burial is equated to O_2 production and organic matter weathering is equated to O_2 consumption.

15.6.2.4   Evidence for oxygenic photosynthesis in the Archean

Fossil evidence for photosynthetic organisms from the same time period can be traced to the existence of stromatolites (Schopf, 1992, 1993; Schopf et al., 2002; Hofmann et al., 1999), although evidence for these early oxygenic photosynthetic communities is debated (Buick, 1990; Brasier et al., 2002). Stromatolites are laminated sediments consistently of alternating organic matter-rich and organic matter-lean layers; the organic matter-rich layers are largely composed of filamentous cyanobacteria. Stromatolites and similar mat-forming cyanobacterial crusts occur today in select restricted shallow marine environments. Fossil evidence from many locales in Archean rocks (Greenland, Australia, South Africa) and Proterozoic rocks (Canada, Australia, South America) coupled with carbon isotope geochemistry provides indirect evidence for the evolution of oxygenic photosynthesis early in Earth's history ($\sim$3.5 Ga).

Other evidence comes from molecular fossils. All cyanobacteria today are characterized by the presence of 2α-methylhopanes in their cell membranes. Brocks et al., (1999) have demonstrated the existence of this taxon-specific biomarker in 2.7 Ga Archean rocks of the Pilbara Craton in NW Australia. Also in these rocks is found a homologous series of C_{27} to C_{30} steranes. Steranes today derive from sterols mainly produced by organisms in the domain Eukarya. Eukaryotes are obligate aerobes that require molecular O_2, and thus Brocks and colleagues argue that the presence of these compounds provides strong evidence for both oxygenic photosynthesis and at least localized utilization of accumulated O_2. Of course, coexistence of two traits within a biological lineage today reveals nothing about which evolved first. It is uncertain whether 2α-methylhopane lipid biosynthesis preceded oxygenic photosynthesis among cyanobacteria. Sterol synthesis certainly occurs in modern prokaryotes (including some anaerobes), but no existing lineage produces the distribution of steranes found in the Brocks et al., (1999) study except eukaryotes.

15.6.3   The Proterozoic Atmosphere

15.6.3.1   Oxygenation of the Proterozoic atmosphere

Although there is evidence for the evolution of oxygenic photosynthesis several hundred million years before the Huronian glaciation (e.g., Brocks *et al.*, 1999) or earlier (Schidlowski, 1988, 2001), high fluxes of UV light reacting with O$_2$ derived from the earliest photosynthetic organisms would have created dangerous reactive oxygen species that severely suppressed widespread development of large populations of these oxygenic photoautotrophs. Cyanobacteria, as photoautotrophs, need be exposed to visible light and have evolved several defense mechanisms to protect cell contents and repair damage. However, two key metabolic pathways (oxygenic photosynthesis and nitrogen fixation) are very sensitive to UV damage. Indisputably cyanobacterial fossil occur in 1,000 Ma rocks, with putative fossils occurring 2,500 Ma and possibly older (Schopf, 1992). Sediment mat-forming cyanobacteria and stromatolites are least ambiguous and oldest. Terrestrial encrusting cyanobacteria are only known in the Phanerozoic. Planktonic forms are not known for the Archean and early Proterozoic. They may not have existed, or they may not be preserved. Molecular evolution, specifically coding for proteins that build gas vesicles necessary for planktonic life, are homologous and conserved in all cyanobacteria. Thus, perhaps planktonic cyanobacteria existed throughout Earth's history since the late Archean.

Today, ozone forms in the stratosphere by reaction of O$_2$ with UV light. This effectively screens much incoming UV radiation. Prior to the accumulation of atmospheric O$_2$, no ozone could form, and thus the UV flux to the Earth's surface would be much greater (with harmful effects on DNA and proteins, which adsorb and are altered by UV). A significant ozone shield could develop at $\sim 10^{-2}$ PAL O$_2$ (Kasting, 1987). However, a fainter young Sun would have emitted somewhat lower UV, mediating the lack of ozone. Although seawater today adsorbs most UV light by 6–25 m (1% transmittance cutoff), seawater with abundant dissolved Fe^{2+} may have provided an effective UV screen in the Archean (Olson and Pierson, 1986). Also, waters rich in humic materials, such as modern coastal oceans, are nearly UV opaque. If Archean seawater contained DOM, this could adsorb some UV. Iron oxidation and BIF at ~ 2.5–1.9 Ga would have removed the UV screen in seawater. Thus, a significant UV stress may have developed at this time. This would be mediated coincidentally by accumulation of atmospheric O$_2$.

In the early Archean before oxygenic photosynthesis evolved, cyanobacteria were limited. Planktonic forms were inhibited, and limited by dissolved iron content of water, and existence of stratified, UV-screen refuges. Sedimentary mat-forming and stromatolite-forming communities were much more abundant. Iron-oxide precipitation and deposition of screen enzymes may have created UV-free colonies under a shield even in shallow waters.

Advent of oxygenic photosynthesis in the Archean generated small oxygen oases (where dissolved O$_2$ could accumulate in the water) within an overall O$_2$-free atmosphere waters containing oxygenic photoautotrophs might have reached 10% air saturation (Kasting, 1992). At this time, both unscreened UV radiation and O$_2$ may have coincided within the water column and sediments. This would lead to increased UV stress for cyanobacteria. At the same time, precipitation of iron from the water would make the environment even more UV-transparent. To survive, cyanobacteria would need to evolve and optimize defense and repair mechanisms for UV damage (Garcia-Pichel, 1998). Perhaps this explains the $\sim$500 Myr gap between origin of cyanobacteria and accumulation of O$_2$ in the atmosphere. For example, the synthesis of scytonemin (a compound found exclusively in cyanobacteria) requires molecular oxygen (implying evolution in an oxic environment); it optimally screens UV-a, the form of UV radiation only abundant in an oxygenated atmosphere.

Oxygenation of the atmosphere at $\sim$2.3–2.0 Ga may derive from at least three separate causes. First, discussed earlier, is the titration of O$_2$ with iron, sulfur, and reduced gases. The other is global rates of photosynthesis and organic carbon burial in sediments. If the rate of atmospheric O$_2$ supply (oxygenic photosynthesis) exceeds all mechanisms of O$_2$ consumption (respiration, chemoautotrophy, reduced mineral oxidation, etc.), then O$_2$ can accumulate in the atmosphere. One means by which this can be evaluated is through seawater carbonate δ^{13}C. Because biological CO$_2$ fixation is associated with significant carbon isotope discrimination, the magnitude of carbon fixation is indicated by the isotopic composition of seawater carbonate. At times of more carbon fixation and burial in sediments, relatively more ^{12}C is removed from the atmosphere+ocean inorganic carbon pool than is supplied through respiration, organic matter oxidation, and carbonate mineral dissolution. Because carbon fixation is dominated by oxygenic photosynthesis (at least since the late Archean), periods of greater carbon fixation and burial of organic matter in sediments (observed as elevated seawater carbonate δ^{13}C) are equated to periods of elevated O$_2$ production through oxygenic photosynthesis. The early Proterozoic Lomagundi event ($\sim$2.3–2.0 Ga) is recorded in the sediment record as a prolonged period of elevated seawater carbonate δ^{13}C, with carbonate δ^{13}C values reaching nearly 10‰ in several sections around the world (Schidlowski, 2001). This represents an extended period of time (perhaps several hundred million years) during which removal of carbon from the "ocean+atmosphere" system as organic matter greatly exceeded supply. By implication, release of O$_2$ through photosynthesis was greatly accelerated during this time.

A third mechanism for oxygenation of the atmosphere at ~2.3 Ga relies on the slow, gradual oxidation of the Earth's crust. Irreversible H_2 escape and basalt-seawater reactions led to a gradual increase in the amount of oxidized and hydrated minerals contained in the Earth's crust and subducted in subduction zones throughout the Archean. Gradually this influenced the oxidation state of volcanic gases derived in part from subducted crustal rocks. Thus, although mantle oxygen fugacity may not have changed since the early Archean, crustal and volcanic gas oxygen fugacity slowly increased as the abundance of oxidized and hydrated crust increased (Holland, 2002; Kasting *et al.*, 1993; Kump *et al.*, 2001). Although slow to develop, Holland (2002) estimates that an increase in f_{O_2} of less than 1 log unit is all that would have been required for transition from an anoxic to an oxic atmosphere, assuming rates of oxygenic photosynthesis consistent with modern systems and the sediment isotope record. Once a threshold volcanic gas f_{O_2} had been reached, O_2 began to accumulate. Oxidative weathering of sulfides released large amounts of sulfate into seawater, facilitating bacterial sulfate reduction.

There are several lines of geochemical evidence that suggest a rise in oxygenation of the atmosphere ~2.3–2.0 Ga, beyond the coincident last occurrence of BIFs (with one late Proterozoic exception) and first occurrence of red beds recognized decades ago, and carbon isotopic evidence suggesting ample burial of sedimentary organic matter (Karhu and Holland, 1996; Bekker *et al.*, 2001; Buick *et al.*, 1998). The Huronian glaciation (~2.3 Ga) is the oldest known glacial episode recorded in the sedimentary record. One interpretation of this glaciation is that the cooler climate was a direct result of the rise of photosynthetically derived O_2. The rise of O_2 scavenged and reacted with the previously high atmospheric concentration of CH_4. Methane concentrations dropped, and the less-effective greenhouse gas product CO_2 could not maintain equable surface temperatures. Kasting *et al.*, (1983) estimate that a rise in p_{O_2} above ~10^{-4} atm resulted in loss of atmospheric CH_4 and onset of glaciation.

Paleosols have also provided evidence for a change in atmosphere oxygenation at some time between 2.3–2.0 Ga. Evidence from rare earth element enrichment patterns and U/Th fractionation suggests a rise in O_2 to ~0.005 bar by the time of formation of the Flin Flon paleosol of Manitoba, Canada, 1.85 Ga (Pan and Stauffer, 2000; Holland *et al.*, 1989; Rye and Holland, 1998). Rye and Holland (1998) examined several early Proterozoic paleosols and observed that negligible iron loss is a consistent feature from soils of Proterozoic age through the present. These authors estimate that a minimum p_{O_2} of >0.03 atm is required to retain iron during soil formation, and thus

atmospheric O_2 concentration has been 0.03 or greater since the early Paleozoic. However, re-evaluation of a paleosol crucial to the argument of iron depletion during soil formation under anoxia, the Hekpoort paleosol dated at 2.2 Ga, has revealed that the iron depletion detected by previous researchers may in fact be the lower zone of a normal oxidized lateritic soil. Upper sections of the paleosol that are not depleted in iron have been eroded away in the exposure examined by Rye and Holland (1998), but have been found in drill core sections. The depletion of iron and occurrence of ferrous-iron minerals in the lower sections of this paleosol have been reinterpreted by Beukes *et al.*, (2002) to indicate an abundant soil surface biomass at the time of deposition that decomposed to generate reducing conditions and iron mobilization during the wet season, and precipitation of iron oxides during the dry season.

Sulfur isotope studies have also provided insights into the transition from Archean low p_{O_2} to higher values in the Proterozoic. In the same studies that revealed extremely low Archean ocean sulfate concentrations, it was found that by ~2.2 Ga, isotopic compositions of sedimentary sulfates and sulfides indicate bacterial sulfate reduction under more elevated seawater sulfate concentrations compared with the sulfate-poor Archean (Habicht *et al.*, 2002; Canfield *et al.*, 2000). As described above, nitrogen isotope ratios in sedimentary kerogens show a large and permanent shift at ~2.0 Ga, consistent with denitrification, significant seawater nitrate concentrations, and thus available atmospheric O_2.

Prior to ~2.2 Ga, low seawater sulfate concentrations would have limited precipitation and subduction of sulfate-bearing minerals. This would maintain a lower oxidation state in volcanic gases derived in part from recycled crust (Holland, 2002). Thus, even while the oxidation state of the mantle has remained constant since ~4.0 Ga (Delano, 2001), the crust and volcanic gases derived from subduction of the crust could only achieve an increase in oxidation state once seawater sulfate concentrations increased.

A strong model for oxygenation of the atmosphere has developed based largely on the sulfur isotope record and innovations in microbial metabolism. The classical interpretation of the disappearance of BIFs relates to the rise of atmospheric O_2, oxygenation of the oceans, and removal of dissolved ferrous-iron by oxidation. However, another interpretation has developed, based largely on the evolving Proterozoic sulfur isotope record. During the oxygenation of the atmosphere at ~2.3–2.0 Ga, the oceans may not have become oxidized, but instead remained anoxic and became strongly sulfidic as well (Anbar and Knoll, 2002; Canfield, 1998 and references therein). Prior to ~2.3 Ga, the oceans were anoxic but not sulfidic. Ferrous-iron was abundant, as was manganese, because both are very soluble in anoxic, sulfide-free waters. The high concentration of dissolved iron and manganese facilitated nitrogen fixation by early cyanobacteria, such that available nitrogen was abundant, and phosphorus became the nutrient limiting biological productivity (Anbar and Knoll, 2002). Oxygenation of the atmosphere at ∼2.3 Ga led to increased oxidative weathering of sulfide minerals on the continents and increased sulfate concentration in seawater. Bacterial sulfate reduction generated ample sulfide, and in spite of limited mixing, the deep oceans would have remained anoxic and now also sulfidic as long as p_{O_2} remained below ∼0.07 atm (Canfield, 1998), assuming reasonable rates of primary production. Both iron and manganese form insoluble sulfides, and thus were effectively scavenged from seawater once the oceans became sulfidic. Thus, the Proterozoic oxygenation led to significant changes in global oxygen balances, with the atmosphere and ocean mixed layer becoming mildly oxygenated (probably <0.01 atm O_2), and the deep oceans becoming strongly sulfidic in direct response to the rise of atmospheric O_2.

In addition to increased oxygen fugacity of volcanic gases, and innovations in biological productivity to include oxygenic photosynthesis, the oxygenation of the Proterozoic atmosphere may be related to large-scale tectonic cycles. There are several periods of maximum deposition of sedimentary rocks rich in organic matter through geologic time. These are ∼2.7 Ga, 2.2 Ga, 1.9 Ga, and 0.6 Ga (Condie *et al.*, 2001). The increased deposition of black shales at 2.7 Ga and 1.9 Ga are associated with superplume events: highly elevated rates of seafloor volcanism, oceanic crust formation. Superplumes lead to increased burial of both organic matter and carbonates (through transgression, increased atmospheric CO_2, accelerated weathering, and nutrient fluxes to the oceans), with no net effect on carbonate isotopic composition. Thus, periods of increased absolute rates of organic matter burial and O_2 production may be masked by a lack of carbon isotopic signature. Breakup of supercontinents may be related to the black shale depositional events at 2.2 Ga and 0.6 Ga. Breakup of supercontinents may lead to more sediment accommodation space on continental shelves, as well as accelerated continental weathering and delivery of nutrients to seawater, fertilizing primary production and increasing organic matter burial. These supercontinent breakup events at 2.2 Ga and 0.6 Ga are clearly observed on the carbonate isotopic record as increases in relative burial of organic matter versus carbonate. Modeling efforts examining the evolution of the carbon, sulfur, and strontium isotope records have shown that gradual growth of the continents during the

Archean and early Proterozoic may in fact play a very large role in controlling the onset of oxygenation of the atmosphere, and that biological innovation may not be directly coupled to atmospheric evolution (Godderís and Veizer, 2000).

15.6.3.2 Atmospheric O₂ during the Mesoproterozoic

The sulfur-isotope-based model of Canfield and colleagues and the implications for limiting nutrient distribution proposed by Anbar and Knoll (2002) suggest that for nearly thousand million years (∼2.2–1.2 Ga), oxygenation of the atmosphere above p_{O_2} ∼0.01 atm was held in stasis. Although oxygenic photosynthesis was active, and an atmospheric ozone shield had developed to protect surface-dwelling organisms from UV radiation, much of the deep ocean was still anoxic and sulfidic. Removal of iron and manganese from seawater as sulfides generated severe nitrogen stress for marine communities, suggesting that productivity may have been limited throughout the entire Mesoproterozoic. The sluggish but consistent primary productivity and organic carbon burial through this time is seen in the carbonate isotope record. For several hundred million years, carbonate isotopic composition varied by no more than ±2‰, revealing very little change in the relative carbonate/organic matter burial in marine sediments. Anbar and Knoll (2002) suggest that this indicates a decoupling of the link between tectonic events and primary production, because although variations in tectonic activity (and associated changes in sedimentation, generation of restricted basins, and continental weathering) occurred during the Mesoproterozoic, these are not observed in the carbonate isotope record. This decoupling is a natural result of a shift in the source of the biological limiting nutrient from phosphorus (derived from continental weathering) to nitrogen (limited by N_2 fixation). Furthermore, the isotopic composition of carbonates throughout the Mesoproterozoic is 1–2‰ depleted relative to average carbonates from the early and late Proterozoic and Phanerozoic. This is consistent with a decrease in the relative proportion of carbon buried as organic matter versus carbonate during this time. It appears that after initial oxygenation of the atmosphere from completely anoxic to low p_{O_2} in the early Proterozoic, further oxygenation was halted for several hundred million years.

Global-scale reinvigoration of primary production, organic matter burial and oxygenation of the atmosphere may be observed in the latest Mesoproterozoic. Shifts in carbonate $\delta^{13}C$ of up to 4‰ are observed in sections around the globe at ∼1.3 Ga (Bartley *et al.*, 2001; Kah *et al.*, 2001). These positive carbon isotope excursions are associated with the formation of the Rodinian supercontinent, which led to increased continental margin length, orogenesis, and greater sedimentation (Bartley *et al.*, 2001). The increased organic matter burial and atmospheric oxygenation associated with this isotope excursion may be related to the first occurrence of laterally extensive $CaSO_4$ evaporites (Kah *et al.*, 2001). Although the rapid 10‰ increase in evaporate $\delta^{34}S$ across these sections is taken to indicate a much reduced seawater sulfate reservoir with much more rapid turnover times than found in the modern ocean, the isotope fractionation between evaporate sulfates and sedimentary sulfides indicates that the oceans were not sulfate-limiting. Atmospheric O_2 was of sufficient concentration to supply ample sulfate for bacterial sulfate reduction throughout the Mesoproterozoic.

15.6.3.3 Neoproterozoic atmospheric O_2

Elevated carbonate isotopic compositions (~4‰) and strong isotope fractionation between sulfate and biogenic sulfide minerals through the early Neoproterozoic indicates a period of several hundred million years of elevated biological productivity and organic matter burial. This may relate in part to the oxygenation of the atmosphere, ammonia oxidation, and increased seawater nitrate availability. Furthermore, oxidative weathering of molybdenum-bearing sulfide minerals, and greater oxygenation of the oceans increased availability of molybdenum necessary for cyanobacterial pathways of N_2-fixation (Anbar and Knoll, 2002). Much of the beginning of the Neoproterozoic, like the late Mesoproterozoic, saw gradual increases in oxygenation of the atmosphere, and possibly of the surface oceans. Limits on the oxygenation of the atmosphere are provided by sulfur isotopic and molecular evidence for the evolution of sulfide oxidation and sulfur disproportionation. The increase in sulfate–sulfide isotope fractionation in the Neoproterozoic (~1.0–0.6 Ga) reflects a shift in sulfur cycling from simple one-step reduction of sulfate to sulfide, to a system in which sulfide was oxidized to sulfur intermediates such as thiosulfate or elemental sulfur, which in turn were disproportionated into sulfate and sulfide (Canfield and Teske, 1996). Sulfur disproportionation is associated with significant isotope effects, generating sulfide that is substantially more ^{34}S-depleted than can be achieved through one-step sulfate reduction. The sulfur isotope record reveals that an increase in sulfate–sulfide isotope fractionation occurred between 1.0 Ga and 0.6 Ga, consistent with evolution of sulfide-oxidation at this time as derived from molecular clock based divergence of non-photosynthetic sulfide-oxidizing bacteria (Canfield and Teske, 1996). These authors estimated that innovation of bacterial sulfide oxidation occurred when much of the coastal shelf sediment (<200 m

water depth) was exposed to water with 13–46 μM O_2, which corresponds to 0.01–0.03 atm p_{O_2}. Thus, after the initial oxygenation of the atmosphere in the early Proterozoic (~2.3–2.0 Ga) to ~0.01 atm, p_{O_2} was constrained to this level until a second oxygenation in the late Proterozoic (~1.0–0.6 Ga) to 0.03 atm or greater.

During the last few hundred million years of the Proterozoic (~0.7–0.5 Ga), fragmentation of the Rodinian supercontinent was associated with at least two widespread glacial episodes (Hoffman *et al.*, 1998; Hoffman and Schrag, 2002). Neoproterozoic glacial deposits are found in Canada, Namibia, Australia, and other locations worldwide (Evans, 2000). In some of these, paleomagnetic evidence suggests that glaciation extended completely from pole to equator (Sumner, 1997; see also Evans (2000) and references therein). The "Snowball Earth", as these events have come to be known, is associated with extreme fluctuations in carbonate isotopic stratigraphy, reoccurrence of BIFs after an ~1.0 Ga hiatus, and precipitation of enigmatic, massive cap carbonate sediments immediately overlying the glacial deposits (Figure 11). It has been proposed that the particular configuration of continents in the latest Neoproterozoic, with land masses localized within the middle–low latitudes, lead to cooling of climate through several mechanisms. These include higher albedo in the subtropics (Kirschvink, 1992), increased silicate weathering as the bulk of continents were located in the warm tropics, resulting in a drawdown of atmospheric CO_2 (Hoffman and Schrag, 2002), possibly accelerated through high rates of OM burial, and reduced meridional Hadley cell heat transport, because tropical air masses were drier due to increased continentality (Hoffman and Schrag, 2002). Growth of initially polar ice caps would have created a positive ice–albedo feedback such that greater than half of Earth's surface became ice covered, global-scale growth of sea ice became inevitable (Pollard and Kasting, 2001; Baum and Crowley, 2001). Kirschvink (1992) recognized that escape from the Snowball Earth becomes possible because during extreme glaciation, the continental hydrologic cycle would be shut down. He proposed that sinks for atmospheric CO_2, namely, photosynthesis and silicated weathering, would have been eliminated during the glaciation. Because volcanic degassing continued during the glaciation, CO_2 in the atmosphere could rise to very high concentrations. It was estimated that an increase in p_{CO_2} to 0.12 bar would be sufficient to induce a strong enough greenhouse effect to begin warming the planet and melting the ice (Caldeira and Kasting, 1992). Once meltback began, it would subsequently accelerate through the positive ice–albedo feedback. Intensely warm temperatures would follow quickly after the glaciation, until accelerated silicate weathering under a reinvigorated

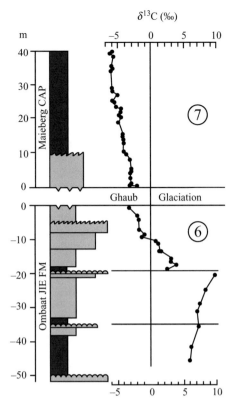

Figure 11 Composite section of carbonate isotope stratigraphy before and after the Ghaub glaciation from Namibia. Enriched carbonates prior to glaciation indicate active primary production and organic matter burial. Successive isotopic depletion indicates a shut down in primary production as the world became covered in ice (source Hoffman *et al.*, 1998).

hydrologic cycle could consume excess CO_2 and restore p_{CO_2} to more equable values. Carbonate mineral precipitation was inhibited during the global glaciation, so seawater became enriched in hydrothermally derived cations. At the end of glaciation, mixing of cation-rich seawater with high alkalinity surface runoff under warming climates led to the precipitation of massive cap carbonates.

Geochemical signatures before and after the Snowball episodes reveal rapid and short-lived changes in global biogeochemical cycles that impacted O_2 concentrations in the atmosphere and oceans. Prior to the Snowball events, the Neoproterozoic ocean experienced strong primary production and organic matter burial, as seen in the carbon isotopic composition of Neoproterozoic seawater in which positive isotope excursions reached 10‰ in some sections (Kaufman and Knoll, 1995). These excursions are roughly coincident with increased oxygenation of the atmosphere in the Neoproterozoic (Canfield and Teske, 1996; Des Marais *et al.*, 1992). Although sulfur isotopic evidence suggests that much of the ocean may have become oxygenated during the Neoproterozoic, it is very likely that this was

a temporary phenomenon. The global-scale glaciations of the late Neoproterozoic would have driven the oceans to complete anoxia through several mechanisms (Kirschvink, 1992). Ice cover inhibited air–sea gas exchange, and thus surface waters and sites of deep-water formation were cut off from atmospheric O_2 supplies. Intense oxidation of organic matter in the water column and sediment would have quickly consumed any available dissolved O_2. Extreme positive sulfur-isotope excursions associated with snowball-succession deposits reflect nearly quantitative, closed-system sulfate reduction in the oceans during glaciation (Hurtgen *et al.*, 2002). As seawater sulfate concentrations were reduced, hydrothermal inputs of ferrous-iron exceeded sulfide supply, allowing BIFs to form (Hoffman and Schrag, 2002). Closed-system sulfate reduction and iron formations require ocean anoxia throughout the water column, although restricted oxygenic photosynthesis beneath thin tropical sea ice may be responsible for ferrous-iron oxidation and precipitation (Hoffman and Schrag, 2002).

Once global ice cover was achieved, it is estimated to have lasted several million years, based on the amount of time required to accumulate 0.12 bar CO_2 at modern rates of volcanic CO_2 emission (Caldeira and Kasting, 1992). This must have placed extreme stress on eukaryotes and other organisms dependent on aerobic respiration. Obviously, oxygenated refugia must have existed, because the rise of many eukaryotes predates the Neoproterozoic Snowball events (Butterfield and Rainbird, 1998; Porter and Knoll, 2000). Nonetheless, carbon isotopic evidence leading up to and through the intense glacial intervals suggests that overall biological productivity was severely repressed during Snowball events. Carbonate isotope compositions fall from extremely enriched values to depletions of ∼(−6‰) at glacial climax (Hoffman *et al.*, 1998; Kaufman *et al.*, 1991; Kaufman and Knoll, 1995). These carbonate isotope values reflect primordial (volcanogenic) carbon inputs, and indicate that effectively zero organic matter was buried in sediments at this time. Sustained lack of organic matter burial and limited if any oxygenic photosynthesis under the glacial ice for several million years would have maintained an anoxic ocean, depleted in sulfate, with low sulfide concentrations due to iron scavenging, and extremely high alkalinity and hydrothermally-derived ion concentrations. Above the ice, the initially mildly oxygenated atmosphere would slowly lose its O_2. Although the hydrologic cycles were suppressed, and thus oxidative weathering of sulfide minerals and organic matter exposed on the continents was inhibited, the emission of reduced volcanic gases would have been enough to fully consume 0.01 bar O_2 within several hundred thousand years.

Thus, it is hypothesized that during Snowball events, the Earth's atmosphere returned to pre-Proterozoic anoxic conditions for at least several million years. Gradual increases in carbonate $\delta^{13}C$ after glaciation and deposition of cap carbonates suggests slow restored increases in primary productivity and organic carbon burial.

Carbon isotopic evidence suggests that reoxygenation, at least of seawater, was extremely gradual throughout the remainder of the Neoproterozoic. Organic matter through the terminal Proterozoic derives largely from bacterial heterotrophs, particularly sulfate reducing bacteria, as opposed to primary producers (Logan *et al.*, 1995). These authors suggested that throughout the terminal Neoproterozoic, anaerobic heterotrophy dominated by sulfate reduction was active throughout the water column, and O_2 penetration from surface waters into the deep ocean was inhibited. Shallow-water oxygen-deficient environments became widespread at the Precambrian–Cambrian boundary (Kimura and Watanabe, 2001), corresponding to negative carbonate $\delta^{13}C$ excursions and significant biological evolution from Ediacaran-type metazoans to emergence of modern metazoan phyla in the Cambrian.

15.6.4 Phanerozoic Atmospheric O_2

15.6.4.1 Constraints on Phanerozoic O_2 variation

Oxygenation of the atmosphere during the latest Neoproterozoic led to a fairly stable and well-oxygenated atmosphere that has persisted through the present day. Although direct measurement or a quantifiable proxy for Phanerozoic paleo-p_{O_2} concentrations have not been reported, multiple lines of evidence point to upper and lower limits on the concentration of O_2 in the atmosphere during the past several hundred million years. Often cited is the nearly continuous record of charcoal in sedimentary rocks since the evolution of terrestrial plants some 350 Ma. The presence of charcoal indicates forest fires throughout much of the Phanerozoic, which are unlikely to have occurred below $p_{O_2} = 0.17$ atm (Cope and Chaloner, 1985). Combustion and sustained fire are difficult to achieve at lower p_{O_2}. The existence of terrestrial plants themselves also provides a crude upper bound on p_{O_2} for two reasons. Above the compensation point of p_{O_2}/p_{CO_2} ratios, photorespiration outcompetes photosynthesis, and plants experience zero or negative net growth (Tolbert *et al.*, 1995). Although plants have developed various adaptive strategies to accommodate low atmospheric p_{CO_2} or aridity (i.e. C_4 and CAM plants), net terrestrial photosynthesis and growth of terrestrial ecosystems are effectively inhibited if p_{O_2} rises too high. This upper bound is difficult to exactly constrain, but is likely to be

~0.3–0.35 atm. Woody tissue is also extremely susceptible to combustion at high p_{O_2}, even if the tissue is wet. Thus, terrestrial ecosystems would be unlikely above some upper limit of p_{O_2} because very frequent reoccurrence of wildfires would effectively wipe out terrestrial plant communities, and there is no evidence of this occurring during the Phanerozoic on a global scale. What this upper p_{O_2} limit is, however, remains disputed. Early experiments used combustion of paper under varying humidity and p_{O_2} (Watson, 1978) but paper may not be the most appropriate analog for inception of fires in woody tissue with greater moisture content and thermal thickness (Robinson, 1989; Berner *et al.*, 2002). Nonetheless, the persistence of terrestrial plant communities from the middle Paleozoic through the present does imply that p_{O_2} concentrations have not risen too high during the past 350 Myr.

Prior to evolution of land plants, even circumstantial constraints on early Paleozoic p_{O_2} are difficult to obtain. Invertebrate metazoans, which have a continous fossil record from the Cambrian on, require some minimum amount of dissolved oxygen to support their aerobic metabolism. The absolute minimum dissolved O_2 concentration able to support aerobic metazoans varies from species to species, and is probably impossible to reconstruct for extinct lineages, but modern infaunal and epifaunal metazoans can accommodate dissolved O_2 dropping to the tens of micromolar in concentration. If these concentrations are extrapolated to equilibrium of the atmosphere with well-mixed cold surface waters, they correspond to ~0.05 atm p_{O_2}. Although local and regional anoxia occurred in the oceans at particular episodes through Phanerozoic time, the continued presence of aerobic metazoans suggests that widespread or total ocean anoxia did not occur during the past ~600 Myr, and that p_{O_2} has been maintained at or above $\sim5\%$ O_2 since the early Paleozoic.

15.6.4.2 Evidence for variations in Phanerozoic O_2

Several researchers have explored possible links between the final oxygenation of the atmosphere and explosion of metazoan diversity at the Precambrian–Cambrian boundary (McMenamin and McMenamin, 1990; Gilbert, 1996).While the Cambrian explosion may not record the origin of these phyla in time, this boundary does record the development of large size and hard skeletons required for fossilization (Thomas, 1997). Indeed, molecular clocks for diverse metazoan lineages trace the origin of these phyla to ~400 Myr before the Cambrian explosion (Doolittle *et al.*, 1996; Wray *et al.*, 1996). The ability of lineages to develop fossilizable hardparts may be linked with increasing oxygenation during the latest

Precambrian. Large body size requires elevated p_{O_2} and dissolved O_2 concentrations, so the diffusion can supply O_2 to internal tissues. Large body size provides several advantages, and may have evolved rather quickly during the earliest Cambrian (Gould, 1995), but large size also requires greater structural support. The synthesis of collagen (a ubiquitous structural protein among all metazoans and possible precursor to inorganic structural components such as carbonate or phosphate biominerals) requires elevated O_2. The threshold for collagen biosynthesis, and associated skeletonization and development of large body size, could not occur until p_{O_2} reached some critical threshold some time in the latest Precambrian (Thomas, 1997).

There is evidence to suggest that the late Paleozoic was a time of very elevated p_{O_2}, to concentrations substantially greater than observed in the modern atmosphere. Coals from the Carboniferous and Permian contain a greater abundance of fusain, a product of woody tissue combustion and charring, than observed for any period of the subsequent geologic past (Robinson, 1989, 1991; Glasspool, 2000), suggesting more abundant forest fires and by implication possibly higher p_{O_2} at this time. Less ambiguous is the biological innovation of gigantism at this time among diverse arthropod lineages (Graham *et al.*, 1995). All arthropods rely on tracheal networks for diffusion of O_2 to support their metabolism; active pumping of O_2 through a vascular system as found in vertebrates does not occur. This sets upper limits on body size for a given p_{O_2} concentration. In comparison with arthropod communities of the Carboniferous and Permian, modern terrestrial arthropods are rather small. Dragonflies at the time reached 70 cm wingspan, mayflies reached 45 cm wingpans, millipedes reached 1 m in length. Even amphibians, which depend in part on diffusion of O_2 through their skin for aerobic respiration, reached gigantic size at this time. These large body sizes could not be supported by today's 21% O_2 atmosphere, and instead require elevated p_{O_2} between 350 Ma and 250 Ma. Insect taxa that were giants in the Carboniferous do not survive past the Permian, suggesting that declining p_{O_2} concentrations in the Permian and Mesozoic led to extinction (Graham *et al.*, 1995; Dudley, 1998).

Increases in atmospheric O_2 affect organisms in several ways. Increased O_2 concentration facilitates aerobic respiration, while elevated p_{O_2} against a constant p_{N_2} increases total atmospheric pressure, with associated changes in atmospheric gas density and viscosity (Dudley, 1998, 2000). In tandem these effects may have played a strong role in the innovation of insect flight in the Carboniferous (Dudley, 1998, 2000), with secondary peaks in the evolution of flight among birds, bats and other insect lineages corresponding to times of high O_2 in the late Mesozoic (Dudley, 1998).

In spite of elevated p_{O_2} in the late Paleozoic, leading up to this time was an extended period of water column anoxia and enhanced burial of organic matter in the Devonian. Widespread deposition of black shales, fine-grained laminated sedimentary rocks rich in organic matter, during the Devonian indicate at least partial stratification of several ocean basins around the globe, with oxygen deficiency throughout the water column. One example of this is from the Holy Cross Mountains of Poland, from which particular molecular markers for green sulfur bacteria have been isolated (Joachimski *et al.*, 2001). These organisms are obligately anaerobic chemophotoautotrophs, and indicate that in the Devonian basin of central Europe, anoxia extended upwards through the water column well into the photic zone. Other black shales that indicate at least episodic anoxia and enhanced organic matter burial during the Devonian are found at several sites around the world, including the Exshaw Shale of Alberta (Canada), the Bakken Shale of the Williston Basin (Canada/USA), the Woodford Shale in Oklahoma (USA), and the many Devonian black shales of the Illinois, Michigan, and Appalachian Basins (USA). Widespread burial of organic matter in the late Devonian has been linked to increased fertilization of the surface waters through accelerated continental weathering due to the rise of terrestrial plant communities (Algeo and Scheckler, 1998). High rates of photosynthesis with relatively small rates of global respiration led to accumulation of organic matter in marine sediments, and the beginnings of a pulse of atmospheric hyperoxia that extended through the Carboniferous into the Permian. Coalescence of continental fragments to form the Pangean supercontinent at this time led to widespread circulation-restricted basins that facilitated organic matter burial and net oxygen release, and later, to extensive infilling to generate nearshore swamps containing terrestrial vegetation which was often buried to form coal deposits during the rapidly fluctuating sea levels of the Carboniferous and Permian.

The largest Phanerozoic extinction occurred at the end of the Permian ($\sim$250 Ma). A noticeable decrease in the burial of organic matter in marine sediments across the Permian–Triassic boundary may be associated with a global decline in primary productivity, and thus, with atmospheric p_{O_2}. The gigantic terrestrial insect lineages, thought to require elevated p_{O_2}, do not survive across this boundary, further suggesting a global drop in p_{O_2}, and the sedimentary and sulfur isotope records indicate an overall increase in sulfate reduction and burial of pyritic shales (Berner, 2002; Beerling and Berner, 2000). Although a long-duration deep-sea anoxic event has been proposed as a cause for the Permian mass extinction, there are competing models to explain exactly how this might have occurred. Hotinski *et al.*, (2001) has shown that while stagnation of the water column to generate deep-water anoxia might at first seem attractive, global thermohaline stagnation would starve the oceans of nutrients, extremely limiting primary productivity, and thus shutting down dissolved O_2 demand in the deep oceans. Large negative excursions in carbon, sulfur, and strontium isotopes during the late Permian may indicate stagnation and reduced ventilation of seawater for extended periods, coupled with large-scale overturn of anoxic waters. Furthermore, sluggish thermohaline circulation at this time could derive from a warmer global climate and warmer water at the sites of high-latitude deep-water formation (Hotinski *et al.*, 2001). The late Permian paleogeography of one supercontinent (Pangea) and one superocean (Panthalassa) was very different from the arrangement of continents and oceans on the modern Earth. Coupled with elevated p_{CO_2} at the time (Berner, 1994; Berner and Kothavala, 2001), GCM models predict warmer climate, weaker wind stress, and low equator to pole temperature gradients. Although polar deep-water formation still occurred, bringing O_2 from the atmosphere to the deep oceans, anoxia was likely to develop at mid-ocean depths (Zhang *et al.*, 2001), and thermohaline circulation oscillations between thermally versus salinity driven modes of circulation were likely to develop. During salinity driven modes, enhanced bottom-water formation in warm, salty low-latitude regions would limit oxygenation of the deep ocean. Thus, although sustained periods of anoxia are unlikely to have developed during the late Permian, reduced oxygenation of deep water through sluggish thermohaline circulation, coupled with episodic anoxia driven by low-latitude warm salty bottom-water formation, may have led to reoccurring episodes of extensive ocean anoxia over period of several million years.

Other researchers have invoked extraterrestrial causes for the End-Permian extinction and anoxia. Fullerenes (cage-like hydrocarbons that effectively trap gases during formation and heating) have been detected in late Permian sediments from southern China. The noble gas complement in these fullerenes indicates an extraterrestrial origin, which has been interpreted by Kaiho *et al.*, (2001) to indicate an unrecognized bolide impact at the Permian–Triassic boundary. The abrupt decrease in $\delta^{34}S$ across this boundary (from 20‰ to 5‰) implies an enormous and rapid release of ^{34}S-depleted sulfur into the ocean–atmosphere system. These authors propose that volatilized bolide- and mantle-derived sulfur ($\sim$0‰) oxidized in air, consumed atmospheric and dissolved O_2, and generated severe oxygen and acid stress in the oceans. Isotope mass balance estimates require $\sim$10^{19} mol sulfur to be released, consuming a similar mass of oxygen. 10^{19} mol O_2 represents some 10–40% of the total available

inventory of atmospheric and dissolved O_2 at this time, removal of which led to immediate anoxia, as these authors propose.

Other episodes of deep ocean anoxia and extensive burial of organic matter are known from the Jurassic, Cretaceous, Miocene, and Pleistocene, although these have not been linked to changes in atmospheric O_2 and instead serve as examples of the decoupling between atmospheric and deep-ocean O_2 concentrations through much of the geologic past. Widespread Jurassic black shale facies in northern Europe (Posidonien Schiefer, Jet Rock, and Kimmeridge Clay) were deposited in a restricted basin on a shallow continental margin. Strong monsoonal circulation led to extensive freshwater discharge and a low-salinity cap on basin waters during summer, and intense evaporation and antiestuarine circulation during winter (Röhl *et al.*, 2001), both of which contributed to water column anoxia and black shale deposition. Several oceanic anoxic events (OAEs) are recognized from the Cretaceous in all major ocean basins, suggesting possible global deep-ocean anoxia. Molecular markers of green sulfur bacteria, indicating photic zone anoxia, have been detected from Cenomanian–Turonian boundary section OAE sediments from the North Atlantic (Sinninghe Damsté and Köster, 1998). The presence of these markers (namely, isorenieratene, a diaromatic carotenoid accessory pigment used during anoxygenic photosynthesis) indicates that the North Atlantic was anoxic and euxinic from the base of the photic zone ($\sim$50–150 m) down to the sediment. High concentrations of trace metals scavenged by sulfide and an absence of bioturbation further confirm anoxia throughout the water column. Because mid-Cretaceous oceans were not highly productive, accelerated dissolved O_2 demand from high rates of respiration and primary production cannot be the prime cause of these OAEs. Most likely, the warm climate of the Cretaceous led to low O_2 bottom waters generated at warm, high salinity regions of low-latitude oceans. External forcing, perhaps through Milankovic-related precession-driven changes in monsoon intensity and strength, influenced the rate of salinity-driven deep-water formation, ocean basin oxygenation, and OAE formation (Wortmann *et al.*, 1999). Sapropels are organic matter-rich layers common to late Cenozoic sediments of the eastern Mediterranean. They are formed through a combination of increased primary production in surface waters, and increased organic matter preservation in the sediment likely to be associated with changes in ventilation and oxygenation of the deep eastern Mediterranean basins (Stratford *et al.*, 2000). The well-developed OMZ located off the coast of Southern California today may have been more extensive in the past. Variations in climate affecting intensity of upwelling and primary production, coupled with tectonic activity altering the depth of basins and height of sills along the California coast, have generated a series of anoxia-facies organic-matter-rich sediments along the west coast of North America, beginning with the Monterey Shale and continuing through to the modern sediments deposited in the Santa Barbara and Santa Monica basins.

As shown by the modeling efforts of Hotinski *et al.*, (2001) and Zhang *et al.*, (2001), extensive deep ocean anoxia is difficult to achieve for extended periods of geologic time during the Phanerozoic when p_{O_2} were at or near modern levels. Thus, while localized anoxic basins are common, special conditions are required to generate widespread, whole ocean anoxia such as observed in the Cretaceous. Deep-water formation in highly saline low-latitude waters was likely in the geologic past when climates were warmer and equator to pole heat gradients were reduced. Low-latitude deep-water formation has a significant effect on deep-water oxygenation, not entirely due to the lower O$_2$ solubility in warmer waters, but also the increased efficiency of nutrient use and recycling in low-latitude surface waters (Herbert and Sarmiento, 1991). If phytoplankton were 100% efficient at using and recycling nutrients, even with modern high-latitude modes of cold deep-water formation, the deep oceans would likely become anoxic.

15.6.4.3 Numerical models of Phanerozoic oxygen concentration

Although photosynthesis is the ultimate source of O$_2$ to the atmosphere, in reality photosynthesis and aerobic respiration rates are very closely coupled. If they were not, major imbalances in atmospheric CO$_2$, O$_2$, and carbon isotopes would result. Only a small fraction of primary production (from photosynthesis) escapes respiration in the water column or sediment to become buried in deep sediments and ultimately sedimentary rocks. This flux of buried organic matter is in effect "net photosynthesis", or total photosynthesis minus respiration. Thus, while over timescales of days to months, dissolved and atmospheric O$_2$ may respond to relative rates of photosynthesis or respiration, on longer timescales it is burial of organic matter in sediments (the "net photosynthesis") that matters. Averaged over hundreds of years or longer, burial of organic matter equates to release of O$_2$ into the "atmosphere + ocean" system:

Photosynthesis:

$$6CO_2 + 6H_2O \rightarrow C_6H_{12}O_6 + 6O_2 \qquad (3)$$

Respiration:

$$C_6H_{12}O_6 + 6O_2 + 6CO_2 + 6H_2O \qquad (4)$$

If for every 1,000 rounds of photosynthesis there are 999 rounds of respiration, the net result is one round of organic matter produced by photosynthesis that is not consumed by aerobic respiration. The burial flux of organic matter in sediments represents this lack of respiration, and as such is a net flux of O$_2$ to the atmosphere. In geochemists' shorthand, we represent this by the reaction

Burial of organic matter in sediments:

$$CO_2 + H_2O \rightarrow O_2 + \text{"CH}_2\text{O"} \qquad (5)$$

where "CH$_2$O" is not formaldehyde, or even any specific carbohydrate, but instead represents sedimentary organic matter. Given the elemental composition of most organic matter in sediments and sedimentary rocks, a more reduced organic matter composition might be more appropriate, i.e., C$_{10}$H$_{12}$O, which would imply release of 12.5 mol O$_2$ for every mole of CO eventually buried as organic matter. However, the simplified stoichiometry of (5) is applied for most geochemical models of C–S–O cycling.

If burial of organic matter equates to O$_2$ release to the atmosphere over long timescales, then the oxidative weathering of ancient organic matter in sedimentary rocks equates to O$_2$ consumption. This process has been called "georespiration" by some authors (Keller and Bacon, 1998). It can be represented by Equation (6), the reverse of (5):

Weathering of organic matter from rocks:

$$O_2 + \text{"CH}_2\text{O"} \rightarrow CO_2 + H_2O \qquad (6)$$

Both Equations (3) and (4) contain terms for addition and removal of O$_2$ from the atmosphere. Thus, if we can reconstruct the rates of burial and weathering of OM into/out of sedimentary rocks through time, we can begin to quantify sources and sinks for atmospheric O$_2$. The physical manifestation of this equation is the reaction of organic matter with O$_2$ during the weathering and erosion of sedimentary rocks. This is most clearly seen in the investigation of the changes in OM abundance and composition in weathering profiles developed on black shales (Petsch *et al.*, 2000, 2001).

In addition to the C–O system, the coupled C–S–O system has a strong impact on atmospheric oxygen (Garrels and Lerman, 1984; Kump and Garrels, 1986; Holland, 1978, 1984). This is through the bacterial reduction of sulfate to sulfide using organic carbon substrates as electron donors. During bacterial sulfate reduction (BSR), OM is oxidized and sulfide is produced from sulfate. Thus, BSR provides a means of resupplying oxidized carbon to the "ocean + atmosphere" system without consuming O$_2$. The net reaction for BSR shows that for every 15 mol of OM consumed, 8 mol sulfate and 4 mol ferric-iron are also reduced to form 4 mol of pyrite (FeS$_2$):

$$4\,Fe(OH)_3 + 8\,SO_4^{2-} + 15\,CH_2O \rightarrow 4\,FeS_2$$
$$+ 15\,HCO_3^- + 13\,H_2O + (OH)^- \qquad (7)$$

Oxidation of sulfate using organic substrates as electron donors provides a means of restoring inorganic carbon to the ocean + atmosphere system without consuming free O_2. Every 4 mol of pyrite derived from BSR buried in sediments represents 15 mol of O_2 produced by photosynthesis to generate organic matter that will not be consumed through aerobic respiration. In effect, pyrite burial equates to net release of O_2 to the atmosphere, as shown by (8), obtained by the addition of Equation (7) to (5):

$$4Fe(OH)_3 + 8SO_4^{2-} + 15CH_2O$$
$$\rightarrow 4FeS_2 + 15HCO_3^- + 13H_2O + (OH)^-$$

$$+15CO_2 + 15H_2O \rightarrow 15O_2 + 15\text{``}CH_2O\text{''}$$

$$4Fe(OH)_3 + 8SO_4^{2-} + 15CO_2 + 2H_2O$$
$$\rightarrow 4FeS_2 + 15HCO_3^- + (OH)^- + 15O_2 \quad (8)$$

Oxidative weathering of sedimentary sulfide minerals during exposure and erosion on the continents results in consumption of O_2 (Equation (9)):

$$4FeS_2 + 15O_2 + 14H_2O \rightarrow 4Fe(OH)_3$$
$$+ 8SO_4^{2-} + 16H^+ \quad (9)$$

Just as for the C–O geochemical system, if we can reconstruct the rates of burial and oxidative weathering of sedimentary sulfide minerals through geologic time, we can use these to estimate additional sources and sinks for atmospheric O_2 beyond organic matter burial and weathering.

In total, then, the general approach taken in modeling efforts of understanding Phanerozoic O_2 variability is to catalog the total sources and sinks for atmospheric O_2, render these in the form of a rate of change equation in a box model, and integrate the changing O_2 mass through time implied by changes in sources and sinks:

$$\begin{aligned} dM_{O_2}/dt &= \sum F_{O_2} \text{ into the atmosphere} \\ &\quad - \sum F_{O_2} \text{ out of the atmosphere} \\ &= F_{\text{burial of organic matter}} \\ &\quad + (15/8)F_{\text{burial of pyrite}} \\ &\quad - F_{\text{weathering of organic matter}} \\ &\quad - (15/8)F_{\text{weathering of pyrite}} \quad (10) \end{aligned}$$

One approach to estimate burial and weathering fluxes of organic matter and sedimentary sulfides through time uses changes in the relative abundance of various sedimentary rock types estimated over Phanerozoic time. Some sedimentary rocks are typically rich in both organic matter and pyrite. These are typically marine shales. In contrast, coal basin sediments contain much organic carbon, but very low amounts of sedimentary sulfides. Non-marine coarse-grained clastic sediments contain very little of either organic matter or sedimentary sulfides. Berner and Canfield (1989) simplified

global sedimentation through time into one of three categories: marine shales + sandstone, coal basin sediments, and non-marine clastic sediments. Using rock abundance estimates derived from the data of Ronov and others (Budyko et al., 1987; Ronov, 1976), these authors estimated burial rates for organic matter and pyrite as a function of time for the past ~600 Myr (Figure 12). Weathering rates for sedimentary organic matter and pyrite were calculated as first order dependents on the total mass of sedimentary organic matter or pyrite, respectively. Although highly simplified, this model provided several new insights into global-scale coupling C–S–O geochemistry. First, the broad-scale features of Phanerozoic O_2 evolution were established. O_2 concentrations in the atmosphere were low in the early Paleozoic, rising to some elevated p_{O_2} levels during the Carboniferous and Permian (probably to a concentration substantially greater than today's 0.21 bar), and then falling through the Mesozoic and Cenozoic to more modern values. This model confirmed the suspicion that in contrast with the Precambrian, Phanerozoic O_2 evolution was a story of relative stability through time, with no great excursions in p_{O_2}. Second, by linking C–S–O cycles with sediments and specifically sedimentation rates, this model helped fortify the idea that a strong control on organic matter burial rates globally, and thus ultimately on release of O_2 to the atmosphere, may be rates of sedimentation in near-shore environments. These authors extended this idea to propose that the close linkage between sedimentation and erosion (i.e., the fact that global rates of sedimentation are matched nearly exactly to global rates of sediment production—in other words, erosion) may in fact be a stabilizing influence on atmospheric O_2 fluctuations. If higher sedimentation rates result in greater burial of organic matter and pyrite, and greater release of O_2 to the atmosphere, at the same time there will be greater rates of erosion on the continents, some of which will involve oxidative weathering of ancient organic matter and/or pyrite.

The other principal approach towards modeling the Phanerozoic evolution of atmospheric O_2 rests on the isotope systematics of the carbon and sulfur geochemical cycles. The significant isotope discriminations associated with biological fixation of CO_2 to generate biomass and with bacterial reduction of sulfate to sulfide have been mentioned several times previously in this chapter. Given a set of simplifications of the exogenic cycles of carbon, sulfur and oxygen, these isotopic discriminations and the isotopic composition of seawater through time ($\delta^{13}C$, $\delta^{34}S$) can be used to estimate global rates of burial and weathering of organic matter, sedimentary carbonates, pyrite, and evaporative sulfates (Figure 13).

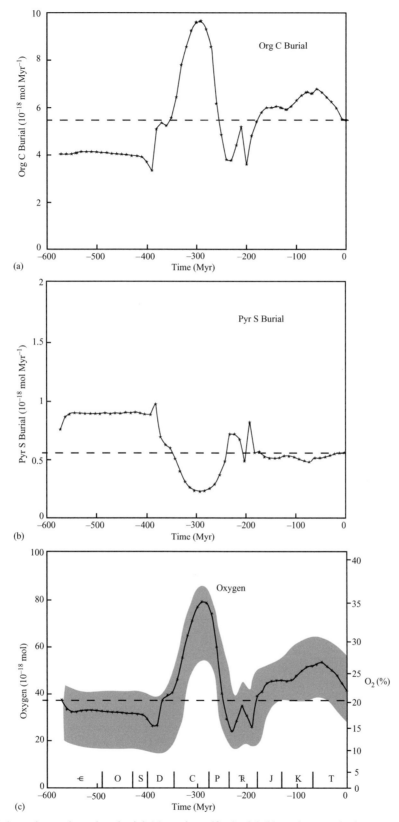

Figure 12 Estimated organic carbon burial (a), pyrite sulfur burial (b), and atmospheric oxygen concentrations (c) through Phanerozoic time, derived from estimates of rock abundance and their relative organic carbon and sulfide content (source Berner and Canfield, 1989).

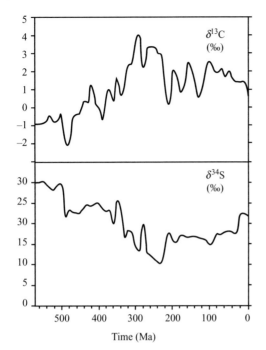

Figure 13 Globally averaged isotopic composition of carbonates (δ^{13}C) and sulfates (δ^{34}S) through Phanerozoic time (source Lindh, 1983).

(i) The first required simplification is that the total mass of exogenic carbon is constant through time (carbon in the oceans, atmosphere, and sedimentary rocks). This of course neglects inputs of carbon and sulfur from volcanic activity and metamorphic degassing, and outputs into the mantle at subduction zones. However, if these fluxes into and out of the exogenic cycle are small enough (or have no effect on bulk crustal carbon or sulfur isotopic composition), then this simplification may be acceptable.

(ii) The second simplification is that the total mass of carbon and sulfur dissolved in seawater plus the small reservoir of atmospheric carbon and sulfur gases remains constant through time. For carbon, this may be a realistic simplification. Dissolved inorganic carbon in seawater is strongly buffered by carbonate mineral precipitation and dissolution, and thus it is unlikely that extensive regions of the world ocean could have become significantly enriched or depleted in inorganic carbon during the geologic past. For sulfur, this assumption may not be completely accurate. Much of the interpretation of sulfur isotope records (with implications for atmospheric O_2 evolution) in the Precambrian depend on varying, but generally low dissolved sulfate concentrations. Unlike carbonate, there is no great buffering reaction maintaining stable sulfate concentrations in seawater. And while the sources of sulfate to seawater have likely varied only minimally, with changes in the sulfate flux to seawater increasing or decreasing smoothly through time as the result of broad-scale

tectonic activity and changes in bulk continental weathering rates, removal of sulfate through excessive BSR or rapid evaporate formation may be much more episodic through time, possibly resulting in fairly extensive shifts in seawater sulfate concentration, even during the Phanerozoic. Nonetheless, using this simplification allows us to establish that for C–S–O geochemical models, total weathering fluxes for carbon and sulfur must equal total burial fluxes for carbon and sulfur, respectively.

Referring to Figure 14, we can see that these simplifications allow us to say that

$$dM_{oc}/dt = F_{wg} + F_{wc} - F_{bg} - F_{bc} = 0 \quad (11)$$

$$dM_{os}/dt = F_{ws} + F_{wp} - F_{bs} - F_{bp} = 0 \quad (12)$$

The full rate equations for each reservoir mass in Figure 14 are as follows:

$$dM_c/dt = F_{bc} - F_{wc} \quad (13)$$

$$dM_g/dt = F_{bg} - F_{wg} \quad (14)$$

$$dM_s/dt = F_{bs} - F_{ws} \quad (15)$$

$$dM_p/dt = F_{bp} - F_{wp} \quad (16)$$

$$dM_{O_2}/dt = F_{bg} - F_{wg} + 15/8(F_{bp} - F_{wp}) \quad (17)$$

This system of equations has four unknowns: two burial fluxes and two weathering fluxes.

(iii) At this point, a third simplification of the carbon and sulfur systems is often applied to the weathering fluxes of sedimentary rocks. As a first approach, it is not unreasonable to guess that the rate of weathering of a given type of rock relates in some sense to the total mass of that rock type available on Earth's surface. If that relation is assumed to be first order with respect to rock mass, an artificial weathering rate constant for each rock reservoir can be derived. Such constants have been derived by assuming that the weathering rate equation has the form $F_{wi} = k_i M_i$. If we can establish the mass of a sedimentary rock reservoir i, and also the average global river flux to the oceans due to weathering of reservoir i, then k_i can easily be calculated. For example, if the total global mass of carbonate in sedimentary rock is 5000×10^{18} mol C, and annually there are 20×10^{12} mol C discharged from rivers to the oceans from carbonate rock weathering, then $k_{carbonate}$ becomes $(20 \times 10^{12}$ mol C yr$^{-1})/$ $(5,000 \times 10^{18}$ mol C), i.e., 4×10^{-3} Myr^{-1}. These simple first-order weathering rate constants have been calculated for each sedimentary rock reservoir in the C–S–O cycle, derived entirely from estimated preanthropogenic carbon and sulfur fluxes from continental weathering. Lack of a true phenomenological relationship relating microscale and outcrop-scale rock weathering reactions to regional- and global-scale carbon and sulfur fluxes remains one of the primary

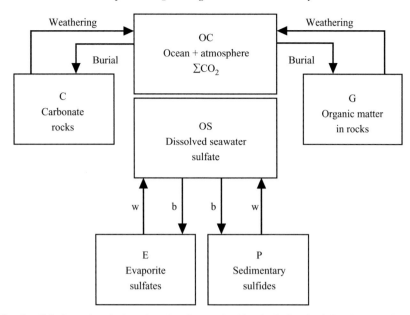

Figure 14 The simplified geochemical cycles of carbon and sulfur, including burial and weathering of sedimentary carbonates, organic matter, evaporites, and sulfides. The relative fluxes of burial and weathering of organic matter and sulfide minerals plays a strong role in controlling the concentration of atmospheric O$_2$.

weaknesses limiting the accuracy of numerical models of the coupled C–S–O geochemical cycles.

If weathering fluxes from eroding sedimentary rocks are independently established, such as through use of mass-dependent weathering fluxes, then all of the weathering fluxes in our system of equations above become effectively known. This leaves only the burial fluxes as unknowns in solving the evolution of the C–S–O system.

The exact isotope discrimination that occurs during photosynthesis and during BSR is dependent on many factors. For carbon, these include cell growth rate, geometry and nutrient availability (Rau *et al.*, 1989, 1992), CO$_2$ availability species-specific effects, modes of CO$_2$ sequestration (i.e., C$_3$, C$_4$, and CAM plants). For sulfur, these can include species–specific effects, the degree of closed-system behavior and sulfate concentration, sulfur oxidation and disproportionation. However, as a simplification in geochemical modeling of the C–S–O cycles, variability in carbon and sulfur isotopic fractionation is limited. In the simplest case, fractionations are constant through all time for all environments. As a result, for example, the isotopic composition of organic matter buried at a given time is set at a constant 25‰ depletion relative to seawater dissolved carbonate at that time, and pyrite isotopic composition is set at a constant 35‰ depletion relative to seawater sulfate. Of course, in reality, α_c and α_s (the isotopic discriminations assigned between inorganic–organic carbon and sulfate–sulfide, respectively) vary greatly in both time and space.

Regardless of how α_c and α_s are set, however, once fractionations have been defined, the mass balance equations given in (13)–(17) above can be supplemented with isotope mass balances as well.

Based on the first simplification listed above, the exogenic cycles of carbon and sulfur are regarded as closed systems. As such, the bulk isotopic composition of average exogenic carbon and sulfur do not vary through time. We can write a rate equation for the rate of change in (mass×isotopic composition) of each reservoir in Figure 14 to reflect this isotope mass balance. For example:

$$
\begin{aligned}
\frac{d(\delta_{oc}M_{oc})}{dt} &\equiv \frac{\delta_{oc}dM_{oc}}{dt} + \frac{M_{oc}d\delta_{oc}}{dt} \\
&= \delta_c F_{wc} + \delta_c F_{wg} - \delta_{oc}F_{bg} \\
&\quad -(\delta_{oc} - \alpha_c)F_{bg}
\end{aligned} \tag{18}
$$

Using the simplification that $dM_{oc}/dt = 0$, Equation (18) reduces to an equation relating the organic matter burial flux F_{bg} in terms of other known entities:

$$
\begin{aligned}
F_{bg} = \left(\frac{1}{\alpha_c}\right) &\left[M_{oc}\left(\frac{d\delta_{oc}}{dt}\right) + F_{wc}(\delta_{oc} - \delta_c) \right. \\
&\left. + F_{wg}(\delta_{oc} - \delta_g) \right]
\end{aligned} \tag{19}
$$

Using (11) above,

$$
F_{bc} = F_{wg} + F_{wc} - F_{bg} \tag{20}
$$

A similar pair of equations can be written for the sulfur system:

$$F_{bp} = \left(\frac{1}{\alpha_s}\right)\left[M_{os}\left(\frac{d\delta_{os}}{dt}\right) + F_{ws}(\delta_{os} - \delta_s)\right.$$

$$\left. + F_{wp}(\delta_{os} - \delta_p)\right] \tag{21}$$

$$F_{bs} = F_{wp} + F_{ws} - F_{bp} \tag{22}$$

Full rate equations for the rate of change in sedimentary rock reservoir isotopic composition can be written as

$$\frac{d\delta_c}{dt} = \frac{F_{bc}(\delta_{oc} - \delta_c)}{M_c} \tag{23}$$

$$\frac{d\delta_g}{dt} = \frac{F_{bg}(\delta_{oc} - \alpha_c - \delta_g)}{M_g} \tag{24}$$

$$\frac{d\delta_s}{dt} = \frac{F_{bw}(\delta_{os} - \delta_s)}{M_s} \tag{25}$$

$$\frac{d\delta_p}{dt} = \frac{F_{bp}(\delta_{os} - \alpha_s - \delta_p)}{M_p} \tag{26}$$

as well as the rate of change in the mass of O_2. The isotopic composition of seawater carbonate and sulfate through time comes directly from the sedimentary rock record. However, it is uncertain how to average globally as well as over substantial period of geologic time. The rates of change in seawater carbonate and sulfate isotopic compositions are fairly small terms, and have been left out of many modeling efforts directed at the geochemical C–S–O system; however, for completeness sake these terms are included here.

One implication of the isotope-driven modeling approach to understanding the C–S–O coupled cycle is that burial fluxes of organic matter and pyrite (which are the sources of atmospheric O_2 in these models) are nearly proportional to seawater isotopic composition of inorganic carbon and sulfate. Recalling the equation for organic matter burial above, it is noted that because sedimentary carbonate and organic matter mass are nearly constant through time, weathering fluxes do not vary greatly through time. Likewise, the average isotopic composition of carbonates and organic matter are also nearly constant, as is the mass of dissolved carbonate and the rate of change in dissolved inorganic carbon isotopic composition. Assuming that F_{wc}, F_{wg}, M_{oc}, $d\delta_{oc}/dt$, δ_c, and δ_g are constant or nearly constant, F_{bg} becomes linearly proportional to the isotopic composition of seawater dissolved inorganic carbon. The same rationale applies for the sulfur system, with pyrite burial becoming linearly proportional to the isotopic composition of seawater sulfate. Although these relationships are not strictly true, even within the constraints of the model simplifications, these relationships provide a useful guide for evaluating changes in organic matter and

pyrite burial fluxes and the impact these have on atmospheric O_2, simply by examining the isotopic records of marine carbonate and sulfate.

Early efforts to model the coupled C–S–O cycles yielded important information. The work of Garrels and Lerman (1984) showed that the exogenic C and S cycles can be treated as closed systems over at least Phanerozoic time, without exchange between sedimentary rocks and the deep "crust + mantle." Furthermore, over timescales of millions of years, the carbon and sulfur cycles were seen to be closely coupled, with increase in sedimentary organic carbon mass matched by loss of sedimentary pyrite (and vice versa). Other models explored the dynamics of the C–S–O system. One important advance was promoted by Kump and Garrels (1986). In these authors' model, a steady-state C–S–O system was generated and perturbed by artificially increasing rates of organic matter burial. These authors tracked the shifts in seawater carbon and sulfur isotopic composition that resulted, and compared these results with the true sedimentary record. Importantly, these authors recognized that although there is a general inverse relationship between seawater $\delta^{13}C$ and $\delta^{34}S$, the exact path along an isotope–isotope plot through time is not a straight line. Instead, because of the vastly different residence times of sulfate versus carbonate in seawater, any changes in the C–S–O system are first expressed through shifts in carbon isotopes, then sulfur isotopes (Figure 15). The authors also pointed out large-scale divisions in C–S isotope coupling through Phanerozoic time. In the Paleozoic, organic matter and pyrite burial were closely coupled (largely because the same types of depositional environment favor burial of marine organic matter and pyrite). During this time, seawater carbonate and sulfur isotopes co-varied positively, indicating concomitant increases (or decreases) in burial of organic matter and pyrite. During the late Paleozoic, Mesozoic, and Cenozoic, terrestrial depositional environments became important settings for burial of organic matter. Because pyrite formation and burial in terrestrial environments is extremely limited, organic matter and pyrite burial became decoupled at this time, and seawater carbonate and sulfur isotopes co-varied negatively, indicating close matching of increased sedimentary organic matter with decreased sedimentary pyrite (and vice versa), perhaps suggesting a net balance in O_2 production and consumption, and maintenance of nearly constant, equable p_{O_2} throughout much of the latter half of the Phanerozoic. The model of Berner (1987) introduced the concept of rapid recycling: the effort to numerically represent the observation that younger sedimentary rocks are more likely to be eroded and weathered than are older sedimentary rocks. Because young sedimentary rocks are likely to

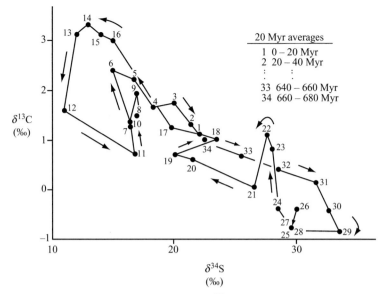

Figure 15 Twenty-million-year average values of seawater δ^{34}S plotted against concomitant carbonate δ^{13}C for the last 700 Myr (source Kump and Garrels, 1986).

be isotopically distinct from older rocks (because they are recording any recent shifts in seawater carbon or sulfur isotopic composition), restoring that isotopically distinct carbon or sulfur more quickly back into seawater provides a type of negative feedback, dampening excessively large or small burial fluxes required for isotope mass balance. This negative feedback serves to reduce calculated fluctuations in organic matter and pyrite burial rates, which in turn reduce fluctuations in release of O_2 to the atmosphere. Results from this study also predict large increases in OM burial fluxes during the Permocarboniferous (~300 Ma) above values present earlier in the Paleozoic. This increase, likely associated with production and burial of refractory terrigenous organic matter (less easily degraded than OM produced by marine organisms), led to elevated concentrations of O_2 ~300 Ma.

One flaw with efforts to model the evolution of Phanerozoic O_2 using the carbon and sulfur isotope records is that unreasonably large fluctuations in organic matter and pyrite burial fluxes (with coincident fluctuations in O_2 production rates) would result. Attempts to model the whole Phanerozoic generated unreasonably low and high O_2 concentrations for several times in the Phanerozoic (Lasaga, 1989), and applications of what seemed to be a realistic feedback based on reality (weathering rates of sedimentary organic matter and pyrite dependent on the concentration of O_2) were shown to actually become positive feedbacks in the isotope-driven C–S–O models (Berner, 1987; Lasaga, 1989).

Phosphorus is a key nutrient limiting primary productivity in many marine environments. If phosphorus supply is increased, primary production and perhaps organic matter burial will also

increase. Degradation and remineralization of OM during transit from surface waters into sediments liberates phosphorus, but most of this is quickly scavenged by adsorption onto the surfaces of iron oxyhydroxides. However, work by Ingall and Jahnke (1994) and Van Cappellen and Ingall (1996) has shown that phosphorus recycling and release into seawater is enhanced under low O_2 or anoxic conditions. This relationship provides a strong negative feedback between primary production, bottom water anoxia, and atmospheric O_2. As atmospheric O_2 rises, phosphorus scavenging on ferric-iron is enhanced, phosphorus recycling back into surface waters is reduced, primary production rates are reduced, and O_2 declines. If O_2 concentrations were to fall, phosphorus scavenging onto ferric-iron would be inhibited, phosphorus recycling back into surface waters would be accelerated, fueling increased primary production and O_2 release to the atmosphere. Van Cappellen and Ingall (1996) applied these ideas to a mathematical model of the C–P–Fe–O cycle to show how O_2 concentrations could be stabilized by phosphorus recycling rates.

Petsch and Berner (1998) expanded the model of Van Cappellen and Ingall (1996) to include the sulfur system, as well as carbon and sulfur isotope effects. This study examined the response of the C–S–O–Fe–P system, and in particular carbon and sulfur isotope ratios, to perturbation in global ocean overturn rates, changes in continental weathering, and shifts in the locus of organic matter burial from marine to terrestrial depocenters. Confirming the idea promoted by Kump and Garrels (1986), these authors showed that perturbations of the exogenic C–S–O cycle result in shifts in seawater carbon and sulfur isotopic composition similar in amplitude and duration to observed isotope excursions in the sedimentary record.

Other proposed feedbacks stabilizing the concentration of atmospheric O_2 over Phanerozoic time include a fire-regulated PO_4 feedback (Kump, 1988, 1989). Terrestrial primary production requires much less phosphate per mole CO_2 fixed during photosynthesis than marine primary production. Thus, for a given supply of PO_4, much more CO_2 can be fixed as biomass and O_2 released from photosynthesis on land versus in the oceans. If terrestrial production proceeds too rapidly, however, p_{O_2} levels may rise slightly and lead to increased forest fires. Highly weatherable, PO_4-rich ash would then be delivered through weathering and erosion to the oceans. Primary production in the oceans would lead to less CO_2 fixed and O_2 released per mole of PO_4.

Hydrothermal reactions between seawater and young oceanic crust have been proposed as an influence on atmospheric O_2 (Walker, 1986; Carpenter and Lohmann, 1999; Hansen and Wallmann, 2002). While specific periods of oceanic anoxia may be associated with accelerated hydrothermal release of mantle sulfide (i.e., the Mid-Cretaceous, see Sinninghe-Damsté and Köster, 1998), long-term sulfur and carbon isotope mass balance precludes substantial inputs of mantle sulfur to the Earth's surface of a different net oxidation state and mass flux than what is subducted at convergent margins (Petsch, 1999; Holland, 2002).

One recent advance in the study of isotope-driven models of the coupled C–S–O cycles is re-evaluation of isotope fractionations. Hayes et al., (1999) published a compilation of the isotopic composition of inorganic and organic carbon for the past 800 Myr. One feature of this dual record is a distinct shift in the net isotopic distance between carbonate and organic carbon, occurring during the past $\sim$100 million years. When carbon isotope distance is compared to estimates of Cenozoic and Mesozoic p_{O_2}, it becomes apparent that there may be some relationship between isotopic fractionation associated with organic matter production and burial and the concentration of O_2 in the atmosphere. The physiological underpinning behind this proposed relationship rests on competition between photosynthesis and photorespiration in the cells of photosynthetic organisms. Because O_2 is a competitive inhibitor of CO_2 for attachment to the active site of Rubisco, as ambient O_2 concentrations rise relative to CO_2, so will rates of photorespiration. Photorespiration is a net consumptive process for plants; previously fixed carbon is consumed with O_2 to produce CO_2 and energy. CO_2 produced through photorespiration may diffuse out of the cell, but it is also likely to be taken up (again) for photosynthesis. Thus, in

cells undergoing fairly high rates of photorespiration in addition to photosynthesis, a significant fraction of total CO_2 available for photosynthesis derives from oxidized, previously fixed organic carbon. The effect of this on cellular carbon isotopic composition is that because each round of photosynthesis results in ^{13}C-depletion in cellular carbon relative to CO_2, cells with high rates of photorespiration will contain more ^{13}C-depleted CO_2 and thus will produce more ^{13}C-depleted organic matter.

In controlled-growth experiments using both higher plants and single-celled marine photosynthetic algae, a relationship between ambient O_2 concentration and net isotope discrimination has been observed (Figure 16) (Berner et al., 2000; Beerling et al., 2002). The functional form of this relationship has been expressed in several ways. The simplest is to allow isotope discrimination to vary linearly with changing atmospheric O_2 mass: $\alpha_c = 25 \times (M_{O_2}/38)$. More complicated relationships have also been derived, based on curve-fitting the available experimental data on isotopic fractionation as a function of $[O_2]$. O_2-dependent isotopic fractionation during photosynthesis has provided the first mathematically robust isotope-driven model of the C–S–O cycle consistent with geologic observations (Berner et al., 2000). Results of this model show that allowing isotope fractionation to respond to changes in ambient O_2 provides a strong negative feedback dampening excessive increases or decreases in organic matter burial rates. Rates of organic matter burial in this model are no longer simply dependent on seawater carbonate $\delta^{13}C$, but now also vary with $1/\alpha_c$. As fractionation becomes greater (through

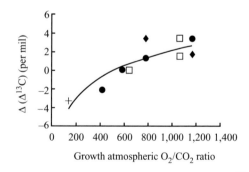

Figure 16 Relationship between change in $\Delta(\Delta^{13}C)$ of vascular land plants determined experimentally in response to growth under different O_2/CO_2 atmospheric mixing ratios. $\Delta(\Delta^{13}C)$ is the change in carbon isotope fractionation relative to fractionation for the controls at present day conditions (21% O_2, 0.036% CO_2). The solid line shows the nonlinear curve fitted to the data, given by $\Delta(\Delta^{13}C) = -19.94 + 3.195 \times \ln(O_2/CO_2)$; ($\bullet$) *Phaseolus vulgaris*; ($\blacklozenge$) *Sinapis alba*; ($\square$) from Berner et al. (2000); (+) from Berry et al. (1972) (Beerling et al., 2002) (reproduced by permission of Elsevier from *Geochem. Cosmochim. Acta* **2002**, *66*, 3757–3767).

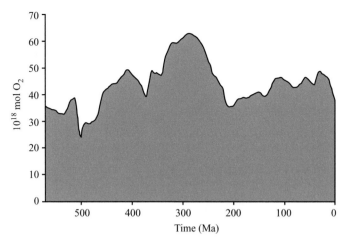

Figure 17 Evolution of the mass of atmospheric O_2 through Phanerozoic time, estimated using an isotope mass balance described in Equations (11)–(26). The model employs the isotope date of Figure 13, and includes new advances in understanding regarding dependence of carbon isotope discrimination during photosynthesis and sulfur isotope discrimination during sulfur disproportionation and organic sulfur formation. The system of coupled differential equations were integrated using an implicit fourth-order Kaps–Rentrop numerical integration algorithm appropriate for this stiff set of equations (sources Petsch, 2000; Berner *et al.*, 2000).

elevated O_2), less of an increase in organic matter burial rates is required to achieve the observed increase in seawater $\delta^{13}C$ than if α_c were constant.

The same mathematical argument can be applied to sulfate–sulfide isotope fractionation during BSR. As O_2 concentrations increase, so does sulfur isotope fractionation, resulting in a strong negative feedback on pyrite burial rates. This is consistent with the broad-scale changes in sulfur isotope dynamics across the Proterozoic, reflecting a large increase in $\Delta^{34}S$ (between sulfate and sulfide) when atmospheric O_2 concentrations were great enough to facilitate bacterial sulfide oxidation and sulfur disproportionation. Perhaps during the Phanerozoic, when O_2 concentrations were greater, sulfur recycling (sulfate to sulfide through BSR, sulfur oxidation, and sulfur disproportionation) was increased, resulting in greater net isotopic distance between sulfate and sulfide. Another means of changing net sulfur isotope discrimination in response to O_2 may be the distribution of reduced sulfur between sulfide minerals and organic matter-associated sulfides. Work by Werne *et al.*, 2000, 2003) has shown that organic sulfur is consistently ~10 ‰ enriched in ^{34}S relative to associated pyrite. This is believed to result from different times and locations of organic sulfur versus pyrite formation. While pyrite may form in shallow sediments or even anoxic portions of the water column, reflecting extreme sulfur isotope depletion due to several cycles of BSR, sulfide oxidation and sulfur disproportion-ation, organic matter is sulfurized within the sediments. Closed, or nearly closed, system behavior of BSR in the sediments results in late-stage sulfide (the source of sulfur in sedimentary organic matter) to be more enriched compared with pyrite in the

same sediments. It is known that burial of sulfide as organic sulfur is facilitated in low O_2 or anoxic waters. If lower atmospheric O_2 in the past encouraged development of more extensive anoxic basins and increased burial of sulfide as organic sulfur instead of pyrite, the 10‰ offset between pyrite and organic sulfur would become effectively a change in net sulfur fractionation in response to O_2.

Applying these newly recognized modifications of carbon and sulfur isotope discrimination in response to changing O_2 availability has allowed development of new numerical models of the evolution of the coupled C–S–O systems and variability of Phanerozoic atmospheric O_2 concentration (Figure 17).

15.7 CONCLUSIONS

Molecular oxygen is generated and consumed by a wide range of processes. The net cycling of O_2 is influenced by physical, chemical, and most importantly, biological processes acting on and beneath the Earth's surface. The exact distribution of O_2 concentrations depends on the specific interplay of these processes in time and space. Large inroads have been made towards understanding the processes that control the concentration of atmospheric O_2, especially regarding O_2 as a component of coupled biogeochemical cycles of many elements, including carbon, sulfur, nitrogen, phosphorus, iron, and others.

Earth's modern oxygenated atmosphere is the product of over four billion years of its history (Figure 18). The early anoxic atmosphere was slowly oxidized (although not oxygenated) as the result of slow H_2 escape. Evolution of oxygenic photosynthesis accelerated the oxidation

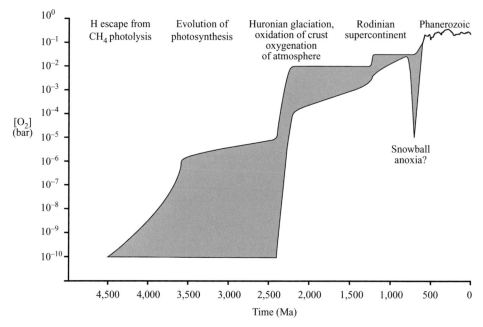

Figure 18 Composite estimate of the evolution of atmospheric oxygen through 4.5 Gyr of Earth's history. Irreversible oxidation of the Earth resulted from CH_4 photolysis and hydrogen escape during early Earth's history. Evolution of oxygenic photosynthesis preceded substantial oxygenation of the atmosphere by several hundred million years. Relative stasis in atmospheric p_{O_2} typified much of the Proterozoic, with a possible pulse of oxygenation associated with formation of the Rodinian supercontinent in the Late Mesoproterozoic, and possible return to anoxia associated with snowball glaciation in the Neoproterozoic (sources Catling *et al.*, 2001; Kasting, 1992; Rye and Holland, 1998; Petsch, 2000; Berner *et al.*, 2000).

of Earth's crust and atmosphere, such that by ~2.2 Ga a small but significant concentration of O_2 was likely present in Earth's atmosphere. Limited primary production and oxygen production compared with the flux of reduced volcanic gases maintained this low p_{O_2} atmosphere for over one billion years until the Neoproterozoic. Rapid oscillations in Earth's carbon and sulfur cycles associated with global Snowball glaciation may also have expression in a return to atmospheric anoxia at this time, but subsequent to the late Proterozoic isotope excursions, oxygenation of the atmosphere to near-modern concentrations developed such that by the Precambrian–Cambrian boundary, O_2 concentrations were high enough to support widespread skeletonized metazoans. Phanerozoic seawater and atmospheric O_2 concentrations have fluctuated in response to tectonic forcings, generating regional-scale anoxia in ocean basins at certain times when biological productivity and ocean circulation facilitate anoxic conditions, but in the atmosphere, O_2 concentrations have remained within ~0.05–0.35 bar p_{O_2} for the past ~600 Myr.

Several outstanding unresolved gaps in our understanding remain, in spite of a well-developed understanding of the general features of the evolution of atmospheric O_2 through time. These gaps represent potentially meaningful directions for future research, including:

(i) assessing the global importance of mineral oxidation as a mechanism of O_2 consumption;

(ii) the flux of reduced gases from volcanoes, metamorphism, and diffuse mantle/lithosphere degassing;

(iii) the true dependence of organic matter oxidation on availability of O_2, in light of the great abundance of microaerophilic and anaerobic microorganisms utilizing carbon respiration as a metabolic pathway, carbon isotopic evidence suggesting continual and essentially constant organic matter oxidation as part of the sedimentary rock cycle during the entire past four billion years, and the inefficiency of organic matter oxidation during continental weathering;

(iv) stasis in the oxygenation of the atmosphere during the Proterozoic;

(v) contrasting biochemical, fossil, and molecular evidence for the antiquity of the innovation of oxygenic photosynthesis; and

(vi) evaluating the relative strength of biological productivity versus chemical evolution of the Earth's crust and mantle in controlling the early stages of oxygenation of the atmosphere.

Thus, study of the global biogeochemical cycle of oxygen, the component of our atmosphere integral and crucial for life as we know it, remains a fruitful direction for Earth science research.

REFERENCES

Algeo T. J. and Scheckler S. E. (1998) Terrestrial-marine teleconnections in the Devonian: links between the evolution of land plants, weathering processes, and marine anoxic events. *Phil. Trans. Roy. Soc. London B* **353**, 113–128.

Anbar A. D. and Knoll A. H. (2002) Proterozoic ocean chemistry and evolution: a bioinorganic bridge? *Science* **297**, 1137–1142.

Arthur M. A. (2000) Volcanic contributions to the carbon and sulfur geochemical cycles and global change. In *Encyclopedia of Volcanoes* (eds. H. Sigurdsson, B. F. Houghton, S. R. McNutt, H. Rymer, and J. Stix). Academic Press, San Diego, pp. 1045–1056.

Bartley J. K., Semikhatov M. A., Kaufman A. J., Knoll A. H., Pope M. C., and Jacobsen S. B. (2001) Global events across the Mesoproterozoic–Neoproterozoic boundary: C and Sr isotopic evidence from Siberia. *Precamb. Res.* **111**, 165–202.

Battle M., Bender M., Sowers R., Tans P. P., Butler J. H., Elkins J. W., Ellis J. T., Conway T., Zhang N., Pang P., and Clarke A. D. (1996) Atmospheric gas concentrations over the past century measured in air from firn at the South Pole. *Nature* **383**, 231–235.

Baum S. K. and Crowley T. J. (2001) GCM response to late Precambrian (~590 Ma) ice-covered continents. *Geophys. Res. Lett.* **28**, 583–586.

Beaumont V. and Robert F. (1999) Nitrogen isotope ratios of kerogens in Precambrian cherts: a record of the evolution of atmospheric chemistry? *Precamb. Res.* **96**, 63–82.

Beerling D. J. and Berner R. A. (2000) Impact of a Permo–Carboniferous high O_2 event on the terrestrial carbon cycle. *Proc. Natl. Acad. Sci.* **97**, 12428–12432.

Beerling D. J., Lake J. A., Berner R. A., Hickey L. J., Taylor D. W., and Royer D. L. (2002) Carbon isotope evidence implying high O_2/CO_2 ratios in the Permo–Carboniferous atmosphere. *Geochim. Cosmochim. Acta* **66**, 3757–3767.

Bekker A., Kaufman A. J., Karhu J. A., Beukes N. J., Quinten S. D., Coetzee L. L., and Kenneth A. E. (2001) Chemostratigraphy of the Paleoproterozoic Duitschland Formation, South Africa. Implications for coupled climate change and carbon cycling. *Am. J. Sci.* **301**, 261–285.

Bender M., Labeyrie L. D., Raynaud D., and Lorius C. (1985) Isotopic composition of atmospheric O_2 in ice linked with deglaciation and global primary productivity. *Nature* **318**, 349–352.

Bender M., Sowers T., and Labeyrie L. (1994a) The Dole effect and its variations during the last 130,000 years as measured in the Vostok ice core. *Global Biogeochem. Cycle* **8**, 363–376.

Bender M. L., Sowers T., Barnola J.-M., and Chappellaz J. (1994b) Changes in the O_2/N_2 ratio of the atmosphere during recent decades reflected in the composition of air in the firn at Vostok Station, Antarctica. *Geophys. Res. Lett.* **21**, 189–192.

Berner R. A. (1987) Models for carbon and sulfur cycles and atmospheric oxygen: application to Paleozoic geologic history. *Am. J. Sci.* **287**, 177–196.

Berner R. A. (1994) GEOCARB II: a revised model of atmospheric CO_2 over Phanerozoic time. *Am. J. Sci.* **294**, 56–91.

Berner R. A. (2002) Examination of hypotheses for the Permo–Triassic boundary extinction by carbon cycle modeling. *Proc. Natl. Acad. Sci.* **99**, 4172–4177.

Berner R. A. and Canfield D. E. (1989) A new model for atmospheric oxygen over Phanerozoic time. *Am. J. Sci.* **289**, 333–361.

Berner R. A. and Kothavala Z. (2001) GEOCARB III: a revised model of atmospheric CO_2 over Phanerozoic time. *Am. J. Sci.* **301**, 182–204.

Berner R. A., Petsch S. T., Lake J. A., Beerling D. J., Popp B. N., Lane R. S., Laws E. A., Westley M. B., Cassar N., Woodward F. I., and Quick W. P. (2000) Isotope fractionation and atmospheric oxygen: Implications for Phanerozoic O_2 evolution. *Science* **287**, 1630–1633.

Berry J. A., Troughton J. H., and Björkman O. (1972) Effect of oxygen concentration during growth on carbon isotope discrimination in C_3 and C_4 species of *Atriplex. Carnegie Inst. Yearbook* **71**, 158–161.

Beukes N. J., Dorland H., Gutzmer J., Nedachi M., and Ohmoto H. (2002) Tropical laterites, life on land, and the history of atmospheric oxygen in the Paleoproterozoic. *Geology* **30**, 491–494.

Brasier M. D., Green O. R., Jephcoat A. P., Kleppe A. K., van Kranendonk M. J., Lindsay J. F., Steele A., and Grassineau N. V. (2002) Questioning the evidence for Earth's earliest fossils. *Nature* **416**, 76–81.

Brocks J. J., Logan G. A., Buick R., and Summons R. E. (1999) Archean molecular fossils and the early rise of eukaryotes. *Science* **285**, 1033–1036.

Budyko M. I., Ronov A. B., and Yanshin A. L. (1987) *History of the Earth's Atmosphere*. Springer, Berlin.

Buick R. (1990) Microfossil recognition in Archean rocks: an appraisal of spheroids and filaments from a 3500 M. Y. old chert-barite unit at North Pole, Western Australia. *Palaios* **5**, 441–459.

Buick I. S., Uken R., Gibson R. L., and Wallmach T. (1998) High-$\delta^{13}C$ Paleoproterozoic carbonates from the Transvaal Supergroup, South Africa. *Geology* **26**, 875–878.

Butterfield N. J. and Rainbird R. H. (1998) Diverse organic-walled fossils, including possible dinoflagellates, from early Neoproterozoic of arctic Canada. *Geology* **26**, 963–966.

Caldeira K. and Kasting J. F. (1992) Susceptibility of the early Earth to irreversible glaciation caused by carbon dioxide clouds. *Nature* **359**, 226–228.

Canfield D. E. (1998) A new model for Proterozoic ocean chemistry. *Nature* **396**, 450–453.

Canfield D. E. and Teske A. (1996) Late Proterozoic rise in atmospheric oxygen concentration inferred from phylogenetic and sulphur-isotope studies. *Nature* **382**, 127–132.

Canfield D. E., Habicht K. S., and Thamdrup B. (2000) The Archean sulfur cycle and the early history of atmospheric oxygen. *Science* **288**, 658–661.

Carpenter S. J. and Lohmann K. C. (1999) Carbon isotope ratios of Phanerozoic marine cements: re-evaluating global carbon and sulfur systems. *Geochim. Cosmochim. Acta* **61**, 4831–4846.

Catling D. C., Zahnle K. J., and McKay C. P. (2001) Biogenic methane, hydrogen escape, and the irreversible oxidation of the early Earth. *Science* **293**, 839–843.

Cloud P. (1972) A working model of the primitive Earth. *Am. J. Sci.* **272**, 537–548.

Cloud P. E. (1976) Beginnings of biospheric evolution and their biogeochemical consequences. *Paleobiol.* **2**, 351–387.

Condie K. C., Des Marais D. J., and Abbott D. (2001) Precambrian superplumes and supercontinents: a record in black shales, carbon isotopes, and paleoclimates? *Precamb. Res.* **106**, 239–260.

Cope M. J. and Chaloner W. G. (1985) Wildfire: an interaction of biological and physical processes. In *Geological Factors and the Evolution of Plants* (ed. B. H. Tiffney). Yale University Press, New Haven, CT, pp. 257–277.

Delano J. W. (2001) Redox history of the Earth's interior since ~3900 Ma: implications for prebiotic molecules. *Origins Life Evol. Biosphere* **31**, 311–341.

Delmelle P. and Stix J. (2000) Volcanic gases. In *Encyclopedia of Volcanoes* (eds. H. Sigurdsson, B. F. Houghton, S. R. McNutt, H. Rymer, and J. Stix). Academic Press, San Diego, pp. 803–816.

Des Marais D. J., Strauss H., Summons R. E., and Hayes J. M. (1992) Carbon isotope evidence for stepwise oxidation of the Proterozoic environment. *Nature* **359**, 605–609.

Deuser W. G. (1975) *Chemical Oceanography*. Academic Press, Orlando, FL, p. 3.

Doolittle R. F., Feng D. F., Tsang S., Cho G., and Little E. (1996) Determining divergence times of the major kingdoms of living organisms with a protein clock. *Science* **271**, 470–477.

Dudley R. (1998) Atmospheric oxygen, giant Paleozoic insects and the evolution of aerial locomotor performance. *J. Exp. Biol.* **201**, 1043–1050.

Dudley R. (2000) The evolutionary physiology of animal flight: Paleobiological and present perspectives. *Ann. Rev. Phys.* **62**, 135–155.

Evans D. A. D. (2000) Stratigraphic, geochronological and paleomagnetic constraints upon the Neoproterozoic climatic paradox. *Am. J. Sci.* **300**, 347–433.

Farquhar J., Bao H., and Thiemens M. (2000) Atmospheric influence of Earth's earliest sulfur cycle. *Science* **289**, 756–758.

Farquhar J., Wing B. A., McKeegan K. D., Harris J. W., Cartigny P., and Thiemens M. H. (2002) Mass-independent sulfur of inclusions in diamond and sulfur recycling on early Earth. *Science* **297**, 2369–2372.

Garcia-Pichel F. (1998) Solar ultraviolet and the evolutionary history of cyanobacteria. *Origins Life Evol. Biosphere* **28**, 321–347.

Garrels R. M. and Lerman A. (1984) Coupling of the sedimentary sulfur and carbon cycles—an improved model. *Am. J. Sci.* **284**, 989–1007.

Gilbert D. L. (1996) Evolutionary aspects of atmospheric oxygen and organisms. In *Environmental Physiology: 2* (eds. M. J. Fregly and C. M. Blatteis). Oxford University Press, Oxford, UK, pp. 1059–1094.

Glasspool I. (2000) A major fire event recorded in the mesofossils and petrology of the late Permian, Lower Whybrow coal seam, Sydney Basin, Australia. *Palaeogeogr. Palaeoclimat. Palaeoecol.* **164**, 373–396.

Godderís Y. and Veizer J. (2000) Tectonic control of chemical and isotopic composition of ancient oceans: the impact of continental growth. *Am. J. Sci.* **300**, 434–461.

Gould S. J. (1995) Of it and not above it. *Nature* **377**, 681–682.

Graham J. B., Dudley R., Anguilar N., and Gans C. (1995) Implications of the late Paleozoic oxygen pulse for physiology and evolution. *Nature* **375**, 117–120.

Habicht K. S., Gade M., Thamdrup B., Berg P., and Canfield D. E. (2002) Calibration of sulfate levels in the Archean ocean. *Science* **298**, 2372–2374.

Haggerty S. E. and Toft P. B. (1985) Native iron in the continental lower crust: petrological and geophysical implications. *Science* **229**, 647–649.

Halmer M. M., Schmincke H.-U., and Graf H.-F. (2002) The annual volcanic gas input into the atmosphere, in particular into the stratosphere: a global data set for the past 100 years. *J. Volcanol. Geotherm. Res.* **115**, 511–528.

Hansen K. W. and Wallmann K. (2003) Cretaceous and Cenozoic evolution of seawater composition, atmospheric O_2 and CO_2: a model perspective. *Am. J. Sci.* **303**, 94–148.

Hartnett H. E., Keil R. G., Hedges J. I., and Devol A. H. (1998) Influence of oxygen exposure time on organic carbon preservation in continental margin sediments. *Nature* **391**, 572–574.

Hayes J. M., Strauss H., and Kaufman A. J. (1999) The abundance of ^{13}C in marine organic matter and isotopic fractionation in the global biogeochemical cycle of carbon during the past 800 Myr. *Chem. Geol.* **161**, 103–125.

Herbert T. D. and Sarmiento J. L. (1991) Ocean nutrient distribution and oxygenation: limits on the formation of warm saline bottom water over the past 91 m.y. *Geology* **19**, 702–705.

Hinkle M. E. (1994) Environmental conditions affecting the concentrations of He, CO_2, O_2 and N_2 in soil gases. *Appl. Geochem.* **9**, 53–63.

Hoehler T. M., Bebout B. M., and Des Marais D. J. (2001) The role of microbial mats in the production of reduced gases on the early Earth. *Nature* **412**, 324–327.

Hoffman P. F. and Schrag D. P. (2002) The snowball Earth hypothesis: testing the limits of global change. *Terra Nova* **14**, 129–155.

Hoffman P. F., Kaufman A. J., Halverson G. P., and Schrag D. P. (1998) A Neoproterozoic snowball Earth. *Science* **281**, 1342–1346.

Hofmann H. J., Gery K., Hickman A. H., and Thorpe R. I. (1999) Origin of 2.45 Ga coniform stromatolites in Warrawoona Group, Western Australia. *Geol. Soc. Am. Bull.* **111**, 1256–1262.

Holland H. D. (1978) *The Chemistry of the Atmosphere and Oceans.* Wiley, New York.

Holland H. D. (1984) *The Chemical Evolution of the Atmosphere and Oceans.* Princeton University Press, Princeton, NJ.

Holland H. D. (2002) Volcanic gases, black smokers and the great oxidation event. *Geochim. Cosmochim. Acta* **66**, 3811–3826.

Holland H. D., Feakes C. R., and Zbinden E. A. (1989) The Flin Flon paleosol and the composition of the atmosphere 1.8 BYBP. *Am. J. Sci.* **289**, 362–389.

Hotinski R. M., Bice K. L., Kump L. R., Najjar R. G., and Arthur M. A. (2001) Ocean stagnation and End-Permian anoxia. *Geology* **29**, 7–10.

Hurtgen M. T., Arthur M. A., Suits N. S., and Kaufman A. J. (2002) The sulfur isotopic composition of Neoproterozoic seawater sulfate: implications for a snowball Earth? *Earth Planet. Sci. Lett.* **203**, 413–429.

Ingall E. and Jahnke R. (1994) Evidence for enhanced phosphorus regeneration from marine sediments overlain by oxygen depleted waters. *Geochim. Cosmochim. Acta* **58**, 2571–2575.

Ingmanson D. E. and Wallace W. J. (1989) *Oceanography: An Introduction.* Wadworth, Belmont, CA, pp. 99.

Joachimski M. M., Ostertag-Henning C., Pancost R. D., Strauss H., Freeman K. H., Littke R., Sinninghe-Damsté J. D., and Racki G. (2001) Water column anoxia, enhanced productivity and concomitant changes in $\delta^{13}C$ and $\delta^{34}S$ across the Frasnian–Famennian boundary (Kowala—Holy Cross Mountains/Poland). *Chem. Geol.* **175**, 109–131.

Kah L. C., Lyons T. W., and Chesley J. T. (2001) Geochemistry of a 1.2 Ga carbonate-evaporite succession, northern Baffin and Bylot Islands: implications for Mesoproterozoic marine evolution. *Precamb. Res.* **111**, 203–234.

Kaiho K., Kajiwara Y., Nakano T., Miura Y., Kawahata H., Tazaki K., Ueshima M., Chen Z., and Shi G. (2001) End-Permian catastrophe by a bolide impact: evidence of a gigantic release of sulfur from the mantle. *Geology* **29**, 815–818.

Karhu J. and Holland H. D. (1996) Carbon isotopes and the rise of atmospheric oxygen. *Geology* **24**, 867–870.

Kasting J. F. (1987) Theoretical constraints on oxygen and carbon dioxide concentrations in the Precambrian atmosphere. *Precamb. Res.* **34**, 205–229.

Kasting J. F. (1992) Models relating to Proterozoic Atmospheric and Oceanic Chemistry. In *The Proterozoic Biosphere* (eds. J. W. Schopf and C. Klein). Cambridge University Press, Cambridge, pp. 1185–1187.

Kasting J. F. and Siefert J. L. (2002) Life and the evolution of Earth's atmosphere. *Science* **296**, 1066–1068.

Kasting J. F., Zahnle K. J., and Walker J. C. G. (1983) Photochemistry of methane in the Earth's early atmosphere. *Precamb. Res.* **20**, 121–148.

Kasting J. F., Eggler D. H., and Raeburn S. P. (1993) Mantle redox evolution and the oxidation state of the Archean atmosphere. *J. Geol.* **101**, 245–257.

Kasting J. F., Pavlov A. A., and Siefert J. L. (2001) A coupled ecosystem-climate model for predicting the methane concentration in the Archean atmosphere. *Orig. Life Evol. Biosph.* **31**, 271–285.

Kaufman A. J. and Knoll A. H. (1995) Neoproterozoic variations in the C-isotopic composition of seawater: stratigraphic and biogeochemical implications. *Precamb. Res.* **73**, 27–49.

Kaufman A. J., Hayes J. M., Knoll A. H., and Germs G. J. B. (1991) Isotopic composition of carbonates and organic carbon from upper Proterozoic successions in Namibia: stratigraphic variation and the effects of diagenesis and metamorphism. *Precamb. Res.* **49**, 301–327.

Keeling R. F. and Shertz S. R. (1992) Seasonal and interannual variations in atmospheric oxygen and implications for the carbon cycle. *Nature* **358**, 723–727.

Keeling R. F., Najjar R. P., Bender M. L., and Tans P. P. (1993) What atmospheric oxygen measurements can tell us about the global carbon cycle. *Global Biogeochem. Cycles* **7**, 37–67.

Keller C. K. and Bacon D. H. (1998) Soil respiration and georespiration distinguished by transport analyses of vadose CO_2, $^{13}CO_2$, and $^{14}CO_2$. *Global Biogeochem. Cycles* **12**, 361–372.

Kiehl J. T. and Dickinson R. E. (1987) A study of the radiative effects of enhanced atmospheric CO_2 and CH_4 on early Earth surface temperatures. *J. Geophys. Res.* **92**, 2991–2998.

Kimura H. and Watanabe Y. (2001) Oceanic anoxia at the Precambrian–Cambrian boundary. *Geology* **29**, 995–998.

Kirschvink J. L. (1992) Late Proterozoic low-latitude global glaciation: the snowball earth. In *The Proterozoic Biosphere* (eds. J. W. Schopf and C. Klein). Cambridge University Press, Cambridge, pp. 51–52.

Kump L. R. (1988) Terrestrial feedback in atmospheric oxygen regulation by fire and phosphorus. *Nature* **335**, 152–154.

Kump L. R. (1989) Chemical stability of the atmosphere and ocean. *Palaeogeogr. Palaeoclimat. Paleoecol.* **75**, 123–136.

Kump L. R. and Garrels R. M. (1986) Modeling atmospheric O_2 in the global sedimentary redox cycle. *Am. J. Sci.* **286**, 337–360.

Kump L. R., Kasting J. F., and Barley M. E. (2001) Rise of atmospheric oxygen and the upside down Archean mantle. *Geochem. Geophys. Geosys.* **2** No. 2000 GC 000114.

Lasaga A. C. (1989) A new approach to isotopic modeling of the variation in atmospheric oxygen through the Phanerozoic. *Am. J. Sci.* **289**, 411–435.

Lasaga A. C. and Ohmoto H. (2002) The oxygen geochemical cycles: dynamics and stability. *Geochim. Cosmochim. Acta* **66**, 361–381.

Lenton T. M. and Watson A. J. (2000) Redfield revisited 2. What regulates the oxygen content of the atmosphere? *Global Biogeochem. Cycles* **14**, 249–268.

Lindh T. B. (1983) Temporal variations in ^{13}C, ^{34}S and global sedimentation during the Phanerozoic. MS Thesis, University of Miami.

Logan G. A., Hayes J. M., Hleshima G. B., and Summons R. E. (1995) Terminal Proterozoic reorganization of biogeochemical cycles. *Nature* **376**, 53–56.

Lovell J. S. (2000) Oxygen and carbon dioxide in soil air. In *Handbook of Exploration Geochemistry*, Geochemical Remote Sensing of the Subsurface (ed. M. Hale). Elsevier, Amsterdam, vol. 7, pp. 451–469.

McMenamin M. A. S. and McMenamin D. L. S. (1990) *The Emergence of Animals—The Cambrian Breakthrough*. Columbia University Press, New York.

Mörner N.-A. and Etiope G. (2002) Carbon degassing from the lithosphere. *Global. Planet. Change* **33**, 185–203.

Murakami T., Utsunomiya S., Imazu Y., and Prasad N. (2001) Direct evidence of late Archean to early Proterozoic anoxic atmosphere from a product of 2.5 Ga old weathering. *Earth Planet. Sci. Lett.* **184**, 523–528.

Najjar R. G. and Keeling R. F. (2000) Mean annual cycle of the air–sea oxygen flux: a global view. *Global Biogeochem. Cycles* **14**, 573–584.

Nutman A. P., Mojzsis S. J., and Friend C. R. L. (1997) Recognition of $\geqq 3850$ Ma water-lain sediments in West Greenland and their significance for the early Archean Earth. *Geochim. Cosmochim. Acta* **61**, 2475–2484.

Ohmoto H. (1996) Evidence in pre-2.2 Ga paleosols for the early evolution of atmospheric oxygen and terrestrial biota. *Geology* **24**, 1135–1138.

Ohmoto H. (1999) Redox state of the Archean atmosphere: evidence from detrital heavy minerals in ca. 3250–2750

Ma sandstones from the Pilbara Craton, Australia: comment and reply. *Geology* **27**, 1151–1152.

Olson J. M. and Pierson B. K. (1986) Photosynthesis 3.5 thousand million years ago. *Photosynth. Res.* **9**, 251–259.

Pan Y. and Stauffer M. R. (2000) Cerium anomaly and Th/U fractionation in the 1.85 Ga Flin Flon Paleosol: clues from REE- and U-rich accessory minerals and implications for paleoatmospheric reconstruction. *Am. Mineral.* **85**, 898–911.

Pavlov A. A., Kasting J. F., Brown L. L., Rages K. A., and Freedman R. (2000) Greenhouse warming by CH_4 in the atmosphere of early Earth. *J. Geophys. Res.* **105**, 11981–11990.

Petsch S. T. (1999) Comment on Carpenter and Lohmann (1999). *Geochim Cosmochim Acta* **63**, 307–310.

Petsch S. T. (2000) A study on the weathering of organic matter in black shales and implications for the geochemical cycles of carbon and oxygen. PhD Dissertation. Yale University.

Petsch S. T. and Berner R. A. (1998) Coupling the geochemical cycles of C, P, Fe, and S: the effect on atmospheric O_2 and the isotopic records of carbon and sulfur. *Am. J. Sci.* **298**, 246–262.

Petsch S. T., Berner R. A., and Eglinton T. I. (2000) A field study of the chemical weathering of ancient sedimentary organic matter. *Org. Geochem.* **31**, 475–487.

Petsch S. T., Smernik R. J., Eglinton T. I., and Oades J. M. (2001) A solid state ^{13}C-NMR study of kerogen degradation during black shale weathering. *Geochim. Cosmochim. Acta* **65**, 1867–1882.

Phoenix V. R., Konhauser K. O., Adams D. G., and Bottrell S. H. (2001) Role of biomineralization as an ultraviolet shield: implications for Archean life. *Geology* **29**, 823–826.

Pollard D. and Kasting J. K. (2001) Coupled GCM-ice sheet simulations of Sturtian (750–720 Ma) glaciation: when in the snowball-earth cycle can tropical glaciation occur? *EOS* **82**, S8.

Porter S. M. and Knoll A. H. (2000) Testate amoebae in the Neoproterozoic Era: evidence from vase-shaped microfossils in the Chuar Group, Grand Canyon. *Paleobiology* **26**, 360–385.

Rasmussen B. and Buick R. (1999) Redox state of the Archean atmosphere: evidence from detrital heavy minerals in ca. 3250–2750 Ma sandstones from the Pilbara Craton, Australia. *Geology* **27**, 115–118.

Rau G. H., Takahashi T., and Des Marais D. J. (1989) Latitudinal variations in plankton $\delta^{13}C$: implications for CO_2 and productivity in past oceans. *Nature* **341**, 516–518.

Rau G. H., Takahashi T., Des Marais D. J., Repeta D. J., and Martin J. H. (1992) The relationship between $d^{13}C$ of organic matter and $[CO_{2(aq)}]$ in ocean surface water: Data from a JGOFS site in the northeast Atlantic Ocean and a model. *Geochim. Cosmochim. Acta* **56**, 1413–1419.

Robinson J. M. (1989) Phanerozoic O_2 variation, fire, and terrestrial ecology. *Palaeogeogr. Palaeoclimat. Palaeoecol. (Global Planet Change)* **75**, 223–240.

Robinson J. M. (1991) Phanerozoic atmospheric reconstructions: a terrestrial perspective. *Palaeogeogr. Palaeoclimat. Palaeoecol.* **97**, 51–62.

Röhl H.-J., Schmid-Röhl A., Oschmann W., Frimmel A., and Schwark L. (2001) The Posidonia Shale (Lower Toarcian) of SW-Germany: an oxygen-depleted ecosystem controlled by sea level and paleoclimate. *Palaeogeogr. Palaeoclimat. Palaeoecol.* **165**, 27–52.

Ronov A. B. (1976) Global carbon geochemistry, volcanism, carbonate accumulation, and life. *Geochem. Int.* **13**, 172–195.

Rye R. and Holland H. D. (1998) Paleosols and the evolution of atmospheric oxygen: a critical review. *Am. J. Sci.* **298**, 621–672.

Rye R. and Holland H. D. (2000) Life associated with a 2.76 Ga ephemeral pond? Evidence from Mount Roe #2 paleosol. *Geology* **28**, 483–486.

Schidlowski M. (1988) A 3,800-million year isotopic record of life from carbon in sedimentary rocks. *Nature* **333**, 313–318.

Schidlowski M. (2001) Carbon isotopes as biogeochemical recorders of life over 3.8 Ga of Earth history: evolution of a concept. *Precamb. Res.* **106**, 117–134.

Schopf J. W. (1992) Paleobiology of the Archean. In *The Proterozoic Biosphere* (eds. J. W. Schopf and C. Klein). Cambridge University Press, pp. 25–39.

Schopf J. W. (1993) Microfossils of the early Archean Apex Chert: new evidence of the antiquity of life. *Science* **260**, 640–646.

Schopf J. W., Kudryavtvev A. B., Agresti D. G., Wdowiak T. J., and Czaja A. D. (2002) Laser-Raman imagery of Earth's earliest fossils. *Nature* **416**, 73–76.

Sinninghe-Damsté J. S. and Köster J. (1998) A euxinic southern North Atlantic Ocean during the Cenomanian/Turonian oceanic anoxic event. *Earth Planet. Sci. Lett.* **158**, 165–173.

Sleep N. H. and Zahnle K. (1998) Refugia from asteroid impacts on early Mars and the early Earth. *J. Geophys. Res. E—Planets.* **103**, 28529–28544.

Sleep N. H., Zahnle K. J., Kasting J. F., and Morowitz H. J. (1989) Annihilation of ecosystems by large asteroid impacts on the early earth. *Nature* **342**, 139–142.

Sleep N. H., Zahnle K. J., and Neuhoff P. S. (2001) Initiation of clement surface conditions on the earliest Earth. *Proc. Natl. Acad. Sci.* **98**, 3666–3672.

Sowers T., Bender M., and Raynaud D. (1989) Elemental and isotopic composition of occluded O_2 and N_2 in polar ice. *J. Geophys. Res.* **94**, 5137–5150.

Stratford K., Williams R. G., and Myers P. G. (2000) Impact of the circulation on sapropel formation in the eastern Mediterranean. *Global Biogeochem. Cycles* **14**, 683–695.

Sumner D. Y. (1997) Carbonate precipitation and oxygen stratification in late Archean seawater as deduced from facies and stratigraphy of the Gamohaan and Frisco Formations, Transvaal Supergroup, South Africa. *Am. J. Sci.* **297**, 455–487.

Thomas A. L. R. (1997) The breath of life—did increased oxygen levels trigger the Cambrian Explosion? *Trends Ecol. Evol.* **12**, 44–45.

Tolbert N. E., Benker C., and Beck E. (1995) The oxygen and carbon dioxide compensation points of C_3 plants: possible role in regulating atmospheric oxygen. *Proc. Natl. Acad. Sci.* **92**, 11230–11233.

Valley J. W., Peck W. H., King E. M., and Wilde S. A. (2002) A cool early Earth. *Geology* **30**, 351–354.

Van Cappellen P. and Ingall E. D. (1996) Redox stabilization of the atmosphere and oceans by phosphorus-limited marine productivity. *Science* **271**, 493–496.

van Zuilen M., Lepland A., and Arrhenius G. (2002) Reassessing the evidence for the earliest traces of life. *Nature* **418**, 627–630.

Vidal R. A., Bahr D., Baragiola R. A., and Peters M. (1997) Oxygen on Ganymede: laboratory studies. *Science* **276**, 1839–1842.

Walker J. C. G. (1979) The early history of oxygen and ozone in the atmosphere. *Pure Appl. Geophys.* **117**, 498–512.

Walker J. C. G. (1986) Global geochemical cycles of carbon, sulfur, and oxygen. *Mar. Geol.* **70**, 159–174.

Watson A. J. (1978) *Consequences for the Biosphere of Grassland and Forest Fires*. PhD Dissertation. Reading University, UK.

Werne J. P., Hollander D. J., Behrens A., Schaeffer P., Albrecht P., and Sinninghe-Damsté J. S. (2000) Timing of early diagenetic sulfurization of organic matter: a precursor-product relationship in Holocene sediments of the anoxic Cariaco Basin, Venezuela. *Geochim. Cosmochim. Acta* **64**, 1741–1751.

Werne J. P., Lyons T. W., Hollander D. J., Formolo M., and Sinninghe-Damsté J. S. (2003) Reduced sulfur in euxinic sediments of the Cariaco Basin: sulfur isotope constraints on organic sulfur formation. *Chem. Geol.* **195**, 159–179.

Wilde S. A., Valley J. W., Peck W. H., and Graham C. M. (2001) Evidence from detrital zircons for the existence of continental crust and oceans on the Earth 4.4 Gyr ago. *Nature* **409**, 175–178.

Wortmann U. G., Hesse R., and Zacher W. (1999) Major-element analysis of cyclic black shales: paleoceanographic implications for the early Cretaceous deep western Tethys. *Paleoceanography* **14**, 525–541.

Wray G. A., Levinton J. S., and Shapiro L. H. (1996) Molecular evidence for deep Precambrian divergences among Metazoan phyla. *Science* **274**, 568–573.

Zhang R., Follows M. J., Grotzinger J. P., and Marshall J. (2001) Could the late Permian deep ocean have been anoxic? *Paleoceanography* **16**, 317–329.

Geochemistry of Earth Surface Systems
ISBN: 978-0-08-096706-6

pp. 511–550

16

The Global Nitrogen Cycle

J. N. Galloway

University of Virginia, Charlottesville, VA, USA

16.1	INTRODUCTION	552
16.2	BIOGEOCHEMICAL REACTIONS	553
	16.2.1 *The Initial Reaction: Nr Creation*	553
	16.2.2 *Atmosphere*	554
	16.2.2.1 *Inorganic reduced nitrogen*	555
	16.2.2.2 *Inorganic oxidized nitrogen*	555
	16.2.2.3 *Reduced organic nitrogen*	555
	16.2.2.4 *Oxidized organic nitrogen*	556
	16.2.3 *Biosphere*	556
16.3	NITROGEN RESERVOIRS AND THEIR EXCHANGES	556
	16.3.1 *Land to Atmosphere*	556
	16.3.2 *Ocean to Atmosphere*	557
	16.3.3 *Atmosphere to Surface*	557
	16.3.4 *Land to Ocean*	557
16.4	Nr CREATION	557
	16.4.1 *Introduction*	557
	16.4.2 *Lightning—Natural*	557
	16.4.3 *Terrestrial BNF—Natural*	557
	16.4.4 *Anthropogenic*	558
	16.4.4.1 *Introduction*	558
	16.4.4.2 *Food production*	558
	16.4.4.3 *Energy production*	560
	16.4.5 *Nr Creation Rates from 1860 to 2000*	560
16.5	GLOBAL TERRESTRIAL NITROGEN BUDGETS	561
	16.5.1 *Introduction*	561
	16.5.2 *Nr Creation*	562
	16.5.3 *Nr Distribution*	563
	16.5.4 *Nr Conversion to N_2*	564
16.6	GLOBAL MARINE NITROGEN BUDGET	565
16.7	REGIONAL NITROGEN BUDGETS	566
16.8	CONSEQUENCES	568
	16.8.1 *Introduction*	568
	16.8.2 *Atmosphere*	569
	16.8.3 *Terrestrial Ecosystems*	570
	16.8.4 *Aquatic Ecosystems*	571
16.9	FUTURE	572
16.10	SUMMARY	573
	ACKNOWLEDGMENTS	574
	REFERENCES	574

16.1 INTRODUCTION

Once upon a time nitrogen did not exist. Today it does. In the intervening time the universe was formed, nitrogen was created, the Earth came into existence, and its atmosphere and oceans were formed! In this analysis of the Earth's nitrogen cycle, I start with an overview of these important events relative to nitrogen and then move on to the more traditional analysis of the nitrogen cycle itself and the role of humans in its alteration.

The universe is $\sim$15 Gyr old. Even after its formation, there was still a period when nitrogen did not exist. It took $\sim$300 thousand years after the big bang for the Universe to cool enough to create atoms; hydrogen and helium formed first. Nitrogen was formed in the stars through the process of nucleosynthesis. When a star's helium mass becomes great enough to reach the necessary pressure and temperature, helium begins to fuse into still heavier elements, including nitrogen.

Approximately 10 Gyr elapsed before Earth was formed ($\sim$4.5 Ga (billion years ago)) by the accumulation of pre-assembled materials in a multistage process. Assuming that N_2 was the predominate nitrogen species in these materials and given that the temperature of space is $-270\,°C$, N_2 was probably a solid when the Earth was formed since its boiling point (b.p.) and melting point (m.p.) are $-196\,°C$ and $-210\,°C$, respectively. Towards the end of the accumulation period, temperatures were probably high enough for significant melting of some of the accumulated material. The volcanic gases emitted by the resulting volcanism strongly influenced the surface environment. Nitrogen was converted from a solid to a gas and emitted as N_2. Carbon and sulfur were probably emitted as CO and H_2S (Holland, 1984). N_2 is still the most common nitrogen volcanic gas emitted today at a rate of $\sim$2 TgN yr^{-1} (Jaffee, 1992).

Once emitted, the gases either remained in the atmosphere or were deposited to the Earth's surface, thus continuing the process of biogeochemical cycling. The rate of transfer depended on the reactivity of the emitted material. At the lower extreme of reactivity are the noble gases, neon and argon. Most neon and argon emitted during the degassing of the newly formed Earth is still in the atmosphere, and essentially none has been transferred to the hydrosphere or crust. At the other extreme are carbon and sulfur. Over 99% of the carbon and sulfur emitted during degassing are no longer in the atmosphere, but reside in the hydrosphere or the crust. Nitrogen is intermediate. Of the $\sim$6$\times$10^6 TgN in the atmosphere, hydrosphere, and crust, $\sim$2/3 is in the atmosphere as N_2 with most of the remainder in the crust. The atmosphere is a large nitrogen reservoir primarily, because the triple bond of the N_2 molecule requires a significant amount of energy to break. In the early atmosphere, the only sources of such energy were solar radiation and electrical discharges.

At this point we had an earth with mostly N_2 and devoid of life. How did we get to an earth with mostly N_2 and teeming with life? First, N_2 had to be converted into reactive N (Nr). (The term reactive nitrogen (Nr) includes all biologically active, photochemically reactive, and radiatively active nitrogen compounds in the atmosphere and biosphere of the Earth. Thus, Nr includes inorganic reduced forms of nitrogen (e.g., NH_3 and NH_4^+), inorganic oxidized forms (e.g., NO_x, HNO_3, N_2O, and NO_3^-), and organic compounds (e.g., urea, amines, and proteins).) The early atmosphere was reducing and had limited NH_3. However, NH_3 was a necessary ingredient in forming early organic matter. One possibility for NH_3 generation was the cycling of seawater through volcanics (Holland, 1984). Under such a process, NH_3 could then be released to the atmosphere where, when combined with CH_4, H_2, H_2O, and electrical energy, organic molecules including amino acids could be formed (Miller, 1953). In essence, electrical discharges and UV radiation can convert mixtures of reduced gases into mixtures of organic molecules that can then be deposited to land surfaces and oceans (Holland, 1984).

To recap, Earth was formed at 4.5 Ga, water condensed at 4 Ga, and organic molecules were formed thereafter. By 3.5 Ga simple organisms (prokaryotes) were able to survive without O_2 and produced NH_3. At about the same time, the first organisms that could create O_2 in photosynthesis (e.g., cyanobacteria) evolved. It was not until 1.5–2.0 Ga that O_2 began to build up in the atmosphere. Up to this time, the O_2 had been consumed by chemical reactions (e.g., iron oxidation). By 0.5 Ga the O_2 concentration of the atmosphere reached the same value found today. As the concentration of O_2 built up, so did the possibility that NO could be formed in the atmosphere during electrical discharges from the reaction of N_2 and O_2.

Today we have an atmosphere with N_2 and there is energy to produce some NO (reaction of N_2 and O_2). Precipitation can transfer Nr to the Earth's surface. Electrical discharges can create nitrogen-containing organic molecules. Simple cells evolved $\sim$3.5 Ga and, over the succeeding years, more complicated forms of life have evolved, including humans. Nature formed nitrogen and created life. By what route did that "life" discover nitrogen?

To address this question, we now jump from 3.5 Ga to $\sim$2.3$\times$10^{-7} Ga. In the 1770s, three scientists—Carl Wilhelm Scheele (Sweden), Daniel Rutherford (Scotland), and Antoine Lavosier (France)—independently discovered the existence of nitrogen. They performed experiments in which an unreactive gas was produced. In 1790, Jean Antoine Claude Chaptal formally named the gas *nitrogène*. This discovery marked the beginning of our understanding of nitrogen and its role in Earth systems.

By the beginning of the second half of the nineteenth century, it was known that nitrogen is a common element in plant and animal tissues, that it is indispensable for plant growth, that there is constant cycling between organic and inorganic compounds, and that it is an effective fertilizer. However, the source of nitrogen was still uncertain. Lightning and atmospheric deposition were thought to be the most important sources. Although the existence of biological nitrogen fixation (BNF) was unknown at that time, in 1838 Boussingault demonstrated that legumes restore Nr to the soil and that somehow they create Nr directly. It took almost 50 more years to solve the puzzle. In 1888, Herman Hellriegel (1831–1895) and Hermann Wilfarth (1853–1904) published their work on microbial communities. They noted that microorganisms associated with legumes have the ability to assimilate atmospheric N_2 (Smil, 2001). They also said that it was necessary for a symbiotic relationship to exist between legumes and microorganisms.

Other important processes that drive the cycle were elucidated in the nineteenth century. In the late 1870s, Theophile Scholesing proved the bacterial origins of nitrification. About a decade later, Serfei Nikolaevich Winogradsky isolated the two nitrifers—*Nitrosomonas* and *Nitrobacter*—and showed that the species of the former genus oxidize ammonia to nitrite and that the species of the latter genus convert nitrite to nitrate. Then in 1885, Ulysse Gayon isolated cultures of two bacteria that convert nitrate to N_2. Although there are only two bacterial genera that can convert N_2 to Nr, several can convert Nr back to N_2, most notably *Pseudomonas, Bacillus,* and *Alcaligenes* (Smil, 2001).

By the end of the nineteenth century, humans had discovered nitrogen and the essential components of the nitrogen cycle. In other words, they then knew that some microorganisms convert N_2 to NH_4^+, other microorganisms convert NH_4^+ to NH_3^-, and yet a third class of microorganisms convert NH_3^- back to N_2, thus completing the cycle.

The following sections of this chapter examine the biogeochemical reactions of Nr, the distribution of Nr in Earth's reservoirs, and the exchanges between the reservoirs. This chapter then discusses Nr creation by natural and anthropogenic processes and nitrogen budgets for the global land mass and for continents and oceans using Galloway and Cowling (2002) and material from Cory Cleveland (University of Colorado) and Douglas Capone (University of Southern California) from a paper in review in Biogeochemistry (Galloway

et al., 2003a). This chapter also presents an overview of the consequences of Nr accumulation in the environment (using Galloway *et al.* (2003b) as a primary reference) and then concludes with estimates of minima and maxima Nr creation rates in 2050.

16.2 BIOGEOCHEMICAL REACTIONS

16.2.1 The Initial Reaction: Nr Creation

In the formation of Earth, most nitrogen was probably in the form of N_2, the most stable molecule and thus the reservoir from which all other nitrogen compounds are formed. It is thus fitting to begin this section on chemical reactions with N_2—with both atoms of the molecule representing the Adam and Eve of nitrogen. N_2 is converted to Nr by either converting N_2 to NH_3 or by converting N_2 to NO. High energy is required to break the triple bond of the N_2 molecule—226 kcal mol^{-1}. In the case of both NH_3 and NO formation, a natural process and an anthropogenic process form Nr from N_2. The next few paragraphs review these processes; a later section discusses rates of Nr formation as a function of time and process.

BNF is a microbially mediated process that occurs in several types of bacteria and blue-green algae. This process uses the enzyme nitrogenase in an anaerobic environment to convert N_2 to NH_3. The microbes can be free-living or in a symbiotic association with the roots of higher plants. Legumes are the best-known example of this type of relationship (Schlesinger, 1997; Mackenzie, 1998):

$$2N_2 + 6H_2O \rightarrow 4NH_3 + 3O_2$$

As will be explained in more detail later, BNF can occur in both unmanaged and managed ecosystems. In the former, natural ecosystems produce Nr. In the latter, the cultivation of legumes enhances BNF and the cultivation of some crops (e.g., wetland rice) creates the necessary anaerobic environment to promote BNF.

Over geologic history, most Nr has been formed by BNF. However, in the last half of the twentieth century, the Haber–Bosch process has replaced BNF as the dominant terrestrial process creating Nr. The Haber–Bosch process was invented and developed commercially in the early 1900s. It uses high temperature and pressure with a metallic catalyst to produce NH_3:

$$N_2 + 3H_2 \rightarrow 2NH_3$$

This process was used extensively during World War I to produce munitions and, since the early 1950s, has become the world's largest source of nitrogenous fertilizer (Smil, 2001). N_2 is taken from the atmosphere, whereas H_2 is produced from a fossil fuel, usually natural gas.

Two processes create Nr in the form of NO from the oxidation of N_2. The natural process is lightning; the anthropogenic process is fossil-fuel combustion. As previously mentioned, at an early point in Earth's history, lightning was an important process in creating Nr from N_2. Although globally much less important now, N_2 is still converted to NO by electrical discharges:

$$N_2 + O_2 + \text{electrical energy} \rightarrow 2NO$$

Lightning formation of NO is most important in areas of deep convective activities such as occurring in tropical continental regions.

The high temperatures and pressures found during fossil-fuel combustion provide the energy to convert N_2 to NO through reaction with O_2:

$$N_2 + O_2 + \text{fossil energy} \rightarrow 2NO$$

Note that, during fossil-fuel combustion, NO can also be formed from the oxidation of fossil organic nitrogen, which is primarily found in coal. This is technically not a creation of new Nr but rather the mobilization of Nr that has been sequestered for millions of years (Socolow, 1999).

Thus, independent of the process, N_2 is transformed to either NO in the atmosphere or a combustion chamber or to NH_3 in an organism or a factory. The rest of this section examines the fate of NO and NH_3 (and their reaction products) in the atmosphere, terrestrial ecosystems, aquatic ecosystems, and agro-ecosystems.

Except in the atmosphere (where nitrogen chemistry is essentially a series of oxidation reactions),

the basic structure of nitrogen reactions is similar in all reservoirs. For soils, freshwaters, coastal waters, and oceans, the chemical reactions of nitrogen are generally self-contained sequences of oxidation and reduction reactions driven by microbial activity and environmental conditions. The central framework of the reactions is constructed through the processes of nitrogen fixation, assimilation, nitrification, decomposition, ammonification, and denitrification (Figure 1). This framework provides nitrogen for amino acid and protein synthesis by primary producers followed perhaps by consumption of the protein by a secondary producer. This framework also provides a mechanism to convert the amino acids/protein back to inorganic nitrogen following excretion from, or the death of, the primary or secondary producer.

16.2.2 Atmosphere

The atmospheric nitrogen cycle is the simplest cycle, because the direct influence of biota is limited and thus chemical and physical processes primarily control transformations. In addition, the cycle of oxidized inorganic nitrogen (NO_y) is, for the most part, decoupled from the cycle of reduced inorganic nitrogen (NH_x). As discussed in the next section, nitrogen cycles in terrestrial and aquatic ecosystems are substantially more complex because of the microbially mediated nitrogen transformations that occur (Figure 1).

The atmospheric chemistry of nitrogen can be divided into four groupings of nitrogen species that are generally independent of each other—inorganic

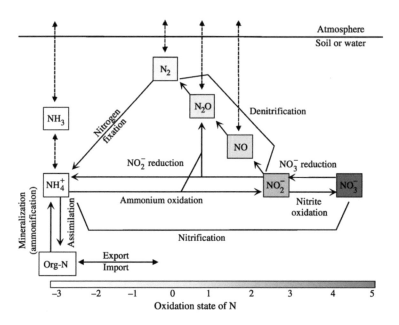

Redrawn from Karl

Figure 1 The processes of nitrogen fixation, assimilation, nitrification, decomposition, ammonification, and denitrification (after Karl, 2002).

reduced nitrogen, inorganic oxidized nitrogen, organic reduced nitrogen, and organic oxidized nitrogen. Reactions are generally within the groupings. For each grouping, I first discuss the emitted species and then its fate in the atmosphere.

16.2.2.1 Inorganic reduced nitrogen

There are two species in this grouping—ammonia (NH_3) and ammonium (NH_4^+) with one valence state ($-III$) (Table 1). The primary species emitted to the atmosphere is NH_3 produced during organic matter decomposition and emitted when the partial pressure in the soil, water, or plant is greater than the partial pressure in the atmosphere. It is the most common atmospheric gaseous base and, once in the atmosphere, can be converted to an aerosol in an acid–base reaction with a gas (e.g., HNO_3) or aerosol (e.g., H_2SO_4):

$$NH_{3(g)} + HNO_{3(g)} \rightarrow NH_4NO_{3(s)}$$

$$NH_{3(g)} + H_2SO_{4(s)} \rightarrow NH_4HSO_{4(s)}$$

$$NH_{3(g)} + NH_4HSO_{4(s)} \rightarrow (NH_4)_2SO_{4(s)}$$

All species are readily removed by atmospheric deposition. NH_3 is primarily removed by dry deposition (often close to its source). Aerosol NH_4^+ is primarily removed by wet deposition; in fact, the hydroscopic aerosol is a cloud-condensation nuclei. If NH_x (NH_3 and NH_4^+) is lifted above the planetary boundary layer (PBL), it can be transported large distances (1,000 km or more). Emissions in one location can impact receptors far downwind.

16.2.2.2 Inorganic oxidized nitrogen

This grouping has many species and valence states (Table 1). Most oxidized nitrogen species in the atmosphere are part of NO_y, which includes $NO + NO_2 + HNO_3 + PAN$ and other trace oxidized species. Within this group, $NO + NO_2$ are referred to as NO_x. All these species are relatively reactive and most have lifetimes of minutes to days in the atmosphere; some (e.g., PAN) can

have longer lifetimes. NO is the most commonly emitted species and has several sources that are usually of two types. As discussed earlier, the first is from the conversion of N_2 to NO by lightning or the high-temperature combustion of fossil fuels. The second is the conversion of one Nr species to NO generally through fire or microbial activity. Once in the atmosphere, NO is quickly oxidized to NO_2 that is then oxidized to HNO_3 and then potentially reacts with NH_3 to form an aerosol:

$$NO + O_3 \rightarrow NO_2 + O_2$$

$$NO_2 + OH \rightarrow HNO_3$$

$$HNO_3 + NH_3 \rightarrow NH_4NO_3$$

However, in this process a significant cycle involving hydrocarbons and ozone has important implications for the oxidation capacity of the atmosphere and human and ecosystem health.

The oxidized inorganic nitrogen species not included in NO_y is N_2O. It is produced during nitrification and denitrification (Figure 1). It has an atmospheric residence time of ~100 yr and, because of this, emissions are globally distributed. Because of its stability, no significant chemical reactions take place in the troposphere but, once in the stratosphere, it is converted to NO by UV radiation:

$$N_2O + O(^1D) \rightarrow 2NO$$

The NO produced will then destroy stratospheric ozone in a reaction that regenerates it:

$$NO + O_3 \rightarrow NO_2 + O_2$$

$$O_3 \rightarrow O + O_2$$

$$NO_2 + O \rightarrow NO + O_2$$

The net reaction is

$$2O_3 \rightarrow 3O_2$$

Increasing concentrations of atmospheric N_2O contribute to two environmental issues of the day. In the troposphere it contributes to the greenhouse potential; in the stratosphere it contributes to ozone destruction.

16.2.2.3 Reduced organic nitrogen

Atmospheric organic nitrogen can occur as bacteria, particulate matter, and soluble species (e.g., amines) (Neff *et al.*, 2002). These materials are emitted to the atmosphere through low-temperature processes such as turbulence and high-temperature processes like biomass burning. The concentrations of bacteria range from ~10 bacteria m^{-3} in the marine environment to >1,000 bacteria m^{-3} in urban environments

Table 1 Valence state of N species.

Grouping	Valence/oxidation state	Species
Inorganic oxidized N	5	NO_3^-, HNO_3
	4	NO_2
	3	NO_2^-
	2	NO
Diatomic N	0	N_2
Inorganic reduced N	-3	NH_3, NH_4^+
Organic reduced N	-3	$R–NH_2$

(Neff *et al.*, 2002). Atmospheric reactions involving bacteria are limited. Particulate organic nitrogen is composed of organic debris and soil matter. Again, there is probably limited atmospheric chemistry of this material. Conversely, soluble reduced organic nitrogen compounds (e.g., urea, free amino acids, and other methylated amines) are quite reactive in the atmosphere. Many species can react quickly with HO and organic nitrates and thus are not transported far from their emission points (Neff *et al.*, 2002).

16.2.2.4 Oxidized organic nitrogen

Atmospheric oxidized organic nitrogen species are generally formed in the atmosphere as the end products of reactions of hydrocarbons with NO_x (Neff *et al.*, 2002). Hydrocarbons can form organic radicals (RO, RO_2) through reaction with light, OH, or ozone. The resulting species can react with NO_2 to form $RONO_2$.

16.2.3 Biosphere

The biosphere is defined as the terrestrial and aquatic ecosystems, including the oceans. As mentioned above, microbial processes have a strong, in many cases controlling, influence on the biogeochemistry of nitrogen in these systems. These processes themselves constitute a cycle (Figure 1). The individual processes are briefly defined below.

Nitrogen fixation is the processes by which N_2 is converted to any nitrogen compound where nitrogen has a nonzero oxidation state. Historically, the most common process has been the biologically driven reduction of N_2 to NH_3 or NH_4^+. However, currently anthropogenically enhanced nitrogen fixation dominates on continents (see Section 16.4).

Ammonia assimilation is the uptake of NH_3 or NH_4^+ by an organism into its biomass in the form of an organic nitrogen compound. For organisms that can directly assimilate reduced inorganic nitrogen, this is an efficient process to incorporate nitrogen into biomass.

Nitrification is the aerobic process by which microorganisms oxidize ammonium to nitrate and derive energy. It is the combination of two bacterial processes: one group of organisms oxidizes ammonia to nitrite (e.g., *Nitrosomonas*), after which a different group oxidizes nitrite to nitrate (e.g., *Nitrobacter*):

$$2NH_4^+ + 3O_2 \rightarrow 2NO_2^- + 2H_2O + 4H$$

$$2NO_2^- + O_2 \rightarrow 2NO_3^-$$

Assimilatory nitrate reduction is the uptake of NO_3^- by an organism and incorporation as biomass through nitrate reduction. It is an important process, because it allows the mobile nitrate ion to be transported to a receptor which can then reduce it to ammonia for subsequent uptake. It is an important input of nitrogen for many plants and organisms.

Ammonification is the primary process that converts reduced organic nitrogen ($R–NH_2$) to reduced inorganic nitrogen (NO_4^+) through the action of microorganisms. This is part of the general process of decomposition where heterotrophic microbes use organic matter for energy and in the process convert organic N to NO_4^+.

Denitrification is the reduction of NO_3^- to any gaseous nitrogen species, normally N_2.

$$5CH_2O + 4H^+ + 4NO_3^- \rightarrow 2N_2 + 5CO_2 + 7H_2O$$

It is an anaerobic process and requires nitrate (NH_x that can be nitrified) and organic matter. Microorganisms use nitrate as an oxidant to obtain energy from organic matter. It is prevalent in waterlogged soils and is the primary process that converts Nr back to N_2.

16.3 NITROGEN RESERVOIRS AND THEIR EXCHANGES

The reservoirs of the Earth contain $\sim 5 \times 10^{21}$ g N, 80% of which is in the atmosphere. Sedimentary rocks contain almost all the remainder with just a trace ($<1\%$ of total) in oceans and living and dead organic matter (Table 2) (Mackenzie, 1998).

The chemical form of the nitrogen depends on the reservoir. In the atmosphere, except for trace amounts of N_2O, NO_y, NH_x, and organic N, it occurs as N_2. In oceans and soils, it primarily occurs as organic nitrogen, nitrate, and ammonium. Atmospheric and hydrologic transport provide the primary paths for exchanges between reservoirs.

16.3.1 Land to Atmosphere

Nitrogen is transferred to the atmosphere by low- and high-temperature processes. The high-temperature processes are biomass combustion

Table 2 N amounts in global reservoirs (TgN yr^{-1}).

Reservoirs	Amount	Percentage of total
Atmosphere, N_2	3,950,000,000	79.5
Sedimentary rocks	999,600,000	20.1
Ocean		
N_2	20,000,000	0.4
NO_3^-	570,000	0.0
Soil organics	190,000	0.0
Land biota	10,000	0.0
Marine biota	500	0.0

Source: Mackenzie (1998) except ocean, N_2 from Schlesigner (1997).

and fossil-fuel combustion; the low-temperature processes are volatilization of gases from soils and waters and turbulent injection of particulate matter into the atmosphere. The gases are generated primarily as a result of microbial activity (e.g., nitrification, denitrification, and ammonification).

16.3.2 Ocean to Atmosphere

When the partial pressure of gas in seawater is greater than the partial pressure of gas in the atmosphere, nitrogen can be emitted from the ocean through volatilization of NH_3, N_2, and N_2O. Nitrogen can also be transferred to the atmosphere via aerosol formed by breaking waves or bubbles.

16.3.3 Atmosphere to Surface

Nitrogen is deposited to land and ocean surfaces by wet and dry deposition. Wet deposition includes rain, snow, and hail; the nitrogen can be inorganic or organic. Dry deposition includes gases and aerosols. The two most important dry-deposited species are gaseous HNO_3 and NH_3. Aerosol dry deposition of nitrate and ammonium does occur, but the fluxes are generally small relative to gaseous deposition.

16.3.4 Land to Ocean

Both inorganic and organic nitrogen compounds are transferred from continents to the coastal ocean via discharge of rivers and groundwater. The former can be soluble (e.g., NO_3^-) or particulate (NH_4^+ adsorbed on surfaces). The latter can also be soluble or particulate. Most nitrogen is transported as either NO_3^- or particulate organic nitrogen (Seitzinger *et al.*, 2002).

16.4 Nr CREATION

16.4.1 Introduction

As discussed earlier, N_2 is converted to Nr by four basic processes—lightning, BNF, combustion, and the Haber–Bosch process. This section details these processes and concludes with an estimate of the trends in Nr creation from 1860 to 2000.

16.4.2 Lightning—Natural

High temperatures in lightning strikes produce NO in the atmosphere from molecular oxygen and nitrogen, primarily over tropical continents. Subsequently this NO is oxidized to NO_2 and then to HNO_3 and quickly (i.e., days) introduces Nr into ecosystems through wet and dry deposition. More current estimates of Nr creation by lightning range

between $3\,TgN\,yr^{-1}$ and $10\,TgN\,yr^{-1}$ (Ehhalt *et al.*, 2001). In this analysis I have used a global estimate of $5.4\,TgN\,yr^{-1}$ (Lelieveld and Dentener, 2000). As will be seen below, although this number is small relative to terrestrial BNF, it can be important for regions that do not have other significant Nr sources. It is also important, because it creates NO_x high in the free troposphere as opposed to NO_x emitted at the Earth's surface. As a result it has a longer atmospheric residence time and is more likely to contribute to tropospheric O_3 formation, which significantly impacts the oxidizing capacity of the atmosphere (E. Holland, personal communication).

16.4.3 Terrestrial BNF—Natural

Before human activity, BNF was the most important process in converting N_2 to Nr. However, quantifying BNF is difficult primarily because of the uncertainty and variability in the existing estimates of BNF rates at the plot scale. In addition, as noted in Cleveland *et al.* (1999), for many large areas where BNF is likely to be important, particularly in the tropical regions of Asia, Africa, and South America, there are virtually no data on natural terrestrial rates of BNF. In a recent compilation of rates of natural BNF by Cleveland *et al.* (1999), symbiotic BNF rates for several biome types are based on a few published rates of symbiotic BNF at the plot scale within each particular biome. For example, based on the few estimates of symbiotic BNF available for tropical rain forests, published estimates indicate that natural BNF in these systems annually represents ~24% of the total global, natural terrestrial BNF (Cleveland *et al.*, 1999).

Difficulties notwithstanding, earlier estimates of BNF in terrestrial ecosystems range from ~$30\,TgN\,yr^{-1}$ to $200\,TgN\,yr^{-1}$ (e.g., Delwiche, 1970; Söderlund and Rosswall, 1982; Stedman and Shetter, 1983; Paul and Clark, 1997; Schlesinger, 1997). Most studies merely present BNF estimates as "global values" and, at best, are broken into a few very broad components (e.g., "forest," "grassland," and "other"; e.g., Paul and Clark (1997)). However, such coarse divisions average enormous land areas that contain significant variation in both BNF data sets and biome types, thus diminishing their usefulness. Many studies also lack a list of data sources from which their estimates are derived (e.g., Söderland and Rosswall, 1982; Stedman and Shetter, 1983; Paul and Clark, 1997; Schlesinger, 1997). In contrast, Cleveland *et al.* (1999) provide a range of estimates of BNF in natural ecosystems from $100\,TgN\,yr^{-1}$ to $290\,TgN\,yr^{-1}$ (with a "best estimate" of $195\,TgN\,yr^{-1}$). These estimates are based on published, data-based rates of BNF in natural ecosystems and differ only in the percentage of cover estimates of symbiotic nitrogen fixers used to scale plot-level estimates to the biome scale.

Although the data-based estimates of Cleveland *et al.* (1999) provide more of a documented, constrained range of terrestrial BNF, there are several compelling reasons to believe that an estimate in the lower portion of the range is more realistic than the higher estimates (Galloway *et al.*, 2003a; C. Cleveland, personal communication). First, rate estimates of BNF in the literature are inherently biased since investigations of BNF are frequently carried out in plot-level studies in areas where BNF is most likely to be important. For example, rates of BNF in temperate forests are based on estimates that include nitrogen inputs from alder and black locust (Cleveland *et al.*, 1999). Although the rates of BNF may be very high within stands dominated by these species (Boring and Swank, 1984; Binkley *et al.*, 1994), these species are certainly not dominant in temperate forests as a whole (Johnson and Mayeux, 1990). Similarly, although the species with high rates of BNF are often common in early successional forests (Vitousek, 1994), they are often rare in mature or late successional forests, especially in the temperate zone (Gorham *et al.*, 1979; Boring and Swank, 1984; Blundon and Dale, 1990). Literature-derived estimates based on reported coverage of nitrogen-fixing species are thus inflated by these inherent biases (Galloway *et al.*, 2003a; C. Cleveland, personal communication).

Therefore, based on the above, it is reasonable to assume that annual global BNF contributed between 100 TgN yr^{-1} and 290 TgN yr^{-1} to natural terrestrial ecosystems before large-scale human disturbances. However, because of the biases inherent in plot-scale studies of nitrogen-fixation rates, the true rate of BNF is probably at the lower end of this range (C. Cleveland, personal communication). Asner *et al.* (2001) use actual evapotranspiration (ET) values generated in the model Terraflux together with the strong, positive relationship between ET and nitrogen fixation (Cleveland *et al.*, 1999) to generate a new, single estimate of BNF before any large-scale human disturbance. This analysis is still based on the relationship between ET and BNF, but uses rates of BNF calculated using the low percentage cover values of symbiotic nitrogen fixers over the landscape (i.e., 5%; Cleveland *et al.*, 1999). This analysis suggests that within the range of 100–290 TgN yr^{-1}, natural BNF in terrestrial ecosystems contributes 128 TgN yr^{-1}. This value is supported by an analysis comparing BNF to nitrogen requirement (by type of biome). Using the Cleveland *et al.* (1999) relationship between ET and BNF, a global nitrogen-fixation value of 128 TgN yr^{-1} would suggest an average of ~15% of the nitrogen requirement across all biome types is met via BNF; higher estimates of BNF would imply that at least 30% of the nitrogen requirement across all biomes is met via natural BNF (Asner *et al.*, 2001).

This estimate (128 Tg N yr^{-1}) represents potential BNF before large-scale human disturbances and does not account for decreases in BNF from land-use change or from any other physical, chemical, or biological factor. To estimate natural terrestrial BNF for 1890 and the early 1990s, BNF is scaled to the extent of altered land at those two times. Of the 1.15×10^4 Mha of natural vegetated land (Mackenzie, 1998), Houghton and Hackler (2002) estimate that in 1860 and 1995, 760 Mha and 2,400 Mha, respectively, had been altered by human action (e.g., cultivation, conversion of forests to pastures). Therefore, in this analysis of BNF in the terrestrial landscape, 128 TgN yr^{-1} is fixed before landscape alteration of the natural world, 120 TgN yr^{-1} for 1860, and 107 Tg N yr^{-1} for the present world (Galloway *et al.*, 2003a). These values are similar to ones of Galloway and Cowling (2002) (which also relied on work of C. Cleveland) for 1890 and 1990, and are also used in this comparison (see below).

16.4.4 Anthropogenic

16.4.4.1 Introduction

As discussed above, in the prehuman world, two processes—BNF and lightning—generated enough energy to break the N_2 triple bond. The former process uses metabolic energy to create Nr on purpose and the latter process uses electrical energy to produce Nr by accident. Humans developed one new process (Haber–Bosch) and enhanced a natural process (BNF) to create Nr on purpose to sustain food production. Humans also create Nr by accident by fossil fuel combustion.

16.4.4.2 Food production

Humans not only need nitrogen to survive but they need it in the form of amino acids because they are not able to synthesize amino acids from inorganic nitrogen. Early hunter gatherers met their nitrogen requirements by consuming naturally occurring amino acids in plants (which can make their own organic nitrogen compounds (e.g., amino acids)) and animals (which eat plants or other animals to get organic nitrogen compounds). Once people established settled communities and started using and reusing the same land, they had to find ways to introduce the harvested Nr back into the earth.

The history of Nr additions can be divided into four stages—recycling of organic matter (e.g., crop residues, manure), C-BNF (e.g., legumes), importing existing Nr (e.g., guano), and creation of mineral nitrogen (e.g., Haber–Bosch). The first two stages are not distinctly separate. Recycling has probably been practiced since the advent of agriculture, and legumes were certainly early candidates for cultivation. Archeological evidence points to legume cultivation over 6.5 ka (Smith, 1995) and perhaps as early as 1.2×10^4 yr BP (Kislev and Baryosef, 1988). Soybeans have been cultivated in China for at least 3,100 yr (Wang, 1987). Rice cultivation, which began in Asia perhaps as early as 7,000 yr ago (Wittwer *et al.*, 1987), also resulted in anthropogenic-induced creation of Nr by creating anaerobic environments where BNF could occur. There are, however, limits to recycling and C-BNF; by the early nineteenth century additional nitrogen sources were required. Guano imported from arid tropical and subtropical islands and from South America became an early source of new Nr with nitrogen contents up to ~30 times higher than most manures (Smil, 2001). From about 1830 to the 1890s, guano was the only source of additional nitrogen. However, by the late 1890s many were concerned that there was not enough nitrogen to provide food for the world's growing population. This concern was highlighted by Sir William Crookes (1832–1919), President of the British Association of Science. In a September 1898 speech he warned that "all civilized nations stand in deadly peril of not having enough to eat" (Crookes, 1871) because of the lack of nitrogen. Crookes went on to encourage chemists to learn how to convert atmospheric N_2 to NH_3 (Smil, 2001). As discussed below, in a very short 15 years, the Haber–Bosch process was developed

exactly for that purpose. However, before that happened, agriculture used two other industrial processes to create nitrogen for fertilizer. When coal is burned or heated in the absence of air, part of the organic nitrogen in the fuel is converted to NH_3 (coke ovens, Table 3). On a global basis, by 1890 trace amounts of NH_3 were captured for agriculture; by 1935 it had overtaken guano and sodium nitrate as the primary nitrogen-fertilizer source (Table 3). Another industrial process was the synthesis of cyanamide—the first method of creating Nr from N_2. In this process, calcium carbide reacts with N_2 to form $CaCN_2$, which is then combined with superheated steam to produce $CaCO_3$ and NH_3. Although this process did create usable Nr, it required a large amount of energy. By the late 1920s, the traditional nitrogen sources, recycling crop residues and manure, were replaced by sodium nitrate, coke-oven gas, and cyanamide synthesis. By 1929, the total amount of nitrogen produced by these processes was 1.2 TgN.

There was, of course, another source of new Nr—the reaction of N_2 and H_2 to produce NH_3. It has long been realized that NH_3 cannot be formed at ambient temperature and pressure. Indeed, the first experiments using higher temperatures began in 1788 (Smil, 2001). It took more than another century for the experiments to bear fruit. In 1904, Fritz Haber began his experiments on ammonia synthesis. By 1909 he had developed an efficient process to create NH_3 from N_2 and H_2 at the laboratory scale using high temperature, high pressure, and a catalyst. For this he received the Nobel prize in Chemistry in 1920. The path from a laboratory synthesis to a factory synthesis took

Table 3 Guano and Chilean nitrate extraction: ammonium sulfate, cyanimide, calcium production from 1850 to 2000 (TgN yr^{-1}).

	$NaNO_3$	Guano	Coke oven	$CaCN_2$	Electric arc	Haber–Bosch	Total
1850	0.01						0.01
1860	0.01	0.07					0.08
1870	0.03	0.07					0.10
1880	0.05	0.03					0.08
1890	0.13	0.02					0.15
1900	0.22	0.02	0.12				0.36
1905	0.25	0.01	0.13				0.39
1910	0.36	0.01	0.23	0.01			0.61
1913	0.41	0.01	0.27	0.03	0.01		0.73
1920	0.41	0.01	0.29	0.07	0.02	0.15	0.95
1929	0.51	0.01	0.43	0.26	0.02	0.93	2.15
1935	0.18	0.01	0.37	0.23		1.30	2.09
1940	0.20	0.01	0.45	0.29		2.15	3.10
1950	0.27		0.50	0.31		3.70	4.78
1960	0.20		0.95	0.30		9.54	10.99
1970	0.12		0.95	0.30		30.23	31.60
1980	0.09		0.97	0.25		59.29	60.60
1990	0.12		0.55	0.11		76.32	77.10
2000	0.12		0.37	0.08		85.13	85.70

Source: Smil (2001).

four years. In 1913, based on the work of Carl Bosch, the first major NH₃-generating plant became operational in Germany. In honor of his work, Bosch received the Nobel prize in Chemistry in 1932. For the first few years, most production went for munitions to sustain Germany during World War I. It was not until the early 1930s that the Haber–Bosch process became the primary source of NH₃ for agricultural use.

In summary, from the advent of agriculture to about 1850, the sole source of nitrogen for food production was BNF naturally occurring in the soil, cultivation-induced BNF and nitrogen added by recycling of organic matter (stage 1). From 1850 to 1890, an additional source was guano and sodium nitrate deposits (stage 2). From 1890 to 1930, the coke oven and $CaCN_2$ processes accounted for up to 40% of the nitrogen fertilizer produced (stage 3). After 1930, the Haber–Bosch process became the dominant source of fertilizer nitrogen (stage 4). Of the early processes, the only ones that are still important sources for some regions are C-BNF and organic matter recycling.

16.4.4.3 Energy production

Combustion of fossil fuels creates energy on purpose and several waste gases by accident, including NO, which is created in the combustion chamber by the reaction of N_2 and O_2. In addition, nitrogen fixed by plants hundreds of millions of years ago (fossil nitrogen) has the potential to become reactive again when fossil fuels are extracted and burned. More than 1% by weight of coal, for example, is nitrogen fixed in geological times. In all, ~50 TgN yr^{-1} of fossil nitrogen are removed from belowground today, primarily

in coal but also in crude oil. When fossil fuel is burned, fossil nitrogen either joins the global pool of Nr in the form of NO_x emissions or is changed back to nonreactive N_2. In a typical coal-fired plant without pollution control, less than half the fossil nitrogen becomes NO_x emissions, a fraction that is much smaller with pollution control (Galloway et al., 2002; Ayres et al., 1994).

16.4.5 Nr Creation Rates from 1860 to 2000

Nr creation rates (Galloway et al., 2003b) are related to global population trends from 1860 to 2000 in Figure 2. For 1920, 1930, and 1940, global anthropogenic fertilizer production (Smil, 2001) was made equivalent to total Nr production by the Haber–Bosch process. For 1950 onwards Nr creation rates by Haber–Bosch were obtained from USGS minerals (Kramer, 1999). The rate in 1900 is estimated to be ~15 TgN yr^{-1} (V. Smil, personal communication). Nr creation rates for 1860, 1870, 1880, and 1890 were estimated from population data using the 1900 data on population and Nr creation. For the period 1961–1999, Nr creation rates were calculated from crop-specific data on harvested areas (FAO, 2000) and fixation rates (Smil, 1999). Decadal data from 1910 to 1950 were interpolated between 1900 and 1961. The data from 1860 to 1990 are from a compilation from Elisabeth Holland and are based on Holland and Lamarque (1997), Müller (1992), and Keeling (1993). These data agree well with those published by van Aardenne et al. (2001) for the decadal time steps from 1890 to 1990. The data for 1991–2000 are estimated by scaling NO_x emissions to increases in fossil-fuel combustion over the same period.

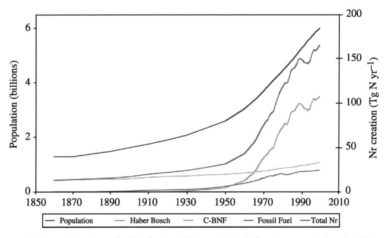

Figure 2 The (purple line) global population from 1860 to 2000 (left axis: population in billions; right axis: Nr creation in TgN yr^{-1}) showing (green line) Nr creation via the Haber–Bosch process, including production of NH₃ for nonfertilizer purposes; (blue line) Nr creation from cultivation of legumes, rice and sugar cane; (brown line) Nr creation from fossil-fuel combustion; and (red line) the sum created by these three processes (source Galloway et al., 2003b).

The global rate of increase in Nr creation by humans was relatively slow from 1860 to 1960. Since 1960, however, the rate of increase has accelerated sharply (Figure 2). Cultivation-induced Nr creation increased from ~15 TgN yr^{-1} in 1860 to ~33 TgN yr^{-1} in 2000. Nr creation by fossil-fuel combustion increased from ~0.3 TgN yr^{-1} in 1860 to ~25 TgN yr^{-1} in 2000. Nr creation from the Haber–Bosch process went from zero before 1910 to >100 TgN yr^{-1} in 2000 with ~85% used in the production of fertilizers. Thus, between 1860 and 2000, the anthropogenic Nr creation rate increased from ~15 TgN yr^{-1} to ~165 TgN yr^{-1} with ~5 times more Nr coming from food production than from energy production and use (Galloway *et al.*, 2002).

16.5 GLOBAL TERRESTRIAL NITROGEN BUDGETS

16.5.1 Introduction

Over the last several decades, there have been numerous compilations of the global nitrogen cycle (Table 4). Most cover the basic fluxes— BNF and denitrification in terrestrial and marine environments, anthropogenic Nr creation, Nr emissions and deposition, and riverine discharge. A better understanding of the anthropogenic processes that create Nr has led to more precise estimates, and spatially defined databases have led to a better understanding of how nitrogen fluxes vary. Although the current estimates reflect

Table 4 Global creation and distribution rates of Nr (TgN yr^{-1}).

	~1970 (Delwiche, 1970)	1970 (Svensson and Söderlund, 1976)	~1980 (Rosswall, 1983[a])	1990 (Galloway et al., 1995)	1990s (Schlesinger, 1997)
Natural Nr creation					
Terrestrial BNF	30	140	44–200	90–130	100
Marine BNF	10	30–130	1–130	40–200	15
Total lightning	7.6	?	0.5–30	3	5
Anthropogenic Nr creation					
Haber–Bosch	30	36	60	78	80
BNF, cultivation	14	89		43	40
Fossil-fuel combustion		19	10–20	21	24
Total terrestrial	74	194		255	249
Total global	174	274		375	264
Atmospheric emission					
NO$_x$, fossil-fuel combustion		19	10–20	21	24
NO$_x$, other		21–89	0–90	14.5	24
Terrestrial NH$_3$		113–244	36–250	53	62
Marine NH$_3$		0		13	13
Total emissions		253		102	123
Atmospheric deposition					
Terrestrial NO$_y$		32–83	110–240	26.5	30
Marine NO$_y$		11–33		12.3	14[b]
Terrestrial NH$_x$	91–186		40–116	52	40
Marine NH$_x$		19–50		17	16[b]
Organic N			10–100		
Total deposition		253	173–496	110	100
Riverine flux to coast	30	13–24	13–40	76[c]	36[d]
Denitrification[e]					
Continental N$_2$O		16–69	16–69	9.1	11.7
Marine N$_2$O		20–80	9–90	2	4
Continental N$_2$	43	91–92	43–390	130–290	13–233
Marine N$_2$	40	5–99	0–330	150–180	110
Total denitrification	83	236		386	249

[a] Deposition values for land plus ocean; NH$_3$ emissions include marine.
[b] Wet.
[c] Total.
[d] Dissolved.
[e] N$_2$O emissions are included with the realization that N$_2$O is also produced during nitrification.

definite improvements, there is still substantial uncertainty about these fluxes.

One of the first global-scale nitrogen cycles was created by Delwiche (1970) (Table 4). He noted that industrial fixation (Haber–Bosch) was the flux with the highest confidence and that all other fluxes "could well be off by a factor of 10." He also stated, and perhaps was the first to do so, that anthropogenic Nr was accumulating in the environment, because denitrification was not keeping up with increased Nr created by humans. Shortly thereafter, Svensson and Söderlund (1976) published a budget based on 1970 data. Their budget was more complete (atmospheric emission and deposition were included). Rosswall (1992) used a number of sources to estimate global-scale nitrogen fluxes for ~1980. In the last several years, several papers have addressed the nitrogen cycle on the global scale (e.g., Ayres et al., 1994; Mackenzie, 1994; Galloway et al., 1995; Vitousek et al., 1997; Galloway, 1998; Seitzinger and Kroeze, 1998; Galloway and Cowling, 2002).

To illustrate the impact of this significant increase in the rate of Nr creation on the global nitrogen cycle, Table 5 contrasts Nr creation and distribution in 1890 with those in 1990 using Galloway and Cowling (2002) as the primary source. Because there was limited Nr created by human

Table 5 Global Nr creation and distribution in 1890 and 1990 (TgN yr^{-1}).

	1890	1990
Natural Nr creation		
BNF, terrestrial	100	89
BNF, marine	120	120
Total lightning	5	5
Anthropogenic Nr creation		
Haber–Bosch	0	85
BNF, cultivation	15	33
Fossil-fuel combustion	0.6	21
Nr creation		
Total terrestrial	121	233
Total global	241	353
Atmospheric emission		
NO_x, fossil-fuel combustion	0.6	21
NO_x, other	6.2	13.0
NH_3, terrestrial	8.7	43.0
NH_3, marine	8	8
Total emissions	24	85
Atmospheric deposition		
NO_y, terrestrial	8	33
NO_y, marine	5	13
NH_x, terrestrial	8	43
NH_x, marine	12	14
Total deposition	33	103
Riverine flux to coast (DIN)	5	20

Source: Galloway and Cowling (2002). N_2O emissions are included here with the realization that N_2O is also produced during nitrification; deposition values are for land plus ocean.

activities in 1890, it is an appropriate point to begin, examining the nitrogen cycle. Although the global population was ~25% its population at the turn of twenty-first century, the world was primarily agrarian and produced only 2% of the energy and 10% of the grain produced today. Most energy (75%) was provided by biomass fuels; coal provided most of the rest (Smil, 1994). Little petroleum or natural gas was produced and what was produced was of little consequence when compared to the global supply of energy and the creation of Nr as NO_x through combustion.

16.5.2 Nr Creation

Fossil-fuel combustion created only ~0.6 Tg N yr^{-1} in 1890 through production of NO_x (Table 5). Crop production was primarily sustained by recycling crop residue and manure on the same land where food was raised. Since the Haber–Bosch process had not yet been invented, the only new Nr created by human activities was by legume and rice cultivation (the latter promotes Nr creation because rice cultivation creates an anaerobic environment that enhances nitrogen fixation). Although estimates are not available for 1890, Smil (1999) estimates that in 1900 cultivation-induced Nr creation was ~15 Tg N yr^{-1}. Additional Nr was mined from guano (~0.02 TgN yr^{-1}) and nitrate deposits (~0.13 TgN yr^{-1}) (Smil, 2000).

Thus, in 1890 the total anthropogenic Nr creation rate was ~15 TgN yr^{-1}, almost all of which was from food production. In contrast, the natural rate of Nr creation was ~220 TgN yr^{-1}. Terrestrial ecosystems created ~100 TgN yr^{-1} and marine ecosystems created ~120 TgN yr^{-1} (within a range of 87–156 TgN yr^{-1}) (D. Capone, personal communication). An additional ~5 TgN yr^{-1} was fixed by lightning. Thus, globally human activities created ~6% of the total Nr fixed and ~13% when only terrestrial systems are considered.

One century later the world's population had increased by a factor of ~3.5, from ~1.5 billion to ~5.3 billion, but the global food and energy production increased approximately sevenfold and 90-fold, respectively. Just as was the case in 1890, in 1990 (and now) food production accounts for most new Nr created. The largest change since 1890 has been the magnitude of Nr created by humans. Smil (1999) estimates that in the mid-1990s cultivation-induced Nr production was ~33 TgN yr^{-1}. The Haber–Bosch process, which did not exist in 1890, created an additional ~85 TgN yr^{-1} in 1990, mostly for fertilizer (~78 TgN yr^{-1}) and the remainder in support of industrial activities such as the manufacture of synthetic fibers, refrigerants, explosives, rocket fuels, nitroparaffins, etc.

From 1890 to 1990, energy production for much of the world was transformed from a bio-fuel to a fossil-fuel economy. The increase in energy production by fossil fuels resulted in increased NO_x emissions—from ~ 0.6 TgN yr^{-1} in 1890 to ~ 21 TgN yr^{-1} in 1990. By 1990 over 90% of the energy produced created new Nr. There was substantial atmospheric dispersal. Thus, in 1990 Nr created by anthropogenic activities was ~ 140 TgN yr^{-1}, an almost ninefold increase over 1890 even though there was only a ~ 3.5-fold increase in global population. With the increase in Nr creation by human activities came a decrease in natural terrestrial nitrogen fixation (from ~ 100 TgN yr^{-1} to ~ 89 TgN yr^{-1}) because of the conversion of natural grasslands and forests to croplands, etc. (C. Cleveland, personal communication).

16.5.3 Nr Distribution

What is the fate of anthropogenic Nr? The immediate fate for the three anthropogenic sources is clear: NO_x from fossil-fuel combustion is emitted directly into the atmosphere; $R-NH_2$ from rice and legume cultivation is incorporated into biomass; NH_3 from the Haber–Bosch process is converted primarily into commercial fertilizer applied to agro-ecosystems to produce food. However, little fertilizer nitrogen actually enters the human mouth in the form of food; in fact, most created Nr is released to environmental systems (Smil, 1999).

In 1890 both the creation and fate of Nr were dominated by natural processes (Figure 3(a)). Only limited Nr was transferred via atmospheric and hydrologic pathways compared to the amount of Nr created. For terrestrial systems, of the ~ 115 TgNr yr^{-1} created, only about ~ 15 TgN yr^{-1} were emitted to the atmosphere as either NH_3 or NO_x. There was limited connection between terrestrial and marine ecosystems; only about 5 TgN yr^{-1} of dissolved inorganic nitrogen were transferred via rivers into coastal ecosystems in 1890 and only about 17 TgN yr^{-1} were deposited to the ocean surface.

By contrast, in 1990 when creation of Nr was dominated by human activities (Figure 3(b)), Nr distribution changed significantly. Increased food production caused NH_3 emissions to increase from ~ 9 TgN yr^{-1} to ~ 43 TgN yr^{-1}, and NO_x emissions increased from ~ 7 TgN yr^{-1} to ~ 34 TgN yr^{-1} because of increased energy and food production. These increased emissions resulted in widespread distribution of Nr to downwind ecosystems (Figure 4). The transfer of Nr to marine systems also increased. By 1990 riverine fluxes of dissolved inorganic nitrogen to the coastal ocean had increased to 20 TgN yr^{-1}, and

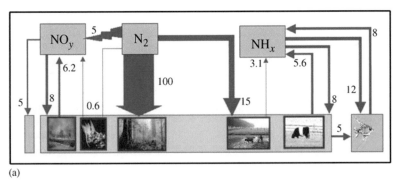

(a)

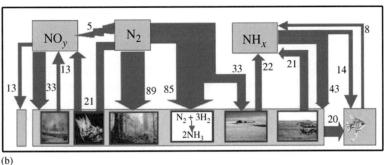

(b)

Figure 3 Global nitrogen budgets for: (a) 1890 and (b) 1990, TgN yr^{-1}. Emissions to the (left) NO_y box from (first from left) vegetation include agricultural and natural soil emissions and combustion of biofuel, biomass (savannah and forests), and agricultural waste and emissions from (second from left) coal reflect fossil-fuel combustion. Emissions to the (right) NH_x box from (third from right) agricultural fields include emissions from agricultural land and combustion of biofuel, biomass (savannah and forests), and agricultural waste, and emissions from (second from right) the cow and feedlot reflect emissions from animal waste. For more details, see text for "global N cycle: past and present" (source Galloway and Cowling, 2002).

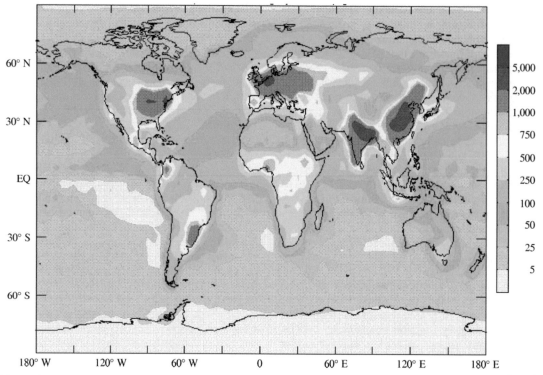

Figure 4 Global atmospheric deposition of Nr to the oceans and continents of the Earth in 1993 (mg N m^{-2} yr^{-1}) (sources F. J. Dentener, personal communication; Lelieveld and Dentener, 2000).

atmospheric nitrogen deposition to marine regions had increased to 27 TgN yr^{-1}. Although evidence suggests that most riverine nitrogen is denitrified in coastal and shelf environments (Seitzinger and Giblin, 1996), most atmospheric flux is deposited directly to the open ocean, with part of the 27 TgN yr^{-1} deposited to coastal ocean and shelf regions with significant ecological consequences (Rabalais, 2002).

Another Nr species emitted to the atmosphere, N$_2$O, bears mention. Produced by nitrification and denitrification, natural emissions of N$_2$O are $\sim$9.6 TgN yr^{-1} primarily from the oceans and tropical soils (Mosier *et al.*, 1998). As compiled in Ehhalt *et al.* (2001), several recent estimates of anthropogenic N$_2$O emissions for the 1990s range from 4.1 TgN yr^{-1} to 8.1 TgN yr^{-1}; the most recent estimate is 6.9 TgN yr^{-1} (Ehhalt *et al.*, 2001). N$_2$O accumulates either in the troposphere (current estimate is 3.8 TgN yr^{-1}) or is destroyed in the stratosphere.

16.5.4 Nr Conversion to N$_2$

Another key component missing from Figure 3 is the ultimate fate of the $\sim$140 TgN yr^{-1} Nr created by human action in 1990. On a global basis, Nr created by human action is either accumulated (stored) or denitrified. Several recent studies estimate denitrification in land and associated freshwaters relative to inputs for large regions. At the

scale of continents, denitrification is estimated as 40% of Nr inputs in Europe (van Egmond *et al.*, 2002) and $\sim$30% in Asia (Zheng *et al.*, 2002). At the scale of large regions, denitrification estimates as a percentage of Nr inputs include 33% for land areas draining to the North Atlantic Ocean (Howarth *et al.*, 1996) and 37% for land areas draining to the Yellow-Bohai Seas (Bashkin *et al.*, 2002). Country-scale estimates of the percentage of Nr inputs denitrified to N$_2$ in soils and waters include $\sim$40% for the Netherlands (Kroeze *et al.*, 2003), 23% for the USA (Howarth *et al.*, 2002a), 15% for China (Xing and Zhu, 2002), and 16% for the Republic of Korea (Bashkin *et al.*, 2002).

On a watershed scale, van Breemen *et al.* (2002) estimate that 47% of total Nr inputs to the collective area of 16 large watersheds in the northeastern US are converted back to N$_2$: 35% in soils and 12% in rivers. Moreover, within the Mississippi River watershed, Burkart and Stoner (2001) divide the basin into six large sub-basins and conclude that soil denitrification losses of N$_2$ range from a maximum of $\sim$10% of total inputs in the Upper Mississippi region to less than 2% in the Tennessee and Arkansas/Red regions. Goolsby *et al.* (1999) estimate denitrification within soils of the entire Mississippi–Atchafalaya watershed to be $\sim$8% but do not quantify additional denitrification losses of Nr inputs in river systems. Relative to inputs, landscape-scale estimates of Nr denitrified to N_2 in terrestrial systems and associated freshwaters are quite variable reflecting major differences in the amount of nitrogen inputs available to be denitrified and in environmental conditions that promote this process. All the above-mentioned studies reporting denitrification at regional scales claim a high degree of uncertainty in their estimates, highlighting the fact that our knowledge of such landscape-level rates of denitrification is quite poor. The few estimates that do exist are subject to enormous uncertainties and must often be derived as the residual after all other terms in a regional nitrogen budget are estimated, terms which themselves are often difficult to constrain (e.g., Erisman *et al.*, 2001; van Breemen *et al.*, 2002).

The wetland/stream/river/estuary/shelf region provides a continuum with substantial capacity for denitrification. Nitrate is commonly found, there is abundant organic matter, and sediments and suspended particulate microsites offer anoxic environments. In this section we discuss denitrification in the stream/river/estuary/shelf continuum. Although there are several specific studies of denitrification in wetlands, the role of wetlands in Nr removal at the watershed scale needs to be better understood.

Seitzinger *et al.* (2002) estimate that, of the Nr that enters the stream/river systems that drain 16 large watersheds in the eastern US, 30–70% can be removed within the stream/river network, primarily by denitrification. In an independent analysis on the same watersheds, van Breemen *et al.* (2002) estimated by difference that the lower end of the Seitzinger *et al.* (2002) range is the more likely estimate. For Nr that enters estuaries, 10–80% can be denitrified, depending primarily on the residence time and depth of water in the estuary (Seitzinger, 1988; Nixon *et al.*, 1996).

Of the Nr that enters the continental shelf environment of the North Atlantic Ocean from continents, Seitzinger and Giblin (1996) estimate that >80% is denitrified. The extent of the denitrification depends, in part, on the size of the continental shelf. The shelf denitrification potential is large on the south and east coasts of Asia and the east coasts of South America and North America. These are also the regions where riverine Nr inputs are the largest. Although most riverine/estuary Nr that enters the shelf region is denitrified, total denitrification in the shelf region is larger than that supplied from the continent and thus Nr advection from the open ocean is required.

The uncertainties about these estimates of denitrification on large scales are great enough that the relative importance of denitrification versus storage is unknown. However, the calculations do support the findings that denitrification is important in all portions of the stream/river/estuary/shelf continuum and that, on a global basis, the continuum is a permanent sink for Nr created by human action.

In summary, the approximate doubling of new Nr added to the world between 1890 and 1990 had a substantial impact on the global nitrogen cycle (Figure 3):

- atmospheric emissions of NO_x increased from $\sim$14 TgN yr^{-1} to 39 TgN yr^{-1};
- atmospheric emissions of NH_3 increased from $\sim$17 TgN yr^{-1} to $\sim$50 TgN yr^{-1};
- terrestrial N_2O emissions increased from 8 Tg N yr^{-1} to 12 TgN yr^{-1};
- riverine discharge of DIN to the coastal zone increased from 5 TgN yr^{-1} to 20 TgN yr^{-1};
- deposition of inorganic nitrogen to continents increased from 16 TgN yr^{-1} to 76 TgN yr^{-1}; and
- deposition of inorganic nitrogen to oceans increased from 17 TgN yr^{-1} to 27 TgN yr^{-1}.

In 1990, human action dominated the addition of new Nr into the landscape. Substantial added Nr is redistributed throughout environmental reservoirs via atmospheric and riverine transport with, as will be demonstrated in a later section, substantial and long-lasting consequences.

16.6 GLOBAL MARINE NITROGEN BUDGET

For the most part, nitrogen cycles within the ocean have the same microbial processes as in the soil (Figure 1). BNF is the dominant source of new nitrogen to the ocean, with other smaller contributions coming from atmospheric deposition and riverine runoff. Ammonification, nitrification, uptake, and decomposition are all critical components. There are two primary removal mechanisms—denitrification and burial in marine sediments. The former is by far the most important.

The pool of Nr in the oceans is $\sim$570 Gt (as $N-NO_3$) (Table 2). Over long periods of time, the inputs and outputs of Nr (and thus the pool size) have to balance; however, over the short term (e.g., centuries), they are probably not in balance given the different controls on the two major fluxes, BNF and denitrification.

Because many ocean regions are nitrogen limited, once Nr is added, it is rapidly used by phytoplankton in the euphotic zone. Zooplankton graze on the phytoplankton. Nr can recycle within the euphotic zone or can be transferred to the subeuphotic zone as total organic nitrogen (TON) where most is mineralized to nitrate and some is deposited to the sediments.

Estimates of marine BNF rates have varied substantially over the last few decades. In the 1970 and early 1980s, estimates ranged from 1 to 130 (Delwiche, 1970; Svensson and Söderlund,

1976). As recently as the late 1980s and early 1990s, global marine BNF rate was ~10–15 TgN yr^{-1} (Capone and Carpenter, 1982; Walsh et al., 1988; Carpenter and Romans, 1991). However, the more recent estimates range from 100 TgN yr^{-1} to 200 TgN yr^{-1} (Karl et al., 2002). In support of this higher range, a recent assessment of marine BNF (D. Capone, personal communication; Galloway et al., 2003a) gives a global range of 87–156 TgN yr^{-1} (mostly in pelagic regions) and states that all ocean basins are important contributors (Table 6).

The other Nr sources to the ocean are riverine injection and atmospheric deposition. Over the last few decades, riverine inputs to the ocean have ranged from 13 TgN yr^{-1} to 76 TgN yr^{-1} with the more recent estimates ~40–50 TgN yr^{-1} (Table 4). These inputs to coastal regions have significantly altered the associated ecosystems (NRC, 2000; Rabalais, 2002).

Using an empirical model that calculates total nitrogen riverine inputs from the inputs of new Nr to a watershed (BNF, nitrogen fertilizer consumption, cultivation-induced BNF, and atmospheric deposition of NO$_y$ from fossil-fuel combustion), Galloway et al., (2003a) quantify riverine Nr export from world regions for 1990. They find that ~59 TgN yr^{-1} is discharged via riverine export, with ~11 TgN yr^{-1} transported to inlands and drylands and ~48 TgN yr^{-1} transported to coastal waters. Asia has the highest rate of Nr transported to inland waters/drylands (5.1 TgN yr^{-1}) and North America the lowest. Interestingly, about twice as much Nr is transported to the inland waters/drylands of Oceania than is transported to the coast. A comparison of Nr riverine transport to coasts shows that the highest transport rate (~16.7 TgN yr^{-1}) is in Asia, with transport in all other regions except Oceania in the range of 6–9 TgN yr^{-1}.

Atmospheric deposition provides ~27 Tg N yr^{-1} of total NO$_y$ and NH$_x$ deposition (Figure 3). The deposition of organic nitrogen is potentially important. On a global basis, Neff et al. (2002) estimate that atmospheric deposition of organic nitrogen ranges from 10 TgN yr^{-1} to 50 TgN yr^{-1}, with highest rates (for marine regions) in the coastal zone. Atmospheric deposition to coastal and shelf regions (defined by the 200 m depth line) is ~8 TgN yr^{-1} compared to ~25 TgN yr^{-1} to the open ocean. Nr inputs to the ocean are balanced primarily by denitrification and, to a much lesser extent, by sediment burial.

River discharge, atmospheric deposition, and BNF introduce Nr to the coastal oceans. Riverine injection is the most important even though, in some regions, atmospheric nitrogen deposition can be as much as 40% of total inputs and has a measurable effect on coastal productivity (Paerl et al., 2002). However, since most Nr that enters the coastal and shelf region is denitrified, there is little continental-based Nr transferred to the open ocean (Nixon et al., 1996; Seitzinger and Giblin, 1996). Although open oceans are essentially chemically decoupled from continents, they are connected through the atmosphere, which deposits ~27 TgN yr^{-1} to the global ocean.

16.7 REGIONAL NITROGEN BUDGETS

An analysis of regional nitrogen budgets is important as it illustrates the differences in Nr creation and distribution as a function of level of development and geographic location. In addition, the short atmospheric lifetimes of NO$_x$ (and most of its reaction products), and NH$_x$ (one day to one week, depending on total burden and altitude) mean that these chemical species vary substantially in both space and time. Atmospheric transport and dynamics, nitrogen emissions, chemical processing, and removal mechanism (dry versus wet deposition) all interact to alter the spatial distribution of reactive nitrogen. Dividing the Earth's land surface by geopolitical region offers a convenient mechanism for examining this spatial variation.

There are several recent studies of the nitrogen cycle on regional scale: Asia—Galloway (2000), Bashkin et al. (2002), and Zheng et al. (2002); North Atlantic Ocean and watershed—Galloway et al. (1996) and Howarth et al. (1996); oceans—Karl (1999) and Capone (2001); Europe—van Egmond et al. (2002); and the US—Howarth et al. (2002a). To give an appreciation of how the regional budgets vary, this section contrasts Nr creation for several regions of the world: Africa, Asia, Europe, Latin America, North America, and Oceania, as defined by the United Nations Food and Agriculture Organization (FAO, 2000).

For the mid-1990s in each region, Galloway and Cowling (2002) estimate the amount of Nr created by fertilizer production, fossil-fuel combustion (F. J. Dentener, personal communication), and rice/legume cultivation (FAO, 2000). Most fertilizer was produced in Asia (40.1 Tg N yr^{-1}), Europe (21.6 TgN yr^{-1}), and North America (18.3 Tg N yr^{-1}), with smaller amounts in Latin America, Africa, and Oceania (3.2 Tg N yr^{-1}, 2.5 Tg N yr^{-1}, and 0.4 TgN yr^{-1}, respectively). The other Nr creation process involved in food production is cultivation-induced BNF. Smil (1999) estimates that ~33 TgN yr^{-1} were globally produced by cultivation in the mid-1990s. The regional breakdown for these BNF rates from cultivation of seed legumes, rice, and sugar cane is 13.7 Tg N yr^{-1} for Asia; ~6.0 Tg N yr^{-1}, ~3.9 Tg N yr^{-1}, and ~5.0 Tg N yr^{-1} for North America, Europe plus the former Soviet Union, and Latin America, respectively; and ~<2 TgN yr^{-1} for Africa and Oceania. Most Nr creation by fossil-fuel

Table 6 N fixation and denitrification by ocean basin (Tg N yr^{-1}).

	N₂ fixation		High N₂ fixation		Denitrification		High denitrification	
North Atlantic								
Pelagic	12	Gruber and Sarmiento (1997)	42	Capone (pers. com.)	0	Christensen et al. (1987)	3	Codispoti et al. (2001)
Shelf	0.38	Capone (1983)	3.7	Howarth et al. (1988)	15	Capone (pers. com.)	75	Devol (1991)
Deep sediments	0		0		1.2		2.4	Bender et al. (1977)
South Atlantic								
Pelagic	5	Capone (pers. com.)	21	Capone (pers. com.)	0	Christensen et al. (1987)	2	Codispoti et al. (2001)
Shelf	0.38	Capone (1983)	3.7	Howarth et al. (1988)	15	Capone (pers. com.)	75	Devol (1991)
Deep sediments	0		0		1.2		2.4	Bender et al. (1977)
North Pacific								
Pelagic	29	Deutsch et al. (2001)	35	Capone (pers. com.)	22	Deutsch et al. (2001)	40	Codispoti et al. (2001)
Shelf	0.28	Capone (1983)	2.8	Howarth et al. (1988)	14	Christensen et al. (1987)	70	Devol (1991)
Deep sediments	0		0		2.1	Capone (pers. com.)	4.2	Bender et al. (1977)
South Pacific								
Pelagic	19	Deutsch et al. (2001)	24	Capone (pers. com.)	26	Deutsch et al. (2001)	40	Codispoti et al. (2001)
Shelf	0.28	Capone (1983)	2.8	Howarth et al. (1988)	7	Christensen et al. (1987)	35	Devol (1991)
Deep sediments	0		0		2.1	Capone (pers. com.)	4.2	Bender et al. (1977)
Indian Ocean								
Pelagic	20	Bange et al. (2000)	19	Capone (pers. com.)	33	Bange et al. (2000)	65	Codispoti et al. (2001)
Shelf	0.19	Capone (1983)	1.9	Howarth et al. (1988)	6.4	Christensen et al. (1987)	32	Devol (1991)
Deep sediments	0		0		1.8	Capone (pers. com.)	3.6	Bender et al. (1977)
All pelagic	85		141		81		150	
All shelves	1.5		14.9		57		287	
All deep sediments	0		0		8.4		17	
Total	87		156		147		454	

Source: Galloway et al. (2003b).

combustion was in North America, Europe plus the former Soviet Union, and Asia ($\sim$7.4 TgN yr^{-1}, $\sim$6.6 Tg N yr^{-1}, and $\sim$6.4 Tg N yr^{-1}, respectively). Latin America, Africa, and Oceania by comparison had more modest Nr creation rates from fossil-fuel combustion with each being <1.5 TgN yr^{-1} (Table 7).

In addition to the regional creation of Nr, Nr is also exchanged among regions. Exports of nitrogen-containing fertilizer, plant material (e.g., grain), and meat from one region can be a source of nitrogen for another region. As reported in Galloway and Cowling (2002), fertilizer is the commodity most often exchanged between regions. In 1995, the global production of nitrogen fertilizers was $\sim$86 TgN yr^{-1}. Of this amount, $\sim$24.9 TgN yr^{-1} was exported to other regions. Over half of the exports were from Europe ($\sim$13.2 TgN yr^{-1}). Other regions with significant exports were Asia ($\sim$10.7 TgN yr^{-1}) and North America ($\sim$5.2 TgN yr^{-1}). The primary receiving regions were Asia ($\sim$7.6 TgN yr^{-1}) and Europe (TgN yr^{-1}). Thus, although $\sim$30% of the fertilizer nitrogen produced was exported, the only region that had a large net loss was Europe ($\sim$6.6 TgN yr^{-1}); the largest net gain of any region was for Asia ($\sim$6.4 TgN yr^{-1}).

The next most frequently exchanged commodity is plant material, mostly cereal grains. Asia, Europe, and Africa had net gains in nitrogen-containing plant material ($\sim$2.2 TgN yr^{-1}, $\sim$1.0 TgN yr^{-1}, $\sim$0.5 TgN yr^{-1}, respectively), whereas North America, Latin America, and Oceania had net losses ($\sim$2.8 TgN yr^{-1}, $\sim$0.8 TgN yr^{-1}, and $\sim$0.2 TgN yr^{-1}, respectively). Net meat exchange was <0.1 TgN yr^{-1} for each region. Summing all three categories, Asia gained the most Nr: $\sim$8.7 TgN yr^{-1}. Although Oceania and Africa also gained nitrogen, the gains were small ($\sim$0.3 TgN yr^{-1} and $\sim$0.2 TgN yr^{-1}, respectively). Europe and North America had net losses of $\sim$5.6 TgN yr^{-1} and $\sim$3.3 TgN yr^{-1}, respectively. Latin America had a small net loss. When these net import/exports are added to the Nr creation in each region (Table 7), North America, with $\sim$5% of the world's population, was responsible for creating $\sim$20% of the

world's Nr. Africa, with $\sim$13% of the world's population, was responsible for creating $\sim$6% of the world's Nr.

Given these regional differences, it is also interesting to express Nr creation and use on a per capita basis (calculated from Table 7 (not including net import/export) and FAO population data (2000)) to illustrate the average amount of Nr mobilized per person, by region (Table 8). At one extreme North America mobilized $\sim$100 kg N person^{-1} yr^{-1}, and at the other extreme Africa mobilized about an order of magnitude less or $\sim$6.8 kg N person^{-1} yr^{-1}. In all regions, food production was larger than energy production, and the primary causes for the regional differences were the amounts of nitrogen mobilized per capita in producing food. In North America, it was $\sim$80 kg N person^{-1} yr^{-1}, and in Africa it was $\sim$6 kg N person^{-1} yr^{-1}. These values show the regions responsible for Nr creation in excess of what is needed for human body sustenance ($\sim$2 kg N capita^{-1} yr^{-1}).

16.8 CONSEQUENCES

16.8.1 Introduction

Much has been written about the impacts of increased Nr creation and its accumulation in environmental reservoirs. The largest impact is, of course, that much of the world's population is sustained by the increased food production made possible by the Haber–Bosch process and by cultivation-induced BNF (Smil, 2000). However, as noted in Galloway et al. (2003b), there are also substantial negative impacts.

- Nr increase leads to the production of tropospheric ozone and aerosols that induce serious respiratory diseases, cancer, and cardiac diseases in humans (Pope et al. 1995; J. R. Follett and R. F. Follett, 2001; Wolfe and Patz, 2002).
- Nr increases and then decreases forest and grassland productivity wherever atmospheric Nr deposition has increased significantly and critical thresholds have been exceeded. Nr additions probably also decrease biodiversity in many natural habitats (Aber et al., 1995).

Table 7 Nr creation rates for various regions of the world in mid-1990s (TgN yr^{-1}).

World regions	Fertilizer production	Cultivation	Combustion	Net import/export	Total
Africa	2.5	1.8	0.8	0.2	5.3
Asia	40.1	13.7	6.4	8.7	68.9
Europe + FSU	21.6	3.9	6.6	−5.6	26.5
Latin America	3.2	5.0	1.4	−0.2	9.4
North America	18.3	6.0	7.4	−3.3	28.4
Oceania	0.4	1.1	0.4	0.3	2.2
World	$\sim$86	$\sim$30	$\sim$23	0.1	$\sim$140

Source: Galloway and Cowling (2002).

- Nr is responsible (together with sulfur) for the acidification and loss in biodiversity of lakes and streams in many regions of the world (Vitousek *et al.*, 1997).
- Nr is responsible for eutrophication, hypoxia, loss of biodiversity, and habitat degradation in coastal ecosystems. Nr is now considered the biggest pollution problem in coastal waters (e.g., Howarth *et al.*, 2000; NRC, 2000; Rabalais, 2002).
- Nr contributes to global climate change and stratospheric ozone depletion, both of which impact human and ecosystem health (e.g., Cowling *et al.*, 1998).

Although most of these nitrogen problems are being studied separately, research is increasingly indicating that these Nr-related problems are linked. For example, the same nitrogen that produces urban air pollution can also contribute to water pollution. These linkages, referred to as the nitrogen cascade, are discussed in detail in Galloway *et al.* (2003b). This section summarizes that discussion. The nitrogen cascade is defined as the sequential transfer of Nr through environmental systems, which results in environmental changes as Nr moves through or is temporarily stored within each system.

Two scenarios in Figure 5 illustrate the nitrogen cascade. The first example shows the fate of NO_x produced during fossil fuel combustion. In the sequence, an atom of nitrogen mobilized as NO_x in the atmosphere first increases ozone concentrations, then decreases atmospheric visibility and increases concentrations of small particles, and then increases precipitation acidity. After that same nitrogen atom is deposited into the terrestrial ecosystem; it may increase soil acidity (if a base cation is lost from the system), decrease biodiversity, and either increase or decrease ecosystem productivity. If discharged to the aquatic ecosystem, the nitrogen atom can increase surface-water acidity in mountain streams and lakes. Following transport to the coast, nitrogen atom can countribute to coastal eutrophication. If the

nitrogen atom is converted to N_2O and emitted back to the atmosphere, it can first increase greenhouse warming potential and then decrease stratospheric ozone.

Example 2 illustrates a similar cascade of effects of Nr from food production. In this case, atmospheric N_2 is converted to NH_3 in the Haber–Bosch process. The NH_3 is primarily used to produce fertilizer. About half the Nr fertilizer applied to global agro-ecosystems is incorporated into crops harvested from fields and used for human food and livestock feed (Smil, 1999, 2001). The other half is transferred to the atmosphere as NH_3, NO, N_2O, or N_2, lost to the aquatic ecosystems, primarily as nitrate or accumulate in the soil nitrogen pool. Once transferred downstream or downwind, the nitrogen atom becomes part of the cascade. As Figure 5 illustrates, Nr can enter the cascade at different places depending on its chemical form. An important characteristic of the cascade is that, once it starts, the source of the Nr (e.g., fossil-fuel combustion and fertilizer production) becomes irrelevant. Nr species can be rapidly interconverted from one Nr to another. Thus, the critical step is the *formation* of Nr.

The simplified conceptual overview of the cascade shown in Figure 5 does not cover some of its important aspects. The internal cycling of Nr within each step of the cascade is ignored; long-term Nr storage is not accounted for and neither are all the multiple pathways that Nr follows as it flows from one step to another. In a recent paper, Galloway *et al.* (2003b) cover these aspects for the atmosphere (troposphere and stratosphere), for terrestrial ecosystems (agro-ecosystems, forests, and grasslands), and for aquatic ecosystems (groundwater, wetlands, streams, lakes, rivers, and marine coastal regions). They evaluate the potential for accumulation and cycling of Nr within the system; the loss of Nr through conversion to N_2 by denitrification; the transfer of Nr to other systems; and the effects of Nr within the system. Galloway *et al.* (2003b) conclude that the cascade of Nr from one system to another will be enhanced if there is a limited potential for Nr to be accumulated or for the loss of N_2 through denitrification within a given system thereby increasing the potential for transfer to the next system. There will be a lag in the cascade if there is a large potential for accumulation within a system. The cascade will be decreased if there is a large potential for denitrification to N_2 within a system. Their analyses are summarized in Table 9, and the remainder of this section presents an overview of their findings (Galloway *et al.*, 2003b).

Table 8 Regional per capita Nr creation during food and energy production in mid-1990s (kg N person^{-1} yr^{-1}).

World regions	Food	Energy	Total
Africa	5.7	1.1	6.8
Asia	15	1.8	17
Europe + FSA	35	9.1	44
Latin America	16	2.8	19
North America	80	24	100
Oceania	50	13	63
World	20	3.9	24

Source: Galloway and Cowling (2002).

16.8.2 Atmosphere

The atmosphere receives Nr mainly as air emissions of NO_x, NH_3, and N_2O from aquatic and

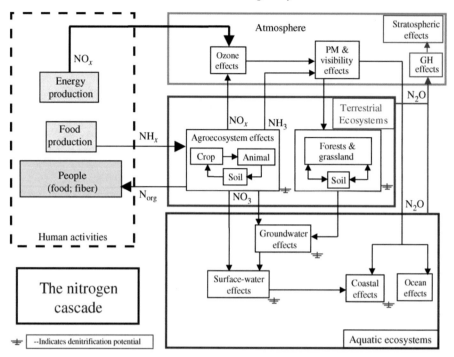

Figure 5 The nitrogen cascade illustrates the sequential effects a single atom of N can have in various reservoirs after it has been converted from a nonreactive to a reactive form (source Galloway *et al.*, 2003b).

terrestrial ecosystems and of NO_x from the combustion of biomass or fossil fuels. NO_x and NH_3 (and their reaction products), can accumulate in the troposphere on a regional scale. However, because of their short residence times in the atmosphere and lack of potential for formation of N_2 by denitrification, almost all Nr emitted as NO_x and NH_3 is transferred back to the Earth's surface within hours to days. There is also an internal cascade of effects. (NO increases the potential first for ozone and then for aerosol formation.) Except for N_2O, there is very limited potential for the long-term storage of Nr (and thus limited lag time), but there are significant effects from Nr while it remains in the atmosphere. There is no potential for denitrification back to N_2 within the troposphere, and a large potential for Nr transfer to the next receptors—terrestrial and aquatic ecosystems.

16.8.3 Terrestrial Ecosystems

Intensively managed agro-ecosystems and, even more so, confined animal feeding operations (CAFOs) are the technical means by which most human needs and dietary preferences are met. About 75% of the Nr created around the world by humans is added to agro-ecosystems to sustain food production. About 70% of that is from the Haber–Bosch process and ~30% from C-BNF (Figure 2). Smil (2001) estimates that ~40% of all the people alive today owe their life to the

production and wide use of fertilizers produced by the Haber–Bosch process.

Most Nr applied to agro-ecosystems is transferred to other systems along the nitrogen cascade; a much smaller portion is denitrified to N_2 (Table 9). Nitrogen-use efficiency in major grain and animal production systems can be improved through the collaboration of ecologists, agronomists, soil scientists, agricultural economists, and politicians. Actual fertilizer nitrogen-use efficiency, nitrogen losses, and loss pathways in major cropping and animal systems must be more accurately measured so that we can: (i) identify opportunities for increased nitrogen-use efficiency through improved crop and soil management; (ii) quantify nitrogen-loss pathways in major food crops, including animal feeding operations; and (iii) improve human understanding of local, regional, and global nitrogen balances and nitrogen losses from major cropping and animal systems.

The residence times (and lag times) of Nr in forests can be years to centuries depending on forest history, forest type, and Nr inputs. The effects of Nr accumulation in forests are numerous, most relate to changes in forest and microbial productivity and function. There is significant potential for Nr to be transferred to the atmosphere as NO and N_2O and especially to surface waters as NO_3^- once Nr additions or availability exceed biotic requirements. Relative to inputs in high Nr deposition areas, there is a limited potential for Nr

Table 9 Characteristics of different systems relevant to the nitrogen cascade.

System	Accumulation potential	Transfer potential	N_2 production potential	Links to systems down the cascade	Effects
Atmosphere	Low	Very high	None	All but groundwater	Human and ecosystem health, climate change
Agroecosystems	Low to moderate	Very high	Low to moderate	All	Human and ecosystem health, climate change
Forests	High	Moderate to high in places	Low	All	Biodiversity, net primary productivity, mortality, groundwater
Grasslands	High	Moderate to high in places	Low	All	Biodiversity, net primary productivity, groundwater
Groundwater	Moderate	Moderate	Moderate	Surface water, atmosphere	Human and ecosystem health, climate change
Wetlands, streams, lakes, and rivers	Low	Very high	Moderate to high	Atmosphere marine costal systems	Biodiversity, ecological structure, fish
Marine coastal regions	Low to moderate	Moderate	High	Atmosphere	Biodiversity, ecological structure, fish, harmful algal blooms

Source: Galloway *et al.* (2003b).

to be removed from the cascade through N_2 formation.

Unmanaged grasslands receive most of their Nr from BNF and atmospheric deposition, with the latter being much more important where deposition rates are large. As with forests, temperate grasslands are potentially major storage reservoirs and short- to long-term sinks within the nitrogen cascade.

Grasslands managed for animal production (e.g., cattle) are much more leaky with respect to loss of added Nr. The addition of fertilizer and/or grazing animals increases the amount of Nr available for loss, especially via the atmosphere (e.g., NH_3 (Sommer and Hutchings, 1997) and N_2O (Fowler *et al.*, 1997)). Thus, the effective residence times for managed ecosystems are potentially less than those for unmanaged systems.

16.8.4 Aquatic Ecosystems

Leaching from agro-ecosystems is the primary source of anthropogenic Nr for groundwater, although in some regions human waste disposal can also be important (Puckett *et al.*, 1999; Nolan and Stoner, 2000). Nitrate is the most common Nr species (Burkart and Stoner, 2001). Nitrogen is lost from groundwater both from denitrification to N_2 and from losses of Nr to surface waters and the atmosphere but the relative mix of these fates is site dependent, as is the residence time of Nr in groundwater reservoirs.

The effects of elevated Nr in groundwater do pose a significant human-health risk through contaminated drinking water. In the human body, nitrate is converted to nitrite, which can cause methemoglobinemia through interference with the ability of hemoglobin to take up O_2. Methemoglobinemia usually occurs after water with high concentrations of nitrate has been consumed; infants and people on kidney dialysis are particularly susceptible (J. R. Follett and R. F. Follett, 2001). Other potential effects associated with elevated concentrations of nitrate in drinking water include respiratory infection, alteration of thyroid metabolism, and cancers induced by conversion of nitrate to N-nitroso compounds in the body (J. R. Follett and R. F. Follett, 2001; Wolfe and Patz, 2002).

Groundwater systems are accumulating Nr, and the effects of contaminated groundwater can be significant. Nr is lost from groundwater through denitrification, from advection of nitrate to surface waters, and from conversion to gaseous Nr forms that diffuse or flow to the atmosphere. Just as there is a lag in Nr release in forests because of the nitrogen-saturation phenomena (Aber *et al.*, 2003), there is also a lag for Nr release from groundwaters to surface waters but this is highly variable among sites (Groffman *et al.*, 1998;

Puckett and Cowdery, 2002). At some time, the Nr introduced into the groundwater will be transferred to surface water if the Nr is not permanently stored in groundwater or denitrified to N_2.

Surface freshwater ecosystems consist of wetlands (e.g., bogs, fens, marshes, swamps, prairie potholes, etc.), streams, lakes (and artificial reservoirs), and rivers. Surface freshwater ecosystems receive most of their Nr from their associated watersheds, from atmospheric deposition, and from BNF within the system. There is limited potential for Nr to accumulate within surface-water ecosystems, because the residence time of Nr within surface waters, like the water itself, is very brief. Residence times may be relatively longer in the sediments associated with wetlands and some larger lakes but are still short when compared to terrestrial ecosystems or the oceans.

However, many headwater streams and lakes are now in highly disturbed landscapes and thus have high nitrate concentrations; this can lead to eutrophication problems either locally or further downstream. In addition, for water in headwater streams and lakes draining poorly buffered soils, increased nitrate concentrations can acidify streams and impact biota.

The potential for Nr accumulation in streams, lakes, rivers, and associated wetlands is small (Table 9). Although changes in Nr inputs to surface waters may significantly alter the internal cycles in these systems, globally such changes are merely extremely efficient avenues for propagating the effects of the nitrogen cascade from higher to lower components. Even though wetlands may delay or prevent this transfer locally (and denitrification may short-circuit some of the Nr transport along the way), surface freshwater ecosystems essentially move Nr from the mountains to the sea and ensure that perturbations at one point in the cascade quickly lead to changes elsewhere.

Most Nr in coastal ecosystems (e.g., estuaries) is from river water and groundwater; direct atmospheric deposition is an important source in some systems, and inputs from the ocean are important in others. These inputs have increased several-fold as a consequence of human activities (Meybeck, 1982; Howarth et al., 1996; Seitzinger and Kroeze, 1998; NRC, 2000; Howarth et al., 2002b). Because of the dynamic nature of coastal ecosystems, there is limited potential for Nr accumulation. In addition, although the potential for Nr to be transferred to the continental shelf regions is large because of the high rates of denitrification (mostly as N_2), transport to the shelf is limited, and the Nr that is transferred is mostly converted to N_2 before being transported to the open ocean.

Although Nr has a short residence time in coastal ecosystems (compared to terrestrial ecosystems), the time that it does spend there can have a profound impact on the ecosystem. Primary production in most coastal rivers, bays, and seas of the temperate zone is limited by Nr supplies (Vitousek and Howarth, 1991; Nixon et al., 1996; NRC, 2000). In the USA, the increased Nr flux is now viewed as the most serious pollution problem in coastal waters (Howarth et al., 2000; NRC, 2000; Rabalais et al., 2002). One-third of the nation's coastal rivers and bays are severely degraded, and another one-third is moderately degraded from nutrient over enrichment (Bricker et al., 1999).

An obvious consequence of increasing Nr inputs to coastal waters over the past few decades has been an increase in the size of water masses that are anoxic (completely devoid of oxygen) or hypoxic (concentrations of oxygen less than 2–3 mg L^{-1}). These "dead zones" can be found in the Gulf of Mexico, Chesapeake Bay, Long Island Sound, Florida Bay, the Baltic Sea, the Adriatic Sea, and many other coastal areas (Diaz and Rosenberg, 1995; NRC, 2000).

Other major effects of Nr increases in coastal regions include the loss of seagrass beds and macro-algal beds and changes in coral reefs (Lapointe and O'Connell, 1989; Valiela et al., 1997; NRC, 2000; Howarth et al., 2000). Thus, Nr inputs to coastal ecosystems have increased significantly over the last few decades. Although most Nr is eventually denitrified to N_2 within the coastal ecosystems and its associated shelf, significant and widespread impacts on various ecosystem components and on human health remain (Table 9).

16.9 FUTURE

In 1995 the rate of global creation of Nr from human activities was $\sim$140 TgN yr^{-1}, which equates to an average per capita Nr creation rate of $\sim$24 kg N person^{-1} yr^{-1}: 20 kg N person^{-1} yr^{-1} attributable to food production and 4 kg N person^{-1} yr^{-1} attributable to energy production. These figures varied greatly in different areas. In Africa, the per capita Nr creation rate in the mid-1990s was 6.8 kg N yr^{-1} (5.7 kg N capita^{-1} yr^{-1} for food production and 1.1 kg N capita^{-1} yr^{-1} for energy production) and in North America it was $\sim$100 kg N capita^{-1} yr^{-1} of which $\sim$80 kg N capita^{-1} yr^{-1} was used to sustain food production (Table 8).

What will it be in 2050? The United Nations (1999) estimates that the world population will be $\sim$8.9 billion. Some of the numerous estimates of future energy and food production scenarios use sophisticated scenario development (e.g., Houghton et al., 2001). For this analysis, I use a simple calculation to set a range of Nr creation rates using projections for population increases and two assumptions of per capita Nr creation rate. To estimate the lower end of the possible range of Nr creation in 2050, I assume that, although the population increases, the Nr per capita creation rate by region will be the same in 2050 as it was in the mid-1990s (Table 8). Multiplying 2050 population rates by 1995 Nr creation rates gives a global total of $\sim$190 TgN yr^{-1} created (compared to $\sim$140 TgN yr^{-1} in 1990), with most of the increase occurring (by definition) in those regions where population increases are the largest (e.g., Asia) (Figure 6). To estimate an upper limit of Nr creation in 2050, I assume that all the world's peoples will have Nr creation rates the same as those in North America in 1990 (100 kg N capita^{-1} yr^{-1}). The resulting global total is $\sim$960 TgN yr^{-1} for 2050 or about a factor of 7 greater than the 1990 rate. Latin America, Africa, and Asia show the largest relative increases and Asia has the largest absolute increase (Figure 6).

The likelihood that the 2050 Nr creation rates at either end of the range will occur is slight. In all probability per capita Nr creation rates will rise to some degree, especially in those areas that will experience the largest population increases; therefore, the lower estimate is probably indeed low. Although the actual level reached in 2050 will probably be larger than the low estimate, it is also very likely to be less (much less?) than the high estimate. This, however, does not preclude the eventuality of the high estimate being reached sometime after 2050, after population growth has leveled off but during a prolonged period of growth in per capita Nr creation.

Given the concerns about Nr, it is perhaps unlikely that the high estimate will be reached. Whatever the final maximum Nr creation rate turns out to be, however, will depend greatly on how the world manages its use of nitrogen for food production and its control of nitrogen in energy production.

There is ample opportunity for a creative management strategy for Nr that will optimize food and energy production as well as environmental health. Several papers in the special issue (**31**) of *Ambio* devoted to the plenary papers of the *Second International Nitrogen Conference* address the issue of Nr management to maximize food and energy production while protecting the health of people and ecosystems: management of energy producing systems (Moomaw, 2002; Bradley and Jones, 2002) and management of food production systems (Cassman *et al.*, 2002; Fixen and West, 2002; Oenema and Pietrzak, 2002; Roy *et al.*, 2002; Smil, 2002). Other recent publications focus more generally on policy issues related to nitrogen (Erisman *et al.*, 2001; Mosier *et al.*, 2001; Melillo and Cowling, 2002).

16.10 SUMMARY

The global cycle of nitrogen has changed over the course of a century from one dominated by natural processes to one dominated by anthropogenic processes. The increased abundance of Nr has positive and negative effects that can occur in sequence as Nr cascades through the environment. A critical topic for research is the fate of anthropogenic Nr. Where is it going? How much is stored? How much is denitrified? A broader question that needs to be answered is how can society optimize nitrogen's beneficial role in sustainable food production and minimize nitrogen's negative effects on human health and the environment resulting from food and energy production.

These critical issues will require a new way of examining how people manage the environment and use the Earth's resources.

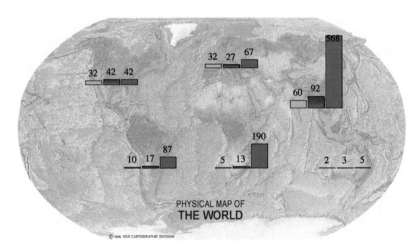

Figure 6 Continental Nr creation rates for (left column) 1995 and low (center column) and high (right column) estimates for 2050, TgN yr^{-1}.

ACKNOWLEDGMENTS

I thank Aaron Mills and Fred Mackenzie for discussions on various aspects of this paper. I thank Mary Scott Kaiser for putting the paper into the correct format and for editing the manuscript. I also thank The Ecosystems Center of the Marine Biological Laboratory and the Woods Hole Oceanographic Institution for providing a sabbatical home to write this paper and to the University of Virginia for the Sesquicentennial Fellowship.

REFERENCES

Aber J. D., Magill A., McNulty S. G., Boone R. D., Nadelhoffer K. J., Downs M., and Hallett R. (1995) Forest biogeochemistry and primary production altered by nitrogen saturation. *Water Air Soil Pollut.* **85**, 1665–1670.

Aber J. D., Goodale C. L., Ollinger S. V., Smith M. -L., Magill A. H., Martin M. E., Hallett R. A., and Stoddard J. L. (2003) Is nitrogen deposition altering the nitrogen status of northeastern forests?. *Bioscience*, **53**, 375–389.

Asner G. P., Townsend A. R., Riley W. J., Matson P. A., Neff J. C., and Cleveland C. C. (2001) Physical and biogeochemical controls over terrestrial ecosystem responses to nitrogen deposition. *Biogeochemistry* **54**, 1–39.

Ayres R. U., Schlesinger W. H., and Socolow R. H. (1994) Human impacts on the carbon and nitrogen cycles. In *Industrial Ecology and Global Change* (eds. R. H. Socolow, C. Andrews, R. Berkhout, and V. Thomas). Cambridge University Press, , pp. 121–155.

Bange H., Rixen T., Johansen A., Siefert R., Ramesh R., Ittekkot V., Hoffmann M., and Andreae M. (2000) A revised nitrogen budget for the Arabian Sea. *Global Biogeochem. Cycles* **14**, 1283–1297.

Bashkin V. N., Park S. U., Choi M. S., and Lee C. B. (2002) Nitrogen budgets for the Republic of Korea and the Yellow Sea region. *Biogeochemistry* **57**, 387–403.

Bender M., Ganning K. A., Froelich P. M., Heath G. R., and Maynard V. (1977) Interstitial nitrate profiles and oxidation of sedimentary organic matter in the Eastern Equatorial Atlantic. *Science* **198**, 605–609.

Binkley D., Cromack K., Jr., and Baker D. D. (1994) Nitrogen fixation by red alder: biology, rates, and controls. In *The Biology and Management of Red Alder* (eds. D. E. Hibbs, D. S. DeBell, and R. F. Tarrant). Oregon State University Press, Corvallis, pp. 57–72.

Blundon D. J. and Dale M. R. T. (1990) Denitrogen fixation (acetylene reduction) in primary succession near Mount Robson, British Columbia, Canada. *Arct. Alp. Res.* 255–263.

Boring L. R. and Swank W. T. (1984) The role of black locust (*Robinia pseudo-acacia*) in forest succession. *J. Ecol.* **72**, 749–766.

Boyer E. W., Goodale C. L., Jaworski N. A., and Howarth R. W. (2002) Anthropogenic nitrogen sources and relationships to riverine nitrogen export in the Northeastern USA. *Biogeochemical* **57**, 137–169.

Bradley M. J. and Jones B. M. (2002) Reducing global NO_x emissions: promoting the development of advanced energy and transportation technologies. *Ambio* **31**, 141–149.

Bricker S. B., Clement C. G., Pirhalla D. E., Orland S. P., and Farrow D. G. G. (1999) *National Estuarine Eutrophication Assessment: A Summary of Conditions, Historical Trends, and Future Outlook*. US Department of Commerce, National Oceanic and Atmospheric Administration, National Ocean Service, Silver Spring, MD.

Burkart M. R. and Stoner J. D. (2001) Nitrogen in groundwater associated with agricultural systems. In *Nitrogen in the Environment: Sources, Problems and Management* (eds. R. F. Follett and J. L. Hatfield). Elsevier, New York, pp. 123–145.

Capone D. G. (1983) Benthic nitrogen fixation. In *Nitrogen in the Marine Environment* (eds. E. J. Carpenter and D. G. Capone). Elsevier, New York, pp. 105–137.

Capone D. (2001) Marine nitrogen fixation: what's the fuss? *Current Opinions Microbiol.* **4**, 341–348.

Capone D. G. and Carpenter E. J. (1982) Nitrogen fixation in the marine environment. *Science* **217**, 1140–1142.

Carpenter E. J. and Romans K. (1991) Major role of the cyanobacterium tricodesmium in nutrient cycling in the North Atlantic Ocean. *Science* **254**, 1356–1358.

Cassman K. G., Dobermann A. D., and Walters D. (2002) Agro-ecosystems, nitrogen-use efficiency, and nitrogen management. *Ambio* **31**, 132–140.

Christensen J. P., Murray J. W., Devol A. H., and Codispoti L. A. (1987) Denitrification in continental shelf sediments has a major impact on the oceanic nitrogen budget. *Global Biogeochem. Cycles* **1**, 97–116.

Cleveland C. C., Townsend A. R., Schimel D. S., Fisher H., Howarth R. W., Hedin L. O., Perakis S. S., Latty E. F., Von Fischer J. C., Elseroad A., and Wasson M. F. (1999) Global patterns of terrestrial biological nitrogen (N_2) fixation in natural ecosystems. *Global Biogeochem. Cycles* **13**, 623–645.

Codispoti L., Brandes J., Christensen J., Devol A., Naqvi S., Paerl H., and Yoshinari T. (2001) The oceanic fixed nitrogen and nitrous oxide budgets: moving targets as we enter the anthropocene? *Scientia Marina* **65**(suppl. 2), 85–105.

Cowling E., Erisman J. W., Smeulders S. M., Holman S. C., and Nicholson B. M. (1998) Optimizing air quality management in Europe and North America: justification for integrated management of both oxidized and reduced forms of nitrogen. *Environ. Pollut.* **102**, 599–608.

Crookes W. Sir (1871) *Select Methods in Chemical Analysis (Chiefly inorganic)*. Longmans, Green, London. (microform, Landmarks of Science, Readex Microprint, 1969, Micro-opaque, New York).

Delwiche C. C. (1970) The nitrogen cycle. *Sci. Am.* **223**, 137–146.

Deutsch C., Gruber N., Key R. M., Sarmiento J. L., and Ganachaud A. (2001) Denitrification and N_2 fixation in the Pacific Ocean. *Global Biogeochem. Cycles* **15**, 483–506.

Devol A. H. (1991) Direct measurement of nitrogen gas fluxes from continental shelf sediments. *Nature* **349**, 319–322.

Diaz R. and Rosenberg R. (1995) Marine benthic hypoxia: a review of its ecological effects and the behavioral responses of benthic macrofauna. *Oceanogr. Mar. Biol. Ann. Rev.* **33**, 245–303.

Ehhalt D., Prather M., Dentener F., Derwent R., Dlugokencky E., Holland E., Isaksen I., Katima J., Kirchhoff V., Matson P., Midgley P., and Wang M. (2001). Atmospheric chemistry and greenhouse gases. In *Climate Change 2001: The Scientific Basis* (eds. J. T. Houghton, Y. Ding, D. J. Griggs, M. Noguer, P. J. van der Linden, X. Da, K. Maskell, and C. A. Johnson). Third Assessment Report, Intergovernmental Panel on Climate change, Cambridge University Press, New York, pp. 239–288 (http://www.grida.no/climate/ipcc_tar/wg1/index.htm).

Erisman J. W., de Vries W., Kros H., Oenema O., van der Eerden L., and Smeulders S. (2001) An outlook for a national integrated nitrogen policy. *Environ. Sci. Policy* **4**, 87–95.

FAO (2000) *FAO Statistical Databases*. http://apps.fao.org.

Fixen P. E. and West F. B. (2002) Nitrogen fertilizers meeting the challenge. *Ambio* **31**, 169–176.

Follett J. R. and Follett R. F. (2001) Utilization and metabolism of nitrogen by humans. In *Nitrogen in the Environment: Sources, Problems and Management* (eds. R. Follett and J. L. Hatfield). Elsevier Science, New York, pp. 65–92.

Fowler D., Skiba U., and Hargreaves K. J. (1997) Emissions of nitrous oxide from grasslands. In *Gaseous Nitrogen*

Emissions from Grasslands (eds. S. C. Jarvis and B. F. Pain). CAB International, Wallingford, UK, pp. 147–164.

Galloway J. N. (1998) The global nitrogen cycle: changes and consequences. *Environ. Pollut.* **102**(S1), 15–24.

Galloway J. N. (2000) Nitrogen mobilization in Asia. *Nutr. Cycl. Agroecosyst.* **57**, 1–12.

Galloway J. N. and Cowling E. B. (2002) Reactive nitrogen and the world: 200 years of change. *Ambio* **31**, 64–71.

Galloway J. N., Schlesinger W. H., Levy H., II, Michaels A., and Schnoor J. L. (1995) Nitrogen fixation: anthropogenic enhancement—environmental response. *Global Biogeochem. Cycles* **9**, 235–252.

Galloway J. N., Howarth R. W., Michaels A. F., Nixon S. W., Prospero J. M., and Dentener F. J. (1996) Nitrogen and phosphorus budgets of the North Atlantic ocean and its watershed. *Biogeochemistry* **35**, 3–25.

Galloway J. N., Cowling E. B., Seitzinger S. J., and Socolow R. (2002) Reactive nitrogen: too much of a good thing? *Ambio* **31**, 60–63.

Galloway J. N., Dentener F. J., Capone D. G., Boyer E. W., Howarth R. W., Seitzinger S. P., Asner G. P., Cleveland C., Green P., Holland E., Karl D. M., Michaels A. F., Porter J. H., Townsend A., and Vörösmary C. (2003a) Nitrogen cycles: past, present, and future. *Biogeochemistry* (submitted).

Galloway J. N., Aber J. D., Erisman J. W., Seitzinger S. P., Howarth R. W., Cowling E. B., and Cosby B. J. (2003b) The nitrogen cascade. *Bioscience* **53**, 1–16.

Goolsby D. A., Battaglin W. A., Lawrence G. B., Artz R. S., Aulenbach B. T., Hooper R. P., Keeney D. R., and Stensland G. J. (1999) Flux and sources of nutrients in the Mississippi–Atchafalaya River Basin: Topic 3. *Report for the Integrated Assessment on Hypoxia in the Gulf of Mexico.* Coastal Ocean Program Decision Analysis Series No. 17, US Dept. of Commerce, National Oceanic and Atmospheric Administration, National Ocean Service, Silver Spring, MD.

Gorham E., Vitousek P. M., and Reiners W. A. (1979) The regulation of chemical budgets over the course of terrestrial ecosystem succession. *Ann. Rev. Ecol. Sys.* **10**, 53–84.

Groffman P. M., Gold A. J., and Jacinthe P. A. (1998) Nitrous oxide production in riparian zones and groundwater. *Nutr. Cycl. Agroecosyst.* **52**, 179–186.

Gruber N. and Sarmiento J. (1997) Global patterns of marine nitrogen fixation and denitrification. *Global Biogeochem. Cycles* **11**, 235–266.

Holland H. D. (1984) *The Chemical Evolution of the Atmosphere and Oceans.* Princeton University, NJ.

Holland E. A. and Lamarque J.-F. (1997) Bio-atmospheric coupling of the nitrogen cycle through NO_x emissions and NO_y deposition. *Nutr. Cycl. Agroecosyst.* **48**, 7–24.

Houghton R. A. and Hackler J. L. (2002) Carbon flux to the atmosphere from land-use changes. In *Trends: A Compendium of Data on Global Change.* Carbon Dioxide Information Analysis Center, Oak Ridge National Laboratory, US Department of Energy, Oak Ridge, TN.

Houghton J. T., Ding Y., Griggs D. J., Noguer M., van der Linden P. J., Da X., Maskell, K., and Johnson C. A. (eds.) (2001) *Climate Change 2001: The Scientific Basis,* Third Assessment Report, Intergovernmental Panel on Climate Change. Cambridge University Press, Cambridge, New York (http://www.grida.no/climate/ipcc_tar/wg1/index.htm).

Howarth R. W., Marino R., and Cole J. J. (1988) Nitrogen fixation in freshwater, estuarine, and marine ecosystems: 2. Biogeochemical controls. *Limnol. Oceanogr.* **33**, 688–701.

Howarth R. W., Billen G., Swaney D., Townsend A., Jaworski N., Lajtha K., Downing J. A., Elmgren R., Caraco N., Jordan T., Berendse F., Freney J., Kudeyarov V., Murdoch P., and Zhao-Liang Z. (1996) Regional nitrogen budgets and riverine N and P fluxes for the drainages to the North Atlantic Ocean: natural and human influences. *Biogeochemistry* **35**, 75–139.

Howarth R. W., Anderson D., Cloern J., Elfring C., Hopkinson C., Lapointe B., Malone T., Marcus N., McGlathery K., Sharpley A., and Walker D. (eds.) (2000) *Nutrient Pollution of Coastal Rivers, Bays, and Seas. Issues in Ecology 7.* Ecological Society of America, Washington, DC. (http://esa.sdsc.edu).

Howarth R. W., Boyer E., Pabich W., and Galloway J. N. (2002a) Nitrogen use in the United States from 1961–2000, and estimates of potential future trends. *Ambio* **31**, 88–96.

Howarth R. W., Sharpley A. W., and Walker D. (2002b) Sources of nutrient pollution to coastal waters in the United States: implications for achieving coastal water quality goals. *Estuaries* **25**, 656–676.

Jaffee D. A. (1992) The global nitrogen cycle. In *Global Biogeochemical Cycles* (eds. S. S. Butcher, G. H. Orians, R. J. Charlson, and G. V. Wolfe). Academic Press, London, pp. 263–284.

Johnson H. B. and Mayeux H. S. (1990) *Prosopis glandulosa* and the nitrogen balance of rangelands: extent and occurrence of nodulation. *Oecologia* **84**, 176–185.

Karl D. M. (1999) A sea of change: biogeochemical variability in the North Pacific Subtropical Gyre. *Ecosystems* **2**, 181–214.

Karl D. M. (2002) Nutrient dynamics in the deep blue sea. *Trends Microbiol.* **10**, 410–418.

Karl D. M., Michaels A., Berman B., Capone D., Carpenter E., Letelier R., Lipschultz F., Paerl H., Sigman D., and Stal L. (2002) Denitrogen fixation in the world's oceans. *Biogeochemistry* **57**, 47–52.

Keeling C. D. (1993) Global historical CO_2 emissions. In *Trends '93: A Compendium of Data on Global Change* (eds. T. A. Boden, D. P. Kaiser, R. J. Sepanski, and F. W. Stoss). ORNL/CDIAC-65. Carbon Dioxide Information Analysis Center, Oak Ridge National Laboratory, Oak Ridge, TN, pp. 501–504.

Kislev M. E. and Baryosef O. (1988) The legumes—the earliest domesticated plants in the Near East. *Current Anthropol.* **29**, 175–179.

Kramer D. A. (1999) *Minerals Yearbook: Nitrogen.* US Geological Survey Minerals Information (http://minerals.usgs.gov/minerals/pubs/commodity/nitrogen/).

Kroeze C., Aerts R., van Breemen N., van Dam D., van der Hoek K., Hofschreuder P., Hoosbeek M., de Klein J., Kros H., van Oene H., Oenema O., Tietema A., van der Veeren R., and de Vries W. (2003) Uncertainties in the fate of nitrogen: I. An overview of sources of uncertainty illustrated with a Dutch case study. *Nutr. Cycl. Agroecosyst.* **66**, 43–69.

Lapointe B. E. and O'Connell J. D. (1989) Nutrient-enhanced productivity of *Cladophroa prolifera* in Harrington Sounds, Bermuda: eutrophication of a confined, phosphorus-limited marine ecosystem. *Est. Coast Shelf Sci.* **28**, 347–360.

Lelieveld J. and Dentener F. (2000) What controls tropospheric ozone? *J. Geophys. Res.* **105**, 3531–3551.

Mackenzie F. T. (1994) Global climatic change: climatically important biogenic gases and feedbacks. In *Biotic Feedbacks in the Global Climatic System: Will the Warming Feed the Warming* (eds. G. M. Woodwell and F. T. Mackenzie). Oxford University Press, New York, pp. 22–46.

Mackenzie F. T. (1998) *Our Changing Planet: An Introduction to Earth System Science and Global Environmental Change,* 2nd edn. Prentice Hall, Upper Saddle River, NJ.

Melillo J. M. and Cowling E. B. (2002) Reactive nitrogen and public policies for environmental protection. *Ambio* **31**, 150–158.

Meybeck M. (1982) Carbon, nitrogen, and phosphorus transport by world rivers. *Am. J. Sci.* **282**, 401–450.

Miller S. L. (1953) A production of amino acids under possible primitive Earth conditions. *Science* **117**, 529–529.

Moomaw W. R. (2002) Energy, industry and nitrogen: strategies for reducing reactive nitrogen emissions. *Ambio* **31**, 184–189.

Mosier A., Kroeze C., Nevison C., Oenema O., Seitzinger S., and van Cleemput O. (1998) Closing the global atmospheric N_2O budget: nitrous oxide emissions through the agricultural nitrogen cycle. *Nutr. Cycl. Agroecosyst.* **52**, 225–248.

Mosier A. R., Bleken M. A., Chaiwanakupt P., Ellis E. C., Freney J. R., Howarth R. B., Matson P. A., Minami K., Naylor R., Weeks J., and Zhu Z.-L. (2001) Policy implications of human accelerated nitrogen cycling. *Biogeochemistry* **52**, 281–320.

Müller J. (1992) Geographical distribution and seasonal variation of surface emissions and deposition velocities of atmospheric trace gases. *J. Geophys. Res.* **97**, 3787–3804.

Neff J. C., Holland E. A., Dentener F. J., McDowell W. H., and Russell K. M. (2002) The origin, composition and rates of organic nitrogen deposition: a missing piece of the nitrogen cycle? *Biogeochemistry* **57**, 99–136.

Nixon S. W., Ammerman J. W., Atkinson L. P., Berounsky V. M., Billen G., Boicourt W. C., Boynton W. R., Church T. M., DiToro D. M., Elmgren R., Garber J. H., Giblin A. E., Jahnke R. A., Owens N. J. P., Pilson M. E. Q., and Seitzinger S. P. (1996) The fate of nitrogen and phosphorus at the land-sea margin of the North Atlantic Ocean. *Biogeochemistry* **35**, 141–180.

Nolan B. T. and Stoner J. D. (2000) Nutrients in groundwaters of the conterminous United States, 1992–1995. *Environ. Sci. Tech.* **34**, 1156–1165.

NRC (National Research Council) (2000) *Clean Coastal Waters: Understanding and Reducing the Effects of Nutrient Pollution.* National Academy Press, Washington, DC.

Oenema O. and Pietrzak S. (2002) Nutrient management in food production: achieving agronomic and environmental targets. *Ambio* **31**, 132–140.

Paerl H. W., Dennis E. L., and Whitall D. D. (2002) Atmospheric deposition of nitrogen: implications for nutrient over-enrichment of coastal waters. *Estuaries* **25**, 677–693.

Paul E. A. and Clark F. E. (1997) *Soil Microbiology and Biochemistry.* Elsevier/Academic Press, New York.

Pope C. A., III, Thun M. J., Namboodiri M. M., Dockery D. W., Evans J. S., Speizer F. E., and Heath C. W. Jr. (1995) Particulate air pollution as a predictor of mortality in a prospective study of US adults. *Am. J. Resp. Crit. Care Med.* **151**, 669–674.

Puckett L. J. and Cowdery T. K. (2002) Transport and fate of nitrate in a glacial outwash aquifer in relation to groundwater age, land use practices and redox processes. *J. Environ. Qual.* **31**, 782–796.

Puckett L. J., Cowdery T. K., Lorenz D. L., and Stoner J. D. (1999) Estimation of nitrate contamination of an agroecosystem outwash aquifer using a nitrogen mass-balance budget. *J. Environ. Qual.* **28**, 2015–2025.

Rabalais N. (2002) Nitrogen in aquatic ecosystems. *Ambio* **31**, 102–112.

Rabalais N. N., Turner R. E., and Scavia D. (2002) Beyond science into policy: gulf of mexico hypoxia and the mississippi river. *Bioscience* **52**, 129–142.

Rosswall T. (1983) The nitrogen cycle. In *The major Biogeochemical Cycles and Their Interactions* (eds. B. Bolin and R. B. Cook). SCOPE, Wiley, Chichester, Vol. 21, pp 46–50.

Rosswall T. (1992) The international geosphere-biosphere program—a study of global change (IGBP). *Environ. Geol. Water Sci.* **20**, 77–78.

Roy R. N., Misra R. V., and Montanez A. (2002) Reduced reliance on mineral nitrogen, yet more food. *Ambio* **31**, 177–183.

Schlesinger W. H. (1997) *Biogeochemistry: An Analysis of Global Change.* Elsevier, New York.

Seitzinger S. P. (1988) Denitrification in freshwater and coastal marine ecosystems: ecological and geochemical importance. *Limnol. Oceanogr.* **33**, 702–724.

Seitzinger S. P. and Giblin A. E. (1996) Estimating denitrification in North Atlantic continental shelf sediments. *Biogeochemistry* **35**, 235–260.

Seitzinger S. P. and Kroeze C. (1998) Global distribution of nitrous oxide production and N inputs in freshwater and coastal marine ecosystems. *Global Biogeochem. Cycles* **12**, 93–113.

Seitzinger S. P., Styles R. V., Boyer E., Alexander R. B., Billen G., Howarth R., Mayer B., and van Breemen N. (2002) Nitrogen retention in rivers: model development and application to watersheds in the Northeastern USA. *Biogeochemistry* **57**, 199–237.

Smil V. (1994) *Energy in World History.* Westview Press, Boulder, CO.

Smil V. (1999) Nitrogen in crop production: an account of global flows. *Global Biogeochem. Cycles* **13**, 647–662.

Smil V. (2000) *Feeding the World.* MIT Press, Cambridge.

Smil V. (2001) *Enriching the Earth.* MIT Press, Cambridge.

Smil V. (2002) Nitrogen and food production: proteins for human diets. *Ambio* **31**, 126–131.

Smith B. D. (1995) *The Emergence of Agriculture.* Scientific American, New York.

Socolow R. (1999) Nitrogen management and the future of food: lessons from the management of energy and carbon. *Proc. Natl. Acad. Sci. USA* **96**, 6001–6008.

Söderlund R. and Rosswall T. (1982) The nitrogen cycles: Topic 1. Report for the integrated assessment on hypoxia in the Gulf of Mexico. In *NOAA Coastal Ocean Program Decision Analysis Series No. 17.* NOAA, Silver Spring, MD.

Sommer S. G. and Hutchings N. J. (1997) Components of ammonia volatilization from cattle and sheep production. In *Gaseous Nitrogen Emissions from Grasslands* (eds. S. C. Jarvis and B. F. Pain). CAB International, UK, pp. 79–93.

Stedman D. H. and Shetter R. (1983) The global budget of atmospheric nitrogen species. In *Trace Atmospheric Constituents: Properties, Transformations and Fates* (ed. S. S. Schwartz). Wiley, New York, pp. 411–454.

Svensson B. H. and Söderlund R. (eds.) (1976) Nitrogen, phosphorus and sulphur—global cycles. Scientific Committee on Problems of the Environment (SCOPE) Report on Biogeochemical Cycles, Royal Swedish Academy of Science.

United Nations (1999) *World Population Prospects: the 1998 Revision.* United Nations, New York.

Valiela I., McClelland J., Hauxwell J., Behr P. J., Hersh D., and Foreman J. (1997) Macroalgal blooms in shallow estuaries: controls and ecophysiological and ecosystem consequences. *Limnol. Oceanogr.* **42**, 1105–1118.

van Aardenne J. A., Dentener F. J., Klijn Goldewijk C. G. M., Lelieveld J., and Olivier J. G. J. (2001) A $1°-1°$ resolution data set of historical anthropogenic trace gas emissions for the period 1890–1990. *Global Biogeochem. Cycles* **15**, 909–928.

van Breemen N., Boyer E. W., Goodale C. L., Jaworski N. A., Paustian K., Seitzinger S., Lajtha L. K., Mayer B., Van Dam D., Howarth R. W., Nadelhoffer K. J., Eve M., and Billen G. (2002) Where did all the nitrogen go? Fate of nitrogen inputs to large watersheds in the Northeastern USA. *Biogeochemistry* **57**, 267–293.

van Egmond N. D., Bresser A. H. M., and Bouwman A. F. (2002) The European nitrogen case. *Ambio* **31**, 72–78.

Vitousek P. M. (1994) Potential nitrogen fixation during primary succession in Hawaii volcanoes national park. *Biotropica* **26**, 234–240.

Vitousek P. M. and Howarth R. W. (1991) Nitrogen limitation on land and in the sea. How can it occur? *Biogeochemistry* **13**, 87–115.

Vitousek P. M., Aber J. D., Howarth R. W., Likens G. E., Matson P. A., Schindler D., Schlesinger W. H., and Tilman G. D. (1997) *Human Alteration of the Global Nitrogen Cycle: Causes and Consequences, Issues in Ecology 1.* Ecological Society of America, Washington, DC, (http://esa.sdsc.edu).

Walsh J. J., Dieterle D. A., and Meyers M. B. (1988) A simulation analysis of the fate of phytoplankton within the mid-Atlantic Bight. *Cont. Shelf Res.* **8**, 757–787.

Wang L. (1987) Soybeans—the miracle bean of China. In *Feeding a Billion: Frontiers of Chinese Agriculture)* (eds. S. Wittwer, L. Wang, and Y. Yu). Michigan State University Press, East Lansing, MI.

Wittwer S., Wang L., and Yu Y. (1987) *Feeding a Billion: Frontiers of Chinese Agriculture*. Michigan State University Press, East Lansing.

Wolfe A. and Patz J. A. (2002) Nitrogen and human health: direct and indirect impacts. *Ambio* **31**, 120–125.

Xing G. X. and Zhu Z. L. (2002) Regional nitrogen budgets for China and its major watersheds. *Biogeochemistry* **57**, 405–427.

Zheng X., Fu C., Xu X., Yan X., Chen G., Han S., Huang Y., and Hu F. (2002) The Asian nitrogen case. *Ambio* **31**, 79–87.

17
Evolution of Sedimentary Rocks

J. Veizer

Ruhr University, Bochum, Germany and University of Ottawa, ON, Canada

and

F. T. Mackenzie

University of Hawaii, Honolulu, HI, USA

17.1	INTRODUCTION	580
17.2	THE EARTH SYSTEM	580
	17.2.1 *Population Dynamics*	580
17.3	GENERATION AND RECYCLING OF THE OCEANIC AND CONTINENTAL CRUST	582
17.4	GLOBAL TECTONIC REALMS AND THEIR RECYCLING RATES	583
17.5	PRESENT-DAY SEDIMENTARY SHELL	584
17.6	TECTONIC SETTINGS AND THEIR SEDIMENTARY PACKAGES	585
17.7	PETROLOGY, MINERALOGY, AND MAJOR ELEMENT COMPOSITION OF CLASTIC SEDIMENTS	586
	17.7.1 *Provenance*	586
	17.7.2 *Transport Sorting*	588
	17.7.3 *Sedimentary Recycling*	588
17.8	TRACE ELEMENT AND ISOTOPIC COMPOSITION OF CLASTIC SEDIMENTS	588
17.9	SECULAR EVOLUTION OF CLASTIC SEDIMENTS	589
	17.9.1 *Tectonic Settings and Lithology*	589
	17.9.2 *Chemistry*	590
	17.9.3 *Isotopes*	591
17.10	SEDIMENTARY RECYCLING	592
17.11	OCEAN/ATMOSPHERE SYSTEM	593
	17.11.1 *The Chemical Composition of Ancient Ocean*	594
	17.11.2 *Isotopic Evolution of Ancient Oceans*	595
	17.11.2.1 *Strontium isotopes*	595
	17.11.2.2 *Osmium isotopes*	598
	17.11.2.3 *Sulfur isotopes*	598
	17.11.2.4 *Carbon isotopes*	600
	17.11.2.5 *Oxygen isotopes*	602
	17.11.2.6 *Isotope tracers in developmental stages*	603
17.12	MAJOR TRENDS IN THE EVOLUTION OF SEDIMENTS DURING GEOLOGIC HISTORY	605
	17.12.1 *Overall Pattern of Lithologic Types*	605
	17.12.2 *Phanerozoic Carbonate Rocks*	606
	17.12.2.1 *Mass-age distribution and recycling rates*	606
	17.12.2.2 *Dolomite/calcite ratios*	607
	17.12.2.3 *Ooids and ironstones*	608
	17.12.2.4 *Calcareous shelly fossils*	609

    17.12.2.5   *The carbonate cycle in the ocean*   610

    17.12.3   *Geochemical Implications of the Phanerozoic Carbonate Record*   610

ACKNOWLEDGMENTS   611

REFERENCES   611

17.1 INTRODUCTION

For almost a century, it has been recognized that the present-day thickness and areal extent of Phanerozoic sedimentary strata increase progressively with decreasing geologic age. This pattern has been interpreted either as reflecting an increase in the rate of sedimentation toward the present (Barrell, 1917; Schuchert, 1931; Ronov, 1976) or as resulting from better preservation of the younger part of the geologic record (Gilluly, 1949; Gregor, 1968; Garrels and Mackenzie, 1971a; Veizer and Jansen, 1979, 1985).

Study of the rocks themselves led to similarly opposing conclusions. The observed secular (=age) variations in relative proportions of lithological types and in chemistry of sedimentary rocks (Daly, 1909; Vinogradov *et al.*, 1952; Nanz, 1953; Engel, 1963; Strakhov, 1964, 1969; Ronov, 1964, 1982) were mostly given an evolutionary interpretation. An opposing, uniformitarian, approach was proposed by Garrels and Mackenzie (1971a). For most isotopes, the consensus favors deviations from the present-day steady state as the likely cause of secular trends.

This chapter attempts to show that recycling and evolution are not opposing, but complementary, concepts. It will concentrate on the lithological and chemical attributes of sediments, but not deal with the evolution of sedimentary mineral deposits (Veizer *et al.*, 1989) and of life (Sepkoski, 1989), both well amenable to the outlined conceptual treatment. The chapter relies heavily on Veizer (1988a) for the sections dealing with general recycling concepts, on Veizer (2003) for the discussion of isotopic evolution of seawater, and on Morse and Mackenzie (1990) and Mackenzie and Morse (1992) for discussion of carbonate rock recycling and environmental attributes.

17.2 THE EARTH SYSTEM

The lithosphere, hydrosphere, atmosphere, and biosphere, or rocks, water, air, and life are all part of the terrestrial exogenic system that is definable by the rules and approaches of general system science theory, with its subsets, such as population dynamics and hierarchical structures.

17.2.1 Population Dynamics

The fundamental parameters essential for quantitative treatment of population dynamics are the population size (A_0) and its recycling rate. A_0 is normalized in the subsequent discussion to one population (or 100%) and the rates of recycling relate to this normalized size. Absolute rates can be established by multiplying this relative recycling rate (parameter b below) by population size.

A steady state natural population, characterized by a continuous generation/destruction (birth/death) cycle, is usually typified by an age structure similar to that in Figure 1, the cumulative curve defining all necessary parameters of a given population. These are its *half-life* τ_{50}, *mean age* τ_{mean}, and *oblivion age* or life expectancy τ_{max}.

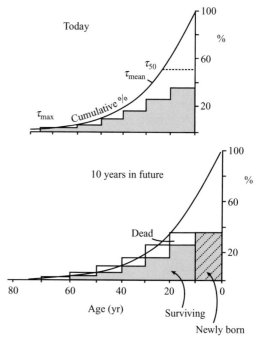

Figure 1 Simplified age distribution pattern for a steady state extant population. In this case, the natality/mortality rate is 35% of the total population for a 10-yr interval $(b = 35 \times 10^{-3}\ yr^{-1})$. τ_{max} is defined as the fifth percentile. In practice τ_{max} is the age at which the resolution of the database becomes indistinguishable from the background. The b value for the same population is inversely proportional to the available time resolution T (Equation (2)). Today's "instantaneous" rates of deposition and erosion of sediments exceed those calculated from the geological record based on time resolution of 10^6–10^7 yr (Sadler, 1981). It is therefore essential to stipulate the resolution T in consideration of rates (after Veizer and Jansen, 1985).

For steady state first-order (=single population) systems, the survival rate of constituent units can be expressed as

$$A_{t^*} = A_0\, e^{-kt^*} \qquad (1)$$

where A_{t^*} is the cumulative fraction of the surviving population older than t^*, $A_0 = 1$ (one population), t^* is age (not time), and k is the rate constant for the recycling process. In the subsequent discussion, the recycling rate is often considered in the form of a *recycling proportionality constant b*, which is related to the above equation through formalism:

$$b = 1 - e^{-kT} \qquad (2)$$

where T is the time resolution or duration of recycling (cf. Veizer and Jansen, 1979, 1985). In general, the larger the b value—i.e., the faster the rate of recycling—the steeper the slope of the cumulative curve and the shorter the τ_{50} and τ_{max} of the population. For a steady state extant population, generation per unit time must equal combined destruction for all age groups during the same time interval. Consequently, the cumulative slope remains the same but propagates into the future (Figure 1).

The above terminology is applicable to internally (cannibalistically) recycling populations. In an external type of recycling, the influx and efflux cause a similar age structure, but the terminology differs. In this case, the average duration an individual unit resides within the population is termed the *residence time* τ_{res}. Mathematically, τ_{res} is similar to the cannibalistic τ_{mean} and it relates to the above parameters as $\tau_{max} \geq \tau_{res} \geq \tau_{50}$. It is this alternative—populations interconnected by external fluxes—that is usually referred to as the familiar box model by natural scientists. Frequently, box models are nothing more than one possible arrangement for propagation of cyclic populations.

In the subsequent discussion, the terminology of the cannibalistic populations is employed. Note, however, that the age distribution patterns and the recycling rates calculated from these patterns are a consequence of both cannibalistic and external recycling. At this stage, we lack the data and the criteria for quantification of their relative significance. Nevertheless, from the point of view of preservation probability, it may be desirable, but not essential, to know whether the constituent units (geologic entities) have been created and destroyed by internal, external, or combined phenomena.

Among natural populations, two major deviations from the ideal pattern are ubiquitous (Pielou, 1977; Lerman, 1979). The first deviation consists of populations with excessive proportions of young units (e.g., planktonic larval stages) because their destruction rate is very high, but

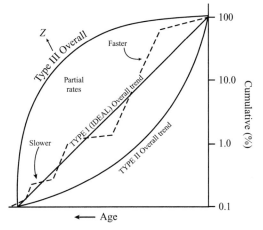

Figure 2 Cumulative age distribution functions for various populations plotted on a semilogarithmic scale (after Veizer, 1988a).

chances for survival improve considerably with maturation (type II in Figure 2). The mathematical formalism for such populations (e.g., Lerman, 1979) is a power-law function,

$$A_{t^*} = A_0(1 + kt^*)^{-z} \qquad (3)$$

where the exponent z increases for populations with progressively larger destruction rates of young units.

The other common exception consists of populations with suppressed destruction of young units (type III, Figure 2). In these instances, the destruction rates increase rapidly as the life span τ_{max} is approached. Mathematical expression for this relation is

$$A_{t^*} = A_0[1 - (kt^*)^z] \qquad (4)$$

where the exponent z increases for populations with progressively larger mortality rates of old units.

These relationships are valid for populations of constant size. For non-steady-state populations with stable age structures—i.e., those with overall rates of growth or decline much slower than the rates of recycling of their constituents—the age distributions approach the pattern of the constant-size populations. The calculated recycling rates are therefore identical. For populations where the overall growth (decline) approaches the rate of recycling of constituent units, independent criteria are required to differentiate recycling from the growth (decline) component.

The above discussion assumed a quasi-continuous generation/destruction process for a first-order system, but natural variability causes oscillations, at a hierarchy of frequencies and amplitudes, around the smooth overall patterns. Furthermore, geological processes are usually discrete and episodic phenomena. Because of all these factors, partial intervals have generation/destruction rates that deviate from the smooth average rates (Figure 2), with the connecting tangents having either shallower or steeper slopes. Note again that a given partial slope may reflect deviations in generation, in destruction, or in their combined effect. Usually, the problem is not resolvable, but the combined effect is the most likely alternative. As the population ages, the magnitude of this higher-order scatter diminishes to the level of uncertainties in the database (Figure 2). Quantitative interpretation of such higher-order features from the preserved record is therefore possible only for a length of time roughly comparable to the life-span τ_{max} of a given population. It would be pointless to attempt quantification of oceanic spreading rates from the fragmentary record of pre-Jurassic ophiolites. Any such quantification must rely on some derivative signal, such as isotopic composition of seawater, which may be preserved in coeval sediments. In contrast to the fast cycling oceanic crust, the higher-order scatter for slowly cycling populations (e.g., continental crust) remains considerable, because it has not yet been smoothed out by the superimposed recycling. Such populations still retain vestiges of ancient episodic events.

For geologic entities (e.g., crustal segments, mineral deposits, tectonic domains, and fossils), the age distribution patterns can be extracted from their stratigraphic and geochronologic assignments. At present, only major features of the record can be quantitatively interpreted, because the database is usually not of the desired reliability.

17.3 GENERATION AND RECYCLING OF THE OCEANIC AND CONTINENTAL CRUST

The concept of global tectonics (Dietz, 1961; Hess, 1962; Morgan, 1968; Le Pichon *et al.*, 1973) combined the earlier proposals of continental drift and seafloor spreading into a unified theory of terrestrial dynamics. It introduced the notion of continual generation and destruction of oceanic crust and implied similar consequences for other tectonic realms.

The present-day age distribution pattern of the *oceanic crust* is well known (Sprague and Pollack, 1980; Sclater *et al.*, 1981; Rowley, 2002), and the plate tectonic concept of ocean floor generation/subduction well established. The age distribution pattern (Figure 3) conforms to systematics with a half-life (τ_{50}) of ~60 Myr, translating to generation/destruction of ~3.5 km^2 (or ~20 km^3) of oceanic crust per year. The maximal life-span τ_{max} (oblivion age) for its tectonic settings and their associated sediments is therefore less than 200 Myr.

The situation for continental crust is more complex and still dogged by the controversy (e.g., Sylvester, 2000) that pits the proponents of its near-complete generation in the early planetary

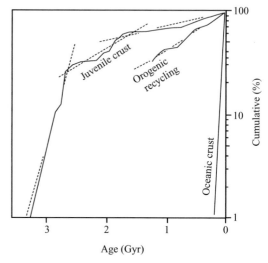

Figure 3 Cumulative age distribution of oceanic and continental crusts. Continental crust based on Condie (2001), oceanic crust after Sclater *et al.* (1981), and Rowley (2002).

history, followed only by crust/mantle recycling (Armstrong, 1981 and the adherents) against those advocating an incremental growth in the course of geologic evolution (see Taylor and McLennan, 1985). The latest summary of the volume/age estimates for the continental crust (Condie, 2001) provides a more definitive constraint for the discussion of the issue. The total volume of the continental crust is estimated at 7.177×10^9 km^3 (Cogley, 1984), and the age distribution pattern of its juvenile component (Figure 3) suggests a tripartite evolution, with ~25% ~4.0–2.6 Gyr old, another 35% added between ~2.6 Gyr and 1.7 Gyr, and the remaining 40% subsequently. The observed growth pattern is of a sinusoidal (logistic) type (Veizer and Jansen, 1979), with commencement of large-scale crustal generation at ~4 Gyr, accelerating growth rate that culminated in major phases of crustal generation and cratonisation during the ~2.6–1.7 Gyr time span, and a declining rate subsequently.

The above crustal generation pattern is only a minimal estimate, based on the assumption that no continental crust was recycled into the mantle. The other limiting alternative can be based on the proposition that the continental crust attained its present-day steady state ~1.75 Gyr ago. If this were the case, today's preserved post 1.75 Gyr crust is only about one-half of that generated originally, with the equivalent amount recycled into the mantle. This recycling may go hand in hand with orogenic activity that has a τ_{50} of ~800 Myr (Figure 3), a value in good agreement with the previous estimates for the low- and high- grade metamorphic reworking rates (τ_{50} of 673 Myr and 987 Myr, respectively) by Veizer and Jansen (1985). Furthermore, in the post-1.6 Ga record,

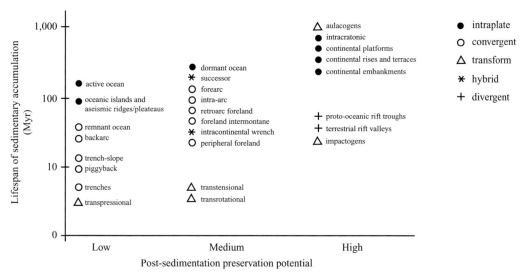

Figure 4 Typical life spans for sedimentary basins versus their post-sedimentation preservation potential. "Preservation potential" refers to average amount of time during which basins will not be uplifted and eroded, or be tectonically destroyed during and following sedimentation. Sedimentary or volcanic fill may be preserved as accretionary complexes during and after basin destruction (true of all strata deposited on oceanic crust). Basins with full circles, particularly intraplate continental margins, are "preserved" in the sense of retaining their basement, but they are likely to be subcreted beneath or within suture belts, and are difficult to recognize in the ancient record in such settings (after Ingersoll and Busby, 1995).

the orogenic segments are composed, on average, of about equal amounts of juvenile $(2.7 \times 10^9 \text{ km}^3)$ and recycled $(2.6 \times 10^9 \text{ km}^3)$ crust (figure 4 in Condie, 2001). This would suggest that each orogenic episode results in incorporation of about one-half of the juvenile crust into the reworked crustal segment, with the other half being subducted into the mantle.

Considering the above scenarios as limiting alternatives, the long-term average rate of continental crust generation would be $\sim 1.7 \pm 0.1 \text{ km}^3 \text{ yr}^{-1}$ $(1.1 \pm 0.5 \text{ km}^3 \text{ yr}^{-1}$ by Reymer and Schubert, 1984) for no mantle recycling and about twice that much for the alternative where about one-half of the juvenile crust contributes to the orogenic buildup of the continents, while the other half is recycled into the mantle.

17.4 GLOBAL TECTONIC REALMS AND THEIR RECYCLING RATES

Tectonic setting is the principal controlling factor of lithology, chemistry, and preservation of sediment accumulations in their depocenters, the sedimentary basins. The latest classification of Ingersoll and Busby (1995) assigns sedimentary basins into five major groups based on their relationship to plate boundaries (Figure 4). It groups together the basins that are associated with the divergent, interplate, convergent, transform, and hybrid settings, and recognizes 26 basin types. This classification, while it cannot take into account the entire complexity of natural systems, implies that basins associated with transform and

convergent boundaries are more prone to destruction than basins associated with divergent and intraplate settings, particularly those developed on continental crust (Figure 4).

This qualitative observation is consistent with the prediction of oblivion ages for specific tectonic realms based on the concept of population dynamics (Veizer and Jansen, 1985). Theoretically, if τ_{max} is taken as the fifth percentile, the oblivion ages for specific tectonic realms should be a factor of ~ 4.5 times the respective half-lives, but empirically, due to deviations from the ideal type I age pattern, the τ_{max} is usually some ~ 3.0–3.5 times τ_{50}. This qualification notwithstanding, the short-lived basins are erased faster from the geologic record and the degree of tectonic diversity must be a function of time. The diversity diminishes as the given segment of the solid earth ages, and the rate of memory loss is inversely proportional to recycling rates of the constituent tectonic realms. For a steady state system, the calculated theoretical preservational probabilities are depicted in Figure 5. This reasoning shows that the realms of the oceanic domain (basins of active margins to immature orogenic belts) should have only $\leq 5\%$ chance of survival in crustal segments older than ~ 100–300 Myr, while the platformal and intracratonic basins can survive for billions of years.

Due to rather poor inventories, the proposed systematics should be viewed as nothing more than a conceptual framework, but it nevertheless helps to visualize the probability of preservation of sedimentary packages in the geologic record.

17.5 PRESENT-DAY SEDIMENTARY SHELL

The present-day mass of global sediments is $\sim 2.7 \times 10^{24}$ g (Ronov, 1982; Hay *et al.*, 2001). Of these, $\sim 72.6\%$ are situated within the confines of the present-day continents (orogenic belts 51.9%, platforms 20.7%), 12.9% at passive margin basins, 5.5% at active margin basins, and the sediments covering the ocean floor account for $\sim 8.3\%$ of the total (Ronov, 1982; Gregor, 1985; Veizer and Jansen, 1985).

The apparent decline of sedimentary thicknesses in progressively older sections (Barrell, 1917; Gilluly, 1949) is reflected also in the latest inventory of mass/age distribution of the global sedimentary mass, which declines exponentially with age (Figure 6). This exponential decline is not clearly discernible from the mass/age distribution of sediments within the confines of the continental crust (Ronov, 1993), but adding the mass of sediments presently associated with the passive margin tectonic settings (Gregor, 1985) and the sediments on the ocean floor (Hay *et al.*, 1988), the pattern clearly emerges. Hence, the preservation of the sedimentary record is a function of tectonic setting, with sediments on continental crust surviving well into the Precambrian, while the continuous record of passive margin sediments ends at ~ 250 Myr ago and that of the ocean floor sediments at ~ 100 Myr ago (Figure 6).

The original concept of recycling, as developed by Garrels and Mackenzie (1971a), was based solely on the "continental" database assembled by the group at the Vernadsky Institute of Geochemistry in Moscow (Ronov, 1949, 1964, 1968, 1976, 1982, 1993). The former authors proposed that the present-day mass/age distribution of global sedimentary mass is consistent with a half-mass age of ~ 600 Myr, resulting in deposition and destruction of about five sedimentary masses over the entire geologic history. Furthermore, the observed temporal relationship of clastics/carbonates/evaporites led Garrels and Mackenzie (1971a) to propose a concept of differential recycling rates for different lithologics based on their susceptibility to chemical weathering, with clastics having half-mass age of ~ 600 Myr, carbonates ~ 200 Myr, and evaporites ~ 100 Myr.

Subsequently, Veizer (1988a) and Hay *et al.* (1988) pointed out that the concept of differential recycling, although valid, can only partially be based on the susceptibility to chemical weathering.

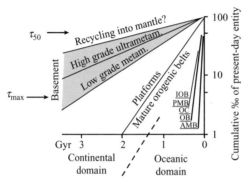

Figure 5 Preservation probabilities for major global tectonic realms. AMB=active margin basins, OB= oceanic intraplate basins, OC=oceanic crust, PMB= passive margin basins, IOB=immature orogenic belts. Rates were derived on the assumption that all deviations from the ideal pattern of the type I age distributions were a consequence of the poor quality of the available database (after Veizer, 1988b).

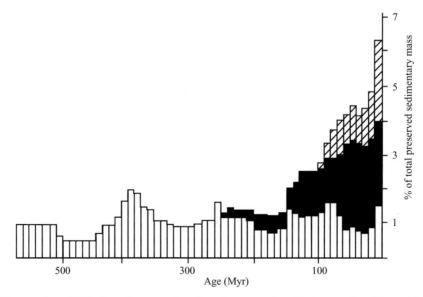

Figure 6 The mass/age distribution of preserved sedimentary mass deposited on continental crust (vertical lines), basin of passive margins (black), and on the oceanic floor (cross-hatched) (courtesy of W. W. Hay).

In a layer cake stratigraphy, the removal of carbonate strata, for instance, would necessarily result in a collapse of all the overlying strata, regardless of their lithology. On a macroscale, therefore, the sediments are removed en-masse, with chemical weathering rates coupled to the physical ones as (Millot *et al.*, 2002)

$$\text{Chem} = 0.39(\text{Phy})^{0.66} \qquad (5)$$

This point of view is supported also by the fact that the particulate load accounts for ~3/4 of the present-day fluvial sediment flux (Garrels and Mackenzie, 1971a).

Indeed, the detailed consideration of temporal lithological trends does not follow the pattern anticipated from recycling based on their susceptibility to chemical weathering. For example, during the Phanerozoic, the relative proportion of carbonates increases, and of clastics decreases, with age (Figure 7). Similarly, the lithological trends of Ronov (1964) that span the entire geological time span, if recast into the present concept (Figure 8), show that it is particularly the most ancient sequences that contain the highly labile immature clastics (arkoses and graywackes). Furthermore, limestones, evaporites, and phosphorites appear to have a similar age distribution pattern, intermediate between that of the passive margin basins and platforms, while dolostones plot on a platformal trend. This suggests that the Ronov (1964) type of secular distribution of lithologies is a reflection of preservation probabilities of different tectonic settings, each having its own type of sediment assemblages.

The above view is clearly supported by the mass/age distribution of lithologies within the same tectonic domain. For example, carbonates, chert, red clay, and terrigeneous sediments on the ocean floor (Hay *et al.*, 1988) all have the same type of age distribution pattern that is controlled by a single variable, the rate of spreading and subduction of the ocean floor. This sedimentary mass also differs lithologically from its "continental" counterpart, because it is comprised of 76 wt. % terrigeneous material, 7% calcium carbonate, 10% opal, and 7% mineral bound water (Plank and Langmuir, 1998). Based on its age distribution pattern (Figure 6), the average rate of pelagic sediment subduction is ~1×10^{21} g Myr^{-1} (Hay *et al.*, 1988), an estimate in good agreement with an upper limit of $1.1 \pm 0.5 \times 10^{21}$ g Myr^{-1} for sediment subduction based on Sm/Nd isotopic constraints (Veizer and Jansen, 1985) discussed later in the text. The estimates based on direct measurements, however, suggest a sediment subduction rate of ~1.4 to 1.8×10^{21} g Myr^{-1} (Rea and Ruff, 1996; Plank and Langmuir, 1998), the difference likely accounted for by material entering the trenches from the adjacent accretionary wedge. This material can be scraped off the subducting slab, uplifted, eroded, and rapidly recycled (Hay *et al.*, 2001).

17.6 TECTONIC SETTINGS AND THEIR SEDIMENTARY PACKAGES

Is the proposition that differential recycling of sediments is controlled by tectonic settings supported by the observational data? Could this explain, for example, the paradox of increasing carbonate/clastic ratio with age during the Phanerozoic (Figure 7)?

The inventories of lithologies for basin types are sparse. Veizer and Ernst (1996) attempted to quantify the sedimentary fill of the North American Phanerozoic basins, and the results are presented in Figure 9. Due to inherent limitations of the database and its interpretation, and particularly

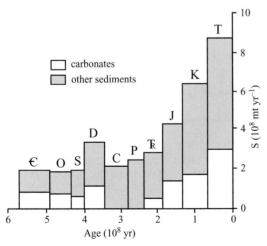

Figure 7 Surviving amounts of epicontinental terrigenous-clastic and marine-carbonates during the Phanerozoic (after Morse and Mackenzie, 1990).

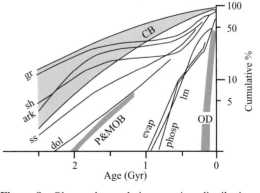

Figure 8 Observed cumulative mass/age distributions of major sedimentary lithological types. Explanation of abbreviations: gr denotes graywackes; sh, shales; ark, arkoses; ss, sandstones; dol, dolostones; evap, evaporites; lm, limestones; phosp, phosphorites; CB, continental basement; P, platforms; MOB, mature orogenic belts; OD, oceanic domain (after Veizer, 1988c).

due to inconsistencies in lithological descriptions, the sedimentary facies were grouped into three categories only: coarse clastics (sandstones, siltstones, conglomerates, and arkoses), fine clastics (shales, graywackes), and "chemical" sediments (carbonates, evaporites, and cherts). These

limitations notwithstanding, the compilation shows that basins associated with immature tectonic settings, such as arc-trench systems, are filled chiefly by clastic sediments, mostly immature fine grained graywackes and shales (forearc). Foreland basins, situated usually on the continental side of the continental-margin/arch-trench system, contain a higher proportion of coarse, often mature, clastics and some chemical sediments. This is even more the case for the passive margin (continental rise, terrace, and embankment) settings. Finally, carbonate sedimentation predominates in the intracratonic settings. Preservation probability of tectonic settings can therefore explain the tendency for the average lithology shifting from clastics towards carbonates with increasing age during the Phanerozoic.

17.7 PETROLOGY, MINERALOGY, AND MAJOR ELEMENT COMPOSITION OF CLASTIC SEDIMENTS

17.7.1 Provenance

The petrology of *coarse clastic* (conglomerate-size) *sediments* is controlled in the first instance by their provenance that, in turn, is a function of tectonic setting. Advancing tectonic stability is accompanied by an increasingly mature composition of the clasts (Figure 10). For first cycle sediments (Cox and Lowe, 1995), the early arc stage of tectonic evolution is dominated by volcanic clasts, from the growing and accreting volcanic pile. Subsequently, plutonic and metamorphic lithologies dominate the orogenic and uplift stages. The post-tectonic conglomerates contain a high proportion of recycled sedimentary clasts. Mineralogically, the evolution is from clasts with abundant plagioclase, to K-feldspar rich, and finally to quartz (chert) dominated clasts.

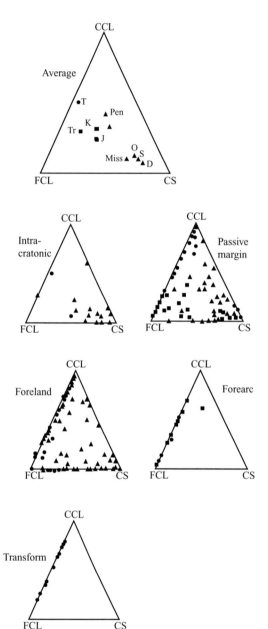

Figure 9 Relative proportions of lithological types within intracratonic to transform basins and the averages for geological periods. Based on the North American cross-sections in Cook and Bally (1975). An independent compilation by Berry and Wilkinson (1994) from the same source, but based on different criteria, yielded comparable temporal patterns. CCL coarse clastics; FCL fine clastics; CS "chemical" sediments. A global review based on the Ronov (1982) database also yields comparable patterns (after Veizer and Ernst, 1996).

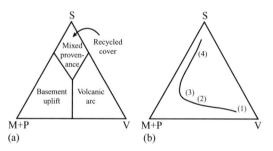

Figure 10 Idealized evolution of conglomerate clast composition. S=sedimentary clasts; M+P= metamorphic+plutonic clasts; V=volcanic clasts: (a) Approximate compositional fields for conglomerate clast populations of differing provenance. (b) Generalized stages in the evolution of conglomerate clast compositions on a crustal block: (1) volcanic arc stage; (2) dissected arc/ accretion orogen stage; (3) post-tectonic granite/basement uplift stage; and (4) sediment-recycling-dominated stage (after Cox and Lowe, 1995).

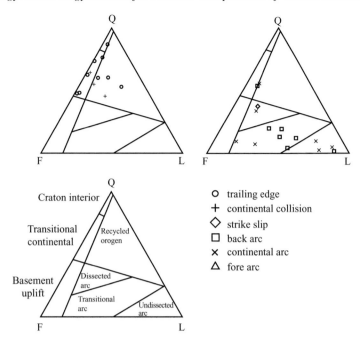

Figure 11 Ternary diagrams of framework quartz (Q)–feldspar (F)–unstable lithic fragments (L) for sands. Provenance fields from Dickinson *et al.* (1983) (after McLennan *et al.*, 1990).

A similar provenance control relates also the *sandstone* petrology to tectonic setting, as expressed in the Q (quartz)—F (feldspar)—L (unstable lithic fragments) ternary diagrams (Figure 11) of Dickinson *et al.* (1983). Again, increasing tectonic maturity shifts the mode of sandstone petrology from L towards the F/Q tangent, terminating in the recycled Q mode.

Fine clastic sediments, mostly *mudrocks*, in contrast to their coarser counterparts, are either derived by first cycle weathering of silicate minerals or glass, or from recycling of older mudrocks. Physical comminution plays only a secondary role. The average shale is composed of ~40–60% clay minerals, 20–30% quartz, 5–10% feldspar and minor iron oxide, carbonate, organic matter, and other components (Yaalon, 1962; Shaw and Weaver, 1965). Granitic source rocks produce shales richer in kaolinite and illite, the mafic ones richer in smectites (Cox and Lowe, 1995).

Geochemical processes associated with weathering and soil formation are dominated by alteration of feldspars (and volcanic glass), feldspars accounting for 70% of the upper crust, if the relatively inert quartz is discounted (Taylor and McLennan, 1985). Advancing weathering leads to a shift towards an aluminum rich composition that can be approximated by the chemical index of alteration (CIA) of Nesbitt and Young (1984)

$$CIA = 100(Al_2O_3/(Al_2O_3 + CaO^* + Na_2O + K_2O))$$

(6)

The suspended sediments of major rivers clearly reflect this alteration trend and plot on a tangent

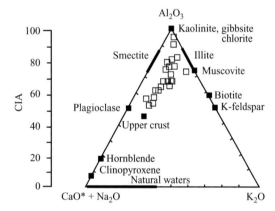

Figure 12 Ternary plots of molecular proportions of Al_2O_3–(Na_2O+CaO^*)–K_2O with the Chemical Index of Alteration (CIA) scale shown on the left. Also plotted are selected idealized igneous and sedimentary minerals and the range of typical natural waters. Squares are suspended sediments from major rivers throughout the world representing a variety of climatic regimes. CaO^* is the silicate bound concentration only (after McLennan *et al.*, 2003).

between the source (upper continental crust (UCC)) and the clays, the end-products of weathering (Figure 12).

The overall outcome is a depletion of the labile Ca–Na plagioclases in the sediments of progressively more stable tectonic settings, the trend being more pronounced in the fine-grained muddy sediments than in their coarser counterparts (Figure 13).

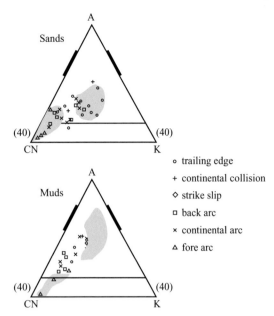

○ trailing edge
+ continental collision
◇ strike slip
□ back arc
× continental arc
△ fore arc

Figure 13 Ternary plots of mole fraction $Al_2O_3(A)$–$CaO+Na_2O$ in silicates (CN)–$K_2O(K)$ (note that the lower part of the ternary diagrams, A<40, is not shown). The plagioclase-K-feldspar join, at A = 50, and illite and smectite compositions (thick bars) as in Figure 12. Fields encompassing data from Fore Arc and Trailing Edge tectonic settings are shown in stippled patterns (after Mc Lennan *et al.*, 1990).

17.7.2 Transport Sorting

Transport processes, involving first cycle as well as recycled components, result in further sorting by grain size and density. For *sandstone* components, the higher stability of quartz, compared to feldspar and lithic grains, results in an increasing SiO_2/Al_2O_3 ratio and a decrease in concentration of trace elements that were associated chiefly with the labile aluminosilicate minerals (McLennan *et al.*, 2003). Simultaneously, the labile nature of plagioclase relative to K-feldspar leads to a rise in the K_2O/Na_2O ratio. As for provenance, the overall shift in major element composition is towards the A–K tangent (Figure 13). More importantly, transport processes are the main factor that separates the sand- and the mud-size fractions. As for sandstones, *mudstones* also evolve towards the A/K tangent, but with increasing maturity they shift more towards the A apex of the ternary diagram (Figure 13).

17.7.3 Sedimentary Recycling

The processes of mechanical weathering and dissolution in recycling systems lead to the diminution of grain size of all mineralogical constituents, as well as to progressively monomineralic quartz-rich sediments (Cox and Lowe, 1995). Sedimentary recycling is particularly effective in redistributing the trace elements, a topic discussed in the next section.

17.8　TRACE ELEMENT AND ISOTOPIC COMPOSITION OF CLASTIC SEDIMENTS

As already pointed out, the suspended load of rivers falls on the tangent connecting the UCC and its clay rich weathering products (Figure 12). The chemical composition of clastic sediments reflects therefore that of the UCC, albeit depleted for those elements that are leached out, and transported as dissolved load, to be eventually concentrated in seawater and precipitated as (bio)chemical sediments. The variable residence times (τ) of these elements in seawater are a reflection of their relative mobility, with the logarithm of τ directly proportional to the logarithm of the ratio of a given element in seawater to its upper crustal abundance (Figure 14). The elements in the upper right corner (sodium, calcium, magnesium, strontium) are rapidly mobilized during sedimentary processes, while those in the lower left corner, such as titanium, zirconium, hafnium, niobium, tantalum, thorium, zirconium, nickel, cobalt, rare-earth elements (REEs) are mostly transferred from the UCC into the clastic sedimentary mass (McLennan *et al.*, 2003). It is the latter assemblage of elements, and even more so their ratios, that are useful for provenance studies. Because of their coherent behavior in geochemical processes, the REE and the elements such as thorium, zirconium, scandium, titanium… (Taylor and McLennan, 1985) are particularly suitable for tracing the ultimate provenence of clastic sediments (McLennan *et al.*, 1990). An example is given in Figure 15, where the modern active margin turbidites reflect directly the composition of the source, with Th/Sc and Zr/Sc ratios increasing in tandem with increasing igneous differentiation of the source rocks. The coherent trends break down in the trailing margin turbidites because these contain mostly recycled components of the older sediments. Repeated sedimentary recycling tends to enrich the sand-size fraction in heavy minerals. Heavy minerals, such as zircon (uranium, hafnium), monazite (thorium), chromite (chromium), titanium-minerals (ilmenite, titanite, rutile), or cassiterite (tin) are the dominant carriers of trace elements in sandstones. In trailing edge turbidites, the recycling tends to concentrate the much more abundant zircon (carrier phase of zirconium) than monazite (thorium), thus increasing the Zr/Sc but not so much the Th/Sc ratio.

Since a number of isotope systematics (U, Th/Pb, Lu/Hf, Sm/Nd) in clastic sediments is

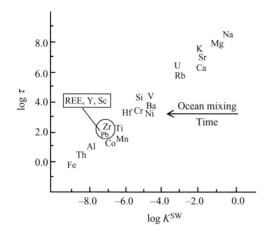

Figure 14 Plot of logτ (residence time in years) versus log K^{SW} (concentration in seawater/concentration in upper continental crust) for selected elements (after McLennan *et al.*, 2003).

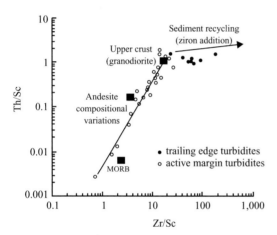

Figure 15 Th/Sc versus Zr/Sc for modern turbidites from active and passive margin setting (after McLennan *et al.*, 2003).

essentially controlled by the heavy mineral fraction, such considerations are of considerable importance for any geological interpretations.

Again, modern turbidites provide a classic example (Taylor and McLennan, 1985). Their Th/Sc ratio in active arc settings straddles the mafic to felsic join, with the bulk of samples reflecting the dominant andesitic component (Figure 16). Their ε_{Nd} of ~+5 is that of modern oceanic crust. The turbidites in progressively more evolved tectonic settings contain increasing proportions of recycled sedimentary components, become more quartzose, have more negative ε_{Nd} and higher Th/Sc ratios. The observation that the more "evolved" tectonic settings incorporate recycled components from progressively older sources is confirmed by the neodymium model ages of these clastic sediments that increase from ~250 Myr from fore arc settings to some 1.8 Gyr for the trailing age settings (Figure 17).

17.9 SECULAR EVOLUTION OF CLASTIC SEDIMENTS

17.9.1 Tectonic Settings and Lithology

Based on the tectonic concept of differential preservational probabilities, the progressively older segments of the continental crust should retain only the remnants of the most stable tectonic settings, that is they should be increasingly composed of basement and its platformal to intracratonic sedimentary cover. This is the case throughout the Phanerozoic and Proterozoic, but not for the oldest segment, the Archean. Compared to the Proterozoic, the Archean contains a disproportionate abundance of the perishable greenstones (e.g., Windley, 1984; Condie, 1989, 2000) with mid-ocean ridge basalt (MORB) and oceanic plateau basalt (OPB) affinities and immature clastic lithologies,

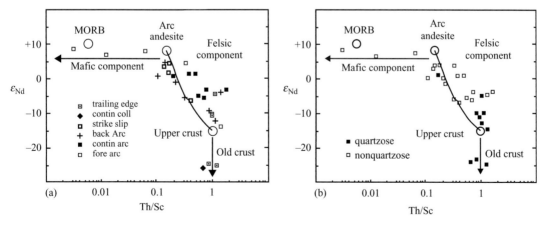

Figure 16 Plot of ε_{Nd} versus Th/Sc ratio for deep-sea turbidites according to tectonic setting of deposition (a) and according to quartz content (b). Also shown are compositions of various geochemical reservoirs (from Taylor and McLennan, 1985) and mixing relationships between average island arc andesite and upper continental crust (after McLennan *et al.*, 1990).

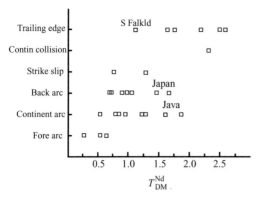

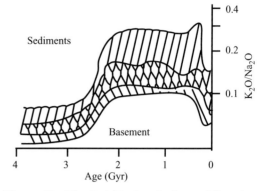

Figure 17 Neodymium model ages (relative to depleted mantle) for recent turbidites from continental margins. The large range in model ages indicates that the rate of addition of new mantle-derived material is highly variable during sedimentary recycling (after McLennan *et al.*, 1990).

Figure 18 The K_2O/Na_2O ratio for undifferentiated "continental" sediments and their basement, based on the data of Engel *et al.* (1974) (after Veizer, 1988a).

particularly graywackes (Figure 8). Regardless of their precise present-day analogue, these greenstones are an expression of the ephemeral oceanic tectonic domain in the sense of Veizer and Jansen (1985) (Figure 5). The temporal distribution of greenstones is, therefore, entirely opposite to that expected from continuous recycling, regardless of its actual rate. The fact that so many of them survived to this day, despite this recycling, argues for their excessive original abundance and entrainment into the growing and stabilizing continents. How is this tectonic and lithological evolution, from an oceanic to continental domain around the Archean/Proterozoic transition, reflected in the chemistry of clastic sediments?

17.9.2 Chemistry

The K_2O/Na_2O ratio of sediments and continental basement increased considerably at about the Archean/Proterozoic transition (Figure 18), as already pointed out by Engel *et al.* (1974). Subsequently, Taylor (1979) and Taylor and McLennan (1985) emphasized that the REE showed similar trends in their overall concentrations, in LREE/HREE and Yb/La ratios, and in the decline of the size of the europium anomaly. The general interpretation of these data was based on the proposition that the compositions of the continental crust, and of "continental" sediments, reflect a major cratonisation event that spanned the Archean/Proterozoic transition and resulted in an UCC of more felsic nature.

This interpretation was questioned by Gibbs *et al.* (1988) and Condie (1993), who argued that the global averages for clastic rocks reflect only the variable proportions of facies associated with the predominant tectonic settings at any given time and not the change in the composition of the continental crust. This is undoubtedly the

case, but—as discussed in the previous section— the types of the Archean tectonic settings are exactly the opposite to that expected from preservation probabilities based on the recycling concept. This feature must therefore reflect the fact that immature tectonic settings and lithologies were the norm in the Archean and they still dominate the preserved record, despite the high rate of recycling. Furthermore, the secular trends are present regardless whether the tectonic facies assemblages are considered together or separately (Figure 19). Combined, or separately, the Th/Sc secular trends show the same features as the K_2O/Na_2O parameter, with a rise in the ratio at the Archean/Proterozoic boundary and a reversal of the trend for the youngest portion of the record. Trends similar to that of Th/Sc one were observed also for Th, U, Th/U ratio (McLennan *et al.*, 2003) and La/Sc ratio (Condie, 1993). This is to a degree true also for the Eu/Eu* anomaly, although the magnitude is smaller than previously believed.

The Archean clastic rocks are also often enriched in chromium, nickel, and cobalt (Danchin, 1967; Condie, 1993), but this feature may only be of regional significance, mostly for the Kaapval Craton and the Pilbara block, where it may reflect the ubiquity of komatiites in the source regions of the sediments. Nevertheless, considering the frequent discrepancies in the anticipated Ni/Cr ratios, in lower than expected MgO content and, at times, high abundance of incompatible elements (Condie, 1993; McLennan *et al.*, 1990), the ultramafic provenance is not an unequivocal explanation and some secondary processes, such as weathering, may have played a role in repartitioning of these elements.

The apparent decline of the K_2O/Na_2O and Th/Sc ratios in the youngest segment of the secular trend is due to the fact that the youngest segments contain mostly the transient immature tectonic settings with their immature clastic assemblages. These are prone to destruction with advancing tectonic maturation (Figure 9) and their preservation into the Paleozoic and Proterozoic is therefore limited.

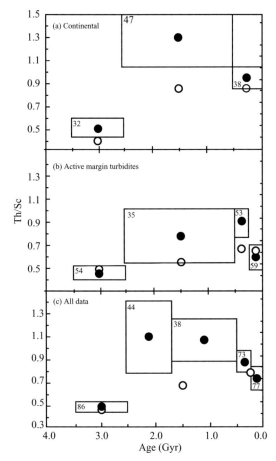

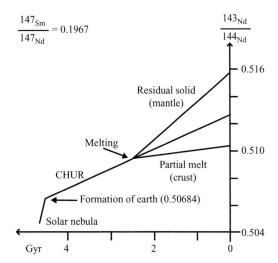

$$\frac{^{147}Sm}{^{147}Nd} = 0.1967$$

Figure 20 Theoretical evolution of Sm/Nd system in the course of planetary evolution.

Figure 19 Secular variation in Th/Sc ratio for sedimentary rocks deposited in stable continental regions (shales) and tectonically active regions (shales/graywackes). Sample numbers are shown in or near boxes. Solid symbols and boxes are arithmetic means and 95% confidence intervals. Open circles are sediment averages reported by Condie (1993) (after McLennan *et al.*, in press).

17.9.3 Isotopes

The discussion of the chemistry of clastic sediments suggested an overall mafic to felsic evolution of global sediments, and presumably of UCC, their ultimate source, in the course of geologic history, with a major evolutionary step across the Archean/Proterozoic transition. The response of isotopes to this evolutionary scenario can best be gauged by consideration of the REE isotope systematics, such as the Sm/Nd and Lu/Hf. The major fractionation of REE is accomplished during igneous differentiation of rocks from the mantle, resulting in lower parent to daughter ratios of the crustal products for both, the $^{147}Sm/^{143}Nd$ (McCulloch and Wasserburg, 1978) and the $^{176}Lu/^{177}Hf$ (White and Patchett, 1982) systematics. Although exceptions do exist (e.g., McLennan *et al.*, 2003; Patchett, 2003), the subsequent igneous, metamorphic, and sedimentary

history of the rocks usually does not affect their inherited parent/daughter ratios. For crustal rocks, including clastic sediments, it is therefore possible to calculate the time when the original material "departed" from the mantle, the latter approximated by the chondritic uniform reservoir (CHUR) evolutionary trend (Figure 20). The intercept with the CHUR is the model age. A similar reasoning applies also to the Lu/Hf systematics, with one exception. In clastic sediments, the usual carrier phase for hafnium is the heavy mineral zircon that tends to be fractionated into the sand fraction by the processes of sedimentary recycling. Mature sands therefore may contain an "excess" of hafnium and low Lu/Hf ratios (Patchett, 2003).

When model ages are plotted against stratigraphic ages of clastic sediments it becomes clear that the former are at best similar to, but mostly higher than their stratigraphic ages (Figure 21). Furthermore, the discrepancy increases from the Archean to today, from an average "excess" of ~250 Myr to ~1.8 Gyr. This is true for both isotope systematics and for muds as well as sands. These excess ages are usually interpreted as "crustal residence ages" (O'Nions *et al.*, 1983; Allègre and Rousseau, 1984), with an implication that they reflect the evolution of the ultimate crystalline crustal source. The average present-day modal excess of ~1.8 Gyr (Figure 21) is indeed a measure of the average ultimate provenance and thus of the mean age of the continental crust. In reality, however, sedimentary recycling is much faster than the metamorphism/erosion of the crystalline basement (Figure 5) and the excess model ages are therefore a consequence of the cannibalistic recycling of the ancient sedimentary mass (Veizer and Jansen, 1979, 1985). These relationships can be utilized for evaluation of the degree of cannibalistic recycling.

17.10 SEDIMENTARY RECYCLING

Sedimentary accumulations were ultimately derived from disintegration of the UCC and the global sedimentary mass should therefore have a

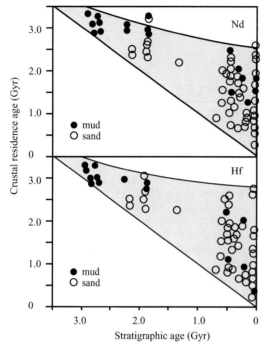

Figure 21 Stratigraphic age versus Nd- and Hf-crustal residence ages. Model ages were calculated using linear ε evolution from 0 to +10 for Nd and 0 to +16 for Hf, from 4.56 Gyr to present. The similarity of the model age systematics underscores the overall coherent behavior of the Sm–Nd and Lu–Hf isotopic systems in the sedimentary environment (after Vervoort *et al.*, 1999).

chemical composition comparable to this part of the crust. This indeed is mostly the case (Figure 22). Compared with the UCC, the composition of present-day average global sediments (AS) for most elements does not deviate more than ±50% from that of the UCC. Exceptions are the enrichments in boron, calcium, vanadium, chromium, iron, cobalt, and nickel, and the depletion in sodium. In addition, the sediments (Goldschmidt, 1933; Rubey, 1951; Vinogradov, 1967; Ronov, 1968) are strongly enriched in excess volatiles and have a higher oxidation state.

The anomalous enrichment in calcium and depletion in sodium are a consequence of hydrothermal exchange between the ocean floor and seawater, processes discussed later in the text. Hydrothermal processes can also account for most of the other elements enriched in sediments. Vanadium, chromium, cobalt, nickel plus tin are even more enriched in the sediments of the ocean floor (global subducting sediment (GLOSS)) than the AS (Figure 22) due to stronger impact of the hydrothermal systems (Plank and Langmuir, 1998). Nevertheless, the general overall absence of large anomalies in the normalized average composition of sediments suggests that the exogenic ± endogenic inputs and sinks for most major elements—except possibly calcium and sodium—were balanced throughout most of geologic history, a proposition supported by the fact that the average composition of the subducting sediments (GLOSS; Plank and Langmuir, 1998) is also approaching that of the continental crust (Figure 22).

Because the continental sedimentary mass accounts for the bulk of the present-day global sediments, the growth of the global sedimentary

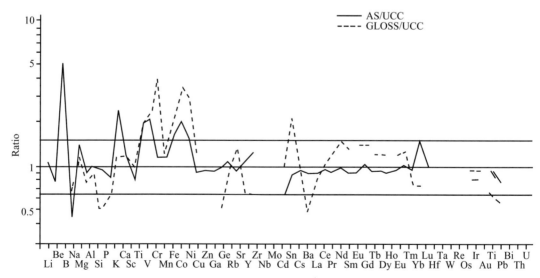

Figure 22 The elemental composition of average global sediment (AS) and global subducting sediment (GLOSS) normalized to upper continental crust (UCC). AS and UCC from McLennan and Murray (1999) and McLennan and Taylor (1999), respectively. Ti, Nb, Cs, and Ta are corrected as proposed in Plank and Langmuir (1998). GLOSS from Plank and Langmuir (1998).

mass should be a function of the growth of the continents. It is feasible, therefore, that at least some of this mass, particularly during the early stages of the earth's history and at the beginning of each tectonomagmatic cycle, evolved on oceanic or intermediate type crust. If so, it would have been derived from a source more mafic than the present-day UCC. In an entirely cannibalistic (closed) recycling system, this composition would have been perpetuated indefinitely regardless of the nature of the later continents.

In reality we must be dealing with a partially open system, because some sediments are being subducted while others are being formed at the expense of primary igneous and metamorphic rocks. Estimates based on Sm/Nd systematics (Figure 23) indicate that the sedimentary cycle is $\sim 90 \pm 5\%$ cannibalistic, attaining its near present-day steady state around the Archean/Proterozoic transition. The *first-order* features of secular trends, such as the K_2O/Na_2O, Th/Sc (Figures 18 and 19), REE, U, Th, and Th/U (Collerson and Kamber, 1999; McLennan *et al.*, 2003) can be explained, provided the Archean was the time when the sedimentary mass was mostly growing by addition of the first-cycle sediments from erosion of contemporaneous young (≤ 250 Myr old) igneous precursors. Subsequent to the large-scale cratonization events, and subsequent to establishment of a substantial global sedimentary mass at $\sim 2.5 \pm 0.5$ Gyr, cannibalistic sediment—sediment recycling became the dominant feature of sedimentary evolution. The general absence of neodymium model ages much in excess of their stratigraphic ages in most Archean sediments (Figure 23) is consistent with the absence of the inherited old detrital components. This observation strongly argues against the presence of large continental landmasses prior to ~ 3 Gyr.

17.11 OCEAN/ATMOSPHERE SYSTEM

The previous discussion dealt with the solid earth component of the exogenic cycle, a system that is recycled on $10^6 - 10^9$ yr timescales. In contrast, the ocean, atmosphere, and life are recycled at much faster rates and the continuity of the past record is lost rapidly. For quantitative evaluations we have to rely therefore on proxy signals embedded in marine (bio)chemical sediments.

Other contributions in this Treatise (Volumes 6–8) deal with the lithological and chemical aspects of evolution of specific types of (bio)-chemical sediments, such as cherts, phosphorites, hydrocarbons, and evaporites, and we will therefore concentrate on the most ubiquitous category, the carbonate rocks (see also Section 17.12). This carrier phase also contains the largest number of chemical and isotopic tracers.

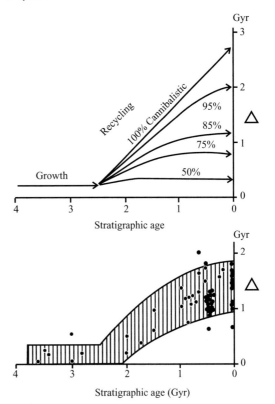

Figure 23 Models of Sm/Nd excess ages for sedimentary rocks. The Sm/Nd systematics dates the time of fractionation from the mantle. Regardless of whether most or only some of the sediments were generated during early terrestrial history, they would inherit Sm/Nd systematics from their igneous precursors. In a cannibalistic sedimentary recycling, these ancient systematics will be perpetuated and, as a consequence, Sm/Nd of all second-stage sediments should exceed their stratigraphic ages (Figure 21) and the Δ (Sm/Nd model age minus stratigraphic age) should increase toward the present, with a 45° slope being an upper limit for a completely closed system. In order to generate the observed smaller Δ, it is necessary to add sediments formed from a young source. The upper figure represents model calculations based on the assumption that prior to 2.5 Ga the sedimentary mass was growing through addition of first-cycle sediments. The post-Archean evolution assumes cannibalistic recycling of the steady state mass, and the slopes represent the degree of cannibalism for this recycling. The bottom part is a collation of experimental data (after Veizer and Jansen, 1985).

As already pointed out in Figure 7, the relative proportion of carbonate rocks within the continental realm generally decreases in the course of the Phanerozoic, and the Mesozoic and Cenozoic "deficiency" of carbonates was attributed to a tectonic cause, the ubiquity of transient immature tectonic settings. Another reason is the general northward drift of continents, which resulted in a progressive decline in the shelf areas that fell within the confines of the tropical climatic belt

(Walker *et al.*, 2002; Bluth and Kump, 1991; Kiessling, 2002). Finally, in the course of the Mesozoic and Cenozoic, the locus of carbonate sedimentation migrated from the shelves to the pelagic realm, mirroring the role that calcareous shells of foraminiferans, pteropods, and cocco-lithophorids commenced to play in the carbonate budget (Kuenen, 1950). This environmental shift may have been accompanied by the deepening of the carbonate compensation depth (Ross and Wilkinson, 1991), which may have enlarged the oceanic areas that were sufficiently shallow for preservation of pelagic carbonates. Another feature of carbonate sedimentation is the relative scarcity of dolostones in Cenozoic and Mesozoic sequences, compared to their Paleozoic and particularly Proterozoic counterparts (Chilingar, 1956; Veizer, 1985). Carbonate sediments as such are mostly associated with low latitude sedimentary environments, but for the Phanerozoic the dolostones/total carbonate ratio within the tropical belt increases polewards (Berry and Wilkinson, 1994), that is towards arid climatic zones. This suggests that the process of dolomitization is a near-surface phenomenon. The more or less consistent offset of the modes of $\delta^{18}O$ between dolostones and limestones, of ~2–3‰, throughout the entire geological history (Shields and Veizer, 2002) is also consistent with such an interpretation, because the observed $\Delta^{18}O_{dol-cal}$ is a near-equilibrium value (Land, 1980). These carbonates are therefore either primary marine precipitates, or more likely they are early diagenetic products of stabilization and dolomitization of carbonate precursors. Pore waters at this stage were still in contact with the overlying seawater and/or contained an appreciable seawater component. The high frequency of dolostones in ancient sequences may again be principally a reflection of the ubiquity of shelf, epicontinental, and platformal tectonic settings preserved from the Paleozoic and Proterozoic times. Changes in seawater chemistry, such as Mg/Ca ratio, saturation state, or SO_4 content, may have been a complementary factor.

17.11.1 The Chemical Composition of Ancient Ocean

Earlier studies (e.g., Holland, 1978, 1984) assumed that the chemical composition of seawater during the Phanerozoic was comparable to the present-day one. Subsequently, experimental data on fluid inclusions in halite (Lowenstein *et al.*, 2001; Horita *et al.*, 1991, 2002) and on carbonate cements (Cicero and Lohmann, 2001) suggested that at least the magnesium, calcium, and strontium, and their ratios, in Phanerozoic seawater may have been variable. Steuber and Veizer (2002) assembled a continuous record of Sr/Ca variations for the Phanerozoic oceans (Figure 24) that covaries positively with the

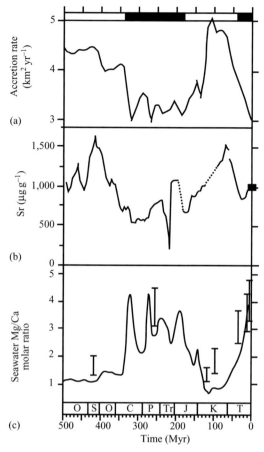

Figure 24 (a) "Accretion rate of oceanic crust" (Gaffin, 1987); (b) trends in Sr concentrations in biological LMC (Steuber and Veizer, 2002; Lear *et al.*, 2003); and (c) Mg/Ca ratio of seawater. Vertical bars are based on experimental data from fluid inclusions in halite (Lowenstein *et al.*, 2001), while the trend line results from model calculations (Stanley and Hardie, 1998). The bars at the top represent "calcitic" (blank) and "aragonitic" (black) seas (Sandberg, 1983) (after Steuber and Veizer, 2002).

"accretion rate of the oceanic crust" (Gaffin, 1987) and negatively with the less well-known Mg/Ca ratio. Such a covariance would suggest that we are dealing with coupled phenomena and they proposed that all these variables are ultimately driven by tectonics, specifically seafloor spreading rates that, in turn, control the associated hydrothermal and low-temperature alteration processes. Since the hydrothermal alteration of young oceanic crust effectively exchanges magnesium for calcium (see Chapter 2), the accretion of the oceanic crust would modulate the Mg/Ca ratio of seawater. At high accretion rates, the low Mg/Ca ratio favors precipitation of calcite and, as a result, higher retention of strontium in seawater. At slow accretion rates, the high Mg/Ca ratio favors aragonite precipitation and a high rate of strontium removal from seawater.

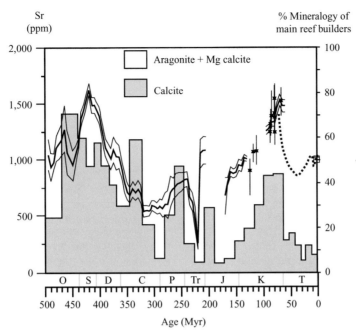

Figure 25 Predominant skeletal mineralogy, calcite (shaded) versus aragonite + high-Mg calcite (blank), in Phanerozoic reefs. Mineralogy from Kiessling (2002), Sr data from Steuber and Veizer (2002) and Lear *et al.* (2003) (see also Figure 24) (courtesy of T. Steuber).

The above scenario is consistent with the observation that calcite was the dominant mineralogy of carbonate skeletal components in the early to mid-Paleozoic and the mid-Jurassic to mid-Tertiary (Figure 25), the times of high Sr/Ca and low Mg/Ca ratio (Figure 24). Aragonite mineralogy, however, dominated the mid-Carboniferous to Jurassic and the Tertiary to Quaternary intervals (Sandberg, 1983; Kiessling, 2002) with opposite chemical attributes. The changing Mg/Ca ratio of seawater can also be at least in part responsible for the general scarcity of $MgSO_4$-bearing potash minerals (Hardie, 1996) in the Paleozoic and Mesozoic marine evaporites.

While all the above trends and their correlations are likely real, the proposed causative mechanism is being questioned lately due to the proposition that seafloor spreading rates have been about constant since at least 180 Myr (Rowley, 2002). If so, the sea-level stands cannot be inverted into "accretion rates of oceanic crust," as done by Gaffin (1987). A causative mechanism for all these covariant phenomena remains therefore enigmatic.

The reconnaissance studies of fluid inclusions (Horita *et al.*, 1991, 2002) suggest also that early Paleozoic seawater was $\sim 2.5 \times$ depleted in SO_4, compared to its present-day counterpart. From model considerations, based on the mineralogy and volume of evaporites, claims have been made also for changes in the potassium concentration of Phanerozoic oceans (Hardie, 1996), and for an increase in the total salinity, from the modern 35 ppt to ~ 50 ppt in the Cambrian (Hay *et al.*, 2001). The

experimental confirmation for all these theoretical assertions is presently not available.

On timescales of billion of years, ancient Precambrian carbonates appear to have been enriched in Fe^{2+} and Mn^{2+}, if compared to their Phanerozoic counterparts (Veizer, 1985). In part, this may be a reflection of diagenetic alteration processes that tend to raise the iron and manganese contents of successor phases (Brand and Veizer, 1980; Veizer, 1983). However, the Archean manganese concentrations, in the 10^3–10^4 ppm range, are likely not explained by diagenetic processes alone. Accepting that the redox state of the Archean and early Proterozoic oceans may have been lower than that of their Phanerozoic counterparts (Cloud, 1976), the high Fe^{2+} and Mn^{2+} content of contemporaneous carbonates may reflect higher concentrations of these elements in the ancient oceans.

17.11.2 Isotopic Evolution of Ancient Oceans

In contrast to chemistry, where the secular trends are still mostly obscured by the natural scatter in the database, the isotope evolution of seawater is better resolved.

17.11.2.1 Strontium isotopes

In modern oceans the concentration of strontium is ~ 8 ppm and its residence time is ~ 4–8 Myr (Holland, 1984). The present isotopic ratio $^{87}Sr/^{86}Sr$ is 0.7092 (McArthur, 1994), controlled essentially by two fluxes, the "mantle" and the "river" flux. The former represents strontium exchanged between seawater and oceanic crust ($^{87}Sr/^{86}Sr \sim 0.703$) in hydrothermal systems on the ocean floor. The latter, reflecting the more fractionated composition of the continental crust, feeds into the oceans more radiogenic strontium, with an average isotope ratio for rivers of ~ 0.711 (Wadleigh *et al.*, 1985; Goldstein and Jacobsen, 1987; Palmer and Edmond, 1989). Note, nevertheless, that the latter may vary from 0.703 to 0.730 or more, depending on whether the river is draining a young volcanic terrane or an old granitic shield. The third input is the flux of strontium from diagenesis of carbonates, which results in expulsion of some strontium from the solid phase during precursor to product (usually aragonite to low-magnesium calcite) recrystallization, but this flux is not large enough to influence the isotopic composition of seawater. A simple balance calculation based on isotopes, therefore, shows that strontium in seawater originates $\sim 3/4$ from the "river" flux and $\sim 1/4$ from the "mantle" flux, generating the modern value of 0.7092. This value is uniform with depth and into marginal seas. Even water bodies such as Hudson Bay, with a salinity $\sim 1/2$ of the open ocean due to large riverine influx, have this same isotope ratio. This is because the rivers are very dilute relative to seawater, with strontium concentrations usually 1,000 times less, and their impact is felt only if the proportion of seawater in the mixtures is less than 10%.

The above considerations show that the strontium isotopic composition of seawater is controlled essentially by tectonic evolution, that is, by relative contributions from weathering processes on continents and from the intensity of submarine hydrothermal systems. Over geological time, however, the isotopic compositions of these

two fluxes have evolved, because ^{87}Sr is a decay product of ^{87}Rb:

$$\left(\frac{^{87}Sr}{^{86}Sr}\right)_p = \left(\frac{^{87}Sr}{^{86}Sr}\right)_o + \left(\frac{^{87}Rb}{^{86}Sr}\right)\left(e^{\lambda t} - 1\right) \quad (7)$$

where p=present, o=initial ratio at the formation of the Earth 4.5 Gyr ago (0.699), λ=decay constant $(1.42 \times 10^{-11} \text{ yr}^{-1})$, and t=time since the beginning (e.g., formation of the Earth 4.5 Gyr ago).

From Equation (7) it is evident that the term $(^{87}Sr/^{86}Sr)_p$ for coeval rocks originating from the same source $(^{87}Sr/^{86}Sr)_o$ depends only on their Rb/Sr ratios. Since this ratio is ~ 6 times larger for the more fractionated continental rocks than for the basalts (~ 0.15 to 0.027; Faure, 1986), the $^{87}Sr/^{86}Sr$ of the continental crust at any given time considerably exceeds that of the mantle and oceanic crust, increasing to the present-day values of ~ 0.730 for the average continental crust, as opposed to ~ 0.703 for the oceanic crust. The rivers draining the continents are less radiogenic (~ 0.711) than the crust itself due to the fact that most riverine strontium originates from the weathering of carbonate rocks rather than from their silicate counterparts. The former, as marine sediments, inherited their strontium from seawater, which—as discussed above—contains also the less radiogenic strontium from hydrothermal sources.

The lower envelope of the strontium isotopic trend during the *Precambrian* (Figure 26) straddles the mantle values until about the Archean/Proterozoic transition, afterwards deviating towards more radiogenic values, reflecting the input from the continental crust. The large spread of values above the lower envelope is a consequence of several factors. First, it reflects the impact of secondary alteration that mostly results in resetting

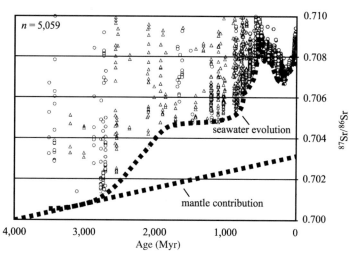

Figure 26 Strontium isotopic composition of sedimentary carbonate rocks during geologic history. Reproduced from Shields and Veizer (2002). Circles and triangles represent samples with good and poor age assignment, respectively.

of the signal towards more radiogenic values (Veizer and Compston, 1974). Second, the scatter also includes higher order oscillations in the strontium isotope ratio of seawater which cannot be as yet resolved for the Precambrian due to poor stratigraphic resolution and inadequate geochronological control.

From the above discussion, it is clear that the primary control on strontium isotopic composition of the first order (billion years trend) for seawater will be exercised by the growth pattern of the continental crust. Two competing hypotheses dominate this debate:

(i) The generation of the entire continental crust was an early event and the present-day scarcity of older remnants is a consequence of their destruction (recycling) by subsequent tectonic processes (Armstrong, 1981; Sylvester *et al.*, 1997).
(ii) The continents were generated episodically over geologic history, with major phases of continent formation in the late Archean and early Proterozoic, and attainment of a near modern extent by ~1.8 Gyr (Veizer and Jansen, 1979; Taylor and McLennan, 1985; McCulloch and Bennett, 1994).

Model calculations by Goddéris and Veizer (2000) of seawater $^{87}Sr/^{86}Sr$ evolution for the two alternatives, and for the coeval mantle, show that the measured experimental data fit much better with the second pattern of continental growth. Note that this scenario is also fully compatible with the evolution of sediments and their chemistry discussed in the preceding sections of this chapter. The Archean oceans were "mantle buffered" (Veizer *et al.*, 1982) by vigorous circulation of seawater via submarine hydrothermal systems. With the exponential decline of internal heat dissipation,

the vigor of the hydrothermal system also declined and at the same time the flux of radiogenic strontium from growing continents brought in by rivers started to assert itself. This tectonically controlled transition from "mantle" to "river buffered" oceans across the Archean/Proterozoic transition is a first order feature of terrestrial evolution, with consequences for other isotopic systematics, for redox state of the ocean/atmosphere system and for other related phenomena.

The resolution of the database is considerably better for the *Phanerozoic* than for the Precambrian due to higher quality of samples and to much better biostratigraphic resolution, with duration of biozones from ~1 Myr in the Cenozoic to ~5 Myr in the early Paleozoic. The first data documenting the strontium isotopic variations in Phanerozoic seawater were published by Peterman *et al.* (1970), with subsequent advances by Veizer and Compston (1974) and Burke *et al.* (1982). The latest version, by Veizer *et al.* (1999), is reproduced in Figure 27. Overall, this second order Phanerozoic trend shows a decline in $^{87}Sr/^{86}Sr$ values from the Cambrian to the Jurassic, followed by a steep rise to today's values, with superimposed third order oscillations at 10^7 yr frequency. Because of the better quality of samples, the experimental data indicate an existence of still higher order $^{87}Sr/^{86}Sr$ oscillations within biozones. However, since these samples often do not originate from the same profile, their relative ages within a biozone are difficult to discern. They have to be treated therefore as coeval and the secular trend thus becomes a band (Figure 28).

In general, it is again tectonics that is the cause of the observed Phanerozoic trend, with the "mantle" input of greater relative importance at times of the troughs and the "river" flux dominating in the

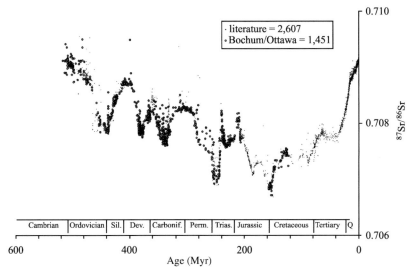

Figure 27 $^{87}Sr/^{86}Sr$ variations for the Phanerozoic based on 4,055 samples of brachiopods, belemnites, and conodonts. Normalized to NBS 987 of 0.710240 (after Veizer *et al.*, 1999).

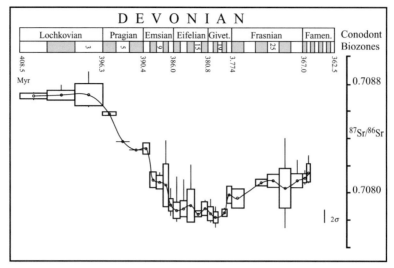

Figure 28 $^{87}Sr/^{86}Sr$ variations during the Devonian based on conodont biozones. Explanations: circle=mean; box=$\pm 1\sigma$; vertical line=minimum and maximum. The 2 σ in the lower right corner is an average 2 σ for the NBS 987 standard. Note that only brachiopods are included in this. Figures 1–37 are conodont biozones (after Veizer *et al.*, 1999).

Tertiary and early Paleozoic. Nevertheless, it is difficult to correlate the overall trend or the superimposed oscillations with specific tectonic events. The problem arises from the fact that model solutions do not produce unique answers. The "river" flux is likely the major reason for the observed $^{87}Sr/^{86}Sr$ oscillations, because the changes in seafloor spreading rates, apart from being disputed (Rowley, 2002), are relatively sluggish and the strontium isotope ratio of the "mantle" flux is relatively constant at ~0.703. The "river" flux, however, may vary widely in both strontium elemental flux and its isotope ratio. For example, the rapid Tertiary rise in $^{87}Sr/^{86}Sr$ (Figure 27) is commonly interpreted as reflecting the uplift of the Himalayas. Accepting this to be the case, it still remains an open question whether the rise is due to higher flux of "river" strontium (increased weathering rate), its more radiogenic nature (unroofing of older core complexes), or both (cf. the contributions in Ruddiman, 1997). For these reasons, it is difficult to utilize the strontium isotopic curve of seawater as a direct proxy for continental weathering rates in model considerations. Nevertheless, it is intriguing that the Phanerozoic seawater strontium isotope curve correlates surprisingly well with the estimated past sediment fluxes (Hay *et al.*, 2001) that were reconstructed by the "population dynamics" approach discussed in the introductory section of this chapter.

17.11.2.2   Osmium isotopes

The isotope ^{187}Os is generated by β decay of ^{187}Re. In many ways the systematics and the presently known secular evolution of $^{187}Os/^{186}Os$ in seawater is similar to that of $^{87}Sr/^{86}Sr$ (Peucker-Ehrenbring and Ravizza, 2000). The present-day $^{187}Os/^{186}Os$ of UCC is ~1.2–1.3, runoff ~1.4 (due likely to preferential weathering of radiogenic black shales), seawater, ~1.06 and that of meteorites and mantle, ~0.13. In contrast to strontium with a seawater residence time of 4–8 Myr, the residence time of osmium is on the order of 10^4 yr due to the effective scavenging of osmium by Fe/Mn crusts and organic-rich sediments. This enables tracing of short-term fluctuations in seawater composition, such as the Quaternary glacial/interglacial cycles, a feat difficult to replicate by the buffered strontium system.

Metalliferous sediments usually have low Re/Os ratios and reflect well the isotopic composition of seawater. Their disadvantage is the slow accumulation rate that limits temporal resolution. Organic rich shales, alternatively, have high Re/Os ratios, requiring age correction by the isochron technique. Corals and carbonate sediments do not appear to preserve the hydrogenous (seawater) $^{187}Os/^{186}Os$ record.

The presently known data for seawater $^{187}Os/^{186}Os$ cover mostly the Cenozoic, with fragmentary results for the Cretaceous and the Jurassic. As for $^{87}Sr/^{86}Sr$, the $^{187}Os/^{186}Os$ declines from present-day value of ~1.06 to 0.15 at the K/T boundary. The sudden drop, from ~0.8 to 0.15 at the K/T boundary likely reflects the cosmic input from the meteoric impact. The $^{187}Os/^{186}Os$ ratio for the Mesozoic oscillates between 0.8 and 0.15, but the details are not yet resolved. For further discussion of Re/Os systematics.

17.11.2.3   Sulfur isotopes

In contrast to strontium and osmium isotopes, the isotopes of sulfur are strongly fractionated by biological processes, particularly during the dissimilatory bacterial reduction of sulfate to sulfide. The laboratory results for this step are anywhere from +4 to −46‰ (CDT), but even larger fractionations have been observed in natural systems (Harrison and Thode, 1958; Chambers and Trudinger, 1979; Habicht and Canfield, 1996).

The geologic record is characterized by a dearth of *Precambrian* evaporitic sequences, including their sulfate facies. Stratiform barites do exist, but they may be, at least in part, of hydrothermal origin. The scarcity of evaporites is partly due to their poor survival rates in the face of tectonic processes, but another reason may be low pO_2 levels in the contemporaneous ocean/atmosphere system, particularly during the Archean (Veizer, 1988a). In addition, most Archean sulfides, such as pyrites, contain $\delta^{34}S$ close to 0‰ CDT, the value typical of the mantle (Figure 29), rather than the expected highly negative ones characteristic of bacterial dissimilatory sulfate reduction. These observations were usually interpreted (e.g., Schidlowski *et al.*, 1983; Hayes *et al.*, 1992) as being due to biological evolution, where it is assumed that the invention of oxygen generating photosynthesis and of bacterial sulfate reduction were only later developments. In that case, most of the sulfur in the Archean host phases would have originated from mantle sources and carried its isotopic signature. Only with the onset of these two biological processes, in about the late Archean or early Proterozoic, was enough oxygen generated to stabilize sulfate in seawater and to

initiate its bacterial reduction to H_2S, the latter eventually forming sulfide minerals, such as pyrite. This development resulted in the burial of large quantities of sulfides depleted in ^{34}S in the sediments, causing the residual sulfate in the ocean to shift towards heavier values. The result is the bifurcation of $\delta^{34}S$ sulfate/sulfide values at the time of "invention" of bacterial dissimilatory sulfate reduction (Figure 29).

The above scenario is appealing, but not mandatory. As shown by Goddéris and Veizer (2000), the same "logistic" scenario of continental growth that generated the strontium isotope trend can also generate the observed $\delta^{34}S$ pattern (Figure 29) and the growth of sulfate in the oceans. This explanation has the advantage that a single scenario generates all these (and other) evolutionary patterns. In short, the early "mantle" buffered oceans (Veizer *et al.*, 1982) had a large consumption of oxygen in the submarine hydrothermal systems (Wolery and Sleep, 1988) because they operated at considerably higher rates than today. The capacity of this sink declined exponentially in the course of geologic history, reflecting the decay in the dissipation of the heat from the core and mantle. As a result, the buffering of the ocean was taken over by the continental "river" flux. In summary, the bacterial dissimilatory sulfate reduction must have been extant at the time of bifurcation of the $\delta^{34}S$ record, but the isotope data do not provide a definitive answer as to the timing of this invention. Tectonic evolution would override its impact even if established much earlier.

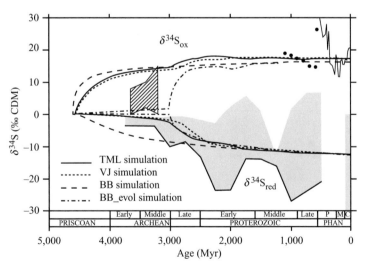

Figure 29 Model sulfur isotopic evolution in the course of geologic history. The lower trend (labeled $\delta^{34}S_{red}$) represents the $\delta^{34}S$ for sulfides. The upper curves (labeled $\delta^{34}S_{ox}$) are the $\delta^{34}S$ of marine sulfates. TML and VJ simulations assume a logistic type of continental growth as proposed by Taylor and McLennan (1985) and Veizer and Jansen (1979), respectively. BB simulation assumes an instantaneous generation of continental crust, BB-evol simulation assumes instantaneous continental generation, but with delayed "invention" of oxygen generating photosynthesis. The Phanerozoic trend as in Figure 30. Dots represent measurements of Precambrian sulfates (Claypool *et al.*, 1980) and the hatched field represents sulfates from Holser *et al.* (1988) (after Goddéris and Veizer, 2000).

The Precambrian δ^{34}S record is spotty for sulfide-S and almost nonexistent for sulfate-S (Canfield and Teske, 1996; Strauss, 1993). A fragmentary record for the latter exists only for the latest Neoproterozoic (Strauss, 1993), suggesting a large shift from ~20 to 33±2‰ at the transition into the Phanerozoic.

The δ^{34}S$_{sulfate}$ variations in *Phanerozoic* oceans, based on evaporites (Holser and Kaplan, 1966; Claypool *et al.*, 1980), form an overall trough-like trend similar to strontium isotopes (Figure 30). Note, however, the large age uncertainties for, and the large gaps between, the studied evaporitic sequences. This is due to their episodic occurrence and uncertain chronology and is part of the reason for the large spread in the coeval δ^{34}S values despite the fact that the δ^{34}S$_{sulfate}$ in seawater is spatially homogeneous (Longinelli, 1989). Another reason for this large spread in the δ^{34}S values is the evolution of sulfur isotopes in the course of the evaporative process, from sulfate to chloride to late salt facies. A recent development of the technique that enabled measurement of δ^{34}S in structurally bound sulfate in carbonates (Kampschulte and Strauss, 1998; Kampschulte, 2001) yielded a Phanerozoic secular curve with much greater temporal resolution (Figure 30).

The δ^{34}S$_{sulfate}$ and δ^{13}C$_{carbonate}$ secular curves correlate negatively (Veizer *et al.*, 1980), suggesting that it is the redox balance (peddling of oxygen between the carbon and sulfur cycles) that controls the δ^{34}S variations in Phanerozoic oceans.

Note, nevertheless, that a physical geological scenario for this coupling is as yet not clarified. If redox balance is indeed a major control mechanism, it would suggest that the withdrawal of ^{32}S due to pyrite burial in sediments was twice as large as today in the early Paleozoic versus about one-half in the late Paleozoic (Kump, 1989).

In addition to the long-term 10^8 yr trends, shorter spikes, on 10^5–10^6 yr scales (e.g., Permian/Triassic transition; Holser, 1977), do exist, but their "catastrophic" geological causes are not as yet resolved.

17.11.2.4 Carbon isotopes

The two dominant exogenic reservoirs of carbon are carbonate rocks and organic matter in sediments. They are linked in the carbon cycle via atmospheric CO_2 and the carbon species dissolved in the hydrosphere. The δ^{13}C for the total dissolved carbon (TDC) in seawater is ~+1±0.5‰ (PDB), with surficial waters generally heavier and deep waters lighter than this average (Kroopnick, 1980; Tan, 1988). Atmospheric CO_2 in equilibrium with TDC of marine surface water has a δ^{13}C of ~−7‰. CO_2 is preferentially utilized by photosynthetic plants for production of organic carbon causing further depletion in ^{13}C (Equation 8):

$$6CO_2 + 6H_2O \rightarrow C_6H_{12}O_6 + 6O_2 \qquad (8)$$

Most land plants utilize the so-called C_3, or Calvin pathway (O'Leary, 1988), that results in tissue with a δ^{13}C$_{org}$ of ~−25‰ to −30‰. The situation

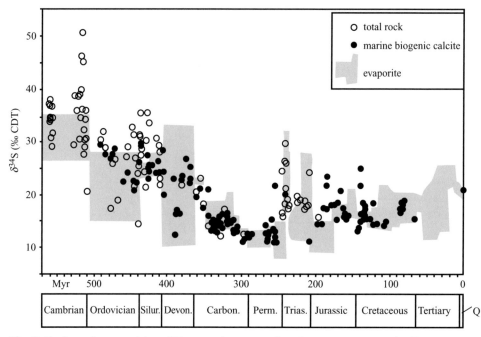

Figure 30 Sulfur isotopic composition of Phanerozoic seawater based on measurements of sulfur structurally bound in calcitic shells as well as evaporites. Note that the shell samples are mostly the same as those of the Sr, O, and C isotopes in Veizer *et al.* (1999) (after Kampschulte, 2001 by courtesy of the author).

for aquatic plants is somewhat different because they utilize dissolved and not gaseous CO_2. Tropical grasses, however, utilize the C_4 (Hatch-Slack or Kranz) pathway and have a $\delta^{13}C$ of some $-10‰$ to $-15‰$. A third group that combines these two pathways, the CAM plants (algae and lichens), has intermediate $\delta^{13}C$ values. In detail, the nature of the discussed variations is far more complex (Deines, 1980; Sackett, 1989) and depends on the type of organic compounds involved. For our purposes, however, it is only essential to realize that C_{org} is strongly depleted in ^{13}C. This organic matter, which is very labile, is easily oxidized into CO_2 that inherits the ^{13}C-depleted signal.

The $\delta^{13}C$ of mantle carbon is $\sim -5‰$ PDB (Schidlowski *et al.*, 1983; Hayes *et al.*, 1992) and in the absence of life and its photosynthetic capabilities, this would also be the isotopic composition of seawater. Yet, as far back as 3.5 Gyr ago, and possibly as far as ~ 4 Gyr ago (Schidlowski *et al.*, 1983), the carbonate rocks ($\sim$seawater) had $\delta^{13}C$ at $\sim 0‰$ PDB (Figure 31). This suggests that a reservoir of reduced organic carbon that accounted for $\sim 1/5$ of the entire exogenic carbon existed already some 4 Gyr ago, "pushing up" the residual 4/5 of carbon, present in the oxidized form in the ocean/atmosphere system, from $-5‰$ to $0‰$ PDB. This is an oxidized/reduced partitioning similar to that we have today. Stated in a simplified manner, life with its photosynthetic capabilities, and possibly of present-day magnitude, can be traced almost as far back as we have a rock record. This photosynthesis may or may not have been generating oxygen as its byproduct, but was essential in order to "lift" the seawater $\delta^{13}C$ to values similar to the present-day ones. In order to sustain seawater $\delta^{13}C$ at this level during the entire geologic history, it is necessary

that the input and output in the carbon cycle have the same $\delta^{13}C$. Since the input from the mantle, via volcanism and hydrothermal systems, has a $\delta^{13}C$ of $-5‰$ and the subducted carbonates are $0‰$, the subduction process must involve also a complementary ^{13}C-depleted component, organic matter. This is possible to contemplate as long as oceanic waters were not fully oxygenated, such as may have been the case in the Archean. This is either because oxygen generating photosynthesis was "invented" as late as the late Archean or early Proterozoic (Cloud, 1976; Holland, 1984), or because tectonic evolution led to a progressive oxygenation of the ocean/atmosphere system due to a switchover from a "mantle"—to a "river"— buffered ocean system (Goddéris and Veizer, 2000). For the latter alternative, and in analogy to sulfur, it is possible to argue that oxygen generating photosynthesis (photosystem 2) may have been extant as far back as we have the geologic record, without necessarily inducing oxygenation of the early ocean/atmosphere system (but see Lasaga and Ohmoto, 2002). Whatever the cause, the oxygenation of the system in the early Proterozoic would have resulted in oxidation of organic matter that was settling down through the water column. Today only $\sim 1\%$ of organic productivity reaches the ocean floor and $\sim 0.1\%$ survives into sedimentary rocks. As a result, the addition of mantle carbon, coupled with the subduction loss of the ^{13}C-enriched limestone carbon, would slowly force the $\delta^{13}C$ of seawater back to mantle values. In order to sustain the near $0‰$ PDB of seawater during the entire geologic history, it is necessary to lower the input of mantle carbon into the ocean/atmosphere system by progressively diminishing the impact of hydrothermal and volcanic activity over geologic time.

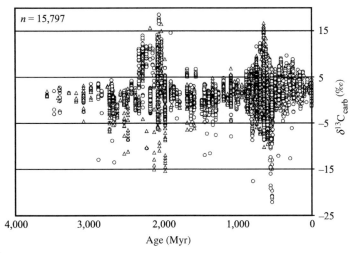

Figure 31 The $\delta^{13}C$ composition of carbonate rocks during geologic history. Reproduced from Shields and Veizer (2002). Triangles—dolostones, circles—limestones and fossil shells. For possible explanation of the large Paleo- and Neoproterozoic spreads, see Rothman *et al.* (2003).

Superimposed on this invariant *Precambrian* $\delta^{13}C$ seawater trend are two intervals with very heavy (and very light) values, at ~2.2 Gyr and in the Neoproterozoic (Figure 31). The former has been interpreted as a result of the invention of oxygen generating photosynthesis that resulted in the sequestration of huge quantities of organic matter (Karhu and Holland, 1996) into coeval sediments and the Neoproterozoic interval was the time of the proposed "snowball earth" (Hoffman et al., 1998). At this stage, the reasons for the high frequency of the anomalous $\delta^{13}C$ values during these two intervals are not well understood, but it is interesting that both were associated with large glaciations, as was the later discussed ^{13}C-enriched Permo/Carboniferous interval.

The sampling density and time resolution in the *Phanerozoic* enabled the delineation of a much better constrained secular curve (Figure 32), with a maximum in the late Permian, but even in this case we are dealing with a band of data, reflecting the fact that the $\delta^{13}C_{DIC}$ of seawater is not uniform in time and space, that organisms can incorporate metabolic carbon into their shells (vital effect), and that some samples may also contain a diagenetic overprint. Superimposed on the overall trend are higher oscillations, at 10^7yr and shorter time-scales, but their meaning is not yet understood.

Frakes et al. (1992) proposed that the $\delta^{13}C_{carbonate}$ (seawater) becomes particularly heavy at times of glaciations, and that such times are also characterized by low CO_2 levels. The coincidences of the $\delta^{13}C$ peaks with the late Ordovician and Permocarboniferous glacial episodes appear to support this proposition, but the Mesozoic/Cenozoic record is divergent. Accepting the

validity of the present $\delta^{13}C$ trend, it is possible to calculate the model p_{CO_2} levels of ancient atmospheres. Three Phanerozoic p_{CO_2} reconstructions exist (Berner and Kothavala, 2001; Berner and Streif, 2001; Rothman, 2002) that are internally inconsistent and not one of them shows any correlation with the paleoclimate deduced from sedimentological criteria (Veizer et al., 2000; Boucot and Gray, 2001; Veizer, 2003; Figure 35). This led Veizer et al. (2000) to conclude that either the estimates of paleo-CO_2 were unreliable or there was no direct relationship between p_{CO_2} levels and climate for most of the Phanerozoic.

Higher order peaks, at a 10^6 yr resolution, have been observed in the geologic record, particularly in deep sea borehole sections and are discussed by Ravizza and Zachos.

17.11.2.5 Oxygen isotopes

The oxygen isotope record of some 10,000 limestones and low-magnesium calcitic fossils (Shields and Veizer, 2002) shows a clear trend of ^{18}O depletion with age of the rocks (Figure 33). This isotope record in ancient marine carbonates (but also cherts and phosphates) is one of the most controversial topics of isotope geochemistry. It centers on the issue of the primary versus post-depositional origin of the secular trend (e.g., Land, 1995 versus Veizer, 1995). Undoubtedly, diagenesis, and other post-depositional phenomena reset the δ^{18}O, usually to more negative values, during stabilization of original metastable phases (e.g., aragonite, high-magnesium calcite), into the more stable phase, diagenetic low-magnesium calcite. Every carbonate rock is subjected to this

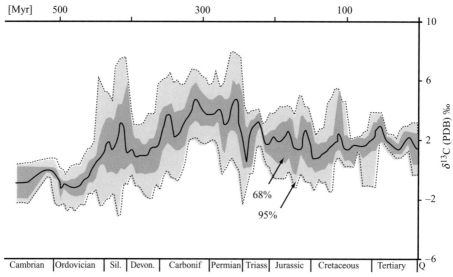

Figure 32 The Phanerozoic $\delta^{13}C$ trend for LMC shells. The running mean for ~4,500 samples is based on a 20 Ma window and 5 Ma forward step. The shaded areas around the running mean include the 68% ($\pm1\sigma$) and 95% ($\pm2\sigma$) of all data (after Veizer et al., 1999).

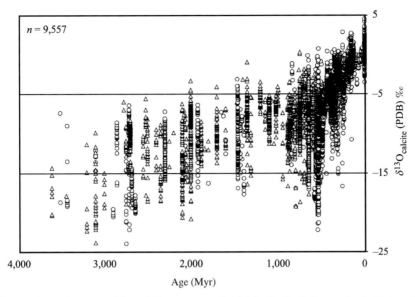

$n = 9,557$

Figure 33 Oxygen isotopic composition of limestones and calcitic shells during geologic history. Triangles and circles represent samples with good and poor age assignment, respectively (after Shields and Veizer, 2002).

stabilization stage and most, if not all, of its internal components are reset. The only exception can be the original low-magnesium calcitic shells of some organisms, such as brachiopods, belemnites, and foraminifera. Yet, the overall bulk rock depletions, relative to these stable phases, are $\sim$2–3‰ (Veizer *et al.*, 1999) and not some 7‰ or more as is the case for the Precambrian limestones (Figure 33). The rocks, once diagenetically stabilized become relatively "inert" to further resetting. The retention of $\Delta^{18}O_{dolomite-calcite}$ of $\sim$3‰ during the entire geologic history (Shields and Veizer, 2002) is also consistent with such an interpretation.

The observed *Precambrian* $\delta^{18}O$ secular trend is therefore real, albeit shifted by 2–3‰ to lighter values, and likely reflects the changing $\delta^{18}O$ of seawater. The exchange of oxygen at $T > 350\,°C$ between percolating seawater and oceanic crust results in ^{18}O enrichment of the water and ultimately oceans. The opposite happens at $T < 350\,°C$ (Muehlenbachs, 1998; Gregory, 1991; Wallmann, 2001). One interpretation could be that over geologic history this "isotopically neutral" crossover point migrated to shallower depths, thus reducing the profile of the low-T alteration relative to the deeper one, perhaps due to blanketing of the ocean floor by pelagic biogenic sediments during the Phanerozoic that sealed the off-ridge oceanic crust from seawater percolation.

The *Phanerozoic* trend (Figure 34) is based on $\sim$4,500 samples of low-magnesium calcitic fossils from about 100 localities worldwide. The reasons for believing that it is essentially a primary trend were discussed in detail by Veizer *et al.* (1999).

The pattern in Figure 34 has considerable implications for our understanding of past climate, but is still subject to debate. The models of

Muehlenbachs (1998) and Gregory (1991) claimed that due to a balance of high and low temperature reactions during interaction of the water cycle with the lithosphere, the $\delta^{18}O$ of the oceans should have been buffered near its present-day value. If seawater always had $\delta^{18}O$ of $\sim$0‰ SMOW (standard mean ocean water), the primary nature of the $\delta^{18}O$ record (Figure 34) would demand cooling oceans in the course of the Phanerozoic. With such an assumption, however, the early- to mid-Paleozoic ocean temperatures would have to have been in excess of 40 °C, even at times of glaciations. This is an unpalatable proposition, not only climatologically, but also in view of the similarity in faunal assemblages, in our case brachiopods, during this entire time span. Accepting that the $\delta^{18}O$ of past seawater was evolving towards ^{18}O-enriched values (Wallmann, 2001) and detrending the data accordingly (Figure 35), the superimposed second order structure of the curve correlates well with the Phanerozoic paleoclimatic record (cf. also Boucot and Gray, 2001). The observed structure, therefore, likely reflects paleotemperatures. If so, this would indicate that global climate swings were not confined to higher latitudes, but involved equatorial regions as well. As already pointed out, neither the $\delta^{18}O$ nor the paleoclimate record correlate with the model p_{CO_2} estimates for the ancient atmosphere.

17.11.2.6 Isotope tracers in developmental stages

The advances in instrumentation and particularly the arrival of the multicollector inductively coupled plasma mass spectrometers (MC-ICPMS)

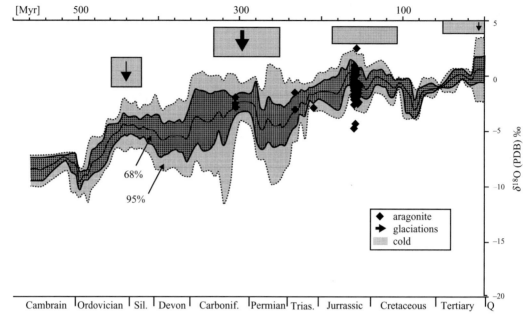

Figure 34 Phanerozoic $\delta^{18}O$ trend for low-Mg calcitic and aragonitic shells. Note that this is a trend for $CaCO_3$, offset by the fractionation factor α from that of seawater. See Figure 32 for further explanations (after Veizer *et al.*, 1999).

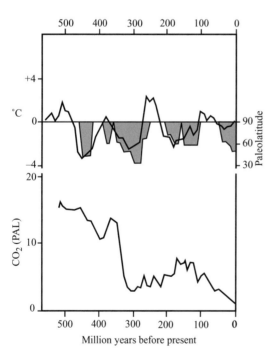

Figure 35 Reconstructed variations in mean temperature of shallow low-latitude seawater during the Phanerozoic based on the data in Figure 34. Note the good agreement of the cooling episodes with the extended latitudinal dispersion of ice rafted debris (shaded histograms). In the subsequent publication, Shaviv and Veizer (2003) showed that the proposed temperature variations correlated with the intensity of cosmic-ray flux reaching the Earth. The p_{CO2} (PAL—present-day atmospheric level) is that for the Geocarb model of Berner (1994).

opened a window for a number of new tracers that were difficult to tackle with the old instrumentation. A pattern for seawater evolution is thus emerging for isotopes of boron and calcium.

The present-day boron concentration in seawater and its $\delta^{11}B$ are uniform at 4.5 ppm and $-39.6‰$, respectively (Lemarchand *et al.*, 2000) and its residence time is ~14 Myr. Boron is present in seawater as $B(OH)_4^-$ and $B(OH)_3$. The relative proportion of these species is a function of pH (Palmer and Swihart, 1996), with $B(OH)_3$ 19.8‰ enriched in ^{11}B relative to $B(OH)_4^-$. Boron incorporation into carbonate skeletons, at concentrations of ~10–60 ppm, is from $B(OH)_4^-$ (Hemming and Hanson, 1992) and their $\delta^{11}B$ can therefore be used for tracing the pH of ancient seawater. Pearson and Palmer (2000), utilizing foraminiferal calcite, argued that the pH of Cenozoic seawater increased from ~7.4 at 60 Ma to its present-day value of 8.1. In contrast, Lemarchand *et al.* (2000) argued that the $\delta^{11}B$ trend in these foraminifera reflects the changing $^{11}B/^{10}B$ composition of seawater at constant pH, a development largely due to scavenging of boron by an increasing flux of clastic sediments.

The residence time of *calcium*, ~1 Myr versus the mixing rates of ocean of $\sim1,000$ yr, means that its isotopes are distributed homogeneously in seawater (Skulan *et al.*, 1997), with $^{40}Ca/^{44}Ca$ equal to 45.143 (Schmitt *et al.*, 2001). Modern carbonate shells show a variation of $\sim4‰$. These shells are $\sim1‰$ enriched relative to magmatic rocks, but compared to seawater they are depleted by 1–3‰, depending on the trophic level of the organism (Skulan *et al.*, 1997). Zhu and MacDougall (1998) showed that calcium isotope composition of the shells was both species and temperature dependent and that river water is depleted by 2‰ relative to seawater. The temperature dependency of calcium isotope fractionation enables this tracer to be a potential paleotemperature proxy (Nägler *et al.*, 2000). The seawater $^{44}Ca/^{40}Ca$ secular variations are indicated by the results of De La Rocha and De Paolo (2000) for the last 160 Myr, and by the data of Zhu (1999) for the entire 3.4 Gyr of earth history. The latter indicate that, in analogy to strontium isotopes (Veizer and Compston, 1976), the Archean samples have $^{44}Ca/^{40}Ca$ ratios similar to the earth mantle, with the crustal-like values first appearing at the Archean/Proterozoic transition. However, in contrast to the strontium isotope trend, the calcium isotope ratios appear to dip towards mantle values also at ~1.6 Gyr ago.

In the near future, the isotopes of *silicon, iron, and magnesium* also will likely develop into useful paleoceanographic tracers, but at this stage their utility for pre-Quaternary studies is limited.

17.12 MAJOR TRENDS IN THE EVOLUTION OF SEDIMENTS DURING GEOLOGIC HISTORY

17.12.1 Overall Pattern of Lithologic Types

The lithologic composition and the relative percentages of sedimentary and volcanic rocks preserved within the confines of present-day continents in crustal segments of various ages (Ronov, 1964; Budyko *et al.*, 1985) are shown in Figure 36. It should be remembered that with increasing geologic age, the total sedimentary rock mass diminishes (Figure 6) and a given

volume percentage of rock 3 Gyr ago represents much less mass than an equal percentage of rocks 200 Myr old. Despite this limitation, some general trends in lithologic rock types agreed on by most investigators are evident in this summary.

The outcrops of very old Archean rocks are few and thus may not be representative of the original sediment compositions deposited. Nevertheless, it appears that carbonate rocks are relatively rare in the Archean. Based on data from the limited outcrops, Veizer (1973) concluded that Archean carbonate rocks are predominantly limestones. During the early Proterozoic, the abundance of carbonates increases markedly, and for most of this Era the preserved carbonate rock mass is typified by the ubiquity of early diagenetic, and perhaps primary, dolostones (Veizer, 1973; Grotzinger and James, 2000). In the Phanerozoic, carbonates constitute ~30% of the total sedimentary mass, with sandstones and shales accounting for the rest. The Phanerozoic record of carbonates will be elaborated upon in the subsequent text.

Other highlights of the lithology-age distribution of Figure 36 are: (i) a marked increase in the abundance of submarine volcanogenic rocks and immature sandstones (graywackes) with increasing age; (ii) a significant percentage of arkoses in the early and middle Proterozoic, and an increase in the importance of mature sandstones (quartz-rich) with decreasing age; red beds are significant rock types of Proterozoic and younger age deposits; (iii) a significant "bulge" in the relative abundance of banded iron-ore formations, chert-iron associations, in the early Proterozoic; and (iv) a lack of evaporitic sulfate (gypsum, anhydrite) and salt (halite, sylvite) deposits in sedimentary rocks older than ~800 Myr. Marine evaporites owe their existence to a unique combination of tectonic, paleogeographic, and sea-level conditions. Seawater bodies must be restricted to some degree, but also must exchange with the open ocean to permit large volumes of seawater to enter these restricted basins and evaporate. Environmental settings of evaporite deposition may occur on cratons or in rifted basins. Figure 37 illustrates that because evaporite deposition requires an unusual combination of circumstances, a "geological accident" (Holser, 1984), the intensity of deposition has varied significantly during geologic time. This conclusion implies that the volume preserved of NaCl and $CaSO_4$ per unit of time in the Phanerozoic reasonably reflects the volume deposited, and because of the lack of any secular trend in evaporite mass per unit time, there has been only minor differential recycling of these rocks relative to other lithologies. Because the oceans are important reservoirs for these components, such large variations in the rate of NaCl and $CaSO_4$ output from the ocean to evaporites imply changes in the salinity and chemistry of seawater (Hay *et al.*, 2001).

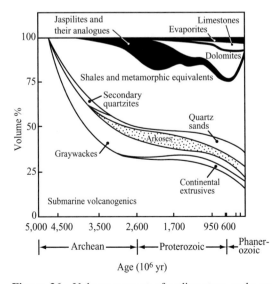

Figure 36 Volume percent of sedimentary rocks as a function of age. Extrapolation beyond ~3 Ga is hypothetical (after Ronov, 1964).

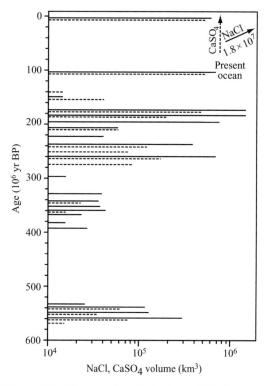

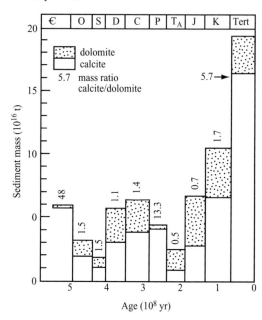

Figure 38 The Phanerozoic sedimentary carbonate mass distribution as a function of geologic age. Period masses of calcite and dolomite and the period mass ratios of calcite/dolomite are also shown (after Mackenzie and Morse, 1992).

Figure 37 Phanerozoic distribution of NaCl (dark line) and CaSO₄ (dashed line) in marine evaporites (after Holser, 1984).

These trends in lithologic features of the sedimentary rock mass are a consequence of evolution of the surface environment of the planet as well as recycling and post-depositional processes, and both secular and cyclic processes have played a role in generating the lithology-age distribution we see today (Veizer, 1973, 1988a; Mackenzie, 1975). For the past 1.5–2.0 Gyr, the Earth has been in a near present-day steady state, and the temporal distribution of rock types since then has been controlled primarily by recycling in response to plate tectonic processes.

Because sedimentary carbonates are important rock types in terms of providing mineralogical, chemical, biological, and isotopic data useful in interpretation of the history of Earth's surface environment, the following sections discuss these rock types in some detail. The discussion is mainly limited to the Phanerozoic because of the more complete database for this Eon than for the Precambrian.

17.12.2 Phanerozoic Carbonate Rocks

17.12.2.1 *Mass-age distribution and recycling rates*

As stated above, carbonate rocks comprise ~30% of the mass of Phanerozoic sediments. Given and Wilkinson (1987) reevaluated all the existing data on Phanerozoic carbonate rocks, their masses, and their relative calcite and dolomite contents (Figure 38). It can be seen that, as with the total sedimentary mass (Garrels and Mackenzie, 1971a,b), the mass of carbonate rock preserved is pushed toward the front of geologic time. The Tertiary, Carboniferous, and Cambrian periods are times of significant carbonate preservation, whereas the preservation of Silurian and Triassic carbonates is minimal.

The survival rates of the carbonate masses for different Phanerozoic systems are shown in Figure 39, together with the Gregor (1985) plot for the total sedimentary mass. The difference between the survival rate of the total carbonate mass and that of dolomite is the mass of limestone surviving per interval of time. The half-life of all the post-Devonian sedimentary mass is 130 Myr, and for a constant mass with a constant probability of destruction, the mean sedimentation rate is $\sim 100 \times 10^{14}$ g yr^{-1}. The modern global erosional flux is $\sim 200 \times 10^{14}$ g yr^{-1}, of which $\sim 15\%$ is particulate and dissolved carbonate. Although the data are less reliable for the survival rate of Phanerozoic carbonate sediments than for the total sedimentary mass, the half-life of the post-Permian carbonate mass is ~ 86 Myr. This gives a mean sedimentation rate of $\sim 35 \times 10^{14}$ g carbonate per year, compared to the present-day carbonate flux of 30×10^{14} g yr^{-1} (Morse and Mackenzie, 1990). The difference in half-lives between the total sedimentary mass, which is principally sandstone and shale, and the carbonate

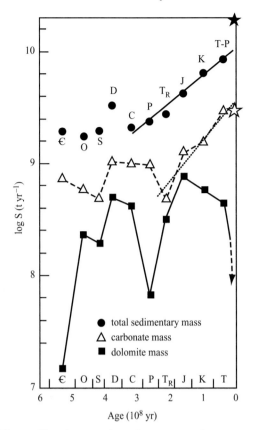

Figure 39 Phanerozoic sedimentary rock mass–age relationships expressed as the logarithm of the survival rate in tons per year versus time. The straight lines are best fits to the total mass data (solid line) and to the carbonate mass data (dotted line) for particular intervals of Phanerozoic time. The difference between the logarithm of S for the carbonate mass and that of the dolomite mass is the survival rate of the calcite mass. Filled star is present total riverine flux to the oceans, whereas open star is carbonate flux (after Mackenzie and Morse, 1992).

mass probably is a consequence of the more rapid recycling of the carbonate mass at a rate ~1.5 times that of the total mass.

This is not an unlikely situation. With the advent of abundant carbonate-secreting, planktonic organisms in the Jurassic, the site of carbonate deposition shifted significantly from shallow-water areas to the deep sea. This gradual shift will increase still further the rate of destruction (by eventual subduction) of the global carbonate mass relative to the total sedimentary mass because the recycling rate of oceanic crust (the "b" values of Veizer and Jansen, 1985; Veizer, 1988a) exceeds that of the "continental" sediments by a factor of 6. Also, Southam and Hay (1981), using a half-life of 100 Myr for pelagic sediment, estimated that as much as 50% of all sedimentary rock formed by weathering of igneous rock may have been lost by subduction during the past 4.5 Gyr.

Thus, it appears, as originally suggested by Garrels and Mackenzie (1969, 1971a, 1972), that the carbonate component of the sedimentary rock mass may have a cycling rate slightly different from that of the total sedimentary mass. These authors argued that the differential recycling rates for different lithologies were related to their resistance to chemical weathering and transport. Evaporites are the most easily soluble, limestones are next, followed by dolostones, and shales and sandstones are the most inert. Although resistance to weathering may play a small role in the selective destruction of sedimentary rocks, it is likely, as argued previously, that differences in the recycling rates of different tectonic regimes in which sediments are deposited are more important.

17.12.2.2 Dolomite/calcite ratios

For several decades it has been assumed that the Mg/Ca ratio of carbonate rocks increases with increasing rock age (see Daly, 1909; Vinogradov and Ronov, 1956a,b; Chilingar, 1956; Figure 40). In these summaries, the magnesium content of North American and Russian Platform carbonates is relatively constant for the latest 100 Myr, and then increases gradually, very close to, if not the same as, the commencement of the general increase in the magnesium content of pelagic limestones (Renard, 1986). The dolomite content in deep sea sediments also increases erratically with increasing age back to ~125 Myr before present (Lumsden, 1985). Thus, the increase in magnesium content of carbonate rocks with increasing age into at least the early Cretaceous appears to be a global phenomenon, and to a first approximation, is not lithofacies related. In the 1980s, the accepted truism that dolomite abundance increases relative to limestone with increasing age has been challenged by Given and Wilkinson (1987). They reevaluated all the existing data and concluded that dolomite abundances do vary significantly throughout the Phanerozoic but may not increase systematically with age (Figure 40). Yet the meaning of these abundance curves, and indeed their actual validity, is still controversial (Zenger, 1989).

Voluminous research on the "dolomite problem" (see Hardie, 1987, for discussion) has shown that the reasons for the high magnesium content of carbonates are diverse and complex. Some dolomitic rocks are primary precipitates; others were deposited as $CaCO_3$ and then converted entirely or partially to dolomite before deposition of a succeeding layer; still others were dolomitized by migrating underground waters tens or hundreds of millions of years after deposition. It is therefore exceedingly important to know the distribution of the calcite/dolomite ratios of carbonate rocks through geologic time. This information

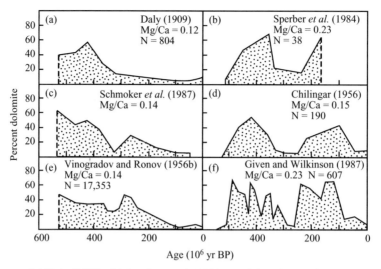

Figure 40 Estimates ((a) Daly (1909), (b) Sperber *et al.* (1984), (c) Schmoker *et al.* (1987), (d) Chilingar (1956), (e) Vinogradov and Ronov (1956b), and (f) Given and Wilkinson (1987)) of percent dolomite in Phanerozoic cratonic carbonate rocks as a function of age. Mg/Ca=average ratio (after Wilkinson and Algeo, 1989).

has a bearing on the origin of dolomite, as well as on the properties of the coeval atmosphere–hydrosphere system (Given and Wilkinson, 1987; Wilkinson and Algeo, 1989; Berner, 1990; Morse and Mackenzie, 1990; Mackenzie and Morse, 1992; Arvidson and Mackenzie, 1999; Arvidson *et al.*, 2000; Holland and Zimmermann, 2000). For example, it could be argued that if the dolomite/calcite ratio progressively increases with age of the rock units (Figure 40), the trend principally reflects enhanced susceptibility of older rock units to processes of dolomitization. Such a trend would then be only a secondary feature of the sedimentary carbonate rock mass due to progressive diagenesis and seawater driven dolomitization (Garrels and Mackenzie, 1971a; Mackenzie, 1975; Holland and Zimmermann, 2000). Alternatively, if the trend in the calcite/dolomite ratio is cyclic in nature, this cyclicity could be interpreted as representing environmental change in the ocean-atmosphere system (Wilkinson and Algeo, 1989). For the discussion here, we accepted the data of Given and Wilkinson (1987) on the Ca/Mg ratio of Phanerozoic sedimentary carbonates (Figure 38) to calculate the mass ratios of these carbonate components as a function of age, but do realize that such data are still a matter of controversy (Zenger, 1989; Holland and Zimmermann, 2000). In reality, it is most likely that both cyclical and secular compositional changes in the ocean-atmosphere-sediment system, as well as diagenesis, contribute to dolomite abundance during geologic time.

The period-averaged mass ratio of calcite to dolomite is anomalously high for the Cambrian and Permian System rocks (Figure 38). For the remainder of the Phanerozoic, it appears to oscillate within the 1.1±0.6 range, except for the limestone-rich Tertiary. Comparison with the generalized Phanerozoic sea-level curves of Vail *et al.* (1977) and Hallam (1984) (Figure 42) hints that dolomites are more abundant at times of higher sea levels. Mackenzie and Agegian (1986, 1989) and Given and Wilkinson (1987) were the first to suggest this possible cyclicity in the calcite/dolomite ratio during the Phanerozoic, and Lumsden (1985) observed a secular decrease in dolomite abundance in deep marine sediments from the Cretaceous to Recent, corresponding to the general fall of sea level during this time interval. These cycles in calcite/dolomite ratios correspond crudely to the Fischer (1984) two Phanerozoic super cycles and to the Mackenzie and Pigott (1981) oscillatory and submergent tectonic modes.

17.12.2.3 Ooids and ironstones

Although still somewhat controversial (e.g., Bates and Brand, 1990), the textures of ooids appear to vary during Phanerozoic time. Sorby (1879) first pointed out the petrographic differences between ancient and modern ooids: ancient ooids commonly exhibit relict textures of a calcitic origin, whereas modern ooids are dominantly made of aragonite. Sandberg (1975) reinforced these observations by study of the textures of some Phanerozoic ooids and a survey of the literature. His approach, and that of others who followed, was to employ the petrographic criteria, among others, of Sorby: i.e., if the microtexture of the ooid is preserved, then the ooid originally had a calcite mineralogy; if the ooid exhibits textural disruption, its original mineralogy was aragonite. The textures of originally aragonitic fossils are usually used as checks to deduce the original mineralogy of the ooids. Sandberg (1975)

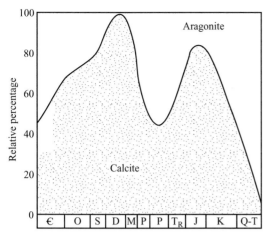

Figure 41 Inferred mineralogy of Phanerozoic ooids (after Morse and Mackenzie, 1990).

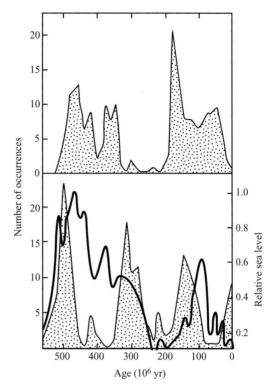

Figure 42 Number of occurrences of Phanerozoic ironstones (upper diagram, source Van Houten and Bhattacharyya, 1982) and oolitic limestones (lower diagram, source Wilkinson *et al.*, 1985) as a function of geologic age. The relative sea-level curve is that of Hallam (1984). Minima in occurrences appear to correlate with times of sea level withdrawal from the continents (after Morse and Mackenzie, 1990).

observed that ooids of inferred calcitic composition are dominant in rocks older than Jurassic.

Following this classical work, Sandberg (1983, 1985) and several other investigators (Mackenzie and Pigott, 1981; Wilkinson *et al.*, 1985; Bates and Brand, 1990) attempted to quantify further this relationship. Figure 41 is a schematic diagram representing a synthesis of the inferred mineralogy of ooids during the Phanerozoic. This diagram is highly tentative, and more data are needed to document the trends. However, it appears that while originally aragonitic ooids are found throughout the Phanerozoic, an oscillatory trend in the relative percentage of calcite versus aragonite ooids may be superimposed on a long-term evolutionary decrease in ooids with an inferred original calcitic mineralogy. Although the correlation is not strong, the two major maxima in the sea-level curves of Vail *et al.* (1977) and Hallam (1984) appear to coincide with times when calcite ooids were important seawater precipitates (Figure 42). Wilkinson *et al.* (1985) found that the best correlation between various data sets representing global eustasy and ooid mineralogy is that of inferred mineralogy with percentage of continental freeboard. Sandberg (1983) further concluded that the cyclic trend in ooid mineralogy correlates with cyclic trends observed for the inferred mineralogy of carbonate cements. Van Houten and Bhattacharyya (1982; and later Wilkinson *et al.*, 1985) showed that the distribution of Phanerozoic ironstones (hematite and chamosite oolitic deposits) also exhibits a definite cyclicity (Figure 42) that too appears to covary with the generalized sea-level curve. Minima appear to coincide with times of sea-level withdrawal from the continents.

17.12.2.4 Calcareous shelly fossils

In some similarity to the trends observed for the inorganic precipitates of ooids and ironstones,

the mineralogy of calcareous fossils during the Phanerozoic also shows a cyclic pattern (Figure 25) with calcite being particularly abundant during high sea levels of the early to mid-Paleozoic and the Cretaceous (Stanley and Hardie, 1998; Kiessling, 2002). Overall (Figure 43) there is a general increase in the diversity of major groups of calcareous organisms such as coccolithophorids, pteropods, hermatypic corals, and coralline algae. It is noteworthy that the major groups of pelagic and benthic organisms contributing to carbonate sediments in today's ocean first appeared in the fossil record during the middle Mesozoic and progressively became more abundant. What is most evident in Figure 43 is a long-term increase in the production of biogenic carbonates dominated by aragonite and magnesian calcite mineralogies. Because organo-detrital carbonates are such an important part of the Phanerozoic carbonate rock record, this increase in metastable mineralogies played an important role in the pathway of diagenesis of carbonate sediments. The ubiquity of low magnesian calcite skeletal organisms for much of the Paleozoic led to production of calcitic organo-detrital sediments

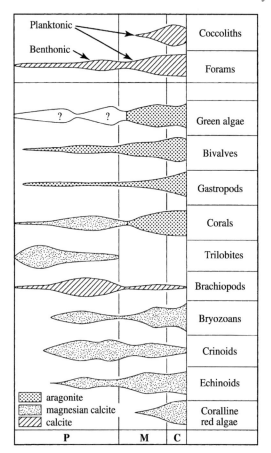

Figure 43   Mineralogical evolution of benthic and planktonic organism diversity during the Phanerozoic based on summaries of Milliken and Pigott (1977) and Wilkinson (1979). P is Paleozoic, M is Mesozoic, and C is Cenozoic.

whose original bulk chemical and mineralogical composition was closer to that of their altered and lithified counterparts of Cenozoic age.

17.12.2.5   The carbonate cycle in the ocean

The partitioning of carbonate burial between shoal-water and deep sea realms has varied in a cyclic pattern through Phanerozoic time (Morse and Mackenzie, 1990). The variation in the magnitudes of the fluxes of $(Ca,Mg)CO_3$ to the two environments through time is difficult to assess; even today's fluxes are probably not known within a factor of 2. To gain some impression of the fluxes involved, a tentative model of the carbonate carbon cycle in the world's oceans is shown in Figure 44. This is a representation of the mean state of the cycle during the most recent glacial to interglacial transition. About 18×10^{12} moles of calcium and magnesium (equivalent to 216×10^6 metric tons of carbon) accumulate yearly as carbonate minerals (Morse and Mackenzie, 1990), mainly as biological precipitates. Of this flux

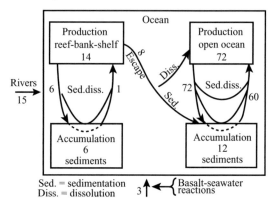

Figure 44   Tentative model of global ocean carbonate cycle. Fluxes are in units of 10^{12} mol C yr^{-1} as $(Ca,Mg)CO_3$ and represent estimates of fluxes averaged over the most recent glacial–interglacial transition (after Mackenzie and Morse, 1992).

$\sim 6 \times 10^{12}$ moles are deposited as calcium and magnesium carbonates in shoal-water areas (Milliman, 1974; Smith, 1978; Wollast, 1993; Milliman *et al.*, 1999), and the remainder accumulates as calcareous oozes in the pelagic realm. The 12×10^{12} moles of carbonate accumulated annually in the deep sea are only $\sim 17\%$ of the annual carbonate production rate of 72×10^{12} mol of the open ocean photic zone. This efficient recycling of carbonate carbon in the open ocean water column and at the sediment-water interface is a well-known feature of the marine carbon cycle (Broecker and Peng, 1982). It is important to note that much shoal-water carbonate production ends up in sediments of reefs, banks, etc., so that, in contrast to the pelagic realm, production rate more closely approximates sedimentation rate. However there is an escape of carbonate sediment from shoal-water areas to the deep sea, where it is deposited or dissolved (Land, 1979; Kiessling, 2002). The magnitude of this flux is poorly known but may affect the chemistry of open-ocean regions owing to dissolution of the carbonate debris (Droxler *et al.*, 1988; Agegian *et al.*, 1988; Sabine and Mackenzie, 1975). Furthermore, its accumulation on the slopes of banks may act as a record of paleoenvironmental change (Droxler *et al.*, 1983).

17.12.3   Geochemical Implications of the Phanerozoic Carbonate Record

The reasons for the mass-age relationships discussed previously are not totally clear. A number of investigators (Mackenzie and Pigott, 1981; Sandberg, 1985; Wilkinson *et al.*, 1985; Wilkinson and Given, 1986; Wilkinson and Algeo, 1989; Mackenzie and Agegian, 1989; Stanley and Hardie, 1998; Arvidson *et al.*, 2000) concluded that these observations are the result of changing atmosphere–hydrosphere environmental conditions through the Phanerozoic. Others argued that, for example, the ooid observations are not statistically significant (Bates and Brand, 1990) or that the Given and Wilkinson (1987) mass-age database is not valid (Holland and Zimmermann, 2000). The latter authors do accept, however, that there is generally lower dolomite abundance in carbonate sediments deposited during the past 200 Myr. These qualifications notwithstanding, a number of previously described parameters (Sr/Ca, Mg/Ca, aragonite/calcite, possibly dolomite/calcite, and frequency of ooids and iron ores) appear to be related in some degree to sea-level stands, the latter at least in part a reflection of plate tectonic activity during the Phanerozoic.

Rowley (2002) in his recent publication argues that the rate of oceanic plate production may not have varied significantly for the latest 180 Myr. If so, this may have major implications on our understanding of the model linkage of tectonics to sea-level change, atmospheric CO_2, seawater chemistry and related phenomena. Nevertheless, while the 30–40% variations in seafloor spreading rates during the latest 100 Myr (Delaney and Boyle, 1986) are probably not justifiable, we cannot at present dismiss entirely the proposition that hydrothermal exchange between seafloor and ocean, and presumably the rate of plate generation, might still have varied somewhat. It may be possible therefore that the first-order changes in sea level can still be driven by the accretion of the ridges, but, at the same time, we should also search for alternative linkages of sea-level stands to the mineralogical and chemical properties of the Phanerozoic sedimentary cycle.

In the standard reasoning, the first-order changes in sea level are driven by the accretion of ridges: high accretion rate, high sea level; low accretion rate, low sea level. Regardless of the tie between oceanic plate production and sea level, extended times of global high sea level may have been times of enhanced atmospheric CO_2 levels (Berner *et al.*, 1983; Lasaga *et al.*, 1985; Berner, 2000), higher temperatures (not necessarily related solely to atmospheric CO_2 concentrations), probably lower seawater Mg/Ca ratios (Lowenstein *et al.*, 2001; Dickson, 2002), different saturation states of seawater (Arvidson *et al.*, 2000), and perhaps different seawater sulfate concentrations than at present. The converse is true for first-order global sea-level low stands. It appears that the environmental conditions for early dolomitization, formation of calcitic ooids and cements, and preponderance of calcitic reef-building organisms are best met during extended times of global high sea levels, with ubiquitous shallow-water and sabkha-like environments (calcite seas; Sandberg, 1983). Dolomitization of precursor calcite and aragonite phases, either in marine waters or in mixed continental-marine waters, would be enhanced under these conditions. Furthermore, the potentially lowered pH of marine waters during times of enhanced atmospheric carbon dioxide would favor syndepositional or later dolomitization in mixed marine-meteoric waters, because the range of seawater-meteoric compositional mixtures over which calcite could be dissolved and dolomite precipitated is expanded (Plummer, 1975). Perhaps superimposed on the hypothesized Phanerozoic cyclic dolomite/calcite ratio is a longer term trend in which dolomite abundance increases with increasing age, particularly in rocks older than 200 Myr, due to favorable environmental conditions as well as to advancing late diagenetic and burial dolomitization. During the past 150 Myr, this magnesium has been transferred out of the dolomite reservoir ("bank" of Holland and Zimmermann, 2000) into the magnesium silicate reservoir by precipitation of silicates and dissolution of dolomite (Garrels and Mackenzie, 1972; Garrels and Perry, 1974; Holland and Zimmermann, 2000) and to a lesser extent into the ocean reservoir, this accounting in part for the increasing Mg/Ca ratio of seawater during the past 150 Myr (Lowenstein *et al.*, 2001).

Thus, it appears that the apparent trends in Phanerozoic carbonate mineralogy are related to changes in atmosphere–hydrosphere conditions that are driven in part by plate tectonic mechanisms. However, we are aware that this tentative proposition requires collection of more data on the detailed chemistry and mineralogy of Phanerozoic carbonate sequences worldwide as well as resolution of the problem related to the past production rates of the oceanic lithosphere.

ACKNOWLEDGMENTS

We would like to dedicate this chapter to Robert M. Garrels.

REFERENCES

Agegian C. R., Mackenzie F. T., Tribble J. S., and Sabine C. (1988) Carbonate production and flux from a mid-depth bank ecosystem. *Natl. Undersea Res. Prog. Res. Report* **88**(1), 5–32.

Allègre C. J. and Rousseau D. (1984) The growth of the continents through geological time studied by Nd isotope analysis of shales. *Earth Planet. Sci. Lett.* **67**, 19–34.

Armstrong R. L. (1981) Radiogenic isotopes: the case for crustal recycling on a near-steady-state no-continental-growth Earth. *Phil. Trans. Roy. Soc. London A Ser.* **301**, 443–472.

Arvidson R. S. and Mackenzie F. T. (1999) The dolomite problem: control of precipitation kinetics by temperature and saturation state. *Am. J. Sci.* **299**, 257–288.

Arvidson R. S., Mackenzie F. T., and Guidry M. W. (2000) Ocean/atmosphere history and carbonate precipitation rates: a solution to the dolomite problem. In *Marine Authigenesis: from Global to Microbial*, SEPM Spec. Publ. No. 65 (eds. C. R. Glenn, L. Prévôt-Lucas, and J. Lucas), SEPM, Tulsa, pp. 1–5.

Bates N. R. and Brand U. (1990) Secular variation of calcium carbonate mineralogy: an evaluation of oöid and micrite chemistries. *Geol. Rundsch.* **79**, 27–46.

Barrell J. (1917) Rhythms and the measurement of geological time. *Geol. Soc. Am. Bull.* **28**, 745–904.

Berner R. A. (1990) Atmospheric carbon dioxide levels over Phanerozoic time. *Science* **249**, 1382–1386.

Berner R. A. (1994) GEOCARB: II. A revised model of atmospheric CO_2 over Phanerozoic time. *Am. J. Sci.* **294**, 56–91.

Berner R. A. (2000) The effect of the rise of land plants on atmospheric CO_2 during the paleozoic. In *Plants Invade the Land: Evolutionary and Environmental Approaches* (eds. P. G. Gensel and D. Edwards). Columbia University Press, New York, pp. 173–178.

Berner R. A. and Kothavala Z. (2001) GEOCARB: III. A revised model of atmospheric CO_2 over phanerozoic time. *Am. J. Sci.* **301**, 182–204.

Berner U. and Streif H. (2001) *Klimafakten, 2001, Der Rückblick—Ein Schlüssel für die Zukunft.* Schweizerbart'sche Verlagsbuchhandlung, Science Publishers, Stuttgart.

Berner R. A., Lasaga A. C., and Garrels R. M. (1983) The carbonate silicate geochemical cycle and its effect on atmospheric carbon dioxide over the past 100 million years. *Am. J. Sci.* **283**, 641–683.

Berry J. P. and Wilkinson B. H. (1994) Paleoclimatic and tectonic control on the accumulation of North American cratonic sediment. *Geol. Soc. Am. Bull.* **106**, 855–865.

Bluth G. J. and Kump L. R. (1991) Phanerozoic paleogeology. *Am. J. Sci.* **291**, 284–308.

Boucot A. J. and Gray J. (2001) A critique of Phanerozoic climatic models involving changes in the CO_2 content of the atmosphere. *Earth Sci. Rev.* **56**, 1–159.

Brand U. and Veizer J. (1980) Chemical diagenesis of a multi-component carbonate system: I. Trace elements. *J. Sedim. Petrol.* **50**, 1219–1236.

Broecker W. S. and Peng T. H. (1982) *Tracers in the Sea.* Eldigio Press, Palisades, NY.

Budyko M. I., Ronov A. B., and Yanshin A. L. (1985) *History of the Earth's Atmosphere.* Springer, Heidelberg.

Burke W. H., Denison R. F., Hetherington E. A., Koepnick R. F., Nelson H. F., and Otto J. B. (1982) Variation of seawater $^{87}Sr/^{86}Sr$ throughout Phanerozoic time. *Geology* **10**, 516–519.

Canfield D. E. and Teske A. (1996) Late Proterozoic rise in atmospheric oxygen concentration inferred from phylogenetic and sulphur-isotope studies. *Nature* **382**, 127–132.

Chambers L. A. and Trudinger P. A. (1979) Microbiological fractionation of stable sulfur isotopes: a review and critique. *Geomicrobiol. J.* **1**, 249–293.

Chilingar G. V. (1956) Relationship between Ca/Mg ratio and geologic age. *Am. Assoc. Petrol. Geol. Bull.* **40**, 2256–2266.

Cicero A. D. and Lohmann K. C. (2001) Sr/Mg variation during rock-water interaction: implications for secular changes in the elemental chemistry of ancient seawater. *Geochim. Cosmochim. Acta* **65**, 741–761.

Claypool G. E., Holser W. T., Kaplan I. R., Sakai H., and Zak I. (1980) The age curves of sulfur and oxygen isotopes in marine sulfate and their mutual interpretation. *Chem. Geol.* **28**, 199–260.

Cloud P. E. (1976) Major features of crustal evolution. *Trans. Geol. Soc. S. Afr.* **79**, 1–32.

Cogley J. G. (1984) Continental margins and the extent and number of continents. *Rev. Geophys. Space Phys.* **22**, 101–122.

Collerson K. D. and Kamber B. S. (1999) Evolution of the continents and the atmosphere inferred from the Th–U–Nb systematics of the depleted mantle. *Science* **283**, 1519–1522.

Condie K. C. (1989) *Plate Tectonics and Crustal Evolution.* Pergamon, London.

Condie K. C. (1993) Chemical composition and evolution of the upper continental crust: contrasting results from surface samples and shales. *Chem. Geol.* **104**, 1–37.

Condie K. C. (2000) Episodic continental growth models: afterthoughts and extensions. *Tectonophysics* **322**, 153–162.

Condie K. C. (2001) Continental growth during formation of Rodinia at 1.35–0.9 Ga. *Gondwana Res.* **4**, 5–16.

Cook T. D. and Bally A. W. (1975) *Shell Atlas: Stratigraphic Atlas of North and Central America.* Princeton University Press, Princeton, NJ.

Cox R. and Lowe D. R. (1995) A conceptual review of regional-scale controls on the composition of clastic sediment and the co-evolution of continental blocks and their sedimentary cover. *J. Sedim. Res.* **A65**, 1–12.

Daly R. A. (1909) First calcareous fossils and evolution of limestones. *Geol. Soc. Am. Bull.* **20**, 153–170.

Danchin R. V. (1967) Chromium and nickel in the Fig Tree shale from South Africa. *Science* **158**, 261–262.

Deines P. (1980) The isotopic composition of reduced organic carbon. In *Handbook of Environment Isotope Gechemistry: Vol. 1. The Terrestrial Environment* (eds. P. Fritz and J. C. Fontes). Elsevier, Amsterdam, pp. 329–406.

Delaney M. L. and Boyle E. A. (1986) Lithium in foraminiferal shells: implications for high-temperature hydrothermal circulation fluxes and oceanic generation rates. *Earth Planet. Sci. Lett.* **80**, 91–105.

De La Rocha C. L. and De Paolo J. (2000) Isotopic evidence for variations in the marine calcium cycle over the Cenozoic. *Science* **289**, 1176–1178.

Dickinson W. R., Beard L. S., Brakenridge G. R., Erjavec J. L., Ferguson R. C., Inman K. F., Knepp R. A., Lindberg F. A., and Ryberg P. T. (1983) Provenance of North American Phanerozoic sandstones in relation to tectonic setting. *Geol. Soc. Am. Bull.* **94**, 222–235.

Dickson J. A. D. (2002) Fossil echinoderms as monitor of the Mg/Ca ratio of Phanerozoic oceans. *Science* **298**, 1222–1224.

Dietz R. S. (1961) Continent and ocean basin evolution by spreading of the seafloor. *Nature* **190**, 854–857.

Droxler A. W., Schlager W., and Wallon C. C. (1983) Quaternary aragonite cycles and oxygen-isotopic records in Bahamian carbonate ooze. *Geology* **11**, 235–239.

Droxler A. W., Morse J. W., and Kornicker W. A. (1988) Controls on carbonate mineral accumulation in Bahamian basins and adjacent Atlantic ocean sediments. *J. Sedim. Petrol.* **58**, 120–130.

Engel A. E. J. (1963) Geologic evolution of North America. *Science* **140**, 143–152.

Engel A. E. J., Itson S. P., Engel C. G., Stickney D. M., and Cray E. J. (1974) Crustal evolution and global tectonics, a petrogenic view. *Bull. Geol. Soc. Am.* **85**, 843–858.

Faure G. (1986) *Principles of Isotope Geology.* Wiley, New York.

Fischer A. G. (1984) The two Phanerozoic super cycles. In *Catastrophes in Earth History* (eds. W. A. Berggren and J. A. Vancouvering). Princeton University Press, NJ, pp. 129–148.

Frakes L. A., Francis J. E., and Syktus J. I. (1992) *Climate Mode of the Phanerozoic: The History of the Earth's Climate over the Past 600 Million Years.* Cambridge University Press, Cambridge, UK.

Gaffin S. (1987) Ridge volume dependence of seafloor generation rate and inversion using long term sea level change. *Am. J. Sci.* **287**, 596–611.

Garrels R. M. and Mackenzie F. T. (1969) Sedimentary rock types: relative proportions as a function of geological time. *Science* **163**, 570–571.

Garrels R. M. and Mackenzie F. T. (1971a) *Evolution of Sedimentary Rocks.* Norton, New York.

Garrels R. M. and Mackenzie F. T. (1971b) Gregor's denudation of the continents. *Nature* **231**, 382–383.

Garrels R. M. and Mackenzie F. T. (1972) A quantitative model for the sedimentary rock cycle. *Mar. Chem.* **1**, 22–41.

Garrels R.M. and Perry E. A. Jr., (1974) Chemical history of the oceans deduced from postdepositional changes in sedimentary rocks. In *Studies in Paleo-Oceanography Special Publication Society Economic Paleontologists and Mineralogists,* **20**, (ed. W. W. Hay). SPEM, Tulsa, OK, pp. 193–204.

Gibbs A. K., Montgomery C. W., O'Day P. A., and Erslev E. A. (1988) Crustal evolution revisited: reply to comments by S. M. McLennan et al., on "The Archean–Proterozoic transition: evidence from the geochemistry of metasedimentary rocks from Guyana and Montana". *Geochim. Cosmochim. Acta* **52**, 793–795.

Gilluly J. (1949) Distribution of mountain building in geologic time. *Geol. Soc. Am. Bull.* **60**, 561–590: **120**, 135–139.

Given R. K. and Wilkinson B. H. (1987) Dolomite abundance and stratigraphic age: constraints on rates and mechanisms of Phanerozoic dolostone formation. *J. Sedim. Petrol.* **57**, 1068–1079.

Goddéris Y. and Veizer J. (2000) Tectonic control of chemical and isotopic composition of ancient oceans: the impact of continental growth. *Am. J. Sci.* **300**, 434–461.

Goldschmidt V. M. (1933) Grundlagen der quantitativen Geochemie. *Fortschr. Mineral. Kristallog. Petrogr.* **17**, 1–112.

Goldstein S. J. and Jacobsen S. B. (1987) The Nd and Sr isotope systematics of river water dissolve material: implications for the source of Nd and Sr in seawater. *Chem. Geol.* **66**, 245–272.

Gregor C. B. (1968) The rate of denudation in post-Algonkian time. *Proc. Koninkl. Ned. Akad. Wetenschap. Ser. B: Phys. Sci.* **71**, 22–30.

Gregor C. B. (1985) The mass-age distribution of Phanerozoic sediments. In *Geochronology and the Geologic Record,* Mem. No. 10 (ed. N. J. Snelling). Geol. Soc. London, London, UK, pp. 284–289.

Gregory R. T. (1991) Oxygen isotope history of seawater revisited: composition of seawater. In *Stable Isotope Geochemistry: a Tribute to Samuel Epstein,* Geochem. Soc. Spec. Publ. 3 (eds. H. P. Tailor, Jr., J. R. O'Neil, and I. R. Kaplan), Min. Soc. America, Washington, DC, pp. 65–76.

Grotzinger J. P. and James N. P. (2000) *Precambrian Carbonates, Evolution and Understanding: Carbonate Sedimentation and Diagenesis in the Evolving Precambrian World.* SEPM Special Publication # 67.

Habicht K. A. and Canfield D. E. (1996) Sulphur isotope fractionation in modern microbial mats and the evolution of the sulphur cycle. *Nature* **382**, 342–343.

Hallam A. (1984) Pre-Quaternary sea-level changes. *Ann. Rev. Earth Planet. Sci.,* **12**, 205–243.

Hardie L. A. (1987) Perspectives on dolomitization: a critical view of some current views. *J. Sedim. Petrol.* **57**, 166–183.

Hardie L. A. (1996) Secular variation in seawater chemistry: an explanation of the coupled secular variations in the mineralogies of marine limestones and potash evaporites over the past 600 Myr. *Geology* **24**, 279–283.

Harrison A. G. and Thode H. G. (1958) Mechanism of the bacterial reduction of sulfate from isotope fractionation studies. *Faraday Soc. Trans.* **54**, 84–92.

Hay W. W., Sloan J. L. II, and Wold C. N. (1988) Mass/age distribution and composition of sediments on the ocean floor and the global rate of sediment subduction. *J. Geophys. Res.* **93**, 14933–14940.

Hay W. W., Wold C. N., Söding E., and Flügel S. (2001) Evolution of sediment fluxes and ocean salinity. In *Geologic Modelling and Simulations: Sedimentary Systems* (eds. D. F. Merriam and J. C. Davis). Kluwer Academic/Plenum, Dordrecht, pp. 153–167.

Hayes J. M., Des Marais D. J., Lambert J. B., Strauss H., and Summons R. E. (1992) Proterozoic biogeochemistry. In *The Proterozoic Biosphere: A Multidisciplinary Study* (eds. J. W. Schopf and C. Klein). Cambridge University Press, Cambridge, UK, pp. 81–134.

Hemming N. G. and Hanson G. N. (1992) Boron isotope composition and concentration in modern marine carbonates. *Geochim. Cosmochim. Acta* **56**, 537–543.

Hess H. H. (1962) History of ocean basins. In *Petrologic Studies* (eds. A. E. J. Engel, H. L. James, and B. F. Leonard). Geological Society of America, Bouldev, co., pp. 599–620.

Hoffman P. F., Kaufman A. J., Halverson G. P., and Schrag D. P. (1998) A Neoproterozoic snowball Earth. *Science* **281**, 1342–1346.

Holland H. D. (1978) *The Chemistry of the Atmosphere and Oceans.* Wiley, New York.

Holland H. D. (1984) *The Chemical Evolution of the Atmosphere and Oceans.* Princeton University Press, Princeton, NJ.

Holland H. D. and Zimmerman H. (2000) The dolomite problem revisited. *Int. Geol. Rev.* **2**, 481.

Holser W. T. (1977) Catastrophic chemical events in the history of the ocean. *Nature* **267**, 403–408.

Holser W. T. (1984) Gradual and abrupt shifts in ocean chemistry during Phanerozoic time. In *Patterns of Change in Earth Evolution* (eds. H. D. Holland and A. F. Trendall). Springer, Heidelberg, pp. 123–143.

Holser W. T. and Kaplan I. R. (1966) Isotope geochemistry of sedimentary sulfates. *Chem. Geol.* **1**, 93–135.

Holser W. T., Schidlowski M., McKenzie F. T., and Maynard J. B. (1988) Geochemical cycles of carbon and sulfur. In *Chemical Cycles in the Evolution of the Earth* (eds. C. B. Gregor, R. M. Garrels, F. T. Mackenzie, and J. B. Maynard). Wiley, New York, pp. 105–173.

Horita J., Friedman T. J., Lazar B., and Holland H. D. (1991) The composition of Permian seawater. *Geochim. Cosmochim. Acta* **55**, 417–432.

Horita J., Zimmermann H., and Holland H. D. (2002) The chemical evolution of seawater during the Phanerozoic: implications from the record of marine evaporites. *Geochim. Cosmochim. Acta* **66**, 3733–3756.

Ingersoll R. V. and Busby C. J. (1995) Tectonics of sedimentary basins. In *Tectonics of Sedimentary Basins* (eds. C. J. Busby and R. V. Ingersoll). Blackwell, Oxford, pp. 1–51.

Kampschulte A. (2001) Schwefelisotopenuntersuchungen an strukturell substituierten Sulfaten in marinen Karbonaten des Phanerozoikums—Implikationen für die geochemische Evolution des Meerwassers und Korrelation verschiedener Stoffkreisläufe. PhD Thesis, Ruhr Universität, Bochum.

Kampschulte A. and Strauss H. (1998) The isotopic composition of trace sulphates in Paleozoic biogenic carbonates: implications for coeval seawater and geochemical cycles. *Min. Mag.* **62A**, 744–745.

Karhu J. A. and Holland H. D. (1996) Carbon isotopes and the rise of atmospheric oxygen. *Geology* **24**, 867–870.

Kiessling W. (2002) Secular variations in the Phanerozoic reef ecosystem. *SEPM Spec. Publ.* **72**, 625–690.

Kroopnick P. (1980) The distribution of ^{13}C in the Atlantic ocean. *Earth Planet. Sci. Lett.* **49**, 469–484.

Kuenen Ph. H. (1950) *Marine Geology.* Wiley, New York.

Kump L. R. (1989) Alternative modeling approaches to the geochemical cycles of carbon, sulfur, and strontium isotopes. *Am. J. Sci.* **289**, 390–410.

Land L. S. (1979) The fate of reef-derived sediment on the north Jamaica island slope. *Mar. Geol.* **29**, 55–71.

Land L. S. (1980) The isotopic and trace element geochemistry of dolomite: the state of the art. *SEPM Spec. Publ.* **28**, 87–110.

Land L. S. (1995) Comment on "Oxygen and carbon isotopic composition of Ordovician brachiopods: implications for coeval seawater" by H. Qing and J. Veizer. *Geochim. Cosmochim. Acta* **59**, 2843–2844.

Lasaga A. C. and Ohmoto H. (2002) The oxygen geochemical cycle: dynamics and stability. *Geochem. Cosmochim. Acta* **66**, 361–381.

Lasaga A. C., Berner R. A., and Garrels R. M. (1985) An improved geochemical model of atmospheric CO_2 fluctuations over the past 100 million years. In *The Carbon Cycle and Atmospheric CO_2: Natural Variations Archean to Present* Geophysical Monograph 32 (eds. E. T. Sundquist and W. S. Broecker). American Geophysical Union, Washington, DC, pp. 397–411.

Lear C. H., Elderfield H., and Wilson P. A. (2003) A Cenozoic seawater Sr/Ca record from benthic foraminiferal calcite and its application in determining global weathering fluxes. *Earth Planet. Sci. Lett.* **208**, 69–84.

Lemarchand D., Gaillardet J., Lewin E., and Allègre J. C. (2000) The influence of rivers on marine boron isotopes and implications for reconstructing past ocean pH. *Nature* **408**, 951–954.

Le Pichon X., Francheteau J., and Bonnin J. (1973) *Plate Tectonics*. Elsevier, Amsterdam.

Lerman A. (1979) *Geochemical Processes: Water and Sediment Environments*. Wiley, New York.

Longinelli A. (1989) Oxygen-18 and sulphur-34 in dissolved oceanic sulphate and phosphate. In *Handbook of Environmental Isotope Geochemistry* (eds. P. Fritz and J. C. Fontes). Elsevier, Amsterdam, pp. 219–255.

Lowenstein T. K., Timofeeft M. N., Brennan S. T., Hardie L. A., and Demicco R. V. (2001) Oscillations in Phanerozoic seawater chemistry: evidence from fluid inclusions. *Science* **294**, 1086–1088.

Lumsden D. N. (1985) Secular variations in dolomite abundance in deep marine sediments. *Geology* **13**, 766–769.

Mackenzie F. T. (1975) Sedimentary cycling and the evolution of seawater. In *Chemical Oceanography,* 2nd edn. (eds. J. P. Riley and G. Skirrow). Academic Press, London, vol. 1, pp. 309–364.

Mackenzie F. T. and Agegian C. (1986) Biomineralization, atmospheric CO_2 and the history of ocean chemistry. In *Proc. 5th Int. Conf. Biomineral. Department of Geology, University Texas, Arlington, 2.*

Mackenzie F. T. and Agegian C. (1989) Biomineralization and tentative links to plate tectonics. In *Origin, Evolution, and Modern Aspects of Biomineralization in Plants and Animals* (ed. R. E. Crick). Plenum, New York, pp. 11–28.

Mackenzie F. T. and Morse J. W. (1992) Sedimentary carbonates through Phanerozoic time. *Geochim. Cosmochim. Acta* **56**, 3281–3295.

Mackenzie F. T. and Pigott J. P. (1981) Tectonic controls of Phanerozoic sedimentary rock cycling. *J. Geol. Soc. London* **138**, 183–196.

McArthur J. M. (1994) Recent trends in strontium isotope stratigraphy. *Terra Nova* **6**, 331–358.

McCulloch M. T. and Bennett V. C. (1994) Progressive growth of the Earth's continental crust and depleted mantle: geochemical constraints. *Geochim. Cosmochim. Acta* **58**, 4717–4738.

McCulloch M. T. and Wasserburg G. J. (1978) Sm–Nd and Rb–Sr chronology of continental crust formation. *Science* **200**, 1003–1011.

McLennan S. M. and Murray R. W. (1999) Geochemistry of sediments. In *Encyclopedia of Geochemistry* (eds. C. P. Marshall and R. W. Fairbridge). Kluwer, Dordrecht, pp. 282–292.

McLennan S. M. and Taylor S. R. (1999) Earth's continental crust. In *Encyclopedia of Geochemistry* (eds. C. P. Marshall and R. W. Fairbridge). Kluwer, Dordrecht, pp. 145–151.

McLennan S. M., Taylor S. R., McCulloch M. T., and Maynard J. B. (1990) Geochemical and Nd–Sr isotopic composition of deep-sea turbidites: crustal evolution and plate tectonic associations. *Geochim. Cosmochim. Acta* **54**, 2015–2050.

McLennan S. M., Bock B., Hemming S. R., Hurowitz J. A., Lev S. M., and McDaniel D. K. (2003) The roles of provenance and sedimentary processes in the geochemistry of sedimentary rocks. In *Geochemistry of Sediments and Sedimentary Rocks: Evolutionary Considerations to Mineral Deposit-Forming Environments.* (ed. D. R. Lentz), Geol. Assoc. Canada GEOtext. St. John's, Nfld, vol. 5, pp. 1–31.

McLennan S. M., Taylor S. R., and Hemming S. R. Composition, differentiation and evolution of continental crust: constraints from sedimentary rocks and heat flow. In *Evolution and Differentiation of Continental Crust* (eds. M. Brown and T. Rushmer). Cambridge University Press (in press).

Milliken K. L. and Pigott J. D. (1977) Variation of oceanic Mg/Ca ratio through time-implications for the calcite sea. *Geol Soc. Am. South-Central Meet. (abstr.)*, 64–65.

Milliman J. D. (1974) *Recent Sedimentary Carbonates: 1. Marine Carbonates.* Springer, Heidelberg.

Milliman J. D., Troy P. J., Balch W. M., Adams A. K., Li Y.-H., and Mackenzie F. T. (1999) Biologically mediated dissolution of calcium carbonate above the chemical lysocline? *Deep Res. I* **46**, 1653–1669.

Millot R., Gaillardet J., Dupre B., and Allègre J. C. (2002) The global control of silicate weathering rates and the coupling with physical erosion: new insights from rivers on the Canadian shield. *Earth Planet. Sci. Lett.* **196**, 83–98.

Morgan W. J. (1968) Rises, trenches, great faults, and crustal blocks. *J. Geophys. Res.* **73**, 1959–1982.

Morse J. W. and Mackenzie F. T. (1990) *Geochemistry of Sedimentary Carbonates.* Elsevier, Amsterdam.

Muehlenbachs K. (1998) The oxygen isotopic composition of the oceans, sediments and the seafloor. *Chem. Geol.* **145**, 263–273.

Nägler T. F., Eisenhauer A., Müller A., Hemleben C., and Kramers J. (2000) The δ^{44}Ca-temperature calibration on fossil and cultured *Globigerinoides sacculifer*: new tool for reconstruction of past sea surface temperatures. *Geochem. Geophys. Geosys.* **1** 20009000091.

Nanz R. H., Jr., (1953) Chemical composition of Precambrian slates with notes on the geochemical evolution of lutites. *J. Geol.* **61**, 51–64.

Nesbitt H. W. and Young G. M. (1984) Predictions of some weathering trends of plutonic and volcanic rocks based on thermodynamic and kinetic considerations. *Geochim. Cosmochim. Acta* **48**, 1523–1534.

O'Leary M. H. (1988) Carbon isotopes in photosynthesis. *Bioscience* **38**, 328–336.

O'Nions R. K., Hamilton P. J., and Hooker P. J. (1983) A Nd isotope investigation of sediments related to crustal development in the British Isles. *Earth Planet. Sci. Lett.* **63**, 329–338.

Palmer M. R. and Edmond J. M. (1989) Strontium isotope budget of the modern ocean. *Earth Planet. Sci. Lett.* **92**, 11–26.

Palmer M. R. and Swihart G. H. (1996) Boron isotope geochemistry: an overview. *Rev. Mineral.* **33**, 709–744.

Patchett P. J. (2003) Provenance and crust-mantle evolution studies based on radiogenic isotopes in sedimentary rocks. In *Geochemistry of Sediments and Sedimentary Rocks: Evolutionary Considerations to Mineral Deposit—Forming Environments,* Geological Association of Canada GEOtext (ed. D. R. Lentz). St. John's, Nfld, vol. 5, pp. 89–97.

Pearson P. N. and Palmer M. R. (2000) Atmospheric carbon dioxide concentrations over the past 60 million years. *Nature* **406**, 695–699.

Peterman Z. E., Hedge C. E., and Tourtelot H. A. (1970) Isotopic composition of strontium in seawater throughout Phanerozoic time. *Geochim. Cosmochim. Acta* **34**, 105–120.

Peucker-Ehrenbring B. and Ravizza G. (2000) The marine osmium isotope record. *Terra Nova* **12**, 205–219.

Pielou E. C. (1977) *Mathematical Ecology.* Wiley, New York.

Plank T. and Langmuir C. H. (1998) The chemical composition of subducting sediment and its consequences for the crust and mantle. *Chem. Geol.* **145**, 325–394.

Plummer L. N. (1975) Mixing of seawater with calcium carbonate ground water. *Geol. Soc. Am. Mem.* **142**, 219–236.

Rea D. K. and Ruff L. J. (1996) Composition and mass flux of sediment entering the world's subduction zones: implications for global sedimentary budgets, great earthquakes, and volcanism. *Earth Planet. Sci. Lett.* **140**, 1–12.

Renard M. (1986) Pelagic carbonate chemostratigraphy (Sr, Mg, ^{18}O, ^{13}C). *Mar. Micropaleontol.* **10**, 117–164.

Reymer A. and Schubert G. (1984) Phanerozoic additions to the continental crust and crustal growth. *Tectonics* **3**, 63–77.

Ronov A. B. (1949) A history of the sedimentation and epeirogenic movements of the European part of the USSR (based on the volumetric method). *AN SSSR Geofiz. Inst. Trudy* **3**, 1–390. (in Russian).

Ronov A. B. (1964) Common tendencies in the chemical evolution of the Earth's crust, ocean and atmosphere. *Geochem. Int.* **1**, 713–737.

Ronov A. B. (1968) Probable changes in the composition of seawater during the course of geological time. *Sedimentology* **10**, 25–43.

Ronov A. B. (1976) Global carbon geochemistry, volcanism, carbonate accumulation, and life. *Geokhimiya* **8**, 1252–1257; *Geochem. Int.* **13**, 175–196.

Ronov A. B. (1982) The Earth's sedimentary shell (quantitative patterns of its structure, compositions, and evolution). *Int. Geol. Rev.* **24**, 1313–1388.

Ronov A. B. (1993) *Stratisphere or Sedimentary Layer of the Earth.* Nauka, Russian.

Ross S. K. and Wilkinson B. M. (1991) Planktogenic/eustatic control on cratonic/oceanic carbonate accumulation. *J. Geol.* **99**, 497–513.

Rothman D. H. (2002) Atmospheric carbon dioxide levels for the last 500 million years. *Proc. Natl. Acad. Sci.* **99**, 4167–4171.

Rothman D. H., Hayes J. M., and Summons R. (2003) Dynamics of the Neoproterozoic carbon cycle. *Proc. Natl. Acad. Sci* **100**, 8124–8129.

Rowley D. B. (2002) Rate of plate creation and destruction: 180 Ma to present. *Geol. Soc. Am. Bull.* **114**, 927–933.

Rubey W. W. (1951) Geologic history of seawater: an attempt to state the problem. *Geol. Soc. Am. Bull.* **62**, 1111–1148.

Ruddiman W. F. (1997) *Tectonic Uplift and Climate Change.* Plenum, New York.

Sabine C. and Mackenzie F. T. (1975) Bank-derived carbonate sediment transport and dissolution in the Hawaiian Archipelago. *Aquat. Geochem.* **1**, 189–230.

Sackett W. M. (1989) Stable carbon isotope studies on organic matter in the marine environment. In *Handbook of Environmental Isotope Geochemistry* (eds. P. Fritz and J. C. Fontes). Elsevier, Heidelberg, vol. 3, pp. 139–169.

Sadler P. M. (1981) Sediment accumulation rates and the completeness of stratigraphic sections. *J. Geol.* **89**, 569–584.

Sandberg P. A. (1975) New interpretation of Great Salt Lake oöids and of ancient non-skeletal carbonate mineralogy. *Sedimentology* **22**, 497–538.

Sandberg P. A. (1983) An oscillating trend in non-skeletal carbonate mineralogy. *Nature* **305**, 19–22.

Sandberg P. A. (1985) Nonskeletal aragonite and pCO_2 in the Phanerozoic and Proterozoic. In *The Carbon Cycle and Atmospheric CO_2: Natural Variations Archean to Present,* Geophys. Monogr. Ser. 32 (eds. E. T. Sundquist and W. S. Broecker). American Geophysical Union, Washington, DC, pp. 585–594.

Schidlowski M., Hayes J. M., and Kaplan I. R. (1983) Isotopic inferences of ancient biochemistries: carbon, sulfur, hydrogen, and nitrogen. In *Earth's Earliest Biosphere* (ed. J. W. Schopf). Princeton University Press, Princeton, NJ, pp. 149–186.

Schmitt A.-D., Bracke G., Stille P., and Kiefel B. (2001) The calcium isotope composition of modern seawater determined by thermal ionisation mass spectrometry. *Geostand. Newslett.* **25**, 267–275.

Schuchert C. (1931) Geochronology of the age of the Earth on the basis of sediments and life. *Natl. Res. Council Bull.* **80**, 10–64.

Sclater J. G., Parsons B., and Jaupart C. (1981) Oceans and continents: similarities and differences in the mechanism of heat loss. *J. Geophys. Res.* **86**, 11535–11552.

Sepkoski J. J., Jr. (1989) Periodicity in extinction and the problem of catastrophism in the history of life. *J. Geol. Soc. London* **146**, 7–19.

Shaviv N. J. and Veizer J. (2003) Celestial driver of Phanerozoic climate? *GSA Today* **13**(7), 4–10.

Shaw D. B. and Weaver C. E. (1965) The mineralogical composition of shales. *J. Sedim. Petrol.* **35**, 213–222.

Shields G. and Veizer J. (2002) The Precambrian marine carbonate isotope database: version 1. *Geochem. Geophys.*

Geosys. **3**(6), June 6, 2002, p. 12 (http://g-cubed.org/gc2002/2001GC000266).

Skulan J., De Paolo D. J., and Owens T. L. (1997) Biological control of calcium isotopic abundences in the global calcium cycle. *Geochim. Cosmochim. Acta* **61**, 2505–2510.

Smith S. V. (1978) Coral reef area and contributions of reefs to processes and resources of the world's oceans. *Nature* **273**, 225–226.

Sorby H. C. (1879) The structure and origin of limestones. *Proc. Geol. Soc. London* **35**, 56–95.

Southam J. R. and Hay W. W. (1981) Global sedimentary mass balance and sea level changes. In *The Oceanic Lithosphere, The Sea* (ed. C. Emiliani). Wiley, New York, vol. 7, pp. 1617–1684.

Sperber C. M., Wilkinson B. H., and Peacor D. R. (1984) Rock composition, dolomite stoichiometry, and rock/water reactions in dolomitic carbonate rocks. *J. Geol.* **92**, 609–622.

Sprague D. and Pollack H. N. (1980) Heat flow in the Mesozoic and Cenozoic. *Nature* **285**, 393–395.

Stanley S. M. and Hardie L. A. (1998) Secular oscillations in the carbonate mineralogy of reef-building and sediment-producing organisms driven by tectonically forced shifts in seawater chemistry. *Palaeogeogr. Palaeoclimat. Palaeoecol.* **144**, 3–19.

Steuber T. and Veizer J. (2002) A Phanerozoic record of plate tectonic control of seawater chemistry and carbonate sedimentation. *Geology* **30**, 1123–1126.

Strakhov N. M. (1964) States and development of the external geosphere and formation of sedimentary rocks in the history of the Earth. *Int. Geol. Rev.* **6**, 1466–1482.

Strakhov N. M. (1969) *Principles of Lithogenesis.* Oliver and Boyd, Edinburgh, Vol. 2.

Strauss H. (1993) The sulfur isotopic record of Precambrian sulfates: new data and a critical evaluation of the existing record. *Precamb. Res.* **63**, 225–246.

Sylvester P. J. (2000) Continental formation, growth and recycling. *Tectonophysics* **322**, 163–190.

Sylvester P. J., Campbell I. H., and Bowyer D. A. (1997) Niobium/uranium evidence for early formation of the continental crust. *Science* **275**, 521–523.

Tan F. C. (1988) Stable carbon isotopes in dissolved inorganic carbon in marine and estuarine environments. In *Handbook of Environmental Isotope Geochemistry* (eds. P. Fritz and J. C. Fontes). Elsevier, Amsterdam, vol. 3, pp. 171–190.

Taylor S. R. (1979) Chemical composition and evolution of continental crust: the rare earth element evidence. In *The Earth: Its Origin, Structure and Evolution* (ed. M. W. McElhinny). Academic Press, New York, pp. 353–376.

Taylor S. R. and McLennan S. M. (1985) *The Continental Crust: its Composition and Evolution.* Blackwell, Oxford, UK.

Vail P. R., Mitchum R. W., and Thompson S. (1977) Seismic stratigraphy and global changes of sea level. 4, Global cycles of relative changes of sea level. *AAPG Mem.* **26**, 83–97.

Van Houten F. B. and Bhattacharyya D. P. (1982) Phanerozoic oölitic ironstones-geologic record and facies. *Ann. Rev. Earth Planet. Sci.* **10**, 441–458.

Veizer J. (1973) Sedimentation in geologic history: recycling versus evolution or recycling with evolution. *Contrib. Mineral. Petrol.* **38**, 261–278.

Veizer J. (1983) Trace element and isotopes in sedimentary carbonates. *Rev. Mineral.* **11**, 265–300.

Veizer J. (1985) Carbonates and ancient oceans: isotopic and chemical record on timescales of 10^7–10^9 years. *Geophys. Monogr. Am. Geophys. Union* **32**, 595–601.

Veizer J. (1988a) The evolving exogenic cycle. In *Chemical Cycles in the Evolution of the Earth* (eds. C. B. Gregor, R. M. Garrels, F. T. Mackenzie, and J. B. Maynard). Wiley, New York, pp. 175–220.

Veizer J. (1988b) Continental growth: comments on "The Archean–Proterozoic transition: evidence from Guyana and Montana" by A. K. Gibbs C. W. Montgomery P. A.

O'Day and E. A. Erslev. *Geochim. Cosmochim. Acta* **52**, 789–792.

Veizer J. (1988c) Solid Earth as a recycling system: temporal dimensions of global tectonics. In *Physical and Chemical Weathering in Geochemical Cycle* (eds. A. Lerman and M. Meyback). Reidel, Dordrecht, pp. 357–372.

Veizer J. (1995) Reply to the comment by L. S. Land on "Oxygen and carbon isotopic composition of Ordovician brachiopods: implications for coeval seawater: discussion". *Geochim. Cosmochim. Acta* **59**, 2845–2846.

Veizer J. (2003) Isotopic evolution of seawater on geological timescales: sedimentological perspective. In *Geochemistry of Sedimentary Rocks: Secular Evolutionary Considerations to Mineral Deposit-forming Environments,* Geological Association of Canda, GEOtext (ed. D. R. Lentz). St. John's, Nfld, vol. 5, pp. 99–114.

Veizer J. and Compston W. (1974) $^{87}Sr/^{86}Sr$ composition of seawater during the Phanerozoic. *Geochim. Cosmochim. Acta* **38**, 1461–1484.

Veizer J. and Compston W. (1976) $^{87}Sr/^{86}Sr$ in Precambrian carbonates as an index of crustal evolution. *Geochim. Cosmochim. Acta* **40**, 905–914.

Veizer J. and Ernst R. E. (1996) Temporal pattern of sedimentation: Phanerozoic of North America. *Geochem. Int.* **33**, 64–76.

Veizer J. and Jansen S. L. (1979) Basement and sedimentary recycling and continental evolution. *J. Geol.* **87**, 341–370.

Veizer J. and Jansen S. L. (1985) Basement and sedimentary recycling: 2. Time dimension to global tectonics. *J. Geol.* **93**, 625–643.

Veizer J., Holser W. T., and Wilgus C. K. (1980) Correlation of $^{13}C/^{12}C$ and $^{34}S/^{32}S$ secular variations. *Geochim. Cosmochim. Acta* **44**, 579–587.

Veizer J., Compston W., Hoefs J., and Nielsen H. (1982) Mantle buffering of the early oceans. *Naturwissenschaften* **69**, 173–180.

Veizer J., Laznicka P., and Jansen S. L. (1989) Mineralization through geologic time: recycling perspective. *Am. J. Sci.* **289**, 484–524.

Veizer J., Ala D., Azmy K., Bruckschen P., Buhl D., Bruhn F., Carden G. A. F., Diener A., Ebneth S., Goddéris Y., Jasper T., Korte C., Pawellek F., Podlaha O. G., and Strauss H. (1999) $^{87}Sr/^{86}Sr$, $\delta^{13}C$ and $\delta^{18}O$ evolution of Phanerozoic seawater. *Chem. Geol.* **161**, 59–88.

Veizer J., Goddéris Y., and François L. M. (2000) Evidence for decoupling of atmospheric CO_2 and global climate during the Phanerozoic eon. *Nature* **408**, 698–701.

Vervoort J. D., Patchett P. J., Blichert-Toft J., and Albarede F. (1999) Relationship between Lu–Hf and Sm–Nd isotopic systems in the global sedimentary system. *Earth Planet. Sci. Lett.* **168**, 79–99.

Vinogradov A. P. (1967) The formation of the oceans. *Izv. Akad. Nauk SSSR, Ser. Geol.* **4**, 3–9.

Vinogradov A. P. and Ronov A. B. (1956a) Composition of the sedimentary rocks of the Russian platform in relation to the history of its tectonic movements. *Geochemistry* **6**, 533–559.

Vinogradov A. P. and Ronov A. B. (1956b) Evolution of the chemical composition of clays of the Russian platform. *Geochemistry* **2**, 123–129.

Vinogradov A. P., Ronov A. B., and Ratynskii V. Y. (1952) Evolution of the chemical composition of carbonate rocks. *Izv. Akad. Nauk SSSR Ser. Geol.* **1**, 33–60.

Wadleigh M. A., Veizer J., and Brooks C. (1985) Strontium and its isotopes in Canadian rivers: fluxes and global implications. *Geochim. Cosmochim. Acta* **49**, 1727–1736.

Walker L. J., Wilkinson B. H., and Ivany L. C. (2002) Continental drift and Phanerozoic carbonate accumulation in shallow-shelf and deep-marine settings. *J. Geol.* **110**, 75–87.

Wallmann K. (2001) The geological water cycle and the evolution of marine $\delta^{18}O$. *Geochim. Cosmochim. Acta* **65**, 2469–2485.

White W. M. and Patchett P. J. (1982) Hf–Nd–Sr and incompatible-element abundances in island arcs: implications for magma origin and crust-mantle evolution. *Earth Planet. Sci. Lett.* **67**, 167–185.

Wilkinson B. H. (1979) Biomineralization, paleoceanography, and the evolution of calcareous marine organisms. *Geology* **7**, 524–527.

Wilkinson B. H. and Algeo T. J. (1989) Sedimentary carbonate record of calcium–magnesium cycling. *Am. J. Sci.* **289**, 1158–1194.

Wilkinson B. H. and Given R. K. (1986) Secular variation in abiotic marine carbonates: constraints on Phanerozoic atmospheric carbon dioxide contents and oceanic Mg/Ca ratios. *J. Geol.* **94**, 321–334.

Wilkinson B. H., Owen R. M., and Carroll A. R. (1985) Submarine hydrothermal weathering, global eustasy and carbonate polymorphism in Phanerozoic marine oolites. *J. Sedim. Petrol.* **55**, 171–183.

Windley B. F. (1984) *The Evolving Continents,* 2nd edn. Wiley, New York.

Wolery T. J. and Sleep N. H. (1988) Interaction of the geochemical cycles with the mantle. In *Chemical Cycles in the Evolution of the Earth* (eds. C. B. Gregor, R. M. Garrels, F. T. Mackenzie, and J. B. Maynard). Wiley, New York, pp. 77–104.

Wollast R. (1993) The relative importance of biomineralization and dissolution of $CaCO_3$ in the global carbon cycle. In *Past and Present Biomineralization Processes.* Bulletin de l'Institut Oceanographie, Monaco No. 13 (ed. Francoise Doumenge), pp. 13–35.

Yaalon D. H. (1962) Mineral composition of average shale. *Clay Mineral Bull.* **5**, 31–36.

Zenger D. H. (1989) Dolomite abundance and stratigraphic age: constraints on rates and mechanisms of Phanerozoic dolostone formation. *J. Sedim. Petrol.* **59**, 162–164.

Zhu P. (1999) *Calcium Isotopes in the Marine Environment.* PhD Thesis, University of California, San Diego.

Zhu P. and MacDougall J. D. (1998) Calcium isotopes in the marine environment and the oceanic calcium cycle. *Geochim. Cosmochim. Acta* **62**, 1691–1698.

18

Generation of Mobile Components during Subduction of Oceanic Crust

M. W. Schmidt

ETH Zürich, Switzerland

and

S. Poli

Universita di Milano, Italy

18.1	INTRODUCTION	617
18.2	SETTING THE SCENE	619
	18.2.1 The Oceanic Lithosphere before Subduction	619
	18.2.2 Continuous versus Discontinuous Reactions	621
	18.2.3 Fluid Production	622
	18.2.4 Fluid Availability versus Multicomponent Fluids	623
	18.2.5 Real World Effects	623
18.3	DEVOLATILIZATION REGIMES IN MORB	623
	18.3.1 High Dehydration Rates and Fluid Production (Typically up to 600 °C and 2.4 GPa)	623
	18.3.2 Low Dehydration Rates and Little Fluid Production (2.4–10 GPa and 500–800 °C)	625
	18.3.3 Melting Regimes (650–950 °C; to 5–6 GPa)	626
	18.3.3.1 Fluid-saturated (flush) melting	626
	18.3.3.2 Fluid-absent melting	626
	18.3.4 Dissolution Regime (>5–6 GPa)	627
18.4	HOW MUCH H_2O SUBDUCTS INTO THE TRANSITION ZONE?	628
18.5	DEVOLATILIZATION IN SEDIMENTS	629
	18.5.1 Pelites	629
	18.5.2 Carbonates	630
	18.5.3 Graywackes and Volcaniclastics	631
	18.5.4 Melting of Sediments Compared to Melting of MORB	631
18.6	SERPENTINIZED PERIDOTITE	632
18.7	IMPLICATIONS FOR TRACE ELEMENTS AND AN INTEGRATED VIEW OF THE OCEANIC LITHOSPHERE	634
	18.7.1 Mobile Phase Production and Trace-element Transfer	634
	18.7.2 Integrating Fluid Flux over the Entire Subducted Oceanic Crust: An Example	635
18.8	CONCLUSIONS AND OUTLOOK	636
	REFERENCES	638

18.1 INTRODUCTION

Subduction zones are the geotectonic settings where the Earth's mantle is refertilized. Whereas the various magmatic geotectonic settings produce oceanic and continental crust (including the supra-subduction arc magmatism), subduction itself consumes oceanic crust. This recycling process replenishes the mantle with most of the element inventory that otherwise would be, with time, strongly depleted in the mantle. Thus, subduction has a major role in maintaining the Earth's magmatic environments and tectonic style over the geological history. For understanding the recycling process it is necessary to understand the reactions that occur during subduction, and within the subduction setting, in particular those that transfer material from the subducting lithosphere to the mantle wedge.

Prograde metamorphism of subducting oceanic crust causes a series of mineralogical reactions that inevitably result in eclogites that may or may not contain hydrous phases and/or carbonates. An alternation of continuous and discontinuous reactions causes devolatilization, i.e., production of a fluid or, more generally, a mobile phase. The mobile phase is either a low-density fluid, a high-density solute-rich fluid, a silicate melt, or a carbonatite melt. This contribution reviews the reaction mechanisms and conditions resulting in the generation of the various mobile phases and also examines the restite(s) that are subducted to great depths.

In general, four different regimes producing a mobile phase can be recognized in the oceanic crust to depth-equivalents of 10 GPa (Figure 1):

(i) *High dehydration rates* at low-to-medium P, low T (<2.5 GPa, <600 °C) where hydrous phases are abundant and dehydration reactions are often perpendicular (in P–T space) to typical subduction geotherms. All subducted lithosphere goes through this first stage.

(ii) *Medium to low dehydration rates* at medium-to-high P, low T (2.5–10 GPa, 500–850 °C) where hydrous phases are already largely reduced in volume and dehydration reactions are often subparallel to possible P–T paths. In this range fluids become increasingly rich in dissolved matter.

(iii) *Melting* where the amount of melt depends mostly on H_2O-availability and the composition of melts is, in addition, strongly pressure sensitive:

(a) *Flush melting* (1–4 GPa, 650–850 °C) at temperatures between the wet granite solidus and

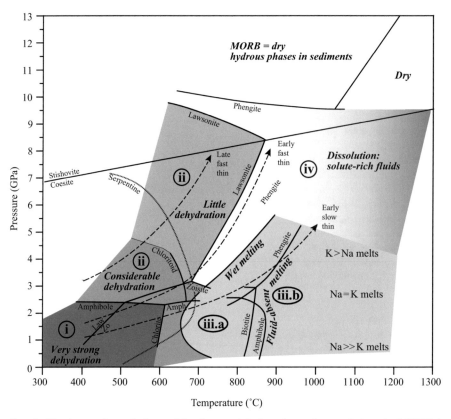

Figure 1 Devolatilization regimes during subduction, mainly based on phase relations in MORB (compare to Figure 3). The P–T region with high dehydration rates is limited by the amphibole stability and the wet solidus, the P–T regions with less dehydration is limited by zoisite and lawsonite stabilities. H_2O-saturated melting takes place in MORB and sediments at the wet solidus, fluid-absent melting to 2.5 GPa is dominated by amphibole (MORB) or biotite (pelite), and above 2.5 GPa by phengite (only in pelites). The classic melting regime is replaced by a continuous dissolution of hydrous phases in a solute-rich fluid at higher pressures (compare to Figure 4). The serpentine stability field in peridotite (dotted line) is given for reference (compare to Figure 5). The stippled arrows represent cold-to-intermediate subduction geotherms (after Kincaid and Sacks, 1997).

the fluid-absent amphibole, biotite, and phengite melting curves. In this case, additional fluid is provided from underlying dehydrating lithologies.

(b) *Fluid-absent* melting at high temperatures (>800–900 °C). At relatively low P, amphibole and biotite (1–2.5 GPa and 1–3 GPa, respectively) are the principal hydrous phases to melt. Adakitic magmas, i.e., slab melts (with Na > K), are likely to occur. At medium P (2.5–5 GPa), phengite is the principal hydrous phase to melt, and melts have K > Na.

(iv) *Dissolution* at high *P*, high *T* (>5 GPa, >800 °C) where the solvus between fluid and melt is closed and a solute-rich fluid might dissolve hydrous phases.

In the following discussion some mechanisms for devolatilization reactions are illustrated, including *real-world* aspects such as chemically heterogeneous protoliths and failure to reach equilibrium (compared to a chemically homogeneous protolith and equilibrated ideal situation). We describe and quantify, as far as possible, the four different dehydration/melting regimes, principally

investigating three bulk compositions, i.e., pelite, MORB, and harzburgite and H_2O as the major volatile component. Subsequently, carbonates and CO_2 transfer are discussed. Phase petrology is then applied to understand the behavior of trace elements. Finally, we argue for the necessity of integrating fluid/melt-producing processes over the entire oceanic lithosphere, as pressure–temperature conditions of the different lithologies within a given column of subducted lithosphere are strictly related to each other.

The purpose of this chapter is to give an overview of the possible mechanisms responsible for fluid or melt transfer into the mantle wedge, rather than to provide a catalog of minerals (see Table 1) and reactions in subducted lithosphere.

18.2 SETTING THE SCENE

18.2.1 The Oceanic Lithosphere before Subduction

In the following, we summarize some aspects that are crucial for the production and quantification

Table 1 Major volatile-carrying phases in subduction zones.

Mineral		Chemical formula	H_2O (wt. %)	Pelite graywackes	Basalts	Mg-gabbros	Peridotite
Phengite	phe	$K(Mg, Fe)_{0.5}Al_2Si_{3.5}O_{10}(OH)_2$	4.3	+++	+	–	–
Biotite/ phlogopite	bt-phl	$K(Mg, Fe)_{2.8}Al_{1.4}Si_{2.8}O_{10}(OH)_2$	4.1	++	+	–	+
Paragonite	par	$NaAl_3Si_3O_{10}(OH)_2$	4.6	+	++	+	–
K-richterite	K-rich	$KCa(Mg,Fe)_4AlSi_8O_{22}(OH)_2$	2.0	–	–	–	+
Glaucophane– barroisite	amp	$NaCa(Mg, Fe)_3Al_3Si_7O_{22}(OH)_2$– $Na_2(Mg, Fe)_3Al_2Si_8O_{22}(OH)_2$	2.2	+	+++	+++	–
Hornblende– pargasite	amp	$Ca_2(Mg, Fe)_4Al_2Si_7O_{22}(OH)_2$– $NaCa_2(Mg, Fe)_4Al_3Si_6O_{22}(OH)_2$	2.2	–	+++	+++	+
Lawsonite	law	$CaAl_2Si_2O_7(OH)_2 \cdot H_2O$	11.2	+	++	++	–
Zoisite/ epidote	zo/epi	$CaAl_2(Al,Fe^{3+})Si_3O_{12}(OH)$	2.0	+	++	++	–
Chloritoid	cld	$(Mg,Fe)_2Al_4Si_2O_{10}(OH)_4$	7.5	++	+	++	–
Chlorite	chl	$(Fe,Mg)_5Al_2Si_3O_{10}(OH)_8$	12.5	++	+++	+++	+
Talc	tc	$(Mg,Fe)_3Si_4O_{10}(OH)_2$	4.8	++	+	++	+
Talc in Si-rich veins							+++
Serpentine	serp	$(Mg,Fe)_{48}Si_{34}O_{85}(OH)_{62}$	12.3	–	–	+	+++
Phase A	"A"	$(Mg,Fe)_7Si_2O_8(OH)_6$	11.8	–	–	?	+++
Phase E	"E"	$(Mg,Fe)_{2.2}Si_{1.1}O_{2.8}(OH)_{3.2}$	11–18	–	–	?	++
10 Å phase	10A	$(Mg,Fe)_4Si_3O_{10}(OH)_2 \cdot H_2O$	8–14	Likely	–	Likely	+
Aragonite/ calcite	ara/cc	$CaCO_3$		+	+	+	–
Dolomite	dol	$CaMg(CO_3)_2$		+	+	+	+
Magnesite	mgs	$MgCO_3$		+	+	+	+

+: <5 vol%, ++: 5–20%, +++: >20%.

of mobile phases in the oceanic lithosphere (Figure 2).

A well-stratified oceanic crust is composed of a sedimentary cover layer (including pelites, carbonates, cherts, and volcaniclastics), a basaltic layer built of pillows and sheeted dikes, and a gabbroic layer with an upper layer chemically close to MORB but a lower part mostly composed of differentiated high-magnesium gabbros, Fe–Ti-gabbros, troctolites, norites, and other cumulates (Nicolas, 1989). The crust is underlain by a partly serpentinized, depleted, mostly harzburgitic peridotite (Snow and Dick, 1995).

All sediments have some natural porosity and begin expelling fluids during compaction. They generally remain fluid saturated during prograde metamorphism, and equilibration of major elements can be generally assumed (Carlson, 2002) as field evidence for prograde disequilibrium is essentially absent. Also, eclogites and blueschists having a basaltic precursor generally appear to be fully hydrated during high-pressure metamorphism, thus assuming fluid saturation and equilibrium during subduction is a reasonable simplification. In contrast, gabbros, which are only partly hydrated (in veins and adjacent alteration zones), may not experience fluid-saturated conditions and may thus fail to reach equilibrium. When pervasive alteration of coarse-grained gabbros takes place, hydration reactions are often limited to grain boundaries. Like the gabbros, serpentinization of peridotite is very heterogeneous. Veins and fractures represent important infiltration pathways for fluids and zones of

intensive serpentinization probably alternate with very weakly altered peridotite. Mammerickx (1989) estimated that 20% of the Pacific Ocean floor is affected by fracturing. For both peridotite and gabbros it remains difficult to provide quantitative estimates of the volume affected by hydration. In addition, the amount of hydration/carbonatization also depends on the spreading velocity at the mid-ocean ridges (MORs); oceanic lithosphere produced at slow-spreading ridges appears to undergo higher amounts of hydration than that produced at fast-spreading ridges.

Alteration of the igneous parts of the oceanic crust also forms veins containing carbonates, and although pervasive carbonatization is rare, carbonate contents might be locally high in veins and their immediate surroundings.

The well-stratified oceanic-crust paradigm applies to the circum-Pacific, where 53% of the total length of present Earth oceanic-crust subduction takes place (own compilation). Fast spreading Pacific-type oceanic crust, however, is not found in the Atlantic and Indian oceans where 4% and 16%, respectively, of present subduction occurs (the remainder being subduction of oceanic crust from comparatively small basins). The slow-spreading Atlantic has an oceanic lithosphere that is more complex structurally, and in which lateral heterogeneity is an important feature (Gente *et al.*, 1995; Cannat, 1993). Large transforms offer a preferential site for the emplacement of ultramafics in the shallowest portions of the lithosphere (Constantin, 1999). In the Atlantic it is common to have serpentinites cropping out on

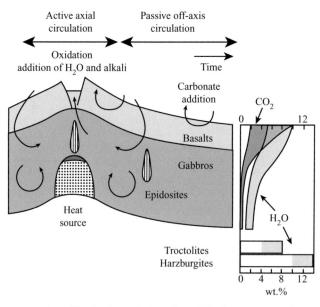

Figure 2 Schematic representation of hydrothermal alteration of the igneous oceanic crust. The inset to the right gives H_2O and CO_2 contents in a depth profile of the oceanic crust, the two horizontal bars indicate typical H_2O contents in troctolites and in intensively serpentinized harzburgite.

the ocean floor; such bodies are rare in the Pacific. The Indian Ocean has a crustal architecture intermediate between the Atlantic and Pacific oceans, it currently subducts in the Macran arc and from Burma to Sunda. Thus, caution is necessary when applying a simple layered oceanic-crust model to infer the geometry of subducted lithologies in the past and in some present subduction zones as, for example, S-Sandwich, Antilles, or Sumatra.

Determining the state of the oceanic lithosphere before subduction is crucial for understanding how the different lithologies evolve. Most of the alteration takes place close to the mid-ocean ridges, however, recent reports show that some hydrothermal activity takes place also in older oceanic crust away from the ridge axis (Kelley *et al.*, 2001). Fracturing at the seafloor and fractures caused by bending of the subducting lithosphere at the trench provide pathways for fluids and hence, further possibilities of hydration. Once the oceanic lithosphere is subducted, continued alteration of the upper layers results from any fluid or melt expelled from deeper levels. Thus, during ongoing subduction the igneous oceanic crust will interact with fluids passing through it from the serpentinized peridotite and the sediments will interact with fluids coming from the basaltic and the serpentinized peridotite layers (Gieskes *et al.*, 1990).

The building blocks of the oceanic crust define the major lithologies involved in devolatilization and melting. To a first approximation, the major players are pelites, basalts, and serpentinized harzburgites. We consider here pelites and carbonates as the main components of the sedimentary layer. From a phase petrological point of view,

graywackes and volcaniclastic sediments of broadly andesitic to dacitic composition can be regarded as intermediate between pelites and basalts. In terms of major elements and dehydration, the sediment layer is thin and volumetrically unimportant (with the exception of K_2O). Its major geochemical significance is as a source of incompatible trace elements, which are concentrated in the sediments, and also in contributing to carbonate cycling.

18.2.2 Continuous versus Discontinuous Reactions

The production of fluids and melts during subduction (the latter occurring in strictly prograde, relatively low-temperature–high-pressure conditions compared to typical crustal anatepis) is dominated by a succession of continuous and discontinuous reactions that are of comparatively equal importance in terms of fluid productivity in both the sedimentary and mafic layers of the oceanic crust. Discontinuous reactions signify the appearance and/or disappearance of phases, whereas continuous reactions only change composition(s) and proportions of the phases already present.

Extensive solid solutions (amphiboles, micas, pyroxenes, garnet) result in continuous reactions that release fluids over several tens of kilometers depths. This can be exemplified by the disappearance reactions of amphibole (Figure 3(a)) within the amphibole-eclogite facies (Poli, 1993). At the minimum pressure necessary for the formation of omphacite in basaltic bulk compositions (1.5 GPa at 650 °C), more than 50% amphibole remains in

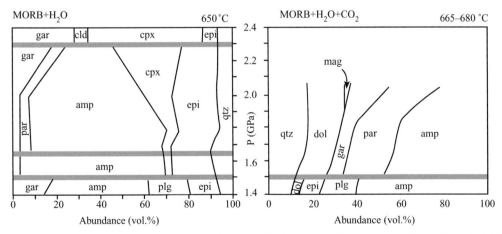

Figure 3 Continuous versus discontinuous reactions contributing to the disappearance of amphibole in (a) H_2O-saturated, CO_2-free MORB (Poli, 1993) and (b) MORB saturated in an H_2O–CO_2-fluid (Molina and Poli, 2000). Modal proportions of minerals in vol.% as a function of pressure, H_2O-contents for each mineral given in Table 1. Note that in both cases continuous reactions produce at least as much fluid (i.e., decompose as much amphibole) as discontinuous reactions (gray bars). In the CO_2-bearing system, amphibole abundances are similar to the CO_2-free system, epidote is replaced by dolomite at 1.5 GPa, and paragonite is more abundant as the appearance of omphacite is retarded to >2.1 GPa. See Table 1 for mineral abbreviations.

the eclogitic assemblage omphacite–garnet–amphibole–epidote–quartz ± paragonite. Within this assemblage, amphibole decomposes progressively, mainly forming omphacite and garnet, until 22% amphibole remains at its upper pressure stability at ~2.3 GPa. Within this pressure interval, amphibole composition changes from calcic and tschermakite-rich to sodic–calcic and barroisite-rich, and the continuous reaction from 1.5 GPa to 2.2 GPa produces more fluid (i.e., 0.7 wt.% H_2O) than the discontinuous terminal amphibole-breakdown at 2.2 GPa (i.e., 0.4 wt.% H_2O), which results in chloritoid as an additional hydrous phase. As in mafic compositions, amphibole decomposition and abundance in peridotites are controlled by continuous reactions that decompose 50–70% of the amphibole present at low pressures before its terminal-pressure breakdown at 2.5–3.0 GPa (e.g., Niida and Green, 1999). These experimental studies demonstrate that continuous reactions and discontinuous terminal breakdown reactions are equally important in fluid generation in subduction zones, in contrast to earlier models (e.g., Tatsumi, 1986), which related fluid flow and the position of the volcanic front within a subduction zone directly to amphibole breaking down exclusively at a given depth.

In sediments and the mafic portion of the subducted crust, all of the reactions involving hydrous phases and carbonates involve solid solutions whose compositions depend on the bulk composition, in addition to pressure and temperature. Different bulk compositions cause different phase compositions and thus cause reactions to shift in *P–T* space, i.e., to start shallower or deeper. In peridotites, amphibole and to some extent chlorite are controlled by continuous and discontinuous reactions; however, the other volumetrically important hydrous phases (e.g., brucite, serpentine, talc, and "phase A") in altered harzburgites display a relatively restricted compositional range, at least compared to those present in mafic eclogites. As a result, breakdown reactions of hydrous phases in harzburgites are dominated (in a first approximation) by discontinuous reactions, and take place over a restricted depth range of only a few kilometers.

18.2.3 Fluid Production

There has been a misconception in the literature that hydrous phases break down in the absence of a free fluid phase and thus the stability fields of hydrous phases are not very relevant, as they decompose when the fluid leaves the rock. This is fundamentally wrong: H_2O and CO_2 are chemical species just like any other species (e.g., SiO_2, Al_2O_3, MgO). The only difference is that the phase corresponding to the composition of such chemical species (i.e., H_2O, CO_2, etc.) happens to have a physical state (i.e., fluid) that is different from other phases on composition of chemical species (e.g., quartz, corundum, periclase). It should be remembered that the thermodynamical treatment of all these phases is identical for the entropy and enthalpy terms and that they only differ for the *P–V–T* (pressure–volume–temperature) relation adopted. Just as SiO_2 saturation is not a prerequisite for the stability of olivine or enstatite, so H_2O saturation is not required for the stability of hydrous phases. In fact, for a given bulk composition it is quite possible (when following a suitable *P–T* trajectory) to pass from a fluid-absent to a fluid-present regime and back again to a fluid-absent regime.

A fluid is only produced if a given rock volume is already completely hydrated (fluid saturated). If fluid saturation is not realized at the beginning of subduction, a number of fluid-absent reactions will take place. These reactions are of the type $A + V_1 = B + V_2$ (where A, B are volatile free phases and V_1, V_2 are hydrous phases or carbonates), involve hydrates and/or carbonates and change the mineralogy of a rock volume according to the stability fields of the minerals, but do not liberate a fluid. Prograde subduction zone metamorphism (as is true for any type of prograde metamorphism) generally reduces the amount of H_2O that can be stored in hydrous minerals with depth. Thus, almost any part of the oceanic crust sooner or later becomes fluid saturated. In an equilibrium situation, the volatile content bound in hydrous phases and carbonates remains constant until fluid saturation occurs. Either continuous or discontinuous reactions may lead to fluid saturation in a rock. The point at which this occurs depends on initial water content, and pressure and temperature, and somewhat counter-intuitively, initial low water contents do not cause early complete dehydration, but delay the onset of fluid production to high pressures.

Due to heterogeneous alteration (and thus varying initial H_2O and CO_2 contents) there is a wide depth range over which different volumes of the oceanic lithosphere become fluid saturated. A second complication arises from the scale at which equilibrium is effective. A few grains may locally form a fluid-saturated environment, but it is questionable whether the fluid produced on a local grain scale is able to escape. Field evidence argues for equilibration of fluids in eclogites on a centimeter to meter scale until they are able to collect in veins where these fluids might escape (Philippot, 1993; Widmer and Thompson, 2001; Zack *et al.*, 2001, possibly through fractures in the overlying lithologies without affecting them much, possibly through pervasive infiltration. Thus, considerable uncertainties regarding the quantification of fluid- or melt-producing processes result from the scale of equilibration and the way fluid migrates (Austrheim and Engvik, 1997) through the overlying layers. The ways fluids migrate also influence the phase equilibria and reactions taking place. Most reactions during subduction occur in response to an increase in pressure and temperature, where fluid is produced and expelled. However, the ascending fluid may change the fluid composition in the overlying layers and thus cause reactions.

18.2.4 Fluid Availability versus Multicomponent Fluids

In calculations of phase equilibria that appear to be H_2O undersaturated, it is commonly assumed that a fluid phase is present and H_2O activity is lowered by CO_2 in the fluid. The latter is not necessarily true, and this section examines the differences between the limited availability of an aqueous fluid compared to the unlimited availability of a mixed H_2O–CO_2 fluid. Note that both cases are described by the term "H_2O undersaturated."

This can be illustrated by a natural example. In the coarse-grained Allanin magnesium-gabbro, infiltration of fluid caused the formation of reaction rims around olivine (Chinner and Dixon, 1973). The succession is olivine $\rightarrow$ anthophyllite (2 wt.% H_2O) $\rightarrow$ talc (4wt.% H_2O + kyanite $\rightarrow$ chloritoid (8wt.% H_2O + talc + kyanite. This reaction rim is H_2O undersaturated, and the succession of mineral assemblages corresponds to an increase of H_2O content towards the rim and can only be modeled by an increase in the availability of water towards the rim. The H_2O-undersaturated character of the inner rim zones does not necessitate (or justify) a CO_2 component in the fluid, but rather reflects limited availability of an H_2O fluid.

Thermodynamic calculations based on measured compositions of solid phases (as commonly performed) result in an evaluation of the chemical potential of H_2O in these phases (μ_{H_2O}). Under equilibrium, the chemical potential of H_2O is equal in all phases and only if some additional constraint implies the presence of a fluid phase (e.g., fluid inclusions), the composition of this fluid phase can be calculated from P, T, and μ_{H_2O}. This can be illustrated in the simple system CaO–Al_2O_3–SiO_2–H_2O–CO_2 where a given chemical potential of H_2O at a given P and T (4 GPa, 600 °C) may correspond to a fluid-absent situation (with lawsonite + zoisite + aragonite + coesite + kyanite present) or to a situation with a mixed fluid phase (with aragonite + coesite + kyanite + fluid present) (Poli and Schmidt, 1998).

The actual consequence of H_2O undersaturation in the context discussed here is that mixed fluid phases, i.e., solutions, will shift phase equilibria in P–T–X space, whereas the presence or absence of a single component fluid (e.g., a purely aqueous fluid) determines whether or not a reaction takes place, without changing the reaction's position in P–T space.

18.2.5 Real World Effects

The compilation of available phase equilibria aims at understanding equilibrium situations in typical (and homogeneous) average bulk compositions. However, different *real-world effects* with amplitudes that may depend on the rock type, are to be expected.

In the real world, the following factors contribute to the continuous character of the devolatilization signal from the downgoing slab: highly variable bulk compositions in the sedimentary and gabbroic layers as well as possibly different degrees of depletion (caused by different amounts of melt extracted) in the hydrated peridotitic layer (Constantin, 1999); heterogeneous distribution of carbonates versus hydrous minerals resulting in an inhomogeneous X_{CO_2} in the fluid phase (Gillis and Robinson, 1990); large temperature gradients within the subducting lithosphere (Kincaid and Sacks, 1997), and finally, possible kinetic effects that inhibit reactions and thus widen reaction zones, which are related both to the effects of fluid availability and deformation history (Austrheim and Engvik, 1997; Molina *et al.*, 2002) and to thermal retardation of sluggish solid–solid transformations (Schmeling *et al.*, 1999).

All of the above effects lead to smearing out the release of a fluid/melt *pulse* over a broader depth range. In contrast, the single focusing mechanism for fluid/melt flow is mechanical: fluids/melts may ascend through channelized flow, hydrofractures or pre-existing fractures, which may focus a broadly distributed fluid into distinct pathways. The relative importance of these mechanisms is largely conjectural, and whether macroscopical fluid/melt focusing is achieved depends on their extent and spacing.

18.3 DEVOLATILIZATION REGIMES IN MORB

Based on phase relations in H_2O-bearing MORBs, four distinct P–T regions with characteristic mobile-phase production mechanisms can be identified and are discussed sequentially.

18.3.1 High Dehydration Rates and Fluid Production (Typically up to 600 °C and 2.4 GPa)

Once the oceanic crust starts subducting, most of its remnant porosity will be immediately lost by compaction and its pore fluids get expelled. At this stage, zeolites, pumpelleyite, and prehnite are the major H_2O-bearing minerals and H_2O contents stored in hydrous minerals amount to 8–9 wt.% H_2O in the bulk rock (Peacock, 1993). Beyond depths of ~15 km the oceanic crust enters into the blueschist facies in which the major hydrous minerals are chlorite, sodium-rich, calcium-poor amphiboles (glaucophane to barroisite), phengite (white mica), lawsonite or zoisite, and paragonite (e.g., Sorensen, 1986). Water contents of the bulk rock at the beginning of the blueschist facies are ~6 wt.% (Figure 4). Initially abundant chlorite has high H_2O contents (12 wt.% H_2O) and decomposes completely in the depth range to 70 km through various continuous and discontinuous reactions. Lawsonite (11 wt.% H_2O) has a maximum abundance of 25 vol.% at the onset of blueschist-facies metamorphism and decreases to ~10 vol.% at the amphibole-out reaction. Anhydrous minerals typically comprise 5–25 vol.% (e.g., Thurston, 1985; Okay, 1980) at 5–10 km depth and some 100 °C, grow in volume to ~50 vol.% just before the amphibole-out reaction, and increase to >70 vol.% at pressures beyond amphibole stability. As a consequence, dehydration rates are considerable within this P–T regime,

where a fully hydrated MORB loses ~4–6 wt.% H_2O when passing through the blueschist stage within the fore-arc region.

At pressures of 2.2–2.4 GPa, i.e., the maximum pressure stability of amphibole in MORB (65–70 km), dehydration reactions are numerous and their orientations in P–T-space are mostly oblique to a typical subduction-type P–T path, resulting in high dehydration rates. Although fully hydrated oceanic crust loses about two-thirds of its initial water content in this interval, this fluid will either serpentinize the cold corner of the mantle wedge or eventually pass, through veins, to the ocean floor in the fore-arc region (e.g., Mariana arc, Fryer *et al.*, 1999). The high dehydration rates in conjunction with a—close to the trench—thin mantle wedge should lead to rapid, full serpentinization of the cold corner (i.e., within 0.8–3.0 Ma) if fluids pass pervasively through the mantle wedge (see also Gerya *et al.*, 2002). As a serpentinized cold corner has a much lower viscosity than anhydrous mantle, the serpentinized cold corner may develop a counter-convection cell and stagnate in its corner, whereas the rest of the mantle wedge undergoes large-scale convection (examples of cold corner counter-flow are found in Davies and Stevenson, 1992).

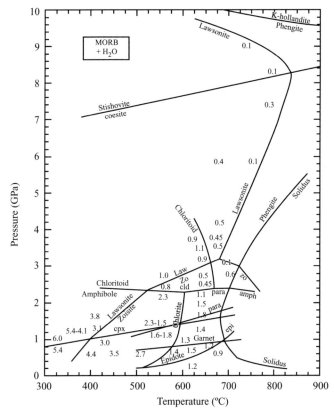

Figure 4   Major phase stability boundaries in H_2O-saturated MORB and H_2O contents (numbers in wt.%) stored in hydrous phases (Schmidt and Poli, 1998). Data below 550 °C are based on natural blueschists and greenschists, all other data are based on experiments (compare with Figure 1).

Most information concerning blueschists is obtained from studying natural occurrences; it is only at and beyond the chlorite- and amphibole-out reactions that reaction rates are sufficiently high to study these systems experimentally. Thermodynamic modeling of phase relations in MORB is generally accurate, however, it becomes difficult when incorporating the amphibole solid solution. Both calculated amounts and stabilities of amphiboles are far from what is observed in nature and experiments (e.g., the pressure stability of amphiboles in MORB is calculated to 4–5 GPa instead of the 2.4–2.8 GPa observed in experiments and deduced from natural eclogites), leaving us to rely almost entirely on natural occurrences to deduce relevant dehydration reactions.

The study of blueschists also demonstrates that many of their minerals contain significant amounts of Fe^{3+} (e.g., Maruyama *et al.*, 1986; Brown and Forbes, 1986). The inhomogeneous degrees of oxidation within the altered oceanic crust add another compositional variable, shift reactions in *P–T*-space, and complicate the geochemistry of trace elements with variable oxidation state (e.g., uranium). The fraction of Fe^{3+}/Fe^{tot} is generally highest at the onset of subduction-zone metamorphism and decreases with increasing grade.

18.3.2 Low Dehydration Rates and Little Fluid Production (2.4–10 GPa and 500–800 °C)

Until the late 1980s it was believed that oceanic crust is fully dehydrated after the pressure-induced amphibole breakdown. However, experiments (Poli, 1993; Pawley and Holloway, 1993; Poli and Schmidt, 1995; Schmidt and Poli, 1998; Okamoto and Maruyama, 1999; Forneris and Holloway, 2001) and natural occurrences of epidote/zoisite, lawsonite, talc, chloritoid, phengite, staurolite, OH-rich topaz, and many other minor hydrous minerals in eclogites that have already lost their amphibole, clearly demonstrate that dehydration continues above 2.4 GPa. In particular, since the early 1990s, the frenetic chase after coesite- and diamond-bearing hydrated eclogites showed that most of the above hydrous minerals are stable at pressures beyond 2.5 GPa (e.g., the coesite-lawsonite bearing eclogites found in xenoliths of Colorado Plateau, Helmstaedt and Schulze, 1988; Usui *et al.*, 2003). Generally, the natural occurrences of coesite- and sometimes diamond-bearing hydrous eclogites (the most famous being Dora Maira, Norway, Dabie Shan, and the Kokchetav massif, see Liou *et al.* (1998) for a concise review) are in good agreement with experimental findings, and the discussion below is based on experimental *P–T* conditions and mineral abundances.

Above 2.4 GPa, a maximum of 1.5 wt.% H_2O remains stored in hydrous phases of the oceanic crust. The major water hosts are lawsonite, zoisite, chloritoid, talc, and phengite. Lawsonite is the most water-rich of these phases (12 wt.% H_2O) and may host >50% of the water present; however it is restricted to relatively low temperatures (Figure 4). Zoisite is stable to ~3.3 GPa and occurs at the wet solidus until 3 GPa. Talc is only a minor phase in MORB but becomes abundant in bulk compositions with high X_{Mg}, i.e., magnesium gabbros. Phengite has an important role, as virtually any K_2O present in the oceanic crust at pressures below 5 GPa will be stored in phengite, irrespective of the bulk composition (Schmidt, 1996). Above this pressure, potassium also enters into clinopyroxene (Okamoto and Maruyama, 1998; Schmidt and Poli, 1998) and the relative amounts of phengite, clinopyroxene, and the coexisting fluid will depend upon the quantity of dissolved potassium and its dissolution rate into the fluid. Above 2.4 GPa continuous reactions dominate over a few discontinuous reactions and most (including the wet solidus) are subparallel to typical subduction *P–T* paths. As a consequence, a particular rock volume crosses reactions over a wide depth interval. Thus, dehydration rates are rather low, and the (post-amphibole) remnant 1.5 wt.% of H_2O is lost over a wide depth range. The pressure–temperature range that defines this "low dehydration rate" regime corresponds to the depth interval of the subarc region.

The hydrous phases in oceanic crust with the widest pressure stability range are lawsonite and phengite, which both reach into the stishovite field. The breakdown of these two phases corresponds to the *end* of the major devolatilization history of oceanic crust.

Because many reactions are parallel to typical *P–T* paths during subduction, effective H_2O contents, mineral assemblages, and mineral compositions are fairly sensitive to temperature. A K_2O-free MORB passing 600–650 °C at a depth of 100 km has 1 wt.% H_2O stored in lawsonite and chloritoid and will arrive at 200 km with ~0.4 wt.% H_2O in lawsonite. Passing 100 km at a temperature of 700 °C causes the loss of the last hydrous phase (zoisite) near this depth and our MORB volume becomes a dry eclogite and thus irrelevant for any further production of mobile phases. Now, if the 100 km depth is achieved at a temperature of 750 °C, a MORB is already above the wet solidus and any fluid infiltrating from below should cause partial melting. As is evident from these considerations, the major difficulty in predicting what type and quantity of mobile phase is produced stems from the uncertainty in temperature distribution in the subducting oceanic crust. Unfortunately, a difference in temperature of 100 °C has a significant impact.

18.3.3 Melting Regimes (650–950 °C; to 5–6 GPa)

Melting of the oceanic crust is evidenced in so-called adakites which form volcanic suites of andesitic to rhyodacitic composition (Kay, 1978; Drummond *et al.*, 1996). These are interpreted to be partial melts of the oceanic crust more or less modified (mainly magnesium enriched) during their ascent through the mantle wedge. Adakitic suites are distinguished from melts originating from a fluid-metasomatized mantle wedge by distinctive trace-element ratios (i.e., high La/Yb and Sr/Y, review by Martin, 1999), by anomalously high X_{Mg} values and MgO concentrations at comparative SiO_2 contents (thought to be acquired when slab melts absorb olivine on their way through the mantle wedge) and by the absence of volcanic rocks more basic than andesite. Melting of subducted crust was probably predominant on the Archean Earth (Martin, 1999) where mantle temperatures were higher and oceanic plates probably were thinner and hotter. There is consensus that on the modern Earth dehydration dominates in subducting lithosphere. However, when young, hot crust is subducted, or when the thermal field within a subduction zone is disturbed at plate boundaries (i.e., at the lateral end of a subducting plate) or due to flat slabs caused by ridge subduction, melting of the oceanic crust may occur.

Melting of subducted crust can occur under fluid-saturated or fluid-absent conditions. The principal hydrous phases involved are phengite, biotite, epidote/zoisite, and amphibole. If melting of the oceanic crust occurs at pressures below 2.5 GPa, the oceanic crust will go through a greenschist and epidote-amphibolite facies stage at low pressures (both stages not discussed above).

The compositions of slab melts principally depend on the bulk composition, the degree of melting and the pressure of melting. However, it is generally true that at low- to moderate-melt ratios, where cpx is a major residual mineral, Na/K ratios mostly depend on the pressure of melting. At any pressure, potassium strongly partitions into the melt. However, $Na^{cpx/melt}$ partition coefficients are small (<0.5) below 2 GPa, close to unity around 3 GPa and increase to 3–5 at 4 GPa. The latter corresponds to ~50–70% jadeite component in residual clinopyroxene, which retains most of the Na_2O, but leaves all of the K_2O in the melt. As a consequence, melts from amphibole-dominated fluid-absent melting have Na > K, whereas melts resulting from fluid-present or fluid-absent phengite melting at higher pressures are strongly potassic peraluminous granites (with K > Na). The latter, which would migrate into the mantle wedge at >70 km depth are not observed at the surface. This is either because they never form or because it is virtually

impossible for them to traverse the thick mantle wedge and retain their peculiar chemistry, which is far out of equilibrium with peridotite. Absorption reactions with mantle minerals would largely modify the major-element composition of these melts, although a sediment-melt contribution is sometimes postulated in arc magmas on the basis of certain trace elements (see Section 18.5).

18.3.3.1 Fluid-saturated (flush) melting

Fluid-saturated melting of basaltic crust begins at temperatures of ~650 °C at 1.5 GPa to ~750 °C at 3 GPa (Figure 5). It should be noted that the wet-solidus temperature is elevated by at least 100–200 °C, if the fluid is in equilibrium with carbonates (an X_{CO_2} of 0.3–0.6 would be expected in the appropriate *P–T* range). Although a small quantity of free fluid (<0.1 vol.%) is likely to be present in any lithology affected by dehydration, this would not be sufficient to produce a significant melt portion through fluid-saturated melting. However, fluid-saturated melting at relatively low temperatures could be achieved in subducted crust through flush melting, i.e., by addition of aqueous fluid from below. In this case, the melt productivity depends on the amount of fluid added to the system. The possibilities to obtain volumetrically significant melt fractions in (i) MORB through flushing with fluid originating from dehydration of serpentinized peridotite situated below the MORB or (ii) sediments through flushing with fluids from underlying MORB or serpentinized peridotites are discussed in Section 18.5.4.

18.3.3.2 Fluid-absent melting

Most adakite suites contain andesites that are consistent with fluid-absent melting of a basaltic source (but not with melting of mica-dominated sediments only). Fluid-absent melting is defined as the production of a silicate melt with low-water contents from an assemblage that contains hydrous phases but no free fluid phase. At pressures below 2.5 GPa, both metapelites and metabasalts contain a pair of hydrous phases with equal water contents (Figure 5): in pelites, phengite (4.3 wt.% H_2O) has a lower modal abundance than biotite (4.1 wt.% H_2O) and also a 100–150 °C lower melting temperature than biotite. In basaltic compositions, epidote/zoisite (2.0 wt.% H_2O, typically 10 vol.%) melts ~100 °C below amphibole (2.2 wt.% H_2O, typically 50–30 vol.%, decreasing with increasing pressure) (Vielzeuf and Schmidt, 2001, and references therein). In both cases, the first fluid-absent melts are produced through the volumetrically less-important phase, i.e., phengite in metapelites and epidote/zoisite in metabasalts. These first melts are thought to be dacitic and

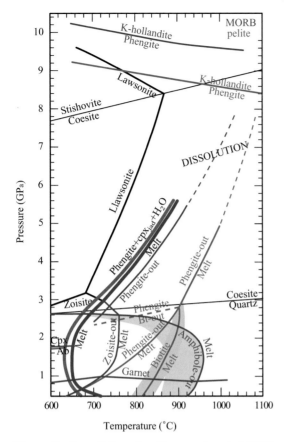

Figure 5 Compilation of melting reactions in MORB and average pelite. Black lines represent reactions with similar *P–T* locations in both bulk compositions, blue lines: MORB, red lines: pelite. The red and blue areas are *P–T* fields of biotite-dominated and amphibole-dominated fluid-absent melting respectively. At subduction-zone melting pressures (i.e., >1 GPa), the first fluid-absent melt in pelite appears at the phengite-out reaction and in MORB at the zoisite/epidote-out reaction. The bold lines represent the H_2O-saturated solidus, which involves amphibole or biotite in addition to phengite + cpx + coesite/quartz below 2.5 GPa and plagioclase instead of cpx below 1.5–1.7 GPa. Upto 5–6 GPa classical melting occurs where a solidus separates a crystal + fluid field from a melt-present field (the fluid having distinctively higher H_2O contents than the melt). A continuous increase of solute in the fluid is observed at high pressures (>5.5 GPa) and at least the K-bearing phases (phengite or at higher pressures K-hollandite) dissolve in a fluid of continuously evolving composition.

probably amount to less than 10 vol.%. In the absence of detailed experimental investigations (at fluid-absent conditions, experimental reaction rates are too sluggish below ~800 °C and equilibration does not take place for the fluid-absent melting of epidote and phengite below 2.5–3 GPa), it is likely that the primary melts of adakites result from 20–35% amphibole-dominated fluid-absent melting of MORB, geochemical modeling is consistent with a garnet + cpx ± amph residue

(Martin, 1999). Nevertheless, the temperature distribution in oceanic crust is such that subducted sediments are at higher temperatures than the igneous oceanic crust. Thus, if amphibolites are melted, the overlying sediments must also melt through mica-dominated fluid-absent melting. This complexity needs to be taken into account in geochemical slab-melting models. Fluid-absent melting of amphibolite or metapelite cannot be fully described in a closed system or a system open only to melt extraction. The temperatures necessary for fluid-absent melting of the MORB layer (800–900 °C) are such that the hydrated peridotite below would reach the serpentine stability limit, thus possibly providing a significant amount of fluid. This fluid would strongly increase the melt fractions in amphibolites and metapelites and might be required to produce a sufficient quantity of primary melts that then give rise to adakite suites.

18.3.4 Dissolution Regime (>5–6 GPa)

The typical concept of low-density H_2O–CO_2 fluids (with small to moderate amounts of solute) in the sub-solidus and high-density silicate liquids (with typically 1–15 wt.% H_2O dissolved) above the solidus does not apply to the subduction environment at pressures above 5–6 GPa. At higher pressures, a chemical continuum between fluids and melts exists (Boettcher and Wyllie, 1969) and, depending on fluid–rock ratios, a continuous dissolution process leaches hydrophile species out of sediments, basalts, and serpentine.

Considerable attention has focused on the role of high-pressure fluids/melts that exist beyond the *second critical end point* (Ricci, 1951). At crustal pressures and temperatures, a large miscibility gap exists between a low-density aqueous (already supercritical) fluid and a high-density hydrous silicate melt. Consequently, a *fluid* and a *melt* have quite distinct compositions and physical properties (e.g., viscosity, compressibility). This is reflected in the common terminology of dehydration/hydration versus melting reactions. However, at higher pressures, the solubilities of silicates or silicate components in fluid and of H_2O in silicate melts increase. Beyond a certain pressure (depending on the chemical system, between 1.5 GPa and >12 GPa), the miscibility gap between classical fluid and melt shrinks, intersects the solidus (this exact locus is defined as the second critical end point) and a chemical continuum between the two extremes, a dry silicate melt and a pure H_2O-fluid, is possible (see Stalder *et al.*, 2000, figure 1). This implies that a continuum exists between the physical properties of former melt and fluid; it does not imply that a mobile phase with melt chemistry and physical properties characteristic of silicate melts may not exist (or with fluid chemistry and properties). The term *supercritical fluid* is not distinct or adequate enough to describe this phase, as any aqueous/carbonic fluid above the first critical end point (at a few hundred bars and degree centigrade) is already supercritical.

At the conditions described above, the wet solidus vanishes, the concept of melting loses its definition, and solid assemblages continuously dissolve in first, a volatile-rich, and with increasing temperature, silicate-rich *nonsolid* phase. The exact pressures (at temperatures relevant for subduction) for the disappearance of the wet solidus are somewhat uncertain and strongly depend on the chemical system. In the simple system SiO_2–H_2O, the second critical end point is only at 1 GPa, 1,100 °C (Kennedy *et al.*, 1962), in albite-H_2O it moves to ~1.5 GPa (Stalder *et al.*, 2000), in CaO–SiO_2–H_2O–CO_2 it occurs at 3.2 GPa, 500 °C (Boettcher and Wyllie, 1969), whereas in the model ultramafic system MgO–SiO_2–H_2O (MSH) the solidus terminates at ~12 GPa, 1,100 °C (Stalder *et al.*, 2001). In potassium-enriched MORB, and in graywackes and pelites, the miscibility gap closes somewhere between 4 GPa and 6.5 GPa (Schmidt and Vielzeuf, 2001). At 4 GPa, a classical sequence of melting reactions and quenched melts are observed in experiments, whereas at 6.5 GPa, initially abundant phengite becomes less and less abundant until it disappears and a mostly K_2O–Al_2O_3–SiO_2-bearing fluid-quench precipitate becomes more and more abundant at grain boundaries. This finding is in agreement with an experimental study on carbonaceous pelites (Domanik and Holloway, 2000), where phengite disappears in a similar fashion at 6.5 GPa and 8 GPa. Thus, there is little doubt that at high pressures the solvus closes and a chemical continuum is realized.

Estimates of the amount of solute in such fluids above 6–7 GPa range from 50 wt.% to 70 wt.% (a molar H_2O : K_2O ratio slightly above unity is in fact sufficient for completely dissolving micas), being close to the solubility of H_2O in water-saturated melts near 3 GPa (solubilities of H_2O in natural silicate melts at higher pressures are unknown). Due to the high solubility of K_2O (and probably other components that are less soluble at low pressures), it is doubtful that much of the subducted potassium and related trace elements could reach great depths. In fact, most of the potassium is likely to be transferred to the overlying mantle wedge, where in a MgO-rich, SiO_2-poor chemical environment it may reprecipitate in phlogopite or potassium-richterite (potassic amphibole) (Sudo and Tatsumi, 1990; Konzett and Ulmer, 1999; Trønnes, 2002) and then be dragged to greater depths in the mantle directly overlying the subducted slab. The little potassium remaining in the oceanic crust after dehydration

and leaching is stored in the anhydrous phase potassium-hollandite ($KAlSi_3O_8$), which is stable to >25 GPa.

18.4   HOW MUCH H_2O SUBDUCTS INTO THE TRANSITION ZONE?

It is well known that refertilization of the mantle takes place through transfer of fluids or melts from the subducting lithosphere to the overlying mantle wedge, as described above. A portion of the elements transferred into the mantle wedge is partitioned into partial melts, which ultimately form arc volcanism and thus do not refertilize the deep mantle. The residue of oceanic crust remaining from this process is subducted to depths >300 km where it may ultimately be mechanically mixed with mantle material (Allegre and Turcotte, 1986). Dixon *et al.* (2002) deduced from the geochemistry of ocean-island basalts, which are widely believed to contain some recycled oceanic crust in their sources, that dehydration is >92% efficient during subduction.

In order to evaluate how much water is subducted to great depths in the mantle, it is necessary to (i) determine the amount of H_2O stored in peridotite after pressure induced decomposition of serpentine and (ii) understand the state of the oceanic crust at pressures just above the phengite breakdown reaction. Any H_2O stored in oceanic peridotites that pass beyond 220 km depth and any oceanic crust that passes beyond 300 km depth is unlikely to be mobilized within the direct subduction context and will participate in mechanical mixing with deep mantle. Only a few hydrous phases may exist beyond 220 km and 300 km depth in peridotite and oceanic crust, respectively, and temperature stabilities for these phases ("phase A," "phase E," and "phase D" in peridotite: Angel *et al.* (2001), Ulmer and Trommsdorff (1999), Frost (1999), and Ohtani *et al.* (2000); and the hydrous aluminosilicates topaz-OH and "phase egg" in aluminous sediments, Ono (1998)) increase faster than any subduction geotherm, at least to 20 GPa. Furthermore, hydrous phases become much less important at pressures beyond 10 GPa: the volumetrically dominant nominally anhydrous phases (NAMs) such as olivine, wadsleyite, garnet, and cpx (3,000 ppm OH in natural cpx from 6 GPa, 1,000 °C; Katayama and Nakashima (2003); up to 3.3 wt.% H_2O in wadsleyite, Kohlstedt *et al.* (1996)) dissolve considerable amounts of hydrogen at these pressures and become the principal hydrogen reservoirs at greater depths.

Whether a significant amount of water is subducted beyond 200 km in peridotitic compositions depends on the exact *P–T* path. As can be seen in Figure 6, any serpentine-bearing peridotite descending along geotherms that are cooler than 580 °C at 6 GPa (termed the *choke point*

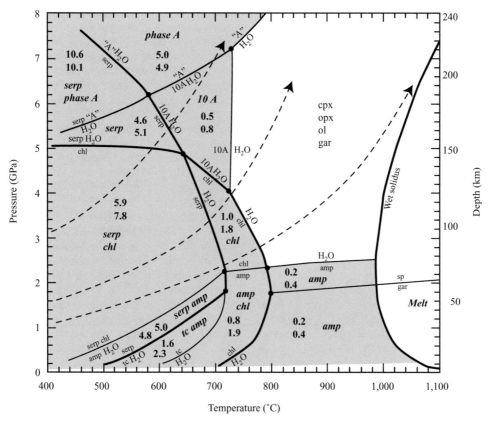

Figure 6 Major phase stability boundaries in H_2O-saturated peridotite and H_2O contents stored in hydrous phases for harzburgite and lherzolite (upper and lower number, respectively; after Schmidt and Poli, 1998 and Fumagalli *et al.*, 2001). The gray field denotes stability fields of hydrous phases. Serpentinized peridotite following the coldest *P–T*-path will not dehydrate up to transition zone depths, as phase boundaries of phase A (and "phase E" at >11 GPa, Frost, 1999) are flatter in *P–T* space than typical subduction *P–T* paths.

by Kawamoto *et al.* (1995)) will conserve H_2O and form phase A (and subsequently phase E and phase D). In oceanic lithosphere subducting along geotherms that pass between 580 °C, 6 GPa and 720 °C, 7 GPa the 10 Å phase forms upon serpentine breakdown (Fumagalli and Poli, 1999; Fumagalli *et al.*, 2001), and holds 0.6 wt.% H_2O in the peridotite. Subsequent entering into the stability field of phase A at greater depths will not lead to a significant increase of the bulk H_2O content as most of the free fluid produced from serpentine breakdown would have escaped already.

18.5 DEVOLATILIZATION IN SEDIMENTS

18.5.1 Pelites

Whereas subduction-zone metamorphism leads to a continuous decrease of H_2O stored in MORB, this is not necessarily true for pelites at pressures up to 2.5–3.0 GPa. By far the most abundant hydrous minerals in metapelites are the potassic micas phengite and biotite. The amount of H_2O

stored in micas (containing 9–10 wt.% K_2O and 4–4.5 wt.% H_2O) is easily calculated from the bulk K_2O content. Other important hydrous phases in subduction zone metapelites are talc, chloritoid, and chlorite (see compilation in Poli and Schmidt, 2002). The stabilities of these phases are strongly dependant on the X_{Mg} which varies widely in oceanic pelagic sediments, from 0.2 to 0.8 (e.g., Plank and Langmuir, 1998). Biotite is expected to transform completely to garnet + phengite-bearing assemblages at pressures above 2.5–3.0 GPa (Poli and Schmidt, 2002) but under high X_{Mg} biotite may persist to a maximum of 4 GPa (Hermann, 2002a). In contrast, phengite is stable to considerably higher pressures (up to 8–9 GPa, Domanik and Holloway, 1996; Ono, 1998).

The absolute amount of hydrous phases, and therefore the maximum amount of H_2O in metapelites, is also strongly controlled by the amount of quartz, which is highly variable in mica schists. An average of 9 metapelites (containing between 12 vol.% and 46 vol.% quartz) from the Nome Blueschist Terrane, Alaska (Thurston, 1985) yields 2.7 wt.% H_2O stored in phengite, chlorite, paragonite, epidote, and glaucophane. The same average bulk composition would contain ~2.0 wt.% H_2O stored in the hypothetical assemblage phengite—lawsonite—chloritoid (+gar + cpx + coesite), which is stable near 3 GPa and 600 °C, and 1.1 wt.% H_2O if phengite is the only hydrous phase at higher pressures. Thus, during subduction from 20 km to 80 km depth, 300 m of mica schist would yield only 2–6% of the mass of H_2O released from the underlying 2 km of MORB. These numbers are fairly approximate, as our current knowledge of phase relationships in metapelites metamorphosed at high-pressure, low-temperature conditions is still fragmentary, and further experimental work is required to explore the significance of talc, chlorite, and chloritoid in the critical pressure range between 2 GPa and 4 GPa. Nevertheless, metapelites contain generally less H_2O than MORB in a low-pressure blueschist stage but more H_2O at pressures beyond 2.5 GPa.

Typically, only a few hundred meters of sediments may be subducted, and the quantity of volatiles stored in pelites is small when compared to MORB and serpentinized peridotite. The importance of pelites to the subduction factory lies in their relatively high concentration of K_2O and other highly incompatible minor and trace elements, which may be concentrated in accessory phases (typically rutile, allanite, zircon, phosphates, ellenbergerite). Thus sediments can impart a strong trace-element signal to the slab fluids (or melts), which is quite distinct from that derived from the igneous oceanic crust. For example, if present, allanite (a phase that may be residual during melting) contains more than 90% of the whole-rock LREE and thorium; rutile contains more than 95% of titanium, niobium, and tantalum; zircon contains 95% of the whole-rock zirconium and hafnium, and phengite, at a modal abundance of 20–35% (Figure 8), incorporates more than 95% of the bulk-rock rubidium, barium, and caesium (see also Hermann, 2002b). Finally, as discussed above, pelites also impact melting, as carbonate-free pelites have the lowest melting temperature of all subducted lithologies (see Section 18.3.4) and, within the oceanic crust, are situated at the highest temperatures.

18.5.2 Carbonates

Most of the discussion above centers on the breakdown of hydrous phases during subduction. The subducted mass of H_2O in most subduction zones is much larger than that of CO_2, and at many trenches, the sediments do not contain any significant carbonates. Near-trench sediment columns have molar $CO_2 : H_2O$ ratios mostly below 1 : 3, and a ratio above 1 : 1 was only found along the middle-American to Peruvian margin (Rea and Ruff, 1996; Plank and Langmuir, 1998; compiled in Poli and Schmidt, 2002). If carbonates are

confined to the sediments, they do not alter phase relations in the igneous part of the crust, as fluid flow is generally buoyancy driven away from the igneous crust. However, some carbonate precipitates during hydrothermal alteration of the igneous crust. In these instances, carbonate is mainly found in veins and is much less pervasively distributed than H_2O. For a more detailed understanding of fluid-producing processes it is thus necessary to evaluate the effect of CO_2 on subduction-related metamorphism and fluid transfer, on hydrous phase and carbonate stabilities, on fluid compositions, and how deep carbonates can subduct.

In general, the succession of carbonate minerals with increasing pressure is calcite → dolomite → magnesite. X_{CO_2} in the fluid decreases in this sequence. For low temperature P–T paths, aragonite replaces calcite. Both calculations (Kerrick and Connolly, 1998, 2001a,b) and experiments (Yaxley and Green, 1994; Molina and Poli, 2000) show that at low to intermediate temperatures (up to 800 °C at 4 GPa), a very small amount of CO_2 saturates the system in carbonate and fluids are buffered to compositions with ≤ 10 mol.% CO_2. Carbon dioxide becomes an important species in the fluid (to X_{CO_2} values of 0.4–0.7) only at higher temperatures and comparatively low pressures in coexistence with calcite. Few experiments are available on the maximum pressure stability of carbonates in natural bulk compositions. Domanik and Holloway (2000) found magnesite in a calcareous pelite from 6.5 GPa to 11 GPa. In synthetic systems it is well known that carbonates are very stable at subsolidus conditions: between 5 GPa and 9 GPa dolomite breaks down along a curved reaction line (Luth, 2001) to magnesite and $CaCO_3$ polymorphs, which are stable down to lower mantle pressures (Biellmann *et al.*, 1993).

Carbonates present within the subducting oceanic crust influence dehydration reactions, but contrary to what is intuitively believed, can enlarge the stability field of hydrous phases. This is especially true for melting reactions in mafic and pelitic compositions where a mixed H_2O–CO_2 fluid shifts melting reactions to higher temperatures and thus enlarges the amphibole and mica-stability fields. Carbonate minerals may remain stable in the presence of a siliceous melt and with increasing temperature decompose forming an immiscible carbonatite melt coexisting with a silicate melt.

Molina and Poli (2000) demonstrated the effect of CO_2 on phase relations in mafic compositions (Figure 3). With increasing pressure, the stable carbonates at 665–730 °C are calcite (<1.4 GPa), dolomite, and dolomite + magnesite (>2 GPa). Fluids coexisting with calcite are CO_2 rich ($X_{CO_2} = 0.4$–0.7) but at higher pressures, fluids coexisting with dolomite have much lower CO_2

contents ($X_{CO_2} = 0.02$–0.2). This implies that carbon tends to fractionate into the solid with increasing pressure. The effect of CO_2 on the stability of hydrous phases is surprising: in the carbonated system, plagioclase disappears at the same pressures as in the pure-H_2O system (~1.5 GPa at 650 °C). However, its breakdown does not cause omphacite formation, which is delayed by ~0.5 GPa. Amphibole breaks down at ~2.5–2.6 GPa (Yaxley and Green, 1994; Crottini *et al.*, 2003) and the fluid saturated solidus is located at ~730 °C at 2.2 GPa, as expected by the presence of H_2O-rich fluids at such conditions.

As X_{CO_2} values are low (<0.15) in fluids produced by subduction zone metamorphism, the only efficient decarbonatization processes are either flushing with aqueous fluids from below or melting at relatively high temperatures. A scenario describing decarbonation reactions caused by fluid-infiltration during subduction would be a layer of carbonaceous sediment overlying hydrated oceanic lithosphere. At any time, the metacarbonate would probably contain a very small amount of equilibrated fluid that would have 5–15 mol.% CO_2 at depths beyond 60 km (the exact CO_2-fraction depending on pressure and temperature, Kerrick and Connolly (2001a,b) and Molina and Poli (2000)). Aqueous fluid produced in the serpentinite and basalt below the carbonate layer will migrate upwards, replacing the CO_2-bearing fluid, and then will locally equilibrate, i.e., consume some carbonates in order to increase the CO_2 content in the fluid. If the aqueous fluid passes pervasively through the limestone, the entire carbonate layer will eventually be dissolved and the fluid migrating into the wedge will transport a significant amount of CO_2 over time. For example, 100 m of average siliceous limestone (15.0 wt.% initial CO_2; Plank and Langmuir (1998)) will be completely decarbonated from reaction with aqueous fluids derived from complete dehydration of 150–400 m of serpentinite (ignoring the basaltic crust) or 900–1,600 m of dehydrating MORB (ignoring the peridotite) at depths beyond 60 km, assuming pervasive infiltration and equilibration of the fluid. The entire CO_2 thus derived will flow out of the carbonaceous sediment and into the mantle wedge. However, if the aqueous fluid is efficiently channeled, a very small fraction of the limestone would be affected (depending on channel-spacing) and >90% of the carbonates may survive complete dehydration of the underlying oceanic lithosphere. In this case, these carbonates will be subducted to depths of the transition zone.

18.5.3 Graywackes and Volcaniclastics

A significant component of sedimentary columns at trenches are graywackes and volcaniclastic sediments of andesitic to dacitic composition (Plank and Langmuir, 1998). Phase diagrams of graywackes and andesites (Vielzeuf and Montel, 1994; Schmidt, 1993) suggest that, for our present purpose, these systems can be viewed as being intermediate between pelites and MORB. The significant K_2O contents in graywackes and volcaniclastics of intermediate composition lead to abundant micas (phengite or biotite), and thus reactions and phases similar to those in pelites. However, the CaO contents are significant enough for the formation of amphiboles, thus leading to phases and reactions similar to those in intrinsically CaO-rich MORB. Obviously, compared to MORB and pelites, reactions are shifted in *P*–*T* space due to different phase compositions, and the amounts of phases are highly variable.

18.5.4 Melting of Sediments Compared to Melting of MORB

In hot subduction zones, melting of the downgoing oceanic crust might be achieved. In order to understand which lithologies might melt under what conditions it is necessary to compare melting relations of sediments and MORB under the various fluid-availability conditions. In this context, it should be noted that of all average sediments, a pelagic clay has the highest melt productivity. However, the initial melting temperature of most carbonate free sediments remains virtually identical as long as mica and quartz saturation is maintained. As for MORB, any carbonate addition to sediments results in a mixed H_2O–CO_2 fluid phase and an increase in melting temperatures by at least 100–200 °C. The melting temperatures for MORB and pelitic sediments at water-saturated conditions are similar, the wet MORB solidus being situated typically 20–50 °C above the wet pelite solidus (Figure 5). As pointed out above, fluid-saturated melting only becomes efficient (in terms of melt productivity) if fluid is added from an external source.

In order to achieve flush-melting of the sediments and concomitant dehydration of the MORB layer, a very particular thermal field is necessary: first, the sediments have to pass 650–800 °C at 1.5–3.0 GPa; secondly, the temperature within the lower part of the MORB layer must be 100–200 °C lower than in the sediments in order to provide a significant quantity of fluid through chlorite- and amphibole-decomposition reactions. Within the temperature range where flush melting would occur in the sediments, and considering a *normal* temperature distribution (i.e., temperature decreasing with depth in the oceanic crust), serpentinized peridotite would remain in the chlorite + serpentine stability field and thus not produce any fluid. For the same reason, flush melting of MORB near the wet solidus by fluids derived from underlying serpentinite is highly unlikely in the depth range of interest, as temperatures in the serpentinized peridotite should remain 50–200 °C lower than in the MORB (compare to Figure 1). Thus, conditions necessary for flush melting might be realized in a particular subduction zone; however, flush melting cannot be treated as the general case and is not expected for the fast subduction zones dominant in the Pacific rim. For these, thermal models predict temperatures far too low for any melting in the 1.5–3.0 GPa depth range (Davies and Stevenson, 1992; Kincaid and Sacks, 1997; Gerya et al., 2002).

Fluid-absent melting involving biotite in sediments or amphibole in MORB produces melts with distinct compositions at different P–T conditions (Figure 5), with different pressure dependencies of the melting reactions. The amphibole-melting reaction bends strongly back around 2 GPa and near the solidus, amphibole disappears at 2.4 GPa (Figure 5). In MORB, epidote/zoisite remains stable at the solidus to ~3 GPa, leaving a possibility for minor fluid-absent melting up to these pressures. Any melting at higher pressures depends on the K_2O content of MORB and the related presence of minor phengite. In average metapelites and graywackes, biotite is completely replaced by phengite at 2.5–2.8 GPa (Auzanneau, unpublished experiments, see also Vielzeuf and Schmidt (2001)) but at high Mg/(Mg+Fe) biotite stability extends to ~4 GPa (Hermann, 2002a).

An interesting feature of melting at pressures above 3 GPa arises from the fact that MORB, pelites, and also intermediate andesite or graywacke compositions all contain garnet, cpx, phengite, and coesite (Schmidt, 1996; Okamoto and Maruyama, 1999; Hermann and Green, 2001; Schmidt and Vielzeuf, 2001). Peraluminous graywackes and pelites have kyanite in addition. Thus, all the lithologies of the oceanic crust contain the same assemblage and fluid-saturated melting occurs through the identical reaction: phengite + coesite + cpx + H_2O = melt. If a significant amount of free water is not available, melting occurs through phengite + cpx = garnet + melt ± kyanite. This reaction takes place ~150–200 °C above the wet solidus (at 3 GPa to at least 5 GPa) and leads to 20–30 wt.% melt in the metasediments and to a few percent melt in the mafic rocks (dependent on the bulk K_2O content). In MORB, phengite is immediately consumed upon melting and the temperature must rise by >100 °C to significantly increase melt fractions through the reaction: cpx = garnet + melt (25–30% melt). Nevertheless, temperatures necessary for such fluid-absent, high-pressure melting are high and unrealistic in most subduction zones.

The most likely melting scenarios for subducted oceanic lithosphere strongly depend on the thermal gradient within the subducted crust.

In the depth range of interest, most thermal models (e.g., Davies and Stevenson, 1992; Furukawa, 1993; Kincaid and Sacks, 1997; Gerya et al., 2002) predict a temperature difference of ≤200 °C from top of the sediments to the bottom of the crust (which at the same time is the top of the serpentinized peridotite layer). With such a gradient, significant fluid-saturated sediment flush melting cannot take place due to fluids originating from the directly adjacent MORB layer. However, the temperatures necessary to obtain fluids from the serpentinized peridotite cause the MORB layer (intercalated between sediments and serpentinite) to be at temperatures above the fluid-saturated solidus. Thus it appears likely to achieve either no significant fluid-saturated melting at all, or flush melting from both sediments and MORB. For depths of 50–80 km, achieving the temperatures necessary for fluid-absent melting of sediments and MORB results in a temperature that causes dehydration in the serpentinized peridotite layer. This again would enhance melt productivities through flush melting.

18.6 SERPENTINIZED PERIDOTITE

The serpentinized peridotite layer situated just below the igneous oceanic crust (or often brought to the surface in slow spreading oceans like the Atlantic) constitutes a major H_2O reservoir in subducted lithosphere, of comparative size to the oceanic crust. It is difficult to estimate the amount of serpentinization within this layer. The only certainty is that the degree of serpentinization is highly variable both on a regional and local scale. Hydrothermal systems near the ridge and transform faults of all dimensions are the primary sites of hydration, which is mostly serpentinization. Recently, hydration has also been suggested to occur in extensional faults that run parallel to the trench and are caused by bending of the subducting plate at the onset of subduction (Peacock, 2001). Our best estimate gives an average of 20% serpentinization to a few kilometer depth (Schmidt and Poli, 1998). Although there might be some localized serpentinization along faults much deeper in the oceanic lithosphere, large-scale serpentinization and the resulting low-density peridotite (2.3–2.5 g cm^{-3}) would cause a buoyancy problem during subduction.

Whereas a number of reactions at intermediate to elevated temperatures are important for hydration of peridotite in the overlying mantle wedge directly adjacent to the top of the oceanic crust (Figures 6 and 7), almost any subduction P–T-path will keep the slab in the serpentine stability field to at least >2 GPa (Ulmer and Trommsdorff, 1995). As a consequence, hydrated peridotite in the downgoing lithosphere will remain as serpentine and chlorite (+ olivine + clinopyroxene) while a multitude of reactions is taking place in

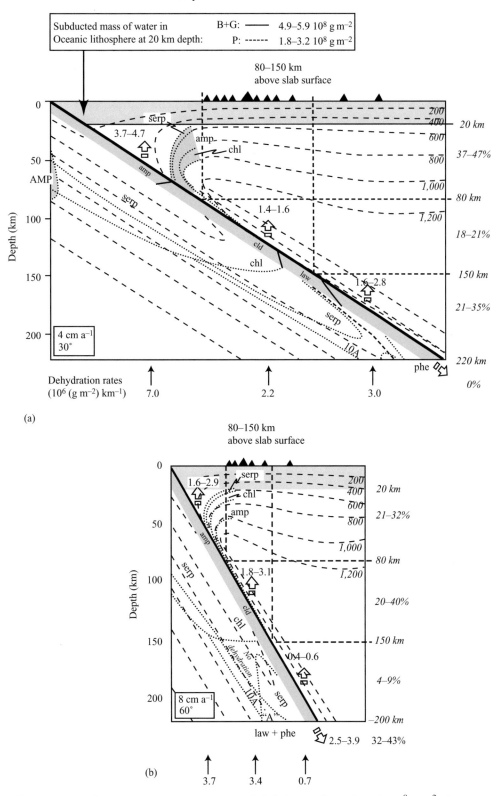

Figure 7 Stabilities of hydrous phases and masses of H_2O in a unit section (in $10^8 \, g \, m^{-2}$) for two typical temperature fields during intra-oceanic subduction (a) for an intermediate burying velocity (4 cm yr^{-1} convergence rate and 30° dip angle of the slab) and (b) a fast and cold subduction zone (8 cm yr^{-1} convergence rate and 60° dip angle; isotherms from Furukawa, 1993). Dark gray: oceanic crust, light gray: lithosphere of the overriding plate in the Furukawa (1993) model. The dotted lines denote maximum stability fields and not effectively hydrated peridotite. In a subduction zone with a fast subduction rate, serpentine and phase A fields overlap, preventing significant

the oceanic crust above. Thus, the serpentinized peridotite layer of the oceanic lithosphere does not produce any significant fluid up to pressures of 3–6 GPa, where, depending on temperature, serpentine breakdown may occur (Figure 6). The H_2O contents of Figure 6 are calculated for average harzburgite. However, oceanic alteration does not only add H_2O, but also removes MgO. Peridotite sitting just above the slab surface experiences alteration by fluids derived from generally quartz-saturated sediments and MORB and thus is likely to become somewhat SiO_2 enriched (Manning, 1994). If the $MgO/(MgO + SiO_2)$ ratio in the peridotite is shifted from lying between olivine and serpentine to a value between serpentine and talc, then talc becomes stable to higher pressures (talc + serpentine reaction; Ulmer and Trommsdorff (1999)). In monomineralic veins, talc might persist to its maximum-pressure stability limit at 4.5–5.0 GPa (Pawley and Wood, 1995).

Subduction zones can be divided into two types: those where the serpentine-out reaction is crossed below 6 GPa (and major dehydration of serpentinized peridotite occurs), and those where the serpentine stability boundary is crossed at >6 GPa. As discussed above, almost no fluid production is expected when serpentine reacts to phase A (see Figures 6 and 7) and the H_2O stored in the hydrated peridotite is expected to subduct deeply. In contrast, if the stabilities of serpentine and phase A do not overlap along a given slab geotherm, a significant flushing zone (over 20–30 km depth) is expected at the transition between serpentine-chlorite   peridotite → chlorite   peridotite → anhydrous garnet peridotite. The width of such a zone is controlled by the actual stability of chlorite, which extends ~100 °C higher than the stability of serpentine in complex systems approaching natural rocks (Fumagalli, 2001).

This picture is somewhat modified when the P–T-path produces the sequence serpentine → 10Å phase → phase A. In this case, a moderate amount (max. 0.8 wt.%) of H_2O subducts to great depth, while >75% of the initially subducted H_2O of the serpentinized peridotite is lost via dehydration.

In a number of arcs (NE Japan–Kuriles–Kamchatka, Aleutians, N. Chile), so-called double seismic zones are observed. Whereas the upper seismic zone correlates with the oceanic crust, it

has been suggested (Seno and Yamanaka, 1996; Peacock, 2001) that the lower seismic zone corresponds to the limit of serpentine stability in the lower part of the oceanic lithosphere (see Figure 7). It has been argued that the lower seismic zone earthquakes are triggered by reactivation of ancient faults through fluid saturation, where the fluids derive from serpentine dehydration.

18.7 IMPLICATIONS FOR TRACE ELEMENTS AND AN INTEGRATED VIEW OF THE OCEANIC LITHOSPHERE

Trace elements and isotopes such as ^{10}Be, B, Li, Ba, Cs, and elements of the U–Th-series provide important information on main element-transfer processes in arcs and are reviewed by Morris. Such studies allow one to distinguish between a sediment and an altered MORB signal in island arc volcanics (Morris *et al.*, 1990), to deduce across-arc variations as a function of slab depth (Ishikawa and Nakamura, 1994; Ryan *et al.*, 1995; Moriguti and Nakamura, 1998), yield time constraints on element transfer (Sigmarsson *et al.*, 1990; see Turner *et al.*, 2003, for review), and allow one to distinguish between dehydration and melting processes (Sigmarsson *et al.*, 1998; Martin, 1999). However, the correct interpretation of their concentrations and spatial distributions rely on major element phase relations and on the understanding of mobile phase production discussed in this chapter.

18.7.1 Mobile Phase Production and Trace-element Transfer

Dehydration, or more generally, devolatilization of the oceanic crust is a process that combines continuous and discontinuous reactions in a variety of heterogeneous bulk compositions. In addition, within a vertical column—the sedimentary, mafic, and serpentinized peridotite layers—each experience a significant thermal gradient. The result is a continuous, but not constant, production of a fluid or melt, with the rate of mobile phase production generally decreasing with depth. Peaks in the volatile flux result from significant discontinuous reactions. However, despite the

---

dehydration of the serpentinized peridotite of the oceanic lithosphere. In such a thermal situation, a thin layer of hydrated peridotite is stable above the oceanic crust and drags some H_2O downwards. Details of the calculations: the input is set to H_2O-amounts at 20 km subduction depth (upper left values), the oceanic lithosphere is composed of 2 km of fully hydrated basalt, 5 km of gabbro hydrated to 20 vol.% and 5 km of serpentinized peridotite with 10–20 vol.% hydrated. Labels on open arrows and percentages at the right side are masses of H_2O (fluxing the mantle wedge) integrated over the depth ranges 20–80 km, 80–150 km, and 150–220 km, and the mass (10^8 g m^{-2} of subducted crust) of H_2O subducting beyond 220 km. The arrows below (a) and (b) give dehydration rates in 10^6 g m^{-2} per (vertical) kilometer of subduction at 50 km, 100 km, and 180 km depth. The highest uncertainty in terms of H_2O masses is in the estimate of initially subducted H_2O in the peridotite of the oceanic lithosphere.

continuous fluid flux, trace elements may not necessarily be released continuously.

All dehydration reactions in oceanic lithosphere take place at temperatures where diffusion rates in most minerals are insignificant compared to the available time span for fluid production in subduction zones. In terms of trace-element partitioning it is thus necessary to distinguish between the mineral mode in a given composition and the reactive volume, which will be much smaller. For example: a MORB at 4 GPa, 700 °C has 48 vol.% garnet, 39 vol.% cpx, 5 vol.% lawsonite, 2 vol.% phengite, and 6 vol.% coesite. The breakdown of lawsonite can be modeled via the reaction lawsonite + clinopyroxene + garnet$_1$ = garnet$_2$ (grossular enriched) + H$_2$O, which produces ~8 vol.% garnet in the rock (Schmidt and Poli, 1998). The garnet that grows from the breakdown of lawsonite is in equilibrium with the fluid. However, diffusion rates at 700 °C for garnet and cpx are so slow that in the available time span (several tens of thousands of years, based on radioactive disequilibrium, see review by Turner *et al.*, 2003) the volumes of the non-reacting garnet and cpx affected by diffusion are negligible (<0.1 vol.%). Thus, the trace elements formerly residing in lawsonite and the 3 vol.% of cpx that decomposed with lawsonite, will be redistributed between the newly formed garnet and fluid. However, >85 vol.% of the rock is not equilibrated and thus should not be included when calculating geochemical residua or the trace-element content of the fluid. As a consequence of slow solid-state diffusion, in most cases the only elements that may be mobilized are those that are hosted in minerals that decompose. It is thus necessary to establish trace-element residence in subduction-zone lithologies (Figure 8).

The following cases may be distinguished:

- A given trace element, for example, boron (Ryan *et al.*, 1995), is extremely soluble in the fluid and any mineral/fluid-partition coefficient is ≪1. Such elements may show a continuously decreasing concentration in the fluid with increasing depth. Concentrations might already be very low in fluids produced at moderate depths due to a shallow effective removal from the subducting crust.
- A given trace element has partition coefficients close to unity and thus its concentration doesn't change in the fluid as subduction progresses.
- A given trace element is strongly partitioned into a particular hydrous phase (e.g., cerium, strontium into lawsonite or epidote; Figure 8) and has only moderate-to-low cpx/fluid and garnet/fluid partition coefficients. These elements will quantitatively enter into the fluid at the breakdown reaction(s) of the given mineral and cause a variation in their

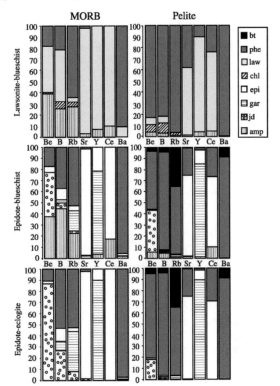

Figure 8 Distribution of Be, B, Rb, Sr, Y, Ce, and Ba between minerals of average MORB and pelite in blueschist and eclogite facies (employing representative mineral modes for natural blueschists and experimental epidote-eclogite, and trace element concentration data mostly from Domanik *et al.*, 1993). At subsolidus temperatures, diffusive equilibration is ineffective (except for micas) and the equilibrating volume that needs to be taken into account for trace-element modeling is defined by the reacting minerals. Thus, a given trace element equilibrates with the fluid only when its host phase(s) break(s) down.

concentrations that is not at all proportional to the fluid-flux.
- A given trace element (e.g., barium, and to some extent beryllium in the sediments) partitions into mica and as mica is dissolved away with increasing depth and temperature, enters into the fluid only at greater depth, its flux at low temperature/pressure being small.
- A given trace element strongly partitions into cpx/garnet and will be returned to replenish the deep-mantle trace-element reservoir.

18.7.2 Integrating Fluid Flux over the Entire Subducted Oceanic Crust: An Example

The layered structure of oceanic lithosphere originating from fast-spreading ridges, combined with temperature gradients within the subducting lithosphere define the relationship between the amount and depth range of fluids and/or melts generated in the sedimentary and mafic layers and fluids generated from the ultramafic layer. The interdependence between sediment and MORB melting and dehydration in serpentinized peridotite was illustrated in Section 18.5.4. Here, an example of the effects on trace-element transfer as a function of fluid : rock ratios is illustrated. The layered structure of the oceanic lithosphere may cause the fluid : rock ratio in the sedimentary layer to be greater than one (as fluids from the underlying mafic and peridotitic layers must rise through the sediments). Thus, some of the trace elements (e.g., beryllium, thorium) commonly considered to be only efficiently mobilized by melts could also be quite effectively mobilized by fluids, if the entire subducted lithosphere is considered.

Recently, an apparent contradiction was put forward to argue for melting of sediments contemporaneously with dehydration of MORB. It was estimated that >30–40% of the subducted beryllium and thorium, which are strongly enriched in sediments, are recycled into the mantle and extracted to the surface via arc volcanism (Johnson and Plank, 1999, and references therein). At the same time, boron, which is strongly enriched in altered MORB, and uranium appear to be effectively recycled into arc magmas by fluids. Based on bulk partition coefficients $^{xtl/fluid}D$ of 2–4.8 and $^{xtl/melt}D$ of 0.7–1.5 for beryllium and thorium, Johnson and Plank (1999) argued that recycling of >30–40% of these elements is only possible when sediments melt while MORB dehydrates (mobilizing boron and uranium). It was then shown by thermal modeling that such a temperature distribution is possible (Van Keken *et al.*, 2002). However, this scenario considers sediments, MORB, and hydrated peridotite as independent systems. Considering that the bulk crystal-fluid partition coefficients for beryllium and thorium are not extremely high, i.e., 2–5, and considering that most of the beryllium partitions into phengite, and that phengite is the only phase where diffusive re-equilibration at sub-solidus temperatures is possible, it follows that an elevated fluid/rock ratio can dissolve >80% of the beryllium (and probably thorium) in the sediments, if pervasive fluid infiltration from the lower layers occurs. Employing the above partition coefficients, the fluid produced from an average thickness MORB layer through dehydration below 3 GPa would be sufficient for leaching 60% of beryllium (and possibly thorium) out of 200 m of pelagic sediments. Thus, sufficient beryllium could be mobilized before typical melting depths are reached. This estimate is conservative because partition coefficients for relatively low-pressure fluids are employed. Solute-rich, high-pressure fluids and continuous dissolution of phengite would greatly facilitate the transfer of beryllium

to the mantle wedge. The trace-element transport capacities of such solute-rich fluids are expected to increase significantly compared to low-pressure, low-density fluids; however, the little experimental data that are available are not enough to suggest this. In fact, significant mobilization of beryllium could simply be taken as evidence that micas dissolve away in a dissolution regime.

18.8 CONCLUSIONS AND OUTLOOK

The complexity of natural processes—i.e., heterogeneous bulk compositions, heterogeneous volatile distribution, equilibrium on different scales, and kinetic effects—may fairly complicate individual subduction zones. Nevertheless, there is no reason for pessimism. The dehydration behavior of the two volumetrically predominant lithologies, basalt and peridotite, are fairly predictable. The resulting mobile component will then have a sediment signal added. This signal can be reasonably well defined when the subducting column of sediments is well known; there is ample geochemical evidence that the efflux in arc magmas is fairly proportional to the influx in terms of subducted sediment component (e.g., Plank and Langmuir, 1993; Morris, 2003; Turner *et al.*, 2003).

The complexity of natural processes in subducting slabs is reflected by the complex distribution of volcanic emissions in subduction zones. First, it should be emphasized that 69% of modern subduction zones on the Earth (40,900 km, unpublished compilation) show active volcanism in the Quaternary. The rest (18,500 km) are not volcanic because of either unfavorable thermomechanical environments (e.g., flat slabs, initiation of subduction, etc.) or possibly the lack of a sufficient amount of volatiles released at depths where melting could take place.

Even though we believe that there is no straightforward relationship between the location of fluid release in the slab and volcanic emissions, the variability of reaction patterns illustrated in this chapter is recorded by variability in the distribution of volcanic arcs on the Earth's surface (Figure 9). The spatial onset of volcanism, the arc, is probably primarily controlled by the thermal structure in the mantle wedge, i.e., when a sufficient thickness and convection intensity in the mantle wedge is reached in order to allow temperatures necessary for the formation of primitive arc magmas (Kushiro, 1987; Schmidt and Poli, 1998). The depth of the slab below the volcanic front is often regarded as a relevant parameter to characterize petrologic processes occurring in subduction zones (Gill, 1981; Tatsumi, 1986). This parameter has often been expressed as single-valued with some sort of standard deviation (128 ± 38 km Gill, 1981; 110 ± 38km Tatsumi,

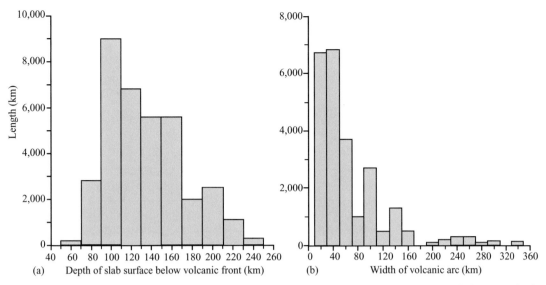

Figure 9 Subduction zone statistics: histograms of depths of (a) the slab surface below the volcanic front and (b) the width of volcanic arcs. The vertical axis denotes arc lengths in km measured at the trench. This is our own compilation (unpublished) based on locations of quaternary volcanoes and slab surfaces from tomography (if available), otherwise earthquake depths.

1986; 108 ± 14 km for the trench side of the chain, 173 ± 12 km for the back-arc side of the chain, Tatsumi and Eggins, 1995) on the basis of a fairly arbitrary *volcano counting* in selected arcs. On the contrary, our compilation is based on recent geophysical acquisitions (slab surface tomography) and on the spatial extent of volcanic activity and shows that, although some maximum is found at ~100 km depth, a continuum in "depth of the slab surface below the volcanic front" is observed (Figure 9(a)). Moreover, despite a majority of volcanic arcs being fairly well focused on the surface (less than 50 km wide), volcanic activity over more than 100 km arc width is not unusual (Figure 9(b)). Such an arc width corresponds to a comparative depth range of the slab surface. In some continental arcs (e.g., Sumatra, the Southern Volcanic Zone in the Andes), a narrow volcanic arc is the result of a position of virtually all volcanoes on a major fault subparallel to the trench. In such a case, it appears likely that the volcano distribution is not directly related to processes taking place below the lithosphere of the overriding plate.

It is beyond the scope of this review to discuss in detail the statistical parameters of subduction zones, and Figure 10 is intended to demonstrate that conditions attained in subducting slabs are highly variable, even for similar convergence parameters, and that interplay between thermomechanical properties and reaction paths are responsible for a complex pattern of fluid release and magma genesis. Furthermore, any correlation of kinematic subduction parameters with volcano location tacitly assumes steady state, which is not necessarily the case.

At present, the resolution of thermal models for the oceanic crust with its large temperature gradient is not sufficient. This is because the thermal field strongly depends on the flow field (and on the degree of mechanical coupling between the subducting slab and the down-dragged mantle wedge), which in turn depends on the viscosity, which is a function of P and T and the materials present. The latter then depend on the chemical reactions taking place, which in turn depend on the temperature field. A temperature distribution model is thus a fairly complex problem and needs input from a large variety of disciplines.

An example of feedback between temperature field and phase petrology is serpentinization in the mantle wedge directly overlying the subducted slab (Gerya *et al.*, 2002). First, the calculated temperature distribution from thermomechanical models results in a pressure–temperature region where serpentine would be potentially stable. Second, dehydration rates and fluid transport mechanisms (pervasive versus channeled flow) allow one to model the amount of serpentine formed. However, the serpentinized peridotite has four to six orders of magnitude lower viscosity than dry peridotite. This strongly influences the coupling of the downgoing slab and the convection of the mantle wedge. It also influences the possible amount of shear heating at the slab surface. If convection patterns change due to a modified rheology caused by phase transformations, the temperature field changes and the permissible region for serpentine changes. We thus need more complex models in which these parameters are varied simultaneously.

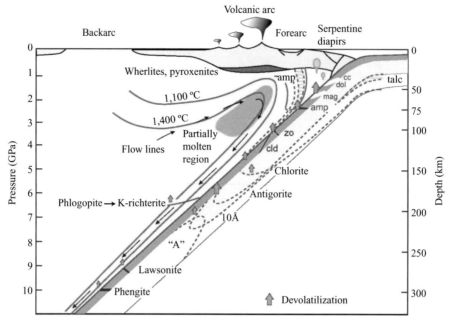

Figure 10 Cartoon of events in a typical cold subduction zone in which melting of the oceanic crust does not occur (after Poli and Schmidt, 2002).

Upwards directed interaction between the different lithologies largely depends on the temperature gradient within the oceanic lithosphere, which is among the most difficult parameters to model. Thus, at present, a purely forward model for mass transfer in a given subduction zone is not reliable, geochemical information on the subduction output is necessary to constrain likely mass-transfer processes.

REFERENCES

Allegre C. J. and Turcotte D. L. (1986) Implications of a 2-component marble-cake mantle. *Nature* **323**, 123–127.

Angel R. J., Frost D. J., Ross N. L., and Hemley R. (2001) Stabilities and equations of state of dense hydrous magnesium silicates. *Phys. Earth Planet. Inter.* **127**, 181–196.

Austrheim H. and Engvik A. (1997) Fluid transport, deformation and metamorphism at depth in a collision zone. In *Fluid Flow and Transport in Rocks* (eds. B. Jamtveit and B. W. D. Yardley). Chapman and Hall, London, pp. 123–135.

Biellmann C., Gillet P., Guyot F., Peyronneau J., and Reynard B. (1993) Experimental evidence for carbonate stability in the Earth's lower mantle. *Earth Planet. Sci. Lett.* **118**, 31–41.

Boettcher A. L. and Wyllie P. J. (1969) The system CaO–SiO₂–CO₂–H₂O: III. Second critical end-point on the melting curve. *Geochim. Cosmochim. Acta* **33**, 611–632.

Brown E. H. and Forbes R. B. (1986) Geology of the shuksan suite, North Cascades, Washington, USA. In *Memoir 164, Blueschists and Eclogites* (eds. B. W. Evans and E. H. Brown). Geological Society of America, , pp. 143–167.

Cannat M. (1993) Emplacement of mantle rocks in the seafloor at mid-ocean ridges. *J. Geophys. Res.* **98**, 4163–4172.

Carlson W. D. (2002) Scales of disequilibrium and rates of equilibration during metamorphism. *Am. Mineral.* **87**, 185–204.

Chinner G. A. and Dixon J. E. (1973) Some high-pressure paragenesis of the allalin gabbro, Valais, Switzerland. *J. Petrol.* **14**, 185–202.

Constantin M. (1999) Gabbroic intrusions and magmatic metasomatism in harzburgites from the Garrett transform fault: implications for the nature of the mantle–crust transition at fast spreading ridges. *Contrib. Min. Petrol.* **136**, 111–130.

Crottini A., Poli S., and Molina J. F. (2003) Carbon cycle in subduction zones: experimental constraints in fluid-saturated MORB eclogites. EGS–AGU–EUG joint assembly, Nice, EAE03-A-12638.

Davies H. J. and Stevenson D. J. (1992) Physical model of source region of subduction zone volcanics. *J. Geophy. Res.* **97**, 2037–2070.

Dixon J. E., Leist L., Langmuir C., and Schilling J. G. (2002) Recycled dehydrated lithosphere observed in plume-influenced mid-ocean-ridge basalt. *Nature* **420**, 385–389.

Domanik K. J. and Holloway J. R. (1996) The stability and composition of phengitic muscovite and associated phases from 5.5 to 11 GPa: implications for deeply subducted sediments. *Geochim. Cosmochim. Acta* **60**, 4133–4150.

Domanik K. J. and Holloway J. R. (2000) Experimental synthesis and phase relations of phengitic muscovite from 6.5 to 11 GPa in a calcareous metapelite from the Dabie mountains. *China. Lithos* **52**, 51–77.

Domanik K. J., Hervig R. L., and Peacock S. M. (1993) Beryllium and boron in subduction zone minerals: an ion microprobe study. *Geochim. Cosmochim. Acta* **57**, 4997–5010.

Drummond M. S., Defant M. J., and Kepezhinskas P. K. (1996) Petrogenesis of slab-derived trondhjemite-tonalite-dacite/adakite magmas. *Trans. Roy. Soc. Edinburgh: Earth Sci.* **87**, 205–215.

Forneris J. and Holloway J. R. (2001) Experimental determination of chloritoid stability in subducting oceanic crust. *EOS, Trans., AGU* **82**(47) (Fall Meet. suppl., abstr. T22D-09).

Frost D. J. (1999) The stability of dense hydrous magnesium silicates in earth's transition zone and lower mantle. In *Special Publication No. 6, Mantle Petrology: Field*

Observations and High-pressure Experimentation (eds. Y. Fei, C. M. Bertka, and B. O. Mysen). Geochemical Society, pp. 283–296.

Fryer P., Wheat C. G., and Mottl M. J. (1999) Mariana blueschist mud volcanism: implications for conditions within the subduction zone. *Geology* **27**, 103–106.

Fumagalli P. (2001) Processi di trasporto e rilascio di H_2O nelle zone di subduzione: uno studio sperimentale su sistemi ultrafemici ad alta pressione. PhD Thesis, University Milano, Milano, 178pp.

Fumagalli P. and Poli S. (1999) Phase relationships in hydrous peridotites at high pressure: preliminary results of multianvil experiments. *Periodico di Mineralogia* **68**, 275–286.

Fumagalli P., Stixrude L., Poli S., and Snyder D. (2001) The 10Å phase: a high-pressure expandable sheet silicate stable during subduction of hydrated lithosphere. *Earth Planet. Sci. Lett.* **186**, 125–141.

Furukawa Y. (1993) Magmatic processes under arcs and formation of the volcanic front. *J. Geophy. Res.* **98**, 8309–8319.

Gente P., Pockalny R. A., Durand C., Deplus C., Maia M., Ceuleneer G., Mevel C., Cannat M., and Laverne C. (1995) Characteristics and evolution of the segmentation of the Mid-Atlantic ridge between 20° N and 24° N during the last 10 million years. *Earth Planet Sci. Lett.* **129**, 55–71.

Gerya T. V., Stöckhert B., and Perchuk A. L. (2002) Exhumation of high-pressure metamorphic rocks in a subduction channel: a numerical simulation. *Tectonics* **21**(6), 1056, doi:10.1029/2002TC001406.

Gieskes J. M., Vrolijk P., and Blanc G. (1990) Hydrogeochemistry of the northern Barbados accretionary complex transect: ocean drilling project leg 110. *J. Geophys. Res.* **95**, 8809–8818.

Gill J. (1981) *Orogenic Andesites and Plate Tectonics.* Springer, New York, 390pp.

Gillis K. M. and Robinson P. T. (1990) Patterns and processes of alteration in the lavas and dykes of the troodos ophiolite, Cyprus. *J. Geophy. Res.* **95**, 21523–21548.

Helmstaedt H. and Schulze D. J. (1988) Eclogite-facies ultramafic xenoliths from Colorado Plateau diatreme breccias: comparison with eclogites in crustal environments, evalation of the subduction hypothesis, and implications for eclogite xenoliths from diamondiferous kimberlites. In *Eclogite and Eclogite Facies Rocks* (ed. D. C. Smith). Elsevier, New York, pp. 387–450.

Hermann J. (2002a) Experimental constraints on phase relations in subduction continental crust. *Contrib. Min. Petrol.* **143**, 219–235.

Hermann J. (2002b) Allanite: thorium and light rare earth element carrier in subducted crust. *Chem. Geol.* **192**, 289–306.

Hermann J. and Green D. H. (2001) Experimental constraints on high pressure melting in subducted crust. *Earth Planet. Sci. Lett.* **188**, 149–168.

Ishikawa T. and Nakamura E. (1994) Origin of the slab component in arc lavas from across-arc variation of B and Pb isotopes. *Nature* **370**, 205–208.

Johnson M. C. and Plank T. (1999) Dehydration and melting experiments constraints the fate of subducted sediments. *Geochem. Geophys. Geosys. Electron. J.* **1**, paper no. 1999GC000014.

Katayama I. and Nakashima S. (2003) Hydroxyl in clinopyroxene from the deep subducted crust: evidence for H_2O transport into the mantle. *Am. Mineral.* **88**, 229–234.

Kawamoto T., Leinenweber K., Hervig R. L., and Holloway J. R. (1995) Stability of hydrous phases in an H_2O-saturated KLB-1 peridotite up to 15 GPa. In *Volatiles in the Earth and Solar System*, pp. 229–239.

Kay R. W. (1978) Aleutian magnesian andesites: melts from subducted pacific ocean crust. *J. Volcanol. Geotherm. Res.* **4**, 117–132.

Kelley D. S., Karson J. A., Blackman D. K., Früh-Green G., Butterfield D. A., Lilley M. D., Olson E. J., Schrenk M. O.,

Roe K. K., Lebon G. T., and Rivizzigno P. (2001) An off-axis hydrothermal vent field near the Mid-Atlantic ridge at 30° N. *Nature* **412**, 145–149.

Kennedy G. C., Wasserburg G. J., Heard H. C., and Newton R. C. (1962) The upper three-phase region in the system SiO_2–H_2O. *Am. J. Sci.* **260**, 501–521.

Kerrick D. M. and Connolly J. A. D. (1998) Subduction of ophicarbonates and recycling of CO_2 and H_2O. *Geology* **26**, 375–378.

Kerrick D. M. and Connolly J. A. D. (2001a) Metamorphic devolatilization of subducted oceanic metabasalts: implications for seismicity, arc magmatism and volatile recycling. *Earth Planet. Sci. Lett.* **189**, 19–29.

Kerrick D. M. and Connolly J. A. D. (2001b) Metamorphic devolatilization of subducted marine sediments and the transport of volatiles into the Earth's mantle. *Nature* **411**, 293–296.

Kincaid C. and Sacks I. S. (1997) Thermal and dynamical evolution of the upper mantle in subduction zones. *J. Geophys. Res.* **102**, 12295–12315.

Kohlstedt D. L., Keppler H., and Rubie D. C. (1996) Solubility of water in the α, β, and γ phases of $(Mg, Fe)_2SiO_4$. *Contrib. Mineral. Petrol.* **123**, 345–357.

Konzett J. and Ulmer P. (1999) The stability of hydrous potassic phases in lherzolite mantle; an experimental study to 9.5 GPa in simplified and natural bulk compositions. *J. Petrol.* **40**, 629–652.

Kushiro I. (1987) A petrological model for the mantle wedge and lower crust in the Japanese island arcs. In *Magmatic Processes: Physicochemical Principle*, Geochemical Society Special Publication 1 (ed. B. O. Mysen), pp. 165–181.

Liou J. G., Zhang R. Y., Ernst W. G., Rumble D., III, and Maruyama S. (1998) High pressure minerals from deeply subducted metamorphic rocks. In *Ultrahigh-pressure Mineralogy: Physics and Chemistry of the Earth's Deep Interior/ Reviews in Mineralogy* (ed. R. J. Hemley). The Mineralogical Society of America, Washington, vol. 37, pp. 33–96.

Luth R. W. (2001) Experimental determination of the reaction aragonite + magnesite = dolomite at 5 to 9 GPa. *Contrib. Mineral. Petrol.* **141**, 222–232.

Mammerickx J. (1989) Large scale undersea features on the north-east pacific. In *The Eastern Pacific Ocean and Hawaii, The Geology of North America* (eds. E. L. Winterer, D. M. Hussong, and R. W. Decker). Geological Society of America, , pp. 5–13.

Manning C. E. (1994) The solubility of quartz in H_2O in the lower crust and upper mantle. *Geochim. Cosmochim Acta* **58**, 4831–4839.

Martin H. (1999) Adakitic magmas: modern analogues of archaean granitoids. *Lithos* **46**, 411–429.

Maruyama S., Cho M., and Liou J. G. (1986) Experimental investigations of blueschist-greenschist transition equilibria: pressure dependence of Al_2O_3 contents in sodic amphiboles—a new geobarometer. In *Memoir 164, Blueschists and Eclogites* (eds. B. W. Evans and E. H. Brown). Geological Society of America, , pp. 1–16.

Molina J. F. and Poli St. (2000) Carbonate stability and fluid composition in subducted oceanic crust: an experimental study on H_2O–CO_2-bearing basalts. *Earth Planet. Sci. Lett.* **176**, 295–310.

Molina J. F., Austrheim H., Glodny J., and Rusin A. (2002) The eclogites of the Marun-Keu complex, Polar Urals (Russia): fluid control on reaction kinetics and metasomatism during high P metamorphism. *Lithos* **61**, 55–78.

Moriguti T. and Nakamura E. (1998) Across-arc variation of Li isotopes in lavas and implications for crust/mantle recycling at subduction zones. *Earth Planet. Sci. Lett.* **163**, 167–174.

Morris J. D., Leeman W. P., and Tera F. (1990) The subducted component in island arc lavas: constraints from Be isotopes and B–Be systematics. *Nature* **344**, 31–36.

Nicolas A. (1989) *Structures of Ophiolites and Dynamics of Oceanic Lithosphere.* Kluwer, Dordrecht.

Niida K. and Green D. H. (1999) Stability and chemical composition of paragasitic amphibole in MORB pyrolite under upper mantle conditions. *Contrib. Mineral. Petrol.* **135**, 18–40.

Ohtani E., Mizobata H., and Yurimoto H. (2000) Stability of dense hydrous magnesium silicate phases in the systems Mg_2SiO_4–H_2O and $MgSiO_3$–H_2O at pressures up to 27 GPa. *Phys. Chem. Mineral.* **27**, 533–544.

Okamoto K. and Maruyama S. (1998) Multi-anvil re-equilibration experiments of a Dabie Shan ultrahigh pressure eclogite within the diamond-stability fields. *The Island Arc* **7**, 52–69.

Okamoto K. and Maruyama S. (1999) The high-pressure synthesis of lawsonite in the MORB + H_2O system. *Am. Mineral.* **84**, 362–373.

Okay A. I. (1980) Mineralogy, petrology, and phase relations of glaucophane-lawsonite zone blueschists from the Tavsanh region, northwest Turkey. *Contrib. Mineral. Petrol.* **72**, 243–255.

Ono S. (1998) Stability limits of hydrous minerals in sediment and mid-ocean ridge basalt compositions: implications for water transport in subduction zones. *J. Geophy. Res.* **103**, 18253–18267.

Pawley A. R. and Holloway J. R. (1993) Water sources for subduction zone volcanism: new experimental constraints. *Science* **260**, 664–667.

Pawley A. R. and Wood B. J. (1995) The high-pressure stability of talc and 10 Å phase: potential storage sites for H_2O in subduction zones. *Am. Mineral.* **80**, 998–1003.

Peacock S. M. (1993) The importance of blueschist to eclogite dehydration reactions in subducting oceanic crust. *Geol. Soc. Am. Bull.* **105**, 684–694.

Peacock S. (2001) Are the lower planes of double seismic zones caused by serpentine dehydration in subducting oceanic mantle? *Geology* **29**, 299–302.

Philippot P. (1993) Fluid-melt-rock interaction in mafic eclogites and coesite bearing sediments: constraints on volatile recycling during subduction. *Chem. Geol.* **108**, 93–112.

Plank T. and Langmuir C. H. (1993) Tracing trace elements from sediment input to volcanic output at subduction zones. *Nature* **362**, 739–743.

Plank T. and Langmuir C. H. (1998) The chemical composition of subducting sediment and its consequences for the crust and mantle. *Chem. Geol.* **145**, 325–394.

Poli S. (1993) The amphibolite-eclogite transformation: an experimental study on basalt. *Am. J. Sci.* **293**, 1061–1107.

Poli S. and Schmidt M. W. (1995) H_2O transport and release in subduction zones-experimental constraints on basaltic and andesitic systems. *J. Geophys. Res.* **100**, 22299–22314.

Poli S. and Schmidt M. W. (1998) The high-pressure stability of zoisite and phase relationships of zoisite-bearing assemblages. *Contrib. Mineral. Petrol.* **130**, 162–175.

Poli S. and Schmidt M. W. (2002) Petrology of subducted slabs. *Ann. Rev. Earth Planet. Sci.* **30**, 207–235.

Rea D. K. and Ruff L. J. (1996) Composition and mass flux of sediment entering the world's subduction zones: implications for global sediment budgets, great earthquakes, and volcanism. *Earth Planet. Sci. Lett.* **140**, 1–12.

Ricci J. E. (1951) *The Phase Rule and Heterogeneous Equilibrium.* Dover Publications, New York, 504pp.

Ryan J. G., Morris J., Tera F., Leeman W. P., and Tsvetkov A. (1995) Cross-arc geochemical variations in the kurile arc as a function of slab depth. *Science* **270**, 625–627.

Schmeling H., Monz R., and Rubie D. C. (1999) The influence of olivine metastability on the dynamics of subduction. *Earth Planet. Sci. Lett.* **165**, 55–66.

Schmidt M. W. (1993) Phase relations and compositions in tonalite as a function of pressure: an experimental study at 650 °C. *Am. J. Sci.* **293**, 1011–1060.

Schmidt M. W. (1996) Experimental constraints of recycling of potassium from subducted oceanic crust. *Science* **272**, 1927–1930.

Schmidt M. W. and Poli S. (1998) Experimentally based water budgets for dehydrating slabs and consequences for arc magma generation. *Earth Planet. Sci. Lett.* **163**, 361–379.

Schmidt M. W. and Vielzeuf D. (2001) How to generate a mobile component in subducting crust: melting vs. dissolution processes. In *11th Ann. V. M. Goldschmidt Conf.*, Abstr. 3366. LPI Contrib. No. 1088, Lunar Planet. Inst., Houston (CD-ROM).

Seno T. and Yamanaka Y. (1996) Double seismic zones, compressional deep trench-outer rise events, and superplumes. In *Subduction; Top to Bottom*, AGU Geophysics Monograph 96 (eds. E. Bebout, D. W. Schol, S. H. Kirby, and J. P. Blatt). AGU, pp. 347–355.

Sigmarsson O., Condomines M., Morris J. D., and Harmon R. S. (1990) Uranium and [10]Be enrichments by fluids in the Andean arc magmas. *Nature* **346**, 163–165.

Sigmarsson O., Martin H., and Knowles J. (1998) Melting of a subducting oceanic crust from U–Th disequilibria in austral Andean lavas. *Nature* **394**, 566–569.

Snow J. E. and Dick H. J. B. (1995) Pervasive magnesium loss by marine weathering of peridotite. *Geochim. Cosmochim. Acta* **59**, 4219–4235.

Sorensen S. S. (1986) Petrologic and geochemical comparison of the blueschist and greenschist units of the catalina schist terrane, southern California. In *Memoir 164, Blueschists and Eclogites* (eds. B. W. Evans and E. H. Brown). Geological Society of America, , pp. 59–75.

Stalder R., Ulmer P., Thompson A. B., and Günther D. (2000) Experimental approach to constrain second critical end points in fluid/silicate systems: near-solidus fluids and melts in the system albite-H_2O. *Am. Mineral.* **85**, 68–77.

Stalder R., Ulmer P., Thompson A. B., and Gunther D. (2001) High pressure fluids in the system MgO–SiO_2–H_2O under upper mantle conditions. *Contrib. Mineral. Petrol.* **140**, 607–618.

Sudo A. and Tatsumi Y. (1990) Phlogopite and K-amphibole in the upper mantle: implication for magma genesis in subduction zones. *Geophy. Res. Lett.* **17**, 29–32.

Tatsumi Y. (1986) Formation of the volcanic front in subduction zones. *Geophys. Res. Lett.* **13**, 717–720.

Tatsumi Y. and Eggins St. (1995) Subduction zone magmatism. In *Frontiers in Earth Sciences*. Blackwell, Cambridge, 211pp.

Thurston S. P. (1985) Structure, petrology, and metamorphic history of the nome group blueschist terrane, Salmon Lake area, Seward Peninsula, Alaska. *Geol. Soc. Am. Bull.* **96**, 600–617.

Trønnes R. G. (2002) Stability range and decomposition of potassic richterite and phlogopite end members at 5–15 GPa. *Mineral. Petrol.* **74**, 129–148.

Turner S., Bourdon B., and Gill J. (2003) Insights into magma genesis at convergent margins from U-series isotopes. *Rev. Mineral. Geochem.* **52**, 255–315.

Ulmer P. and Trommsdorff V. (1995) Serpentine stability to mantle depths and subduction-related magmatism. *Science* **268**, 858–861.

Ulmer P. and Trommsdorff V. (1999) Phase relations of hydrous mantle subducting to 300 km. In *Mantle Petrology: Field Observations and High Pressure Experimentation: A Tribute to Francis R. (Joe) Boyd.* (eds. Y. W. Fei, C. Bertka, and B. O. Mysen), Geochemical Society Special Publication No. 6, pp. 259–281.

Usui T., Nakamura E., Kobayashi K., Maruyama S., and Helmstaedt H. (2003) Fate of the subducted Farallon plate inferred from eclogite xenoliths in the Colorado Plateau. *Geology* **31**(7), 589–592.

Van Keken P. E., Kiefer B., and Peacock S. (2002) High-resolution models of subduction zones: implications for mineral dehydration reactions and the transport of water into the deep mantle. *Geochem. Geophys. Geosys. Electron. J.* **3**(10), 1056, doi: 10.1029/2001GC000256.

Vielzeuf D. and Montel J. M. (1994) Partial melting of Al-metagraywackes: Part I. Fluid-absent experiments and phase relationships. *Contrib. Mineral. Petrol.* **117**, 375–393.

Vielzeuf D. and Schmidt M. W. (2001) Melting relations in hydrous systems revisited: application to metapelites, metagreywackes, and metabasalts. *Contrib. Mineral. Petrol.* **141**, 251–267.

Widmer T. and Thompson A. B. (2001) Local origin of high pressure vein material in eclogite facies rocks of the zermatt-saas zone, Switzerland. *Am. J. Sci.* **301**, 627–656.

Yaxley G. M. and Green D. H. (1994) Experimental demonstration of refractory carbonate-bearing eclogite and siliceous melt in the subduction regime. *Earth Planet. Sci. Lett.* **128**, 313–325.

Zack T., Rivers T., and Foley S. F. (2001) Cs–Rb–Ba systematics in phenite and amphibole: an assesment of fluid mobility at 2.0 GPa in eclogites from Tescolmen (Central Alps). *Contrib. Mineral. Petrol.* **140**, 651–669.

APPENDIX 1. Periodic Table of the Elements.

Atomic number
Element symbol
Atomic mass

1 H 1.00794																	2 He 4.00260
3 Li 6.941	4 Be 9.01218											5 B 10.811	6 C 12.011	7 N 14.0067	8 O 15.9994	9 F 18.9984	10 Ne 20.1797
11 Na 22.9898	12 Mg 24.3050											13 Al 26.9815	14 Si 28.0855	15 P 30.9738	16 S 32.066	17 Cl 35.4527	18 Ar 39.948
19 K 39.0983	20 Ca 40.078	21 Sc 44.9559	22 Ti 47.88	23 V 50.9415	24 Cr 51.9961	25 Mn 54.9380	26 Fe 55.847	27 Co 58.9332	28 Ni 58.69	29 Cu 63.546	30 Zn 65.39	31 Ga 69.723	32 Ge 72.61	33 As 74.9216	34 Se 78.96	35 Br 79.904	36 Kr 83.80
37 Rb 85.4678	38 Sr 87.62	39 Y 88.9059	40 Zr 91.224	41 Nb 92.9064	42 Mo 95.94	43 Tc (98)	44 Ru 101.07	45 Rh 102.906	46 Pd 106.42	47 Ag 107.868	48 Cd 112.411	49 In 114.82	50 Sn 118.710	51 Sb 121.75	52 Te 127.60	53 I 126.905	54 Xe 131.29
55 Cs 132.905	56 Ba 137.327	57 La 138.906 ★	72 Hf 178.49	73 Ta 180.948	74 W 183.85	75 Re 186.207	76 Os 190.2	77 Ir 192.22	78 Pt 195.08	79 Au 196.967	80 Hg 200.59	81 Tl 204.383	82 Pb 207.2	83 Bi 208.980	84 Po (209)	85 At (210)	86 Rn (222)
87 Fr (223)	88 Ra 226.025	89 Ac 227.028 ◀	104 (261)	105 (262)	106 (263)	107 (262)	108 (265)	109 (267)									

★ Lanthanides

58 Ce 140.115	59 Pr 140.908	60 Nd 144.24	61 Pm (145)	62 Sm 150.36	63 Eu 151.965	64 Gd 157.25	65 Tb 158.925	66 Dy 162.50	67 Ho 164.930	68 Er 167.26	69 Tm 168.934	70 Yb 173.04	71 Lu 174.967

◀ Actinides

90 Th 232.038	91 Pa 231.036	92 U 238.029	93 Np 237.048	94 Pu (244)	95 Am (243)	96 Cm (247)	97 Bk (247)	98 Cf (251)	99 Es (252)	100 Fm (257)	101 Md (258)	102 No (259)	103 Lr (260)

APPENDIX 2.   Table of Isotopes[a].

A	Element	Abundance/half-life	Source
1	H	99.985%	
2	H	0.015%	
3	H	12.33 yr	C, B
3	He	0.000137%	
4	He	99.999863%	
6	Li	7.5%	
7	Li	92.5%	
7	Be	53.12 d	C
9	Be	100%	
10	Be	1.51e+6 yr	C
10	B	19.9%	
11	B	80.1%	
12	C	98.90%	
13	C	1.10%	
14	C	5,730 yr	C, B
14	N	99.634%	
15	N	0.366%	
16	O	99.762%	
17	O	0.038%	
18	O	0.200%	
19	F	100%	
20	Ne	90.48%	
21	Ne	0.27%	
22	Ne	9.25%	
22	Na	2.6019 yr	C
23	Na	100%	
24	Mg	78.99%	
25	Mg	10.00%	
26	Mg	11.01%	
26	Al	7.17e+5 yr	C
27	Al	100%	
28	Si	92.23%	
29	Si	4.67%	
30	Si	3.10%	
32	Si	150 yr	C
31	P	100%	
32	P	14.262 d	C
33	P	25.34 d	C
32	S	95.02%	
33	S	0.75%	
34	S	4.21%	
35	S	87.32 d	C
36	S	0.02%	
35	Cl	75.77%	
36	Cl	3.01e+5 yr	C, B
37	Cl	24.23%	
36	Ar	0.337%	
37	Ar	35.04 d	
38	Ar	0.063%	
39	Ar	269 yr	C
40	Ar	99.600%	

APPENDIX 2.   (Continued).

A	Element	Abundance/half-life	Source
39	K	93.2581%	
40	K	1.277e+9 yr	
		0.0117%	
41	K	6.7302%	
40	Ca	96.941%	
41	Ca	1.03e+5 yr	C
42	Ca	0.647%	
43	Ca	0.135%	
44	Ca	2.086%	
46	Ca	0.004%	
48	Ca	6e+18 yr	
		0.187%	
45	Sc	100%	
46	Ti	8.0%	
47	Ti	7.3%	
48	Ti	73.8%	
49	Ti	5.5%	
50	Ti	5.4%	
50	V	1.4e+17 yr	
		0.250%	
51	V	99.750%	
50	Cr	1.8e+17 yr	
		4.345%	
51	Cr	27.7025 d	B
52	Cr	83.789%	
53	Cr	9.501%	
54	Cr	2.365%	
53	Mn	3.74e+6 yr	E
54	Mn	312.3 d	C
55	Mn	100%	
54	Fe	5.8%	
56	Fe	91.72%	
57	Fe	2.2%	
58	Fe	0.28%	
60	Fe	1.5e+6 yr	E
59	Co	100	
60	Co	5.2714 yr	B
58	Ni	68.077%	
59	Ni	7.6e+4 yr	C
60	Ni	26.223%	
61	Ni	1.140%	
62	Ni	3.634%	
63	Cu	69.17%	
65	Cu	30.83%	
64	Zn	48.6%	
65	Zn	244.26 d	B
66	Zn	27.9%	
67	Zn	4.1%	
68	Zn	18.8%	
69	Ga	60.108%	
71	Ga	39.892%	
70	Ge	21.23%	
72	Ge	27.66%	
73	Ge	7.73%	

(Continued)

(Continued)

APPENDIX 2.   (Continued).

A	Element	Abundance/half-life	Source
74	Ge	35.94%	
76	Ge	7.44%	
75	As	100%	
74	Se	0.89%	
76	Se	9.36%	
77	Se	7.63%	
78	Se	23.78%	
79	Br	50.69%	
81	Br	49.31%	
78	Kr	0.35%	
80	Kr	2.25%	
81	Kr	2.29e+5 yr	C
82	Kr	11.6%	
83	Kr	11.5%	
84	Kr	57.0%	
85	Kr	10.756 yr	B
85	Rb	72.165%	
87	Rb	4.75e+10 yr	
		27.835%	
84	Sr	0.56%	
86	Sr	9.86%	
87	Sr	7.00%	
88	Sr	82.58%	
90	Sr	28.79 yr	B
89	Y	100%	
90	Zr	51.45%	
91	Zr	11.22%	
92	Zr	17.15%	
93	Zr	1.53e+6 yr	
94	Zr	17.38%	
96	Zr	3.8e+19 yr	
		2.80 2%	
93	Nb	100%	
92	Mo	14.84%	
94	Mo	9.25%	
95	Mo	15.92%	
96	Mo	16.68%	
97	Mo	9.55%	
98	Mo	24.13%	
100	Mo	1.00e+19 yr	
		9.63%	
99	Tc	2.111e+5 yr	E
98	Ru	1.88%	
99	Ru	12.7%	
100	Ru	12.6%	
101	Ru	17.0%	
102	Ru	31.6%	
104	Ru	18.7%	
103	Rh	100%	
102	Pd	1.02%	
104	Pd	11.14%	
105	Pd	22.33%	
106	Pd	27.33%	

APPENDIX 2.   (Continued).

A	Element	Abundance/half-life	Source
107	Pd	6.5e+6 yr	E
108	Pd	26.46%	
110	Pd	11.72%	
107	Ag	51.839%	
109	Ag	48.161%	
106	Cd	1.25%	
108	Cd	0.89%	
110	Cd	12.49%	
111	Cd	12.80%	
112	Cd	24.13%	
113	Cd	7.7e+15 yr	
		12.22%	
114	Cd	28.73%	
113	In	4.3%	
115	In	4.41e+14 yr	
		95.7%	
112	Sn	0.97%	
114	Sn	0.65%	
115	Sn	0.34%	
116	Sn	14.53%	
117	Sn	7.68%	
118	Sn	24.23%	
119	Sn	8.59%	
120	Sn	32.59%	
122	Sn	4.63%	
124	Sn	5.79%	
121	Sb	57.36%	
123	Sb	42.64%	
120	Te	0.096%	
122	Te	2.603%	
123	Te	1e+13 yr	
		0.908%	
124	Te	4.816%	
125	Te	7.139%	
126	Te	18.95%	
128	Te	2.2e+24 yr	
		31.69%	
130	Te	7.9e+20 yr	
		33.80%	
127	I	100%	
129	I	1.57e+7 yr	E,C,B
124	Xe	1.6e+14 yr	
		0.10%	
126	Xe	0.09%	
128	Xe	1.91%	
129	Xe	26.4%	
130	Xe	4.1%	
131	Xe	21.2%	
132	Xe	26.9%	
134	Xe	10.4%	
136	Xe	2.36e+21 yr	
		8.9%	
133	Cs	100%	
134	Cs	2.0648 yr	B

(Continued)

(Continued)

APPENDIX 2. (Continued).

A	Element	Abundance/half-life	Source
137	Cs	30.07 yr	B
130	Ba	0.106%	
132	Ba	0.101%	
134	Ba	2.417%	
135	Ba	6.592%	
136	Ba	7.854%	
137	Ba	11.23%	
138	Ba	71.70%	
138	La	1.05e+11 yr 0.0902%	
139	La	99.9098%	
138	Ce	0.25%	
140	Ce	88.48%	
142	Ce	5e+16 yr 11.08%	
141	Pr	100%	
142	Nd	27.13%	
143	Nd	12.18%	
144	Nd	2.29e+15 yr 23.80%	
145	Nd	8.30%	
146	Nd	17.19%	
148	Nd	5.76%	
150	Nd	1.1e+19 yr 5.64%	
–	Pm	no stable or long-lived isotope	
144	Sm	3.1%	
146	Sm	1.03e+8 yr	E
147	Sm	1.06e+11 yr 15.0%	
148	Sm	7e+15 yr 11.3%	
149	Sm	2e+15 yr 13.8%	
150	Sm	7.4%	
152	Sm	26.7%	
154	Sm	22.7%	
151	Eu	47.8%	
153	Eu	52.2%	
152	Gd	1.08e+14 yr 0.20%	
154	Gd	2.18%	
155	Gd	14.80%	
156	Gd	20.47%	
157	Gd	15.65%	
158	Gd	24.84%	
160	Gd	21.86%	
159	Tb	100%	
156	Dy	0.06%	
158	Dy	0.10%	
160	Dy	2.34%	
161	Dy	18.9%	
162	Dy	25.5%	
163	Dy	24.9%	

APPENDIX 2. (Continued).

A	Element	Abundance/half-life	Source
164	Dy	28.2%	
165	Ho	100%	
162	Er	0.14%	
164	Er	1.61%	
166	Er	33.6%	
167	Er	22.95%	
168	Er	26.8%	
170	Er	14.9%	
169	Tm	100%	
168	Yb	3.05%	
171	Yb	14.3%	
172	Yb	21.9%	
173	Yb	16.12%	
174	Yb	31.8%	
176	Yb	12.7%	
175	Lu	97.41%	
176	Lu	3.78e+10 yr 2.59%	
174	Hf	2.0e+15 yr 0.162%	
176	Hf	5.206%	
177	Hf	18.606%	
178	Hf	27.297%	
179	Hf	13.629%	
180	Hf	35.100%	
180	Ta	1.2e+15 yr 0.012%	
181	Ta	99.988%	
180	W	0.13%	
182	W	26.3%	
183	W	1.1e+17 yr 14.3%	
184	W	3e+17 yr 30.67%	
186	W	28.6%	
185	Re	37.40%	
187	Re	4.35e+10 yr 62.60%	
184	Os	5.6e+13 yr 0.02%	
186	Os	2.0e+15 yr 1.58%	
187	Os	1.6%	
188	Os	13.3%	
189	Os	16.1%	
190	Os	26.4%	
192	Os	41.0%	
191	Ir	37.3%	
193	Ir	62.7%	
190	Pt	6.5e+11 yr 0.01%	
192	Pt	0.79%	
194	Pt	32.9%	
195	Pt	33.8%	
196	Pt	25.3%	

APPENDIX 2. (Continued).

A	Element	Abundance/half-life	Source
198	Pt	7.2%	
197	Au	100%	
196	Hg	0.15%	
198	Hg	9.97%	
199	Hg	16.87%	
200	Hg	23.10%	
201	Hg	13.18%	
202	Hg	29.86%	
204	Hg	6.87%	
203	Tl	29.524%	
205	Tl	70.476%	
206	Tl	4.199 min	U238
207	Tl	4.77 min	U235
208	Tl	3.053 min	Th232
210	Tl	1.30 min	U238
204	Pb	1.4e+17 yr	
		1.4%	
207	Pb	22.1%	
208	Pb	52.4%	
210	Pb	22.3 yr	U238
211	Pb	36.1 min	U235
212	Pb	10.64 h	Th232
214	Pb	26.8 min	U238
209	Bi	100%	
210	Bi	5.013 d	U238
211	Bi	2.14 min	U235
212	Bi	60.55 min	Th232
214	Bi	19.9 min	U238
215	Bi	7.6 min	U235
210	Po	138.376 d	U238
211	Po	0.516 s	U235
212	Po	0.299 μs	Th232
214	Po	164.3 μs	U238
215	Po	1.781 ms	U235

(Continued)

APPENDIX 2. (Continued).

A	Element	Abundance/half-life	Source
216	Po	0.145 s	Th232
218	Po	3.10 min	U238
215	At	0.10 ms	U235
218	At	1.5 s	U238
219	Rn	3.96 s	U235
220	Rn	55.6 s	Th232
222	Rn	3.8235 d	U238
223	Fr	21.8 min	U235
223	Ra	11.435 d	U235
224	Ra	3.66 d	Th232
226	Ra	1600 yr	U238
228	Ra	5.75 yr	Th232
227	Ac	21.773 yr	U235
228	Ac	6.15 h	Th232
228	Th	1.9116 yr	Th232
230	Th	7.538e+4 yr	U238
232	Th	1.405e+10 yr	
		100%	
234	Th	24.10 d	U238
231	Pa	32760 yr	U235
234	Pa	6.70 h	U238
234	U	2.455e+5 yr	
		0.0055%	
235	U	7.038e+8 yr	
		0.7200%	
238	U	4.468e+9 yr	
		99.2745%	

Sources of short-lived radionuclides: B, bomb or reactor sources; C, cosmogenic; E, extinct radioactivities; U235, U238, Th232—nuclides in respective decay chains.

Note: the symbol e indicates that the number following is that raised to the power of 10.

[a] Modified from: Lawrence Berkeley Laboratory web site: http://ie.lbl.gov/education/isotopes.htm

APPENDIX 3. The Geologic Timescale.

Eon	Era	Period	Epoch	Millions of years ago
Phanerozoic	Cenozoic	(Quaternary)	Holocene	
				0.011
			Pleistocene	
				1.82
		(Tertiary)	Pliocene	
				5.32
			Miocene	
				23
			Oligocene	
				33.7
			Eocene	
				55
			Paleocene	
				65
	Mesozoic	Cretaceous		
				144
		Jurassic		
				200
		Triassic		
				250
	Paleozoic	Permian		
				295
		Carboniferous Pennsylvanian		
				320
		Mississippian		
				355
		Devonian		
				410
		Silurian		
				440
		Ordovician		
				500
		Cambrian		
				543
Proterozoic				
				2,500
Archean		Oldest rock		4,400
		Age of the solar system		4,550

APPENDIX 4. Useful Values.

Molecular mass of dry air, $m_a = 28.966$
Molecular mass of water, $m_w = 18.016$
Universal gas constant, $R = 8.31436 \, \text{J mol}^{-1} \, \text{K}^{-1}$
Gas constant for dry air, $R_a = R/m_a = 287.04 \, \text{J kg}^{-1} \, \text{K}^{-1}$
Gas constant for water vapor, $R_v = R/m_w = 461.50 \, \text{J kg}^{-1} \, \text{K}^{-1}$
Molecular weight ratio $\varepsilon \equiv m_w/m_a = R_a/R_v = 0.62197$
Stefan's constant $\sigma = 5.67 \times 10^{-8} \, \text{W m}^{-2} \, \text{K}^{-4}$
Acceleration due to gravity, g (m s^{-2}) as a function of latitude φ and height z (m)

$$g = (9.78032 + 0.005172 \sin^2\varphi - 0.00006 \sin^2 2\varphi)(1 + z/a)^{-2}$$

Mean surface value, $\bar{g} = \int_0^{\pi/2} g \cos\varphi \, d\varphi = 9.7976$
Radius of sphere having the same volume as the Earth, $a = 6{,}371$ km (equatorial radius = 6,378 km, polar radius = 6,357 km)
Rotation rate of Earth, $\Omega = 7.292 \times 10^{-5} \, \text{s}^{-1}$
Mass of Earth = 5.977×10^{24} kg
Mass of atmosphere = 5.3×10^{18} kg
Mass of ocean = 1400×10^{18} kg
Mass of groundwater = 15.3×10^{18} kg
Mass of ice caps and glaciers = 43.4×10^{18} kg
Mass of water in lakes and rivers = 0.1267×10^{18} kg
Mass of water vapor in atmosphere = 0.0155×10^{18} kg
Area of Earth = $5.10 \times 10^{14} \, \text{m}^2$
Area of ocean = $3.61 \times 10^{14} \, \text{m}^2$
Area of land = $1.49 \times 10^{14} \, \text{m}^2$
Area of ice sheets and glaciers = $1.62 \times 10^{13} \, \text{m}^2$
Area of sea ice = $1.9 \times 10^{13} \, \text{m}^2$ in March and $2.9 \times 10^{13} \, \text{m}^2$ in September (averaged between 1979 and 1987)

INDEX

NOTES:

Page numbers suffixed by *t* and *f* refer to Tables and Figures respectively. vs. indicates a comparison.

A

Abitibi Belt, 172–173
accretion
  solar nebula, 513*f*
  volatile elements, 135
acidification
  acid rain
    anthropogenic impacts, 156–157
    forest ecosystems, 113*t*, 114
acid mine drainage
  sulfur (S) concentrations, 157–158
acylpolysaccharides, 400–402, 402*f*, 403*f*, 404–405, 404*f*
adenosine triphosphate (ATP)
  aerobic cellular respiration, 518
  photosynthesis, 516*f*
  sulfur biomolecules, 142
aerosols
  cloud condensation nuclei (CCN), 154–155
  organosulfur emissions, 156
  sea salt
    ice core analysis, 360*f*
  sulfur (S)
    aerosol sulfates, 154–155
    sulfuric acid (H_2SO_4), 158–159
  volcanic emissions, 24*t*, 28, 29*f*
agriculture
  biological pump
    C:N:P ratio variations, 446
    runoff effects, 446
    silica limitations, 446
air-sea exchange
  biological pump, 455*f*
  calcium carbonate ($CaCO_3$) production, 94
  carbon cycle
    carbon dioxide (CO_2)
      dissolved inorganic carbon (DIC), 92–94
    gas fluxes
      molecular oxygen (O_2), 512–515, 514*f*,
        515*f*, 517
      radon (^{222}Rn), 94
  nitrogen cycle, 557
  oceanic tracers, 97
  sulfates, 143
algae
  nitrogen fixation, 435–436
alkalinity
  hydrothermal vents, 60*t*
  oceans
    alkalinity pump, 364, 455*f*
    biological pump, 456–457
aluminum (Al)
  atmospheric emissions, 27*t*
  chemical weathering
    elemental mobility, 210*t*, 211*f*

  complexation, 317*f*
  elemental abundances
    clastic sediments, 588–589, 589*f*
    Congo River, 308*f*
    hydrothermal vents, 60*t*
    rivers, 296*t*, 307*f*, 308*f*, 313*f*
    sedimentary environments, 592*f*
  hydrolysis, 316*f*
  phytoplankton photosynthesis, 428*t*
  solubility, 325, 326*f*
  ultrafiltration techniques, 320*f*
Amazon, Brazil, 115, 116*f*
Amazon River
  adsorption processes, 330–331, 331*f*
  neodymium (Nd) concentrations, 324*f*
  organic carbon cycle, 394*f*, 395
  rare earth elements (REEs), 323*f*
  trace elements, 296*t*, 310, 311*f*, 312*f*
amides, 401, 402*f*
amino acids
  oceanic carbon cycle, 401–402, 401*f*, 402*f*
ammonia (NH_3)
  biological pump, 435–436, 435*f*, 440–441
amphiboles
  weatherability, 224
amphibolites
  river basin lithology, 282*t*
Amu Darya, Central Asia, 290*f*
andesites
  arc magmatism, 6–8
  turbidites, 589*f*
  viscosity, 5*f*
  volatile elements, 7*t*, 12*t*
anhydrite ($CaSO_4$)
  hydrothermal vents, 52*f*, 68, 68*f*
anorthosites
  river basin lithology, 282*t*
Antarctica
  Antarctic Cold Reversal, 365–366
  Antarctic sea ice cover, 468–469, 471*f*
  biological pump, 7*f*, 17*f*, 478–481, 483*f*
  carbon cycle, 355–359, 358*f*, 362–365, 365–367
  global warming, 365, 366*f*
  ice cores, 455–456, 457*f*
  temperature variations, 363*f*
Antarctic Bottom Water (AABW), 17*f*, 78, 391
Antarctic Intermediate Water (AAIW), 17*f*, 479
antimony (Sb)
  elemental abundances
    rivers, 296*t*, 307*f*, 308*f*, 314–315
  hydrolysis, 316*f*
aquifers
  hydrologic properties, 216–217, 217*f*
  sulfur cycle, 146–147

Arabian Sea
  biological pump, 476
  molecular oxygen (O_2) concentrations, 513–514
  oceanic carbon cycle, 416f
  organic carbon cycle, 411, 412f
arabinose, 400–401, 401f
aragonite ($CaCO_3$)
  Phanerozoic
    calcareous fossils, 595f, 609–610, 610f
    oöids, 608–609, 609f
  solubility in seawater, 595f
Aral Sea
  dust, 158
Archaea
  organic carbon cycle, 414–415
Archaean
  atmospheric evolution
    carbon ($\delta^{13}C$) isotope concentrations, 527–528, 527f
    carbon dioxide (CO_2) concentrations, 527–528, 527f
    molecular oxygen (O_2) concentrations, 524–526
    oxygenic photosynthesis, 526–527, 528
  banded iron formations (BIFs), 522–524, 523f
  carbon cycle, 375–376
  continental crust
    clastic sediments, 590
    evolutionary trends, 605–606
  geological record
    Isua Supercrustal Belt, Greenland, 527–528
  isotopic evolution, 595–598, 596f
  nitrogen (N) isotope concentrations, 525–526
  paleosols, 524
  seawater
    strontium (Sr) isotopic evolution, 595–598, 596f, 597f
  sulfur (S) concentrations, 135–137
  sulfur (S) isotope concentrations, 525–526
Archer, David, 496–497
arc magmatism *see* magmatic arcs
Arctic
  Arctic Ocean, 63–64
argon (Ar)
  elemental abundances
    Earth's crust, 188–189
    surface inventories, 169t
  isotopes
    atmospheric composition, 170t
    degassing models, 180–181, 186
    mantle (Earth), 171–173, 172f, 180–181, 186
    mid-ocean ridge basalts (MORBs), 171–173, 172f, 174–175, 175f
    nuclear processes, 169t
    ocean island basalts (OIBs), 171–173
    potassium/argon (^{40}K-^{40}Ar) isotopic budget, 178, 188–189
    subduction fluxes, 177–178
  volcanic emissions, 22t
arkoses, 585f, 605
arsenic (As)
  atmospheric emissions, 27t
  biological pump, 437–440
  elemental abundances
    hydrothermal vents, 73–74, 77, 77t
    rivers, 296t, 307f, 308f, 314–315
  hydrolysis, 316f
Atlantic Ocean
  biogenic opal (SiO_2) production, 478–481
  biological pump efficiency, 17f, 467
  calcium carbonate ($CaCO_3$) abundance, 493–495, 504–508
  helium (He) isotopes, 171f
  hydrothermal plumes, 75–76, 77, 77t
  hydrothermal vents, 46–47, 46f, 63–64
  molecular oxygen (O_2) concentrations, 512–515, 514f
  nitrogen cycle, 565–566, 567t
  organic carbon cycle, 408–409, 409f

atmospheric chemistry
  carbon cycle, 89–90, 89f, 89t, 343f, 346t
  carbon dioxide (CO_2) concentrations
    anthropogenic emissions, 93f, 95–97, 96t, 101–106, 102f, 103t
    biological pump, 453–490
    calcium carbonate ($CaCO_3$) production, 457
  Mars
    argon (Ar) isotopes, 194
    degassing, 194–196
    radiogenic isotopes, 194
    xenon (Xe) isotopes, 194–195
  methane (CH_4) carbon subcycle, 344f, 348–350, 353t
  nitrogen cycle
    inorganic oxidized nitrogen, 555, 555t
    inorganic reduced nitrogen, 555, 555t
    nitrogen cascade, 569–570, 570f, 571t
    oxidized organic nitrogen, 556
    reactive nitrogen (Nr), 561t, 562t
    reduced organic nitrogen, 555, 555t
    valence states, 555t
  noble gases
    iodine/xenon (I/Xe) isotopic ratios, 178–179
    planetary degassing, 167–204
    plutonium/xenon (Pu/Xe) isotopic ratios, 178–179
    potassium/argon (^{40}K-^{40}Ar) isotopic budget, 178
  oxygenation, 528–531
  oxygen (O)
    molecular oxygen (O_2) concentrations, 102, 102f, 512
  planetary degassing, 167–204
  radiogenic isotopes, 178
  sulfur cycle
    aerosol sulfates, 154–155
    carbon disulfide (CS_2), 150
    carbonyl sulfide (COS), 150, 151f
    deposition processes, 155
    dimethyl sulfide (DMS), 150–151, 152f
    dimethylsulfoxide (DMSO), 151–152
    elemental sulfur, 152
    global budget, 149–150, 149f
    hydrogen sulfide (H_2S), 150, 150f
    methanesulfonic acid, 151–152
    sulfur dioxide (SO_2), 152–154, 153f
    volcanic emissions, 25t
  volatile elements
    composition, 170t
    surface inventories, 169, 169t
  see also stratosphere; troposphere
atmospheric p_{CO2} signature, 456–457, 464
Augustine Volcano, Alaska, 26–27, 27t

B

bacteria
  anoxygenic photosynthesis, 136
  hydrothermal plumes, 76
  organic carbon cycle, 414–415
Baitoushan volcano, 25t
banded iron formations (BIFs)
  Archaean, 522–524, 523f
barite
  Archaean sedimentary environment, 135–136, 136f
  hydrothermal vents, 68f
  marine sediments, 472
  sulfur (S) concentrations, 137
barium (Ba)
  biological pump, 437–438, 438f, 472
  complexation, 317f
  elemental abundances
    clastic sediments, 588–589, 589f
    Congo River, 308f
    hydrothermal vents, 60t
    rivers, 296t, 307f, 308f, 312f, 313f

sedimentary environments, 592*f*
hydrolysis, 316*f*
phytoplankton photosynthesis, 428*t*
basalts
  arc magmatism
    turbidites, 589*f*
    volatile elements, 7*t*, 12*t*
  ocean island basalts (OIBs)
    carbon (C) abundances, 191
    degassing models, 181–183, 182*f*
    isotopic composition
      argon (Ar) isotopes, 171–173
      helium (He) isotopes, 169–170
      neon (Ne) isotopes, 170–171
      xenon (Xe) isotopes, 173–174
    noble gases, 176–177
    volatile elements, 7*t*, 12*t*
  river basin lithology, 282*t*
  viscosity, 5*f*
  volatile elements, 12*t*
  weathering rates, 231*f*, 232*f*
  see also mid-ocean ridge basalts (MORBs)
Beatton River, 296*t*
benthic foraminifera
  boric acid (B(OH)$_3$) concentrations, 499, 502–503
  deep-sea carbonate-rich sedimentary environments,
    495–496, 495*f*, 497–498, 499–500, 499*f*
  deep-water oxygen isotope record, 457*f*
  interglacial periods, 502–503
  oceanic carbon cycle, 408*f*, 409*f*, 456–457
  radiogenic isotopes, 498–499, 503–504
  shell weights
    carbonate ion concentrations, 492*f*, 499*f*, 501–502,
      505, 508
    wall thickness, 501*f*
    water depth, 506*f*, 508*f*
  skeletal mineralogy, 595*f*, 609–610, 610*f*
  zinc/cadmium (Zn/Cd) ratios, 499, 503–504
Bermuda Rise, 408–409, 409*f*
beryllium (Be)
  biological pump, 437–440
  complexation, 317*f*
  elemental abundances
    hydrothermal vents, 60*t*, 77, 77*t*
    rivers, 296*t*, 307*f*, 308*f*
    sedimentary environments, 592*f*
  hydrolysis, 316*f*
  isotopes
    rivers, 331, 332*f*
biogenic opal (SiO$_2$)
  biological pump, 25*f*, 456–457
  Pacific Ocean, 481, 483*f*
biological pump, 425–452
  Antarctic sea ice cover, 468–469, 471*f*
  Antarctic stratification, 7*f*, 17*f*, 467, 478–481, 482
  basic concept
    aggregation processes, 430–431
    bacterial hydrolysis, 432–433
    burial processes, 433
    cycling processes, 427–434
    dissolved organic matter (DOM), 433–434
    exopolymer particles, 430–431
    flocculation, 430–431
    marine snow, 430
    new production, 434
    nutrient limitations, 430
    nutrient uptake, 428–430
    particle decomposition, 431–433
    particulate organic matter (POM), 431
    photosynthesis, 428–430
    phytoplankton, 428–430
    primary production levels, 429, 429*f*
    schematic diagram, 427*f*

sedimentation, 433
sinking fluxes, 431–433, 432*f*
sinking rates, 430–431
zooplankton grazing, 432
biogenic opal (SiO$_2$), 25*f*, 456–457, 481, 483*f*
calcium carbonate (CaCO$_3$) production, 94, 462, 462*t*
carbon cycle
  atmospheric carbon dioxide concentrations, 453–490
  carbon dioxide (CO$_2$) concentrations, 92–93
  carbon sequestration, 454*f*
  Holocene, 455–456, 455*f*
  nested cycles, 454, 456*f*
CYCLOPS model, 17*f*
dissolved inorganic compounds, 427*f*
efficiency
  carbon:nitrogen:phosphorus (C:N:P) ratios,
    444–445, 461–462
  community composition changes, 444
  high-nutrient, low chlorophyll (HNLC) areas,
    443–444, 443*f*
  nutrient concentrations, 17*f*, 465–469, 466*f*, 468*f*
  particulate organic matter (POM), 444–445
  transport enhancements, 445
export production, 25*f*, 471–472, 475–478
future trends
  agricultural runoff, 446
  carbon:nitrogen:phosphorus (C:N:P) ratios, 446
  carbon sequestration, 446–447
  increased carbon dioxide (CO$_2$) levels, 445–446
  silica limitations, 446
general discussion, 426–427
glacial/interglacial effects, 17*f*, 25*f*, 455–456, 457*f*, 471*f*,
  475–478, 478–481
high-nutrient, low chlorophyll (HNLC) areas, 426, 440,
  443–444, 443*f*
impact processes
  macronutrients
    dissolved inorganic carbon (DIC), 427*f*, 434–435
    nitrogen cycle, 435–436, 435*f*, 439*f*, 440–441
    nitrogen fixation, 435–436
    phosphorus cycle, 436, 439*f*
    silica cycle, 436–437, 437*f*, 439*f*
    silica/nitrate ratios, 437, 438*f*
    trace elements, 437–440, 438*f*, 439*f*
latitude controls
  equilibrium processes, 460–465, 466*f*
  high-latitude ocean, 465–469, 478–481
  low-latitude ocean, 460–465, 466*f*, 475–478
  mid-latitude ocean, 460–465, 466*f*, 475–478
  phytoplankton production, 7*f*, 458–469
nutrient status
  cadmium/calcium (Cd/Ca) ratio, 472–473, 481
  carbon (C) isotope distribution, 474
  molecular oxygen (O$_2$) concentrations, 474–475
  nitrate assimilation, 473, 480–481
  phasing, 475
  phytoplankton, 7*f*, 472–473
  silicate concentrations, 457*f*, 473–474, 480–481
organic carbon cycle, 25*f*, 453–458, 454*f*, 466*f*, 471–472
phytoplankton
  calcium carbonate (CaCO$_3$) production, 462, 462*t*
  latitude controls, 7*f*, 458–469, 478–481
  nitrogen fixation, 463
  nutrient status, 7*f*, 472–473
  nutrient supply, 454*f*, 458–469, 466*f*
  phosphorus cycle, 461
  silicate/nitrate uptake, 480–481
process steps
  Redfield ratio, 461–462
  sinking rates, 93
quantification studies
  general discussion, 440–443
  new production measurements, 440–441

biological pump (*continued*)
    oxygen utilization rates, 442–443
    particle flux measurements, 441–443
    particle-reactive nuclides, 442, 442*f*
    sediment traps, 441–442
    thorium/uranium (Th/U) isotopic ratios, 472
biomarkers
    oceanic carbon cycle, 416*f*
biomass
    biomass burning
        fire
            forest fires, 534–545
            methane (CH_4) carbon subcycle, 353*t*
biomineralization
    processes
        carbonate biomineralization
            coccoliths, 445
        silica biomineralization
            biological pump, 436–437, 437*f*
biotite
    chemical weathering, 215–216
    mineral fluxes, 223*t*
    sedimentary environments, 587*f*
    weatherability, 225, 225*f*
    weathering reactions, 208–209, 212*t*
Bishop Tuff, 7*t*
bismuth (Bi)
    elemental abundances
        sedimentary environments, 592*f*
    hydrolysis, 316*f*
black carbon, 397–398, 399*f*
Black Sea
    anaerobic methane oxidation, 416
    hydrogen sulfide (H_2S), 144
    molecular oxygen (O_2) concentrations, 513–514
Bölling Allerod warm period
    carbon cycle, 365–367
    methane (CH_4) carbon subcycle, 365–366, 366*f*
bomb radiocarbon (^{14}C)
    oceanic tracers, 97, 98
    riverine organic matter, 394–395
boron (B)
    boric acid ($B(OH)_3$)
        oceanic concentrations, 502–503
        rivers, 331
    elemental abundances
        hydrothermal vents, 60*t*
        rivers, 296*t*, 307*f*, 308*f*
        sedimentary environments, 592*f*
    hydrolysis, 316*f*
    isotopes
        benthic foraminifera, 373, 499, 502–503
        paleo pH proxy, 373, 502–503
Bosch, Carl, 559–560
Brahmaputra River, 296*t*
Brogger, Waldemar, 156–157
bromine (Br)
    elemental abundances
        hydrothermal vents, 60*t*
    hydrogen bromide (HBr), 26–27, 26*t*
    hydrothermal vents, 54–56, 60*t*
Byrd, Antarctica, 362*f*
Byrd ice core, Antarctica, 353*t*

C

cadmium (Cd)
    benthic foraminifera, 499, 503–504
    biological pump, 439–440, 439*f*
    complexation, 317*f*
    elemental abundances
        hydrothermal vents, 60*t*
        rivers, 296*t*, 307*f*, 308*f*, 314–315

    sedimentary environments, 592*f*
    hydrolysis, 316*f*
    phytoplankton photosynthesis, 428*t*
calcite ($CaCO_3$)
    deep-sea carbonate-rich sedimentary environments, 492, 492*f*, 493*f*
    magnesium/calcium (Mg/Ca) ratio, 607–608, 608*f*
    Phanerozoic
        calcareous fossils, 595*f*, 609–610, 610*f*
        oöids, 608–609, 609*f*
        sedimentary environments, 606–607, 606*f*, 607*f*
    solubility
        seawater, 595*f*
calcium (Ca)
    chemical weathering
        elemental mobility, 210*t*, 211*f*
    complexation, 317*f*
    elemental abundances
        clastic sediments, 588–589, 589*f*
        Congo River, 308*f*
        hydrothermal vents, 60*t*
        rivers, 308*f*, 313*f*
        sedimentary environments, 592*f*
    groundwater concentrations
        distribution, 217–218
    hydrolysis, 316*f*
    ionic concentrations
        anthropogenic sources, 290–291, 290*t*
        idealized model, 284–286, 285*f*
        pristine river distributions, 279*t*, 282*t*
        river sources/sinks, 280*t*
        runoff, 283–284, 284*f*, 286–287, 286*f*, 286*t*
        weathering rates, 289*t*
    phytoplankton photosynthesis, 428*t*
calcium carbonate ($CaCO_3$)
    air-sea exchange, 94
    biological pump, 453–490
    carbon cycle, 343*f*, 346*t*, 367–368
    dissolution mechanisms
        saturation as function of water depth, 492–493, 493*f*, 496–498, 497*f*, 499*f*, 500*f*, 502*t*
    oceans
        boric acid ($B(OH)_3$) concentrations, 502–503
        burial process, 457
        carbonate cycle, 610, 610*f*
        carbonate pump, 455*f*, 457
        dissolution mechanisms, 496–498, 496*f*, 498*f*, 504–508, 505*f*
        fossil fuels, 509
        glacial periods, 508
        interglacial periods, 504–508
        ionic concentrations, 491–510
        oceanic carbon cycle, 364
        preservation events, 504–508, 505*f*, 506*f*
        rain rate, 365, 455*f*
        terrestrial biomass, 504–505
Cambrian
    molecular oxygen (O_2), 534–535
    sulfide minerals, 137
Cameroon River, 323*f*
Camp Century ice core, 353*t*
Cape Verde Plateau
    calcium carbonate ($CaCO_3$) dissolution, 498*f*
carbonatites, 12*t*
carbon (C)
    bicarbonate (HCO_3)
        anthropogenic sources, 290–291, 290*t*
        pristine river distributions, 279*f*, 279*t*, 282*t*, 283*f*
        riverine model, 284–286, 285*f*
        runoff, 283–284, 284*f*, 286–287, 286*t*, 287*f*
        weathering rates, 289*t*
    biological pump
        carbon dioxide (CO_2) concentrations

carbon sequestration, 446–447
C:N:P ratio variations, 444–445
community composition changes, 444
general discussion, 426–427
high-nutrient, low chlorophyll (HNLC) areas,
443–444, 443*f*
increased carbon dioxide (CO_2) levels, 445–446
particle decomposition, 431–433
particulate organic matter (POM), 431, 444–445
photosynthesis, 428–430
sinking fluxes, 431–433, 432*f*
dissolved inorganic carbon (DIC), 427*f*, 434–435
carbonates
carbonate biomineralization
coccoliths, 445
Phanerozoic, 527–528, 527*f*, 531–532, 540*f*
rivers, 280*t*
carbon (δ^{13}C) isotopes
paleo-ocean, 600–602, 601*f*, 602*f*
carbon dioxide (CO_2) concentrations
anthropogenic sources, 93*f*, 95–97, 96*t*, 101–106, 102*f*
atmospheric concentrations, 89–90, 91–94, 93*f*
basaltic melts, 8, 8*f*
biological pump
carbon sequestration, 446–447
C:N:P ratio variations, 444–445
community composition changes, 444
dissolved inorganic carbon (DIC), 427*f*, 434–435
general discussion, 426–427
high-nutrient, low chlorophyll (HNLC) areas,
443–444, 443*f*
increased carbon dioxide (CO_2) levels, 445–446
particle decomposition, 431–433
particulate organic matter (POM), 431, 444–445
photosynthesis, 428–430
sinking fluxes, 431–433, 432*f*
greenhouse gases, 88, 96
human health effects, 35–36
hydrothermal vents, 60*t*, 61
magma abundances, 7*t*
noneruptive emissions, 15–16, 15*f*
organic carbon cycle, 391*f*
volcanic emissions, 12*t*, 22*t*, 23, 26, 35–36
carbon disulfide (CS_2)
occurrences, 133
seawater, 144
tropospheric concentrations, 150
carbon monoxide (CO)
atmospheric concentrations, 90, 97
volcanic emissions, 12*t*, 22*t*
carbon:nitrogen (C:N) ratios, 401
carbon:nitrogen:phosphorus (C:N:P) ratios, 461–462
carbon:phosphorus (C:P) ratios, 401
carbon transfer, 281–283, 283*t*
carbonyl sulfide (COS)
occurrences, 133
seawater, 144, 146
tropospheric concentrations, 150, 151*f*
volcanic emissions, 12*t*
dissolved inorganic carbon (DIC)
biological pump, 427*f*
elemental abundances
rivers, 308*f*
surface inventories, 169*t*
hydrolysis, 316*f*
isotopes
atmospheric chemistry, 170*t*
Late Paleocene Thermal Maximum, 374–375, 374*f*
oceanic waters, 497, 497*f*
paleo-ocean, 600–602, 601*f*, 602*f*
organic carbon cycle, 389–424
rivers
carbon transfer, 281–283, 283*t*

dissolved inorganic carbon (DIC), 281–283, 283*t*
dissolved organic carbon (DOC), 283*t*
particulate inorganic carbon (PIC), 283*t*
particulate organic carbon (POC), 283*t*
volatile cycling, 190–192
see also net primary production (NPP); radiocarbon (^{14}C)
carbon cycle
abrupt change mechanisms, 374–375
Antarctica, 355–359, 358*f*, 362–365
anthropogenic emissions
air-land exchange, 96*t*, 98–99, 99*f*, 103*t*
bomb radiocarbon (^{14}C), 97
carbon dioxide (CO_2) concentrations, 93*f*, 95–97, 96*t*,
101–106, 102*f*, 103*t*
chlorofluorocarbons (CFCs), 98
fossil fuels, 94–95, 95*f*, 96*t*, 103*t*
land use changes, 96*t*, 99–100, 103*t*, 105, 108*t*
molecular oxygen (O_2) concentrations, 102, 102*f*
oceanic carbon uptake, 96*t*, 97–98, 103*t*
period 1850-2000, 93*f*, 94–101, 95*f*, 96*t*
period 1980-2000, 101–111
radiocarbon (^{14}C), 97
regional distributions, 106–108, 107*t*, 108*t*
residual fluxes, 96*t*, 99*f*, 103*t*
atmospheric chemistry, 89–90, 89*f*, 89*t*, 343*f*, 346*t*,
348–350, 353*t*
biological pump
calcium carbonate ($CaCO_3$) production, 462, 462*t*
carbon dioxide (CO_2), 453–458, 462, 462*t*
carbon sequestration, 454*f*
export production, 471–472
Holocene, 455–456, 455*f*
nested cycles, 454, 456*f*
calcium carbonate ($CaCO_3$), 343*f*, 346*t*, 364, 367–368
carbon (C) abundances, 191
carbon dioxide (CO_2)
air-land exchange, 91–92
air-sea exchange, 92–94
anthropogenic emissions, 93*f*, 95–97, 96*t*, 101–106, 102*f*,
103*t*, 342–348
atmospheric concentrations, 89–90, 89*t*, 91*t*, 93*f*, 95–97
atmospheric evolution, 527–528, 527*f*
biological pump, 453–490, 462*t*
carbon (δ^{13}C) isotope concentrations, 103, 527–528, 527*f*
carbon dioxide fertilization, 111*f*, 112–117, 112*t*, 113*t*
fluxes, 91–94, 93*f*, 105*t*, 108–110, 342–348
fossil fuels, 94–95, 95*f*, 96*t*, 103*t*
GEOCARB model, 371, 372*f*
geologic evidence, 372*f*, 376
glacial/interglacial variations, 359–367, 360*f*, 363*f*
global carbon budget, 101–106, 102*f*
Holocene variations, 355–359, 356*f*, 357*f*, 358*f*
ice core gas analysis, 351–367, 353*t*
land use changes, 100–101
oceanic concentrations, 91*t*, 97–98, 118–120
reservoirs, 342–348, 343*f*
sea-land exchange, 94
silicate weathering, 368–369
source reduction, 120–121
sources/sinks, 104*f*
terrestrial ecosystems, 101*t*
carbon pools, 103
continental crust
methane (CH_4) carbon subcycle, 344*f*, 353*t*
preindustrial cycle, 343*f*, 346*t*
dissolved inorganic carbon (DIC)
glacial periods, 345–346
oceanic carbon uptake, 97–98, 460–465, 466*f*
oceans, 90–91, 91*t*
pristine river distributions, 283*f*
rivers, 281–283, 283*t*
river sources/sinks, 280*t*
dissolved organic carbon (DOC)

carbon cycle (*continued*)
  acylpolysaccharides, 400–402, 402*f*, 403*f*, 404–405, 404*f*
  amides, 401, 402*f*
  background information, 398–399
  biopolymerization versus geopolymerization, 399–404
  gel polymers, 404–405, 404*f*, 405*f*
  glacial periods, 345–346
  high molecular weight dissolved organic matter (HMW DOM), 399–404
  marine sediments, 391
  monosaccharides, 400–402, 401*f*
  NMR spectra, 400–402, 400*f*, 406*f*
  organic carbon cycle, 391*f*
  proteins, 402–404
  rivers, 283*t*, 394–396
  water column fluxes, 392–393
 fluxes
  air-land exchange, 91–92, 96*t*, 99–100, 103*t*
  air-sea exchange, 92–94, 455*f*
  alkalinity pump, 364, 455*f*
  biological pump, 92–93
  carbon dioxide (CO$_2$) concentrations, 91–94, 93*f*
  estimation methods, 101*t*
  glacial periods, 346–347
  measurement techniques, 108–109
  methane (CH$_4$) carbon subcycle, 342–350, 344*f*, 353*t*
  organic carbon cycle, 391*f*, 392–393
  oxidation processes, 91–92
  preindustrial cycle, 342–350, 343*f*, 346*t*
  residual fluxes, 96*t*, 99*f*, 100–101, 103*t*
  sea-land exchange, 94
  solubility pump, 92–93, 363–364, 455*f*
  terrestrial ecosystems, 101*t*
 forest ecosystems
  acid rain, 113*t*, 114
  Amazon, Brazil, 115, 116*f*
  carbon dioxide fertilization, 112–117, 112*t*, 113*t*
  climate effects, 113*t*, 114
  disturbance mechanisms, 101*t*, 117–118
  nitrogen fertilization, 113–114, 113*t*
  northern mid-latitudes, 106–108, 108*t*, 111
  regional distributions, 107*t*
  reservoirs, 90, 90*t*
  sources/sinks, 103, 106, 111–118, 111*f*, 113*t*
  synergistic mechanisms, 116
  terrestrial carbon models, 117
  tropics, 108–110, 110*t*, 111
  tropospheric ozone concentrations, 113*t*, 114
 fossil fuels
  anthropogenic emissions, 88, 94–95, 95*f*
  carbon dioxide (CO$_2$) concentrations, 96*t*, 103*t*
  contemporary carbon cycle, 89*f*, 89*t*
  preindustrial cycle, 343*f*, 346*t*
  reservoirs, 91
 future research, 121
 geologic history, 341–388
 glacial terminations, 365–367
 global carbon budget, 96*t*, 101–106, 102*f*, 103*t*
 gradual change mechanisms, 367–371
 gradual change simulations, 375–376
 Great Oxidation Event, 375–376
 greenhouse gases, 88, 89*t*, 96
 ice core gas analysis
  carbon dioxide (CO$_2$) concentrations, 455–456, 457*f*
  glacial/interglacial variations, 359–367, 360*f*, 363*f*
  Holocene variations, 355–359, 356*f*, 357*f*, 358*f*
  polar ice core data, 353*t*
  preindustrial cycle, 351–367
 importance, 88
 Late Paleocene Thermal Maximum, 374–375, 374*f*
 mantle (Earth), 343*f*, 346*t*
 marine sediments, 89*f*, 89*t*, 343*f*, 346*t*, 347, 353*t*
 methane (CH$_4$) carbon subcycle

  atmospheric concentrations, 90, 96–97, 344*f*, 348–350, 353*t*
  continental crust, 344*f*, 353*t*
  fluxes, 348–350, 353*t*
  geological timescales, 342–350, 343*f*, 350–351
  glacial/interglacial variations, 359–367, 360*f*, 362*f*
  glucose production, 369
  greenhouse gases, 88, 96–97
  Holocene variations, 355–359, 356*f*, 357*f*, 358*f*
  ice core gas analysis, 351–367, 353*t*
  marine sediments, 353*t*
  methane hydrates, 350, 351*f*, 353*t*
  microbial processes, 369
  natural gas, 344*f*, 353*t*
  oceans, 344*f*, 348, 353*t*
  reservoirs, 344–345, 348–350, 353*t*
  soils, 344*f*, 353*t*
  termites, 353*t*
  volcanism, 353*t*
  wetlands, 344*f*, 353*t*, 359
  wildfires, 353*t*
 microbes
  abundances, 416*f*
  anaerobic methane oxidation, 415–416
  crenarchaeota, 415, 416*f*
  planktonic archaea, 415, 416*f*
  prokaryotes, 414–415
 oceans
  air-sea exchange, 92–94
  contemporary carbon cycle, 89*f*, 89*t*
  dissolved inorganic carbon (DIC), 90–91, 91*t*
  glaciation, 533
  methane (CH$_4$) carbon subcycle, 344*f*, 348, 353*t*
  preindustrial cycle, 343*f*, 346*t*
  sources/sinks, 119–120, 120–121
 organic carbon cycle
  aeolian dust, 392
  black carbon, 397–398, 399*f*
  burial process, 537–545, 539*f*
  fluxes, 391*f*, 392–393
  future research, 417–418
  global trends, 370, 370*f*
  glucose production, 374–375
  hydrothermal plumes, 76
  marine sediments, 396–398, 396*f*
  microbes, 414–416
  preindustrial cycle, 343*f*
  preservation, 406–414
  reservoirs, 390–392, 391*f*
  rivers, 283*t*, 343*f*, 346*t*, 392, 394–396
  water column fluxes, 392–393
  weathering sources, 392
 paleoclimatology, 361–362, 362–365, 365–367
 paleosol formation, 524
 particulate inorganic carbon (PIC), 283*t*
 particulate organic carbon (POC)
  composition, 407*f*
  compositional transformations, 406–409
  fluxes, 407*f*
  rivers, 394–396
  seawater, 393
 Phanerozoic, 367–375, 372*f*, 376
 Precambrian, 375–376
 preindustrial cycle
  atmospheric chemistry, 343*f*, 346*t*
  calcium carbonate (CaCO$_3$), 343*f*, 346*t*
  coal, 346*t*, 349*t*
  continental crust, 343*f*, 346*t*
  dolomite weathering, 367–368
  fluxes, 342–350, 343*f*, 346*t*
  fossil fuels, 346*t*
  geological timescales, 342–350, 343*f*, 350–351
  geologic evidence, 372*f*, 376

glacial/interglacial variations, 359–367, 360*f*, 363*f*
glacial periods, 345–346
Holocene variations, 355–359, 356*f*, 357*f*, 358*f*
ice core gas analysis, 351–367, 353*t*
mantle (Earth), 343*f*, 346*t*
marine sediments, 343*f*, 346*t*, 347
natural gas, 346*t*, 349*t*
oceans, 343*f*, 346*t*
reservoirs, 342–350, 343*f*, 346*t*
seawater, 346*t*
silicate weathering, 368–369
soils, 346*t*
vegetation, 346*t*
volcanism, 346*t*
preservation
anoxia, 411–412, 412*f*
background information, 406
chemical protection, 412–414
compositional transformations, 406–409
particulate organic carbon (POC), 407*f*
physical protection, 410–411, 410*f*
radiocarbon (^{14}C) concentrations, 407, 408*f*, 409*f*
refractory biomolecules, 412
stable isotope compositions, 407, 408*f*, 409*f*
sulfurization, 413, 413*f*, 414*f*
prokaryotes, 414–415
Quaternary, 351–367
reservoirs
contemporary carbon cycle, 88–91, 89*f*, 89*t*
methane (CH_4) carbon subcycle, 344*f*, 348–350, 353*t*
organic carbon cycle, 390–392, 391*f*
preindustrial cycle, 342–350, 343*f*, 346*t*
sedimentary environments, 89*f*, 89*t*
soils
contemporary carbon cycle, 89*f*, 89*t*
methane (CH_4) carbon subcycle, 353*t*
preindustrial cycle, 343*f*, 346*t*
reservoirs, 90, 90*t*
sources/sinks
acid rain, 113*t*, 114
Amazon, Brazil, 115, 116*f*
biological processes, 119–120
carbon dioxide fertilization, 112–117, 112*t*, 113*t*
climate effects, 113*t*, 114
disturbance mechanisms, 101*t*, 117–118
geological formations, 121
global carbon budget, 101–106, 102*f*, 103*t*, 104*f*
nitrogen fertilization, 113–114, 113*t*
northern mid-latitudes, 106–108, 108*t*, 111
oceanic mechanisms, 118–120, 120–121
source reduction, 120–121
synergistic mechanisms, 116
terrestrial carbon models, 117
terrestrial ecosystems, 98–99, 105*t*, 107*t*, 110–111
terrestrial mechanisms, 111–118, 111*f*, 113*t*, 120
tropics, 108–110, 110*t*, 111
tropospheric ozone concentrations, 113*t*, 114
terrigenous fluxes
background information, 393
black carbon, 397–398, 399*f*
deltaic sediments, 395*f*
dust, 392
estuaries, 395
land use changes, 105, 105*t*, 108*t*
marine sediments, 396–398, 396*f*
radiocarbon (^{14}C) concentrations, 394–395, 394*f*
regional distributions, 106–108, 107*t*, 108*t*
rivers, 392, 394–396
sources/sinks, 101*t*
weathering, 392
vascular plants
molecular oxygen (O_2), 544–545, 544*f*
vegetation

carbon dioxide fertilization, 112–117, 113*t*
contemporary carbon cycle, 89*f*, 89*t*
preindustrial cycle, 343*f*, 346*t*
reservoirs, 90, 90*t*
weathering, 538, 541*f*
Carboniferous
carbon cycle models, 371
molecular oxygen (O_2), 535
sulfide minerals, 137
Cariaco Basin
molecular oxygen (O_2) concentrations, 513–514
sulfurization of organic matter, 413*f*, 414*f*
Caroni River, 296*t*, 323*f*, 324*f*
Ceara Rise
calcium carbonate ($CaCO_3$) dissolution, 496–497, 497*f*,
    498, 498*f*
foraminifera, 492*f*, 499*f*, 501–502, 508*f*
ionic concentrations, 501–502
celestite/celestine ($SrSO_4$), 137
Cenomanian-Turonian Boundary (CTB) extinction
    event, 536
Cenozoic
carbon cycle models, 371
ocean chemistry, 593–605
cerium (Ce)
elemental abundances
Congo River, 308*f*
hydrothermal vents, 58*f*, 77*t*
rivers, 296*t*, 307*f*, 308*f*, 312*f*, 313*f*, 323*f*
sedimentary environments, 592*f*
cesium (Cs)
elemental abundances
Congo River, 308*f*
hydrothermal vents, 60*t*
rivers, 296*t*, 307*f*, 308*f*, 313*f*
sedimentary environments, 592*f*
hydrolysis, 316*f*
CFCs
air-sea exchange, 98
chalcopyrite ($CuFeS_2$)
metalliferous ridge and basal sediments, 77–78
occurrences, 137
Changjiang River, 296*t*, 313*f*, 331*f*
Chaptal, Jean Antoine Claude, 552–553
chemical weathering
erosion processes, 232–233
field studies
glacial sediments, 216–217
rivers, 312–314, 313*f*
sedimentary environments, 588
silicate weathering
Archaean, 525–526
etch pits, 210–211, 214*f*
fluid transport, 214–215
groundwater solutes, 216–218
hydrologic effects, 227–229
hydrolysis reactions, 207–208
mass changes, 207–220
mineral fluxes, 223*t*
morphological changes, 210–211
pore water concentrations, 215–216, 215*f*
pore water flow, 216*f*
preindustrial carbon cycle, 368–369
rivers, 280*t*
solute compositions, 211–220, 226–227, 226*f*
surface area parameters, 220–221, 222*f*, 223–224
surface-water solutes, 218–220
weatherability, 224–226
weathering rates, 207, 212*t*, 222–223
weathering rates
Bowen's reaction series, 224–226
climate effects, 229–231
contemporary/geologic comparisons, 220–221

chemical weathering (*continued*)
   definition, 207
   elemental fluxes, 222
   etch pits, 210–211, 214*f*
   fluid transport, 214–215
   geochemical models, 233–234, 234*f*
   groundwater solutes, 216–218
   hydrologic effects, 227–229
   mass changes, 207–220
   mineral fluxes, 223*t*
   morphological changes, 210–211
   permeabilities, 227–229, 230*f*
   pore waters, 208, 208*f*, 215–216, 215*f*, 216*f*
   precipitation effects, 230–231, 232*f*
   riverine chemistry, 277–292
   runoff, 230–231, 232*f*
   silicate weathering, 207, 212*t*, 222–223
   silicon fluxes, 231, 232*f*
   soil chronosequences, 220
   solute compositions, 211–220, 226–227, 226*f*
   surface area parameters, 220–221, 222*f*, 223–224
   surface-water solutes, 218–220
   temperature effects, 229–230, 231*f*
   time factors, 220–221
   watersheds, 218–220, 219*f*, 222, 223*t*
   weatherability, 224–226
chemolithoautotrophy
   carbon (δ^{13}C) isotope concentrations, 527–528
   hydrothermal plumes, 76
chemolithotrophy
   redox processes, 519
chlorine (Cl)
   elemental abundances
      hydrothermal vents, 60*t*
      magmas, 7*t*
      rivers, 308*f*
   halite (NaCl)
      hydrothermal vents, 53–54, 55*f*
      volcanic emissions, 27*t*
   hydrochloric acid (HCl)
      volcanic emissions, 12*t*, 22–23, 22*t*, 27*t*
   hydrothermal vents, 53, 55*f*, 58*f*, 60*t*
   ionic concentrations
      anthropogenic sources, 290–291, 290*f*, 290*t*
      idealized model, 284–286, 285*f*
      pristine river distributions, 279*t*, 282*t*
      river sources/sinks, 280*t*
      runoff, 283–284, 284*f*, 286–287, 286*t*, 287*f*
      weathering rates, 289*t*
   iron dichloride (FeCl$_2$), 27*t*
   potassium chloride (KCl), 27*t*
   solubility, 9
   see also CFCs
chlorite
   sedimentary environments, 587*f*
chlorocarbons *see* CFCs
chlorofluorocarbons (CFCs) *see* CFCs
chlorophyll
   photosynthesis, 517
   phytoplankton abundance, 7*f*, 458–469
chromium (Cr)
   elemental abundances
      clastic sediments, 588–589, 589*f*
      Congo River, 308*f*
      hydrothermal vents, 73–74, 77, 77*t*
      rivers, 296*t*, 307*f*, 308*f*, 313*f*, 314–315
      sedimentary environments, 592*f*
   hydrolysis, 316*f*
   phytoplankton photosynthesis, 428*t*
cinnabar (HgS)
   occurrences, 137
clay minerals
   sedimentary environments, 587, 587*f*

clinopyroxenes
   sedimentary environments, 587*f*
clouds
   cloud condensation nuclei (CCN)
      sulfates, 154–155
coal
   organosulfur compounds, 147, 148*f*, 156
   preindustrial carbon cycle, 346*t*, 349*t*
   sulfur cycle, 147
   sulfur emissions, 155–156
cobalt (Co)
   atmospheric emissions, 27*t*
   complexation, 317*f*
   elemental abundances
      clastic sediments, 588–589, 589*f*
      Congo River, 308*f*
      hydrothermal vents, 60*t*
      rivers, 296*t*, 307*f*, 308*f*, 312*f*, 313*f*, 314–315
      sedimentary environments, 592*f*
   hydrolysis, 316*f*
coccolithophorids
   calcium carbonate (CaCO$_3$) production, 456–457
   deep-sea carbonate-rich sedimentary environments,
      495–496, 495*f*
   mineralogical evolution, 610*f*
coccoliths
   biomineralization, 428
   carbonate biomineralization, 445
   nitrogen fixation, 435–436
colloids
   aquatic systems
      analytical techniques, 321–322
      characteristics, 318–319, 319*f*
      colloid dynamics, 324–325
      dissolved organic carbon (DOC), 319–320, 320*f*,
         324–325, 325*f*
      general discussion, 318–325
      metal-humate stability constants, 320–321, 321*f*
      neodymium (Nd) concentrations, 324*f*
      nonorganic colloidal pool, 321–322
      rare earth elements (REEs) fractionation, 322–324, 323*f*
      trace element speciation, 319–321
      ultrafiltration techniques, 319–321, 320*f*
   high molecular weight dissolved organic matter (HMW
      DOM), 404–405
Columbia River, 296*t*, 397
condensation processes
   fractionation, 135
Congo River, 296*t*, 308*f*
Connecticut River, 296*t*, 323*f*, 330
continental crust
   age distribution, 582–583, 582*f*
   carbon cycle, 343*f*, 346*t*
   composition
      sedimentary environments, 592–593, 592*f*
   formation conditions
      recycling, 582–583
   isotopic evolution
      boron (B) isotopes, 604
      calcium (Ca) isotopes, 604–605
      carbon (δ^{13}C) isotope concentrations, 600–602,
         601*f*, 602*f*
      isotopic tracers, 603–605
      osmium (Os) isotopes, 598
      oxygen (δ^{18}O) isotope concentrations, 602–603, 603*f*
      strontium (Sr) isotopes, 595–598, 596*f*, 597*f*, 598*f*
      sulfur (S) isotopes, 598–600, 600*f*
   mass/age distribution, 584–585, 585*f*
   methane (CH$_4$) carbon subcycle, 344*f*, 353*t*
   as trace element source, 312–315, 313*f*
   upper continental crust
      elemental abundances, 588–589, 589*f*
      turbidites, 589*f*

copper (Cu)
  biological pump, 437–440, 439*f*
  complexation, 317*f*
  elemental abundances
    hydrothermal vents, 60*t*, 73–74, 74*f*
    rivers, 296*t*, 307*f*, 308*f*, 313*f*, 314–315
    sedimentary environments, 592*f*
  hydrolysis, 316*f*
  phytoplankton photosynthesis, 428*t*
corals
  skeletal mineralogy, 595*f*, 610*f*
cosmogenic nuclides
  chemical weathering, 220–221
crater lakes, 15*f*, 16, 140–141
Cretaceous
  silica concentrations, 436–437, 437*f*
cyanobacteria
  nitrogen fixation, 435–436
  photosynthesis
    Calvin cycle, 516–517
    evolution, 528–531
    oxygenic photosynthesis, 528
CYCLOPS model, 17*f*

D

D47 ice core, 353*t*
D57 ice core, 353*t*
dacites
  viscosity, 5*f*
  volatile elements, 7*t*, 12*t*
Dahl, Knut, 156–157
Dansgaard-Oeschger events, 362–363
dating techniques
  ice core gas analysis, 351–367
  lutetium/hafnium (Lu/Hf) isotopic ratios
    clastic sediments, 591
  radiometric dating techniques
    samarium/neodymium (Sm/Nd) isotopic ratios
      chondritic uniform reservoir (CHUR)
        trend, 591, 591*f*
      clastic sediments, 591, 593*f*
  see also cosmogenic nuclides; long-lived radioisotopes;
        radiocarbon (^{14}C)
deuterium (^{2}H)
  hydrogen (δ^2H) isotopes
    ice core analysis, 457*f*
Devonian
  black shale deposits, 535
  carbon cycle models, 371
  molecular oxygen (O_2), 535
  strontium (Sr) isotopic evolution, 598*f*
  sulfide minerals, 137
diamond (C)
  sulfide inclusions, 136
diatoms
  abundances, 7*f*
  biogenic opal (SiO_2) production, 478–481
  biological pump, 429, 436–437, 437*f*
  nitrogen fixation, 435–436
dimethyl sulfide (DMS)
  global fluxes, 152*f*
  seawater, 144–146, 145*f*
  tropospheric concentrations, 150
dinoflagellates, 435–436
dissolved organic matter (DOM)
  biological pump, 433–434
  dissolved organic carbon (DOC)
    aquatic systems, 319–320, 320*f*, 324–325, 325*f*,
        332–333
dolomite ($CaMg(CO_3)_2$)
  magnesium/calcium (Mg/Ca) ratio, 607–608, 608*f*
  mass/age distribution, 585*f*, 605*f*

Phanerozoic sediments, 606–607, 606*f*, 607*f*
  river basin lithology, 282*t*
  weathering, 367–368
Dome C ice core, 353*t*, 356–357, 366*f*
Don River, 290*f*
Douna River, 296*t*
dust
  ice core analysis, 360*f*
  organic carbon cycle, 392
  sulfur (S) concentrations, 158
Dye 3 ice core, 353*t*
dysprosium (Dy)
  elemental abundances
    hydrothermal vents, 58*f*
    rivers, 296*t*, 307*f*, 308*f*, 313*f*, 323*f*
    sedimentary environments, 592*f*

E

Earth
  age distribution, 580–582, 580*f*, 581*f*
  Archaean
    sulfur (S) concentrations, 135–137
  crust
    carbon (C) abundances, 190
    degassing models, 188–190
    formation, 189
    nitrogen (N) abundances, 192
    noble gases, 177–178
    potassium/argon (^{40}K-^{40}Ar) isotopic budget, 188–189
    subduction fluxes, 177–178
    water (H_2O), 193–194
  degassing models, 167–204
  early history
    molecular oxygen (O_2) concentrations, 522–524, 523*f*
    origins of life, 141
    RNA molecules, 141
    sulfur (S) concentrations, 135–137
    volcanic emissions, 526
  formation conditions
    nitrogen (N) compounds, 552
  inner core
    photosynthesis evolution
      anoxygenic photosynthesis, 136
    population dynamics, 580–582
    volatile elements, 190–194
  see also mantle (Earth); net primary production (NPP)
East Pacific Rise (EPR)
  hydrothermal vents, 46–47, 48–49, 64–66, 72*f*, 78
ecosystem dynamics *see* net primary
        production (NPP)
Eel River Basin, 416
El Chichón volcano, 12–13, 24*t*, 27*t*, 29
elements
  volatile elements
    accretion, 135
    atmospheric composition, 170*t*
    elemental abundances, 190–194
    magmas
      abundances, 6–8, 7*t*
      human health effects, 4–6
      magmatic volatile phase (MVP), 4
      partitioning, 4
      seawater, 6–8
      solubility, 8–9
      sources, 6–8, 7*t*
      speciation, 8–9
    mantle (Earth)
      carbonatites, 12*t*
      volcanism, 3
    noble gases, 167–204
    surface inventories, 169, 169*t*
Emiliania huxleyi, 445

Eocene
  silica concentrations, 436–437, 437*f*
EQ3/6 codes, 233
erbium (Er)
  elemental abundances
    hydrothermal vents, 58*f*
    rivers, 296*t*, 307*f*, 308*f*, 313*f*, 323*f*
    sedimentary environments, 592*f*
Erta 'Ale, Ethiopia, 12*t*, 15*f*, 23
estuaries
  nitrogen cycle, 564–565, 571–572
  organic carbon cycle, 395
  sulfur cycle, 146
eukaryotes
  anoxygenic photosynthesis, 136
  glycolysis, 518, 518*f*
  Krebs cycle, 518, 518*f*
  oxidative phosphorylation, 518, 518*f*
  photosynthesis, 516, 528
  respiration pathways, 518–522, 518*f*
europium (Eu)
  elemental abundances
    hydrothermal vents, 58*f*
    rivers, 296*t*, 308*f*, 312*f*, 323*f*
    sedimentary environments, 592*f*
evaporation processes
  seawater, 137
evaporite minerals
  carbon cycle, 538, 541*f*
  deposition rates, 138, 138*f*
  evolutionary trends, 605, 606*f*
  mass/age distribution, 585, 585*f*, 605*f*
  Phanerozoic
    sediments, 606*f*
  river basin lithology, 282*t*
  sulfur cycle, 137, 538, 541*f*

F

feldspar
  hydrolysis reactions, 207–208
  mineral fluxes, 223*t*
  sandstones, 587*f*
  sedimentary environments, 588*f*
  weathering rates
    characteristics, 212*t*
    dissolution rates, 227
    fluid fluxes, 229*f*
    reaction times, 224*f*
fertilizers
  carbon dioxide fertilization, 112–117, 113*t*
  nitrogen fertilization, 113–114, 113*t*
  sulfur (S) concentrations, 158
Flin Flon Belt, 530
fluorine (F)
  difluorooxosilane (SiOF$_2$), 27*t*
  elemental abundances
    hydrothermal vents, 60*t*
    magmas, 7*t*
  hydrofluoric acid (HF)
    animal health, 35
    volcanic emissions, 12*t*, 26*t*, 35
  silicon tetrafluoride (SiF$_4$), 27*t*
  solubility, 9
forest ecosystems
  carbon cycle
    acid rain, 113*t*, 114
    Amazon, Brazil, 115, 116*f*
    carbon dioxide fertilization, 112–117, 112*t*, 113*t*
    climate effects, 113*t*, 114
    disturbance mechanisms, 101*t*, 117–118
    nitrogen fertilization, 113–114, 113*t*
    northern mid-latitudes, 106–108, 108*t*, 111

    regional distributions, 107*t*
    reservoirs, 90, 90*t*
    sources/sinks, 103, 106, 111–118, 111*f*, 113*t*
    synergistic mechanisms, 116
    terrestrial carbon models, 117
    tropics, 108–110, 110*t*, 111
    tropospheric ozone concentrations, 113*t*, 114
  deforestation
    carbon cycle, 113*t*, 115, 116*f*
    tropics, 110, 110*t*
fossil fuels
  anthropogenic emissions
    annual observations, 95*f*
    carbon dioxide (CO$_2$) concentrations, 94–95, 103*t*
    global carbon budget, 96*t*, 101–106
    sources/sinks, 102*f*
  calcium carbonate (CaCO$_3$) dissolution, 509
  carbon cycle
    anthropogenic emissions, 88
    contemporary carbon cycle, 89*f*, 89*t*
    preindustrial cycle, 343*f*, 346*t*
    reservoirs, 91
  molecular oxygen (O$_2$) concentrations, 512, 513*f*
  nitrogen cycle, 560, 560*f*, 568–569, 570*f*
Frasch process, 139
Fraser River, 296*t*
fructose, 400–401, 401*f*
fullerenes, 536
fumaroles
  gas emissions, 14–15, 15*f*
  sulfur cycle, 140
  volatile elements, 7*t*
fumigants *see* pesticides
fungicides *see* pesticides

G

gadolinium (Gd)
  elemental abundances
    hydrothermal vents, 58*f*
    rivers, 296*t*, 308*f*, 313*f*, 323*f*
    sedimentary environments, 592*f*
galactose, 400–401, 401*f*
Galapagos Spreading Center, 46–47, 49, 69
galena (PbS)
  occurrences, 137
gallium (Ga)
  elemental abundances
    rivers, 296*t*, 307*f*, 308*f*
    sedimentary environments, 592*f*
  hydrolysis, 316*f*
Ganges River, India
  trace elements, 296*t*
Garonne River, 296*t*
Gayon, Ulysse, 553
gel polymers, 404–405, 404*f*, 405*f*
GEOCARB model, 371, 372*f*
geochemical modeling
  carbon cycle
    GEOCARB model, 371, 372*f*
    gradual change simulations, 375–376
    ocean carbon models, 98
    terrestrial carbon models, 117
  chemical weathering rates, 233–234
  CYCLOPS model, 17*f*
  EQ3/EQ6 code, 233
  geochemical codes
    The Geochemist's Workbench™ (GWB), 233
  GIMRT code, 233
  IDREAT code, 233, 234*f*
  MAGIC code, 233
  mantle (Earth)
    mantle degassing, 179–188

oceans
  ocean carbon models, 98
PATHARC.94 code, 233
Phanerozoic atmospheric evolution,
    537–545, 539*f*
PHREEQC code, 233
PROFILE code, 233
riverine ionic concentrations, 284–286, 285*f*
SAFE code, 233
SMART code, 233
UNSATCHEM code, 233
The Geochemist's Workbench™ (GWB), 233
geological timescale
  carbon cycle, 342–350, 343*f*
  methane (CH_4) carbon subcycle, 342–350, 344*f*
germanium (Ge)
  biological pump, 437–440
  elemental abundances
    rivers, 296*t*, 307*f*, 308*f*
    sedimentary environments, 592*f*
  hydrolysis, 316*f*
geysers, 14–15, 15*f*
gibbsite
  sedimentary environments, 587*f*
  solubility, 326*f*
GIMRT code, 233
GISP2 ice core, 20, 20*f*, 353*t*, 358
glaciation
  atmospheric carbon dioxide concentrations, 17*f*, 468*f*
  biological pump, 17*f*, 25*f*, 455–456, 457*f*, 471*f*,
      475–478, 478–481
  Dansgaard-Oeschger events, 362–363
  foraminifera analysis, 502–503
  glacial sediments, 216–218
  glaciated terrain
    carbonate hydrolysis
      carbonate isotope stratigraphy, 532–533, 533*f*
    interglacial periods, 502–503, 504*f*
    last glacial maximum (LGM)
      biological pump, 475–478
    Little Ice Age, 356
    Medieval Warm Period, 356
    methane (CH_4) carbon subcycle, 350
    Milankovitch cycles, 361–362, 456
    ocean chemistry, 502–503, 508
    paleoclimatology, 361–362, 362–365, 365–367
    preindustrial carbon cycle, 345–346, 359–367, 360*f*,
        372*f*, 373
    seawater composition, 500, 502–503
    Snowball Earth, 3, 532–533, 546*f*
    termination events, 365, 366*f*
glauberite
  sulfur (S) concentrations, 137
global warming
  Antarctica, 365, 366*f*
  ocean uptake, 119
glucose, 374–375, 400–401, 401*f*
gneiss
  river basin lithology, 282*t*
gold (Au)
  elemental abundances
    sedimentary environments, 592*f*
  hydrolysis, 316*f*
Goldschmidt, V. M., 130
granites
  river basin lithology, 282*t*
  weathering rates
    fluid fluxes, 227–228
    runoff, 232*f*
    silicate weathering, 227, 228*f*
graywacke
  evolutionary trends, 605
  mass/age distribution, 585*f*, 605*f*

Great Oxidation Event, 375–376
greenhouse gases
  carbon cycle, 88, 96, 455–456
  methane (CH_4), 88, 96–97
  preindustrial levels, 361
  volcanism, 3
  see also methane (CH_4)
Greenland
  climate effects, 362–365, 365–367
  Isua Supercrustal Belt
    carbon ($\delta^{13}C$) isotope concentrations, 527–528
GRIP ice core, 18*f*, 20, 353*t*, 358, 362*f*
groundwater
  aquifers
    calcium distributions, 217–218
    hydrologic properties, 216–217, 217*f*
    silicon distributions, 217–218
    sulfur cycle, 146–147
  chemical weathering effects, 216–218
  molecular oxygen (O_2) concentrations, 516
  nitrogen cycle, 570*f*, 571–572, 571*t*
  sulfur cycle, 146–147
Guaymas Basin, 61, 65–66
Gulf of Mexico
  carbon cycle, 396, 396*f*
gypsum ($CaSO_4$ $2H_2O$)
  rivers, 280*t*, 282*t*
  sulfur (S) concentrations, 137

H

Haber, Fritz, 559–560
hafnium (Hf)
  elemental abundances
    clastic sediments, 588–589, 589*f*
    Congo River, 308*f*
    rivers, 296*t*, 307*f*, 308*f*, 313*f*
    sedimentary environments, 592*f*
  hydrolysis, 316*f*
Hales, Burke, 496–497
halite (NaCl)
  hydrothermal vents, 53–54, 55*f*
  rivers, 280*t*
  volcanic emissions, 27*t*
halogen chemistry
  volcanic emissions, 26–27, 27*t*
halons *see* CFCs
Harz Mountains, Germany, 296*t*, 315
Hawaiian Islands
  helium (He) isotopes, 171*f*
HCFCs *see* CFCs
helium (He)
  elemental abundances
    surface inventories, 169*t*
  isotopes
    degassing models, 181–183, 182*f*, 186
    hydrothermal vents, 46, 61, 70–71, 72*f*
    mantle (Earth), 169–170, 171*f*, 181–183, 182*f*, 186
    mantle fluxes, 175–176, 176–177
    mid-ocean ridge basalts (MORBs), 169–170, 171*f*
    nuclear processes, 169*t*
    ocean island basalts (OIBs), 169–170
  isotopic abundances
    atmospheric composition, 170*t*
    volcanic emissions, 22*t*
Helland, Amund, 156–157
Hellriegel, Herman, 553
heterotrophs, 76
holmium (Ho)
  elemental abundances
    hydrothermal vents, 58*f*
    rivers, 296*t*, 307*f*, 308*f*, 313*f*, 323*f*
    sedimentary environments, 592*f*

Holocene
  biological pump, 455–456, 455*f*
  calcium carbonate (CaCO₃) concentrations, 505, 506*f*
  ice core gas analysis, 351–367, 353*t*
  methane (CH₄) carbon subcycle, 348–350
  preindustrial carbon cycle, 342–348
hornblende
  chemical weathering, 210–211, 214*f*
  mineral fluxes, 223*t*
  sedimentary environments, 587*f*
  weathering rates, 212*t*
hot springs, 14–15
Huanghe River, 296*t*, 313*f*
Huaynaputina volcano, 25*t*
Hudson River, 296*t*, 323*f*, 394*f*
hydrocarbons
  natural gas
    methane (CH₄) carbon subcycle, 344*f*, 353*t*
    preindustrial carbon cycle, 346*t*, 349*t*
  sulfur emissions, 155–156
hydrochlorofluorocarbons (HCFCs) *see* CFCs
hydrogen (H)
  hydrothermal vents, 71–72
  isotopes
    atmospheric chemistry, 170*t*
  molecular hydrogen (H₂)
    hydrothermal vents, 60*t*
    volcanic emissions, 12*t*, 22*t*
hydrothermal processes
  hydrothermal circulation
    discharge zones, 67–68
    reaction stages, 52*f*
    reaction zones, 67–68
    recharge zones, 67
  hydrothermal fluxes
    heat and water volume, 51*t*
    heat flow, 46, 46*f*
    importance, 50–51
  hydrothermal plumes
    biogeochemical interactions, 76, 76*f*
    buoyancy, 69–70, 70*f*
    chemical composition, 70–75, 71*f*
    dispersion processes, 75–76
    dynamics, 69–70, 70*f*
    helium (He) isotopes, 70–71, 72*f*
    hydrogen sulfide (H₂S), 71–72
    iron (Fe) concentrations, 72–73
    iron (Fe) oxyhydroxides, 74–75, 79
    iron (Fe) uptake, 73–74, 74*f*
    manganese (Mn) concentrations, 72–73
    metalliferous ridge and basal sediments, 78
    methane (CH₄), 71–72
    molecular hydrogen (H₂), 71–72
    noble gases, 70–71
    ocean chemistry, 77, 77*t*
    radon (²²²Rn), 70–71
    rare earth elements (REEs), 74–75
    redox processes, 72–73
    trace elements, 73–74, 74*f*
    trace metals, 73–74
    tracers, 74–75
  hydrothermal vents
    chemical composition, 51–66
    East Pacific Rise, 46–47, 48–49, 72*f*, 78
    Galapagos Spreading Center, 46–47, 49, 69
    global distribution, 47*f*, 48–49
    Guaymas Basin, 65–66
    Juan de Fuca Ridge, 48*f*, 65
    metalliferous ridge and basal sediments, 46, 47*f*, 69, 77–79
    molecular oxygen (O₂) consumption, 521
    origins of life, 141

  RNA molecules, 141
    sulfur cycle, 139
    vent fluids, 48, 48*f*
  near-vent deposits
    formation conditions, 68, 68*f*
    Lost City vent site, 69
    metalliferous ridge and basal sediments, 77–78
    mineralization, 67–68, 68–69
    sulfide mounds, 67*f*, 68–69
  net impact, 66–67
  vent fluids
    alkalinity, 60*t*
    biological processes, 56
    carbon dioxide (CO₂), 60*t*, 61
    chemical composition, 48, 51–66, 58*f*, 60*t*, 70–75, 71*f*
    chlorinity, 53, 55*f*, 58*f*
    East Pacific Rise, 64–66
    geographic variations, 62–64
    helium (He) isotopes, 61
    hydrogen sulfide (H₂S), 60*t*, 61
    hydrothermal vents, 48*f*
    magmatic degassing, 56
    major elements, 56–59, 59*f*, 60*t*
    methane (CH₄), 60*t*, 61
    noble gases, 61
    nutrient concentrations, 61
    organic geochemistry, 62
    pH, 58, 60*t*
    phase separation, 52–56, 52*f*, 55*f*, 58*f*
    radon (²²²Rn), 61
    rare earth elements (REEs), 56, 58*f*, 79
    spatial variations, 64
    spreading rates, 63–64
    steady-state conditions, 56
    substrate compositions, 62
    sulfate concentrations, 53
    sulfide minerals, 58–59, 60*t*
    temperature/pressure processes, 62–63
    temporal variations, 64–66
    time-series measurements, 58*f*
    trace metals, 59–61
    water-rock interactions, 52–56, 52*f*, 55*f*
  volcanism, 64–66
hydroxyl radical (OH)
  methane (CH₄), 350

I

ice cores
  Antarctica, 455–456, 457*f*
  carbon cycle
    carbon dioxide (CO₂) concentrations, 95–96, 455–456, 457*f*
    glacial/interglacial variations, 359–367, 360*f*, 362*f*, 455–456
    Holocene variations, 355–359, 356*f*, 357*f*, 358*f*
    ice core gas analysis, 351–367, 353*t*
  hydrogen (δ²H) isotopes, 457*f*
  molecular oxygen (O₂) concentrations, 512, 513*f*
  timescale calibration, 351–367
  volatile measurement techniques, 18*f*, 20–21, 20*f*
Idel River, 296*t*
IDREAT code, 233, 234*f*
Illinois River, 296*t*
illite
  sedimentary environments, 587, 587*f*
Indian Ocean
  biological pump, 476
  calcium carbonate (CaCO₃) abundance, 493–495, 508
  hydrothermal vents, 46–47, 46*f*, 63–64
  nitrogen cycle, 565–566, 567*t*

Indin River, 296*t*
indium (In)
  complexation, 317*f*
  hydrolysis, 316*f*
Indus River, 296*t*
insecticides *see* pesticides
interstellar medium (ISM)
  elemental abundances, 135
iodine (I)
  isotopes
    iodine/xenon (I/Xe) isotopic ratios, 178–179
iridium (Ir)
  elemental abundances
    sedimentary environments, 592*f*
iron (Fe)
  biological pump, 437–440, 443–444, 443*f*
  chemical weathering, 210*t*, 211*f*
  complexation, 317*f*
  elemental abundances
    clastic sediments, 588–589, 589*f*
    Congo River, 308*f*
    hydrothermal vents, 60*t*
    rivers, 296*t*, 307*f*, 308*f*, 312*f*, 313*f*
    sedimentary environments, 592*f*
  Great Oxidation Event, 375–376
  hydrolysis, 316*f*
  hydrothermal plumes, 73–74, 74*f*
  ionic concentrations
    hydrothermal plumes, 72–73
  iron dichloride (FeCl$_2$), 27*t*
  phytoplankton photosynthesis, 428*t*
  solubility, 325, 326*f*
  trace metal concentrations
    uptake, 73–74, 74*f*
isotopes
  long-lived radioisotopes
    particle flux measurements, 442, 442*f*

J

jaspilites, 605*f*
Juan de Fuca Ridge
  hydrothermal vents, 48*f*, 65
Junge layer, 158–159
Jupiter
  Europa
    sulfur cycle, 160–161, 161*f*
  Io
    sulfur cycle, 160
  sulfur cycle, 160
Jurassic
  anoxic environments, 536
  black shale deposits, 536

K

Kaapvaal craton, South Africa
  clastic sediments, 590
Kalix River, 296*t*, 310, 323*f*, 324–325
kaolinite
  hydrolysis reactions, 207–208
  sedimentary environments, 587, 587*f*
  solubility, 326*f*
Kasai River, 296*t*
Katmai volcano, 25*t*
Kawah Ijen, Indonesia, 15*f*
Kerguelen Igneous Province
  argon (Ar) isotopes, 172*f*
kerogen
  organic carbon cycle, 390–391, 391*f*
Kilauea Caldera
  carbon dioxide (CO$_2$) emissions, 26
  environmental impacts, 35

gas emissions, 11–12, 12*t*
  sulfur emissions, 24*t*
  volatile elements, 7*t*
Krakatau volcano
  halogen compounds, 27*t*
  sulfur emissions, 25*t*
Krebs cycle, 518, 518*f*
krypton (Kr)
  elemental abundances
    surface inventories, 169*t*
  isotopic abundances
    atmospheric composition, 170*t*
Kudriavy volcano, Kurile Islands, 6–8, 7*t*
Kuwae volcano, 25*t*
kyeserite, 137

L

lakes
  nitrogen cycle, 571–572
  rivers, 280*t*
  sulfur cycle, 146–147
Laki fissure, Iceland, 33, 35
lanthanum (La)
  complexation, 317*f*
  elemental abundances
    Congo River, 308*f*
    hydrothermal vents, 58*f*
    rivers, 296*t*, 307*f*, 308*f*, 312*f*, 313*f*, 323*f*
    sedimentary environments, 592*f*
  hydrolysis, 316*f*
Lanyang Hsi River, 394*f*
Late Paleocene Thermal Maximum, 374–375, 374*f*
Lavoisier, Antoine, 552–553
Law Dome ice core, 353*t*, 355, 356*f*
lead (Pb)
  complexation, 317*f*
  elemental abundances
    clastic sediments, 588–589, 589*f*
    hydrothermal vents, 60*t*, 73–74
    rivers, 296*t*, 307*f*, 308*f*, 312*f*, 313*f*, 314–315
    sedimentary environments, 592*f*
  hydrolysis, 316*f*
Lena River, 296*t*
life
  beginnings, 141
  photosynthesis evolution
    anoxygenic photosynthesis, 136
    RNA molecules, 141
limestones
  mass/age distribution, 585, 585*f*, 605*f*
  river basin lithology, 282*t*
Lindgren, Waldemar, 130
lithium (Li)
  elemental abundances
    hydrothermal vents, 60*t*
    rivers, 296*t*, 307*f*, 308*f*
    sedimentary environments, 592*f*
Little Ice Age, 356
Lohmann, Pat, 501–502
Loihi Seamount
  argon (Ar) isotopes, 172, 172*f*
  helium (He) isotopes, 169–170, 176
  xenon (Xe) isotopes, 174*f*
long-lived radioisotopes
  particle flux measurements, 442, 442*f*
Lost City Vent Field, 69
Luce River, 332
lutetium (Lu)
  complexation, 317*f*
  elemental abundances
    hydrothermal vents, 77*t*

lutetium (Lu) (*continued*)
    rivers, 296*t*, 307*f*, 308*f*, 313*f*, 323*f*
    sedimentary environments, 592*f*
  hydrolysis, 316*f*

M

Mackenzie River
  organic carbon cycle, 394*f*
  trace elements, 296*t*, 313*f*
Madeira River, 296*t*, 323*f*, 324*f*
mafic rocks *see* subduction zones
MAGIC code, 233
magmas
  degassing
    bubble coalescence, 10–11
    bubble growth, 10
    chemical composition, 12*t*, 16
    excess degassing, 12–13
    exsolution, 9
    fragmentation, 12
    gas separation, 11–12, 12*t*
    gas speciation, 22*t*
    nucleation, 10
    saturation, 9–10
    supersaturation, 10
    vesiculation, 4, 9–13
    viscosity, 10–11
    volatile measurements, 16–23
  magmatic-hydrothermal systems, 21, 21*f*
  magmatic volatile phase (MVP), 4
  sulfur (S) compounds, 139–140
  viscosity, 4, 5*f*
  volatile elements
    abundances, 6–8, 7*t*
    human health effects, 4–6
    measurement techniques, 16–23
    partitioning, 4
    solubility, 8–9
    sources, 6–8, 7*t*
    speciation, 8–9
  volatile measurement techniques
    geochemical surveillance methods, 21–23
    ice cores, 18*f*, 20–21, 20*f*
    petrological methods, 18*f*, 19–20
    portable remote sensing systems, 18–19, 18*f*
    satellite remote sensing systems, 18*f*, 19
    in situ measurements, 17–18, 18*f*
    sulfate markers, 20–21, 20*f*
magmatic arcs
  andesites, 6–8
  volatile elements, 6–8, 7*t*
magnesium (Mg)
  chemical weathering
    elemental mobility, 210*t*, 211*f*
    watersheds, 219*f*
  complexation, 317*f*
  elemental abundances
    clastic sediments, 588–589, 589*f*
    hydrothermal vents, 60*t*
    rivers, 308*f*, 313*f*
    sedimentary environments, 592*f*
  hydrolysis, 316*f*
  hydrothermal vents, 58*f*
  ionic concentrations
    anthropogenic sources, 290–291, 290*t*
    idealized model, 284–286, 285*f*
    pristine river distributions, 279*t*, 282*t*
    river sources/sinks, 280*t*
    runoff, 283–284, 284*f*, 286–287, 286*t*, 287*f*
    weathering rates, 289*t*
  phytoplankton photosynthesis, 428*t*
Mammoth Mountain, United States, 15*f*

Mamore River, 296*t*
manganese (Mn)
  complexation, 317*f*
  elemental abundances
    clastic sediments, 588–589, 589*f*
    hydrothermal vents, 60*t*
    rivers, 296*t*, 307*f*, 308*f*, 311, 311*f*
    sedimentary environments, 592*f*
  hydrolysis, 316*f*
  ionic concentrations
    hydrothermal plumes, 72–73
  phytoplankton photosynthesis, 428*t*
mannose, 400–401, 401*f*
mantle (Earth)
  carbon cycle, 343*f*, 346*t*
  composition
    volatile elements, 2–4
  degassing
    core/mantle boundary, 187
    early earth models, 180
    interacting reservoirs, 183–185
    multiple reservoir model, 181–183, 182*f*
    open-system models, 185–187
    single reservoir model, 180–181
  noble gases
    abundances, 174–175, 175*f*
    argon (Ar) isotopes, 171–173, 172*f*
    degassing models, 180–181, 181–183, 182*f*,
      183–185, 185–187
    fluxes, 175–176, 176–177
    helium (He) isotopes, 169–170, 171*f*
    neon (Ne) isotopes, 170–171, 171*f*
    xenon (Xe) isotopes, 173–174, 174*f*
  upper mantle
    carbon (C) abundances, 190–191
    degassing models, 181–183, 182*f*, 183–185,
      185–187
    nitrogen (N) abundances, 192
    water (H_2O), 193
  see also mid-ocean ridge basalts (MORBs); peridotites
marble, 282*t*
marcasite (FeS_2)
  hydrothermal vents, 68*f*
marine sediments
  aerobic respiration, 519–520
  barium accumulations, 472
  carbon cycle
    contemporary carbon cycle, 89*f*, 89*t*
    dissolved organic carbon (DOC), 391
    methane (CH_4) carbon subcycle, 353*t*
    preindustrial carbon cycle, 343*f*, 346*t*, 347
    terrigenous fluxes, 396–398, 396*f*
  continental margin sediments
    sulfur cycle, 146–147
  deep-sea carbonate-rich sedimentary environments
    calcium carbonate ($CaCO_3$), 491–510
  mass/age distribution, 585*f*
  molecular oxygen (O_2) concentrations, 515, 519–520
  organic carbon cycle, 389–424
  organic matter decomposition
    burial rate, 454, 455*f*, 456*f*
    heterotrophic respiration, 454, 456*f*
  sedimentary environments
    black carbon, 397–398, 399*f*
    contemporary carbon cycle, 89*f*, 89*t*
    methane (CH_4) carbon subcycle, 353*t*
    preindustrial carbon cycle, 343*f*, 346*t*, 347
    preservation, 406–414
    terrigenous fluxes, 396–398, 396*f*
  siliciclastic sediments
    carbon-sulfur (C-S) relationships, 542–543, 543*f*
  sulfur (S)
    sulfur cycle, 146–147

Mars
  atmospheric chemistry
    argon (Ar) isotopes, 194
    degassing, 194–196
    radiogenic isotopes, 194
    xenon (Xe) isotopes, 194–195
  mantle
    noble gases, 195
Masaya, Nicaragua, 15*f*, 24*t*, 26, 27–36
mass extinction
  Cenomanian-Turonian Boundary (CTB) extinction
      event, 536
  molecular oxygen (O$_2$), 535–536
  Permian-Triassic (P-Tr) boundary
    molecular oxygen (O$_2$), 535–536
Mauna Loa volcano, 24*t*, 95–97, 115–116
Mbam River, 324*f*
McCorkle, Dan, 497
Medieval Warm Period, 356
Mekong River, 296*t*
Mengong River, 296*t*, 324*f*
Merapi volcano, Indonesia, 24*t*
mercury (Hg)
  complexation, 317*f*
  elemental abundances
    rivers, 314–315
  hydrolysis, 316*f*
Mesoproterozoic
  atmospheric evolution, 531–532, 546*f*
Mesozoic
  carbon cycle models, 371
  ocean chemistry, 593–605
metabolism
  C$_1$ metabolism, 519
  inorganic metabolism, 519
  molecular oxygen (O$_2$), 534
methane (CH$_4$)
  atmospheric concentrations, 90
  atmospheric lifetime, 348–350
  carbon cycle
    atmospheric concentrations, 90, 344*f*, 348–350, 353*t*
    continental crust, 344*f*, 353*t*
    fluxes, 344*f*, 348–350, 353*t*
    geological timescales, 342–350, 343*f*, 350–351
    glacial/interglacial variations, 350, 359–367, 360*f*, 362*f*
    greenhouse gases, 88, 96–97
    Holocene variations, 355–359, 356*f*, 357*f*, 358*f*
    ice core gas analysis, 351–367, 353*t*
    marine sediments, 353*t*
    methane hydrates, 353*t*
    microbial processes, 369
    natural gas, 344*f*, 353*t*
    oceans, 344*f*, 348, 353*t*
    reservoirs, 344–345, 344*f*, 348–350, 353*t*
    soils, 344*f*, 353*t*
    termites, 353*t*
    volcanism, 353*t*
    wetlands, 344*f*, 353*t*, 359
    wildfires, 353*t*
  greenhouse gases, 88, 96–97
  hydrothermal vents, 60*t*, 61, 71–72
  hydroxyl radical (OH), 350
  methane hydrates, 350, 351*f*
  methanogenesis
    Archaean, 526–527
  tropospheric concentrations, 350
  volcanic emissions, 22*t*
Meuse River, 394*f*
microcline
  chemical weathering, 210–211, 214*f*
mid-ocean ridge basalts (MORBs)
  argon (Ar) isotopes, 171–173, 172*f*, 174–175
  carbon (C) abundances, 190–191

degassing models, 180–181
helium (He) isotopes, 169–170, 171*f*
neon (Ne) isotopes, 170–171, 171*f*
nitrogen (N) abundances, 192
noble gases
  abundances, 174–175, 175*f*
  fluxes, 175–176, 176–177
  helium (He) isotopes, 169–170, 171*f*
  neon (Ne) isotopes, 170–171, 171*f*, 174–175, 175*f*
  xenon (Xe) isotopes, 173–174, 174–175, 174*f*, 175*f*,
      183–185, 184*f*
noble gas fluxes, 175–176, 176–177
turbidites, 589*f*
volatile elements, 7*t*, 12*t*
water content, 193
water (H$_2$O), 193
xenon (Xe) isotopes, 173–174, 174*f*, 183–185, 184*f*
Milankovitch cycles, 361–362, 456
mineral dissolution
  feldspar
    dissolution rates, 227
  plagioclase
    dissolution rates, 227, 228*f*
  sedimentary environments, 588
mining
  acid mine drainage
    sulfur (S) concentrations, 157–158
  sulfur mineral extraction, 138–139, 139*f*
Mississippi River
  adsorption processes, 330–331, 331*f*
  boric acid (B(OH)$_3$), 331
  organic carbon cycle, 394*f*, 396, 396*f*
  rare earth elements (REEs), 330
  trace elements, 296*t*, 310
Missouri River, 296*t*
Mistassini River, 296*t*
modeling *see* geochemical modeling
mofettes, 15–16
molecular oxygen (O$_2$)
  air-sea exchange, 512–515, 514*f*, 515*f*, 517
  atmospheric concentrations, 102, 102*f*, 512
  atmospheric evolution, 522–545, 545*f*, 546*f*
  carbon-sulfur (C-S) relationships, 542–543, 543*f*
  consumption mechanisms
    abiotic organic matter oxidation, 521–522
    aerobic cellular respiration, 518, 518*f*
    aerobic respiration, 519–520
    C$_1$ metabolism, 519
    chemolithoautotrophy, 519
    glycolysis, 518, 518*f*
    hydrothermal vents, 521
    inorganic metabolism, 519
    iron oxidation, 521
    Krebs cycle, 518, 518*f*
    mineral oxidation, 520
    oxidative phosphorylation, 518, 518*f*
    photorespiration, 518–519
    sulfur oxidation, 521
    volcanic emissions, 520
  deep water concentrations, 474–475, 514*f*
  formation conditions, 552
  fossil fuels, 512, 513*f*
  freshwater ecosystems, 515
  global oxygen budgets, 522, 523*f*
  historical background
    Archaean, 522–524, 523*f*, 524–528
    banded iron formations (BIFs), 522–524, 523*f*
    glaciation, 532–533, 533*f*
    Mesoproterozoic, 531–532
    Neoproterozoic, 532–534
    oxygenic photosynthesis, 528–531
    Phanerozoic, 534–545, 545*f*
    Proterozoic, 528–534

molecular oxygen (O$_2$) (*continued*)
    hydrothermal processes, 544
    marine sediments, 515
    net primary production (NPP), 517, 523*f*
    ocean chemistry, 464
    oxygen cycle, 511–550
    phosphorus (P), 543
    photosynthesis, 512–513, 516–517, 516*f*, 528
    pore waters, 515
    production mechanisms
        photosynthesis, 516–517, 516*f*
        water vapor photolysis, 517–518
    pyritization, 545
    reservoirs
        atmosphere, 512
        freshwater ecosystems, 515
        groundwater, 516
        oceans, 512–515
        soils, 516
        thermohaline circulation, 512–513, 514*f*
    Rubisco, 544
    seasonality, 512, 513*f*
    seawater concentrations, 512–515, 514*f*
    surface inventories, 169*t*
molybdenum (Mo)
    atmospheric emissions, 27*t*
    elemental abundances
        hydrothermal vents, 77*t*
        rivers, 296*t*, 307*f*, 308*f*, 314–315
        sedimentary environments, 592*f*
    hydrolysis, 316*f*
Momotombo, Nicaragua, 12*t*
monosaccharides, 400–402, 401*f*
Mount Etna volcano
    aerosol emissions, 18*f*, 19, 24*t*
    environmental impacts, 27–36
    excess degassing, 12–13
    sulfur emissions, 24*t*, 25–26
Mount Logan, 353*t*
Mount St. Augustine, 12*t*
Mount St. Helens
    eruptive emissions, 16
    excess degassing, 13
    gas emissions, 12*t*
    halogen compounds, 27*t*
    sulfur emissions, 140*f*
    sulfur fluxes, 23
    volatile elements, 7*t*
Murray River, Australia, 323*f*
muscovite
    sedimentary environments, 587*f*
    weatherability, 225, 225*f*

N

N-acetyl galactosamine, 400–401, 401*f*
N-acetyl glucosamine, 400–401, 401*f*, 403–404
Negro River, 296*t*
Neman River, 290*f*
neodymium (Nd)
    clastic sediments, 589, 590*f*
    elemental abundances
        Congo River, 308*f*
        hydrothermal vents, 58*f*, 74*f*, 77*t*
        rivers, 296*t*, 307*f*, 308*f*, 312*f*, 313*f*, 323*f*
        sedimentary environments, 592*f*
        turbidites, 589*f*
    rivers, 324*f*
neon (Ne)
    elemental abundances
        surface inventories, 169*t*
    isotopes
        degassing models, 186

    mantle (Earth), 170–171, 171*f*, 186
        mid-ocean ridge basalts (MORBs), 170–171, 171*f*,
            174–175, 175*f*
        nuclear processes, 169*t*
        ocean island basalts (OIBs), 170–171
    isotopic abundances
        atmospheric composition, 170*t*
Neoproterozoic
    atmospheric evolution, 3, 532–534
net primary production (NPP)
    biological pump, 7*f*, 428–430, 454, 456*f*, 458–469
    carbon dioxide fertilization, 112–117, 113*t*
    environmental impacts, 113*t*, 114
    global distribution, 429*f*
    gross primary production (GPP), 92
    nutrients
        limitations, 430
        phytoplankton, 428–430
    organic carbon cycle, 391*f*
    photosynthesis, 92, 94, 94*t*
nickel (Ni)
    biological pump, 437–440, 439*f*
    complexation, 317*f*
    elemental abundances
        clastic sediments, 588–589, 589*f*
        Congo River, 308*f*
        hydrothermal vents, 60*t*
        rivers, 296*t*, 307*f*, 308*f*, 311, 311*f*, 312*f*, 313*f*, 314–315
        sedimentary environments, 592*f*
    hydrolysis, 316*f*
    phytoplankton photosynthesis, 428*t*
Niger River, 296*t*
niobium (Nb)
    chemical weathering, 209, 210*f*
    elemental abundances
        rivers, 296*t*, 307*f*, 308*f*, 313*f*
        sedimentary environments, 592*f*
    hydrolysis, 316*f*
Nitrobacter, 441
nitrogen cycle
    ammonia assimilation, 556
    ammonification, 556
    assimilatory nitrate reduction, 556
    atmospheric chemistry
        inorganic oxidized nitrogen, 555, 555*t*
        inorganic reduced nitrogen, 555, 555*t*
        nitrogen cascade, 569–570, 570*f*, 571*t*
        oxidized organic nitrogen, 556
        reactive nitrogen (Nr), 561*t*, 562*t*
        reduced organic nitrogen, 555–556, 555*t*
        valence states, 555*t*
    biosphere, 554*f*, 556
    chemical reactions, 553–554, 554*f*
    coastal regions, 571–572
    denitrification
        biological pump, 463
        continental shelf, 565
        marine ecosystems, 567*t*
        processes, 554*f*, 556
        reactive nitrogen (Nr), 561*t*, 562*t*
    estuaries, 564–565, 571–572
    fertilizers, 113–114, 113*t*
    groundwater, 570*f*, 571–572, 571*t*
    historical background, 552–553
    lakes, 571–572
    mantle (Earth), 192
    marine ecosystems
        biological nitrogen fixation (BNF), 565–566
        biological pump, 463
        denitrification, 567*t*
        global marine budgets, 565–566
        nitrogen cascade, 570*f*, 571–572, 571*t*
        nitrogen fixation, 567*t*

nitrogen reservoirs, 556–557
reactive nitrogen (Nr), 561*t*, 562*t*
nitrification, 554*f*, 556
nitrogen fixation
biosphere, 556
chemical reactions, 554*f*
legume cultivation, 553
marine ecosystems, 567*t*
phytoplankton, 463
nitrogen (N) abundances, 192
nitrogen reservoirs
air-land exchange, 556–557
air-sea exchange, 557
deposition processes, 557
marine ecosystems, 556–557
sea-land exchange, 557
pollution, 568–572
reactive nitrogen (Nr)
anthropogenic sources, 558–560, 561*t*, 562*t*
biological nitrogen fixation (BNF), 553, 557–558,
561*t*, 562*t*
chemical reactions, 553–554
creation rates, 560–561, 560*f*, 561*t*, 562*t*
cyanimide synthesis, 558–559, 559*t*
distribution, 563–564, 563*f*, 564*f*
environmental impacts, 568–572, 570*f*, 571*t*
food production, 558–560, 568–569, 570*f*
formation conditions, 552–553, 557–561
fossil fuel combustion, 560, 560*f*, 561*t*, 562*t*,
568–569, 570*f*
future esimates, 572–573, 573*f*
guano, 558–559, 559*t*
Haber-Bosch process, 553, 558–559, 559*t*, 560*f*,
561*t*, 562*t*
legume cultivation, 553, 558–559, 560*f*, 561*t*,
562–563, 562*t*
lightning, 557, 561*t*, 562*t*
molecular nitrogen (N$_2$) conversion, 564–565
nitrate extraction, 559*t*
nitrogen cascade, 568–572, 570*f*, 571*t*
regional nitrogen budgets, 566–568, 569*t*
redox processes, 554*f*
rivers, 561*t*, 562*t*, 564–565, 570*f*, 571–572
soils, 563–564, 564–565, 570*f*
terrestrial ecosystems
biological nitrogen fixation (BNF), 553, 557–558
global terrestrial budgets, 561–565, 561*t*, 562*t*, 563*f*
molecular nitrogen (N$_2$) conversion, 564–565
nitrogen cascade, 570–571, 570*f*, 571*t*
nitrogen reservoirs, 556–557
processes, 554*f*
reactive nitrogen (Nr), 561*t*, 562*t*
nitrogen (N)
ammonia (NH$_3$)
ammonia assimilation, 556
ammonification, 556
emissions, 563–564, 563*f*
formation conditions, 552, 554*f*
hydrothermal vents, 60*t*
biosphere, 554*f*, 556
carbon:nitrogen (C:N) ratios, 401
carbon:nitrogen:phosphorus (C:N:P) ratios, 461–462
elemental abundances
surface inventories, 169*t*
isotopes
Archaean, 525–526
atmospheric chemistry, 170*t*
geologic evolution, 525*f*
marine ecosystems, 556–557
molecular nitrogen (N$_2$)
chemical reactions, 553–554
emissions, 563–564, 563*f*
formation conditions, 552–553

volcanic emissions, 22*t*
nitrates (HNO$_3$)
biological pump, 7*f*, 454*f*, 458–469
depth profiles, 427*f*
high-nutrient, low chlorophyll (HNLC) areas, 443*f*
new production measurements, 440–441
nitrogen cycle, 435–436, 435*f*, 439*f*, 440–441
silica/nitrate ratios, 437, 438*f*
nitrogen fixation, 435–436
nitrification, 554*f*
nitrogen fertilization, 113–114, 113*t*
nitrogen oxides (NO$_x$)
emissions, 563–564, 563*f*
photosynthesis, 428*t*
reactive nitrogen (Nr)
anthropogenic sources, 558–560, 561*t*, 562*t*
biological nitrogen fixation (BNF), 557–558, 561*t*, 562*t*
chemical reactions, 553–554
creation rates, 560–561, 560*f*, 561*t*, 562*t*
cyanimide synthesis, 558–559, 559*t*
distribution, 563–564, 563*f*, 564*f*
environmental impacts, 568–572, 570*f*, 571*t*
food production, 558–560, 568–569, 570*f*
formation conditions, 552–553, 557–561
fossil fuel combustion, 560, 560*f*, 561*t*, 562*t*,
568–569, 570*f*
future esimates, 572–573, 573*f*
guano, 558–559, 559*t*
Haber-Bosch process, 553, 558–559, 559*t*, 560*f*,
561*t*, 562*t*
legume cultivation, 553, 558–559, 560*f*, 561*t*,
562–563, 562*t*
lightning, 557, 561*t*, 562*t*
molecular nitrogen (N$_2$) conversion, 564–565
nitrate extraction, 559*t*
nitrogen cascade, 568–572, 570*f*, 571*t*
nitrogen fixation, 553
regional nitrogen budgets, 566–568, 569*t*
volatile cycling, 192–193
Nitrosomonas, 441
Nitrospira, 441
noble gases
hydrothermal vents, 61, 70–71
iodine/xenon (I/Xe) isotopic ratios, 178–179
mantle (Earth)
abundances, 174–175, 175*f*
argon (Ar) isotopes, 171–173, 172*f*, 180–181, 186
core/mantle boundary, 187
degassing models, 180–181, 181–183, 182*f*,
183–185, 185–187
fluxes, 175–176, 176–177
helium (He) isotopes, 169–170, 171*f*, 181–183, 182*f*, 186
neon (Ne) isotopes, 170–171, 171*f*, 186
xenon (Xe) isotopes, 183–185, 184*f*, 186
Mars, 195
mid-ocean ridge basalts (MORBs), 174–175, 175–176,
175*f*, 176–177
ocean island basalts (OIBs), 170–171, 171–173,
173–174, 176–177
planetary degassing, 167–204
plutonium/xenon (Pu/Xe) isotopic ratios, 178–179
potassium/argon (^{40}K-^{40}Ar) isotopic budget, 178
radiogenic gases, 178
subduction fluxes, 177–178
terrestrial planets
atmospheric composition, 170*t*
planetary degassing, 167–204
surface inventories, 169, 169*t*
North Atlantic Deep Water (NADW)
biological pump efficiency, 17*f*
calcium carbonate (CaCO$_3$) abundance, 493–495, 493*f*
hydrothermal vents, 78
organic carbon cycle, 391

Ntalla River, 324*f*
Nyong River, 296*t*, 319–320

O

Ob River, 296*t*, 324–325
oceanic crust
  accretion rate, 594–595, 594*f*
  age distribution, 582–583, 582*f*
  boron (B) isotopes, 604
  calcium (Ca) isotopes, 604–605
  formation conditions, 582–583
  hydrothermal fluxes
    circulation, 71*f*
    heat and water volume, 51*t*
    heat flow, 46, 46*f*
    hydrothermal plumes, 77, 77*t*
    importance, 50–51
    mineralization processes, 67–68
  isotopes
    carbon (δ^{13}C) isotope concentrations, 600–602, 601*f*, 602*f*
    isotopic evolution, 595–605
    isotopic tracers, 603–605
    osmium (Os) isotopic evolution, 598
    oxygen (δ^{18}O) isotope concentrations, 602–603, 604*f*
    strontium (Sr) isotopic evolution, 595–598, 596*f*, 597*f*, 598*f*
    sulfur (S) isotopic evolution, 598–600, 600*f*
  magnesium/calcium (Mg/Ca) ratio, 594–595, 594*f*, 595*f*
  ocean ridges, 46–48, 47*f*, 48–49
  preservation potentials, 584*f*
oceans
  alkalinity
    alkalinity pump, 455*f*
    biological pump, 456–457
  anoxic environments, 536
  biological fluxes, 119–120
  biological pump, 425–452
    basic concept
      aggregation processes, 430–431
      bacterial hydrolysis, 432–433
      burial processes, 433
      cycling processes, 427–434
      dissolved organic matter (DOM), 433–434
      exopolymer particles, 430–431
      flocculation, 430–431
      marine snow, 430
      new production, 434
      nutrient limitations, 430
      nutrient uptake, 428–430
      particle decomposition, 431–433
      particulate organic matter (POM), 431
      photosynthesis, 428–430
      phytoplankton, 428–430
      primary production levels, 429, 429*f*
      schematic diagram, 427*f*
      sedimentation, 433
      sinking fluxes, 431–433, 432*f*
      sinking rates, 430–431
      zooplankton grazing, 432
    calcium carbonate ($CaCO_3$) production, 455*f*
    dissolved inorganic compounds, 427*f*
    efficiency
      C:N:P ratio variations, 444–445
      community composition changes, 444
      high-nutrient, low chlorophyll (HNLC) areas, 443–444, 443*f*
      particulate organic matter (POM), 444–445
      transport enhancements, 445
    export production, 471–472
    future trends
      agricultural runoff, 446
      carbon sequestration, 446–447

      C:N:P ratio variations, 446
        increased carbon dioxide (CO_2) levels, 445–446
        silica limitations, 446
    general discussion, 426–427
    high-latitude ocean, 465–469
    high-nutrient, low chlorophyll (HNLC) areas, 426, 440, 443–444, 443*f*
    impact processes
      carbon nutrients, 427*f*, 434–435
      nitrogen cycle, 435–436, 435*f*, 439*f*, 440–441
      phosphorus cycle, 436, 439*f*
      silica cycle, 436–437, 437*f*, 439*f*
      silica/nitrate ratios, 437, 438*f*
      trace elements, 437–440, 438*f*, 439*f*
    latitude controls, 458–469, 466*f*
    low-latitude ocean, 460–465, 466*f*
    mid-latitude ocean, 460–465, 466*f*
    nutrient concentrations, 454*f*, 458–469, 466*f*
    phytoplankton production, 458–469, 466*f*
    quantification studies
      general discussion, 440–443
      new production measurements, 440–441
      oxygen utilization rates, 442–443
      particle flux measurements, 441–443
      particle-reactive nuclides, 442, 442*f*
      sediment traps, 441–442
    Redfield ratios, 461–462
  calcium carbonate ($CaCO_3$)
    accumulation rate, 492*f*
    biological pump, 455*f*
    boric acid ($B(OH)_3$) concentrations, 502–503
    dissolution mechanisms, 496–498, 496*f*, 498*f*, 504–508, 505*f*
    fossil fuels, 509
    glacial periods, 508
    interglacial periods, 504–508
    ionic concentrations, 491–510
    marine organisms, 492–493, 493*f*
    oceanic carbon cycle, 364
    pore waters, 497*f*, 498–499, 498*f*
    preservation events, 504–508, 505*f*, 506*f*
    rain rate, 491–492, 492*f*
    saturation as function of water depth, 492–493, 493*f*, 496–498, 497*f*, 499*f*, 500*f*, 502*t*
    saturation horizon depth, 495–496, 495*f*
    terrestrial biomass, 504–505
    transition zone, 492–493
    water column redissolution, 496–498
  carbonate sediments
    boron (B) isotopes, 604
    calcareous fossils, 595*f*, 609–610, 610*f*
    calcium (Ca) isotopes, 604–605
    carbonate cycle, 610, 610*f*
    carbon (δ^{13}C) isotope concentrations, 600–602, 601*f*, 602*f*
    evolutionary trends, 605
    general discussion, 593–605
    geochemical implications, 610–611
    ironstones, 609, 609*f*
    isotopic evolution, 595–605
    isotopic tracers, 603–605
    magnesium/calcium (Mg/Ca) ratio, 594–595, 594*f*, 595*f*, 607–608, 608*f*
    mass/age distribution, 606–607, 606*f*
    oöids, 608–609, 609*f*
    osmium (Os) isotopic evolution, 598
    oxygen (δ^{18}O) isotope concentrations, 602–603, 603*f*
    recycling rates, 606–607
    strontium (Sr) isotopic evolution, 595–598, 596*f*, 597*f*, 598*f*
    sulfur (S) isotopic evolution, 598–600, 600*f*
    survival rates, 607*f*
  carbon cycle
    air-sea exchange, 92–94

carbon (δ^{13}C) isotope concentrations, 392–393
  contemporary carbon cycle, 89*f*, 89*t*
  dissolved inorganic carbon (DIC), 90–91, 91*t*
  glaciation, 533
  methane (CH$_4$) carbon subcycle, 344*f*, 353*t*
  photosynthesis, 392–393
  preindustrial cycle, 343*f*, 346*t*
  sources/sinks, 119–120, 120–121
deep-sea carbonate-rich sedimentary environments
  foraminifera, 499–500
  measurement techniques, 499–500
deep water
  biological pump, 455*f*
  carbon cycle, 363
  molecular oxygen (O$_2$) concentrations, 464,
    474–475, 514*f*
magnesium/calcium (Mg/Ca) ratio
  oceanic crust, 594–595, 594*f*, 595*f*
methane (CH$_4$)
  methane (CH$_4$) carbon subcycle, 344*f*, 353*t*
net primary production (NPP)
  photosynthesis, 94, 94*t*
nitrogen cycle
  denitrification, 567*t*
  global marine budgets, 565–566
  nitrogen cascade, 570*f*, 571*t*
  nitrogen fixation, 463, 567*t*
  nitrogen reservoirs, 556–557
  reactive nitrogen (Nr), 561*t*, 562*t*
nutrient production
  biological pump, 460–465
  carbon cycle, 119–120
ocean chemistry
  atmospheric oxygenation, 530–531
  boric acid (B(OH)$_3$) concentrations, 502–503
  boron (B) isotope concentrations, 604
  calcium (Ca) isotopes, 604–605
  carbon (δ^{13}C) isotope concentrations, 600–602, 601*f*, 602*f*
  carbon uptake, 96*t*, 97–98, 103*t*
  hydrothermal plumes, 77, 77*t*
  isotopic evolution, 595–605
  isotopic tracers, 603–605
  molecular oxygen (O$_2$) concentrations, 464,
    512–515, 514*f*
  osmium (Os) isotopes, 598
  oxygen (δ^{18}O) isotope concentrations, 602–603, 604*f*
  pore waters, 495–496
  strontium (Sr) isotopes, 595–598, 596*f*, 597*f*, 598*f*
  sulfur (S) isotopes, 598–600, 600*f*
ocean mixing
  biological pump, 455*f*, 456*f*
  bomb radiocarbon (^{14}C), 97
  carbonate pump, 455*f*
  radiocarbon (^{14}C), 97
  tracers, 97
organic carbon cycle, 370, 370*f*, 389–424
siliciclastic sediments
  carbon-sulfur (C-S) relationships, 542–543, 543*f*
thermohaline circulation, 512–513, 514*f*, 535–536
tracers
  bomb radiocarbon (^{14}C), 98
  chlorofluorocarbons (CFCs), 98
  ocean carbon models, 98
  radiocarbon (^{14}C), 97
uptake
  buffer factor, 118–119
  carbon dioxide (CO$_2$), 97–98, 118–120
  global warming, 119
  nutrient production, 119–120
  ocean carbon models, 98
  vertical mixing, 119
  see also mid-ocean ridge basalts (MORBs)
Ohio River, 296*t*

Old Faithful geyser, Yellowstone National Park,
  United States, 15*f*
Oldoinyo Lengai, Tanzania, 12*t*, 18*f*
olivine
  weatherability, 225
Ontong Java plateau (OJP)
  calcium carbonate (CaCO$_3$) dissolution, 496–497, 496*f*,
    498, 498*f*
  carbonate ion concentrations, 505, 506*f*
  foraminifera, 499*f*, 501–502, 505, 506*f*, 508*f*
  ionic concentrations, 501–502
  radiocarbon (^{14}C) concentrations, 498, 500*f*
oöids, 608–609, 609*f*
opal
  biomineralization, 428, 436–437
ophiolites
  hydrothermal vents, 69
Ordovician
  sulfide minerals, 137
organic matter
  carbon cycle, 538, 541*f*
  dissolved organic carbon (DOC)
    acylpolysaccharides, 400–402, 402*f*, 403*f*, 404–405, 404*f*
    amides, 401, 402*f*
    background information, 398–399
    biopolymerization versus geopolymerization, 399–404
    gel polymers, 404–405, 404*f*, 405*f*
    high molecular weight dissolved organic matter (HMW
      DOM), 399–404
    marine sediments, 391
    monosaccharides, 400–402, 401*f*
    NMR spectra, 400–402, 400*f*, 406*f*
    organic carbon cycle, 391*f*
    proteins, 402–404
    rivers, 283*t*, 394–396
    water column fluxes, 392–393
  microbes
    abundances, 416*f*
    burial process, 537–545, 539*f*
    crenarchaeota, 415, 416*f*
    planktonic archaea, 415, 416*f*
    prokaryotes, 414–415
  molecular oxygen (O$_2$)
    consumption mechanisms, 521–522
  organic carbon cycle
    aeolian dust, 392
    biological pump, 25*f*, 453–458, 454*f*, 466*f*, 471–472
    black carbon, 397–398, 399*f*
    fluxes, 391*f*, 392–393
    future research, 417–418
    marine sediments, 396–398, 396*f*
    microbes, 414–416
    preservation, 406–414
    reservoirs, 390–392, 391*f*
    rivers, 283*t*, 392, 394–396
    water column fluxes, 392–393
    weathering sources, 392
  particulate organic carbon (POC)
    composition, 407*f*
    compositional transformations, 406–409
    fluxes, 407*f*
    seawater, 393
  particulate organic material (POM)
    rivers, 394–396
  Permian-Triassic (P-Tr) boundary, 535–536
  preservation
    anoxia, 411–412, 412*f*
    background information, 406
    chemical protection, 412–414
    compositional transformations, 406–409
    particulate organic carbon (POC), 407*f*
    physical protection, 410–411, 410*f*
    radiocarbon (^{14}C) concentrations, 407, 408*f*, 409*f*

organic matter (*continued*)
  refractory biomolecules, 412
  stable isotope compositions, 407, 408*f*, 409*f*
  sulfurization, 413, 413*f*, 414*f*
  prokaryotes, 414–415
  river concentrations, 318–325
  sulfur cycle, 538, 541*f*
  terrigenous fluxes
    background information, 393
    black carbon, 397–398, 399*f*
    deltaic sediments, 395*f*
    dust, 392
    estuaries, 395
    marine sediments, 396–398, 396*f*
    radiocarbon (^{14}C) concentrations, 394–395, 394*f*
    rivers, 392, 394–396
    weathering, 392
Orinoco River, 296*t*, 323*f*, 324*f*, 330–331, 331*f*, 332*f*
osmium (Os)
  elemental abundances
    rivers, 296*t*, 307*f*, 308*f*
    sedimentary environments, 592*f*
  isotopes
    paleo-ocean, 598
Ottawa River, 296*t*
Oubangui River, 296*t*
oxygen (O)
  biological pump, 442–443
  Great Oxidation Event, 375–376
  isotopes
    oxygen (δ^{18}O) isotope concentrations
      ice core analysis, 360*f*
      oceans, 602–603, 603*f*
      paleo-ocean, 602–603, 603*f*
ozone (O$_3$)
  stratosphere
    aerosol sulfates, 159
  troposphere
    forest ecosystems, 113*t*, 114

P

Pacific Ocean
  biogenic opal (SiO$_2$), 25*f*, 481, 483*f*
  biological pump, 25*f*, 475–478, 481
  calcium carbonate (CaCO$_3$)
    abundances, 493–495, 493*f*, 508
  hydrothermal processes
    hydrothermal plumes, 75–76, 77, 77*t*
    hydrothermal vents, 46, 46*f*, 63–64
  isotopic composition
    helium (He) isotope concentrations, 171*f*
  molecular oxygen (O$_2$) concentrations, 512–515, 514*f*
  nitrogen cycle, 565–566, 567*f*
  organic carbon cycle, 407*f*, 408*f*
Paleocene
  Late Paleocene Thermal Maximum, 374–375, 374*f*
  silica concentrations, 436–437, 437*f*
paleoclimatology, 361–362, 362–365, 365–367
paleo-ocean
  boron (B) isotopes, 604
  calcium (Ca) isotopes, 604–605
  carbon (δ^{13}C) isotope concentrations, 600–602,
    601*f*, 602*f*
  isotopic evolution, 595–605
  isotopic tracers, 603–605
  magnesium/calcium (Mg/Ca) ratio, 594–595, 594*f*, 595*f*
  metalliferous ridge and basal sediments, 78–79
  osmium (Os) isotopic evolution, 598
  oxygen (δ^{18}O) isotope concentrations, 602–603, 604*f*
  strontium (Sr) isotopic evolution, 595–598, 596*f*,
    597*f*, 598*f*
  sulfur (S) isotopic evolution, 598–600, 600*f*

Paleoproterozoic
  atmospheric evolution, 3
paleosols
  Archaean, 524
  atmospheric oxygenation, 530
  carbon cycle, 524
Paleozoic
  carbon cycle models, 371
  molecular oxygen (O$_2$), 535
  ocean chemistry, 594–595
palladium (Pd)
  elemental abundances
    rivers, 296*t*, 307*f*, 308*f*
Parker River, 394*f*
particulate organic matter (POM), 431, 444–445
partitioning
  beryllium (Be), 331, 332*f*
  magma volatiles, 4
PATHARC.94 code, 233
Peel River, 296*t*
peridotites
  river basin lithology, 282*t*
Permian
  molecular oxygen (O$_2$), 535
pesticides
  sulfur (S) concentrations, 158
petroleum
  natural gas
    methane (CH$_4$) carbon subcycle, 344*f*, 353*t*
    preindustrial carbon cycle, 346*t*, 349*t*
  sulfur emissions, 156
pH
  aquatic systems, 324–325, 325*f*, 329–330, 329*f*
  boron (B) isotope concentrations
    pH proxy, 502–503, 503*f*
    foraminifera, 502–503
    hydrothermal vents, 58, 60*t*
  ocean chemistry
    boric acid (B(OH)$_3$) concentrations, 502–503
Phanerozoic
  atmospheric evolution, 534–545, 545*f*
  carbonate sediments
    calcareous fossils, 595*f*, 609–610, 610*f*
    carbonate cycle, 610, 610*f*
    geochemical implications, 610–611
    ironstones, 609, 609*f*
    magnesium/calcium (Mg/Ca) ratio, 607–608, 608*f*
    mass/age distribution, 606*f*, 608–609
    oöids, 608–609, 609*f*
    recycling rates, 606–607
    survival rates, 607*f*
  carbon cycle, 367–375, 372*f*, 376
  molecular oxygen (O$_2$)
    carbon (δ^{13}C) isotope concentrations, 540*f*
    carbon-sulfur (C-S) relationships, 542–543, 543*f*
    constraints, 534–537
    metazoan development, 534–537
    numerical models, 537–545, 539*f*
    phosphorus (P), 543
    variability, 534–537
  oceans
    carbon (δ^{13}C) isotope concentrations, 600–602,
      601*f*, 602*f*
    ocean chemistry, 594–595, 595*f*, 597*f*, 598*f*
    oxygen (δ^{18}O) isotope concentrations, 602–603, 604*f*
    sulfur isotopic record, 598–600, 600*f*
  organic carbon cycle, 370, 370*f*
  sedimentary environments
    evaporites, 606*f*
    evolutionary trends, 605–606, 606*f*
    red beds, 523*f*
    sedimentary basins, 585–586, 585*f*
  silica concentrations, 436–437, 437*f*

phosphorites
  marine sediments, 585, 585*f*
phosphorus cycle
  biological pump, 461
phosphorus (P)
  biological pump, 427*f*, 436, 439*f*, 443*f*
  carbon:nitrogen:phosphorus (C:N:P) ratios, 461–462
  carbon:phosphorus (C:P) ratios, 401
  elemental abundances
    hydrothermal vents, 77, 77*t*
    rivers, 296*t*
    sedimentary environments, 592*f*
  hydrolysis, 316*f*
  metalliferous ridge and basal sediments, 78–79
  phosphates
    biological pump, 7*f*, 454*f*, 458–469
  phytoplankton photosynthesis, 428*t*
photosynthesis
  anoxygenic photosynthesis, 516*f*
  biological pump, 427*f*, 428–430
  Calvin cycle, 516–517, 516*f*
  carbon cycle
    carbon dioxide (CO$_2$) concentrations, 373
  chlorophyll, 517
  evolution
    anoxygenic photosynthesis, 136
    geological record, 546*f*
  glucose production, 374–375
  gross primary production (GPP), 92
  molecular oxygen (O$_2$) production, 512–513,
    516–517, 516*f*
  net primary production (NPP)
    carbon dioxide fertilization, 112–117, 113*t*
    climate effects, 113*t*, 114
    molecular oxygen (O$_2$), 517, 523*f*
    respiration, 92
  oceanic carbon cycle, 392–393
  organic matter, 392–393
  oxygenic photosynthesis
    eukaryotes, 516, 528
    fossil evidence, 528
    molecular oxygen (O$_2$) production, 516*f*
  Phanerozoic, 537–545
  photorespiration, 518–519
  phytoplankton, 517, 523*f*
  Rubisco, 516–517, 518–519, 544
  seasonality, 517
  volcanic emissions, 34–35
PHREEQC code, 233
physical erosion *see* chemical weathering
phytoplankton
  biological pump
    calcium carbonate (CaCO$_3$) production, 462, 462*t*
    dissolved organic matter (DOM), 433–434
    latitude controls, 7*f*, 458–469, 478–481
    nitrogen fixation, 463
    nutrient status, 7*f*, 472–473
    nutrient supply, 454*f*, 458–469, 466*f*
    phosphorus cycle, 461
    photosynthesis, 428–430
    primary production levels, 429
    silicate/nitrate uptake, 480–481
    sinking rates, 456–457
  cadmium (Cd) concentrations, 439–440
  chlorophyll supplies, 7*f*, 458–469
  iron (Fe) concentrations, 440
  oceanic carbon cycle, 392–393, 408*f*
  photosynthesis, 517, 523*f*
  trace element concentrations, 428*t*
  zinc (Zn) concentrations, 438–439
Pilbara Craton, Western Australia
  clastic sediments, 590
  oxygenic photosynthesis, 528

Pinatubo volcano
  climate effects, 4, 31–33
  environmental impacts
    general discussion, 28–31
    optical/radiative effects, 30–31
    ozone (O$_3$) concentrations, 28–30
    stratospheric sulfate aerosols, 28, 29*f*
  eruptive emissions, 16
  excess degassing, 12–13
  halogen compounds, 27*t*
  sulfur emissions, 24*t*, 25*t*
  sulfur fluxes, 22–23
  volatile elements, 7*t*
plagioclase
  mineral fluxes, 223*t*
  sedimentary environments, 587, 587*f*
  weathering rates
    characteristics, 212*t*, 223
    dissolution rates, 227, 228*f*
    etch pits, 210–211
    fluid fluxes, 229*f*
    reaction times, 224*f*
    weathering gradients, 208
Pleistocene
  Milankovitch cycles, 361–362
  preindustrial carbon cycle, 361
plutonium (Pu)
  isotopes
    plutonium/xenon (Pu/Xe) isotopic
      ratios, 178–179
polar ice core data, 353*t*, 355–359
pollution
  acid rain, 156–157
  coastal pollution, 158
  nitrogen cycle, 568–572
  sulfur emissions, 155–156
  volcanic emissions, 27–36
Popocat, 22–23, 35
pore waters
  chemical weathering
    pore water concentrations, 215–216, 215*f*
    pore water flow, 216*f*
  molecular oxygen (O$_2$) concentrations, 515
  oceans
    calcium carbonate (CaCO$_3$), 495–496, 497*f*,
      498–499, 498*f*
  sodium distributions, 208, 208*f*
potassium (K)
  chemical weathering
    elemental mobility, 210*t*, 211*f*
  complexation, 317*f*
  elemental abundances
    clastic sediments, 588–589, 589*f*
    Congo River, 308*f*
    Earth's crust, 188–189
    hydrothermal vents, 60*t*, 73–74
    rivers, 308*f*, 312*f*, 313*f*
    sedimentary environments, 592*f*
  hydrolysis, 316*f*
  hydrothermal vents, 58*f*
  ionic concentrations
    anthropogenic sources, 290–291, 290*t*
    pristine river distributions, 279*f*, 279*t*, 282*t*
    river sources/sinks, 280*t*
    runoff, 283–284, 284*f*, 286–287, 286*t*, 287*f*
    weathering rates, 289*t*
  isotopes
    potassium/argon (^{40}K-^{40}Ar) isotopic budget,
      178, 188–189
  potassium chloride (KCl), 27*t*
praseodymium (Pr)
  elemental abundances
    hydrothermal vents, 58*f*

praseodymium (Pr) (*continued*)
  rivers, 296*t*, 308*f*, 313*f*, 323*f*
  sedimentary environments, 592*f*
Precambrian
  carbon cycle, 375–376
  molecular oxygen (O$_2$), 534–535
  oceans
    carbon (δ^{13}C) isotope concentrations, 600–602,
      601*f*, 602*f*
    oxygen (δ^{18}O) isotope concentrations,
      602–603, 603*f*
    sulfur isotopic record, 598–600, 600*f*
precipitation
  chemical weathering, 230–231, 232*f*
primary production
  net primary production (NPP)
    biological pump, 7*f*, 454, 456*f*, 458–469
    carbon dioxide fertilization, 112–117, 113*t*
    climate effects, 113*t*, 114
    gross primary production (GPP), 92
    molecular oxygen (O$_2$), 517, 523*f*
    organic carbon cycle, 391*f*
    photosynthesis, 94, 94*t*
    respiration, 92
PROFILE code, 233
prokaryotes
  molecular oxygen (O$_2$) consumption, 519
  organic carbon cycle, 414–415
  photosynthesis, 516
protactinium (Pa), 442
  hydrothermal vents, 77
  isotopes
    biological pump, 472
Proteobacteria, 441
Proterozoic
  atmospheric evolution
    Mesoproterozoic, 531–532, 546*f*
    Neoproterozoic, 532–534
    oxygenic photosynthesis, 528–531
  carbon cycle, 375–376
  continental crust
    clastic sediments, 590
  red beds, 523*f*
pyrite (FeS$_2$)
  hydrothermal vents, 68–69, 68*f*, 77–78
  metalliferous ridge and basal sediments, 77–78
  molecular oxygen (O$_2$), 545
  occurrences, 137
  Phanerozoic, 537–545, 539*f*
  rivers, 280*t*
pyroxene
  weatherability, 224
pyrrhotite (Fe$_7$S$_8$)
  hydrothermal vents, 68*f*

Q

quartzite
  mass/age distribution, 605*f*
quartz (SiO$_2$)
  hydrothermal vents, 68–69
  sandstones, 587*f*
  weathering rates, 212*t*
Quaternary
  carbon cycle, 351–367
  ice core gas analysis, 351–367, 353*t*

R

radiocarbon (^{14}C)
  bomb radiocarbon (^{14}C)
    oceanic tracers, 97, 98
    riverine organic matter, 394–395

oceans
  oceanic uptake, 97
  ocean mixing, 97
  organic carbon cycle, 394–395, 394*f*, 407, 408*f*, 409*f*
radiogenic isotopes
  planetary degassing, 178–179
  see also long-lived radioisotopes
radiolaria
  biological pump, 436–437, 437*f*
radium (Ra)
  biological pump, 437–440
  rivers, 296*t*, 307*f*, 308*f*
radon (^{222}Rn)
  air-sea exchange, 94
  hydrothermal vents, 61, 70–71
rainwater, 156–157
rare earth elements (REEs)
  biological pump, 437–440
  chemical weathering, 209
  clastic sediments, 588–589, 589*f*, 591
  fractionation, 322–324, 323*f*
  hydrothermal vents, 56, 58*f*, 74–75, 79
  metalliferous ridge and basal sediments, 78
  rivers, 309–310, 330
Redfield ratio
  biological pump, 461–462
  biomineralization, 428
redox processes
  chemolithotrophy, 519
  hydrothermal plumes, 72–73
  nitrogen cycle, 554*f*
Red Sea
  hydrothermal vents, 46
  metalliferous ridge and basal sediments, 77–79
regolith
  elemental mobility, 209–210, 210*t*, 211*f*
  erosion, 232–233
  weathering rates
    bulk compositional changes, 209–210, 210*f*
    etch pits, 210–211, 214*f*
    fluid transport, 214–215, 227–228
    groundwater solutes, 216–218
    niobium (Nb) concentrations, 209, 210*f*
    permeabilities, 227–229, 230*f*
    physical development, 232–233
    pore water concentrations, 215–216, 215*f*
    pore water flow, 216*f*
    small-scale compositional changes, 210–211
    solute compositions, 211–220
    surface-water solutes, 218–220
    titanium (Ti) concentrations, 209, 210*f*
    weathering gradients, 208, 208*f*
    zirconium (Zr) concentrations, 209, 210*f*
rhamnose, 400–401, 401*f*
rhenium (Re)
  elemental abundances
    rivers, 296*t*, 307*f*, 308*f*
    sedimentary environments, 592*f*
  hydrolysis, 316*f*
Rhine River, Europe
  colloids, 318–319
  organic carbon cycle, 394*f*
  river chemistry alterations, 290*f*
  trace elements, 296*t*, 314–315
rhyolite
  river basin lithology, 282*t*
  viscosity, 5*f*
  volatile elements, 7*t*
Rio Beni, 296*t*
Rio Negro Basin, Amazon, 312*f*, 323*f*, 324*f*
river systems
  carbon (C)
    bicarbonate (HCO$_3$), 279*f*, 279*t*, 282*t*, 283*f*

carbon transfer, 281–283, 283*t*
dissolved inorganic carbon (DIC), 280*t*,
    281–283, 283*t*
dissolved organic carbon (DOC), 392, 394–396
dissolved organic matter (DOM), 283*t*
particulate inorganic carbon (PIC), 283*t*
particulate organic carbon (POC), 283*t*
particulate organic material (POM), 394–396
preindustrial carbon cycle, 343*f*, 346*t*
ionic concentrations
    anthropogenic sources, 280*t*, 290–291, 290*f*, 290*t*
    background information, 277–278
    climate effects, 283–284
    data sources, 278
    global budget, 287–289, 289*t*
    global land distribution, 287–289, 288*t*
    global range, 278–279, 279*f*, 279*t*, 283*f*
    idealized model, 284–286, 285*f*
    ionic yields, 286–287, 286*t*, 287*f*, 289*t*
    lithology, 281, 282*t*
    river fluxes, 287–289, 288*f*, 288*t*, 289*t*
    runoff, 283–284, 284*f*, 286–287, 286*t*, 287*f*
    sources/sinks, 280*t*
    tectonics, 280–281
    weathering intensities, 286–287
nitrogen cycle, 561*t*, 562*t*, 564–565, 570*f*, 571–572
sulfur cycle, 146–147
trace elements
    abundances, 295–312, 296*t*, 307*f*
    adsorption processes, 330–331, 331*f*
    adsorption properties, 327–328, 328*f*
    Africa, 295, 296*t*
    aluminum solubility, 326*f*
    anion adsorption, 331–332
    anthropogenic sources, 314
    aqueous speciation, 315–318, 315*f*, 316*f*, 317*f*
    Asia, 296*t*, 306
    atmospheric sources, 314
    Born analysis, 316*f*
    colloids, 318–325
    concentration ranges, 306–307
    concentrations, 308*f*
    conservative behavior, 311, 311*f*
    crustal concentrations, 307–308, 308*f*
    dissolved organic carbon (DOC), 332–333
    equilibrium solubility, 325–327
    Europe, 295, 296*t*
    general discussion, 333–335
    hydrous oxides, 328–330, 328*f*
    isotope decoupling, 309–310
    metal-humate stability constants, 320–321, 321*f*
    neodymium (Nd) concentrations, 309–310, 309*f*
    North America, 295, 296*t*
    occurrences, 294–295
    organic matter, 332–333
    particle dynamics, 333
    pH, 329–330, 329*f*, 330–331
    rare earth elements (REEs), 309–310, 322–324,
        323*f*, 330
    rock weathering, 312–314, 313*f*
    seasonality, 330
    sources, 312–315, 313*f*
    South America, 295–306, 296*t*
    surface reactions, 327
    temporal variations, 310–311
    trace metal adsorption, 328–330, 329*f*
    transport mechanisms, 311–312, 312*f*
    ultrafiltration techniques, 319–321
    uranium solubility, 326*f*
RNA molecules, 141
rodenticides *see* pesticides
Ross Sea
    calcium carbonate (CaCO₃) abundance, 494

rubidium (Rb)
    elemental abundances
        clastic sediments, 588–589, 589*f*
        Congo River, 308*f*
        hydrothermal vents, 60*t*
        rivers, 296*t*, 307*f*, 308*f*, 312*f*, 313*f*
        sedimentary environments, 592*f*
    hydrolysis, 316*f*
Rubisco
    molecular oxygen (O₂), 544
    photosynthesis, 516–517, 518–519
runoff
    chemical weathering, 230–231
    ionic concentrations, 284*f*, 286–287, 286*f*, 286*t*
    rivers, 283–284
Rutherford, Daniel, 552–553

S

SAFE code, 233
Saint Lawrence River, 296*t*
salinity
    rivers, 284–286, 285*f*
    seawater, 143
    sulfates, 143
salinization
    seawater
        evaporation processes, 137
samarium (Sm)
    elemental abundances
        Congo River, 308*f*
        hydrothermal vents, 58*f*
        rivers, 296*t*, 307*f*, 308*f*, 312*f*, 313*f*, 323*f*
        sedimentary environments, 592*f*
    ultrafiltration techniques, 320*f*
Sanaga River, 296*t*, 324*f*
Santa Clara River, 394*f*
Santa Maria volcano, 25*t*
saprolites
    chemical weathering, 209
    elemental mobility, 210*t*
    mica weatherability, 225*f*
    pore water concentrations, 216*f*
    weathering rates, 212*t*, 220
    see also chemical weathering; soils
scandium (Sc)
    biological pump, 437–440
    complexation, 317*f*
    elemental abundances
        clastic sediments, 588–589, 589*f*
        Congo River, 308*f*
        rivers, 296*t*, 307*f*
        sedimentary environments, 592*f*
        turbidites, 589, 589*f*
    hydrolysis, 316*f*
Scheele, Carl Wilhelm, 552–553
schists
    river basin lithology, 282*t*
Scholesing, Theophile, 553
seawater
    aragonite solubility, 594–595, 595*f*
    biological pump, 453–490
    boron (B)
        boron (B) isotopes, 604
        pH proxy, 502–503, 503*f*
    calcite solubility, 595*f*
    calcium (Ca) isotopes, 604–605
    carbon cycle
        biological processes, 119–120
        carbon (δ¹³C) isotope concentrations, 600–602, 601*f*, 602*f*
        carbon dioxide (CO₂) concentrations, 118–120
        carbon uptake, 96*t*, 97–98, 103*t*
        dissolved organic carbon (DOC), 392–393

seawater (*continued*)
    methane (CH_4) carbon subcycle, 344*f*
    preindustrial carbon cycle, 343*f*, 346*t*
  elemental abundances, 588–589, 589*f*
  evaporation processes, 137
  glaciation effects, 502–503
  hydrothermal vents, 52–56, 52*f*, 55*f*, 71*f*
  isotopic composition
    isotopic evolution, 595–605
    isotopic tracers, 603–605
  magma volatiles, 6–8
  magnesium/calcium (Mg/Ca) ratio
    oceanic crust, 594–595, 594*f*, 595*f*
  osmium (Os)
    isotopic evolution, 598
  oxygen (O)
    molecular oxygen (O_2) concentrations, 512–515, 514*f*
    oxygen ($\delta^{18}O$) isotope concentrations, 602–603, 604*f*
  strontium (Sr)
    isotopic evolution, 595–598, 596*f*, 597*f*, 598*f*
  sulfur (S)
    carbon disulfide (CS_2), 144
    carbonyl sulfide (COS), 144, 146
    coastal marshes, 146, 146*f*
    coastal pollution, 158
    concentrations, 135–136
    dimethyl sulfide (DMS), 144–146, 145*f*
    dimethylsulfoniopropionate (DMSP), 144–146, 145*f*
    dimethylsulfoxide (DMSO), 145
    estuaries, 146
    hydrogen sulfide (H_2S), 143–144, 146
    isotopic evolution, 598–600, 600*f*
    organosulfur compounds, 144–146
    sulfates, 143, 530
  temperature variations, 604*f*
sedimentary basins
  classification, 583, 583*f*
  lithologies, 585–586, 586*f*
  preservation potentials, 583, 583*f*, 584*f*
  turbidites, 589*f*
sedimentary environments
  age distribution, 580–582, 580*f*, 581*f*, 582*f*, 605*f*
  carbonate sediments
    boron (B) isotopes, 604
    calcareous fossils, 595*f*, 609–610, 610*f*
    calcium (Ca) isotopes, 604–605
    carbonate cycle, 610, 610*f*
    evolutionary trends, 605, 606–610
    general discussion, 593–605
    geochemical implications, 610–611
    ironstones, 609, 609*f*
    isotopic evolution, 595–605
    isotopic tracers, 603–605
    magnesium/calcium (Mg/Ca) ratio, 594–595, 594*f*, 595*f*, 607–608, 608*f*
    mass/age distribution, 606–607, 606*f*
    oöids, 608–609, 609*f*
    osmium (Os) isotopes, 598
    oxygen ($\delta^{18}O$) isotope concentrations, 602–603, 604*f*
    recycling rates, 606–607
    strontium (Sr) isotopes, 595–598, 596*f*, 597*f*, 598*f*
    sulfur (S) isotopes, 598–600, 600*f*
    survival rates, 607*f*
  carbon cycle, 89*f*, 89*t*
  carbon ($\delta^{13}C$) isotope concentrations
    atmospheric evolution, 527*f*
    paleo-ocean, 600–602, 601*f*, 602*f*
    Phanerozoic, 527–528, 531–532, 540*f*
  Chemical Index of Alteration (CIA), 587*f*
  clastic sediments
    chemical analysis, 590
    elemental abundances, 588–589, 589*f*

    isotope systematics, 591
    isotopic abundances, 588–589
    lithologies, 589–590
    lutetium/hafnium (Lu/Hf) isotopic ratios, 591
    potassium oxide (K_2O)/sodium oxide (Na_2O) ratio, 590, 590*f*
    provenance analysis, 586–587, 586*f*, 587*f*
    rare earth elements (REEs), 591
    samarium/neodymium (Sm/Nd) isotopic ratios, 591, 592*f*, 593*f*
    secular evolution, 589–591, 591*f*
    stratigraphic age, 591, 592*f*
    tectonics, 586–587, 586*f*, 588*f*, 589–590
    thorium/scandium (Th/Sc) isotopic ratios, 589, 589*f*, 591*f*
    turbidites, 589*f*
  elemental composition, 592–593, 592*f*
  evolutionary trends
    carbonate sediments, 605, 606–610
    evaporites, 605, 606*f*
    lithologies, 605–606
    Phanerozoic, 606–610
  marine sediments
    black carbon, 397–398, 399*f*
    contemporary carbon cycle, 89*f*, 89*t*
    mass/age distribution, 585*f*
    methane (CH_4) carbon subcycle, 353*t*
    molecular oxygen (O_2) concentrations, 515
    organic carbon cycle, 389–424
    preindustrial carbon cycle, 343*f*, 346*t*, 347
    preservation, 406–414
    terrigenous fluxes, 396–398, 396*f*
  mass/age distribution, 584–585, 585*f*
  mineral dissolution, 588
  mudstones
    clastic sediments, 587
    transport mechanisms, 588
  oceanic crust
    boron (B) isotopes, 604
    calcium (Ca) isotopes, 604–605
    carbonate sediments, 593–605
    carbon ($\delta^{13}C$) isotope concentrations, 600–602
    chemical composition, 594–595, 594*f*, 595*f*, 597*f*, 598*f*
    isotopic evolution, 595–605
    isotopic tracers, 603–605
    osmium (Os) isotopes, 598
    oxygen ($\delta^{18}O$) isotope concentrations, 602–603
    strontium (Sr) isotopes, 595–598, 596*f*, 597*f*, 598*f*
    sulfur (S) isotopes, 598–600, 600*f*
  organic carbon cycle, 391*f*
  Phanerozoic, 537–545
  population dynamics, 580–582
  recycling, 588, 590*f*, 592–593
  red beds, 523*f*
  river basin lithology, 282*t*
  sandstones
    evolutionary trends, 605–606
    mass/age distribution, 585*f*
    provenance analysis, 587, 587*f*
    ternary petrologic diagrams, 587*f*
    transport mechanisms, 588
  silicate minerals, 588*f*
  siliciclastic sediments
    carbon-sulfur (C-S) relationships, 542–543, 543*f*
  sulfur (S)
    sulfur (S) concentrations, 135–136, 136*f*
  tectonics
    clastic sediments, 586–587, 586*f*, 587*f*, 589–590
    lithologies, 585–586, 586*f*
    preservation potentials, 583, 584*f*
  volcanogenic deposits, 605, 605*f*
  weathering, 588

sediment traps
    foraminifera, 409*f*
    phytoplankton, 407*f*, 408*f*
Seine River, 296*t*, 314–315
selenium (Se)
    atmospheric emissions, 27*t*
    biological pump, 437–440
    elemental abundances
        rivers, 296*t*, 307*f*, 308*f*, 314–315
    hydrolysis, 316*f*
serpentinite, 282*t*
shales
    Devonian, 535
    Jurassic Period, 536
    mass/age distribution, 585*f*, 605*f*
    river basin lithology, 282*t*
Shinano River, 296*t*
siliciclastic sediments
    proxy data
        carbon-sulfur (C-S) relationships, 542–543, 543*f*
silicon (Si)
    biogenic opal (SiO_2)
        biological pump, 456–457
        Pacific Ocean, 481, 483*f*
    biological pump
        depth profiles, 427*f*, 439*f*
        silica cycle, 436–437, 437*f*
        silica/nitrate ratios, 437, 438*f*
    biomineralization
        opal, 428, 436–437
    chemical weathering
        elemental mobility, 210*t*, 211*f*
        silicon fluxes, 231, 232*f*
        watersheds, 219*f*
    elemental abundances
        clastic sediments, 588–589, 589*f*
        hydrothermal vents, 60*t*
        rivers, 308*f*, 313*f*
        sedimentary environments, 592*f*
    hydrolysis, 316*f*
    opal
        biomineralization, 428, 436–437
        phytoplankton photosynthesis, 428*t*
    silicic acid ($Si(OH)_4$)
        depth profiles, 427*f*, 439*f*
        silica cycle, 436–437, 437*f*
        silica/nitrate ratios, 437
    silicon dioxide (SiO_2)
        anthropogenic sources, 290–291, 290*t*
        pristine river distributions, 279*f*, 279*t*, 282*t*
        river sources/sinks, 280*t*
        runoff, 284*f*, 286–287, 286*f*, 286*t*
        ultrafiltration techniques, 320*f*
        weathering rates, 289*t*
    silicon tetrafluoride (SiF_4), 27*t*
    weathering rates, 207
    see also quartz (SiO_2)
silver (Ag)
    complexation, 317*f*
    elemental abundances
        hydrothermal vents, 60*t*
        rivers, 296*t*
    hydrolysis, 316*f*
Siple Dome ice core, 353*t*, 355
Skeena River, 296*t*
SMART code, 233
smectite
    sedimentary environments, 587, 587*f*
Snowball Earth hypothesis, 3, 532–533, 546*f*
sodium (Na)
    chemical weathering
        elemental mobility, 210*t*, 211*f*
    complexation, 317*f*

elemental abundances
    clastic sediments, 588–589, 589*f*
    Congo River, 308*f*
    hydrothermal vents, 60*t*
    rivers, 308*f*, 312*f*, 313*f*
    sedimentary environments, 592*f*
halite (NaCl)
    hydrothermal vents, 53–54, 55*f*
    volcanic emissions, 27*t*
ice core analysis, 360*f*
ionic concentrations
    anthropogenic sources, 290–291, 290*t*
    idealized model, 284–286, 285*f*
    pristine river distributions, 279*f*, 279*t*, 282*t*
    river sources/sinks, 280*t*
    runoff, 283–284, 284*f*, 286–287, 286*t*, 287*f*
    weathering rates, 289*t*
phytoplankton photosynthesis, 428*t*
pore water distributions, 208, 208*f*
soils
    carbon cycle
        contemporary carbon cycle, 89*f*, 89*t*
        methane (CH_4) carbon subcycle, 344*f*, 353*t*
        preindustrial cycle, 343*f*, 346*t*
        reservoirs, 90, 90*t*
    chemical weathering
        weathering rates, 212*t*, 227–229
    fertilizers, 158
    methane (CH_4)
        methane (CH_4) carbon subcycle, 353*t*
    molecular oxygen (O_2) concentrations, 516
    nitrogen cycle, 563–564, 564–565, 570*f*
    organic carbon (C)
        organic carbon cycle, 391*f*
    permeabilities, 227–229, 230*f*
    rivers, 280*t*
    sulfur cycle, 147–149, 148*f*, 149*t*
    volcanic emissions, 34–35
    see also net primary production (NPP); paleosols
solar system
    elemental abundances
        origins, 134–135
solar wind
    xenon (Xe) isotope concentrations, 184*f*
solfataras, 3*f*, 14–15
Solimoes River, 296*t*, 312*f*, 323*f*, 324*f*
solubility pump, 92–93, 455*f*
Soufrière Hills volcano, 4, 7*t*, 11, 15*f*, 22–23
South China Sea, 476
Southern Ocean
    Antarctic stratification, 7*f*, 17*f*, 467, 478–481, 482
    biogenic opal (SiO_2) production, 25*f*
    biological pump, 25*f*
    biological pump efficiency, 17*f*, 465–469, 466*f*, 468*f*
    black carbon, 397–398, 399*f*
    calcium carbonate ($CaCO_3$) abundance,
        493–495, 506–508
    hydrothermal vents, 46–47
    molecular oxygen (O_2) concentrations, 514*f*
    preindustrial carbon cycle, 361
South Pole, 353*t*
sphalerite
    hydrothermal vents, 68–69, 68*f*
    metalliferous ridge and basal sediments, 77–78
sponges
    biological pump, 436–437, 437*f*
stable isotopes
    organic carbon cycle, 407, 408*f*, 409*f*
star formation
    elemental abundances, 134–135
stratosphere
    sulfur cycle
        aerosols, 158–159

stratosphere (*continued*)
    aircraft emissions, 159–160
    ozone (O_3), 159
    sulfur (S) concentrations, 158–159
    volcanic emissions, 158–159
  volcanic emissions, 27–36
stromatolites, 528
Stromboli
  emissions style, 15*f*
  excess degassing, 12–13
  sulfur dioxide (SO_2) emissions, 24*t*
strontium (Sr)
  complexation, 317*f*
  elemental abundances
    clastic sediments, 588–589, 589*f*
    Congo River, 308*f*
    hydrothermal vents, 60*t*
    rivers, 296*t*, 307*f*, 308*f*, 312*f*, 313*f*
    sedimentary environments, 592*f*
  hydrolysis, 316*f*
  isotopes
    paleo-ocean, 595–598
  phytoplankton photosynthesis, 428*t*
  ultrafiltration techniques, 320*f*
Stuiver, M *see* radiocarbon (^{14}C)
subduction zones
  carbon (C) abundances, 191
  nitrogen (N) abundances, 192
  noble gases, 177–178
  water (H_2O), 193
sulfur cycle
  amino acids, 141–142, 142*f*
  anthropogenic impacts
    acid rain, 156–157
    agricultural/industrial pollutants, 157–158
    coastal pollution, 158
    combustion emissions, 155–156, 156*f*
    organosulfur emissions, 156
  aquifers, 146–147
  atmospheric chemistry
    aerosol sulfates, 154–155
    carbon disulfide (CS_2), 150
    carbonyl sulfide (COS), 150, 151*f*
    deposition processes, 155
    dimethyl sulfide (DMS), 150–151, 152*f*
    dimethylsulfoxide (DMSO), 151–152
    elemental sulfur, 152
    global budget, 149–150, 149*f*
    hydrogen sulfide (H_2S), 150, 150*f*
    methanesulfonic acid, 151–152
    sulfur dioxide (SO_2), 152–154, 153*f*
  biochemistry
    origins of life, 141
    sulfur biomolecules, 141–142
    sulfur uptake, 142–143, 142*f*
  crater lakes, 140–141
  cysteine, 142, 142*f*
  environmental impacts, 141
  estuaries, 146
  Europa, 160–161, 161*f*
  evaporation processes, 137
  fluxes, 138
  fumaroles, 140
  geological timescales, 137–138, 138*f*
  global fluxes, 161*f*
  glutathione, 142, 142*f*
  groundwater, 146–147
  hydrothermal vents, 139, 141
  Io, 160
  Jupiter, 160
  lakes, 146–147
  rivers, 146–147
  seawater

carbon disulfide (CS_2), 144
carbonyl sulfide (COS), 144, 146
coastal marshes, 146, 146*f*
dimethyl sulfide (DMS), 144–146, 145*f*
dimethylsulfoniopropionate (DMSP), 144–146, 145*f*
dimethylsulfoxide (DMSO), 145
estuaries, 146
hydrogen sulfide (H_2S), 143–144, 146
organosulfur compounds, 144–146
sulfates, 143
soils, 147–149, 148*f*, 149*t*
stratosphere
  aerosols, 158–159
  ozone (O_3), 159
  sulfur (S) concentrations, 158–159
  volcanic emissions, 158–159
sulfur mineral extraction, 138–139, 139*f*
troposphere
  aerosol sulfates, 154–155
  carbon disulfide (CS_2), 150
  carbonyl sulfide (COS), 150, 151*f*
  deposition processes, 155
  dimethyl sulfide (DMS), 150–151, 152*f*
  dimethylsulfoxide (DMSO), 151–152
  elemental sulfur, 152
  global budget, 149–150, 149*f*
  hydrogen sulfide (H_2S), 150, 150*f*
  methanesulfonic acid, 151–152
  sulfur dioxide (SO_2), 152–154, 153*f*
vegetation, 147–149, 148*f*, 149*t*, 155
Venus, 160
volcanism, 136, 136*f*, 139–141, 140*f*, 141*f*
weathering, 538, 541*f*
sulfur (S)
  agricultural importance, 142
  aircraft emissions, 159–160
  alloptropes, 131, 132*t*
  carbon disulfide (CS_2)
    occurrences, 133
    seawater, 144
    tropospheric concentrations, 150
  carbonyl sulfide (COS)
    occurrences, 133
    seawater, 144, 146
    tropospheric concentrations, 150, 151*f*
    volcanic emissions, 12*t*
  chemical reactions, 132–134
  condensation processes, 135
  dimethyl sulfide (DMS)
    global fluxes, 152*f*
    seawater, 144–146, 145*f*
    tropospheric concentrations, 150
  dimethylsulfoniopropionate (DMSP), 144–146, 145*f*
  dimethylsulfoxide (DMSO), 145, 151–152
  elemental abundances
    core composition
      Earth, 134, 134*t*
      elemental abundances, 134, 134*t*
      interstellar medium (ISM), 135
      magmas, 7*t*
      rivers, 308*f*
      solar system, 134–135
    elemental sulfur, 137, 138–139, 152
    evaporite minerals, 137
    geochemical cycle, 131*f*
    historical background, 130
    hydrogen sulfide (H_2S)
      characteristics, 132
      chemolithoautotrophy, 519
      environmental impacts, 141
      human health effects, 132
      hydrothermal vents, 58–59, 60*t*
      seawater, 143–144, 146

sulfur extraction, 139
  tropospheric concentrations, 150, 150*f*
  volcanic emissions, 12*t*, 22*t*
marine sediments
  sulfur cycle, 146–147
metal sulfide structure, 132–133, 133*f*
methanesulfonic acid, 151–152
methanethiol (MSH), 144
mineral extraction and utilization, 138–139, 139*f*
occurrences, 137–139
organosulfur compounds, 133–134, 144–146, 147, 148*f*, 156
oxo compounds, 134
oxyacids, 133
isotopes
  Archaean, 525–526
  atmospheric oxygenation, 530
  geologic evolution, 525*f*
  paleo-ocean, 598–600, 600*f*
  radioactive isotopes, 131, 131*t*
  stable isotopes, 130–131, 131*t*
solubility, 9
stratosphere, 159–160
sulfates
  aerosols, 154–155
  anthropogenic sources, 290–291, 290*f*, 290*t*
  coastal pollution, 158
  geological timescales, 137–138
  hydrothermal vents, 53
  occurrences, 133
  Phanerozoic, 540*f*
  pristine river distributions, 279*t*, 282*t*
  riverine model, 284–286, 285*f*
  river sources/sinks, 280*t*
  runoff, 283–284, 284*f*, 286–287, 286*t*, 287*f*
  seawater
    atmospheric oxygenation, 530
    salinity, 143
  volatile measurement techniques, 20–21, 20*f*
  volcanic emissions, 28, 29*f*
  weathering rates, 289*t*
  Western Australia, 135–136
sulfides
  geologic evolution, 525*f*
  hydrothermal vents, 58–59
  metalliferous ridge and basal sediments, 77–78
  Neoproterozoic, 532–534
  occurrences, 137
sulfonic acids
  occurrences, 134
sulfur dioxide (SO$_2$)
  characteristics, 133
  environmental impacts, 141
  human health effects, 35
  tropospheric concentrations, 152–154, 153*f*
  uptake, 142–143
  volcanic emissions, 12*t*, 21–22, 22*t*, 35, 140
sulfuric acid (H$_2$SO$_4$)
  characteristics, 133
  stratospheric chemistry, 158–159
sulfur trioxide (SO$_3$)
  characteristics, 133
  tropospheric concentrations, 153
vapor pressure, 131–132, 132*f*
volcanic emissions
  Archaean cycle, 136*f*
  atmospheric releases, 25*t*
  continuously erupting volcanoes, 24*t*
  gas vents, 12*t*
  general discussion, 24–26
  geochemistry, 139–141
  stratospheric releases, 24*t*
  yield estimates, 25*t*

T

Takahashi, Taro, 493–494
Tambora volcano
  climate effects, 32
  halogen compounds, 27*t*
  sulfur emissions, 25*t*, 29, 32
tantalum (Ta)
  elemental abundances
    rivers, 296*t*, 307*f*, 308*f*, 313*f*
    sedimentary environments, 592*f*
  hydrolysis, 316*f*
Tapajos River, 296*t*, 324*f*
Taupo volcano, 25*t*
Taylor Dome ice core, 353*t*, 356–357, 358*f*, 363*f*
terbium (Tb)
  elemental abundances
    Congo River, 308*f*
    hydrothermal vents, 58*f*
    rivers, 296*t*, 308*f*, 312*f*, 313*f*, 323*f*
    sedimentary environments, 592*f*
termites, 353*t*
terrestrial ecosystems
  aerobic respiration, 519–520
  carbon sinks, 101*t*
  molecular oxygen (O$_2$), 519–520
  Phanerozoic, 534–545
Tertiary
  strontium (Sr) isotopic evolution, 597*f*
thallium (Tl)
  elemental abundances
    rivers, 296*t*, 307*f*, 313*f*
    sedimentary environments, 592*f*
  hydrolysis, 316*f*
thermoclines
  biological pump, 454*f*
thorium (Th)
  complexation, 317*f*
  elemental abundances
    clastic sediments, 588–589, 589*f*
    Congo River, 308*f*
    hydrothermal vents, 77
    rivers, 296*t*, 307*f*, 308*f*, 312*f*, 313*f*
    sedimentary environments, 592*f*
    turbidites, 589, 589*f*
  hydrolysis, 316*f*
  isotopes
    biological pump, 472
    particle flux measurements, 442, 442*f*
thulium (Tm)
  elemental abundances
    rivers, 296*t*, 307*f*, 308*f*, 313*f*, 323*f*
    sedimentary environments, 592*f*
tin (Sn)
  elemental abundances
    rivers, 307*f*
    sedimentary environments, 592*f*
  hydrolysis, 316*f*
titanium (Ti)
  biological pump, 437–440
  chemical weathering, 209, 210*f*
  elemental abundances
    clastic sediments, 588–589, 589*f*
    rivers, 296*t*, 307*f*, 308*f*, 313*f*
    sedimentary environments, 592*f*
  hydrolysis, 316*f*
  phytoplankton photosynthesis, 428*t*
Toba volcano, 27*t*, 31, 32
Torgersen, Haakon, 156–157
trace elements
  atmospheric emissions
    anthropogenic sources, 27*t*
    volcanic emissions, 27, 27*t*

trace elements (*continued*)
  biological pump, 437–440, 438*f*, 439*f*
  hydrothermal plumes, 73–74, 74*f*
  hydrothermal vents, 59–61, 73–74
  phytoplankton
    photosynthesis, 428–430
  rivers
    abundances, 295–312, 296*t*, 307*f*
    adsorption processes, 330–331, 331*f*
    adsorption properties, 327–328, 328*f*
    Africa, 295, 296*t*
    aluminum solubility, 326*f*
    anion adsorption, 331–332
    anthropogenic sources, 314–315
    aqueous speciation, 315–318, 315*f*, 316*f*, 317*f*
    Asia, 296*t*, 306
    atmospheric sources, 314
    Born analysis, 316*f*
    colloids, 318–325
    concentration ranges, 306–307
    concentrations, 308*f*
    conservative behavior, 311, 311*f*
    crustal concentrations, 307–308, 308*f*
    dissolved organic carbon (DOC), 332–333
    equilibrium solubility, 325–327
    Europe, 295, 296*t*
    general discussion, 333–335
    hydrous oxides, 328–330, 328*f*
    isotope decoupling, 309–310
    metal-humate stability constants, 320–321, 321*f*
    neodymium (Nd) concentrations, 309–310, 309*f*
    North America, 295, 296*t*
    occurrences, 294–295
    organic matter, 332–333
    particle dynamics, 333
    pH, 329–330, 329*f*, 330–331
    rare earth elements (REEs), 309–310, 322–324, 323*f*, 330
    rock weathering, 312–314, 313*f*
    seasonality, 330
    sources, 312–315, 313*f*
    South America, 295–306, 296*t*
    surface reactions, 327
    temporal variations, 310–311
    trace metal adsorption, 328–330, 329*f*
    transport mechanisms, 311–312, 312*f*
    ultrafiltration techniques, 319–321
    uranium solubility, 326*f*
  volcanic emissions, 27, 27*t*
tracers
  air-sea exchange, 97
  bomb radiocarbon (^{14}C), 97, 98
  hydrothermal plumes, 74–75
  oceans
    air-sea exchange, 97
    bomb radiocarbon (^{14}C), 98
    chlorofluorocarbons (CFCs), 98
    ocean carbon models, 98
    ocean mixing, 97
    radiocarbon (^{14}C), 97
  radiocarbon (^{14}C)
    bomb radiocarbon (^{14}C), 97, 98
  tritium (^{3}H)
    oceanic tracers, 97
trachy-andesite, 282*t*
Trichodesmium, 430, 435–436
tritium (^{3}H)
  tracers
    oceanic tracers, 97
Trompetas River, 296*t*, 323*f*, 324*f*
troposphere
  methane (CH_4), 350
  ozone (O_3) concentrations
    forest ecosystems, 113*t*, 114

sulfur cycle
  aerosol sulfates, 154–155
  carbon disulfide (CS_2), 150
  carbonyl sulfide (COS), 150, 151*f*
  deposition processes, 155
  dimethyl sulfide (DMS), 150–151, 152*f*
  dimethylsulfoxide (DMSO), 151–152
  elemental sulfur, 152
  global budget, 149–150, 149*f*
  hydrogen sulfide (H_2S), 150, 150*f*
  methanesulfonic acid, 151–152
  sulfur dioxide (SO_2), 152–154, 153*f*
  volcanic emissions, 31–33, 33–34
tungsten (W)
  elemental abundances
    rivers, 296*t*, 307*f*, 308*f*
    sedimentary environments, 592*f*
  hydrolysis, 316*f*
Tungurahua, Ecuador, 22–23
turbidites, 589, 589*f*

U

ultramafic rocks *see* peridotites; subduction zones
ultraviolet (UV) radiation
  photosynthesis, 528–531
UNSATCHEM code, 233
Upper Yukon River, 296*t*
uranium (U)
  elemental abundances
    clastic sediments, 588–589, 589*f*
    Congo River, 308*f*
    hydrothermal vents, 73–74, 77, 77*t*
    rivers, 296*t*, 307*f*, 308*f*, 312*f*, 313*f*
    sedimentary environments, 592*f*
  hydrolysis, 316*f*
  isotopes
    biological pump, 472
    particle flux measurements, 442, 442*f*
    uranium/xenon (U/Xe) composition, 184*f*
  solubility, 326, 326*f*

V

vanadium (V)
  elemental abundances
    clastic sediments, 588–589, 589*f*
    hydrothermal vents, 73–74, 74*f*, 77, 77*t*
    rivers, 296*t*, 307*f*, 308*f*, 311, 311*f*, 314–315
    sedimentary environments, 592*f*
  hydrolysis, 316*f*
  metalliferous ridge and basal sediments, 78–79
vegetation
  carbon cycle
    carbon dioxide (CO_2), 373
    carbon dioxide fertilization, 112–117, 113*t*
    contemporary carbon cycle, 89*f*, 89*t*
    preindustrial cycle, 343*f*, 346*t*
    reservoirs, 90, 90*t*
  net primary production (NPP), 92
  organic carbon cycle, 391*f*
  Phanerozoic, 534–545
  plant biomass, 112–117, 113*t*
  rivers, 280*t*
  sulfur cycle, 147–149, 148*f*, 149*t*, 155
  sulfur uptake, 142–143
  vascular plants
    molecular oxygen (O_2), 544–545, 544*f*
  volcanic emissions, 34–35
Venus
  atmospheric chemistry
    degassing, 196
    noble gas abundances, 196
  sulfur cycle, 160

volatile elements
  accretion, 135
  atmospheric composition, 170*t*
  elemental abundances, 190–194
  magmas
    abundances, 6–8, 7*t*
    human health effects, 4–6
    magmatic volatile phase (MVP), 4
    partitioning, 4
    seawater, 6–8
    solubility, 8–9
    sources, 6–8, 7*t*
    speciation, 8–9
  mantle (Earth)
    carbonatites, 12*t*
    volcanism, 3
  noble gases, 167–204
  surface inventories, 169, 169*t*
volcanism
  atmospheric oxygenation, 531
  crater lakes, 140–141
  degassing
    bubble coalescence, 10–11
    bubble growth, 10
    excess degassing, 12–13
    exsolution, 9
    fragmentation, 12
    gas separation, 11–12, 12*t*
    nucleation, 10
    saturation, 9–10
    supersaturation, 10
    vesiculation, 4, 9–13
    viscosity, 10–11
  emissions
    air quality, 35
    carbon dioxide (CO_2) degassing, 15–16, 15*f*
    carbon dioxide (CO_2) fluxes, 26
    chemical composition, 12*t*, 16, 21–23, 22*t*
    climate effects, 4, 31–33
    early Earth history, 526
    emission styles, 13–16
    future research, 36
    gas speciation, 22*t*
    halogen compounds, 26–27, 27*t*
    hazards, 4–6, 35–36
    historic volcanoes, 25*t*
    human health effects, 4–6, 35–36
    mafic eruptions, 33
    molecular oxygen (O_2), 520
    silicic eruptions, 31–33
    solfataras, 3*f*
    sulfur dioxide (SO_2), 35
    sulfur fluxes, 24–26, 24*t*, 25*t*
    trace metals, 27, 27*t*
    troposphere, 31–33, 33–34
    volatile measurements, 16–23
    water (H_2O), 26
  emission styles
    crater lakes, 15*f*, 16, 140–141
    eruptive emissions, 15*f*, 16
    fumaroles, 14–15, 15*f*
    geysers, 14–15, 15*f*
    hot springs, 14–15
    mofettes, 15–16
    noneruptive emissions, 14–16, 15*f*
    open-vent degassing, 15*f*, 16
    solfataras, 3*f*, 14–15
  environmental impacts
    animal health, 35–36
    general discussion, 28–31
    human health effects, 35–36
    ozone (O_3) concentrations, 28–30
    soils, 34–35
    stratospheric sulfate aerosols, 28, 29*f*
    vegetation, 34–35
  historical background, 2–4
  hot spots
    carbon (C) abundances, 191
    nitrogen (N) abundances, 192
    water (H_2O), 193
  methane (CH_4) carbon subcycle, 353*t*
  molecular oxygen (O_2) consumption, 520
  preindustrial carbon cycle, 346*t*
  sulfide minerals, 137
  sulfur emissions
    Archaean cycle, 136*f*
    geochemistry, 139–141
    magma degassing, 139–140
    Mount St. Helens, 140*f*
    stratospheric chemistry, 159
    variability, 141*f*
  volatile measurement techniques
    geochemical surveillance methods, 21–23
    ice cores, 18*f*, 20–21, 20*f*
    petrological methods, 18*f*, 19–20
    portable remote sensing systems, 18–19, 18*f*
    satellite remote sensing systems, 18*f*, 19
    in situ measurements, 17–18, 18*f*
    sulfate markers, 20–21, 20*f*
  volcanic plumes, 33–34, 34*f*, 36
Volga River, 290*f*
Vosges River, 296*t*
Vostok Station ice core
  carbon dioxide (CO_2) concentrations, 457*f*
  glacial/interglacial variations, 353*t*, 359,
    360*f*, 362*f*

W

water (H_2O)
  abundances
    Earth, 193–194
    magmas, 7*t*
  basaltic melts, 8, 8*f*
  photolysis, 517–518
  volcanic emissions, 12*t*, 22*t*, 26
  water cycle, 193
watersheds
  chemical weathering
    surface-water solutes, 218–220
    weathering rates, 219*f*, 230–231, 232*f*
  nitrogen cycle, 564–565
weathering processes
  calcium carbonate ($CaCO_3$), 367–368, 455*f*
  carbon cycle, 538, 541*f*
  organic carbon cycle, 392
  rivers, 277–292, 312–314, 313*f*
  sedimentary environments, 588
  sulfur cycle, 538, 541*f*
Weddell Sea
  calcium carbonate ($CaCO_3$) abundance, 494
wetlands
  glaciation, 362–365
  methane (CH_4)
    methane (CH_4) carbon subcycle, 344*f*,
    353*t*, 359
Wilfarth, Hermann, 553
Winogradsky, Sergei Nikolaevich, 553

X

xenon (Xe)
  elemental abundances
    surface inventories, 169*t*
  isotopes
    degassing models, 183–185, 184*f*, 186
    iodine/xenon (I/Xe) isotopic ratios, 178–179
    mantle (Earth), 183–185, 184*f*, 186

xenon (Xe) (*continued*)
  mid-ocean ridge basalts (MORBs), 173–174,
    174–175, 174*f*, 175*f*, 183–185, 184*f*
  nuclear processes, 169*t*
  ocean island basalts (OIBs), 173–174
  plutonium/xenon (Pu/Xe) isotopic
    ratios, 178–179
  subduction fluxes, 177–178
  uranium/xenon (U/Xe) composition, 184*f*
  isotopic abundances
    atmospheric composition, 170*t*
Xijiang River, 296*t*
xylose, 400–401, 401*f*

Y

Yangtze River, 330–331
Yenisei River, 296*t*, 324–325
York River, 394*f*
Younger Dryas Event
  methane (CH$_4$) carbon subcycle, 365–366, 366*f*
ytterbium (Yb)
  elemental abundances
    Congo River, 308*f*
    hydrothermal vents, 77
    rivers, 296*t*, 307*f*, 308*f*, 312*f*, 313*f*, 323*f*
    sedimentary environments, 592*f*
yttrium (Y)
  biological pump, 437–440
  complexation, 317*f*
  elemental abundances

clastic sediments, 588–589, 589*f*
    rivers, 296*t*, 307*f*, 308*f*, 313*f*
    sedimentary environments, 592*f*
  hydrolysis, 316*f*

Z

Zaire River, 296*t*
zinc (Zn)
  benthic foraminifera, 499, 503–504
  biological pump, 438–439, 439*f*
  complexation, 317*f*
  elemental abundances
    hydrothermal vents, 60*t*, 73–74
    rivers, 296*t*, 307*f*, 308*f*, 314–315
    sedimentary environments, 592*f*
  hydrolysis, 316*f*
  phytoplankton photosynthesis, 428*t*
zirconium (Zr)
  biological pump, 437–440
  chemical weathering, 209, 210*f*
  elemental abundances
    clastic sediments, 588–589, 589*f*
    rivers, 296*t*, 307*f*, 308*f*, 313*f*
    sedimentary environments, 592*f*
    turbidites, 589, 589*f*
  hydrolysis, 316*f*
zooplankton
  biological pump, 432, 441–442, 454
  heterotrophic respiration, 454, 456*f*
  oceanic carbon cycle, 392–393, 408*f*, 456–457

Printed in the United States
By Bookmasters